Environmental Science and Engineering

Environmental Science

Series Editors

Rod Allan
Ulrich Förstner
Wim Salomons

For further volumes:
http://www.springer.com/series/3234

Khan M. G. Mostofa · Takahito Yoshioka
M. Abdul Mottaleb · Davide Vione
Editors

Photobiogeochemistry of Organic Matter

Principles and Practices in Water Environments

Springer

Editors
Khan M. G. Mostofa
Institute of Geochemistry
Chinese Academy of Sciences
Guiyang, Guizhou
People's Republic of China

M. Abdul Mottaleb
Department of Chemistry and Physics
Northwest Missouri State University
Missouri
USA

Takahito Yoshioka
Field Science Education
Kyoto University
Kyoto
Japan

Davide Vione
Department of Analytical Chemistry
University of Turin
Turin
Italy

ISSN 1431-6250
ISBN 978-3-642-32222-8 ISBN 978-3-642-32223-5 (eBook)
DOI 10.1007/978-3-642-32223-5
Springer Heidelberg New York Dordrecht London

Library of Congress Control Number: 2012952610

Printed on acid-free paper

Springer is part of Springer Science+Business Media (www.springer.com)

Contents

Dissolved Organic Matter in Natural Waters 1
Khan M. G. Mostofa, Cong-qiang Liu, M. Abdul Mottaleb,
Guojiang Wan, Hiroshi Ogawa, Davide Vione, Takahito Yoshioka
and Fengchang Wu

Photoinduced and Microbial Generation of Hydrogen Peroxide and Organic Peroxides in Natural Waters 139
Khan M. G. Mostofa, Cong-qiang Liu, Hiroshi Sakugawa,
Davide Vione, Daisuke Minakata and Fengchang Wu

Photoinduced Generation of Hydroxyl Radical in Natural Waters 209
Khan M. G. Mostofa, Cong-qiang Liu, Hiroshi Sakugawa,
Davide Vione, Daisuke Minakata, M. Saquib and M. Abdul Mottaleb

Photoinduced and Microbial Degradation of Dissolved Organic Matter in Natural Waters 273
Khan M. G. Mostofa, Cong-qiang Liu, Daisuke Minakata,
Fengchang Wu, Davide Vione, M. Abdul Mottaleb,
Takahito Yoshioka and Hiroshi Sakugawa

Colored and Chromophoric Dissolved Organic Matter in Natural Waters 365
Khan M. G. Mostofa, Cong-qiang Liu, Davide Vione,
M. Abdul Mottaleb, Hiroshi Ogawa, Shafi M. Tareq and
Takahito Yoshioka

Fluorescent Dissolved Organic Matter in Natural Waters 429
Khan M. G. Mostofa, Cong-qiang Liu, Takahito Yoshioka,
Davide Vione, Yunlin Zhang and Hiroshi Sakugawa

Photosynthesis in Nature: A New Look 561
Khan M. G. Mostofa, Cong-qiang Liu, Xiangliang Pan, Takahito Yoshioka, Davide Vione, Daisuke Minakata, Kunshan Gao, Hiroshi Sakugawa and Gennady G. Komissarov

Chlorophylls and their Degradation in Nature 687
Khan M. G. Mostofa, Cong-qiang Liu, Xiangliang Pan, Davide Vione, Kazuhide Hayakawa, Takahito Yoshioka and Gennady G. Komissarov

Complexation of Dissolved Organic Matter with Trace Metal Ions in Natural Waters 769
Khan M. G. Mostofa, Cong-qiang Liu, Xinbin Feng, Takahito Yoshioka, Davide Vione, Xiangliang Pan and Fengchang Wu

Impacts of Global Warming on Biogeochemical Cycles in Natural Waters 851
Khan M. G. Mostofa, Cong-qiang Liu, Kunshan Gao, Shijie Li, Davide Vione and M. Abdul Mottaleb

Editors Biography 915

Dissolved Organic Matter in Natural Waters

Khan M. G. Mostofa, Cong-qiang Liu, M. Abdul Mottaleb, Guojiang Wan, Hiroshi Ogawa, Davide Vione, Takahito Yoshioka and Fengchang Wu

1 Introduction

Organic matter (OM) in water is composed of two major fractions: dissolved and non-dissolved, defined on the basis of the isolation technique using filters (0.1–0.7 μm). Dissolved organic matter (DOM) is the fraction of organic substances that passes the filter, while particulate organic matter (POM) remains on the filter (Danielsson 1982; Kennedy et al. 1974; Liu et al. 2007; Mostofa et al. 2009a). DOM is generally originated from three major sources: (i) allochthonous

K. M. G. Mostofa (✉) · C. Q. Liu · G. Wan
State Key Laboratory of Environmental Geochemistry, Institute of Geochemistry, Chinese Academy of Sciences, Guiyang 550002, China
e-mail: mostofa@vip.gyig.ac.cn

M. A. Mottaleb
Center for Innovation and Entrepreneurship (CIE), Department of Chemistry/Physics, Northwest Missouri State University, 800 University Drive, Maryville, MO 64468, USA

H. Ogawa
Atmospheric and Ocean Research Institute, The University of Tokyo, 1-15-1, Minamidai, Nakano, 164-8639 Tokyo, Japan

D. Vione
Dipartimento di Chimica Analitica, University of Turin, I-10125 Turin, Italy

Centro Interdipartimentale NatRisk, I-10095 Grugliasco (TO), Italy

T. Yoshioka
Field Science Education and Research Center, Kyoto University, Kitashirakawa Oiwake-cho, Sakyo-ku, Kyoto 606-8502, Japan

F. C. Wu
State Environmental Protection Key Laboratory of Lake Pollution Control, Chinese Research Academy of Environmental Sciences, Chaoyang 100012, China

K. M. G. Mostofa et al. (eds.), *Photobiogeochemistry of Organic Matter*,
Environmental Science and Engineering, DOI: 10.1007/978-3-642-32223-5_1,

or terrestrial material from soils, (ii) autochthonous or surface water-derived of algal or phytoplankton origin, and (iii) syhthetic organic substances of man-made or industrial origin. DOM in natural waters is composed of a heterogeneous mixture of organic compounds with molecular weights ranging from less than 100 to over 300,000 Daltons (Hayase and Tsubota 1985; Thurman 1985a; Ma and Ali 2009). On the other hand, POM is composed of plant debris, algae, phytoplankton cell, bacteria, and so on (Mostofa et al. 2009a). Humic substances (fulvic and humic acids) of terrestrial origin are the dominant DOM fractions in freshwater and coastal seawater (Mostofa et al. 2009a). On the other hand, autochthonous fulvic acids (or marine humic-like) of algal or phytoplankton and bacterial origin are the key DOM fractions in lakes and oceans (Mostofa et al. 2009a, b; Coble 1996, 2007; Parlanti et al. 2000; Amado et al. 2007; Zhang et al. 2009). In addition, among the major classes of DOM components there are carbohydrates, proteins, amino acids, lipids, phenols, alcohols, organic acids and sterols (Mostofa et al. 2009a).

DOM can display physical properties such as the absorption of energy from ultraviolet (UV) and photosynthetically available radiation (PAR) (Kirk 1976; Morris et al. 1995; Siegel and Michaels 1996; Morris and Hargreaves 1997; Tranvik 1998; Bertilsson and Tranvik 2000; Laurion et al. 2000; Markager and Vincent 2000; Huovinen et al. 2003; Sommaruga and Augustin 2006; Hayakawa and Sugiyama 2008; Effler et al. 2010), chemical properties such as complex formation with trace metal ions (Mostofa et al. 2009a, 2011; Lead et al. 1999; Wang and Guo 2000; Koukal et al. 2003; Mylon et al. 2003; Wu et al. 2004; Lamelas and Slaveykova 2007; Lamelas et al. 2009; Fletcher et al. 2010; Reiller and Brevet 2010; Sachs et al. 2010; Da Costa et al. 2011), the ability to maintain acidity and alkalinity (Mostofa et al. 2009a; Oliver et al. 1983; Wigington et al. 1996; Pace and Cole 2002; Hudson et al. 2003; Kopáćek et al. 2003), the occurrence of redox and photo-Fenton reactions (Voelker and Sulzberger 1996; Voelker et al. 1997, 2000; Kwan and Voelker 2002; Jeong and Yoon 2004; Wu et al. 2005; Vione et al. 2006; Nakatani et al. 2007), as well as the ability to control the cycling of nutrients such as NH_4^+, NO_3^+, and PO_4^{3-} in natural waters (Bronk 2002; Zhang et al. 2004, 2008; Kim et al. 2006; Vähätalo and Järvinen 2007; Li et al. 2008).

DOM can photolytically generate strong oxidants such as superoxide radical ($O_2^{\bullet -}$), hydrogen peroxide (H_2O_2), and hydroxyl radical ($HO^\bullet$), which also play a role in its photoinduced decomposition in natural waters (Mostofa and Sakugawa 2009; Vione et al. 2006, 2010; Zellner et al. 1990; Zepp et al. 1992; Moran et al. 2000; Farias et al. 2007; Mostofa et al. 2007a; Minakata et al. 2009). Correspondingly, DOM can undergo photoinduced and microbial degradation processes, which can produce a number of degradation products such as dissolved inorganic carbon (DIC), CO_2, CH_4, CO, low molecular weight (LMW) DOM, organic acids. These compounds are very important in the aquatic environments (Jones and Amador 1993; Miller and Zepp 1995; Lovley and Chapelle 1995; Lovley et al. 1996; Moran and Zepp 1997; Miller 1998; Conrad 1999; Johannessen and Miller 2001; Ma and Green 2004; Xie et al. 2004; Johannessen et al. 2007; Yoshioka et al. 2007; Brandt et al. 2009; Rutledge et al. 2010; Omar et al. 2010; Ballaré et al. 2011; Zepp et al. 2011). DOM with its degradation products can extensively influence photosynthesis, thereby playing a key role in global carbon cycle processes (Mostofa et al. 2009a; Mostofa and Sakugawa 2009; Ma and Green 2004; Johannessen et al. 2007; Palenik and Morel 1988; Fujiwara et al. 1993; Komissarov 1994, 1995, 2003; Miller and Moran 1997; Meriläinen et al. 2001; Malkin et al. 2008). DOM also plays important roles in regulating drinking water quality, complexing behavior with metal ions, water photochemistry, biological activity, photosyhthesis, and finally global warming.

This chapter will provide an overview on the origin of DOM, its contents and sources in natural waters, the contribution of organic substances to DOM, the biogeochemical functions of DOM, its physical and chemical properties, as well as its molecular size distribution. It comprehensively discusses the controlling factors and their effects on the distribution of DOM in natural waters, the emerging contaminants and their sources, transportation and impacts, as well as methodologies and techniques for the detection of pharmaceuticals in fish tissue. Finally, it is discussed how DOM acts as energy source for living organisms and aquatic ecosystems.

2 What is Dissolved Organic Matter?

DOM is conventionally defined as any organic material that passes through a given filter (0.1–0.7 μm). The organic material that is retained on the filter is termed 'particulate organic matter (POM) (Mostofa et al. 2009a). The permeate from ultrafiltration (<10 kiloDaltons or kDa) is often defined as the truly dissolved organic carbon fraction and the filter-passing fraction between >10 kDa and <0.4 or 0.7 μm as the total dissolved organic carbon fraction in aqueous solution. Colloids are operationally defined as particles between 1 nm and 1 μm in size, and the 'dissolved' fraction can include a subset of the colloidal materials (Sharp 1973; Vold and Vold 1983; Koike et al. 1990; Benner et al. 1992; Buesseler et al. 1996; Wells 2002). These types of colloidal particles are not entirely retained by filters with pore sizes between 0.2 and 0.7 μm. DOM can be in the size range of tens to hundreds of nm when they are associated with other colloidal materials in water (Lead and Wilkinson 2006). It has been shown that colloids make up a significant fraction, approximately 10–40 %, of the marine DOM pool.

DOM in natural waters is composed of a heterogeneous mixture of numerous allochthonous and autochthonous organic compounds containing low molecular weight substances (e.g. organic acids) and macromolecules such as fulvic and humic acids (humic substances), with molecular weight ranging from less than 100 to over 300,000 Daltons (Thurman 1985a, 1986; Ma and Ali 2009; Rashid and King 1969; MacFarlane 1978; Hayase and Tsubota 1983; Amy et al. 1987; Wagoner et al. 1997; Jerry and Jean-Philippe 2003; Xiao and Wu 2011). DOM found in natural ground and surface waters are also referred as natural organic matter (NOM). The most common organic substances are humic substances (fulvic and humic acids) of terrestrial origin, autochthonous fulvic acids of phytoplankton or algal origin, carbohydrates, sugars, amino acids, proteins, lipids, organic acids, phenols, alcohols, acetylated amino sugars, and so on. On the other hand, POM includes plant debris, detritus, living organisms, bacteria, algae, phytoplankton, corals, coral reefs, and so on. DOM is considered as the larger pool of organic matter in a variety of waters, which can include more than 90 % of total organic matter (Thurman 1986; Kececioglu et al. 1997).

2.1 Biogeochemical Functions of OM (DOM and POM)

DOM of both allochthonous and autochthonous origin can play multiple functions in photoinduced, chemical, microbial and geochemical processes in natural waters. They can be classified as follows:

(1) Photoinduced functions of DOM. Irradiated DOM can produce H_2O_2 (Mostofa and Sakugawa 2009), which in turn can produce the strong oxidizing agent hydroxyl radical ($HO^•$), either directly by photoinduced dissociation ($H_2O_2 + hv \rightarrow HO^•$) or by the photo-Fenton reaction. These processes are involved in the photoinduced degradation of organic compounds (Vione et al. 2006, 2010; Zellner et al. 1990; Zepp et al. 1992; Farias et al. 2007). DOM undergoes rapid photoinduced decomposition by natural sunlight, and this process is less efficient in waters with high contents of DOM and more efficient with high DOM concentrations (Moran et al. 2000; Ma and Green 2004; Vähätalo et al. 2000; Mostofa et al. 2007b; Vione et al. 2009). DOM can thus control redox and photo-Fenton reactions in natural waters (Voelker and Sulzberger 1996; Voelker et al. 1997, 2000; Kwan and Voelker 2002; Jeong and Yoon 2004; Wu et al. 2005; Vione et al. 2006; Nakatani et al. 2007). The biogeochemical functions of H_2O_2 and $HO^•$ are discussed in details in chapters "Photoinduced and Microbial Generation of Hydrogen Peroxide and Organic Peroxides in Natural Waters", "Photoinduced Generation of Hydroxyl Radical in Natural Waters".

(2) Microbial functions of OM (DOM and POM). DOM and POM are decomposed biologically by microorganisms in natural waters (Moran et al. 2000; Mostofa et al. 2007a; Ma and Green 2004; Lovley and Chapelle 1995; Hopkinson et al. 2002; Coble 2007; Koschorreck et al. 2008; Lønborg et al. 2009a, b; Lønborg and Søndergaard 2009). This process can produce new autochthonous DOM or nutrients in water (Mostofa et al. 2009b; Zhang et al. 2009; Kim et al. 2006; Weiss et al. 1991; Harvey et al. 1995; Yamashita and Jaffé 2008; Fu et al. 2010; Li et al. 2011), so that DOM is responsible for the maintenance of the microbial loop in natural waters (utilization of DOC by bacteria, consumption and decomposition of bacteria by protozoans and release of dissolved organic compounds and CO_2) (Sherr and Sherr 1989; Carrick et al. 1991; Jones 1992; Tranvik 1992). Bioavailable carbon

substrates produced from DOM and OM either photolytically or biologically can enhance biological productivity in waters (Mostofa et al. 2009a; Bertilsson and Tranvik 1998, 2000; Vähätalo and Järvinen 2007; Lovley et al. 1996; Komissarov 2003; Tranvik 1992; Norrman et al. 1995; Wetzel et al. 1995). Production of nutrients and DIC through photoinduced and microbial degradation of DOM or POM can control the food-chains for microorganisms (Mostofa et al. 2009a; Miller and Zepp 1995; Ma and Green 2004; Tranvik 1992; Salonen et al. 1992; Kirchman et al. 1995; Wheeler et al. 1997; Guildford and Hecky 2000; Rosenstock et al. 2005). The biogeochemical functions of microbial processes are discussed in details in "Photoinduced and Microbial Degradation of Dissolved Organic Matter in Natural Waters".

(3) Optical (or physical) functions of DOM: a fraction of DOM is named as either colored and chromophoric dissolved organic matter (CDOM) based on the absorption of ultraviolet (UV) and photosynthetically available radiation (PAR), or fluorescent DOM (FDOM) based on the emission of fluorescence photons after radiation absorption. DOM generally controls the downward irradiance flux through the water column of UV-B (280–320 nm), UV-A (320–400 nm), total UV (280–400 nm) as well as photosynthetically available radiation (PAR, 400–700 nm) (Kirk 1976; Morris et al. 1995; Siegel and Michaels 1996; Morris and Hargreaves 1997; Tranvik 1998; Bertilsson and Tranvik 2000; Laurion et al. 2000; Markager and Vincent 2000; Huovinen et al. 2003; Sommaruga and Augustin 2006; Hayakawa and Sugiyama 2008; Effler et al. 2010). DOM is responsible for water color, water transparency, occurrence of the euphotic zone and thermal stratification in the surface waters of lakes and oceans because it affects (decreases) the penetration of solar radiation (Laurion et al. 2000; Effler et al. 2010; Hudson et al. 2003; Eloranta 1978; Jones and Arvola 1984; Howell and Pollock 1986; Perez-Fuentetaja et al. 1999; Snicins and Gunn 2000; Watts et al. 2001; Mostofa et al. 2005a). Biogeochemical functions of CDOM and FDOM are discussed in detail in the respective chapters (see chapters "Colored and Chromophoric Dissolved Organic Matter (CDOM) in Natural Waters" and "Fluorescent Dissolved Organic Matter in Natural Waters").

(4) Cycling of nutrients (NH_4^+, NO_3^+, and PO_4^{3-}) by DOM and POM. Nutrients are produced by degradation of DOM and typically derive from dissolved organic nitrogen (DON) and dissolved organic phosphorus (DOP) in DOM molecular structure (Bronk 2002; Zhang et al. 2004, 2008; Kim et al. 2006; Vähätalo and Järvinen 2007; Li et al. 2008). Nutrients are mostly released during the photoinduced and microbial respiration (or assimilation) of POM (e.g. algae or phytoplankton biomass), as shown by in situ experiments conducted under light and dark incubations (Kim et al. 2006; Li et al. 2008; Yamashita and Jaffé 2008; Carrillo et al. 2002; Kopáček et al. 2004; Fu et al. 2005; Mostofa KMG et al. unpublished data). NO_3^- and NO_2^- can be produced by oxidation of ammonia in nitrification ($NH_4^+ + 2O_2 \rightarrow NO_3^- + 2H^+ + H_2O$) and of DON in lake waters (Ma and

Green 2004; Kopáček et al. 2004; Lehmann et al. 2004; Minero et al. 2007). Nutrients produced by DOM and OM can fuel new primary and secondary production in natural waters. Total contents of DOM in lake waters are responsible for variation of the trophic level, due to eutrophication/oligotrop hication processes. The latter are a major driver of change for chemical variables such as major ions, nutrients (phosphorus and nitrogen compounds, silica) and the chemical nature of DOM.

(5) DOM can control photosynthesis in natural waters. DOC can limit productivity (Jackson and Hecky 1980; Carpenter et al. 1998) and affect epilimnetic (Hanson et al. 2003) and hypolimnetic respiration (Houser et al. 2003). Photoinduced and microbial oxidation of DOM is responsible for the simultaneous generation of H_2O_2, CO_2 and DIC (Mostofa and Sakugawa 2009; Ma and Green 2004; Johannessen et al. 2007; Palenik and Morel 1988; Fujiwara et al. 1993; Miller and Moran 1997; Meriläinen et al. 2001; Malkin et al. 2008). Such compounds could favor the occurrence of photosynthesis in natural waters. Some studies show that H_2O_2 could be involved as reactant in photosynthesis ($xCO_2 + yH_2O_{2(H2O)} + h\upsilon \rightarrow C_x(H_2O)_y + O_2 +$ ene rgy; and $2H_2O_2 + h\upsilon \rightarrow 2H_2O + O_2$) (Mostofa et al. 2009a; Komissarov 1994, 1995, 2003; Miller and Moran 1997). Nutrients (PO_4^{3-} and NH_4^+) released by DOM and POM might also favor the occurrence of photosynthesis and subsequently enhance the cyanobacterial or algal blooms in natural waters (Zhang et al. 2008, 2009; Kim et al. 2006; Li et al. 2008; Lehmann et al. 2004; Huszar et al. 2006; Nõges et al. 2008; McCarthy et al. 2009; Mohlin and Wulff 2009). High chlorophyll *a* concentrations are often detected in waters with high contents of DOM, and the reverse happens in low-DOM waters (Meriläinen et al. 2001; Malkin et al. 2008; Fu et al. 2010; Guildford and Hecky 2000; Mostofa et al. 2005a, Mostofa KMG et al., unpublished data; Satoh et al. 2006; Yacobi 2006; Komatsu et al. 2007).

(6) Chemical functions of OM (DOM and POM). DOM and POM are composed of various functional groups in their molecular structures, which can form complexes with trace metal ions (M) in aqueous solution via strong π-electron bonding systems (Mostofa et al. 2009a, 2011; Lead et al. 1999; Wang and Guo 2000; Koukal et al. 2003; Mylon et al. 2003; Wu et al. 2004; Lamelas and Slaveykova 2007; Lamelas et al. 2009; Fletcher et al. 2010; Reiller and Brevet 2010; Sachs et al. 2010; Da Costa et al. 2011). These studies imply that the M-DOM complexation is important for speciation, bioavailability, transport and ultimate fate of trace metal ions in the water environment. The detailed functions of M-DOM complexes are discussed in Complexation of Dissolved Organic Matter With Trace Metal Ions in Natural Waters. DOM can also influence the cycling of aluminum and iron oxides in natural waters (McKnight et al. 1992).

(7) Maintenance of the drinking water quality by DOM and POM in waters (Mostofa et al. 2009a). The production of POM is significantly dependent on the DOM contents in natural waters, and POM can produce new autochthonous DOM and nutrients under both irradiation and microbial respiration

or assimilation (Mostofa et al. 2005a, 2009b; Zhang et al. 2009; Kim et al. 2006; Li et al. 2008; Yamashita and Jaffé 2008; Carrillo et al. 2002; Kopáček et al. 2004; Fu et al. 2005). Simultaneously, DOM can release nutrients upon exposure to natural sunlight in waters (Bronk 2002; Zhang et al. 2004, 2008; Kim et al. 2006; Vähätalo and Järvinen 2007; Li et al. 2008). Increases in nutrients and autochthonous DOM severely deteriorate the drinking water quality, but DOM can also balance acidity and alkalinity through its photoinduced or microbial decomposition (Mostofa et al. 2009a; Oliver et al. 1983; Wigington et al. 1996; Pace and Cole 2002; Hudson et al. 2003; Kopáćek et al. 2003).

(8) OM can maintain global carbon cycle processes through production, distribution, transportation and decomposition of carbon compounds in the biosphere (Mostofa et al. 2009a; Brandt et al. 2009; Rutledge et al. 2010; Omar et al. 2010; Ballaré et al. 2011; Zepp et al. 2011; Hedges 1992; Amon and Benner 1994; Ogawa and Tanoue 2003; Freeman et al. 2004; Lavoie et al. 2005; Fenner et al. 2007a, b; Wolf et al. 2007). The photoinduced and microbial decomposition of DOM and POM yields CO_2, CO, CH_4, DIC (DIC is defined jointly as dissolved CO_2, H_2CO_3, HCO_3^-, and CO_3^{2-}), low molecular weight DOM and other inorganic ions (Jones and Amador 1993; Miller and Zepp 1995; Lovley and Chapelle 1995; Lovley et al. 1996; Moran and Zepp 1997; Miller 1998; Conrad 1999; Johannessen and Miller 2001; Ma and Green 2004; Xie et al. 2004; Johannessen et al. 2007; Yoshioka et al. 2007; Brandt et al. 2009; Rutledge et al. 2010; Omar et al. 2010; Ballaré et al. 2011; Zepp et al. 2011). The produced CO_2 and CH_4 increase the atmospheric green house gases and contribute to the global carbon cycle (Davidson and Janssens 2006; Porcal et al. 2009). Elevated atmospheric CO_2 can enhance DOC supply, particularly in peat soils. This is attributed to elevated net primary productivity of plants and increased root exudation of DOC in soil environments, which ultimately leach into the aquatic ecosystem (Freeman et al. 2004; Lavoie et al. 2005; Fenner et al. 2007a, b; Wolf et al. 2007; Kang et al. 2001; Pastor et al. 2003).

(9) Character and energy functions of OM in the water ecosystem. DOM and POM can provide a major source of energy, in the form of C and N, which are essential to all living organisms in natural waters (Mostofa et al. 2009a; Tranvik 1992; Salonen et al. 1992; Wetzel 1984, 1992). Thermal energy produced during the photoinduced and microbial degradation of DOM and organic matter, photoinduced redox reactions, microbial loop, as well as photosynthesis are key drivers in aquatic ecosystems (Mostofa et al. 2009a; Komissarov 1994, 1995, 2003; Miller and Moran 1997; Sherr and Sherr 1989; Carrick et al. 1991; Jones 1992; Tranvik 1992; Salonen et al. 1992; Wetzel 1984, 1992; Hedges et al. 2000). DOM itself can provide energy and matter for the growth of bacterial films on the surface of drinking-water pipes, a process that involves also fulvic and humic acids (humic substances) depending on their occurrence in groundwater in developing and developed countries (Mostofa et al. 2009a).

3 Origin of DOM in Natural Waters

DOM is generally originated from three major sources in natural waters: (i) DOM derived from terrestrial soils, termed allochthonous DOM; (ii) DOM derived from in situ production in natural surface waters, termed autochthonous DOM, and (iii) DOM derived from human activities (e.g. industrial synthesis), termed anthropogenic DOM.

3.1 Origin of Allochthonous DOM in Soil Ecosystems

DOM including fulvic and humic acids (humic substances) originates from the decomposition of vascular plant material, root exudates and animal residues in terrestrial soil. Origin of allochthonous DOM from vascular plant materials or particulate detrital pools is significantly varied in different regions (tropical, temperate and boreal), which is regulated by the occurrence of three key factors or functions (Mostofa et al. 2009a; Wetzel 1983, 1990, 1992; Malcolm 1985; Dai et al. 1996; Nakane et al. 1997; Wershaw 1999; Jaramillo and Dilcher 2000; Kalbitz et al. 2000; Trumbore 2000; Uchida et al. 1998, 2000; Moore et al. 2008; Braakhekke et al. 2011; Spence et al. 2011; Tu et al. 2011): (i) Physical functions that include temperature and moisture; (ii) Chemical functions that include nutrient availability, amount of available free oxygen and redox activity, and (iii) Microbial processes that include microfloral succession patterns and availability of microorganisms (aerobic or anaerobic).

It is suggested that microorganisms can alter sugars, starch, proteins, cellulose and other carbon compounds bound to organic matter of plant or animal origin during their metabolic processes. These processes can transform the aromatic and lipid plant components into amphiphilic molecules including humic substances, i.e., molecules that consist of separate hydrophobic (non-polar) and hydrophilic (polar) parts (Wershaw 1999). The non-polar parts of the molecules are composed of relatively unaltered segments of plant polymers, while the polar parts include carboxylic acid groups (Wershaw 1999). Aerobic microorganisms can decompose organic matter at a faster rate than anaerobic ones, depending on the availability of free oxygen. Compositional changes of DOM occur with soil depth, leading to a decrease of aromatic compounds and carbohydrates whilst alkyl, methoxy and carbonyl moieties increase with depth (Dai et al. 1996). The increase in alkyl and carboxylic C with depth are the result of biodegradation of forest litter and oxidation of lignin side chains, respectively (Zech et al. 1985; Kogel-Knabner et al. 1988; Kogel-Knabner 1992).

The origin of allochthonous DOM from microbial processes can be judged from significant variations in respired organic carbon in different soil environments. The mean age of soil respired organic carbon determined using ^{14}C tracer is lowest (1 year) in tropical forest soils (eastern Amazonia, Brazil), relatively low (3 years) in temperate forest soils (central Massachusetts, USA), and highest (16 years) in boreal forest soils (Manitoba, Canada) (Trumbore 2000). Experimental studies using $\delta^{13}C$ or ^{14}C to track sources and turnover of DOC indicate that DOM, which is transported over decimetres or metres down into subsoil, mainly represents highly altered residues of organic matter processing (Schiff et al. 1997; Flessa et al. 2000; Hagedorn et al. 2004; Fröberg et al. 2007). Note that allochthonous DOM is mostly derived, in zero to a few decimeter depth from the decomposition of plant material by microbial processes in soils and shallow groundwater (Uchida et al. 1998, 2000; Fröberg et al. 2007; IPCC 1996; Buckau et al. 2000).

DOC leached from soil is partly retained in the vadose zone before reaching aquifers (Siemens and Kaupenjohann 2003; Mikutta et al. 2007; Kalbitz and Kaiser 2008; Scheel et al. 2008). For the range of groundwater recharge of 95–652 mm yr^{-1}, it is shown that a constant flux of DOC from soil into surface waters often takes place (Kindler et al. 2011). Therefore, allochthonous DOM is partly discharged through hydrological processes directly into streams or riverbeds or surrounding water bodies, which ultimately flux to lake or oceanic environments as final water reservoir.

3.2 *Origin of Autochthonous DOM in Natural Waters*

Production of autochthonous DOM is generally observed at the epilimnion (upper water layers) compared to the hypolimnion (deeper layers) during the summer stratification period, particularly in lakes and oceans. A rough estimation by comparing the upper with the deeper layers demonstrates that the contribution of autochthonous DOM is largely varied in lakes and oceans: it reaches 0–55 % in Lake Hongfeng (181–250 μM C at 0–6 m and 161–223 μM C at 22–25 m depth, respectively, during March–September), 3–47 % in Lake Baihua (183–264 μM C at 0–3 m and 157–206 μM C at 14–15 m during March-September), 6–35 % in Lake Baikal (93–142 μM C at 0–100 m and 88–105 μM C at 600–720 m during August–September in 1995, 1998, 1999), 3–82 % in Lake Biwa (93–183 μM C at 2.5–10 m and 78–101 μM C at 70 m during May–September in 1999–2002), 21–49 % in Lake Ashino in Japan (99–111 μM C at 0–10 m and 74–84 μM C at 30–38 m in September 1997), 81–102 % in Lake Ikeda in Japan (101–112 μM C at 0–10 m and 55–56 μM C at 200–233 m for site I1; at 41 m for site I2 in October 1997), 52 % in Lake Suwa in Japan (216 μM C at 0 m in September and 142 μM C at 0 m in December 1997), 61–81 % in Lake Inawashiro in Japan (42–47 μM C at 0–10 m and 26 μM C at 70 m), 13–29 % in Lake Fuxian (123–135 μM C at 0–10 m and 95–105 μM C at 50–140 m in June 2001), 19 % in Lake Hovsgol (95 μM C at 0 m and 80 μM C at 50–200 m in July 1999), 0–88 % in Lake Kinneret (270–485 μM C at 0–10 m and 258–368 μM C at 38 m during the summer period in 2004), 17–41 % in Lake Peter (data not shown), 11–29 % (biological production) in Lake Bret, 0–104 % in Middle Atlantic Bight (82–98 μM C at 0 m and 48–90 μM C at 90–2600 m in June 2001), 16–77 % in Western North Pacific (85–117 μM C at 0 m and 66–73 μM C at 150 m), 0–194 % in Atlantic Ocean (50–97 μM C at <100 m and 33–59 μM C at >1000 m), 0–165 % in Pacific Ocean (40–90 μM C at <100 m and 34–45 μM C at >1000 m), 28–121 % in Indian Ocean and Arabian Sea (55–95 μM C at <100 m and 43 μM C at >1000 m), 0–121 in Antarctic Ocean (38–75 μM C at <100 m and 34–60 μM C at >1000 m), as well as 0–118 % in Arctic Ocean (34–107 μM C at <100 m and 49–54 μM C at >1000 m) (Mostofa et al. 2005a, 2009a; Fu et al. 2010; Ogawa and Tanoue 2003; Ogawa and Ogura 1992; Wilkinson et al. 1997; Mitra et al. 2000; Yoshioka et al. 2002a; Hayakawa et al. 2003, 2004; Annual Report 2004; Bade 2004; Sugiyama et al. 2004).

The contribution of extracellular release of photosynthetically-derived DOM varies from 5 to 70 % in natural waters (Lancelot 1979; Fogg 1983; Connolly et al. 1992). The autochthonous production is significantly higher in oceans with a high water temperature (WT) than in those with a low water temperature, particularly in the Arctic Ocean. The key contributors to autochthonous DOM in natural waters as well as in sediment pore waters are considered to be phytoplankton or algal biomass, bacteria, coral, coral reef, submerged aquatic vegetation, krill (shrimp-like marine crustaceans), seagrass, and marsh- and mangrove forest (Mostofa et al. 2009a, b; Zhang et al. 2009; Li et al. 2011; McKnight et al. 1991, 1993, 1994, 2001; Tanoue et al. 1995, 1996; Fukuda et al. 1998; Nelson et al. 1998, 2004; Tanoue 2000; Kahru and Mitchell 2001; Ogawa et al. 2001; Hata et al. 2002; Rochelle-Newall and Fisher 2002a, b; Burdige et al. 2004; Cammack et al. 2004; Steinberg et al. 2004; Wild et al. 2004; Yamashita and Tanoue 2004; Biers et al. 2007; Chen et al. 2007; Vantrepotte et al. 2007; Wada et al. 2007; Wang et al. 2007; Hanamachi et al. 2008; Henderson et al. 2008; Tanaka et al. 2008; Tzortziou et al. 2008; Ortega-Retuerta et al. 2009; Tranvik et al. 2009). These studies demonstrate that autochthonous DOM is produced from POM by several processes such as photoinduced and microbial respiration (or assimilation), zooplankton grazing, bacterial release and uptake, viral interactions, and complex microbial processes in sediment pore waters.

3.2.1 Respiration or Assimilation of Algae or Phytoplankton Species and Bacteria

Algae or phytoplankton biomass and bacteria can release new DOM in natural waters by two key processes: first, photoinduced respiration or assimilation of algae or phytoplankton biomass and bacteria, which can produce new DOM (Mostofa et al. 2005a, 2009b, 2011; Rochelle-Newall and Fisher 2002a; Varela et al. 2003; Aoki et al. 2008; Biddanda and Benner 1997; Hulatt et al. 2009). Second, microbial respiration or assimilation of algae or phytoplankton and bacteria, which can release the new DOM in natural waters (Mostofa et al. 2009a, b, 2011; Parlanti et al. 2000; Zhang et al. 2009; Fu et al. 2010; McKnight et al. 1991, 1994, 2001; Nelson et al. 2004; Rochelle-Newall and Fisher 2002a; Cammack et al. 2004; Yamashita and Tanoue 2004, 2008; Wada et al. 2007; Hanamachi et al. 2008; Ortega-Retuerta et al. 2009; Aoki et al. 2008; Biddanda and Benner 1997; Hulatt et al. 2009; Bertilsson and Jones 2003; Chen and Gardner 2004; Stedmon and Markager 2005a; Stedmon et al. 2007a, b; Wetz and Wheeler 2007; Zhao et al. 2009).

Re-suspension of algae or phytoplankton in ultrapure water (Milli-Q), artificial sea water and natural waters can release new organic compounds, either under irradiation or under dark incubation. These organic substances, produced either under irradiation (Fig. 1a) or in the dark (Fig. 1b, c) show fluorescence (excitation-emission matrix, EEM) properties. The EEM spectra of autochthonous DOM (Fig. 1a, b) are roughly similar to those of allochthonous fulvic acid and show two fluorescence peaks at peak C- and A-regions (Fig. 1d). In contrast, they are different from allochthonous humic acids that show more than two peaks at peak C-region (Fig. 1f). Based on the similarities of the EEM spectra, the key component of autochthonous fluorescent DOM is defined as "autochthonous fulvic acid (C-like)" of algal or phytoplankton origin. The other component (Fig. 1c) is defined as "autochthonous fulvic acid (M-like)" of algal or phytoplankton origin, based on the similarities with the marine humic-like component (Coble 1996, 2007). Identification of autochthonous DOM of algal or phytoplankton origin is

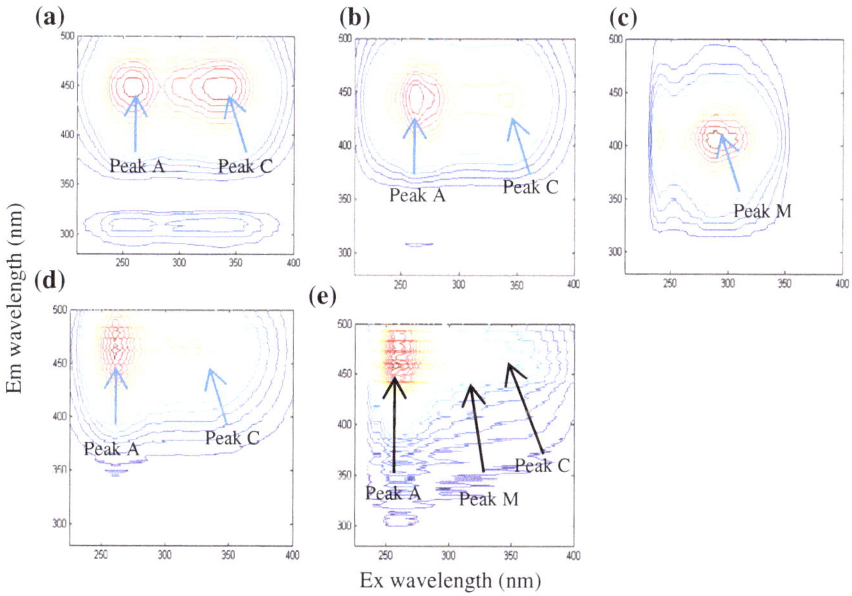

Fig. 1 Comparison of the fluorescent components of autochthonous fulvic acid (C-like) produced under microbial respiration of lake algae (**a**), autochthonous fulvic acid (C-like) under photorespiration or assimilation of algal biomass (**b**) and autochthonous fulvic acid (M-like) under microbial respiration of algae (**c**) with aqueous samples of standard Suwannee River Fulvic Acid (**d**) and Suwannee River Humic Acid (**e**) identified using PARAFAC modeling on the EEM spectra of their respective samples. *Data source* Mostofa KMG et al., (unpublished data)

discussed extensively in the FDOM chapter (see chapter "Fluorescent Dissolved Organic Matter in Natural Waters"). Note that "autochthonous fulvic acids" of algal or phytoplankton origin are newly termed in this study for mostly two reasons: first, to distinguish and generalize between all freshwaters and marine waters; second, because of the confusion in different studies that use several names such as marine humic-like (Coble 1996, 2007), sedimentary fulvic acids (Hayase and Tsubota 1983), microbially derived fulvic acids or marine fulvic acids (McKnight et al. 1991, 1994; Harvey and Boran 1985; Meyers-Schulte and Hedges 1986).

DOM is produced significantly by eleven species of intertidal and sub-tidal macroalgae when they are illuminated, providing evidence for a light-driven exudation mechanism (Hulatt et al. 2009). The contribution of the released DOC has been detected as 6.4 and 17.3 % of the total organic carbon in cultures of *Chlorella vulgaris* and *Dunaliella tertiolecta*, respectively, upon light exposure (Hulatt and Thomas 2010). DOM can support a significant growth of bacterial biomass, representing a further loss of algal assimilated carbon in water (Hulatt and Thomas 2010). Dissolved combined amino acids, middle-reach peaks of particulate amino acids and non-protein amino acids are often decreased in downstream rivers, which is likely the result of photoinduced degradation of DOM and algae (Duan and Bianchi 2007).

On the other hand, the key processes of autochthonous DOM release by microbial respiration of algae or phytoplankton biomass in waters are presumably the extracellular release by living cells, cell death and lysis, or herbivore grazing that may occur in the deeper waters of rivers, lakes and oceans (Mostofa et al. 2009a; Tanoue 2000; Tranvik et al. 2009; Hulatt et al. 2009). In fact, bacteria play a specific role in subsequent processing of the DOM released by algae in natural water (Nelson et al. 1998, 2004; Rochelle-Newall and Fisher 2002a; Cammack et al. 2004; Biers et al. 2007; Ortega-Retuerta et al. 2009). Cultivation of three kinds of phytoplankton (green algae *Microcystis aeruginosa* and *Staurastrum dorcidentiferum* and dark-brown whip-hair algae *Cryptomonas ovata* collected from lake waters) shows that fulvic acid-like and protein-like fluorescent components are released when they are cultivated under a 12:12 h light/dark cycle in a MA medium and an improved VT medium at 20 °C (Aoki et al. 2008). This study implies that the increase of the refractory organic matter in lake waters may be attributed to a change of the predominant phytoplankton. Similarly, cultivation of three kinds of phytoplankton (*Prorocentrum donghaiense, Heterosigma akashiwo* and *Skeletonema costatum* collected from sea water) can produce visible humic-like (C-like and M-like) and protein-like or tyrosine-like components in waters (Zhao et al. 2006a, 2009).

Releases of DOM by eleven species of intertidal and sub-tidal macroalgae in the dark account for 63.7 % of that in the light in the UV-B band (Hulatt et al. 2009). Some brown algae can produce considerably less DOM (e.g. *Pelvetia canaliculata*), which are more comparable to the green and red species (Hulatt et al. 2009). It is shown that thin, subsurface DOM maxima are observed below the plume during the highly stratified summer period but are absent in the spring, which is the strong evidence of significant in situ biological production of CDOM in natural waters (Chen and Gardner 2004).

Incubation of coastal seawater in the presence of model (DON: amino sugars and amino acids) and DIN compounds shows that net biological DOM formation occurs upon addition of amino sugars (formation of fluorescent, mostly labile DOM) and tryptophan (formation of non-fluorescent, refractory DOM) (Biers et al. 2007). Similarly, natural assemblages of marine bacteria can rapidly produce refractory material (in <48 h) utilizing labile compounds (glucose, glutamate), as observed in a laboratory experiment (Ogawa et al. 2001). On the other hand, photoinduced formation of DOM is only detected when tryptophan is added to the water (Biers et al. 2007). This CDOM is highly fluorescent, with excitation-emission matrices (EEMs) resembling those of terrestrial, humic-like fluorophores (Biers et al. 2007). The bulk particulate organic carbon (POC) during the decomposition process of freshwater or marine algae and phytoplankton is significantly decreased during the first few days. It subsequently remains almost constant (Zhang et al. 2009; Hanamachi et al. 2008; Matsunaga 1981; Fukami et al. 1985; Osinga et al. 1997; Fujii et al. 2002). The carbohydrate contents of both the particulate and dissolved pools are increased during the phytoplankton growth cycle, accounting for 18–45 % and 26–80 % of total organic carbon (TOC), respectively, in natural waters (Biddanda and Benner 1997). Photoreactions driven by UV-B can reduce the microbial availability of certain organic substrates such as peptone and algal exudates (Morris and Hargreaves 1997; Thomas and Lara 1995; Naganuma et al. 1996). This phenomenon can be caused by light-induced cross-linking between DOM and algal exudates (Morris and Hargreaves 1997).

LMW organic acids are presumably formed by four major processes (Lovley et al. 1996; Xiao and Wu 2011; Wetzel et al. 1995; Smith and Oremland 1983; Kieber et al. 1990; Corin et al. 1996; Janczarek et al. 1997; Evans 1998; Bertilsson et al. 1999; Tedetti et al. 2006; Lu et al. 2007; Xiao et al. 2009, 2011): first, photoinduced decomposition of allochthonous and autochthonous DOM in surface waters; second, photoinduced and microbial respiration or assimilation of algae or phytoplankton biomass in natural waters; third, conversion of anaerobic organic substances (carbohydrates, fats, proteins, etc.) into CH_4 and CO_2 in pore waters or soil ecosystems; and fourth, root exudations of plants or plant–microbe associations (e.g. Rhizobium symbiosis with leguminous roots).

A number of factors can influence the DOC release by algae or phytoplankton and bacteria in waters, which can be distinguished as: (i) occurrence of the phytoplankton species and their contents; (ii) water quality; (iii) presence of nutrients; (iv) effect of UV and PAR; (v) water temperature; (vi) occurrence of microbes; (vii) metabolic abilities or inabilities and so on (Norrman et al. 1995; Mostofa KMG et al., unpublished data; Lancelot 1979; Fogg 1983; Nelson et al. 1998, 2004; Rochelle-Newall and Fisher 2002a, b; Cammack et al. 2004; Biers et al. 2007; Ortega-Retuerta et al. 2009; Hulatt et al. 2009; Zhao et al. 2006a, 2009; Williams 1990, 1995; Obernosterer and Herndl 1995; Anderson and Williams 1998; McCallister et al. 2004).

3.2.2 Photosynthesis

Photosynthesis is the key process for the formation of organic carbon or OM (e.g. algae or cyanobacteria, phytoplankton, etc.) through light-stimulated inorganic carbon acquisition in surface waters (Mostofa et al. 2009a; Komissarov 1994, 1995, 2003; Li et al. 2011; Li 1994; Zubkov and Tarran 2008; Beardall et al. 2009a, b; Wu and Gao 2009; Liu et al. 2010). Photosynthetic organisms are then able to produce autochthonous DOM via photoinduced respiration (or photoinduced assimilation) and microbial respiration or assimilation in natural waters (Mostofa et al. 2009b; Zhang et al. 2009; Conrad 1999; Weiss et al. 1991; Harvey et al. 1995; Fu et al. 2010; Thomas and Lara 1995; Druon et al. 2010; Yamashita et al. 2008). A new hypothesis on photosynthesis also considers that H_2O_2 might be involved in the occurrence of oxygenic photosynthesis in both higher plants (Komissarov 1994, 1995, 2003; Miller and Moran 1997) and natural water organisms (Mostofa et al. 2009a, b). Occurrence of photosynthesis in natural waters includes two facts: the first is the generation of numerous chemical species from DOM, which may proceed as follows: (i) photoinduced degradation of DOM can produce many photoproducts, such as H_2O_2, CO_2, DIC, CO, LMW DOM, and so on in upper surface waters (Mostofa and Sakugawa 2009; Miller and Zepp 1995; Miller 1998; Johannessen and Miller 2001; Ma and Green 2004; Xie et al. 2004; Johannessen et al. 2007; Salonen and Vähätalo 1994; Amon and Benner 1996; Granéli et al. 1996; Remington et al. 2011; Zepp et al. 1998; Cai et al. 1999; Gennings et al. 2001; Clark et al. 2004; Fichot and Miller 2010; White et al. 2010; Cai 2011); (ii) microbial degradation of DOM including DON and DOP can produce compounds such as H_2O_2, CO_2, DIC, PO_4^{3-}, NH_4^+, CH_4, LMW DOM and so on (Mostofa and Sakugawa 2009; Zhang et al. 2004; Vähätalo and Järvinen 2007; Lovley et al. 1996; Ma and Green 2004; Palenik and Morel 1988; Li et al. 2011; Zinder 1990; Kotsyurbenko et al. 2001; Zagarese et al. 2001; Semiletov et al. 2007). Many of these compounds can favor the occurrence of photosynthesis either directly or indirectly and lead to fixation of organic carbon or OM from inorganic carbon in surface waters (Mostofa et al. 2009a; Komissarov 1994, 1995, 2003; Miller and Moran 1997; Li et al. 2011; Ortega-Retuerta et al. 2009; Li 1994; Zubkov and Tarran 2008; Beardall et al. 2009a, b; Wu and Gao 2009; Liu et al. 2010).

A general scheme for the photoinduced (Eq. 3.1) and microbial or biological (Eq. 3.2) degradation of DOM can be expressed as follows (Mostofa et al. 2009a, b):

$$\mathrm{DOM} + h\upsilon \rightarrow H_2O_2 + CO_2 + \mathrm{DIC} + \mathrm{CO} + \mathrm{LMW\ DOM} \quad (3.1)$$

$$\mathrm{DOM} + \mathrm{microbes} \rightarrow CO_2 + \mathrm{DIC} + PO_4^{3-} + NH_4^+ + CH_4 + \mathrm{LMW\ DOM} \quad (3.2)$$

The second fact is that H_2O_2 and CO_2, produced by either photoinduced or microbial degradation of DOM and POM can take part to photosynthesis, to form new OM or carbohydrate-type compounds (Mostofa et al. 2009a, b):

$$xCO_{2(H_2O)} + yH_2O_{2(H_2O)} + h\upsilon \rightarrow C_x(H_2O)_y + O_2 + E(\pm) \quad (3.3)$$

$$2H_2O_2 + h\upsilon \rightarrow 2H_2O + O_2 \quad (3.4)$$

where $C_x(H_2O)_y$ represents a generic carbohydrate (Eq. 3.3). According to this hypothesis, H_2O_2 acts together with carbon dioxide (CO_2) to form carbohydrates and oxygen (Eq. 3.3). The formation of oxygen occurs via H_2O_2 disproportionation (Eq. 3.4) that is a common conversion reaction of H_2O_2 in water ecosystems and the atmosphere (see the photosynthesis chapter for detailed description for these reactions) (Liang et al. 2006; Buick 2008). In Eq. (3.3), E (±) is the energy produced during photosynthesis.

Currently, model results imply that the progressive release of DON in the ocean's upper layer during summer increases the primary production by 30–300 %. This will in turn enhance the DOC production mostly from phytoplankton exudation in the upper layer and the solubilization of POM deeper in the water column (Druon et al. 2010). Experimental studies observe that both the quantity and the spectral quality of DOM produced by bacteria can be influenced by the presence of photoproducts in aqueous media (Ortega-Retuerta et al. 2009). Photosynthetically produced POM (algae or phytoplankton) and their photo- and microbial respirations are significantly influenced by several key factors, such as chemical nature and contents of DOM (Jones 1992; Hessen 1985; Tranvik and Hafle 1987; Tranvik 1989); high precipitation (Freeman et al. 2001a; Tranvik and Jasson 2002; Hejzlar et al. 2003; Zhang et al. 2010); land use changes that cause high transport of DOC from catchments to adjacent surface waters (Worrall et al. 2004a; Raymond and Oh 2007); nitrogen deposition (Pregitzer et al. 2004; Findlay 2005); sulfate (SO_4^{2-}) deposition (Zhang et al. 2010; Evans et al. 2006; Monteith et al. 2007); droughts and alteration of hydrologic pathways (Hongve et al. 2004; Worrall and Burt 2008); changes in total solar UV radiation or an increase in temperature due to global warming (Freeman et al. 2001a; Zhang et al. 2010; Sinha et al. 2001; Sobek et al. 2007; Rastogi et al. 2010).

Finally, H_2O_2 can react with CO_2 under abiotic conditions to produce various organic substances (CH_2O, HCOOH, CH_3OH, CH_4, $C_6H_{12}O_6$; Eqs. 3.5–3.9, respectively) in aqueous solution (Lobanov et al. 2004). The reactions between H_2O_2 and CO_2 as well as their thermodynamic parameters such as enthalphy changes (ΔH^0) and the Gibbs free energy changes (ΔG^0) are as follows (Lobanov et al. 2004):

$$H_2O_2 + CO_2 \rightarrow CH_2O + 3/2O_2 \quad (3.5)$$

$$\Delta H^0 = 465\text{kJ}, \ \Delta G^0 = 402\text{kJ}$$

$$H_2O_2 + CO_2 \rightarrow HCOOH + O_2 \quad (3.6)$$

$$\Delta H^0 = 172\text{kJ}, \ \Delta G^0 = 166\text{kJ}$$

$$2H_2O_2 + CO_2 \rightarrow CH_3OH + 5/2O_2 \quad (3.7)$$

$$\Delta H^0 = 530\,kJ\,,\ \Delta G^0 = 464\,kJ$$

$$2H_2O_2 + CO_2 \rightarrow CH_4 + 3O_2 \quad (3.8)$$

$$\Delta H^0 = 694\,kJ\,,\ \Delta G^0 = 580\,kJ$$

$$H_2O_2 + CO_2 \rightarrow 1/6C_6H_{12}O_6 + 3/2O_2 \quad (3.9)$$

$$\Delta H^0 = 426\,kJ$$

3.3 DOM Derived from Anthropogenic and Human Activities

Organic pollutants derived from sewerage and from domestic, agricultural and industrial effluents significantly contribute to increase the concentration levels of DOM in natural waters (Fu et al. 2010; McCalley et al. 1981; Silberhorn et al. 1990; Kramer et al. 1996; Mudge and Bebianno 1997; Manoli and Samara 1999; Abril et al. 2002; Newton et al. 2003; Mostofa et al. 2005b, 2010; Richardson 2003, 2007; Mottaleb et al. 2005, 2009; Mudge and Duce 2005; Richardson and Ternes 2005, 2011; Buser et al. 2006; Field et al. 2006; Lishman et al. 2006; Rudel et al. 2006; Xia et al. 2006; Brown et al. 2007; Schmid et al. 2007; Farré et al. 2008; Kinney et al. 2008; Guo et al. 2009; Ramirez et al. 2009; Citulski and Farahbakhsh 2010; Kumar and Xagoraraki 2010; Pal et al. 2010; Yoon et al. 2010; Kleywegt et al. 2011; Yu et al. 2011). The organic matter pollution is an important problem in both developed and developing countries through input of untreated sewerage and industrial effluents into natural waters. However, its impacts may be much worse in developing countries due to the lack of sewerage treatment and of industrial effluent treatment plants. The occurrence of DOM derived from anthropogenic and human activities is gradually increasing because of the increasing diffusion of domestic, agricultural and industrial activities. Some components of sewerage-impacted DOM are made up of detergents or fluorescent whitening agents (FWAs), including mostly diaminostilbene type (DAS1) and distyryl biphenyl (DSBP), protein-like components, sterols, and unknown organics (McCalley et al. 1981; Mudge and Bebianno 1997; Mostofa et al. 2005b, 2010; Mudge and Duce 2005). The organic components originating from agricultural wastes are pesticides, herbicides, dichlorodiphenyltrichloroethane (DDT) and their degradation products (Richardson 2007; Guo et al. 2009; Derbalah et al. 2003; Medana et al. 2005).

Recent studies show that emerging organic contaminants such as pharmaceuticals and personal care products (PPCPs) are a ubiquitous class of organic chemicals of considerable concern for natural waters, and will be discussed in details later. Wastewater-derived organic compounds can produce three major types of toxic byproducts such as trihalomethanes (THMs), N-nitrosodimethylamine (NDMA) and organic chloramines. These compounds may be formed either upon chlorination or in conventional and advanced wastewater treatment plants (Scully et al. 1988; Jensen and Helz 1998; Jameel and Helz 1999; Mitch et al. 2003).

4 Contribution of Organic Substances to DOM in Natural Water

The contributions of major organic substances in streams and rivers to the total DOM pool are 20–85 % of humic substances, of which 15–80 % fulvic acid and 5–29 % humic acid (the ratio of fulvic acid to humic acid is 9:1 for lower stream DOC and it decreases to 4:1 or less for higher stream DOC), 10–30 % of carbohydrates, 2–48 % of dissolved amino acids, organic acids or hydrophilic acids (9–25 %), autochthonous fulvic acids of phytoplankton or algal origin (or marine humic-like: see Sect. 3.2 and also FDOM chapter for detailed description), organic acids, organic peroxides (ROOHs), sterols; organic contaminants of anthropogenic origin and so on (Mostofa et al. 2009a; Malcolm 1985, 1990; Bertilsson et al. 1999; Lu et al. 2007; Wetzel and Manny 1972; Meybeck 1982; Meyer and Tate 1983; Ittekkot et al. 1985; Thurman 1985b; Meyer 1986; Tipping et al. 1988; Lewis and Saunders 1989; Peuravuori 1992; Hedges et al. 1994; Eatherall 1996; Volk et al. 1997; Kusel and Drake 1999; Peuravuori and Pihlaja 1999; Alberts and Takács 1999; Ma et al. 2001; Raymond and Bauer 2001a; van Hees et al. 2002; Nagai et al. 2005; Mostofa 2005; Guéguen et al. 2006). Hydrophilic acids generally include amino acids, proteins, carbohydrates and free sugars. The contribution of humic substances (hydrophobic acids) in groundwater is approximately 12–98 % (1–80 % of fulvic acid and 2–97 % of humic acid), and the contribution of hydrophilic fractions is 1–82 % (Buckau et al. 2000; Bertilsson et al. 1999; Peuravuori and Pihlaja 1999; Leenheer et al. 1974; Thurman 1985c; Ford and Naiman 1989; Schiff et al. 1990; Wassenaar et al. 1990; Malcolm 1991; Grøn et al. 1996; Christensen et al. 1998; McIntyre et al. 2005; Mladenov et al. 2008). These studies observe high variation in the contribution of humic substances from stream (source) to the end of river mouths. The main reasons are the mixing up of various sources of water in the downstream locations as well as the photoinduced and microbial changes during transportation.

In lakes the contributions of humic substances (fulvic and humic acids) account for 14–90 % of total DOM (14–70 % of fulvic acid and 0–22 % of humic acid); the DOM pool is also made up of ~12–60 % of autochthonous fulvic acids (see FDOM chapter for detailed description) of algal or phytoplankton origin; of carbohydrates for 1–65 %; of amino acids, proteins and organic acids that together account for 10–33 % of total DOM; of organic acids (2.5–7.5 %, but 0–11 % in pore water); sterols; algal toxins, organic contaminants of anthropogenic origin and so on (Mostofa et al. 2009a, b; Parlanti et al. 2000; Xiao and Wu 2011; Wilkinson et al. 1997; McKnight et al. 1991, 1994, 1997; Xiao et al. 2009, 2011; Thurman 1985b; Peuravuori 1992; Peuravuori and Pihlaja 1999; Ma et al. 2001; Nagai et al. 2005; Schiff et al. 1990; Steinberg and Muenster 1985; Hama and Handa 1987; Baron et al. 1991; Søndergaard and Middelboe 1995; Reitner et al. 1997; Malcolm and MacCarthy 1992; Imai et al. 1998; Rosenstock and Simon 2001; Frimmel 2004; Hayakawa 2004; Sugiyama et al. 2005). Biomolecules (e.g. carbohydrates and proteins) as well as organic acids account for approximately 70 % of high molecular weight (HMW) DOM, and only for approximately 2 % of (LMW) DOM in lake water (Hama and Handa 1992). These studies also show that allochthonous fulvic acids in lakes are largely varied during the summer and winter season, with winter maxima and summer minima. Their total content is also low in algal-dominated lakes.

The percentages of major organic substances in bulk DOM in shelf, coastal and open ocean are: 1–75 % of allochthonous fulvic acids of terrestrial origin; 5–10 % of autochthonous fulvic acids (or marine humic-like: see Sect. 3.2 and also FDOM chapter for detailed description) of algal or phytoplankton origin; 10–80 % of carbohydrates (~25 % in deeper layers); 10–28 % of amino acids, proteins and lipids taken together (amino acids alone account for 7 %); organic acids; organic peroxides (ROOH); sterols; algal toxins, and so on (Mostofa et al. 2009a, b; Coble 1996, 2007; Zhang et al. 2009; Bronk 2002; Ogawa and Tanoue 2003; Ogawa et al. 2001; Biddanda and Benner 1997; Harvey and Boran 1985; Meyers-Schulte and Hedges 1986; Druon et al. 2010; Richardson 2007; Thurman 1985b; Alberts and Takács 1999; Ma et al. 2001; Beck et al. 1974; Stuermer and Harvey 1977; Gagosian and Stuermer 1977; Burney et al. 1982; Thurman and Malcolm 1983; Romankevich 1984; Williams and Druffel 1987; Moran et al. 1991; Moran and Hodson 1994; Pakulski and Benner 1994; McCarthy et al. 1996; Opsahl and Benner 1997; Gašparovic et al. 1998; Kirchman et al. 2001; Aluwihare et al. 2002; Benner and Kaiser 2003; Yamashita and Tanoue 2003). The contributions of allochthonous humic substances in shelf seawater are 11–75 %, of which around 38 % of marsh origin and 62 % of river origin (Moran and Hodson 1994). Carbohydrates can comprise 10–70 % of the organic matter in the plankton cell (Romankevich 1984) and are presumably released directly to the water column by algae or phytoplankton under photo- and microbial respiration (Mostofa et al. 2009b; Zhang et al. 2009; Hellebust 1965; Ittekkot et al. 1981; Mopper et al. 1995; Cowie and Hedges 1994, 1996). Carbohydrates (originally polysaccharides) make up approximately 15–60 % of marine HMW DOM (Druon et al. 2010; Burney et al. 1982; Romankevich 1984; Pakulski and Benner 1994; McCarthy et al. 1996). Carbohydrates also account for ~5–20 % of particulate material in seawater (Pakulski and Benner 1994; Tanoue and Handa 1987; Hernes et al. 1996; Panagiotopoulos et al. 2002). Autochthonously produced carbohydrates, proteins and lipids are vital biochemical organic groups that together constitute approximately 10–80 % of organic carbon and 15–50 % of the nitrogen assimilated during photosynthesis by phytoplankton in natural waters (Sundh 1992; Bronk et al. 1994; Braven et al. 1995; Malinsky-Rushansky and Legrand 1996; Wakeham et al. 1997; Slawyk et al. 1998).

The main organic substances in rainwater are hydrophobic DOM (major fraction; ~<50 %), including allochthonous humic substances (fulvic and humic acids) or marine humic-like substances, hydrophilic DOM (major fraction; ~>50 %), including organic acids (~14–40 %) such as acetic and formic acid, dicarboxylic acids (~<6 %, including oxalic, succinic, malonic and maleic acids), pyruvic acid (~<1 %), amino acids (~2 %) including tryptophan-like and tyrosine-like components, formaldehyde (~2–8 %), acetaldehyde (~5 %), organic peroxides (ROOHs: see chapter "Photoinduced and Microbial Generation of Hydrogen Peroxide and Organic Peroxides in Natural Waters" for detailed description) (McDowell and Likens 1988; Hellpointner and Gäb 1989; Hewitt and Kok 1991; Guggenberger and Zech 1993; Sakugawa et al. 1993; Sempéré and Kawamura 1994; Chebbi and Carlier 1996; Williams et al. 1997; Willey et al. 2000, 2006; Ciglasch et al. 2004; Avery et al. 2006; Kieber et al. 2006; Muller et al. 2008; Miller et al. 2008, 2009; Santos et al. 2009a, b; Southwell et al. 2010; Zhang et al. 2011; Nichols and Espey 1991; Brassell et al. 1980; Sargent et al. 1981). These studies also show that rainwater mostly consists of low molecular weight organic substances, having MW < 1000 Dalton. Note that factors such as wind speed, storm trajectory and rainwater volume can influence DOM contents in rainwater. The relative importance of these factors depends on the sources of the rainwater constituents (Miller et al. 2008).

The contribution of allochthonous fulvic and humic acids is significantly high in source waters (streams and rivers), then their contributions decrease during the flow into the downward water ecosystem (lakes, estuaries and oceans) because of three major processes: first, photoinduced and microbial degradation; second, dilution of the source waters with other water bodies; third, high contents of autochthonous DOM can decrease the relative contribution of allochthonous fulvic and humic acids in stagnant waters, particularly in lakes, estuaries and oceans.

On the other hand, the contribution of autochthonous DOM including autochthonous fulvic acids of algal or phytoplankton origin, carbohydrates, proteins, amino acids, lipids, organic acids etc. is relatively low in source waters, but significantly high in lakes and oceans. Autochthonous production of DOM is typically detected in the epilimnion of lake and ocean during the stratification period. A rough estimate shows that the contribution of autochthonous DOM is 0–102 % in lakes and 0–194 % in oceans, which has been discussed in earlier section (Mostofa et al. 2009a; Wigington et al. 1996; Fu et al. 2010; Ogawa and Tanoue 2003; Ogawa and Ogura 1992; Mitra et al. 2000; Yoshioka et al. 2002a; Hayakawa et al. 2003, 2004; Annual Report 2004; Bade 2004; Sugiyama et al. 2004).

The sterol biomarkers used for identifying DOM sources in water are terrestrial (b-sitosterol and ergosterol), sewage (5b-coprostanol and epi-coprostanol), phytoplankton (cholest-5,22-dien-ol, brassicasterol, dinosterol), and marine markers (cholesterol) (McCalley et al. 1981; Mudge and Bebianno 1997; Mudge and Duce 2005; Nichols and Espey 1991). Long-chain C22-C30 alkanols are generally considered to originate from terrestrial plants, while short-chain alkanols have unspecified marine, terrestrial and bacterial origins (Brassell et al. 1980; Sargent et al. 1981). From the above contributions to the DOM composition in various sources of waters, it is evidenced that, on average, approximately 80–90 % of bulk DOM in streams, rivers, lakes and oceans is specifically identified as allochthonous fulvic and humic acids, autochthonous fulvic acids, carbohydrates, proteins, lipids, amino acids, fatty acids, sterols, and organic acids.

4.1 Physical and Chemical Properties of DOM

Naturally-originated organic compounds such as humic substances (fulvic and humic acids) of terrestrial plant origin, autochthonous DOM of algal or phytoplankton origin, proteins, amino acids, peptides and polysaccharides exhibit, to varying degrees, several major properties (Mostofa et al. 2009a, b; Malcolm 1985; Xue et al. 1995; Mandal et al. 1999; Filella 2008). They are: (i) physically heterogeneous; (ii) polyfunctional, due to the existence of a variety of functional groups and the presence of a broad range of functional reactivity; (iii) polyelectrolytical, with high electric charge density due to the presence of a large number of dissociated functional groups; (iv) structurally labile, because of their capacity to associate intermolecularly and to change molecular conformation in response to changes in pH, pE, ionic strength, trace metal binding, and so on; (v) polydisperse in size.

Water Color:

The yellow color in natural waters is due to the occurrence of humic substances (fulvic and humic acid) and of autochthonous fulvic acids (C- and M-like) of algal or phytoplankton origin, which absorb light in the blue and ultraviolet (Kalle 1966; Jerlov 1968). These substances were formerly referred to collectively as yellow substances or gelbstoff (Kirk 1976; Kalle 1966). Water color is generally related to the occurrence and contents of these substances in natural waters (Eloranta 1978; Jones and Arvola 1984). Ocean color is an important feature of water that was recently determined using remote sensing applications (Hopkinson et al. 2002; Morel et al. 2007; Morel and Gentili 2009; Van der Woerd et al. 2011; Volpe et al. 2011; Son et al. 2011). It is mostly due to the effect of autochthonous fulvic acids of algal or phytoplankton origin as well as partly to allochthonous fulvic and humic acids (humic substances). A recent study has shown that autochthonous fulvic acids (C-like and M-like) of lake algal origin under dark incubation can exhibit yellow color (Mostofa et al. 2009b). Note that autochthonous fulvic acids (C-like and M-like) are characterized based on their similar fluorescence properties to allochthonous fulvic acids (C-like and M-like), which will be discussed in detail in the FDOM chapter (see chapter "Fluorescent Dissolved Organic Matter in Natural Waters").

Attenuation of Spectral UV Irradiance

DOM is the key factor that controls the downward irradiance flux through the water column of UV-B (280–320 nm), UV-A (320–400 nm), total UV (280–400 nm) and photosynthetically available radiation (PAR, 400–700 nm) (Kirk 1976; Morris et al. 1995; Siegel and Michaels 1996; Morris and Hargreaves 1997; Tranvik 1998; Bertilsson and Tranvik 2000; Laurion et al. 2000; Markager and Vincent 2000; Huovinen et al. 2003; Sommaruga and Augustin 2006; Hayakawa and Sugiyama 2008; Effler et al. 2010). These studies show that UV-B penetration depths vary from only a few centimeters in highly humic lakes to dozens of meters in the oceans, due to variation in DOM contents. It is also observed that 99 % of the UV-B radiation is attenuated in an approximately 0.5-m water column in the clearest lake for DOC ranging from 408 to 725 μM C and for chlorophyll *a* ranging from 1.6 to 16 μg L^{-1} (Huovinen et al. 2003). In the UV-A region at 380 nm, the corresponding attenuation is limited to the upper one meter.

The absorption coefficients predict that, in a small humic lake (DOC 1100–1242 μM C), UV-B radiation is attenuated to 1 % of the subsurface irradiance within the top 10 cm water column, whereas UV-A radiation (at 380 nm) penetrates more than twice as deep (maximum 25 cm) (Huovinen et al. 2003). However, in clear lakes with low DOC concentration the contribution of phytoplankton to UV attenuation can be significant (Sommaruga and Psenner 1997). Any enhancement of photoinduced degradation of DOC by UV radiation and acidification can substantially increase the UV transparency in lakes (Morris and Hargreaves 1997; Vione et al. 2009; Schindler et al. 1996; Yan et al. 1996; Scully et al. 1997). The consequence is an enhanced penetration of UV radiation into the water column, which can significantly damage aquatic biota. DOM is thus responsible for UV attenuation in the water column and for the related protection of aquatic organisms in natural waters.

Aggregation of DOM

Aggregation of fulvic and humic acid (humic substances) can occur at the intramolecular (involving a single polymer molecule) or intermolecular (involving multiple chains) levels in aqueous solution (Wershaw 1999; Engebretson and von Wandruszka 1996; Lippold et al. 2008). The interior of the resulting aggregates is relatively hydrophobic, whilst the exterior is more hydrophilic. They can exist in a pseudomicellar form or as micelle-like aggregates in solution, and as membrane-like aggregates on mineral surfaces (Wershaw 1999; Sutton and Sposito 2005; Piccolo et al. 2001). The results of the chemical analysis of humic acids isolated from natural environments (water, soil, peat, sediments, and sludge from wastewater treatment facilities) demonstrate that the percentage elemental composition, the contents of carboxylic groups and of aromatic phenolic groups is very variable. They range from 33.2 (river) to 60.7 % (Aldrich) of C; 2.25 (river) to 5.4 % (soil) of H; 0.65 (river) to 3.7 % (peat) of N; 34.1 (Aldrich) to 63.8 (river) of O; 0.06 (soil) to 0.10 % (sewage sludge) of S; 1.0 (river) to 8.1 mmol g^{-1} (peat) of carboxylic groups (-COOH), and from 0.36 (bog peat) to 4.4 mmol g^{-1} of phenolic moieties (ArOH) (Klavins and Purmalis 2010).

Humic acids behave like surface-active substances when they are added to solutions, which depend on their origin and molecular properties. Therefore, their surface tension decreases as their concentration increases (Lippold et al. 2008; Klavins and Purmalis 2010; Wershaw 1993; Engebretson and von Wandruszka 1994; Terashima et al. 2004). Humic acids can be significantly modified in their functional groups such as the benzene ring in phenolic structures with the addition of hydrophilic sulfonic, hydroxyl or trimethylammonium functional groups (Klavins and Purmalis 2010). This effect can be used for the development of biopolymers with surfactant properties (Klavins and Purmalis 2010; Heinze and Liebert 2001). Humic substances might influence plankton food chains in lakes in two ways (Jones 1992): (i) By altering the physical or chemical environment and thus modifying autotrophic primary production and the dependent food chains; and (ii) By acting as a direct carbon/energy source for food chains.

4.1.1 Redox Behavior of Fulvic and Humic Acids

Fulvic and humic acids (humic substances) can act as reductants and oxidants in aqueous media (Lovley et al. 1996; Richardson 2007; Wilson and Weber 1979; Nash et al. 1981; Skogerboe and Wilson 1981; Österberg and Shirshova 1997; Scott et al. 1998; André and Choppin 2000; Steelink 2002; Kuczewski et al. 2003; Shcherbina et al. 2007). They are capable of reducing Fe^{3+}, Sn^{4+}, V^{5+} and Cr^{4+}. The +IV oxidation states of the redox-sensitive actinides (e.g. Pa, Np, U, Pu) are stabilized by complexation with fulvic and humic acids. Fulvic and humic acids are thus capable of detoxifying surface water and soils contaminated with toxic organic and inorganic chemicals. Some examples are (i) reduction of metals from toxic valence states to non-toxic states, such as Cr^{4+} to Cr^{3+}, V^{5+} to V^{4+}, or $UO_2{}^{2+}$ and UO_2OH^+ to U^{4+} (Steelink 2002; Wittbrodt and Palmer 1995; Markich 2002; Freyer et al. 2009); (ii) reductive cleavage of halogenated hydrocarbons such as trichloroethylene, a common pollutant in soil and groundwater, which can be degraded to ethylene and hydrochloric acid (Steelink 2002); (iii) abiotic reduction of mercury in the presence of a competing ion as well as methylation of the carboxylic groups of humic and fulvic acids, which can consume methylmercury (Allard and Arsenie 1991); and (iv) reduction of organic nitro groups to amines. For instance, trinitrotoluene (TNT) is reduced to compounds such as aminodinitrotoluene that can form complexes with fulvic and humic acids (Steelink 2002). Note that TNT is an explosive that can migrate to and pollute groundwater.

On the other hand, it has also been observed that the functional groups in fulvic and humic acids can be oxidized, as is the case of catechol moieties (oxidized to quinones), aldehydes (to carboxylic acids), alcohols (to aldehydes or carboxylic acids) and so on (Steelink 2002). These redox processes account for the presence of intermediates such as semiquinones in fulvic and humic acids. A typical redox process involving fulvic acids (FA) and humic acids (HA) can be depicted as below (Wilson and Weber 1979; Skogerboe and Wilson 1981):

$$FA_{(ox)} + e^- + H^+ = FA_{(red)},\ E^\circ = 500\ \text{mV (FA) and } 700 - 794\ \text{mV (HA)} \tag{2.1}$$

For instance, SRHA has standard reduction potential $E^\circ = 760 \pm 6$ at pH 5–7. The E° values are variable depending on the pH (Wilson and Weber 1979;

Skogerboe and Wilson 1981; Matthiesen 1994; Struyk and Sposito 2001). Some studies also suggest that functional groups such as quinone or quinone-like moieties in fulvic and humic acids are largely responsible for the observed reversible redox behavior in natural waters (Scott et al. 1998; Tratnyek and Macalady 1989; Schwarzenbach et al. 1990; Nurmi and GTratnyek 2002; Cory and McKnight 2005; Macalady and Walton-Day 2009). In addition, fulvic and humic acids can donate electrons photolytically in aqueous media, which can induce the production of oxidizing agents such as superoxide ion ($O_2^{\bullet -}$) and hydrogen peroxide (H_2O_2) (see detailed description in chapter "Photoinduced and Microbial Generation of Hydrogen Peroxide and Organic Peroxides in Natural Waters") (Mostofa and Sakugawa 2009; Fujiwara et al. 1993; Baxter and Carey 1983).

The presence of diverse functional groups in the molecular structure of fulvic and humic acids is responsible for their redox behavior in waters. The redox behavior of humic acids depends on the redox potential of the aqueous solutions as well as on the complexation capacity with multicharged cations in water (Österberg and Shirshova 1997; Struyk and Sposito 2001; Kerndorff and Schnitzer 1980; Zauzig et al. 1993).

4.1.2 Definition and Chemical Nature of Allochthonous Fulvic and Humic Acids

Allochthonous DOM of vascular plant origin is primarily composed of humic substances (fulvic and humic acids), which are also termed as hydrophobic acids. Stream fulvic and humic acid are therefore vital to understand the nature of the allochthonous DOM, because the chemical composition and optical properties of these substances are greatly altered photolytically and microbially during their transportation after leaching from soil into rivers, lakes or oceans.

Allochthonous Fulvic Acids

Allochthonous fulvic acids can be defined as molecularly heterogeneous and supramolecular, with molecular weight ranging from less than 100 to over 300,000 Daltons and with the largest fractions ranging less than 50,000. They are optically active, typically refractory to microbial degradation, photolytically reactive, biogenic, and yellow-colored. They are also soluble under all pH conditions in water (Ma and Ali 2009; Mostofa et al. 2005b, 2007a; MacFarlane 1978; Dai et al. 1996; McKnight et al. 1988, 2001; Hayase and Tsubota 1983; Frimmel 2004; Aiken et al. 1985; Aiken and Malcolm 1987; Aiken and Gillam 1989; Amador et al. 1989; David and Vance 1991; Allard et al. 1994; Hummel 1997; Fimmen et al. 2007). Allochthonous fulvic acids in surface waters have relatively low contents of organic N compared to organic C, i.e. a high C:N ratio. This ratio is in the range ~45–202, and standard SRFA (1S101F and 2S101F) have values of 73–78. Allochthonous fulvic acids also have relatively high contents of O and organic P, low contents of S, relatively low aromaticity (17–30 % of total C) and high aliphatic C (63 %) (Malcolm 1985; Wetzel 1983; McKnight et al. 2001;

Meyers-Schulte and Hedges 1986; Ma et al. 2001; McIntyre et al. 2005; Frimmel 2004; Aiken and Malcolm 1987; Abbt-Braun and Frimmel 1990; Abbt-Braun et al. 1991; IHSS 2011; Senesi 1990).

Allochthonous fulvic acids are supramolecular structures composed of a variety of functional groups or components such as benzene-containing carboxyl groups, ketones, methoxylate and phenolic groups (catechol-type), carboxylic and di-carboxylic groups, ethers, esters, amides, aliphatic OH, carbohydrate OH, –C = C–, hydroxycoumarin-like structures, chromone, xanthone, quinones, flavones, O, N, S, and P-atom-containing functional groups attached to aromatic and aliphatic C, indole groups, degraded lignins, and so on (Malcolm 1985; Dai et al. 1996; Frimmel 2004; Allard et al. 1994; McKnight et al. 1988; Leenheer et al. 1995, 1998, 2001; Brown and Rice 2000; Haiber et al. 2001; Kujawinski et al. 2002; Lambert and Lankes 2002; Cook et al. 2003; Stenson et al. 2003; Leenheer and Croué 2003; Leenheer 2007; Killops and Killops 1993). Lignins are complex, high-mass, primarily ether-linked phenylpropanoid biopolymers including only C, H, and O atoms in their molecular structure. They are mostly found in wood cells, whereas the main building blocks for the phenyl portion of lignins are coumaryl, coniferyl, and sinapyl alcohols that vary from plant to plant (Helm 2000; Filley et al. 2002; Lewis and Yamamoto 1990; Christman and Oglesby 1971). The lignin biopolymer is degraded by fungi and eventually bacteria through different pathways that include depolymerization, demethylation, side-chain oxidation, and aromatic ring cleavage (Lewis and Yamamoto 1990; Nelson et al. 1987; Grushnikov and Antropova 1975; Higuchi 1993; Radnoti de Lipthay et al. 1999; Leonowicz et al. 2001; Lowe and Bustin 1989).

In humic substances, 60–90 % of the acid groups are carboxylic and the remainder are phenolic (Leenheer et al. 1995). S-XANES have shown that sulphur is present in humic substances in many different oxidation states: organic sulfides (R–S–R), thiol (–SH), di– and polysulfides (R–S–S–R), sulfoxide (R–SO–R), sulfone S compounds ($R\text{-}SO_2\text{-}R$), sulfonate ($HSO_3\text{-}R$), and sulfate esters ($HSO_4\text{-}R$) (Frimmel 2004; McKnight et al. 1988; Morra et al. 1997; Xia et al. 1998, 1999; Schnitzer and Khan 1978).

Depending on the major elemental composition of C, H, O and N disregarding S, an average empirical formula for fulvic acid has been considered as $C_{12}H_{12}O_9N$ (Steelink 2002; Leenheer et al. 1998; Paciolla et al. 1998; Schnitzer 1985). Based on accurate mass measurements, molecular formulas have been assigned to 4626 individual Suwannee River fulvic acids with molecular masses between 316 and 1098 Da, which led to plausible structures consistent with degraded lignin (Leenheer and Croué 2003).

Hummel (Fimmen et al. 2007) has shown that a fulvic molecule (i) contains on average 5.5 mmoles of carboxyl groups per gram, which corresponds to one carboxylic group per six carbon atoms, or one group per aromatic ring if distributed evenly; (ii) has an average phenolic group content of 1.2 mol per gram, which means one phenolic group per 30 carbon atoms, or only two phenolic groups per fulvic molecule; and (iii) has hydroxyl and carbonyl groups that, put together, are as abundant as carboxyl groups (5–7 mmol g^{-1}). Therefore, an average fulvic acid molecule (molecular weight 2,000 g mol^{-1}) would have one carboxylic, hydroxyl or carbonyl group every three carbon atoms. Amino acids, amino sugars, ammonium (NH_4^+) and nucleic acid bases make up 45–59 % of fulvic acid-N (Smith and Epstein 1971).

The stable carbon isotope ($\delta^{13}C = {}^{13}C/{}^{12}C$) fractionation of standard SRFA is –27.6 ‰, while other isolated allochthonous fulvic acids in rivers have [–(25.6–26.4 ‰)] and in lakes have [−(23.02–33.13) ‰]. These data indicate that SRFA are most likely derived from higher plant matter (Thurman 1985a; McIntyre et al. 2005; Senesi 1990; Simpson et al. 2002; Caraco et al. 1998). Note that standard FAs of Elliot Soil I have $\delta^{13}C = -25.4$ ‰, Elliot Soil II have $\delta^{13}C = -25.6$ ‰, Pahikee peat I have $\delta^{13}C = -25.8$ ‰. Reference FA of Suwannee River have $\delta^{13}C = -27.9$ ‰, Pahikee peat I have $\delta^{13}C = -26.1$ ‰, Nordic Lake have $\delta^{13}C = -27.8$ ‰ (Senesi 1990).

Terrestrial DOM from groundwater, streams, rivers, lakes and sea water (0 salinity) is confined to a narrow range of $\delta^{13}C$ (from –25.3 ‰ to –28.6 ‰), with 80 % of the values falling within 0.5 ‰ of –27.0 ‰ (Schiff et al. 1997; McCallister et al. 2004; McIntyre et al. 2005; Elder et al. 2000; Nagao et al. 2011; Fry and Sherr 1984). Note that the $\delta^{13}C$ is largely different for fresh deciduous leaves (–30.4 ‰), it increases in the top soil (–28.9 ‰) and then from –27.8 to –26.4 ‰ in soil. Plant leaves with C3 photosynthesis have $\delta^{13}C$ = –(25.9–29.2 ‰) and soil profiles have $\delta^{13}C$ = –(23.8–25.9 ‰). $\delta^{13}C$ has lower values in litter-rich soil DOC [−(26.6–27.7 ‰)] than in litter-lacking soil DOC [approximately −(23–27 ‰)] or terrigenous soil with surface/forest litter [−(23–27 ‰)], terrestrial leaf OM (−27 ‰), terrigenous vascular plant [−(26–30 ‰)], yellow soil profile [−(21.1–24.8 ‰)] or limestone soil profile [−(23.0–24.1 ‰)] (Tu et al. 2011; McCallister et al. 2004; Elder et al. 2000; Trumbore et al. 1992; Deegan and Garritt 1997; Stevenson 1982; Richter et al. 1999; Raymond and Bauer 2001b; Cloern et al. 2002; Zhu and Liu 2006; Stenson et al. 2002). Therefore, the origin of allochthonous DOM is significantly dependent on the types and nature of terrestrial vegetation in soil environments.

The combination of flow path analysis and ^{14}C content of DOC suggests that the DOC in upland streams is composed of two pools (Schiff et al. 1997). First, the DOC pool is carried to the stream by discharging groundwater. This DOC has been extensively recycled in the soil zone, has low ^{14}C content and probably has a low proportion of labile functional groups. Although groundwater contributions to stream flow are high even during storm events, groundwater DOC concentrations are low. The relative contribution of this older recalcitrant pool is limited by the amount of soluble carbon which elutes through the overlying soil column. The second pool is composed of recently fixed and potentially more microbially labile DOC leached from the A horizon or litter layer. The potential contribution of this second pool is very high especially after leaffall.

Allochthonous Humic Acids

Allochthonous humic acids in surface waters can be defined as molecularly heterogeneous and supramolecular, with molecular weight ranging from less than 500 to over 300,000 Daltons. The largest fraction is found in the range larger than 300,000 Daltons. They are optically active, typically refractory to microbial degradation, photolytically reactive, biogenic, and yellow-colored organic acids. They are insoluble and form precipitates at $pH < 2$ (MacFarlane 1978; Hayase and Tsubota 1983; Sutton and Sposito 2005; Steelink 2002; Aiken and Malcolm 1987; Aiken and Gillam 1989; Schulten and Schnitzer 1998). Allochthonous humic acids of various origin (soil, bog peat, sewerage sludge) have relatively high contents of organic N to organic C, i.e. they have relatively low C:N atomic ratio (8–51). Standard SRHA (1S1011H and 2S101H) have C:N = 44–45. Allochthonous humic acids also have relatively low contents of O and organic P, high contents of S, relatively high aromaticity (30–40 % of total C) and relatively low contents of aliphatic C (~30–47 %) compared to fulvic acids (Malcolm 1985; Wetzel 1983; McKnight et al. 2001; Meyers-Schulte and Hedges 1986; Ma et al. 2001; McIntyre et al. 2005; Frimmel 2004; Aiken and Malcolm 1987; Abbt-Braun and Frimmel 1990; Abbt-Braun et al. 1991; IHSS 2011; Senesi 1990). It has been shown that the contents of aromatic and other functional groups are very variable depending on the different sources of humic acids and their photobiogeochemical changes in natural waters. The aromaticity of humic acids is very low (~15 %) in marine waters (Malcolm 1990).

Allochthonous humic acids have a supramolecular structure composed of a variety of functional groups (or fluorophores), such as aromatic carboxylic and dicarboxylic acids, aromatic OH groups including phenols (or catechols) and phenolic acids, aliphatic or carbohydrate OH, aldehyde or aliphatic ketones, amide/amino groups, peptides, esters (COOR) or benzene-containing methoxylates, polymethylenes ($-CH_2-$), hydroxycoumarin-like structures, chromone, xanthone, quinone, O, N, S, and P-atom-containing functional groups attached to aromatic and aliphatic carbon, methylated forms of para-coumaric, ferrulic, vanillic and syringic acids, pyrrole, indole, imidazole and pyridine groups (Malcolm 1985; Sutton and Sposito 2005; Steelink 2002; Lambert and Lankes 2002; Leenheer and Croué 2003; Stevenson 1982; Schulten and Schnitzer 1998; Laane 1984; Mao et al. 1998; Hu et al. 2000; Mahieu et al. 2000, 2002; Zang et al. 2000; Kujawinski et al. 2009; Piccolo 2002; Vairavamurthy and Wang 2002; Abe and Watanabe 2004; Schmidt-Rohr et al. 2004; Guignard et al. 2005; Fiorentino et al. 2006). A typical humic acid containing 0.2 % reduced sulphur has only 63 $\mu mol\ g^{-1}$ of thiol sites (Bloom et al. 2001). Amino acids, amino sugars, ammonium (NH_4^+) and nucleic acid bases make up 46–53 % of the N associated with humic acids (Schnitzer 1985). Depending on the elemental compositions of C, H, O, and N, an empirical formula for humic acids has been proposed as $C_{10}H_{12}O_5N$ and a representative molecular formula as $C_{72}H_{72}O_{30}N_4{\cdot}8H_2O$ (Steelink 2002; Schnitzer and Khan 1978; Paciolla et al. 1998).

The stable carbon isotope ($\delta^{13}C$) fractionation of standard SRHA is –27.7 ‰, which indicates that they are most likely derived from higher plant matter (IHSS

2011). Note that Standard HAs of Elliot Soil have $\delta^{13}C = -22.6$ ‰; Pahikee peat have $\delta^{13}C = -26.0$ ‰, and Leonardite have $\delta^{13}C = -23.8$ ‰. Reference HAs of Suwannee River have $\delta^{13}C = -28.2$ ‰, Pahikee peat have $\delta^{13}C = -26.3$ ‰, Nordic Lake have $\delta^{13}C = -27.8$ ‰, and Summit Hill soil have $\delta^{13}C = -26.3$ ‰ (IHSS 2011). In addition, carbon isotope composition of dissolved humic and fulvic acids shows that the $\Delta^{14}C$ values are ranged from −247 to +26 ‰ whilst the average values are -170 ± 79 ‰ for humic acid and -44 ± 73 ‰ for fulvic acid (Nagao et al. 2011). This suggests that the residence time of fulvic acid in the watershed is being shorter than that of humic acid (Nagao et al. 2011).

4.1.3 Definition of Autochthonous Fulvic Acids and Chemical Nature of Autochthonous DOM

The key autochthonously produced biochemical organic groups or substances (Mostofa et al. 2009a) identified in natural waters can be classified as: autochthonous fulvic acids (C-like and M-like) of algal (cyanobacteriam) or phytoplankton origin; carbohydrates such as uranic acids, amino sugars and neutral sugars including free mono-, oligo-. lipopoly-, exopoly-, homopoly-, and heteropolysaccharides; nitrogen-containing organic compounds including amino acids, proteins, amines, amides, urea, purines, pyrimidines, peptides, polypeptides, pyrrole, and indole; lipids, including saturated, monounsaturated, polyunsaturated, branched-chain and odd-chain fatty acids (mostly composed of oleic acid, arachidonic acid, eicosapentanoic acid, linoleic acid, docosahexaenoic acid, cis-vaccenic acid, iso- and anteiso-C_{15} and C_{17} fatty acids, polyunsaturated C_{22} and C_{20} fatty acids, high molecular-weight, straight-chain (C_{24}, C_{26}, C_{28},C_{30}) fatty acids; organic acids including mono-, di- and tri-carboxylic acids, glycollate, and hydroxamate; allelopathic compounds. There are also steroidal alcohols (sterols) such as 24-methyl-cholesta-5,24(28)-dien-3*β*-ol, 24-ethylcholest-5-en-3*β*-ol, cholesta-5,22*E*-dien-3*β*-ol, cholest-5-en-3*β*-ol, cholesta-5,22-dien-3*β*-ol, 27-Nor-24-methylcholesta-5,22-dien-3*β*-ol, 4α,23,24-trimethyl-5α-cholest-22*E*-en-3*β*-ol (dinosterol), 24-methylcholesta-5,22-dien-3*β*-ol, 24-ethylcholesta-5,22*E*-dien 3*β*-ol, 24-ethylcholesta-5-en-3*β*-ol, 24-ethylcholesta-5,24(28)*E*-dien-3*β*-ol, 24-n-propylcholesta-5,24(28)*E*-dien-3*β*-ol, 3-methyllidene-7,11,15-trimethylhexadecan-1,2-diol (phytyldiol); vanillyl and syringyl phenols including vanillin, acetovanillone, vanillic acid, syringaldehyde, acetosyringone and syringic acid from lignin-derived oxidation products; bisnorhopane and various alkenones such as four polyunsaturated C_{37} and C_{38} methyl- and ethyl- alkenones, 6,10,14-trimethylpentadecan-2-one; pigments including melanin, mycosporine-like amino acids (shinorine, palythine, porphyra-334, palythene and usujirene); carotenoids (diadinoxanthin, zeaxanthin, myxoxanthophyll, and echinenone); algal toxins (mostly cyanobacterial toxins produced from blue–green algae) including microccystins, nodularins, anatoxins, cylindrospermopsin, and saxitoxins; red tide toxins including brevetoxins (Parlanti et al. 2000; Mostofa et al. 2009b; Zhang et al. 2009; Xiao and Wu 2011; Coble 2007; Norrman et al. 1995; Hanamachi et al. 2008;

Richardson 2007; Singh and Singa 2002; Miller et al. 2002; Hama et al. 2004; McCallister et al. 2006; Prince et al. 2008). Most of these autochthonous substances have been extensively discussed in earlier studies (Mostofa et al. 2009a).

"Autochthonous fulvic acids" of algal or phytoplankton origin are molecularly heterogeneous, with molecular weight ranging from less than 100 to over 1,898 Daltons. They are optically active, biogenic, highly photoreactive, microbially refractory and yellow-colored organic acids (Mostofa et al. 2009b, Mostofa KMG et al., unpublished data; Zhang et al. 2009; Johannessen et al. 2007; Amon and Benner 1994; McKnight et al. 1991, 1994; Ogawa et al. 2001; Aoki et al. 2004, 2008; Nagai et al. 2005; Williams and Druffel 1987; Fimmen et al. 2007; Barber 1968; Ogura 1972). Autochthonous fulvic acids or DOM in freshwater and seawater have relatively high contents of dissolved organic N compared to organic C, i.e. low C:N atomic ratios (ca. 8–36, but lower in surface waters and higher in deeper waters). They are rich in S, highly aliphatic in nature and have low contents of aromatic carbon (ca. 5–21 % of total carbon) (Wetzel 1983; McKnight et al. 1991, 1994, 1997, 2001; Ogawa et al. 1999, 2001; Meyers-Schulte and Hedges 1986; Aluwihare et al. 2002; Fimmen et al. 2007; McCallister et al. 2006; Nissenbaum and Kaplan 1972; Carder et al. 1989; Karl et al. 1991; Midorikawa and Tanoue 1996, 1998; McCarthy et al. 1997; Engel and Passow 2001; Carlson et al. 2000; Church et al. 2002). Autochthonous fulvic acids have higher nitrogen content (C:N = 8–36) than allochthonous standard fulvic and humic acids (C:N = 44–78). This may indicate that autochthonous fulvic acids are less refractory than allochthonous fulvic and humic acids, probably because autochthonous DOM has fewer aromatic compounds and relatively more proteins and lipids, which decreases its carbon to nitrogen ratio compared to allochthonous DOM (McCallister et al. 2006). Cyanobacteria may contain significant quantities of lipids (fats and oil) which are esters of fatty acids and alcohols that comprise a large group of structurally distinct organic compounds including fats, waxes, phospholipids, glycolipids etc. (Singh and Singa 2002). The lipids of some cyanobacterial species are also rich in essential fatty acids such as the C_{18} linoleic (18:2ω6) and y-linolenic (18:3ω3) acids and their C_{20} derivatives, eicosapentaenoic acids (20:5ω3) and arachidonic acid (20:4ω6) (Singh and Singa 2002). These fatty acids are essential components of the diet of humans and animals and are becoming important feed additives in aquaculture (Borowitzka 1988).

Spectroscopic studies of isolated autochthonous fulvic acids show that they are composed of methylated isomers of hydroxy-benzenes and hydroxy-benzoic acids, aliphatic acids, carbohydrate OH, protein amide and amine groups; they also contain Schiff-base derivatives (–N = C–C = C–N–), fatty acid methyl esters (heptanedioic acid, octanedioic acid, nonanedioic acid, methyl tetradecanoate, 12-methyl-tetradecanoic acid, 7-hexadecenoic acid, and hexadecanoic acid), N- and S-containing amino and sulfidic functional groups. The latter include 3-(methylthio)-propanoic acid; dimethyl sulfone; N,N-dimethyl-2-butanamine, N-methyl proline; N-methyl aniline; 3-piperidinemethanol; 1-methyl-2,5-pyrrolidinedione; 1-methyl-2-piperidinone; caprolactam; 3-ethyl-1,3-dimethyl-2,5-pyrrolidinedione; 2-amino-5,6-dihydro-4,4,6-trimethyl-4 H-1,3-oxazine; 3-ethyl-2,6-piperidinedione; 1,3,5-trimethyl-1,3,5-triazine-2,4,6-trione; 1,3-dimethyl-2,4-pyrimidinedione; 2-methyl-isoindole-1,3-dione; 5-methoxy-2-methyl-indole; 1,3,5-trimethyl-2,4-pyrimidinedione; and 3,3-dimethyl-4-[(2-methoxycarbonyl)ethyl]-2,5-dione-pyrrolidine (McKnight et al. 1997; Fimmen et al. 2007; Laane 1984; Borowitzka 1988; Wershaw 1992; Xue and Sigg 1993; Xue et al. 1995). The aromatic compounds present in autochthonous DOM originate from intracellular quinones in the chloroplasts and mitochondria of algae and bacteria (McKnight et al. 1997; McKnight and Aiken 1998; Klapper et al. 2002).

Algal toxins such as as microcystins and nodularins have high molecular weight and cyclic peptide structures and are hepatotoxic; anatoxins, cylindrospermopsin and saxitoxins have heterocyclic alkaloid structures. Anatoxins and saxitoxins are neurotoxic, while cylindrospermopsin is hepatotoxic (Richardson 2007). On the other hand, red tide toxins such as brevetoxins have heterocyclic polyether structures and are neurotoxic. Note that bacteria, algae and their exudates also consist of a mosaic of functional groups such as amino, phosphoryl, sulfhydryl and carboxylic groups. The net charge on the cell wall depends on the pH of the medium (Filella 2008). Algae and bacteria have no lignin-like components in their molecular structure (McKnight et al. 1997; McKnight and Aiken 1998; Opsahl and Benner 1998), thus the low aromaticity of autochthonous fulvic acids can reflect the lower content of moieties with sp^2-hybridized carbon in cell wall material and in other components of microbial cells (McKnight et al. 1994).

Algal- or phytoplankton-derived autochthonous fulvic acids can absorb light to a lesser extent (by approximately 3–5 times) than allochthonous fulvic acids. They show a progressive increase in absorbance with decreasing wavelength that is typical of fulvic acids (McKnight et al. 1991, 1994). However, the autochthonous fulvic acids (C-like and M-like) of algae or phytoplankton origin can exhibit higher fluorescence intensity at peak C-region than at peak A-region, which is an opposite behavior compared to allochthonous fulvic acids (C-like and M-like) of terrestrial plant origin (Fig. 1; McKnight et al. 2001; Mostofa et al. 2009b). Autochthonous fulvic acids can persist with ages up to 3,000 yr in the desert lakes in Antarctica (McKnight et al. 1991, 1994).

The stable carbon isotope ($\delta^{13}C$) fractionation of autochthonous DOM of algal or phytoplankton origin ranges from −17.2 to 23.7 ‰ in lake and marine environments (Thurman 1985a; Raymond and Bauer 2001a; Nissenbaum and Kaplan 1972). The $\delta^{13}C$ values of algae or phytoplankton shows high variation in freshwater [−(18.3–34.6 ‰)] and sea water [−(18–24.2 ‰)] (Mostofa KMG et al., unpublished data; McCallister et al. 2004; McKnight et al. 1997; Fry and Sherr 1984; Anderson and Arthur 1983; Sigleo and Macko 1985; Yoshioka et al. 1989; Currin et al. 1995; Yoshioka 1997; Lehmann et al. 2004). In addition, $\delta^{13}C$ shows high variations between benthic microalgae [−(12–18 ‰)]; benthic marsh microalgae [−(23.7–27.7 ‰)]; C-4 salt marsh plants [−(12–14 ‰)]; C-3 freshwater/brackish marsh plants [−(23–26 ‰)]; submerged macrophytes [−(21.7–22.2 ‰)]; emergent macrophytes (−26 ‰); marsh macrophytes [−(23.3–28.9 ‰)]; marsh OM [−(22.3–26.4 ‰)]; and freshwater grass leachate such as *Peltandra virginica* [−(29.6 ‰)] (McCallister et al. 2004; Raymond and Bauer 2001a, c; Caraco et al. 1998; Fry and Sherr 1984; Currin et al. 1995; Sullivan and Moncreiff 1990). Depending on the origin of DOM from these algae and plants, there can be found variable carbon isotope ratios for DOM in natural waters.

The autochthonous DOM of algal or phytoplankton origin is usually very suitable for bacterial use, as suggested by the pattern of increased bacterial production with increased primary production (Cole et al. 1988). Autochthonous DOM is in fact relatively labile (Søndergaard and Middelboe 1995; Kirchman et al. 1991). However, autochthonously derived DOC may become persistent over time (Ogawa et al. 2001; Fry et al. 1996; Tranvik and Kokalj 1998). Laboratory studies have shown that natural assemblages of marine bacteria become rapidly able (in <48 h) to utilize labile compounds (glucose, glutamate) and produce refractory DOM that can persist for more than a year (Ogawa et al. 2001). It has also been shown that only 10–15 % of the bacterially derived DOM is identified as hydrolysable amino acids and sugars, which is a characteristic nature of marine DOM (Ogawa et al. 2001). Moreover, the higher concentrations of DON observed in total DOM during the summer period than in winter (Fellman et al. 2009; Vazquez et al. 2011) are most likely accounted for by the produced autochthonous DOM in natural waters.

4.2 *Molecular Size Distribution of DOM*

The molecular size distribution of DOM is significantly variable in natural waters (Table 1). One of the techniques for isolating DOM in natural waters is tangential flow ultrafiltration (also called cross-flow ultrafiltration). The results show that the contributions of the various fractions to total DOC are 21–65 % for the fraction <1 kDa, 44–68 % for <5 kDa, 57–65 % for <10–12 kDa. Moreover, they are 41 % for 1–30 kDa, 32–56 % for 1 kDa–0.1 μm, 67–84 % for 1 kDa–0.45 μm, and 0.1–16 % for 0.1–0.45 or 0.1-GF/F μm in rivers (Table 1) (Yoshioka et al. 2007; Guéguen et al. 2002, 2006; Martin et al. 1995; Mannino and Harvey 2000; de Zarruk et al. 2007; Wu and Tanoue 2001; Wu et al. 2003; Waiser and Robarts 2000; Huguet et al. 2010; Carlson et al. 1985). In lakes, the relative abundances of various DOM fractions are 42–73 % for <1 kDa, 54–79 % for <5 kDa, 21–43 % for 5 kDa–0.1 μm, and 0–2 % for 0.1–0.45 μm (Table 1) (Yoshioka et al. 2007; Guéguen et al. 2002; Wu and Tanoue 2001; Wu et al. 2003; Waiser and Robarts 2000). In estuaries or lagoons, the contributions are 26–98 % for <1 kDa, 11–25 % for 1–3 kDa, 63–75 % for <10 kDa, 25–31 % for 1–30 kDa, 2–45 % for 1 kDa–0.2 μm, 22–48 % for 3 kDa–0.2 μm, 14–20 % for 30 kDa–0.2 μm, and 1–2 % for 30 kDa–0.2 μm (Table 1) (Hagedorn et al. 2004; Mannino and Harvey 2000; Guéguen et al. 2002; Waiser and Robarts 2000; Huguet et al. 2010). In coastal and open oceans, the contributions of the relative DOM fractions are 30–85 % for <1 kDa (30–70 % in coastal waters, 49–85 % in the open ocean), 23–53 % for the fraction between 1 kDa and 10 kDa, 3–19 % for the fraction between 10 kDa and 0.1–0.2 μm, 15–70 % for the fraction between 1 kDa and 0.2 μm, 85 % for the fraction between 1.8 kDa and 0.2 μm (Table 1) (Buesseler et al. 1996; Druon

Table 1 Molecular size distribution of the fractionated DOM in natural waters

Samples	Contribution % of molecular size distribution of DOM							References
	<1 kDa	<3–3.5 kDa	<5 kDa	<10–12 kDa	<30 kDa	<0.1 μm	<0.2–0.45 μm or 0.1-GF/F	
Soils								
Soil DOM, collected French agriculture	–	~50	–	23	–	–	–	de Zarruk et al. (2007)
Rivers								
Rivers, Lake Biwa watershed	–	–	58–68	–	–	32–40	0.2–16.0	Yoshioka et al. (2007)
Rivers, Lake Baikal watershed	–	–	44–56	–	–	43–56	0.1–0.5	Yoshioka et al. (2007)
Delaware river	51	–	–	–	41	–	2	Mannino and Harvey (2000)
Pearl River, Guangzhou section, China	65	–	–	–	–	–	–	
Vistula river, Poland	21–64	–	–	–	–	–	36–86	Guéguen et al. (2002)
Channel fresh water, Venice Lagoons, Itali (n = 3)	–	–	–	57–65	–	–	13–26	Martin et al. (1995)*
Yukon river, Canada	–	–	–	–	–	–	67–84 (>1 kDa)	Guéguen et al. (2006)
Lakes								
Lake Biwa (2.5 m, n = 2)	–	–	68–77	–	–	22–28	0.8–1.1	Yoshioka et al. (2007)
Lake Biwa (70 m, n = 2)	–	–	68–76	–	–	23–32	0.2–0.4	Yoshioka et al. (2007)
Lake Biwa (2.5 and 20 m depth: n = 3)	–	–	54–59	–	–	40–43	2	Wu and Tanoue (2001)
Lake Biwa (2.5 depth: n = 2)	–	–	55–58	–	–	40–43	2	Wu et al. (2003)
Lake Biwa (70 m depth: n = 1)	–	–	69	–	–	30	1	Wu and Tanoue (2001)
Lake Biwa (70 m depth: n = 1)	–	–	69	–	–	30	1	Wu et al. (2003)
Lake Baikal (2 m, n = 2)	–	–	70–77	–	–	23–29	0–0.2	Yoshioka et al. (2007)
Lake Baikal (200 and 1400 m, n = 4)	–	–	72–79	–	–	21–27	0.1–0.5	Yoshioka et al. (2007)

(continued)

Table 1 (continued)

Samples	Contribution % of molecular size distribution of DOM							References
	<1 kDa	<3–3.5 kDa	<5 kDa	<10–12 kDa	<30 kDa	<0.1 μm	<0.2–0.45 μm or 0.1-GF/F	
Redberry lake	73	–	–	–	–	–	–	Waiser and Robarts (2000)
Creeks (Oscar and Trout pond)	55–61	–	–	–	–	–	–	Waiser and Robarts (2000)
Lake Geneva, France	42–64	–	–	–	–	–	36–58	Guéguen et al. (2002)
Estuaries or Lagoons								
Venice Lagoons, Itali (n = 5)	–	–	–	63–75	–	–	14–20	Martin et al. (1995)
Delaware Estuary	71–74	–	–	–	25–31	–	1–2	Mannino and Harvey (2000)
Gironde Estuary: surface water, French Atlantic coast	41–47	13–17	–	–	–	–	22–32	Huguet et al. (2010)
Gironde Estuary: deep water, French Atlantic coast	41–47	11–19	–	–	–	–	31–43	Huguet et al. (2010)
Seine Estuary: surface water, French Atlantic coast	26–56	12–24	–	–	–	–	22–48	Huguet et al. (2010)
Seine Estuary: deep water, French Atlantic coast	35–48	20–25	–	–	–	–	22–34	Huguet et al. (2010)
Adour Estuary, France	55–98	–	–	–	–	–	2–45	Guéguen et al. (2002)
Oceans								
Galveston Bay	30–37	–	–	–	–	–	63–70	Santschi et al. (1995)
Chesapeake Bay and Galveston Bay	39–41	–	–	46–53	–	–	7–11	Guo and Santschi (1997a)
Middle Atlantic Bight	65–70	–	–	23–30			3–11	Guo et al. (1996)
Middle Atlantic Bight	51–59	–	–	–	–	–	41–49	Santschi et al. (1995)
Gulf of Mexico and Middle Atlantic Bight	55–65	~24	–	7–14	–	–	4–7	Guo et al. (1995)

(continued)

Table 1 (continued)

Samples	Contribution % of molecular size distribution of DOM							References
	<1 kDa	<3–3.5 kDa	<5 kDa	<10–12 kDa	<30 kDa	<0.1 μm	<0.2–0.45 μm or 0.1-GF/F	
Gulf of Mexico	47–66	–	–	–	–	–	34–53	Santschi et al. (1995)
Gulf of Mexico	55	–	–	35	–	–	10	Guo et al. (1994)
Open Ocean deep water (off Hawaii: ~600 m deep)	85	–	–	–	–	–	15	Mopper et al. (1996)
North Pacific Ocean, subarctic region (8–59 m)	49–62	–	–	26–33		10–19	–	Midorikawa and Tanoue (1998)
Open North Pacific Ocean (22°45′N, 158°00′W)	67–78	–	–	–	–	–	22–33	Benner et al. (1992)
North Atlantic surface waters	50–70	–	–	–	–	–	34	Carlson et al. (1985)
Northwestern Pacific Ocean surface and deep waters	–	–	–	–	–	–	85 (>1.8 kDa)	Sugimura and Suzuki (1988)
North Pacific Ocean: deep waters	65	–	–	–	–	–	–	Guo and Santschi (1996)
North Pacific Ocean: deep waters	57	–	–	–	–	–	–	Buesseler et al. (1996)

et al. 2010; Midorikawa and Tanoue 1998; Carlson et al. 1985; Sugimura and Suzuki 1988; Guo et al. 1994, 1995, 1996; Santschi et al. 1995; Guo and Santschi 1996, 1997a; Mopper et al. 1996). These results demonstrate that the contribution of the lower MW fraction (<1–10 kDa) is relatively low in rivers and that it significantly increases in lakes, coastal waters and the open ocean. Comparison of molecular fractions between surface (epilimnion) and deep (hypolimnion) waters shows that the molecular size fraction of <1–5 kDa in deep water is often more important than in the surface waters of lakes and oceans (Table 1) (Yoshioka et al. 2007; Wu and Tanoue 2001; Wu et al. 2003; Mopper et al. 1996). It is suggested that either microbial degradation of DOM or new releases of DOM from microbial respiration of organic matter in deeper waters are responsible for the high contents of the low molecular size fractions of DOM in natural waters. An additional implication is that significant microbial or biological degradation of DOM and organic matter occurs in deep waters. The high percentage of colloidal DOC or colloidal organic carbon included in the >1 kDa to 0.45 μm range suggests that colloids are the predominant phase in bulk DOC transported by rivers (Guéguen et al. 2006; Benner and Hedges 1993; Guo and Santschi 1997b; Guéguen and Dominik 2003).

The optical and chemical characteristics of the molecular size fractions of DOM show that truly dissolved DOM (<1–10 kDa) includes fulvic acid (59–96 % on the basis of fluorescence), total hydrolyzed amino acids (51–63 %), tryptophan (free tryptophan has a molecular weight of 0.2 kDa) and total dissolved carbohydrates (10–20 %). In contrast, the DOM fraction between >1–10 kDa and 0.2–0.45 μm or 0.1-GF/F includes fulvic acid (5–22 % on the basis of fluorescence), total dissolved carbohydrates (80–90 %) and total hydrolyzed amino acids (29–42 %). The DOM fraction of 0.1 μm-GF/F (0.45–0.7 μm) includes protein-like or tryptophan-like or bacterial cells or phytoplankton cells, total hydrolyzed amino acids (7–11 %) and fulvic acid (2–8 % on the basis of fluorescence) (Liu et al. 2007; Guéguen et al. 2006; McCarthy et al. 1996; Midorikawa and Tanoue 1998; Wu and Tanoue 2001; Wu et al. 2003; Pakulski and Benner 1992; Skoog and Benner 1997; Boehme and Wells 2006). The contributions to the molecular size fractions of sedimentary fulvic acid extracted from Tokyo Bay sediment samples are 44.8 % for <1 kDa, 3.5 % for 10 kDa, 31.8 % for 50 kDa, 14.6 % for 100 kDa and 5.3 % for 300 kDa. The corresponding contributions of humic acid are 2.4 % for <1 kDa, 0.8 % for 10 kDa, 5.3 % for 50 kDa, 16.1 % for 100 kDa and 75.4 % for 300 kDa (Hayase and Tsubota 1983, 1985). This suggests that allochthonous fulvic acid is mostly composed of low molecular size fractions (<1–10 kDa) whilst allochthonous humic acid is mostly composed of high molecular size fractions, >300 kDa (Hayase and Tsubota 1983, 1985; Rashid and King 1969; MacFarlane 1978). Therefore, molecular size fractions could be a useful indicator to distinguish between fulvic and humic acids in DOM in a variety of natural waters.

These results also imply that allochthonous fulvic acid of terrestrial origin or the autochthonous fulvic acid (C-like) of algal or phytoplankton origin can primarily undergo photoinduced and microbial in situ degradation, which can decrease the molecular size and increase as a consequence the low molecular size fraction of DOM (Yoshioka et al. 2007; Amon and Benner 1994; Corin et al. 1996; Amador et al. 1989; Leenheer and Croué 2003; Opsahl and Benner 1998; Boehme and Wells 2006; Mopper et al. 1991; Senesi et al. 1991; Allard et al. 1994; Benner and Biddanda 1998; Mopper and Kieber 2002). The autochthonous fulvic acid (C-like) of algal or phytoplankton origin can show the fluorescence excitation-emission (Ex/Em) maxima of peak C in a longer wavelength region (Ex/Em = 340–370/434–480 nm), whilst the autochthonous fulvic acid (M-like) can show its Ex/Em maxima in a shorter wavelength region (290–330/358–434 nm) compared to allochthonous fulvic acids (standard SRFA at Ex/Em = 325–345/442–462 nm in Milli-Q and Seawater) (Parlanti et al. 2000; Mostofa et al. 2009b; Zhang et al. 2009; Vähätalo and Järvinen 2007; Yamashita and Jaffé 2008; Nakajima 2006; Murphy et al. 2008; Balcarczyk et al. 2009). Note that autochthonous fulvic acids (C-like and M-like) are defined on the basis of the similarity with the fluorescence properties of allochthonous fulvic acids (C-like and M-like) for both freshwater and marine environments (for a detailed explanation see the FDOM chapter: "Fluorescent Dissolved Organic Matter in Natural Waters").

Humic-like fluorescence is a key component in DOM size fractions between ~15 and 150 kDa. A bathychromic shift (blue shift) of the humic fluorescence peak is often detected with decreasing molecular size, and interestingly the maximum in humic fluorescence moves to lower excitation and emission wavelengths in estuarine waters (Boehme and Wells 2006). Blue-shift phenomena are generally observed in field studies (Coble 1996; Mostofa et al. 2005a, b, 2007a, b; Moran et al. 2000; Burdige et al. 2004; de Souza-Sierra et al. 1994; Komada et al. 2002).

The molecular size distribution of DOM plays significant roles in various kinds of physical, photoinduced and biological processes in natural waters. They are listed below.

(i) The bioreactivity of POM and DOM decreases along a continuum of larger to smaller sizes. Diagenetic processes lead to the formation of structurally complex LMW compounds that are more resistant to biodegradation (Amon and Benner 1994, 1996; Hama et al. 2004; Mannino and Harvey 2000; Harvey and Mannino 2001; Benner 2002; Loh et al. 2004; Zou et al. 2004; Seitzinger et al. 2005; Kaiser and Benner 2009). This hypothesis is termed as size-reactivity continuum model and is based on the results of size-fractionation experiments that demonstrate that bacterial utilization of (HMW) DOM is typically higher compared to (LMW) DOM (Amon and Benner 1994). It has also been shown that neutral sugars and amino sugars are considerably more bioreactive than amino acids in all organic matter size fractions of DOM in deep mesopelagic waters (Kaiser and Benner 2009). Furthermore, nonspecific enzyme reactions can lead to secondary products that are resistant to degradation (Ogawa et al. 2001). Products of such enzymatic degradations may not resemble the structure of the original compounds, thereby reducing enzymatic recognition and further biodegradation. In addition, size can affect the bioreactivity of individual organic matter fractions. Colloidal organic matter, which is part of HMW DOM, is much less accessible to bacteria than particles larger than a few μm because it occupies a minimum between two different transport regimes (Kaiser and Benner 2009; Kepkay 1994; Wells and Goldberg 1993). In fact, Brownian motion dominates transport of smaller colloids to bacteria, whilst larger particles are primarily transported to bacteria by turbulent shear (Kepkay 1994).

(ii) Photoinduced and microbial processes that involve DOM, including fulvic and humic acids, can produce biologically labile LMW organic substances (e.g. organic acids) in natural waters (Moran and Zepp 1997; Carrick et al. 1991; Kieber et al. 1989, 1990; Corin et al. 1996; Mopper et al. 1991; Allard et al. 1994; Mopper and Stahovec 1986; Backlund 1992). These LMW organic compounds are important intermediates of the conversion of organic substances such as carbohydrates, fats and proteins into CH_4 and CO_2 in aqueous media (Smith and Oremland 1983; Evans 1998; Xiao et al. 2009; Wellsbury and Parkes 1995).

(iii) The absorption of natural sunlight is greatly dependent on the molecular size of DOM and has a high biogeochemical importance in natural waters. For example, fulvic and humic acids (humic substances) can absorb both visible and UV radiation (Kieber et al. 1990; Kramer et al. 1996; Sadtler 1968; Strome and Miller 1978). Many low molecular weight organic acids photogenerated from large CDOM or FDOM can only absorb in the UV-C range, with no absorption of UV-B, UV-A or visible radiation (Carrick et al. 1991; Kieber et al. 1990; Mopper et al. 1991; Sadtler 1968). Further details are provided in the DOM degradation chapter (see chapter "Photoinduced and Microbial Degradation of Dissolved Organic Matter in Natural Waters").

4.3 Autochthonous Fulvic Acids and their Differences with Allochthonous Fulvic Acids

The key component of autochthonous DOM is variously termed as marine humic-like substances (Coble 1996), sedimentary fulvic acid (Hayase et al. 1987, 1988) or marine fulvic acids (Malcolm 1990), which is contradictory. However, recent studies show that the two fluorescent components are primarily produced under either photoinduced or microbial respiration (or assimilation) of algal (phytoplankton) biomass (Mostofa et al. 2009b; Zhang et al. 2009; Stedmon and Markager 2005a). PARAFAC modeling of EEM spectra of algal-originated DOM suggests that the fluorescence peaks and the images of the first fluorescent component are similar to those of allochthonous fulvic acid (Fig. 1). On the other hand, the fluorescence peaks and the images of the second fluorescent component are similar to those of marine humic-like substances (Coble 1996). However, the fluorescence intensity and the peak positions of the first fluorescent component are quite different from EEM spectra of standard fulvic acid, which justifies their being denoted with a new name. To avoid the difficulties of indicating the two algal-originated fluorescent components and considering the similarities of their EEM images with allochthonous fulvic acid, it is suggested to denote the first and the second fluorescent component as 'autochthonous fulvic acid (key component)' and 'autochthonous fulvic acid (minor component)', respectively. These names could be useful to denote the two fluorescent components originated from algae or phytoplankton in fresh- and marine waters in future research studies. The differences between autochthonous fulvic acids and allocthonous fulvic acids and their identification are extensively discussed in the next chapter, 'Fluorescent dissolved Organic Matter in Natural Waters'.

5 Measurement, Distribution and Sources of DOM in Natural Waters

Measurement of DOM

DOM is generally determined as dissolved organic carbon (DOC) concentration, because of the predominant presence of organic carbon in all dissolved organic substances included in bulk DOM. The amount of DOC in natural waters is determined using a high-temperature catalytic oxidation (HTCO) method developed by Sugimura and Suzuki (Sugimura and Suzuki 1988). This technique is very precise and rapid for the determination of non-volatile DOM in concentrations between 0 and 2000 μM, compared to conventional wet chemical oxidation methods (Menzel and Vaccaro 1964; Jonathan 1973). In the HTCO method (Sugimura and Suzuki 1988), the oxidation of DOM in water is carried out on a platinum catalyst at 680 °C under an oxygen atmosphere after the sample has been freed of inorganic carbon. The concentration of CO_2 generated is measured with a non-dispersive IR gas analyzer. The determination can be carried out with a precision of ±2 % using a sample volume of 100–200 μl.

Methodology for HTCO (Sugimura and Suzuki 1988): After collection of water samples using polycarbonate bottles, water is filtered with precombusted (450 °C) glass-fiber or any other filters (0.1–0.7 μm size). Triplicate samples (15 ml) are stored in brown glass bottles (30 ml in volume). 25 μl of 6 N HCl solution is added to remove dissolved inorganic carbon (DIC). These bottles are sealed with Teflon-coated butyl-rubber stoppers and aluminum caps and stored in a freezer (−40 °C). There is need to analyze the samples as soon as possible. For sample measurement, DIC is firstly removed by bubbling the brown bottles with pure air for approximately 15 min. After removing DIC, 200 μl of the water sample is injected into a TOC analyzer (e.g. TOC-5000A, Shimadzu, Kyoto, Japan). Note that analytical blanks for the DOC measurement originating from the instrument (system blank) and from pure water (e.g. Milli Q TOC, Millipore) are on average in the range of 2–4 μM C and 6 μM C, respectively (Yoshioka et al. 2002a). The system blank is determined during sample measurement according to the instrument software of the TOC 5000A. The system blank is generally used for the correction of DOC concentration for samples. Potassium hydrogen phthalate is generally used as a standard for calibration.

5.1 Distribution and Sources of DOM in Natural Waters

DOM Contents in Stream, Rivers, Groundwater and Rainwater

DOC concentrations are very variable in different upstream locations of the world (Table 2). Relatively low values such as 7–970 μM C are found in Asia (Mostofa et al. 2005a, b, 2010, Mostofa KMG et al., unpublished data; Hayakawa et al. 2004; Konohira and Yoshioka 2005) and 17–3300 μM C in North America (Table 2) (McKnight et al. 1993, 2001; Volk et al. 1997; David and Vance 1991; Fellman et al. 2009; Eckhardt and Moore 1990; Dosskey and Bertsch 1994; Wahl et al. 1997; Cory et al. 2004; Meier et al. 2004; Fahey et al. 2005; Raymond and Saiers 2010). Value found in Europe are a bit higher (21–6250 μM C) (Stedmon et al. 2007b; Worrall et al. 2004a; Evans et al. 2006; Chapman et al. 2001; Monteith and Evans 2005; Gielen et al. 2011). Stream DOM is mostly released from the leaching of ground water in high mountain areas that in Asia are densely shaded by coniferous-mixed forests or typical grassland. In Europe-North America, the major sources of stream DOM are riparian vegetation, woodland streams (major sources of detritus), wetlands, swamps, and peat-land.

DOC concentrations in rivers vary in different locations of the world (Table 2). It has been found 32–2429 μM C in Asia (Mostofa et al. 2005b, 2007a, 2010, Mostofa KMG et al., unpublished data; Yoshioka et al. 2002b, 2007; Ittekkot et al. 1985; Safiullah et al. 1987; Cauwet and Mackenzie 1993; Tao 1996; Zhang 1996; Kao and Liu 1997; Zhang et al. 1999; Gao et al. 2002; Nagao et al. 2003; Ishikawa et al. 2006; Yue et al. 2006; He et al. 2010); 83–833 μM C in Africa (Martins and Probst 1991); ~50–3917 μM C in Europe (Vazquez et al. 2011; Eisma et al. 1982; Cadée 1987; Meybeck et al. 1988; Rostan and Cellot 1995; Elbe 1997; Lara et al. 1998; Veyssy 1998; Duff et al. 1999; Abril et al. 2000, 2002; Baker 2001, 2002; Brodnjak-Vončina et al. 2002; Guéguen et al. 2002; Baker and Spencer 2004; Kaiser et al. 2004; Romani et al. 2004); 40–4167 μM C in North America (Wu et al. 2005; Xie et al. 2004; McKnight et al. 2001; Alberts and Takács 1999; Guéguen et al. 2006; Morel and Gentili 2009; Raymond and Bauer 2001b; Haines 1979; Newbern et al. 1981; Spiker 1981; Alberts et al. 1984; Findlay et al. 1991; Perry and Perry 1991; Prahl et al. 1998; Crandall et al. 1999; Davis et al. 2001; Biddanda and Cotner 2002; Repeta et al. 2002; Wang et al. 2004; Zanardi-Lamardo et al. 2004; Schwede-Thomas et al. 2005; See and Bronk 2005; Stepanauskas et al. 2005; Osburn et al. 2009); and ~108–7500 μM C in Latin America (Raymond and Bauer 2001b; Richey et al. 1990; Depetris and Kempe 1993; Daniel et al. 2002). These results generally show that DOC concentrations are relatively low in Asian and African Rivers and relatively high in Europe, North and South America Rivers. The major sources of DOC in Asian Rivers are natural ones such as leaching of groundwater in mountainous areas covered by coniferous-mixed forests, deciduous conifer forest, grassland, irrigated grassland, and swamps, but also anthropogenic sources such as urban sewerage, industrial and agricultural activities. In African Rivers, the major sources of DOC are mostly from the typical rain forest belt, leaching and heterotrophic processes of

Table 2 Distribution of dissolved organic carbon (DOC) concentrations in a variety of natural waters

Study sites	Country/ latitudes	Length/ watershed area[a] km or (km^2)[a]	DOC Surface (Depth)[c] (μM C)	DOC Deeper (Depth)[b]	Reference
Streams: Asia					
Upstreams, Lake Biwa watershed (n = 8)	Japan	(3174)	7–110	–	Mostofa et al. (2005a)
Upstreams, Lake Biwa watershed, Okutama and Uryu regions (n = 35)	Japan	–	12–280	–	Konohira and Yoshioka (2005)
Upstream, Kurose River, Hiroshima Prefecture (n = 2)	Japan	–	47–239	–	Mostofa et al. (2005b)
Upstream, Ohta River, Hiroshima Prefecture (n = 1)	Japan	–	59–67	–	Mostofa et al. (2005b)
Upstream, Lake Xingyun basin (n = 1)	24°N	–	~50	–	Hayakawa et al. (2004)
Upstream sites, Nanming River	26°N	–	50–100	–	Mostofa et al. (2010)
Branches of Upper Region, Yellow River (Huang He) (n = 13)	China	–	174–970	–	Mostofa KMG et al., (unpublished data)
Europe					
Upstreams (Warkworth and Afon Hafren) (n = 2)	UK	–	21–1000	–	Worrall et al. (2004a)
Forest streams, Denmark	55°N	–	274–1051	–	Stedmon and Markager (2005b)
Fuirosos, forested stream	Spain	–	150	–	Vazquez et al. (2011)
De Inslag forest, Brasschaat, Belgian Campine region	Belgian (51°N)	–	~1000–6250	–	Gielen et al. (2011)
Upstreams, The United Kingdom Acid Waters Monitoring Network (n = 11)	UK	(2.1–14.1)	100–1075	–	Monteith and Evans (2005), Evans et al. (2006)
Streams, Scotland	Scotland	–	400–933	–	Chapman et al. (2001)
Stream, boreal watershed, in northern Sweden	Sweden	(2940)	1308–2192	–	Bertilsson et al. (1999)
North America					
Upstream rivers (Hubbard Brook)	USA	–	158–717	–	Fahey et al. (2005)
Upstream Rivers (National Park)	USA	–	17–367	–	Cory et al. (2004)

(continued)

Table 2 (continued)

Study sites	Country/ latitudes	Length/ watershed area[a] km or (km^2)[a]	DOC Surface (Depth)[c] (μM C)	DOC Deeper (Depth)[b]	Reference
Streams, central Maine (n = 11)	USA	–	125–1141	–	David and Vance (1991)
Stream, Fourmile Branch watershed, South Carolina	USA	(12.6)	442	–	Dosskey and Bertsch (1994)
Streams, Colorado Rocky Mountains	USA	–	108–433	–	McKnight et al. (1993, 2001)
Streams, Dog Creek and Oyster Creek, forested watershed (n = 2)	USA	–	933 ± 167–2208 ± 333	–	Wahl et al. (1997)
Tongass National Forest, stream	USA	–	275–358	–	Fellman et al. (2009)
Upstream, White Clay Creek	USA	–	67–867	–	Volk et al. (1997)
Upstreams, forest land watershed (n = 30)	USA	(1.9–226)	75–358	–	Raymond and Saiers (2010)
Stream water, New Jersey Pine Barrens	USA	–	2100	–	Meier et al. (2004)
Upstreams, south-central Ontario (n = 11)	Canada	–	225–900	–	Wu et al. (2005)
Streams (n = 42)	Canada	–	290–3300	–	Eckhardt and Moore (1990)
Groundwaters					
Groundwater (Tubewell waters)	Bangladesh	–	17–424	–	Anawar et al. (2002)
Groundwater, BaiCheng City and main west Liaohe river, North-east China (n = 2)	China	–	49–371	–	Mostofa KMG et al., (unpublished data)
Groundwater, Lake Biwa basin, Japan	Japan	–	16–328 (40–100 m)	–	Mostofa et al. (2007a)
Groundwater, Tomago sand beds, Newcastle	Australia	–	183–842	–	McIntyre et al. (2005)
Groundwater, Germany	Germany	–	42–1400 (1–24 m)	–	Buckau et al. (2000)
Groundwater, Germany	Germany	–	6117–15333 (35–137 m)	–	Buckau et al. (2000)
Soil water, De Inslag forest, Brasschaat, Belgian Campine region	Belgian (51°N)	–	~1000–6250	–	Gielen et al. (2011)
Groundwater, boreal watershed, in northern Sweden	Sweden	–	200–850	–	Bertilsson et al. (1999)

(continued)

Table 2 (continued)

Study sites	Country/ latitudes	Length/ watershed area[a] km or (km^2)[a]	DOC Surface (Depth)[c] (μM C)	Deeper (Depth)[b]	Reference
Groundwater (Shallow and deep wells; groundwater spring), Norton-in-Hales	UK	(38)	467 ± 375–2708 ± 1558	–	Bradley et al. (2007)
Groundwater, Suwannee River basin	USA	–	8–2333	–	Crandall et al. (1999), Schwede-Thomas et al. (2005)
Shallow groundwater (1.6 m), New Jersey Pine Barrens	USA	–	1558–2583	–	Meier et al. (2004)
Groundwater, various types of aquifers	USA	–	42–8333	–	Thurman (1985a)
Groundwater, Cape Cod, USA	USA	–	<167		Pabich et al. (2001)
Groundwater, Amazon Basin	Brazil	–	100–3000	–	Richey et al. (2002)
Groundwater, Okavango Delta	Botswana	–	1108 ± 217–14167 ± 6333	–	Mladenov et al. (2007)
Soil water (upper soil horizons)	North America; Europe	–	1667–7500	–	Michalzik et al. (2001)
Soil water (lower soil horizons)	North America; Europe	–	167–2917	–	Michalzik et al. (2001)
Rainwater					
Rainwater, upstream regions of Yellow River (n = 2)	China	–	93–1784	–	Mostofa KMG et al., (unpublished data)
Rainwater (n = 483), Northern China	China	–	25–3675	–	Pan et al. (2010)
Rainwater (n = 13), Guangzhou city	China	–	78–694	–	Xu et al. (2008)
Rainwater, Brazilian savanna	Brazil	–	217	–	Ciglasch et al. (2004)
Rainwater (summer), Westwood, Los Angeles	USA	–	<1908	–	Sakugawa et al. (1993)
Rainwater (winter), Westwood, Los Angeles	USA	–	17–758	–	Sakugawa et al. (1993)
Rainwater (n = 120), University of North Carolina, Wilmington campus	USA	–	4–379	–	Kieber et al. (2006)

(continued)

Table 2 (continued)

Study sites	Country/ latitudes	Length/ watershed area[a] km or (km^2)[a]	DOC Surface (Depth)[c] (μM C)	DOC Deeper (Depth)[b]	Reference
Rainwater (n = 4), University of North Carolina, Wilmington campus	USA	–	32–105	–	Kieber et al. (2007)
Rainwater, mixed forests in New Hampshire and New York	USA	–	92–158	–	Likens et al. (1983)
Rainwater (n = 18), University of North Carolina, Wilmington	USA	–	77 ± 20	–	Likens et al. (1983)
Rainwater (n = 13), Wilmington, North Carolina, USA	USA	–	12–461	–	Southwell et al. (2010)
Rainwater, UNCW campus, Southeastern, North Carolina	USA	–	81	–	Avery et al. (2006)
Rainwater, Wilmington, NC	USA	–	5–238	–	Miller et al. (2008)
Rainwater, Fichtelgebirge	Germany	–	333–633	–	Guggenberger and Zech (1993)
Rainwater, Meteorological Station, University of Aveiro	Portugal	–	649–1078	–	Santos et al. (2009a)
Rainwater (cold season), town of Aveiro	Portugal	–	28–157	–	Santos et al. (2009b)
Continental rainwater	–	–	161	–	Willey et al. (2000)
Marine rainwater	–	–	23	–	Willey et al. (2000)
Rivers, Tributaries and Channels: Asia					
Ganges River, Bangladesh portion	Bangladesh, India	~2500 (975,000)	108–317	–	Ittekkot et al. (1985)
Brahmaputra River, Bangladesh Portion	Bangladesh, India, China	~2900 (580,000)	83–500	–	Safiullah et al. (1987)
Yellow River (Huang He), mainstream in Upper Region	China	~5400 (745,000)	95–503	–	Mostofa KMG et al., (unpublished data)

(continued)

Table 2 (continued)

Study sites	Country/ latitudes	Length/ watershed area[a] km or (km^2)[a]	DOC Surface (Depth)[c] (μM C)	DOC Deeper (Depth)[b]	Reference
Yellow river (Huang He)	China	~5400 (745,000)	167–333	–	Cauwet and Mackenzie (1993), Zhang et al. (1999)
Yangtze River (or Cháng Jiang)	China	~6300 (1,950,000)	142–400	–	Cauwet and Mackenzie (1993), Zhang (1996)
Xijiang River, South China	China	–	84–495	–	Gao et al. (2002)
Nunming River, Guizhou Proviance, South-west China	China	118 (–)	82–463	–	Mostofa et al. (2010), Yue et al. (2006)
NenJiang River (or Nen River), Heilongjiang Proviance and Inner Mongolia, North-east China	China	1370 (–)	346–1286	–	Mostofa KMG et al., (unpublished data)
The Second Song Hua Jiang River, Jilin and Zhejiang proviance, North-east China	China	849 (73,000)	362–1352	–	Mostofa KMG et al., (unpublished data)
LiaoHe River, Heibei, Jilin, Inner Mongolia, Jilin, and Liaoning Proviances, North-east China	China	1394 (201,600)	169–1048	–	Mostofa KMG et al., (unpublished data)
DaLingHe River, Inner Mongolia, North-east China	China	435? (13,000)	90–283	–	Mostofa KMG et al., (unpublished data)
Rivers, Lake Biwa watershed (n = 10)	Japan	20–60	32–375	–	Mostofa et al. (2005b)
Kurose River, Hiroshima Prefecture	Japan	43 (250)	123–385	–	Mostofa et al. (2005b)
Ohta River, Hiroshima Prefecture	Japan	103 (1,700)	45–164	–	Mostofa et al. (2005b)
Rivers, Lake Fuxian and Lake Xingyun basin (n = 9)	24°N	–	37–428	–	Hayakawa et al. (2004)
Yinluan River, Tianjin	China	–	396	–	Tao (1996)
Bang Nara River; Saiburi River, Thailand	1–10°N	–	125–625	–	Yoshioka et al. (2002b)
Kahayan River and Rungan River, Indonesia	10°N–10°S	–	558–1042	–	Ishikawa et al. (2006)
Lanyang His River	Taiwan	–	42–667	–	Kao and Liu (1997)
3 Rivers, Lake Hovsgol basin	50–52°N	–	400–1400	–	Hayakawa et al. (2003)
5 Major Rivers, Lake Baikal basin	Russia	–	43–542	–	Yoshioka et al. (2002a, 2007)

(continued)

Table 2 (continued)

Study sites	Country/ latitudes	Length/ watershed area[a] km or (km^2)[a]	DOC Surface (Depth)[c] (μM C)	DOC Deeper (Depth)[b]	Reference
Kuji River & 3 Tributaries, North Kanto	Japan	–	92–175	–	Nagao et al. (2003)
Tributaries (n = 8)	30–35°N	–	122–271	–	Mostofa et al. (2007a)
Dongjiang tributary, South China Sea	China	–	186–227	–	He et al. (2010)
Tributaries of NenJiang River	North-East China	–	177–2429	–	Mostofa KMG et al., (unpublished data)
Tributary of LiaoHe River (n = 2)	North-East China	–	631–640	–	Mostofa KMG et al., (unpublished data)
Irrigation Channels, Japan (n = 19)	30–35°N	3–5	65–210	–	Mostofa et al. (2007a)
Africa					
Zaïre (or Congo) River (Angola, Burundi, Cameroon, Congo, Gabon, Rwanda, Tanzania, Zambia)	Africa	4700 (40,00,000)	250–833	–	Martins and Probst (1991)
Niger River (Guinea, Mali, Niger, Benin, Nigeria)	Africa	4180 (11,25,000)	167–500	–	Martins and Probst (1991)
Gambia River (Republic of Guinea, The Gambia)	West Africa	1130 (42,000)	83–333	–	Martins and Probst (1991)
Europe					
Lena River, East Siberia	Russia	4472 (2,500,000)	309–1042	–	Lara et al. (1998)
Rivers (n = 3), Arctic Russia	66–71°N		142–3917	–	Duff et al. (1999)
Mura River, Slovenia	Slovenia	438 (13,824)	150–708	–	Brodnjak-Vončina et al. (2002)
River Tyne, UK	55°N	(2,935)	275–1275	–	Baker and Spencer (2004)
Rhône River	Switzerland-France	(98,000)	94–818	–	Rostan and Cellot (1995)
Scheldt River	France, Belgium, Netherlands	~350 (21,600)	567	–	Abril et al. (2000)
Rhine River (Germany, Italy, Austria, Switzerland, France, Netherlands)	–	~1233 (224,000)	242–442	–	Eisma et al. (1982), Abril et al. (2002)
Gironde River	France, Spain	~602 (71,000)	258	–	Veyssy (1998)

(continued)

Table 2 (continued)

Study sites	Country/ latitudes	Length/ watershed area[a] km or (km^2)[a]	DOC Surface (Depth)[c] (μM C)	DOC Deeper (Depth)[b]	Reference
Thames River	UK	~346 (14,000)	483	–	Abril et al. (2002)
Elbe River	Czech Republic, Germany	~1091 (145,800)	342–383	–	Elbe (1997), Abril et al. (2002)
Ems River	Germany, Netherlands	~371 (9,000)	567–675	–	Cadée (1987), Abril et al. (2002)
Sado River	Portugal	~175 (7,600)	558	–	Abril et al. (2002)
Douro River	Portugal, Spain	~897 (115,320)	208	–	Abril et al. (2002)
Loire River	France	~1012 (115,000)	325	–	Meybeck et al. (1988), Abril et al. (2002)
Tagliamento River, Italy	46°N	~178 (2,900)	32–95	–	Kaiser et al. (2004)
Vistula River (Poland, Ukraine, Belarus, Slovakia)	–	~1000 (194,000)	397–653	–	Guéguen et al. (2002)
Fuirosos, downstream rivers (summer season)	Spain	–	417–2750	–	Vazquez et al. (2011)
Ebro River, NE Iberian Peninsula	Spain	~900 (80, 000)	183–195	–	Romani et al. (2004)
Öre River catchement, boreal watershed, in northern Sweden	Sweden	(2,940)	600–1558	–	Bertilsson et al. (1999)
Latin America					
Upstreams and Rivers	Brazil		242–7500	–	Daniel et al. (2002)
Paraná River, Argentina	Argentina, Brazil, Paraguay	~4800 (2,582,000)	108–458	–	Depetris and Kempe (1993)
Amazon River, Brazil	Brazil, Colombia, Ecuador, Peru	~6800 (6,300,000)	235–1000	–	Raymond and Bauer (2001b), Richey et al. (1990)
North America					
Hudson River (New York, and New Jersey)	USA	507 (21,000)	196	–	Raymond and Bauer (2001b)
York River (Virginia)	USA	55 (4350)	390–701	–	Raymond and Bauer (2001b)

(continued)

Table 2 (continued)

Study sites	Country/ latitudes	Length/ watershed area[a] km or (km^2)[a]	DOC Surface (Depth)[c] (μM C)	DOC Deeper (Depth)[b]	Reference
New River (North Carolina, Virginia and West Virginia)	USA	515	83–4167	–	Newbern et al. (1981)
Parker River (Massachusetts)	USA	~37 (609)	986	–	Raymond and Bauer (2001b)
Suwannee River (Florida and Georgia)	USA	396	775–2583	–	Crandall et al. (1999), Schwede-Thomas et al. (2005)
Ogeechee River, Georgia	USA	~470 (14,000)	267–525	–	See and Bronk (2005)
Almaha River, Georgia	USA	~220 (36,000)	300–475	–	See and Bronk (2005)
Satilla River, Georgia	USA	–	525–3133	–	See and Bronk (2005)
Savannah River, Geogia	USA	–	225–417	–	See and Bronk (2005)
Sapelo Island River, Geogia	USA	–	~50–2333	–	Alberts and Takács (1999), Haines (1979)
Satilla River and Altamaha River, Georgia	USA	~378; ~220	1545–1620; 1430–1442	–	Xie et al. (2004)
Columbia River (Washington, D. C., Oregon and British Columbia)	USA, Canada	2000	117–158	–	Prahl et al. (1998)
4 Major Rivers (Muskegon, Grand, St. Joseph and Kalamazoo), Lake Michigan basin	USA	–	430–1870	–	Biddanda and Cotner (2002)
Taylor River (Chennel), southern Everglades National Park, Florida	USA	–	708–1533	–	Davis et al. (2001)
Rivers (n = 8), southeastern United States	USA	–	417–2333	–	Alberts and Takács (1999)
Hudson River & Mohawk River (New York)	USA	507; 225	375–508	–	Findlay et al. (1991)
Potomac River (Virginia, Maryland, Washington, D. C.)	USA	616	364	–	Spiker (1981)
Susquehanna River, Maryland, Pennsylvania, New York	USA	715	292	–	Spiker (1981)

(continued)

Table 2 (continued)

Study sites	Country/ latitudes	Length/ watershed area[a] km or (km^2)[a]	DOC Surface (Depth)[c] (μM C)	DOC Deeper (Depth)[b]	Reference
Rappahannock River, Virginia	USA	314	125	–	Spiker (1981)
James River, Virginia	USA	560	167	–	Spiker (1981)
Missouri River; Ohio River	USA	~3700 (1,371,000); 1570 (490,000)	283–308	–	McKnight et al. (2001)
Mississippi River; and Atchafalaya River	28–29°N	~3734 (3,267,000); 220 (7,600)	275–446	–	Wang et al. (2004)
Delaware, Eel, and Mississippi River	USA	~480 (36,500); 300 (9,500); -	214, 80, 181	–	Repeta et al. (2002)
3 Rivers (Shark, Broad and Caloosahatchee)	25–26°N	–	170–1178	–	Zanardi-Lamardo et al. (2004)
2 Rivers (Flathead and Kootenai), Montana (n = 2)	Canada, USA	~250 (22,700); 780 (50,000)	40–667	–	Perry and Perry (1991)
Yukon River	Canada, Alaska	~3185 (800,000)	508–2835	–	Guéguen et al. (2006)
Rivers, South-central Ontario	Canada	–	258–900	–	Wu et al. (2005)
Mackenzie River, Arctic region	Canada	~1700 (1,810)	312–576	–	Osburn et al. (2009)
Lakes, Wetlands and Swamps: Asia					
Lake Biwa, Japan	30–35°N	(3,174)	88–183 (2.5–20 m)	76–101 (40–80 m)	Mostofa et al. (2005a), Sugiyama et al. (2004)
Lake Ashino, Japan (A1 and A2)	Japan	–	99–111 (0–10 m)	74–84 (30–38 m)	Sugiyama et al. (2004)
Lake Ikeda, Japan ((I1 and I2)	Japan	–	101–112 (0–10 m)	55–56 (200–233 m)	Sugiyama et al. (2004)
Lake Suwa, Japan (Center)	Japan	–	142–216 (0 m)	–	Sugiyama et al. (2004)
Lake Inawashiro, Japan	Japan	–	42–47 (0–10 m)	26 (70 m)	Sugiyama et al. (2004)
Lake Hongfeng & Lake Baihua, China	26°N	–	169–330 (0–8 m)	161–194 (20–25 m)	Fu et al. (2010)
Lakes, China (n = 3)	24–25°N	–	98–614	105–123 (30–50 m)	Sugiyama et al. (2004)

(continued)

Table 2 (continued)

Study sites	Country/ latitudes	Length/ watershed area[a] km or (km^2)[a]	DOC Surface (Depth)[c] (μM C)	DOC Deeper (Depth)[b]	Reference
Lake Ponds, (WangHua and TLH), North-east China (n = 2)	China	–	2075–3152	–	Mostofa KMG et al., (unpublished data)
Lake Fuxian, China	24°N	–	116	95–100 (50–140 m)	Hayakawa et al. (2004)
Lake Xingyun, China	24°N	–	629–658	–	Hayakawa et al. (2004)
8 Lakes (Batu, Tehang, Bunter, Bajawak, Pahewan, Hampapak, Rengas & Takapan)	Indonesia (10°N–10°S)	–	458–1300	–	Ishikawa et al. (2006)
Lake Dapur and Lake Hurung)	Indonesia (10°N–10°S)	–	2042, 4133	–	Ishikawa et al. (2006)
Lake Hovsgol, Middle site, Mongolia	50–52°N	–	85–100	85–90 (50–200 m)	Hayakawa et al. (2003)
Lake shore site toward middle of Lake Hövsgöl, Mongolia	50–52°N	–	93–550	–	Hayakawa et al. (2003)
Lake Baikal (North, Central & South basin)	51–55°N	(556,000)	92–142 (0–100 m)	88–105 (500–1620 m)	Yoshioka et al. (2002a, 2007), Sugiyama et al. (2004)
Lakes, New Zealand (n = 11)	New Zealand	(3.4–352)	25–833	–	Rae et al. (2001)
Swamps, Bang Nara River basin, Thailand	1–10°N	–	475–10333	–	Yoshioka et al. (2002b)
Middle East					
Lake Kinnerret	Israel	–	270–485 (0–10 m)	258–368 (38 m)	Annual Report (2004)
Latin America					
12 Lakes, Barciloche region	Argentina	–	21–222	–	Morris et al. (1995)
Lake Barata, Brazil	1°N	–	420–710	–	Farjalla et al. (2006)
Europe		–		–	
Lake Great Dun Fell	UK	–	333–3750	–	Worrall et al. (2004a)
Lake Geneva, Switzerland–France	Switzerland–France	–	59–128	–	Guéguen et al. (2002)

(continued)

Table 2 (continued)

Study sites	Country/ latitudes	Length/ watershed area[a] km or (km^2)[a]	DOC Surface (Depth)[c] (μM C)	DOC Deeper (Depth)[b]	Reference
Lakes (n = 38), Sweden	55–71°N	–	167–1833	–	Bertilsson and Tranvik (2000)
Lakes, The United Kingdom Acid Waters Monitoring Network (n = 11)	UK	(0.42–14.7)	67–550	–	Monteith and Evans (2005), Evans et al. (2006)
Lakes, Sweden (n = 5)	Sweden	–	325–1617 (1.5–7.5 m)	–	Granéli et al. (1996)
Lakes (n = 1000), Norway	Norway	–	167 (median)	–	Larsen et al. (2011)
North America					
Lake Michigan	USA	–	124–216	–	Biddanda and Cotner (2002)
Lake Superior	USA	–	110–119	–	Biddanda et al. (2001)
Lakes (Christmas, Turtle, Minnetonka, Owasso, and Round)	USA	–	537–712	–	Biddanda et al. (2001)
Lakes (Josephine, Johanna, Eagle, Medicine, and Mitchell)	USA	–	545–785	–	Biddanda et al. (2001)
Pony Lake and Lake Fryxel	USA	–	1789–4908	–	Schwede-Thomas et al. (2005)
Lakes, Michigan (n = 20)	46°N	–	120–1764	–	Pace and Cole (2002)
Banks Lake, and Okefenokee Swamp	25–30°N	–	950, 2425	–	Alberts et al. (1984)
Laramie River; Chimney Park wetland, USA	41°N	–	592, 1250	–	Brooks et al. (2007)
Lakes, Northeast USA, Colorado, Alaska (n = 47)	USA	–	67–1958	–	Morris et al. (1995)
Lake Superior and Nobska Pond	47, 41°N	–	117, 550	–	Repeta et al. (2002)
Upper Great Lakes, North American temperate forest (n = 188)	USA	–	133–2417	–	Xenopoulos et al. (2003)
Seepage Lakes, central Maine (n = 17)	USA	–	132–1492	–	David and Vance (1991)
Lakes (Sky Pond and The Loch), Loch Vale Watershed, Colorado	USA	–	67–308	–	Baron et al. (1991)

(continued)

Table 2 (continued)

Study sites	Country/ latitudes	Length/ watershed area[a] km or (km^2)[a]	DOC Surface (Depth)[c] (μM C)	DOC Deeper (Depth)[b]	Reference
Redberry Lake and its three inflows, Canada	52°N	–	933–2917	–	Waiser and Robarts (2000)
Lakes (n = 59), southwestern Québec, Canada	51°N–74°W	–	58–2350	–	Houle et al. (1995)
Lakes (n = 20), Québec, Canada	45°N	–	208–900	–	Cammack et al. (2004)
Lakes: Saskatchewan lakes (n = 26), Canada	51–53°N	–	342–13017	–	Arts et al. (2000)
Ponds and 8 Wetlands (n = 10), Canada	52°N	–	1283–6675	–	Arts et al. (2000)
Prairie wetlands, Saskatchewan, Canada	52°N	–	2933–10000	–	Waiser and Robarts (2004)
Lakes, South-central Ontario (n = 4)	Canada	–	225–675	–	Wu et al. (2005)
Freshwater Lakes (n = 23), Alberta	Canada	–	1833–5833	–	Curtis and Prepas (1993), Curtis and Adams (1995)
Saline Lakes (n = 35), Alberta	Canada	–	2667–27500	–	Curtis and Prepas (1993)
Saline Lakes (n = 37), Prairies	Canada	–	4983–10000 (median)	–	Molot et al. (2004)
Non-saline Lakes (n = 32), Prairies	Canada	–	3250–3833 (median)	–	Molot et al. (2004)
Lakes (n = 45 & 11), Atlantic Maritime, Nova Scotia and New Brunswick	Canada	–	483, 283 (median)	–	Molot et al. (2004)
Lakes (n = 60), Pacific Maritime	Canada	–	267 (median)	–	Molot et al. (2004)
Boreal lakes (n = 9), Ontario	Canada	–	150–425	–	Hudson et al. (2003)
Lakes (n = 30), Boreal Plains	Canada	–	3000 (mean)	–	Molot et al. (2004)
Lakes (n = 42 & 16), Boreal Shield, Quebec and Newfoundland	Canada	–	508 (median)	–	Molot et al. (2004)
Lakes (n = 42 & 16), Boreal Shield, Quebec and Newfoundland	Canada	–	725, 358 (median)	–	Molot et al. (2004)
Lakes (n = 24), Boreal Cordillera	Canada	–	1225 (median)	–	Molot et al. (2004)
Lakes (n = 49), Taiga Shield	Canada	–	475–2042 (median)	–	Molot et al. (2004)

(continued)

Table 2 (continued)

Study sites	Country/ latitudes	Length/ watershed area[a] km or (km^2)[a]	DOC Surface (Depth)[c] (μM C)	DOC Deeper (Depth)[b]	Reference
Lakes (n = 177), Montane Cordillera	Canada	–	125–1025 (median)	–	Molot et al. (2004)
Lakes (n = 38), Hudson Plains	Canada	–	783–1083 (median)	–	Molot et al. (2004)
Lakes (n = 37 & 18), Southern Arctic	Canada	–	525, 192 (median)	–	Molot et al. (2004)
Lakes (n = 12 & 6), Axel Heiberg and Victoria Islands, Southern Arctic	Canada	–	192, 92 (median)	–	Molot et al. (2004)
Lakes (n = 9 & 1), Banks and Prince Patrick Islands, Southern Arctic	Canada	–	283, 292 (median)	–	Molot et al. (2004)
Precambrian Shield lakes	Canada	–	357–1142	–	Curtis and Schindler (1997)
Lakes (n = 12), Northwest Ontario, Canada	51°N	(0.024–80000)	149–816	–	Kelly et al. (2001)
Lakes (n = 3), Loch Vale Watershed, Rocky Mountain National Park	Canada	(204–660)	31–117	–	McKnight et al. (1997)
Estuaries					
Pearl River Estuary, upper to lower reaches	China	–	84–473	–	He et al. (2010), Chen et al. (2004)
Caeté Estuary, mangroves forest	Brazil	–	283–558	–	Dittmar and Lara (2001)
Satilla Estuary, Geogia	USA	–	1972–2046	–	Moran et al. (2000)
Hudson River Esturay	USA	–	217–539	–	Hummel and Findlay (2006)
York River Estuary	USA	–	254–713	–	Raymond and Bauer (2001a)
Coastal Estuaries (n = 2), San Francisco Bay and Chesapeake Bay, USA	37–39°N	–	193 ± 4–222 ± 3	–	Boyd and Osburn (2004)
Arctic Estuary, the Beaufort Sea	Canada	–	133–453	–	Osburn et al. (2009)
Two Esturies, UK	55°N	–	427–1427	–	Baker and Spencer (2004)

(continued)

Table 2 (continued)

Study sites	Country/ latitudes	Length/ watershed area[a] km or (km^2)[a]	DOC Surface (Depth)[c] (μM C)	DOC Deeper (Depth)[b]	Reference
Estuary, Denmark	55°N	–	149–426	–	Stedmon et al. (2003)
Adour Estuary, France	France	–	76–316	–	Guéguen et al. (2002)
Gironde Estuary, French Atlantic coast	France	–	98–542	–	Huguet et al. (2010)
Seine Estuary, French Atlantic coast	France	–	133–554	–	Huguet et al. (2010)
Scheldt Estuary	Europe	–	183–517	–	Abril et al. (2002)
Rhine Estuary	Europe	–	142–258	–	Abril et al. (2002)
Gironde Estuary	Europe	–	92–208	–	Abril et al. (2002)
Thames Estuary	Europe	–	217–417	–	Abril et al. (2002)
Elbe Estuary	Europe	–	258–367	–	Abril et al. (2002)
Ems Estuary	Europe	–	425–592	–	Abril et al. (2002)
Sado Estuary	Europe	–	300–525	–	Abril et al. (2002)
Douro Estuary	Europe	–	158–208	–	Abril et al. (2002)
Loire Estuary	Europe	–	200–292	–	Abril et al. (2002)
Coastal and open Oceans					
Hiroshima Bay; Seto Inland Sea, Japan	32–34°N	–	83–135 (0–30 m)	71–92 (50–300 m)	Mostofa KMG et al., (unpublished data)
Barguzin Bay; Coastal areas, Lake Baikal	51–55°N	–	105–363	–	Yoshioka et al. (2002a, 2007)
Mississippi River Plume	28–29°N	–	54–124	–	Chen et al. (2004)
Orinoco River Plume	USA	–	70–276	–	Blough et al. (1993), del Castillo et al. (1999)
Chesapeake Bay	35–41°N	–	118–215	–	Mitra et al. (2000)
Florida Bay	24–25°N	–	139–147	–	Zanardi-Lamardo et al. (2004)
West Florida Shelf	USA	–	89–305	–	del Castillo et al. (2000)
Middle Atlantic Bay	USA	–	70–150	–	Vodacek et al. (1995)

(continued)

Table 2 (continued)

Study sites	Country/ latitudes	Length/ watershed area[a] km or (km^2)[a]	DOC Surface (Depth)[c] (μM C)	Deeper (Depth)[b]	Reference
Kara Sea, an estuary of Arctic Ocean	72–73°N	–	423–537	–	Opsahl et al. (1999)
Middle Atlantic Bight (MAB)	USA	–	82–98	48–90 (90–2600 m)	Mitra et al. (2000)
Arctic shelf, the Beaufort Sea	Canada	–	97–229	–	Osburn et al. (2009)
Arctic Gulf, the Beaufort Sea	Canada	–	70–126	–	Osburn et al. (2009)
South Baltic Sea	53–66°N	–	474–616	–	Ferrari et al. (1996), Ferrari and Dowell (1998)
Northern Gulf of Maxico	35–41°N	–	50–100	–	Wang et al. (2004)
North Pacific Ocean	31°N	–	87	38–55 (200–6000 m)	Williams and Druffel (1987)
Western North Pacific	35°N	–	85–117	66–73 (150 m)	Ogawa and Ogura (1992)
Arctic Ocean	74–81°N	–	58–85 (50–200 m)	49–54 (>1000 m)	Opsahl et al. (1999)
Atlantic Ocean	3–75°N	–	50–97 (<100 m)	33–59 (>1000 m)	Ogawa and Tanoue (2003)
Pacific Ocean	0–58°N, 1267°S	–	40–90	34–45	Ogawa and Tanoue (2003)
Indian Ocean; Arabian Sea	5–20°N	–	55–95	43	Ogawa and Tanoue (2003)
Antarctis Ocean	42–77°S	–	38–75	34–60	Ogawa and Tanoue (2003)
Arctic Ocean	70–85°N	–	34–107	49–54	Ogawa and Tanoue (2003)

"n" indicates the number of upstreams, rivers and lakes whereas DOC levels are mentioned for ranges or average of the samples
[a]means the watershed area (km^2) applicable to lakes and the values in parentheses are river watershed area
[b]means the water depth in lakes and oceans and values are mentioned within the bracket
[c]means the surface layer and depth belongs within 0–50 m

newly flooded terrestrial vegetation, and leaching of organic matter from floodplain soils. In Latin American Rivers, the major sources of DOC are the dumping of untreated urban sewage into the rivers, the leaching of DOM from humid evergreen and deciduous forests, floodplains of ox-bows and ponds, seasonally humid savannas and shrublands, as well as swamps. The sources of DOC in European Rivers may be natural such as wetlands, swamps and peat-land or anthropogenic such as industrial activities and sewerage. The major sources in North America that account for the high DOC levels are natural (wetlands, swamps, marshlands, ground water of mountainous areas covered by riparian vegetation ecosystems) and anthropogenic (agricultural, sewerage and industrial effluents).

In groundwater, DOC concentrations are 16–424 μM C in Asia, 183–842 μM C in Australia, 42–15333 μM C in Europe, 8–8333 μM C in North America, 100–3000 in Brazil, 1108 ± 217–14167 ± 6333 μM C in Botswana, and finally 167–7500 μM C in soil solution in North America and Europe (Table 2) (Thurman 1985a; Mostofa et al. 2007a, Mostofa KMG et al., unpublished data; Buckau et al. 2000; Bertilsson et al. 1999; McIntyre et al. 2005; Mladenov et al. 2008; Meier et al. 2004; Gielen et al. 2011; Crandall et al. 1999; Schwede-Thomas et al. 2005; Pabich et al. 2001; Michalzik et al. 2001; Anawar et al. 2002; Richey et al. 2002; Bradley et al. 2007). The results of the available studies show that DOC concentrations in groundwater are significantly higher in Europe, North America, South America, and African regions than in Asia. High ground-water DOC concentrations are directly related to the high DOC concentrations in surface waters and high infiltration rates (Mladenov et al. 2007).

In rainwater, DOC concentrations are substantially varied among different countries, such as 25–3675 μM C in China, 217 μM C in Brazil, 4–1908 μM C in USA, 28–1078 in Europe, 161 μM C in a continental location, and 23 μM C in a marine location (Table 2) (Likens et al. 1983; Guggenberger and Zech 1993; Sakugawa et al. 1993; Willey et al. 2000; Ciglasch et al. 2004; Avery et al. 2006; Kieber et al. 2006; Santos et al. 2009a, b; Southwell et al. 2010; Miller et al. 2008; Mostofa KMG et al., unpublished data; Xu et al. 2008; Pan et al. 2010). Rainwater DOC concentrations are significantly high in China and Europe, but relatively low in USA, and lower in marine locations than in continental ones. These results suggest that DOC concentrations in rainwater are mostly affected by the atmospheric organic contaminants, which generally derive from the surrounding environments.

DOM Contents in Lakes, Wetland and Swamps

DOC concentrations are significantly variable in lakes situated among different continents: 80–4133 μM C in Asia (Table 2) (Yoshioka et al. 2002a, 2007; Fu et al. 2010; Mostofa et al. 2005a, Mostofa KMG et al., unpublished data; Hayakawa et al. 2003, 2004; Sugiyama et al. 2004; Ishikawa et al. 2006; Rae et al. 2001); approximately 21–710 μM C in Latin America (Morris et al. 1995; Farjalla et al. 2006); approximately 59–3750 μM C in Europe (Bertilsson and Tranvik 2000; Lu et al. 2007; Worrall et al. 2004a; Evans et al. 2006; Monteith and Evans 2005; Guéguen et al. 2002; Larsen et al. 2011); and approximately 70–27500 μM

C in North America (Morris et al. 1995; Pace and Cole 2002; Hudson et al. 2003; Wu et al. 2005; Cammack et al. 2004; McKnight et al. 1997; Allard et al. 1994; Waiser and Robarts 2000; Biddanda and Cotner 2002; Repeta et al. 2002; Schwede-Thomas et al. 2005; Curtis and Prepas 1993; Curtis and Schindler 1997; Arts et al. 2000; Biddanda et al. 2001; Kelly et al. 2001; Xenopoulos et al. 2003; Molot et al. 2004; Brooks et al. 2007). The general results of DOC concentrations studies in lakes show some characteristic phenomena (Table 2): (i) relatively low DOC concentrations have been found in Asia, but lakes situated Indonesia and North-East China shows high contents of DOC. (ii) Lakes situated in Europe and North America show relatively high DOC values. DOC concentrations in Saline lakes are significantly higher than in non-Saline lakes situated in the Prairies and Alberta in Canada. Boreal lakes often show high contents of DOC. This is interesting because the boreal region contains roughly 30 % of the global lakes and their water is rich in OM (Molot and Dillon 1996; Downing et al. 2006; Benoy et al. 2007).

A large lake database (7,514 lakes from 6 continents) shows that mean DOC concentrations are 632 ± 16 μM C, ranging from 8 μM C to 27667 μM C (Sobek et al. 2007). In 87 % of the lakes DOC was between 83 and 1667 μM C, whilst 8.3 % of the lakes had concentrations lower than 83 μM C. Lakes between 1667 and 3333 μM C are relatively few (4.2 %), and only 0.4 % of the lakes had DOC concentrations above 3333 μM C (Fig. 2a). In 55 % of the lakes DOC concentration was above 417 μM C, which is suggested to be a threshold value for the transition between net autotrophy and net heterotrophy in lakes (Sobek et al. 2007; Jansson et al. 2000; Prairie et al. 2002). Several hypotheses are considered based on DOC concentration and terrestrial vegetation (Sobek et al. 2007). First, arctic lakes are generally characterized by low DOC concentrations where land-cover types are wooded tundra to bare desert and correspond to an annual mean temperature of <−4 °C (Fig. 2b). Second, boreal lakes display a tendency toward higher DOC concentration if the land-cover types are deciduous and mixed boreal forest to deciduous conifer forest, which corresponds to an annual mean temperature of 0.5– 4 °C (Fig. 3b). Third, lakes on the northern Great Plains in Saskatchewan and Canada show the highest DOC concentrations (~80 to ~10500 μM C) and the land-cover types are cool grasses, shrubs, cool crops and towns (Fig. 3c). Fourth, Fig. 2b shows that DOC concentrations for lakes situated in the warmer climate zones (average mean annual temperature >17.4 and <22.3 °C) reach relatively high values (approximately 80–3300 μM C) for land-cover types including broadleaf crops, corn and cropland and conifer forests (Fig. 2b). Correspondingly, DOC concentrations for lakes situated in the low warmer climate zones (average mean annual temperature >8.0 and <13.3 °C) are relatively low (approximately 50–2000 μM C). Here typical land-cover types are crops, mixed and deciduous broadleaf forests, cool and cool rain forests, grass, and shrubs. Fifth, Fig. 2b suggests that DOC concentrations for lakes that experience relatively low annual temperature (>1.5 and <4.3 °C) are relatively high (approximately 50–2100 μM C), and the land-cover types are mostly woods, deciduous and mixed boreal forest, cool mixed forest, conifer boreal forest, narrow conifers and deciduous conifer forests.

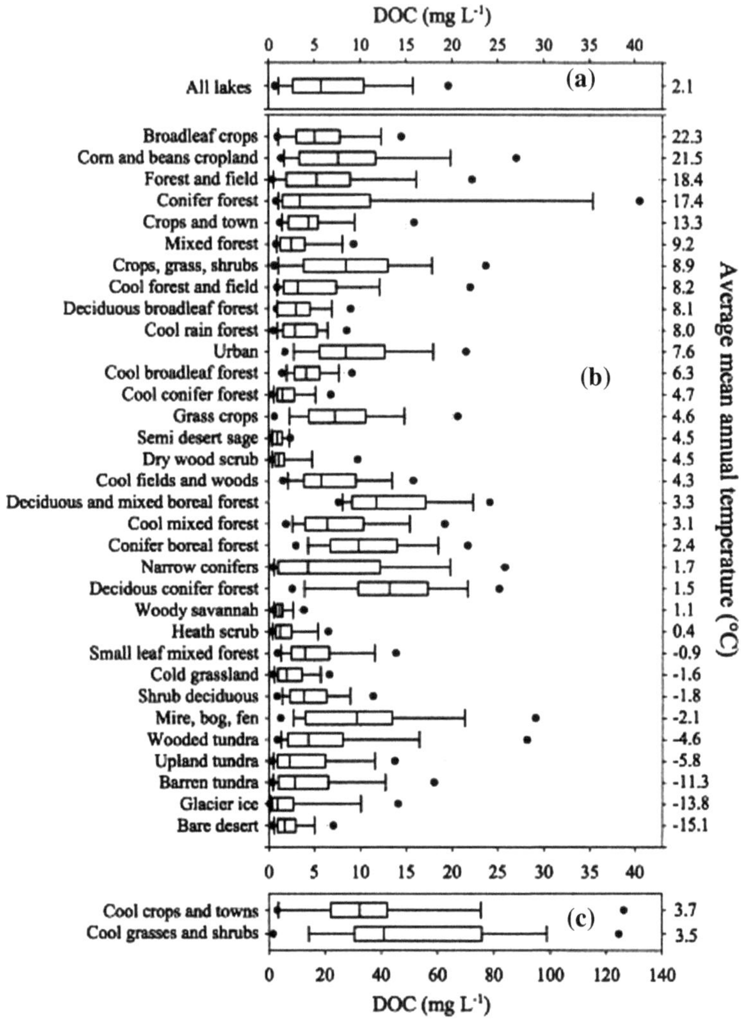

Fig. 2 Box and whisker-plot of the distribution of DOC concentration (**a**) for all lakes, (**b**) for lakes divided into different land-cover types, and (**c**) for lakes in Saskatchewan, Canada. In (**b**), the land-cover types have been sorted according to the average mean annual temperature. The boxes display the median and the quartiles, the whiskers represent the 10 and 90 % percentiles, and the points represent the 5 and 95 % percentiles. Only land-cover types containing 10 or more lakes are shown. The land-cover types "inland water" and "sea water" are omitted from the plot because they are not indicative of climate or geography. *Data source* Sobek et al. (2007)

DOM Contents in Estuaries

DOC concentrations in estuaries are often lower than in lakes all over the world. They reach 84–473 μM C in China (Chen and Gardner 2004; He et al. 2010); 283–558 μM C in Brazil (Dittmar and Lara 2001); 190–2046 μM C in the USA

(Moran et al. 2000; Raymond and Bauer 2001a; Boyd and Osburn 2004; Hummel and Findlay 2006); 133–453 μM C in Canada (Osburn et al. 2009); 76–1427 μM C in Europe (Abril et al. 2002; Guéguen et al. 2002; Huguet et al. 2010; Baker and Spencer 2004; Stedmon et al. 2003). DOC in estuaries is originated from both autochthonous sources (algae or phytoplankton) and allochthonous sources such as material of terrestrial plant origin.

DOM Contents in Coastal and Open Oceans

DOC concentrations are substantially higher (50–616 μM C) in coastal seawaters than in the open ocean (40–117 μM C) (Table 2) (Ogawa and Tanoue 2003; Mitra et al. 2000; Chen and Gardner 2004; Williams and Druffel 1987; Opsahl and Benner 1998; Wang et al. 2004; Zanardi-Lamardo et al. 2004; Osburn et al. 2009; Blough et al. 1993; Vodacek et al. 1995; Ferrari et al. 1996; Ferrari and Dowell 1998; Del Castillo et al. 1999, 2000; Mostofa KMG et al., unpublished data). DOC concentrations in coastal waters are generally regulated by terrestrial or riverine input, zooplankton feeding and algal or phytoplankton production (Lee and Wakeham 1988; Lee and Henrichs 1993; Mann and Wetzel 1995; Hedges et al. 1997). Increased biomass of primary producers also plays a major role in regulating the DOM contents in coastal areas (Ittekkot 1982; Billen and Fontigny 1987; Bronk et al. 1998). DOC concentrations in open oceans are relatively low and are included in the range of 40–117 μM C at epilimnion and 30–90 μM C at hypolimnion (Table 2) (Ogawa and Tanoue 2003). DOC values appear to be relatively uniform in most of the oceans, whilst in coastal areas the DOM pool becomes much more heterogeneous because of terrestrial inputs. In open oceans, the DOC concentrations in epilimnetic layers follow the order of Arctic Ocean (70–107 μM C) > subtropical zone (~80 μM C) > tropical (equatorial) and temperate zones (60–70 μM C) > subarctic and subantarctic regions (50–60 μM C) > Antarctic region (40–60 μM C) (Ogawa and Tanoue 2003). There are two main sources of DOC in coastal oceans: allochthonous DOM of terrestrial origin and autochthonous DOM of algal or phytoplankton origin. In open oceans, allochthonous DOM contents are gradually decreased photolytically depending on the distance from coastal areas, whilst autochthonous DOM is substantially increased (see the contribution of DOM for detailed discussion). The three major sources of DOC in the Arctic Ocean are in situ production (56 %), river run-off (25 %), and Pacific water (19 %) (Kirchman et al. 1995).

6 Factors Affecting DOM in Natural Waters

DOM contents and its dynamics are mostly dependent on two issues: origin and/or input, as well as its consequent mineralization by various environmental factors that are associated with the watershed activities of natural waters. The river-groundwater interface can act as a source or sink for DOM, depending on the volume and direction of flow, DOC concentrations and biotic activity (Brunke and Gonser 1997). The dynamics of lake DOM is greatly affected by specific and regional factors such as pH of lake water, air temperature, solar radiation, precipitation, sulfate deposition, DOC contents in the adjacent rivers, vegetation of the terrestrial ecosystem, and southern oscillation index (SOI) (Mostofa et al. 2005a, 2009b; Hudson et al. 2003). A complex balance of abiotic and biotic processes controls the molecular composition of marine DOM to produce signatures that are characteristic of different environments (Kujawinski et al. 2009). Therefore, the controlling factors affecting the origin and dynamics of both allochthonous and autochthonous DOM in natural waters can be distinguished as: (i) Types and nature of terrestrial plant material in soil; (ii) Land management and natural effects (precipitation, flood and drought); (iii) Effect of temperature; (iv) Microbial processes; (v) Photoinduced processes; (vi) Photosynthesis in natural waters; (vii) Metal ions complexation and salinity; and (viii) Global warming.

6.1 Types and Nature of Terrestrial Plant Material in Soil

Allochthonous DOM is originated in soil by microbial degradation of leachable organic carbon, which varies depending on the types and nature of terrestrial plant communities, soil types and other regional effects (Mostofa et al. 2009a; Nakane et al. 1997; Uchida et al. 1998, 2000; Moore et al. 2008; Tu et al. 2011; Kindler et al. 2011; Duff et al. 1999; Michalzik et al. 2001; Rae et al. 2001; Cronan and Aiken 1985; Frost et al. 2006; Johnson et al. 2006). The litter-rich surface soils have relatively higher DOC concentration than the litter-lacking ones, which can be distinguished because in the former case the $\delta^{13}C$ values of DOC are closer to the $\delta^{13}C$ of litter than to the $\delta^{13}C$ of organic carbon in forest soil (Tu et al. 2011). In most temperate and boreal landscapes the DOC concentrations in inland waters are regulated by a wide variety of watershed characteristics, including the quantity and type of vegetation, watershed slope, and particularly the extent and nature of wetlands (Kindler et al. 2011; Allard et al. 1994; Rae et al. 2001; Xenopoulos et al. 2003; Frost et al. 2006; Engstrom 1987; Williamson et al. 2001; Rice 2002; Canham et al. 2004; Winn et al. 2009).

DOC leaching from topsoils in the presence of different vegetation is largely variable. Therefore, different values have been observed in the presence of grasslands (range 158–1425 μM C and mean: 667 μM C), croplands (range: 325–1442 μM C and mean: 1000 μM C) and forests (range: 592–3592 μM C and mean: 1917 μM C), but large variations have also been observed within land use classes (Kindler et al. 2011). Under Cerrado vegetation, total organic C (TOC) concentrations (filtered < ca. 1 μm) found in the soil solution (ca. 417 μM C) between 15 and 200 cm depth were lower than those usually found in the soil of temperate forests (833–1667 μM C) (Michalzik et al. 2001; Lilienfein et al. 2001). TOC concentrations in the soil solution under *Pinus* are lower than under Cerrado (Lilienfein et al. 2001). In uplands, soils derived from coniferous forests are much richer in DOC than those from hardwood stands or grass lands (Rae et al. 2001; Cronan and Aiken 1985). Lakes with catchments containing 55–65 % natural grassland and <30 % forest have low DOC concentration (<83 μM C), whilst lakes in moderately forested (50–60 %) catchments have DOC concentrations of 83–208 μM C, and those in densely forested (>70 %) catchments have DOC concentrations of 667 μM C (Cronan and Aiken 1985). Forested wetlands, particularly those with coniferous trees, are positively related to lake DOC, open-water wetlands including lakes are inversely related to DOC, and scrub-shrub and emergent wetlands are not related to DOC (Xenopoulos et al. 2003). Upstream rivers covered by coniferous, deciduous or moxed-type forests have generally low (~<200 μM C) DOC concentrations (Table 2) (Mostofa et al. 2005a, b; Sugiyama et al. 2005). A model analysis has shown that the terrestrial land cover such as conifer boreal forest and barren tundra strongly affects DOC in lakes (Sobek et al. 2007). The land cover type "conifer boreal forest" is positively related with lake DOC, while "cool conifer forest" is negatively related to DOC. However, cool conifer forest is confined to high altitude areas such as the Rocky Mountains and the Alps, which may explain the relatively low DOC concentrations found in these lakes (Sobek et al. 2007).

In polar desert lakes, DOM is generated autochthonously by microbial processes in water, since there is no catchment vegetation (Rae et al. 2001). This DOM thus differs from that of temperate latitudes by having a reduced ratio of aromatic to aliphatic residues (McKnight et al. 1994). It has also been shown that DOC from predominantly grassland catchments is qualitatively different in terms of its UVR attenuation properties than DOC from a mainly forested catchment (Rae et al. 2001). Therefore, the type and amount of terrestrial vegetation surrounding a catchment plays a significant role in defining the concentration levels of DOC in the catchment water. It is hypothesized that high ground-water DOC concentrations are directly related to high DOC concentrations in surface waters (Mladenov et al. 2007).

6.2 Land Management and Natural Effects (Precipitation, Flood and Drought)

Land management and natural effects (precipitation, flood and drought) are important factors for controlling DOM release from soil environments to natural water catchments (Mostofa et al. 2005b, 2007a; Watts et al. 2001; Ittekkot et al. 1985; Safiullah et al. 1987; Newbern et al. 1981; Richey et al. 1990; Depetris and Kempe 1993; Shaw 1979; Worrall et al. 2003; Worrall and Burt 2004; Yallop and Clutterbuck 2009; Clutterbuck and Yallop 2010; Yallop et al. 2010). These processes include several phenomena:

(i) DOC is largely released from soil into water during agricultural activities, particularly in plantation and growing seasons of rice plants as well as other plantations through rainfall or water overflow (Mostofa et al. 2005a, 2007a). Rapid photo- and microbial respiration or assimilation of soil OM might be responsible for high releases of DOC into water during agricultural activities.

(ii) DOC concentrations are increased with afforestation in catchments (Ciglasch et al. 2004; Neal et al. 1998, 2004). In addition, deforestation can reduce evaporation and increase surface temperature. Changes in land-surface cover can enhance the degradation of soil DOM and OM by both photoinduced and microbial processes (Brandt et al. 2009; Rutledge et al. 2010; Raich and Schlesinger 1992; Borges et al. 2008). Therefore, either afforestation or deforestation in soil environments can contribute to the rapid washout of allochthonous DOM and OM to water catchments by precipitation or runoff.

(iii) Controlled heather burning as a management tool for red grouse (Lagopus lagopus) husbandry in peat surface is often (but not always) identified as a highly significant driver of spatial variance in DOC concentration in drainage water (Yallop and Clutterbuck 2009; Yallop et al. 2006, 2008; Ward et al. 2007; Worrall et al. 2007).

(iv) Increased precipitation and runoff can lead to higher DOC export from the catchment into natural surface waters (Pace and Cole 2002; Zhang et al. 2010; Monteith et al. 2007; Hongve et al. 2004; Sobek et al. 2007; Ciglasch et al. 2004; Gielen et al. 2011; Anderson et al. 1997; Evans et al. 1999). Total solar radiation and precipitation can account for 49–84 % of the variation in the long-term DOC patterns in various catchments (Zhang et al. 2010). The DOC concentrations in Swedish lakes and streams have substantially increased during the 1970–1980s, mostly due to higher precipitation (Tranvik and Jasson 2002). DOC concentrations vary from 4 μM C to 3675 μM C in rainwater, which may largely affect the natural surface waters (Table 2) (Likens et al. 1983; McDowell and Likens 1988; Guggenberger and Zech 1993; Chebbi and Carlier 1996; Willey et al. 2000, 2006; Ciglasch et al. 2004; Avery et al. 2006; Kieber et al. 2006, 2007; Miller et al. 2009; Santos et al. 2009a, b; Southwell et al. 2010; Pan et al. 2010). Factors affecting the variation in DOC concentrations are rainwater volume, season (winter, spring or summer), location of the rain events, wind speed, storm trajectory, and the air mass pathways during precipitation. During highly rainy seasons, DOC concentrations are significantly increased through flushing of organic-rich waters from upper soil horizons, agricultural and forest runoff, primary production and subsequent flooding of the ox-bow or flood-plain lakes, through lake out-flowing into the nearby catchment waters (Mostofa et al. 2005b; Ittekkot et al. 1985; Safiullah et al. 1987; Ishikawa et al. 2006; Newbern et al. 1981; Richey et al. 1990; Depetris and Kempe 1993; Mulholland 2003). Leaching and heterotrophic processing of newly flooded terrestrial vegetation, leaching of organics from floodplain soil and catchment areas can also affect the DOC concentrations (Wetzel 1992; Duff et al. 1999; Mladenov et al. 2007; Reche et al. 1999). The flow path of water in the catchment after precipitation is an important factor that can reduce soil erosion (Sobek et al. 2007), thereby reducing the rapid washout of allochthonous DOM, POM and nutrients to water catchments.

Many studies observe that DOC export is positively correlated with runoff, and two issues are involved (Sobek et al. 2007): First, the carbon budget of the studied landscapes is not in steady state, i.e., increased runoff exports more DOC than is produced in soil, which implies that observed increases are temporary. Second, changes in runoff are concomitant to changes in leachable organic carbon stocks in soil. High runoff indicates a high water table, which favors DOC leaching and hampers microbial degradation due to anoxia and humification. In addition, the negative relationship between runoff and lake DOC concentration indicates that when high runoff prevails over extended periods of time, the leachable soil organic carbon pool will eventually be reduced (Sobek et al. 2007).

(v) Severe drought seasons can either greatly decrease or increase the DOC levels in natural water (Watts et al. 2001; Meier et al. 2004; Shaw 1979; Worrall et al. 2003, 2005, 2006; Worrall and Burt 2004; Ward et al. 2007). DOC concentrations in shallow groundwater are very low (1558 μM C) under drought condition compared to spring samples (2583 μM C). Correspondingly, the properties of DOM are largely different (Meier et al. 2004). Such differences may be attributed to biogeochemical changes in the DOM pool over the summer and fall seasons under drought conditions. Studies show that runoff characteristics and flow-paths within peat soils change as a result of severe drought, which could increase DOC concentrations in runoff water (Evans et al. 1999; Holden and Burt 2002).

In addition, releases of allochthonous DOM are largely dependent on several catchment properties such as drainage ratio (catchment : lake area), proportion of wetlands, proportion of upstream lakes, watershed slope, altitude, catchment area, % peat cover, water area, wetland cover, and soil C:N ratio in the catchment (Sobek et al. 2007; Eckhardt and Moore 1990; Xenopoulos et al. 2003; Rasmussen et al. 1989; Kortelainen 1993; Hope et al. 1997; Aitkenhead and McDowell 2000). The export of DOC from catchments is often related to the organic carbon stocks in the catchment soils (Hope et al. 1994; Aitkenhead et al. 1999). Studies show that DOC fluxes are small: 0.8 ± 0.2 % relative to gross primary productivity, 1.0 ± 0.3 % relative to ecosystem respiration, and (2.4 ± 0.4 %) relative to soil respiration, when the DOC fluxes are considered relative to the gross ecosystem carbon fluxes in a specific catchment (Gielen et al. 2011).

6.3 *Effect of Temperature*

Temperature can affect DOM in two ways: First, water temperature (WT), linked with solar radiation, is one of the most important variables in the production of autochthonous DOM in natural waters because it affects the physical, photoinduced, microbial and ecological processes (Mostofa et al. 2009a; Sobek et al. 2007; Gielen et al. 2011; Gudasz et al. 2010). The mineralization of organic carbon in lake sediments exhibits a strong positive relationship with temperature, which suggests that increasing temperature would lead to increased mineralization of OM in natural waters (Gudasz et al. 2010). As already seen, the distribution of DOC concentrations for various lakes (7500 lakes from 6 continents; Fig. 2) (Sobek et al. 2007) does not show a simple relationship between DOC and temperature, because the DOC values are affected by both temperature and the surrounding vegetation. However, autochthonous DOM is often higher in lake waters where the water temperature (WT) is higher. For example, WT in the surface water of Lake Baikal is generally lower (4–16 °C: summer period) compared to Lake Biwa (10–28.7 °C: summer period), although the DOC levels are almost similar: 88–114 μM C at 0–1400 m depth and 76–135 μM C at 0–80 m at central basins, respectively, during the summer stratification period (Weiss et al. 1991; Mostofa et al. 2005a; Yoshioka et al. 2002a; Goldman et al. 1996). However, autochthonous production in Lake Biwa is significantly higher (3–82 %) than in Lake Baikal (6–35 %) (Table 2) (Mostofa et al. 2005a; Yoshioka et al. 2002a; Sugiyama et al. 2004). Autochthonous production is not observed in a region where the WT is very low (ca. $\leq$0 °C) and, at the same time, chlorophyll *a* (Chl *a*) production does not occur in the upper water column (Bussmann and Kattner 2000) However, a little increase in WT may produce a little amount of Chl *a* with a corresponding increase in autochthonous DOM in natural waters (Wheeler et al. 1996, 1997; Bussmann and Kattner 2000; Melnikov and Pavlov 1978; Tremblay et al. 2006). Low WT may affect the DOM contents by several pathways (Mostofa et al. 2009a): (i) Photoinduced degradation of surface DOM is less effective due to low solar effects at low WT and air temperature. This may result in low contents of photoproducts, such as DIC, CO_2, H_2O_2, LMW organic substances and so on, which subsequently decreases photosynthesis and primary production. The result is a decrease of autochthonous DOM production in natural waters. (ii) Mineralization of DOM by photoinduced degradation becomes significantly low at low WT, which may preserve the DOM in natural waters and lead to increased allochthonous DOM contents.

An increase in air temperature can significantly enhance DOC export from soil to surface water by increasing soil respiration and mineralization of plant organic material (Mostofa et al. 2005a, b; Monteith et al. 2007; Raymond and Saiers 2010; Gielen et al. 2011; Evans et al. 1999, 2002; Gudasz et al. 2010; Newson et al. 2001). This can lead to DOM leaching from groundwater to stream or riverbeds or lakes, and the DOC concentrations are linearly increased with increasing temperature in natural waters. Coherently, the DOC leaching from forest catchments to streams is significantly enhanced during the summer season. It has been shown that releases of DOM in upstream waters of forest mountainous origin are much higher (28–84 %) during summer than in winter season. This has been estimated during monthly samplings in four upstream rivers and the releases were highest (52–84 %) in forest soils at upstream sites of Kurose River than in forest granite mountain (28–31 %) in Lake Biwa watershed (Mostofa et al. 2005a, b). It is suggested that increased temperature during the summer season can lead to higher microbial activity and enhanced decomposition of organic matter or peat, which increases production of DOC. Increases in temperature can also induce higher drawdown of water tables in summer, increasing the depth of the zone where oxidation and production of DOC take place (Evans et al. 1999). In the UK, the latter effects would probably be exacerbated by the decreased summer rainfall over the last 40 years (Burt et al. 1998). It is suggested that the effect of increased temperature on water tables can account for between 10 and 20 % of the increase in DOC concentration (Worrall et al. 2004b, 2007; Cole et al. 2002).

6.4 Microbial Processes

Microbial processes have two important effects on OM (DOM and POM). First microbial respiration or assimilation of OM into algal or phytoplankton biomass or bacterial biomass can release autochthonous DOM in deep water (Mostofa et al. 2009a, b, 2011; Zhang et al. 2009; Yamashita and Jaffé 2008; Fu et al. 2010; Rochelle-Newall and Fisher 2002a; Yamashita and Tanoue 2004; Aoki et al. 2008; Stedmon and Markager 2005a; Stedmon et al. 2007a). This process can give an important contribution to autochthonous DOM in natural waters. Second, microbial processes can change the molecular structure of DOM components and their optical properties, either absorption properties of CDOM or fluorescence properties of FDOM (Moran et al. 2000; Mostofa et al. 2007a; Hur 2011). Such properties will be discussed in chapters "Photoinduced and Microbial Degradation of Dissolved Organic Matter in Natural Waters", "Colored and Chromophoric Dissolved Organic Matter in Natural Waters", and "Fluorescent Dissolved Organic Matter in Natural Waters". Microbial degradation can mineralize DOC by approximately 0–85 % in natural waters (see chapter "Colored and Chromophoric Dissolved Organic Matter in Natural Water"). High molecular weight protein-like structures in plant-derived DOM are degraded primarily through physical–chemical and microbial processes (Scully et al. 2004). Microbial activity is significantly stimulated by the photoproducts of readily assimilable nitrogen compounds such as ammonium and amino acids (Bushaw et al. 1996; Jørgensen et al. 1998). Under N-limiting conditions, nitrogenous photoproducts can significantly increase the rates of bacterial growth in natural waters (Bushaw et al. 1996). Microbial degradation depends on several key factors, such as occurrence and nature of microbes; sources of DOM and the quantity of its fermentation products; temperature; pH; and sediment depth (see chapter "Photoinduced and Microbial Degradation of Dissolved Organic Matter in Natural Waters").

6.5 Photoinduced Processes

Photoinduced processes have two effects on OM (DOM and POM): First, photo-respiration or assimilation of POM can release autochthonous DOM in surface waters (Mostofa et al. 2009b, 2011; Harvey et al. 1995; Fu et al. 2010; Rochelle-Newall and Fisher 2002a; Hiriart-Baer and Smiith 2005). This process can give a significant contribution to autochthonous DOM in natural waters. Second, solar radiation causes changes in the molecular structure of DOM and decomposes its functional groups. This effect can be detected chemically as mineralization of DOC, by approximately 0–54 % during irradiation times ranging from hours to months. It can also be detected optically as alteration of either chromophoric dissolved organic matter (CDOM) or fluorescent dissolved organic matter (FDOM) (see chapters "Photoinduced and Microbial Degradation of Dissolved Organic Matter in Natural Waters", "Colored and Chromophoric Dissolved Organic Matter in Natural Waters", and "Fluorescent Dissolved Organic Matter in Natural Waters"). Photoinduced degradation can also reduce the mean molecular size of the high molecular weight DOM (Lovley and Chapelle 1995; Lovley et al. 1996; Yoshioka et al. 2007; Amador et al. 1989), which subsequently produces low molecular weight (LMW) intermediates (Lovley et al. 1996; Wetzel et al. 1995; Amon and Benner 1994; Dahlén et al. 1996). This process ultimately ends up in mineralization with formation of e.g. COS, CO, CO_2, DIC, ammonium and gaseous hydrocarbons (Miller and Zepp 1995; Miller 1998; Johannessen and Miller 2001; Ma and Green 2004; Xie et al. 2004; Johannessen et al. 2007; Gennings et al. 2001; Clark et al. 2004). Photoinduced degradation generally occurs in the mixing zone and decreases with an increase in water depth (Bertilsson and Tranvik 2000; Ma and Green 2004; Vähätalo et al. 2000; Mostofa et al. 2005a; Granéli et al. 1996). The photoreactivity of fluorescent DOM is greatly decreased when passing from freshwater to marine waters, but deep waters in lakes or marine environments are often more sensitive to photoinduced degradation processes than surface waters (Mostofa et al. 2011). Similar effects have been observed as far as photomineralization is concerned (Vione et al. 2009). Photoinduced degradation is significantly affected by several key factors, such as solar radiation, water temperature, effects of total dissolved Fe and photo-Fenton reaction, occurrence and quantity of NO_2^- and NO_3^- ions, molecular nature of DOM, water pH and alkalinity, dissolved oxygen (O_2), water depth, physical mixing in the surface mixing zone, increased UV-radiation during ozone hole events, global warming, and salinity (see chapter "Photoinduced and Microbial Degradation of Dissolved Organic Matter in Natural Waters").

6.6 Photosynthesis in Natural Waters

Autochthonous production of DOM in natural waters is mostly accompanied by photosynthesis (Takahashi et al. 1995; Hamanaka et al. 2002; Marañòn et al. 2004). Photosynthetically produced POM (mostly algae or phytoplankton) and the related release of new DOM are significantly influenced by several key factors, such as high precipitation (Freeman et al. 2001a; Tranvik and Jasson 2002; Hejzlar et al. 2003; Zhang et al. 2010), nitrogen deposition (Pregitzer et al. 2004; Findlay 2005), sulfate (SO_4^{2-}) deposition (Zhang et al. 2010; Evans et al. 2006; Monteith et al. 2007), and changes in total solar UV radiation or an increase in temperature due to global warming (Freeman et al. 2001a; Zhang et al. 2010; Sinha et al. 2001; Sobek et al. 2007; Rastogi et al. 2010). The increase in temperature driven by solar radiation is effective in inducing photoinduced and microbial processes of OM (including DOM and POM) as well as in enhancing photosynthesis. This is consistent with data from the Central England Temperature Record (Parker et al. 1992), showing that mean summer temperatures across England were 0.66 °C higher during the 1990s than in the preceding 30 years. Model studies predict that the production of new DOM due to photosynthetic processes from winter to summer would vary from 6 to 60 %, due to a large seasonal variation in light intensity (Anderson and Williams 1998; Bratback and Thingstad 1985). The factors affecting the photosynthesis in natural waters are discussed in detail in chapter "Photosynthesis in Nature: A New Look".

6.7 Metal Ions Complexation and Salinity

Metal ions can complex the DOM functional groups (fulvic and humic acids of vascular plant origin, autochthonous fulvic acids of algal or phytoplankton origin, tryptophan, protein, algae and so on) and can induce structural changes (e.g. molecular conformation or rigidity) and formation of aggregates. Complexation would thus change the outer appearance of the molecule and its optical properties, such as absorption properties of CDOM and fluorescence properties of FDOM, either increasing or decreasing them (Mostofa et al. 2009a, 2011; Lead et al. 1999; Wang and Guo 2000; Koukal et al. 2003; Mylon et al. 2003; Wu et al. 2004; Lamelas and Slaveykova 2007; Lamelas et al. 2009; Fletcher et al. 2010; Reiller and Brevet 2010; Sachs et al. 2010; Da Costa et al. 2011). Correspondingly, salinity can also affect the DOM components in seawater, both structurally and optically, modifying them in comparison to freshwater (Nakajima 2006; Blough et al. 1993; del Vecchio and Blough 2002; Boyd et al. 2010). Complexation of metal ions and the effect of salinity are extensively discussed in Chapters "Photoinduced and Microbial Degradation of Dissolved Organic Matter in Natural Waters", "Fluorescent Dissolved Organic Matter in Natural Waters", and "Complexation of Dissolved Organic Matter with Trace Metal Ions in Natural Waters".

6.8 Global Warming

Global warming may affect DOM in two ways: First, global warming could further enhance atmospheric CO_2, because of elevated net primary productivity and increases root exudation of DOC in soil environments (Freeman et al. 2001b, 2004; Lavoie et al. 2005; Fenner et al. 2007a, b; Wolf et al. 2007; Kang et al. 2001; Tranvik and Jasson 2002; Monteith et al. 2007; Evans et al. 2002; Dorodnikov et al. 2011). This process ultimately leaches allochthonous DOM into the aquatic ecosystem. Second, global warming may accelerate the photoinduced and microbial decomposition of DOM to produce compounds such as H_2O_2, CO_2, DIC, NO_3^-, PO_4^{3-}, NH_4^+, LMWDOM and so on (Mostofa and Sakugawa 2009; Johannessen and Miller 2001; Ma and Green 2004; Xie et al. 2004; Johannessen et al. 2007; Palenik and Morel 1988; Kotsyurbenko et al. 2001; Lovley 2006). The availability of these compounds can enhance photosynthesis and ultimately increase the primary and secondary production. These processes can induce the formation of autochthonous DOM and are usually expected to deteriorate the quality of natural waters. All these processes are extensively discussed in chapter "Impacts of Global Warming on Biogeochemical Cycles in Natural Waters".

7 Possible Mechanisms for Increased and Declined DOM Contents in Surface Waters

Production of autochthonous DOM is a well-known phenomenon in stagnant surface waters, particularly in lakes and oceans. The corresponding increase in autochthonous DOC is, on average, 0–102 % in lakes and 0–194 % in the oceans' epilimnion compared to the hypolimnion during the summer stratification period (Table 2). Increased concentrations of DOC in surface waters are a commonly observed phenomenon in North America and in North and Central Europe including UK, the Czech Republic, Finland, Norway, Canada, USA and so on (Hejzlar et al. 2003; Worrall et al. 2004a, 2007; Evans et al. 2005, 2006; Monteith et al. 2007; Hongve et al. 2004; Monteith and Evans 2005; Larsen et al. 2011; Clutterbuck and Yallop 2010; Yallop et al. 2010; Freeman et al. 2001b; Bouchard 1997; Skjelkvåle et al. 2001, 2005; Driscoll et al. 2003; Stoddard et al. 2003; Vuorenmaa et al. 2006). The increase in DOC export also enhances the export of humic DOC from upland peat catchments (Yallop et al. 2010). It is estimated that the increase in mean DOC concentrations between the first and last 5 years of monitoring in UK's Acid Waters Monitoring Network (AWMN) streams and lakes are 32–135 % in 11 streams and 31–140 % in 11 lakes (Evans et al. 2006; Monteith and Evans 2005). DOC in UK rivers arises from a number of sources including: decomposition of deep peat if present (McDonald et al. 1991), sewage (Eatherall et al. 2000), industrial point-source effluents (Tipping et al. 1997) and products of early stages of plant decomposition (Palmer et al. 2001). The long-term trends of increasing or decreasing DOC concentration are not evident in various lakes except at the Experimental Lakes Area, where an increase in DOC is correlated with a decrease in summer total solar radiation and an increase in summer precipitation (Zhang et al. 2010). The initial DOM contents are also important to enhance the autochthonous production of DOC in aquatic ecosystems. Moreover, several mechanisms have been suggested to explain the enhancement of aquatic DOC including: increased terrestrial vegetation cover in response to climate change (Larsen et al. 2011; Worrall et al. 2003; Freeman et al. 2001b; Stoddard et al. 2003; Evans et al. 2005) and associated increases in enchytraeid worm activity (Cole et al. 2002; Carrera et al. 2009); increasing CO_2-mediated stimulation of primary productivity (Freeman et al. 2004); hydrological change (Hongve et al. 2004; Evans et al. 2005); artificial peat drainage (Worrall et al. 2003); the occurrence of severe drought events (Watts et al. 2001; Worrall and Burt 2004); and the removal of decomposition-inhibiting phenolic compounds following prolonged water table drawdown (Freeman et al. 2001a). However, these mechanisms are not sufficiently well documented yet to understand the increased DOC concentrations in natural waters.

One of the possible mechanisms leading to increased DOC concentrations in surface waters is the enhancement of photosynthesis. Two different processes are involved, depending on the sources of DOM.

(i) The first issue is that increased soil respiration may increase the decomposition rates of soil OM due to the effect of global warming. Furthermore, elevated CO_2 enhances DOC supply in peat soils because of elevated net primary productivity and increased root exudation of DOC in soil environments (Freeman et al. 2001b, 2004; Lavoie et al. 2005; Fenner et al. 2007a, b; Wolf et al. 2007; Kang et al. 2001; Tranvik and Jasson 2002; Monteith et al. 2007; Evans et al. 2002; Dorodnikov et al. 2011). This process ultimately leaches allochthonous DOM into the aquatic ecosystem. The increased activity of enchytraeid worms (the dominant invertebrates in upland peats) at higher temperature increases the microbial activity in peat and enhances nutrient mineralization (Cole et al. 2002). The mineralization of C- and N-containing compounds would increase the losses of nitrate and DOC (Cole et al. 2002).

(ii) The second issue is that the allochthonous DOM that is increasingly released into surface waters can undergo photoinduced decomposition to generate H_2O_2, CO_2 and DIC (dissolved CO_2, H_2CO_3, HCO_3^-, and CO_3^{2-}), or microbial degradation with production of H_2O_2, CO_2, DIC, CH_4, PO_4^{3-}, NH_4^+ and so on (Lovley et al. 1996; Johannessen and Miller 2001; Ma and Green 2004; Xie et al. 2004; Johannessen et al. 2007; Palenik and Morel 1988; Clark et al. 2004; Kotsyurbenko et al. 2001). Many of these compounds are able to enhance photosynthesis (Mostofa et al. 2009a; Komissarov 1994, 1995, 2003; Li et al. 2011; Li 1994; Zubkov and Tarran 2008; Beardall et al. 2009a, b; Wu and Gao 2009; Liu et al. 2010). This process can fuel primary and secondary production, thereby leading to enhanced aquatic OM and DOM. In fact, algae and phytoplankton can produce autochthonous DOM under both photoinduced and microbial respiration or assimilation, which contributes to increasing DOM in natural waters. Under elevated carbon dioxide levels, the proportion of DOM derived from recently assimilated CO_2 is ten times higher compared to the control cases (Freeman et al. 2004). In addition, new DOC release is far more sensitive to environmental drivers that affect net primary productivity compared to decomposition alone (Freeman et al. 2004).

> Moreover, photosynthesis depends on several key factors that have already been discussed and that could thus indirectly affect the occurrence of DOM in natural waters (see also chapter "Photosynthesis in Nature: A New Look").

On the other hand, declined concentrations of DOC have been observed in several surface waters including south west of England, northern Scandanavia and Italy (Worrall et al. 2004a, 2007; Schindler et al. 1996; Skjelkvåle et al. 2001; Bertoni et al. 2010; Minella et al. 2011). Of the 315 catchments examined in the UK, 18 % (55 catchments) have shown significant decreases in DOC concentration over the last 10 years (Worrall et al. 2007). DOC concentrations in the epilimnion have decreased from 119 to 57 μM C (average values during 1980–1984 and 2000–2007, respectively). In lake Maggiore (Italy), chlorophyll *a* concentrations averaged 5.9 $\mu g\ l^{-1}$ in the period 1980–1990, decreased to 4.0 $\mu g\ l^{-1}$ in the following decade (1990–2000) and underwent a further decrease (to 2.0 $\mu g\ l^{-1}$) in the period 2000–2007 (Bertoni et al. 2010). The observed DOC decline is presumably caused by a decrease of total phosphorus and of the organic loadings to the lake, because of a decrease of the anthropic impact. The consequences are a decrease of in-lake productivity and pronounced changes in phytoplankton composition, including higher biodiversity, reduced biovolume and lower average community cell size (Bertoni et al. 1998, 2008; Callieri and Piscia 2002; Morabito and OggioniA 2003; Salmaso et al. 2003; Rogora 2007).

The decline of DOC concentrations in surface waters would be linked to lower photosynthesis and often to relatively low contents of DOM. The latter may also be the result of low precipitation, which generally decreases to input of soil allochthonous DOM to natural waters. Waters with low contents of DOM would produce low amounts of photoinduced or microbial end products, which may significantly decrease the primary and secondary production with a subsequent decline of autochthonous DOM (see also chapter "Photosynthesis in Nature: A New Look"). Low production of autochthonous DOM would further contribute to the decline of DOC in natural waters. In fact, production of autochthonous DOM may sometimes offset the DOM decomposition by natural sunlight. For the same reason, soil inputs of allochthonous humic substances to surface waters during the summer stratification period may enhance photosynthesis and increase the autochthonous DOM in natural waters.

In some cases, drought can increase the DOM levels (Freeman et al. 2001a; Worrall and Burt 2008; Vazquez et al. 2011; Evans et al. 1999; Worrall et al. 2006; Holden and Burt 2002, 2003). The mechanism behind this phenomenon is that waters with high contents of DOM may undergo high photoinduced and microbial DOM degradation under drought conditions. The related production of photoinduced and microbial end products may be responsible for enhancement of high photosynthesis and, therefore, of high primary and secondary production. The latter phenomenon would ultimately lead to increased DOM contents.

The autochthonous production of DOM depends on several factors in natural waters, and particularly in lakes and oceans (Table 2) (Mostofa et al. 2005a, 2009a; Fu et al. 2010; Ogawa and Ogura 1992; Mitra et al. 2000; Yoshioka et al.

2002a; Hayakawa et al. 2003, 2004; Anderson and Williams 1998; Bushaw et al. 1996; Takahashi et al. 1995; Marañòn et al. 2004). These factors can be summarized as follows (Mostofa et al. 2009a): seasonal terrestrial riverine input; acidity or alkalinity, conductivity and pH, the variation of which indicates major differences in water chemistry that could influence photoreaction rates as well as the structure and speciation of organic matter. Other important factors are: anthropogenic activities; water transparency and the related light penetration through the water column; standing stocks of carbon; stratification of the water column; stirring/mixing of water by strong wind; photosynthetically active radiation (PAR) and water temperature; microbial degradation of DOM; ecosystem metabolism and bacterial growth; release of large amounts of dissolved organic compounds by wetland and littoral macrophytes; vertical mixing of the water column due to temperature effects; habitat structure and diversity.

8 Emerging Contaminants in Natural Waters

Emerging contaminants are generally detected in soil, sediment, air, water, aquatic biota including fish, wildlife, terrestrial earthworms, and humans (Richardson 2003, 2007; Mottaleb et al. 2005, 2009; Richardson and Ternes 2005, 2011; Buser et al. 2006; Schmid et al. 2007; Farré et al. 2008; Kinney et al. 2008; Guo et al. 2009; Ramirez et al. 2009; Citulski and Farahbakhsh 2010; Kumar and Xagoraraki 2010; Pal et al. 2010; Yoon et al. 2010; Kleywegt et al. 2011; Yu et al. 2011; Daughton and Ternes 1999; Keith et al. 2001; Heberer 2002; Balmer et al. 2004; Brooks et al. 2005; Duedahl-Olesen et al. 2005). According to these studies, emerging contaminants (usually emerging *organic* contaminants) are typically defined as organic substances that occur in very small amount (usually at concentration levels of nanograms to micrograms per liter), are persistent and have potential health effects on organisms including humans, fish and wildlife, or other adverse ecological effects. Such contaminants are either of anthropic origin (e.g. municipal, industrial, agricultural and human activities and waste water treatment processes) or naturally produced, e.g. during the algal (or phytoplankton) blooms in surface water.

Emerging contaminants include a diverse group of organic compounds that can be classified as pharmaceuticals, personal care products (PCPs), endocrine-disrupting compounds (EDCs), steroids and hormones, drinking water disinfection byproducts (DBPs), perfluorinated compounds (PFCs), brominated flame retardants (including polybrominated diphenyl ethers), sucralose and other artificial sweeteners, benzotriazoles, naphthenic acids, antimony, siloxanes, sunscreens/UV filters, musks, algal toxins, perchlorate, dioxane, pesticides, ionic liquids or organic salts, nanomaterials, gasoline additives and their transformation products, as well as microorganisms (Richardson 2003, 2007; Mottaleb et al. 2005, 2009; Richardson and Ternes 2005, 2011; Buser et al. 2006; Schmid et al. 2007; Farré et al. 2008; Kinney et al. 2008; Guo et al. 2009; Ramirez et al. 2009; Citulski and Farahbakhsh 2010; Kumar and Xagoraraki 2010; Pal et al. 2010; Yoon et al. 2010; Kleywegt et al. 2011; Yu et al. 2011; Daughton and Ternes 1999; Keith et al. 2001; Heberer 2002; Balmer et al. 2004; Brooks et al. 2005; Duedahl-Olesen et al. 2005).

It is estimated that approximately 3000 different substances are used as pharmaceutical ingredients, including painkillers, antibiotics, antidiabetics, beta-blockers, contraceptives, lipid regulators, antidepressants and impotence drugs (Richardson and Ternes 2011). Pharmaceuticals that arise concern for possible chronic toxicity are salicylic acid, diclofenac, propranolol, clofibric acid, carbamazepine, atenolol, bezafibrate, cyclophosphamide, ciprofloxacin, furosemide, hydrochlorothiazide, ibuprofen, lincomycin, ofloxacin, ranitidine, salbutamol, sulfamethoxazole, diltiazem, acetaminophen, chloramphenicol, florfenicol, thiamphenicol and fluoxetine (Pal et al. 2010; Richardson and Ternes 2011; Hoeger et al. 2005; Carlsson et al. 2006; Pomati et al. 2006; Kim et al. 2007; Lai et al. 2009).

Occurences of various hormones in natural waters as priority drinking water contaminants are estriol [E3], estrone, progesterone, 17α-ethinylestradiol [EE2], 17α-estradiol, 17β-estradiol [E2], testosterone, androstenedione, equilenin, equilin, mestranol, and norethindrone (Pal et al. 2010; Richardson and Ternes 2011). Synthetic musk compounds have diverse chemical structures, such as nitroaromatic groups including musk xylene (1-tert-butyl-3,5-dimethyl-2,4,6-trinitrobenzene) and musk ketone (4-tert-butyl-2,6-dimethyl-3,5-dinitroacetophenone); polycyclic structures including 7-acetyl-1,1,3,4,4,6-hexamethyl-1,2,3,4-tetrahydronaphthalene (AHTN; trade name, tonalide), 1,3,4,6,7,8-hexahydro-4,6,6,7,8,8-hexamethylcyclopenta-(g)-2-benzopyran (HHCB; trade name, galaxolide), 4-acetyl-6-tert-butyl-1,1-dimethylindan (ADBI; trade name, celestolide), dihydropentamethylindanone (DPMI; trade name, cashmeran), and 5-acetyl-1,1,2,3,3,6-hexamethylindan (AHMI, trade name phantolide). They are studied in waters, waste sludge and air (Richardson and Ternes 2011; Gomez et al. 2009; Wombacher and Hornbuckle 2009; Clara et al. 2011; Ramirez et al. 2011). It has been shown that galaxolide is the most abundant musk detected in wastewater, reaching up to 2069 and 1432 ng L^{-1} in influents and effluents, respectively.

Siloxanes include cyclic siloxanes, octamethylcyclotetrasiloxane (D4), decamethylcyclopentasiloxane (D5), dodecamethylcyclohexasiloxane (D6), tetradecamethylcycloheptasiloxane (D7) and linear siloxanes (Richardson and Ternes 2011; Kierkegaard et al. 2011). Approximately 600 different pesticides are applied annually in the US, whilst in Japan more than 450 active products are distributed annually among 5,400 commercial products. The two countries are key pesticide users in the world (Chen et al. 2007; Guo et al. 2009; Majewski et al. 2000; Derbalah et al. 2003; Qiu et al. 2005). Several pesticide degradation products are also of concern, such as: alachlor ethanesulfonic acid (ESA), alachlor oxanilic acid (OA), acetochlor ESA, acetochlor OA, metolachlor ESA, metolachlor OA, 3-hydroxycarbofuran, terbufos sulfone, alachlor ESA and OA, acetochlor ESA and OA, metolachlor ESA and OA, thiophenol and phenyl disulfide from dyfonate hydrolysis; 4-chloro-2-methylphenol and 4-chloro-2-methyl-6-nitrophenol from [(4-chloro-2-methylphenoxy)acetic acid] (MCPA) phototransformation; desphenyl-chloridazon (DPC) and methylated-DPC of N-chloridazon degradation (Richardson and Ternes 2011; Buttiglieri et al. 2009; Chiron et al. 2009; Wang et al. 2010).

Benzotriazoles and other benzo-related contaminants are detected in water, the most common being benzotriazole, tolyltriazole, benzothiazoles, and benzosulfonamides in waters (Richardson and Ternes 2011; Jover et al. 2009; van Leerdam et al. 2009; Matamoros et al. 2010). Perfluorinated compounds (PFCs) include perfluorooctane sulfonate (PFOS), perfluorooctanoic acid (PFOA) and a number of structurally-related compounds such as fluorinated telomer alcohols (FTOHs), perfluorobutanoic acid (PFBA), perfluorohexanoic acid (PFHxS), Perfluorobutanesulfonate (PFBS), perfluoropropane sulfonate (PFPrS), perfluoroethane sulfonate (PFEtS), perfluorooctane sulfonamide (PFOSA), N-ethyl perfluorooctane sulfonamide acetate (N-EtFOSAA), perfluorododecanoic acid (PFDoDa), perfluoroundecanoic acid (PFUnDa), perfluorodecanoic acid (PFDA), perfluorononanoic acid (PFNA), perfluoroheptanoic acid (PFHpA), perfluorohexanoic acid (PFHxA), perfluoropentanoic acid (PFPeA), and perfluoropropanoic acid (PFPrA). They are mostly volatile and subject to metabolism or degradation that leads to the formation of their persistent sulfonate and carboxylic acid forms (Richardson and Ternes 2011; Andersen et al. 2008; Farré et al. 2008; Shi et al. 2008; Mak et al. 2009).

Perchlorate is mostly found as an impurity of sodium hypochlorite (liquid bleach) and as ammonium perchlorate (Richardson and Ternes 2011). Ionic liquids are composed of cationic or anionic polar headgroups with accompanying alkyl side chains. The cationic head groups include imidazolium, pyridinium, pyrrolidinium, morpholinium, piperidium, quinolinium, quaternary ammonium, and quaternary phosphonium moieties. The anionic head groups include tetrafluoroborate (BF_4^-), hexafluorophosphate (PF_6^-), bis(trifluoromethylsulfonyl)-imide [$(CF_3SO_2)2\ N^-$], dicyanamide [$(CN)_2N^-$], chloride, and bromide (Richardson and Ternes 2011; Pham et al. 2010).

The chemical structures of nanomaterials are highly varied, including fullerenes, nanotubes, quantum dots, metal oxanes, TiO_2 nanoparticles, nanosilver, and zerovalent iron nanoparticles (Richardson and Ternes 2011). Artificial sweeteners in natural waters mainly include sucralose, acesulfame, cyclamate, saccharin, aspartame, neotame, and neohesperidin dihydrochalcone (NHDEC) (Richardson and Ternes 2011; Scheurer et al. 2009).

Toxins of blue–green algal origin commonly occur as microcystins, nodularins, anatoxins, cylindrospermopsin, and saxitoxins, while red tide toxins are detected as brevetoxins in natural waters (Richardson and Ternes 2011; dos Anjos et al. 2006; Wood et al. 2006; Zhao et al. 2006b; Smith et al. 2011). Saxitoxin variants recorded in cyanobacteria include decarbamoyl derivatives (dc), gonyautoxins (GTX); neosaxitoxin (neoSTX), N-sulphonocarbamoyl toxins (C-toxins), saxitoxin (STX) and a class of toxins produced by *Lyngbya wollei* (Humpage et al. 2010). Furthermore, chlorination of microcystin-LR and of cylindrospermopsin can give several byproducts, which are identified as chloro-microcystin, chloro-dihydroxy-microcystin, dichloro-dihydroxy-microcystin, trichlorohydroxy-microcystin, and several dihydroxy-microcystins (Merel et al. 2009).

Endocrine disrupting compounds or chemicals (EDCs) can disrupt the development of the endocrine system and of organs that respond to endocrine signals in organisms. These organisms can be indirectly exposed during prenatal and/or early postnatal life and the effects of exposure during development are permanent and irreversible (Colborn 1993). In addition, transgenerational exposure can result from the exposure of the mother to a chemical at any time throughout her life before producing offspring, due to persistence of EDCs in body fat. EDCs can then be mobilized during egg laying or pregnancy and lactation (Colborn 1993). EDCs include pesticides and their metabolites, DDT and its metabolites, pentachlorophenol (PCP), alkylphenols (e.g. penta- to nonylphenols, 4-tert-octylphenol) polychlorinated compounds (e.g. polychlorinated dibenzo-*p*-dioxins, polychlorinated dibenzofurans, and polychlorinated biphenyls: PCBs), bisphenolic compounds (phenylphenol and bisphenol A), polybrominated diphenyl ethers, nematocides, naphthenic acids, pharmaceuticals (diethylstilbestrol), triclosan (2,4,4′-trichloro-2′ 4′-trichloro-2′-hydroxydiphenyl ether), phthalate esters (e.g. benzylbutyl phthalate, diethylhexyl phthalate, diphenylphthalate), steroids (e.g. ethynyl estradiol, 17β-estradiol, diethylstilbestrol), natural estrogens (e.g. 17*β*-sitosterol, estriol, estrone), natural androgens (e.g. testosterone), natural phytoestrogens (soy, alfalfa and clover), naturally occurring compounds (lignans, coumestans, isofavones, mycotoxins), parabenes (hydroxybenzoate derivatives), organotins, and inorganic metal ions (cadmium, lead, mercury, antimony, uranium) (Citulski and Farahbakhsh 2010; Richardson and Ternes 2011; Colborn 1993; Bolger et al. 1998; Servos 1999; Petrović et al. 2001; Rhind 2002; Montgomery-Brown et al. 2003; Rudel et al. 2003; Ishibashi 2004; Kimura et al. 2004; Auriol et al. 2006; Diamanti-Kandarakis et al. 2009; Xu et al. 2011).

Microorganisms of concern are *Cryptosporidium*, *Giardia*, *E. coli*, pathogens, aeromonas, coliphages, viruses, total coliforms, *Helicobacter pylori*, enterococci, *E. coli O157:H7* and H1N1 (swine flu) (Liu et al. 2009, 2010; Richardson and Ternes 2011; Yañez et al. 2009; Li et al. 2010; Vikesland and Wigginton 2010; Wildeboer et al. 2010).

8.1 Sources of Emerging Contaminants in the Aquatic Environment

Emerging contaminants are commonly derived from three major sources (Richardson 2003, 2007; Mottaleb et al. 2005, 2009; Richardson and Ternes 2005, 2011; Buser et al. 2006; Schmid et al. 2007; Farré et al. 2008; Kinney et al. 2008; Guo et al. 2009; Ramirez et al. 2009; Citulski and Farahbakhsh 2010; Kumar and Xagoraraki 2010; Pal et al. 2010; Yoon et al. 2010; Kleywegt et al. 2011; Yu et al. 2011; Daughton and Ternes 1999; Keith et al. 2001; Heberer 2002; Balmer et al. 2004; Brooks et al. 2005; Duedahl-Olesen et al. 2005): (i) anthropogenic sources including atmospheric deposition but very often involving the effluents released by municipal, industrial, agricultural, and human activities, waste-water treatment processes, and so on; (ii) natural sources, including most notably the algal blooms in surface water; and (iii) photoinduced and microbial alteration of organic substances during transport from rivers to lakes, oceans or other water sources.

The key point sources of pharmaceuticals are discharge of wastes and drugs from hospitals; discharge of expired and consumed drugs from household;

wastewater and solid wastes discharge from pharmaceutical industries; hormones and antibiotics used in aquacultures; hormones and drugs used in livestock; compounds excreted from the human body in the form of non-metabolized parent molecules or as metabolites after ingestion and subsequent excretion, as well as the disposal of unused or expired medicinal products (Pal et al. 2010; Hernández et al. 2007; Pérez and Barceló 2007a, b; Nakada et al. 2008). The percentage of parent compound from the human body is 6–39 or $\geq$70 % for various antibiotics, 6–39 or $\leq$5 % for analgesics and anti-inflammatory drugs, $\leq$5 % for antiepileptic drugs (e.g. Carbamazepine), <0.5 or 50–90 % for beta-blockers, and 40–69 % for blood lipid regulators such as bezafibrate (Pal et al. 2010; Mompelat et al. 2009). The key entry route for pharmaceutical contaminants into natural waters is the point-source release from wastewater treatment plants (Daughton and Ternes 1999; Heberer 2002). Pharmaceuticals can also enter surface waters by run-off from fields treated with digested sludge (Farré et al. 2008). Veterinary drugs and their metabolites are transported through leaching or run-off from livestock slurries when liquid manure is sprayed on agricultural field waters (Farré et al. 2008).

Personal care products (PCPs) (e.g. fragrances) can be discharged into aquatic ecosystems through shower waste and finally from waste water treatment plants (Farré et al. 2008; Rimkus and Wolf 1996; Käfferlein et al. 1998; Smital et al. 2004; Peck 2006). UV filters used in sunscreens, cosmetics, and other PCPs are persistent in chlorinated water. Several halogenated by-products have been identified, which can cause endocrine and developmental toxicity and estrogenicity (Kunz and Fent 2006; Negreira et al. 2008; Schmitt et al. 2008). Synthetic musk compounds are widely used as fragrance additives in many personal care products, such as cleaning agents, air fresheners, house-hold products, perfumes, lotions, sunscreens, and laundry detergents (Richardson and Ternes 2011; Rimkus and Wolf 1996; Käfferlein et al. 1998; Smital et al. 2004). Steroids are excreted in urine of humans as more hydrophilic glucuronides and sulfates, and free steroids and conjugates are detected in sewage influent and effluent (Ascenzo et al. 2003; Reddy et al. 2005). Livestock wastes are potential sources of endocrine disrupting compounds and of steroidal estrogen hormones such as estradiol, estrone and estriol in natural waters (Raman et al. 2001; Hanselman et al. 2003; Furuichi et al. 2006). Steroids were detected in more than 86 % of water samples from creeks where the cattle had direct access to the water (Kolodziej and Sedlak 2007).

Sources and pathways of xenohormone uptake by humans are mostly inhalation (e.g., from indoor air), dermal absorption (e.g., from personal care products), and ingestion of food (Wagner and Oehlmann 2009). Another source of xenobiotics in foodstuff is the substances migrating from the packaging material, which can accumulate in the foodstuff. A variety of additives, such as stabilizers, antioxidants, coupling agents and pigments are used to optimize the properties of packaging materials, which include for instance durability, elasticity and color (Lau and Wong 2000; Casajuana and Lacorte 2003; Zygoura et al. 2005; Fankhauser-Noti et al. 2006).

Dioxane is a high production chemical that is used as solvent stabilizer in the manufacture and processing of paper, cotton, textile products, automotive coolants, cosmetics and shampoos, as well as a stabilizer in 1,1,1-trichloroethane (TCA), a popular degreasing solvent (Richardson 2007; Richardson and Ternes 2011; Lee et al. 2011). Siloxanes are widely used in PCPs and in a number of household products, such as cosmetics, deodorants, soaps, hair conditioners, hair dyes, car waxes, baby pacifiers, cookware, cleaners, furniture polishes, and water-repellent windshield coatings (Richardson and Ternes 2011; Kierkegaard et al. 2011).

Ammonium perchlorate is used in solid propellants for rockets, missiles, fireworks and highway flares. It can also be added in drinking water treatment processes as an impurity of sodium hypochlorite (liquid bleach), and is present as naturally-occuring perchlorate in fertilizers (e.g., Chilean nitrate). Formation of perchlorate can also take place upon reaction of chlorine radicals with ozone in the troposphere during the summer periods (Richardson and Ternes 2011; Parker 2009; Furdui and Tomassini 2010).

Benzotriazoles are complexing agents that are mostly used as anticorrosives or corrosion inhibitors (e.g., in engine coolants, aircraft deicers and antifreezing liquids), as UV-light stabilizers for plastics, for silver protection in dish-washing liquids, as anti-foggants in photography, and in aircraft de-icing/anti-icing fluids (ADAFs). These compounds are responsible for acute Microtox activity (Richardson 2007; Lovley 2006; Cancilla et al. 1997). They are soluble in water, resistant to biodegradation, and are only partially removed in wastewater treatment (Richardson 2007).

Naphthenic acids are a complex mixture of alkylsubstituted acyclic and cycloaliphatic carboxylic acids that dissolve in water at neutral or alkaline pH and have surfactant-like properties (Richardson and Ternes 2011). The key sources of naphthenic acids are the residual tailing water left over from the extraction of crude oil from oil sands, coal deposits, and petroleum industries (Richardson 2007; Richardson and Ternes 2011; Headley et al. 2009; Scott et al. 2009). PFCs are widely used in fabrics and carpets, paints, adhesives, waxes, polishes, metals, electronics, fire-fighting foams and caulks, as well as grease-proof coatings for food packaging such as microwave popcorn bags, French fry boxes, hamburger wrappers, and so on (Richardson and Ternes 2011).

Pesticides are generally released from agricultural fields to rivers or nearby waters by surface runoff, induced by either atmospheric precipitation or overflow and drainage of agricultural field waters (Wang et al. 2007; Guo et al. 2009; Richards and Baker 1993; Majewski et al. 2000; Derbalah et al. 2003; Qiu et al. 2005). However, the new sources of dichlorodiphenyltrichloroethane (DDT) are mostly connected with continuing illegal applications of technical DDT, use of technical DDT-containing antifouling paint in commercial fishing boat maintenance, and presence of DDT residues in dicofol, although its use is internationally forbidden (Wang et al. 2007; Guo et al. 2009; Qiu et al. 2005). Nonylphenol polyethoxylates (NPEOs) and alkylphenol ethoxylates (AEOs) are non-ionic surfactants, which are widely used in household, cleaning products, paints, pesticides, and industrial processes such as paper and petroleum production (Farré et al. 2008; Fairchild et al. 1999; Strynar and Lindstrom 2008).

Ionic liquids have unique properties including tunable viscosity, miscibility, and electrolytic conductivity. These properties make them useful for many applications, such as organic synthesis and catalysis, production of fuel cells, batteries, coatings, oils, nanoparticles, as well as other chemical engineering and biotechnology applications (Richardson and Ternes 2011).

Algal toxins are produced during algal or cyanobacterial blooms in natural waters (Richardson and Ternes 2011; dos Anjos et al. 2006; Wood et al. 2006; Zhao et al. 2006b; Smith et al. 2011).

Emerging pollutants can be altered in the environment by direct and indirect photolysis, hydrolysis, other chemical processes, biodegradation, sorption, volatilization and dispersion, or by a combination of these processes. Environmental transformation can either contribute to the complete removal of the organic contaminants, or produce transformation intermediates that can sometimes occur in the environment at higher levels than the parent compound and that can be as toxic or more toxic (Scully et al. 1988; Jensen and Helz 1998; Jameel and Helz 1999; Mitch et al. 2003; Strynar and Lindstrom 2008; Boxall et al. 2004; Gurr and Reinhard 2006; Jahan et al. 2008).

8.2 Transport of Emerging Contaminants in the Aquatic Environment

Once released into the environment, emerging contaminants are transported into different aquatic organisms, sediments and plants, depending on the emission routes as well as their physico-chemical properties such as water solubility, vapor pressure and polarity (Guo et al. 2009; Richardson and Ternes 2011; Daughton and Ternes 1999; Farré et al. 2008; Epel and Smital 2001). Emerging contaminants are generally persistent, have a wide range of hydrophilicity/hydrophobicity, and many of them are liable to bioaccumulation and biomagnification in organisms and plants when present in the aquatic environment (Guo et al. 2009; Richardson and Ternes 2011; Daughton and Ternes 1999; Farré et al. 2008; Epel and Smital 2001). Aquatic organisms including fish can accumulate emerging contaminants in certain body tissues. This phenomenon can take place either directly by bioaccumulation and biomagnification from water or by uptake of food such as OM (e.g. algae), sediments in water bed, small aquatic plants and so on, which have come in contact with the contaminants.

Emerging contaminants are mostly transmitted to humans through food consumption, particularly fish and seafood (Wong et al. 2002; Meng et al. 2007). Synthetic musks are potential candidates as substrates or inhibitors of multixeno biotic resistance (MXR) transporters (Daughton and Ternes 1999; Epel and Smital 2001). The multixenobiotic resistance (MXR) in aquatic organisms is mediated by the transport activity of transmembrane proteins belonging to the ATP-binding cassette (ABC) superfamily. These proteins are primarily involved in the active, ATP-dependent transport of biological molecules across plasma membranes (Smital et al. 2004; Higgins et al. 1988; Dean et al. 2001). The P-glycoprotein (P-gp) detected in ABC can transport drugs, xenobiotic compounds, anticancer agents including the vinca alkaloids and anthracyclines, drugs against human immunodeficiency virus (HIV), fluorophores as well as typical environmental pollutants (Smital et al. 2004; Danø 1972; Juliano and Ling 1976; Smital and Kurelec 1998; Bard 2000; Litman et al. 2001). Various transmembrane transport proteins can thus cause a rapid efflux of a wide variety of potentially toxic xenobiotics out of the cells of aquatic organisms. This is a 'first line of defense' against endogenous and exogenous toxicants (Smital et al. 2004; Kurelec 1992; Epel 1998).

However, some environmental chemicals act as specific MXR inhibitors and have the potential to block the active efflux of xenobiotics, thereby causing a significant increase of their intracellular accumulation. The main consequence of inhibition is an increase in chemosensitivity of aquatic organisms toward the many xenobiotics that are typically present in aquatic environments (Smital et al. 2004). Based on these considerations, otherwise innocuous environmental chemicals can be seen as a new class of environmentally hazardous chemicals that are termed as MXR inhibitors or chemosensitizers (Smital et al. 2004).

Perchlorate is a very water-soluble and environmentally stable anion, which can accumulate in plants (including lettuce, wheat, and alfalfa) and can thus contribute to exposure in humans and animals (Richardson and Ternes 2011).

8.2.1 Toxicological Impacts of Emerging Contaminants

Emerging contaminants and their transformation byproducts have adverse effects on the health of aquatic organisms (including algae, bacteria and fish), of animals and humans, as well as aquatic ecological effects (Guo et al. 2009; Kumar and Xagoraraki 2010; Pal et al. 2010; Derbalah et al. 2003; Scully et al. 1988; Jensen and Helz 1998; Jameel and Helz 1999; Mitch et al. 2003; Pomati et al. 2006; Farré et al. 2008; Fairchild et al. 1999; Boxall et al. 2004; Jahan et al. 2008; McLeese et al. 1981; Ahel et al. 1987; Tyler et al. 1998; Scott and Jones 2000; Oberdorster and Cheek 2001; Cleuvers 2004; Ferrari et al. 2004; Bedner and MacCrehan 2006; Owen et al. 2007). Their effects can be characterized using seven attributes: prevalence, frequency of detection, removal, bioaccumulation, ecotoxicity (for fish, daphnid, and algae aquatic indicator species), pregnancy effects, and health effects. The latter attribute was characterized using seven sub-attributes: carcinogenicity, mutagenicity, impairment of fertility, central nervous system action, endocrine effects, immunotoxicity, and developmental effects (Kumar and Xagoraraki 2010).

Production of byproducts such as trihalomethanes (THMs), N-nitrosodimethylamine (NDMA), and organic chloramines in conventional and advanced wastewater treatment plants arises considerable concern. These compounds are in fact extremely toxic and carcinogenic to human beings and aquatic organisms, and have been found in drinking and natural waters (Scully et al. 1988; Jensen and Helz 1998; Jameel and Helz 1999; Mitch et al. 2003; Farré et al. 2008). Transformation products of some organics are often more persistent than the corresponding parent compounds, and can cause greater toxicity (Boxall et al. 2004). For example, the major biodegradation product of nonylphenol ethoxylates, nonylphenol, is much more persistent than the parent compounds and has estrogenic properties (Jahan et al. 2008). Pharmaceuticals and their transformation byproducts show acute toxicity to bacteria, algae, invertebrates, fish, mussels, and human embryonic cells (Guo et al. 2009; Kumar and Xagoraraki 2010; Pal et al.

2010; Derbalah et al. 2003; Scully et al. 1988; Jensen and Helz 1998; Jameel and Helz 1999; Mitch et al. 2003; Pomati et al. 2006; Farré et al. 2008; Fairchild et al. 1999; Boxall et al. 2004; Jahan et al. 2008; McLeese et al. 1981; Ahel et al. 1987; Tyler et al. 1998; Scott and Jones 2000; Oberdorster and Cheek 2001; Cleuvers 2004; Ferrari et al. 2004; Bedner and MacCrehan 2006; Owen et al. 2007). It has been shown that low part per trillion (10–100 ng L^{-1}) concentrations of steroidal estrogen hormones can adversely affect the reproductive biology of aquatic wildlife such as fish, turtles and frogs, by disrupting the normal function of their endocrine systems (Tyler et al. 1998; Oberdorster and Cheek 2001). The sex hormones (mainly estrogens and androgens) are of very high potential concern, followed by cardiovascular drugs, antibiotics and antineoplastics, the latter being used to cure abnormal tissue growth (neoplasms) (Sanderson et al. 2004).

Ethylene dibromide (EDB) is among the most commonly detected contaminants in groundwater. It is classified as a probable human carcinogen and is highly persistent in water (Richardson 2007). 1,4-Dioxane is also a widespread contaminant in groundwater and is a probable human carcinogen (Richardson 2007). The transformation intermediates of nonylphenol ethoxylates (NPEOs) and alcohol ethoxylates (AEOs), in addition to the endocrine disrupting properties, are highly toxic and refractory and can cause hazards to aquatic ecosystems (Derbalah et al. 2003; Fairchild et al. 1999; McLeese et al. 1981; Ahel et al. 1987; Scott and Jones 2000). DDT and its metabolites can damage the nervous system, reproductive system, and liver. It is also a potential human carcinogen that can cause liver cancer (Guo et al. 2009). Ionic liquids are toxic, and their toxicity can widely vary among organisms and trophic levels (Pham et al. 2010).

EDCs have potential effects on organisms (microorganisms, wildlife, animals, and humans) including: androgenic, estrogenic, anti-estrogenic and anti-androgenic properties; disruption of the development of vital systems such as the endrocrine, reproductive, immune, and thyroid functions; sexual differentiation of the brain during fetal development; cognitive and motor function. Many of them are also suspected carcinogens (Richardson and Ternes 2011; Colborn 1993; Rhind 2002; Jansen et al. 1993; Nimrod and Benson 1996; Hansen 1998; Langer et al. 1998; Helleday et al. 1999; Vine et al. 2000; Moore et al. 2001; Fenton 2006). EDCs can have transient and persistent effects on mammary gland development depending on dose, exposure parameters and on whether exposure occurs during critical periods of gland growth or differentiation (Fenton 2006). Adverse effects from these abnormal developmental patterns include the presence of carcinogen-sensitive structures in the gland, in greater numbers or for longer periods. Inhibited functional differentiation can also be observed, leading to malnutrition or increased mortality of the offspring (Fenton 2006). Individually, adverse effects of EDCs exposure are detected on sperm production in rats and humans and reductions in Sertoli cell number in sheep (Carlsen et al. 1995; Toppari et al. 1996; Lee et al. 1999; Sweeney et al. 2000). Reductions in embryo survival and consequent effects on the reproductive rate in females are observed for many mammalian and bird species (IEH) (IEH 1999). Finally, human health is adversely affected by consumption of food contaminated by EDCs.

Currently, algal toxins or red tide toxins produced during algal blooms in lakes, estuaries and oceans are responsible for adverse effects, including the increasing incidence of loss of phytoplankton competitor motility, inhibition of photosynthesis and of enzymes, membrane damage, large fish kills, shellfish poisoning, deaths of livestock and wildlife, as well as illness and death in humans associated with the consumption of contaminated shellfish (Richardson 2007; Prince et al. 2008; Negri et al. 1995; Landsberg 2002; Legrand et al. 2003; Llewellyn 2006; Etheridge 2010). It has been shown that saxitoxin and its analogues are the only neurotoxins identified in *Anabaena circinalis* from the Murray Darling River. There an extensive *A. circinalis* bloom in 1991 resulted in the death of over 1600 stock (Humpage et al. 1994; Bowling and Baker 1996; Steffensen et al. 1999). The mechanisms behind the effects of harmful algal blooms on organisms will be discussed in chapter "Photosynthesis in Nature: A New Look". Finally, microorganisms are responsible for outbreaks of waterborne illness that have killed millions of people over the last few decades all over the world (Richardson and Ternes 2011).

8.3 Methodologies and Techniques of Emerging Contaminants (Pharmaceuticals and Personal Care Products) Detection

The liquid chromatography-tandem mass spectrometry (LC–MS/MS) screening method described here has been developed to target 23 pharmaceuticals and 2 metabolites with differing physicochemical properties in fish tissue by Ramirez and his colleagues (Ramirez et al. 2007). In this method, analysis of pharmaceuticals and their metabolites in fish tissue is conducted using reversed-phase separation of target compounds with a C18 column and a nonlinear gradient, consisting of 0.1 % (v/v) formic acid and methanol. A 1:1 mixture of 0.1 M aqueous acetic acid (pH 4) and methanol is identified as optimal, resulting in extraction recoveries for 24 of 25 compounds exceeding 60 % among 10 solvents tested. Eluted analytes are then introduced into the mass analyzer using positive or negative electrospray ionization. Note that moderate-polarity solvents are generally observed to be most effective at removing target analytes from tissue.

Sample collection and preservation (Ramirez et al. 2007): Fish (*Lepomis* sp.) were sampled from Pecan Creek (impacted by pharmaceuticals) and Clear Creek (not impacted by contaminants) streams to serve as test and reference specimens, respectively. Lateral fillets were dissected from fish collected at both sites and homogenized using a Tissuemiser (Fisher Scientific, Fair Lawn, NJ) set to rotate at 30,000 rpm. Pecan creek homogenates were stored individually, while Clear Creek homogenates were composited into a single sample. All tissues were stored at −20 °C prior to analysis.

Preparation of analytical sample (Ramirez et al. 2007): Approximately 1.0 g of tissue was combined with 8 mL of extraction solvent [a 1:1 mixture of

0.1 M aqueous acetic acid (pH 4) and methanol] in a 20mL borosilicate glass vial (Wheaton; VWR Scientific, Rockwood, TN), and the mixture was homogenized using a Tissuemiser (Fisher Scientific) at 30,000 rpm. Five surrogates (100.0 μg/mL in acetonitrile) were added to each sample: acetaminophen-d_4 (454 ng), fluoxetine-d_6 (636 ng), diphenhydramine-d_3 (8.9 ng), carbamazepine-d_{10} (38.5 ng), and ibuprofen-$^{13}C_3$ (789 ng). Samples were shaken vigorously and mixed on a rotary extractor for 5 min. Following extraction, samples were rinsed into 50-mL polypropylene copolymer round-bottomed centrifuge tubes (Nalge Co.; Nalgene Brand Products, Rochester, New York) using 1 mL of extraction solvent and centrifuged at 16,000 rpm for 40 min at 4 °C. The supernatant was decanted into 18-mL disposable borosilicate glass culture tubes (VWR Scientific), and the solvent was evaporated to dryness under a stream of nitrogen at 45 °C using a Zymark Turbovap LC concentration workstation (Zymark Corp., Hopkinton, MA). Samples were reconstituted in 1 mL of mobile phase, and a constant amount of the internal standards 7-aminoflunitrazapam-d_7 (100 ng) and meclofenamic acid (1000 ng) was added. Prior to analysis, samples were sonicated for 1 min and filtered using Pall Acrodisc hydrophobic Teflon Supor membrane syringe filters (13 mm diameter; 0.2 μm pore size; VWR Scientific, Suwanee, GA).

LC–MS/MS detection (Ramirez et al. 2007): A Varian ProStar model 210 binary pump equipped with a model 410 autosampler was used to detect the analytes, which were separated on a 15 cm × 2.1 mm (5 μm, 80 Å) Extend-C18 column (Agilent Technologies, Palo Alto, CA) connected with an Extend-C18 guard cartridge 12.5 mm × 2.1 mm (5 μm, 80 Å) (Agilent Technologies). A binary gradient consisting of 0.1 % (v/v) formic acid in water and 100 % methanol was employed to achieve chromatographic separation, whereas the time-scheduled elution program was as follows (min): 0, 2, 7, 12, 21 28, 34, 45, 50, 51, 65. The mobile-phase composition for 0.1 % formic acid was 93, 93, 85, 85, 52, 52, 41, 2, 2, 93, 93 and for methanol was 7, 7, 15, 15, 48, 48, 59, 98, 98, 7, 7, respectively. Additional chromatographic parameters were as follows: injection volume, 10 μL; column temperature, 30 °C; flow rate, 350 μL min^{-1}. Eluted analytes were monitored by MS/MS using a Varian model 1200L triple-quadrupole mass analyzer equipped with an electrospray interface (ESI).

Each compound was infused individually into the mass spectrometer at a concentration of 1 μg mL^{-1} in aqueous 0.1 % (v/v) formic acid at a flow rate of 10 μL min^{-1} for determining the best ionization mode (ESI + or −) and optimal MS/MS transitions for target analytes. All analytes were initially tested using both positive and negative ionization modes while the first quadrupole was scanned from *m/z* 50 to [M + 100]. This can enable identification of the optimal source polarity and the most intense precursor ion for each compound. Once these parameters have been defined, the energy at the collision cell was varied, while the third quadrupole was scanned to identify and optimize the intensity of product ions for each compound. Additional instrumental parameters held constant for all analytes were as follows: nebulizing gas, N_2 at 60 psi; drying gas, N_2 at 19 psi; temperature, 300 °C; needle voltage, 5000 V ESI+ , 4500 V ESI-; declustering potential, 40 V; collision gas, argon at 2.0 mTorr.

Extraction Recoveries (Ramirez et al. 2007): All samples were analyzed by LC–MS/MS, and individual analyte recoveries were calculated using the following equation:

$$\text{Recovery} = (A_{x1}/A_{IS1}) / (A_{x2}/A_{IS2}) \times 100\%$$

where A_{X1}, A_{IS1}, A_{X2}, and A_{IS2} represent peak areas for the analyte (X) and internal standard (IS) in groups 1 and 2 samples, respectively.

Identification of Pharmaceuticals Using LC–MS/MS (Ramirez et al. 2007):

A LC–MS/MS total ion chromatogram resulting from analysis of clean tissue (non-affected by contaminants) spiked with a mixture of standard pharmaceuticals is depicted in Fig. 3. Peak identifications for pharmaceuticals in the chromatogram are as follows: (1) acetaminophen-d_4, (2) acetaminophen, (3) atenolol, (4) cimetidine, (5) codeine, (6) 1,7-dimethylxanthine, (7) lincomycin, (8) trimethoprim, (9) thiabendazole, (10) caffeine, (11) sulfamethoxazole, (12) 7-aminoflunitrazepam-d_7 (+IS), (13) metoprolol, (14) propranolol, (15) diphenhydramine-d_3, (16) diphenhydramine, (17) diltiazem, (18) carbamazepine-d_{10}, (19) carbamazepine, (20) tylosin, (21) fluoxetine-d_6, (22) fluoxetine, (23) norfluoxetine, (24) sertraline, (25) erythromycin, (26) clofibric acid, (27) warfarin, (28) miconazole, (29) ibuprofen-$^{13}C_3$, (30) ibuprofen, (31) meclofenamic acid (-IS), and (32) gemfibrozil. Three factors were presumably considered in selecting the target analytes (Table 3): First, number of prescriptions dispensed in the United States during 2005 (RxList 2005). Second, variability in structure, physicochemical properties, and therapeutic use. Third, relative frequency of occurrence in soils, sediments, and biosolids. The frequency of detection of various PPCPs in analyzed sediment, soil, and biosolid samples (64–100 %) is typically much higher than in water (5 %). This may be due to variation in physicochemical properties favoring compound partitioning from water to solid environmental matrixes. Compounds residing in sediment may then be taken up by aquatic organisms via ingestion (Furlong et al. 2004; Brooks et al. 2005; Ramirez et al. 2007).

Optimized MS/MS transitions and collision energies employed for detection and quantitation of each analyte are presented in Table 3, along with the molecular structure and most common therapeutic use for each analyte. With the exception of erythromycin, selected precursors represent the molecular ion $[M + H]^+$ or $[M - H]^-$ for each analyte. The most abundant precursor for erythromycin was found to be the $[M + H - H_2O]^+$ ion at *m/z* 716. Selected product ions generally represent the most abundant fragment observed for each precursor at the noted collision energy. Once suitable MS/MS transitions have been identified for each analyte, an aqueous mixture of reference standards was employed to optimize chromatographic parameters. A nonlinear gradient consisting of 0.1 % (v/v) formic acid and methanol resulted in near baseline resolution of the majority of analytes in ~50 min (Fig. 3). A 15-min

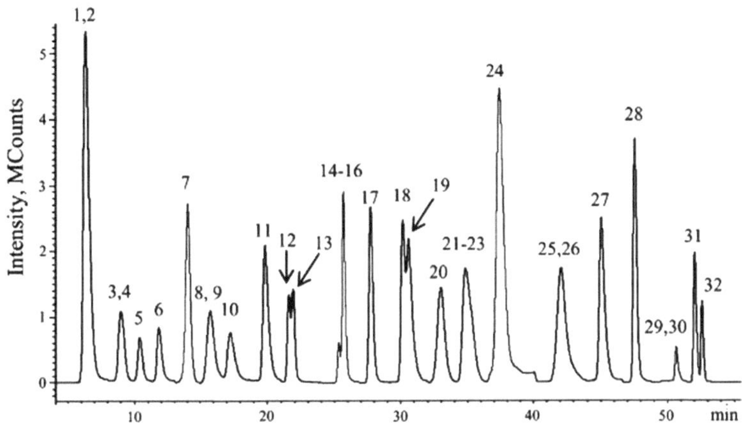

Fig. 3 LC-MS/MS total ion chromatogram resulting from analysis of clean tissue spiked with a mixture of pharmaceutical standards. Peak identifications are as follows: (1) acetaminophen-d4, (2) acetaminophen, (3) atenolol, (4) cimetidine, (5) codeine, (6) 1,7-dimethylxanthine, (7) lincomycin, (8) trimethoprim, (9) thiabendazole, (10) caffeine, (11) sulfamethoxazole, (12) 7 aminoflunitrazepam-d_7 (+IS), (13) metoprolol, (14) propranolol, (15) diphenhydramine-d_3, (16) diphenhydramine, (17) diltiazem, (18) carbamazepine-d_{10}, (19) carbamazepine, (20) tylosin, (21) fluoxetine-d_6, (22) fluoxetine, (23) norfluoxetine, (24) sertraline, (25) erythromycin, (26) clofibric acid, (27) warfarin, (28) miconazole, (29) ibuprofen-$^{13}C_3$, (30) ibuprofen, (31) meclofenamic acid (-IS), and (32) gemfibrozil. *Data source* Ramirez et al. (2007)

isocratic hold (93:7 formic acid–methanol) was added to the end of each run to allow for column equilibration between injections. While the majority of analytes were eluted as single peaks, erythromycin was consistently eluted as two partially resolved peaks, which are attributed to differing retention characteristics for presumed Stereoisomers (Vanderford et al. 2003; Yang and Carlson 2004). In addition, isotope effects on retention behavior were often observed for carbamazepine-d_{10} and fluoxetine-d_6 (peaks 18 and 19, Fig. 3). The observed retention time for carbamazepine-d_{10} (30.08 min) was shorter than that observed for carbamazepine (30.53 min) by almost 30 s. Correspondingly, a 20-s difference in retention time was observed for fluoxetine-d_6 (34.58 min) relative to that observed for fluoxetine (34.93 min), although it is not evident in Fig. 3 due to coelution of norfluoxetine (35.13 min). Finally, isotope effects were not observed for acetaminophen (peaks 1 and 2) and diphenhydramine (peaks 15 and 16, Fig. 3) due to a lower degree of deuterium substitution and decreased resolution at shorter retention times. Four compounds such as diphenhydramine, diltiazem, carbamazepine, and norfluoxetine were detected in fish environmental samples (affected by contaminants; Fig. 4a), which were confirmed by comparing the results of the fish samples unaffected by contaminants and spiked with known amounts of their respective standards (Fig. 4b). The concentrations of these pharmaceuticals were 0.66–1.32 ng g^{-1} for diphenhydramine, 0.11–0.27 ng g^{-1} for diltiazem, 0.83–1.44 ng g^{-1} for carbamazepine, and 3.49–5.14 ng g^{-1} for norfluoxetine detected in 11 of 11 contaminated environmental fish samples.

Table 3 Analyte-dependent mass spectrometry parameters for target compounds (data source Ramirez et al. 2007; RxList (The Internet Drug Index) 2005)

Compound	Use	Structure	Precursor ion	Collision energy (eV)	Product ion	pKa[a]
ESI positive analytes						
Acetaminophen	Analgesic	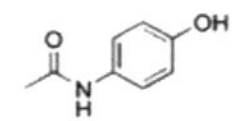	152 $[M+H]^+$	−11.0	110	9.86
Atenolol	Anti-hypertension	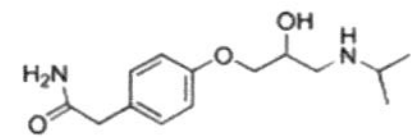	267 $[M+H]^+$	−21.5	145	9.16
Cimetidine	Anti-acid reflux	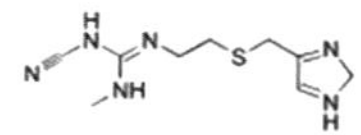	253 $[M+H]^+$	−13.5	159	7.07
Codeine	Analgesic	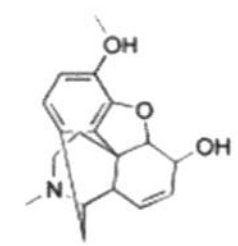	300 $[M+H]^+$	−38.0	215	8.25
1,7-dimethylxanthine	Caffeine metabolite	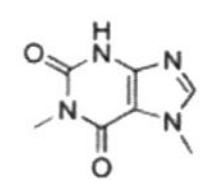	181 $[M+H]^+$	−15.5	124	8.50
Liltcomycin	Antibiotic	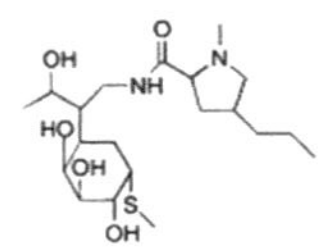	407 $[M+H]^+$	−15.5	359	8.78

(continued)

Table 3 (continued)

Compound	Use	Structure	Precursor ion	Collision energy (eV)	Product ion	pKa[a]
Trimethoprim	Antibiotic	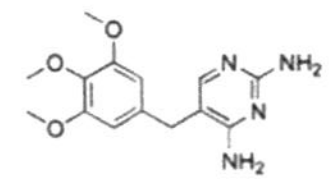	291 $[M + H]^+$	−17.5	261	7.20
Thiabendazole	Antibiotic	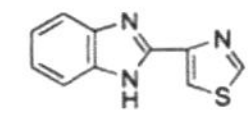	202 $[M + H]^+$	−23.0	175	
Caffeine	Stimulant	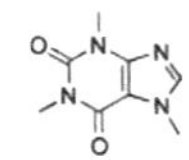	195 $[M + H]^+$	−16.0	138	
Sulfamethoxazole	Antibiotic	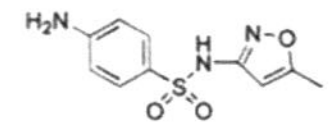	254 $[M + H]^+$	−13.0	156	5.81
Metoprelol	Anti-hypertension	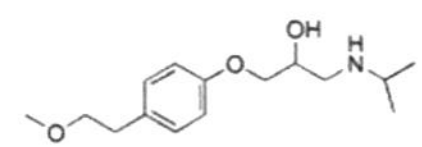	268 $[M + H]^+$	−15.5	191	9.17
Propranolol	Anti-hypertension	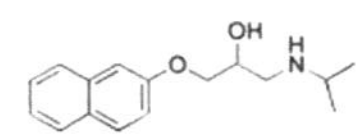	260 $[M + H]^+$	−11.0	116	9.14
Diphenhydramine	Antihistamine	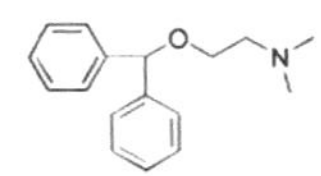	256 $[M + H]^+$	−11.5	167	8.76
Diltiazem	Anti-hypertension	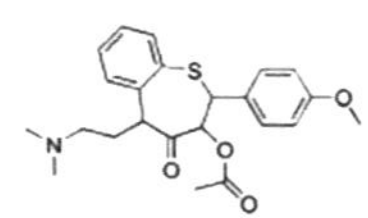	415 $[M + H]^+$	−22.0	178	8.94

(continued)

Table 3 (continued)

Compound	Use	Structure	Precursor ion	Collision energy (eV)	Product ion	pKa[a]
Carbamazepine	Anti-seizure	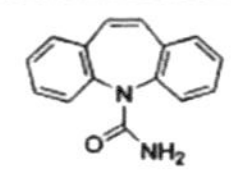	237 $[M+H]^+$	−13.5	194	
Tylosin	Antibiotic	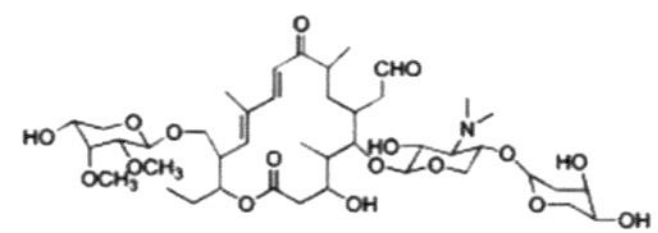	916 $[M+H]^+$	−31.5	174	7.39
Fluoxetine	Anti depressant	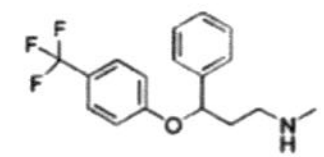	310 $[M+H]^+$	−6.0	148	10.1
Norfluoxetine	Fluoxetine metabolite	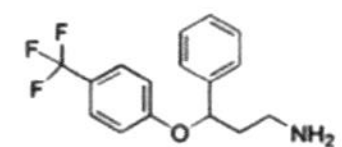	296 $[M+H]^+$	−45	134	9.05
Sertraline	Antidepressant	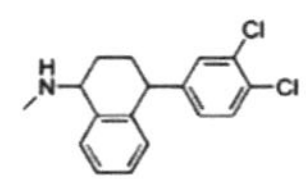	306 $[M+H]^+$	−11.0	275	9.47
Erythromycin	Antibiotic	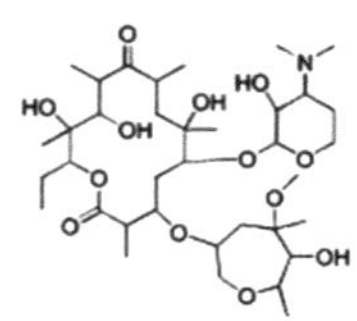	716 $[M+H-H_2O]^+$	−18.0	558	8.16

(continued)

Table 3 (continued)

Compound	Use	Structure	Precursor ion	Collision energy (eV)	Product ion	pKa[a]
Warfarin	Anli-coaulant	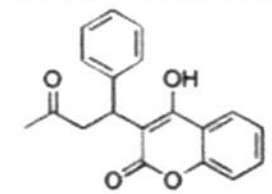	309 $[M + H]^+$	−14.0	163	4.50
Miconazole	Antibiotic	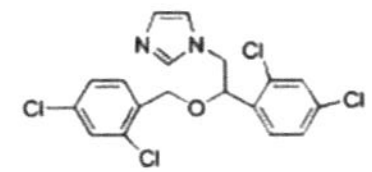	417 $[M + H]^+$	−27.5	161	6.67
ESI Negative Analytes						
Clofibric Acid	Antilipemi	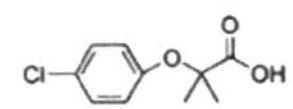	213 $[M - H]^-$	15.4	127	3.18
Ibuprofen	Analgesic	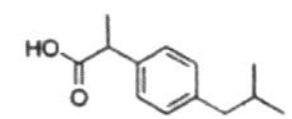	205 $[M - H]^-$	7.0	161	4.41
Gemfibrozil	Antilipemie	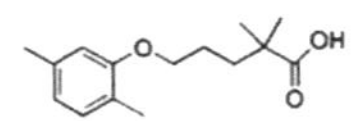	249 $[M - H]^-$	13.0	121	4.75

[a]Calculated values obtained from SciFinder database (@ 2006 American Chemical Society)

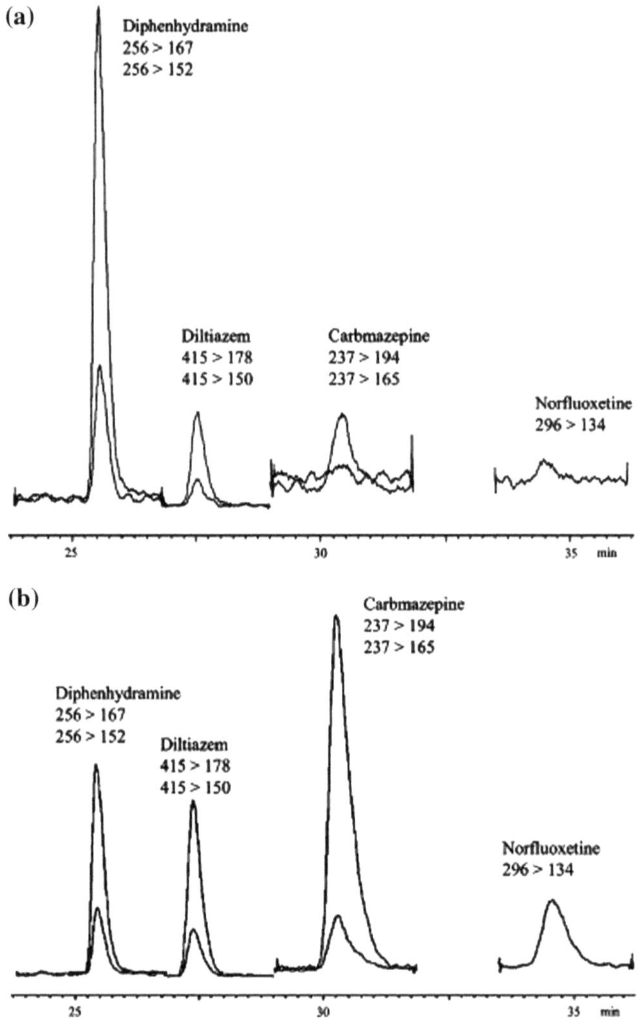

Fig. 4 LC-MS/MS reconstituted ion chromatograms displaying analyte-specific quantitation and qualifier ions monitored for (**a**) a tissue extract from a fish (Lepomis sp.) that is affected by contaminants and (**b**) an extract from fish tissue (not affected by contaminants) spiked with known amounts of diphenhydramine (1.6 ng g^{-1}), diltiazem (2.4 ng g^{-1}), carbamazepine (16 ng g^{-1}), and norfluoxetine (80 ng g^{-1}). The higher m/z fragment is more intense in all cases. Data source Ramirez et al. (2007)

Identification of Personal Care Products Using GC–MS/MS (Mottaleb et al. 2009)

Two gas chromatography-mass spectrometry (GC–MS) methods have been described for simultaneous determination in fish of ten extensively used personal care products (PCPs) such as benzophenone, 4-methylbenxylidine camphor

(4-MBC), m-toluamide, galaxolide, tonalide, musk xylene, musk ketone, celestolide, triclosan, octocrylene and two alkylphenol surfactants such as p-octylphenol and p-nonylphenol. These methods consisted of extraction, clean-up, derivatization and analysis by gas chromatography–mass spectrometry with selected ion monitoring (GC–SIM–MS) or gas chromatography–tandem mass spectrometry (GC–MS/MS) techniques (Mottaleb et al. 2009). To assess recovery of target compounds from 1-g tissue homogenates, acetone was selected as optimal solvent for extracting compounds with dissimilar physicochemical properties from fish tissue. Initial experiments confirmed that GC–SIM–MS could be applied for analysis of lean fillet tissue (<1 % lipid) without gel-permeation chromatography (GPC), and this approach was applied to assess the presence of target analytes in fish fillets collected from a regional effluent-dominated stream in Texas, USA. Benzophenone, galaxolide, tonalide, and triclosan were detected in 11 of 11 environmental samples at concentrations ranging from 37 to 90, 234 to 970, 26 to 97, and 17 to 31 ng g^{-1}, respectively. However, performance of this analytical approach declined appreciably with increasing lipid content of analyzed tissues. Successful analysis of samples with increased lipid content was enabled by adding GPC to the sample preparation protocol and monitoring analytes with tandem mass spectrometry. Both analytical approaches were validated using fortified fillet tissue collected from locations expected to be minimally impacted by anthropogenic influences. Average analyte recoveries ranged from 87 % to 114 % with RSDs <11 % and from 54 % to 107 % with RSDs <20 % for fish tissue containing <1 % and 4.9 % lipid, respectively. Statistically derived method detection limits (MDLs) for GC–SIM–MS and GC–MS/MS methodologies ranged from 2.4 to 16 ng g^{-1}, and from 5.1 to 397 ng g^{-1}, respectively (Mottaleb et al. 2009). In a following study, improvement of the MDL has been observed between 12 and 38 ng g^{-1} by the GC–MS/MS methodology for the same PCPs using 2.0–2.5 g of fish (Subedi et al. 2011).

9 Does DOM Act as Energy Source for Living Organisms and Aquatic Ecosystem?

The concentration levels of DOC in groundwater are very variable: they reach 16–424 μM C in Asia, 42–15333 μM C in Europe, 8–2333 μM C in North America, 1108 ± 217–14167 ± 6333 μM C in Botswana (Africa), 100–3000 μM C in Brazil (South America) (Table 2) (Mostofa et al. 2007a, Mostofa KMG et al., unpublished data; Buckau et al. 2000; Bertilsson et al. 1999; McIntyre et al. 2005; Meier et al. 2004; Crandall et al. 1999; Schwede-Thomas et al. 2005; Pabich et al. 2001; Michalzik et al. 2001; Anawar et al. 2002; Richey et al. 2002; Bradley et al. 2007). Groundwater is the main source of drinking water for many developing and developed countries, including the USA. Groundwater has the advantage over surface water of being usually free of suspended solids, bacteria and other disease-causing microorganisms (Mostofa et al. 2009a). Interestingly, upland areas make up 30 %

of the surface of Great Britain, but supply over 70 % of its drinking water (Watts et al. 2001). Therefore, all the people uptake a certain amount of DOC everyday from drinking water. According to the level of DOC in groundwater and considering an average water intake of 2 liters per day for adults (which can rise to ~5 liters for manual labor at high temperature), on average, every person intakes per day ~50–800 μM C in Asia, ~100–30000 μM C in Europe and 20–5000 μM C in the U.S.A. The interesting question that arises is that these DOC contents are significant energy sources for human beings and for the other living organisms. Before addressing this question, it is important to examine which substances make up DOM in natural waters.

The contribution of humic substances (hydrophobic acids) in groundwater is very variable in different countries, and is approximately included in the range of 12–98 % (1–80 % of fulvic acid and 2–97 % of humic acid). The contribution of hydrophilic fractions is 1–82 % (Buckau et al. 2000; Bertilsson et al. 1999; Peuravuori and Pihlaja 1999; Leenheer et al. 1974; Thurman 1985c; Ford and Naiman 1989; Schiff et al. 1990; Wassenaar et al. 1990; Malcolm 1991; Grøn et al. 1996; Christensen et al. 1998; McIntyre et al. 2005; Mladenov et al. 2008). Along with the humic substances, hydrophilic compounds (acidic, basic and neutral) and carbohydrates (mainly polysaccharides, ~1–10 %) are also present in groundwater (Thurman 1985a; Peuravuori and Pihlaja 1999; Artinger et al. 2000). The intake of DOC by every person is approximately 20–30000 μM C, or 0.2–360 mg C L^{-1} per day, for the average hydration of a human body in the case of groundwater.

It is generally well-known that carbohydrates can produce energy for all living organisms. The sources of carbohydrates and humic substances are the same vascular plant material. DOM with its content of organic C and N is a thermodynamic anomaly that provides a major source of energy to drive aquatic and terrestrial ecosystems (Tranvik 1992; Salonen et al. 1992; Wetzel 1984, 1992; Hedges et al. 2000; Berner 1989). Energy changes (±) such as supply (+) or consumption (−) of energy in the aquatic environment generally occur during the photoinduced and microbial degradation of DOM and organic matter, during the microbial loop and the photosynthesis (Mostofa et al. 2009a; Komissarov 1994, 1995, 2003; Miller and Moran 1997; Li et al. 2011; Sherr and Sherr 1989; Carrick et al. 1991; Jones 1992; Tranvik 1992; Wetzel 1984, 1992). In addition, terrestrial DOM represents a source of allochthonous energy for heterotrophs in receiving lakes, rivers, reservoirs, estuaries and coastal oceans (Mostofa et al. 2009a; Wetzel 1992; Smith and Hollibaugh 1993; Kemp et al. 1997; Pace et al. 2004; Aller and Blair 2006). It has been shown that DOM makes up 47 % of the energy which enters and 70 % of the energy which leaves the groundwater ecosystem (Fisher and Likens 1973). It has also been shown that undisturbed groundwater basins export only small amounts of energy (~1 %) from the upland regions, while the ramaining 99 % of forest production is consumed terrestrially (Fisher and Likens 1972, 1973). It is therefore concluded that the DOM including humic substances can act as energy source and are vital for all living organisms (Mostofa et al. 2009a). Note that DOM in drinking water can play a negligible energetic role for humans, due to the uptake of a substantially lower amount of organic carbon compared to foods (e.g. boiled rice, vegetables, fish, meat and so on) and beverages (e.g. fruit juices, alcohol, etc.).

Humic substances (humic and fuvic acids) are extensively applied as biomedicines to decrease the gastric damage induced by ethanol, to protect organisms against cell-wall disruption, to maintain antibacterial and antiviral properties, decrease viral respiratory illness, and to protect against cancer and related cancer-causing viruses (Brzozowski et al. 1994; Klöcking et al. 2002; Peña-Méndez et al. 2005). On the other hand, humic acid is a toxic factor for many mammalian cells and can be involved in the so-called humic acid-induced cytotoxicity (Peña-Méndez et al. 2005; Ho et al. 2003).

10 Scope of the Future Research

After the development of an effective method for TOC analysis in 1988, DOM has been mostly determined in developed countries since 1990 to date, but fewer studies have been carried out in developing countries. Considering the importance of DOM, it is important to determine its levels in natural water in developing countries, also considering that the DOC concentrations in many watersheds have changed (either increased or decreased) over the last few decades. Moreover, emerging contaminants and their transformation byproducts are extensively examined currently, but only limited information is available on their ecotoxicological impacts on the aquatic environments.

Some important research demands for future challenges are the following: (i) Determination of concentration levels of DOM in important rivers and lakes in developing countries. (ii) Extraction of autochthonous fulvic acids from algae or phytoplankton under both photorespiration by natural sunlight or artificial light, and microbial respiration or assimilation under dark incubation. (iii) Characterization of the extracted autochthonous fulvic acids to examine the presence of functional groups, elemental composition, and possible molecular structure with reference to standard Suwannee River Fulvic Acid and Humic Acid. (vi) Investigation on lakes having reduced DOC contents, using incorporation of terrestrial soils in lake surface waters. (v) Investigation on lakes having increased DOC contents, trying to reduce photosynthesis and primary production in the lake surface waters. (vi) Joint chemical and toxicological evaluation of emerging contaminants and their transformation byproducts, for important end points and target organs and effects such as mutagenicity, carcinogenicity, hepatotoxicity, nephrotoxicity, immunotoxicity, neurotoxicity, developmental neurotoxicity and pharmacokinetics (Farré et al. 2008).

Nomenclature

ABC	ATP binding cassette
ADAFs	Aircraft deicing/antiicing fluids
AEOs	Alkylphenol ethoxylates
BF_4^-	Tetrafluoroborate
CDOM	Colored and chromopheric dissolved organic matter

$(CF_3SO_2)2N^-$	Bis(trifluoromethylsulfonyl)-imide
$(CN)_2N^-$	Dicyanamide
DBPs	Disinfection byproducts (DBPs)
DDT	Dichlorodiphenyltrichloroethane
DIC	Dissolved inorganic carbon (DIC is defined jointly as dis solved CO_2, H_2CO_3, HCO_3^-, and CO_3^{2-})
DOC	Dissolved organic carbon
DOM	Dissolved organic matter
DON	Dissolved organic nitrogen
DOP	Dissolved organic phosphorus
EDB	Ethylene dibromide
EDC	Endocrine-disrupting compounds
EEM	Excitation-emission matrix
FDOM	Fluorescent dissolved organic matter
FTOHs	Fluorinated telomer alcohols
HIV	Human immunodeficiency virus
H_2O_2	Hydrogen peroxide
IHSS	International humic substances society
LMW	Low molecular weight
MXR	Multixenobiotic resistance
OM	Organic matter
$O_2^{\bullet-}$	Superoxide radical
$HO^{\bullet}$	Hydroxyl radical
NDMA	N-nitrosodimethylamine
N-EtFOSAA	N-ethyl perfluorooctane sulfonamide acetate
NHDEC	Neohesperidin dihydrochalcone
NPEOs	Nonylphenol polyethoxylates
PAR	Photosynthetically available radiation
PCBs	Polychlorinated biphenyls
PF_6^-	Hexafluorophosphate
PFBA	Perfluorobutanoic acid
PFCs	Perfluorinated compounds
PFEtS	Perfluoroethane sulfonate
PFHxA	Perfluorohexanoic acid
PFOA	Perfluorooctanoic acid
PFOS	Perfluorooctane sulfonate
PFOSA	Perfluorooctane sulfonamide
PFPeA	Perfluoropentanoic acid
PFPrA	Perfluoropropanoic acid
PFPrS	Perfluoropropane sulfonate
POM	Particulate organic matter
PPCPs	Pharmaceuticals, personal care products
PCPs	Personal care products
SRFA	Suwannee River Fulvic Acid
SRHA	Suwannee River Humic Acid

THMs	Trihalomethanes
TOC	Total organic carbon
UV	Ultraviolet
1 mg L^{-1}	$(1 \times 1000)/12 = 83$ μM C

Problems

(1) Define the dissolved organic matter (DOM) and explain how does it differ from organic matter?
(2) What are the major sources of DOM in natural waters?
(3) Explain the DOM functions shortly.
(4) Explain the origin of allochthonous DOM in soil and autochthonous DOM in natural waters.
(5) What are the contributions of humic substances (fulvic and humic acids) in groundwater, rivers, lakes and oceans?
(6) Explain the redox behavior of fulvic and humic acids.
(7) Define the allochthonous fulvic and humic acids, and the autochthonous fulvic acids. What are the chemical differences among these classes of humic substances?
(8) Why does the molecular size of DOM decrease from rivers to lakes and from lakes to oceans?
(9) What are the controlling factors that affect the DOM contents in natural waters? Explain the two most important factors that affect DOM in natural waters.
(10) Explain the possible mechanisms for the increased or declined DOM contents in surface waters.
(11) What are the emerging contaminants? Explain the sources, transportation and toxicological effects of these contaminants in the aquatic environments.
(12) How does DOM act as energy source for living organisms and aquatic ecosystems?

Acknowledgments We are particularly grateful to Dr. Liu Cong-Qiang, Professor and Academician; Dr. Hu Ruizhong, Professor and Director General; Dr. Wang Shijie, Professor and Vice-director; Dr. Feng Xin Bin, Professor and Vice Director; Prof. Yun Liu, Prof. Xiao Tangfu, Dr. Li Xiao-Dong, Dr. Ding Hu of Institute of Geochemistry, Chinese Academy of Sciences; Mrs. Asma Mostofa and Mrs. Rafia Khatun for their kind assistances, constant support and inspiration during the preparations of the primary and final draft of the manuscripts. This work was financially supported by the Institute of Geochemistry, the Chinese Academy of Sciences, China. This work was partly supported by Center for Innovation and Entrepreneurship, Northwest Missouri State University, USA; Atmospheric and Ocean Research Institute, The University of Tokyo, Japan; University of Turin and Centro Interdipartimentale NatRisk, I-10095 Grugliasco (TO), Italy; Kyoto University, Japan; and Chinese Research Academy of Environmental Sciences, China. This chapter acknowledges Ramirez AJ, Mottaleb MA, Brooks BW, Chambliss CK, 2007, Analysis of Pharmaceuticals in Fish Using Liquid Chromatography-Tandem Mass Spectrometry, Analytical Chemistry, 79 (8), 3155–3163. Copyright (2007) Americal Chemical Society; reprinted from Journal of Chromatography A, 1216 (5), Mottaleb MA, Usenko S, O'Donnell JG, Ramirez AJ, Brooks BW, Chambliss CK, Gas chromatography–mass spectrometry screening methods for select UV-filters, synthetic musks, alkylphenols, an antimicrobial agent, and an insect repellent in fish, 815–823, Copyright (2009), with permission from Elsevier; and copyright (2007) by the Association for the Sciences of Limnology and Oceanography, Inc.

References

Abbt-Braun G, Frimmel FH (1990) Restmetallgehalte isolierter Huminstoffe aus Erde, Moor und Deponie. Acta Hydrochim Hydrobiol 18:649–656

Abbt-Braun G, Frimmel FH, Lipp P (1991) Isolation of organic substances from aquatic and terrestial systems—Comparison of some methods. Z Wasser Abwasser Forsch 24:285–292

Abe T, Watanabe A (2004) X-ray photoelectron spectroscopy of nitrogen functional groups in soil humic acids. Soil Sci 169:35–43

Abril G, Etcheber H, Borges AV, Frankignoulle M (2000) Excess atmospheric crbon dioxide transported by rivers into the Scheldt Estuary. Comptes Rendus l'Académie des Sciences Série IIa 330:761–768

Abril G, Nogueira M, Etcheber H, Cabeçadas G, Lemaire E, Brogueira MJ (2002) Behaviour of organic carbon in nine contrasting European estuaries. Estuar Coast Shelf Sci 54:241–262

Ahel M, Conrad T, Giger W (1987) Persistent organic chemicals in sewage effluents. 3. Determination of nonylphenoxy carboxylic acids by high-resolution gas chromatography/mass spectrometry and high-performance liquid chromatography. Environ Sci Technol 21:697–703

Aiken GR, Gillam AH (1989) Determination of molecular weights of humic substances by ultracentrifugation. In: Hayes MB, MacCarthy P, Malcolm RL (eds) Humic substances II. In search of structure. Wiley, New York, pp 515–544

Aiken GR, Malcolm RL (1987) Molecular weight of aquatic fulvic acids by vapor pressure osmometry. Geochem Cosmochim Acta 51:2177

Aiken GR, McKnight DM, Wershal RL, MacCarthy P (1985) Humic substances in soil, sediment and water. Wiley, New York

Aitkenhead JA, McDowell WH (2000) Soil C:N ratios as a predictor of annual riverine DOC flux at local and global scales. Global Biogeochem Cycles 14:127–138

Aitkenhead JA, Hope D, Billett MF (1999) The relationship between dissolved organic carbon in stream water and soil organic carbon pools at different spatial scales. Hydrol Process 13:1289–1302

Alberts JJ, Takács M (1999) Importance of humic substances for carbon and nitrogen transport into southeastern United States estuaries. Org Geochem 30:385–395

Alberts JJ, Giesy JP, Evans DW (1984) Distribution of dissolved organic carbon and metal-binding capacity among ultrafilterable fractions isolated from selected surface waters of the southeastern United States. Environ Geol Water Sci 6:91–101

Allard B, Arsenie I (1991) Abiotic reduction of mercury by humic substances in aquatic system—an important process for the mercury cycle. Water Air Soil Pollut 56:457–464

Allard B, Borén H, Pattersson C, Zhang G (1994) Degradation of humic substances by UV irradiation. Environ Int 20:97–101

Aller RC, Blair NE (2006) Carbon remineralization in the Amazon-Guianas tropical mobile mudbelt: a sedimentary incinerator. Cont Shelf Res 26:2241–2259

Aluwihare LI, Repeta DJ, Chen RF (2002) Chemical composition and cycling of dissolved organic matter in the Mid-Atlantic Bight. Deep Sea Res Part II Top Stud Oceanogr 49:4421–4437

Amado AM, Cotner JB, Suhett AL, de Assis Esteves F, Reinaldo Luiz Bozelli RL, Farjalla VF (2007) Contrasting interactions mediate dissolved organic matter decomposition in tropical aquatic ecosystems. Aquat Microb Ecol 49:25–34

Amador JA, Alexander M, Zika RG (1989) Sequential photochemical and microbial degradation of organic molecules bound to humic acid. App Environ Microb 55:2843–2849

Amon RMW, Benner R (1994) Rapid cycling of high-molecular-weight dissolved organic matter in the ocean. Nature 369:549–552
Amon RMW, Benner R (1996) Bacterial utilization of different size classes of dissolved organic matter. Limnol Oceanogr 41:41–51
Amy GL, Collins MR, Kuo CJ, King PH (1987) Comparing gelpermeation chromatography and ultrafiltration for the molecularweight characterization of aquatic organic-matter. J Am Water Works Assoc 79:43–49
Anawar HM, Akai J, Mostofa KMG, Safiullah S, Tareq SM (2002) Arsenic Poisoning in groundwater: health risk and geochemical sources in Bangladesh. Environ Int 27:597–604
Andersen ME, Butenhoff JL, Chang SC, Farrar DG, Kennedy GL Jr, Lau C, Olsen GW, Seed J, Wallace KB (2008) Perfluoroalkyl acids and related chemistries—Toxicokinetics and modes of action. Toxicol Sci 102:3–14
Anderson TF, Arthur MA (1983) Stable isotopes of oxygen and carbon and their application to sedimentologic and paleoenvironmental problems. In: Arthur MA, Anderson TF, Kaplan IR, Veizer J, Land LS (eds) Stable isotopes in sedimentary geology. Society of Economic Paleontologists and Mineralogists, Tulsa, pp 1–151
Anderson TR, Williams PJleB (1998) Modelling the seasonal cycle of dissolved organic carbon at station E1 in the English Channel. Estuar Coastal Shelf Sci 46:93–109
Anderson SP, Dietrich WE, Torres R, Montgomery DR, Loague K (1997) Concentration-discharge relationships in runoff from a steep, unchanneled catchment. Water Resour Res 33:211–225
André C, Choppin GR (2000) Reduction of Pu(V) by humic acid. Radiochim Acta 88:613–616
Annual Report 2004 (2005) Monitoring and research in Lake Kinneret Yigal Allon Kinneret. Limnological Laboratory IOLR report T7/2005, pp 75–76
Aoki S, Fuse Y, Yamada E (2004) Determinations of humic substances and other dissolved organic matter and their effects on the increase of COD in Lake Biwa. Anal Sci 20:159–164
Aoki S, Ohara S, Kimura K, Mizuguchi H, Fuse Y, Yamada E (2008) Characterization of fluorophores released from three kinds of lake phytoplankton using gel chromatography and fluorescence spectrophotometry. Anal Sci 24:1461–1467
Artinger R, Buckau G, Geyer S, Fritz P, Wolf M, Kim JI (2000) Characterization of groundwater humic substances: influence of sedimentary organic carbon. Appl Geochem 15:97–116
Arts MT, Robarts RD, Kasai F, Waiser MJ, Tumber VP, Plante AJ, Rai H, De Lange HJ (2000) The attenuation of ultraviolet radiation in high dissolved organic carbon waters of wetlands and lakes on the northern Great Plains. Limnol Oceanogr 45:292–299
Ascenzo GD, Di Corcia a, Gentili A, Mancini R, Mastropasqua R, Nazzari M, Samperi R (2003) Fate of natural estrogen conjugates in municipal sewerage transport and treatment facilities. Sci Total Environ 302:199–209
Auriol M, Filali-Meknassi Y, Tyagi RD, Adams CD, Surampalli RY (2006) Endocrine disrupting compounds removal from wastewater, a new challenge. Process Biochem 41:525–539
Avery GB, Kieber RJ, Witt M, Willey JD (2006) Rainwater monocarboxylic and dicarboxylic acid concentrations in southeastern North Carolina, USA as a function of air mass back trajectory. Atmos Environ 40:1683–1693
Backlund P (1992) Degradation of aquatic humic material by ultraviolet light. Chemosphere 25:1869–1878
Bade DL (2004) Ecosystem carbon cycles: whole-lake fluxes estimated with multiple isotopes. University of Wisconsin, Thesis, Madison
Baker A (2001) Fluorescence excitation-emission matrix characterization of some sewage impacted rivers. Environ Sci Technol 35:948–953
Baker A (2002) Fluorescence excitation-emission matrix characterization of some farm wastes: implications for water quality monitoring. Water Res 36:189–194
Baker A, Spencer RGM (2004) Characterization of dissolved organic matter from source to sea using fluorescence and absorbance spectroscopy. Sci Total Environ 333:217–232
Balcarczyk KL, Jones JB Jr, Rudolf Jaffe′ R, Maie N (2009) Stream dissolved organic matter bioavailability and composition in watersheds underlain with discontinuous permafrost. Biogeochemistry 94:255–270

Ballaré CL, Caldwell MM, Flint SD, Robinson SA, Bornman JF (2011) Effects of solar ultraviolet radiation on terrestrial ecosystems Patterns, mechanisms, and interactions with climate change. Photochem Photobiol Sci 10:226–241
Balmer ME, Poiger T, Droz C, Romanin K, Bergqvist P-A, Muller MD, Buser H-A (2004) Occurrence of methyl triclosan, a transformation product of the bacericide triclosan, in fish and from various lakes in Switzerland. Environ Sci Technol 38:390–395
Barber RT (1968) Dissoved organic carbon from deep waters resists microbial oxidation. Nature 220:274
Bard SM (2000) Multixenobiotic resistance as a cellular defense mechanism in aquatic organisms. Aquat Toxicol 48:357–389
Baron J, McKnight DM, Denning AS (1991) Sources of dissolved and particulate organic material in Loch ValeWatershed, Rocky Mountain National Park, Colorado, USA. Biogeochemistry 15:89–110
Baxter RM, Carey JH (1983) Evidence for photochemical generation of superoxide ion in humic waters. Nature 306:575–576
Beardall J, Sobrino C, Stojkovic S (2009a) Interactions between the impacts of ultraviolet radiation, elevated CO2, and nutrient limitation on marine primary producers. Photochem Photobiol Sci 8:1257–1265
Beardall J, Stojkovic S, Larsen S (2009b) Living in a high CO2 world: impacts of global climate change on marine phytoplankton. Plant Ecol Divers 2:191–205
Beck KC, Reuter JH, Perdue EM (1974) Organic and inOrg Geochemof some coastal plain rivers of the southeastern United States. Geochim Cosmochim Acta 38:341–364
Bedner M, MacCrehan WA (2006) Transformation of acetaminophen by chlorination produces the toxicants 1,4-benzoquinone and N-acetyl-p-benzoquinone imine. Environ Sci Technol 40:516–522
Benner R (2002) Chemical composition and reactivity. In: Hansell DA, Carlson CA (eds) Biogeochemistry of marine dissolved organic matter. Academic Press, Amsterdam, pp 59–85
Benner R, Biddanda B (1998) photochemical transformations of surface and deep marine dissolved organic matter: effects on bacterial growth. Limnol Oceanogr 43:1373–1378
Benner R, Hedges JI (1993) A test of the accuracy of freshwater DOC measurements by high temperature catalytic oxidation and UV-promoted persulfate oxidation. Mar Chem 41:161–166
Benner R, Kaiser K (2003) Abundance of amino sugars and peptidoglycan in marine particulate and dissolved organic matter. Limnol Oceanogr 48:118–128
Benner R, Pakulski JD, McCarthy M, Hedges JI, Hatcher PG (1992) Bulk chemical characteristics of dissolved organic matter in the ocean. Science 255(5051):1561–1564
Benoy G, Cash K, McCauley E, Wrona F (2007) Carbon dynamics in lakes of the boreal forest under a changing climate. Environ Rev 15:175–189
Berner RA (1989) Biogeochemical cycles of carbon and sulfur and their effect on atmospheric oxygen over Phanerozoic time. Palaeogeogr Palaeoclimatol Palaeoecol 73:97–122
Bertilsson S, Jones JB (2003) Supply of dissolved organic matter to aquatic ecosystems: autochthonous sources. In: Findlay SEG, Sinsabaugh RL (eds) Aquatic ecosystems: interactivity of dissolved organic matter. Academic Press, San Diego, pp 3–24
Bertilsson S, Tranvik LJ (1998) Photochemically produced carboxylic acids as substrates for freshwater bacterioplankton. Limnol Oceanogr 43:885–895
Bertilsson S, Tranvik LJ (2000) Photochemically transformation of dissolved organic matter in lakes. Limnol Oceanogr 45:753–762
Bertilsson S, Stepanauskas R, Cuadros-Hansson R, Granéli W, Wikner J, Tranvik LJ (1999) Photochemically induced changes in bioavailable carbon and nitrogen pools in a boreal watershed. Aquat Microb Ecol 19:47–56
Bertoni R, Callieri C, Morabito G, Pinolini ML, Pugnetti A (1998) Quali-quantitative changes in organic carbon production during the oligotrophication of Lake Maggiore, Italy. Verh Int Ver Limnol 26:300–304

Bertoni R, Callieri C, Caravati E, Contesini M, Corno G, Manca D (2008) Indagini sull'ambiente pelagico Carbonio organico e popolamenti batterici eterotrofi In CNR-ISE Ricerche sull'evoluzione del Lago Maggiore Aspetti Limnologici Programma quinquennale 2003–2007 Campagna 2007 e Rapporto quinquennale 2003–2007. Commissione Internazionale per la protezione delle acque italo-svizzere: 67–72

Bertoni R, Callieri C, Corno G, Rasconi S, Caravati E, Contesini M (2010) Long-term trends of epilimnetic and hypolimnetic bacteria and organic carbon in a deep holo-oligomictic lake. Hydrobiologia 644:279–287

Biddanda B, Benner R (1997) Carbon, nitrogen, and carbohydrate fluxes during the production of particulate and dissolved organic matter by marine phytoplankton. Limnol Oceanogr 42:506–518

Biddanda BA, Cotner JB (2002) Love handles in aquatic ecosystems: the role of dissolved organic carbon drawdown, resuspended sediments, and terrigenous inputs in the carbon balance of Lake Michigan. Ecosystems 5:431–445

Biddanda B, Ogdahl M, Cotner J (2001) Dominance of bacterial metabolism in oligotrophic relative to eutrophic waters. Limnol Oceanogr 46:730–773

Biers EJ, Zepp RG, Moran MA (2007) The role of nitrogen in chromophoric and fluorescent dissolved organic matter formation. Mar Chem 103:46–60

Billen G, Fontigny A (1987) Dynamics of a Phaeocystis dominated bloom in Belgian coastal waters. 2. Bacterioplankton dynamics. Mar Ecol Prog Ser 37:249–257

Bloom PR, Bleam WF, Xia K (2001) X-ray spectroscopy applications for the study of humic substances. In: Clapp CE, Hayes MHB, Senesi N, Bloom PR, Jardine PM (eds) Humic substances and chemical contaminants. Soil Science Society of America, Madison, p 317

Blough NV, Zafiriou OC, Bonilla J (1993) Optical absorption spectra of waters from the Orinoco River outflow: terrestrial input of colored organic matter to the Caribbean. J Geophys Res 98:2271–2278

Boehme J, Wells M (2006) Fluorescence variability of marine and terrestrial colloids: examining size fractions of chromophoric dissolved organic matter in the Damariscotta River estuary. Mar Chem 101:95–103

Bolger R, Wiese TE, Ervin K, Nestich S, Checovich W (1998) Rapid screening of environmental chemicals for estrogen receptor binding capacity. Environ Health Perspect 106:551–557

Borges AV, Ruddick K, Schiettecatte L-S, Delille B (2008) Net ecosystem production and carbon dioxide fluxes in the Scheldt estuarine plume BMC. Ecology 8:15. doi:101186/1472-6785-8-15

Borowitzka MA (1988) Fats, oils and hydrocarbons. In: Borowitzka MA, Borowitzka LJ (eds) Micro-algal biotechnology. Cambridge University Press, Cambridge, pp 257–287

Bouchard A (1997) Recent lake acidification and recovery trends in southern Quebec, Canada. Water Air Soil Pollut 94:225–245

Bowling L, Baker P (1996) Major cyanobacterial bloom in the Barwon- Darling River, Australia, in 1991, and underlying limnological conditions. Aust J Mar Freshw Res 47:643–657

Boxall ABA, Sinclair CJ, Fenner K, Kolpin D, Maund SJ (2004) When synthetic chemicals degrade in the environment. Environ Sci Technol 38:368A–375A

Boyd TJ, Osburn CL (2004) Changes in CDOM fluorescence from allochthonous and autochthonous sources during tidal mixing and bacterial degradation in two coastal estuaries. Mar Chem 89:189–210

Boyd TJ, Barham BP, Hall GJ, Schumann BS, Paerl RW, Osburn CL (2010) Variation in ultrafiltered and LMW organic matter fluorescence properties under simulated estuarine mixing transects: 2. Mixing with photoexposure. J Geophys Res 115:G00F14. doi:101029/2009JG000994

Braakhekke MC, Beer C, Hoosbeek MR, Reichstein M, Kruijt B, Schrumpf M, Kabat P (2011) SOMPROF: a vertically explicit soil organic matter model. Ecol Model 222:1712–1730

Bradley C, A Baker A, Cumberland S, Boomer I, Morrissey I (2007) Dynamics of water movement and trends in dissolved carbon in a headwater wetland in a permeable catchment. Wetlands 27:1066–1080

Brandt LA, Bohnet C, King JY (2009) Photochemically induced carbon dioxide production as a mechanism for carbon loss from plant litter in arid ecosystems. J Geophys Res 114(G2):G02004

Brassell SC, Comet PA, Eglinton G, Isaacson PJ, McEvoy J, Maxwell JR, Thompson ID, Tibbetts PJC, Volkman JK (1980) The origin and fate of lipids in the Japan Trench In: Douglas AG, Maxwell JR (eds) Advances in Organic Geochemistry. Pergamon, Oxford, pp 375–392

Bratback G, Thingstad TF (1985) Phytoplankton-bacteria interactions: an apparent paradox? analysis of a model system with both competition and commensalism. Mar Ecol Prog Ser 25:23–30

Braven J, Butler EI, Chapman J, Evens R (1995) Changes in dissolved free amino acid composition in seawater associated with phytoplankton populations. Sci Total Environ 172:145–150

Brodnjak-Vončina D, Dobenik D, Novie M, Zupan J (2002) Chemometrics characterisation of the quality of river water. Analyt Chim Acta 462:87–100

Bronk DA (2002) Dynamics of DON. In: Hansell DA, Carlson CA (eds) Biogeochemistry of marine dissolved organic matter. Academic Press, San Diego, pp 153–249

Bronk DA, Glibert PM, Ward BB (1994) Nitrogen uptake, dissolved organic nitrogen release, and new production. Science 265:1843–1846

Bronk DA, Glibert PM, Malone TC, Banahan S, Sahlsten E (1998) Inorganic and organic nitrogen cycling in Chesapeake Bay: autotrophic versus heterotrophic processes and relationships to carbon flux. Aquat Microb Ecol 15:177–189

Brooks BW, Chambliss CK, Stanley JK, Ramirez AJ, Banks KE, Johnson RD, Lewis RJ (2005) Determination of select antidipressants in fish from an effluent-dominated stream. Environ Toxicol Chem 24:464–469

Brooks ML, Meyer JS, McKnight DM (2007) Photooxidation of wetland and riverine dissolved organic matter: altered copper complexation and organic composition. Hydrobiologia 579:95–113

Brown TL, Rice JA (2000) Effect of experimental parameters on the ESI FT-ICR mass spectrum of fulvic acid. Anal Chem 72:384–390

Brown JN, Paxe′us N, Förlin L, Larsson JDG (2007) Variations in bioconcentration of human pharmaceuticals from sewage effluents into fish blood plasma. Environ Toxicol Pharmacol 24:267–274

Brunke M, Gonser TOM (1997) The ecological significance of exchange processes between rivers and groundwater freshwater. Biology 37:1–33

Brzozowski T, Dembinski A, Konturek S (1994) Influence of Tolpa Peat Preparation on gastroprotection and on gastric and duodenal ulcers. Acta Pol Pharm 51:103–107

Buckau G, Antinger R, Geyer S, Wolf M, Fritz P, Kim JI (2000) Groundwater in situ generation of aquatic humic and fulvic acid and the mineralization of sedimentary organic carbon. Appl Geochem 15:819–832

Buesseler KO, Bauer JE, Chen RF, Eglinton TI, Gustafsson O, Landing W, Mopper K, Moran SB, Santschi PH, VernonClark R, Wells ML (1996) An intercomparison of cross-flow filtration techniques used for sampling marine colloids: overview and organic carbon results. Mar Chem 55:1–31

Buick R (2008) When did oxygenic photosynthesis evolve? When did oxygenic photosynthesis evolve? Phil Trans R Soc B 363:2731–2743

Burdige DJ, Kline SW, Chen WH (2004) Fluorescent dissolved organic matter in marine sediment pore waters. Mar Chem 89:289–311

Burney CM, Davis PG, Johnson KM, Sieburth JMcN (1982) Diel relationships of microbial tropic groups and in situ dissolved carbohydrate dynamics in the Caribbean Sea. Mar Biol 67:311–322

Burt TP, Adamson JK, Lane AMJ (1998) Long-term rainfall and streamflow records for north central England: putting the Environmental Change Network site at Moor House, Upper Teesdale, in context. Hydrol Sci J 43:775–787

Buser H-R, Balmer ME, Schmid P, Kohler M (2006) Occurrence of UV-filters 4-methylbenzylidene camphor and octocrylene in fish from various Swiss rivers with inputs from wastewater treatment plants. Environ Sci Technol 40:1427–1431

Bushaw KL, Zepp RG, Tarr MT, Schulz-Jander D, Bourbonniere RA, Hodson RE, Miller WL, Bronk DA, Moran MA (1996) Photochemical release of biologically available nitrogen from aquatic dissolved organic matter. Nature 381:404–407

Bussmann I, Kattner G (2000) Distribution of dissolved organic carbon in the central Arctic Ocean: the influence of physical and biological properties. J Mar Syst 27:209–219

Buttiglieri G, Peschka M, Fromel T, Muller J, Malpei F, Seel P, Knepper TP (2009) Environmental occurrence and degradation of the herbicide n-chloridazon. Water Res 43:2865–2873

Cadée GC (1987) Organic carbon in the Ems River and estuary: a comparison of summer and winter conditions In: Degens ET, Kempe S, Wei-Bin G (eds) Transport of carbon and minerals in major world rivers, SCOPE/UNEP Sonderbd 64 Mitt Geological-Palaeontological Institute and Museum of the University of Hamburg, pp 359–374

Cai W-J (2011) Estuarine and coastal ocean carbon paradox: CO2 sinks or sites of terrestrial carbon incineration? Annu Rev Mar Sci 3:123–145

Cai WJ, Pomeroy LR, Moran MA, Wang YC (1999) Oxygen and carbon dioxide mass balance for the estuarine-intertidal marsh complex of five rivers in the southeastern US. Limnol Oceanogr 44:639–649

Callieri C, Piscia R (2002) Photosynthetic efficiency and seasonality of autotrophic picoplankton in Lago Maggiore after its recovery. Freshw Biol 47:941–956

Cammack WKL, Kalff J, Prairie YT, Smith EM (2004) Fluorescent dissolved organic matter in lakes: relationships with heterotrophic metabolism. Limnol Oceanogr 49:2034–2045

Cancilla DA, Holtkamp A, Matassa L, Fang X (1997) Isolation and characterization of Microtox-active components from aircraft deicing/anti-icing fluids. Environ Toxicol Chem 16:430–434

Canham CD, Pace ML, Papaik MJ, Primack AGB, Roy KM, Maranger RJ, Curran RP, Spada DM (2004) A spatially explicit watershed-scale analysis of dissolved organic carbon in Adirondack lakes. Ecol Appl 14:839–854. doi:101890/02-5271

Caraco NF, Lapman G, Cole JJ, Limburg KE, Pace ML, D Fischer D (1998) Microbial assimilation of DIN in a nitrogen rich estuary: implications for food quality and isotope studies. Mar Ecol Prog Ser 167:59–71

Carder KL, Steward RG, Harvey GR, Ortner PB (1989) Marine humic and fulvic acids: their effects on remote sensing of ocean chlorophyll. Limnol Oceanogr 34:68–81

Carlsen E, Giwercman A, Keiding N, Skakkebaek NE (1995) Declining semen quality and increasing incidence of testicular cancer: is there a common cause? Environ Health Perspect 103(Suppl 7):137–139

Carlson DJ, Brann ML, Mague TH, Mayer LM (1985) Molecular weight distribution of dissolved organic materials in seawater determined by ultrafiltration: a re-examination. Mar Chem 16:155–171

Carlson CA, Hansell DA, Peltzer ET, Smith WO Jr (2000) Stocks and dynamics of dissolved and particulate organic matter in the southern Ross Sea, Antarctica. Deep Sea Res I 47:3201–3225

Carlsson C, Johansson AK, Alvan G, Bergman K, Kuhler T (2006) Are pharmaceuticals potent environmental pollutants? Part I: environmental risk assessments of selected active pharmaceutical ingredients. Sci Total Environ 364:67–87

Carpenter SR, Cole JJ, Kitchell JF, Pace ML (1998) Impact of dissolved organic carbon, phosphorus, and grazing on phytoplankton biomass and production in experimental lakes. Limnol Oceanogr 43:73–80

Carrera N, Barreal ME, Gallego PP, Briones MJI (2009) Soil invertebrates control peat land C fluxes in response to warming. Funct Ecol 23:637–648

Carrick HJ, Fahnenstiel GL, Stoermer EF, Wetzel RG (1991) Protozoan growth rates and trophic couplings in Lake Michigan. Limnol Oceanogr 36:133–1345

Carrillo P, Medina-Sánchez JM, Villar-Argaiz M (2002) The interaction of phytoplankton and bacteria in a high mountain lake: importance of the spectral composition of solar radiation. Limnol Oceanogr 47:1294–1306

Casajuana N, Lacorte S (2003) Presence and release of phthalic esters and other endocrine disrupting compounds in drinking water. Chromatographia 57(9):649–655
Cauwet G, Mackenzie FT (1993) Carbon inputs and distribution in estuaries of turbid rivers: the Yang Tze and Yellow rivers (China). Mar Chem 43:235–246
Chapman PJ, Edwards AC, Cresser MS (2001) The nitrogen composition of streams in upland Scotland: some regional and seasonal differences. Sci Total Environ 265:65–83
Chebbi A, Carlier P (1996) Carboxylic acids in the troposphere, occurence, sources, and sinks: a review. Atmos Environ 30:4233–4249
Chen RF, Gardner GB (2004) High-resolution measurements of chromophoric dissolved organic matter in the Mississippi and Atchafalaya River plume regions. Mar Chem 89:103–125
Chen Z, Li Y, Pan JM (2004) Distributions of colored dissolved organic matter and dissolved organic carbon in the Pearl River Estuary, China. Cont Shelf Res 24:1845–1856
Chen G, Huang L, Tan Y, Yin J, Wang H, Huang H, Zou K, Li R (2007) Antibacterial substance from mucus of a scleractinian coral, Symphyllia gigantea. Acta Oceanol Sinica 26:140–143
Chiron S, Comoretto L, Rinaldi E, Maurino V, Minero C, Vione D (2009) Pesticide By-Products in the Rhône Delta (Southern France) The Case of 4-Chloro-2-methylphenol and of its Nitroderivative. Chemosphere 74:599–604
Christensen JB, Jensen DL, Grøn C, Filip Z, Christensen TH (1998) Characterization of the dissolved organic carbon in landfill leachate-polluted groundwater. Water Res 32:125–135
Christman RF, Oglesby RT (1971) Microbiol degradation and the formation of humus. In: Sarkanen KV, Ludwig CH (eds) Lignins: occurrence, Formation, structure and reactions. Wiley-Interscience, New York, pp 769–796
Church MJ, Ducklow HW, Karl DM (2002) Multiyear increases in dissolved organic matter inventories at Station ALOHA in the North Pacific Gyre. Limnol Oceanogr 47:1–10
Ciglasch H, Lilienfein J, Kaiser K, Wilcke W (2004) Dissolved organic matter under native Cerrado and Pinus caribaea plantations in the Brazilian savanna. Biogeochemistry 67:157–182
Citulski JA, Farahbakhsh K (2010) Fate of endocrine-active compounds during municipal biosolids treatment: a Review. Environ Sci Technol 44:8367–8376
Clara M, Gans O, Windhofer G, Krenn U, Hartl W, Braun K, Scharf S, Scheffknecht C (2011) Occurrence of polycyclic musks in wastewater and receiving water bodies and fate during wastewater treatment. Chemosphere 82:1116–1123
Clark CD, Hiscock WT, Millero FJ, Hitchcock G, Brand L, Miller WL, Ziolkowski L, Chen RF, Zika RG (2004) CDOM distribution and CO2 production on the Southwest Florida Shelf. Mar Chem 89:145–167
Cleuvers M (2004) Mixture toxicity of the anti-inflammatory drugs diclofenac, ibuprofen, naproxen and acetylsalicylic acid. Ecotoxicol Environ Saf 59:309–315
Cloern JE, Canuel EA, Harris D (2002) Stable carbon and nitrogen isotope composition of aquatic and terrestrial plants of the San Francisco Bay estuarine system. Limnol Oceanogr 47:713–729
Clutterbuck B, Yallop AR (2010) Land management as a factor controlling dissolved organic carbon release from upland peat soils 2: changes in DOC productivity over four decades. Sci Total Environ 408:6179–6191
Coble PG (1996) Characterization of marine and terrestrial DOM in sea water using excitation-emission matrix spectroscopy. Mar Chem 52:325–336
Coble PG (2007) Marine optical biogeochemistry: the chemistry of ocean color. Chem Rev 107:402–418
Colborn T, vom Saal FS, Soto AM (1993) Developmental effects of endocrine-disrupting chemicals in wildlife and humans. Environ Health Pespect 101:378–384
Cole JJ, Findlay S, Pace ML (1988) Bacterial production in fresh and saltwater ecosystems: a cross system overview. Mar Ecol Prog Ser 43:1–10
Cole L, Bardgett RD, Ineson P, Adamson JK (2002) Relationships between enchytraeid worms (Oligochaeta), climate change, and the release of dissolved organic carbon from blanket peat in northern England. Soil Biol Biochem 34:599–607
Connolly JP, Coffin RB, Landeck RE (1992) Modeling carbon utilization by bacteria in natural water systems. In: Hurst CJ (ed) Modelling the metabolic and physiologic activities of microorganisms. Wiley, Chichester, pp 249–276

Conrad R (1999) Contribution of hydrogen to methane production and control of hydrogen concentrations in methanogenic soils and sediments. FEMS Microbiol Ecol 28:193–202
Cook RL, McIntyre DD, Langford CH, Vogel HJ (2003) A comprehensive liquid-state heteronuclear and multidimensional NMR study of Laurentian fulvic acid. Environ Sci Technol 37:3935–3944
Corin N, Backlund P, Kulovaara M (1996) Degradation products formed during UV-irradiation of humic waters. Chemosphere 33:245–255
Cory RM, McKnight DM (2005) Fluorescence spectroscopy reveals ubiquitous presence of oxidized and reduced quinines in dissolved organic matter. Environ Sci Technol 39:8142–8149
Cory RM, Green SA, Pregitzer KS (2004) Dissolved organic matter concentration and composition in the forests and streams of Olympic National Park, WA. Biogeochemistry 67:269–288
Cowie GL, Hedges JI (1994) Biochemical indicators of diagenetic alteration in natural organic matter mixtures. Nature 369:304–307
Cowie GL, Hedges JI (1996) Digestion and alteration of the biochemical constituents of a diatom (Thalassiosira weissflogii) ingested by an herbivorous zooplankton (Calanus pacificus). Limnol Oceanogr 41:581–594
Crandall CA, Katz BG, Hirten JJ (1999) Hydrochemical evidence for mixing of river water and groundwater during high-flow conditions, lower Suwannee River basin, Florida, USA. Hydrogeol J 7:454–467
Cronan CS, Aiken GR (1985) Chemistry and transport of soluble humic substances in forested watersheds of the Adirondack Park, New York. Geochim Cosmochim Acta 49:1697–1705
Currin CA, Newell SY, Pearl HW (1995) The role of standing dead Spartina alterniflora and benthic microalgae in salt marsh food webs: considerations based on multiple stable isotope analysis. Mar Ecol Prog Ser 121:99–116
Curtis PJ, Adams HE (1995) Dissolved organic matter quantity and quality from freshwater and saline lakes in east-central Alberta. Biogeochemistry 30:59–76
Curtis PJ, Prepas EE (1993) Trends of dissolved organic carbon (DOC) concentrations from freshwater to saltwater. Verh Int Ver Limnol 25:298–301
Curtis PJ, Schindler DW (1997) Hydrologic control of dissolved organic matter in low-order Precambrian Shield lakes. Biogeochemistry 36:125–138
Da Costa F, Lubes G, Rodríguez M, Lubes V (2011) Study of the ternary complex formation between vanadium(III), dipicolinic acid and small blood serum bioligands. J Solution Chem 40:106–117
Dahlén J, Bertilsson S, Pettersson C (1996) Effects of UV-A irradiation on dissolved organic matter in humic surface waters. Environ Int 22:501–506
Dai KO, David M, Vance G (1996) Characterization of solid and dissolved carbon in a spruce-fir Spodosol. Biogeochemistry 35:339–365
Daniel MHB, Montebelo AA, Bernardes MC, Ometto JPHB, De Camargo PB, Krusche AV, Ballester MV, Victoria RL, Martinelli LA (2002) Effects of urban sewage on dissolved oxygen, dissolved inorganic and organic carbon, and electrical conductivity of small streams along gradient of urbanization in the Piracicaba River basin. Water Air Soil Poll 136:189–206
Danielsson LG (1982) On the use of filters for distinguishing between dissolved and particulate fractions in natural waters. Water Res 16:179–182
Danø K (1972) Cross resistance between vinca alkaloids and anthracyclines in Ehrlich ascites tumor in vivo, Cancer. Chemother Rep 56:701–708
Daughton CG, Ternes TA (1999) Pharmaceuticals and personal care products in the environment: agents of subtle change? Environ Health Perspect 107:907–938
David MB, Vance GF (1991) Chemical character and origin of organic acids in streams and seepage lakes of central Maine. Biogeochemistry 12:17–41. doi:101007/BF00002624
Davidson EA, Janssens IA (2006) Temperature sensitivity of soil carbon decomposition and feedbacks to climate change. Nature 440:165–173
Davis SE III, Childers DL, Day JW Jr, Rudnick DT, Sklar FH (2001) Nutrient dynamics in vegetated and unvegetated areas of a southern everglades mangrove creek. Estuar Coast Shelf Sci 52:753–765

de Souza-Sierra MM, Donard OFX, Lamotte M, Belin C, Ewald M (1994) Fluorescence spectroscopy of coastal and marine waters. Mar Chem 47:127–144

de Zarruk KK, Scholer G, Dudal Y (2007) Fluorescence fingerprints and Cu2+-complexing ability of individual molecular size fractions in soil- and waste-borne DOM. Chemosphere 69:540–548

Dean M, Rzhetsky A, Allikmets R (2001) The human ATP-binding cassette (ABC) transporter superfamily. Genome Res 11:1156–1166

Deegan LA, Garritt RH (1997) Evidence for spatial variability in estuarine food webs. Mar Ecol Prog Ser 147:31–47

del Castillo CE, Coble PG, Morell JM, Lopez JM, Corredor JE (1999) Analysis of the optical properties of the Orinoco River Plume by absorption and fluorescence spectroscopy. Mar Chem 66:35–51

del Castillo CE, Gilbes F, Coble PG, Muller-Karger FE (2000) On the dispersal of riverine colored dissolved organic matter over the West Florida Shelf. Limnol Oceanogr 45:1425–1432

del Vecchio R, Blough NV (2002) Photobleaching of chromophoric dissolved organic matter in natural waters: kinetics and modeling. Mar Chem 78:231–253

Depetris PJ, Kempe S (1993) Carbon dynamics and sources in the Paraná River. Limnol Oceanogr 38:382–395

Derbalah ASH, Nakatani N, Sakugawa H (2003) Distribution, seasonal pattern, flux and contamination source of pesticides and nonylphenol residues in Kurose River water, Higashi-Hiroshima, Japan. Geochem J 37:217–232

Diamanti-Kandarakis E, Bourguignon JP, Giudice L et al (2009) Endocrine-disrupting chemicals: an endocrine society scientific statement. Endocr Rev 30:293–342

Dittmar T, Lara RJ (2001) Driving forces behind nutrient and organic matter dynamics in a mangrove tidal creek in north Brazil. Coast Shelf Sci 52:249–259

Dorodnikov M, Kuzyakov Y, Fangmeier A, Wiesenberg GLB (2011) C and N in soil organic matter density fractions under elevated atmospheric CO2: turnover vs stabilization. Soil Biol Biochem 43:579–589

dos Anjos FM, Bittencourt-Oliveira MD, Zajac MP, Hiller S, Christian B, Erler K, Luckas B, Pinto E (2006) Detection of harmful cyanobacteria and their toxins by both PCR amplification and LC-MS during a bloom event. Toxicon 48:239–245

Dosskey MG, Bertsch PM (1994) Forest sources and pathways of organic matter transport to a blackwater stream: a hydrologic approach. Biogeochemistry 24:1–19

Downing JA, Prairie YT, Cole JJ, Duarte CM, Tranvik LJ, Striegl RG, McDowell WH, Kortelainen P, Caraco NF, Melack JM, Middelburg JJ (2006) The global abundance and size distribution of lakes, ponds, and impoundments. Limnol Oceanogr 51:2388–2397

Driscoll CT, Driscoll KM, Roy KM et al (2003) Chemical response of lakes in the Adirondack Region of New York to declines in acidic deposition. Environ Sci Technol 37:2036–2042

Druon JN, Mannino A, Signorini S, McClain C, Friedrichs M, Wilkin J, Fennel K (2010) Modeling the dynamics and export of dissolved organic matter in the Northeastern US continental shelf. Estuar Coast Shelf Sci 88:488–507

Duan S, Bianchi TS (2007) Particulate and dissolved amino acids in the lower Mississippi and Pearl Rivers (USA). Mar Chem 107:214–229

Duedahl-Olesen L, Cederberg T, Pedersen KH, Højgård A (2005) Synthetic musk fragrances in trout from Danish fish farms and human milk. Chemosphere 61:422–431

Duff KE, Laing TE, Smol JP, Lean DRS (1999) Limnological characteristics of lakes located across arctic treeline in northern Russia. Hydrobiol 391:205–222

Eatherall AE (1996) The role of carbon in river basins. LOIS working note no 7 institute of hydrology Wallingford, p 53

Eatherall A, Warwick MS, Tolchard S (2000) Identifying sources of dissolved organic carbon on the River Swale, Yorkshire. Sci Total Environ 251(252):173–190

Eckhardt BW, Moore TR (1990) Controls on dissolved organic carbon concentrations in streams, southern Quebec. Can J Fish Aquat Sci 47:1537–1544

Effler SW, Perkins M, Peng F, Strait C, Weidemann AD, Auer MT (2010) Light-absorbing components in Lake Superior. J Great Lakes Res 36:656–665
Eisma D, Cadée GC, Laane R (1982) Supply of suspended matter and particulate and dissolved organic carbon from the Rhine to the Coastal North Sea. In: Degens ET (ed) Transport of carbon and minerals in major world rivers, SCOPE/UNEP Sonderbd 52 Mitt Geological-Palaeontological Institute and Museum of the University of Hamburg, pp 483–506
Elbe ARGE (1997) Wassergütedaten der Elbe. Jährliche Zahlentafeln, Hamburg, p 95
Elder J, Rybicki N, Carter V, Weintraub V (2000) Sources and yields of dissolved carbon in northern Wisconsin stream catchments with differing amounts of peatland. Wetlands 20:113–125
Eloranta P (1978) Light penetration in different types of lakes in Central Finland. Holarct Ecol 1:362–366
Engebretson RR, von Wandruszka R (1994) Microorganization in dissolved humic acids. Environ Sci Technol 28:1934–1941. doi:101021/es00060a026
Engebretson RR, von Wandruszka R (1996) Quantitative approach to humic acid associations. Environ Sci Technol 30:990–997. doi:101021/es950478g
Engel A, Passow U (2001) Carbon and nitrogen content of transparent exopolymer particles (TEP) in relation to their Alcian Blue adsorption. Mar Ecol Prog Ser 219:1–10
Engstrom DR (1987) Influence of vegetation and hydrology on the humus budgets of Labrador lakes. Can J Fish Aquat Sci 44:1306–1314. doi:101139/f87-154
Epel D (1998) Use of multidrug transporters as first lines of defense against toxins in aquatic organisms. Comp Biochem Physiol A 120:23–28
Epel D, Smital T (2001) Multidrug-multixenobiotic transporters and their significance with respect to environmental levels of pharmaceuticals and personal care products. In: Daughton CG, Jones-Lepp TL (eds) Pharmaceuticals and personal care products in the environment scientific and regulatory issues, vol 791. American Chemical Society, Washington, pp 244–263
Etheridge SM (2010) Paralytic shellfish poisoning: seafood safety and human health perspectives. Toxicon 56:108–122
Evans AJ (1998) Biodegradation of 14C-labeled low molecular organic acids using three biometer methods. J Geochem Explor 65:17–25
Evans MG, Burt TP, Holden J, Adamson JK (1999) Runoff generation and water table fluctuations in blanket peat: evidence from UK data spanning the dry summer of 1995. J Hydrol 221:141–160
Evans CD, Freeman C, Monteith DT, Reynolds B, Fenner N (2002) Climate change—terrestrial export of organic carbon: communication arising. Nature 415:862
Evans CD, Monteith DT, Cooper DM (2005) Long-term increases in surface water dissolved organic carbon: observations, possible causes and environmental impacts. Environ Pollut 137:55–71
Evans CD, Chapman PJ, Clark JM, Monteith DT, Cresser MS (2006) Alternative explanations for rising dissolved organic carbon export from organic soils. Glob Chang Biol 12:2044–2053
Fahey TJ, Siccama TG, Driscoll CT, Likens GE, Campbell J, Johnson CE, Battles JJ, Aber JD, Cole JJ, Fisk MC, Groffman PM, Hamburg SP, Holmes RT, Schwarz PA, Yanai RD (2005) The biogeochemistry of carbon at Hubbard Brook. Biogeochemistry 75:109–176
Fairchild WL, Swansburg OE, Arsenault JT, Brown SB (1999) Does an association between pesticide use and subsequent declines in catch of Atlantic salmon (Salmo salar) represent a case of endocrine disruption? Environ Health Perspect 107:349–358
Fankhauser-Noti A, Biedermann-Brem S, Grob K (2006) PVC plasticizers/additives migrating from the gaskets of metal closures into oily food: Swiss market survey June 2005. Eur Food Res Technol 223:447–453
Farias J, Rossetti GH, Albizzati ED, Alfano OM (2007) Solar degradation of formic acid: temperature effects on the photo-Fenton reaction. Ind Eng Chem Res 46:7580–7586
Farjalla VF, Azevedo DA, Esteves FA, Bozelli RL, Roland F, Enrich-Prast A (2006) Influence of hydrological pulse on bacterial growth and DOC uptake in a clear-water Amazonian Lake. Microb Ecol 52:334–344
Farré Ml, Pérez S, Kantiani L, Barceló D (2008) Fate and toxicity of emerging pollutants, their metabolites and transformation products in the aquatic environment TrAC. Trends Anal Chem 27:991–1007

Fellman JB, Hood E, Edwards RT, Jones JB (2009) Uptake of allochthonous dissolved organic matter from soil and salmon in coastal temperate rainforest streams. Ecosystems 12:747–759. doi:101007/s10021-009-9254-4
Fenner N, Freeman C, Lock MA, Harmens H, Reynolds B, Sparks T (2007a) Interactions between elevated CO2 and warming could amplify DOC Exports from Peatland Catchments. Environ Sci Tech 41:3146–3152
Fenner N, Ostle NJ, McNamara N, Sparks T, Harmens H, Reynolds B, Freeman C (2007b) Elevated CO2 effects on peatland plant community carbon dynamics and DOC production. Ecosystems 10:635–647
Fenton SE (2006) Endocrine-disrupting compounds and mammary gland development: early exposure and later life consequences. Endocrinology 147:18–24
Ferrari GM, Dowell MD (1998) CDOM absorption characteristic with relation to fluorescence and salinity in coastal areas of the Southern Baltic Sea. Estuar Coast Shelf Sci 47:91–105
Ferrari GM, Dowell MD, Grossi S, Targa C (1996) The relationship between the optical properties of chromophoric dissolved organic matter and total concentration of dissolved organic carbon in the southern Baltic Sea region. Mar Chem 55:299–316
Ferrari B, Mons R, Vollat B, Fraysse B, Paxéus N, Giudice RL, Pollio A, Garric J (2004) Environmental Risk Assessment of six human pharmaceuticals: are the current environmental risk assessment procedures sufficient for the protection of the aquatic environment? Environ Toxical Chem 23:1344–1354
Fichot CG, Miller WL (2010) An approach to quantify depth-resolved marine photochemical fluxes using remote sensing: application to carbon monoxide (CO) photoproduction. Remote Sens Environ 114:1363–1377
Field JA, Johnson CA, Rose JB (2006) What is emerging? Environ Sci Technol 40:7105
Filella M (2008) NOM site binding heterogeneity in natural waters: discrete approaches. J Mol Liq 143:42–51
Filley TR, Cody GD, Goodell B, Jellison J, Noser C, Ostrofsky A (2002) Lignin demethylation and polysaccharide decomposition in spruce sap wood degraded by brown rot fungi. Org Geochem 33:111–124
Fimmen RL, Cory RM, Chin Y-P, Trouts TD, McKnight DM (2007) Probing the oxidation-reduction properties of terrestrially and microbially derived dissolved organic matter. Geochim Cosmoch Acta 71:3003–3015
Findlay SEG (2005) Increased carbon transport in the Hudson River: unexpected consequence of nitrogen deposition? Front Ecol Environ 3:133–137
Findlay S, Pace M, Lints D (1991) Variability and transport of suspended sediment, particulate and dissolved organic carbon in the tidal freshwater Hudson River. Biogeochemistry 12:149–169
Fiorentino G, Spaccini R, Piccolo A (2006) Separation of molecular constituents from a humic acid by solid-phase extraction following a transesterification reaction. Talanta 68:1135–1142
Fisher SG, Likens GE (1972) Stream ecosystem: organic energy budget. Bioscience 22:33–35
Fisher SG, Likens GE (1973) Energy flow in Bear Brook, New Hampshire: an integrative approach to stream ecosystem metabolism. Ecol Monogr 43:421–439
Flessa H, Ludwig B, Heil B, Merbach W (2000) The origin of soil organic C, dissolved organic C and respiration in a long-term maize experiment in Halle, germany, determined by C-13 natural abundance. J Plant Nutr Soil Sci Z Pflanzenernahr Bodenkd 163(2):157–163
Fletcher KE, Boyanov MI, Homas SH, Wu Q, Kemner KM, Löffler FE (2010) U(VI) reduction to mononuclear U(IV) by Desulfitobacterium species. Environ Sci Technol 44:4705–4709
Fogg GE (1983) The ecological significance of extracellular products of phytoplankton photosynthesis. Bot Mar 26:3–14
Ford TE, Naiman RJ (1989) Groundwater-surface water relationships in boreal forest watersheds: dissolved organic carbon and inorganic nutrient dynamics. Can J Fish Aquat Sci 46:41–49
Freeman C, Ostle N, Kang H (2001a) An enzymic 'latch' on a global carbon store—a shortage of oxygen locks up carbon in peatlands by restraining a single enzyme. Nature 409:149

Freeman C, Evans CD, Monteith DT, Reynolds B, Fenner N (2001b) Export of organic carbon from peat soils. Nature 412:785
Freeman C, Fenner N, Ostle NJ, Kang H, Dowrick DJ, Reynolds B, Lock MA, Sleep D, Hughes S, Hudson J (2004) Export of dissolved organic carbon from peatlands under elevated carbon dioxide levels. Nature 430:195–198
Freyer M, Walther C, Stumpf T, Buckau G, Fanghänel T (2009) Formation of Cm humate complexes in aqueous solution at pHc 3 to 55: the role of fast interchange. Radiochim Acta 97:547–558
Abbt-Braun G, Lankes U, Frimmel, F (2004) Structural characterization of aquatic humic substances—the need for a multiple method approach. Aquat Sci Res Across Bound 66(2):151–170
Fröberg M, Berggren Kleja D, Hagedorn F (2007) The contribution of fresh litter to dissolved organic carbon leaching from a coniferous forest floor. Eur J Soil Sci 58:108–114
Frost PC, Mack A, Larson JH, Bridgham SD, Lamberti GA (2006) Environmental controls of UV-B radiation in forested streams of Northern Michigan. Photochem Photobiol 82:781–786. doi:101562/2005-07-22-RA-619
Fry B, Sherr EB (1984) 13C Measurements as indicators of carbon flow in marine and freshwater ecosystems. Contrib Mar Sci 27:13–47
Fry B, Hopkinson CS, Nolin A (1996) Long-term decomposition of DOC from experimental diatom blooms. Limnol Oceanogr 41:1344–1347
Fu F-X, Zhang Y, Leblanc K, Sañudo-Wilhelmy SA, Hutchins DA (2005) The biological and biogeochemical consequences of phosphate scavenging onto phytoplankton cell surfaces. Limnol Oceanogr 50:1459–1472
Fu P, Mostofa KMG, Wu FC, Liu CQ, Li W, Liao H, Wang L, Wang J, Mei Y (2010) Excitation-emission matrix characterization of dissolved organic matter sources in two eutrophic lakes (Southwestern China Plateau). Geochem J 44:99–112
Fujii M, Murashige S, Ohnishi Y, Yuzawa A, Miyasaka H, Suzuki Y, Komiyama H (2002) Decomposition of phytoplankton in seawater Part 1: kinetic analysis of the effect of organic matter concentration. J Oceanogr 58:433–438
Fujiwara K, Ushiroda T, Takeda K, Kumamoto Y, Tsubota H (1993) Diurnal and seasonal distribution of hydrogen peroxide in seawater of Seto Inland Sea. Geochem J 27:103–115
Fukami K, Simidu U, Taga N (1985) Microbial decomposition of phyto- and zooplankton in seawater I Changes in organic matter. Mar Ecol Prog Ser 21:1–5
Fukuda R, Ogawa H, Nagata T, Koike I (1998) Direct determination of carbon and nitrogen contents of natural bacterial assemblages in marine environments. Appl Environ Microbiol 64:3352–3358
Furdui VI, Tomassini F (2010) Trends and Sources of Perchlorate in Arctic Snow. Environ Sci Technol 44:588–592
Furlong ET, Kinney CA, Ferrer I, Werner SL, Cahill JD, Ratterman G (2004) Pharmaceuticals and personal-care products in solids: Analysis and field results for sediment, soil and biosolid samples. Extended Abstracts of Papers, 228th ACS National Meeting, Division of Environmental Chemistry 44:1320–1324
Furuichi T, Kannan K, Suzuki K, Tanaka S, Giesy JP, Masunaga S (2006) Occurrence of estrogenic compounds in and removal by a swine farm waste treatment plant. Environ Sci Technol 40:7896–7902
Gagosian RB, Stuermer DH (1977) Cycling of biogenic compounds and their diagenetically transformed products in seawater. Mar Chem 5:605–632
Gao Q, Tao Z, Shen C, Sun Y, Yi W, Xing C (2002) Riverine organic carbon in the Xijiang River (South China): seasonal variation in content and flux budget. Environ Geo 41:826–832
Gašparovic B, Cosovic B, Vojvodic V (1998) Contribution of organic acids to the pool of surface active substances in model and marine samples using o-nitrophenol as an electrochemical probe. Org Geochem 29:1025–1032
Gennings C, Molot LA, Dillon PJ (2001) Enhanced photochemical loss of DOC in acidic waters. Biogeochemistry 52:339–354
Gielen B, Neirynck J, Luyssaert S, Janssens IA (2011) The importance of dissolved organic carbon fluxes for the carbon balance of a temperate Scots pine forest. Agric For Meteorol 151:270–278

Goldman CR, Elser JJ, Richards RC, Reuters JE, Priscu JC, Levin AL (1996) Thermal stratification, nutrient dynamics, and phytoplankton productivity during the onset of spring phytoplankton growth in Lake Baikal, Russia. Hydrobiologia 331:9–24
Gomez MJ, Gomez-Ramos MM, Aguera A, Mezcua M, Herrera S, Fernandez-Alba AR (2009) A new gas chromatography/mass spectrometry method for the simultaneous analysis of target and non-target organic contaminants in waters. J Chromatogr A 1216:4071–4082
Granéli W, Lindell M, Tranvik L (1996) Photo-oxidative production of dissolved inorganic carbon in lakes of different humic content. Limnol Oceanogr 41:698–706
Grøn C, Wassenaar L, Krogh M (1996) Origin and structures of groundwater humic substances from three Danish aquifers. Environ Int 22:519–534
Grushnikov OP, Antropova ON (1975) Microbiological degradation of lignin. Russ Chem Rev 44:431. doi:10.1070/RC1975v044n05ABEH002352
Gudasz C, Bastviken D, Steger K, Premke K, Sobek S, Tranvik LJ (2010) Temperature-controlled organic carbon mineralization in lake sediments. Nature 466:478–481
Guéguen C, Dominik J (2003) Partitioning of trace metals between particulate, colloidal and truly dissolved fractions in a polluted river: the Upper Vistula River (Poland). Appl Geochem 18:457–470
Guéguen C, Belin C, Dominik J (2002) Organic colloid separation in contrasting aquatic environments with tangential flow filtration. Water Res 36:1677–1684
Guéguen C, Guo L, Wang D, Tanaka N, Hung C-C (2006) Chemical characteristics and origin of dissolved organic matter in the Yukon River. Biogeochemistry 77:139–155
Guggenberger G, Zech W (1993) Dissolved organic carbon control in acid forest soils of the Fichtelgebirge (Germany) as revealed by distribution patterns and structural composition analysis. Geoderma 59:109–129
Guignard C, Lemée L, Amblés A (2005) Lipid constituents of peat humic acids and humin Distinction from directly extractable bitumen components using TMAH and TEAAc thermochemolysis. Org Geochem 36:287–297
Guildford SJ, Hecky RE (2000) Total nitrogen, total phosphorus, and nutrient limitation in lakes and oceans: is there a common relationship? Limnol Oceanogr 45:1213–1223
Guo L, Santschi PH (1996) A critical evaluation of the cross-flow ultrafiltration technique for sampling colloidal organic carbon in seawater. Mar Chem 55:113–127
Guo L, Santschi PH (1997a) Isotopic and elemental characterization of colloidal organic matter from the Chesapeake Bay and Galveston Bay. Mar Chem 59:1–15
Guo L, Santschi PH (1997b) Composition and cycling of colloids in marine environments. Rev Geophys 35:17–40
Guo L, Coleman CH Jr, Santschi PH (1994) The distribution of colloidal and dissolved organic carbon in the Gulf of Mexico. Mar Chem 45:105–119
Guo L, Santschi PH, Warnken KW (1995) Dynamics of dissolved organic carbon (DOC) in oceanic environments. Limnol Oceanogr 40:1392–1403
Guo L, Santschi PH, Cifuentes LA, Trumbore SE, Southon J (1996) Cycling of high-molecular-weight dissolved organic matter in the Middle Atlantic Bight as revealed by carbon isotopic (13C and 14C) signatures. Limnol Oceanogr 41:1242–1252
Guo Y, Yu H-Y, Zeng EY (2009) Occurrence, source diagnosis, and biological effect assessment of DDT and its metabolites in various environmental compartments of the Pearl River Delta, South China: a review. Environ Pollut 157:1753–1763
Gurr CJ, Reinhard M (2006) Harnessing Natural Attenuation of pharmaceuticals and hormones in rivers. Environ Sci Technol 40:2872–2876
Hagedorn F, Saurer M, Blaser P (2004) A C-13 tracer study to identify the origin of dissolved organic carbon in forested mineral soils. Eur J Soil Sci 55(1):91–100
Haiber S, Herzog H, Burba P, Gosciniak B, Lambert J (2001) Quantification of carbohydrate structures in size fractionated aquatic humic substances by two-dimensional nuclear magnetic resonance. Environ Sci Technol 35:4289–4294
Haines EB (1979) Nitrogen pools in Georgia coastal waters. Estuaries 2:34–39
Hama T, Handa N (1987) Pattern of organic matter production by natural phytoplankton population in a eutrophic lake 1 Intracellular products. Archiv Für Hydrobiologie 109:107–120

Hama T, Handa N (1992) Diel variation of water-extractable carbohydrate composition of natural phytoplankton populations in Kirnu-ura Bay. J Exp Mar Biol Ecol 162:159–176
Hama T, Yanagi K, Hama J (2004) Decrease in molecular weight of photosynthetic products of marine phytoplankton during early diagenesis. Limnol Oceanogr 49:471–481
Hamanaka J, Tanoue E, Hama T, Handa N (2002) Production and export of particulate fatty acids, carbohydrates and combined amino acids in the euphotic zone. Mar Chem 77:55–69
Hanamachi Y, Hama T, Yanai T (2008) Decomposition process of organic matter derived from freshwater phytoplankton. Limnology 9:57–69
Hanselman TA, Graetz DA, Wilkie ANNC (2003) Manure-borne estrogens as potential environmental contaminants: a review. Environ Sci Technol 37:5471–5478
Hansen L (1998) Stepping backward to improve assessment of PCB congener toxicities. Environ Health Perspect 106:171–189
Hanson PC, Bade DL, Carpenter SR, Kratz TK (2003) Lake metabolism: relationships with dissolved organic carbon and phosphorus. Limnol Oceanogr 48:1112–1119
Harvey GR, Boran DA (1985) The geochemistry of humic substances in seawater. In: Aiken GR et al (eds) Humic substances in soil, sediment and water. Wiley, New York, pp 233–247
Harvey HR, Mannino A (2001) The chemical composition and cycling of particulate and macromolecular dissolved organic matter in temperate estuaries as revealed by molecular organic tracers. Org Geochem 32:527–542
Harvey HR, Tuttle JH, Bell JT (1995) Kinetics of phytoplankton decay during simulated sedimentation: changes in biochemical composition and microbial activity under oxic and anoxic conditions. Geochim Cosmochim Acta 59:3367–3377
Hata H, Kudo S, Yamano H, Kurano N, Kayanne H (2002) Organic carbon flux in Shiraho coral reef (Ishigaki Island, Japan). Mar Ecol Prog Ser 232:129–140
Hayakawa K (2004) Seasonal variations and dynamics of dissolved carbohydrates in Lake Biwa. Org Geochem 35:169–179
Hayakawa K, Sugiyama Y (2008) Spatial and seasonal variations in attenuation of solar ultraviolet radiation in Lake Biwa, Japan. J Photochem Photobiol B Biol 90:121–133
Hayakawa K, Sekino T, Yoshioka T, Maruo M, Kumagai M (2003) Dissolved organic carbon and fluorescence in Lake Hovsgol: factors reducing humic content of the lake water. Limnology 4:25–33
Hayakawa K, Sakamoto M, Kumagai M, Jiao C, Song X, Zhang Z (2004) Fluorescence spectroscopic characterization of dissolved organic matter in the water of Lake Fuxian and adjacent rivers in Yunnan, China. Limnology 5:155–163
Hayase K, Tsubota H (1983) Sedimentary humic acid and fulvic acid as surface active substances. Geochim Cosmochim Acta 47:947–952
Hayase K, Tsubota H (1985) Sedimentary humic and fulvic acids as fluorescent organic materials. Geochim Cosmochim Acta 49:159–163
Hayase K, Yamamoto M, Nakazawa I, Tsubota H (1987) Behavior of natural fluorescence in Sagami Bay and Tokyo Bay, Japan—vertical and lateral distributions. Mar Chem 20:265–276
Hayase K, Tsubota H, Sunada I, Goda S, Yamazaki H (1988) Vertical distribution of fluorescent organic matter in the North Pacific. Mar Chem 25:373–381
He B, Dai MH, Zhai WD, Wang LF, Wang KJ, Chen JH, Lin JR, Han A, Xu YP (2010) Distribution, degradation and dynamics of dissolved organic carbon and its major compound classes in the Pearl River estuary. China Mar Chem 119:52–64
Headley JV, Peru KM, Barrow MP (2009) Mass spectrometric characterization of naphthenic acids in environmental samples: a review. Mass Spectrom Rev 28:121–134
Heberer T (2002) Occurrence, fate, and removal of pharmaceutical residues in the aquatic environment: a review of recent research data. Toxicol Lett 131:5–17
Hedges JI (1992) Global biogeochemical cycles: progress and problems. Mar Chem 39:67–93
Hedges JI, Cowie GL, Richey JE, Quay PD, Benner R, Strom M, Forshey BR (1994) Origins and processing of organic matter in the Amazon River as indicated by carbohydrates and amino acids. Limnol Oceanogr 39:743–761
Hedges JI, Keil RG, Benner R (1997) What happens to terrestrial organic matter in the ocean? Org Geochem 27:195–212

Hedges JI, Eglinton G, Hatcher PG, Kirchman DL, Arnosti C, Dereenne S, Evershed RP, Kögel-Knabner I, de Leeuw JW, Littke R, Michaelis W, Rullkötter J (2000) The molecularly-uncharacterized component of nonliving organic matter in natural environments. Org Geochem 31:945–958
Heinze T, Liebert T (2001) Unconventional methods in cellulose functionalization. Prog Polym Sci 26:1689–1762. doi:101016/S0079-6700(01)00022-3
Hejzlar J, Dubrovsky M, Buchtele J, Ruzicka M (2003) The apparent and potential effects of climate change on the inferred concentration of dissolved organic matter in a temperate stream (the Malse River, South Bohemia). Sci Total Environ 310:143–152
Hellebust JA (1965) Excretion of some organic compounds by marine phytoplankton. Limnol Oceanogr 10:192–206
Helleday T, Tuominen L-L, Bergman A, Jenssen D (1999) Brominated flame retardants induce intragenic recombination in mammalian cells. Mutat Res 439:137–147
Hellpointner E, Gäb S (1989) Detection of methyl, hydroxymethyl and hydroxyethyl hydroperoxides in air and precipitation. Nature 337:631–634
Helm RF (2000) In: Glasser WG, Northey RA, Schultz TP (eds) Lignin: historical, biological, and materials perspective. Oxford University Press, Washington, pp 161–171
Henderson RK, Baker A, Parsons SA, Jefferson B (2008) Characterisation of algogenic organic matter extracted from cyanobacteria, green algae and diatoms. Water Res 42:3435–3445
Hernández F, Sancho JV, Ibáñez M, Guerrero C (2007) Antibiotic residue determination in environmental waters by LC-MS TrAC. Trends Anal Chem 26:466–485
Hernes PJ, Hedges JI, Peterson ML, Wakeham SG, Lee C (1996) Neutral carbohydrate geochemistry of particulate material in the central equatorial Pacific. Deep Sea Res II 43:1181–1204
Hessen DO (1985) The relation between bacterial carbon and dissolved humic compounds in oligotrophic lakes. FEMS Microbiol Ecol 31:215–223
Hewitt CN, Kok GL (1991) Formation and occurrences of organic hydroperoxides in the troposphere: laboratory and field observations. J Atmos Chem 12:181–194
Higgins CF, Gallagher MP, Mimmack MM, Pearce SR (1988) A family of closely related ATP-binding subunits from prokaryotic and eukaryotic cells. Bioessays 8:111–116
Higuchi TJ (1993) Biodegradation mechanism of lignin by white-rot basidiomycetes. J Biotechnol 30:1–8
Hiriart-Baer VP, Smiith REH (2005) The effect of ultraviolet radiation on freshwater planktonic primary production: the role of recovery and mixing processes. Limnol Oceanogr 50:1352–1361
Ho KJ, Liu TK, Huang TS, Lu FJ (2003) Humic acid mediates iron release from ferritin and promotes lipid peroxidation in vitro: a possible mechanism for humic acid-induced cytotoxicity. Arch Toxicol 77:100–109
Hoeger B, Kollner B, Dietrich DR, Hitzfeld B (2005) Water-borne diclofenac affects kidney and gill integrity and selected immune parameters in brown trout (Salmo trutta f fario). Aquat Toxicol 75:53–64
Holden J, Burt TP (2002) Laboratory experiments on drought and runoff in blanket peat European. J Soil Sci 53:675–689
Holden J, Burt TP (2003) Hydrological studies on blanket peat: the significance of the acrotelm–catotelm model. J Ecol 91:86–102
Hongve D, Riise G, Kristiansen JF (2004) Increased colour and organic acid concentrations in Norwegian forest lakes and drinking water—a result of increased precipitation? Aquat Sci 66:231–238
Hope D, Billett MF, Cresser MS (1994) A review of the export of carbon in river water: fluxes and processes. Environ Pollut 84:301–324
Hope D, Billett MF, Cresser MS (1997) Exports of organic carbon in two river systems in NE Scotland. J Hydrol 193:61–82
Hopkinson CS, Vallino JJ, Nolin A (2002) Decomposition of dissolved organic matter from the continental margin. Deep Sea Res Part II Top Stud Oceanogr 49:4461–4478

Houle D, Carignan R, Lachance M, Dupont J (1995) Dissolved organic carbon and sulfur in southwestern Quebec lakes: relationships with catchment and lake properties. Limnol Oceanogr 40:710–717

Houser JN, Bade DL, Cole JJ, Pace ML (2003) The dual influences of dissolved organic carbon on hypolimnetic metabolism: organic substrate and photosynthetic reduction. Biogeochemistry 64:247–269

Howell GD, Pollock TL (1986) Sulphate, water colour and dissolved organic carbon relationships in organic waters of Atlantic Canada. In: El-Shaarawi AH, Kwiatkowski RE (eds) Developments in water science, vol 27. Elsevier, Amsterdam, pp 53–63

Hu W-G, Mao J-D, Xing B, Schmidt-Rohr K (2000) Poly(methylene) crystallites in humic substances detected by nuclear magnetic resonance. Environ Sci Technol 34:530–534

Hudson JJ, Dillon PJ, Somers KM (2003) Long-term patterns in dissolved organic carbon in boreal lakes: the role of incident radiation, precipitation, air temperature, southern oscillation and acid deposition. Hydrol Earth Syst Sci 7:390–398

Huguet A, Vacher L, Saubusse S, Etcheber H, Abril G, Relexans S, Ibalot F, Parlanti E (2010) New insights into the size distribution of fluorescent dissolved organic matter in estuarine waters. Org Geochem 41:595–610

Hulatt CJ, Thomas DN (2010) Dissolved organic matter (DOM) in microalgal photobioreactors: a potential loss in solar energy conversion? Bioresour Technol 101:8690–8697

Hulatt CJ, Thomas DN, Bowers DG, Norman L, Zhang C (2009) Exudation and decomposition of chromophoric dissolved organic matter (CDOM) from some temperate macroalgae. Estuar Coast Shelf Sci 84:147–153

Hummel W (1997) Binding models for humic substances. In: Grenthe I, Puigdomenech I (eds) Modelling in aquatic chemistry. Nuclear Energy Agency, Paris, pp 153–201

Hummel M, Findlay S (2006) Effects of water chestnut (Trapa natans) beds on water chemistry in the tidal freshwater Hudson River. Hydrobiologia 559:169–181

Humpage A, Rositano J, Bretag A, Brown R, Baker P, Nicholson B, Steffensen D (1994) Paralytic shellfish poisons from Australian cyanobacterial blooms. Aust J Mar Freshw Res 45:761–771

Humpage AR, Magalhaes V, Froscio SM (2010) Comparison of analytical tools and biological assays for detection of paralytic shellfish poisoning toxins. Anal Bioanal Chem 397:1655–1671

Huovinen PS, Penttilä H, Soimasuo MR (2003) Spectral attenuation of solar ultraviolet radiation in humic lakes in Central Finland. Chemosphere 51:205–214

Hur J (2011) Microbial changes in selected operational descriptors of dissolved organic matters from various sources in a watershed. Water Air soil Soil Pollut 215:465–476

Huszar VLM, Caraco NF, Roland F, Cole J (2006) Nutrient–chlorophyll relationships in tropical–subtropical lakes: do temperate models fit? Biogeochemistry 79:239–250

IEH (1999) IEH assessment of the ecological significance of endocrine disruption: effects on reproductive function and consequences for natural populations (assessment A4). MRC Institute for Environment and Health, Leicester

IHSS (2011) Elemental compositions and stable isotopic ratios of IHSS samples. http://ihssgatec hedu/ihss2/

Imai A, Fukushima T, Matsushige K, Inoue T, Ishibashi T (1998) Fractionation of dissolved organic carbon from the waters of Lake Biwa and its inflowing rivers. Jpn J Limnol 59:53–68 (in Japanese)

IPCC (1996). In: Houghton JT, Filho LGM, Calander BA, Harris N, Kattenberg A, Maskell K (eds) Climate change 1995: the science of climate change. Cambridge University Press, New York, pp 572–573

Ishibashi H, Matsumura N, Hirano M, Matsuoka M, Shiratsuchi H, Ishibashi Y, Takao Y, Arizono K (2004) Effects of triclosan on the early life stages and reproduction of medaka Oryzias latipes and induction of hepatic vitellogenin. Aquat Toxicol 67:167–179

Ishikawa T, Trisliana Yurenfrie, Ardianor Gumiri S (2006) Dissolved organic carbon concentration of a natural water body and its relationship to water color in Central Kalimantan, Indonesia. Limnology 7:143–146

Ittekkot V (1982) Variations of dissolved organic matter during a plankton bloom: qualitative aspects based on sugar and amino acid analysis. Mar Chem 11:143–158
Ittekkot V, Brockman U, Michaelis W, Degens ET (1981) Dissolved free and combined carbohydrates during a phytoplankton bloom in the northern North Sea. Mar Ecol Prog Ser 4:299–305
Ittekkot V, Safiullah S, Mycke B, Seifert R (1985) Seasonal variability and geochemical significance of organic matter in the River Ganges, Bangladesh. Nature 317:800–802
Jackson TA, Hecky RE (1980) Depression of primary productivity by humic matter in lake and reservoir waters of the boreal forest zone. Can J Fish Aquat Sci 37:2300–2317
Jahan K, Ordóñez R, Ramachandran R, Balzer S, Stern M (2008) Modeling biodegradation of nonylphenol. Water Air Soil Pollut 8:395–404
Jameel RH, Helz GR (1999) Organic Chloramines in disinfected wastewaters: rates of reduction by sulfite and toxicity. Environ Toxicol Chem 18:1899–1904
Janczarek M, Urbanik-Sypniewska T, Skorupaka A (1997) Effect of authentic flavonoids and the exudates of clover root on growth rate and inducing ability of nod genes of Rhizobium leguminosarum bv Trifolii. Microbiol Res 152:93–98
Jansen HT, Cooke PS, Porcelli J, Tsuei-Chu L, Hansen LG (1993) Estrogenic and antiestrogenic actions of PCBs in the female rat: in vitro and in vivo studies. Reprod Toxicol 7:237–248
Jansson M, Bergström A-K, Blomqvist P, Drakare S (2000) Allochthonous organic carbon and phytoplankton/bacterioplankton production relationships in lakes. Ecology 81:3250–3255
Jaramillo CA, Dilcher DL (2000) Microfloral diversity patterns of the late Paleocene–Eocene interval in Colombia, northern South America. Geology 9:815–818
Jensen JS, Helz GR (1998) Rates of reduction of N-chlorinated peptides by sulfite: relevance to incomplete dechlorination of wastewaters. Environ Sci Technol 32:516–522
Jeong J, Yoon J (2004) Dual roles of CO_2^- for degrading synthetic organic chemicals in the photo/ferrioxalate system. Water Res 38:3531–3540
Jerlov NG (1968) Optical oceanography. Elsevier, Amsterdam
Jerry AL, Jean-Philippe C (2003) Characterizing aquatic dissolved organic matter. Environ Sci Technol A 37:18A–26A
Johannessen SC, Miller WL (2001) Quantum yield for the photochemical production of dissolved inorganic carbon in seawater. Mar Chem 76:271–283
Johannessen SC, Peña MA, Quenneville ML (2007) Photochemical production of carbon dioxide during a coastal phytoplankton bloom. Estuar Coast Shelf Sci 73:236–242
Johnson MS, Lehmann J, Selva EC, Abdo M, Riha S, Couto EG (2006) Organic carbon fluxes within and stream water exports from headwater catchments in the southern Amazon. Hydrol Process 20:2599–2614. doi:101002/hyp6218
Jonathan HS (1973) Total organic carbon in seawater—comparison of measurements using persulfate oxidation and high temperature combustion. Mar Chem 1:211–229
Jones RI (1992) The influence of humic substances on lacustrine planktonic food chains. Hydrobiologia 229:73–91
Jones RD, Amador JA (1993) Methane and carbon monoxide production, oxidation, and turnover times in the Caribbean Sea as influenced by the Orinoco River. J Geophys Res 98:2353–2359
Jones RI, Arvola L (1984) Light penetration and some related characteristics in small forest lakes in southern Finland. Verh Int Ver Limnol 22:811–816
Jørgensen NOG, Tranvik L, Edling H, Granéli W, Lindell M (1998) Effects of sunlight on occurrence and bacterial turnover of specific carbon and nitrogen compounds in lake water. FEMS Microbiol Ecol 25:217–227
Jover E, Matamoros V, Bayona JM (2009) Characterization of benzothiazoles, benzotriazoles and benzosulfonamides in aqueous matrixes by solid-phase extraction followed by comprehensive two-dimensional gas chromatography coupled to time-of-flight mass spectrometry. J Chromatogr A 1216:4013–4019
Juliano RL, Ling V (1976) A surface glycoprotein modulating drug permeability in Chinese hamster ovary cell mutants. Biochim Biophys Acta 455:152–162
Käfferlein HU, Goen T, Angerer JJ (1998) Musk xylene: analysis, occurrence, kinetics, and toxicology. Crit Rev Toxicol 28:431–476

Kahru M, Mitchell BG (2001) Seasonal and nonseasonal variability of satellite-derived chlorophyll and dissolved organic matter concentration in the California current. J Geophys Res 106:2517–2529
Kaiser K, Benner R (2009) Biochemical composition and size distribution of organic matter at the Pacific and Atlantic time-series stations. Mar Chem 113:63–77
Kaiser E, Arscott DB, Tockner K, Sulzberger B (2004) Sources and distribution of organic carbon and nitrogen in the Tagliamento River, Italy. Aquat Sci 66:103–116
Kalbitz K, Kaiser K (2008) Contribution of dissolved organic matter to carbon storage in forest mineral soils. J Plant Nutr Soil Sci 171:52–60
Kalbitz K, Solinger S, Park J-H, Michalzik B, Matzner E (2000) Controls on the dynamics of dissolved organic matter in soils: a review. Soil Sci 165:277–304
Kalle K (1966) The problem of the gelbstoff in the sea. Oceanog Mar Biol Annu Rev 4:91–104
Kang H, Freeman C, Ashendon TW (2001) Effects of elevated CO2 on fen peat biogeochemistry. Sci Total Environ 279:45–50
Kao SJ, Liu KK (1997) Fluxes of dissolved and nonfossil particulate organic carbon from an Oceania small river (Lanyang Hsi) in Taiwan. Biogeochemistry 39:255–269
Karl DM, Tilbrook BA, Tien G (1991) Seasonal coupling of organic matter production and particle flux in the western Bransfield Strait, Antarctica. Deep Sea Res I 38:1097–1126
Kececioglu J, Ming L, Tromp J, Benner R, Biddanda B, Black B, McCarthy M (1997) Abundance, size distribution, and stable carbon and nitrogen isotopic compositions of marine organic matter isolated by tangential-flow ultrafiltration. Mar Chem 57:243–263
Keith TL, Snyder SA, Naylor CG, Staples CA, Summer C, Kannan K, Giesy JP (2001) Identification and quantitation of nonylphenol ethoxylates and nonylphenol in fish tissues from Michigan. Environ Sci Technol 35:10–13
Kelly CA, Fee E, Ramlal PS, Rudd JWM, Hesslein RH, Anema C, Schindler EU (2001) Natural variability of carbon dioxide and net epilimnetic production in the surface waters of boreal lakes of different sizes. Limnol Oceanogr 46(5):1054–1064
Kemp WM, Smith EM et al (1997) Organic carbon balance and net ecosystem metabolism in Chesapeake Bay. Mar Ecol Prog Ser 150:229–248
Kennedy VC, Zellweger GW, Jones BP (1974) Filter pore-size effects on the analysis of Al, Fe, Mn, and Ti in water. Water Resour Res 10:785–790
Kepkay PE (1994) Particle aggregation and the biological reactivity of colloids. Mar Ecol Prog Ser 109:293–304
Kerndorff H, Schnitzer M (1980) Sorption of metals on humic acid. Geochim Cosmochim Acta 44:1701–1708
Kieber DJ, McDaniel J, Mopper K (1989) Photochemical source of biological substrates in sea water: implications for carbon cycling. Nature 341:637–639
Kieber RJ, Zhou X, Mopper K (1990) Formation of carbonyl compounds from UV-induced photodegradation of humic substances in natural waters: fate of riverine carbon in the sea. Limnol Oceanogr 35:1503–1515
Kieber RJ, Whitehead RF, Willey JD, Reid S, Seaton PJ (2006) Chromophoric dissolved organic matter (CDOM) in rainwater collected in southeastern North Carolina, USA. J Atmos Chem 54:21–41
Kieber RJ, Willey JD, Whitehead RF, Reid SN (2007) Photobleaching of chromophoric dissolved organic matter (CDOM) in rainwater. J Atmos Chem 58:219–235
Kierkegaard A, van Egmond R, McLachlan MS (2011) Cyclic volatile methylsiloxane bioaccumulation in flounder and ragworm in the Humber Estuary. Environ Sci Technol 45:5936–5942
Killops SD, Killops VJ (eds) (1993) An introduction to organic geochemistry. Wiley, New York
Kim C, Nishimura Y, Nagata T (2006) Role of dissolved organic matter in hypolimnetic mineralization of carbon and nitrogen in a large, monomictic lake. Limnol Oceanogr 51:70–78
Kim Y, Choi K, Jung J, Park S, Kim P-G, Park J (2007) Aquatic toxicity of acetaminophen, carbamazepine, cimetidine, diltiazem and six major sulfonamides, and their potential ecological risks in Korea. Environ Int 33:370–375

Kimura K, Toshima S, Amy G, Watanabe Y (2004) Rejection of neutral endocrine disrupting compounds (EDCs) and pharmaceutical active compounds (PhACs) by RO membranes. J Membr Sci 245:71–78

Kindler R, Siemens JAN, Kaiser K, Walmsley DC, Bernhofer C, Buchmann N, Cellier P, Eugster W, Gleixner G, GrŨNwald T, Heim A, Ibrom A, Jones SK, Jones M, Klumpp K, Kutsch W, Larsen KS, Lehuger S, Loubet B, McKenzie R, Moors E, Osborne B, Pilegaard KIM, Rebmann C, Saunders M, Schmidt MWI, Schrumpf M, Seyfferth J, Skiba UTE, Soussana J-F, Sutton MA, Tefs C, Vowinckel B, Zeeman MJ, Kaupenjohann M (2011) Dissolved carbon leaching from soil is a crucial component of the net ecosystem carbon balance. Glob Change Biol 17:1167–1185

Kinney CA, Furlong ET, Kolpin DW, Burkhardt MR, Zaugg SD, Werner SL, Bossio JP, Benotti MJ (2008) Bioaccumulation of pharmaceuticals and other anthropogenic waste indicators in earthworms from agricultural soil amended with biosolid or swine manure. Environ Sci Technol 42:1863–1870

Kirchman DL, Suzuki Y, Garside C, Ducklow HW (1991) High turnover rates of dissolved organic carbon during a spring phytoplankton bloom. Nature 352:612–614

Kirchman DL, Rich JH, Barber RT (1995) Biomass and biomass production of heterotrophic bacteria along 140°W in the equatorial Pacific: effect of temperature on the microbial loop. Deep Sea Res II 42:603–619

Kirchman DL, Meon B, Ducklow HW, Carlson CA, Hansell DA, Steward GF (2001) Glucose fluxes and concentrations of dissolved combined neutral sugars (polysaccharides) in the Ross Sea and Polar Front Zone, Antarctica. Deep Sea Res II 48:4179–4197

Kirk JTO (1976) Yellow substance (gelbstoff) and its contribution to the attenuation of photosynthetically active radiation in some inland and coastal south-eastern Australian waters. Aust J Mar Freshw Res 27:61–71

Klapper L, McKnight DM, Fulton JR, Blunt-Harris EL, Nevin KP, Lovley DR, Hatcher PG (2002) Fulvic acid oxidation state detection using fluorescence spectroscopy. Environ Sci Technol 36:3170–3175

Klavins M, Purmalis O (2010) Humic substances as surfactants. Environ Chem Lett 8:349–354

Kleywegt S, Pileggi V, Yang P, Hao C, Zhao X, Rocks C, Thach S, Cheung P, Whitehead B (2011) Pharmaceuticals, hormones and bisphenol A in untreated source and finished drinking water in Ontario, Canada—occurrence and treatment efficiency. Sci Total Environ 409:1481–1488

Klöcking R, Helbig B, Schotz G, Schacke M, Wutzler P (2002) Anti-HSV-1 activity of ssynthetic humic acid-like polymers derived from p-diphenolic starting compounds. Antivir Chem Chemother 13:241–249

Kogel-Knabner I (1992) Forest soil organic matter: structure and formation Bayreuther Bodenkundliche Berichte. University of Bayreuth, Germany

Kogel-Knabner I, Zech W, Hatcher PG (1988) Chemical composition of the organic matter in forest soils: III the humus layer. Z Pflanzenemrnhr Bodenk 151:331–340

Koike I, Hara S, Terauchi K, Kogure K (1990) Role of sub-micrometre particles in the ocean. Nature 345:242–244

Kolodziej EP, Sedlak DL (2007) Rangeland grazing as a source of steroid hormones to surface waters. Environ Sci Technol 41:3514–3520

Komada T, Schofield OME, Reimers CE (2002) Fluorescence characteristics of organic matter released from coastal sediments during resuspension. Mar Chem 79:81–97

Komatsu E, Fukushima T, Harasawa H (2007) A modeling approach to forecast the effect of long-term climate change on lake water quality. Ecol Model 209:351–366

Komissarov GG (1994) Photosynthesis: a new look. Sci Russ 5:52–55

Komissarov GG (1995) Photosynthesis as a physical process. Chem Phys Rep 14:1723–1732

Komissarov GG (2003) Photosynthesis: the physical-chemical approach. J Adv Chem Phys 2:28–61

Konohira E, Yoshioka T (2005) Dissolved organic carbon and nitrate concentrations in streams: a useful index indicating carbon and nitrogen availability in catchments. Ecol Res 20:359–365

Kopáćek J, Hejzlar J, Kaňa J, Porcal P, Klementová S (2003) Photochemical, Chemical, and biological transformations of dissolved organic carbon and its effect on alkalinity production in acidified lakes. Limnol Oceanogr 48:106–117

Kopáček J, Brzáková M, Hejzlar J, Nedoma J, Porcal P, Vrba J (2004) Nutrient cycling in a strongly acidified mesotrophic lake. Limnol Oceanogr 49:1202–1213

Kortelainen P (1993) Content of total organic carbon in Finnish lakes and its relationship to catchment characteristics. Can J Fish Aquat Sci 50:1477–1483

Koschorreck M, Wendt-Potthoff K, Scharf B, Richnow HH (2008) Methanogenesis in the sediment of the acidic Lake Caviahue in Argentina. J Volcanol Geotherm Res 178:197–204

Kotsyurbenko OR, Glagolev MV, Nozhevnikova AN, Conrad R (2001) Competition between homoacetogenic bacteria and methanogenic archaea for hydrogen at low temperature. FEMS Microbiol Ecol 38:153–159

Koukal B, Guéguen C, Pardos M, Dominik J (2003) Influence of humic substances on the toxic effects of cadmium and zinc to the green alga Pseudokirchneriella subcapitata. Chemosphere 53:953–961

Kramer JB, Canonica S, Hoigne J, Kaschig J (1996) Degradation of fluorescent whitening agents in sunlit natural waters. Environ Sci Technol 30:2227–2234

Kuczewski B, Marquardt CM, Seibert A, Geckeis H, Kratz JV, Trautmann N (2003) Separation of plutonium and neptunium species by capillary electrophoresis-inductively coupled plasma-mass spectrometry and application to natural groundwater samples. Anal Chem 75:6769–6774

Kujawinski EB, Freitas MA, Zang X, Hatcher PG, Green-Church KB, Jones RB (2002) The application of electrospray ionization mass spectrometry (ESI MS) to the structural characterization of natural organic matter. Org Geochem 33:171–180

Kujawinski EB, Longnecker K, Blough NV, del Vecchio R, Finlay L, Kitner JB, Giovannoni SJ (2009) Identification of possible source markers in marine dissolved organic matter using ultrahigh resolution mass spectrometry. Geochim Cosmochim Acta 73:4384–4399

Kumar A, Xagoraraki I (2010) Pharmaceuticals, personal care products and endocrine-disrupting chemicals in US surface and finished drinking waters: a proposed ranking system. Sci Total Environ 408:5972–5989

Kunz PY, Fent K (2006) Estrogenic activity of UV filter mixtures. Toxicol Appl Pharmacol 217:86–99

Kurelec B (1992) The multixenobiotic resistance mechanism in aquatic organisms. Crit Rev Toxicol 22:23–43

Kusel K, Drake HL (1999) Microbial turnover of low molecular weight organic acids during leaf litter decomposition. Soil Biol Biochem 31:107–118

Kwan WP, Voelker BM (2002) Decomposition of hydrogen peroxide and organic compounds in the presence of dissolved iron and ferrihydrite. Environ Sci Technol 36:1467–1476

Laane RWPM (1984) Comment on the structure of marine fulvic and humic acids. Mar Chem 15:85–87

Lai HT, Hou JH, Su CI, Chen CL (2009) Effects of chloramphenicol, florfenicol, and thiamphenicol on growth of algae Chlorella pyrenoidosa, Isochrysis galbana, and Tetraselmis chui. Ecotoxicol Environ Safe 72:329–334

Lambert J, Lankes U (2002) Application of nuclear magnetic resonance spectroscopy to structural investigations of refractory organic substances: principles and definitions. In: Frimmel FH, Abbt-Braun G, Heumann KG, Hock B, Lüdemann H-D, Spiteller M (eds) Refractory organic substances (ROS) in the environment. Wiley-VCH, Weinheim, pp 89–95

Lamelas C, Slaveykova VI (2007) Comparison of Cd(II), Cu(II), and Pb(II) biouptake by green algae in the presence of humic acid. Environ Sci Technol 41:4172–4178

Lamelas C, Pinheiro JP, Slaveykova VI (2009) Effect of humic acid on Cd(II), Cu(II), and Pb(II) uptake by freshwater algae: kinetic and cell wall speciation considerations. Environ Sci Technol 43:730–735

Lancelot C (1979) Gross excretion rates of natural marine phytoplankton and heterotrophic uptake of excreted products in the southern North Sea, as determined by short-term experiments. Mar Ecol Prog Ser 1:179–186

Landsberg JH (2002) The effects of harmful algal blooms on aquatic organisms. Rev Fish Sci 10:113–390
Langer P, Tajtakova M, Fodor G, Kocan A, Bohov P, Michalek J, Kreze A (1998) Increased thyroid volume and prevalence of thyroid disorders in an area heavily polluted by polychlorinated biphenyls. Eur J Endocrinol 139:402–409
Lara RJ, Rachold V, Kattner G, Hubberten HW, Guggenberger G, Skoog A, Thomas DN (1998) Dissolved organic matter and nutrients in the Lena River, Siberian Arctic: characteristics and distribution. Mar Chem 59:301–309
Larsen S, Andersen TOM, Hessen DO (2011) Climate change predicted to cause severe increase of organic carbon in lakes. Glob Change Biol 17:1186–1192
Lau OW, Wong SK (2000) Contamination in food from packaging material J Chromatogr A 882:255–270 Casajuana N, Lacorte S (2003) Presence and release of phthalic esters and other endocrine disrupting compounds in drinking water. Chromatographia 57:649–655
Laurion I, Ventura M, Catalan J, Psenner R, Sommarruga R (2000) Attenuation of untraviolet radiation in mountain lakes: factors controlling the among- and within-lake variability. Limnol Oceanogr 45:1274–1288
Lavoie M, Paré D, Bergeron Y (2005) Impact of global change and forest management on carbon sequestration in northern forested peatlands. Environ Rev 13:199–240
Lead JR, Wilkinson KJ (2006) Natural aquatic colloids: current knowledge and future trends. Environ Chem 3:159–171
Lead JR, Davison W, Hamilton-Taylor J, Harper M (1999) Trace metal sorption by natural particles and coarse colloids. Geochim Cosmochim Acta 63:1661–1670
Lee C, Henrichs SM (1993) How the nature of dissolved organic matter might affect the analysis of dissolved organic carbon. Mar Chem 41:105–120
Lee C, Wakeham SG (1988) Organic matter in seawater: biogeochemical processes. In: Riley JP (ed) Chemical oceanography, vol 9. Academic Press, New York, p 1–51
Lee PC, Arndt P, Nickels C (1999) Testicular abnormalities in male rats after lactational exposure to nonylphenols. Endocrine 11:61–68
Lee I-S, Sim W-J, Kim C-W, Chang Y-S, Oh J-E (2011) Characteristic occurrence patterns of micropollutants and their removal efficiencies in industrial wastewater treatment plants. J Environ Monit 13:391–397
Leenheer JA (2007) Progression from model structures to molecular structures of natural organic matter components. Ann Environ Sci 1:57–68
Leenheer JA, Croué JP (2003) Characterizing aquatic dissolved organic matter. Environ Sci Technol 37:18–26
Leenheer JA, Malcolm RL, McKinley PW, Eccles LA (1974) Occurrence of dissolved organic carbon in selected ground-water samples in the United States. J Res 2:361–369
Leenheer JA, Wershaw RL, Reddy MM (1995) Strong-acid, carboxyl-group structures in fulvic acid from the Suwannee River, Georgia 1 Minor structures. Environ Sci Technol 29:393–398
Leenheer JA, Brown GK, MacCarthy P, Cabaniss SE (1998) Models of metal binding structures in fulvic acid from the Suwannee river, Georgia. Environ Sci Technol 32:2410–2416
Leenheer JA, Rostad CE, Gates PM, Furlong ET, Ferrer I (2001) Molecular resolution and fragmentation of fulvic acid by electrospray ionization/multistage tandem mass spectrometry. Anal Chem 73:1461–1471
Legrand C, Rengefors K, Fistarol GO, Graneli E (2003) Allelopathy in phytoplankton—biochemical, ecological and evolutionary aspects. Phycologia 42:406–419
Lehmann MF, Bernasconi SM, Mckenzie JA, Barbieri A, Simona M, Veronesi M (2004) Seasonal variation of the δ13C and δ15N of particulate and dissolved carbon and nitrogen in Lake Lugano: constraints on biogeochemical cycling in a eutrophic lake. Limnol Oceanogr 49:415–429
Leonowicz A, Cho N-S, Luterek J, Wilkolazka A, Wojtas-Wasilewska M, Matuszewska A, Hofrichter M, Wesenberg D, Rogalski J (2001) Fungal laccase: properties and activity on lignin. J Basic Microbiol 41:185–227

Lewis WM, Saunders JF III (1989) Concentration and transport of dissolved and suspended substances in the Orinoco River. Biogeochemistry 7:203–240

Lewis NG, Yamamoto E (1990) Lignin: occurrence, biogenesis and biodegradation. Annu rev plant Physiol Plant Mol Biol 41(1):455–496

Li WKW (1994) Primary production of prochlorophytes, cyanobacteria, and eukaryotic ultraphytoplankton—measurements from flow cytometric sorting. Limnol Oceanogr 39:169–175

Li W, Wu FC, Liu CQ, Fu PQ, Wang J, Mei Y, Wang L, Guo J (2008) Temporal and spatial distributions of dissolved organic carbon and nitrogen in two small lakes on the Southwestern China. Plateau Limnol 9:163–171

Li F, Zhao Q, Wang CA, Lu XF, Li XF, Le XC (2010) Detection of Escherichia coli O157:H7 using gold nanoparticle labeling and inductively coupled plasma-mass spectrometry. Anal Chem 82:3399–3403

Li G, Gao K, Gao G (2011) Differential impacts of solar UV radiation on photosynthetic carbon fixation from the coastal to offshore surface waters in the South China Sea. Photochem Photobiol 87:329–334

Liang M-C, H Hartman H, Kopp RE, Kirschvink JL, Yung YL (2006) Production of hydrogen peroxide in the atmosphere of a Snowball Earth and the origin of oxygenic photosynthesis. PNAS 103:18896–18899

Likens GE, Edgerton RS, Galloway JN (1983) The composition and deposition of organic carbon in precipitation. Tellus 35B:16–24

Lilienfein J, Wilcke W, Thomas R, Vilela L, do Carmo Lima S, Zech W (2001) Effects of Pinus caribaea plantations on the C, N, P, and S status of Brazilian savanna oxisols. For Ecol Manag 147:171–182

Lippold H, Gottschalch U, Kupsch H (2008) Joint influence of surfactants and humic matter on PAH solubility are mixed micelles formed? Chemosphere 70:1979–1986. doi:101016/jchemosphere200709040

Lishman L, Smyth SA, Sarafin K, Kleywegt S, Toito J, Peart T, Lee B, Servos M, Beland M, Seto P (2006) Occurrence and reductions of pharmaceuticals and personal care products and estrogens by municipal wastewater treatment plants in Ontario, Canada. Sci Total Environ 367:544–558

Litman T, Druley TE, Stein WD, Bates SE (2001) From MDR to MXR: new understanding of multidrug resistance systems, their properties and clinical significance. Cell Mol Life Sci 58:931–959

Liu R, Lead JR, Baker A (2007) Fluorescence characterization of cross flow ultrafiltration derived freshwater colloidal and dissolved organic matter. Chemosphere 68:1304–1311

Liu YM, Wang CA, Tyrrell G, Hrudey SE, Li XF (2009) Induction of Escherichia coli O157:H7 into the viable but non-culturable state by chloraminated water and river water, and subsequent resuscitation. Environ Microbiol Rep 1:155–161

Liu YM, Wang C, Fung C, Li XF (2010) Quantification of Viable but Nonculturable Escherichia coli O157:H7 by Targeting the rpoS mRNA. Anal Chem 82:2612–2615

Llewellyn LE (2006) Saxitoxin, a toxic marine natural product that targets a multitude of receptors. Nat Prod Rep 23:200–222

Lobanov AV, Kholuiskaya SN, Komissarov GG (2004) Photocatalytic synthesis of formaldehyde from CO2 and H2O2. Doklady Phys Chem Part I 399:266–268

Loh AN, Bauer JE, Druffel ERM (2004) Variable ageing and storage of dissolved organic components in the open ocean. Nature 430:877–881

Lønborg C, Søndergaard M (2009) Microbial availability and degradation of dissolved organic carbon and nitrogen in two coastal areas. Estuar Coast Shelf Sci 81:513–520

Lønborg C, Álvarez-Salgado XA, Davidson K, Miller AEJ (2009a) Production of bioavailable and refractory dissolved organic matter by coastal heterotrophic microbial populations. Estuar Coast Shelf Sci 82:682–688

Lønborg C, Davidson K, Álvarez-Salgado XA, Miller AEJ (2009b) Bioavailability and bacterial degradation rates of dissolved organic matter in a temperate coastal area during an annual cycle. Mar Chem 113:219–226

Lovley DR (2006) Bug juice: harvesting electricity with microorganisms. Nature Rev Microbiol 4:497–508
Lovley DR, Chapelle FH (1995) Deep subsurface microbial processes. Rev Geophys 33:365–381
Lovley DR, Coates JD, Blunt-Harris EL, Phillips EJP, Woodward JC (1996) Humic substances as electron acceptors for microbial respiration. Nature 382:445–448
Lowe LE, Bustin RM (1989) Forms and hydrolytic behavior of sulfur in humic acid and residue fractions of four peats from the Fraser Lowland. Can J Soil Sci 69:287–293
Lu H, Yan C, Liu J (2007) Low-molecular-weight organic acids exuded by Mangrove (Kandelia candel (L) Druce) roots and their effect on cadmium species change in the rhizosphere. Environ Exp Bot 61:159–166
Ma X, Ali N (2009) Detection of a DNA-like materials in Suwannee River fulvic acid. In: Wu FC, Xing B (eds) Natural organic matter and its significance in the environment. Science Press, Beijing, pp 66–89
Ma X, Green SA (2004) Photochemical transformation of dissolved organic carbon in Lake Superior-an in situ experiment. J Great Lakes Res 30(suppl 1):97–112
Ma H, Allen E, Yin Y (2001) Characterization of isolated fractions of dissolved organic matter from natural waters and a wastewater effluent. Water Res 35:985–996
Macalady DL, Walton-Day K (2009) New light on a dark subject: on the use of fluorescence data to deduce redox states of natural organic matter (NOM). Aquat Sci. doi:101007/s00027-009-9174-6
MacFarlane RB (1978) Molecular weight distribution of humic and fulvic acids of sediments from a north Florida estuary. Geochim Cosmochim Acta 42:1579–1582
Mahieu N, Olk DC, Randall EW (2000) Accumulation of heterocyclic nitrogen in humified organic matter: a 15 N NMR study of lowland rice soils. Eur J Soil Sci 51:379–389
Mahieu N, Olk DC, Randall EW (2002) Multinuclear magnetic resonance analysis of two humic acid fractions from lowland rice soils. J Environ Qual 31:421–430
Majewski MS, Foreman WT, Goolsby DA (2000) Pesticides in the atmosphere of the Mississippi River Valley, part I—rain. Sci Total Environ 248:201–212
Mak YL, Taniyasu S, Yeung LWY, Lu G, Jin L, Yang Y, Lam PKS, Kannan K, Yamashita N (2009) Perfluorinated compounds in tap water from China and several other countries. Environ Sci Technol 43:4824–4829
Malcolm RL (1985) Geochemistry of stream fulvic and humic substances. In: Aiken GR, McKnight DM, Wershaw RL, MacCarthy P (eds) Humic substances in soil, sediment, and water: geochemistry, isolation and characterization. Wiley, New York, pp 181–209
Malcolm RL (1990) The uniqueness of humic substances in each of soil, stream and marine environments. Anal Chim Acta 232:19–30
Malcolm RL (1991) Factors to be considered in the isolation and characterization of aquatic humic substances In: Allard B, Borén H, Grimvall A (eds) Humic substances in the aquatic and terrestrial environments. Lecture notes in earth sciences, vol 33. Springer, Berlin, pp 9–36
Malcolm RL, MacCarthy P (1992) Quantitative evaluation of XAD-8 and XAD-4 resins used in tandem for removing organic solutes from water. Environ Int 18:597–607
Malinsky-Rushansky NZ, Legrand C (1996) Excretion of dissolved organic carbon by phytoplankton of different sizes and subsequent bacterial uptake. Mar Ecol Prog Ser 132:249–255
Malkin SY, Guildford SJ, Hecky RE (2008) Modeling the growth response of Cladophora in a Laurentian Great Lake to the exotic invader Dreissena and to lake warming. Limnol Oceanogr 53:1111–1124
Mandal R, Sekaly ALR, Murimboh J, Hassan NM, Chakrabarti CL, Back MH, Grégoire DC, Schroeder WH (1999) Effect of the competition of copper and cobalt on the lability of Ni(II)-organic ligand complexes Part I In model solutions containing Ni(II) and a well-characterized fulvic acid. Anal Chim Acta 395:323
Mann CJ, Wetzel RG (1995) Dissolved organic carbon and its utilization in a riverine wetland ecosystem. Biogeochemistry 31:99–120

Mannino A, Harvey HR (2000) Biochemical composition of particles and dissolved organic matter along an estuarine gradient: sources and implications for DOM reactivity. Limnol Oceanogr 45:775–788

Manoli E, Samara C (1999) Polycyclic aromatic hydrocarbons in natural waters: sources, occurrence and analysis TrAC. Trends Anal Chem 18:417–428

Mao J, Hu W, Schmidt-Rohr K, Davies G, Ghabbour EA, Xing B (1998) In: Davies G, Ghabbour EA (eds) Humic Substances: structures, Properties, and Uses. Royal Society of Chemistry, Cambridge, p 83

Marañòn E, Cermeño P, Fernández E, Rodrìguez J, Zabala L (2004) Significance and mechanisms of photosynthetic production of dissolved organic carbon in a coastal eutrophic ecosystem. Limnol Oceanogr 49:1652–1666

Markager S, Vincent WF (2000) Spectral light attenuation and the absorption of UV and blue light in natural waters. Limnol Oceanogr 45:642–650

Markich SJ (2002) Uranium speciation and bioavailability in aquatic systems: an overview. Sci World J 2:707–729

Martin JM, Dai MH, Cauwet G (1995) Significance of colloids in the biogeochemical cycling of organic carbon and trace metals in the Venice Lagoon (Italy). Limnol Oceanogr 40:119–131

Martins O, Probst JL (1991) Biogeochemistry of major African Rivers: carbon and mineral transport. In: Degens ET, Kempe S, Richey JE (eds) Biogeochemistry of major world rivers, SCOPE 42, Hamburg, Ch 6

Matamoros V, Jover E, Bayona JM (2010) Occurrence and fate of benzothiazoles and benzotriazoles in constructed wetlands. Water Sci Technol 61:191–198

Matsunaga K (1981) Studies on the decompositive processes of phytoplanktonic organic matter. Jap J Limnol 42:220–229

Matthiesen A (1994) Evaluating the redox capacity and the redox potential of humic acids by redox titrations. In: Senesi N, Miano TM (eds) Humic substances in the global environment and applications for human health. Elsevier, Amsterdam, pp 187–192

McCalley DV, Cooke M, Nickless G (1981) Effect of sewagetreatment on fecal sterols. Water Res 15:1019–1025

McCallister SL, Bauer JE, Cherrier JE, Ducklow HW (2004) Assessing sources and ages of organic matter supporting river and estuarine bacterial production: a multiple-isotope (Δ14C, δ13C, and δ15 N) approach. Limnol Oceanogr 49:1687–1702

McCallister SL, Bauer JE, Canuel EA (2006) Bioreactivity of estuarine dissolved organic matter: a combined geochemical and microbiological approach. Limnol Oceanogr 51:94–100

McCarthy M, Hedges J, Benner R (1996) Major biochemical composition of dissolved high molecular weight organic matter in seawater. Mar Chem 55:281–297

McCarthy M, Pratum T, Hedges J, Benner R (1997) Chemical composition of dissolved organic nitrogen in the ocean. Nature 390:150–154

McCarthy MJ, James RT, Chen Y, East TL, Gardner WS (2009) Nutrient ratios and phytoplankton community structure in the large, shallow, eutrophic, subtropical Lakes Okeechobee (Florida, USA) and Taihu (China). Limnology 10:215–227. doi:101007/s10201-009-0277-5

McDonald AT, Mitchell GN, Naden PS, Martin DSJ (1991) Discoloured water investigations. Report to Yorkshire Water, University of Leeds

McDowell WH, Likens GE (1988) Origin, composition, and flux of dissolved organic carbon in the Hubbard Brook Valley. Ecol Monogr 58:177–195

McIntyre C, McRae C, Batts BD, Piccolo A (2005) Structural characterisation of groundwater hydrophobic acids isolated from the Tomago Sand Beds Australia. Org Geochem 36:385–397

McKnight DM, Aiken GR (1998) Sources and age of aquatic humus. In: Hessen DO, Tranvik LJ (eds) Aquatic humic substances. Springer, New York, pp 9–39

McKnight DM, Thorn KA, Wershaw RL, Bracewell JM, Robertson GW (1988) Rapid changes in dissolved humic substances in Spirit Lake and South Fork Castle Lake, Washington. Limnol Oceanogr 33:1527–1541

McKnight DM, Aiken GR, Smith RL (1991) Aquatic fulvic acids in microbially based ecosystems-results from two desert lakes in Antarctica. Limnol Oceanogr 36:998–1006

McKnight DM, Bencala KE et al (1992) Sorption of dissolved organic-carbon by hydrous aluminum and iron-oxides occuring at the confluence of Deer Creek with the Snake River, Summit County, Colorado. Environ Sci Technol 26:1388–1396
McKnight DM, Smith RL, Harnish RA, Miller CL, Bencala KE (1993) Seasonal relationships between planktonic microorganisms and dissolved organic material in an alpine stream. Biogeochemistry 21:39–59
McKnight DM, Andrews ED, Spaulding SA, Aiken GR (1994) Aquatic fulvic acids in algal-rich Antarctic ponds. Limnol Oceanogr 39:1972–1979
McKnight DM, Harnish R, Wershaw RL, Schiff S, Baron JS (1997) Chemical characteristics of particulate, colloidal, and dissolved organic material in Loch Vale Watershed, Rocky Mountain National Park. Biogeochemistry 36:99–124
McKnight DM, Boyer EW, Westerhoff PK, Doran PT, Kulbe T, Andersen DT (2001) Spectrofluorometric characterization of dissolved organic matter for indication of precursor organic material and aromaticity. Limnol Oceanogr 46:38–48
McLeese DW, Zito DW, Sergeant DB, Burridge L, Metcalfe CD (1981) Lethality and accumulation of alkylphenols in aquatic fauna. Chemosphere 10:723–730
Medana C, Calza P, Baiocchi C, Pelizzetti E (2005) Liquid chromatography tandem mass spectroscopy as tool to investigate pesticides and their degradation products. Curr Org Chem 9:859–873
Meier M, Chin Y-P, Maurice P (2004) Variations in the composition and adsorption behavior of dissolved organic matter at a small, forested watershed. Biogeochemistry 67:39–56
Melnikov IA, Pavlov GL (1978) Characteristics of organic carbon distribution in the waters and ice of the Arctic Basin. Oceanology 18:163–167
Meng XZ, Zeng EY, Yu LP, Guo Y, Mai BX (2007) Assessment of human exposure to polybrominated diphenyl ethers in China via fish consumption and inhalation. Environ Sci Technol 41:4883–4887
Menzel DW, Vaccaro RF (1964) The measurement of dissolved and particulate carbon in seawater. Limnol Oceanogr 9:138–142
Merel S, Le Bot B, Clement M, Seux R, Thomas O (2009) Ms identification of microcystin-LR chlorination by-products. Chemosphere 74:832–839
Meriläinen JJ, Hynynen J, Palomäki A, Veijola H, Witick A, Mäntykoski K, Granberg K, Lehtinen K (2001) Pulp and paper mill pollution and subsequent ecosystem recovery of a large boreal lake in Finland: a palaeolimnological analysis. J Paleolimnol 26:11–35
Meybeck M (1982) Carbon, nitrogen, and phosphorus transport by world rivers. Am J Sci 282:401–450
Meybeck M, Cauwet G, Dessery S, Somville M, Gouleau D, Billen G (1988) Nutrients (Organic C, P, N, Si) in the eutrophic river Loire and its estuary. Estuar Coast Shelf Sci 27:595–624
Meyer JL (1986) Dissolved organic carbon dynamics in two subtropical blackwater rivers. Archiv Hydrobiol 108:119–134
Meyer JL, Tate CM (1983) The effects of watershed disturbance on dissolved organic carbon dynamics of a stream. Ecology 64:33–44
Meyers-Schulte KJ, Hedges J (1986) Molecular evidence for a terrestrial component of organic matter dissolved in ocean water. Nature 321:61–63
Michalzik B, Kalbitz K, Park JH, Solinger S, Matzner E (2001) Fluxes and concentrations of dissolved organic carbon and nitrogen—a synthesis for temperate forests. Biogeochemistry 52:173–205
Midorikawa T, Tanoue E (1996) Extraction and characterization of organic ligands from oceanic water columns by immobilized metal ion affinity chromatography. Mar Chem 52:157–171
Midorikawa T, Tanoue E (1998) Molecular masses and chromophoric properties of dissolved organic ligands for copper(II) in oceanic water. Mar Chem 62:219–239
Mikutta R, Mikutta C, Kalbitz K, Scheel T, Kaiser K, Jahn R (2007) Biodegradation of forest floor organic matter bound to minerals via different binding mechanisms. Geochim Cosmochim Acta 71:2569–2590
Miller WL (1998) Effects of UV radiation on aquatic humus: photochemical principles and experimental considerations. Ecol Stud 133:125–143

Miller WL, Moran MA (1997) Interaction of photochemical and microbial processes in the degradation of refractory dissolved organic matter from a coastal marine environment. Limnol Oceanogr 42:1317–1324

Miller WL, Zepp RG (1995) Photochemical production of dissolved inorganic carbon from terrestrial organic matter: significance to the oceanic organic carbon cycle. Geophys Res Lett 22:417–420

Miller WL, Moran MA, Sheldon WM, Zepp RG, Opsahl S (2002) Determination of apparent quantum yield spectra for the formation of biologically labile photoproducts. Limnol Oceanogr 47:343–352

Miller C, Willey JD, Kieber R (2008) Changes in rainwater composition in Wilmington, NC during tropical storm Ernesto. Atmos Environ 42:846–855

Miller C, Gordon KG, Kieber RJ, Willey JD, Seaton PJ (2009) Chemical characteristics of chromophoric dissolved organic matter in rainwater. Atmos Environ 43:2497–2502

Minakata D, Li K, Westerhoff P, Crittenden J (2009) Development of a group contribution method to predict aqueous phase hydroxyl radical (HO•) reaction rate constants. Environ Sci Technol 43:6220–6227

Minella M, Rogora M, Vione D, Maurino V, Minero C (2011) A model approach to assess the long-term trends of indirect photochemistry in lake water the case of Lake Maggiore (NW Italy). Sci Total Environ 409:3463–3471

Minero C, Chiron S, Falletti G, Maurino V, Pelizzetti E, Ajassa R, Carlotti ME, Vione D (2007) Photochemical processes involving nitrite in surface water samples. Aquat Sci 69:71–85

Mitch WA, Sharp JO, Trussell RR, Valentine RL, Alvarez-Cohen L, Sedlak DL (2003) N-Nitrosodimethylamine as a drinking water contaminant: a review. Environ Eng Sci 20:389–404

Mitra S, Bianchi TS, Guo L, Santschi PH (2000) Terrestrially derived dissolved organic matter in the Chesapeake Bay and the Middle Atlantic Bight. Geochim Cosmochim Acta 64:3547–3557

Mladenov N, McKnight DM, Wolski P, Murray-Hudson M (2007) Simulation of DOM fluxes in a seasonal floodplain of the Okavango Delta, Botswana. Ecol Model 205:181–195

Mladenov N, Huntsman-Mapila P, Wolski P, Masamba WRL, McKnight DM (2008) Dissolved organic matter accumulation, reactivity, and redox state in ground water of a recharge wetland. Wetlands 28:747–759

Mohlin M, Wulff A (2009) Interaction effects of ambient UV radiation and nutrient limitation on the toxic Cyanobacterium Nodularia spumigena. Microb Ecol 57:675–686

Molot LA, Dillon PJ (1996) Storage of terrestrial carbon in boreal lake sediments and evasion to the atmosphere. Glob Biogeochem Cycles 10:483–492

Molot LA, Keller W, Leavitt PR, Robarts RD, Waiser MJ, Arts MT, Clair TA, Pienitz R, Yan ND, McNicol DK, Prairie Y, Dillon PJ, Macrae M, Bello R, Nordin RN, Curtis PJ, Smol JP, Douglas MSV (2004) Risk analysis of dissolved organic mattermediated ultraviolet B exposure in Canadian inland waters. Can J Fish Aquat Sci 61:2511–2521

Mompelat S, LeBot B, Thomas O (2009) Occurrence and fate of pharmaceutical products and by-products, from resource to drinking water. Environ Int 35:803–814

Monteith DT, Evans CD (2005) The United Kingdom Acid Waters Monitoring Network: a review of the first 15 years and introduction to the special issue. Environ Pollut 137:3–13

Monteith DT, Stoddard JL, Evans CD, de Wit HA, Forsius M, Hogasen T, Wilander A, Skjelkvale BL, Jeffries DS, Vuorenmaa J, Keller B, Kopacek J, Vesely J (2007) Dissolved organic carbon trends resulting from changes in atmospheric deposition chemistry. Nature 450:537–540

Montgomery-Brown J, Drewes JE, Fox P, Reinhard M (2003) Behavior of alkylphenol polyethoxylate metabolites during soil aquifer treatment. Water Res 37:3672–3681

Moore RW, Rudy TA, Lin T-M, Ko K, Peterson RE (2001) Abnormalities of sexual development in male rats with in utero and lactational exposure to the antiandrogenic plasticizer di-(2-ethylhexyl)phthalate. Environ Health Perspect 109:229–237

Moore TR, Paré D, Boutin R (2008) Production of Dissolved organic carbon in Canadian forest soils. Ecosystems 11:740–751

Mopper K, Kieber DJ (2002) Photochemistry of carbon, sulfur, nitrogen and phosphorus. In: Hansell DA, Carlson CA (eds) Biogeochemistry of marine dissolved organic matter. Academic Press, Amsterdam, pp 59–85

Mopper K, Stahovec WL (1986) Sources and sinks of low molecular weight organic carbonyl compounds in seawater. Mar Chem 19:305–321

Mopper K, Zhou X, Kieber RJ, Kieber DJ, Sikorski RJ, Jones RD (1991) Photochemical degradation of dissolved organic carbon and its impact on the oceanic carbon cycle. Nature 353:60–62

Mopper K, Zhou J, Ramana KS, Passow Dam HG, Drapeau DT (1995) The role of surface-active carbohydrates in the flocculation of a diatom bloom in a mesocosm. Deep Sea Res 42:47–73

Mopper K, Feng Z, Bentjen SB, Chen RF (1996) Effects of cross-flow filtration on the absorption and fluorescence properties of seawater. Mar Chem 55:53–74

Morabito G, OggioniA Panzani P (2003) Phytoplankton assemblage at equilibrium in large and deep subalpine lakes: a case study from Lago Maggiore (N Italy). Hydrobiologia 502:37–48

Moran MA, Hodson RE (1994) Dissolved humic substances of vascular plant origin in a coastal marine-environment. Limnol Oceanogr 39:762–771

Moran MA, Zepp RG (1997) Role of photoreactions in the formation of biologically labile compounds from dissolved organic matter. Limnol Oceanogr 42:1307–1316

Moran MA, Pomeroy LR, Sheppard ES, Atkinson LP, Hodson RE (1991) Distribution of terrestrially derived dissolved organic matter on the southeastern US continental shelf. Limnol Oceanogr 36:1134–1149

Moran MA Jr, Sheldon WM, Zepp RG (2000) Carbon loss and optical property changes during long-term photochemical and biological degradation of estuarine dissolved organic matter. Limnol Oceanogr 45:1254–1264

Morel A, Gentili B (2009) A simple band ratio technique to quantify the colored dissolved and detrital organic material from ocean color remotely sensed data. Remote Sens Environ 113:998–1011. doi:101016/jrse200901008

Morel A, Claustre H, Antoine D, Gentili B (2007) Natural variability of bio-optical properties in case 1 waters: attenuation and reflectance within the visible and near-UV spectral domains as observed in South Pacific and Mediterranean waters. Biogeosciences 4:2147–2178

Morra MJ, Fendorf SE, Brown PD (1997) Speciation of sulfur in humic and fulvic acids using X-ray absorption near-edge structure (XANES) spectroscopy. Geochim Cosmochim Acta 61:683–688

Morris DP, Hargreaves BR (1997) The role of photochemical degradation of dissolved organic carbon in regulating the UV transparency of three lakes on the Pocono Plateau. Limnol Oceanogr 42:239–249

Morris DP, Zagarese H, Williamson CE, Balseiro EG, Hargreaves BR, Modenutti B, Moeller R, Queimalinos C (1995) The attenuation of solar UV radiation in lakes and the role of dissolved organic carbon. Limnol Oceanogr 40:1381–1391

Mostofa KMG (2005) Dynamics, characteristics and photochemical processes of fluorescent dissolved organic matter and peroxides in river water. Ph.D. Thesis, September 2005, Hiroshima University, Japan

Mostofa KMG, Sakugawa H (2009) Spatial and temporal variations and factors controlling the concentrations of hydrogen peroxide and organic peroxides in rivers. Environ Chem 6:524–534

Mostofa KMG, Yoshioka T, Konohira E, Tanoue E, Hayakawa K, Takahashi M (2005a) Three-dimensional fluorescence as a tool for investigating the dynamics of dissolved organic matter in the Lake Biwa watershed. Limnology 6:101–115

Mostofa KMG, Honda Y, Sakugawa H (2005b) Dynamics and optical nature of fluorescent dissolved organic matter in river waters in Hiroshima prefecture, Japan. Geochem J 39:257–271

Mostofa KMG, Yoshioka T, Konohira E, Tanoue E (2007a) Dynamics and characteristics of fluorescent dissolved organic matter in the groundwater, river and lake water. Water Air Soil Pollut 184:157–176

Mostofa KMG, Yoshioka T, Konohira E, Tanoue E (2007b) Photodegradation of fluorescent dissolved organic matters in river waters. Geochem J 41:323–331

Mostofa KMG, Wu FC, Yoshioka T, Sakugawa H, Tanoue E (2009a) Dissolved organic matter in the aquatic environments. In: Wu FC, Xing B (eds) Natural organic matter and its significance in the environment. Science Press, Beijing, pp 3–66

Mostofa KMG, Liu CQ, Wu FC, Fu PQ, Ying WL, Yuan J (2009b) Overview of key biogeochemical functions in lake ecosystem: impacts of organic matter pollution and global warming. In: Keynote Speech, Proceedings of the 13 th World Lake Conference, 1–5 November 2009. Wuhan, China, pp 59–60

Mostofa KMG, Wu FC, Liu CQ, Fang WL, Yuan J, Ying WL, Li W, Yi M (2010) Characterization of Nanming River (Southwestern China) impacted by sewerage pollution using excitation-emission matrix and PARAFAC. Limnology 11:217–231

Mostofa KMG, Wu FC, Liu CQ, Yoshioka T, Sakugawa H, Tanoue E (2011) Photochemical, microbial and metal complexation behavior of fluorescent dissolved organic matter in the aquatic environments (invited review). Geochem J 45:235–254

Mottaleb MA, Brumley WC, Curtis LR, Sovocool GW (2005) Nitro musk adducts of rainbow trout hemoglobin: dose–response and toxicokinetics determination by GC-NICI-MS for a sentinel species. Am Biotech Lab 23:24–29

Mottaleb MA, Usenko S, O'Donnell JG, Ramirez AJ, Brooks BW, Chambliss CK (2009) Gas chromatography–mass spectrometry screening methods for select UV-filters, synthetic musks, alkylphenols, an antimicrobial agent, and an insect repellent in fish. J Chromatogr A 1216:815–823

Mudge SM, Bebianno MJ (1997) Sewage contamination following an accidental spillage in the Ria Formosa, Portugal. Mar Pollut Bull 34:163–170

Mudge SM, Duce CE (2005) Identifying the source, transport path and sinks of sewage derived organic matter. Environ Pollut 136:209–220

Mulholland PJ (2003) Large-scale patterns in dissolved organic carbon concentration, flux, and sources. In: Findlay S, Sinsabaugh RL (eds) Aquatic ecosystems—interactivity of dissolved organic matter. Academic Press, New York, pp 139–160

Muller CL, Baker A, Hutchinson R, Fairchild IJ, Kidd C (2008) Analysis of rainwater dissolved organic carbon compounds using fluorescence spectrophotometry. Atmos Environ 42:8036–8045

Murphy KR, Stedmon CA, Waite TD, Ruiz GM (2008) Distinguishing between terrestrial and autochthonous organic matter sources in marine environments using fluorescence spectroscopy. Mar Chem 108:40–58

Mylon SE, Twining BS, Fisher NS, Benoit G (2003) Relating the speciation of Cd, Cu, and Pb in two connecticut rivers with their uptake in algae. Environ Sci Technol 37:1261–1267

Nagai K, Aoki S, Fuse Y, Yamada E (2005) Fractionation of dissolved organic matter (DOM) as precursors of trihalomethane in Lake Biwa and Yodo Rivers. Bunseki Kagaku 54:923–928

Naganuma T, Konishi T, Inoue T, Nakane T, Sukizaki S (1996) Photodegradation or photoalteration? Microbial assay of dissolved organic matter. Mar Ecol Progr Ser 135:309–310

Nagao S, Matsunaga T, Suzuki Y, Ueno T, Amano H (2003) Characteristics of humic substances in the Kuji River as determined by high-performance size exclusion chromatography with fluorescence detection. Water Res 37:4159–4170

Nagao S, Kodama H, Aramaki T, Fujitake N, Uchida M, Shibata Y (2011) Carbon isotope composition of dissolved humic and fulvic acids in the Tokachi river system. Radiat Prot Dosimetry 146:322–325

Nakada N, Kiri K, Shinohara H, Harada A, Kuroda K, Takizawa S et al (2008) Evaluation of pharmaceuticals and personal care products as water-soluble molecular markers of sewage. Environ Sci Technol 42:6347–6353

Nakajima H (2006) Studies on photochemical degradation processes of dissolved organic matter in seawater. MS Thesis, Hiroshima University, pp 1–173

Nakane K, Kohno T Horikoshi T, Nakatsubo T (1997) Soil carbon cycling at a black spruce (Picea mariana) forest stand in Saskatchewan, Canada. J Geophy Res 102(D24):28,785–28,793

Nakatani N, Ueda M, Shindo H, Takeda K, Sakugawa H (2007) Contribution of the photo-Fenton reaction to hydroxyl radical formation rates in river and rain water samples. Anal Sci 23:1137–1142

Nash KL, Fried S, Friedman AM, Sullivan JC (1981) Redox behavior, complexing, and adsorption of hexavalent actinides by humic acid and selected clays Stiring marine disposal of high-level radioactive waste. Environ Sci Technol 15:834–837
Neal C, Reynolds B, Wilkinson J, Hill T, Neal M, Hill S, Harrow M (1998) The impacts of conifer harvesting on runoff water quality: a regional survey for Wales. Hydrol Earth Syst Sci 2:323–344
Neal C, Reynolds B, Neal M, Wickham H, Hill L, Williams B (2004) The impact of conifer harvesting on stream water quality: the Afon Hafren, mid-Wales. Hydrol Earth Syst Sci 8:503–520
Negreira N, Canosa P, Rodriguez I, Ramil M, Rubi E, Cela R (2008) Study of some UV filters stability in chlorinated water and identification of halogenated by-products by gas chromatography-mass spectrometry. J Chromatogr A 1178:20–214
Negri AP, Jones GJ, Hindmarsh M (1995) Sheep mortality associated with paralytic shellfish poisons from the cyanobacterium. Anabaena circinalis Toxicon 33:1321–1329
Nelson MJK, Montgomery SO, Maha¤ey WR, Pritchard PH (1987) Biodegradation of tricholoroethylene and involvement of an aromatic biodegradative pathway. Appl Environ Microbiol 53:949–954
Nelson NB, Siegel DA, Michaels AF (1998) Seasonal dynamics of colored dissolved material in the Sargasso Sea. Deep Sea Res I 45:931–957
Nelson NB, Carlson CA, Steinberg DK (2004) Production of chromophoric dissolved organic matter by Sargasso Sea microbes. Mar Chem 89:273–287
Newbern LA, Webster JR, Benefield EF, Kennedy JH (1981) Organic matter transport in an Appalachian Mountain river in Virginia, USA. Hydrobiologia 83:73–83
Newson M, Baker A, Mounsey S (2001) The potential role of freshwater luminescence measurements in exploring runoff pathways in upland catchments. Hydrol Process 15:989–1002
Newton A, Icely JD, Falcao M, Nobre A, Nunes JP, Ferreira JG, Vale C (2003) Evaluation of eutrophication in the Ria Formosa coastal lagoon, Portugal. Cont Shelf Res 23:1945–1961
Nichols PD, Espey QI (1991) Characterization of organic-matter at the air sea interface, in subsurface water, and in bottom sediments near the Malabar sewage outfall in Sydney coastal region Australian. J Mar FreshW Res 42:327–348
Nimrod AC, Benson WH (1996) Environmental estrogenic effects of alkylphenol ethoxylates. Crit Rev Toxicol 26:335–364
Nissenbaum A, Kaplan IR (1972) Chemical and isotopic evidence for the in situ origin of marine humic substances. Limnol Oceanogr 17:570–582
Nõges T, Laugaste R, Nõges P, Tõnno I (2008) Critical N:P ratio for cyanobacteria and N-fixing species in the large shallow temperate lakes Peipsi and Võrtsjärv, North-East Europe. Hydrobiologia 599:77–86
Norrman B, Zweifel UL, Hopkinson CS Jr, Fry B (1995) Production and utilisation of dissolved organic carbon during an experimental diatom bloom. Limnol Oceanogr 40:898–907
Nurmi JT, GTratnyek P (2002) Electrochemical properties of natural organic matter (NOM), fractions of NOM and model biogeochemical electron shuttles. Environ Sci Technol 36:617–624
Oberdorster E, Cheek AO (2001) Gender benders at the beach: endocrine disruption in marine and estuarine organisms. Environ Toxicol Chem 20:23–36
Obernosterer I, Herndl GJ (1995) Phytoplankton extracellular release and bacterial growth: dependence on the inorganic N:P ratio. Mar Ecol Prog Ser 116:247–257
Ogawa H, Ogura N (1992) Comparison of two methods for measuring dissolved organic carbon in sea water. Nature 356:696–698
Ogawa H, Tanoue E (2003) Dissolved organic matter in oceanic waters. J Oceanogr (Rev) 59:129–147
Ogawa H, Fukuda R, Koike I (1999) Vertical distributions of dissolved organic carbon and nitrogen in the Southern Ocean. Deep Sea Res I 46:1809–1826
Ogawa H, Amagai Y, Koike I, Kaiser K, Benner R (2001) Production of refractory dissolved organic matter by bacteria. Science 292:917–920
Ogura N (1972) Rate and extend of decomposition of dissolved organic-matter in surface seawater. Mar Biol 13:89

Oliver BG, Thurman EM, Malcolm RL (1983) The contribution of humic substances to the acidity of colored natural waters. Geochim Cosmochim Acta 47:2031–2035
Omar AM, Olsen A, Johannessen T, Hoppema M, Thomas H, Borges AV (2010) Spatiotemporal variations of fCO2 in the North Sea. Ocean Sci 6:77–89
Opsahl S, Benner R (1997) Distribution and cycling of terrigenous dissolved organic matter in the ocean. Nature 386:480–482
Opsahl S, Benner R (1998) Photochemical reactivity of dissolved lignin in river and ocean waters. Limnol Oceanogr 43:1297–1304
Opsahl S, Benner R, Amon RMW (1999) Major flux of terrigenous dissolved organic matter through the Arctic Ocean. Limnol Oceanogr 44:2017–2023
Ortega-Retuerta E, Frazer TK, Duarte CM, Ruiz-Halpern S, Tovar-Sánchez A, Arrieta JM, Reche I (2009) Biogeneration of chromophoric dissolved organicmatter by bacteria and krill in the Southern Ocean. Limnol Oceanogr 54:1941–1950
Osburn CL, Retamal L, Vincent WF (2009) Photoreactivity of chromophoric dissolved organic matter transported by the Mackenzie River to the Beaufort Sea. Mar Chem 115:10–20
Osinga R, de Vries KA, Lewis WE, van Raaphorst W, Dijkhuizen L, van Duyl FC (1997) Aerobic degradation of phytoplankton debris dominated by Phaeocystis sp in different physiological stages of growth. Aquat Microb Ecol 12:11–19
Österberg R, Shirshova L (1997) Oscillating, nonequilibrium redox properties of humic acids. Geochim Cosmochim Acta 61:4599–4604
Owen SF, Giltrow E, Huggett DB, Hutchinson TH, Saye J, Winter MJ, Sumpter JP (2007) Comparative physiology, pharmacology and toxicology of β-blockers: mammals versus fish. Aquat Toxicol 82:145–162
Pabich WJ, Valiela I, Hemond HF (2001) Relationship between DOC concentration and vadose zone thickness and depth below water table in groundwater of Cape Cod, USA. Biogeochemistry 55(3):247–268
Pace ML, Cole JJ (2002) Synchronous variation of dissolved organic carbon and color in lakes. Limnol Oceanogr 47:333–342
Pace ML, Cole JJ et al (2004) Whole-lake carbon-13 additions reveal terrestrial support of aquatic food webs. Nature 427:240–243
Paciolla MD et al (1998) In: Davies G, Ghabbour EA (eds) Humic substances: structures, properties and uses. Royal Society of Chemistry, Cambridge, pp 20–214
Pakulski JD, Benner R (1992) An improved method for the hydrolysis and MBTH analysis of dissolved and particulate carbohydrates in seawater. Mar Chem 40:143–160
Pakulski J, Benner R (1994) Abundance and distribution of carbohydrates in the ocean. Limnol Oceanogr 39:930–940
Pal A, Gin KY-H, Lin AY-C, Reinhard M (2010) Impacts of emerging organic contaminants on freshwater resources: review of recent occurrences, sources, fate and effects. Sci Total Environ 408:6062–6069
Palenik B, Morel FMM (1988) Dark production of H2O2 in the Sargasso Sea. Limnol Oceanogr 33:1606–1611
Palmer SM, Hope D, Billett MF, Dawson JJC, Bryant CL (2001) Sources of organic and inorganic carbon in a headwater stream: evidence from carbon isotope studies. Biogeochemistry 52:321–338
Pan Y, Wang Y, Xin J, Tang G, Song T, Wang Y, Li X, Wu F (2010) Study on dissolved organic carbon in precipitation in Northern China. Atmos Environ 44:2350–2357
Panagiotopoulos C, Sempéré R, Obernosterer I, Striby L, Goutx M, Van Wambeke F, Gautier S, Lafont R (2002) Bacterial degradation of large particles in the southern Indian Ocean using in vitro incubation experiments. Org Geochem 33:985–1000
Parker DR (2009) Perchlorate in the environment: the emerging emphasis on natural occurrence. Environ Chem 6:10–27
Parker DE, Legg TP, Folland CK (1992) A new daily Central England Temperature series, 1772–1991. Int J Climatol 12:317–342

Parlanti E, Worz K, Geoffroy L, Lamotte M (2000) Dissolved organic matter fluorescence spectroscopy as a tool to estimate biological activity in a coastal zone submitted to anthropogenic inputs. Org Geochem 31:1765–1781
Pastor J, Solin J, Bridgham SD, Updegraff K, Harth C, Weishampel P, Dewey B (2003) Global warming and the export of dissolved organic carbon from boreal peatlands. Oikos 100:380–386
Peck AM (2006) Analytical methods for the determination of persistent ingredients of personal care products in environmental matrices. Anal Bioanal Chem 386:907–939
Peña-Méndez EM, Havel J, Patočka J (2005) Humic substances-compounds of still unknown structure: applications in agriculture, industry, environment, and biomedicine. J Appl Biomed (Rev) 3:13–24
Pérez S, Barceló D (2007a) Application of advanced MS techniques to analysis and identification of human and microbial metabolites of pharmaceuticals in the aquatic environment. Trends Anal Chem 26:494–514
Pérez S, Barceló D (2007b) In: Aga DS (Ed) Fate of pharmaceuticals in the environment and in water treatment systems, 1st edn. CRC Press, Boca Raton, ch 2
Perez-Fuentetaja A, Dillon PJ, Yan ND, McQueen D (1999) Significance of dissolved organic carbon in the prediction of thermocline depthin small Canadian shield lakes. Aquat Ecol 33:127–133
Perry SA, Perry WB (1991) Organic carbon dynamics in two regulated rivers in northwestern Montana, USA. Hydrobiologia 218:193–203
Petrović M, Eljarrat E, López de Alda MJ, Barceló D (2001) Analysis and environmental levels of endocrine-disrupting compounds in freshwater sediments TrAC. Trends Anal Chem 20:637–648
Peuravuori J (1992) Isolation, fractionation and characterization of aquatic humic substances. Does a distinct humic molecule exist?. Academic dissertation, University of Turku, vol 4. Finland Finnish Humus News, pp 1–99
Peuravuori J, Pihlaja K (1999) Some approaches for modelling of dissolved aquatic organic matter In: Keskitalo J, Eloranta P (eds) Limnology of humic waters. Backhuy Publishers, Leiden, pp 11–39
Pham TPT, Cho C-W, Yun Y-S (2010) Environmental fate and toxicity of ionic liquids: a review. Water Res 44:352–372
Piccolo A (2002) The supramolecular structure of humic substances: a novel understanding of humus chemistry and implications in soil science. Adv Agron 75:57–134 (Academic Press)
Piccolo A, Conte P, Cozzolino A (2001) Chromatographic and spectrophotometric properties of dissolved humic substances compared with macromolecular polymers. Soil Sci 166:174–185. doi:101097/00010694-200103000-00003
Pomati F, Castiglioni S, Zuccato E, Fanelli R, Vigetti D, Rossetti C et al (2006) Effects of a complex mixture of therapeutic drugs at environmental levels on human embryonic cells. Environ Sci Technol 40:2442–2447
Porcal P, Koprivnjak J-F, Molot LA, Dillon PJ (2009) Humic substances—part 7: the biogeochemistry of dissolved organic carbon and its interactions with climate change. Environ Sci Pollut Res 16:714–726
Prahl FF, Small LF, Sullivan BA (1998) Biogeochemical gradients in the lower Columbia River. Hydrobiologia 361:37–52
Prairie YT, Bird DF, Cole JJ (2002) The summer metabolic balance in the epilimnion of southeastern Quebec lakes. Limnol Oceanogr 47:316–321
Pregitzer K, Zak DR, Burton AJ, Ashby JA, MacDonald NW (2004) Chronic nitrate additions dramatically increase the export of carbon and nitrogen from northern hardwood ecosystems. Biogeochemistry 68:179–197
Prince EK, Myers TL, Kubanek J (2008) Effects of harmful algal blooms on competitors: allelopathic mechanisms of the red tide. Limnol Oceanogr 53:531–541
Qiu X, Zhu T, Yao B, Hu J, Hu S (2005) Contribution of dicofol to the current DDT pollution in China. Environ Sci Technol 39:4385–4390

Radnoti de Lipthay J, Barkay T, Vekova J, Sorensen SJ (1999) Utilization of phenoxyacetica acid, by strains using either the ortho or meta cleavage of catechol during phenol degradation, after conjugal transfer of oftLtl, the gene encoding a 2,4-dichlorophenoxyacetic acid12-oxoglutarate dioxygenase. Appl Microbiol Biotechnol 51:207–214

Rae R, Howard-Williams C, Hawes I, Schwarz A-M, Vincent WF (2001) Penetration of solar ultraviolet radiation into New Zealand lakes: influence of dissolved organic carbon and catchment vegetation. Limnology 2:79–89

Raich JW, Schlesinger WH (1992) The global carbon dioxide flux in soil respiration and its relationship to vegetation and climate. Tellus B 44:81–99

Raman DR, Layton AC, Moody LB, Easter JP, Sayler GS, Burns RT, Mullen MD (2001) Degradation of estrogens in dairy waste solids: effects of acidification and temperature. Trans ASAE 44:1881–1888

Ramirez AJ, Mottaleb MA, Brooks BW, Chambliss CK (2007) Analysis of pharmaceuticals in fish using liquid chromatography-tandem mass spectrometry. Ana Chem 79:3155–3163

Ramirez AJ, Brain RA, Usenko S, Mottaleb MA, O'Donnell JG, Stahl LL, Wathen JB, Snyder BD, Pitt JL, Perez-Hurtado P, Dobbins LL, Brooks BW, Chambliss CK (2009) Occurrence of pharmaceuticals and personal care products in fish: results of a national pilot study in the United States. Environ Toxicol Chem 28:2587–2597

Ramirez N, Marce RM, Borrull F (2011) Development of a stir bar sorptive extraction and thermal desorption-gas chromatography-mass spectrometry method for determining synthetic musks in water samples. J Chromatogr A 1218:156–161

Rashid MA, King LH (1969) Molecular weight distribution measurements on humic and fulvic acid fractions from marine clays on the Scotian Shelf. Geochim Cosmochim Acta 33:147–151

Rasmussen JB, Godbout L, Schallenberg M (1989) The humic content of lake water and its relationship to watershed and lake morphometry. Limnol Oceanogr 34:1336–1343

Rastogi RP, Richa Sinha RP, Singh SP, Häder D-P (2010) Photoprotective compounds from marine organisms. J Ind Microbiol Biotechnol 37:537–558

Raymond PA, Bauer JE (2001a) DOC cycling in a temperate estuary: a mass balance approach using natural 14C and 13C isotopes. Limnol Oceanogr 46:655–667

Raymond PA, Bauer JE (2001b) Riverine export of aged terrestrial organic matter to the North Atlantic Ocean. Nature 409:497–500

Raymond PA, Bauer JE (2001c) Use of 14C and 13C natural abundances for evaluating riverine, estuarine, and coastal DOC and POC sources and cycling: a review and synthesis. Org Geochem 23:469–485

Raymond PA, Oh NH (2007) An empirical study of climatic controls on riverine C export from three major US watersheds. Global Biogeochem Cycles 21:GB2022

Raymond P, Saiers J (2010) Event controlled DOC export from forested watersheds. Biogeochemistry 100:197–209

Reche I, Pace M, Cole JJ (1999) Relationship of trophic and chemical conditions to photobleaching of dissolved organic matter in lake ecosystems. Biogeochemistry 44:259–280

Reddy S, Iden CR, Brownawell BJ (2005) Analysis of steroid estrogen conjugates in municipal waste waters by liquid chromatography-tandem mass spectrometry. Anal Chem 77:7032–7038

Reiller PE, Brevet J (2010) Bi-exponential decay of Eu(III) complexed by Suwannee River humic substances: spectroscopic evidence of two different excited species. Spectrochim Acta Part A Mol Biomol Spectrosc 75:629–636

Reitner B, Herndl G, Herzig A (1997) Role of ultraviolet-B radiation on photochemical and microbial oxygen consumption in a humic-rich shallow lake. Limnol Oceanogr 42:950–960

Remington S, Krusche A, Richey J (2011) Effects of DOM photochemistry on bacterial metabolism and CO 2 evasion during falling water in a humic and a whitewater river in the Brazilian Amazon. Biogeochemistry 105(1):185–200

Repeta DJ, Quan TM, Aluwihare LI, Accardi A (2002) Chemical characterization of high molecular weight dissolved organic matter in fresh and marine waters. Geochim Cosmochim Acta 66:955–962

Rhind SM (2002) Endocrine disrupting compounds and farm animals: their properties, actions and routes of exposure. Domest Anim Endocrinol 23:179–187
Rice CW (2002) Storing carbon in soil: why and how? Geotimes. American Geological Institute, Alexandria. http://wwwgeotimesorg/jan02/feature_carbonhtml
Richards RP, Baker DB (1993) Pesticides concentration patterns in agricultural drainage networks in the lake Eire basin. Environ Toxicol Chem 12:13–26
Richardson SD (2003) Water analysis: emerging contaminants and current issues. Anal Chem 75:2831–2857
Richardson SD (2007) Water analysis: emerging contaminants and current issues. Anal Chem 79:4295–4324
Richardson SD, Ternes TA (2005) Water analysis: emerging contaminants and current issues. Anal Chem 77:3807–3838
Richardson SD, Ternes TA (2011) Water analysis: emerging contaminants and current issues. Anal Chem 83:4614–4648
Richey JE, Hedges JI, Devol AH, Quay PD (1990) Biogeochemistry of carbon in the Amazone River. Limnol Oceanogr 35:352–371
Richey JE, Melack JM, Aufdenkampe AK, Ballester VM, Hess LL (2002) Outgassing from Amazonian rivers and wetlands as a large tropical source of atmospheric CO2. Nature 416:617–620
Richter DD, Markewitz D, Trumbore SE, Wells CG (1999) Rapid accumulation and turnover of soil carbon in a re-establishing forest. Nature 400:56–58
Rimkus GG, Wolf M (1996) Polycyclic musk fragrances in human adipose tissue and human milk. Chemosphere 33:2033–2043
Rochelle-Newall EJ, Fisher TR (2002a) Production of chromophoric dissolved organic matter fluorescence in marine and estuarine environments: an investigation into the role of phytoplankton. Mar Chem 77:7–21
Rochelle-Newall EJ, Fisher TR (2002b) Chromophoric dissolved organic matter and dissolved organic carbon in Chesapeake Bay. Mar Chem 77:23–41
Rogora M (2007) Considerazioni generali sull'evoluzione del chimismo delle acque lacustri e tributarie In CNR-ISE Ricerche sull'evoluzione del Lago Maggiore Aspetti limnologici Programma quinquennale 2003–2007 Campagna 2007 e Rapporto quinquennale 2003–2007 Commissione Internazionale per la protezione delle acque italo svizzere: 89–97
Romani AM, Guasch H, Muñoz I, Ruana J, Vilalta E, Schwartz T, Emtiazi F, Sabater S (2004) Biofilm structure and function and possible implications for riverine DOC dynamics. Microb Ecol 47:316–328
Romankevich EA (1984) Geochemistry of organic matter in the ocean. Springer, Berlin, p 274
Rosenstock B, Simon M (2001) Sources and sinks of dissolved free amino acids and protein in a large and deep mesotrophic lake. Limnol Oaceanogr 50:90–101
Rosenstock B, Zwisler W, Simon M (2005) Bacterial consumption of humic and non-humic low and high molecular weight DOM and the effect of solar irradiation on the turnover of labile DOM in the Southern Ocean. Microb Ecol 50:90–101
Rostan JC, Cellot B (1995) On the use of UV spectrophotometry to assess dissolved organic carbon origin variations in the upper Rhône River. Aquat Sci 57:70–81
Rudel RA, Camann DE, Spengler JD, Korn LR, Brody JG (2003) Phthalates, alkylphenols, pesticides, polybrominated diphenyl ethers, and other endocrine-disrupting compounds in indoor air and dust. Environ Sci Technol 37:4543–4553
Rudel H, Böhmer W, Schröter-Kermani C (2006) Retrospective monitoring of synthetic musk compounds in aquatic biota from German rivers and coastal areas. J Environ Monit 8:812–823
Rutledge S, Campbell DI, Baldocchi D, Schipper LA (2010) Photodegradation leads to increased carbon dioxide losses from terrestrial organic matter. Glob Change Biol 16:3065–3074
RxList (The Internet Drug Index) (2005) Top 300 prescriptions for 2005 by number of US prescriptions dispensed. http://wwwrxlistcom/top200htm
Sachs S, Reich T, Bernhard G (2010) Study of the role of sulfur functionalities in humic acids for uranium(VI) complexation. Radiochim Acta 98:467–477

Sadtler (1968) The Sadtler standard spectra. Sadtler Research Lab, Inc, Philadelphia

Safiullah S, Mofizuddin M, Iqbal Ali SM, Enamul Kabir S (1987) Biogeochemical cycles of carbon in the rivers of Bangladesh. In: Degens ET, Kempe S, Weibin G (eds) Transport of carbon and minerals in major world rivers. Pt4 mitt Geological-Palaeontological Institute and Museum of the University of Hamburg, SCOPE/UNEP Sonderbd 64, pp 435–442

Sakugawa H, Kaplan IR, Shepard LS (1993) Measurements of H2O2, aldehydes and organic acids in Los Angeles rainwater: their sources and deposit rates. Atmos Environ 27B:203–219

Salmaso N, Morabito G, Mosello R, Garibaldi L, Simona M, Buzzi F, Ruggiu D (2003) A synoptic study of phytoplankton in the deep lakes south of the Alps (lakes Garda, Iseo, Como, Lugano and Maggiore). J Limnol 62:207–227

Salonen K, Vähätalo A (1994) Photochemical mineralization of dissolved organic matter in Lake Skjervatjern. Environ Int 20:307–312

Salonen K, Kairesalo T, Jones RI (1992) Dissolved organic matter in lacustrine ecosystems: energy-source and system regulator. Hydrobiologia 229:1–291

Sanderson H, Brain RA, Johnson DJ, Wilson CJ, Solomon KR (2004) Toxicity classification and evaluation of four pharmaceuticals classes: antibiotics, antineoplastics, cardiovascular, and sex hormones. Toxicology 203:27–40

Santos PSM, Otero M, Duarte RMBO, Duarte AC (2009a) Spectroscopic characterization of dissolved organic matter isolated from rainwater. Chemosphere 74:1053–1061

Santos PSM, Duarte RMBO, Duarte AC (2009b) Absorption and fluorescence properties of rainwater during the cold season at a town in Western Portugal. J Atmos Chem 62:45–57

Santschi PH, Guo L, Baskaran M, Trumbore S, Southon J, Bianchi TS, Honeyman B, Cifuentes L (1995) Isotopic evidence for the contemporary origin of high-molecular weight organic matter in oceanic environments. Geochim Cosmochim Acta 59:625–631

Sargent JR, Gatten RR, Henderson RJ (1981) Marine wax esters. Pure Appl Chem 53:867–871

Satoh Y, Katano T, Satoh T, Mitamura O, Anbutsu K, Nakano S-I, Ueno H, Kihira M, Drucker V, Tanaka Y, Mimura T, Watanabe Y, Sugiyama M (2006) Nutrient limitation of the primary production of phytoplankton in Lake Baikal. Limnology 7:225–229

Scheel T, Jansen B, van Wijk AJ, Verstraeten JM, Kalbitz K (2008) Stabilization of dissolved organic matter by aluminium: a toxic effect or stabilization through precipitation? Eur J Soil Sci 59:1122–1132

Scheurer M, Brauch H-J, Lange FT (2009) Analysis and occurrence of seven artificial sweeteners in German waste water and surface water and in soil aquifer treatment (SAT). Anal Bioanal Chem 394:1585–1594

Schiff SL, Arvena R, Trumbore SE, Dillon PJ (1990) Dissolved organic carbon cycling in forested watersheds: a carbon isotope approach. Water Res 26:2949–2957

Schiff SL, Aravena R, Trumbore SE, Hinton MJ, Elgood R, Dillon PJ (1997) Export of DOC from forested catchments on the Precambrian Shield of Central Ontario: clues from 13C and 14C. Biogeochemistry 36:43–65

Schindler DW, Curtis PJ, Parker BR, Stainton MP (1996) Consequences of climate warming and lake acidification for UV-B penetration in North American boreal lakes. Nature 379:705–708

Schmid P, Martin Kohler M, Gujer E, Zennegg M, Lanfranchi M (2007) Persistent organic pollutants, brominated flame retardants and synthetic musks in fish from remote alpine lakes in Switzerland. Chemosphere 67:S16–S21

Schmidt-Rohr K, Mao J-D, Olk DC (2004) Nitrogen-bonded aromatics in soil organic matter and their implications for a yield decline in intensive rice cropping. Proc Natl Acad Sci U S A 101:6351–6354

Schmitt C, Oetken M, Dittberner O, Wagner M, Oehlmann J (2008) Endocrine modulation and toxic effects of two commonly used UV screens on the aquatic invertebrates Potamopyrgus antipodarum and Lumbriculus variegates. Environ Pollut 152:322–329

Schnitzer M (1985) Nature of nitrogen in humic substances. In: Aiken GR, McKnight DM, Wershaw RL, MacCarthy P (eds) Humic Substances in Soil, Sediment, and Water. Wiley, New York, pp 303–328

Schnitzer M, Khan SU (1978) Soil organic matter. Elsevier, New York
Schulten H-R, Schnitzer M (1998) The chemistry of soil organic nitrogen: a review. Biol Fertil Soils 26:1–15
Schwarzenbach RP, Stierli R, Lanz K, Zeyer J (1990) Quinone and iron porphyrin mediated reduction of nitroaromatic compounds in homogenous aqueous solution. Environ Sci Technol 24:1566–1574
Schwede-Thomas SB, Chin Y, Dria KJ, Hatcher P, Kaiser E, Sulzberger B (2005) Characterizing the properties of dissolved organic matter isolated by XAD and C-18 solid phase extraction and ultrafiltration. Aquat Sci 67:61–71
Scott MJ, Jones MN (2000) The biodegradation of surfactants in the environment. Bioc Biophy Acta 1508:235–251
Scott DT, McKnight DM, Blunt-Harris EL, Kolesar SE, Lovley DR (1998) Quinone moieties act as electron acceptors in the reduction of humic substances by humicsreducing microorganisms. Environ Sci Technol 32:2984–2989
Scott AC, Whittal RM, Fedorak PM (2009) Coal is a potential source of naphthenic acids in groundwater. Sci Total Environ 407:2451–2459
Scully FE, Howell GD, Kravitz R, Jewell JT, Hahn V, Speed M (1988) Proteins in natural-waters and their relation to the formation of chlorinated organics during water disinfection. Environ Sci Technol 22:537–542
Scully NM, Vincent WF, Lean DRS, Cooper WJ (1997) Implications of ozone depletion for surface-water photochemistry: sensitivity of clear lakes. Aquat Sci 59:260–274
Scully NM, Maie N, Dailey SK, BoyerJN Jones RD, Jaffé R (2004) Early diagenesis of plant-derived dissolved organic matter along a wetland, mangrove, estuary ecotone. Limnol Oceanogr 49:1667–1678
See JH, Bronk DA (2005) Changes in C:N ratios and chemical structures of estuarine humic substances during aging. Mar Chem 97:334–346
Seitzinger SP, Hartnett H, Lauck R, Mazurek M, Minegishi T, Spyres G, Styles R (2005) Molecular-level chemical characterization and bioavailability of dissolved organic matter in stream water using electrospray-ionization mass spectrometry. Limnol Oceanogr 50:1–12
Semiletov IP, Pipko II, Repina I, Shakhova NE (2007) Carbonate chemistry dynamics and carbon dioxide fluxes across the atmosphere-ice-water interfaces in the Arctic Ocean: Pacific sector of the Arctic. J Mar Syst 66:204–226
Sempéré R, Kawamura K (1994) Comparative distributions of dicarboxylic acids and related polar compounds in snow, rain, and aerosols from urban atmosphere. Atmos Environ 28:449–459
Senesi N (1990) Molecular and quantitative aspects of the chemistry of fulvic acid and its interactions with metal ions and organic chemicals. Part II. The fluorescence spectroscopy approach. Anal Chim Acta 232:77–106
Senesi N, Miano T, Provenzano M, Burnett G (1991) Characterization, differentiation, and classification of humic substances by fluorescence spectroscopy. Soil Sci 152:259–271
Servos MR (1999) Review of the aquatic toxicity, estrogenic responses and bioaccumulation of alkylphenols and alkylphenol polyethoxylates. Water Qual Res J Can 34:123–177
Sharp JH (1973) Size classes of organic carbon in seawater. Limnol Oceanogr 18:441–447
Shaw EM (1979) 1975–76 drought in England and Wales in perspective. Disasters 3:103–110
Shcherbina NS, Perminova IV, Kalmykov SN, Kovalenko AN, Haire RG, Novikov AP (2007) Redox and complexation interactions of neptunium(V) with quinonoid-enriched humic derivatives. Environ Sci Technol 41:7010–7015
Sherr B, Sherr E (1989) Trophic impacts phagotrophic Protozoa in pelagic foodwebs In: Hattori T, Ishida Y, Maruyama Y, Morita RY, Uchida A (eds) Recent advances in microbial ecology. Japan Scientific Society Press,Tokyo, pp 388–393
Shi X, Du Y, Lam PKS, Wu RSS, Zhou B (2008) Developmental toxicity and alteration of gene expression in zebrafish embryos exposed to PFOS. Toxicol Appl Pharmacol 230:23–32
Siegel DA, Michaels AF (1996) Quantification of the non-algal light attenuation in the Sargasso Sea: implications for biogeochemistry and remote sensing. Deep Sea Res 43:321–345

Siemens J, Kaupenjohann M (2003) Dissolved organic matter induced denitrification in subsoils and aquifers? Geoderma 113:253–271
Sigleo AC, Macko SA (1985) Stable isotope and amino acid composition of estuarine dissolved colloidal material. In: Sigleo AC, Hattori A (eds) Marine and estuarine geochemistry. Lewis, Boca Raton, pp 29–46
Silberhorn EM, Glauert HP, Robertson LW (1990) Carcinogenicity of polyhalogenated biphenyls: PCBs and PBBs. Crit Rev Toxicol 20:440–496
Simpson AJ, Kingery WL, Hatcher PG (2002) The Identification of Plant Derived Structures in Humic Materials Using Three-Dimensional NMR Spectroscopy. Environ Sci Technol 37(2):337–342
Singh SC, Singa RP (2002) Häder D-P (2002) Role of lipids and fatty acids in stress tolerance in cyanobacteria. Acta Protozool 41:297–308
Sinha R, Klisch M, Gröniger A, Häder D-P (2001) Responses of aquatic algae and cyanobacteria to solar UV-B. Plant Ecol 154:219–236
Skjelkvåle BL, Mannio J, Wilander A, Andersen T (2001) Recovery from acidification of lakes in Finland, Norway and Sweden 1990–1999 Hydrology and Earth System. Science 5:327–338
Skjelkvåle BL, Stoddard J, Jeffries D et al (2005) Regional scale evidence for improvements in surface water chemistry 1990–2001. Environ Pollut 137:165–176
Skogerboe RK, Wilson SA (1981) Reduction of ionic species by fulvic acid. Anal Chem 53:228–232
Skoog A, Benner R (1997) Aldoses in various size fractions of marine organic matter: implications for carbon cycling. Limnol Oceanogr 42:1803–1813
Slawyk G, Raimbault B, Garcia N (1998) Measuring gross uptake of 15 N-labeled nitrogen by marine phytoplankton without particulate matter collection: evidence of low 15 N losses to the dissolved organic nitrogen pool. Limnol Oceanogr 43:1734–1739
Smital T, Kurelec B (1998) The chemosensitizers of multixenobiotic resistance mechanism in aquatic invertebrates: a new class of pollutants. Mutat Res 399:43–53
Smital T, Luckenbach T, Sauerborn R, Hamdoun AM, Vega RL, Epel D (2004) Emerging contaminants–pesticides, PPCPs, microbial degradation products and natural substances as inhibitors of multixenobiotic defense in aquatic organisms. Mutat Res/Fundam Mol Mech Mutagen 552:101–117
Smith BN, Epstein S (1971) Two categories of 13C:12C ratios for higher plants. Plant Physiol 47:380–384
Smith SV, Hollibaugh JT (1993) Coastal metabolism and the organic-carbon balance. Rev Geophys 31:75–89
Smith RL, Oremland RS (1983) Anaerobic oxalate degradation: widespread natural occurrence in aquatic sediments. Appl Environ Microb 46:106–113
Smith FMJ, Wood SA, van Ginkel R, Broady PA, Gaw S (2011) First report of saxitoxin production by a species of the freshwater benthic cyanobacterium. Scytonema Agardh Toxicon 57:566–573
Snicins E, Gunn J (2000) Interannual variation in the thermal structure of clear and colored lakes. Limnol Oceanogr 45:1647–1654
Sobek S, Tranvik LJ, Prairie YT, Kortelainen P, Cole JJ (2007) Patterns and regulation of dissolved organic carbon: an analysis of 7,500 widely distributed lakes. Limnol Oceanogr 52:1208–1219
Sommaruga R, Augustin G (2006) Seasonality in UV transparency of an alpine lake is associated to changes in phytoplankton biomass. Aquat Sci 68:129–141. doi:101007/s00027-006-0836-3
Sommaruga R, Psenner R (1997) Ultraviolet radiation in a high mountain lake of the Austrian Alps: air and underwater measurements. Photochem Photobiol 65:957–963
Son S, Wang M, Shon J-K (2011) Satellite observations of optical and biological properties in the Korean dump site of the Yellow Sea. Remote Sens Environ 115:562–572
Søndergaard M, Middelboe M (1995) A cross-system analysis of labile dissolved organic carbon. Mar Ecol Prog Ser 118:283–294
Southwell MW, Smith JD, Avery GB, Kieber RJ, Willey JD (2010) Seasonal variability of formaldehyde production from photolysis of rainwater dissolved organic carbon. Atmos Environ 44:3638–3643

Spence A, Simpson AJ, McNally DJ, Moran BW, McCaul MV, Hart K, Paull B, Kelleher BP (2011) The degradation characteristics of microbial biomass in soil. Geochim Cosmochim Acta 75:2571–2581
Spiker E (1981) Flux of organic carbon by rivers to the oceans report CONF-8009140, US Department of Energy, Springfield, pp 79–109
Stedmon CA, Markager S (2005a) Tracing the production and degradation of autochthonous fractions of dissolved organic matter by fluorescence analysis. Limnol Oceanogr 50:1415–1426
Stedmon CA, Markager S (2005b) Resolving the variability in dissolved organic matter fluorescence in a temperate estuary and its catchment using PARAFAC analysis. Limnol Oceanogr 50:686–697
Stedmon CA, Markager S, Bro R (2003) Tracing dissolved organic matter in aquatic environments using a new approach to fluorescence spectroscopy. Mar Chem 82:239–254
Stedmon CA, Markager S, Tranvik L, Kronberg L, Slätis T, Martinsen W (2007a) Photochemical production of ammonium and transformation of dissolved organic matter in the Baltic Sea. Mar Chem 104:227–240
Stedmon CA, Thomas DN, Granskog M, Kaartokallio H, Papaditriou S, Kuosa H (2007b) Characteristics of dissolved organic matter in Baltic coastal sea ice: allochthonous or autochthonous origins? Environ Sci Technol 41:7273–7279
Steelink C (2002) Investigating humic acids in soils. Anal Chem 74:328A–333A
Steffensen D, Burch M, Nicholson B, Drikas M, Baker P (1999) Management of toxic blue-green algae (Cyanobacteria) in Australia. Environ Toxicol 14:183–195
Steinberg C, Muenster U (1985) Geochemistry and ecological role of humic substances in lake water. In: Aiken GR, McKnight DM, Wershaw RL, MacCarthy P (eds) Humic substances in soil, sediment, and water: geochemistry, isolation, and characterization. Wiley, New York, pp 105–145
Steinberg DK, Nelson NB, Carlson CA, Prusak AC (2004) Production of chromophoric dissolved organic matter (CDOM) in the open ocean by zooplankton and the colonial cyanobacterium Trichodesmium spp. Mar Ecol Prog Ser 267:45–56
Stenson AC, Landing WM, Marshall AG, Cooper WT (2002) Ionization and fragmentation of humic substances in electrospray ionization Fourier transform-ion cyclotron resonance mass spectrometry. Ana Chem 74(17):4397–4409
Stenson AC, Marshall AG, Cooper WT (2003) Exact masses and chemical formulas of individual Suwannee river fulvic acids from ultrahigh resolution electrospray ionization Fourier transform ion cyclotron resonance mass spectra. Anal Chem 75:1275–1284
Stepanauskas R, Moran MA, Bergamaschi BA, Hollibaugh JT (2005) Sources, bioavailability, and photoreactivity of dissolved organic carbon in the Sacramento–San Joaquin River Delta. Biogeochemistry 74(2):131–149
Stevenson FJ (1982) Humus chemistry: genesis-composition-reactions. Wiley, New York, pp 221–243
Stoddard JL, Kahl JS, Deviney FA, DeWalle DR, Driscoll CT, Herlihy AT, Kellogg JH, Murdoch PS, WebbJR, Webster KE (2003) Response of surface water chemistry to the Clean Air Act Amendments of 1990 Corvallis. US Environmental Protection Agency EPA/620/R-03/001, Oregon
Strome DJ, Miller MC (1978) Photolytic changes in dissolved humic substances. Int Ver Theor Angew Limnol 20:1248–1254
Struyk Z, Sposito G (2001) Redox properties of standard humic acids. Geoderma 102:329–346
Strynar MJ, Lindstrom AB (2008) Perfluorinated compound in house dust from Ohio and North Carolina, USA. Environ Sci Technol 42:3751–3756
Stuermer DH, Harvey GR (1977) Isolation of humic substances and alcohol-soluble organic-matter from seawater. Deep Sea Res 24(3):303–309
Subedi B, Mottaleb MA, Chamblish CK, Usenko S (2011) Simultaneous analysis of select pharmaceuticalsand personal care products in fish tissue using pressurized liquid extraction combined with silica gel cleanup. J Chromatogr A 1218:6278–6284

Sugimura Y, Suzuki Y (1988) A high-temperature catalytic oxidation method for the determination of non-volatile dissolved organic carbon in seawater by direct injection of a liquid sample. Mar Chem 24:105–131

Sugiyama Y, Anegawa A, Kumagai T, Harita Y-N, Hori T, Sugiyama M (2004) Distribution of dissolved organic carbon in lakes of different trophic types. Limnology 5:165–176

Sugiyama Y, Anegawa A, Inokuchi H, Kumagai T (2005) Distribution of dissolved organic carbon and dissolved fulvic acid in mesotrophic Lake Biwa, Japan. Limnology 6:161–168

Sullivan MJ, Moncreiff CA (1990) Edaphic algae are an important component of salt marsh food webs: evidence from multiple stable isotope analyses. Mar Ecol Prog Ser 62:149–159

Sundh I (1992) Biochemical composition of dissolved organic carbon released from natural communities of lake phytoplankton. Arch Hydrobiol 125:347–369

Sutton R, Sposito G (2005) Molecular structure in soil humic substances: the new view. Environ Sci Technol 39:9010–9015

Sweeney T, Nicol L, Roche JF, Brooks AN (2000) Maternal exposure to octylphenol suppresses ovine fetal follicle-stimulating hormone secretion, testis size, and Sertoli cell number. Endocrinology 141:2667–2673

Takahashi M, Hama T, Matsunaga K, Handa N (1995) Photosynthetic organic carbon production and respiratory organic carbon consumption in the trophogenic layer of Lake Biwa. J Plankton Res 17:1017–1025

Tanaka Y, Miyajima T, Koike I, Hayashibara T, Ogawa H (2008) Production of dissolved and particulate organic matter by the reef-building corals Porites cylindrica and Acropora pulchra. Bull Mar Sci 82:237–245

Tanoue E (2000) Proteins in the sea-synthesis. In: Handa N, Tanoue E, Hama T (eds) Dynamics and characterization of marine organic matter. TERRAPUB/Kluwer, Tokyo, pp 383–463

Tanoue E, Handa N (1987) Monosaccharide composition of marine particles and sediments from the Bering Sea and northern North Pacific. Oceanol Acta 10:91–100

Tanoue E, Nishiyama S, Kamo M, Tsugita A (1995) Bacterial membranes: possible source of a major dissolved protein in seawater. Geochim Cosmochim Acta 59:2643–2648

Tanoue E, Ishii M, Midorikawa T (1996) Discrete dissolved and particulate proteins in oceanic waters. Limnol Oceanogr 41:1334–1343

Tao S (1996) Fractionation and chlorination of organic carbon in water from Yinluan River, Tianjin, China. Geo J 40:213–217

Tedetti M, Kawamura K, Charriere B, Chevalier N, Sempere R (2006) Determination of low molecular weight dicarboxylic and ketocarboxylic acids in seawater samples. Anal Chem 78:6012–6018

Terashima M, Fukushima M, Tanaka S (2004) Influence of pH on the surface activity of humic acid: micelle-like aggregate formation and interfacial adsorption. Colloids Surf A Physicochem Eng Asp 247:77–83. doi:101016/jcolsurfa200408028

Thomas DN, Lara RJ (1995) Photodegradation of algal derived dissolved organic carbon. Mar Ecol Prog Ser 116:309–310

Thurman EM (1985a) Humic substances in groundwater. In: Aiken GR, McKnight DM, Wershaw RL, MacCarthy P (eds) Humic substances in soil, sediment, and water: geochemistry, isolation, and characterization. Wiley, New York, pp 87–103

Thurman EM (1985b) Organic geochemistry of natural waters. Martinus Nijhoff/Dr W Junk Publishers, Dordrecht

Thurman EM (1985c) Humic substances in groundwater. In: Aiken GR, McKnight DM, Wershaw RL, MacCarthy P (eds) Humic substances in soil, sediment, and water: geochemistry, isolation, and characterization. Wiley, New York, pp 87–103

Thurman EM (1986) Organic geochemistry of natural waters, 2nd edn. Martinus Nijhoff/Dr W Junk Publishers, The Netherlands

Thurman EM, Malcolm RL (1983) Structural study of humic substances: new approaches and methods. In: Christman RF, Yu ZG (eds) Aquatic and terrestrial humic materials. Ann Arbor Science, Ann Arbor, pp 1–23

Tipping E, Hilton J, James B (1988) Dissolved organic matter in Cumbrian lakes and streams. Freshw Biol 19:371–378

Tipping E, Marker AFH, Butterwick C, Collett GD et al (1997) Organic carbon in the Humber rivers. Sci Total Environ 194/195:345–355

Toppari J, Larsen JC, Christiansen P, Giwercman A, Grandjean P, Guillette LJ, Jegou B, Jensen TK, Jouannet P, Keiding N, Leffers H, McLachlan JA, Meyer O, Muller J, Meyts ER, Scheike T, Sharpe R, Sumpter J, Skakkebaek NE (1996) Male reproductive health and environmental xenoestrogens. Environ Health Perspect 104(Suppl 4):741–803

Tranvik LJ (1989) Bacterioplankton growth, grazing mortality and quantitative relationship to primary production in a humic and a clearwater lake. J Plankton Res 11:985–1000

Tranvik LJ (1992) Allochthonous dissolved organic matter as an energy source for pelagic bacteria and the concept of the microbial loop. Hydrobiologia 229:107–114

Tranvik L (1998) Degradation of dissolved organic matter in humic wates by bacteria. In: Hessen DO, Tranvik L (eds) Aquatic humic substances: ecology and biogeochemistry. Springer, Berlin, pp 259–283

Tranvik LJ, Hafle MG (1987) Bacterial growth in mixed cultures on dissolved organic carbon from humic and clear waters. Appl envir Microbiol 53:482–488

Tranvik LJ, Jasson M (2002) Climate change—terrestrial export of organic carbon. Nature 415:861–862

Tranvik LJ, Kokalj S (1998) Decreased biodegradability of algal DOC due to interactive effects of UV radiation and humic matter. Aquat Microb Ecol 14:301–307

Tranvik LJ, Downing JA, Cotner JB, Loiselle SA, Striegl RG, Ballatore TJ, Dillon P, Finlay K, Fortino K, Knoll LB, Kortelainen PL, Kutser T, Larsen S, Laurion I, Leech DM, McCallister SL, McKnight DM, Melack JM, Overholt E, Porter JA, Prairie Y, Renwick WH, Roland F, Sherman BS, Schindler DW, Tremblay SSA, Vanni MJ, Verschoor AM, von Wachenfeldt E, Weyhenmeyer GA (2009) Lakes and reservoirs as regulators of carbon cycling and climate. Limnol Oceanogr 54:2298–2314

Tratnyek PG, Macalady DL (1989) Abiotic reduction of nitro aromatic pesticides in anaerobic laboratory systems. J Agric Food Chem 37:248–254

Tremblay J, Michel C, Hobson KA, Gosselin M, Price NM (2006) Bloom dynamics in early opening waters of the Arctic Ocean. Limnol Oceanogr 51:900–912

Trumbore S (2000) Age of soil organic matter and soil respiration: radiocarbon constraints on below ground dynamics. Ecol Appl 10:399–411

Trumbore SE, Schiff SL, Aravena R, Elgood R (1992) Sources and transformation of dissolved organic carbon in the Harp Lake forested catchment: the role of soils. Radiocarbon 34:626–635

Tu C-L, Liu C-Q, Lu X-H, Yuan J, Lang Y-C (2011) Sources of dissolved organic carbon in forest soils: evidences from the differences of organic carbon concentration and isotope composition studies. Environ Earth Sci 63:723–730

Tyler CR, Jobling S, Sumpter JP (1998) Endocrine disruption in wildlife: a critical review of the evidence. Crit Rev Toxicol 28:319–361

Tzortziou M, Neale PJ, Osburn CL, Megonigal JP, Maie N, Jaffé R (2008) Tidal marshes as a source of optically and chemically distinctive colored dissolved organic matter in the Chesapeake Bay. Limnol Oceanogr 53:148–159

Uchida M, Nakatsubo T, Horikoshi T, Nakane K (1998) Contribution of micro-organisms to the carbon dynamics in black spruce (Picea mariana) forest soil in Canada. Ecol Res 13:17–26

Uchida M, Nakatsubo T, Kasai Y, Nakane K, Horikoshi T (2000) Altitude differences in organic matter mass loss and fungal biomass in a subalpine coniferous forest, Mr Fuji, Japan. Arctic Antarctic Alp Res 32:262–269

Vähätalo AV, Järvinen M (2007) Photochemically produced bioavailable nitrogen from biologically recalcitrant dissolved organic matter stimulates production of a nitrogen-limited microbial food web in the Baltic Sea. Limnol Oceanogr 52:132–143

Vähätalo AV, Salkinoja-Saonen M, Taalas P, Salnen K (2000) Spectrum of the quantum yield for photochemical mineralization of dissolved organic carbon in a humic lake. Limnol Oceanogr 45:664–676

Vairavamurthy A, Wang S (2002) Organic nitrogen in geomacromolecules: insights on speciation and transformation with K-edge XANES spectroscopy. Environ Sci Technol 36:3050–3056

Van der Woerd HJ, Blauw A, Peperzak L, Pasterkamp R, Peters S (2011) Analysis of the spatial evolution of the 2003 algal bloom in the Voordelta (North Sea). J Sea Res 65(2):195–204

van Hees PAW, Jones DL, Godbold DL (2002) Biodegradation of low molecular weight organic acids in coniferous forest podzolic soil. Soil Biol Biochem 34:1261–1272

van Leerdam JA, Hogenboom AC, van der Kooi MME, de Voogt P (2009) Determination of polar 1H-benzotriazoles and benzothiazoles in water sample by solid-phase extraction and liquid chromatography LTQ FT Orbitrap massspectrometry. Int J Mass Spectrom 282:99–107

Vantrepotte V, Brunet C, Mériaux X, Lécuyer E, Vellucci V, Santer R (2007) Bio-optical properties of coastal waters in the Eastern English Channel. Estuar Coast Shelf Sci 72:201–212

Varela MM, Barquero S, Bode A, Fernández E, González N, Teira E, Varela M (2003) Microplanktonic regeneration of ammonium and dissolved organic nitrogen in the upwelling area of the NW of Spain: Relationships with dissolved organic carbon production and phytoplankton size-structure. J Plankton Res 25(7):719–736

Vazquez E, Amalfitano A, Fazi S, Butturini A (2011) Dissolved organic matter composition in a fragmented Mediterranean fluvial system under severe drought conditions. Biogeochemistry 102:59–72. doi:101007/s10533-010-9421-x

Veyssy E (1998) Transferts de matie`res organiques des bassins versants aux estuaries de la Gironde et de l'Adour. Ph.D. Thesis, University of Bordeaux 1, p 262

Vikesland PJ, Wigginton KR (2010) Nanomaterial enabled biosensors for pathogen monitoring—a review. Environ Sci Technol 44:3656–3669

Vine MF, Stein L, Weigle K, Schroeder J, Degnan D, Chiu-Kit JT, Hanchette C, Backer L (2000) Effects on the immune system associated with living near a pesticide dump site. Environ Health Perspect 108:1113–1124

Vione D, Falletti G, Maurino V, Minero C, Pelizzetti E, Malandrino M, Ajassa R, Olariu R-I, Arsene C (2006) Sources and sinks of hydroxyl radicals upon irradiation of natural water samples. Environ Sci Technol 40:3775–3781

Vione D, Lauri V, Minero C, Maurino V, Malandrino M, Carlotti ME, Olariu RI, Arsene C (2009) Photostability and photolability of dissolved organic matter upon irradiation of natural water samples under simulated sunlight. Aquat Sci 71:34–45

Vione D, Ponzo M, Bagnus D, Maurino V, Minero C, Carlotti ME (2010) Comparison of different probe molecules for the quantification of hydroxyl raidcals in aqueous solution. Environ Chem Lett 8:95–100

Vodacek A, Hoge FE, Swift RN, Yungel JK, Peltzer ET, Blough NV (1995) The use of in situ and airborne fluorescence measurements to determine UV absorption coefficients and DOC concentrations in surface waters. Limnol Oceanogr 40:411–415

Voelker BM, Sulzberger B (1996) Effects of fulvic acid on Fe(II) oxidation by hydrogen peroxide. Environ Sci Technol 30:1106–1114

Voelker BM, Morel FMM, Sulzberger B (1997) Iron redox cycling in surface waters: effects of humic substances and light. Environ Sci Technol 31:1004–1011

Voelker BM, Sedlak DL, Zafiriou OC (2000) Chemistry of superoxide radical in seawater: reactions with organic Cu complexes. Environ Sci Technol 34:1036–1042

Vold RD, Vold MJ (1983) Colloid and interface chemistry. Addison-Wesley, Reading

Volk CJ, Volk CB, Kaplan LA (1997) Chemical composition of biodegradable dissolved organic matter in streamwater. Limnol Oceanogr 42:39–44

Volpe V, Silvestri S, Marani M (2011) Remote sensing retrieval of suspended sediment concentration in shallow waters. Remote Sensing Environ 115:44–54

Vuorenmaa J, Forsius M, Mannio J (2006) Increasing trends of total organic carbon concentrations in small forest lakes in Finland from 1987 to 2003. Sci Total Environ 365:47–65

Wada S, Aoki MN, Tsuchiya Y, Sato T, Shinagawa H, Hama T (2007) Quantitative and qualitative analyses of dissolved organic matter released from Ecklonia cava Kjellman, in Oura Bay, Shimoda, Izu Peninsula, Japan. J Exp Mar Biol Ecol 349:344–358

Wagner M, Oehlmann J (2009) Endocrine disruptors in bottled mineral water: total estrogenic burden and migration from plastic bottles. Environ Sci Pollut Res. doi:101007/s11356-009-0107-7

Wagoner DB, Christman RF, Cauchon G, Paulson R (1997) Molar mass and size of Suwannee River natural organic matter using multiangle laser light scattering. Environ Sci Technol 31:937–941
Wahl MH, McKellar HN, Williams TM (1997) Patterns of nutrient loading in forested and urbanized coastal streams. J Exp Mar Biol Ecol 213:111–131
Waiser MJ, Robarts RD (2000) Changes in composition and reactivity of allochthonous DOM in a prairie saline lake. Limnol Oceanogr 45:763–774
Waiser MJ, Robarts RD (2004) Photodegradation of DOC in a shallow prairie wetland: evidence from seasonal changes in DOC optical properties and chemical characteristics. Biogeochemistry 69(2):263–284
Wakeham SG, Lee C, Hedges JI, Hernes PJ, Peterson ML (1997) Molecular indicators of diagenetic status in marine organic matter. Geochim Cosmochim Acta 61:5363–5369
Wang WX, Guo L (2000) Influences of natural colloids on metal bioavailability to two marine bivalves. Environ Sci Technol 34:4571–4576
Wang XC, Chen RF, Gardner GB (2004) Sources and transport of dissolved and particulate organic carbon in the Mississippi River estuary and adjacent coastal waters of the northern Gulf of Mexico. Mar Chem 89(1):241–256
Wang XC, Litz L, Chen RF, Huang W, Feng P, Altabet MA (2007) Release of dissolved organic matter during oxic and anoxic decomposition of salt marsh cordgrass. Mar Chem 105:309–321
Wang TW, Chamberlain E, Shi HL, Adams CD, Ma YF (2010) Identification of hydrolytic metabolites of dyfonate in alkaline aqueous solutions by using high performance liquid chromatography—UV detection and gas chromatography-mass spectrometry. Int J Environ Anal Chem 90:948–961
Ward SE, Bardgett RD, McNamara NP, Adamson JK, Ostle NJ (2007) Long-term consequences of grazing and burning on northern peatland carbon dynamics. Ecosystems 10:1069–1083
Wassenaar L, Hendry J, Aravena R, Fritz P, Barker J (1990) Isotopic composition (13C, 14C, 2H) and geochemistry of aquatic humic substances from groundwater. Org Geochem 15:383–396
Watts CD, Naden PS, Machall J, Banks J (2001) Long term variation in water colour from Yorkshire catchments. Sci Total Environ 278:57–72
Weiss RF, Carmack EC, Koropalov VM (1991) Deep-water renewal and biological production in Lake Baikal. Nature 349:665–669
Wells ML (2002) Marine colloids and trace metals. In: Hansell D, Carlson C (eds) Biogeochemistry of marine dissolved organic matter. Academic Press, San Diego, pp 367–397
Wells ML, Goldberg ED (1993) Colloid aggregation in seawater. Mar Chem 41:353–358
Wellsbury P, Parkes RJ (1995) Acetate bioavailability and turnover in estuarine sediment. FEMS Microbiol Ecol 17:85–94
Wershaw RL (1992) Membrane-micelle model for humus in soils and sediments and its relation to humification. US geological survey open-file report 91-513
Wershaw RL (1993) Model for humus in soils and sediments. Environ Sci Technol 27:814–816. doi:101021/es00042a603
Wershaw RL (1999) Molecular aggregation of humic substances. Soil Sci 164:803–813
Wetz MS, Wheeler PA (2007) Release of dissolved organic matter by coastal diatoms. Limnol Oceanogr 52:798–807
Wetzel RG (1983) Limnology, 2nd edn. Saunders College Publ, Philadelphia, p 860
Wetzel RG (1984) Detrital dissolved and particulate organic carbon functions in aquatic ecosystems. Bull Mar Sci 35:503–509
Wetzel RG (1990) Land-water interfaces: metabolic and limnological regulators Edgardo Baldi Memorial Lecture. Verh Int Ver Limnol 24:6–24
Wetzel RG (1992) Gradient-dominated ecosystems: sources and regulatory functions of dissolved organic matter in freshwater ecosystems. Hydrobiologia 229:181–198
Wetzel RG, Manny BA (1972) Decomposition of dissolved organic carbon and nitrogen compounds from leaves in an experimental hard-water stream. Limnol Oceanogr 17:927–931

Wetzel RG, Hatcher PG, Bianchi TS (1995) Natural photolysis by ultraviolet irradiance of recalcitrant dissolved organic matter to simple substrates for rapid bacterial metabolism. Limnol Oceanogr 40:1369–1380

Wheeler PA, Gosselin M, Sherr E, Thibault D, Kirchman DL, Benner R, Whitledge TE (1996) Active cycling of organic carbon in the central Arctic Ocean. Nature 380:697–699

Wheeler PA, Watkins JM, Hansing RL (1997) Nutrients, organic carbon and organic nitrogen in the upper water column of the Arctic Ocean: implications for the sources of dissolved organic carbon. Deep Sea Res II 44:1571–1592

White EM, Kieber DJ, Sherrard J, Miller WL, Mopper K (2010) Carbon dioxide and carbon monoxide photoproduction quantum yields in the Delaware Estuary. Mar Chem 118:11–21

Wigington PJ, DeWalle DR et al (1996) Episodic acidification of small streams in the northeastern United States: ionic controls of episodes. Ecol Appl 6:389–407

Wild C, Rasheed M, Werner U, Franke U, Johnstone R, Huettel M (2004) Degradation and mineralization of coral mucus in reef environments. Mar Ecol Prog Ser 267:159–171

Wildeboer D, Amirat L, Price RG, Abuknesha RA (2010) Rapid detection of E coli in water using a hand-held fluorescence reader. Water Res 44:2621–2628

Wilkinson KJ, Negre JC, Buffle J (1997) Coagulation of colloidal material in surface waters: the role of natural organic matter. J Contam Hydrol 26(1–4):229–243

Willey JD, Kieber RJ, Eyman MS, Avery GB (2000) Rainwater dissolved organic carbon: concentrations and global flux. Global Biogeochem Cycles 14:139–148

Willey JD, Kieber RJ, Avery GB (2006) Changing chemical composition of precipitation in Wilmington, North Carolina, USA: implications for the continental USA. Environ Sci Technol 40:5675–5680

Williams PJleB (1990) The importance of losses during microbial growth: commentary on the physiology, measurement and ecology of the release of dissolved organic material. Mar Microb Food Webs 4:175–206

Williams PJleB (1995) Seasonal accumulation of carbon-rich dissolved organic material, its scale in comparison with changes in particulate material and the consequential effect on net C/Nassimilation ratios. Mar Chem 51:17–29

Williams PM, Druffel ERM (1987) Radiocarbon in dissolved organic matter in the central North Pacific Ocean. Nature 330:246–248

Williams MR, Fisher TR, Melack JM (1997) Chemical composition and deposition of rain in the central Amazon, Brazil. Atmos Environ 31:207–217

Williamson CE, Olson OG, Lott SE, Walker ND, Engstrom DR, Hargreaves BR (2001) Ultraviolet radiation and zooplankton community structure following deglaciation in Glacier Bay, Alaska. Ecology 82:1748–1760

Wilson SA, Weber JH (1979) An EPR study of the reduction of vanadium (V) to vanadium (IV) by fulvic acids. Chem Geol 26:345–354

Winn N, Williamson CE, Abbitt R, Rose K, Renwick W, Henry M, Saros J (2009) Modeling dissolved organic carbon in subalpine and alpine lakes with GIS and remote sensing. Landscape Ecol. doi:101007/s10980-009-9359-3

Wittbrodt PR, Palmer CD (1995) Reduction of Cr(VI) in the Presence of Excess Soil Fulvic Acid. Environ Sci Technol 29:255–263

Wolf AA, Drake BG, Erickson JE, Megonigal JP (2007) An oxygenmediated positive feedback between elevated carbon dioxide and soil organic matter decomposition in a simulated anaerobic wetland. Global Change Biol 13:2036–2044

Wombacher WD, Hornbuckle KC (2009) Synthetic musk fragrances in a conventional drinking water treatment plant with lime softening. J Environ Eng ASCE 135:1192–1198

Wong CKC, Leung KM, Poon BHT, Lan CY, Wong MH (2002) Organochlorine hydrocarbons in human breast milk collected in Hong Kong and Guangzhou. Arch Environ Contam Toxicol 43:364–372

Wood SA, Holland PT, Stirling DJ, Briggs LR, Sprosen J, Ruck JG, Wear RG (2006) Survey of cyantotoxins in New Zealand water bodies between 2001 and 2004. NZ J Mar Freshw Res 40:585–597

Worrall F, Burt TP (2004) Time series analysis of long term river DOC records. Hydrol Process 18:893–911
Worrall F, Burt T (2008) The effect of severe drought on the dissolved organic carbon (DOC) concentration and flux from British rivers. J Hydrol 361:262–274
Worrall F, Burt TP, Shedden R (2003) Long terms records of riverine carbon flux. Biogeochemistry 64:165–178
Worrall F, Harriman R, Evans CR, Watts CD, Adamson J, Neal C, Tipping E, Burt T, Grieve I, Monteith D, Naden PS, Nisbet T, Reynolds B, Stevens P (2004a) Trends in dissolved organic carbon in UK rivers and lakes. Biogeochemistry 70:369–402
Worrall F, Burt TP, Adamson JK (2004b) Can climate change explain increases in DOC flux from upland peat catchments? Sci Total Environ 326:95–112
Worrall F, Burt TP, Adamson JK (2005) Fluxes of dissolved carbon dioxide and inorganic carbon from an upland peat catchment: implications for soil respiration. Biogeochemistry 73:515–539
Worrall F, Burt TP, Adamson JK (2006) Trends in drought frequency—the fate of northern peatlands. Clim Chang 76:339–359
Worrall F, Armstrong A, Adamson J (2007) The effects of burning and sheep-grazing on water table depth and soil water quality in an upland peat. J Hydrol 339:1–14
Wu H, Gao K (2009) UV radiation-stimulated activity of extracellular carbonic anhydrase in the marine diatom Skeletonema costatum Funct. Plant Biol 36:137–143
Wu FC, Tanoue E (2001) Molecular mass distribution and fl uorescence characteristics of dissolved organic ligands for copper(II) in Lake Biwa, Japan. Org Geochem 32:11–20
Wu FC, Tanoue E, Liu CQ (2003) Fluorescence and amino acid characteristics of molecular size fractions of DOM in the waters of Lake Biwa. Biogeochemistry 65:245–257
Wu FC, Mills RB, Evans RD, Dillon PJ (2004) Kinetics of metal-fulvic acid complexation using a stopped-flow technique and three-dimensional excitation emission fluorescence spectrometer. Anal Chem 76:110–113
Wu FC, Mills RB, Evans RD, Dillon PJ (2005) Photodegradation-induced changes in dissolved organic matter in acidic waters. Can J Fish Aquat Sci 62:1019–1027
Xenopoulos MA, Lodge DM, Frentress J, Kreps TA, Bridgham SD, Grossman E, Jackson CJ (2003) Regional comparisons of watershed determinants of dissolved organic carbon in temperate lakes from the Upper Great Lakes region and selected regions globally. Limnol Oceanogr 48:2321–2334
Xia K, Weesner F, Bleam WF, Bloom PR, Skyllberg UL, Helmke PA (1998) XANES studies of oxidation states of sulfur in aquatic and soil humic substances. Soil Sci Soc Am J 62:1240–1246
Xia K, Skyllberg UL, Bleam WF, Bloom PR, Nater EA, Helmke PA (1999) X-ray absorption spectroscopic evidence for the complexation of Hg(II) by reduced sulfur in soil humic substances. Environ Sci Technol 33(2):257–261
Xia XH, Yu H, Yang ZF, Huang GH (2006) Biodegradation of polycyclic aromatic hydrocarbons in the natural waters of the Yellow River: effects of high sediment content on biodegradation. Chemosphere 65:457–466
Xiao M, Wu FC (2011) Biogeochemical characteristics and environmental effects of low-molecularweight organic acids in lacustrine ecosystem. In: Goldsmith Conference, pp 2190
Xiao M, Wu FC, Liao HQ, Li W, Lee XQ, Huang RS (2009) Vertical profiles of low molecular weight organic acids in sediment porewaters of six Chinese lakes. J Hydrol 365:37–45
Xiao M, Wu FC, Zhang R, Wang L, Li X, Huang R (2011) Temporal and spatial variations of low-molecular-weight organic acids in Dianchi Lake, China. J Environ Sci 23:1249–1256
Xie HX, Zafiriou OC, Cai WJ, Zepp RG, Wang YC (2004) Photooxidation and its effects on the carboxyl content of dissolved organic matter in two coastal rivers in the Southeastern United States. Environ Sci Technol 38:4113–4119
Xu T, Song Z, Liu J, Wang C, Wei J, Chen H (2008) Organic composition in the dry season rainwater of Guangzhou, China. Environ Geochem Health 30(1):53–65
Xu Y, Luo F, Pal A, Gin KY-H, Reinhard M (2011) Occurrence of emerging organic contaminants in a tropical urban catchment in Singapore. Chemosphere 83:963–969

Xue H-B, Sigg L (1993) Free cupric ion concentration and Cu(II) speciation in a eutrophic lake. Limnol Oceanogr 38:1200–1213
Xue H-B, Kistler D, Sigg L (1995) Competition of copper and zinc for strong ligands in a eutrophic lake. Limnol Oceanogr 40:1142–1152
Yacobi YZ (2006) Temporal and vertical variation of chlorophyll a concentration, phytoplankton photosynthetic activity and light attenuation in Lake Kinneret: possibilities and limitations for simulation by remote sensing. J Plankton Res 28:725–736
Yallop AR, Clutterbuck B (2009) Land management as a factor controlling dissolved organic carbon release from upland peat soils 1: spatial variation in DOC productivity. Sci Total Environ 407:3803–3813
Yallop AR, Thacker JI, Thomas G, Stephens M, Clutterbuck B, Brewer T et al (2006) The extent and intensity of management burning in the English uplands. J Appl Ecol 43:1138–1148
Yallop AR, White SM, Clutterbuck B (2008) Evidence for a mechanism driving recent observed trends in dissolved organic carbon release from upland peat soils. Asp Appl Biol 85:127–132
Yallop AR, Clutterbuck B, Thacker J (2010) Increases in humic dissolved organic carbon export from upland peat catchments: the role of temperature, declining sulphur deposition and changes in land management. Clim Res 45:43–56
Yamashita Y, Jaffé R (2008) Characterizing the interactions between trace metals and dissolved organic matter using excitation-emission matrix and parallel factor analysis. Environ Sci Technol 42:7374–7379
Yamashita Y, Tanoue E (2003) Distribution and alteration of amino acids in bulk DOM along a transect from bay to oceanic waters. Mar Chem 82:145–160
Yamashita Y, Tanoue E (2004) In situ production of chromophoric dissolved organic matter in coastal environments. Geophys Res Lett 31:L14302. doi:101029/2004GL019734
Yamashita Y, Tanoue E (2008) Production of bio-refractory fluorescent dissolved organic matter in the ocean interior. Nature Geosci 1:579–582. doi:101038/ngeo279
Yamashita Y, Jaffé R, Maie N, Tanoue E (2008) Assessing the dynamics of dissolved organic matter (DOM) in coastal environments by excitation emission matrix fluorescence and parallel factor analysis (EEM-PARAFAC). Limnol Oceanogr 53:1900–1908
Yan ND, Keller W, Scully NM, Lean DRS, Dillon PJ (1996) Increased UV-B penetration in a lake owing to drought-induced acidification. Nature 381:141–143
Yañez M, Barbera V, Soria E, Catalan V (2009) Quantitative detection of Helicobacter pylori in water samples by real–time PCR amplification of the cag pathogenicity island gene, cage. J Appl Microbiol 107:416–424
Yoon Y, Ryu J, Oh J, Choi BG, Snyder SA (2010) Occurrence of endocrine disrupting compounds, pharmaceuticals, and personal care products in the Han River (Seoul, South Korea). Sci Total Environ 408:636–643
Yoshioka T (1997) Phytoplanktonic carbon isotope fractionation: equations accounting for CO2-concentrating mechanisms. J Plankton Res 19:1455–1476
Yoshioka T, Wada E, Hayashi H (1989) Seasonal variations of carbon and nitrogen isotope ratios of plankton and sinking particles in Lake Kizaki. Jpn J Limnol 50:313–320
Yoshioka T, Ueda S, Khodzher T, Bashenkhaeva N, Korovyakova I, Sorokovikova L, Gorbunova L (2002a) Distribution of dissolved organic carbon in Lake Baikal and its watershed. Limnology 3:159–168
Yoshioka T, Ueda S, Miyajima T, Wada E, Yoshida N, Sugimoto A, Vijarnsorn P, Boonprakub S (2002b) Biogeochemical properties of a tropical swamp forest ecosystem in southern Thailand. Limnology 3:51–59
Yoshioka T, Mostofa KMG, Konohira E, Tanoue E, Hayakawa K, Takahashi M, Ueda S, Katsuyama M, Khodzher T, Bashenkhaeva N, Korovyakova I, Sorokovikova L, Gorbunova L (2007) Distribution and characteristics of molecular size fractions of freshwater dissolved organic matter in watershed environments: its implication to degradation. Limnology 8:39–42
Yu Y, Huang Q, Cui J, Zhang K, Tang C, Peng X (2011) Determination of pharmaceuticals, steroid hormones, and endocrine-disrupting personal care products in sewage sludge by

ultra-high-performance liquid chromatography-tandem mass spectrometry. Anal Bioanal Chem 399:891–902

Yue L, Wu FC, Liu C, Li W, Fu P, Bai Y, Wang L, Yin Z, Lü Z (2006) Relationship between fluorescence characteristics and molecular weight distribution of natural dissolved organic matter in Lake Hongfeng and Lake Baihua, China. Chin Sci Bull 51:89–96

Zagarese HE, Diaz M, Pedrozo F, Ferraro M, Cravero W, Tartarotti B (2001) Photodegradation of natural organic matter exposed to fluctuating levels of solar radiation. J Photochem Photobiol B Biol 61:35–45

Zanardi-Lamardo E, Moore CA, Zika RG (2004) Seasonal variation in molecular mass and optical properties of chromophoric dissolved organic material in coastal waters of southwest Florida. Mar Chem 89:37–54

Zang X, van Heemst JDH, Dria KJ, Hatcher PG (2000) Encapsulation of protein in humic acid from a histosol as an explanation for the occurrence of organic nitrogen in soil and sediment. Org Geochem 31:679–695

Zauzig J, Stepniewski W, Horn R (1993) Oxygen concentration and redox potential gradient in unsaturated model soil aggregates. Soil Sci Soc Am J 57:908–916

Zech W, Kogel I, Zucker A, Alt H (1985) CP-MAS-'3C-NMR spektren organischer Lagen einer Tangelrendzina. Z Pflanzenemahr Bodenk 150:262–265

Zellner R, Exner M, Herrmann H (1990) Absolute OH quantum yields in the laser photolysis of nitrate, nitrite and dissolved H2O2 at 308 and 351 nm in the temperature range 278–353 K. J Atmos Chem 10:411–425

Zepp RG, Faust BC, Hoigné J (1992) Hydroxyl radical formation in aqueous reactions (pH 3–8) of iron(II) with hydrogen peroxide: the photo-Fenton reaction. Environ Sci Technol 26:313–319

Zepp RG, Callaghan TV, Erickson DJ (1998) Effects of enhanced solar ultraviolet radiation on biogeochemical cycles. J Photochem Photobiol B Biol 46:69–82

Zepp RG, Erickson DJ, Paul ND, Sulzberger B (2011) Effects of solar UV radiation and climate change on biogeochemical cycling: interactions and feedbacks. Photochem Photobiol Sci 10:261–279

Zhang J (1996) Nutrient elements in large Chinese estuaries. Cont Shelf Res 16:1023–1045

Zhang J, Yu ZG, Wang JT, Ren JL, Chen HT, Xiong H, Dong LX, Xu WY (1999) The subtropical Zhujiang (Pearl River) estuary: nutrient, trace species and their relationship to photosynthesis. Estuar Coast Shelf Sci 49:385–400

Zhang Y, Zhu L, Zeng X, Lin Y (2004) The biogeochemical cycling of phosphorus in the upper ocean of the East China Sea. Estuar Coast Shelf Sci 60:369–379

Zhang R, Wu FC, Liu CQ, Fu PQ, Li W, Wang LY, Liao HQ, Guo JY (2008) Characteristics of organic phosphorus fractions in different trophic sediments of lakes from the middle and lower reaches of Yangtze River region and Southwestern Plateau, China. Environ Pollut 152:366–372

Zhang Y, van Dijk MA, Liu M, Zhu G, Qin B (2009) The contribution of phytoplankton degradation to chromophoric dissolved organic matter (CDOM) in eutrophic shallow lakes: field and experimental evidence. Water Res 43:4685–4697

Zhang J, Hudson J, Neal R, Sereda J, Clair T, Turner M, Jeffries D, Dillon P, Molot L, Somers K, Hesslein R (2010) Long-term patterns of dissolved organic carbon in lakes across eastern Canada: evidence of a pronounced climate effect. Limnol Oceanogr 55:30–42

Zhang YL, Lee XQ, Cao F (2011) Chemical characteristics and sources of organic acids in precipitation at a semi-urban site in Southwest China. Atmos Environ 45:413–419

Zhao WH, Wang JT, Cui X, Ji NY (2006a) Research on fluorescence excitation and emission matrix spectra of dissolved organic matter in phytoplankton growth process. Chin High Technol Lett 16:425–430 (in Chinese with English abstract)

Zhao Z-Y, Gu J-D, Fan X-J, Li H-B (2006b) Molecular size distribution of dissolved organic matter in water of the Pearl River and trihalomethane formation characteristics with chlorine and chlorine dioxide treatments. J Hazard Mater 134:60–66

Zhao W, Wang J, Chen M (2009) Three-dimensional fluorescence characteristics of dissolved organic matter produced by Prorocentrum donghaiense Lu Chinese. J Oceanol Limnol 27:564–569. doi:101007/s00343-009-9141-z

Zhu S, Liu C-Q (2006) Vertical patterns of stable carbon isotope in soils and particle-size fractions of karst areas, Southwest China. Environ Geol 50:1119–1127

Zinder SH (1990) Conversion of acetic acid to methane by thermophiles. FEMS Microbiol Lett 75:125–137

Zou L, Wang XC, Callahan J, Culp RA, Chen RF, Altabet MA, Sun MY (2004) Bacterial roles in the formation of high-molecular-weight dissolved organic matter in estuarine and coastal waters: evidence from lipids and the compound-specific isotopic ratios. Limnol Oceanogr 49:297–302

Zubkov MV, Tarran GA (2008) High bacterivory by the smallest phytoplankton in the North Atlantic Ocean. Nature 455:224–226. doi:101038/nature07236

Zygoura PD, Paleologos EK, Riganakos KA, Kontominas MG (2005) Determination of diethylhexyladipate and acetyltributylcitrate in aqueous extracts after cloud point extraction coupled with microwave assisted back extraction and gas chromatographic separation. J Chromatogr A 1093:29–35

Photoinduced and Microbial Generation of Hydrogen Peroxide and Organic Peroxides in Natural Waters

Khan M. G. Mostofa, Cong-qiang Liu, Hiroshi Sakugawa, Davide Vione, Daisuke Minakata and Fengchang Wu

1 Introduction

The concentration of hydrogen peroxide (H_2O_2) in natural waters has been determined for the first time in 1925 by Harvey (Harvey 1925), who studied inshore and offshore water from the English Channel. The concentration of H_2O_2 has been determined in seawater in the 1970's (van Baalen and Marler 1966) and in some Russian freshwaters in the 1980's (Sinel'nikov 1971; Sinel'nikov and Demina 1974). In the same period the occurrence and concentration of H_2O_2 was being studied in air (Penkett et al. 1979; Lazrus et al. 1986; Sakugawa and Kaplan 1987), rain and cloud water, freshwater and coastal and open ocean waters (Cooper and Zika 1983; Draper and Crosby 1983; Helz and Kieber 1985; Lazrus et al. 1985;

K. M. G. Mostofa (✉) · C. Q. Liu
State Key Laboratory of Environmental Geochemistry, Institute of Geochemistry, Chinese Academy of Sciences, Guiyang 550002, China
e-mail: mostofa@vip.gyig.ac.cn

H. Sakugawa
Department of Environmental Dynamics and Management, Graduate School of Biosphere Science, Hiroshima University, 1-7-1, Kagamiyama, Higashi-Hiroshima 739-8521, Japan

D. Vione
Dipartimento Chim Analit, University Turin, I-10125 Turin, Italy

Centro Interdipartimentale NatRisk, I-10095 Grugliasco, (TO), Italy

D. Minakata
School of Civil and Environmental Engineering, Brook Byers Institute for Sustainable Systems, Georgia Institute of Technology, 828 West Peachtree Street, Suite 320, Atlanta, GA 30332, USA

F. C. Wu
State Environmental Protection Key Laboratory of Lake Pollution Control, Chinese Research Academy of Environmental Sciences, Chaoyang 100012, China

K. M. G. Mostofa et al. (eds.), *Photobiogeochemistry of Organic Matter*, Environmental Science and Engineering, DOI: 10.1007/978-3-642-32223-5_2,

Zika et al. 1985a, b; Moffett and Zika 1987a; Palenic and Morel 1988; Cooper and Lean 1989; Hellpointner and Gäb 1989; Johnson et al. 1989). Starting from the 1980's, organic peroxides (ROOH) have been detected in air (Sakugawa and Kaplan 1987; Lazrus et al. 1985; Hellpointner and Gäb 1989; Sauer et al. 2001), cloudwater and rain (Kelley and Reddy 1986). The ROOH concentrations have also been determined in freshwater (Mostofa 2005; Sakugawa et al. 2006; Mostofa and Sakugawa 2009) and seawater (Sakugawa et al. 2000; Gerringa et al. 2004).

Recent studies have demonstrated that natural sunlight or solar radiation is a key factor for the generation of H_2O_2 and ROOH in the atmosphere and in natural waters. Microbial processes can produce small amounts of both H_2O_2 and ROOH in living organisms (Kim and Portis 2004; Boveris et al. 2006; Grivennikova et al. 2008; Roy and Atreja 2008) as well as in the deeper water layers (i.e., under dark conditions) of river, lake and marine environments (Komissarov 2003). H_2O_2 is found to link with the occurrence of oxygenic photosynthesis in both higher plants (Komissarov 1994, 1995, 2003) and natural waters (Mostofa et al. 2009a, b). Therefore, H_2O_2 generated mostly by solar radiation and microbial processes could simultaneously be important for the occurrence of photosynthesis in terrestrial higher plants and for the production of organic matter (ca. algae, cyanobacteria, etc.) in water environments. There is evidence that the microbial processing of vascular-plant spoils in the terrestrial soil environment can produce humic substances (fulvic and humic acids), which are then released into river, lake and marine waters (Mostofa et al. 2009a). The action of sunlight on fulvic and humic acids correspondingly produces H_2O_2 that, by favoring photosynthesis in the surface layer of rivers, lakes and oceans, would induce the generation of algae and other aquatic organisms. These organisms are then able to produce autochthonous DOM via photorespiration (or photo-assimilations) and microbial respiration or processes (Mostofa et al. 2009b; Collen et al. 1995; McCarthy et al. 1997; Rosenstock and Simon 2001; Medina-Sánchez et al. 2006; Nieto-Cid et al. 2006; Zhang et al. 2009; Fu et al. 2010). The photoinduced reactions of autochthonous DOM also yield H_2O_2 in natural waters. The production of H_2O_2 would mostly depend on the amount of DOM and on solar irradiance. Global warming with the associated increase in water temperature would enhance the production of H_2O_2, simultaneously affecting both the photodegradation of DOM and the photosynthesis (Mostofa et al. 2009b). Photosynthesis in higher plants and in natural waters can be significantly increased by rain, also because of the elevated concentration of H_2O_2 and ROOH in rainwater. Therefore, the photoinduced and microbial generation of H_2O_2 is a key factor for the occurrence of many photoinduced, biological, physical and geochemical processes. Such processes include the production of hydroxyl radical and other free radical species, photosynthesis, production of chlorophyll and of autochthonous DOM, photodegradation of DOM, CDOM and FDOM, and complexation of DOM with trace elements in natural water environments. On the other hand, production of ROOH could be a marker of microbial modification of bulk organic matter and of DOM under dark conditions. A few studies have previously been conducted to examine the photoinduced and microbial production of ROOH, their chemical nature and relationships with DOM.

Despite the universal and unique functions that H_2O_2 and ROOH may play in water ecosystems, their roles on some key biogeochemical functions in natural waters have hardly been investigated. This chapter will provide a general overview on the biogeochemical functions of H_2O_2 and ROOH, their production mechanisms and the controlling factors for formation and decay, as well as their significance and impact in natural waters.

1.1 Hydrogen Peroxide and its Biogeochemical Functions

Hydrogen peroxide (H_2O_2) is a simple chemical compound (H–O–O–H) that appears like water (H–O–H) in its chemical formula, with an additional oxygen atom. Hydrogen peroxide can undergo dismutation into water and oxygen:

$$2H_2O_2 \rightarrow 2H_2O + O_2$$

H_2O_2 is a universal constituent of the hydrosphere and occurs in freshwater, seawater, mineral water, rain, dew, cloud, snow, air, and in all living organisms. H_2O_2 also finds effective application in experiments as well as in treatment processes. It acts as an useful indicator for a variety of photoinduced, biological and abiotic processes in the aquatic environment.

The various biogeochemical functions of H_2O_2 can be classified as follows: (i) H_2O_2 is the most stable reactive oxygen species (ROS) and is used as an indicator of photoinduced activity, because it is for instance photolytically generated through irradiation of various dissolved organic matter (DOM) components in natural waters (Cooper and Zika 1983; Zika et al. 1985a, b; Mostofa and Sakugawa 2009; Obernosterer et al. 2001; Fujiwara et al. 1993; Moore et al. 1993; Scully et al. 1996). (ii) H_2O_2 and its precursor superoxide ($O_2^{\bullet -}$) can be both oxidising and reducing agents and are, therefore, potentially important for a number of redox reactions in natural waters (Moffett and Zika 1987a, b; Petasne and Zika 1987; Moffett and Zafiriou 1990; Zafiriou 1990; Zepp et al. 1992; Zafiriou et al. 1998; Voelker et al. 2000; Jeong and Yoon 2005). (iii) H_2O_2 is a natural tracer of the surface-water mixing zone or of stratification processes in lake and marine environments (Johnson et al. 1989; Sikorsky and Zika 1993a, b; Sarthou et al. 1997; Scully and Vincent 1997). (iv) H_2O_2 is an indicator of the photodegradation of dissolved organic matter (DOM) and of organic pollutants in surface natural waters (Gao and Zepp 1998; Westerhoff et al. 1999; Southworth and Voelker 2003). (v) H_2O_2 is involved in oxidative stress in biota/living cells, because of its elevated reactivity by both oxidation and reduction (Berlett and Stadtman 1997; Paradies et al. 2000; Blokhina et al. 2003; Richard et al. 2007). (vi) H_2O_2 can be helpful in the identification of biological activity, in particular in coastal waters where higher biological activity with rapid decay of H_2O_2 is commonly observed compared to the open oceans (Fujiwara et al. 1993; Moffett and Zafiriou 1990; Cooper and Zepp 1990; Petasne and Zika 1997). (vii) H_2O_2 is a useful tracer of the vertical advection transport or the convective overturn, which is usually caused by nocturnal cooling in the upper ocean and can transport significant amounts of H_2O_2 to deep waters (Johnson et al. 1989; Sarthou et al. 1997; Scully and Vincent 1997; Yuan and Shiller 2001). (viii) H_2O_2 is thought to play an important role in the occurrence of photosynthesis in higher plants (Komissarov 1994, 1995; 2003) and in natural waters (Mostofa et al. 2009a, b), by which effect it can induce the production of autochthonous DOM in the aqueous environment. (ix) H_2O_2 can react with CO_2 under irradiation to produce various organic substances in aqueous solution (Lobanov et al. 2004), with a potentially significant role in biogeochemical processes in natural waters. (x) H_2O_2 plays an important role in controlling the physiology of plants, including the activity of some enzymes and the photophosphorylation and photorespiration rates; it is also responsible for fungitoxicity of the leaf surface (Lobanov et al. 2008). (xi) H_2O_2 is generated inside cells by peroxisomes and mitochondria; the formation of H_2O_2 is caused by the reduction of O_2 absorbed in intracellular fluid during the photorespiration (Komissarov 2003; Lobanov et al. 2008). (xii) H_2O_2 acts as an oxidant in the conversion of SO_2 to SO_4^{2-} in rainwater, thereby contributing to the acid rain phenomenon that is a harmful threat which damages plant tissues and contributes to forest decline worldwide (Calvert et al. 1985; Sakugawa et al. 1990, 1993). (xiii) The environmental concentration of H_2O_2 is influenced by algae, which simultaneously cause its decay and induce its photoinduced production by exposure of algal suspensions to sunlight (Zepp et al. 1987). (xiv) The photoinduced generation of H_2O_2 from algal suspensions plays a key role in the oxidation of anilines; the latter are able to decrease H_2O_2 production, possibly by consuming it on the surface of algal cells (Zepp et al. 1987; Zepp and Schlotzhauer 1983). (xv) Elevated levels of H_2O_2 induce damage and cell lysis in microorganisms (Gonzalez-Flecha and Demple 1997; Weinbauer and Suttle 1999); H_2O_2 is also implicated as a cause of mortality of fecal indicator bacteria in marine sewage fields (Mitchell and Chamberlin 1975; Clark et al. 2008). (xvi) Bioelectrochemical oxidation of wastewater organic matter can effectively produce H_2O_2 on an industrial scale, with an overall 83 % efficiency that could be useful for industrial purposes (Rozendal et al. 2009). (xvii) H_2O_2 produced from DOM may contribute approximately 1–50 % of hydroxyl radical ($HO^\bullet$), a strong oxidizing agent, which is responsible for indirect photoinduced changes in the DOM components in natural waters (Mostofa and Sakugawa 2009; Takeda et al. 2004; Nakatani et al. 2007; Page et al. 2011).

1.2 Organic Peroxides (ROOH) and Their Biogeochemical Functions

Organic peroxides (ROOH) are organic compounds containing the peroxide functional group (–O–O–), and may be considered as derivatives of hydrogen peroxide (H–O–O–H) where one or both of the hydrogen atoms have been replaced by organic radicals. Organic peroxides can commonly be denoted as ROOH, where

R can be CH_3–, CH_3–CH_2–, etc. and H can be H or R. The organic peroxides are ubiquitously distributed in air, cloud, dew, rain, mineral water, freshwater and seawater (Sakugawa and Kaplan 1987; Lazrus et al. 1985; Hellpointner and Gäb 1989; Sauer et al. 2001; Kelley and Reddy 1986; Mostofa 2005; Sakugawa et al. 2006; Mostofa and Sakugawa 2009; Sakugawa et al. 2000; Gerringa et al. 2004).

The major ROOH compounds identified in the aquatic environments are methyl hydroperoxide (CH_3OOH), hydroxymethyl hydroperoxide ($HOCH_2OOH$), ethyl hydroperoxide (CH_3CH_2OOH), 1-hydroxyethyl hydroperoxide ($CH_3CH(OH)OOH$), 2-hydroxyethyl hydroperoxide ($CH_2(OH)CH_2OOH$), 1-hydroxypropyl hydroperoxide ($CH_3CH_2CH(OH)OOH$), 2-hydroxypropyl hydroperoxide ($CH_3CH(OH)CH_2OOH$), 3-hydroxypropyl hydroperoxide ($CH_2(OH)CH_2CH_2OOH$), and bis(hydroxymethyl) peroxide ($HOCH_2OOCH_2OH$) (Hellpointner and Gäb 1989; Hewitt and Kok 1991). The concentration levels of ROOH compounds are commonly low (~<390 nM) in natural waters, and their concentrations are also low when they are generated in photoexperiments conducted on natural waters or on aqueous solutions of standard DOM components.

The various biogeochemical functions of ROOH can be categorized as follows: (i) Production of ROOH compounds would be a marker of microbial changes in bulk organic matter or DOM under dark conditions, which are usually occurring in deeper layers of lake or seawater (Sakugawa et al. 1995, 2000; Hayase and Shinozuka 1995; Mostofa et al. 2005). (ii) ROOH compounds are readily decomposed and correspondingly generated, so that they reach a steady-state concentration in natural waters. (iii) ROOH compounds might be important transformation intermediates of DOM and may be chemically converted into stable DOM components in natural waters. (iv) The photoinduced and thermal decomposition of organic peroxides generally yields organic peroxide radicals; they may combine with other organic substances to form new compounds, or can form polymeric compounds in aqueous solution (Mageli and Kolczynski 1966; Mill et al. 1980; Kieber and Blough 1990; Faust and Allen 1992). Future research is expected to further highlight the importance of ROOH in natural waters.

1.3 *Nature and Characteristics of H_2O_2 and ROOH*

In natural waters, H_2O_2 shows several characteristic properties that can be listed as follows: (i) The photoinduced generation of H_2O_2 follows a regular trend of increasing concentration with increasing irradiation time, in photoexperiments conducted under a solar simulator (Fig. 1a, b). It suggests that the formation rate is higher than the transformation one. (ii) Photogenerated H_2O_2 is gradually consumed in aqueous media in the absence of solar radiation (Fig. 2a). It suggests that H_2O_2 in aqueous solution is presumably decomposed by chemical and/or enzymatic reactions. (iii) The rate of H_2O_2 photoproduction is higher in filtered than in unfiltered natural waters samples (Fig. 2a), suggesting that particulate matter may rapidly consume H_2O_2 in aqueous solution. (iv) The photoinduced generation of

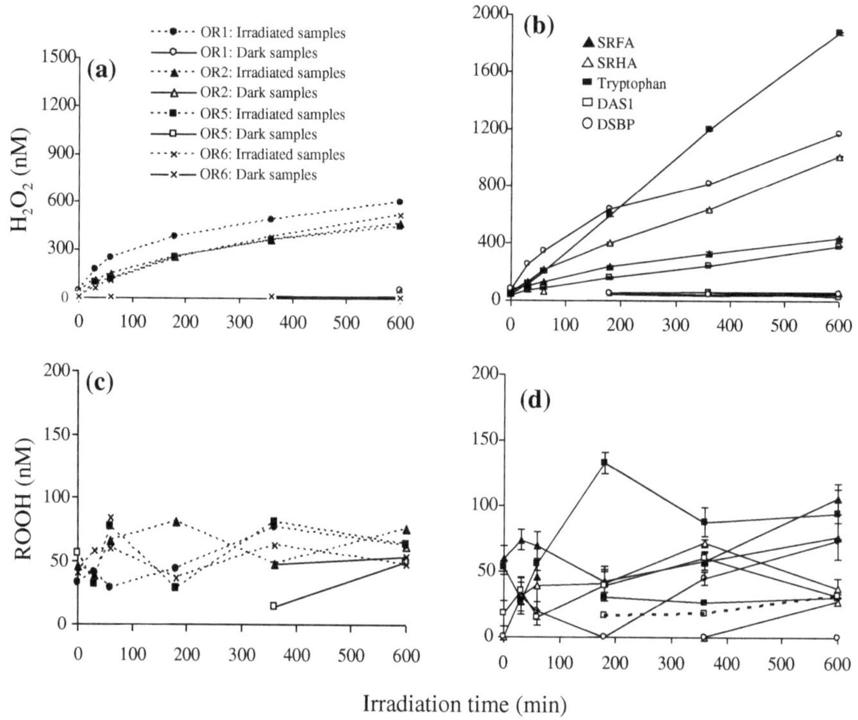

Fig. 1 Production of H_2O_2 (**a**, **b**) and ROOH (**c**, **d**) as a result of solar irradiation on the Ohta River waters (sites OR1, OR2, OR5, and OR6) and on various standard substances, respectively, in photo-experiments conducted using a solar simulator. *Data source* Mostofa and Sakugawa (2009)

H_2O_2 is highly variable in the presence of various standard organic substances in aqueous media (Fig. 1b), which suggests that the concentration of H_2O_2 depends on the nature of the DOM components. (v) The photoinduced generation of H_2O_2 increases with an increase in the contents of fulvic acid in photo-irradiated samples under a solar simulator (Fig. 3), which suggests that H_2O_2 production depends on the DOM amount. (vi) When photogenerated H_2O_2 in unfiltered river water is incubated in the dark, it is entirely decomposed in the first day of incubation and it is not produced further during the incubation period (Fig. 2b). Therefore, microbial reactions may be more effective in consuming than in producing H_2O_2 in river water.

ROOH compounds typically show the following features in natural waters: (i) The photoinduced generation of ROOH does not follow a regular trend of increasing concentration with increasing irradiation time, in photoexperiments conducted using a solar simulator; in contrast, produced ROOH is very low and fluctuates heavily without any observable trends (Fig. 1c, d). It is suggested that ROOH compounds are readily decomposed in aqueous solution. (ii) The photoinduced generation of ROOH compounds is typically higher in filtered than in

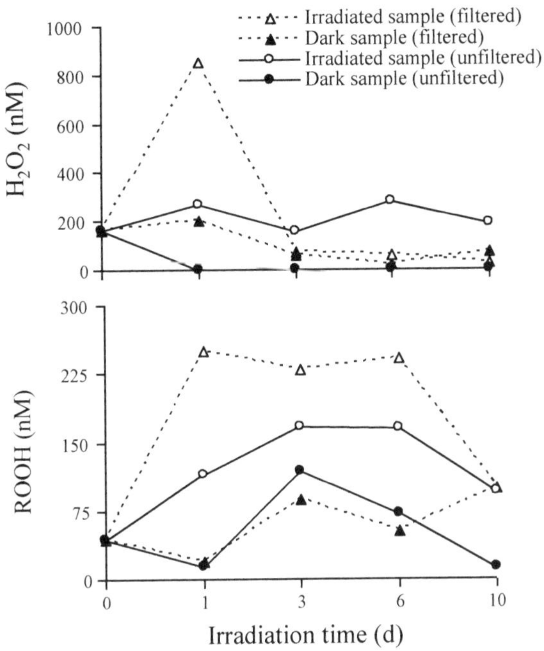

Fig. 2 Production of H_2O_2 and ROOH as a result of photoinduced and microbial incubation on filtered and unfiltered river waters. *Data source* Mostofa et al. (Manuscript in preparation)

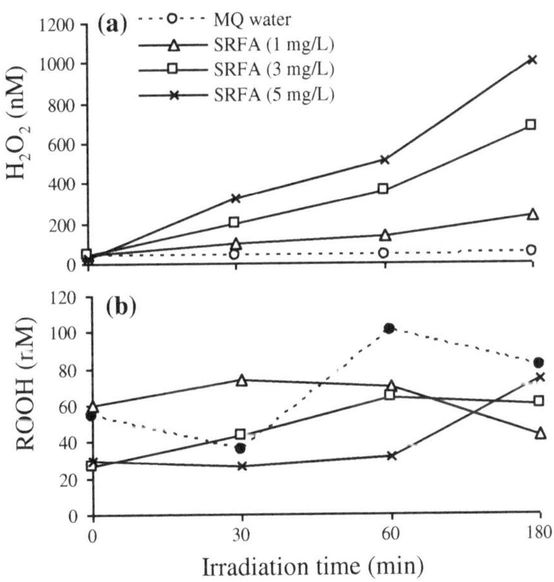

Fig. 3 Production of H_2O_2 and ROOH as a result of solar irradiation on the aqueous solutions of fulvic acid in photoexperiments conducted using a solar simulator

unfiltered samples of river water (Fig. 2b), which suggests that particulate matter (or microbes) in unfiltered river water are susceptible to rapidly degrade ROOH. (iii) ROOH compounds were frequently generated under dark incubation (which followed irradiation) in unfiltered and filtered river waters (Fig. 1b), which indicates that dark production pathways of ROOH are operational in natural waters. (iv) The photoinduced generation of ROOH compounds is typically higher for low concentration of fulvic acid (FA, 1 mg L^{-1}), and decreases with increasing FA concentration (3 and 5 mg L^{-1}, Fig. 3b). This finding suggests that the formation of ROOH compounds does not depend on DOM concentration which, on the contrary, might favor ROOH decomposition. These results indicate that ROOH compounds are quickly decomposed, which might be due to their inherently unstable chemical nature. ROOH compounds are sensitive to acid, alkali, redox and light in aqueous solution (Mostofa and Sakugawa 2009).

1.4 Steady State Concentration and Half-Life of H_2O_2 and ROOH

The concentration levels of H_2O_2 and ROOH are often measured in natural waters or in irradiated aqueous solutions, and they are often in a steady state. Steady-state concentrations of H_2O_2 and ROOH compounds in natural waters are mostly dependent on three major phenomena. First, enzymes (catalase, peroxidase and superoxide dismutase) in microbes, phytoplankton and algae present in natural waters are active agents for the rapid decay of peroxides (Mostofa 2005; Fujiwara et al. 1993; Moffett and Zafiriou 1990; Petasne and Zika 1997). These processes limit the occurrence of organic peroxides in natural waters. Second, the incident solar irradiance may be involved into the production of peroxides in waters (Cooper and Zika 1983; Moore et al. 1993; Baxter and Carey 1983; Mostofa and Sakugawa 2003). Third, the organic peroxides may take part to the generation of free radicals ($HO^•$ or $RO^•$) by direct photolysis or photo-Fenton reactions in natural waters (Zepp et al. 1992; Jeong and Yoon 2005; Southworth and Voelker 2003; Voelker et al. 1997). The free radicals then cause the photodegradation of DOM (Gao and Zepp 1998; Brezonik and Fulkerson-Brekken 1998; Goldstone et al. 2002). A general scheme for the steady-state concentration of H_2O_2 and ROOH in aqueous media can be expressed as follows (Fig. 4):

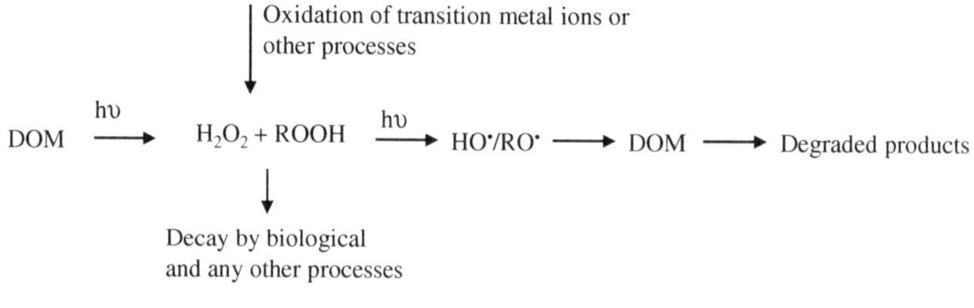

Fig. 4 A schematic diagram of steady state concentration of photoinduced generation of H_2O_2 and ROOH from DOM in natural waters

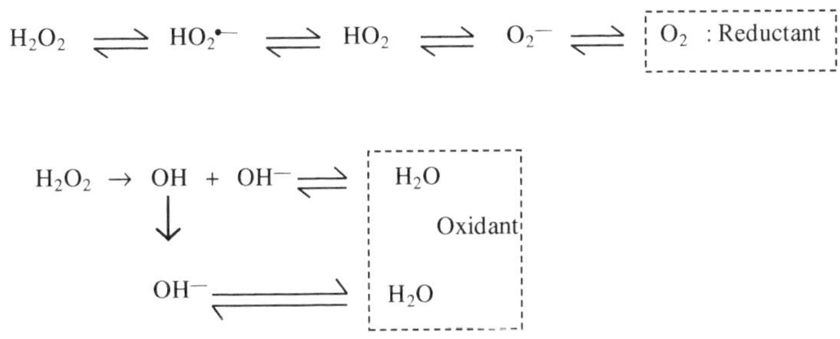

Fig. 5 Electron transfer and proton transfer reactions in the reduction of O_2 from H_2O_2 to H_2O, demonstrating the intermediates involved *Data source* Moffett and Zafiriou (1990)

More simply, "Peroxides$_{SSC}$ = produced peroxides—(decay by microbles and any other processes + contribution to DOM photo degradation)", where SSC = Steady-State Concentration. Therefore, enzymes might be an important factor in regulating the occurrence of H_2O_2 and ROOH compounds in natural waters.

The decay rates of H_2O_2 and ROOH, expressed as half-life times ($t_{1/2}$), are hours to days depending on the presence of enzymes in natural waters (Harvey 1925; Mostofa 2005; Richard et al. 2007; Cooper and Zepp 1990; Cooper and Lean 1992). For example, the half-life of H_2O_2 is gradually increased from unfiltered to filtered lake waters, from 4.4 h for unfiltered water to 4.7 h for 64 μm filtered water (zooplakton removed), 6.4 h for 12 μm filtered water (large algae removed), 19.1 h for 1.0 μm filtered water (small algae removed), and 58.7 h for 0.2 μm filtered water (bacteria removed) (Cooper and Lean 1992). Similarly, the half-lives are approximately 3 h or less for highly biologically productive coastal waters or freshwaters, and hundreds of hours for oligotrophic unfiltered waters (Mostofa 2005; Fujiwara et al. 1993; Moore et al. 1993; Richard et al. 2007).

1.5 H_2O_2 Acts as a Reductant and Oxidant-REDOX

H_2O_2 acts as a reductant and oxidant (REDOX) in many reactions occurring in natural waters (Moffett and Zika 1987a; b; Moffett and Zafiriou 1990; Zepp et al. 1992; Jeong and Yoon 2005). When H_2O_2 acts as a reductant, O from H_2O_2 is transformed into O_2. When H_2O_2 acts as an oxidant, O from H_2O_2 is converted into H_2O (Moffett and Zafiriou 1990). The chain reactions of H_2O_2 as reductant and oxidant are schematically depicted below (Fig. 5) (Moffett and Zafiriou 1990).

1.6 Concentration Levels of H_2O_2 and ROOH Compounds in Natural Water

The levels of H_2O_2 and ROOH are greatly variable for a variety of natural waters (Table 1) (van Baalen and Marler 1966; Sinel'nikov 1971; Sinel'nikov and Demina 1974; Cooper and Zika 1983; Helz and Kieber 1985; Lazrus et al. 1985;

Zika et al. 1985a, b; Moffett and Zika 1987a; Palenic and Morel 1988; Cooper and Lean 1989; Johnson et al. 1989; Sakugawa et al. 2000, 2006; Mostofa and Sakugawa 2009; Gerringa et al. 2004; Obernosterer et al. 2001; Fujiwara et al. 1993; Moore et al. 1993; Sikorsky and Zika 1993a, b; Sarthou et al. 1997; Richard et al. 2007; Petasne and Zika 1997; Lobanov et al. 2008; Sakugawa et al. 1995; Cooper and Lean 1992; Moffett and Zika 1983; Szymczak and Waite 1991; Resing et al. 1993; Miller and Kester 1994; Amouroux and Donard 1995; Fujiwara et al. 1995; Kieber and Heltz 1995; Herut et al. 1998; Cooper et al. 2000; Akane et al. 2004, 2005; Avery et al. 2005; Croot et al. 2005; Miller et al. 2005; O'Sullivan et al. 2005; Olasehinde et al. 2008; Boehm et al. 2009; Clark et al. 2010a, b; Rusak et al. 2010). H_2O_2 concentrations in surface freshwater are 6–68 nM in upstream rivers and 9–501 nM in rivers in Japan, 1300–3200 nM in rivers and 700–1300 nM in reservoirs in Russia, 88–320 nM in rivers in the USA, and 10–1300 nM in several lakes in USA and Canada (Table 1). H_2O_2 concentrations in surface seawater are 11–440 nM in estuaries in USA and Japan, 0–496 nM in coastal Bay and coastal seawaters in Japan, 25–360 nM in Amazon and Orinoco River plume, 3–1700 nM in Chesapeake Bay, 22–256 nM in Bay of Biscay (Atlantic Ocean), 124–275 nM in Biscayne Bay and Gulf Stream, <200 nM in Port Aransas seawater, <150 nM in Florida west coast, 8–50 nM in Peru upwelling area (Coastal and offshore), 8–100 nM in the Mediterranean (Israeli coastal waters) and the Red Sea (Gulf of Aqaba), 20–80 nM in Baltic Sea (German Coastal waters), 15–110 nM in Great Barrier Reef seawater (Australia), 120–280 nM in Gulf of Mexico, 50–420 nM in Caribbean Sea, 95–175 nM in Sargasso Sea and Western Mediterranean, 16–220 nM in Atlantic Ocean, and 5–25 nM in Southern Ocean in Antarctic regions (Table 1). H_2O_2 concentrations are remarkably higher in Russian rivers and reservoir (700–3200 nM) than in other rivers (6–501 nM) and lakes (10–1300 nM) in the freshwater environments. High concentrations (0–420 nM) are commonly observed in estuaries, bays and coastal seawaters, and an exceptionally high concentration (1700 nM) was detected in Cheasapeake Bay. H_2O_2 concentrations are apparently lowest in the Southern Ocean, Antactic (5–25 nM). On the other hand, the occurrence of ROOH compounds is not often studied in natural waters (Table 1). ROOH concentrations are 9–73 nM in upstreams, 0–200 nM in rivers, 32–389 nM in coastal seawaters, and 1–6 nM in the eastern Atlantic Ocean (Table 1).

1.7 Production Rates and Sources of H_2O_2

Production rates of H_2O_2 are greatly variable among upstreams (245–903 nM h^{-1}), groundwater (0–4800 nM h^{-1}), rivers (390–7400 nM h^{-1}), lakes (81–2400 nM h^{-1}), coastal waters (4536–35640 nM h^{-1}), and seawaters (0–161 nM h^{-1}) (Table 2) (Mostofa and Sakugawa 2009; Obernosterer et al. 2001; Scully et al. 1996; Richard et al. 2007; Miller and Kester 1994; Cooper et al. 1988; Moffett and Zafiriou 1993; Yocis et al. 2000; Clark et al. 2009; Mostofa KMG and Sakugawa H, unpublished;

Table 1 Hydrogen peroxide (H_2O_2) and organic peroxides (ROOH) concentrations reported in natural waters (rivers, lakes, seawaters), and rainwater

Sampling sites/Regions	Latitude/ Country	Sample type /Time/Period	H_2O_2 (nM)	ROOH	Solar intensity (MJm^{-2})	DOC (μM C)	References
Rivers							
Upstream waters (3 sites), Hiroshima Prefecture	Japan	Monthly (12 months)	6.0–68.0	9.0–73.0	0.50–2.88	43–146	Mostofa and Sakugawa (2009)
Ohta Rivers (2 sites), Hiroshima Prefecture	Japan	Monthly (12 months)	38–171	1.0–80.0	0.24–3.19	40–164	Mostofa and Sakugawa (2009)
Kurose Rivers (2 sites), Hiroshima Prefecture	Japan	Monthly (12 months)	9–213	0.0–67.0	0.48–3.13	130–383	Mostofa and Sakugawa (2009)
Upstream waters, Hiroshima Prefecture	Japan	Diel (August)	9.0–43.0	–	0–2.74	118–239	Mostofa and Sakugawa (2009)
Kurose downstream waters, Hiroshima Prefecture	Japan	Diel (September)	4.0–69.0	–	0–2.84	326–384	Mostofa and Sakugawa (2009)
Kurose Rivers (2 sites), Hiroshima Prefecture	Japan	Summer (August)	345–501	74–78	2.70	299–329	Mostofa and Sakugawa (2009)
Ohta River, Hiroshima Prefecture	Japan	December and June	66–107	33–200	–	–	Sakugawa et al. (2006)
Stream, water of Leith	New Zealand	October	15–491	–	–	–	Richard et al. (2007)
Stream, water of Leith	New Zealand	September 2003–March 2006	688 ± 8.9–72 ± 2.7	–	–	–	Rusak et al. (2010)
Ashida River, Fukuyama	Japan	December and June	91–169	80–178	–	–	Sakugawa et al. (2006)
Fuji River, Onomichi	Japan	December and June	98–301	87–125	–	–	Sakugawa et al. (2006)
Nuta River, Mihara	Japan	December and June	78–249	78–105	–	–	Sakugawa et al. (2006)
Gohno River, Miyoshi	Japan	December and June	17–101	55–69	–	–	Sakugawa et al. (2006)
Saijo River, Shoubara	Japan	December	72	65	–	–	Sakugawa et al. (2006)
Volga River	Russia	Surface water	1300–3200	–	–	–	Sinel'nikov (1971)
Chechessee River, S·C.	USA	Surface water	88	–	–	183	Cooper and Zika (1983)
Combahee River, S.C.	USA	Surface water	160	–	–	1225	Cooper and Zika (1983)

(continued)

Table 1 (continued)

Sampling sites/Regions	Latitude/ Country	Sample type /Time/Period	H_2O_2 (nM)	ROOH	Solar intensity (MJm^{-2})	DOC (μM C)	References
Newman River, Fla.	USA	Surface water	170	–	–	1392	Cooper and Zika (1983)
Peacock River, Ga.	USA	Surface water	320	–	–	1483	Cooper and Zika (1983)
Tamiami Canal, Fla.	USA	Surface water	90	–	–	1033	Cooper and Zika (1983)
Lakes and reservoir							
Jacks Lake, Ontario	Canada	Surface water	10–800	–	–	–	Cooper and Lean (1992)
Lake Erie	USA	Surface water	50–200	–	–	–	Cooper and Lean (1992)
Lake Ontario	USA	Surface water	100	–	–	–	Cooper and Lean (1992)
Jacks Lake, Ontario	44°N	Diel	10–800	–	–		Cooper and Lean (1989)
VH Pond, Miami, Fla.	USA	Surface water	140	–	–	–	Cooper and Zika (1983)
Reservoir	Russia	Surface water	700–1300	–	–	–	Sinel'nikov and Demina (1974)
Estuaries							
Patuxent Estuary	38–39°N	Diel (Feb: 14:30–10:00)	25–61	–	–	–	Kieber and Heltz (1995)
Patuxent Estuary	38–39°N	Diel (Aug: 09:30–07:30)	177–350	–	–	–	Kieber and Heltz (1995)
Patuxent Estuary	38–39°N	Diel (Sept: 11:40–09:00)	11–194	–	–	–	Kieber and Heltz (1995)
Patuxent Estuary	38–39°N	Diel (Nov: 09:00–08:30)	39–95	–	–	–	Kieber and Heltz (1995)
Estuary of Ohta River, Japan	34°N	Summer	<400	–	–	–	Fujiwara et al. (1995)
Estuary of Ohta River, Japan	34°N	Winter	60–140	–	–	–	Fujiwara et al. (1995)
Coastal seawater off Rhode Island	USA	Surface waters	60–280	–	–	–	Miller et al. (2005)
Estuarine, Chesapeake Bay	USA	Surface waters	440	–	–	–	O'Sullivan et al. (2005)
Coastal waters	USA	Surface waters	110–260	–	–	–	Cooper et al. (2000)

(continued)

Table 1 (continued)

Sampling sites/Regions	Latitude/ Country	Sample type /Time/Period	H_2O_2 (nM)	ROOH	Solar intensity (MJm^{-2})	DOC (μM C)	References
Gironde estuary, France	France	Surface waters	22–256	–	–	–	Amouroux and Donard (1995)
Sea beaches, southern California	USA	Surf zone	49–175	–	–	–	Clark et al. (2010)
Sea Beach, southern California (n = 4)	USA	Diel	25–200	–	–	–	Clark et al. (2010)
Sea beach, Santa Catalina Island	USA	Surface waters	93–329	–	–	–	Boehm et al. (2009)
Seawaters							
Hiroshima Bay, Japan	34°N	Surface water	143–348	–	–	–	Olasehinde et al. (2008)
Hiroshima Bay, Japan	34°N	Surface water (0–20 m)	39–496	–	–	78–212	Akane et al. (2004)
Hiroshima Bay, Japan	34°N	Diel and surface water (5:00–19:00)	143–448	–	–	–	Akane et al. (2004)
Hiroshima Bay, Japan	34°N	Diel and surface water (20:00–4:00)	85–259	–	–	–	Akane et al. (2004)
Hiroshima Bay, Japan	34°N	Diel and surface water (20:00–4:00)	0–195	32–389	–	100–150	Sakugawa et al. (2000)
Iyo-Nada Bay, Japan	33°N	Diel and surface water (20:00–4:00)	7–146	38–296			Sakugawa et al. (2000)
Seto Inland Sea, Japan	33°N	Diel and surface water (20:00–4:00)	40–191	–	–	–	Sakugawa et al. (1995)
Tokyo Bay	35°N	Diel and surface water	20–207	50–130	–	–	Sakugawa et al. (1995)
Sagami Bay	35°N	Diel and surface water	40–80	40–90	–	–	Sakugawa et al. (1995)
Seto Inland Sea, Japan	34°N	Surface water	60–400	–	–	–	Fujiwara et al. (1993)
Hiroshima Bay, Japan	34°N	Surface water (day time)	143–448				Fujiwara et al. (1993)
Hiroshima Bay, Japan	34°N	Surface water (night time)	85–259				Fujiwara et al. (1993)

(continued)

Table 1 (continued)

Sampling sites/Regions	Latitude/ Country	Sample type /Time/Period	H_2O_2 (nM)	ROOH	Solar intensity (MJm^{-2})	DOC (μM C)	References
Taira Bay, Japan	26°N	Surface (diel, red-soil polluted)	40–160	–	–	73–118	Arakaki et al. (2005)
Sesoko Island Bay, Japan	26°N	Surface (diel, red-soil polluted)	30–110	–	–	70–118	Arakaki et al. (2005)
Amazon plume	10°S–40°N	Surface water	25–165	–	–	–	Yuan and Shiller (2001)
Bay of Biscay, France	France	Surface waters	138–186	–	–	–	Amouroux and Donard (1995)
Grizzly Bay, California	USA	Surface waters	37	–	–	–	O'Sullivan et al. (2005)
Plume of Orinoco River	9–11°N	Surface water (fall and spring)	~75–360	–	–	–	Sikorsky and Zika (1993)
Biscayne Bay & Gulf Stream	9–11°N	Surface (late afternoon)	124–275	–	–	–	Petasne and Zika (1997)
Peru upwelling area (Coastal and offshore)	9–11°N	Surface (late afternoon)	8.0–50.0	–	–	–	Zika et al. (1985b)
Chesapeake Bay	9–11°N	Surface (late afternoon)	3–1700	–	–	–	Helz and Kieber (1985)
Marine bathing water, Southern California	USA	Surf zone waters (noon)	49–175				Clark et al. (2010a)
Marine bathing water, Southern California	USA	Diel	20–200	–	–	–	Clark et al. (2010b)
Florida west coast	25°N	Surface water (April)	<150	–	–	–	Moffett and Zika (1987a)
Port Aransas seawater	USA	Surface water	<200	–	–	–	van Baalen and Marler (1966)
Mediterian, Israeli coastal waters	32–33°N	Diel	10.0–80.0	–	–	–	Herut et al. (1998)
Red Sea, Gulf of Aqaba	29°N	Diel	8–100	–	–	–	Herut et al. (1998)
Baltic Sea, German coastal waters	54°N	Surface water	20–80	–	–	–	Herut et al. (1998)

(continued)

Table 1 (continued)

Sampling sites/Regions	Latitude/ Country	Sample type /Time/Period	H_2O_2 (nM)	ROOH	Solar intensity (MJm^{-2})	DOC (μM C)	References
Bay of Biscay, Atlantic Ocean	France	Surface water (May)	24–256	–	–	–	Amouroux and Donard (1995)
Bay of Biscay, Atlantic Ocean	France	Surface water (December)	22–69	–	–	–	Amouroux and Donard (1995)
Great Barrier Reef seawater	Australia	Surface water (December)	15–110	–	–	–	Szymczak and Waite (1991)
Gulf of Mexico	22–30°N	Surface water	150–197	–	–	–	van Baalen and Marler (1966)
Gulf of Mexico	22–30°N	Surface water	120–140	–	–	–	Moffett and Zika (1983)
Gulf of Mexico	22–30°N	Surface water	180–280	–	–	–	Zika et al. (1985)
Caribbean Sea	10°N	Surface water	95–420	–	–	–	Moore et al. (1993)
Eastern Caribbean Sea	10°N	Near surface	60–120	–	–	–	Moore et al. (1993)
Eastern Caribbean Sea	15–19°N	Surface water (fall & spring)	~75–150	–	–	–	Sikorsky and Zika (1993)
Eastern Caribbean Sea	12–15°N	Surface water (fall & spring)	~50–180	–	–	–	Sikorsky and Zika (1993)
Sargasso Sea	32°N	Surface water (fall & spring)	95–175	–	–	–	Miller and Kester (1994)
Saragasso Sea	32°N	Surface water (June)	<150	–	–	–	Palenic and Morel (1988)
Western Mediterranean	35–38°N	Surface water (May)	100–140	–	–	–	Johnson et al. (1989)
Subtropical Atlantic Ocean	12–34°N	Surface water	75–220	–	–	77–91 (50 m)	Obernosterer et al. (2001)
Subtropical Atlantic Ocean	12–34°N	Deeper water	5.0–10.0	–	–	57–71 (150 m)	Obernosterer et al. (2001)

(continued)

Table 1 (continued)

Sampling sites/Regions	Latitude/ Country	Sample type /Time/Period	H_2O_2 (nM)	ROOH	Solar intensity (MJm^{-2})	DOC (μM C)	References
South and Tropical Atlantic	10°S–40°N	Surface water	16–68	–	–	70–110	Yuan and Shiller (2001)
Atlantic Ocean	10–20°N	Surface water	37–48	–	–	–	Yuan and Shiller (2001)
Eastern Atlantic Ocean	–	Surface water	20–80	1.0–6.0	–	–	Gerringa et al. (2004)
Lagrangian, Atlantic Ocean	10°S–40°N	Diel	23–55	–	–	70–110	Yuan and Shiller (2001)
Underway, Atlantic Ocean	10°S–40°N	Diel	27–47	–	–	70–111	Yuan and Shiller (2001)
Bermuda Atlantic Time Series Station	Bermuda	Diel (6:00–14:00)	25–84	–	–	–	Avery et al. (2005)
Southern Ocean: Coasta & Cintinental Shelf Zone	61–70°N	Surface water	13–20	–	–	–	Sarthou et al. (1997)
Southern Ocean: Seasonal Ice Zone	61°N	Surface water	7.0–11.0	–	–	–	Sarthou et al. (1997)
Southern Ocean: Permanently Open Ocean Zone	55°N	Surface water	7.0–10.0	–	–	–	Sarthou et al. (1997)
Southern Ocean: Polar Front Zone	48–57°N	Surface water	~5	–	–	–	Sarthou et al. (1997)
LTER-6000, transect, Antarctic	64°N	Surface water	12.0–21.0	–	–	–	Resing et al. (1993)
Paradise Harbor, Antarctic	Antarctic	Surface water	8.5–25.0	–	–	–	[996]
Southern Ocean	48°N	Surface water (20 m)	18–25	–	–	–	Croot et al. (2005)
Rainwaters							
Rainwater, Freising/Munich	Germany	Diel (March)	2300–8600	400–1100	–	–	Hellpointner and Gäb (1989)
Rainwater, Freising/Munich	Germany	Diel (May)	9000–110600	1400–1600	–	–	Hellpointner and Gäb (1989)
Rainwater, Central Europe	Europe	Summer	500–71000	–	–	–	Sakugawa et al. (1990)
Rainwater, Central Europe	Europe	Winter	10–200	–	–	–	Sakugawa et al. (1990)

(continued)

Table 1 (continued)

Sampling sites/Regions	Latitude/ Country	Sample type /Time/Period	H_2O_2 (nM)	ROOH	Solar intensity (MJm^{-2})	DOC (μM C)	References
Rainwater, Dortmund, W. Germany	Germany	Summer	700–65000	–	–	–	Sakugawa et al. (1990)
Rainwater, Dortmund, W. Germany	Germany	Winter	0– 8500	–	–	–	Sakugawa et al. (1990)
Rainwater, The Netherlands	Netherland	–	<8200	–	–	–	Sakugawa et al. (1990)
Rainwater, Salvador area, Bahia	Brazil	March–April	17000–199000	–	–	–	Sakugawa et al. (1990)
Rainwater, Research Triangle Park, NC	USA	Summer	<5800	–	–	–	Sakugawa et al. (1990)
Rainwater, Research Triangle Park, NC	USA	Winter	60–240	–	–	–	Sakugawa et al. (1990)
Rainwater, Claremont, Los Angeles Basin	USA	–	30–46800	–	–	–	Sakugawa et al. (1990)
Rainwater, Southern Florida	USA	–	10000–70000	–	–	–	Sakugawa et al. (1990)
Rainwater, Eastern U.S.	USA	–	100–63000	–	–	–	Sakugawa et al. (1990)
Rainwater, Long Island, NY	USA	April–June	<120000	–	–	–	Sakugawa et al. (1990)
Rainwater, Summit of Whitetop Mountain, VA	USA	Spring-fall	40–39800	–	–	–	Sakugawa et al. (1990)
Rainwater, Westwood, Los Angeles Basin	USA	–	100–95000	–	–	–	Sakugawa et al. (1990)
Rainwater, Philadelphia	USA	Spring	500–5000	–	–	–	Sakugawa et al. (1990)
Rainwater, Northwestern New York state	USA	Winter	100–50000	–	–	–	Sakugawa et al. (1990)

(continued)

Table 1 (continued)

Sampling sites/Regions	Latitude/ Country	Sample type /Time/Period	H_2O_2 (nM)	ROOH	Solar intensity (MJm^{-2})	DOC (μM C)	References
Rainwater, New York	USA	No indication	2900–28800	–	–	–	Lazrus et al. (1985)
Rainwater, Niwot Ridge, Colorado	USA	10:30–17:00	16100–52500	1780–5820	–	–	Hewitt and Kok (1991)
Rainwater, Niwot Ridge, Colorado	USA	8:00–10:00	300–1300	60–80	–	–	Hewitt and Kok (1991)
Rainwater, Westwood, Los Angeles	USA	Summer	43000 (mean, n = 9)	<6500	–	<1908	Sakugawa et al. (1993)
Rainwater, Westwood, Los Angeles	USA	Winter	4300 (mean, n = 53)	–	–	17–758	Sakugawa et al. (1993)
Rainwater, Wilmington, NC	USA	Aug–Sept: 11:00–3.30	1200–11600			5–238	Miller et al. (2008)
Rainwater, North Bay, Ontario	Canada	Jan–Feb	500–5000	–	–	–	Sakugawa et al. (1990)
Rainwater, Jacks Lake, Ontario	44°N	Diurnal (no lightening)	4400–29600				Cooper and Lean (1989)
Rainwater, Jacks Lake, Ontario	44°N	One sample (lightening)	34000				Cooper and Lean (1989)
Jacks Lake	Canada	–	1300–34000	–	–	–	Cooper and Lean (1992)
Rainwater, Kanagawa	Japan	–	10300–25300	–	–	–	Sakugawa et al. (1990)
Rainwater, Tokyo	Japan	–	200–31300	–	–	–	Sakugawa et al. (1990)
Rainwater, Higashi-Hiroshima	Japan	Monthly (Jul–Jan)	39–56400	–	–	–	Sakugawa et al. (2006)
Rainwater, Mt. Gokurakuji (site 1)	Japan	Monthly (Aug–Nov)	24–1050	–	–	–	Sakugawa et al. (2006)

(continued)

Table 1 (continued)

Sampling sites/Regions	Latitude/ Country	Sample type /Time/Period	H_2O_2 (nM)	ROOH	Solar intensity (MJm^{-2})	DOC (μM C)	References
Rainwater, Mt. Gokurakuji (site 3)	Japan	Monthly (Aug–Nov)	189–10100	–	–	–	Sakugawa et al. (2006)
Rainwater, South and Central Atlantic Ocean	10–11°S	8:30–12:30	3500–9200	–	–	–	Yuan and Shiller (2000)
Rainwater, South and Central Atlantic Ocean	3.4°N	16:30–17:10	46200–49300	–	–	–	Yuan and Shiller (2000)
Rainwater, South and Central Atlantic Ocean	4.7°N	23:00– 0.30	14000–14200	–	–	–	Yuan and Shiller (2000)
Rainwater, South and Central Atlantic Ocean	6.79°N	15:30–16:15	6000–12400	–	–	–	Yuan and Shiller (2000)
Rainwater, South and Central Atlantic Ocean	8°N	14:30–23:00	46000–70900	–	–	–	Yuan and Shiller (2000)
Rainwater, Gulf of Mexico	Marine areas	–	11400–82000	–	–	–	Cooper and Lean (1992)
Rainwater, Western Atlantic	Marine areas	–	8400–20600	–	–	–	Cooper and Lean (1992)
Rainwater, Florida Keys	Marine areas	–	24300–31900	–	–	–	Cooper and Lean (1992)

Vermilyea et al. 2010).Variations in production rates of H_2O_2 are likely to be caused by the amount and the molecular nature of DOM (Table 2). This fact can be easily understood from a significant difference in the production rates of H_2O_2 estimated in the presence of various standard organic substances (Table 2). The major source of H_2O_2 in river water is fulvic acid, which contributed 23–61 % in upstream rivers, 28–63 % in polluted Kurose waters, and 67–70 % in clean Ohta river waters (Mostofa and Sakugawa 2009). Tryptophan-like substances are a minor source of H_2O_2 (~1 %) in all river waters. The contribution of the fluorescent whitening agents (DAS + DSBP) to H_2O_2 production was minor (2 %), although they were dominant FDOM components in the downstream waters of the Kurose river. The 4-biphenyl carboxaldehyde (4BCA), one photoproduct of DSBP, showed that the percent contribution to total H_2O_2 production was 2.0–5.0 % in the downstream waters of the Kurose river (Mostofa and Sakugawa 2009). Unknown sources of H_2O_2 (other than fulvic acid-like and tryptophan-like substances or FWAs) accounted for 34–68 % of H_2O_2 in the upstream waters of the Kurose, 35–67 % in the upstream areas of the Ohta, 14–15 % in the downstream sites of the Ohta, and 51–70 % in the downstream sites of the Kurose (Mostofa and Sakugawa 2009). The unknown sources of H_2O_2 may be other fluorescent and non-fluorescent substances (Kramer et al. 1996), which can originate from forest ecosystems in the upstream regions of a river and from various anthropogenic sources affecting the downstream regions. The production rate of H_2O_2 for Suwannee River Fulvic Acid (SRFA) is relatively low (344 nM h^{-1}) compared to DSBP (1073 nM h^{-1}), tryptophan (648 nM h^{-1}), and Suwannee River Humic Acid, SRHA, (644 nM h^{-1}, Table 2). However, fulvic acids may be important H_2O_2 sources due to their significant occurrence (30–80 % of total DOM) in the aquatic environments (Mostofa et al. 2009; Malcolm 1985; Peuravuori and Pihlaja 1999).

1.8 Diurnal Cycle or Diel Variation of H_2O_2 and its Controlling Factors in Natural Waters

A diurnal cycle is a regular and ubiquitous phenomenon of H_2O_2 production and decay. H_2O_2 concentration in natural waters gradually increases as incident solar radiation increases during the period from dawn to noon. The solar radiation reaches a peak at noon time and then the concentration gradually decreases with the decrease of sunlight intensity (Fig. 6). The amplitude of the H_2O_2 diurnal cycle (highest concentration at noon time minus concentration during the period before sunrise) was 35 nM in upstream and 65 nM in Kurose River (Fig. 6) (Mostofa and Sakugawa 2009), 790 nM in Jacks Lake (Cooper and Lean 1989), 36 nM (February), 173 nM (August), 183 nM (September), and 56 nM (November) in Patuxent Estuary (Kieber and Heltz 1995), 187 nM in Seto Inland Sea (Sakugawa et al. 1995), 305 nM in Hiroshima Bay (Akane et al. 2004), 120 nM in Taira Bay and 80 nM in Sesoko Island Bay (Arakaki et al. 2005), 70 nM in Mediterranean (Israeli) coastal waters, 92 nM in Red Sea in Gulf

Table 2 Production rates of H_2O_2 reported from natural waters, and standard fluorescent dissolved organic substances

Type of samples/substances	Source of light/ wavelength (nm)	Production rate of		References
		H_2O_2 (nM h^{-1})	DOC (μM C)	
Rivers				
Upstream waters (Ohta River, OR1 & OR2: Aug), Japan	Xe lamp[a]	400–768	88–101	Mostofa and Sakugawa (2009) and Mostofa KMG and Sakugawa H (unpublished)
Upstream waters (Kurose River, KR1: May), Japan	Xe lamp[a]	342	111	Mostofa and Sakugawa (2009) and Mostofa KMG and Sakugawa H (unpublished)
Upstream waters (Kurose River, KR1: Aug), Japan	Xe lamp[a]	903	152	Mostofa and Sakugawa (2009) and Mostofa KMG and Sakugawa H (unpublished)
Upstream waters (Kurose River, KR2: May), Japan	Xe lamp[a]	723	134	Mostofa and Sakugawa (2009) and Mostofa KMG and Sakugawa H (unpublished)
Upstream waters (Kurose River, KR2: Aug), Japan	Xe lamp[a]	761	106	Mostofa and Sakugawa (2009) and Mostofa KMG and Sakugawa H (unpublished)
Surface stream, Water of Leith, New Zealand	Full solar spectrum	245–444	–	Richard et al. (2007)
Ohta River (midstreams waters, OR3 & OR4: Aug), Japan	Xe lamp[a]	390–485	112–116	Mostofa and Sakugawa (2009) and Mostofa KMG and Sakugawa H (unpublished)
Ohta River (downstream waters, OR5 & OR6: Aug), Japan	Xe lamp[a]	427–468	115–124	Mostofa and Sakugawa (2009) and Mostofa KMG and Sakugawa H (unpublished)

(continued)

Table 2 (continued)

Type of samples/substances	Source of light/ wavelength (nm)	Production rate of H_2O_2 (nM h^{-1})	DOC (μM C)	References
Kurose River (Izumi): polluted site (KR5: May), Japan	Xe lamp[a]	1931	505	Mostofa and Sakugawa (2009) and Mostofa KMG and Sakugawa H (unpublished)
Kurose River (Izumi): polluted site (KR5: Aug), Japan	Xe lamp[a]	1401	310	Mostofa and Sakugawa (2009) and Mostofa KMG and Sakugawa H (unpublished)
Kurose River (Hinotsume): polluted site (KR6: May), Japan	Xe lamp[a]	1429	445	Mostofa and Sakugawa (2009) and Mostofa KMG and Sakugawa H (unpublished)
Kurose River (Hinotsume) polluted site (KR6: Aug), Japan	Xe lamp[a]	1363	276	Mostofa and Sakugawa (2009) and Mostofa KMG and Sakugawa H (unpublished)
Kurose River (Machida): downstream (KR7: May), japan	Xe lamp[a]	545	368	Mostofa and Sakugawa (2009) and Mostofa KMG and Sakugawa H (unpublished)
Kurose River (Shinkeiji): downstream (KR8: May), Japan	Xe lamp[a]	739	392	Mostofa and Sakugawa (2009) and Mostofa KMG and Sakugawa H (unpublished)
Kurose River (Shinkeiji): downstream (KR8: Aug), Japan	Xe lamp[a]	623	299	Mostofa and Sakugawa (2009) and Mostofa KMG and Sakugawa H (unpublished)
Eastern Caribbean, Orinoco River	Xe lamp[a]	33		Moffett and Zafiriou (1993)
Chechesse River, SC (USA)	Sunlight	830	183	Cooper et al. (1988)
VH Pond, Miami, FL	Sunlight	1600	575	Cooper et al. (1988)
VH Pond, Miami, FL (unfiltered)	Sunlight	1400	575	Cooper et al. (1988)
Tamiami Canal, Miami, FL	Sunlight	3800	1033	Cooper et al. (1988)
Tamiami Canal, Miami, FL (unfiltered)	Sunlight	2700	1033	Cooper et al. (1988)

(continued)

Table 2 (continued)

Type of samples/substances	Source of light/ wavelength (nm)	Production rate of		References
		H_2O_2 (nM h^{-1})	DOC (μM C)	
Combahee River, SC	Sunlight	4400	1225	Cooper et al. (1988)
Peacock River, GA	Sunlight	7400	1483	Cooper et al. (1988)
Ground water				
Tucson, Ariz	Sunlight	0	18	Cooper et al. (1988)
Spring water, Coudersport, PA	Sunlight	270	44	Cooper et al. (1988)
Spring water, Sodus, NY	Sunlight	0	78	Cooper et al. (1988)
Well 18, Miami, FL	Sunlight	590	242	Cooper et al. (1988)
Preston Well 5, Miami, FL	Sunlight	1600	517	Cooper et al. (1988)
Well 23, Miami, FL	Sunlight	1700	858	Cooper et al. (1988)
Northwest Well 5, Miami, FL	Sunlight	4800	1100	Cooper et al. (1988)
Northwest Well 1, Miami, FL	Sunlight	4600	1467	Cooper et al. (1988)
Lake				
Amituk (75°N)	Quartz Halogen lamp	81	133	Scully et al. (1996)
Small (74°N)	Quartz Halogen lamp	413	167	Scully et al. (1996)
Char (74°N)	Quartz Halogen lamp	96	117	Scully et al. (1996)
Muretta (74°N)	Quartz Halogen lamp	249	167	Scully et al. (1996)
Drinking Water (55°N)	Quartz Halogen lamp	587	858	Scully et al. (1996)
West Twin (52°N)	Quartz Halogen lamp	451	342	Scully et al. (1996)
Boulder (45°N)	Quartz Halogen lamp	925	525	Scully et al. (1996)
Bat Bog (45°N)	Quartz Halogen lamp	–	700	Scully et al. (1996)
Spruce Bog (45°N)	Quartz Halogen lamp	–	1125	Scully et al. (1996)
Wolf Howl Bog (45°N)	Quartz Halogen lamp	–	1117	Scully et al. (1996)
Cromwell (45°N)	Quartz Halogen lamp	2120	650	Scully et al. (1996)
Croche (45°N)	Quartz Halogen lamp	1222	558	Scully et al. (1996)
Deer Fen (45°N)	Quartz Halogen lamp	–	1667	Scully et al. (1996)

(continued)

Table 2 (continued)

Type of samples/substances	Source of light/ wavelength (nm)	Production rate of		References
		H_2O_2 (nM h^{-1})	DOC (μM C)	
Vernon (45°N)	Quartz Halogen lamp	1322	408	Scully et al. (1996)
Dawson Bog (44°N)	Quartz Halogen lamp	1620	633	Scully et al. (1996)
Sharpes Bay (44°N)	Quartz Halogen lamp	695	492	Scully et al. (1996)
Brookes Bay (44°N)	Quartz Halogen lamp	1315	633	Scully et al. (1996)
Bay of Quinte (44°N)	Quartz Halogen lamp	1473	633	Scully et al. (1996)
Lake Ontario 401 (43°N)	Quartz Halogen lamp	193	242	Scully et al. (1996)
Lake Ontario 403 (43°N)	Quartz Halogen lamp	161	292	Scully et al. (1996)
Lake Ontario 007 (43°N)	Quartz Halogen lamp	175	233	Scully et al. (1996)
Lake Ontario 206 (43°N)	Quartz Halogen lamp	234	225	Scully et al. (1996)
Hamilton Harbor (43°N)	Quartz Halogen lamp	790	325	Scully et al. (1996)
Newnans Lake, Gainesville, FL	Sunlight	2400	967	Cooper et al. (1988)
Coastal waters				
Source waters: San Juan Creek outlet (33°N)	Xe lamp[a]	35640	1500 < 1 kDa	Clark et al. (2009)
Source waters: Upper Newport Back Bay (33°N)	Xe lamp[a]	10260	1400 < 1 kDa	Clark et al. (2009)
Source waters: Talbert Marsh (33°N)	Xe lamp[a]	4536	600 < 1 kDa	Clark et al. (2009)
Surf zone waters: Huntington Beach pier (33°N)	Xe lamp[a]	35640	400 < 1 kDa	Clark et al. (2009)
Surf zone waters: Newport Beach pier (33°N)	Xe lamp[a]	28800	400 < 1 kDa	Clark et al. (2009)
Surf zone waters: San Clemente Beach pier (33°N)	Xe lamp[a]	22320	500 < 1 kDa	Clark et al. (2009)

(continued)

Table 2 (continued)

Type of samples/substances	Source of light/ wavelength (nm)	Production rate of H_2O_2 (nM h^{-1})	Production rate of DOC (μM C)	References
Seawater				
Seto Inland Sea (Near Coastal area: site 2)	Xe lamp[a]	161	107	Mostofa and Sakugawa (2009) and Mostofa KMG and Sakugawa H (unpublished)
Seto Inland Sea (Far from Coastal area: site 11)	Xe lamp[a]	54	89	Mostofa and Sakugawa (2009) and Mostofa KMG and Sakugawa H (unpublished)
Seto Inland Sea (Near open ocean: site 23)	Xe lamp[a]	58	99	Mostofa and Sakugawa (2009) and Mostofa KMG and Sakugawa H (unpublished)
Gulf of Alaska	Xe lamp[a]	0.5–8	–	Vermilyea et al. (2010)
Eastern Caribbean, coastal	Xe lamp[a]	17	–	Moffett and Zafiriou (1993)
Eastern Caribbean, estuarine	Xe lamp[a]	25	–	Moffett and Zafiriou (1993)
Sargasso Sea	–	4 ± 1	–	Miller and Kester (1994)
Antactic waters	Xe lamp[a]	2.1–9.6	–	Yocis et al. (2000)
Subtropical Atlantic Ocean (14°N): 5 m	Xe lamp[a]	5.5	90.6	Obernosterer et al. (2001)
10 m	Xe lamp[a]	3.6	90.6	Obernosterer et al. (2001)
20 m	Xe lamp[a]	1.3	90.6	Obernosterer et al. (2001)
30 m	Xe lamp[a]	1.7	90.6	Obernosterer et al. (2001)
40 m	Xe lamp[a]	1.0	90.6	Obernosterer et al. (2001)
50 m	Xe lamp[a]	0.0	90.6	Obernosterer et al. (2001)
Subtropical Atlantic Ocean (23°N): 5 m	Xe lamp[a]	5.2	80.5	Obernosterer et al. (2001)
10 m	Xe lamp[a]	2.1	80.5	Obernosterer et al. (2001)
20 m	Xe lamp[a]	2.5	80.5	Obernosterer et al. (2001)
30 m	Xe lamp[a]	1.3	80.5	Obernosterer et al. (2001)
40 m	Xe lamp[a]	1.5	80.5	Obernosterer et al. (2001)

(continued)

Table 2 (continued)

Type of samples/substances	Source of light/ wavelength (nm)	Production rate of		References
		H_2O_2 (nM h^{-1})	DOC (μM C)	
50 m	Xe lamp[a]	0.0	80.5	Obernosterer et al. (2001)
Subtropical Atlantic Ocean (34°N): 5 m	Xe lamp[a]	4.8	78	Obernosterer et al. (2001)
10 m	Xe lamp[a]	4.1	78	Obernosterer et al. (2001)
Subtropical Atlantic Ocean (34°N): 5 m	Xe lamp[a]	3.6	81	Obernosterer et al. (2001)
10 m	Xe lamp[a]	1.8	81	Obernosterer et al. (2001)
Standard fluorescent organic substances				
Suwannee River Fulvic Acid (SRFA)	Xe lamp[a]	344	49	Mostofa and Sakugawa 2009 and Mostofa KMG and Sakugawa H (unpublished)
Suwannee River Humic Acid (SRHA)	Xe lamp[a]	644	49	Mostofa and Sakugawa (2009) and Mostofa KMG and Sakugawa H (unpublished)
Tryptophan	Xe lamp[a]	648	69	Mostofa and Sakugawa (2009) and Mostofa KMG and Sakugawa H (unpublished)
Fluorescent whitening agents, DAS1	Xe lamp[a]	190	38	Mostofa and Sakugawa (2009) and Mostofa KMG and Sakugawa H (unpublished)
Fluorescent whitening agents, DSBP	Xe lamp[a]	1073	34	Mostofa and Sakugawa (2009) and Mostofa KMG and Sakugawa H (unpublished)
Tyrosine	Xe lamp[a]	275	33	Mostofa and Sakugawa (2009) and Mostofa KMG and Sakugawa H (unpublished)

(continued)

Table 2 (continued)

Type of samples/substances	Source of light/ wavelength (nm)	Production rate of		References
		H_2O_2 (nM h^{-1})	DOC (μM C)	
Phenylalanine	Xe lamp[a]	39	47	Mostofa and Sakugawa (2009) and Mostofa KMG and Sakugawa H (unpublished)
Phenol	Xe lamp[a]	69	21	Mostofa and Sakugawa (2009) and Mostofa KMG and Sakugawa H (unpublished)
4-Biphenyl carboxaldehyde (4BAC)	Xe lamp[a]	224	74	Mostofa and Sakugawa (2009) and Mostofa KMG and Sakugawa H (unpublished)
2-Sulfonic acid benzaldehyde (2SAB)	Xe lamp[a]	153	80	Mostofa and Sakugawa (2009) and Mostofa KMG and Sakugawa H (unpublished)

[a]production rate is normalized to sunlight intensity (noon time) at the Campus of Hiroshima University, Japan

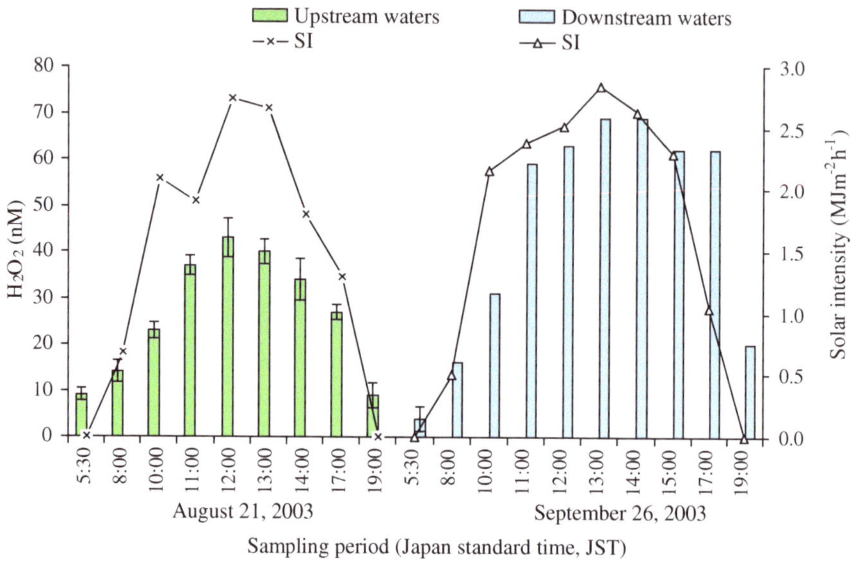

Fig. 6 Diurnal variations of H_2O_2 concentrations in the upstream waters (site KR2) on 21August 2003 and in the downstream waters (site KR6) on 26 September 2003, in the Kurose River. *Data source* Mostofa and Sakugawa (2009)

of Aqaba (Herut et al. 1998), 32 nM in Lagrangian-Atlantic Ocean, 20 nM in Underway-Atlantic Ocean (Yuan and Shiller 2001), 59 nM in Bermuda, Atlantic Time Series Station (Avery et al. 2005), 491 in a shallow freshwater stream (Richard et al. 2007), and 365 nM in marine bathing waters at Huntington State Beach (Clark et al. 2010).

The magnitude of the diurnal cycle of H_2O_2 production shows seasonal and spatial variations in natural waters, depending on several factors. First, the solar intensity varies greatly among tropical, sub-tropical, Arctic and Antarctic regions. The diurnal cycle of H_2O_2 is in fact the best paradigm for the dependence of its production on solar intensity. Second, the contents and nature of DOM components are widely different for a variety of waters and cause correspondingly variable production rates of H_2O_2. For example, H_2O_2 concentration is almost doubled in waters having high DOC concentration (326–384 μM C) than in waters with low DOC (118–239 μM C), even in the presence of similar solar irradiance (Mostofa and Sakugawa 2009). A third factor is the presence of catalase and peroxidase enzymes associated with microbes or algae. Biological processes are widely variable for a variety of natural waters and can control the steady-state concentration by rapidly decomposing H_2O_2 (Fujiwara et al. 1993; Petasne and Zika 1987; Moffett and Zafiriou 1990; Mostofa (Manuscript in preparation). Fourth, iron (Fe) can reduce the steady-state H_2O_2 concentration by producing $HO^{\bullet}$ through the photo-Fenton or other photoinduced reactions in natural waters (Moffett and Zafiriou 1990; Zepp et al. 1992; Southworth and Voelker 2003).

2 Fluorometric Method for Determining H_2O_2 and ROOH in Natural Waters

Theory: The fluorometric method described here has been developed by Fujiwara et al. (1993) and Sakugawa et al. (2000) by implementation of earlier methods (Lazrus et al. 1986; Guilbault et al. 1968; Miller and Kester 1988). The compounds H′OOH (where H′ = H or CH_3–, –OCH_3, etc.) react with p-hydroxyphenyl acetic acid in the presence of peroxidase, to produce the 6,6′-dihydroxy-3,3′-biphenyldiacetic acid (POPHA dimer: Eq. 2.1). The latter is detected using a fluorometer at excitation/emission = 320/400 nm.

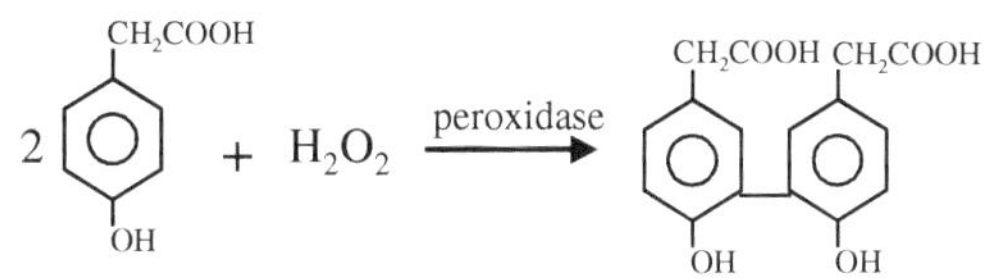

(2.1)

$$2H_2O_2 \xrightarrow{\text{catalase}} 2H_2O + O_2 \tag{2.2}$$

A sodium hydroxide solution is used to increase the pH to approximately 12, which largely enhances the fluorescence intensity of the POPHA dimer. In this way it is possible to detect few nano molar (nM) levels of H_2O_2 in natural waters. To make the analytical blanks and to distinguish H_2O_2 from ROOH, one should add catalase to the samples, which causes the rapid decomposition of H_2O_2 (Eq. 2.2).

2.1 *Chemicals Preparation*

Note: ultrapure water should be used throughout. It should be kept in the dark for 3 days before use to allow for the decomposition to undetectable levels of H_2O_2 and ROOH, which could possibly be present.

Preparation of *p*-hydroxyphenyl acetic acid solution:

(i) Take potassium hydrogen phthalate (71.48 g) in approximately 650 mL water in a 1-L beaker, and dissolve it at approximately 40 °C under gentle stirring.
(ii) Dissolve 12 g NaOH in approximately 50 mL water in a 100 mL beaker.
(iii) The pH of the solution (i) is adjusted to 5.5 upon addition of solution (ii) under constant stirring.
(iv) Add 18.62 g of di-sodium dihydrogen ethylenediamine tetraacetate dehydrate (EDTA) to the solution (iii) under constant stirring. The EDTA is added to eliminate the effect of metal ions, particularly Fe^{2+}, and to prevent the formation of a $Mg(OH)_2$ precipitate after addition of NaOH to seawater samples (Fujiwara et al. 1993). It can be noted that without EDTA, 1 mg/mL Fe^{2+} can reduce the signal intensity by 80 % (Fujiwara et al. 1993).
(v) Add 0.304 g of *p*-hydroxyphenyl acetic acid to the solution (iv) under constant stirring, then adjust the total solution to 1-L in a volumetric flask.

Preparation of the catalase solution: For 50,000 units of catalase solution, add 5 mg of catalase to 2 mL water in a 10 mL glass bottle, then mix up by shaking gently. This solution can be used for one week by keeping it in a refrigerator. For 500 units of catalase solution, add 100 μL of 50,000 units catalase solution to 10 mL water. Such a solution must be freshly prepared each time.

Preparation of peroxidase solution: Add 0.022 mg of peroxidase to 5 mL water in a 10 mL glass bottle, then mix up by shaking gently. This solution can be used for two weeks by keeping it in a refrigerator. Add 250 μL of the peroxidase solution to approximately 100 mL of *p*-hydroxyphenyl acetic acid solution.

NaOH solution: Prepare a fresh 0.6 M NaOH solution.

Preparation of standard H_2O_2 solution: Original H_2O_2 (30 %; KANTO Chemical Co., Japan) was considered as 10 M, then 1 mL of that H_2O_2 solution was used to prepare 100 mM H_2O_2. The 100 mM H_2O_2 solution was then diluted to concentrations of 0, 100, 200, 300, 500, and 1000 nM, as standards for H_2O_2 determination.

Preparation of standard ROOH solution: Original peracetic acid (9 % in diluted acetic acid; KANTO Chemical Co., Japan) was considered as 1 M, then 10 mL of that peracetic acid solution was used to prepare 100 mM ROOH. The 100 mM solution was then diluted to concentrations of 0, 100, 200, 300, 500, and 1000 nM as standards for ROOH determination.

2.2 Analytical Procedure

A flow injection apparatus should be used, of which a scheme is provided in Fig. 6 (Sakugawa et al. 2000; Fujiwara et al. 1993). The instrument shown consists of an auto sampler (TOSOH, model AS8020), fluorescence detector (Shimadzu: RF-10AXL), plunger pump (Sanuki Ind. Co., model 4P2U-4016), and recorder (Shimadzu: C-R5A Chromatopac) (Fig. 7) (Fujiwara et al. 1993).

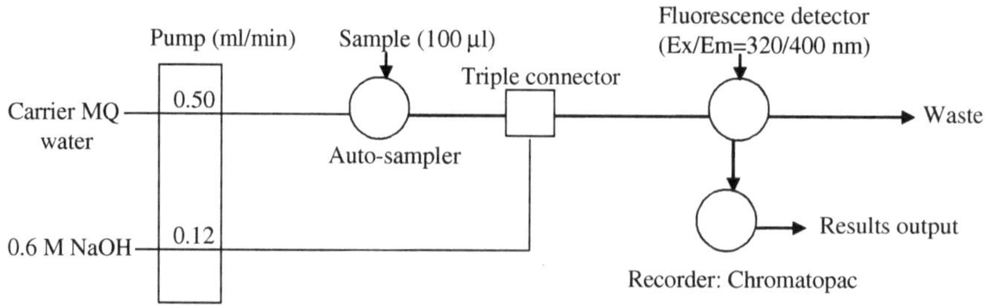

Fig. 7 Modified flow diagram for measuring H_2O_2 and ROOH concentrations in natural waters. *Data source* Fujiwara et al. (1993)

The flow lines were made of Teflon tubing (i.d. = 0.5 mm). After filling up with carrier ultrapure water and 0.6 M NaOH solution, all flow lines should be freed from air bubbles before starting. The fluorescence detector should be set at Ex/Em = 320/400 nm, and the zero level of fluorescence recorded. After completion of the baseline one should set again the fluorescence level to zero, then the analysis can be started. After completion of the measurements, before turning off the plunger pump, one should wash the flow lines. In particular, the NaOH line should be flushed with water and the outgoing flow should be checked for pH until neutrality.

In sample preparation, 1 mL sample in a Teflon or glass container is first treated with catalase (20 μL, 500 units mL^{-1}) in order to decompose all the H_2O_2 present (Eq. 2.2), shaking well for a few seconds and keeping still for six minutes. This solution can be used as a blank. Moreover, 1 mL of the same sample where catalase is replaced with 20 μL of ultrapure water is used to obtain the signal from H_2O_2. Fluorescence can be induced upon addition (300 μL) of peroxidase mixed with *p*-hydroxyphenylacetic acid. The difference in the fluorescence values (Ex/Em = 320/400 nm) between samples treated with catalase and those without the enzyme will provide the estimate of H_2O_2 concentration. Calibration can be carried out by use of the external standards already described (Fig. 8a). A typical example of calibration curves for standard H_2O_2 and peracetic acid

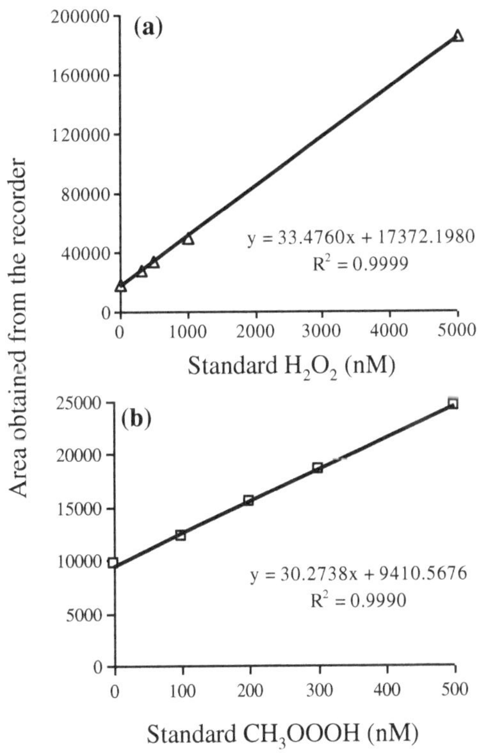

Fig. 8 A typical example of calibration curve for aqueous solutions of standards H_2O_2 (**a**) and peracetic acid (CH_3OOOH) (**b**) measured using this fluorometric method

(CH_3OOOH) is reported in earlier studies (Fig. 8). For ROOH measurement, 50,000 units mL^{-1} catalase solution was used to decompose nearly all of the ROOH in the samples during the same six minute reaction. In this way it is possible to provide only the signal of the background DOM or water fluorescence. The fluorescence-developing reagent is peroxidase mixed with *p*-hydroxyphenylacetic acid also in this case. The difference between the fluorescence measurements using 500 and 50,000 units mL^{-1} of catalase (decomposition of H_2O_2 alone and of H_2O_2 and ROOH, respectively) provides an estimate of the ROOH concentrations in the samples. Also in this case it is possible to use the external standards for calibration (Fig. 8b).

The production of H_2O_2 and ROOH in water samples is normalized as a function of natural sunlight using the following (Eq. 2.3) (Mostofa and Sakugawa 2009):

$$r_{(H_2O_2,Is)} = \frac{D_{(2-NB,Is)} \times r_{(H_2O_2,Ixe)}}{D_{(2-NB,Ixe)}} \tag{2.3}$$

where $r_{(H_2O_2,Is)}$ is the rate of H_2O_2 production, corrected for the intensity of natural sunlight (at noon under clear-sky conditions, on 6 July 2004 at Hiroshima University Campus), in natural water samples and standard DOM materials, $D_{(2\text{-}NB,Is)}$ and $D_{(2\text{-}NB,Ixe)}$ are the degradation rates of 2-NB (2-nitro-benzaldehyde) estimated using the intensity of natural sunlight and the adopted irradiation device, respectively, and $r_{(H_2O_2,Ixe)}$ is the observed H_2O_2 production rate under the adopted irradiation device.

The production rate of H_2O_2 in irradiated water samples can be determined from the net production of H_2O_2 (final concentration minus initial concentration) measured for the initial 60 min of the irradiation period. The rate of H_2O_2 generation is then normalised to sunlight intensity with (Eq. 2.3). The normalised rate of H_2O_2 production of a specific fluorescent DOM component (identified by parallel factor modeling on DOM) is estimated on the basis of its fluorescence intensity observed in waters and can be determined using (Eq. 2.4) (Mostofa and Sakugawa 2009):

$$r_{Fi(DOM)} = \frac{FI_{Fi(DOM)} \times r_{RS}}{FI_{RS}} \tag{2.4}$$

where $r_{Fi\ (DOM)}$ is the normalised production rate of H_2O_2 of an identified fluorescent DOM component in natural waters, $FI_{Fi(DOM)}$ is the fluorescence intensity of the identified fluorescent DOM component in natural waters, FI_{RS} is the fluorescence intensity of the relevant standard substance in the aqueous solution, and r_{RS} is the normalised production rate of H_2O_2 of the relevant standard substance in solution. Finally, percentages of each identified DOM component contributing to the rate of production of H_2O_2 are calculated using the following (Eq. 2.5) (Mostofa and Sakugawa 2009):

$$F_{i(DOM)} = \frac{r_{Fi(DOM)} \times 100}{r_{net(DOM)}} \tag{2.5}$$

where $F_{i(DOM)}$ is the contribution percentage to the normalised net H_2O_2 production rate in the water (%) by each identified fluorescent DOM component, $r_{Fi(DOM)}$ is the normalised H_2O_2 production rate generated by each identified DOM component, and $r_{net(DOM)}$ is the whole, normalised net H_2O_2 production rate in the water samples. The percent contributions of unknown sources of H_2O_2 in the water samples are estimated using a simple formula: $F_{unknown} = 100$— $(F_{FA} + F_{TRYP} + F_{OTHERS})$. In the formula, the sum of the normalized H_2O_2 production rate of FA-like substances (F_{FA}), tryptophan-like substances (F_{TRYP}), and other organic substances if any (F_{OTHERS}) is subtracted from the normalised, net H_2O_2 production rate that is assumed as 100 %.

2.3 *Advanced Analytical Method for H_2O_2 Determination in Natural Waters*

Theory: This method is based on the Fenton reaction, where H_2O_2 reacts with Fe^{2+} in acidic solution to yield $HO^\bullet$. The latter is scavenged by an aromatic compound (e.g. benzene) to produce the respective phenolic compound (e.g. phenol) according to the following reactions (Eqs. 2.6, 2.7) (Olasehinde et al. 2008; Lee et al. 1994; Liu et al. 2003):

$$H_2O_2 + Fe^{2+} \rightarrow Fe^{3+} + HO^\bullet + OH^- \tag{2.6}$$

$$HO^\bullet + C_6H_6 \rightarrow C_6H_5OH \tag{2.7}$$

where the rate constant of the first reaction (Eq. 2.6) is $k = 63$ at pH 3, 1.2×10^2 at pH 4 and 5.7×10^2 M^{-1} s^{-1} at pH 5, respectively (Kwan and Voelker 2002). Phenol produced by the second reaction (Eq. 2.7) is determined by high performance liquid chromatography (HPLC) with fluorescence detector (Olasehinde et al. 2008). The amount of phenol produced is directly proportional to the H_2O_2 concentration present in the sample solution.

Based on this theoretical framework, Olasehinde and his co-workers (Olasehinde et al. 2008) developed a new method for the measurement of H_2O_2 in the aqueous solution, which is highly sensitive and simpler than any other enzymatic process applied earlier to natural waters. The chemicals preparation, analytical procedure and HPLC instrumentation for this method are depicted below (Olasehinde et al. 2008):

Chemicals preparation

Benzene stock solution: 2×10^{-2} M benzene solution is prepared by adding 88.8 μL of 99.7 % benzene in 50 mL of ultrapure water.

Fe^{2+} solution: A 0.1 M Fe^{2+} solution is prepared by dissolving 1.39 g ferrous sulphate pentahydrate into 50 mL of 0.07 M H_2SO_4 solution.

H_2SO_4 solution: A 3.0 M sulphuric acid stock solution is prepared by diluting 16.3 mL of 98 % H_2SO_4 to 100 mL with ultrapure water.

H_2O_2 standard solution: A 1×10^{-2} M H_2O_2 standard stock solution is prepared by diluting 1.0 mL of 30 % H_2O_2 to 100 mL with ultrapure water. The concentration of H_2O_2 is determined based on the molar extinction coefficient at 240 nm ($\varepsilon = 38.1$ L mol^{-1} cm^{-1}) (Miller and Kester 1988).

HPLC system: An HPLC-fluorescence system is adopted. The separation is carried out on a RP-C18 column with acetonitrile–water mixture as eluent. Tentative elution conditions are (CH_3CN/H_2O 40/60 v/v) at a flow rate of 1 mL min^{-1} (note: optimal conditions may vary depending on the actual system adopted). For the detection of phenol, the fluorescence detector is operated at 270 and 298 nm for excitation and emission, respectively.

Analytical procedure: 3.0 mL of water sample (natural water or standard H_2O_2) is first treated with 200 μL of 2×10^{-2} M benzene in a 5 mL amber vial and then mixed by gently shaking. It is then added 50 μL of 0.1 M Fe^{2+} in 0.07 M H_2SO_4 solution, waiting 5 min at room temperature for completion of the Fenton reaction. The final pH of the solution should be adjusted to ca. 4 with addition of sulphuric acid solution. It can be noted that the rate constant of the Fenton reaction is much higher at pH 4 to 5 than at pH 3, thus the reaction can be conducted in these pH ranges. An aliquot of the solution (e.g. 150 μL) is then injected into the HPLC system for analysis. Phenol and benzene are separated by reverse-phase chromatography. The standard phenol and H_2O_2 concentrations might be 0, 100, 200, 300, 500 and 1000 nM, and can be prepared freshly by diluting their stock solutions. The H_2O_2 concentration is determined by calibration of the peak areas of phenol produced in each standard solution against the H_2O_2 concentration of the sample. It can be noted that the addition of 10 μM NO_2^- to the water samples shows no significant interference on the fluorescence intensity of phenol. In contrast, addition of 50 μM NO_2^- to the samples decreases the fluorescence intensity signal of phenol by almost 40 %.

3 Mechanism of Production of H_2O_2 and ROOH in Natural Waters

3.1 *Photoinduced Formation of H_2O_2 and ROOH*

H_2O_2 and ROOH are photolytically produced by several pathways in the aquatic environments. First, H_2O_2 and ROOH are photogenerated by chromophoric or fluorescent dissolved organic matter (CDOM or FDOM) in aqueous media (Cooper and Zika 1983; Mostofa and Sakugawa 2009; Moore et al. 1993; Richard et al. 2007; Baxter and Carey 1983; Clark et al. 2009; Cooper et al. 1989a, b; Dalrymple et al. 2010). A second pathway is linked with the redox cycling of transition metal ions in aqueous media (Moffett and Zika 1983; Moffett and Zika 1987a, b). An additional process is the intracellular H_2O_2 formation in chloropigments in aquatic organisms (Lobanov et al. 2008; Hong et al. 1987; Bazanov et al. 1999). Finally, various chemical reactions cause the production of H_2O_2 and ROOH in the gaseous and aqueous phases in the atmosphere.

In the gas phase, H_2O_2 and ROOH compounds are formed through several chain reactions (Eqs. 3.1–3.5) as a combined effect of solar radiation on organic substances, nitrogen oxides (NO_x), and oxygen (O_2) (Sakugawa et al. 1990, 1993; Zuo and Hoigné 1992, 1993). The relevant processes are shown below:

$$RCHO + h\nu \xrightarrow{O_2} 2RO^{\bullet} + CO \ (R = H, \text{methyl}, \text{alkyl}, \text{etc}) \tag{3.1}$$

$$NO_3^- + HCHO \rightarrow HNO_3 + HO_2^{\bullet} + CO \tag{3.2}$$

$$HO_2^{\bullet} + RO_2^{\bullet} \rightarrow H_2O_2 + O_2 \tag{3.3}$$

$$HO_2^{\bullet} + HO_2^{\bullet} \rightarrow H_2O_2 + O_2 \tag{3.4}$$

$$HO_2^{\bullet} + RO_2^{\bullet} \rightarrow ROOH + O_2 \tag{3.5}$$

In atmospheric waters the formation and decomposition of H_2O_2 is mechanistically different compared to the gas-phase reactions (Eqs. 3.6–3.9). A general scheme can be expressed as follows below (Sakugawa et al. 1990):

$$HO_2^{\bullet}(aq) + O_2 \xrightarrow{H_2O} H_2O_2(aq) + O_2 + OH^- \tag{3.6}$$

$$HSO_3^- + H_2O_2(aq) + H^+ \rightarrow SO_4^{2-} + 2H^+ + H_2O \tag{3.7}$$

$$H_2O_2(aq) + HO^{\bullet}(aq) \rightarrow H_2O + HO_2^{\bullet}(aq) \tag{3.8}$$

$$H_2O_2(aq) + h\nu \rightarrow 2HO^{\bullet}(aq) \tag{3.9}$$

H_2O_2 is also formed by photodecomposition of Fe(III) complexes with oxalic, glyoxalic and pyruvic acids, under the typical acidic conditions that can be found in atmospheric waters (Zuo and Hoigné 1992, 1993; Faust et al. 1993. A general mechanism for the formation of H_2O_2 via this route is reported below (Eqs. 3.10–3.12) (Sakugawa et al. 1990; Kim et al. 2003):

$$\text{Fe(III)-L} + h\nu \rightarrow \text{Fe(II)} + L^{\bullet} \tag{3.10}$$

$$L^{\bullet} + O_2 \rightarrow O_2^{\bullet -} + \text{oxidized L} \tag{3.11}$$

$$O_2^{\bullet -} + 2H^+ \rightarrow H_2O_2 + O_2 \tag{3.12}$$

In (Eqs. 3.10, 3.11), Fe(III)-L is a complex of Fe(III) with an organic ligand, $h\upsilon$ is the energy of a photon, and $L^{\bullet}$ is the organic radical of L. Superoxide ion ($O_2^{\bullet -}$) is a major intermediate in many O_2-mediated oxidations, such as the well-known Haber–Weiss mechanism of iron oxidation (Haber and Weiss 1934).

In natural waters, the main sources of H_2O_2 are fulvic acid (FA), humic acid, tryptophan amino acid, fluorescent whitening agents (DSBP and DAS1) and their photoproducts, as well as various unknown organic substances belonging to DOM (Mostofa and Sakugawa 2009). There is evidence that H_2O_2 may be a photoproduct of reaction chains involving dissolved organic matter (DOM) components in the presence of dissolved oxygen under natural sunlight (Eqs. 3.13–3.18) (Mostofa and Sakugawa 2009; Moore et al. 1993; Richard et al. 2007; O'Sullivan et al. 2005; Cooper et al. 1988; Clark et al. 2009; Fischer et al. 1985; Fischer et al. 1987; Power et al. 1987; Cabelli 1997). In these chain reactions, the functional groups of DOM absorb photons and are promoted to the singlet excited states ($^1DOM^*$). The latter can undergo intersystem crossing (ISC) and be converted into the triplet states ($^3DOM^*$) (Eq. 3.13). The reaction of oxygen with photo-excited DOM might generate the superoxide radical anion ($O_2^{\bullet -}$) (Eq. 3.14) in equilibrium with its conjugate acid perhydroxyl radical ($HO_2^{\bullet}$) (Eq. 3.15). Both $O_2^{\bullet -}$ and $HO_2^{\bullet}$ disproportionate to form H_2O_2 (Eqs. 3.17 and 3.18, respectively). The scheme of the reaction chain is reported below:

$$DOM + h\nu \rightarrow {}^1DOM^* \overset{ISC}{\rightarrow} {}^3DOM^* \quad (3.13)$$

$$^3DOM^* + O_2 \rightarrow DOM^{\bullet} + O_2^{\bullet -} \quad (3.14)$$

$$O_2^{\bullet -} + H^+ \rightarrow HO_2^{\bullet} \quad pK_a = 4.8 \quad (3.15)$$

$$2O_2^{\bullet -} \rightarrow O_2^{2-} + O_2 \quad pK_a = < 0.35\,M^{-1}\,s^{-1} \quad (3.16)$$

$$HO_2^{\bullet} + HO_2^{\bullet} \rightarrow H_2O_2 + O_2 \quad k = 8.6 \times 10^5\,M^{-1}\,s^{-1} \quad (3.17)$$

$$HO_2^{\bullet} + O_2^{\bullet -} + H_2O \rightarrow H_2O_2 + O_2 + OH^- \quad k = 1.0 \times 10^8\,M^{-1}\,s^{-} \quad (3.18)$$

The reaction of $HO_2^{\bullet}$ with $O_2^{\bullet -}$ (Eq. 3.28) is faster than that of $HO_2^{\bullet}$ with $HO_2^{\bullet}$ (Eq. 3.17), and the termination reaction of two $O_2^{\bullet -}$ radicals is too slow to be significant (Clark et al. 2009). The acidic constant of $HO_2^{\bullet}$ ($pK_a = 4.8$) supports the generation of the perhydroxyl radical ($HO_2^{\bullet}$) in coastal waters (Clark et al. 2009; Cabelli 1997). Therefore, the steady-state concentrations of $O_2^{\bullet -}$ and H_2O_2 (Eq. 3.18) are the result of the photoinduced activity of DOM components in sunlit surface freshwater and oceanic environments, as well as in other aqueous media (Inoue et al. 1982; Cooper et al. 1994; Millington and Maurdev 2004). $DOM^{\bullet +}$ is susceptible to further photoinduced degradation by photoinduced generation of hydroxyl radical, and the relevant pathways are depicted in the DOM degradation chapter (see chapter "Photoinduced and Microbial Degradation of Dissolved Organic Matter in Natural Waters"). It can be noted that the excitation of DOM would involve its functional groups (chromophores or fluorophores) that are the easiest to be excited. Therefore, the reactivity of DOM toward H_2O_2 production will often resemble that of simple photoactive organic molecules. Recent evidence highlights that DOM can form complexes with trace elements by a strong π–electron bonding system (Mostofa et al. 2009a, b). The metal-DOM complexes are susceptible to undergoing rapid photoinduced excitation that would finally result into the production of H_2O_2.

In studies mimicking the process of intracellular H_2O_2 formation, it has been found that the synthetic analogues of chlorophyll, metal complexes of porphyrins and phthalocyanines, act as photoactive species that produce H_2O_2 under irradiation in aqueous solutions saturated with dioxygen (Lobanov et al. 2008;

Hong et al. 1987; Bazanov et al. 1999). The highest photoinduced activity has been reported for porphyrin and phthalocyanine complexes with metals such as Mg, Zn, Al, and Cd (Komissarov 2003; Vedeneeva et al. 2005), which can typically produce long-lived triplet excited states (lifetimes up to 1 ms) with a high quantum yield (60–90 %) (Parmon 1985). Photosynthetically produced organic matter (e.g. algae) can enhance the production of H_2O_2 by natural sunlight in aquatic ecosystems (Zepp et al. 1987). It can be hypothesized that the photoinduced and microbial assimilation of algae produce autochthonous fulvic acid and other DOM components (Mostofa et al. 2009b; Fu et al. 2010; Mostofa et al. (Manuscript In preparation), which may induce H_2O_2 photoproduction by the pathways already described for DOM.

In natural waters, ROOH compounds are formed upon photodegradation of DOM (including both CDOM and FDOM) via pathways that also induce the production of H_2O_2 (Mostofa and Sakugawa 2009; Sakugawa et al. 1990; Faust and Hoigne 1987; Perkowski et al. 2006). A generalized chain-reaction scheme for the formation of ROOH from DOM in natural waters can be depicted as follows (Eqs. 3.19–3.24):

$$H_2O_2 + h\nu \rightarrow 2HO^\bullet \tag{3.19}$$

$$DOM^{\bullet+} + HO^\bullet \rightarrow R^\bullet + H^\bullet \tag{3.20}$$

$$R^\bullet + O_2 \rightarrow RO_2{}^\bullet \tag{3.21}$$

$$RO_2{}^\bullet + O_2{}^{\bullet-} + H^+ \rightarrow ROOH + O_2 \tag{3.22}$$

$$RO_2{}^\bullet + R^\bullet \rightarrow ROOR \tag{3.23}$$

$$RO_2{}^\bullet + RO_2{}^\bullet \rightarrow ROOR + O_2 \tag{3.24}$$

First, the photodecomposition of H_2O_2 generates the hydroxyl radical, $HO^\bullet$ (Eq. 3.19), which subsequently oxidizes DOM or $DOM^{\bullet+}$ (the latter is formed by $^3DOM^*$ and O_2, see Eq. 3.20) to form the organic radical $R^\bullet$ (Eq. 3.20) (Mostofa and Sakugawa 2009). Afterwards, $R^\bullet$ reacts with O_2 to form the organo peroxide radical $RO_2{}^\bullet$ (Eq. 3.21). The reduction of $RO_2{}^\bullet$, e.g. by $O_2{}^{\bullet-}$, can form ROOH in natural waters (Eq. 3.22) whereas $O_2{}^{\bullet-}$ is formed using (Eq. 3.14). Organic radicals ($R^\bullet$ and $RO_2{}^\bullet$) can rapidly associate with one another (Eq. 3.23), and organo peroxide radicals can combine (Eq. 3.24) to terminate the chain reactions. The termination reactions (Eqs. 3.23, 3.24) are competitive with (3.21, 3.22), which leads to complicated reaction kinetics (Perkowski et al. 2006).

Oxidation–reduction of transition metal ions is an important pathway for the formation of organic peroxides in natural waters. A general mechanistic scheme for these oxidation–reduction chain reactions (Eqs. 3.25–3.27) can be expressed as follows:

First, oxidation of the metal ions (M^{n+}) forms the superoxide radical anion ($O_2{}^{\bullet-}$) (Eq. 3.25). $O_2{}^{\bullet-}$ then combines with H^+ or with an alkyl ion (R^+=H^+, positively charged alkyl group, etc.) to form an hydro-peroxide or organo-peroxide radical ($RO_2{}^\bullet$, R = H or alkyl group, Eq. 3.26). $RO_2{}^\bullet$ can then associate with H^+ or a metal ion ($M^{(n+1)+}$), to form ROOH (where R = H or an alkyl group) and a further oxidized $M^{(n+2)+}$ ion (Eq. 3.27). Therefore, formation of $O_2^{\bullet-}$ is an important step in the generation of organic peroxides in natural waters.

$$M^{n+} + O_2 \rightarrow M^{(n+1)} + O_2^{\bullet-} (M = Fe^{2+}, Cu^{+}, etc) \quad (3.25)$$

$$O_2^{\bullet-} + R^{+} \rightarrow RO_2^{\bullet} \ (R = H^{+}/CH_3^{+}, etc) \quad (3.26)$$

$$RO_2^{\bullet} + M^{(n+1)+} + H^{+} \rightarrow ROOH + M^{(n+2)+} \ (R = H^{+}/CH_3^{+}) \quad (3.27)$$

3.2 Microbial Formation of H_2O_2 and ROOH

H_2O_2 and ROOH compounds are typically produced under dark incubation by microbial activity in natural waters (Fig. 2) (Palenic and Morel 1988; Moffett and Zafiriou 1990; Vermilyea et al. 2010a, b). They are susceptible to be formed by several biological processes. Biota is thought to be the main source of dark H_2O_2 and ROOH production in natural waters (Fig. 2b) (Paradies et al. 2000; Forman and Boveris 1982). For instance, dark production of H_2O_2 in seawater is particle-dependent and the production rates are in the range of 0.8–2.4 nM h^{-1} (Moffett and Zafiriou 1990). Recent studies demonstrate the high dark production rate (29–122 nM h^{-1}) of H_2O_2 in several lake waters (Vermilyea et al. 2010). Moreover, H_2O_2 and ROOH may be formed extracellularly by marine phytoplankton or cyanobacteria (Palenic and Morel 1988; Zepp et al. 1986). Extracellular H_2O_2 can be produced under dark conditions by enzymatic reduction of oxygen at the cell surface (Palenic et al. 1987) and upon oxygen reduction by other electron transport chains. The latter include the mitochondrial reduction of oxygen followed by H_2O_2 diffusion out of the cell (Forman and Boveris 1982; Frimer et al. 1983). Also the autooxidation of organic material may produce H_2O_2 and ROOH in the aquatic environment (Stevens et al. 1973). In seawater, H_2O_2 may be produced by particle-dependent and light-independent microbial processes (Moffett and Zafiriou 1990). For example, a net H_2O_2 production (dark production minus dark consumption) of 1–3 nM h^{-1} has been observed at 40–60 m in an in situ experiment conducted in the Sargasso Sea (Palenic and Morel 1988). Finally, ROOH compounds are produced in bulk natural-water DOM by light-independent microbial processes (Fig. 2) (Sakugawa et al. 2000). For example, net ROOH production has been observed in both filtered and unfiltered river waters (2b), while H_2O_2 is merely produced in filtered river waters (Fig. 2a). ROOH compounds are typically more concentrated in deep seawaters than in surface waters (Sakugawa et al. 2000).

4 Factors Controlling the Production and Decay of H_2O_2 and ROOH in Natural Waters

Concentration levels of H_2O_2 and ROOH as well as production rates of H_2O_2 differ in a variety of natural waters (Table 1). The magnitude of the H_2O_2 production decreases from coastal waters to open oceans (Zika et al. 1985a, b; Fujiwara et al.

1993). The influence of riverine fluxes having high DOM plays an important role in the production of H_2O_2 in coastal seawaters. The lowest H_2O_2 concentration was seasonally detected in southern oceans (5–25 nM), which was 10 to 20 times lower compared to other oceanic environments (Table 1). The major factors behind the low H_2O_2 concentration in the southern ocean are thought to be: (i) Low incident solar intensity and penetration efficiency in the surface water layer (Zika et al. 1985), solar irradiance being a major factor for the photoinduced formation of H_2O_2 in natural waters. (ii) Water temperature that is normally below <5 °C in the southern ocean. (iii) Vertical mixing (Johnson et al. 1989). (iv) DOC concentration (Zika et al. 1985a, b). (v) Distinct latitude or solar zenith angle, considering that H_2O_2 photoproduction decreases with increasing apparent-noon solar zenith angle (Sikorsky and Zika 1993a, b).

Therefore, the production and decay of H_2O_2 and ROOH and their lifetimes in the aquatic environment (Table 1) generally depend upon a complex set of factors, which can be distinguished as: (1) Effects and variation of solar radiation; (2) Contents and molecular nature of DOM; (3) Presence of phytoplankton, algae and microbes; (4) Photodegradation; (5) Photosynthesis; (6) Photolytic and chemical processes; (7) Physical processes; and (8) Precipitation (e.g. rain).

4.1 Effects and Variation of Solar Radiation

Solar radiation is one of the key factors in the generation of H_2O_2 and ROOH in natural waters (Mostofa and Sakugawa 2009; Obernosterer et al. 2001; Richard et al. 2007; Rusak et al. 2010; Holm-Hansen et al. 1993). The diurnal cycle of H_2O_2, where an increase of solar radiation intensity increases the production of H_2O_2 and vice versa, is a typical example of the strong dependence between solar intensity and H_2O_2 generation (Fig. 6). It has been estimated that the production of H_2O_2 and ROOH is usually higher by several times in the summer season than in the winter one. Production of H_2O_2 is higher in summer by 55–79 % in upstream waters, 162–364 % in polluted waters, and 137–146 % in clean river waters. In the case of ROOH the summer production is higher by 116–240 %, 521–1322 %, and 244–550 %, respectively, compared to the winter one. Such effects are mostly considered to be the effect of variation in solar intensity, which is much higher in the summer season (by 84 %, 32 %, and 216 %, respectively) compared to winter during a sampling day (Fig. 9) (Mostofa and Sakugawa 2009). Therefore, an increase in solar intensity would enhance the production of H_2O_2 in aqueous solution (Mostofa and Sakugawa 2009).

The solar intensity is highly variable in different regions. In the subtropical zone, ultraviolet (UV) B radiation (280–320 nm) is stable, but it is much higher (≈ten fold) than that in the Antarctica (Holm-Hansen et al. 1993). Depletion of the stratospheric ozone layer increases the ground-level UV B radiation in the polar regions (Crutzen 1992) as well as at temperate latitudes (Stolarski et al. 1992). H_2O_2 formation is largely dependent on the radiation wavelengths (Obernosterer

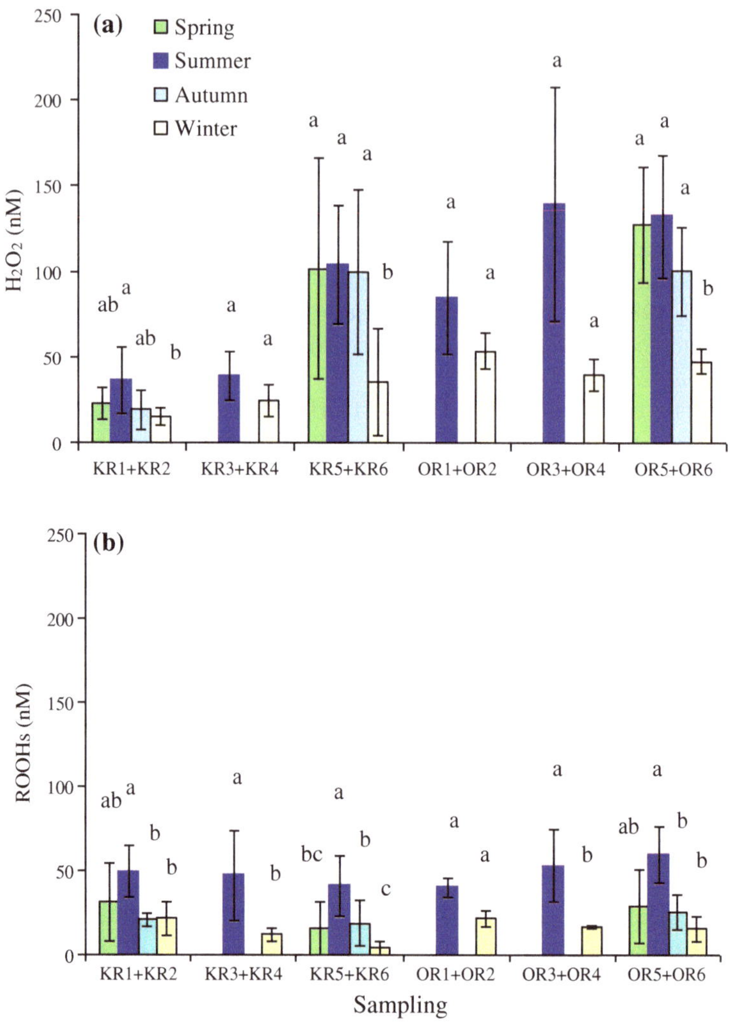

Fig. 9 Seasonal variations of the H_2O_2 (**a**) and ROOH (**b**) concentrations in the waters of the Kurose River and Ohta River in the Hiroshima prefecture, Japan. The error bar indicates the standard deviation of seasonal average value of peroxides. Mean values labelled with different letters are significantly different at $P < 0.05$ (Fisher's l.s.d. analysis). *Data source* Mostofa and Sakugawa (2009)

et al. 2001; Richard et al. 2007), and the contribution of UV-B, UV-A and photosynthetically active radiation (PAR) is 40, 33 and 27 %, respectively (Richard et al. 2007). Production of H_2O_2 at vertical depths depends on the penetration of solar radiation, and decreases with an increase in depth of lakes or oceans (Obernosterer et al. 2001). A model study on a freshwater stream shows that the H_2O_2 concentrations over time significantly depend on photoinduced production rates from ultraviolet-B (UVB), UVA and photosynthetically active radiation (PAR), and loss rates from temperature-dependent and temperature-independent processes (Rusak et al. 2010). The retrieved model terms confirm that H_2O_2 is produced by both UVB and UVA radiation. These results demonstrate that changes in solar radiation reaching the study site are closely correlated with the observed seasonal pattern in H_2O_2 concentrations in the water (Rusak et al. 2010).

4.2 Production and Decay Affected by Contents and Molecular Nature of DOM

The production and decay of H_2O_2 and ROOH in natural waters are significantly affected by the total contents and molecular nature of DOM (Mostofa and Sakugawa 2009; Scully et al. 1995). An increase in standard Suwannee River Fulvic Acid (SRFA) contents in aqueous solution increases the photoinduced production of H_2O_2, but the production of ROOH decreases with an increase in SRFA concentration. It is suggested that the photoinduced generation of H_2O_2 depends on the total contents of DOM components in natural waters. It is demonstrated that the production rates of H_2O_2 are greatly different for a variety of waters, and the production rates for various standard organic substances are also widely variable (Table 2). The photoinduced generation of H_2O_2 by natural waters and standard organic substances is generally much higher at short irradiation times (60 min), after which it often decreases. Such an effect has been observed in upstream waters as well as in aqueous solutions of Suwannee River Fulvic Acid (SRFA), Suwannee River Humic Acid (SRHA), tryptophan, DSBP and DAS1, during photo experiments carried out with a solar simulator (Fig. 1a, b) (Mostofa and Sakugawa 2009). The production of H_2O_2 and its disappearance for prolonged irradiation times has suggested two important phenomena. Firstly, H_2O_2 is initially generated as a consequence of the excitation of highly efficient functional groups of organic substances. These groups are effectively excited and transformed by solar radiation, after which the effectiveness of the functional groups to yield H_2O_2 gradually decreases. This effect, combined with consumption processes, causes a decrease of H_2O_2 concentration at the end of the long-term irradiation period. Secondly, H_2O_2 produced upon irradiation is photolytically converted to $HO^{\bullet}$, which can degrade the parent organic substances and yields a variety of photoproducts in the aqueous solution (Southworth and Voelker 2003; Kramer et al. 1996; Legrini et al. 1993; von Sonntag et al. 1993; Corin et al. 1996; Schmitt-Kopplin et al. 1998; Wang et al. 2001; Leenheer and Croué 2003). These results suggest that the photoinduced generation of H_2O_2 and ROOH depends on the molecular nature of DOM components in natural waters. The relationship between DOC concentration and production rates of H_2O_2 (Fig. 10) shows that the rate is higher for upstreams and groundwater, and increases non-linearly with an increase of DOC concentration in rivers (Fig. 10) and lakes (Scully et al. 1996). It can be considered that the highly reactive DOM is photolytically and rapidly degraded into photoproducts in

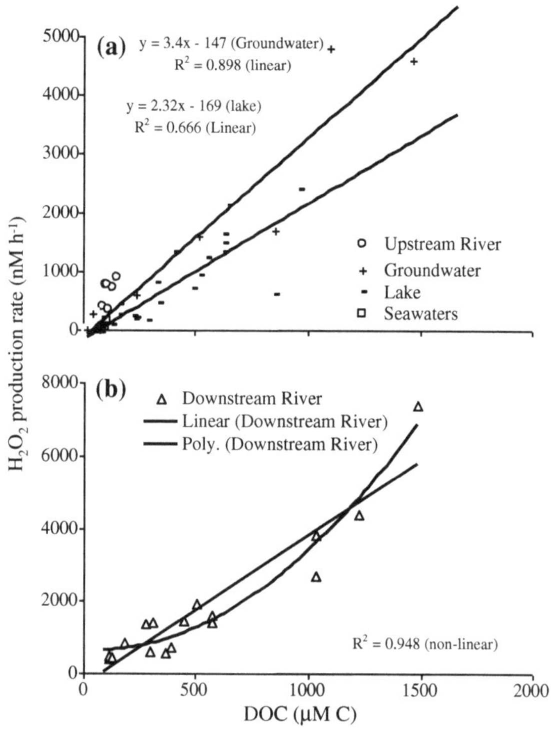

Fig. 10 Relationship between DOC concentration and production rates of H_2O_2 generated from photoexperiments conducted on upstream river, groundwater, lake and seawaters (**a**); as well as on the waters of the downstream river (**b**)

stagnant lake or seawaters or during the transportation of water from the source to the lake or ocean (Moran et al. 2000; Mostofa et al. 2005a, 2007a, b; Wu et al. 2005). Indeed, previous photoprocessing is a likely reason for the photostability of DOM sampled in surface lake environments (Vione et al. 2009). H_2O_2 production is less efficient in the presence of DOM from lake or seawater, which suggests that the generation of hydrogen peroxide depends also on the nature and not only on the total amount of DOM in natural waters. Therefore, H_2O_2 production follows the order: upstreams > groundwater > rivers > lake > coastal sea > open ocean.

Fluorescent DOM (FDOM) or chromophoric DOM (CDOM) plays an active role in the generation of H_2O_2 and ROOH in natural waters (Table 1) (Mostofa and Sakugawa 2009; Obernosterer et al. 2001; Fujiwara et al. 1993; Moore et al. 1993; O'Sullivan et al. 2005). It can be noted that CDOM or FDOM moieties undergo rather efficient photoionization under sunlight (Wu et al. 2005; Senesi 1990). For example, a significant correlation has been observed between fluorescence intensity (FI) of fulvic acid and the photoproduction of hydrated electrons (Fujiwara et al. 1993). Similarly, the production rates of H_2O_2 are highly correlated with the fluorescence of fulvic acid present in river (Mostofa 2005) and lake waters, rather than with DOC concentrations (Scully et al. 1996). Moreover, the production of H_2O_2 by a variety of river waters is highly different due to a variation in their DOM components such as fulvic acid, fluorescent whitening agents and tryptophan-like compounds (Mostofa and Sakugawa 2009). Thus, production of H_2O_2 and ROOH significantly depends on the molecular nature and composition of FDOM or CDOM rather than on DOC concentration.

4.3 Production and Decay Affected by Phytoplankton, Algae and Microbes

Production and decay of H_2O_2 and ROOH are greatly influenced by marine phytoplankton, algae and microbes. Two phenomena are involved. First, marine phytoplankton, algae and microbes may produce autochthonous DOM, which is then involved into the photoinduced or microbiological (the latter being highlighted under dark incubation) generation of H_2O_2 and ROOH compounds in natural waters. Second, the decay of H_2O_2 and ROOH compounds may be caused by catalase, peroxidase and superoxide dismutase produced by phytoplankton, algae and microbes.

A variety of marine organisms or phytoplankton can produce or excrete organic compounds such as riboflavin (Dunlap and Susic 1985; Mopper and Zika 1987), amino acids including tryptophan, proteins, carbohydrates and saturated and unsaturated fatty acids (McCarthy et al. 1997; Rosenstock and Simon 2001; Nieto-Cid et al. 2006). All of these organic compounds are photolytically reactive. For example, 1 nM riboflavin added to seawater can produce approximately 10 nM H_2O_2 (Mopper and Zika 1987), and tryptophan can produce H_2O_2 at a rate of 648 nM h^{-1} in aqueous media (Table 2). The organisms, 10^5 coccolithophorid cells L^{-1}, can produce H_2O_2 at a rate of 1–2 nM h^{-1} in oligotrophic waters (Palenic et al. 1987). Production of H_2O_2 by the eukaryotic phytoplankton species *Hymenomonas carrterae* is induced by amino acid oxidation by cell-surface enzymes (Palenic et al. 1987). The photorespiration cycle of phytoplankton involves production of H_2O_2 during glycolate oxidation (Lehninger 1970), which can be expressed as follows (Eq. 4.1):

$$CH_2OHCOOH + O_2 \xrightarrow{\text{glycolate oxidase}} CHOCOOH + H_2O_2 \tag{4.1}$$

The rate of photorespiration increases with high light intensity, possibly as a way to dissipate the excess light energy (Harris 1979), but its exact role is unclear (Ogren 1984).

The exposure of algae suspensions to sunlight can produce H_2O_2 ((Johnson et al. 1989; Collen et al. 1995; Zepp et al. 1987), possibly by photoinduced excitation of DOM released under photo- and microbial assimilation of algae (Mostofa et al. 2009b; Medina-Sánchez et al. 2006; Fu et al. 2010; Takahashi et al. 1995; Marañòn et al. 2004). This hypothesis is supported by the fact that the H_2O_2 production from algal suspensions is low in the initial two hours of irradiation, and is greatly increased with further irradiation (Zepp et al. 1987). It can be assumed that the high production of H_2O_2 after two hours occurs because of the photodegradation of organic substances newly released from algal suspensions in the reaction media during the initial irradiation period. For example, the production rates of

H_2O_2 due to sunlight effects on algae are 0.04–1.7 × 10^6 M h^{-1} for five algae at a concentration of 0.097–1.0 × 10^{-3} mg m^{-3} Chl *a* (Zepp et al. 1987).

4.3.1 Mechanism of Microbial Decomposition of H_2O_2 and ROOH

Decay of peroxides (H_2O_2 and ROOH) by phytoplankton, algae and microbes is a reverse effect of peroxide production in natural waters. Peroxides (H′OOH, H′ = H or R) may be decomposed by catalase, peroxidase and superoxide dismutase, produced by phytoplankton, algae and microbes to generate energy for their growth and to eliminate excessive intracellular levels of H_2O_2 and $O_2^{\bullet-}$ (Fujiwara et al. 1993; Moffett and Zafiriou 1990; Zepp et al. 1987; Mostofa et al. (Manuscript in preparation); Wong et al. 2003). Such a decomposition effect induced by phytoplankton, algae and microbes would usually occur constantly, until the concentration of peroxides reaches a minimum level that would afford inefficient further decomposition. Catalase enzymatically activates the peroxides (H′OOH*) to use them as oxidants (electron acceptors) and reductants (electron donors). Afterwards, disproportionation of activated H′OOH* converts them into water or alcohols and oxygen. A reaction scheme (Eqs. 4.2, 4.3) for the decomposition of peroxides by catalase can be generalized as follows (Moffett and Zafiriou 1990):

$$\mathrm{H'OOH} + \mathrm{Catalase} \rightarrow \mathrm{H'OOH}^* + \mathrm{Catalase}^\# \quad (4.2)$$

$$2\mathrm{H'OOH}^* + \mathrm{Catalase}^\# \rightarrow \mathrm{H'-O-H} + \mathrm{O_2} + \mathrm{Catalase} \quad (4.3)$$

where Catalase$^\#$ is the activated state of catalase.

Peroxidase enzymatically activates the peroxides (H′OOH*) to detoxify them to H_2O or any other end product. As reducing species it uses organic compounds (H_2R) other than H′OOH. A reaction scheme (Eqs. 4.4, 4.5) for the decomposition of peroxides is presented below (Moffett and Zafiriou 1990):

$$\mathrm{H'OOH} + \mathrm{Peroxidase} \rightarrow \mathrm{H'OOH}^* + \mathrm{Peroxidase}^\# \quad (4.4)$$

$$\mathrm{H'OOH}^* + \mathrm{H_2R} + \mathrm{Peroxidase}^\# \rightarrow \mathrm{H'-O-H} + \mathrm{H-O-H} + \mathrm{R} + \mathrm{Peroxidase} \quad (4.5)$$

where Peroxidase$^\#$ is the activated state of peroxidase. It has been shown that the percentage decay of H_2O_2 was 65–80 % by catalase and 20–35 % by peroxidase, as estimated by isotopic measurements in seawater (Moffett and Zafiriou 1990). The sources of catalase and peroxidase in natural waters are bacteria and marine phytoplankton (Kim and Zobell 1974), but these enzymes are also part of the dissolved organic matter (Serban and Nissenbaum 1986). Similarly, chloroplasts have a peroxidase-mediated H_2O_2 scavenging system (Tanaka et al. 1985). Natural marine peroxidases are also capable of catalyzing H_2O_2-mediated halogenation reactions in the oceanic environments (Theiler et al. 1978; Baden and Corbett 1980). The decay of H_2O_2 is usually low (12 % after 5 h incubation) in upstream waters due to the presence of few bacteria (some 10^5 cells mL^{-1}), and much higher in polluted

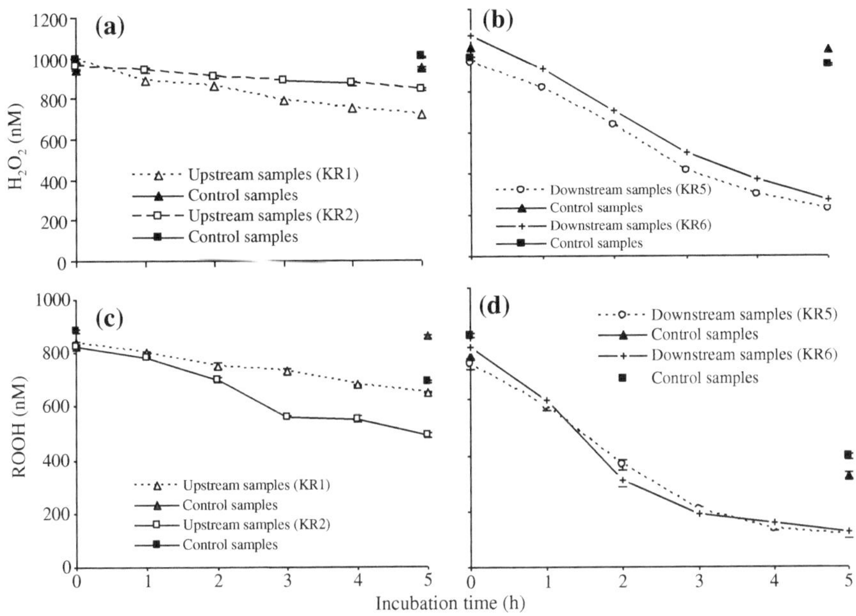

Fig. 11 The decay of peroxides by the occurrence of bacterial incidences in upstream and polluted river waters with an addition of standard 1,000 nM of H_2O_2 (**a**) and 1,000 nM of peracetic acid (**b**) under dark incubation in NK system BIOTRON at controlled temperature (21 °C). Controlled or sterilized samples (addition of 2 % solution of $HgCl_2$) conducted under the same condition and same incubation period. *Data source* Mostofa et al. (Manuscript in preparation)

rivers (74 %) where the bacteria are more numerous (of order 10^6 cells mL^{-1}) (Fig. 11a). Similarly, the decay of peracetic acid (ROOH) was 40 % and 85 %, respectively (Fig. 11b). The initial degradation rate is roughly double for ROOH (peracetic acid) than for H_2O_2, thus the concentrations of ROOH found in rivers are generally lower than those of H_2O_2. It is suggested that ROOH compounds are chemically unstable and more reactive than H_2O_2 in natural waters (Mostofa and Sakugawa 2009). Therefore, enzymatic or microbial degradation of peroxides is a rapid process that may control the steady-state concentrations of both H_2O_2 and ROOH compounds in natural waters (Fujiwara et al. 1993; Cooper and Zepp 1990; Zepp et al. 1987; Serban and Nissenbaum 1986; Tanaka et al. 1985).

It has been shown that the algal-catalyzed decomposition of H_2O_2 under dark conditions is second-order overall, first-order with respect to H_2O_2 and first-order with respect to the algal biomass (Petasne and Zika 1997; Zepp et al. 1987; Cooper and Lean 1992). The median second-order rate constant for nine algae is approximately 4×10^{-3} m^3 (mg Chl a)$^{-1}$ h^{-1}. Natural levels of the blue-green *Cyanobacterium sp.* can greatly increase the decay rates of H_2O_2, which follow a second-order rate constant of 3.5×10^{-10} L $cell^{-1}$ h^{-1} (Petasne and Zika 1997). Similar kinetics has been observed for *Vibrio alginolyticus*, in which case the decay of H_2O_2 was second-order overall, and first-order in both H_2O_2 concentration and number of bacterial cells (Cooper and Lean 1992). Such a kinetic can be expressed as follows:

$$\text{Rate} = -d\,[H_2O_2]\,/dt = k_2\,[H_2O_2]\;[\text{Number of bacterial cells}] \quad (4.6)$$

where $k_2 = 1.6 \times 10^{-9}$ mL cell^{-1}min^{-1}. The freshwater bacterium *Enterobacter cloaceae* showed a similar rate constant, $k_2 = 1.5 \times 10^{-9}$ mL cell^{-1} min^{-1}.

4.4 Production and Decay by DOM Photochemistry

Photodegradation of DOM depends on the incident light intensity, which is directly linked to the production of H_2O_2 and ROOH through photoinduced reactions in natural waters (Cooper and Zika 1983; Moore et al. 1993; Baxter and Carey 1983). For example, H_2O_2 concentration gradually increases with irradiation time in natural waters as well as in aqueous solutions of standard organic substances (Fig. 3) (Obernosterer et al. 2001; Cooper et al. 1988). Similarly, a 10–20 times lower H_2O_2 production was observed in river waters during the cold season compared to summer, and in the Southern Ocean in Antarctic regions (5–25 nM) compared to other oceanic environments, respectively (Fig. 9; Table 1). The photodegradation of DOM is greatly influenced by the stratospheric ozone hole events, particularly in Antarctic waters. The ozone hole can increase the fluxes of solar ultraviolet radiation (UVR, 280–400 nm), which may substantially enhance the photoinduced generation of reactive species (H_2O_2, ROOH, $HO^{\bullet}$, etc.) in natural waters (Yocis et al. 2000; Rex et al. 1997; Qian et al. 2001). For example, a decrease in stratospheric ozone from 336 to 151 Dobson units during an ozone hole event resulted in an increase by 19–42 % in the production of H_2O_2 at the surface of Antarctic waters (Yocis et al. 2000). An increase in ozone hole events can thus cause a higher degree of DOM photodegradation upon generation of highly reactive free radicals.

4.5 Production and Decay by Photosynthesis

As a result of photodegradation of DOM, along with the production of H_2O_2 and ROOH compounds, several other photoproducts such as CO_2, CO or other forms of dissolved inorganic carbon (DIC = sum of dissolved CO_2, H_2CO_3, HCO_3^-, and CO_3^{2-}), low molecular weight (LMW) DOM, and thermal energy, E (±) are simultaneously produced in natural waters (Mostofa et al. 2009; Wu et al. 2005; Amador et al. 1989; Moran and Zepp 1997; Wang et al. 2009; Xie and Zafiriou 2009. A general scheme (Eq. 4.7) for the photodegradation of DOM can be expressed as follows (Mostofa et al. 2009a, b):

$$\text{DOM} + h\upsilon \rightarrow H_2O_2 + CO_2/CO/DIC + \text{LMW DOM} + E\,(\pm) \quad (4.7)$$

H_2O_2 and CO_2 that are simultaneously produced by DOM photodegradation can take part to photosynthesis, to form carbohydrate-type compounds (Eq. 4.8) (Mostofa et al. 2009a, b):

$$xCO_{2(H_2O)} + yH_2O_{2(H_2O)} + h\upsilon \rightarrow C_x(H_2O)_y + O_2 + E(\pm) \qquad (4.8)$$

$$2H_2O_2 + \text{photo}(h\upsilon) \text{ or biological processes} \rightarrow O_2 + 2H_2O \text{ or unknown oxidants} \qquad (4.9)$$

where $C_x(H_2O)_y$ (Eq. 4.8) represents a generic carbohydrate. In natural waters, H_2O_2 acts as a key component together with carbon dioxide (CO_2) to form carbohydrates and oxygen through photosynthesis (Eq. 4.8). The formation of oxygen in the photosynthesis process might also occur via either H_2O_2 disproportionation or biological processes (Eq. 4.9) (Komissarov 2003; Moffett and Zafiriou 1990; Liang et al. 2006; Buick 2008). Note that the contribution of H_2O_2 decay is 65–80 % by catalase enzyme and 20–35 % by peroxidase enzyme, respectively, as estimated by isotopic measurements in seawater (Moffett and Zafiriou 1990). E (±) is the energy produced during photosynthesis. The new concept of photosynthesis was firstly hypothesized in plants by Komissarov (1994, 2003). He proposed that interaction of CO_2 in air and H_2O_2, instead of H_2O, may form carbohydrate in plants. It is interesting to note that during the diurnal cycle, H_2O_2 production is the highest at noon time, thereby simultaneously causing the maximum production of CO_2 or DIC due to photodegradation by H_2O_2 or photoinduced generation of $HO^•$. The new reaction mechanim for photosynthesis (Eq. 4.2) will be discussed in details in photosynthesis chapter “Photosynthesis in Nature: A New Look”.

It is demonstrated that microbial consumption is the dominant sink of oceanic carbon monoxide (CO), and that the rate constant (k_{CO}) of microbial CO consumption is positively correlated with chlorophyll *a* (Chl *a*). It is suggested that Chl *a* concentration can be used as an indicator of CO-consuming bacterial activity in natural waters (Xie et al. 2005). Photodegradation and photosynthesis may be important in natural waters with high contents of DOM; photodegradation induces the production of CO_2 and peroxides, which would in turn favor photosysnthesis in the aquatic environments. This would lead to the multiplication of algae, small aquatic plants and phytoplankton. For example, high algal production is operational in some Chinese Lakes during the summer season, which might also be an effect of high DOM photodegradation that favor photosynthesis in lake ecosystems (Mostofa et al. 2009b).

4.6 Production and Decay by Photolytic and Chemical Processes

Production of H_2O_2 and ROOH by photolytic processes may involve their photoinduced formation from DOM under natural sunlight, as explained earlier. The decay of peroxides by photolytic processes (Moffett and Zafiriou 1990; 1993;

Petasne and Zika 1997) may follow two pathways. First, photolytic decomposition of H_2O_2 can occur in seawater (e.g., filtered Vineyard Sound waters) to yield O_2. The photodecomposition was approximately 5 % of the corresponding photoproduction (Moffett and Zafiriou 1990). However, H_2O_2 decomposition typically does not occur in oligotrophic seawater after 2 h of irradiation. This suggests that the contaminants associated with H_2O_2 synthesis in Vineyard Sound samples might be susceptible to the photolytic decomposition of H_2O_2 (Moffett and Zafiriou 1990). Second, H_2O_2 and ROOH can photolytically form free radicals ($R'OOH + h\upsilon \rightarrow RO^{\bullet\prime} + HO^{\bullet}$ where $R' = H$ or R). For example, ROOH compounds are lower in surface seawater than in the deeper layers (Sakugawa et al. 2000). The ROOH compounds are negatively correlated with solar intensity (Sakugawa et al. 2000). This suggests that ROOH may be decomposed by photolytic processes in surface seawater. This result can be justified by the observation of a significant correlation between H_2O_2 and $HO^{\bullet}$ generated photolytically in experiments conducted on river waters, standard Suwannee River Fulvic Acid and DAS1 using a solar simulator Mostofa KMG and Sakugawa H (unpublished), which indicates the photoinduced formation of $HO^{\bullet}$ from H_2O_2. Therefore, decay of peroxides by photolytic processes is a typical phenomenon that may significantly occur in natural waters.

Formation of H_2O_2 and ROOH by chemical processes may include several chain-reactions among various reactant species (Eqs. (3.2–3.5, 3.10–3.12, 3.27). The decomposition of peroxides by chemical processes may involve the Fenton reaction ($H_2O_2 + Fe^{2+} \rightarrow Fe^{3+} + HO^{\bullet} + OH^-$) (Fenton 1894), photo-Fenton reaction ($H_2O_2 + Fe^{2+} + h\nu \rightarrow Fe^{3+} + HO^{\bullet} + OH^-$) (Zepp et al. 1992), photo-ferrioxalate reaction ($Fe^{II}(C_2O_4) + H_2O_2 + h\nu \rightarrow Fe^{III}(C_2O_4) + HO^{\bullet} + OH^-$) (Safazadeh-Amiri et al. 1997) and other chain reactions (Eqs. 3.7, 3.8, 3.16). Free radical oxidation of H_2O_2 by transition metal ions is one of the most important chemical decomposition processes of H_2O_2 in natural waters (Jeong and Yoon 2005; Fenton 1894; Millero and Sotolongo 1989).

4.7 Physical Mixing Processes

The rates of production and decay of peroxides may be influenced by physical processes, such as the mixing by strong waves in the surface mixing zone (Mostofa KMG and Sakugawa H, unpublished; Scully et al. 1998). Physical mixing by strong waves can facilitate the contact of the reactants and increase the reaction rates. For example, the production rate of H_2O_2 was increased by mechanical stirring during irradiation of seawater (86 nM h^{-1}) and standard Suwannee River Fulvic Acid (445 nM h^{-1}) samples, compared to the same samples that were not stirred (51 and 211 nM h^{-1}, respectively). The photoexperiments on site were carried out with a solar simulator Mostofa KMG and Sakugawa H (unpublished). Mixing phenomena can contribute to the relatively elevated H_2O_2 concentration that is often observed in the mixing zone or in the upper surface layers of lake or seawater (Johnson et al. 1989; Sakugawa et al. 2000; Sikorsky and Zika 1993a, b; Scully et al. 1998). Similarly the vertical convective overturn, which is usually caused by nocturnal cooling in the upper lake or ocean, may greatly decrease the surface H_2O_2 concentration through distribution in the whole water column (Johnson et al. 1989; Sarthou et al. 1997; Yuan and Shiller 2001).

4.8 Salinity Effect on Production of H_2O_2

The photoproduction of H_2O_2 significantly increases with salinity in natural waters (Osburn et al. 2009). The generation of H_2O_2 upon irradiation of ultrafiltered river DOM substantially increases from 15 to 368 nM h^{-1} with increasing salinity at circumneutral pH values (Osburn et al. 2009). The increase in H_2O_2 production with salinity has a linear trend (Eq. 4.10) (Osburn et al. 2009):

$$H_2O_2\,(\text{nM}) = 83.15 \times \text{ salinity } - 69.16\left(r^2 = 0:99, p = 0:001, n = 10\right) \tag{4.10}$$

The apparent quantum yield of H_2O_2 photoproduction from ultrafiltered river DOM, *Qhp*, also increases with salinity from 1.64×10^{-4} to 37.02×10^{-4} (Osburn et al. 2009).

The mechanism of high production of H_2O_2 with salinity is not well documented in ealier studies. It is hypothesiszed that hydrated electrons ($e_{aq}{}^-$) are considerably formed in ionic (saline) solution under irradiation. This phenomenon can substantially increase the production of superoxide radical ($O_2{}^{\bullet-}$) and, through disproportionation, of H_2O_2 in aqueous solution. This is evidenced by the photoinduced formation of aqueous electrons ($e_{aq}{}^-$) from organic substances and by their high production in NaCl-mixed solutions compared to pure (Milli-Q) water (Fujiwara et al. 1993; Gopinathan et al. 1972; Zepp et al. 1987b; Nakanishi et al. 2002; Assel et al. 1998; Richard and Canonica 2005). In the presence of high salinity it was also observed a significant increase of CDOM loss (10–40 %) and high photoelectrochemical degradation of methyl orange (~48 % increase in 0.5 M NaCl) (Osburn et al. 2009; Zhang et al. 2010). The mechanisms behind the high photoinduced reactivity of DOM with salinity are discussed in details in other chapters (see chapters "Colored and Chromophoric Dissolved Organic Matter in Natural Waters" and "Fluorescent Dissolved Organic Matter in Natural Waters").

4.9 Production Affected by Precipitation

Precipitation in the form of e.g. rain greatly increases the peroxide concentrations in natural waters (Sakugawa et al. 1995; Avery et al. 2005; Cooper et al. 1987; Yuan and Shiller 2000). This might be caused by the mixing of highly concentrated H_2O_2 in rainwater, where the measured levels are 0–110,600 nM in Europe, 17,000–199,000 nM in Brazil, 30–120,000 nM in the USA, 500–34,000 nM in Canada, 24–56,400 nM in Japan and 3,500–82,000 nM in marine areas (Table 1) (Lazrus et al. 1985; Cooper and Lean 1989; Hellpointner and Gäb 1989; Sakugawa et al. 1990, 1993, 2006; Hewitt and Kok 1991; Cooper and Lean 1992; Yuan and Shiller 2000; Miller et al. 2008). ROOH concentrations in rainwater are 400–1600 nM in Europe and 60–6500 nM in the USA (Table 1) (Hellpointner and Gäb 1989; Sakugawa et al. 1993; Hewitt and Kok 1991). The levels of H_2O_2 and ROOH in rainwater (Table 1) usually show some common trends. First, there are strong diel variations with highest concentrations in the afternoon and lowest ones in the night time and in the early morning. Second, high variations are observed between summer and winter, which are presumably caused by high light intensity in summer that induces elevated H_2O_2 production. Rain drops may scavenge H_2O_2 and ROOH generated in the gas phase or within cloud droplets. Because of the observed diel trend, daytime precipitation might be a more important source of peroxides to natural waters compared to the nighttime one.

5 Significance of H_2O_2 and ROOH in the Aquatic Environment

H_2O_2 and ROOH compounds are uncharged, non-radical active oxygen species that may act as oxidants and reductants in natural waters. These features of peroxides are also of importance for their use in chemical reactions and in our daily life. The main effects of H_2O_2 and ROOH can be distinguished as: (1) Natural purifiers in natural waters; (2) Photo-Fenton reaction for the decomposition of organic pollutants; (3) Indicators of microbial changes in bulk DOM; (4) Function as a redox agents in aqueous solution, (5) Medical treatment and commercial uses; (6) Growth of terrestrial vegetation by rainwater H_2O_2 and ROOH; and (7) Oxygen evolution in photosynthesis.

5.1 Natural Purification in Aquatic Ecosystems

H_2O_2 and ROOH compounds are powerful oxidants, which can directly oxidize the DOM or other reactants in natural waters (Draper and Crosby 1984; Ho 1986; Samuilov et al. 2001). Peroxides are formed photolytically from DOM in natural water, and their productions reach maximum at noon time. The photoinduced generation of $HO^•$ from peroxides can degrade organic pollutants or DOM (Gao and Zepp 1998; Brezonik and Fulkerson-Brekken 1998; Goldstone et al. 2002), which accounts for the role of H_2O_2 and ROOH as purifiers in natural waters.

5.2 Photo-Fenton Type Reaction for Decomposing Organic Pollutants

One of the key applications of H_2O_2 is its use in the degradation of organic pollutants in the wastewater treatment industry by means of Fenton's reaction (Fe^{2+} and H_2O_2), photo-Fenton reaction (UV/Visible-Fe^{2+}/H_2O_2, $\lambda < 580$ nm), UV/Visible-ferrioxalate/H_2O_2 reaction and ozone with H_2O_2 (Zepp et al. 1992; Voelker et al. 1997; Fenton 1894; Safazadeh-Amiri et al. 1997; Glaze and Kang 1989; Tizaoui et al. 2007). Among many other applied technologies, these four are major commercialized technologies.

5.3 Indicators for Microbial Modification of Bulk DOM

ROOH compounds are significantly produced in natural waters under dark conditions (Figs. 1 and 2) and are more concentrated in deep seawater than in the surface layer (Sakugawa et al. 2000). Net ROOH formation (dark production minus dark consumption) is observed in both filtered and unfiltered river waters (Fig. 2). In contrast, net H_2O_2 formation is only observed in filtered waters. The microbial modification of bulk DOM can yield ROOH compounds in natural waters. Microbially-induced changes in the bulk DOM composition are in agreement with the observation of a red shift of the fulvic acid-like fluorescence (peak C) with an increase in fluorescence in deeper lake or seawaters (Hayase and Shinozuka 1995; Mostofa et al. 2005; Moran et al. 2000). Therefore, dark production of organic peroxides could be a useful indicator for the microbial modification of bulk DOM in aquatic environments.

5.4 Function of H_2O_2 as an Oxidizing-Reducing Agent in Aqueous Solution

On the basis of the reduction potential V, the oxidizing agents in aqueous solution can be classified according to the following order: Fluorine (V = 3.0) > Hydroxyl radical ($HO^{\bullet}$) (V = 2.8) > Ozone (V = 2.1) > Peracetic acid (ROOH) (V = 1.8) > H_2O_2 (V = 1.8) > Potassium permanganate (V = 1.7) > Hypochlorite (V = 1.5) > Chlorine dioxide (V = 1.5) > Chlorine (V = 1.4) (Buettner 1993).

H_2O_2 and ROOH compounds act as intermediates in the reduction of oxygen in natural waters. They can act as oxidants or reductants in their reactions with metal ions (Moffett and Zika 1987a, b). For example, H_2O_2 and ROOH compounds can oxidize Cu(I) and Fe(II) in natural waters (Moffett and Zika 1987a, b), a process that can be schematically generalized as follows:

$$M^{n+} + R'OOH \rightarrow M^{(n+1)+} + R'O^{\bullet} + OH^{-} \ (R' = \text{H or R}) \qquad (5.1)$$

$$M^{n+} + R'O^{\bullet} + H^{+} \rightarrow M^{(n+1)+} + R'OH \quad (5.2)$$

On the other hand, the reduction of Cu(II) and Fe(III) by H_2O_2 and ROOH compounds (Moffett and Zika 1987a; Moffett and Zika 1987) can be generalized in the following scheme:

$$R'OOH \rightleftharpoons H^{+} + R'O_2^{-} \; (R' = H \text{ or } R) \quad (5.3)$$

$$M^{(n+1)+} + R'O_2^{-} \rightarrow M^{n+} + R'O_2^{\bullet} \quad (5.4)$$

$$HO_2^{\bullet} \rightleftharpoons H^{+} + O_2^{\bullet -} \quad (5.5)$$

$$M^{(n+1)+} + O_2^{\bullet -} \rightarrow M^{n+} + O_2 \quad (5.6)$$

These reactions have already been verified for various chemical and biochemical processes in natural waters.

5.5 Medical Treatment and Commercial Uses of H_2O_2

H_2O_2 therapy is commonly used in bio-medical sciences. The singlet oxygen atoms produced from H_2O_2 in the human body ($H_2O_2 \rightarrow H_2O + O_1$) can kill or severely inhibit the growth of anaerobic organisms (bacteria and viruses that use carbon dioxide for fuel and leave oxygen as a by-product) (Gorren et al. 1986). Bacteria and viruses do not have an enzyme coating, thus they are easily oxidized by O_1. Application of H_2O_2 is particularly effective for asthma, leukemia, multiple sclerosis, degenerative spinal disc disease, high blood pressure and wound care (Gorren et al. 1986; Nathan and Cohn 1981). In addition, H_2O_2 is widely used to bleach textiles and paper products, in processing foods, minerals, petrochemicals, consumer products (detergents), and in some daily uses such as cleaning and sanitizing the kitchen, soaking the toothbrush to prevent transfer of germs, cleaning vegetables and fruits for freshness and good taste.

5.6 Growth of Terrestrial Vegetation by Rainwater's H_2O_2 and ROOH

High concentrations of H_2O_2 (0–199000 nM) and ROOH (60–6500 nM) in rainwater (Table 1) should be able to promote photosynthesis in plants and algae (Komissarov 1995, 2003; Mostofa et al. 2009a, b). The detailed mechanism in that regard has been discussed in photosynthesis chapter (see chapter "Photosynthesis in Nature: A New Look"). The occurrence of H_2O_2 and ROOH in rainwater could thus contribute to the good health and efficient growth of plants. However, high concentrations of H_2O_2 (50–100 μM) in the presence of iron (Fe) and oxalate can generate $HO^•$ that would decrease the plant productivity and growth (Kobayashi et al. 2002). Furthermore, the ability of H_2O_2 and ROOH compounds to act as anti-bacterial and anti-fungal agents additionally suggests that an optimal level of peroxides could play a positive role toward good health and efficient growth of earth's plants.

5.7 *Role of H_2O_2 in Oxygen Production by Photosynthesis*

Photosynthetic O_2 evolution involves different stages that carry out a gradual accumulation of oxidizing equivalents in the Mn-containing water-oxidizing complex (WOC) (Samuilov et al. 2001). The WOC can exist in different oxidation states (S_n, where high n indicates the most oxidised states), which can be probed by addition of different redox-active molecules. The interaction of H_2O_2 with the S states of the WOC is depicted in the scheme below (Fig. 12) (Samuilov et al. 2001):

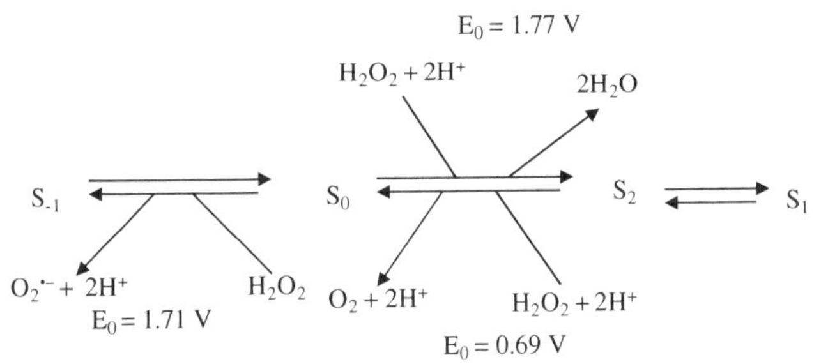

Fig. 12 Different oxidation states of H_2O_2 and its interaction with the S states of the water-oxidizing complex. *Data source* Samuilov et al. (2001)

6 Impacts of H_2O_2 and ROOH in Natural Waters

H_2O_2 and ROOH compounds are uncharged and non-radical active oxygen species, and capable of acting as oxidants and reductants in chemical reactions in natural waters. These properties have some impact on the aquatic organisms, which can be listed as follows: (1) Acid rain; (2) Inhibition of photosynthetic electron transport in cells of cyanobacteria; (3) Effect of H_2O_2 on bacterial growth in waters; and (4) Impact of H_2O_2 on microbial quality of recreational bathing waters.

6.1 Acid Rain

H_2O_2 and ROOH compounds are key components in the conversion of dissolved sulfur dioxide (SO_2) to sulfate (SO_4^{2-}) or sulfuric acid (H_2SO_4) in atmospheric clouds (Sakugawa et al. 1990; Zuo and Hoigné 1993). Sulfuric acid (SO_4^{2-}) can be formed in cloud drops by reaction of HSO_3^- with H_2O_2 (Eq. 3.7) and is a major contributor to acid precipitation to the earth surface.

6.2 Inhibition of Photosynthetic Electron Transport in Cells of Cyanobacteria

H_2O_2 can control a large number of stages of cell metabolism, including those involved in the induction of programmed cell death (Samuilov et al. 2001). H_2O_2 can inhibit growth at concentrations as low as 10^{-5}–10^{-4} M under the conditions of a dialysis culture (Samuilov et al. 2001). H_2O_2 can inhibit the photosynthetic electron transport in cells of cyanobacteria (Samuilov et al. 2001, 2004). It can also destroy the function of the oxygen-evolving complex (OEC) in some chloroplasts and photosystem II preparations, causing release of manganese from the cyanobacterial cells, which inhibits the OEC activity.

6.3 Impact of H_2O_2 on Bacterial Growth in Aquatic Ecosystems

Bacterial growth has a seasonal variability, reaching the maximum in spring to early summer and greatly decreasing in summer, e.g. when water temperature in lakes becomes higher than 25.5 °C (Zhao et al. 2003). Sunlight inactivates bacteria in seawater (Fujioka et al. 1981), which suggests that some photoinduced processes may be involved. The bacterial abundance is commonly affected by water temperature (Zhao et al. 2003; Darakas 2002), but the latter is directly connected with solar radiation that can generate strong oxidizing agents such as peroxides (H_2O_2 and ROOH), $O_2^{\bullet -}$ and $HO^{\bullet}$. These reactive species can reduce the activity of the catalase, peroxidase and superoxide dismutase enzymes present in bacterial cells, DOM, algae and phytoplankton. Bacterial cells protect themselves from the oxidizing species (H_2O_2, $O_2^{\bullet -}$ and $HO^{\bullet}$) by adjusting the level of their enzymes (Chance et al. 1979). An experimental study conducted on marine invertebrates suggests that H_2O_2-scavenging enzymes can protect against external photodynamic effects and internal respiratory by-products (Dykens 1984). It can be assumed that the activity of the enzymes in dealing with the external effects would decrease their ability to scavenge the internal by-products, with harmful effects for the organisms. Low levels of H_2O_2 (~100 nM) affect oxidative stress to bacteria in coastal waters by increasing the concentration of the catalase enzyme. The diurnal periodicity of catalase activity matched the diurnal changes of H_2O_2 (Clark et al. 2008; Angel et al. 1999). The effects of H_2O_2 and peroxides can be particularly important during the summer season when their levels are higher. Moreover, ozone hole events in Antarctic waters may greatly increase photodegradation processes that can generate reactive free radicals and peroxides, with a damaging influence on biogeochemical cycles in Antarctic waters (Diffey 1991; Smith et al. 1992; Randall et al. 2005).

6.4 Impacts of H_2O_2 on Microbial Quality of Recreational Bathing Waters

Microbial water quality is assessed from the concentration of fecal indicator bacteria (FIB) because of their adverse health effects (Cabelli et al. 1979; US Environmental Protection Agency 2000; Wade et al. 2003. Frequent FIB contamination episodes in the surf zone resulted in multiple beach closures in the USA (Boehm et al. 2002). It is shown that elevated levels of H_2O_2, ROOH, superoxide ($O_2^{\bullet-}$) and hydroxyl radical, photolytically produced, can cause damage and cell lysis in microorganisms. This may result into high FIB mortality in recreational bathing waters (Gonzalez-Flecha and Demple 1997; Weinbauer and Suttle 1999; Mitchell and Chamberlin 1975; Clark et al. 2008). It is estimated that approximately ~100 nM of H_2O_2 can cause oxidative stress to bacteria in waters (Angel et al. 1999). Diurnal cycles of FIB mortality in the surf zone (Clark et al. 2008; Boehm et al. 2002), which well resemble the diurnal cycle of H_2O_2, suggest that the FIB mortality may be linked to the photoinduced generation of H_2O_2 and ROOH in sunlit surface waters.

7 Role of H_2O_2 in the Origin of Autochthonous DOM and of other Oxidising Agents

H_2O_2 can contribute to the production of autochthonous DOM by different important processes. First, it is involved in the photosysthesis process that is a major source of organic matter (e.g. algae) (Mostofa et al. 2009a, b). The photoinduced and microbial assimilation of organic matter, including algae, can produce autochthonous DOM in natural waters (Mostofa et al. 2009b; Fu et al. 2010; Harvey et al. 1995; Carrillo et al. 2002; Coble 2007; Yamashita and Tanoue 2004; Yamashita and Tanoue 2008). Coherently, a correlation has been observed between production of organic carbon and concentration of photolytically formed H_2O_2 (Anesio et al. 2005). The autochthonous production of DOM (Mostofa et al. 2005; Yoshioka et al. 2002) is typically observed during the summer season, and a major DOM component that is produced is autochthonous fulvic acid, often termed sedimentary fulvic acid (Hayase and Tsubota 1985). Other produced compounds include marine humic substances (Coble 1996, 2007), carbohydrates and unknown substances (Fu et al. 2010; Mostofa et al. (Manuscript In preparation); Hamanaka et al. 2002; Hayakawa 2004; Farjalla et al. 2006).

Second, H_2O_2, formed photolytically from water using UV radiation, can react with CO_2 under abiogenic conditions to produce various organic substances (CH_2O, HCOOH, CH_3OH, CH_4, and $C_6H_{12}O_6$; Eqs. 7.1–7.5, respectively) in the aqueous solutions (Lobanov et al. 2004). The reactions between H_2O_2 and CO_2 as well as their thermodynamic parameters such as enthalphy changes (ΔH^0) and the Gibbs free energy changes (ΔG^0) are mentioned as follows (Lobanov et al. 2004):

$$H_2O_2 + CO_2 \rightarrow CH_2O + 3/2O_2 \quad (7.1)$$

$$\Delta H^0 = 465\,\text{kJ}, \Delta G^0 = 402\,\text{kJ}$$

$$H_2O_2 + CO_2 \rightarrow HCOOH + O_2 \quad (7.2)$$

$$\Delta H^0 = 172\,\text{kJ}, \Delta G^0 = 166\,\text{kJ}$$

$$2H_2O_2 + CO_2 \rightarrow CH_3OH + 5/2O_2 \quad (7.3)$$

$$\Delta H^0 = 530\,\text{kJ}, \Delta G^0 = 464\,\text{kJ}$$

$$2H_2O_2 + CO_2 \rightarrow CH_4 + 3O_2 \quad (7.4)$$

$$\Delta H^0 = 649\,\text{kJ}, \Delta G^0 = 580\,\text{kJ}$$

$$H_2O_2 + CO_2 \rightarrow 1/6C_6H_{12}O_6 + 3/2O_2 \quad (7.5)$$

$$\Delta H^0 = 426\,\text{kJ}$$

Therefore, these organic substances produced photolytically may play an important role in biogeochemical processes in natural waters.

Third, H_2O_2 can react with nitrous acid to yield peroxynitrous acid, a powerful nitrating agent and an important intermediate in atmospheric chemistry (Vione et al. 2003). The kinetics of the reaction is compatible with a rate-determining step involving either $H_3O_2^+$ and HNO_2, with rate constant $179.6 \pm 1.4\ M^{-1}\ s^{-1}$, or H_2O_2 and protonated nitrous acid, with rate constant $1.68 \pm 0.01 \times 10^{10}\ M^{-1}\ s^{-1}$ (diffusion-controlled reaction) (Vione et al. 2003). Thus, H_2O_2 might be a key environmental factor in atmospheric oxidative chemistry.

8 Scope of the Future Challenges

The determination of H_2O_2 and ROOH as well as their spatial–temporal variations, sources, production and decay mechanisms have been examined in natural waters. Compared to H_2O_2, relatively little attention is paid to the determination of ROOH compounds and their concentrations in natural waters. Investigations based on the detection of ROOH would be crucial to improve the understanding of the photoinduced processes along with H_2O_2 generation in natural waters. Limited attention is also devoted to what fractions of DOM are most involved in the photoinduced production of peroxides in freshwater and marine environments. Other important research demands for future challenges are the following: (i) Identification of the DOM components involved into the production of H_2O_2 in freshwater and marine water. (ii) Elucidation of the temperature and pH effect on the production of H_2O_2 and ROOH compounds by aquatic DOM components and standard organic substances. (iii) Clarification of the correlation between diurnal variations of H_2O_2 and ROOH levels in natural waters and DOM concentration. (iv) Investigation of the role of the photo-Fenton reaction in the production of peroxides in iron-rich waters. (v) Elucidation of the relationship between peroxides and free radicals. (vi) Study of the dark production of H_2O_2 and ROOH by phytoplankton, algae and bacteria in fresh and marine waters. (vii) Effect of variable diurnal levels of H_2O_2 on bacteria in DOM-rich waters. (viii) Detection of ROOH compounds generated photochemically by standard organic substances in aqueous solution and by DOM components in natural waters.

Problems

(1) Explain the nature and biogeochemical function of H_2O_2 and ROOH in natural waters.
(2) Discuss the steady-state concentration of H_2O_2 and ROOH in natural waters
(3) Explain how H_2O_2 acts as REDOX agent.
(4) Explain the mechanisms of photoinduced generation of H_2O_2 and ROOH in the gas phase and in natural waters.
(5) How does fulvic acid photolytically produce H_2O_2 and ROOH in natural waters?
(6) What are the sources of H_2O_2 in natural waters?
(7) What factors are involved in the diurnal cycle of H_2O_2 production in natural waters?
(8) Explain the controlling factors for the decay of H_2O_2 and its decay mechanism by biological processes in natural waters.
(9) What is the importance of H_2O_2 and ROOH?
(10) Which is the impact of H_2O_2 on organisms?
(11) What is the link between H_2O_2, photosynthesis and the autochthonous production of DOM in natural waters?
(12) How does H_2O_2 differ chemically from ROOH?
(13) What is the principle of H_2O_2 and ROOH measurement by the fluorometric method?
(14) In a diurnal cycle in river water, the concentration of H_2O_2 gradually increased from 4 to 69 nM in the period from before sunrise to noon and

then gradually decreased to 20 nM. What is the concentration of H_2O_2 accounted for by diurnal effects in the river waters?

(15) What is the steady state concentration of H_2O_2 in natural waters? Find out the production of H_2O_2 in a natural water when its steady-state concentration is 350 nM, microbial degradation 20 nM, and consumption by DOM photodegradation 30 nM.

Acknowledgments We thank Dr. Li Wen of China University of Geosciences, Wuhan for her generous help during the manuscript preparation. This work was financially supported jointly by the National Natural Science Foundation of China (Grant Nos. 1314765) and Institute of geochemistry, Chinese Academy of Sciences, China. This study was also partly supported by Hiroshima University, Japan; University Turin, Italy; Brook Byers Institute for Sustainable Systems at Georgia Institute of Technology, the United States; and Chinese Research Academy of Environmental Sciences, China. This study acknowledges the Copyright (1990) by the Association for the Sciences of Limnology and Oceanography, Inc.; copyright (1993) by The Geochemical Society of Japan; reprinted from Analytica Chimica Acta, 627(2), Olasehinde EF, Makino S, Kondo H, Takeda K, Sakugawa H, Application of Fenton reaction for nanomolar determination of hydrogen peroxide in seawater, 270–276. Copyright (2008) with permission from Elsevier; Copyright (2009) CSIRO; reprinted (adapted) with permission from Sakugawa H, Kaplan IR, Tsai W, Cohen Y, Atmospheric hydrogen peroxide, Environ Sci Technol, 24(10), 1452–1462. Copyright (1990) American Chemical Society; Springer and the original Biochemistry (Moscow), 66, 2001, 640–645, Hydrogen peroxide inhibits photosynthetic electron transport in cells of cyanobacteria. Biochem (Moscow), Samuilov VD, Bezryadnov DB, Gusev MV, Kitashov AV, Fedorenko TA, with kind permission from Springer Science and Business Media; and Original Russian Text Copyright (2004) by Lobanov AV, Kholuiskaya SN, GG Komissarov.

References

Akane S, Makino S, Hashimoto N, Yatsuzuka Y, Kawai Yu, Takeda K, Sakugawa H (2004) Hydrogen peroxide in the sea water of Hiroshima Bay Japan. Oceanogr Jpn 13:185–196 (In Japanese)

Amador JA, Alexander M, Zika RG (1989) Sequential photochemical and microbial degradation of organic molecules bound to humic acid. App Environ Microb 55:2843–2849

Amouroux D, Donard OFX (1995) Hydrogen peroxide determination in estuarine and marine waters by flow injection with fluorescence detection. Oceanol Acta 18:353–361

Anesio AM, Granéli W, Aiken GR, Kieber DJ, Mopper K (2005) Effect of humic substance photodegradation on bacterial growth and respiration in lake water. Appl Environ Microb 71:6267–6275

Angel DL, Fiedler U, Eden N, Kress N, Adelung D, Herut B (1999) Catalase activity in macro- and micro-organisms as an indicator of biotic stress in coastal waters of the eastern Mediterranean Sea. Helgoland Mar Res 53:209–218

Arakaki T, Fujimura H, Hamdun AM, Okada K, Kondo H, Oomori T, Tanahara A, Taira H (2005) Simultaneous measurement of hydrogen peroxide and Fe species (Fe(II) and Fe(tot) in Okinawa island seawater: impacts of red soil pollution. J Oceanogr 61:561–568

Assel M, Laenen R, Laubereau A (1998) Ultrafast electron trapping in an aqueous NaCl-solution. Chem Phys Lett 289:267–274

Avery GB Jr, Cooper WJ, Kieber RJ, Willey JD (2005) Hydrogen peroxide at the Bermuda Atlantic time series station: temporal variability of seawater hydrogen peroxide. Mar Chem 97:236–244

Baden DG, Corbett MD (1980) Bromoperoxidases from penicillus capitatus lamourouxii, and rhipocepholus. Biochem J 187:205–211

Baxter RM, Carey JH (1983) Evidence for superoxide ion in natural waters. Nature 306:575–576

Bazanov MI, Berezin BD, Berezin DB et al. (1999) Uspekhi khimii porfirinov (Progress in the Chemistry of Porphyrins), St. Petersburg: NII khimii SPbGU

Berlett BS, Stadtman ER (1997) Protein oxidation in aging, disease, and oxidative stress. J Biol Chem 272:20313–20316

Blokhina O, Virolainen E, Fagerstedt KV (2003) Antioxidants, oxidative damage and oxygen deprivation stress: a review. Ann Bot 91:179–194

Boehm AB, Grant SB, Kim JH, Mowbray SL, McGee CD, Clark CD, Foley DM, Wellman DE (2002) Decadal and shorter period variability of surf-zone water quality at Huntington Beach California. Environ Sci Technol 36:3885

Boehm AB, Yamahara KM, Love DC, Peterson BM, McNeill K, Nelson KL (2009) Covariation and photoinactivation of traditional and novel indicator organisms and human viruses at a sewage-impacted marine beach. Environ Sci Technol. doi:10.1021/es9015124

Boveris A, Valdez LB, Zaobornyj T, Bustamante J (2006) Mitochondrial metabolic states regulate nitric oxide and hydrogen peroxide diffusion to the cytosol. Biochim Biophys Acta 1757:535–542

Brezonik PL, Fulkerson-Brekken J (1998) Nitrate-induced photolysis in natural waters: controls on concentrations of hydroxyl radical photo-intermediates by natural scavenging agents. Environ Sci Technol 32:3004–3010

Buettner GR (1993) The pecking order of free radicals and antioxidants: lipid peroxidation, α-tocopherol, and ascorbate. Arch Biochem Biophys 300:535–543

Buick R (2008) When did oxygenic photosynthesis evolve? Phil Trans R Soc B 363:2731–2743. doi:10.1098/rstb.2008.0041

Cabelli DE (1997) The reactions of HO_2/O_2^- radicals in aqueous solution. In: Alfassi ZB (ed) Peroxyl radicals. Wiley, New York, pp 407–437

Cabelli V, Dufour AP, Levin MA, McCabe LJ, Harberman PW (1979) Relationship of microbial indicators to health effects at marine bathing beaches. Am J Public Health 69:690–695

Calvert JG, Lazrus A, Kok GL, Heikes BG, Walega JG, Lind J, Cantrell CA (1985) Chemical mechanisms of acid generation in the troposphere. Nature 317:27–35

Carrillo P, Medina-Sánchezv JM, Villar-Argaiz M (2002) The interaction of phytoplankton and bacteria in a high mountain lake: importance of the spectral composition of solar radiation. Limnol Oceanogr 47:1294–1306

Chance B, Sies H, Boveris A (1979) Hydroperoxide metabolism in mammalian organs. Physiol Rev 59:527–605

Clark CD, Bruyn WJ, Jakubowski SD, Grant SB (2008) Hydrogen peroxide production in marine bathing waters: implications for fecal indicator bacteria mortality. Mar Pollut Bull 56:397–401

Clark CD, De Bruyn WJ, Jones JG (2009) Photochemical production of hydrogen peroxide in size-fractionated Southern California coastal waters. Chemosphere 76:141–146

Clark CD, De Bruyn WJ, Hirsch CM, Jakubowski SD (2010a) Hydrogen peroxide measurements in recreational marine bathing waters in Southern California USA. Water Res 44:2203–2210

Clark CD, De Bruyn WJ, Hirsch CM, Aiona P (2010b) Diel cycles of hydrogen peroxide in marine bathing waters in Southern California, USA: In situ surf zone measurements. Mar Pollut Bull 60:2284–2288

Coble PG (1996) Characterization of marine and terrestrial DOM in sea water using excitation-emission matrix spectroscopy. Mar Chem 52:325–336

Coble PG (2007) Marine optical biogeochemistry: the chemistry of ocean color. Chem Rev 107:402–418

Collen J, Del Rio MJ, Garcia-Reina G, Pedersen M (1995) Photosynthetic production of hydrogen peroxide by Ulva rigida C. Ag. (Chlorophyta). Planta 196:225–230

Cooper WJ, Lean DRS (1989) Hydrogen peroxide concentration in a Northern lake: photochemical formation and diel variability. Environ Sci Technol 23:1425–1428

Cooper WJ, Lean DRS (1992) Hydrogen peroxide dynamics in marine and fresh water systems. Encyclopedia of Earth system science. Academic Press Inc 2:527–535
Cooper WJ, Zepp RG (1990) Hydrogen peroxide decay in waters with suspended soils: evidence for biologically mediated processes. Can J Fish Aquat Sci 47:888–893
Cooper WJ, Zika RG (1983) Photochemical formation of hydrogen peroxide in surface and ground waters exposed to sunlight. Science 220:711–712
Cooper WJ, Saltzman ES, Zika RG (1987) The contribution of rainwater to variability in surface ocean hydrogen peroxide. J Geophys Res 92:2970–2980
Cooper WJ, Zika RG, Petasne RG, Plane JMC (1988) Photochemical formation of H_2O_2 in natural waters exposed to sunlight. Environ Sci Technol 22:1156–1160
Cooper WJ, Zika RG, Petasne RG, Fisher AM (1989a) Aquatic humic substances: influence on fate and treatment of pollutants. In: Suffet IH, MacCarthy P (Ed), Advances in chemistry series 219. Am Chem Soc, Washington
Cooper WJ, Lean DRS, Carey J (1989b) Spatial and temporal patterns of hydrogen peroxide in lake waters. Can J Fish Aquat Sci 46:1227–1231
Cooper WJ, Shao C, Lean DRS, Gordon AS, Scully FE Jr (1994) Factors affecting the distribution of H_2O_2 in surface waters. In: Baker LA (ed) Environmental chemistry of lakes and reservoirs, Adv Chem Ser 237. Am Chem Soc, Washington, pp 391–422
Cooper WJ, Moegling JK, Kieber RJ, Kiddle JJ (2000) A chemiluminescence method for the analysis of H_2O_2 in natural waters. Mar Chem 70:191–200
Corin N, Backlund P, Kulovaara M (1996) Degradation products formed during UV-irradiation of humic waters. Chemosphere 33:245–255
Croot PL, Laan P, Nishioka J, Strass V, Cisewski B, Boye M, Timmermans KR, Bellerby RG, Goldson L, Nightingale P, Baar HJW (2005) Spatial and temporal variation distribution of Fe(II) and H_2O_2 during EisenEx, an open ocean mescoscale iron enrichment. Mar Chem 95:65–88
Crutzen PJ (1992) Ultraviolet on the increase. Nature 356:104–105
Dalrymple RM, Carfagno AK, Sharpless CM (2010) Correlations between dissolved organic matter optical properties and quantum yields of singlet oxygen and hydrogen peroxide. Environ Sci Technol 44:5824–5829
Darakas E (2002) E. Coli kinetics—effect of temperature on the maintenance and respectively the decay phase. Environ Monit Assess 78:101–110
Diffey BL (1991) Solar ultraviolet radiation effects on biological systems. Phys Med Biol 36:299–328
Draper WM, Crosby DG (1983) The photochemical generation of hydrogen peroxide in natural waters. Arch Environ Contam Toxicol 12:121–126
Draper WM, Crosby DG (1984) Solar photooxidation of pesticides in diluted hydrogen-peroxide. J Agric Food Chem 32:231–237
Dunlap WC, Susic M (1985) Determination of pteridines and flavins by reverse-phase, high-performance liquid chromatography with fluorimetric detection. Mar Chem 17:185–198
Dykens JA (1984) Enzymic defenses against oxygen toxicity in marine cnidarians containing endosymbiotic algae. Mar Biol Lett 5:291–301
Farjalla VF, Azevedo DA, Esteves FA, Bozelli RL, Roland F, Enrich-Prast A (2006) Influence of hydrological pulse on bacterial growth and DOC uptake in a clear-water Amazonian Lake. Microb Ecol 52:334–344
Faust BC, Allen JM (1992) Aqueous-phase photochemical sources of peroxyl radicals and singlet molecular-oxygen in clouds and fog. J Geophys Res-Atmospheres 97(D12):12913-12926
Faust BC, Hoigne J (1987) Sensitized photooxidation of phenols by fulvic acid and in natural waters. Environ Sci Technol 21:957–964
Faust BC, Anastasio C, Allen JM, Arakaki T (1993) Aqueous-phase photochemical formation of peroxides in authentic cloud and fog waters. Science 260:73–75
Fenton HJ (1894) Oxidation of tartaric acid in presence of iron. J Chem Soc 65:899–910
Fischer AM, Kliger DS, Winterle JS, Mill T (1985) Direct observations of phototransients in natural waters. Chemosphere 14:1299–1306

Fischer AM, Winterle JS, Mill T (1987) Direct observations of primary photochemical processes in photolysis mediated by humic substances. In: Zika RG and Cooper RW (Ed) Photochemistry of environmental aquatic systems, ACS Symposium Series 327, Am Chem Soc, pp 141–156
Forman HJ, Boveris A (1982) Superoxide radical and hydrogen peroxide in mitochodria. In: Pryor W (Ed), Free radicals in biology, Vol 5, Academic Publishers, pp 65–90
Frimer AA, Forman A, Borg DC (1983) H_2O_2 diffusion through liposomes. Israel J Chem 23:442–445
Fu P, Mostofa KMG, Wu FC, Liu CQ, Li W, Liao H, Wang L, Wang J, Mei Y (2010) Excitation-emission matrix characterization of dissolved organic matter sources in two eutrophic lakes (Southwestern China Plateau). Geochem J 44:99–112
Fujioka RS, Hashimoto HH, Siwak EB, Young RHF (1981) Effect of sunlight on survival of indicator bacteria in seawater. Appl Environ Microb 41:690–696
Fujiwara K, Ushiroda T, Takeda K, Kumamoto Y, Tsubota H (1993) Diurnal and seasonal distribution of hydrogen peroxide in seawater of Seto Inland Sea. Geochem J 27:103–115
Fujiwara K, Takeda K, Kumamoto Y (1995) Generations of carbonyl sulfides and hydrogen peroxide in the Seto Inland Sea-Photochemical reactions progressing in the coastal seawater. In: Sakai H, Nozaki Y (Ed), Biogeochemical Processes and Ocean flux in the Western Pacific, TERRAPUB, Tokyo, pp 101–127
Gao H, Zepp RG (1998) Factors influencing photoreactions of dissolved organic matter in a coastal river of the southern United States. Environ Sci Technol 32:2940–2946
Gerringa LJA, Rjkenberg MJA, Timmermans KR, Buma AGJ (2004) The influence of solar ultraviolet radiation on the photochemical production of H_2O_2 in the equatorial Atlantic Ocean. J Sea Res 51:3–10
Glaze WH, Kang JW (1989) Advanced oxidation processes. Description of a kinetic model for the oxidation of hazardous materials in aqueous media with ozone and hydrogen peroxide in a semibatch reactor. Ind Eng Chem Res 28(11):1573–1580
Goldstone JV, Pullin MJ, Bertilsson S, Voelker BM (2002) Reactions of hydroxyl radical with humic substances: bleaching, mineralization, and production of bioavailable carbon substrates. Environ Sci Technol 36:364–372
Gonzalez-Flecha B, Demple B (1997) Homeostatic regulation of intracellular hydrogen peroxide concentration in aerobically growing Escherichia coli. J Bacteriol 47:382–388
Gopinathan C, Damle PS, Hart EJ (1972) Gamma-ray irradiated sodium chloride as a source of hydrated electrons. J Phys Chem 76:3694–3698
Gorren AC, Dekker H, Wever R (1986) Kinetic investigations of the reaction of cytochrome C oxidase by hydrogen peroxide. Biochem Biophys Acta 852:81–92
Grivennikova VG, Cecchini G, Vinogradov AD (2008) Ammonium-dependent hydrogen peroxide production by mitochondria. FEBS Lett 582:2719–2724
Guilbault GG, Brignac P Jr, Juneau M (1968) New substrate for the fluorometric determination of oxidative enzymes. Anal Chem 40:1256–1263
Haber F, Weiss J (1934) The catalytic decomposition of hydrogen peroxide by iron salts. Proc Roy Soc Lond, Ser A:332–351
Hamanaka J, Tanoue E, Hama T, Handa N (2002) Production and export of particulate fatty acids, carbohydrates and combined amino acids in the euphotic zone. Mar Chem 77:55–69
Harris GP (1979) Photosyntheis, productivity and growth: the physiological ecology of phytoplankton. Arch fur Hydrobiol 16:1–191
Harvey HW (1925) Oxidation in seawater. J Mar Biol Assoc UK 13:953–969
Harvey HR, Tuttle JH, Bell JT (1995) Kinetics of phytoplankton decay during simulated sedimentation: changes in biochemical composition and microbial activity under oxic and anoxic conditions. Geochim Cosmochim Acta 59:3367–3377
Hayakawa K (2004) Seasonal variations and dynamics of dissolved carbohydrates in Lake Biwa. Org Geochem 35:169–179
Hayase K, Shinozuka N (1995) Vertical distribution of fluorescent organic matter along with AOU and nutrients in the Equatorial pacific. Mar Chem 48:282–290

Hayase K, Tsubota H (1985) Sedimentary humic and fulvic acids as fluorescent organic materials. Geochim Cosmochim Acta 49:159–163
Hellpointner E, Gäb S (1989) Detection of methyl, hydroxymethyl and hydroxyethyl hydroperoxides in air and precipitation. Nature 337:631–634
Helz GR, Kieber RJ (1985) Water chlorination: Chem Environ. Impact Health Eff Proc Conf 5th, 1033–1040
Herut B, Shoham-Frider E, Kress N, Fiedler U (1998) Hydrogen peroxide production rates in clean and polluted coastal marine waters of the Mediterranean, red and Baltic Seas. Mar Poll Bull 36:994–1003
Hewitt CN, Kok GL (1991) Formation and occurrences of organic hydroperoxides in the troposphere: laboratory and field observations. J Atmos Chem 12:181–194
Ho P (1986) Photooxidation of 2,4 dinitrotoluene in aqueous solution in the presence of H_2O_2. Environ Sci Technol 20:260–267
Holm-Hansen O, Lubin D, Helbling EW (1993) Ultraviolet radiation and its effects on organisms in aquatic environments. In: Young AR (Ed), Environmental UV photobiology, Plenum, pp 379–425
Hong AP, Bahnemann DW, Hoffmann MR (1987) J Phys Chem 91:2109–2117
Inoue K, Matsuur T, Saito I (1982) Photogeneration of superoxide ion and hydrogen peroxide from tryptophan and its photooxidation products: The role of 3a-hydroperoxypyrrolidinoindole. Photochem Photobiol 35:133–139
Jeong J, Yoon J (2005) pH effect on OH radical production in photo/ferrioxalate system. Water Res 39:2893–2900
Johnson KS, Willason SW, Wiesenburg DA, Lohrenz SE, Arnone RA (1989) Hydrogen peroxide in the western Mediterranean Sea: a tracer for vertical advection. Deep-Sea Res 36:241–254
Kelley RL, Reddy CA (1986) Identification of glucose oxidase activity as the primary source of hydrogen peroxide production in ligninolytic cultures of *Phanerochaete Chrysosporium*. Arch Microbiol 144:248–253
Kieber DJ, Blough NV (1990) Determination of carboncentered radicals in aqueous solution by liquid chromatography with fluorescence detection. Anal Chem 62:2275–2283
Kieber RJ, Heltz GR (1995) Temporal and seasonal variations of hydrogen peroxide levels in estuarine waters. Estuar Coast Shelf Sci 46:645–656
Kim K, Portis AR Jr (2004) Oxygen-dependent H_2O_2 production by Rubisco. FEBS Lett 571:124–128
Kim J, Zobell CE (1974) Occurrence an activities of cell-free enzymes in oceanic environments. In: Colwell RR, Mortia RY (Ed), Effect of the ocean environment on microbial activities, University Park Press, University Park, pp 368–385
Kim DH, Takeda K, Sakugawa H, Lee J-S (2003) The photochemical reactions of iron species in rain and snow in Higashi-Hiroshima, Japan. Anal Sci Technol 16:466–474
Kobayashi T, Natanani N, Hirakawa T, Suzuki M, Miyake T, Chiwa M, Yuhara T, Hashimoto N, Inoue K, Yamamura K, Agus N, Sinogaya JR, Nakane K, Kume A, Arakaki T, Sakugawa H (2002) Variation in CO_2 assimilation rate induced by simulated dew waters with different sources of hydroxyl radical ($^{\cdot}OH$) on the needle surfaces of Japanese red pine (*Pinus densiflora* Sieb. et Zucc.). Environ Pollut 118:383–391
Komissarov GG (1994) Photosynthesis: a new look. Sci Russia 5:52–55
Komissarov GG (1995) Photosynthesis as a physical process. Chem Phys Reports 14(11):1723–1732
Komissarov GG (2003) Photosynthesis: the physical-chemical approach. J Adv Chem Phys 2(1):28–61
Kramer JB, Canonica S, Hoigne J, Kaschig J (1996) Degradation of fluorescent whitening agents in sunlit natural waters. Environ Sci Technol 30:2227–2234
Kwan WP, Voelker BM (2002) Decomposition of hydrogen peroxide and organic compounds in the presence of dissolved iron and ferrihydrite. Environ Sci Technol 36:1467–1476
Lazrus AL, Kok GL, Gitlin SN, Lind JA, Mclaren SE (1985) Automated fluorometric method for hydrogen peroxide in atmospheric precipitation. Anal Chem 57:917–922

Lazrus AL, Kok GL, Lind JA, Gitlin SN, Heikes BG, Shetter RE (1986) Automated fluorometric method for hydrogen peroxide in air. Anal Chem 58:594–597

Lee JH, Tang IN, Weinstein-Lloyd B, Halper EB (1994) Improved nonenzymatic method for the determination of gas-phase peroxides. Environ Sci Technol 28:1180–1185

Leenheer JA, Croué JP (2003) Characterizing aquatic dissolved organic matter. Environ Sci Technol 37:18–26

Legrini O, Oliveros E, Braun AM (1993) Photochemical processes for water treatment. Chem Rev 93:671–698

Lehninger AL (1970) Biochemistry. Worth, New york, p 478

Liang M-C, Hartman H, Kopp RE, Kirschvink JL, Yung YL (2006) Production of hydrogen peroxide in the atmosphere of a Snowball Earth and the origin of oxygenic photosynthesis. PNAS 103:18896–18899

Liu J, Steinberg SM, Johnson BJ (2003) A high performance liquid chromatography method for determination of gas-phase hydrogen peroxide in ambient air using Fenton's chemistry. Chemosphere 52:815–823

Lobanov AV, Kholuiskaya SN, Komissarov GG (2004) Photocatalytic synthesis of formaldehyde from CO_2 and H_2O_2 Doklady. Phys Chem Part I 399:266–268

Lobanov AV, Rubtsova NA, Vedeneeva YA, Komissarov GG (2008) Photocatalytic activity of chlorophyll in hydrogen peroxide generation in water. Doklady Chem Part 2(421):190–193

Mageli OL, Kolczynski JR (1966) Organic peroxides. Ind Eng Chem 58:25–32

Malcolm RL (1985) Geochemistry of stream fulvic and humic substances. In: Aiken GR, McKnight DM, Wershaw RL, MacCarthy P (Ed), Humic substances in soil, sediment, and water: geochemistry. Isolation and Characterization, Wiley, pp 181–209

Marañòn E, Cermeño P, Fernåndez E, Rodrìguez J, Zabala L (2004) Significance and mechanisms of photosynthetic production of dissolved organic carbon in a coastal eutrophic ecosystem. Limnol Oceanogr 49:1652–1666

McCarthy M, Pratum T, Hedges J, Benner R (1997) Chemical composition of dissolved organic nitrogen in the ocean. Nature 390:150–154

Medina-Sánchez Manuel J, Villar-Argaiz M, Carrillo P (2006) Solar radiation-nutrient interaction enhances the resource and predation algal control on bacterioplankton: a short-term experimental study. Limnol Oceanogr 51:913–924

Mill T, Hendry DG, Richardson H (1980) Free-radical oxidants in natural waters. Science 207:886–887

Miller WL, Kester DR (1988) Hydrogen peroxide measurement in seawater by *p*-hydroxyphenyl acetic acid dimerization. Anal Chem 60:2711–2715

Miller WL, Kester DR (1994) Peroxide variations in the Sargasso Sea. Mar Chem 48:17–29

Miller GW, Morgan CA, Kieber DJ, Whitney King D, Snow JA, Heikes BG, Mopper K, Kiddle JJ (2005) Hydrogen peroxide method intercomparison study in seawater. Mar Chem 97:4–13

Miller C, Willey JD, Kieber R (2008) Changes in rainwater composition in Wilmington, NC during tropical storm Ernesto. Atmos Environ 42:846–855

Millero FJ, Sotolongo S (1989) The oxidation of Fe(II) with H_2O_2 in seawater. Geochim Cosmochim Acta 53:1867–1873

Millington KR, Maurdev G (2004) The generation of superoxide and hydrogen peroxide by exposure of fluorescent whitening agents to UVA radiation and its relevance to the rapid photoyellowing of whitened wool. J Photochem Photobiol A Chem 165:177–185

Mitchell R, Chamberlin C (1975) Factors influencing the survival of enteric microorganisms in the sea: an overview In: Gameson ALH (Ed), Proceedings of the international symposium on discharge of Sewage from Ocean outfalls Pergamon Press, London, pp 237–251

Moffett JW, Zafiriou OC (1990) An Investigation of hydrogen peroxide chemistry in surface waters of Vineyard sound with $H_2{}^{18}O_2$ and $^{18}O_2$. Limnol Oceanogr 35:1221–1229

Moffett JW, Zafiriou OC (1993) The photochemical decomposition of hydrogen peroxide in surface waters of the eastern Caribbean and Orinoco River. J Geophys Res 98(C2):2307–2313

Moffett JW, Zika RG (1983) Oxidation kinetics of Cu(I) in seawater: implications for its existence in the marine environment. Mar Chem 13:239–251

Moffett JW, Zika RG (1987a) Photochemistry of a copper complexes in sea water In: Zika RG, Cooper WJ (Ed), Photochemistry of environmental aquatic systems, ACS Sym Ser 327, Am Chem Soc, Washington pp 116–130
Moffett JW, Zika RG (1987b) Reaction kinetics of hydrogen peroxide with copper and iron in seawater. Environ Sci Technol 21:804–810
Moore CA, Farmer CT, Zika RG (1993) Influence of the Orinoko river on hydrogen peroxide distribution and production in the Eastern Caribean. J Geophys Res 98(C2):2289–2298
Mopper K, Zika RG (1987) Natural photosensitizers in sea water: riboflavin and its breakdown products In: Zika RG, Cooper WJ (Ed), Photochemistry of environmental aquatic systems, Am Chem Soc, Washington, pp 174–190
Moran MA, Zepp RG (1997) Role of photoreactions in the formation of biologically labile compounds from dissolved organic matter. Limnol Oceanogr 42(6):1307–1316
Moran MA Jr, Sheldon WM, Zepp RG (2000) Carbon loss and optical property changes during long-term photochemical and biological degradation of estuarine dissolved organic matter. Limnol Oceanogr 45:1254–1264
Mostofa KMG (2005) Dynamics, characteristics and photochemical processes of fluorescent dissolved organic matter and peroxides in river water. Ph D Thesis, September 2005, Hiroshima University, Japan
Mostofa KMG, Akane S, Sakugawa H Role of microbial function in controlling the concentrations of hydrogen peroxide and organic peroxides in rivers. (Manuscript in preparation)
Mostofa KMG, Sakugawa H (2003) Spatial and temporal variation of hydrogen peroxide in stream and river waters: effect of photo-bio-physio-chemical processes of aquatic matter Abstracts of the 13th Annual VM Goldschmidt Conference, Kurashiki, Japan. Geochim Cosmochim Acta 67(18S), p A309
Mostofa KMG, Sakugawa H (2009) Spatial and temporal variations and factors controlling the concentrations of hydrogen peroxide and organic peroxides in rivers. Environ Chem 6:524–534
Mostofa KMG, Yoshioka T, Konohira E, Tanoue E, Hayakawa K, Takahashi M (2005a) Three-dimensional fluorescence as a tool for investigating the dynamics of dissolved organic matter in the Lake Biwa watershed. Limnology 6:101–115
Mostofa KMG, Honda Y, Sakugawa H (2005b) Dynamics and optical nature of fluorescent dissolved organic matter in river waters in Hiroshima prefecture, Japan. Geochem J 39:257–271
Mostofa KMG, Yoshioka T, Konohira E, Tanoue E (2007a) Dynamics and characteristics of fluorescent dissolved organic matter in the groundwater, river and lake water. Water Air Soil Pollut 184:157–176
Mostofa KMG, Yoshioka T, Konohira E, Tanoue E (2007b) Photodegradation of fluorescent dissolved organic matters in river waters. Geochem J 41:323–331
Mostofa KMG, Wu FC, Yoshioka T, Sakugawa H, Tanoue E (2009a) Dissolved organic matter in the aquatic environments In: Wu FC, Xing B (Ed), Natural organic matter and its significance in the environment, Science Press, Beijing, pp 3–66
Mostofa KMG, Liu CQ, Wu FC, Fu PQ, Ying WL, Yuan J (2009b) Overview of key biogeochemical functions in lake ecosystem: impacts of organic matter pollution and global warming. Keynote Speech. In: Proceedings of the 13th World Lake Conference Wuhan, China, 1–5 Nov 2009, pp 59–60
Nakanishi I, Fukuzumi S, Konishi T, Ohkubo K, Fujitsuka M, Ito O, Miyata N (2002) DNA cleavage via superoxide anion formed in photoinduced electron transfer from NADH to γ-Cyclodextrin-Bicapped C60 in an oxygen-saturated aqueous solution. J Phys Chem B 106:2372–2380
Nakatani N, Ueda M, Shindo H, Takeda K, Sakugawa H (2007) Contribution of the photo-Fenton reaction to hydroxyl radical formation rates in river and rain water samples. Anal Sci 23:1137–1142
Nathan CF, Cohn ZA (1981) Antitumer effects of hydrogen peroxide in vivo. J Exp Med 154:1539–1553
Nieto-Cid M, Álvarez-Salgado A, Pérez FF (2006) Microbial and photochemical reactivity of fluorescent dissolved organic matter in a coastal upwelling system. Limnol Oceanogr 51:1391–1400

O'Sullivan DW, Neale PJ, Coffin RB, Boyd TJ, Osburn CL (2005) Photochemical production of hydrogen peroxide and methylhydroperoxide in coastal waters. Mar Chem 97:14–33
Obernosterer I, Ruardij P, Herndl GJ (2001) Spatial and diurnal dynamics of dissolved organic matter (DOM) fluorescence and H_2O_2 and the photochemical oxygen demand of surface water DOM across the subtropical Atlantic Ocean. Limnol Oceanogr 46:632–643
Ogren WL (1984) Photorespiration: pathways, regulation and modification. Ann Rev Plant Physiol 35:415–442
Olasehinde EF, Makino S, Kondo H, Takeda K, Sakugawa H (2008) Application of Fenton reaction for nanomolar determination of hydrogen peroxide in seawater. Analyt Chim Acta 627:270–276
Osburn CL, O'Sullivan DW, Boyd TJ (2009) Increases in the longwave photobleaching of chromophoric dissolved organic matter in coastal waters. Limnol Oceanogr 54:145–159
Page SE, Arnold WA, McNeill K (2011) Assessing the contribution of free hydroxyl radical in organic matter-sensitized photohydroxylation reactions. Environ Sci Technol 45:2818–2825
Palenic B, Morel FMM (1988) Dark production of H_2O_2 in the Sargasso Sea. Limnol Oceanogr 33:1606–1611
Palenic B, Zafiriou OC, Morel FMM (1987) Hydrogen peroxide production by a marine phytoplankton. Limnol Oceanogr 32:1365–1369
Paradies G, Petrosillo G, Pistolese M, Ruggiero FM (2000) The effect of reactive oxygen species generated from the mitochondrial electron transport chain on the cytochrme *C* oxidase activity and on the cardilipin content in bovine heart submitochondrial particles. FEBS Lett 466:323–326
Parmon VN (1985) in Fotokataliticheskoe preobrazovanie solnechnoi energii, Ch 2 Molekulyarnye sistemy dlya razlozheniya vody (Photocatalytic Sunlight Conversion, part 2: Molecular systems for water decomposition). Nauka, Novosibirsk
Penkett SA, Jones BMR, Brice KA, Eggleton AEJ (1979) The importance of atmospheric ozone and hydrogen peroxide in oxidizing sulphur dioxide in cloud and rainwater. Atmos Environ 13:123–137
Perkowski J, Jóźwiak W, Kos L, Stajszczyk P (2006) Applications of Fenton's reagent in detergent separation in highly concentrated water solutions. Fibres Textile Eastern Europe 14:114–119
Petasne RG, Zika RG (1987) The fate of superoxide in coastal seawater. Nature 325:516–518
Petasne RG, Zika RG (1997) Hydrogen peroxide lifetimes in south Florida coastal and offshore waters. Mar Chem 56:215–225
Peuravuori J, Pihlaja K (1999) Some approaches for modelling of dissolved aquatic organic matter. In: Keskitalo J, Eloranta P (Ed), Limnology of humic waters, Backhuy Publishers, Leiden, pp 11–39
Power JF, Sharma DK, Langford CH, Bonneau R, Joussot-Dubein J (1987) Laser flash photolytic studies of a well-characterized soil humic substances. In: Zika RG, Cooper WJ (Ed), Photochemistry of environmental aquatic systems, ACS Symposium Series 327, Am Chem Soc, Washington, pp 17–173
Qian J, Mopper K, Kieber DJ (2001) Photochemical production of the hydroxyl radical in Antarctic waters. Deep-Sea Res I 48:741–759
Randall CE, Harvey VL, Manney GL, Orsolini Y, Codrescu M, Sioris C, Brohede S, Haley CS, Gordley LL, Zawdony JM, Russell JM (2005) Stratospheric effects of energetic particle precipitation in 2003–2004. Geophys Res Lett LO5082 doi:101029/2004GL022003
Resing J, Tien G, letelier R, Karl DM (1993) Palmer LTER: hydrogen peroxide in the Palmer LTER region: II Water column distribution Antarctic. J US 227–229
Rex M, Harris NRP, der Gathen P, Lehman R, Braathen GO, Reimer E, Beck A, Chipperfield MP, Alfier R, Allaart M, O'Conner F, Dier H, Dorokhov V, Fast H, Gil M, Kyro E, Litynska Z, Mikkelsen IB, Molyneux MG, Nakane H, Notholt J, Rummukainen M, Viatte P, Wenger J (1997) Prolonged stratospheric ozone loss in the 1995–96 Arctic winter. Nature 389:835–838
Richard C, Canonica S (2005) Aquatic phototransformation of organic contaminants induced by coloured dissolved natural organic matter. Hdb Env Chem 2(Part M):299–323

Richard LE, Peake BM, Rusak SA, Cooper WJ, Burritt DJ (2007) Production and decomposition dynamics of hydrogen peroxide in freshwater. Environ Chem 4:49–54. doi:101071/EN06068
Rosenstock B, Simon M (2001) Sources and sinks of dissolved free amino acids and protein in a large and deep mesotrophic lake. Limnol Oaceanogr 50:90–101
Roy SC, Atreja SK (2008) Production of superoxide anion and hydrogen peroxide by capacitating buffalo (*Bubalus bubalis*) spermatozoa animal reproduction. Science 103:260–270
Rozendal RA, Leone E, Keller J, Rabaey K (2009) Efficient hydrogen peroxide generation frm organic matter in a bioelectrochemical system. Electrochem Commun 11:1752–1755
Rusak SA, Richard LE, Peake BM, Cooper WJ, Bodeker GE (2010) The influence of solar radiation on hydrogen peroxide concentrations in freshwater. Mar Freshwater Res 61:1147–1153
Safazadeh-Amiri A, Bolton JR, Cater SR (1997) Ferrioxalate-mediated photodegradation of organic pollutants in contaminated water. Water Res 31:2079–2085
Sakugawa H, Kaplan IR (1987) Atmospheric H_2O_2 measurement: comparison of cold trap method with impinger bubbling method. Atmos Environ 21:1791–1798
Sakugawa H, Kaplan IR, Tsai W, Cohen Y (1990) Atmospheric hydrogen peroxide. Environ Sci Technol 24:1452–1462
Sakugawa H, Kaplan IR, Shepard LS (1993) Measurements of H_2O_2, aldehydes and organic acids in Los Angeles rainwater: their sources and deposit rates. Atmos Environ 27B:203–219
Sakugawa H, Yamashita T, Fujiwara K (1995) Determination of hydrogen peroxide and organic peroxides in seawater. In: Tsunogai S, Iseki K, Koike I, Oba T (Ed), Global fluxes of carbon and its related substances in the Coastal Sea-Ocean-atmosphere system. M & J Intern pp 452–457
Sakugawa H, Takami A, Kawai H, Takeda K, Fujiwara K, Hirata S (2000) The occurrence of organic peroxide in seawater. In: Handa N, Tanoue E, Hama T (eds) Dynamics and characterization of marine organic matter. TERRAPUB/Kluwer, Tokyo, pp 231–240
Sakugawa H, Yamashita T, Kwai H, Masuda N, Hashimoto N, Makino S, Nakatani N, Takeda K (2006) Measurements, and production and decomposition mechanisms of hydroperoxides in air, rain, dew, river and drinking waters, Hiroshima prefecture Japan. Geochem 40:47–63 (In Japanese)
Samuilov VD, Bezryadnov DB, Gusev MV, Kitashov AV, Fedorenko TA (2001) Hydrogen peroxide inhibits photosynthetic electron transport in cells of cyanobacteria. Biochem (Moscow) 66:640–645
Samuilov VD, Timofeev KN, Sinitsyn SV, Bezryadnov DB (2004) H_2O_2-induced inhibition of photosynthetic O_2 evolution by Anabaena variabilis cells. Biochem (Moscow) 69:926–933
Sarthou G, Jeandel C, Brisset L, Amouroux D, Besson T, Donard OFX (1997) Fe and H_2O_2 distributions in the upper water column in the Indian sector of the Southern Ocean Earth. Planetary Sci Lett 147:83–92
Sauer F, Beck J, Schuster G, Moortgat GK (2001) Hydrogen peroxide, organic peroxides and organic acids in forested area during FIELDVOC'94 Chemosphere-Global changes. Science 3:309–326
Schmitt-Kopplin P, Hertkorn N, Schulten H-R, Kettrup A (1998) Structural changes in a dissolved soil humic acid during photochemical degradation processes under O_2 and N_2 atmosphere. Environ Sci Technol 32:2531–2541
Scully NM, Vincent WF (1997) Hydrogen peroxide: a natural tracer of stratification and mixing processes in subarctic lakes. Arch Hydrobiol 139:1–15
Scully NM, Lean DRS, McQueen DJ, Cooper WJ (1995) Photochemical formation of hydrogen peroxide in lakes: effects of dissolved organic carbon and ultraviolet radiation. Can J Fish Aquat Sci 52:2675–2681
Scully NM, McQueen DJ, Lean DRS, Cooper WJ (1996) Hydrogen peroxide formation: the interaction of ultraviolet radiation and dissolved organic carbon in lake waters along a 43–75°N gradient. Limnol Oceanogr 41:540–548
Scully NM, Vincent WF, Lean DRS, MacIntyre S (1998) Hydrogen peroxide as a natural tracer of mixing in surface layers. Aquat Sci 60:169–186

Senesi N (1990) Molecular and quantitative aspects of the chemistry of fulvic acid and its interactions with metal ions and organic chemicals: Part II The fluorescence spectroscopy approach. Anal Chim Acta 232:77–106

Serban A, Nissenbaum A (1986) Humic acid association with peroxidase and catalase. Soil Biol Biochem 18:41–44

Sikorsky RJ, Zika RG (1993a) Modeling mixed-layer photochemistry of H_2O_2: optical and chemical modeling of production. J Geophys Res 98:2315–2328

Sikorsky RJ, Zika RG (1993b) Modeling mixed-layer photochemistry of H_2O_2: physical and chemical modeling of distribution. J Geophys Res 98:2329–2340

Sinel'nikov VE (1971) Hydrogen peroxide level in river water, and methods for detecting it. Gibrobiol Zh 7:115–119 (Chem Abst 75:25016a, 1971)

Sinel'nikov VE, Demina VI (1974) Origin of hydrogen peroxide contained in the water of open reservoirs. Gidrokhim Mater 60:30–40 (Chem Abst 83:151980j, 1975)

Smith RC, Prezelin BB, Baker KS, Bidigare RR, Boucher NP, Coley T, Karentz D, MacIntyre S, Matlick HA, Menzies D et al (1992) Ozone depletion: ultraviolet radiation and phytoplankton biology in Antarctic waters. Science 255:952–959

Southworth BA, Voelker BM (2003) Hydroxyl radical production via the photo-Fenton reaction in the presence of fulvic acid. Environ Sci Technol 37:1130–1136

Stevens SE Jr, Patterson COP, Myers J (1973) The production of hydrogen peroxide by blue-green algae: A survey. J Phycol 9:427–430

Stolarski R, Bojkov R, Bishop L, Zereros C, Staehelin J, Zawodny J (1992) Measured trends in stratospheric ozone. Science 256:342–349

Szymczak R, Waite TD (1991) Photochemical activity in waters of the Great Barrier Reef. Estuar Coastal Shelf Sci 33:605–622

Takahashi M, Hama T, Matsunaga K, Handa N (1995) Photosynthetic organic carbon production and respiratory organic carbon consumption in the trophogenic layer of Lake Biwa. J Plankton Res 17:1017–1025

Takeda K, Takedoi H, Yamaji S, Ohta K, Sakugawa H (2004) Determination of hydroxyl radical photoproduction rates in natural waters. Anal Sci 20:153–158

Tanaka K, Suda Y, Kondo N, Sugahara K (1985) Ozone tolerance and the ascorbate-dependent hydrogen peroxide decomposing system in chloroplasts. Plant Cell Physiol 26:1425–1431

Theiler R, Cook JC, Hager LP, Siuda JF (1978) Halocarbon synthesis by bromoperoxidases. Science 202:1094–1096

Tizaoui C, Bouselmi L, Mansouri L, Ghrabi A (2007) Landfill leachate treatment with ozone and ozone/hydrogen peroxide systems. J Hazard Mater 140(1–2):316–324

US Environmental Protection Agency (2000) Improved enumeration methods for the recreational water quality indicators enterococci and escherichia coli, US EPA office of science and technology, 20460 Washington, March 2000, EPA/821/R-97/004

van Baalen C, Marler JE (1966) Occurrence of hydrogen peroxide in sea water. Nature 211:951

vedeneeva YA, Lobanov AV, Kholulskaya SN, Komissarov GG (2005) Abstracts of papers, VI all-Russia conference 'Molecular Modeling', Moscow, p 54

Vermilyea AW, Hansard SP, Voelker BM (2010a) Dark production of hydrogen peroxide in the Gulf of Alaska. Limnol Oceanogr 55:580–588

Vermilyea AW, Dixon TC, Voelker BM (2010b) Use of $H_2{}^{18}O_2$ to measure absolute rates of dark H_2O_2 production in freshwater systems. Environ Sci Technol 44:3066–3072

Vione D, Maurino V, Minero C, Borghesi D, Lucchiari M, Pelizzetti E (2003) New processes in the environmental chemistry of nitrite 2. The role of hydrogen peroxide. Environ Sci Technol 37:4635–4641

Vione D, Lauri V, Minero C, Maurino M, Malandrino M, Carlotti ME, Olariu RI, Arsene C (2009) Photostability and photolability of dissolved organic matter upon irradiation of natural water samples under simulated sunlight. Aquat Sci 71:34–45

Voelker BM, Morel FMM, Sulzberger B (1997) Iron redox cycling in surface waters: effects of humic substances and light. Environ Sci Technol 31:1004–1011

Voelker BM, Sedlak DL, Zafiriou OC (2000) Chemistry of superoxide radical in seawater: reactions with organic Cu complexes. Environ Sci Technol 34:1036–1042

von Sonntag C, Mark G, Mertens R, Schuchmann MN, Schuchmann H-P (1993) UV-radiation and/or oxidants in water pollution control. J Water Supply Res Technol—Aqua 42:201–211

Wade TJ, Pai N, Eisenberg JS, Colford JM (2003) Do US environmental protection agency water quality guidelines for recreational water prevent gastrointestinal illness? A systematic review and meta-analysis. Environ Health Perspect 111:1102–1109

Wang GS, Liao CH, Wu FJ (2001) Photodegradation of humic acids in the presence of hydrogen peroxide. Chemosphere 42:379–387

Wang W, Johnson CG, Takeda K, Zafiriou OC (2009) Measuring the photochemical production of carbon dioxide from marine dissolved organic matter by Pool isotope exchange. Environ Sci Technol 43:8604–8609

Weinbauer MG, Suttle CA (1999) Lysogeny and prophage induction in coastal and offshore bacterial communities. Aquat Microb Ecol 18:217–225

Westerhoff P, Aiken G, Army G, Debroux J (1999) Relationships between the structure of natural organic matter and its reactivity towards molecular ozone and hydroxyl radicals. Water Res 33:2265–2276

Wong GTF, Dunstan WM, Kim D-B (2003) The decomposition of hydrogen peroxide by marine phytoplankton La décomposition du peroxyde d'hydrogène par le phytoplancton marin. Oceanol Acta 26:191–198

Wu FC, Mills RB, Evans RD, Dillon PJ (2005) Photodegradation-induced changes in dissolved organic matter in acidic waters. Can J Fish Aqua Sci 62:1019–1027

Xie HX, Zafiriou OC (2009) Evidence for significant photochemical production of carbon monoxide by particles in coastal and oligotrophic marine waters. Geophys Res Lett 36:L23606. doi:101029/2009GL041158

Xie HX, Zafiriou OC, Umile TP, Kieber DJ (2005) Biological consumption of carbon monoxide in Delaware Bay, NW Atlantic and Beaufort Sea. Mar Ecol Prog Ser 290:1–14

Yamashita Y, Tanoue E (2004) In situ production of chromophoric dissolved organic matter in coastal environments. Geophys Res Lett 31 [doi: 101029/2004GL019734]

Yamashita Y, Tanoue E (2008) Production of bio-refractory fluorescent dissolved organic matter in the ocean interior. Nature Geosci 579-582 doi:101038/ngeo279

Yocis BH, Kieber DJ, Mopper K (2000) Photochemical production of hydrogen peroxide in Antarctic waters. Deep Sea Res Part I: Oceanogr Res Pap 47:1077–1099

Yoshioka T, Ueda S, Khodzher T, Bashenkhaeva N, Korovyakova I, Sorokovikova L, Gorbunova L (2002) Distribution of dissolved organic carbon in Lake Baikal and its watershed. Limnology 3:159–168

Yuan J, Shiller AM (2000) The variation of hydrogen peroxide in rainwater over the South and central Atlantic Ocean. Atmos Envriron 34:3973–3980

Yuan J, Shiller AM (2001) The distribution of hydrogen peroxide in the southern and central Atlantic ocean. Deep-Sea Res II 48:2947–2970

Zafiriou OC (1990) Chemistry of superoxide ion-radical ($O_2^{-}\cdot$) in seawater. I. *pK*asw(HOO) and uncatalyzed dismutation kinetics studied by pulse radiolysis. Mar Chem 30:31–43

Zafiriou OC, Voelker BM, Sedlak DL (1998) Chemistry of the superoxide radical ($O_2^{-}\cdot$) in seawater: reactions with inorganic copper complexes. J Phys Chem A 102:5693–5700

Zepp RG, Schlotzhauer PF (1983) Influence of algae on photolysis rates of chemicals in water. Environ Sci Technol 17:462–468

Zepp RG, Skurlatov YI, Pierce JT (1986) Algal-induced decay and formation of hydrogen peroxide in water. ACS Symp Ser 327, Am Chem Soc, pp 215–224

Zepp RG, Skurlatov YI, Pierce JT (1987a) Algal-induced decay and formation of hydrogen peroxide in water: its possible role in oxidation of anilines by algae. In: Zika RG and Cooper WJ (Ed), Photochemistry of environmental aquatic systems, ACS Symposium Series No. 327, Am Chem Soc, Washington, pp 213–224

Zepp RG, Braun AM, Hoigne J, Leenheer JA (1987b) Photoproduction of hydrated electrons from natural organic solutes in aquatic environments. Environ Sci Technol 21:485–490

Zepp RG, Faust BC, Hoigné J (1992) Hydroxyl radical formation in aqueous reactions (pH 3–8) of iron(II) with hydrogen peroxide: the photo-fenton reaction. Environ Sci Technol 26:313–319

Zhang Y, van Dijk MA, Liu M, Zhu G, Qin B (2009) The contribution of phytoplankton degradation to chromophoric dissolved organic matter (CDOM) in eutrophic shallow lakes: field and experimental evidence. Water Res 43:4685–4697

Zhang WJ, Yu Y, Wang XX (2010) Photocatralytic degradation of methyl orange in TiO_2 suspension-Ti electrode system. Abstrct in Bioinformatrics and Biomedical engineering 2010 4th International conference, 18–20 June 2010, Chengdu, China

Zhao Y, Yu Y, Feng W, Shen Y (2003) Growth and production of free-living heterotrophic nanoflagellates in a eutrophic lake-Lake Donghu, Wuhan, China. Hydrobiol 498:85–95

Zika RG, Moffett W, Petasne RG, Cooper WJ, Saltzman ES (1985a) Spatial and temporal variations of hydrogen peroxide in Gulf of Mexico waters. Geochim Cosmochim Acta 49:1173–1184

Zika RG, Saltzman ES, Cooper WJ (1985b) Hydrogen peroxide concentrations in the Peru upwelling area. Mar Chem 17:265–275

Zuo Y, Hoigné J (1992) Formation of hydrogen peroxide and depletion of oxalic acid in atmospheric water by photolysis of iron(III)-oxalato complexes. Environ Sci Technol 26:1014–1022

Zuo Y, Hoigné J (1993) Evidence for photochemical formation of H_2O_2 and oxidation of SO_2 in authentic fog water. Science 260:71–73

Photoinduced Generation of Hydroxyl Radical in Natural Waters

Khan M. G. Mostofa, Cong-qiang Liu, Hiroshi Sakugawa, Davide Vione, Daisuke Minakata, M. Saquib and M. Abdul Mottaleb

1 Introduction

Hydroxyl radical ($HO^{\bullet}$) is a short-lived free radical, and it is the most potent oxidizing transient among the reactive oxygen species. It is an effective, nonselective and strong oxidant that is ubiquitously formed in natural sunlit surface waters (rivers, lakes and seawater and so on), rain, dew, cloud, fog, snow, aerosol, and in all living organisms. The $HO^{\bullet}$ is photolytically formed from a variety of sources in natural waters. The first experimental report of a reaction that is now known to produce $HO^{\bullet}$ dates back to Henry John Horstman Fenton, who described the oxidation

K. M. G. Mostofa (✉) · C. Q. Liu
State Key Laboratory of Environmental Geochemistry, Institute of Geochemistry, Chinese Academy of Sciences, Guiyang 550002, China
e-mail: mostofa@vip.gyig.ac.cn

H. Sakugawa
Department of Environmental Dynamics and Management, Graduate School of Biosphere Science, Hiroshima University, 1-7-1, Kagamiyama, Higashi-Hiroshima 739-8521, Japan

D. Vione
Dipartimento di Chimica Analitica, University of Turin, I-10125 Turin, Italy

Centro Interdipartimentale NatRisk, I-10095 Grugliasco, (TO), Italy

D. Minakata
School of Civil and Environmental Engineering, Georgia Institute of Technology, 828 West Peachtree Street, Suite 320, Atlanta, GA 30332, USA

M. Saquib
Department of Chemistry, Aligarh Muslim University, Aligarh 202002, Uttar Pradesh, India

M. A. Mottaleb
Department of Chemistry/Physics Northwest Missouri State University, Center for Innovation and Entrepreneurship (CIE), 800 University Drive, Maryville, MO 64468, USA

K. M. G. Mostofa et al. (eds.), *Photobiogeochemistry of Organic Matter*, Environmental Science and Engineering, DOI: 10.1007/978-3-642-32223-5_3,

of Fe(II) with H_2O_2 in aqueous media (Fenton 1894). The Fenton's reaction has been studied by several researchers afterwards (Haber and Weiss 1934; Barb et al. 1951; Hardwick 1957; Wells and Salam 1967, 1968; Po and Sutin 1968; Skinner et al. 1980; Rush and Bielski 1985; Moffett and Zika 1987a, b; Lloyd et al. 1997; Kremer 1999; Lindsey and Tarr 2000). Haber and Weiss in 1934 firstly postulated that the reactivity of the Fenton's reagent is due to the generation of $HO^{\cdot}$ in aqueous solution, and that Fe(II) acts as a catalyst for the decomposition of H_2O_2 into $HO^{\bullet}$. The Fenton's reaction can be used to promote the oxidation of organic compounds (Walling 1975) and has been widely studied to this purpose in the last 25 years (Sychev and Isak 1995; Chen and Pignatello 1997; Gallard et al. 1998; Barbeni et al. 1987; Lindsey and Tarr 2000; Kang et al. 2002; Pignatello et al. 2006).

Hydroxyl radical is also a photo-product of many photolysis reactions that occur in natural waters (Zafiriou 1974; Zafiriou and True 1979a, b; Mill et al. 1980; Draper and Crosby 1981; Russi et al. 1982; Zafiriou et al. 1984; Cooper et al. 1988; Mopper and Zhou 1990; Gjessing and Källqvist 1991; Dister and Zafiriou 1993; Takeda et al. 2004; Vione et al. 2006, 2009a, b; al Housari et al. 2010). In particular, $HO^{\bullet}$ can be produced photolytically from NO_2^- and NO_3^- (Zafiriou and True 1979a, b; Russi et al. 1982; Takeda et al. 2004; Zafiriou and Bonneau 1987; Zepp et al. 1987; Zellner et al. 1990; Brezonik and Fulkerson-Brekken 1998; Mack and Bolton 1999) and upon irradiation of various dissolved organic matter (DOM) components (Mill et al. 1980; Mopper and Zhou 1990; Vaughn and Blough 1998; Holder-Sandvik et al. 2000). Hydroxyl radical can be experimentally determined by use of selective probe molecules such as cumene (isopropylbenzene) and pyridine in dilute solution, benzene, terephthalic acid and and *para*-chlorobenzoic acid (*p*CBA) (Mill et al. 1980; Takeda et al. 2004; Fang et al. 1996). The rate of $HO^{\bullet}$ production mostly depends on the quantity and quality of DOM, on the concentration of other chemical species such as nitrate and nitrite, and on the pH of natural waters.

The chemical reactivity of the Fenton's reaction (Fe^{2+} and H_2O_2) is significantly increased by UV/Visible irradiation ($\lambda < 580$ nm), which has for instance been shown to enhance the mineralization of organic pollutants (Haag and Hoigné 1985; Cooper et al. 1991; Zepp et al. 1992; Ruppert et al. 1993; Faust 1994; Voelker et al. 1997; Arakaki et al. 1998; Bossmann et al. 1998; Rossetti et al. 2002; Zepp 2002; Southworth and Voelker 2003; White et al. 2003). Similarly, the H_2O_2/UV process can produce $HO^{\bullet}$ that can decompose organic substances in aqueous solution (Draper and Crosby 1981; Zellner et al. 1990; Hunt and Taube 1952; Baxendale and Wilson 1956; Volman and Chen 1959; Ho 1986; Vel Leitner and Dore 1996; Berger et al. 1999; Wang et al. 2001; Goldstein and Rabani 2008) as well as in ice (Chu and Anastasio 2005). An advanced process that exploits the photo-Fenton system is the photo-ferrioxalate/H_2O_2 reaction, where UV/visible irradiation ($\lambda < 550$ nm) is combined with the presence of excess oxalate (Huston and Pignatello 1996; Safazadeh-Amiri et al. 1996, 1997; Wu et al. 1999; Arslan et al. 2000; Nogueira and Guimaraes 2000; Emilio et al. 2002; Lee et al. 2003; Hislop and Bolton 1999; Jeong and Yoon 2005). The $HO^{\bullet}$ radical can also be generated in aqueous suspensions of

TiO_2, which plays a key role in the heterogeneous photocatalytic degradation of organic contaminants (Sun and Bolton 1996; Ullah et al. 1998; Konstantinou and Albanis 2004). However, an important difference between TiO_2 photocatalysis and the other processes of $HO^•$ generation described before is that the irradiation of TiO_2 mainly causes the production of surface-bound $HO^•$ groups, which are somewhat less reactive than homogeneous $HO^•$ (Serpone and Pelizzetti 1989). The hydroxyl radical has been detected in rainwater, dew, cloud and fog (Arakaki et al. 1998, 1999a, b; Arakaki and Faust 1998; Nakatani et al. 2001; Kobayashi et al. 2002), snow (Chu and Anastasio 2005; Anastasio et al. 2007; Matykiewiczová et al. 2007), aerosols (Anastasio and Jordan 2004), in aqueous extracts of cigarette tar (Zang et al. 1995), and in living organisms (Buettner et al. 1978; Buettner 1987; Miller et al. 1990; Buettner and Jurkiewicz 1996; Cadet et al. 1999; Bourdat et al. 2000; Paradies et al. 2000; Blokhina et al. 2003; Li et al. 2008). The $HO^•$ is rapidly consumed in natural waters by the subsequent reactions with dissolved organic compounds (Schuchmann and von Sonntag 1979; Neta et al. 1988; Westerhoff et al. 1999; Goldstone et al. 2002; Miller and Chin 2002; Miller et al. 2002; Ervens et al. 2003) and several inorganic species (Zafiriou et al. 1984, 1987; Brezonik and Fulkerson-Brekken 1998; Neta et al. 1988; Song et al. 1996).

The generation of $HO^•$ and its interaction with the dynamics of DOM and nutrient as well as with aquatic organisms are very important in natural waters. There are a number of factors that can control the production and consumption of $HO^•$ in that ecosystem. However, there is no general overview published on $HO^•$ in natural waters. A short review by von Sonntag (2007) covers the formation of free radicals and their reactions in aqueous solution.

This review will provide a general overview on sources, production mechanisms, steady state concentration and biogeochemical functions of $HO^•$ in water environment. This paper also discusses the analytical methods that can be adopted to measure the photoinduced generation of $HO^•$, the factors controlling its production and decay, as well as the significance and impact of $HO^•$ in the aquatic ecosystems. It is shown how the production of $HO^•$ differs among DOM components, as well as between freshwaters and marine environments.

2 Hydroxyl Radical ($HO^•$) and Other Free Radical Species

The hydroxyl radical ($HO^•$) is the most powerful oxidizing agent among the photolytically generated ones. It is a short-lived, highly reactive and nonselective transient, able to oxidize dissolved organic substances and other chemical species in natural waters. The oxidation potentials for a series of common oxidants in surface waters is as follows: Fluorine ($E = 3.03$ V) > $HO^•$ (2.80 V) > Atomic oxygen (2.42 V) > Ozone (2.07 V) > Peracetic acid (ROOH) (1.80 V) > H_2O_2 (1.78 V) > Perhydroxyl radical (1.70 V) > Potassium permanganate (1.68 V) > Chlorine dioxide (1.57 V) > Hypochlorous acid (1.49 V) > Chlorine

Table 1 Oxidation potentials of major oxidants

Free radicals	Oxidation potentials (E°) (V)
Fluorine	3.03
Hydroxyl radical	2.80
Atomic oxygen	2.42
Ozone	2.07
Hydrogen peroxide	1.78
Perhydroxyl radical	1.7
Permanganate	1.68
Chlorine dioxide	1.57
Hypochlorous acid	1.49
Chlorine	1.36

Data source Sun et al. (1997)

(1.36 V) whilst one (Table 1) (Sun et al. 1997). The oxidizing capacity of the hydroxyl radical can be described in terms of its reduction potential (E), which allows the comparison with other powerful oxidants (Buettner and Jurkiewicz 1996; Buettner 1993; Ross et al. 1994). One-electron reduction potentials at pH 7.0 for selected radical couples are 2.31 V for $HO^•$, H^+/H_2O; 1.60 V for $RO^•$, H^+/ROH (aliphatic alkoxyl radical); 1.00 V for $ROO^•$, H^+/ROOH (alkyl peroxyl radical); 0.92 V for $GS^•/GS^-$ (glutathione); 0.60 V for $PUFA^•$, H^+/PUFA-H) (*bis*-allylic-H); 0.59 V for $HU^{•-}$, H^+/UH^{2-} (urate); 0.48 V for $TO^•$, H^+/TOH (tocopherol); 0.32 V for H_2O_2, H^+/H_2O, $HO^•$; 0.28 V for ascorbate$^{•-}$, H^+/ascorbate monoanion; 0.12 V for Fe(III)EDTA/Fe(II)EDTA; and −3.30 V for $O_2/O_2^•$ (Buettner and Jurkiewicz 1996; Buettner 1993). The $HO^•$ reacts with organic compounds at close to diffusion-limited rate constants, which are the fastest after equilibrium reactions and the rate constants (k_{obs}) for the reaction of the equilibrium mixture of ascorbic acid species ($AscH_2/AscH^-/Asc^{2-}$ at pH 7.4) are 1.1×10^{10} M^{-1} s^{-1} for $HO^•$; 1.6×10^9 M^{-1} s^{-1} for tert-Butyl alkoxyl radical ($RO^•$); $1–2 \times 10^6$ M^{-1} s^{-1} for Alkyl peroxyl radical, e.g. $CH_3OO^•$ ($ROO^•$); 1.8×10^8 M^{-1} $s^{-1}$for $ClCOO^•$; 6×10^8 M^{-1} s^{-1} for glutathiol radical ($GS^•$); 1×10^6 M^{-1} s^{-1} for urate radical ($HU^{•-}$); 2×10^5 M^{-1} s^{-1} for tocopheroxyl radical ($TO^•$); 2×10^5 M^{-1} s^{-1} for dismutation ($Asc^{•-}$); 1.4×10^9 M^{-1} s^{-1} for chlorpromazine radical action ($CPZ^{+•}$); $\approx 10^2$ M^{-1} s^{-1} for Fe^{III}-EDTA/ Fe^{II}-EDTA; and 1×10^5 M^{-1} s^{-1} for $O_2^{•-}/HO_2^•$ (Buettner and Jurkiewicz 1996; Buettner 1988; Ross et al. 1994). The $HO^•$ radical is formed by a variety of sources such as NO_2^- and NO_3^- under UV irradiation, the Fenton and the photo-Fenton reaction, the photo-ferrioxalate/H_2O_2 system and so on (Legrini et al. 1993). It is directly responsible for a number of important biogeochemical functions in natural waters.

Among other radical species present in natural waters, organic peroxy radicals ($ROO^•$) are intermediates formed photolytically and thermally from organic peroxides, or directly from the degradation of dissolved organic matter. These radicals are short-lived and highly reactive transients. An important process that involves $ROO^•$ is the formation of new organic compounds upon rapid combination of peroxide radicals with organic substances in aqueous solution (Mill et al. 1980; Mageli and Kolczynski 1966; Faust and Hoigne 1987; Blough 1988; Kieber and Blough 1990; Sakugawa et al. 1990; Faust and Allen 1992; Mostofa and Sakugawa 2009). Furthermore, the thermal decomposition of organic peroxides can initiate the polymerization of vinyl monomers or induce cross-linking of a polymeric substrate upon formation of free radical sites on the polymer (Mageli and Kolczynski 1966). The overall (unspeciated) photostationary-state concentration of peroxyl radicals in sunlit cloud and fog waters is around 1–30 nM (Faust and Allen 1992).

The superoxide radical anion ($O_2^{\bullet-}$) is the one-electron reduction product of molecular oxygen. It is an early photoinduced and short-lived intermediate that is formed in chemical reactions occurring in natural waters, where oxygen acts as the ultimate electron acceptor (Jeong and Yoon 2005; Bielski et al. 1985; Petasne and Zika 1987; Zafiriou 1990; Micinski et al. 1993; Zafiriou et al. 1998; Millington and Maurdev 2004). It has been shown that the photoinduced superoxide production rates are 0.1–6.0 nM min^{-1} under full-sun irradiation in spring, and 0.2–8.0 nM min^{-1} in fall in a variety of Eastern Caribbean waters (Micinski et al. 1993). A key reaction of $O_2^{\bullet-}$ is the production of H_2O_2 by dismutation; hydrogen peroxide is then able to generate $HO^{\bullet}$ by direct photolysis or upon photo-Fenton type reactions in sunlit aqueous solutions (Cooper et al. 1988; Micinski et al. 1993; Fischer et al. 1985). Interestingly, the organic complexes of Cu as well as the copper-catalyzed dismutation (involving Cu^{+} and Cu^{2+}) can be significant sinks of photoproduced $O_2^{\bullet-}$ in seawater (Zafiriou et al. 1998; Voelker et al. 2000).

The carbon dioxide radical anion ($CO_2^{\bullet-}$) is a short-lived and highly reactive intermediate that is photolytically formed in the ferrioxalate reaction system. It is a strong oxidizing agent that is able to oxidize metals and other chemical species in aqueous solution. The $CO_2^{\bullet-}$ is formed photolytically ($C_2O_4^{\bullet-} \rightarrow CO_2^{\bullet-} + CO_2$; $k = 2 \times 10^6$ s^{-1}) upon decomposition of the oxalyl radical anion ($C_2O_4^{\bullet-}$). The latter is produced by the photoinduced decomposition of the highly photosensitive ferrioxalate complex $[Fe(C_2O_4)_3]^{3-}$ in aqueous solution (Hislop and Bolton 1999; Jeong and Yoon 2004, 2005; Mulazzani et al. 1986). A key role played by $CO_2^{\bullet-}$ is its ability to oxidize the metal ions, therefore affecting the biogeochemical cycling of metal-containing species. These processes can have an impact on the generation of $HO^{\bullet}$ and of the superoxide radical anion in natural waters (Hislop and Bolton 1999; Jeong and Yoon 2004, 2005; Wang et al. 2010). Another potentially important process is the transformation of organic substances induced by $CO_2^{\bullet-}$, which is formed photolytically from ferrioxalate complexes in the aqueous solution (Huston and Pignatello 1996).

In addition, it has been reported that quinones photolytically produce species capable of hydroxylation (Alegria et al. 1997; Pochon et al. 2002; Gan et al. 2008; Maurino et al. 2008; Maddigapu et al. 2010; Page et al. 2011). Some of these quinone-derived hydroxylating species exhibit reactivity that is around one order of magnitude lower than free $HO^{\bullet}$ (Pochon et al. 2002; Gan et al. 2008). It is hypothesized that quinone-derived hydroxylating species may contribute at least in part to the photoinduced $HO^{\bullet}$ production by DOM (Vaughn and Blough 1998; Page et al. 2011).

2.1 Sources of $HO^\bullet$ in Natural Waters

The $HO^\bullet$ radical is formed photolytically from various sources in natural waters. In rivers, contributions to $HO^\bullet$ photoproduction are 1–89 % from NO_2^-, 2–70 % from NO_3^-, 1–50 % from H_2O_2, and 2–70 % from the photo-Fenton reaction and/or irradiated CDOM (Takeda et al. 2004; Vione et al. 2006; White et al. 2003; Page et al. 2011; Nakatani et al. 2007; Mostofa KMG and Sakugawa H, unpublished data). Experimental studies show that DOM isolates from rivers may contribute up to 50 % of the hydroxylation through production of H_2O_2 (Page et al. 2011). The results demonstrate that NO_2^- is a key contributor (48–80 %) for $HO^\bullet$ production in sewerage-polluted river waters, but NO_3^- can be a major contributor (16–49 %) in clean river waters. In seawater the major sources of $HO^\bullet$ are 7–75 % from NO_2^-, 1–8 % from NO_3^-, 0–1 % from H_2O_2, and 24–93 % from unknown sources. These data were obtained from a study carried out in Seto Inland and the Yellow Sea (Takeda et al. 2004). The formation of $HO^\bullet$ from different sources in natural waters can be distinguished as: (i) the photolysis of nitrite and nitrate in the aqueous solution (Mopper and Zhou 1990; Takeda et al. 2004; Zepp et al. 1987); (ii) the irradiation of CDOM components via formation of H_2O_2 in the aqueous solution. In this case the production of $HO^\bullet$ depends on the nature of the CDOM components (Fig. 1) (White et al. 2003; Mostofa and Sakugawa 2009; Mostofa KMG and Sakugawa H, unpublished data), but a useful correlation has been found between the formation rate of $HO^\bullet$ and the content of dissolved organic carbon in different lake water samples (Vione et al. 2006); (iii) the Fenton reaction (Fenton 1894; Walling 1975; Kang et al. 2002), the photo-Fenton reaction (Zepp et al. 1992; Arakaki et al. 1998; Southworth and Voelker 2003) as well as the photo-ferrioxalate/H_2O_2 system in natural waters (Southworth and Voelker 2003; Safazadeh-Amiri et al. 1997; Hislop and Bolton 1999); (iv) the direct photolysis of hydrogen peroxide, i.e. UV/H_2O_2 processes in aqueous solution (Draper and Crosby 1981; Wang et al. 2001). The UV irradiation of natural waters can produce H_2O_2 that further yields $HO^\bullet$ (Gjessing and Källqvist 1991; Cooper et al. 1996); (v) the reaction of hydroperoxide radical ($HO_2^\bullet$) with NO ($HO_2^\bullet + NO \rightarrow HO^\bullet + NO_2$) (Sakugawa et al. 1990); (vi) the photolysis of dimeric $\left[Fe_2(OH)_2(H_2O)_8\right]^{4+}$ species in aqueous solution (Langford and Carey 1975); (vii) the photolysis of $Fe^{III}(OH)^{2+}$ in aqueous solution. The generation of $HO^\bullet$ upon photolysis of $Fe^{III}(OH)^{2+}$ is very efficient (quantum yield ~0.2), but the Fe(III) hydroxocomplex is present in significant concentration only at strongly acidic pH values that have little environmental significance(Jeong and Yoon 2005; Pozdnyakov et al. 2000); (viii) the generation of singlet states of oxygen atoms ($^1O_1$) by ozonolysis, followed by reaction with H_2O to form $HO^\bullet$ (Hoigné and Bader 1978, 1979; Staehelin and Hoigné 1985; Takahashi et al. 1995); (ix) the reaction of O_3 with H_2O_2 (peroxone process), which generates $HO^\bullet$ ($H_2O_2 + 2O_3 \rightarrow 2HO^\bullet + 3O_2$) (Hoigné 1998); (x) the production of $HO^\bullet$ by auto-oxidation of cytotoxic agents (Cohen and Heikkila 1974); (xi) chemical effects of ultrasound, which can generate $HO^\bullet$ in aqueous solution (Makino et al. 1983); (xii) ultrasound-induced cavitation in aqueous solution, yielding $HO^\bullet$ upon water splitting (H_2O + ultrasound $\rightarrow HO^\bullet$,

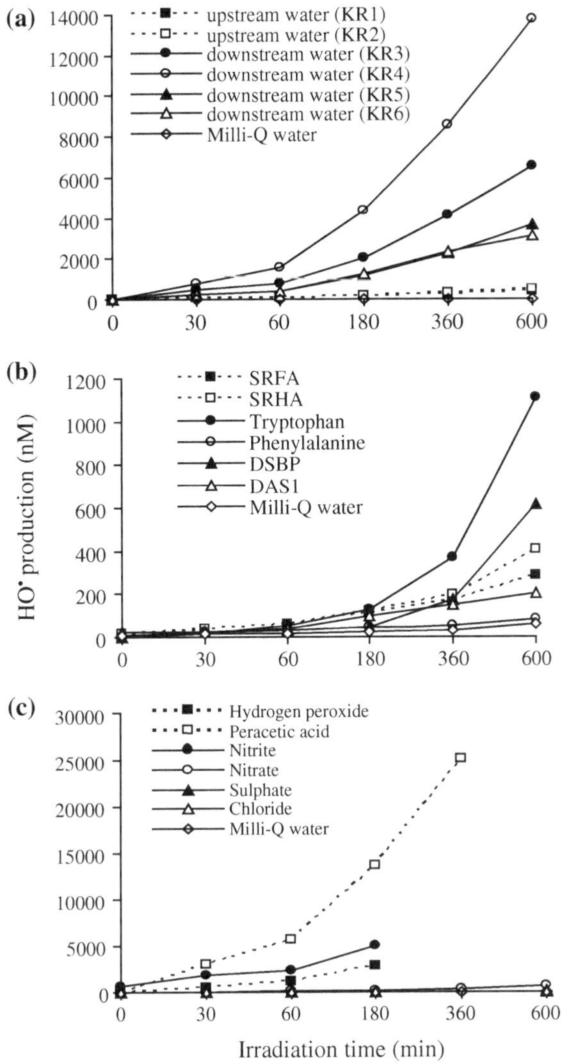

Fig. 1 Photoinduced generation of HO• from river waters (**a**), various standard organic substances (**b**) and various (inorganic and organic) chemical species (**c**) in photoexperiments conducted using a solar simulator. Aqueous solutions (1 mg L^{-1}) of standard all organic substances are used for production of HO radicals in (**b**) and all chemical species in (**c**) are adjusted to 100 μM. All data depicted in these figures are calibrated for natural sunlight on 6 July 2004 at Hiroshima University Campus at noon under clear sky conditions. *Data source* Mostofa KMG and Sakugawa H (unpublished data)

H_2O_2, H_2) (Henglein 1987); (xiii) autooxidation of aqueous extracts of cigarette tar (ACT), giving HO• in air-saturated, buffered aqueous solutions. It is thought that the process is caused by the autooxidation of hydroquinone- and catechol-related species in ACT (Zang et al. 1995); (xiv) photoinduced HO• production from aqueous suspensions of algae(Li et al. 2008); and (xv) photoinduced HO• production can occur from DOM, the reactive triplet states of which could be involved in oxidation of water and/or OH^- and in the production of lower energy hydroxylating species that simulate DOM reactivity (Alegria et al. 1997; Pochon et al. 2002; Gan et al. 2008; Maurino et al. 2008; Maddigapu et al. 2010; Page et al. 2011; Maddigapu et al. 2011; Brigante et al. 2010; Sur et al. 2011).

2.2 Biogeochemical Functions of HO• in Natural Waters

The HO• is responsible for the occurrence of many important biogeochemical functions in natural waters: (i) photoinduced decomposition of DOM, which causes the production of a number of low molecular weight (LMW) photoproducts. The latter are microbiologically labile and constitute a significant source of carbon and energy to the microbial food chains, as well as an important pathway for DOM turnover in natural waters (Zhou and Mopper 1990; Blough and Zepp 1995; Tranvik 1992; Moran and Zepp 1997; Bertilsson and Tranvik 1998; Mopper and Kieber 2000; Mostofa et al. 2009a, b). However, despite the major role played by HO• in the mineralization processes of organic pollutants in the framework of the AOPs, the HO• is expected to contribute to a minor extent to the photomineralization of natural DOM in surface waters (Vione et al. 2009); (ii) photoinduced production of low-molecular weight chemical species such as H_2O_2 and CO_2 (both dissolved and gaseous forms). These processes play some role in the occurrence of photosynthesis, which produces algal biomass that is involved into the generation of autochthonous DOM in natural waters (Mostofa et al. 2009a, b; Komissarov 2003; Fu et al. 2010). However, the importance of such reactions is limited by the relatively low generation rate of HO• in surface waters (Brezonik and Fulkerson-Brekken 1998); (iii) photo-bleaching of DOM induced by solar radiation in waters (Moran et al. 2000; del Vecchio and Blough 2002; Mostofa et al. 2005, 2007); (iv) photodegradation of persistent organic pollutants, which are usually recalcitrant to biological, chemical, and direct photodegradation in water (Brezonik and Fulkerson-Brekken 1998; Haag and Yao 1992; Grannas et al. 2006; Vione et al. 2009); (v) cycling of transition metal ions that can be oxidized by HO• (Jeong and Yoon 2004; Faust and Zepp 1993; Kwan and Voelker 2002); (vi) use of HO• in water treatment processes such as the Advanced Oxidation Technology (AOT), to purify sewerage or industrial wastewater effluents, with the purpose of controlling the organic pollution for sustainable development (Safazadeh-Amiri et al. 1996, 1997; Kang et al. 2000); (vii) damage to macromolecules such as DNA, proteins and lipids, membrane leakage, breakdown of the cellular metabolism, and finally of tissues in biological systems. These processes can be induced by the HO•, alkoxyl (RO•) and peroxyl (ROO•) radicals, which may be produced by the autooxidation of biomolecules such as ascorbate, catecholamines or thiols in organisms (Paradies et al. 2000; Blokhina et al. 2003; Berlett and Stadtman 1997; von Sonntag 2006).

2.3 Steady-State Concentration and Life-Time of HO• in Natural Waters

The steady state concentration of HO• can be determined on the basis of its major sources, which control the total photoinduced formation rate constants, and on sinks or scavengers of HO• that control the total consumption rate constants in natural waters (al Housari et al. 2010; Brezonik and Fulkerson-Brekken 1998; Arakaki et al. 1999b; Hoigné et al. 1989; Schwarzenbach et al. 1993; Nakatani et al. 2004). DOM and carbonate are the major scavengers or sinks of $HO^{\bullet}$ in freshwaters (White et al. 2003; Nakatani et al. 2004), but in seawater the bromide (Br^-) ions are actually the main scavengers (Song et al. 1996; Nakatani et al. 2004; Zafiriou et al. 1987). The steady-state concentration of $HO^{\bullet}$ shows a large variability in natural waters. Examples of concentration values reported in the literature are $(3.0\text{–}8.5) \times 10^{-16}$ M in rivers (Brezonik and Fulkerson-Brekken 1998; Arakaki et al. 1999b; Nakatani et al. 2004), $(9.41 \pm 0.12) \times 10^{-17}$ M to $(1.72 \pm 0.01) \times 10^{-16}$ M in estuarine waters (al Housari et al. 2010), 12×10^{-18} M in coastal surface seawater and 1.1×10^{-18} M in the open ocean (Mopper and Zhou 1990). In Antarctic waters the steady-state concentrations have been determined as 4.3×10^{-19} M in coastal waters and 2.6×10^{-19} M in the open ocean (Qian et al. 2001). Elevated $HO^{\bullet}$ concentration values (from 6.7×10^{-15} to 4.0×10^{-12} M) have been described in surface stream waters contaminated with acidic mine drainage (AMD). These waters have pH ~2.1–3.4, are highly rich of iron (6–1203 mg L^{-1}) and have a high concentration of NO_3^- (5.9×10^{-6}–5.8×10^{-3} M) (Allen et al. 1996). The reported data suggest that the steady-state concentration of $HO^{\bullet}$ can be very variable in different water systems. A major caveat that should be considered while comparing different studies is that the irradiation conditions are usually unequal, which accounts for at least part of the variability. However, variations in the steady-state $HO^{\bullet}$ concentration have also been observed with waters of different origin under the same irradiation conditions. The major factors that account for the variation of the steady-state concentration of $HO^{\bullet}$ in the aquatic environments are: (i) presence of elevated concentrations of NO_2^- and NO_3^- ions; (ii) presence of the elevated amounts of Fe^{3+}-containing complexes; (iii) occurrence of photo Fenton-type reactions that take place between H_2O_2 and reduced transition metal ions; (iv) amount and nature of the dissolved organic matter (DOM).

The reciprocal of the consumption rate constant allows the assessment of the life-time of $HO^{\bullet}$, which is $(2.6\text{–}6.0) \times 10^{-6}$ s in river, dew and cloud water (Arakaki and Faust 1998; Arakaki et al. 1999b; Nakatani et al. 2004) and several times higher ($3.0\text{–}66.0 \times 10^{-6}$ s) in remote polluted clouds, as estimated from a modeling study (Jakob 1986).

2.4 An HPLC Method for Measuring $HO^{\bullet}$ in Irradiated Natural Waters, Based on Benzene as Probe Molecule

This section reports a detailed description of a possible method that can be adopted for the determination of $HO^{\bullet}$, based on benzene as a probe. The description is very detailed to enable the reader easily reproducing a similar experimental set-up. Note that other probe molecules can also be used for $HO^{\bullet}$ determination, e.g. cumene (isopropylbenzene), pyridine and terephthalic acid (Mill et al. 1980;

Fang et al. 1996). Other substrates, such as benzoic acid and nitrobenzene are less suitable as $HO^\bullet$ probe molecules (Vione et al. 2010). The photoinduced generation of $HO^\bullet$ can be quantitatively determined by measuring the phenol (Eq. 2.1) that is produced photolytically from benzene + $HO^\bullet$, when natural waters under illumination are added with benzene. For irradiation it is possible to adopt a quartz cell under a Xe lamp or sunlight (Takeda et al. 2004; Mostofa KMG and Sakugawa H, unpublished data). Note that the light of a xenon lamp should be equipped with special glass filters to filter out the radiation below 300 nm, if one wants to simulate sunlight irradiation.

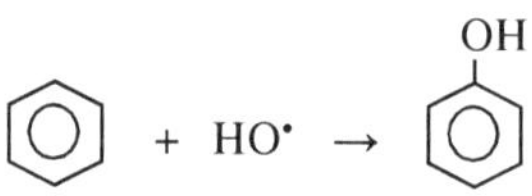

(2.1)

The phenol concentration can be determined by a HPLC method.

Experimental Details

The benzene solutions (e.g. a ~1 mM stock solution) should be prepared by benzene addition to water, followed by gentle shaking. The solution should be kept for 24-h under dark conditions to mix up benzene well with the water sample. The light intensity of the artificial Xe lamp or sunlight can be determined by measuring the photo-degradation rate of 8-μM standard aqueous solution of 2-nitrobenzaldehyde (2-NB) after illumination in a quartz cell (e.g. 60 mL). The illumination time of 2-NB should usually be kept short, e.g. up to 5 min for an irradiation intensity comparable to that of sunlight. In the case of the transformation reaction of benzene into phenol the irradiation time should be longer (up to 10–20 h), except for samples with unusually elevated $HO^\bullet$ production rate. 2-NB can be measured by HPLC–UV. Elution can be carried out with a C18 reverse-phase column. Upon adoption of $H_2O:CH_3CN = 40:60$ as isocratic eluent at a flow rate of 1 ml min^{-1}, if one employs a (5 μm, 4.6 × 250 mm) column the retention time of 2-NB could be something around 5 min. The recommended detection wavelength is 260 nm. It is also recommended to remove the air from the eluent before use, by 20–30 min sonication or by magnetic stirring under vacuum.

Phenol can be determined under the same elution conditions, setting the UV detection wavelength at 210 nm or adopting a fluorescence detector. In the latter case, recommended wavelengths are 270 nm for excitation and 297 nm for emission.

Calculations: The concentration of phenol that are produced can be estimated by comparison with a standard. The photo-formation rate of $HO^\bullet$ (r_{HO}) can be calculated from Eq. 2.2 (Takeda et al. 2004; Nakataniet al. 2004):

$$r_{HO} = \frac{r_P}{F_{B,HO} \times Y_P} \tag{2.2}$$

where r_P is the photo-formation rate of phenol obtained experimentally ($M\ s^{-1}$), $F_{B,HO}$ is the fraction of HO radicals that react with benzene [i.e., $k_{HO/Ph} \times C_{HO} \times C_{Ph}/(k_{HO/Ph} \times C_{HO} \times C_{Ph} + \sum k_i\ C_{HO}\ C_i)$, where i is scavenger], and Y_P is the yield of phenol formed per benzene oxidized by $HO^•$. It is $Y_P = 0.75 \pm 0.07$ in natural waters (Arakaki and Faust 1998). The $F_{B,OH}$ values are much variable for a variety of natural waters, also depending on the concentration of added benzene and on the amount of the natural $HO^•$ scavengers. For addition of 1.2 mM benzene it has been found $F_{B,HO} = 0.94$ for cloud waters (Arakaki and Faust 1998), 0.92 and 0.99 for rivers, and 0.68 for seawaters (Takeda et al. 2004). The high values of $F_{B,HO}$ in rivers and cloud suggest that most of the HO radicals formed photolytically reacts with benzene. In contrast, the low value in seawater samples suggests the 32–34 % of photolytically formed HO radicals reacts with various scavengers of HO radical other than benzene. There are many scavengers, DOM components, HCO_3^-, CO_3^{2-}, NO_2^-, halides (X^-, but chloride only in acidic medium) etc. that can interact with $HO^•$ in aqueous solution (Zafiriou 1974; Mopper and Zhou 1990; Vione et al. 2006; Zepp et al. 1987; Voelker and Sulzberger 1996; Minakata et al. 2009). In seawater, the bromide ion (Br^-) alone scavenges approximately 93 % of photo-generated $HO^•$ (Mopper and Zhou 1990).

To mathematically derive the terms in Eq. 2.2 for any added benzene concentration, one requires a kinetic model where the scavengers and benzene simultaneously react with photogenerated $HO^•$ at the rates R_{SC} and R_B, respectively (Takeda et al. 2004). Under the steady-state condition, the formation rate of $HO^•$ is equal to the consumption rate as follows (Takeda et al. 2004):

$$r_{HO} = r_{SC} + r_B \tag{2.3}$$

$$r_{SC} = k_{SC}[SC][HO]_{SS} = k'_{SC}[HO]_{SS} \tag{2.4}$$

$$r_S = k_B[B][HO]_{SS} \tag{2.5}$$

where k_{SC} is the reaction rate constant of HO radicals with various scavengers in the water sample, $k'_{SC} = k_{SC}$ *[SC]* is the apparent scavenging rate constant of the HO radical, k_B is the reaction rate constant of $HO^•$ with benzene (i.e., $7.8 \times 10^9\ M^{-1}\ s^{-1}$), [B] is the concentration of benzene added to the water sample (e.g. ~1 mM), and $[HO]_{SS}$ is the steady-state concentration of $HO^•$. Under conditions where benzene is in excess (i.e., $r_B >> r_{SC}$), most of the photo-generated $HO^•$ react with benzene, thus $F_{B,HO} \approx 1$. In contrast, in the most general case it is:

$$F_{B,HO} = \frac{r_B}{r_B + R_{SC}} = \frac{[B]}{[B] + k'_{SC}/k_B} \tag{2.6}$$

From Eqs. 2.3–2.6, the phenol formation rate (r_P) can be expressed as:

$$\frac{1}{r_P} = \frac{1}{r_{HO} + Y_P} + \frac{[k'_{sc}]}{r_{HO} + Y_P \times k_B[B]} \tag{2.7}$$

According to Eq. 2.7, the plot of $1/r_P$ versus $1/[B]$ should be straight line (Takeda et al. 2004) and k'_{SC}/k_B can be calculated from the slope and intercept of the plot. Thus, $F_{B,HO}$ can be calculated with Eq. 2.6 using the values of k'_{SC}/k_B and $[B]$.

2.5 Levels of Photoinduced Generation of $HO^•$ in Natural Waters

The production rates of $HO^•$ that have been estimated in a variety of waters, in the presence of standard chemical species (NO_2^-, NO_3^- and H_2O_2) or of standard organic substances under sunlight are summarized in Table 2 (Mopper and Zhou 1990; Takeda et al. 2004; Zepp et al. 1987; Haag and Hoigné 1985; White et al. 2003; Arakaki and Faust 1998; Nakatani et al. 2007; Mostofa KMG and Sakugawa H, unpublished data; Nakatani et al. 2004; Qian et al. 2001; Allen et al. 1996; Mabury 1993; Grannas et al. 2006; Anastasio and Newberg 2007). The rates are typically varied in a range from 10^{-7} to 10^{-10} M s^{-1} in aqueous solution (Table 2). Production rates in rivers are (0.6–7.5) × 10^{-11} M s^{-1} in upstream waters, (0.4–7.4) × 10^{-8} M s^{-1} in upstream waters contaminated with AMD, (1.0–2.9) × 10^{-11} M s^{-1} in non-polluted river waters, 2.4 × 10^{-11} M s^{-1} in Ogeechee River, (2.0–6.0) × 10^{-10} M s^{-1} in Wetland on Lake Erie and Artificial Agricultural wetland, 6.4 × 10^{-11} M s^{-1} in Rice field water, (2.0–17.0) × 10^{-10} M s^{-1} in Satilla River and Pine Barrens that have iron-rich waters (Table 2). It is noticeable that the production rates of $HO^•$ are higher by two to five orders of magnitude in stream waters contaminated with AMD (Allen et al. 1996) than in typical river waters. Such an effect might be caused by the photo-Fenton reaction that is considerably favored in the presence of elevated iron contents (Allen et al. 1996; McKnight et al. 1988). Similarly, high production rates of $HO^•$ have been observed in Satilla River water (White et al. 2003), where more than 70 % of the total $HO^•$ production is accounted for by the photo-Fenton reaction. Therefore, the latter process is expected to be the main contributor to $HO^•$ photo-production in iron-rich waters. In contrast, upstream waters mainly contain DOM components (mostly fulvic and humic acids) that are the major contributors to $HO^•$ photo-production in these systems. A possible pathway that yields $HO^•$ from DOM is the photoinduced formation of H_2O_2 (Eqs. 3.13–3.18, see chapter "Photoinduced and Microbial Generation of Hydrogen Peroxide and Organic Peroxides in Natural Waters"), which could induce the photo-Fenton reaction in the presence of Fe or produce $HO^•$ by direct photolysis (Nakatani et al. 2007; Mostofa KMG and Sakugawa H, unpublished data). An alternative explanation for the production of $HO^•$ from DOM is the oxidation of water by the excited triplet states ($^3DOM^*$) (Brigante et al. 2010).

In lake water the production rates of $HO^•$ are very variable, ranging from 1.8 × 10^{-13} to 4.6 × 10^{-11} M s^{-1} (Table 2). The $HO^•$ photo-production depends on the irradiation wavelength. For instance, the formation rate of $HO^•$ observed on extracted lake DOM under sunlight is higher [(1.6–1.8) × 10^{-10} Ms^{-1} at 308 nm]

Table 2 Production rates of hydroxyl radical ($HO^•$) reported from natural waters, standard nitrite (NO_2^-) nitrate (NO_3^-) and various standard organic substances

Type of samples/substances	Sample	Source of light/ wavelength (nm)	Production rate of $HO^• \times 10^{-11}$ (Ms^{-1})	References
Rivers				
Stream, contaminated with acid mine drainage	Freshwater-polluted	Sunlight	400–7,400	Allen et al. (1996)
Kurose River (upstream): 1 site	Freshwater	Xe lamp[b]	1.8	Takeda et al. (2004)
Kurose River (downstream regions): 7 sites	Freshwater-sewage polluted	Xe lamp[b]	17–89	Takeda et al. (2004)
Ohta River (upstream regions): 4 sites	Freshwater	Xe lamp[b]	0.6–1.1	Takeda et al. (2004)
Ohta River (downstream regions): 4 sites	Freshwater	Xe lamp[b]	1.0–2.6	Takeda et al. (2004)
Kurose River (upstream regions): 2 sites	Freshwater	Xe lamp[b]	7.0–0.5	Nakatani et al. (2004)
Kurose River (downstream): 2 sites	Freshwater-sewage polluted	Xe lamp[b]	17–33	Nakatani et al. (2007)
River waters: upstream & downstream: 33 sites	Freshwater-sewage polluted	Xe lamp[b]	0.78–530	Nakatani et al. (2007)
Kurose River (upstream regions): 2 sites	Freshwater	Xe lamp[b]	0.9–2.5	Mostofa KMG and Sakugawa H (unpublished data)[a]
Kurose River (downstream regions): 4 sites	Freshwater-sewage polluted	Xe lamp[b]	8.0–570	Mostofa KMG and Sakugawa H (unpublished data)[a]
Ohta River (upstream regions): 3 sites	Freshwater	Xe lamp[b]	1.3–1.8	Mostofa KMG and Sakugawa H (unpublished data)[a]
Ohta River (downstream regions): 3 sites	Freshwater	Xe lamp[b]	1.7–2.9	Mostofa KMG and Sakugawa H (unpublished data)[a]
Ogeechee River	Freshwater	Xe lamp[b]	24.0	White et al. (2003)
Satilla River	Freshwater	Xe lamp[b]	45.0	White et al. (2003)
Satilla River (bleached)	Freshwater	Xe lamp[b]	20.0	White et al. (2003)
Pine Barrens	Freshwater	Xe lamp[b]	170.0	White et al. (2003)
Wetland on Lake Erie	Freshwater	Xe lamp[b]	20–26	White et al. (2003)
Artificial agricultural wetland	Freshwater	Xe lamp[b]	23.0	White et al. (2003)
Rice Field water	Freshwater	Xe lamp[b]	64.0	Mabury (1993)

(continued)

Table 2 (continued)

Type of samples/substances	Sample	Source of light/ wavelength (nm)	Production rate of $HO^{\bullet} \times 10^{-11}$ (Ms^{-1})	References
Lakes				
Lake Greifensee, Switzerland	Freshwater	Xe lamp[b]	1.0	Haag and Hoigné (1985)
Lake Greifensee, Switzerland	Freshwater	Xe lamp[b]	2.5	Zepp et al. (1987)
Clear Lake	Freshwater	Xe lamp[b]	4.6	Mabury (1993)
Lake Tahoe	Freshwater	Xe lamp[b]	2.2	Mabury (1993)
Nitrate-rich shallow water body	Freshwater	Xe lamp[b]	0.018	Zepp et al. (1987)
DOM extracted-XAD, Toolik Lake	Freshwater	308 nm	18	Grannas et al. (2006)
DOM extracted-XAD, Toolik Lake	Freshwater	330 nm	8.1	Grannas et al. (2006)
DOM extracted-XAD, Toolik Lake	Freshwater	355 nm	5.6	Grannas et al. (2006)
DOM extracted-C-18, Toolik Lake	Freshwater	308 nm	18	Grannas et al. (2006)
DOM extracted-C-18, Toolik Lake	Freshwater	330 nm	6.4	Grannas et al. (2006)
DOM extracted-C-18, Toolik Lake	Freshwater	355 nm	6.1	Grannas et al. (2006)
DOM extracted-UF, Toolik Lake	Freshwater	308 nm	16	Grannas et al. (2006)
DOM extracted-UF, Toolik Lake	Freshwater	330 nm	8.1	Grannas et al. (2006)
DOM extracted-UF, Toolik Lake	Freshwater	355 nm	4.7	Grannas et al. (2006)
Seawaters				
Everglades	Seawater	Sunlight	42.0	Mopper and Zhou (1990)
Biscayne Bay	Seawater	Sunlight	2.9	Mopper and Zhou (1990)
Vineyard Sound	Seawater	Sunlight	2.7	Mopper and Zhou (1990)
Mississippi River Plume	Seawater	Xe lamp[b]	6.2	White et al. (2003)
Florida Keys	Seawater	Xe lamp[b]	0.43	White et al. (2003)
Seto Inland Sea: 0 m depth	Seawater	Xe lamp[b]	1.5	Takeda et al. (2004)
Seto Inland Sea: 5 m depth	Seawater	Xe lamp[b]	0.57	Takeda et al. (2004)

(continued)

Table 2 (continued)

Type of samples/substances	Sample	Source of light/ wavelength (nm)	Production rate of $HO^{\bullet} \times 10^{-11}$ (Ms^{-1})	References
Seto Inland Sea: 10 m depth	Seawater	Xe lamp[b]	0.91	Takeda et al. (2004)
Seto Inland Sea: 15 m depth	Seawater	Xe lamp[b]	1.0	Takeda et al. (2004)
Seto Inland Sea: 20 m depth	Seawater	Xe lamp[b]	2.4	Takeda et al. (2004)
Seto Inland Sea: 30 m depth	Seawater	Xe lamp[b]	4.9	Takeda et al. (2004)
Yellow Sea: 0 m depth	Seawater	Xe lamp[b]	0.39	Takeda et al. (2004)
Yellow Sea: 20 m depth	Seawater	Xe lamp[b]	0.36	Takeda et al. (2004)
Yellow Sea: 30 m depth	Seawater	Xe lamp[b]	0.49	Takeda et al. (2004)
Yellow Sea: 40 m depth	Seawater	Xe lamp[b]	0.82	Takeda et al. (2004)
Yellow Sea: 50 m depth	Seawater	Xe lamp[b]	0.73	Takeda et al. (2004)
Yellow Sea: 60 m depth	Seawater	Xe lamp[b]	0.72	Takeda et al. (2004)
Gulf Stream	Seawater	Sunlight	0.31	Mopper and Zhou (1990)
Sargasso Sea	Seawater	Sunlight	0.28	Mopper and Zhou (1990)
Weddel Sea, Antarctic	Seawater	Sunlight	0.10	Qian et al. (2001)
Crystal Sound, Antarctic	Seawater	Sunlight	0.10	Qian et al. (2001)
Paradise Harbor, Antarctic	Seawater	Sunlight	0.01	Qian et al. (2001)
Standard substances				
NO_2^-	Seawater	Sunlight	2.30	Mopper and Zhou (1990)
NO_2^-	Milli-Q water	Sunlight	2.80	Arakaki and Faust (1998)
NO_2^-	Milli-Q water	Sunlight	2.30	Takeda et al. (2004)
NO_2^-	Milli-Q water	Xe lamp[b]	2.20	Mostofa KMG and Sakugawa H (unpublished data)[a]
NO_3^-	Fresh water	Sunlight	0.03	Zepp et al. (1987)
NO_3^-	Seawater	Sunlight	0.03	Mopper and Zhou (1990)
NO_3^-	Milli-Q water	Sunlight	0.02	Arakaki and Faust (1998)
NO_3^-	Milli-Q water	Sunlight	0.02	Takeda et al. (2004)

(continued)

Table 2 (continued)

Type of samples/substances	Sample	Source of light/ wavelength (nm)	Production rate of $HO^{\bullet} \times 10^{-11}$ (Ms^{-1})	References
NO_3^-	Milli-Q water	Xe lamp[b]	0.01	Mostofa KMG and Sakugawa H (unpublished data)[a]
Suwannee River Fulvic Acid (SRFA)	Milli-Q water	Xe lamp[b]	1.20	Mostofa KMG and Sakugawa H (unpublished data)[a]
SRFA	Milli-Q water	Xe lamp[b]	6.00	White et al. (2003)
Tryptophan	Milli-Q water	Xe lamp[b]	1.70	Mostofa KMG and Sakugawa H (unpublished data)[a]
Suwannee River Humic Acid (SRHA)	Milli-Q water	Xe lamp[b]	1.10	Mostofa KMG and Sakugawa H (unpublished data)[a]
Phenylalanine	Milli-Q water	Xe lamp[b]	0.40	Mostofa KMG and Sakugawa H (unpublished data)[a]
DAS1	Milli-Q water	Xe lamp[b]	0.31	Mostofa KMG and Sakugawa H (unpublished data)[a]
DSBP	Milli-Q water	Xe lamp[b]	0.30	Mostofa KMG and Sakugawa H (unpublished data)[a]
H_2O_2	Milli-Q water	Xe lamp[b]	0.26	Mostofa KMG and Sakugawa H (unpublished data)[a]
Peracetic acid	Milli-Q water	Xe lamp[b]	0.25	Mostofa KMG and Sakugawa H (unpublished data)[a]
Sea-salt particulate matter, extracted from coastal waters	Seawater	summer, midday, sunlight	~2778–27778	Anastasio and Newberg (2007)

[a]Production rates (Mostofa and Sakugawa) calculated for initial 60 min irradiation and normalized to sunlight intensity (noon time) using 2 nitro benzaldehyde solution (8 mM) at the Campus of Hiroshima University, Japan as well as the rates for standard substances mentioned here is calculated after deduction of rate of MQ water

Rates for standard organic substances adjusted for 1 mg L^{-1} of each substances

[b]Xe light intensity is normalized to sunlight intensity

for shorter wavelength than for longer one [(4.7–6.1) × 10^{-11} Ms^{-1} at 355 nm] in a variety of lake waters (Table 2) (Grannas et al. 2006). Thus, the production of $HO^{\bullet}$ greatly depends on the light wavelength.

In seawater, the $HO^{\bullet}$ production rate is higher [(4.3–42.0) × 10^{-12} M s^{-1}] in coastal waters than in the open ocean [(2.8–3.1) × 10^{-12} M s^{-1}], and it is very low [(1.04–10.3) × 10^{-13} M s^{-1}] in Antarctic waters (Table 2). The production rates of $HO^{\bullet}$ are lower in surface seawater and gradually increase with increasing depth (Takeda et al. 2004; Zafiriou and Dister 1991). For example, the production rate of $HO^{\bullet}$ is (5.7–15.0) × 10^{-12} M s^{-1} at 0–20 m depth, 4.9 × 10^{-11} M s^{-1} at 30 m, and 7.2 × 10^{-11} M s^{-1} at 60 m in Seto Inland Sea and the Yellow Sea (Takeda et al. 2004), obviously under the same irradiation conditions that is not the case of the real environment. The lower production rates at the surface compared to the deeper layers may be caused by the fact that the natural solar radiation is active in surface waters where it produces $HO^{\bullet}$ and other reactive transients. These species can lead to the photo-degradation of DOM (e.g. photo-bleaching that reduces the ability of DOM to absorb sunlight), a process to which the direct photolysis could also contribute. Therefore, the sources of $HO^{\bullet}$ in surface waters can be reduced, e.g. by decreasing the concentration of NO_2^- and the capability of DOM to photo-generate $HO^{\bullet}$. The intensity of solar radiation that reaches the deeper layers is quite limited. Therefore, the deep water has the double feature of generating the highest potentiality to produce $HO^{\bullet}$, but also of being involved to a very limited extent into $HO^{\bullet}$ photo-production in the real natural environment. An important exception could be represented by the sites where the deep oceanic water emerges to the surface.

Studies observe that the sea-salt particulate matter (SS PM) extracted from coastal seawaters can demonstrate substantially high $HO^{\bullet}$ production (rate: ~2778–27778 M s^{-1}), approximately 3–4 orders of magnitude greater than $HO^{\bullet}$ photoformation rates in surface seawater (Anastasio and Newberg 2007). The results show that photolysis of nitrate is a dominant source of $HO^{\bullet}$ (on average 59±25 %) in the SS PM whilst other source is presumably considered the organic compounds. The fact behind the other phenomenon is that irradiated organic compounds or DOM can induce photoinduced production of H_2O_2 that is a $HO^{\bullet}$ source via photolysis or the Fenton reaction, and the photoinduced generation of H_2O_2 is enhanced by salinity. Salinity or NaCl solutions are capable of generating high production of aqueous electrons (e_{aq}^-) photolytically in aqueous media (Gopinathan et al. 1972; Assel et al. 1998) that may enhance the H_2O_2 production from DOM components in waters (Mostofa and Sakugawa 2009; Moore et al. 1993) (see also chapter "Photoinduced and Microbial Generation of Hydrogen Peroxide and Organic Peroxides in Natural Waters"). In fact, photogeneration of H_2O_2 from river DOM was substantially increased with salinity, from 15 to 368 nM h^{-1} at circumneutral pH that may enhance the H_2O_2 production from DOM components in waters (Osburn et al. 2009). Salinity effect on irradiated CDOM might be another most important source of high photoproduction of $HO^{\bullet}$ in sea-salt particulate matter in seawaters.

3 Mechanisms for Photoinduced Generation of $HO^{\bullet}$ in Natural Waters

The photoinduced generation of $HO^{\bullet}$ significantly depends on several important factors such as the presence of NO_2^- and NO_3^- and their concentration, the chemical nature of DOM and its quantity, and finally the total content of Fe. The most important mechanisms for $HO^{\bullet}$ formation in natural waters are discussed below.

3.1 In situ Generation of $HO^{\bullet}$ from DOM

One of the main $HO^{\bullet}$ sources in natural waters (Table 2) (Vione et al. 2006; Mostofa KMG and Sakugawa H, unpublished data) is the photoinduced generation of $HO^{\bullet}$ from DOM components (either Fluorescent Dissolved Organic Matter-FDOM or Colored Dissolved Organic Matter-CDOM). This process can be accounted for either by the oxidation of water by the triplet states $^3DOM^*$, or by the generation of H_2O_2 upon DOM irradiation (reaction 3.1) and its detailed mechanisms are discussed in earlier chapter (see chapter "Photoinduced and Microbial Generation of Hydrogen Peroxide and Organic Peroxides in Natural Waters"). In the latter case, $HO^{\bullet}$ could be produced upon photolysis of H_2O_2 (Eq. 3.2) (Legrini et al. 1993; von Sonntag et al. 1993).

$$DOM + O_2 + H_2O + H^+ \xrightarrow{h\upsilon} H_2O_2 + DOM^{\bullet +} + O_2 + HO^- \quad (3.1)$$

$$H_2O_2 \xrightarrow{h\upsilon} 2HO^{\bullet} \quad (3.2)$$

The quantum yield of reaction 3.2 has been determined as 0.5 under UVC irradiation (Legrini et al. 1993; von Sonntag et al. 1993). The quantum yield varies with wavelength, but it also depends on the band that absorbs radiation. In the case of H_2O_2, the same band is responsible for radiation absorption and photolysis in both the UVC and UVB regions. The hypothesis that the formation of $HO^{\bullet}$ by irradiated DOM is accounted for by H_2O_2 photoproduction. It is consistent with the observed, gradual and parallel increase of H_2O_2 concentration and of phenol formation from benzene, upon irradiation of natural river waters and of relevant standard organic substances (Fig. 2). A similar, parallel trend of both phenol (its formation being used as $HO^{\bullet}$ probe) and H_2O_2 has been observed upon irradiation of upstream DOM mostly containing fulvic acid (Fig. 2a), of standard Suwannee River Fulvic Acid (SRFA) (Fig. 2c) and of diaminostilbene type (DAS1) (Fig. 2d). The same trend has not been observed in sewage polluted river waters (Fig. 2b), which might be the effect of additional production from other $HO^{\bullet}$ sources such as the NO_2^- and NO_3^- ions, present in high amount (Takeda

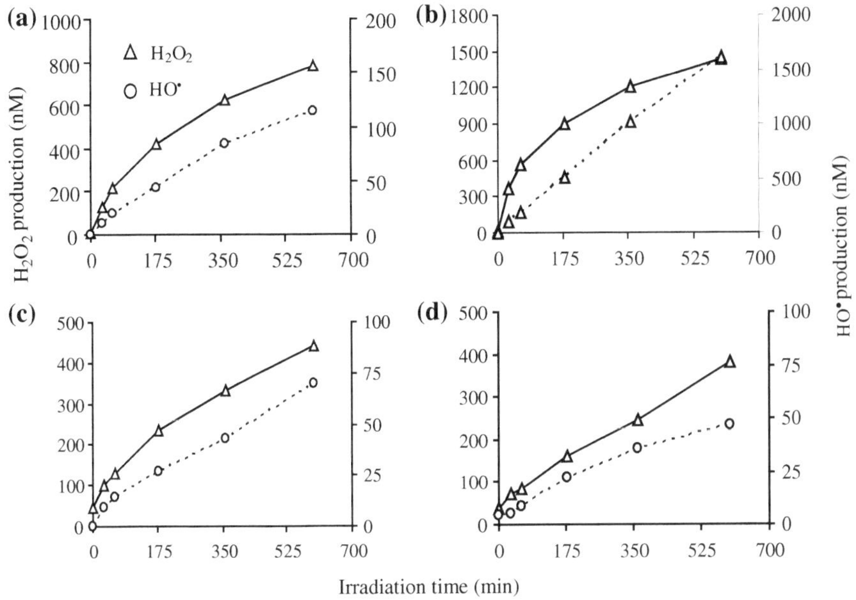

Fig. 2 In-situ generation of H_2O_2 and HO$^•$ for river waters and standard organic substances during the 10 h of irradiation period in photoexperiments conducted using a solar simulator. Upstream DOM having mostly fulvic acid (**a**); polluted river waters, mostly affected by mixture of sewage effluents and upstream DOM (**b**); standard Suwannee River Fulvic Acid (**c**); and standard diaminostilbene (DAS1) (**d**). *Data source* Mostofa KMG and Sakugawa H (unpublished data)

et al. 2004; Mostofa KMG and Sakugawa H, unpublished data; Nakatani et al. 2004). Therefore, the generation of H_2O_2 by DOM could account for most of the production of HO$^•$ by unpolluted water samples, with a relatively elevated content of fulvic acid in DOM (Fig. 3) and a relatively low concentration of other HO$^•$ sources, such as nitrate, nitrite and Fe.

3.2 Direct Photolysis of Nitrate and Nitrite

The direct photolysis of nitrite and nitrate induces HO$^•$ photoproduction (Zafiriou and True 1979a, b; Takeda et al. 2004; Zepp et al. 1987; Mack and Bolton 1999). There is evidence that irradiation in the 200–400 nm wavelength region can convert NO_2^- into NO$^•$ and O$^{•-}$ (Eqs. 3.3, 3.4) (Zepp et al. 1987; Mack and Bolton 1999):

$$NO_2^- + h\upsilon \rightarrow [NO_2^-]^* \quad (3.3)$$

$$[NO_2^-]^* \rightarrow NO^• + O^{•-} \quad (3.4)$$

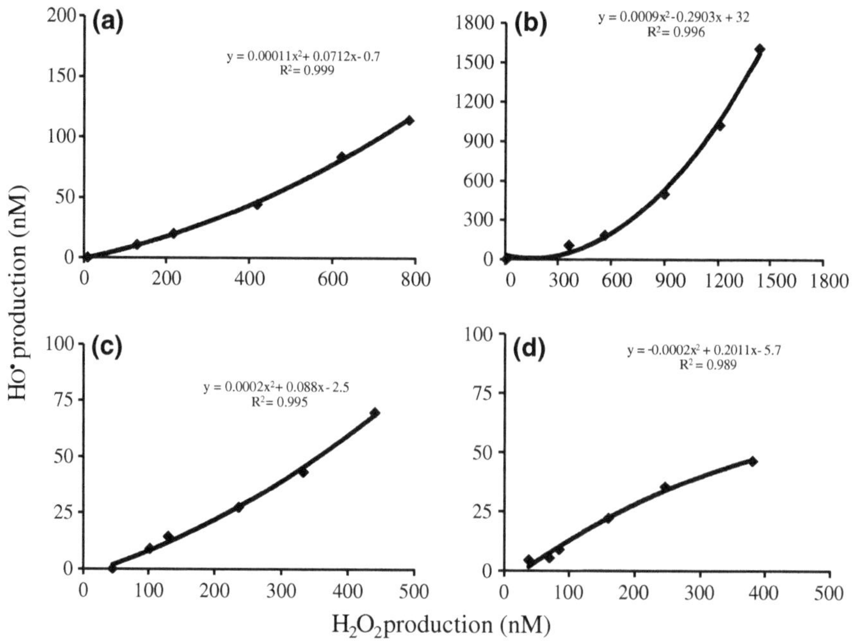

Fig. 3 Relationship between H_2O_2 and $HO^•$ in situ produced from river waters and standard organic substance during the 10 h of irradiation period in photoexperiments conducted using a solar simulator. The relationships of the (**a**, **b**, **c** and **d**) are the same samples of Fig. 1. *Data source* Mostofa KMG and Sakugawa H (unpublished data)

At pH < 12 in aqueous solution, $O^{•-}$ is protonated to form $HO^•$:

$$O^{\bullet -} + H_2O \underset{k_{-3.5}}{\overset{k_{3.5}}{\rightleftharpoons}} HO^{\bullet} + HO^{-} \tag{3.5}$$

where $k_{3.5} = 1.7 \times 10^6\ M^{-1}\ s^{-1}$ for the $HO^•$ formation reaction and $k_{-3.5} = 1.2 \times 10^{10}\ M^{-1}\ s^{-1}$ for the reverse reaction. The radical $HO^•$ can significantly recombine with $NO^•$ and NO_2^{-}; such reactions are very fast (diffusion-controlled) in aqueous media (Mack and Bolton 1999):

$$HO^{\bullet} + NO^{\bullet} \rightarrow HNO_2 \tag{3.6}$$

where $k_{3.6} = 1.0 \times 10^6\ M^{-1}\ s^{-1}$.

$$HO^{\cdot} + NO_2^{-} \rightarrow NO_2^{\cdot} + OH^{-} \tag{3.7}$$

where $k_{3.7} = 1.0 \times 10^{10}\ M^{-1}\ s^{-1}$. These reactions can limit the steady-state concentration of $HO^•$ and, therefore, the ability of the hydroxyl radical to take part in photooxidation reactions of organic compounds in natural waters. Note, however, that the main $HO^•$ scavengers are DOM in freshwater and bromide in seawater (Takeda et al. 2004).

As far as NO_3^- photolysis is concerned (Eq. 3.8), the process can be expressed as follows (Zepp et al. 1987; Mack and Bolton 1999):

$$NO_3^- + h\upsilon \rightarrow NO_2^- + 1/2O_2 \quad (3.8)$$

This stoichiometry can be followed in the absence of $HO^\bullet$ scavengers, over the entire pH range and at irradiation wavelength around 200 nm (Shuali et al. 1969; Wagner et al. 1980). However, irradiation above 280 nm results into two primary photoinduced pathways (Zepp et al. 1987; Mack and Bolton 1999):

$$NO_3^- + h\upsilon \rightarrow [NO_3^-]^* \quad (3.9)$$

$$[NO_3^-]^* \rightarrow NO_2^- + O(^3P) \quad (3.10)$$

$$[NO_3^-]^* \rightarrow NO_2^\bullet + O^{\bullet -} \xrightarrow{H_2O} HO^\bullet + HO^- + NO_2^\bullet \quad (3.11)$$

In this mechanism, NO_3^- absorbs a UVB photon yielding an excited state, $[NO_3^-]^*$ (Eq. 3.9), which undergoes disintegration following two pathways: the first one produces the nitrite ion (NO_2^-) and atomic oxygen, $O(^3P)$ (Eq. 3.10). The second pathway produces nitrogen dioxide ($NO_2^\bullet$) and $O^{\bullet -}$. The latter is rapidly protonated to form $HO^\bullet$ (Eq. 3.11). The formation of NO_2^- in Eq. 3.10 can be followed by nitrite photolysis to give $HO^\bullet$, as shown in (Eqs. 3.3–3.5). It can be noted that $HO^\bullet$ is a strong oxidant that can react with DOM more quickly than does atomic oxygen, $O(^3P)$. Indeed, the main fate of $O(^3P)$ (Eq. 3.4) would be the reaction with oxygen to form ozone, which is rapidly consumed in natural waters by reaction with NO_2^- or decomposition to $HO^\bullet$ (Zepp et al. 1987).

3 *$HO^\bullet$ Production from the Fenton Reaction*

The ferrous ions (Fe^{2+}) catalyzes the formation of $HO^\bullet$ in the presence of H_2O_2 (Fenton 1894). An aqueous solution of H_2O_2 and ferrous or ferric salts is termed as Fenton's reagent. The oxidation efficiency of the Fenton reaction is the highest at pH values ranging from 2 to 5 and at a 1:1 molar ratio of H_2O_2 and Fe^{2+} (Walling 1975). The reactivity of the Fenton's reagent is the effect of the generation of $HO^\bullet$ in the reaction media (Haber and Weiss 1934). The mechanism for the chain Fenton reaction was initially depicted as follows (Eqs. 3.12–3.16) (Barb et al. 1951):

$$Fe^{2+} + H_2O_2 \rightarrow Fe^{3+} + HO^\bullet + HO^- \quad (3.12)$$

$$Fe^{2+} + HO^{\bullet} \rightarrow Fe^{3+} + HO^{-} \quad (3.13)$$

$$HO^{\bullet} + H_2O_2 \rightarrow HO_2^{\bullet} + H_2O \quad (3.14)$$

$$Fe^{2+} + HO_2^{\bullet} \rightarrow Fe^{3+} + HO_2^{\bullet-} \quad (3.15)$$

$$Fe^{3+} + O^{2-} \rightarrow Fe^{2+} + O_2 \quad (3.16)$$

Recently, the reaction rate constants of the Fenton system have been measured at various pH values, an issue that will be discussed later (Kwan and Voelker 2002; Duesterberg et al. 2008). At low $[H_2O_2]/[Fe^{2+}]$ ratios in both reactions (3.12) and (3.13), only the reactions are important, and the overall process is second-order with respect to the reactants. At high $[H_2O_2]/[Fe^{2+}]$ ratios, there is also an important contribution from the competitive reactions (3.15, 3.16) (Rush and Bielski 1985). The earlier studies have examined the Fenton reaction at low pH (<1.0) with the reactants at mM levels. The production of $HO^{\bullet}$ from $Fe^{2+} + H_2O_2$ is also important in AOTs, as it allows the use of transition metals to catalyze the oxidation of organic compounds.

Recent studies demonstrate that the formation of $HO^{\bullet}$ increases linearly with the H_2O_2 concentration (Lindsey and Tarr 2000). Experiments carried out using ESR spin trapping together with water labeled with ^{17}O suggest that $HO^{\bullet}$ is derived exclusively from hydrogen peroxide, and that there is no exchange of oxygen atoms between H_2O_2 and the water solvent (Lloyd et al. 1997). It has been demonstrated that fulvic and humic acid reduce the $HO^{\bullet}$ formation in the Fenton reaction under most conditions, but fulvic acid increases $HO^{\bullet}$ formation at certain pH values (Lindsey and Tarr 2000; Voelker and Sulzberger 1996). Fulvic acid can inhibit the degradation of dissolved aromatic compounds (e.g. phenol, fluorene and phenanthrene) by the Fenton reagent in aqueous solution (Lindsey and Tarr 2000). Accordingly, natural organic matter could inhibit the remediation of pollutants by the Fenton process in water and soil environments. However, it has also been shown that humic acids are able to enhance the degradation of phenol in the second step of the Fenton process. Indeed, after the reaction between Fe^{2+} and H_2O_2 is completed, further degradation of the organic substrate can be directed by the reduction of Fe(III) to Fe^{2+}. The latter process is enhanced by humic substances (Vione et al. 2004).

Indeed, the ferric ion (Fe^{3+}) catalyzes the decomposition of H_2O_2 into $HO^{\bullet}$ and Fe^{2+} ($Fe^{3+} + H_2O_2 \rightarrow Fe^{2+} + HO_2^{\bullet} + H^{+}$) (Barb et al. 1951; Walling and Weil 1974; Lee et al. 2003; Lee and Sedlak 2009). The rate of the Fenton process is greatly enhanced when the temperature is raised from 10 to 50 °C, because of the high activation energy ($\approx$126 kJ mol^{-1}) of the reaction (Lee et al. 2003). A high production of Fe^{2+} causes a correspondingly high production of $HO^{\bullet}$ in the reaction system (Eq. 3.12).

The effective catalytic oxidation of organic compounds by the system Fe^{3+}/H_2O_2 is usually limited to the acidic pH region, because of the low solubility of Fe^{3+} and of the low efficiency of the oxidant generation at neutral pH values (Lee and Sedlak 2009). The addition of polyoxometalate ions (POM) greatly increases the yield of oxidant species in the reaction. Under acidic conditions, POM can mediate the electron transfer from nanoparticle zerovalent iron or Fe(II) to oxygen, thereby increasing the production of H_2O_2. The latter is subsequently converted to $HO^{\bullet}$ through the Fenton reaction (Lee et al. 2008).

3.4 $HO^{\bullet}$ Production from the Photo-Fenton Reaction

The reactivity of the Fenton's reagent, and in particular of $Fe^{3+} + H_2O_2$ is greatly enhanced by UV/Visible irradiation ($\lambda < 580$ nm), which can for instance increase the extent of mineralization of organic pollutants (Zepp et al. 1992; Voelker et al. 1997; Southworth and Voelker 2003; Nakatani et al. 2007; Vermilyea and Voelker 2009). The photo-Fenton reaction is defined as the reaction of photoproduced Fe^{2+} with H_2O_2 to form the highly reactive $HO^{\bullet}$ (Eqs. 3.17, 3.18). The main chemical reactions occurring in the photo-Fenton system are the following:

$$Fe^{3+} + h\upsilon \rightarrow Fe^{2+} + O_2 + H^{+} \tag{3.17}$$

$$Fe^{2+} + H_2O_2 \xrightarrow{h\upsilon} Fe^{3+} + HO^{\bullet} + HO^{-} \tag{3.18}$$

Irradiation of Fe(III) species causes the production of Fe^{2+} and possibly of $HO^{\bullet}$ (e.g. in the photolysis of the complex $FeOH^{2+}$). In the presence of H_2O_2, further $HO^{\bullet}$ is produced by the Fenton process between H_2O_2 itself and the photogenerated Fe^{2+}. The latter reaction yields again Fe(III), but continuing irradiation may recycle Fe(III) to Fe(II) in the reaction media. Therefore, it is possible to continuously generate $HO^{\bullet}$ under irradiation without any net consumption of Fe(II), which significantly accelerates the overall reaction rate in the photo-Fenton system. The $HO^{\bullet}$ formation in the photo-Fenton system is limited only by the availability of radiation and by the content of H_2O_2 in the reaction medium. The photo-Fenton reaction is observed in open ocean whereas vertical profiles of Fe(II) show maxima consistent with the plume of the iron infusion whilst H_2O_2 profiles demonstrate a corresponding minima showing the effect of oxidation of Fe(II) by H_2O_2 (Croot et al. 2005). Observations also show that the detectable Fe(II) concentrations can exist for up to 8 days after an iron infusion (Croot et al. 2005).

3.5 $HO^{\bullet}$ Production from Photo-Ferrioxalate/H_2O_2 Reaction

The photo-ferrioxalate/H_2O_2 reaction is an advanced modification of the photo-Fenton reaction and an effective technique of generating Fe(II) in the reaction media. The addition of oxalate to the photo-Fenton system significantly accelerates the $HO^{\bullet}$ production under UV/visible irradiation ($\lambda < 550$ nm) (Eqs. 3.19–3.25; Table 3) (Hislop and Bolton 1999; Jeong and Yoon 2005;

Table 3 The reaction rate constants for the production of $HO^{\bullet}$ in photo-ferrioxalate/H_2O_2 reaction

Reaction type	Rate constant (*k*) ($M^{-1}\,s^{-1}$)	Equations	References
$Fe^{III}(C_2O_4)^{3-2n} + h\upsilon \rightarrow Fe^{2+} + (n\text{-}1)C_2O_4^{2-} + C_2O_4^{\bullet -}$	–	3.19	Jeong and Yoon (2005)
$C_2O_4^{\bullet -} \rightarrow CO_2 + CO_2^{\bullet -}$	2×10^{12}	3.20	Mulazzani et al. (1986)
$Fe^{III}(C_2O_4)_n^{3-2n} + CO_2^{\bullet -} \rightarrow CO_2 + Fe^{II}(C_2O_4)_n^{2-2n}$	$\approx 8 \times 10^{9}$	3.21	Jeong and Yoon (2004)
$CO_2^{\bullet -} + O_2 \rightarrow CO_2 + O_2^{\leftarrow \leftarrow \bullet}$	2.4×10^{9}	3.22	Hislop and Bolton (1999)
$Fe^{III}(C_2O_4)_n^{3-2n} + O_2^{\bullet -} \rightarrow O_2 + Fe^{II}(C_2O_4)_n^{2-2n}$	$<1 \times 10^{6}$	3.23	Sedlak and Hoigné (1993)
$Fe^{III}(C_2O_4)_n^{3-2n} + HO_2^{\bullet} \rightarrow O_2 + Fe^{II}(C_2O_4)_n^{2-2n}$	$<1.2 \times 10^{5}$	3.24	Sedlak and Hoigné (1993)
$Fe^{II}(C_2O_4) + H_2O_2 \rightarrow Fe^{III}(C_2O_4)^{+} + HO^{\bullet} + OH^{-}$	3.1×10^{4}	3.25	Sedlak and Hoigné (1993)
$Fe^{2+} + HO\bullet \rightarrow Fe^{III}(OH)^{2+}$	4.3×10^{8}	3.26	Zuo and Hoigné (1992)
$Fe^{III}(OH)^{2+} + h\upsilon \rightarrow Fe^{2+} + HO\bullet$	Negligible	3.27	Jeong and Yoon (2005)
$Fe^{3+} + O_2^{\bullet -} \rightarrow Fe^{2+} + O_2$	1.5×10^{8}	3.28	Balmer and Sulzberger (1999)
$C_2O_4^{2-} + HO\bullet \rightarrow CO_2 + CO_2^{\bullet -} + OH^{-}$	7.7×10^{6}	3.29	Buxton et al. (1988)
$HC_2O_4^{-} + HO\bullet \rightarrow CO_2 + CO_2^{\bullet -} + H_2O$	4.7×10^{7}	3.30	Buxton et al. (1988)
$CPAA^{a} + HO\bullet \rightarrow$ products	3.0×10^{9}	3.31	Pignatello (1992)
Equilibrium reactions	*Equilibrium constants* (M^{-1})		
$Fe^{3+} + C_2O_4^{2-} \leftrightarrow Fe^{III}(C_2O_4)^{+}$	2.5×10^{9}	3.32	Faust and Zepp (1993)
$Fe^{III}(C_2O_4)^{+} + C_2O_4^{2-} \leftrightarrow Fe^{III}(C_2O_4)_2^{-}$	6.3×10^{6}	3.33	Faust and Zepp (1993)
$Fe^{III}(C_2O_4)^{-} + C_2O_4^{2-} \leftrightarrow Fe^{III}(C_2O_4)_3^{3-}$	3.8×10^{4}	3.34	Faust and Zepp (1993)
$Fe^{2+} + C_2O_4^{2-} \leftrightarrow Fe^{II}(C_2O_4)$	2.0×10^{5}	3.35	Faust and Zepp (1993)
$Fe^{II}(C_2O_4) + C_2O_4^{2-} \leftrightarrow Fe^{II}(C_2O_4)_2^{2-}$	1.2×10^{2}	3.36	Faust and Zepp (1993)
$HC_2O_4^{+} \leftrightarrow C_2O_4^{2-} + H^{+}$	6.2×10^{1}	3.37	Zuo and Hoigné (1992)
$HO_2^{\bullet} \leftrightarrow O_2^{\bullet -} + H^{+}$	6.2×10^{1}	3.38	Bielski et al. (1985)

[a] CPAA means the 2,4-dichlorophenoxyacetic acid

Bielski et al. 1985; Mulazzani et al. 1986; Jeong and Yoon 2004; Buxton et al. 1988; Pignatello 1992; Zuo and Hoigné 1992; Sedlak and Hoigné 1993; Balmer and Sulzberger 1999). In the photo-Fenton reaction, the $HO^{\bullet}$ radical is formed photolytically from $Fe(OH)^{2+}$ (Eq. 3.27). The relevant reaction mainly takes place at pH 2.5–5 (Eq. 3.26), but its quantum yield is relatively low: $\varphi_{Fe(II)} = 0.14 \pm 0.04$ at 313 nm and $\varphi_{HO} = 0.195 \pm 0.03$ at 310 nm (Hislop and Bolton 1999). When Fe(III) is complexed with a carboxylic anion (e.g. oxalate), the quantum yield of Fe(II) production ($\varphi_{Fe(II)}$) is significantly increased to $\varphi_{Fe(II)} = 1.24$, at 300 nm, pH = ~2 and 6 mM ferrioxalate (Murov et al. 1993). This result is accounted for by the considerable photosensitive nature of the ferrioxalate complex $[Fe(C_2O_4)_3]^{3-}$, which combines elevated absorption of visible radiation with a very high quantum yield of Fe^{2+} photoproduction. Interestingly, the photolysis of the ferrioxalate complex generates an additional reactive radical species, the carbon dioxide radical anion ($CO_2^{\bullet-}$) (Eqs. 3.19–3.21; Table 3).

The radical $CO_2^{\bullet-}$ can produce Fe(II) via reaction (Eq. 3.21) and by other reaction pathways (Eqs. 3.23, 3.24). The kinetic and equilibrium constants for the photo-ferrioxalate/H_2O_2 reaction are shown in Table 3. $CO_2^{\bullet-}$ can react with oxygen to form the superoxide anion ($O_2^{\bullet-}$) (Eq. 3.22), which can further enhance the quantum yield for the generation of Fe^{2+} (Eqs. 3.23, 3.24, 3.28) and contributes to the production of H_2O_2 (Eq. 3.17). When ferrioxalate is irradiated in the presence of H_2O_2 under ideal conditions, a radical $HO^{\bullet}$ is produced by the Fenton reaction per every Fe(II) generated (Eq 3.25, Table 3). In the reaction media (Eq. 3.25), both uncoordinated Fe^{2+} and $Fe^{II}(C_2O_4)$ can react with H_2O_2. Therefore one gets an overall, apparent second-order rate constant for the reaction between Fe(II) and H_2O_2. In the presence of excess oxalate, Fe(III) will be coordinated with either two or three oxalate ligands. Fe(III) is recycled to Fe(II) in both the photo-Fenton and the photo-ferrioxalate/H_2O_2 reaction. In the latter case the formation of $HO^{\bullet}$ depends on the availability of radiation, H_2O_2 and oxalate, the latter two components being consumed by the reaction. The enhancement of $HO^{\bullet}$ photoproduction that is observed upon addition of oxalate depends on the very high photolysis quantum yield of the Fe(III)-oxalate complex(es), which largely compensates for the facts that the photolysis of Fe(III)-oxalate, unlike that of $FeOH^{2+}$, does not yield $HO^{\bullet}$, and that oxalate is a $HO^{\bullet}$ scavenger.

3.6 $HO^{\bullet}$ Production from Photocatalytic Metal Oxide (TiO_2) Suspensions

Titanium dioxide is the most frequently used metal oxide photocatalyst, which undergoes excitation at near-UV wavelengths. The irradiation by sunlight of aqueous suspensions of TiO_2 can induce very significant generation of $HO^{\bullet}$ in aqueous solution. Below it is reported a general scheme of $HO^{\bullet}$ photo-production, proposed in early studies to describe the behavior of aqueous suspensions of TiO_2 in the presence of DOM (Konstantinou and Albanis 2004; Murov et al. 1993; Tseng and Haung 1991; Fox 1993; Chen et al. 2001; Han et al. 2009) (Eqs. 3.13–3.18, see chapter "Photoinduced and Microbial Generation of Hydrogen Peroxide and Organic Peroxides in Natural Waters"):

$$\mathrm{DOM} + h\upsilon \rightarrow {}^{1}\mathrm{DOM}^{*} \xrightarrow{\mathrm{ISC}} {}^{3}\mathrm{DOM}^{*} \tag{3.39}$$

$$\mathrm{T_iO_2} + h\upsilon \rightarrow \mathrm{T_iO_2^{*}} \rightarrow \mathrm{T_iO_2}\,(e^{-} + h^{+}) \tag{3.40}$$

$${}^{3}\mathrm{DOM}^{*} + \mathrm{O_2} \rightarrow \mathrm{DOM}^{*+} + \mathrm{O_2}^{\bullet-} \tag{3.41}$$

$$\mathrm{TiO_2}(e^{-} + h^{+}) + \mathrm{O_2} \rightarrow \mathrm{TiO_2}(h^{+}) + \mathrm{O_2}^{\bullet-} \tag{3.42}$$

$$\mathrm{O_2^{\bullet-}} + \mathrm{H^+} \rightarrow \mathrm{HO_2}^{\bullet} \quad \mathrm{p}K_a = 4.8 \tag{3.43}$$

$$2\mathrm{O_2}^{\bullet-} \rightarrow \mathrm{O_2^{2-}} + \mathrm{O_2} \quad \mathrm{p}K_a = < 0.35\,\mathrm{M^{-1}s^{-1}} \tag{3.44}$$

$$\mathrm{HO_2}^{\bullet} + \mathrm{HO_2}^{\bullet} \rightarrow \mathrm{H_2O_2} + \mathrm{O_2} \quad k = 8.6 \times 10^{5}\,\mathrm{M^{-1}s^{-1}} \tag{3.45}$$

$$\mathrm{HO_2}^{\bullet} + \mathrm{O_2}^{\bullet-} + \mathrm{H_2O} \rightarrow \mathrm{H_2O_2} + \mathrm{O_2} + \mathrm{OH^-} \quad k = 1.0 \times 10^{8}\,\mathrm{M^{-1}s^{-1}} \tag{3.46}$$

$$\mathrm{H_2O_2} + h\upsilon \rightarrow 2\mathrm{HO}^{\bullet} \tag{3.47}$$

$$\mathrm{H_2O_2} + \mathrm{HO}^{\bullet} \rightarrow \mathrm{HO_2}^{\bullet} + \mathrm{HO_2} \quad k = 3.0 \times 10^{7}\,\mathrm{M^{-1}s^{-1}} \tag{3.48}$$

$$\mathrm{DOM}^{\bullet+} + \mathrm{HO}^{\bullet} \rightarrow \text{the photo degraded products} \tag{3.49}$$

Major oxidants arising upon UV irradiation of TiO_2 are the valence-band holes (h^+), which can be trapped by surface Ti–OH^- groups to yield Ti-$HO^\bullet$ radical species (surface-bound $HO^\bullet$ or $HO^\bullet{}_{ads}$). In contrast, the conduction-band electrons (e^-) can be trapped to form surface and sub-surface Ti^{III} species: The latter can react with dissolved oxygen to give hydrogen peroxide radical ($HO_2{}^\bullet$), which disproportionates to H_2O_2 that is photolyzed to give $HO^\bullet$ in solution (Murov et al. 1993; Tseng and Haung 1991; Fox 1993).

As already reported, a semiconductor photocatalyst generates electron/hole pairs upon irradiation, with free electrons produced in a nearly empty conduction band and positive holes remaining in the valence band (Bard 1979; Schiavello 1987). The lifetime of an electron/hole pair can vary from a few nano-seconds to a few hours depending on the ambient conditions (Pleskov and Gurevich 1986). Once formed the $HO^\bullet$ radical (surface-bound or in solution) reacts with organic compounds, which results into the heterogeneous photocatalytic degradation of organic contaminants in solution (Sun and Bolton 1996; Ullah et al. 1998; Konstantinou and Albanis 2004; Tseng and Haung 1991; Fox 1993; Han et al. 2009; Prousek 1996; Vione et al. 2001; Muñoz et al. 2006; Saquib et al. 2008).

3.7 Photoinduced Generation of HO• from Algae

Recent studies demonstrate that chlorophyll *a* and algae can photolytically produce HO• in natural waters (Li et al. 2008; Dykens et al. 1992; Oda et al. 1992). It is hypothesized that HO• is produced by H_2O_2 (Eq. 3.2), which is initially formed in irradiated aqueous suspensions of algae. The photo formation of H_2O_2 from algae is a well-known phenomenon in natural waters (Lehninger 1970; Zepp et al. 1987; Collen et al. 1995), which has been explained in detail in the second chapter (see chapter "Photoinduced and Microbial Generation of Hydrogen Peroxide and Organic Peroxides in Natural Waters"). As found by in vitro studies carried out in aerated water–acetone mixtures, the generation of H_2O_2 proceeds through the photo-formation of superoxide ion by chlorophyll *a* (You and Fong 1986). It has been shown that the HO• photoproduction increases with increasing algal concentration and irradiation time (Li et al. 2008). It has also been shown that for the marine microalga *Dunaliella salina*, the HO• production is significantly enhanced by the addition of Pb(II), or by Pb(II) and methylmercury. However, the HO• production is decreased by addition of methylmercury only, suggesting a complex effect of metal pollution on the HO• production from algae suspensions (Li et al. 2008).

4 Factors Controlling the Production and Decay of HO• in Natural Waters

Major sources of HO• in irradiated natural waters are NO_2^-, NO_3^-, H_2O_2, photo-Fenton reaction (which depends on the concentrations of Fe and H_2O_2), humic substances (fulvic and humic acids), tryptophan amino acid, autochthonous and even unknown DOM components as well as anthropogenic DOM such as components of household detergents or fluorescent whitening agents (FWAs) such as (DAS1) and distyryl biphenyl (DSBP) (Table 2). The concentration values and, in the case of DOM, also the nature of these species are very variable among rivers, lakes and oceans, which significantly affects the HO• sources in natural waters (Takeda et al. 2004; White et al. 2003; Nakatani et al. 2007; Mostofa KMG and Sakugawa H, unpublished data; Mostofa et al. 2005; Zika et al. 1993). Interestingly, the concentration values of NO_2^- and NO_3^- in surface seawaters are lowered by solar irradiation that induces transformation reactions. This is a key factor that limits the role of nitrite and nitrate as HO• sources in the surface seawater layer. Unknown or little known photoinduced HO• sources in natural waters are still poorly characterized DOM components such as humic substances (fulvic and humic acid), amino acids, proteins, components of detergents or FWAs, as well as their daughter photoproducts, unidentified organic compounds and photo-Fenton reactions. Therefore, the production of HO• in natural waters depends on several factors, which can be listed as: (1) Wavelength spectrum of solar radiation; (2) Nature and contents of DOM; (3) Consumption of HO• by DOM and other chemical species; (4) Nitrite and nitrate: Effect of wavelength spectrum, temperature, and pH; (5) H_2O_2: Effect of wavelength spectrum and temperature; (6) Fenton reaction: Effect of pH, temperature and salinity; (7) Photo-Fenton reaction; and (8) Photo-ferrioxalate/H_2O_2 reaction: Dependence of pH and reactants. These issues will be discussed below.

4.1 Wavelength Spectrum of Solar Radiation

The production of $HO^•$ significantly depends on the spectrum of the incident radiation in the ultraviolet (UVC = 200–280 nm; UVB = 280–320 nm; UVA = 320–400 nm) and visible (400–700 nm) regions. Note that UVC radiation is not present in the sunlight spectrum in the troposphere. The maximum $HO^•$ photo-production has been observed upon UVB irradiation of three DOM fractions isolated from lake water, and it decreased at higher wavelengths (Grannas et al. 2006). A rough estimation shows that the $HO^•$ production is approximately 191–247 % and 103–178 % higher at 308 nm than at 355 and 330 nm, respectively. The photoproduction of $HO^•$ by nitrate is only induced by UVB radiation, while the production of $HO^•$ by nitrite and DOM can take place under both UVB and UVA. Note that UVA radiation is able to penetrate more deeply than UVB into the water bodies (Mopper and Zhou 1990; Zafiriou and Bonneau 1987; Zellner et al. 1990; Qian et al. 2001). H_2O_2 generates $HO^•$ (Eq. 3.2) mostly by non-environmental UVC with quantum yield around 0.5, but the radiation absorption by H_2O_2 is extended to the UVB region and the photolysis quantum yield is expected to be similar (Legrini et al. 1993; von Sonntag et al. 1993). During non-ozone hole conditions, $HO^•$ formation from nitrate contributed ~33 % to the total $HO^•$ photoinduced generation in open oceanic surface waters, while the role of DOM plus nitrite was ~67 % (Qian et al. 2001). During an ozone hole event (e.g., 151 Dobson units) the corresponding results were ~40 and 60 % for nitrate and DOM plus nitrite, respectively, because of an increase in UVB irradiance that enhanced the photolysis of nitrate. A model estimation of $HO^•$ photoinduced generation in Antarctic seawater during an ozone hole (151 Dobson units) shows that the production is enhanced by at least 20 %, mostly from nitrate photolysis and to a minor extent from DOM photoinduced reactions (Qian et al. 2001). Similar results have been obtained for Arctic water (Rex et al. 1997; Randall et al. 2005).

4.2 Nature and Amount of DOM Components

The radical $HO^•$ is generated photolytically from various organic substances in natural waters (Table 2) (Mill et al. 1980; Mopper and Zhou 1990; Vaughn and Blough 1998; Holder-Sandvik et al. 2000; Page et al. 2011; Grannas et al. 2006). The most common dissolved organic compounds are humic substances (fulvic and humic acid), amino acids, fluorescence whitening agents (FWAs) or components of detergents such as diaminostilbene (DAS1) and distyryl biphenyl (DSBP), phenolic compounds, various kinds of autochthonous DOM components and unknown ones (Page et al. 2011; Mostofa et al. 2011). The $HO^{\bullet}$ production greatly depends on the DOM components, an increase of which can enhance the $HO^{\bullet}$ formation rate in the aqueous solution (Fig. 1) (Page et al. 2011; Mostofa KMG and Sakugawa H, unpublished data). Study shows that DOM isolates from the Suwannee River photolytically produce free $HO^{\bullet}$, while Elliot Soil Humic Acid (ESHA), and Pony Lake Fulvic Acid (PLFA) hydroxylate arenes, at least in part, through a lower-energy hydroxylating species (Page et al. 2011). The photoinduced generation of $HO^{\bullet}$ from various standard organic substances (1 mg L^{-1} aqueous solutions) has been studied using a solar simulator (Table 2; Fig. 1) (Mostofa KMG and Sakugawa H unpublished data). The differences in production rates among different substances are attributed to the variation of the fluorophores or functional groups on the highly unsaturated aliphatic carbon chains present in various DOM components (FDOM or CDOM) (Mostofa et al. 2011; Senesi 1990). The electronic transitions involving illuminated DOM can lead to the release of free electrons, which can induce the production of H_2O_2 in natural waters. It would then follow $HO^{\bullet}$ production upon H_2O_2 photolysis, and the photogenerated hydroxyl radicals can contribute to the transformation of DOM or of the pollutants present in natural waters. That would lead to photoinduced self-transformation of DOM via photogenerated $HO^{\bullet}$. Table 2 reports the main $HO^{\bullet}$ sources that are operational in the natural environment.

In polluted sewage waters, the $HO^{\bullet}$ formation is greatly enhanced after 60 min of irradiation (Fig. 3a). This finding could be compatible with $HO^{\bullet}$ production by photogenerated H_2O_2, which would not be present in the system before irradiation and would undergo accumulation at earlier irradiation times. The $HO^{\bullet}$ generation being proportional to $[H_2O_2]$, the accumulation of hydrogen peroxide would lead to an increase of the formation rate of the $HO^{\bullet}$.

The photogenerated $HO^{\bullet}$ is rapidly consumed in natural waters upon reaction with dissolved organic compounds (Brezonik and Fulkerson-Brekken 1998; Southworth and Voelker 2003; Westerhoff et al. 1999; Goldstone et al. 2002; Miller and Chin 2002; Miller et al. 2002; Moran and Zepp 1997). A rough estimation showed that the consumption of $HO^{\bullet}$ by DOM is 12–56 % in rivers (Nakatani et al. 2004). Considering the DOM as a major sink, the maximum steady-state concentration of $HO^{\bullet}$ is equal to the production rate divided by the decay rate constant, and can be depicted by (Eq. 4.1) (Schwarzenbach et al. 1993):

$$[\mathrm{HO}]_{\mathrm{SS,Fenton}} = \sum_{\lambda} \frac{I(\lambda)\varphi(\lambda)[1 - 10^{-\varepsilon(\lambda)Cz}]}{zK_{\mathrm{DOM}}C} \tag{4.1}$$

where the $1-10^{-\epsilon(\lambda)Cz}$ is the light attenuation at the depth z (cm), φ is the apparent quantum yield for the generation of $HO^{\bullet}$ from DOM (mole Einstein^{-1}), ϵ is the absorption coefficient of DOM [(mg C $L^{-1})^{-1}cm^{-1}$], C its concentration of DOM (mg C L^{-1}), and k_{DOM} (M^{-1} s^{-1}) is the second-order reaction rate constant between $HO^{\bullet}$ and DOM. The rate constant for the reaction of $HO^{\bullet}$ with

Suwannee River Fulvic Acid is (3.8–4.4) $\times$ 10^4 (mg C)$^{-1}$ L s^{-1} (Southworth and Voelker 2003; Westerhoff et al. 2007). Rate constants in the same ranges, (1–7) $\times$ 10^4 (mg C)$^{-1}$ L s^{-1}, have been determined for the reaction between $HO^\bullet$ and DOM present in natural water samples or extracted from them (Vione et al. 2006; Westerhoff et al. 1999; Goldstone et al. 2002; Gao and Zepp 1998). The half-life of a model pollutant can be estimated as $t_{1/2} = \ln 2\ (k_{HO}[HO^\bullet]_{SS})^{-1}$, where k_{HO} is the second-order reaction rate constant with $HO^\bullet$, and $[HO^\bullet]_{ss}$ is given by Eq. (4.1). Depending on the mixed layer depths that influence $[HO^\bullet]_{ss}$, and for k_{HO} values of the order of 10^9–10^{10} M^{-1} s^{-1}, $t_{1/2}$ can vary from some days to some months.

4.2.1 Fulvic Acid as a Producer and Scavenger of $HO^\bullet$ in Natural Waters

Fulvic acid (FA) can produce $HO^\bullet$ photolytically in aqueous solution (Table 2) (Vaughn and Blough 1998; Goldstone et al. 2002). FA can account for approximately 23–70 % of H_2O_2 production in rivers (Mostofa and Sakugawa 2009). A general reaction of FA that leads to the formation of H_2O_2 (Eq. 4.2) can be depicted on the basis of Eqs. (3.13–3.18) (see chapter "Photoinduced and Microbial Generation of Hydrogen Peroxide and Organic Peroxides in Natural Waters"):

$$FA + O_2 + H_2O \xrightarrow{h\upsilon} FA^{\bullet +} + H_2O_2 + O_2 + OH^- \quad (4.2)$$

The generation of H_2O_2 from FA can lead to $HO^\bullet$ ($H_2O_2 + h\upsilon \rightarrow 2HO^\bullet$) that could further react with FA, at the same time being consumed and causing transformation of FA (Voelker and Sulzberger 1996). Recent experimental studies indicate that at least 50 % of the hydroxylation reactions photosensitized by DOM isolates would be a result of a pathway that is independent of hydrogen peroxide (Page et al. 2011). Recently, the photo-degradation of various functional groups in DOM by $HO^\bullet$ has been observed, and the rates determined in aqueous solution (Minakata et al. 2009). The results suggest that DOM or FA is important scavengers of photolytically generated $HO^\bullet$ in aqueous solution.

4.3 Other Chemical Species or Processes as $HO^\bullet$ Sinks

There are several processes that can inhibit $HO^\bullet$ formation or consume these radicals in the aquatic environments, which can be distinguished as:

(1) Decrease in light intensity in deeper waters, which reduces the formation rate of H_2O_2 and of Fe(II). Photo-generated H_2O_2 and Fe(II) at the surface could be moved downward through vertical mixing processes, thereby reducing their concentration in the surface layer (Southworth and Voelker 2003; Pullin et al. 2004). Such an effect can greatly decrease the $HO^\bullet$ production in water.

(2) Upon oxidation of carbonate and bicarbonate by $HO^\bullet$, the carbonate radical ($CO_3^{\bullet-}$) is produced (Eqs. 4.3, 4.4) in surface waters (Minella et al. 2011) and DOM is the key component to sink $CO_3^{\bullet-}$ by the reaction between them (Eq. 4.5) in surface waters (Buxton et al. 1988; Canonica et al. 2005).

$$HO^\bullet + CO_3^{2-} \rightarrow OH^- + CO_3^{\bullet-} \left[k_{4.3} = 3.9 \times 10^8\,M^{-1}\,s^{-1}\right] \quad (4.3)$$

$$HO^\bullet + HCO_3^- \rightarrow H_2O + CO_3^{\bullet-} \left[k_{4.4} = 8.5 \times 10^6\,M^{-1}\,s^{-1}\right] \quad (4.4)$$

$$DOM + CO_3^{\bullet-} \rightarrow DOM^{+\bullet} + CO_3^{2-} \quad \left[k_{4.5} = 10^2\,(mg\,C)^{-1}\,s^{-1}\right] \quad (4.5)$$

These reactions are all dependent on pH since pK_a of carbonate and bicarbonate has 10.3 and 6.3 (Tossell 2005). At around neutral pH, bicarbonate is dominant over carbonate so HO radical scavenging is not so significant as to higher pH (i.e., carbonate is more dominant).

(3) In seawater, $HO^\bullet$ is mainly consumed by Br^- ions (Zafiriou et al. 1984; Neta et al. 1988; Goldstone et al. 2002; Song et al. 1996; Zafiriou et al. 1987; Das et al. 2009). The reaction of $HO^\bullet$ with Br^- generates the oxidized bromine radical species ($BrOH^{\bullet-}$) (Eq. 4.6), which can be transformed into Br and $Br_2^{\bullet-}$ (Eq. 4.7) (Goldstone et al. 2002; Song et al. 1996; Zafiriou et al. 1987; von Gunten and Oliveras 1997; Grebel et al. 2009).

$$HO^\bullet + Br^- \rightarrow BrOH^{\bullet-} \quad (4.6)$$

$$BrOH^- + H^+ \rightarrow Br^\bullet + H_2O \xrightarrow{Br^-} Br_2^{\bullet-} + H_2O \quad (4.7)$$

Moreover, the bromine radical species could also react with DOM and induce transformation reactions. The half-life of $BrOH^{\bullet-}$ is less than a few seconds, for a H_2O_2 concentration of 0.1 mg L^{-1} in cloud water at pH 8. However, at pH 5 the half-life of $BrOH^{\bullet-}$ is several hours for the same H_2O_2 concentration (von Gunten and Oliveras 1997). A recent study shows that bromide scavenges the $HO^\bullet$ radicals formed upon photolysis of nitrate, before they leave the solvent cage, thereby inhibiting the geminate recombination between the photogenerated $HO^\bullet$ and $^\bullet NO_2$. The result is that the formation rate of $HO^\bullet + Br_2^{\bullet-}$ with nitrate + bromide is higher than that of $HO^\bullet$ with nitrate alone. Such an effect compensates for the lower reactivity of $Br_2^{\bullet-}$ compared to $HO^\bullet$ toward certain organic substrates, e.g. phenol and tryptophan (Das et al. 2009). It has also been shown that the radical $Br_2^{\bullet-}$ is an effective brominating agent for phenol (Vione et al. 2008).

(4) In stimulated polymorphonuclear leukocyte systems, a water extract from green tea and green tea polyphenols had the strongest scavenging effect on $HO^\bullet$, much stronger than vitamin C or vitamin E. Moreover, rosemary antioxidants and curcumin have weaker scavenging effects than vitamin C, but a stronger one than vitamin E (Zhao et al. 1989).

4.4 Dependence on NO_2^- and NO_3^- Photolysis: Effect of Wavelength Spectrum, Temperature, and pH

The photoinduced generation of $HO^•$ largely depends on the presence of NO_2^- and NO_3^- in natural waters (Zafiriou and True 1979a, b; Takeda et al. 2004; Zafiriou and Bonneau 1987; Zepp et al. 1987; Zellner et al. 1990; Brezonik and Fulkerson-Brekken 1998; Mack and Bolton 1999). The quantum yield of $HO^•$ (Φ_{HO}) for NO_2^- and NO_3^- photolysis significantly depends on the temperature, wavelength and pH in the aqueous solution (Tables 4 and 5) (Zepp et al. 1987; Zellner et al. 1990; Goldstein and Rabani 2008; Matykiewiczová et al. 2007; Treinin and Hayon 1970; Strehlow and Wagner 1982; Warneck and Wurzinger 1988; Mark et al. 1996; Vione et al. 2001; Chu and Anastasio 2003). The Φ_{HO} of NO_2^- photolysis has been determined for a NO_2^- concentration of 3×10^{-3} M, for $4 \leq pH \leq 11$ and temperatures between 278–353 K at 308 nm (Table 4) (Zellner et al. 1990). The absorption coefficient of NO_2^- at 308 nm is $\varepsilon_{308} = 4.1$ L mol^{-1} cm^{-1} (Strickler and Kasha 1963). For the same solution, Φ_{HO} has also been determined in the temperature range of 278–358 K, for pH 8 and under 351 nm irradiation (Table 4). Note that the absorption coefficient of NO_2^- is $\varepsilon_{351} = 9.4$ L mol^{-1} cm^{-1} (Strickler and Kasha 1963). The results have demonstrated that Φ_{HO} generally increases with temperature and decreases with pH (Fig. 4).

The temperature dependence of Φ_{HO} is mathematically derived, based on the results of NO_2^- photolysis at 308 (Eq. 4.8) and 351 nm (Eq. 4.9) at room temperature (Table 4) (Zellner et al. 1990). It can be expressed as:

$$\Phi_{HO}(298\text{ K}) = 0.07 \pm 0.01 \qquad \text{for } 7 \leq \text{pH} \leq 9 \tag{4.8}$$

$$\Phi_{HO}(298\text{ K}) = 0.046 \pm 0.009 \qquad \text{for pH} = 8 \tag{4.9}$$

Using the averaged data over the range $7 \leq pH \leq 9$, the temperature dependence of Φ_{HO} at 308 nm (Eq. 4.10) becomes:

$$\Phi_{HO}(T) = \Phi_{HO}(298\text{ K}) \exp\left[(1560 \pm 360)\left(\frac{1}{298} - \frac{1}{T}\right)\right] \tag{4.10}$$

where the apparent activation energy is $E_A = 13 \pm 3$ kJ mol^{-1}. This equation is in good agreement with previous data (Zafiriou and Bonneau 1987), where it has been obtained $E_A = 12 \pm 6$ kJ mol^{-1} at 298.5 nm and in the temperature range of 296–322 K.

The temperature dependence of Φ_{HO} for NO_2^- photolysis at 351 nm can be expressed as (Eq. 4.11):

$$\Phi_{HO}(T) = \Phi_{HO}(298\text{ K}) \exp\left[(1800 \pm 400)\left(\frac{1}{298} - \frac{1}{T}\right)\right] \tag{4.11}$$

Table 4 The quantum yield of $HO^{\bullet}$ (Φ_{HO}) of nitrite photolysis as a function of pH and temperature (T)

T (K)	pH							
	4	5	6	7	8	9	10	11
For wavelength 308 nm								
278	0.047 ± 0.004	0.037 ± 0.002	0.039 ± 0.005	0.040 ± 0.005	0.040 ± 0.002	0.039 ± 0.001	0.034 ± 0.002	0.030 ± 0.003
298	0.091 ± 0.011	0.080 ± 0.007	0.068 ± 0.004	0.068 ± 0.009	0.071 ± 0.009	0.068 ± 0.008	0.060 ± 0.007	0.046 ± 0.006
318	0.100 ± 0.011	0.085 ± 0.005	0.081 ± 0.006	0.082 ± 0.011	0.079 ± 0.003	0.080 ± 0.003	0.076 ± 0.003	0.069 ± 0.005
353	0.160 ± 0.011	0.159 ± 0.008	0.148 ± 0.006	0.141 ± 0.007	0.143 ± 0.003	0.132 ± 0.005	0.125 ± 0.003	0.115 ± 0.004
For wavelength 351 nm								
278	nd	nd	nd	nd	0.027 ± 0.003	nd	nd	nd
288	nd	nd	nd	nd	0.038 ± 0.004	nd	nd	nd
298	nd	nd	nd	nd	0.046 ± 0.003	nd	nd	nd
308	nd	nd	nd	nd	0.058 ± 0.002	nd	nd	nd
318	nd	nd	nd	nd	0.078 ± 0.003	nd	nd	nd
328	nd	nd	nd	nd	0.091 ± 0.010	nd	nd	nd
338	nd	nd	nd	nd	0.096 ± 0.003	nd	nd	nd
348	nd	nd	nd	nd	0.118 ± 0.008	nd	nd	nd
358	nd	nd	nd	nd	0.153 ± 0.009	nd	nd	nd

nd means ‘not detected’
Data source Zellner et al. (1990)

Table 5 The quantum yield of $HO^{\bullet}$ (Φ_{HO}) of nitrate photolysis as a function of pH and temperature (T)

T (K)	pH							
	4	5	6	7	8	9	10	11
For wavelength 308 nm								
278	0.008 ± 0.001	0.007 ± 0.001	0.008 ± 0.001	0.007 ± 0.001	0.008 ± 0.001	0.007 ± 0.001	0.007 ± 0.001	0.004 ± 0.001
298	0.016 ± 0.001	0.017 ± 0.001	0.017 ± 0.001	0.016 ± 0.001	0.017 ± 0.001	0.016 ± 0.001	0.014 ± 0.001	0.009 ± 0.001
318	0.028 ± 0.002	0.026 ± 0.003	0.027 ± 0.002	0.028 ± 0.001	0.027 ± 0.002	0.024 ± 0.002	0.015 ± 0.002	0.011 ± 0.001
353	0.038 ± 0.004	0.034 ± 0.003	0.036 ± 0.004	0.035 ± 0.002	0.036 ± 0.003	0.030 ± 0.005	0.017 ± 0.003	0.012 ± 0.001

Data source (Zellner et al. 1990)

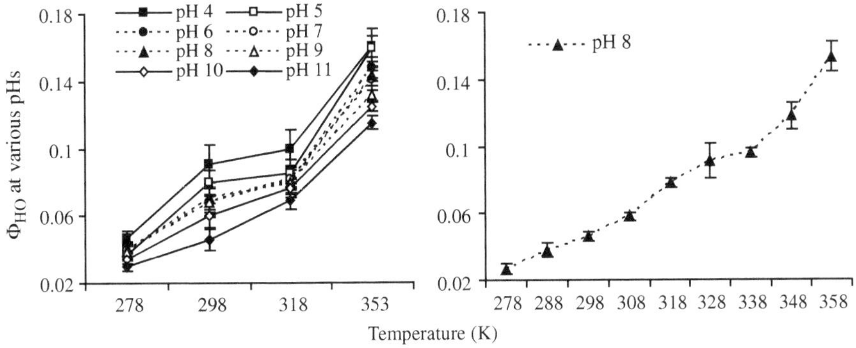

Fig. 4 Quantum yields of $HO^{\bullet}$ of NO_2^- photolysis with variation in temperature at a specific wavelength at 308 nm for various pH (**a**) and at 351 nm (**b**) at pH 8. *Data source* Zellner et al. (1990)

These results allow the hypothesis that the temperature dependence of Φ_{HO} for NO_2^- photolysis is independent of wavelength, despite a decline of the absolute Φ_{HO} value with increasing wavelength (Fig. 4) (Zellner et al. 1990).

On the other hand, the Φ_{HO} of NO_3^- photolysis has been determined for a NO_3^- concentration of 3×10^{-3} M in the range $4 \leq pH \leq 11$, at temperature between 278–353 K and 308 nm irradiation (Table 5) (Zellner et al. 1990). The absorption coefficient of NO_3^- at 308 nm is $\varepsilon_{308} = 6.5$ L mol^{-1} cm^{-1} (Meyerstein and Treinin 1961). It has been shown that the quantum yields at each temperature are generally independent of pH in the range $4 \leq pH \leq 9$. At higher pH's, the quantum yield is typically decreased (Fig. 5).

Fig. 5 Quantum yields of $HO^{\bullet}$ of NO_3^- photolysis with variation in temperature at a specific wavelength (308 nm) for various pH. *Data source* Zellner et al. (1990)

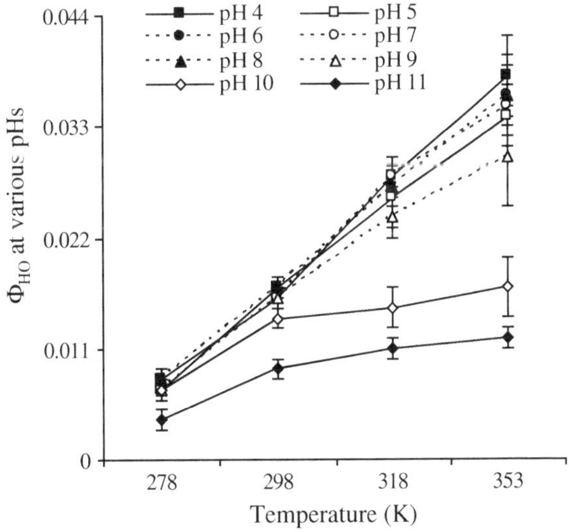

The Φ_{HO} of NO_3^- photolysis at room temperature (298 K) is (Eq. 4.12)

$$\Phi_{HO}(298\text{ K}) = 0.017 \pm 0.03 \text{ for } 4 \leq \text{pH} \leq 9 \quad (4.12)$$

The temperature dependence of Φ_{HO} is mathematically derived from the results of NO_3^- photolysis at 308 nm, in the range of $4 \leq \text{pH} \leq 9$ (Table 5). It can be written as (Eq. 4.13) (Zellner et al. 1990):

$$\Phi_{HO}(\text{T}) = \Phi_{HO}(298\text{ K})\exp\left[(1800 \pm 480)\left(\frac{1}{298} - \frac{1}{T}\right)\right] \quad (4.13)$$

where the apparent activation energy $E_A = 15 \pm 4$ kJ mol^{-1}. The effect of temperature on the $HO^\bullet$ quantum yield is relatively stronger than the effect of pH: Φ_{HO} increases by almost a factor of 5 over the temperature range 278–353 K, for $4 \leq \text{pH} \leq 9$ (Zellner et al. 1990). Because of the low molar absorption coefficient of NO_3^- at 351 nm ($\varepsilon351 \approx 0.05$ L mol^{-1} cm^{-1}) (Meyerstein and Treinin 1961), nitrate is a less effective $HO^\bullet$ source at 351 than at 308 nm. However, the results indicate that the $HO^\bullet$ quantum yield and the temperature dependence are constant at these wavelength (Zellner et al. 1990). There is an overall consistency between the results of different studies into the temperature and wavelength dependence of the $HO^\bullet$ quantum yield for NO_3^- photolysis (e.g., obtained $\Phi_{HO} = 0.015 \pm 0.003$ at 313 nm) (Zepp et al. 1987).

4.5 Dependence on H_2O_2 Photolysis: Effect of Wavelength Spectrum and Temperature

The photoinduced generation of $HO^\bullet$ from H_2O_2 ($H_2O_2 + h\upsilon \rightarrow 2\ HO^\bullet$) depends on the wavelength spectrum of UV-B radiation and on water temperature (Zellner et al. 1990; Hunt and Taube 1952; Baxendale and Wilson 1956; Volman and Chen 1959; Goldstein and Rabani 2008; Chu and Anastasio 2005; Goldstein et al. 2007). The quantum yield of $HO^\bullet$ (Φ_{HO}) has been determined for 3×10^{-3} M H_2O_2 at pH 7, at the wavelengths of 308 and 351 nm (Zellner et al. 1990). The H_2O_2 absorption coefficients for these wavelengths are $\varepsilon_{308} = 1.10$ and $\varepsilon_{351} = 0.11$ L mol^{-1} cm^{-1}, respectively (Taylor and Cross 1949). It has also been shown that the $HO^\bullet$ production gradually increases with an increase in temperature at pH 7 (Fig. 6; Table 6) (Zellner et al. 1990). Zellner and his co-workers derived the equations for the temperature dependence of Φ_{HO} upon H_2O_2 photolysis.

For the photolysis at 308 nm, Φ_{HO} can be expressed as (Eq. 4.14):

$$\Phi_{HO}(298\text{ K}) = 0.98 \pm 0.03 \quad (4.14)$$

The temperature dependence of Φ_{HO} can be expressed as (Eq. 4.15):

$$\Phi_{HO}(T) = \Phi_{HO}(298\text{ K})\exp\left[(660 \pm 190)\left(\frac{1}{298} - \frac{1}{T}\right)\right] \quad (4.15)$$

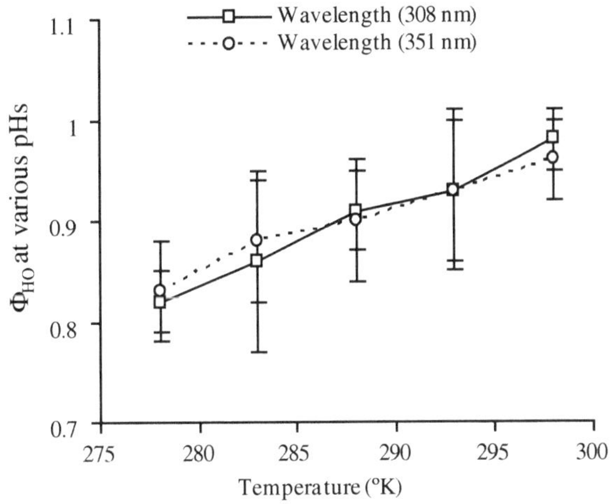

Fig. 6 Quantum yields of $HO^{\bullet}$ of H_2O_2 photolysis with variation in temperature at specific wavelengths (308 and 351 nm, respectively) at pH 7. *Data source* Zellner et al. (1990)

Table 6 Quantum yield of $HO^{\bullet}$ production (Φ_{HO}) from H_2O_2 photolysis at pH 7 as a function of temperature (T)

T (°K)	Φ_{HO} at pH 7	
	308 nm	351 nm
278	0.82 ± 0.03	0.83 ± 0.05
283	0.86 ± 0.09	0.88 ± 0.06
288	0.91 ± 0.04	0.90 ± 0.06
293	0.93 ± 0.07	0.93 ± 0.08
298	0.98 ± 0.03	0.96 ± 0.04

Data source Zellner et al. (1990)

where the activation energy is 5.5 ± 1.6 kJ mol^{-1}.

These values are essentially constant for H_2O_2 photolysis at 351 nm (Eqs. 4.16, 4.17):

$$\Phi_{HO}(298\ \text{K}) = 0.96 \pm 0.04 \tag{4.16}$$

$$\Phi_{HO}(T) = \Phi_{HO}(298\ \text{K}) \exp\left[(580 \pm 160)\left(\frac{1}{298} - \frac{1}{T}\right)\right] \tag{4.17}$$

A recent study has shown that the quantum yield $\Phi_{HO} = 1.11 \pm 0.07$ in the excitation range of 205–280 nm for the photolysis of H_2O_2. This is in agreement with earlier studies (Goldstein et al. 2007). Therefore, the available data suggest that the photolysis of aqueous H_2O_2 at wavelengths ≥ 300 nm generates $HO^{\bullet}$ with a quantum yield Φ_{HO} ~1 at room temperature (25 °C). Φ_{OH} decreases to approximately 0.82 at 5 °C (278 K). It can be noted that the effective quantum yield of H_2O_2 dissociation is approximately 0.5, but the photolysis of H_2O_2 yields two

Table 7 The reaction rate constants for generation of HO• in Fenton reaction at various pH ranges in the absence of light

Reaction type	pH 3	pH 4	pH 5	Equations
	Reaction rate constant (k) ($M^{-1}s^{-1}$)			
$Fe^{(II)} + H_2O_2 \rightarrow Fe^{(III)} + HO^{\bullet} + OH^-$	63	1.2×10^2	5.7×10^2	4.18
$Fe^{(III)} + H_2O_2 \rightarrow Fe^{(II)} + HO_2^{\bullet}/O_2^- + H^+$	2×10^{-3}	2.5×10^{-3}	2.6×10^{-3}	4.19
$H_2O_2 + HO^{\bullet} \rightarrow HO_2^{\bullet}/O_2^- + H_2O$	3.7×10^7	3.3×10^7	3.3×10^7	4.20
$Fe^{(III)} + HO_2^{\bullet}/O_2^- \rightarrow Fe^{(II)} + O_2 + H^+$	7.8×10^5	6.8×10^6	3.1×10^7	4.21
$Fe^{(II)} + HO^{\bullet} \rightarrow Fe^{(III)} + OH^-$	3.2×10^8	3.2×10^8	3.2×10^8	4.22
$Fe^{(II)} + HO_2^{\bullet}/O_2^- \rightarrow Fe^{(III)} + HO_2^{\bullet}$	1.3×10^6	2.4×10^6	6.6×10^6	4.23
$HO\bullet + HO^{\bullet} \rightarrow H_2O_2$	5.2×10^9	5.2×10^9	5.2×10^9	4.24
$HO_2^{\bullet}/O_2^- + HO_2^{\bullet}/O_2^- \rightarrow H_2O_2$	2.3×10^6	1.2×10^7	2.3×10^7	4.25
$HO^{\bullet} + HO_2^{\bullet}/O_2^- \rightarrow H_2O + O_2$	7.1×10^9	7.5×10^9	8.9×10^9	4.26

Data source (Kwan and Voelker 2002)

$HO^{\bullet}$. Therefore, the quantum yield of $HO^{\bullet}$ photoinduced generation from H_2O_2 is double compared to that of H_2O_2 photolysis (Hunt and Taube 1952; Baxendale and Wilson 1956; Volman and Chen 1959).

4.6 Fenton Reaction: Effect of pH, Temperature and Salinity

The Fenton reaction depends on the presence of Fe^{2+} (or Fe^{3+}) and H_2O_2 in natural waters. The oxidation of Fe(II) with H_2O_2 in seawater depends on pH (2–8.5), temperature (5–45 °C) and salinity (0–35 g L^{-1}) (Wells and Salam 1968; Moffett and Zika 1987a, b; Gallard et al. 1998; Bossmann et al. 1998; Duesterberg et al. 2008; Millero and Sotolongo 1989; de Laat and Gallard 1999; Duesterberg and Waite 2006; Duesterberg et al. 2005; Farias et al. 2007; Jung et al. 2009). The rate constants of the chain Fenton reactions and the relevant dependence on pH are presented in Table 7 (Kwan and Voelker 2002). The reaction rate constant between Fe^{2+} and H_2O_2 significantly increases with pH in the range from 3 to 5 (Eq. 4.18; Table 7), and a similar effect is observed with some of the chain reactions (Eqs. 4.19–4.26; Table 7).

A recent study that has been carried out in the pH range 2.5–4.0, both in the presence and absence of a target organic substance (formic acid), also highlights the importance of the Fenton system in the catalytic redox cycling of iron (Duesterberg et al. 2008). Supply of oxygen can enhance the efficiency of the Fenton oxidation, which is understandably attenuated by competition with the organic intermediates in the reaction media (Sychev and Isak 1995; Duesterberg et al. 2005).

It is shown that the addition of phosphotungstate ($PW_{12}O_{40}{}^{3-}$), a polyoxometalate, extends the working pH range of the Fenton system (Fe^{3+}/H_2O_2) up to pH 8.5. This is because $PW_{12}O_{40}{}^{3-}$ forms a soluble complex with iron that converts H_2O_2 into $HO^{\bullet}$ (Lee and Sedlak 2009). The Fenton reaction for the catalytic oxidation of organic compounds is usually limited to the acidic pH region, due to the low solubility of Fe(III) and the low efficiency of oxidant production at neutral pH values. The coordination of Fe^{2+} by $PW_{12}O_{40}{}^{3-}$ can alter the mechanism of the reaction of Fe(II) with H_2O_2 at neutral pH, yielding a reactive species capable of oxidizing aromatic compounds (Lee and Sedlak 2009).

An increase in temperature significantly enhances the efficiency of the Fenton reaction towards the degradation of organic pollutants in aqueous solution (Fig. 7) (Farias et al. 2007). From the results of Farias et al. (2007) it is possible to estimate a HCOOH degradation enhancement of 164–191 % at 313 K and 370–561 % at 328 K compared to 298 K, after 20 min of reaction time, with H_2O_2: formic acid molar ratios between 3 and 8 (Fig. 7). The results also show that the conversion of the organic pollutant is the highest at 328 K, with a low H_2O_2: formic acid molar ratio. Conversion gradually decreases when the H_2O_2: formic acid molar ratio is increased from 3 to 8 at the same temperature (328 K) (Fig. 7). However, at 313 K the conversion of the organic pollutant gradually increases when the H_2O_2: formic acid molar ratio is increased from 3 to 8 (Fig. 7).

The overall Fenton reaction ($Fe^{2+} + H_2O_2$) is first-order with respect to the concentration of total Fe(II) and H_2O_2. It can be expressed as follows (Eq. 4.27) (Moffett and Zika 1987a, b; Millero and Sotolongo 1989):

$$\mathrm{d}\,[\mathrm{Fe}\,(\mathrm{II})]\,/\mathrm{dt} = -k\,\{[\mathrm{Fe}\,(\mathrm{II})]\,[\mathrm{H_2O_2}]\} \tag{4.27}$$

where the brackets denote the total molar concentrations. With an excess of $[H_2O_2]$, the overall reaction becomes pseudo-first order and can be depicted as follows (Eq. 4.28):

$$\mathrm{d}\,[\mathrm{Fe}\,(\mathrm{II})]\,/\mathrm{dt} = -k_1'\,[\mathrm{Fe}\,(\mathrm{II})] \tag{4.28}$$

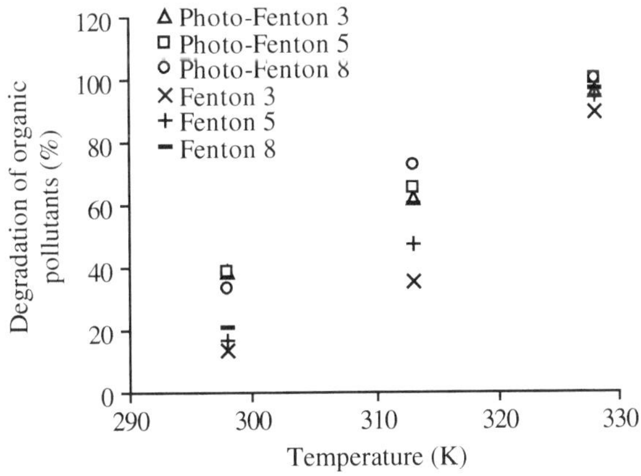

Fig. 7 The percentage degradation of organic pollutants by photo-Fenton and Fenton reactions in a stirred tank laboratory photoreactor. 3, 5 and 8 indicates the molar ratio of the reactant H_2O_2 to formic acid in the aqueous solutions. *Data Source* Farias et al. (2007)

where $k_1' = k[H_2O_2]$. The reaction is quite fast at high temperature and pH, and it becomes quite difficult to be examined at 25 °C and pH 8. The reaction at 5 °C and pH 3.5 followed first-order kinetics with respect to Fe(II) for seawater samples, giving $k' = 0.0385 \pm 0.0009$ min^{-1} and log $k = 1.50 \pm 0.02$ (k in M^{-1} s^{-1}) (Millero and Sotolongo 1989). At higher pH and temperature, the second-order reaction rate constants have been determined at the stoichiometric ratio [Fe(II)]/ $[H_2O_2] = 2$, following Eq. (4.29) (Benson 1960)

$$1/[\mathrm{Fe(II)}] = 1/[\mathrm{Fe(II)}]_o + (k/2)t \quad (4.29)$$

where $[\mathrm{Fe(II)}]_o$ is the initial concentration of Fe(II).

The rate constant k (M^{-1} s^{-1}) for Eq. (4.29) is independent of pH below pH 4, and increases significantly at high pH values. It is a linear function of $[H^+]$ or $[HO^-]$ from pH 6 to 8 in seawater. The effect of pH is not affected by a variation of the temperature

The effects of temperature (T) and ionic strength (I) on k at pH 6 can be expressed as (Eq. 4.30) (Millero and Sotolongo 1989):

$$\log k = 13.73 - 2{,}948/T - 1.70I^{1/2} + 1.20I \quad (4.30)$$

The reaction rates can be depicted as (Eq. 4.31) (Millero and Sotolongo 1989):

$$d[\mathrm{Fe(II)}]/dt = -k_2[\mathrm{Fe(II)}][H_2O_2][HO^-] \quad (4.31)$$

where the values of k_2 are independent of temperature. This is attributed to the fact that the activation energy is of the same order of magnitude as the heat of ionization of water (ΔH_w^*). The effect of the ionic strength on log k_2 can be depicted as (Eq. 4.32) (Millero and Sotolongo 1989):

$$\log k_2 = 11.72 - 2.14I^{1/2} + 1.38I \quad (4.32)$$

The overall reaction rate k over the entire range of pH, temperature and ionic strength can be expressed by (Eq. 4.33) (Millero and Sotolongo 1989):

$$k = k_o\alpha_{\mathrm{Fe}} + k_1\alpha_{\mathrm{FeOH}} \quad (4.33)$$

where α_{Fe} and α_{FeOH}, k_o and k_1 are the fractions of the two Fe(II) species and the rate constants for the oxidation of Fe^{2+} and $FeOH^+$, respectively. The values of k_o and k_1 can be expressed by (Eqs. 4.34, 4.35) (Millero and Sotolongo 1989):

$$\log k_0 = 8.37 - 1{,}866/\mathrm{T} \quad (4.34)$$

$$\log k_1 = 17.26 - 2.948/T - 1.70I^{1/2} + 1.20\mathrm{I} \quad (4.35)$$

The addition of HCO_3^- at constant pH linearly increases the reaction rate, independently of the temperature and salinity. This result can be attributed to $FeCO_3^0$ reacting faster than $FeOH^+$ with H_2O_2. At a given pH and ionic strength, the reaction rates in seawater are almost the same as in NaCl. These results can explain the differences in the reaction rates observed between estuarine waters of various alkalinity and seawater diluted with water (Millero and Sotolongo 1989).

A recent study demonstrates that the rates of the Fenton reaction do not vary in the presence of chloride, nitrate and perchlorate. However, in the presence of sulfate the rate of Fe(II) oxidation is higher and depends on pH and the concentration of sulfate. This result has been obtained under dark conditions at pH < 3, 25 ± 0.5 °C and controlled ionic strength (≤1 M) (Le Truong et al. 2004). It has also been shown that H_2O_2 is more stable in Fenton-like than in Fenton's systems, and that the lifetime of H_2O_2 is highly affected by the solution pH. Indeed, pH-buffered acidic conditions are preferred to ensure sufficient H_2O_2 lifetime, which is an important factor in the feasibility of Fenton's and Fenton-like reactions with haematite and magnetite for soil and groundwater remediation (Jung et al. 2009).

4.7 Photo-Fenton Reaction

The formation of $HO^•$ in the photo-Fenton reaction (Fe(III) + H_2O_2 + hν) significantly depends on light intensity, H_2O_2 and Fe(III) concentration, and pH (Zepp et al. 1992; Voelker et al. 1997; Southworth and Voelker 2003; Vermilyea and Voelker 2009). Studies on the ferrioxalate system (air- and argon-saturated) suggest that the Fe(II) photoproduction rates are not affected by the reducing intermediates ($CO_2^{•-}$, $O_2^{•-}$, scavenger-derived radicals) produced in the aqueous solution (Zepp et al. 1992). The results demonstrate that the photolytically generated Fe(II) can efficiently react with H_2O_2 to produce $HO^•$ in water at pH 3–8 (Zepp et al. 1992). It has been shown that the Fe(II) concentration in upstream waters gradually increased from zero (0530 h JST, Japan standard time) to 3.8 μM (1300 h JST), and decreased to zero again after sunset (1900 h JST) (Fig. 8a) (Mostofa KMG and Sakugawa H unpublished data). In the meantime, pH varied from 7.0 to 7.6 (Mostofa and Sakugawa 2009). In downstream waters, the Fe(II) concentration increases from 2.3 (0530 JST) to 4.0 μM (1200 JST) and then decreased to 2.2 μM (1900 JST) after sunset (Fig. 8b) (Mostofa KMG and Sakugawa H unpublished data). The pH varied from 7.0 to 7.3 (Mostofa and Sakugawa 2009). The production of Fe(II) was the highest at noon, and it was almost the same in clean upstream river (3.8 μM, 7.3 % of the total Fe) and polluted river waters (4.0 μM, 1.9 % of the total Fe). In a similar way, the H_2O_2 production was also maximum at noon and reached 40 and 63 nM, respectively (Fig. 8, "Photoinduced and Microbial Generation of Hydrogen Peroxide and Organic Peroxides in Natural Waters") (Mostofa and Sakugawa 2009). Similarly, Fe^{2+} has been observed to reach up to 0.9 nM in a Swiss Lake at pH 8.0–8.5; 15 nM in a low-land lake in the United Kingdom at pH 7.0–7.5; and approximately 0–145 μM in a salt marsh at Skidaway Island in the USA (Viollier et al. 2000; Aldrich et al. 2001; Emmenegger et al. 2001). The peak concentration of Fe(II) ranged from 4 to 8 % of the total iron concentration at pH 8 in waters of Narragansett Bay. It was also observed an increase of Fe(II) concentration by lowering the pH over the entire

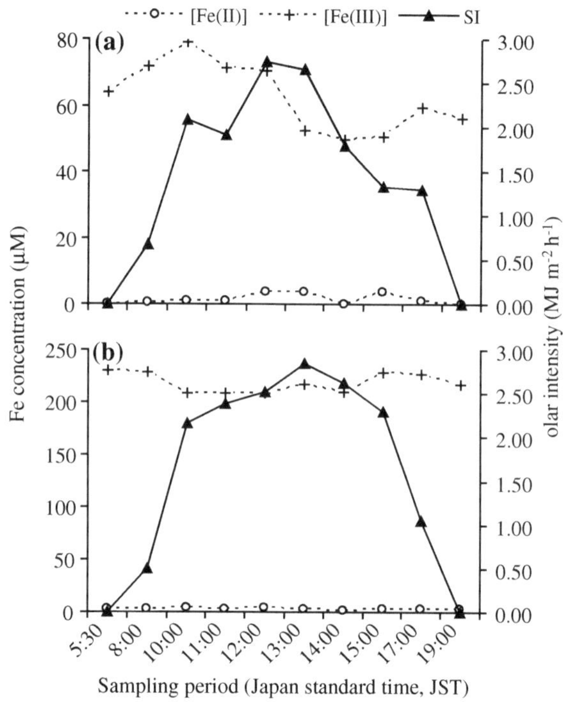

Fig. 8 Diurnal variations of Fe(II) and Fe(III) in the upstream waters (Site KR2, Shouriki) on 21 August 2003 and in the downstream waters (Site KR6, Hinotsume) on 26 September 2003, in the Kurose River, Japan. *Data Source* Mostofa and Sakugawa (unpublished data)

course of the experiment (Miller et al. 1995). All the reported results suggest that the photo-Fenton reaction under sunlight proceeds at the highest rate at noon in natural waters, in correspondence with the peak values of Fe^{2+} and H_2O_2. The diurnal changes in the concentrations of Fe(II) and H_2O_2 are strongly correlated to the losses in the DOM fluorescence: the latter was 28 % lower at noon than before sunrise in the river waters (Mostofa et al. 2005). However, the contribution of the photo-Fenton reaction to the production of $HO^{\bullet}$ was minor (2–29 %) as compared to NO_3^- (3–70 %) and NO_2^- (1–89 %) upon irradiation of river water samples from Japan (pH 6.7–9.0) (Nakatani et al. 2007). Interestingly, it has also been shown that the oxidation of Fe^{2+} by H_2O_2 is the key reaction step in the presence of high concentrations of Fe^{2+}; in contrast, the back reduction of Fe(III) by superoxide is important at low initial Fe^{2+} concentrations and high pH (Pham and Waite 2008). It has also been suggested that precipitation of Fe(III) has a marked effect on the overall Fe(II) oxidation, particularly at high pH. A recent study has shown that the inhibition of the photo-Fenton degradation of organic material (both synthetic phenol wastewater and an aqueous extract of Brazilian gasoline) in the presence of chloride ions can be circumvented, by maintaining the pH of the medium at or slightly above 3 throughout the process. In this way, it is possible to limit inhibition of the oxidation reactions even in the presence of 0.5 M chloride (Machulek et al. 2007).

4.7.1 Kinetics of the Photo-Fenton Reaction

The kinetics of the photo-Fenton reaction can be determined as a function of pH, based on the yield of $HO^{\bullet}$ formed per Fe(II) oxidized by H_2O_2, and considering the photoreactions of aqueous organic substrates (Zepp et al. 1992; Hoigné et al. 1988). Under illumination with constant irradiance of a diluted probe compound (P) that reacts with $HO^{\bullet}$ (Eq. 4.36), the hydroxyl radical would rapidly reach a steady-state concentration. In the presence of P and of other $HO^{\bullet}$ scavengers (S), the decay of $HO^{\bullet}$ can be expressed as follows (Eqs. 4.36, 4.37) (Zepp et al. 1992):

$$HO^{\bullet} + P \xrightarrow{K_P,\,[P]} \text{rection products} \tag{4.36}$$

$$HO^{\bullet} + S \xrightarrow{\sum K_s,\,[S]} \text{scavenging products} \tag{4.37}$$

where k_p is the second-order rate constant ($M^{-1}\ s^{-1}$) for the reaction of $HO^{\bullet}$ with the probe P, and $\sum k_s[S]$ is the pseudo-first order rate constant (s^{-1}) for $HO^{\bullet}$ scavenging by all the components present in the reaction medium, except the probe compound.

The scavenging rate of $HO^{\bullet}$ can be expressed as (Eq. 4.38) (Zepp et al. 1992):

$$r_S = \{k_P\,[P] + \sum k_s\,[S\,]\}[\,HO^{\bullet}] \tag{4.38}$$

If the concentration of P or the reaction rate for the P is sufficiently low (i.e., $\sum k_s[S] \gg k_P[P]$), it is $r_S = (\sum k_s[S])[HO^{\bullet}]$. Under the steady-state condition the rate of generation of $HO^{\bullet}$ is $r_{OH} = r_S$, from which the hydroxyl radical concentration becomes (Eq. 4..39) (Zepp et al. 1992):

$$\left[HO^{\bullet}\right]_{ss} = r_{HO}/\left(\sum k_s\,[S]\right) \tag{4.39}$$

The oxidation rate (Ms^{-1}) of the probe compound in an irradiated system (conversion per unit time) is described as (Eq. 4.40) (Zepp et al. 1992):

$$-d\,[P]\,/dt\ = k_P\left[HO^{\bullet}\right]_{ss}[P]\ = k\,[P] \tag{4.40}$$

If the concentrations of the photoactive Fe(III) species, H_2O_2, and the scavengers show a negligible variation as compared to [P], both r_{OH} and $\sum k_s[S]$ (and $[HO^{\bullet}]_{ss}$ as a consequence) would be about constant. That would give a pseudo-first order reaction with rate constant k. If the second-order rate constant, k_P and the scavenging rate constant, $\sum k_s[S]$ are known, then r_{OH} can be determined from k by the following equation (Eq. 4.41) (Zepp et al. 1992):

$$r_{HO} = k\left(\sum k_s\,[S]\right)/(k_P) \tag{4.41}$$

This steady-state approach has been successfully applied to examine the thermal production of $HO^{\bullet}$ in ozone-treated natural waters, as well as the photoinduced generation of $HO^{\bullet}$ upon irradiation of natural waters and of nitrate ions (Zepp et al. 1987, 1992; Haag and Hoigné 1985).

In natural waters for a thin layer at the surface of a water body, the photoinduced production rate of the reactive species (Schwarzenbach et al. 1993) can be expressed as (Eq. 4.42):

$$r_p = 2.3 \sum_{\lambda} I(\lambda) \times \varepsilon(\lambda) \times \Phi(\lambda) \times C \tag{4.42}$$

where r_P is the production rate (M s^{-1}), I is the incident light intensity (mEinstein cm^{-2} s^{-1}), φ is either the quantum yield (mol $Einstein^{-1}$) or the apparent quantum yield, ϵ and C are the absorption coefficient and the concentration of the relevant light-absorbing reactive species, respectively, and λ is the wavelength. Thus, it is possible to determine the near-surface production rate of $HO^{\bullet}$ from NO_3^- photolysis from Eq. (4.39), from which one gets (Eq. 4.43) (Southworth and Voelker 2003):

$$r_{NO3} = \left(2 \times 10^{-7}\right) [NO_3^-] \tag{4.43}$$

To obtain (Eq. 4.43), the light intensity values were integrated over wavelength for a solar declination of 20° (24-h averaged) (Schwarzenbach et al. 1993), adopting a quantum yield of 0.015 for the $HO^{\bullet}$ photoproduction upon nitrate irradiation at 25 °C (Zepp et al. 1987).

The degradation rate of formic acid in the photo-Fenton reaction increases with temperature (Fig. 7) (Farias et al. 2007). From the cited results it can be estimated that the conversion of HCOOH after 20 min of reaction time is increased approximately by 70–120 % at 313 K and 160–202 % at 328 K compared to the initial temperature of 298 K, with H_2O_2:HCOOH molar ratios in the range from 3 to 8 (Fig. 7). It is also observed that irradiation in the photo-Fenton system enhances degradation, compared to the corresponding dark Fenton system at equal temperature. However, the effect of irradiation is decreased dramatically as temperature increases, so that at 328 K there is little advantage in irradiating the system.

4.8 Photo-Ferrioxalate/H_2O_2 Reaction: Dependence on pH and Reactants

Without addition of H_2O_2 to the photo-ferrioxalate system, the reaction rate gradually increases with increasing pH as can be measured from the degradation of specific organic compounds (Jeong and Yoon 2005; Balmer and Sulzberger 1999). The pH effect is thought to involve two phenomena (Jeong and Yoon 2005; Balmer and Sulzberger 1999). First, an increase in pH may enhance the conversion of dominant ferric complexes such as $[Fe^{III}(C_2O_4)]^+$ and $[Fe^{III}(C_2O_4)_3]^{3-}$ (Eq. 3.19). It subsequently enhances the overall reaction rates through the chain reactions of $CO_2^{\bullet -}$ that form H_2O_2 and, through formation of Fe^{2+} (Eqs. 3.21, 3.23, 3.24), produce $HO^{\bullet}$ as a consequence (Eqs. 3.22, 3.43, 3.45, 4.23, 4.25). Second, the formation of the $Fe^{II}(C_2O_4)$ complex increases with pH and subsequently enhances the Fenton reaction, because $Fe^{II}(C_2O_4)$ can react with H_2O_2 at a faster rate (Eq. 3.25) than Fe^{2+} (Eq. 4.18). Thus, without addition of H_2O_2 the rate-determining step for the production of $HO^{\bullet}$ is the formation of H_2O_2. The latter is formed upon reaction of Fe(II) with $O_2^{\bullet -}$ or $HO_2^{\bullet}$, or from $O_2^- + H^+$ and chain propagation within $HO_2^{\bullet}$. $O_2^{\bullet -}$ is formed from $CO_2^{\bullet -}$ or from the reaction of O_2 with photoinduced electron (e^-), emitted upon photo-ionization of organic compounds. Therefore, an increase in pH will favor the occurrence of Fe(II), $CO_2^{\bullet -}$ and $O_2^{\bullet -}$, which leads to higher amounts of H_2O_2 and of $Fe^{II}(C_2O_4)$. Then, the reaction between H_2O_2 and $Fe^{II}(C_2O_4)$ favors the formation of $HO^{\bullet}$. It is obvious that higher $HO^{\bullet}$ photoproduction causes faster degradation of the dissolved organic substrates.

With addition of H_2O_2 to the photo-ferrioxalate system, the formation of $HO^{\bullet}$ depends on the concentration of H_2O_2 (Hislop and Bolton 1999; Jeong and Yoon 2005). With H_2O_2 above 10 mM the reaction rate may decrease, but the addition of H_2O_2 from 0.1 to 1 mM may enhance the degradation of organic substances. Therefore, a high concentration of H_2O_2 is a major factor for decreasing the overall formation rate of $HO^{\bullet}$. First of all, excess H_2O_2 can contribute significantly to the $HO^{\bullet}$ scavenging, consuming hydroxyl radicals and producing $HO_2^{\bullet}/O_2^{\bullet -}$. The latter species are then able to oxidize Fe(II), which might be kept low so that the formation rate of $HO^{\bullet}$ in the Fenton process is decreased. Lower $HO^{\bullet}$ formation and higher $HO^{\bullet}$ consumption by H_2O_2 can inhibit the degradation of dissolved organic compounds; the inhibition would be higher at higher pH, where the oxidation of Fe(II) by $HO_2^{\bullet}/O_2^{\bullet -}$ is favored.

On the other hand, relatively low levels of H_2O_2 (0.1 to 1 mM) can enhance degradation, because the addition of hydrogen peroxide would by-pass the slow step of H_2O_2 formation in the photo-ferrioxalate system without H_2O_2. Moreover, low H_2O_2 levels would not be able to scavenge $HO^{\bullet}$ significantly, nor to cause Fe(II) oxidation.

5 Significance of $HO^{\bullet}$ in Natural Waters

The $HO^{\bullet}$ radical is the most reactive transient among the reactive oxygen species (ROS) that can be present in natural waters. It is an effective, nonselective and strong oxidant in natural waters for the following reasons:

(i) Photoinduced transformation of DOM by $HO^{\bullet}$ into bioavailable compounds. An example is the degradation of persistent organic substances, which are otherwise recalcitrant to biological and chemical degradation in natural waters (Zepp et al. 1987; Miller and Chin 2002; Haag and Yao 1992; Vione et al. 2009; Grannas et al. 2006; Pullin et al. 2004; Draper and Crosby 1984; Ollis et al. 1991; Sun and Pignatello 1993; Zimbron and Reardon 2005).

(ii) Degradation of organic pollutants or DOM in natural waters (Brezonik and Fulkerson-Brekken 1998; Southworth and Voelker 2003; White et al. 2003; Westerhoff et al. 1999; Goldstone et al. 2002; Kang et al. 2000; Gao and Zepp 1998; Arslan et al. 1999; Katsumata et al. 2006; Muñoz et al. 2006; Farias et al. 2010). Suwannee River Fulvic Acid (SRFA) can produce H_2O_2, which can give $HO^\bullet$ by direct photolysis or through the Fenton reactions (Fig. 9) (Southworth and Voelker 2003). The Fenton process can also be exploited from a technological point of view. Recently, a new pilot-plant solar reactor for the photo-Fenton treatment of waters containing toxic organic substances has been developed. In this reactor, the combined photoinduced and thermal effects of sunlight can degrade 98.2 % of the initial pollutant, for a reaction time of 180 min and a relatively low iron concentration (Farias et al. 2010).

(iii) A great interest is presently devoted to the utilization of $HO^\bullet$ in the Advanced Oxidation Technology (AOT) for treatment of sewerage or industrial effluents, as a help to achieve sustainable development (Safazadeh-Amiri et al. 1996, 1997; Arslan et al. 2000; Venkatadri and Peters 1993).

(iv) The cycling of metals occurs through various redox chemical reactions in natural waters, to which $HO^\bullet$ gives a contribution (Faust 1994; Voelker et al. 1997; Jeong and Yoon 2005; Faust and Zepp 1993; Kwan and Voelker 2002; Emmenegger et al. 2001).

(v) Photoinduced generation of $HO^\bullet$ upon nitrite and nitrate photolysis in natural waters can cause hydroxylation, nitration and nitrosation reactions of many aromatic compounds or organic contaminants (Matykiewiczová et al. 2007; Torrents et al. 1997; Vione et al. 2003a, b, 2004). Some of the reaction intermediates are mutagenic or carcinogenic and are, therefore, of immense environmental concern in the water bodies.

(vi) The $HO^\bullet$ radical is for instance involved in the ultrasonically induced oxidation of arsenite, which plays a key role in the conversion of As(III) in aqueous media and may be applicable as a pretreatment step for the removal of arsenic from water (Xu et al. 2005).

6 Impact of Free Radicals on Biota (Proteins and Living Cells) and Plants

Free radicals are an unavoidable by-product in biological systems and can arise from foods containing unnecessary fat, smoking, alcohol, H_2O_2, ozone, toxins including the carcinogenic ones, ionization, environmental pollutants etc. The major sources of free radicals such as the superoxide ion (O_2^-) and the hydroperoxyl radical ($HO_2^•$) are modest leakages from the electron transport chains of mitochondria, chloroplasts and the endoplasmic reticulum (Miller et al. 1990; Paradies et al. 2000; Blokhina et al. 2003). The $HO^•$, alkoxyl radical ($RO^•$), peroxyl radical ($ROO^•$), and H_2O_2 produced from the autooxidation of biomolecules such as ascorbate, catecholamines or thiols, can damage the macromolecules such as DNA, proteins and lipids in biological systems (Miller et al. 1990; Buettner and Jurkiewicz 1996; Cadet et al. 1999; Blokhina et al. 2003; Berlett and Stadtman 1997; Morse et al. 1977; Radtke et al. 1992; Abele-Oeschger et al. 1994). These events have implications for numerous human health problems. Autoxidation reactions would mostly be catalyzed by transition metal ions (Fe^{2+}, Cu^+), and by semiquinones which can act as electron (e^-) donors (Buettner and Jurkiewicz 1996; Blokhina et al. 2003). Four-electron reduction of oxygen in the respiratory electron transport chain (ETC.) is generally associated with a partial one- to three-electron reduction, yielding reactive oxygen species such as $O_2^{•-}$, $HO^•$, H_2O_2, singlet oxygen ($^1O_2$) and O_3 (Blokhina et al. 2003). Both $O_2^{•-}$ and $HO_2^•$ undergo spontaneous dismutation to produce H_2O_2. In the presence of reduced transition metals such as Fe^{2+} or its complexes that are quite common in biological systems, the $HO^•$ radical can be produced by the Fenton reaction. A mechanistic scheme for the generation of $HO^•$ in biological systems can be depicted as follows (Eqs. 6.1–6.3) (Buettner et al. 1978; Buettner 1987; Buettner and Jurkiewicz 1996; Blokhina et al. 2003):

$$Fe^{3+}\text{-chelate} + O_2^{•-}/AscH^• \rightarrow Fe^{2+}\text{-chelate} + O_2/Asc^{•-} \tag{6.1}$$

$$Fe^{2+}\text{-chelate} + H_2O_2 \rightarrow Fe^{3+}\text{-chelate} + OH^- + HO^• \tag{6.2}$$

$$\text{Biomolecules} + HO^• + O_2 \rightarrow Fe^{3+}\text{-chelate} + OH^- + HO^• \tag{6.3}$$

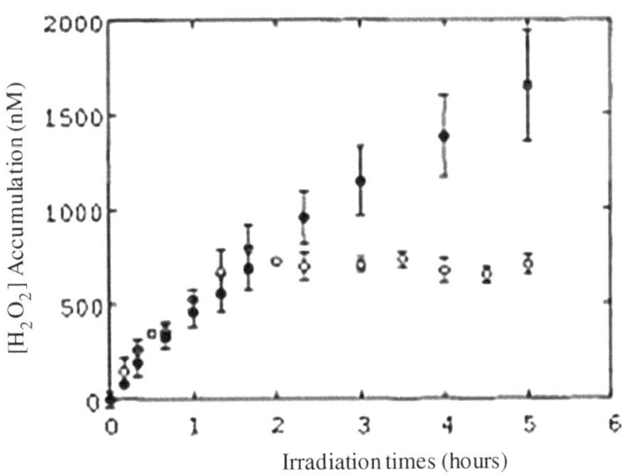

Fig. 9 H_2O_2 concentration during photoirradiation of Suwannee River Fulvic Acid (10 mg L^{-1}) solutions in the absence (●, average of five experiments) and the presence (○, average of three experiments) of Fe (10 μM) at pH 6.0. *Data source* Southworth and Voelker (2003)

where $O_2^{\bullet-}$ and ascorbate ($AscH^-$) can reduce the Fe^{3+}-chelate to form Fe^{2+}-chelate (Eq. 6.1), which subsequently reacts with H_2O_2 to form $HO^\bullet$ in the Fenton reaction (Eq. 6.2). $HO^\bullet$ then induces the oxidation of biomolecules (Eq. 6.3) in biological systems.

The presence of metals that can act as catalysts in biological systems is caused by the fact that many biomolecules bind transition metals, especially protein moieties containing oxygen, nitrogen or sulfur atoms. The transition metals are coordinated with biomolecules through the *d*-orbitals that can also permit the simultaneous binding of a biomolecule and dioxygen, thus providing a bridge between O_2 and the biomolecule (Miller et al. 1990; Buettner and Jurkiewicz 1996; Khan and Martell 1967; Valentine 1973). The major free radicals with the highest reduction potential are $HO^\bullet$, $RO^\bullet$, $LOO^\bullet$, $GS^\bullet$, urate, and even the tocopheroxyl radical ($TO^\bullet$). Ascorbate itself acts as an antioxidant by replacing a potentially very damaging radical ($X^\bullet$), through the following reaction (Buettner and Jurkiewicz 1996; Buettner 1993, 1988):

$$AscH^\bullet + X^\bullet \rightarrow Asc^{\bullet-} + XH \quad (6.4)$$

where $Asc^{\bullet-}$ is the ascorbate radical, which has low reduction potential and does not give additional reaction with O_2 to form further oxidizing species. The kinetics of these electron/hydrogen atom transfer reactions are very fast, as is observed for the equilibrium mixture of $AscH_2/AscH^-/Asc^{2-}$ at pH 7.4 which has been mentioned in earlier section (Buettner and Jurkiewicz 1996; Buettner 1988; Ross et al. 1994). Therefore, ascorbate is an excellent antioxidant from both a thermodynamic and kinetic point of view, but Asc^{2-} or $Asc^{\bullet-}$ can produce low levels of superoxide. The removal of $O_2^{\bullet-}$ by superoxide dismutase can prevent further free radical oxidation in biological systems (Buettner and Jurkiewicz 1996; Williams and Yandell 1982; Scarpa et al. 1983; Winterbourn 1993).

One of the most important impacts of harmful solar UV radiation in biological systems is the skin cancer, which is generally induced by the photoinduced production of free radicals. Formation of $^1O_2$ within cell membranes is caused by the photodynamic action of the photosensitizers photofrin and merocyanine 540 (Buettner and Jurkiewicz 1996). The $^1O_2$ reacts with the membrane lipids to form lipid hydroperoxides (LOOH). Skin is a significant iron excretion site (Green et al. 1968), and Fe(II) can react with LOOH to form highly oxidizing lipid alkoxyl radicals ($LO^\bullet$) by a Fenton-like reaction (Eq. 6.5):

$$Fe^{2+} + LOOH \rightarrow Fe^{3+} + LO^\bullet + OH^- \quad (6.5)$$

Applications of iron chelators to the skin can give photoprotection in the case of chronic exposure to UV radiation, by reducing the formation of free radicals (Bissett et al. 1991).

Second, another most important impact of $HO^\bullet$ is the declining of plants by the effect of $HO^\bullet$, which is formed in dew waters of plants by several sources such as NO_2^-, NO_3^- and H_2O_2 + Fe + oxalate system present in the plants

(Arakaki et al. 1998, 1999a, b; Arakaki and Faust 1998; Nakatani et al. 2001; Kobayashi et al. 2002). It is considered that the $HO^\bullet$ formed in the liquid phase on the needle surfaces of Japanese red pine, which are frequently present in the dew on sunny mornings in the warm-temperate region, are the cause of ecophysiological disorders in plants (Kobayashi et al. 2002).These effects subsequently affect to decrease in the maximum CO_2 assimilation rate, stomatal conductance, minimal fluorescence and needle lifespan (Kume et al. 2000).

7 Summary and Scope of the Future Challenges

The $HO^\bullet$ radical is the most reactive among the many oxygen transient species produced photolytically in natural waters. However, a few studies on $HO^\bullet$ production have been conducted on fresh- and seawater as a whole. The $HO^\bullet$ production in acidic lake waters is not well investigated, although it is known that acidification enhances the formation of $HO^\bullet$ (Vione et al. 2009). The $HO^\bullet$ production varies with the contents of DOM, which may be a key factor to understand the photoinduced processes in a variety of natural waters. Moreover, there is no study conducted about production of alkoxide radicals ($RO^\bullet$) and their role in the photodegradation kinetics of DOM along with $HO^\bullet$. An example of key possible research on free radicals needed for the future challenges are: (i) Effect of pH and temperature on the production of $HO^\bullet$ for a variation in quality and quantity of DOM in natural waters. (ii) Investigation on the sources of free radicals ($HO^\bullet$ and $RO^\bullet$), particularly from fluorescent dissolved organic matter (FDOM) for a variety of waters. (iii) Photoinduced generation of alkoxide radicals ($RO^\bullet$) and their relative reaction kinetics with respect to $HO^\bullet$ and other solution components. (iv) Impacts of free radicals on specific biota in the aquatic environment.

Problems

(1) List the various sources of $HO^\bullet$ and their role in natural waters.
(2) List three important free radical species except $HO^\bullet$ and their importance in natural waters.
(3) How is the steady state concentration of $HO^\bullet$ defined, and why does the steady state concentration vary for a variety of natural waters?
(4) Explain why the potential of $HO^\bullet$ formation is low in surface waters compared to those of deep lakes and the sea.
(5) Mention the important processes of $HO^\bullet$ formation in natural waters and explain the mechanism of in situ generation of $HO^\bullet$ from DOM via H_2O_2.
(6) Explain why the photo-Fenton reaction is more suitable than the Fenton reaction in the degradation of organic pollutants in aqueous solution.

(7) Explain how the photo-ferrioxalate/H_2O_2 reaction system degrades the organic pollutants in the aqueous solution.
(8) Explain how photocatalytic TiO_2 suspensions degrade the organic pollutants in aqueous solutions.
(9) What are the controlling factors for the production and decay of $HO^•$? Explain how fulvic acid plays a dual role in production and decay of $HO^•$.
(10) Explain the effect of wavelength spectrum, temperature, and pH on the formation of $HO^•$ from nitrite and nitrate photolysis in aqueous solution.
(11) Explain the effect of wavelength spectrum and temperature on H_2O_2 photolysis in aqueous solution.
(12) Explain how the Fenton reaction is affected by pH, temperature and salinity in natural waters.
(13) Explain what the photo-Fenton reaction is and what is its kinetics.
(14) What is the photo-ferrioxalate/H_2O_2 reaction? How is this reaction system affected by variations in pH and reactants in aqueous solution?
(15) The quantum yield (Φ_{HO}) for the UV photoproduction of $HO^•$ by nitrite photolysis at 308 nm is 0.07 at room temperature (298 K). Calculate Φ_{HO} at the temperatures of 288 and 328 K.
(16) The Φ_{HO} for the UV photoproduction of $HO^•$ by nitrate photolysis at 308 nm is 0.017 at room temperature (298 K). Calculate the Φ_{HO} at the temperatures of 278 and 318 K.
(17) If the photolysis of aqueous H_2O_2 at 308 nm generates $HO^•$ with $\Phi_{HO} = 1$ at room temperature (298 K), then calculate Φ_{HO} at 288 and 303 K.
(18) Explain shortly the significance of $HO^•$ formation in natural waters.
(19) Explain how $HO^•$ impacts on biota in natural waters.

Acknowledgments This work was financially supported jointly by the National Natural Science Foundation of China and the Chinese Academy of Sciences. This work was partly supported by Hiroshima University, Japan; PNRA—Progetto Antartide, University of Turin, Italy; Brook Byers Institute for Sustainable Systems at Georgia Institute of Technology, the United States; Aligarh Muslim University, India; and Northwest Missouri State University, USA. This chapter acknowledges the reprinted from reprinted (adapted) with permission from Kwan WP, Voelker BM, Decomposition of hydrogen peroxide and organic compounds in the presence of dissolved iron and ferrihydrite, Environmental Science & Technology, 36 (7), 1467–1476. Copyright (2002) Americal Chemical Society; reprinted (adapted) with permission from Farias J, Rossetti GH, Albizzati ED, Alfano OM, Solar degradation of formic acid: temperature effects on the photo-Fenton reaction. Industrial & Engineering Chemistry Research, 46(23):7580–7586). Copyright (2007) American Chemical Society; reprinted from Journal of Photochemistry and Photobiology A: Chemistry, 128(1–3), Mack J, Bolton JR, Photochemistry of nitrite and nitrate in aqueous solution: a review, 1–13. Copyright (1999), with permission from Elsevier; reprinted from Geochimica et Cosmochimica Acta, 53(8), Millero FJ, Sotolongo S, The oxidation of Fe(II) with H_2O_2 in seawater, 1867–1873. Copyright (1989), with permission from Elsevier; reprinted (adapted) with permission from Southworth BA, Voelker BM, Hydroxyl radical production via the photo-Fenton reaction in the presence of fulvic acid, Environmental Science & Technology, 37(6), 1130–1136. Copyright (2003) American Chemical Society; reprinted with permission from Zepp RG, Faust BC, Hoigné J, Hydroxyl radical formation in aqueous reactions (pH 3–8) of iron(II) with hydrogen peroxide: The Photo-Fenton reaction, Environmental Science & Technology, 26 (2), 313–319. Copyright

(1992) American Chemical Society; and reprinted from Water Research, 39(13), Jeong J, Yoon J, pH effect on OH radical production in photo/ferrioxalate system, 2893–2900. Copyright (2005), with permission from Elsevier; reprinted (adapted) with permission from Balmer ME, Sulzberger B, Atrazine degradation in irradiated iron/oxalate system: effects of pH and oxalate, Environmental Science & Technology, 33 (14), 2418–2424. Copyright (1999) American Chemical Society; Springer and the original Journal of Atmospheric Chemistry, 10, 1990, 411–425, Absolute OH quantum yields in the laser photolysis of nitrate, nitrite and dissolved H_2O_2 at 308 and 351 nm in the temperature range 278–353 K, Journal of Atmospheric Chemistry, Zellner R, Exner M, Herrmann H, with kind permission from Springer Science and Business Media; reprinted from Journal of Electrostatics, 39(3), Sun B, Sato M, Sid Clements J, Optical study of active species produced by a pulsed streamer corona discharge in water, 189–202, Copyright (1997), with permission from Elsevier; and Copyright (2004) by The Japan Society for Analytical Chemistry.

References

Abele-Oeschger D, Oeschger R, Theede H (1994) Biochemical adaptations of Nereis diversicolor (Polychaeta) to temporarily increased hydrogen peroxide levels in intertidal sandflats. Mar Ecol Prog Ser 106:101–110

al Housari F, Vione D, Chiron S, Barbati S (2010) Reactive photoinduced species in estuarine waters Characterization of hydroxyl radical, singlet oxygen and dissolved organic matter triplet state in natural oxidation processes. Photochem Photobiol Sci 9:78–86

Aldrich AP, van Berg den CMG, Thies H, Nickus U (2001) The redox speciation of iron in two lakes. Mar Freshw Res 52:885–890

Alegria AE, Ferrer A, Sepulveda E (1997) Photochemistry of water-soluble quinones production of a water-derived spin adduct. Photochem Photobiol 66:436–442

Allen JM, Lucas S, Allen SK (1996) Formation of hydroxyl radical in illuminated surface waters contaminated with acid mine drainage. Environ Sci Technol 15:107–113

Anastasio C, Jordan AL (2004) Photoformation of hydroxyl radical and hydrogen peroxide in aerosol particles from Alert, Nunavut: implications for aerosol and snowpack chemistry in the Arctic. Atmos Environ 38:1153–1166

Anastasio C, Newberg JT (2007) Sources and sinks of hydroxyl radical in sea-salt particles. J Geophys Res 112:D10306. doi:101029/2006JD008061

Anastasio C, Galbavy ES, Hutterli MA, Burkhart JF, Friel DK (2007) Photoformation of hydroxyl radical on snow grains at Summit Greenland. Atmos Environ 41:5110–5121

Arakaki T, Faust BC (1998) Sources, sinks, and mechanisms of hydroxyl radical (OH) photoproduction and consumption in authentic acidic continental cloud waters from Whiteface Mountain, New York: the role of the Fe(r) (r = II, III) photochemical cycle. J Geophys Res 103(D3):3487–3504

Arakaki T, Miyake T, Shibata M, Sakugawa H (1998) Measurement of photolytically formed hydroxyl radical in rain and dew waters. Nippon Kagaku Kaishi 9:619–625

Arakaki T, Miyake T, Hirakawa T, Sakugawa H (1999a) pH dependent photoformation of hydroxyl radical and absorbance of aqueous-phase N(III) (HNO2 and NO2-). Environ Sci Technol 33:2561–2565

Arakaki T, Miyake T, Shibata M, Sakugawa H (1999b) Photochemical formation and scavenging of hydroxyl radical in rain and dew waters. Nippon Kagaku Kaishi 5:335–340 (in Japanese)

Arslan I, Barcioglu A, Tuhkanen T (1999) Oxidative treatment of simulated dyehouse effluent by UV and near-UV light assisted Fenton's reagent. Chemosphere 39:2767–2783

Arslan I, Balcioglu IA, Bahnemann DW (2000) Advanced chemical oxidation of reactive dyes in simulated dyehouse effluents by ferrioxalate-Fenton/UV-A and TiO_2/UV-A processes. Dyes Pigm 47:207–218

Assel M, Laenen R, Laubereau A (1998) Ultrafast electron trapping in an aqueous NaCl-solution. Chem Phys Lett 289:267–274

Balmer ME, Sulzberger B (1999) Atrazine degradation in irradiated iron/oxalate system: effects of pH and oxalate. Environ Sci Technol 33:2418–2424

Barb WG, Boxendale JH, George P, Hargrove KR (1951) Reactions of ferrous and ferric ions with hydrogen peroxide, part II The ferric ion reaction. Trans Faraday Soc 47:591–616

Barbeni M, Minero C, Pelizzetti E (1987) Chemical degradation of chlorophenols with Fenton's reagent. Chemosphere 16:2225–2237

Bard AJ (1979) Photoelectro chemistry and heterogeneous photocatalysis at semiconductors. J Photochem 10:59–75

Baxendale JH, Wilson JA (1956) The photolysis of hydrogen peroxide at high light intensities. Trans Faraday Soc 53:344–356

Benson SW (1960) The foundation of chemical kinetics Ch 15. McGraw-Hill, New York

Berger P, Leitner N Karpel Vel, Doré M, Legube B (1999) Ozone and hydroxyl radicals induced oxidation of glycine. Water Res 33:433–441

Berlett BS, Stadtman ER (1997) Protein oxidation in aging, disease, and oxidative stress. J Biol Chem 272:20313–20316

Bertilsson S, Tranvik LJ (1998) Photolytically produced carboxylic acids as substrates for freshwater bacterioplankton. Limnol Oceanogr 43:885–895

Bielski BHJ, Cabelli DE, Arudi RL, Ross AB (1985) Reactivity of $HO2^{\bullet}/O2^{\bullet-}$ radicals in aqueous solution. J Phys Chem Ref Data 14:1041–1100

Bissett DL, Chatterjee R, Hannon DP (1991) Chronic ultraviolet radiation-induced increase in skin iron and the photoprotective effects of topically applied iron chelators. Photochem Photobiol 54:215–223

Blokhina O, Virolainen E, Fagerstedt KV (2003) Antioxidants, oxidative damage and oxygen deprivation stress: a review. Ann Bot 91:179–194

Blough NV (1988) Electron paramagnetic resonance measurements of photochemical radical production in humic substances: 1 Effects of O_2 and charge on radical scavenging by nitroxides. Environ Sci Technol 22:77–82

Blough NV, Zepp RG (1995) Reactive oxygen species in natural waters. In: Foote CS, Valentine JS (eds) Active oxygen in chemistry. Blackie Academic and Professional, New York, pp 280–333

Bossmann SH, Oliveros E, Gob S, Siegwart S, Dahlen EP, Payawan L, Straub M, Worner M, Braun AM (1998) New evidence against hydroxyl radicals as reactive intermediates in the thermal and photolytically enhanced Fenton reactions. J Phys Chem A 102:5542–5550

Bourdat A-G, Douki T, Frelon S, Gasparutto D, Cadet J (2000) Tandem base lesions are generated by hydroxyl radical within isolated DNA in aerated aqueous solution. J Am Chem Soc 122:4549–4556

Brezonik PL, Fulkerson-Brekken J (1998) Nitrate-induced photolysis in natural waters: controls on concentrations of hydroxyl radical photo-intermediates by natural scavenging agents. Environ Sci Technol 32:3004–3010

Brigante M, Charbouillot T, Vione D, Mailhot G (2010a) Photochemistry of 1-nitronaphthalene: a potential source of singlet oxygen and radical species in atmospheric waters. J Phys Chem A 114:2830–2836

Brigante M, Charbouillot T, Vione D, Mailhot G (2010b) Photochemistry of 1-nitronaphthalene: a potential source of singlet oxygen and radical species in atmospheric waters. J Phys Chem A 114:2830–2836

Buettner GR (1987) Activation of oxygen by metal complexes and its relevance to autoxidative processes in living systems. Bioelectochem Bioenerg 18:29–36

Buettner GR (1988) In the absence of catalytic metals ascorbate does not autoxidize at pH 7: ascorbate as a test for catalytic metals. J Biochem Biophys Methods 16:27–40

Buettner GR (1993) The pecking order of free radicals and antioxidants: Lipid peroxidation, α-tocopherol, and ascorbate. Arch Biochem Biophys 300:535–543

Buettner GR, Jurkiewicz BA (1996) Catalytic metals, ascorbate and free radicals: combinations to avoid. Radiat Res 145:532–541

Buettner GR, Oberley LW, Leuthauser SWHC (1978) The effect of iron on the distribution of superoxide and hydroxyl radicals as seen by spin trapping and on the superoxide dismutase assay. Photochem Photobiol 28:693–695

Buxton GV, Greenstock CL, Helman WP, Ross AB (1988) Critical review of rate constants for reaction of hydrated electrons, hydrogen atoms and hydroxyl radicals (OH/O^-) in aqueous solution. J Phys Chem Ref Data 17:513–886

Cadet J, Delatour T, Douki T, Gasparutto D, Pouget J-P, Ravanat J-L, Sauvaigo S (1999) Hydroxyl radicals and DNA base damage. Mutat Res Fundam Mol Mech Mutagen 424:9–21

Canonica S, Kohn T, Mac M, Real FJ, Wirz J, von Gunten U (2005) Photosensitizer method to determine rate constants for the reaction of carbonate radical with organic compounds. Environ Sci Technol 39:9182–9188

Chen R, Pignatello JJ (1997) Role of quinone intermediates as electron shuttles in Fenton and photoassisted Fenton oxidations of aromatic compounds. Environ Sci Technol 31:2399–2406

Chen C, Li X, Ma W, Zhao J, Hidaka H, Serpone N (2001) Effect of transition metal ions on the TiO_2-assisted photodegradation of dyes under visible irradiation: a probe for the interfacial electron transfer process and reaction mechanism. J Phys Chem B 106:318–324

Chu L, Anastasio C (2003) Quantum yields of hydroxyl radical and nitrogen dioxide from the photolysis of nitrate on ice. J Phys Chem A 107:9594–9602

Chu L, Anastasio C (2005) Formation of hydroxyl radical from the photolysis of frozen hydrogen peroxide. J Phys Chem A 109:6264–6271

Cohen G, Heikkila E (1974) The Generation of hydrogen peroxide, superoxide radical, and hydroxyl radical by 6-hydroxydopamine, dialuric acid, and related cytotoxic agents. J Biol Chem 249:2447–2452

Collen J, del Rio MJ, Garcia-Reina G, Pedersen M (1995) Photosynthetic production of hydrogen peroxide by Ulva rigida C Ag (Chlorophyta). Planta 196:225–230

Cooper WJ, Zika RG, Petasne RG, Fischer AM (1988) Sunlight-induced photochemistry of humic substances in natural waters: major reactive species In: Suffett IH, MacCarthy P (eds) Aquatic humic substances. American Chemical Society, Washington, pp 333–362

Cooper WJ, Nickelson MG, Waite TD, Kurucz CN (1991) High energy electron beam irradiation: an advanced oxidation process for the treatment of aqueous based organic hazardous wastes. J Environ Sci Health A27:219

Cooper WJ, Sawal KL, Hoogland YS, Slifker R, Nickelsen MG, Kurucz CN, Waite TD (1996) Disinfection by-product precursor removal from natural waters using gamma radiation to stimulate an innovative water treatment process. In: Minear RA, Amy GL (eds) Disinfection bi-products in water treatment. CRC Press, Inc, Boca Raton, pp 151–162

Croot PL, Laan P, Nishioka J, Strass V, Cisewski B, Boye M, Timmermans KR, Bellerby RG, Goldson L, Nightingale P, de Baar HJW (2005) Spatial and temporal distribution of Fe(II) and H_2O_2 during EisenEx, an open ocean mescoscale iron enrichment. Mar Chem 95:65–88

Das R, Dutta BK, Maurino V, Vione D, Minero C (2009) Suppression of inhibition of substrate photodegradation by scavengers of hydroxyl radicals: the solvent-cage effect of bromide on nitrate photolysis. Environ Chem Lett 7:337–342

de Laat J, Gallard H (1999) Catalytic decomposition of hydrogen peroxide by Fe(III) in homogeneous aqueous solution: mechanism and kinetic modeling. Environ Sci Technol 33:2726–2732

del Vecchio R, Blough NV (2002) Photobleaching of chromophoric dissolved organic matter in natural waters: kinetics and modeling. Mar Chem 78:231–253

Dister B, Zafiriou OC (1993) Photochemical free-radical production-rates in the eastern Caribbean. J Geophys Res Oceans 98(C2):2341–2352

Draper WM, Crosby DG (1981) Hydrogen peroxide and hydroxyl radical intermediates in indirect photolysis reactions in water. J Agric Food Chem 32:231–237

Draper WM, Crosby DG (1984) Solar photooxidation of pesticides in dilute H_2O_2. J Agric Food Chem 32:231–237
Duesterberg CK, Waite TD (2006) Process optimization of Fenton oxidation using kinetic modeling. Environ Sci Technol 40:4189–4195
Duesterberg CK, Cooper WJ, Waite TD (2005) Fenton-mediated oxidation in the presence and absence of oxygen. Environ Sci Technol 39:5052–5058
Duesterberg CK, Mylon SE, Waite TD (2008) pH effects on iron-catalyzed oxidation using Fenton's reagent. Envion Sci Technol 42:8522–8527
Dykens JA, Shick JM, Benoit C, Buettner GR, Winston GW (1992) Oxygen radical production in the sea anemone Anthopleura Elegantissima and its endosymbiotic algae. J Exp Biol 168:219–241
Emilio CA, Jardim WF, Littera MI, Mansilla HD (2002) EDTA destruction using the solar ferrioxalate AOT comparison with solar photo-Fenton. J Photochem Photobiol A Chem 151:121–127
Emmenegger L, Schwarzenbach R, Sigg L, Sulzberger B (2001) Light-induced redox cycling of iron in circumneutral lakes. Limnol Oceanogr 46:49–61
Ervens B, Gligorovski B, Herrmann H (2003) Temperature-dependent rate constants for hydroxyl radical reactions with organic compounds in aqueous solutions. Phys Chem Chem Phys 5:1811–1824
Fang X, Mark G, von Sonntag C (1996) OH radical formation by ultrasound in aqueous solutions part I: the chemistry underlying the terephthalate dosimeter. Ultrason Sonochem 3:57–63
Farias J, Rossetti GH, Albizzati ED, Alfano OM (2007) Solar degradation of formic acid: temperature effects on the photo-Fenton reaction. Ind Eng Chem Res 46:7580–7586
Farias J, Albizzati ED, Alfano OM (2010) New pilot-plant photo-Fenton solar reactor for water decontamination. Ind Eng Chem Res 49:1265–1273
Faust BC (1994) A review of the photochemical redox reactions of iron species in atmosphere, oceanic, and surface waters: influences of geochemical cycles and oxidant formation. Helz GR, Zepp RG, Crosby DG (eds) Aquatic and surface photochemistry. Lewis Publishers, Boca Raton, pp 3–38
Faust BC, Allen JM (1992) Aqueous-phase photochemical sources of peroxyl radicals and singlet molecular-oxygen in clouds and fog. J Geophys Res Atmos 97(D12):12913–12926
Faust BC, Hoigne J (1987) Sensitized photooxidation of phenols by fulvic acid and in natural waters. Environ Sci Technol 21:957–964
Faust BC, Zepp RG (1993) Photochemistry of aqueous iron(III)-polycarboxylate complexes: roles in the chemistry of atmospheric and surface waters. Environ Sci Technol 27:2517–2522
Fenton HJ (1894) Oxidation of tartaric acid in presence of iron. J Chem Soc 65:899–910
Fischer AM, Kliger DS, Winterle JS, Mill T (1985) Direct observations of phototransients in natural waters. Chemosphere 14:1299–1306
Fox MA (1993) The role of hydroxyl radicals in the photocatalyzed detoxification of organic pollutants—pulse-radiolysis and time-resolved diffuse-reflectance measurements. In: Ollis DF, Alekabi H (eds) Trace metals in the environment, 3, pp 163–167
Fu P, Mostofa KMG, Wu FC, Liu CQ, Li W, Liao H, Wang L, Wang J, Mei Y (2010) Excitation-emission matrix characterization of dissolved organic matter sources in two eutrophic lakes (Southwestern China Plateau). Geochem J 44:99–112
Gallard H, De Laat J Legube B (1998) Effect of pH on the oxidation rate of organic compounds by $FeII/H_2O_2$ mechanisms and simulation. New J Chem 22:263–268
Gan D, Jia M, Vaughan PP, Falvey DE, Blough NV (2008) Aqueous photochemistry of methylbenzoquinone. J Phys Chem A 112:2803–2812
Gao H, Zepp RG (1998) Factors influencing photoreactions of dissolved organic matter in a coastal river of the southern United States. Environ Sci Technol 32:2940–2946
Gjessing ET, Källqvist T (1991) Algicidal and chemical effect of uv-radiation of water containing humic substances. Water Res 25:491–494
Goldstein S, Rabani J (2008) Polychromatic UV photon irradiance measurements using chemical actinometers based on NO_3^- and H_2O_2 excitation: applications for industrial photoreactors. Environ Sci Technol 42:3248–3253

Goldstein S, Aschengrau D, Diamant Y, Rabani J (2007) Photolysis of aqueous H_2O_2: quantum yield and applications for polychromatic UV actinometry in photoreactors. Environ Sci Technol 41:7486–7490
Goldstone JV, Pullin MJ, Bertilsson S, Voelker BM (2002) Reactions of hydroxyl radical with humic substances: bleaching, mineralization, and production of bioavailable carbon substrates. Environ Sci Technol 36:364–372
Gopinathan C, Damle PS, Hart EJ (1972) Gamma-Ray irradiated sodium chloride as a source of hydrated electrons. J Phys Chem 76:3694–3698
Grannas AM, Martin CB, Chin Y, Platz M (2006a) Hydroxyl radical production from irradiated Arctic dissolved organic matter. Biogeochemistry 78:51–66
Grannas AM, Martin CB, Chin Y, Platz M (2006b) Hydroxyl radical production from irradiated Arctic dissolved organic matter. Biogeochemistry 78:51–66
Grebel JE, Pignatello JJ, Song W, Cooper WJ, Mitch WA (2009) Impact of halides on the photobleaching of dissolved organic matter. Mar Chem 115:134–144
Green R, Charlton R, Seftel H, Bothwell T, Mayet F, Adams B, Finch C, Layrisse M (1968) Body iron excretion in man: a collaborative study. Am J Med 45:336–353
Haag WR, Hoigné J (1985) Photo-sensitized oxidation in natural water via OH radicals. Chemosphere 14:1659–1671
Haag WR, Yao CCD (1992) Rate constants for reaction of hydroxyl radicals with several drinking water contaminants. Environ Sci Technol 26:1005–1013
Haber F, Weiss J (1934) The catalytic decomposition of hydrogen peroxide by iron salts. Proc R Soc Lond Ser A 147:332–351
Han F, Kambala VSR, Srinivasan M, Rajarathnam D, Naidu R (2009) Tailored titanium dioxide photocatalysts for the degradation of organic dyes in wastewater treatment: a review. Appl Catal A Gen 359:25–40
Hardwick TJ (1957) The rate constant of the reaction between ferrous ions and hydrogen peroxide in acid solution. Can J Chem 35:428–436
Henglein A (1987) Sonochemistry: historical developments and modern aspects. Ultrasonics 25:6–16
Hislop KA, Bolton JR (1999) The photochemical generation of hydroxyl radicals in the UV-vis/ferrioxalate/H_2O_2 system. Environ Sci Technol 33:3119–3126
Ho P (1986) Photooxidation of 2,4 dinitrotoluene in aqueous solution in the presence of H_2O_2. Environ Sci Technol 20:260–267
Hoigné J, Bader H (1978) Ozone and hydroxyl radical-initiated oxidations of organic and organometallic trace impurities in water. In: Brinkman FE, Bellama JM (eds) Organometals and organometalloids occurrence and fate in the environment. American Chemical Society, Washington, pp 292–313
Hoigné J, Bader H (1979) Ozonation of water: oxidation-competition values of different types of waters used in Switzerland. Ozone Sci Eng 1:357–372
Hoigné J, Faust BC, Haag WR, Zepp RG (1988) In influence of aquatic humic substances on fate and treatment of pollutants. In: MacCarthy P, Suffet IH (eds) ACS Symposium Series 219 American Chemical Society, Washington, pp 363–383
Hoigné J, Faust BC, Haag WR, Scully FE, Zepp RG (1989) Aquatic humic substances as sources and sinks of photolytically produced transient reactants. In: Suffett IH, MacCarthy P (eds) Aquatic humic substances: influence on fate and treatment of pollutants. American Chemical Society, Washington, pp 363–381
Hoigne′ J (1998) Chemistry of aqueous ozone and transformation of pollutants by ozonation and advanced oxidation processes. In: Hrubec J (ed) The handbook of environmental chemistry. Springer Verlag, Heidelberg, pp 83–141
Holder-Sandvik SL, Bilski P, Pakulski JD, Chignell CF, Coffin RB (2000) Photogeneration of singlet oxygen and free radicals in dissolved organic matter isolated from the Mississippi and Atchafalaya River plumes. Mar Chem 69:139–152
Hunt JP, Taube H (1952) The photochemical decomposition of hydrogen peroxide quantum yields, tracer and fractionation effects. J Am Chem Soc 74:5999–6002

Huston PL, Pignatello JJ (1996) Reduction of perchloroalkanes by ferrioxalate-generated carboxylate radical preceding mineralization by the photo Fenton reaction. Environ Sci Technol 30:3457–3463

Jakob DJ (1986) Chemistry of OH in remote clouds and its role in the production of formic acid and peroxymonosulfate. J Geophys Res 91:9807–9826

Jeong J, Yoon J (2004) Dual roles of $CO2^-$ for degrading synthetic organic chemicals in the photo/ferrioxalate system. Water Res 38:3531–3540

Jeong J, Yoon J (2005) pH effect on OH radical production in photo/ferrioxalate system. Water Res 39:2893–2900

Jung YS, Lim WT, Park JY, Kim YH (2009) Effect of pH on Fenton and Fenton-like oxidation. Environ Technol 30:183–190

Kang SF, Liao CH, Po ST (2000) Decolorization of textile wastewater by photo-Fenton oxidation technology. Chemosphere 41:1287–1294

Kang NG, Lee D, Yoon J (2002) Kinetic modeling of Fenton oxidation of phenol and monochlorophenols. Chemosphere 47:915–924

Katsumata H, Kaneco S, Suzuki T, Ohta K, Yobico Y (2006) Photo-Fenton degradation of alachlor in the presence of citrate solution. J Photochem Photobiol A Chem 180:38–45

Khan MMT, Martell AE (1967) Metal ion and metal chelate catalyzed oxidation of ascorbic acid by molecular oxygen I cupric and ferric ion catalyzed oxidation. J Am Chem Soc 89:4176–4185

Kieber DJ, Blough NV (1990) Determination of carboncentered radicals in aqueous solution by liquid chromatography with fluorescence detection. Anal Chem 62:2275–2283

Kobayashi T, Nakatani N, Hirakawa T, Suzuki M, Miyake T, Chiwa M, Yuhara T, Hashimoto N, Inoue K, Yamamura K, Agus N, Sinogaya JR, Nakane K, Kume A, Arakaki T, Sakugawa H (2002) Variation in CO2 assimilation rate induced by simulated dew waters with different sources of hydroxyl radical ($^\bullet$OH) on the needle surfaces of Japanese red pine (*Pinus densiflora* Sieb Et Zucc.). Environ Pollut 118:383–391

Komissarov GG (2003) Photosynthesis: the physical-chemical approach. J Adv Chem Phys 2:28–61

Konstantinou IK, Albanis TA (2004) TiO2-assisted photocatalytic degradation of azo dyes in aqueous solution: kinetic and mechanistic investigations: a review. Appl Catal B Environ 49:1–14

Kremer ML (1999) Mechanism of the Fenton reaction evidence for a new intermediate. Phys Chem Chem Phys 1:3595–3605

Kume A, Tsuboi N, Satomura T, Suzuki M, Chiwa M, Nakane K, Sakurai N, Horikoshi T, Sakugawa H (2000) Physiological characteristics of Japanese red pine, Pinus densiflora Sieb et Zucc, in declined forests at Mt Gokurakuji in Hiroshima Prefecture, Japan. Trees 14:305–311

Kwan WP, Voelker BM (2002) Decomposition of hydrogen peroxide and organic compounds in the presence of dissolved iron and ferrihydrite. Environ Sci Technol 36:1467–1476

Langford JH, Carey CH (1975) Outer-sphere oxidations of alcohols and formic acid by charge transfer excited states of iron(III) species. Can J Chem 53:2436–2440

Le Truong G, De Laat J, Legube B (2004) Effects of chloride and sulfate on the rate of oxidation of ferrous ion by H_2O_2. Water Res 38:2384–2394

Lee C, Sedlak DL (2009) A novel homogeneous Fenton-like system with Fe(III)-phosphotungstate for oxidation of organic compounds at neutral pH values. J Mol Catal A Chem 311:1–6

Lee Y, Jeong J, Lee C, Yoon J (2003a) Influence of various reaction parameters on 2,4-D removal in photo/ferrioxalate/H_2O_2 process. Chemosphere 51:901–912

Lee Y, Lee C, Yoon J (2003b) High temperature dependence of 2,4-dichlorophenoxyacetic acid degradation by Fe^{3+}/H_2O_2 system. Chemosphere 51:963–971

Lee C, Keenan CR, Sedlak DL (2008) Polyoxometalate-enhanced oxidation of organic compounds by nanoparticulate zero-valent iron and ferrous ion in the presence of oxygen. Environ Sci Technol 42:4921–4926

Legrini O, Oliveros E, Braun AM (1993) Photochemical processes for water treatment. Chem Rev 93:671–698

Lehninger AL (1970) Biochemistry. Worth, New York, p 478
Li SX, Hong HS, Zheng FY, Deng NS (2008) Effects of metal pollution and macronutrient enrichment on the photoproduction of hydroxyl radicals in seawater by the alga Dunaliella salina. Mar Chem 108:207–214
Lindsey ME, Tarr MA (2000a) Quantitation of hydroxyl radical during Fenton oxidation following a single addition of iron and peroxide. Chemosphere 41:409–417
Lindsey ME, Tarr MA (2000b) Inhibited hydroxyl radical degradation of aromatic hydrocarbons in the presence of dissolved fulvic acid. Water Res 34:2385–2389
Lloyd RV, Hanna PM, Mason RP (1997) The origin of the hydroxyl radical oxygen in the Fenton reaction. Free Radic Biol Med 22:885–888
Mabury SA (1993) Hydroxyl radical in natural waters. Ph D dissertation, University of California, Davis, California
Machulek A, Moraes JEF, Vautier-Giongo C, Silverio CA, Friedrich LC, Nascimento CAO, Gonzalez MC, Quina FH (2007) Abatement of the inhibitory effect of chloride anions on the photo-Fenton process. Environ Sci Technol 41:8459–8463
Mack J, Bolton JR (1999) Photochemistry of nitrite and nitrate in aqueous solution: a review. J Photochem Photobiol A Chem 128:1–13
Maddigapu PR, Bedini A, Minero C, Maurino V, Vione D, Brigante M, Mailhot G, Sarakha M (2010) The pH-dependent photochemistry of anthraquinone-2-sulfonate. Photochem Photobiol Sci 9:323–330
Maddigapu PR, Minero C, Maurino V, Vione D, Brigante M, Charbouillot T, Sarakha M, Mailhot G (2011) Photoinduced and photosensitised reactions involving 1-nitronaphthalene and nitrite in aqueous solution. Photochem Photobiol Sci 10:601–609. doi:http:dxdoiorg/101039/C0PP00311E
Mageli OL, Kolczynski JR (1966) Organic peroxides. Ind Eng Chem 58:25–32
Makino K, Mossoba MM, Riesz P (1983) Chemical effects of ultrasound on aqueous solutions formation of hydroxyl radicals and hydrogen atoms. J Phys Chem 87:1369–1377
Mark G, Korth H-G, Schuchmann H-P, von Sonntag C (1996) The photochemistry of aqueous nitrate ion revisited. J Photochem Photobiol A Chem 101:89–103
Matykiewiczová N, Kurková R, Klánová J, Klán P (2007) Photolytically induced nitration and hydroxylation of organic aromatic compounds in the presence of nitrate or nitrite in ice. J Photochem Photobiol A Chem 187:24–32
Maurino V, Borghesi D, Vione D, Minero C (2008) Transformation of phenolic compounds upon UVA irradiation of anthraquinone-2-sulfonate. Photochem Photobiol Sci 7:321–327
McKnight DM, Kimball BA, Bencala KE (1988) Iron photoreduction and oxidation in an acidic mountain stream. Science 240:637–640
Meyerstein D, Treinin A (1961) Absorption spectra of NO_3^- in solution. Tram Faraday Soc 57:2104–2112
Micinski E, Ball LA, Zafiriou OC (1993) Photochemical oxygen activation: Superoxide radical detection and production rates in the eastern Caribbean. J Geophys Res Oceans 98(C2):2299–2306
Mill T, Hendry DG, Richardson H (1980) Free-radical oxidants in natural waters. Science 207:886–887
Miller PL, Chin YP (2002) Photoinduced degradation of carbaryl in wetland surface water. J Agric Food Chem 50:6758–6765
Miller DM, Buettner GR, Aust SD (1990) Transition metals as a catalysts of "autoxidation" reactions. Free Radic Biol Med 8:95–108
Miller WL, King DW, Lin J, Kester DR (1995) Photochemical redox cycling of iron in coastal seawater. Mar Chem 50:63–77
Miller WL, Moran MA, Sheldon WM, Zepp RG, Opsahl S (2002) Determination of apparent quantum yield spectra for the formation of biologically labile photoproducts. Limnol Oceanogr 47:343–352
Millero FJ, Sotolongo S (1989) The oxidation of Fe(II) with H_2O_2 in seawater. Geochim Cosmochim Acta 53:1867–1873

Millington KR, Maurdev G (2004) The generation of superoxide and hydrogen peroxide by exposure of fluorescent whitening agents to UVA radiation and its relevance to the rapid photoyellowing of whitened wool. J Photochem Photobiol A Chem 165:177–185
Minakata D, Li K, Westerhoff P, Crittenden J (2009) Development of a group contribution method to predict aqueous phase hydroxyl radical (HO•) reaction rate constants. Environ Sci Technol 43:6220–6227
Minella M, Rogora M, Vione D, Maurino V, Minero C (2011) A model approach to assess the long-term trends of indirect photochemistry in lake water. The case of Lake Maggiore (NW Italy). Sci Total Environ 409:3463–3471
Moffett JW, Zika RG (1987a) Reaction kinetics of hydrogen peroxide with copper and iron in seawater. Environ Sci Technol 21:804–810
Moffett JW, Zika RG (1987b) Photochemistry of a copper complexes in sea water In: Zika RG, Cooper WJ (eds) Photochemistry of environmental aquatic systems, ACS symposium Ser 327. American Chemical Society, Washington, pp 116–130
Moore CA, Farmer CT, Zika RG (1993) Influence of the Orinoko River on hydrogen peroxide distribution and production in the Eastern Caribean. J Geophys Res 98(C2):2289–2298
Mopper K, Kieber DJ (2000) Marine photochemistry and its impact on carbon cycling. In: de Mora S, Demers S, Vernet M (eds) The effects of UV radiation in the marine environment. Cambridge University Press, Cambridge, pp 101–129
Mopper K, Zhou X (1990) Hydroxyl radical photoproduction in the sea and its potential impact on marine processes. Science 250:661–664
Moran MA, Zepp RG (1997) Role of photoreactions in the formation of biologically labile compounds from dissolved organic matter. Limnol Oceanogr 42:1307–1316
Moran MA Jr, Sheldon WM, Zepp RG (2000) Carbon loss and optical property changes during long-term photochemical and biological degradation of estuarine dissolved organic matter. Limnol Oceanogr 45:1254–1264
Morse DE, Duncan H, Hooker N, Morse A (1977) Hydrogen peroxide induces spawning in mollusks, with activation of prostaglandin endoperoxide synthetase. Science 196:298–300
Mostofa KMG, Sakugawa H (2009) Spatial and temporal variations and factors controlling the concentrations of hydrogen peroxide and organic peroxides in rivers. Environ Chem 6:524–534
Mostofa KMG, Honda Y, Sakugawa H (2005) Dynamics and optical nature of fluorescent dissolved organic matter in river waters in Hiroshima prefecture Japan. Geochem J 39:257–271
Mostofa KMG, Yoshioka T, Konohira E, Tanoue E (2007) Photodegradation of fluorescent dissolved organic matters in river waters. Geochem J 41:323–331
Mostofa KMG, Wu FC, Yoshioka T, Sakugawa H, Tanoue E (2009a) Dissolved organic matter in the aquatic environments. In: Wu FC, Xing B (eds) Natural organic matter and its significance in the environment. Science Press, Beijing, pp 3–66
Mostofa KMG, Liu CQ, Wu FC, Fu PQ, Ying WL, Yuan J (2009b) Overview of key biogeochemical functions in lake ecosystem: impacts of organic matter pollution and global warming keynote speech. In: Proceedings of the 13th world lake conference Wuhan, China, 1–5 Nov 2009, pp 59–60
Mostofa KMG, Wu FC, Liu CQ, Yoshioka T, Sakugawa H, Tanoue E (2011) Photochemical, microbial and metal complexation behavior of fluorescent dissolved organic matter in the aquatic environments (Invited review). Geochem J 45:235–254
Mulazzani QG, D'Angelantonio M, Venturi M, Hoffmann MZ, Rodgers MA (1986) Interaction of formate and oxalate ions with radiation-generated radicals in aqueous solution Methylviologen as a mechanistic probe. J Phys Chem 90:5352–5437
Muñoz I, Rieradevall J, Torrades F, Peral J, Dome`nech X (2006a) Environmental assessment of different advanced oxidation processes applied to a bleaching kraft mill effluent. Chemosphere 62:9–16
Muñoz I, Rieradevall J, Torrades F, Peral J, Dome`nech X (2006b) Environmental assessment of different advanced oxidation processes applied to a bleaching kraft mill effluent. Chemosphere 62:9–16

Murov SL, Carmichael I, Hug GL (1993) Handbook of photochemistry, 2nd edn. Marcel Dekker, New York, pp 299–305

Nakatani N, Miyake T, Chiwa M, Hashimoto M, Arakaki T, Sakugawa H (2001) Photochemical formation of OH radicals in dew formed on the pine needles at Mt Gokurakuji. Water Air Soil Pollut 130:397–402

Nakatani N, Hashimoto N, Sakugawa H (2004) An evaluation of hydroxyl radical formation in river water and the potential for photodegradation of bisphenol A. In: Hill RJ, Leventhal J, Aizenshtat Z, Baedecker MJ, Claypool G, Eganhouse R, Goldhaber M, Peters K (eds) The geochemical society special publication series 9, Geochemical investigations in earth and space science: a tribute to Isaac R Kaplan. Elsevier, Amsterdam, pp 233–242

Nakatani N, Ueda M, Shindo H, Takeda K, Sakugawa H (2007) Contribution of the photo-Fenton reaction to hydroxyl radical formation rates in river and rain water samples. Anal Sci 23:1137–1142

Neta P, Huie RE, Ross AB (1988) Rate constants for reactions of inorganic radicals in aqueous solution. J Phys Chem Ref Data 17:1027–1284

Nogueira RP, Guimaraes JR (2000) Photodegradation of dichloroacetic acid and 2,4-dichlorophenol by ferrioxalate/H_2O_2 system. Water Res 34:895–901

Oda T, Akaike T, Sato K, Ishimatsu A, Takeshita S, Muramatsu T, Maeda H (1992) Hydroxyl radical generation by red tide algae. Arch Biochem Biophys 294:38–43

Ollis DF, Pellizetti E, Serpone N (1991) Photocatalytic destruction of water contaminants. Environ Sci Technol 25:1522–1529

Osburn CL, O'Sullivan DW, Boyd TJ (2009) Increases in the longwave photobleaching of chromophoric dissolved organic matter in coastal waters. Limnol Oceanogr 54:145–159

Page SE, Arnold WA, McNeill K (2011) Assessing the contribution of free hydroxyl radical in organic matter-sensitized photohydroxylation reactions. Environ Sci Technol 45:2818–2825

Paradies G, Petrosillo G, Pistolese M, Ruggiero FM (2000) The effect of reactive oxygen species generated from the mitochondrial electron transport chain on the cytochrome C oxidase activity and on the cardilipin content in bovine heart submitochondrial particles. FEBS Lett 466:323–326

Petasne RG, Zika RG (1987) Fate of superoxide in coastal sea water. Nature 325:516–518

Pham AN, Waite TD (2008) Oxygenation of Fe(II) in natural waters revisited: Kinetic modeling approaches, rate constant estimation and the importance of various reaction pathways. Geochim Cosmochim Acta 72:3616–3630

Pignatello JJ (1992) Dark and photoassisted Fe^{3+}-catalyzed degradation of chlorophenoxy herbicides by hydrogen peroxide. Environ Sci Technol 26:944–951

Pignatello JJ, Oliveros E, MacKay A (2006) Advanced oxidation processes for organic contaminant destruction based on the Fenton reaction and related chemistry. Crit Rev Environ Sci Technol 36:1–84

Pleskov YV, Gurevich YY (1986) Semiconductor photoelectron chemistry. Consultants Bureau, New York, p 29

Po HN, Sutin N (1968) The stability constant of the monochloro complex of iron (II). Inorg Chem 7:621–624

Pochon A, Vaughan PP, Gan DQ, Vath P, Blough NV, Falvey DE (2002) Photochemical oxidation of water by 2-methyl-1,4-benzoquinone: evidence against the formation of free hydroxyl radical. J Phys Chem A 106:2889–2894

Pozdnyakov IP, Glebov EM, Plyusnin VF, Grivin VP, Ivanov YV, Vorobyev DY, Bazhin NM (2000) Hydroxyl radical formation upon photolysis of the $Fe(OH)^{2+}$ complex in aqueous solution. Mendeleev Commun 10:185–186

Prousek J (1996) Advanced oxidation processes for water treatment photochemical processes. Chem Listy 90:307–315

Pullin MJ, Bertilsson S, Goldstone JV, Voelker BM (2004) Effects of sunlight and hydroxyl radical on dissolved organic matter: bacterial growth efficiency and production of carboxylic acids and other substrates. Limnol Oceanogr 49:2011–2022

Qian J, Mopper K, Kieber DJ (2001) Photochemical production of the hydroxyl radical in Antarctic waters. Deep-Sea Res I 48:741–759

Radtke K, Byrnes RW, Kerrigan P, Antholine WE, Petering DH (1992) Requirement of endogenous iron for cytotoxicity caused by hydrogen peroxide in euglena gracilis. Mar Environ Res 34:339–343

Randall CE, Harvey VL, Manney GL, Orsolini Y, Codrescu M, Sioris C, Brohede S, Haley CS, Gordley LL, Zawdony JM, Russell JM (2005) Stratospheric effects of energetic particle precipitation in 2003–2004. Geophys Res Lett LO5082 doi:101029/2004GL022003

Rex M, Harris NRP, der Gathen P, Lehman R, Braathen GO, Reimer E, Beck A, Chipperfield MP, Alfier R, Allaart M, O'Conner F, Dier H, Dorokhov V, Fast H, Gil M, Kyro E, Litynska Z, Mikkelsen IB, Molyneux MG, Nakane H, Notholt J, Rummukainen M, Viatte P, Wenger J (1997) Prolonged stratospheric ozone loss in the 1995–96 Arctic winter. Nature 389:835–838

Ross AB, Mallard WG, Helman WP, Buxton, Huie RE, Neta P (1994) NDRL-NIST Solution Kinetics Database: -Ver 20. National Institute for Standards and Technology, Gaithersburg

Rossetti GH, Albizzati ED, Alfano OM (2002) Decomposition of formic acid in a water solution employing the photo-Fenton reaction. Ind Eng Chem Res 41:1436–1444

Ruppert G, Bauer R, Heisler GJ (1993) The photo-Fenton reaction- an effective photochemical wastewater treatment process. J Photochem Photobiol A Chem 73:75–78

Rush JD, Bielski GHJ (1985) Pulse radiolytic studies of the reactions of HO2/O2-with Fe(II)/Fe(III) ions. The reactivity of HO2/O2-with ferric ions and its implications on the occurrence of the Haber-Weiss reaction. J Phys Chem 89:5062–5066

Russi H, Kotzias D, Korte F (1982) Photoinduzierte hydroxylierungsreaktionen organischer chemikalien in naturlichen gewassern: nitrate als potentielle OH-radikalquellen. Chemosphere 11:1041–1048

Safazadeh-Amiri A, Bolton JR, Cater SR (1996) Ferrioxalate-mediated solar degradation of organic contaminants in water. Sol Energy 56:439–443

Safazadeh-Amiri A, Bolton JR, Cater SR (1997) Ferrioxalate-mediated photodegradation of organic pollutants in contaminated water. Water Res 31:2079–2085

Sakugawa H, Kaplan IR, Tsai W, Cohen Y (1990) Atmospheric hydrogen peroxide. Environ Sci Technol 24:1452–1462

Saquib M, Tariq MA, Haque MM, Muneer M (2008) Photocatalytic degradation of disperse blue 1 using UV/TiO_2/H_2O_2 process. J Environ Manag 88:300–306

Scarpa M, Stevanato R, Viglino P, Rigo A (1983) Superoxide ion as active intermediate in the autooxidation of ascorbate by molecular oxygen. J Biol Chem 258:6695–6697

Schiavello M (1987) Basic concepts in photocatalysis. In: Schiavello M (ed) Photocatalysis and environmental trends and applications. Kluwer Academic Publishers, The Netherlands, pp 351–360

Schuchmann MN, von Sonntag C (1979) Hydroxyl radical-induced oxidation of 2-methyl-2-propanol in oxygenated aqueous solution: a product and pulse radiolysis study. J Phys Chem 83:780–784

Schwarzenbach RP, Gschwend PM, Imboden DM (1993) Environmental organic chemistry. Wiley, New York, pp 436–484

Sedlak DL, Hoigné J (1993) The role of copper and oxalate in the redox cycling of iron in atmospheric waters. Atmos Environ 27A:2173–2185

Senesi N (1990) Molecular and quantitative aspects of the chemistry of fulvic acid and its interactions with metal ions and organic chemicals Part II The fluorescence spectroscopy approach. Anal Chim Acta 232:77–106

Serpone N, Pelizzetti E (1989) Photocatalysis: fundamentals and applications. Wiley, New York, p 650

Shuali U, Ottolenghi M, Rabani J, Yelin Z (1969) On photochemistry of aqueous nitrate solutions excited in 195-nm band. J Phys Chem 73:3445–3451

Skinner JF, Glasel A, Hsu L-C, Funt BL (1980) Rotating ring disk electrode study of the hydrogen peroxide oxidation of Fe(II) and Cu(I) in hydrochloric acid. J Electrochem Soc 127:315–324

Song RG, Westerhoff P, Minear RA, Amy GL (1996) Interaction between bromine and natural organic matter. In: Minear RA, Amy GL (eds) Water disinfection and natural organic matter. American Chemical Society, Washington, pp 298–321

Southworth BA, Voelker BM (2003) Hydroxyl radical production via the photo-Fenton reaction in the presence of fulvic acid. Environ Sci Technol 37:1130–1136

Staehelin J, Hoigné J (1985) Decomposition of ozone in water in the presence of organic solutes acting as promoters and inhibitors of radical chain reactions. Environ Sci Technol 19:1206–1213

Strehlow H, Wagner I (1982) Flash photolysis of nitrite ions in aqueous solutions. Z Phys Chem Neue Folge 132:151–160

Strickler SJ, Kasha M (1963) Solvent effects on the electronic absorption spectrum of nitrite ion. J Am Chem Soc 85:2899–2901

Sun L, Bolton JR (1996) Determination of the quantum yield for the photochemical generation of hydroxyl radicals in TiO_2 suspensions. J Phys Chem 100:4127–4134

Sun Y, Pignatello J (1993) Photochemical reactions involved in the total mineralization of 2,4-D by Fe^{3+}/H_2O_2/UV. Environ Sci Technol 27:304–310

Sun B, Sato M, Sid Clements J (1997) Optical study of active species produced by a pulsed streamer corona discharge in water. J Electrostat 39(3):189–202

Sur B, Rolle M, Minero C, Maurino V, Vione D, Brigante M, Mailhot G (2011) Formation of hydroxyl radicals by irradiated 1-nitronaphthalene (1NN): oxidation of hydroxyl ions and water by the 1NN triplet state. Photochem Photobiol Sci 10:1817–1824

Sychev AY, Isak VG (1995) Iron compounds and the mechanisms of the homogeneous catalysis of the activation of O_2 and H_2O_2 and of the oxidation of organic substrates. Russ Chem Rev 64:1105–1129

Takahashi N, Nakai T, Satoh Y, Katoh Y (1995) Ozonolysis of humic acid and its effect on decoloration and biodegradability. Ozone Sci Eng 17:511–525

Takeda K, Takedoi H, Yamaji S, Ohta K, Sakugawa H (2004) Determination of hydroxyl radical photoproduction rates in natural waters. Anal Sci 20:153–158

Taylor RC, Cross PC (1949) Light absorption of aqueous hydrogen peroxide solutions in the near ultraviolet region. J Am Chem Soc 71:2266–2268

Torrents A, Anderson BG, Bilboulian S, Johnson WE, Hapeman CJ (1997) Atrazine photolysis: mechanistic investigations of direct and nitrate mediated hydroxy radical processes and the influence of dissolved organic carbon from the Chesapeake Bay. Environ Sci Technol 31:1476–1482

Tossell JA (2005) Calculation of the interaction of bicarbonate ion with arsenites in aqueous solution and with the surfaces of Al hydroxide minerals. ACS Symp Ser, Chapter 9, 915:118–130

Tranvik LJ (1992) Allochthonous dissolved organic matter as an energy source for pelagic bacteria and the concept of the microbial loop. Hydrobiologia 229:107–114

Treinin A, Hayon E (1970) Absorption spectra and reaction kinetics of NO_2, N_2O_3, and N_2O_4 in aqueous solution. J Am Chem Soc 92:5821–5828

Tseng JM, Haung CP (1991) Removal of chlorophenols from water by photocatalytic oxidation Water Sci Technol 23:377–387

Ullah SS, Khan MGM, Rahman ABMS (1998) Photocatalytic decomposition of phenols by titanium dioxide under sunlight and UV. J Bang Acad Sci 22:29–37

Valentine JS (1973) The dioxygen ligand in mononuclear group VIII transition metal complexes. Chem Rev 73:235–345

Vaughn PP, Blough NV (1998) Photochemical formation of hydroxyl radical by constituents of natural waters. Environ Sci Technol 32:2947–2953

Vel Leitner NK, Dore M (1996) Hydroxyl radical induced decomposition of aliphatic acids in oxygenated and deoxygenated aqueous solutions. J Photochem Photobiol A Chem 99:137–143

Venkatadri R, Peters R (1993) Chemical oxidation technologies. Hazard Waste Hazard Mater 10:107–149

Vermilyea AW, Voelker BM (2009) Photo-Fenton reaction at near neutral pH. Environ Sci Technol 43:6927–6933

Viollier E, Inglett PW, Hunter K, Roychoudhury AN, van Cappellen P (2000) The ferrozine method revisited: Fe(II)/Fe(III) determination in natural waters. Appl Geochem 15:785–790
Vione D, Maurino V, Minero C, Pelizzetti E (2001a) Phenol photonitration upon UV irradiation of nitrite in aqueous solution II: effects of pH and TiO_2. Chemosphere 45:903–910
Vione D, Maurino V, Minero C, Pelizzetti E (2001b) Phenol photonitration upon UV irradiation of nitrite in aqueous solution II: effects of pH and TiO_2. Chemosphere 45:903–910
Vione D, Maurino V, Minero C, Vincenti M, Pelizzetti E (2003a) Aromatic photonitration in homogeneous and heterogeneous aqueous systems. Environ Sci Pollut Res 10:321–324
Vione D, Maurino V, Minero C, Borghesi D, Lucchiari M, Pelizzetti E (2003b) New processes in the environmental chemistry of nitrite 2 the role of hydrogen peroxide. Environ Sci Technol 37:4635–4641
Vione D, Merlo F, Maurino V, Minero C (2004a) Effect of humic acids on the fenton degradation of phenol. Environ Chem Lett 2:129–133
Vione D, Maurino V, Pelizzetti E, Minero C (2004b) Phenol photonitration and photonitrosation upon nitrite photolysis in basic solution. Int J Environ Anal Chem 84:493–504
Vione D, Falletti G, Maurino V, Minero C, Pelizzetti E, Malandrino M, Ajassa R, Olariu R-I, Arsene C (2006) Sources and sinks of hydroxyl radicals upon irradiation of natural water samples. Environ Sci Technol 40:3775–3781
Vione D, Maurino V, Cucu Man S, Khanra S, Arsene C, Olariu RI, Minero C (2008) Formation of organobrominated compounds in the presence of bromide under simulated atmospheric aerosol conditions. ChemSusChem 1:197–204
Vione D, Khanra S, Man SC, Maddigapu PR, Das R, Arsene C, Olariu RI, Maurino V, Minero C (2009a) Inhibition vs enhancement of the nitrate-induced phototransformation of organic substrates by the OH scavengers bicarbonate and carbonate. Water Res 43:4718–4728
Vione D, Maurino V, Minero C, Carlotti ME, Chiron S, Barbati S (2009b) Modelling the occurrence and reactivity of the carbonate radical in surface freshwater. Comptes Rendus Chimie 12:865–871
Vione D, Lauri V, Minero C, Maurino V, Malandrino M, Carlotti ME, Olariu RI, Arsene C (2009c) Photostability and photolability of dissolved organic matter upon irradiation of natural water samples under simulated sunlight. Aquat Sci 71:34–45
Vione D, Casanova I, Minero C, Duncianu M, Olariu RI, Arsene C (2009d) Assessing the potentiality of Romanian surface waters to produce hydroxyl and nitrite radicals. Revista De Chimie 60:123–126
Vione D, Ponzo M, Bagnus D, Maurino V, Minero C, Carlotti ME (2010) Comparison of different probe molecules for the quantification of hydroxyl raidcals in aqueous solution. Environ Chem Lett 8:95–100
Voelker BM, Sulzberger B (1996) Effects of fulvic acid on Fe(II) oxidation by hydrogen peroxide. Environ Sci Technol 30:1106–1114
Voelker BM, Morel FMM, Sulzberger B (1997) Iron redox cycling in surface waters: effects of humic substances and light. Environ Sci Technol 31:1004–1011
Voelker BM, Sedlak DL, Zafiriou OC (2000) Chemistry of superoxide radical in seawater: Reactions with organic Cu complexes. Environ Sci Technol 34:1036–1042
Volman DH, Chen JC (1959) The photochemical decomposition of hydrogen peroxide in aqueous solutions of allyl alcohol at 2537 A. J Am Chem Soc 81:4141–4144
von Gunten U, Oliveras Y (1997) Kinetics of the reaction between hydrogen peroxide and hypobromous acid: implication on water treatment and natural systems. Water Res 31:900–906
von Sonntag C (2006) Free-radical-induced DNA damage and its repair a chemical perspective. Springer Verlag, Berlin, pp 359–482
von Sonntag C (2007) The basics of oxidants in water treatment Part A: OH radical reactions. Water Sci Technol 55:19–23
von Sonntag C, Mark G, Mertens R, Schuchmann MN, Schuchmann H-P (1993) UV-radiation and/or oxidants in water pollution control. J Water Supply Res Technol Aquat 42:201–211
Wagner I, Strehlow H, Busse G (1980) Flash-photolysis of nitrate ions in aqueous-solution. Z Phys Chem 123:1–33

Walling C (1975) Fenton's reagent revisited. Acc Chem Res 8:125–131

Walling C, Weil T (1974) The ferric ion catalyzed decomposition of hydrogen peroxide in perchloric acid solution. Int J Chem Kinet 6:507–516

Wang GS, Liao CH, Wu FJ (2001) Photodegradation of humic acids in the presence of hydrogen peroxide. Chemosphere 42:379–387

Wang Z, Chen X, Ji H, Ma W, Chen C, Zhao J (2010) Photochemical cycling of iron mediated by dicarboxylates: special effect of malonate. Environ Sci Technol 44:263–268

Warneck P, Wurzinger C (1988) Product quantum yields for the 305-nm photodecomposition of nitrate in aqueous solution. J Phys Chem 92:6278–6283

Wells CF, Salam MA (1967) Complex formation between Fe(II) and inorganic anions. Trans Faraday Soc 63:620–629

Wells CF, Salam MA (1968) The effect of pH on the kinetics of the reaction of iron (II) with hydrogen peroxide in perchlorate media. J Chem Soc (A):24–29

Westerhoff P, Aiken G, Army G, Debroux J (1999) Relationships between the structure of natural organic matter and its reactivity towards molecular ozone and hydroxyl radicals. Water Res 33:2265–2276

Westerhoff P, Mezyk SP, Cooper WJ, Minakata D (2007) Electron pulse radiolysis determination of hydroxyl radical rate constants with Suwannee river fulvic acid and other dissolved organic matter isolates. Environ Sci Technol 41:4610–4646

White EM, Vaughan PP, Zepp RG (2003) Role of photo-Fenton reaction in the production of hydroxyl radicals and photobleaching of coloured dissolved organic matter in a coastal river of the southern United States. Aquat Sci 65:402–414

Williams NH, Yandell JK (1982) Outer-sphere electron-transfer reaction of ascorbate anions. Aust J Chem 35:1133–1144

Winterbourn CC (1993) Superoxide as an intracellular radical sink. Free Radic Biol Med 14:85–90

Wu F, Deng N, Zuo Y (1999) Discoloration of dye solutions induced by solar photolysis of ferrioxalate in aqueous solutions. Chemosphere 39:2079–2085

Xu T, Cai Y, Mezyk SP, O'Shea KE (2005) The roles of hydroxyl radical, superoxide anion radical, and hydrogen peroxide in the oxidation of arsenite by ultrasonic irradiation advances in arsenic research. American Chemical Society, Washington, pp 333–343

You J-L, Fong FK (1986) Superoxide photogeneration by chlorophyll A in water/acetone Electron spin resonance studies of radical intermediates in chlorophyll A photoreaction in vitro. Biochem Biophys Res Commun 139:1124–1129

Zafiriou OC (1974) Sources and reactions of OH and daughter radicals in seawater. J Geophys Res 79:4491–4497

Zafiriou OC (1990) Chemistry of superoxide ion-radical (O2-) in seawater: I pK^*_{asw} (HOO) and uncatalyzed dismutation kinetics studies by pulse radiolysis. Mar Chem 30:31–43

Zafiriou OC, Bonneau R (1987) Wavelength-dependent quantum yield of OH radical formation from photolysis of nitrite ion in water. Photochem Photobiol 15:723–727

Zafiriou OC, Dister B (1991) Photochemical free-radical production-rates Guld of marine and Woods-Hole Miami transect. J Geophys Res Oceans 96(C3):4939–4945

Zafiriou OC, True MB (1979a) Nitrite photolysis in seawater by sunlight. Mar Chem 8:9–32

Zafiriou OC, True MB (1979b) Nitrate photolysis in seawater by sunlight. Mar Chem 8:33–42

Zafiriou OC, Joussot-Dubien J, Zepp RG, Zika RG (1984) Photochemistry of natural waters. Environ Sci Technol 18:356A–371A

Zafiriou OC, True Mary B, Hayon E (1987) Consequences of OH radical reaction in sea water: formation and decay of Br2- ion radical, Photochemistry of environmental aquatic systems. American Chemical Society, Washington, pp 89–105

Zafiriou OC, True Mary B, Hayon E (1987) Consequences of OH radical reaction in sea water: formation and decay of Br^{2-} ion radical, Photochemistry of environmental aquatic systems. American Chemical Society, Washington, pp 89–105

Zafiriou OC, Voelker BM, Sedlak DL (1998) Chemistry of the superoxide radical ($O2^-$) in seawater: Reactions with inorganic copper complexes. J Phys Chem A 102:5693–5700

Zang L-Y, Stone K, Pryor WA (1995) Detection of free radicals in aqueous extracts of cigarette tar by electron spin resonance. Free Radic Biol Med 19(2):161–167

Zellner R, Exner M, Herrmann H (1990) Absolute OH quantum yields in the laser photolysis of nitrate, nitrite and dissolved H_2O_2 at 308 and 351 nm in the temperature range 278–353 K. J Atmos Chem 10:411–425

Zepp RG (2002) Solar ultraviolet radiation and aquatic carbon, nitrogen, sulfur and metals cycles In: Helbling EW, Zagarese H (eds) UV effects in aquatic organisms and ecosystems. Royal Society of Chemistry, Cambridge, pp 137–183

Zepp RG, Hoigné J, Bader H (1987a) Nitrate-induced photooxidation of trace organic chemicals in water. Environ Sci Technol 21:443–450

Zepp RG, Skurlatov YI, Pierce JT (1987) Algal-induced decay and formation of hydrogen peroxide in water: its possible role in oxidation of anilines by algae. In: Zika RG, Cooper WJ (eds) Photochemistry of environmental aquatic systems, ACS Symp Ser 327. American Chemical Society, Washington, pp 213–224

Zepp RG, Faust BC, Hoigné J (1992) Hydroxyl radical formation in aqueous reactions (pH 3–8) of iron(II) with hydrogen peroxide: the Photo-Fenton reaction. Environ Sci Technol 26:313–319

Zhao B, Li X, He R, Cheng S, Wenjuan X (1989) Scavenging effect of extracts of green tea and natural antioxidants on active oxygen radicals. Cell Biochem Biophys 14:175–185

Zhou X, Mopper K (1990) Determination of photolytically produced hydroxyl radicals in seawater and freshwater. Mar Chem 30:71–88

Zika RG, Milne PJ, Zafiriou OC (1993) Photochemical studies of the eastern Caribbean—an introductory overview. J Geophys Res Oceans 98(C2):2223–2232

Zimbron JA, Reardon KF (2005) Hydroxyl free radical reactivity toward aqueous chlorinated phenols. Water Res 39:865–869

Zuo Y, Hoigné J (1992) Formation of hydrogen peroxide and depletion of oxalic acid in atmospheric water by photolysis of iron(III)-oxalato complexes. Environ Sci Technol 26:1014–1022

Photoinduced and Microbial Degradation of Dissolved Organic Matter in Natural Waters

Khan M. G. Mostofa, Cong-qiang Liu, Daisuke Minakata, Fengchang Wu, Davide Vione, M. Abdul Mottaleb, Takahito Yoshioka and Hiroshi Sakugawa

1 Background

Solar radiation is a universal and regular phenomenon in biosphere that is vital for all life in the Earth's crust. It maintains all the physical, chemical, photoinduced and biological processes of organic matter and dissolved organic matter

K. M. G. Mostofa (✉) · C. Q. Liu
State Key Laboratory of Environmental Geochemistry, Institute of Geochemistry, Chinese Academy of Sciences, Guiyang 550002, China
e-mail: mostofa@vip.gyig.ac.cn

D. Minakata
School of Civil and Environmental Engineering, Brook Byers Institute for Sustainable Systems, Georgia Institute of Technology, 828 West Peachtree Street, Suite 320, Atlanta, GA 30332, USA

F. C. Wu
State Environmental Protection Key Laboratory of Lake Pollution Control, Chinese Research Academy of Environmental Sciences, Chaoyang 100012, China

D. Vione
Dipartimento Chim Analit, University Turin, I-10125 Turin, Italy

Centro Interdipartimentale NatRisk, I-10095 Grugliasco, TO, Italy

M. A. Mottaleb
Department of Chemistry/Physics, Center for Innovation and Entrepreneurship (CIE), Northwest Missouri State University, 800 University Drive, Maryville, MO 64468, USA

T. Yoshioka
Field Science Education and Research Center, Kyoto University, Kitashirakawa Oiwake-cho, Sakyo-ku, Kyoto 606-8502, Japan

H. Sakugawa
Department of Environmental Dynamics and Management, Graduate School of Biosphere Science, Hiroshima University, 1-7-1, Kagamiyama, Higashi-Hiroshima 739-8521, Japan

Note that 'Photoinduced and photolytically' has been used instead of "photochemical and photochemically" in this book.

K. M. G. Mostofa et al. (eds.), *Photobiogeochemistry of Organic Matter*, Environmental Science and Engineering, DOI: 10.1007/978-3-642-32223-5_4,

(DOM) in natural waters. Photoinduced or photolytic processes can maintain the acidity-alkalinity, water transparency, thermal stratification, redox reactions, production of bioavailable carbon substrates to enhance biological productivity, nutrient concentrations, production of dissolved inorganic carbon (DIC), autochthonous production of DOM, photosynthesis, formation of surface chlorophyll *a* maxima (SCM) and so on (Harvey et al. 1995; Moran and Zepp 1997; Laurion et al. 2000; Kopacek et al. 2003; Barbiero and Tuchman 2004; Huisman et al. 2006; Mostofa et al. 2009a, b). Finally, solar radiation is a major source of energy that is of essential importance in natural water ecosystems. The Photoinduced degradation of DOM and its consequences on natural waters are significantly dependent on the spectral range of sunlight under consideration, namely the UV-A (315–400 nm), UV-B (280–315 nm) or visible light (400–700 nm). Depending on the wavelength, there are significant variations as far as sunlight penetration in the water column is concerned (Scully et al. 1996; Morris and Hargreaves 1997; Reche et al. 1999). DOM is typically able to absorb UV radiation in sea and lake water (Kirk 1994; Morris et al. 1995), thereby controlling the penetration of UV in the deep water layers. The penetration depths of UV radiation in natural waters are greatly varied, with typical penetration in clear ocean water of ~20 m for UV-B and ~50 m for UV-A radiation. In oligotrophic marine waters, penetration of UV-B radiation is 5–10 m and 0.5–3 m in freshwater (Kirk 1994; Smith and Baker 1981; Waiser and Robarts 2000). Therefore, any changes in the radiation wavelengths or an increase in global temperature can greatly impact on the various biogeochemical processes mentioned earlier. However, researchers do not pay much attention to photoinduced processes to assess the biogeochemical processes in natural waters.

The microbial process is a well-known observable fact that is typically responsible for the in situ generation of DOM, cycling of nutrients, occurrence of deep chlorophyll *a* maxima (DCM), photosynthesis, thermal energy and degradation of organic matter in soil or sediment porewaters, which are vital to water environments (Mostofa et al. 2009a, b; Conrad 1999; Guildford and Hecky 2000; Rochelle-Newall and Fisher 2002; Roberts et al. 2004; Lovley 2006; Yamashita and Tanoue 2008). Microbial process is generally controlled by bacterial cells and microorganisms, both autotrophs (plants, algae, bacteria) and heterotrophs (animals, fungi, bacteria) in the aquatic environments. Microbial activity is often taking place at the hypolimnion in natural waters as well as in sediment porewaters, and it is most significant in temperate, Arctic and Antarctic regions when the lake epilimnion is covered by ice. There is also no photoinduced degradation taking place at the epilimnion during the ice-covered period. The overall photoinduced and microbial changes in organic matter and DOM components in waters play a significant role in the global carbon cycle and in the biogeochemical processes in the aquatic environment.

This review will provide a common overview on biogeochemical functions of DOM for photoinduced and microbial processes, DOM degradation for a variety of waters, theoretical model and mechanisms for photoinduced and microbial degradation of organic matter and DOM components, reaction rate constants by functional group contribution method, kinetics of photoinduced degradation of DOM, and interactions between photoinduced and microbial degradation in waters. This study also discusses the key factors that may significantly affect the photoinduced and microbial activities in waters. Finally, this study summarizes the various photoproducts of DOM and depicts their importance on biogeochemical cycles in the aquatic environments.

1.1 Biogeochemical Functions of DOM for Photoinduced Processes in Natural Waters

Photoinduced reactions caused by solar radiation can produce various biogeo chemical alterations of DOM that can be listed as follows (Mostofa et al. 2011):

(1) Photoinduced generation of free radicals, which are susceptible to induce photoinduced degradation of DOM in aqueous media. The free radicals sources include: (i) Photolysis of NO_2^- and NO_3^- ions inducing production of hydroxyl radical ($HO^\bullet$) in waters (Zafiriou and True 1979a, b; Takeda et al. 2004; Vione et al. 2006; Minero et al. 2007) (ii) Generation of free radical species such as superoxide ion ($O_2^{\bullet-}$), hydrogen peroxide (H_2O_2), organic peroxides (ROOH) and $HO^\bullet$ by photolysis of CDOM or FDOM in waters (Cooper et al. 1988; Moore et al. 1993; O'Sullivan et al. 2005; Mostofa and Sakugawa 2009); (iii) Photoinduced production of $HO^\bullet$ by the photo-Fenton (Zepp et al. 1992; White et al. 2003) as well as the photo-ferrioxalate/H_2O_2 reactions (Safarzadeh-Amiri et al. 1996, 1997; Southworth and Voelker 2003).

(2) Production of new organic substances by photorespiration or assimilation of particulate organic matter (POM: ca. algae or phytoplankton) and high molecular weight DOM. These processes have a deep impact on the carbon cycling and include: (i) Photo-respiration or assimilations of algae that can produce autochthonous fulvic acid and other organic substances as well as nutrients in the aquatic environments (Mostofa et al. 2009b; Thomas and Lara 1995; Fu et al. 2010). (ii) Photoinduced degradation of chlorophyll *a* with production of new organic substances; this process is typically occurring in the photic layer of natural lake and seawater (Rontani 2001; Cuny et al. 2002). (iii) Photoinduced assimilation or degradation of algal biomass in surface waters under natural sunlight, which may produce new DOM or FDOM species in the aquatic environments (Rochelle-Newall and Fisher 2002; Thomas and Lara 1995; Fu et al. 2010; Henrichs and Doyle 1986; Biddanda and Benner 1997; Carrillo et al. 2002; Mostofa KMG et al. unpublished). (iv) Microbial assimilation or degradation of algal biomass or phytoplankton; in vitro experiments have shown that under dark incubation these processes may produce new DOM or FDOM (Rochelle-Newall and Fisher 2002; Yamashita and Tanoue 2008; Fu et al. 2010; Biddanda and Benner 1997; Mostofa KMG et al. unpublished; Yamashita and Tanoue 2004; Stedmon and Markager 2005). (v) Photoinduced transformation of high-molecular weight DOM into low-molecular weight organic substances; in some cases the process can lead to complete mineralization (Moran and Zepp 1997; Dahlén et al. 1996; Ma and Green 2004; Vähätalo and Järvinen 2007; Vione et al. 2009).

(3) Photoinduced degradation can regulate water-quality parameters. In particular: (i) Photoinduced degradation changes the physical, chemical and optical properties of water (Kopacek et al. 2003; Vahatalo et al. 2000; Twardowski and Donaghay 2002; Mostofa et al. 2005; Mostofa et al. 2007a, b; Moran et al. 2000), the structure of DOM (Kramer et al. 1996; Kulovaara 1996; Bertilsson and Allard 1996) as well as its molecular weight (Twardowski and Donaghay 2002; Allard et al. 1994; Kaiser and Sulzberger 2004; Yoshioka et al. 2007); (ii) Photoinduced degradation processes can affect the acidity-alkalinity balance and the consumption of dissolved oxygen at the epilimnion level in both lacustrine and oceanic environments (Kopacek et al. 2003; Amon and Benner 1996; Gao and Zepp 1998); (iii) Photoinduced degradation decreases the absorbance of CDOM (and/or FDOM), which can result in water discoloration (Reche et al. 1999) and increases the water-column transparency. A notable consequence is the increased penetration of photosynthetically active radiation (PAR, 400–700 nm) but also of damaging UV radiation (280–400 nm) in the water column (Laurion et al. 2000). These effects have also an influence on the photoinduced degradation of deep-water DOM (Morris and Hargreaves 1997; Siegel and Michaels 1996); (iv) Photoinduced processes can induce the degradation of organic pollutants or contaminants. A wide variety of photogenerated transients is involved ($HO^{\bullet}$, $CO_3^{\bullet-}$, $^1O_2$, $^3CDOM^*$) (Maddigapu et al. 2011; Minella et al. 2011; Arsene C 2011), but the hydroxyl radical is the reactive species that is less likely to produce secondary toxic pollutants. Therefore, $HO^{\bullet}$-induced processes are most likely to achieve efficient decontamination. The photo-Fenton reaction or photo-ferrioxalate/H_2O_2 reactions are particularly effective to this purpose (Safarzadeh-Amiri et al. 1997; Southworth and Voelker 2003; Brezonik and Fulkerson-Brekken 1998); (v) Photoinduced degradation of DOM can interact with eutrophication by increasing the phosphate concentration upon decomposition of organic phosphorus present in DOM (Reche et al. 1999; Carpenter et al. 1998; Kim and Kim 2006; Li et al. 2008); (vi) Production of CO_2 as well as other dissolved inorganic carbon (DIC) species upon photoinduced degradation of DOM can potentially influence the carbon cycling, and may have an impact on climate change (Salonen and Vähätalo 1994; Graneli et al. 1996, 1998); (vii) The decomposition of DOM affects directly or indirectly the distribution of trace elements in natural waters (Kopacek et al. 2003; Kieber et al. 1989).

(4) Photoinduced degradation of DOM can be beneficial to the water ecosystem and provides energy for microbial loops. Its effects include: (i) Supply of nutrients, which are naturally important for plankton productivity in natural waters (Kim and Kim 2006; Kirchman et al. 1991; Salonen et al. 1992; Wetzel 1992); (ii) Increase in the pool of bioavailable carbon substrates, which are essential foods for microorganisms (Bertilsson and Allard 1996; Lindell et al. 1996; Wetzel et al. 1995; Benner and Biddanda 1998; Bertilsson and Tranvik 1998; Bertilsson et al. 1999); (iii) Photo-production of reactive species by CDOM or FDOM, such as hydrogen peroxide (H_2O_2), organic peroxides (ROOH) and $HO^{\bullet}$. These species can contribute damage to macromolecules such as DNA, proteins and lipids (O'Sullivan et al. 2005; Samuilov et al. 2001; Blokhina et al. 2003; Zhao et al. 2003). (iv) The simultaneous generation of H_2O_2, CO_2 and DIC from DOM

could favor the occurrence of photosynthesis in natural waters, where H_2O_2 instead of H_2O would be involved as reactant in photosynthesis ($xCO_2 + yH_2O_{2(H_2O)} + h\upsilon \rightarrow C_x(H_2O)_y + O_2 +$ energy; and $2H_2O_2 + h\upsilon \rightarrow 2H_2O + O_2$) (Mostofa et al. 2009a; Komissarov 1994, 1995, 2003). (v) Production of CO_2 and DIC by photoinduced degradation of DOM and particulate organic matter (POM: ca. algae) can potentially influence carbon cycling and climate change (Ma and Green 2004; Salonen and Vähätalo 1994; Graneli et al. 1998; Clark et al. 2004; Xie et al. 2004; Borges et al. 2008; Kujawinski et al. 2009; Tranvik et al. 2009; Omar et al. 2010; Ballaré et al. 2011; Zepp et al. 2011).

1.2 *Biogeochemical Functions of DOM for Microbial Processes in Natural Waters*

The major changes in organic materials and DOM by microbial processes can be listed as follows:

(i) Microbial degradation of vascular plant material is the only anaerobic process that, according to physical (temperature, moisture), chemical (redox, nutrient availability) and microbial (microfloral successional patterns, availability of microorganisms) factors can diagenetically produce the humic substances (mostly fulvic and humic acids), proteins, carbohydrates, unidentified organic substances and nutrients in soil or sediment pore water environments (Mostofa et al. 2009a; Conrad 1999; Lovley 2006; Wetzel 1992; Malcolm 1985; Nakane et al. 1997; Uchida et al. 1998, 2000).

(ii) Microbial respiration or assimilation of POM (ca. algae or phytoplankton) can produce autochthonous fulvic acid-like substances (C- and M-like) (see the FDOM chapter for detailed description), various DOM components as well as nutrients at different rates in aquatic environments (Harvey et al. 1995; Mostofa et al. 2009b; Conrad 1999; Lovley 2006; Yamashita and Tanoue 2008; Fu et al. 2010; Weiss et al. 1991; Lehmann et al. 2002; Zhang et al. 2009). The microbial origin of these autochthonous organic substances and nutrients in association with photoinduced production are susceptible to control the organic carbon and nutrients dynamics in natural waters.

(iii) Microbial processes may generally affect the chemical composition of aliphatic carbon (e.g. carbohydrates) in macromolecules such as humic substances (fulvic and humic acids) of vascular plant origin, or autochthonous fulvic acids of algal or phytoplankton origin. Chemical alterations have been observed either experimentally under dark incubation or in sediment pore waters (Mostofa et al. 2007a, b, 2009b; Conrad 1999; Moran et al. 2000; Deshmukh et al. 2002; Li W et al., unpublished data). The changes in pore-water autochthonous fulvic acid of algal origin can be identified by the differences in excitation-emission matrix (EEM) spectra and by an increase with depth of fluorescence intensity in sediment pore waters. Studies observe that the relative increase in fluorescence intensity of peak C (relative to humic-like fluorophores) is 43 % in 20–40 cm and 88 % in 42–60 cm depth samples, compared to the average fluorescence intensity at 1–20 cm (Li W et al., unpublished data). Similarly, microbial processes can degrade the protein-like and tryptophan-like components, which are labile to microbial degradation. Microbial processing of these components can be monitored by a decrease in their fluorescence intensity, experimentally under dark incubation and in field observations in deeper layers of natural waters (Mostofa et al. 2005, 2010, 2011; Baker and Inverarity 2004).

(iv) Methanogenesis caused by microorganisms (methanogens and acetogens) is an important anaerobic process that can produce CH_4 and CO_2 by converting either acetate (and formate) or H_2/CO_2 in anaerobic environments (Conrad 1999; Zinder 1993; Lovley et al. 1996; Kotsyurbenko et al. 2001).

(v) Microbial changes either in organic substances (e.g. glucose) or in the functional groups of macromolecules such as fulvic and humic acids of vascular plant origin, as well as autochthonous fulvic acids of algal or phytoplankton origin, may occur with the release of a variety of byproducts such as CH_4, CO_2, DIC (sum of dissolved $CO_2 + H_2CO_3 + HCO_3^- + CO_3^{2-}$), PO_4^{3-}, NH_4^+, H_2O_2 and organic peroxides in waters (Conrad 1999; Mostofa and Sakugawa 2009; Fu et al. 2010; Ma and Green 2004; Li et al. 2008; Tranvik et al. 2009; Zhang et al. 2009; Li W et al., unpublished data; Lovley et al. 1996; Palenik and Morel 1988; Zhang et al. 2004; Kim et al. 2006).

(vi) The *Nitrospira* genus and *Nitrobacter* species are the key nitrite-oxidizing bacteria (NOB) in nitrifying waste water treatment plants, which are likely to depend mostly on nitrite concentration (Kim and Kim 2006). In addition, extracellular polymeric substances (EPSs), biologically produced by most bacteria, are composed of a mixture of polysaccharides, mucopolysacharides and proteins (Arundhati Pal 2008). EPSs produced by anaerobic sludge under sulfate-reducing conditions are capable of biosorption of heavy metals and can remove them from the waste water treatment plant (Chen et al. 1995; Zhang et al. 2010).

(vii) A new metabolic class of microorganisms demonstrates that a wide diversity of organic compounds can be effectively converted to electricity in self- sustaining microbial fuel cells with long-term stability (Lovley 2006; Chaudhuri and Lovley 2003; Logan and Regan 2006; Cheng and Logan 2007; Li and Fang 2007; Rozendal et al. 2007; Call and Logan 2008; Lee et al. 2008; Lee and Rittmann 2009, 2010; Premier et al. 2012). These organisms, known as electricigens, can completely oxidize organic compounds to carbon dioxide, with direct quantitative electron transfer to electrodes that serve as the sole electron acceptor (Lovley 2006).

2 Theoretical Model for DOM Degradation in Natural Waters

Solar radiation causes the sequential degradation of functional groups in DOM, which can be optically detected either as chromophoric dissolved organic matter (CDOM, Fig. 1a–c) or as fluorescent dissolved organic matter (FDOM, Fig. 1d–f).

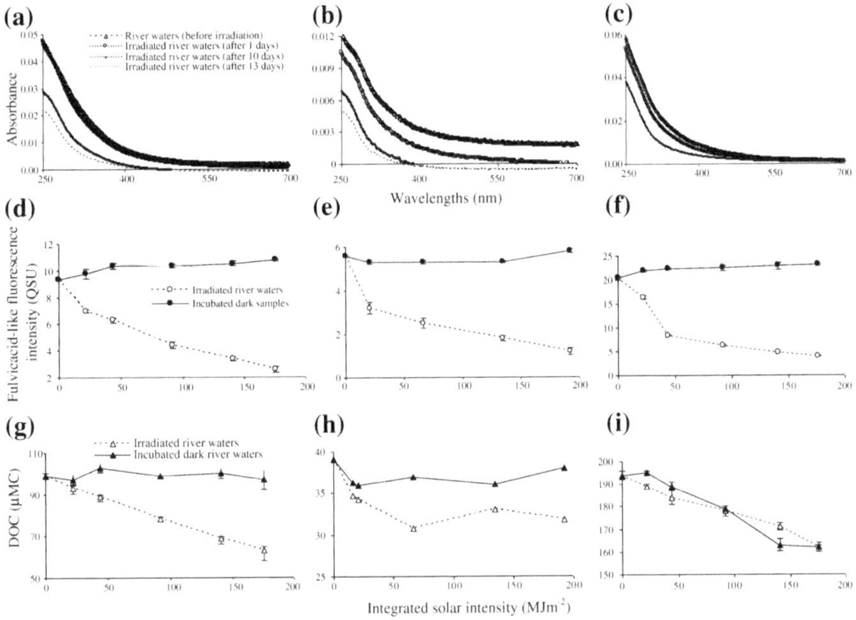

Fig. 1 Photoinduced and/or microbial mineralization of chromophoric dissolved organic matter (CDOM) (**a**, **b**, **c**), fulvic acid-like fluorescence intensity at peak C (**d**, **e**, **f**), and dissolved organic carbon (DOC) concentration (**g**, **h**, **i**) in the upstream waters of Kago and Nishi-Mataya as well as in the downstream waters of Yasu River, respectively. Photoexperiments are conducted under natural sunlight with integrated solar intensity (0, 22, 44, 92, 141, and 176 MJ m^{-2}) during the irradiation period (0, 1, 4, 7, 10, and 13 days, respectively) and microbial experiments are conducted on the filtered samples under dark condition. The quinine sulfate unit (QSU) is estimated using the fluorescence of standard quinine sulfate solution for 1 $\mu g\ L^{-1}$ = 1 QSU. The error bar indicates the standard deviation between triplicate samples. *Data source* Mostofa et al. (2007); Mostofa KMG et al. (unpublished)

Degradation takes place with simultaneous mineralization of dissolved organic carbon (DOC) (Fig. 1g–i) in natural waters (Fig. 1) (Vione et al. 2009; Vahatalo et al. 2000; Mostofa et al. 2005, Mostofa et al. 2007; Moran et al. 2000; Bertilsson and Allard 1996; Allard et al. 1994; Mostofa et al. 2005; Amador et al. 1989; Vähatalo and Wetzel 2004; del Vecchio and Blough 2002). The functional groups in stream humic substances (fulvic and humic acids), with ^{14}C age = 0 (Malcolm 1990), are highly photosensitive/photoreactive, particularly in rivers (Mostofa et al. 2005, 2007; Wu et al. 2005). Photosensitivity sequentially decreases during water transportation from source (stream water) to rivers and then to lakes or coastal or marine waters on the basis of water residence time (Mostofa et al. 2005a, b, 2007; Malcolm 1990; Kieber et al. 1990; Mopper et al. 1991; Miller and Zepp 1995). The functional groups in DOM (either chromophores in CDOM or fluorophores in FDOM) can efficiently absorb solar radiation in natural waters (Mostofa and Sakugawa 2009; Wu et al. 2005; Zafiriou et al. 1984; Mopper and Zhou 1990;

Cooper et al. 1989; Senesi 1990). This can subsequently lead to the decomposition of those functional groups in DOM, thereby causing either losses of absorbance in the UV and visible wavelength regions(Fig. 1a–c) (Vahatalo et al. 2000; Vähätalo and Wetzel 2004; del Vecchio and Blough 2002; Blough and del Vecchio 2002) or losses in fluorescence intensity of FDOM in natural waters (Fig. 1d–f) (Mostofa et al. 2005a, b, 2007; Moran et al. 2000). It can be noted that photoinduced degradation is generally occurring in the mixing zone and decreases with an increase in water depth in natural waters (Vahatalo et al. 2000; Graneli et al. 1996; Mostofa et al. 2005; Bertilsson and Tranvik 2000). Photoinduced degradation can reduce the mean molecular size of the high molecular weight DOM (Moran and Zepp 1997; Yoshioka et al. 2007; Amador et al. 1989; Amon and Benner 1994), which subsequently produces low molecular weight (LMW) intermediate substances (Moran and Zepp 1997; Dahlén et al. 1996; Bertilsson and Tranvik 1998; Mopper et al. 1991). This process ultimately ends up in mineralization with formation of e.g. COS, CO, CO_2, DIC, ammonium, gaseous hydrocarbons and so on in natural waters (Moran and Zepp 1997; Ma and Green 2004; Gao and Zepp 1998; Graneli et al. 1996, 1998; Clark et al. 2004; Xie et al. 2004; Borges et al. 2008; Kujawinski et al. 2009; Tranvik et al. 2009; Omar et al. 2010; Ballaré et al. 2011; Zepp et al. 2011; Mopper et al. 1991; Miller and Zepp 1995; Bertilsson and Tranvik 2000; Chen et al. 1978; Fujiwara et al. 1995; Bushaw et al. 1996; Miller and Moran 1997; Stiller and Nissenbaum 1999; White et al. 2010; Cai 2011).

The rate of photoinduced mineralization of DOM at the depth z (pm_z, mol C m^{-3} d^{-1}), modified by Vähätalo et al. (2000) from Schwarzenbach 1993) and Miller (1998), can be expressed as follows:

$$pm_z = \int_{\lambda_{min}}^{\lambda_{max}} \varphi_\lambda Q_{s,z,\lambda} a_{CDOM,\lambda} \, d\lambda \tag{2.1}$$

where φ_λ is the spectrum of the apparent quantum yield for photoinduced mineralization (mol produced DIC/mol absorbed photons), $Q_{s,z,\lambda}$ is the scalar photon flux density spectrum at a depth z (also referred to as actinic flux, mol photons m^{-2} d^{-1}), and $a_{CDOM,\lambda}$ is the absorption spectrum of CDOM (m^{-1}). CDOM or FDOM is the part of DOM that can absorb solar radiation. The parameters λ_{max} and λ_{min} are the minimum and maximum wavelengths contributing to photoinduced mineralization.

In the whole water column the rate of photoinduced mineralization, modified by Vähätalo et al. (2000) from Miller (1998), can be expressed as follows:

$$pm = \int_{\lambda_{min}}^{\lambda_{max}} \varphi\lambda Q_{a,\lambda} (a_{CDOM,\lambda}/a_{tot,\lambda}) \, d\lambda \tag{2.2}$$

where $Q_{a,\lambda}$ represents the photons absorbed by the water column (mol photons m^{-2} d^{-1}) and the $a_{CDOM,\lambda}/a_{tot,\lambda}$ ratio expresses how much CDOM contributes to the total absorption. In infinitely deep waters, $Q_{a,\lambda}$ roughly equals the downward vector photon flux density just below the surface $Q_{d,v,0-\lambda}$, (Sikorski and Zika 1993a, b).

The quantum yields related to DOM decrease exponentially with increasing wavelength (Moran and Zepp 1997; Vahatalo et al. 2000; Gao and Zepp 1998; Sikorski and Zika 1993; Ratte et al. 1998). A generalized equation to find an exponential relation between quantum yield and wavelength (Vahatalo et al. 2000) can be expressed below:

$$\varphi_\lambda = c \times 10^{-d\lambda} \quad (2.3)$$

where c (dimensionless) and d (nm^{-1}) are positive constants and λ is a wavelength (nm). Different combinations of c and d can cover a wide range of exponential relationships between quantum yield and wavelength.

2.1 Photoinduced Degradation of DOM in Natural Waters

Photoinduced degradation can decompose the DOM, estimated as dissolved organic carbon (DOC) concentration, in natural waters. This has been verified in photoexperiments conducted on waters by using direct natural sunlight or artificial UV radiation in laboratory. Photoirradiation of the samples can gradually decrease the DOC concentration as a function of integrated solar intensity (MJ m^{-2}) during the irradiation period (Fig. 1g–i). The initial DOC concentration, the changes in DOC changes including relative percentages (%), as well as experimental conditions in photoinduced degradation experiments are summarized in Table 1 (Morris and Hargreaves 1997; Ma and Green 2004; Mostofa et al. 2007; Bertilsson and Allard 1996; Amon and Benner 1996; Mostofa et al. 2005; Vähätalo and Wetzel 2004; Miller and Moran 1997; Mostofa K et al., unpublished; Mostofa and Sakugawa unpublished data; Borisover et al. 2011; Pullin et al. 2004; Shiller et al. 2006; Brooks et al. 2007; Corin et al. 1996; Winter et al. 2007; Skoog et al. 1996). It is shown that DOC concentration photolytically decreases by approximately 21–36 % in stream waters during 12–13 days, 2–54 % in downstream river waters during hours to 10 days, 6–41 % in lake waters during hours to 70 days, 31–36 % in Estuarine waters during 71 days, 3–7 % in seawater during hours to 6 days of irradiation (Table 1). Photoirradiation can decompose 35 % of Nordic Reference humic acid (NoHA) and 24 % of Nordic Reference fulvic acid (NoFA) extracted from humus-rich pond water (Table 1).

It is shown that DOC concentration in rivers photolytically decreases a little, approximately 1–2 % in the surface layer (0 m), during 5.5–15.5 h of sunlight irradiation. The DOC losses decrease in deeper layers (<1 % at 6.5 and 24 m), as has been observed in an in situ photo experiment conducted on rivers submerging at different vertical depths (0, 6.5 and 24 m) in Lake Superior (Table 1) (Ma and Green 2004). In lake waters the DOC mineralization is 22–23 % at the surface layer (0 m), 23–24 % at 6.5 m and 4–9 % at 24 m depth, respectively, during the 5.5–15.5 h irradiation period. The results show that DOM mineralization gradually decreases from surface (0 m) to deeper layers because of declining solar radiation that penetrates into lake water.

Table 1 Photoinduced and microbial changes of dissolved organic carbon (DOC) concentration during photoirradiation experiments conducted on natural waters as reported elsewhere

Type of samples/Locations	Filtration size/type (μm)	Irradiation time (h, or, day)	Solar intensity (MJm^{-2})	DOC Before irradiation	Changes in DOC		Changes in DOC		References
					Photoirradiation	Microbial	Photoirradiation	Microbial	
				(μM C)	(μM C)		%		
Kago upstream (July 2000), Japan (35°N)	0.45	12 (irradiated)	192	119	−29	na	−24	na	Mostofa et al. (2005b)
Kago upstream (July 2000), Japan (35°N)	0.45	12 (dark)	192	119	na	1[a]	na	1[a]	Mostofa et al. (2005b)
Kago upstream (June 2001), Japan (35°N)	0.45	13 (irradiated)	176	99	−36	na	−36	na	Mostofa et al. (2007a)
Kago upstream (June 2001), Japan (35°N)	0.45	13 (dark)	176	99	na	−2	na	−2	Mostofa et al. (2007a)
Nishi-Mataya upstream, Japan (35°N)	0.45	13 (irradiated)	176	38	−12	na	−32	na	Mostofa et al. (2007a)
Nishi-Mataya upstream, Japan (35°N)	0.45	13 (dark)	176	38	na	−3	na	−8	Mostofa K et al. (2007a)
Nishi-Mataya upstream, Japan (35°N)	0.45	12 (irradiated)	192	39	−8	na	−21	na	Mostofa K et al. (unpublished data)
Nishi-Mataya upstream, Japan (35°N)	0.45	12 (dark)	192	39	na	−1	na	−3	Mostofa K et al. (unpublished data)
Yasu River, Japan (35°N)	0.45	13 (irradiated)	176	194	−32	na	−16	na	Mostofa et al. (2007a)
Yasu River, Japan (35°N)	0.45	13 (dark)	176	194	na	−32	na	−16	Mostofa et al. (2007a)
Kurose River (Izumi site), 34°N	0.20	6 (irradiated)	118.5	406	−220	na	−54	na	Mostofa K et al. (unpublished data)

(continued)

Table 1 (continued)

Type of samples/Locations	Filtration size/type (μm)	Irradiation time (h, or, day)	Solar intensity (MJm^{-2})	DOC Before irradiation	Changes in DOC		Changes in DOC		References
					Photoirradiation	Microbial	Photoirradiation	Microbial	
				(μM C)	(μM C)		%		
Kurose River (Izumi site), 34°N	0.45	6 (dark)	118.5	433	na	−235	na	−82	Mostofa K et al. (unpublished data)
Kurose River (Izumi site), 34°N	unfiltered[b]	6 (irradiated)	118.5	406	81[a]	na	20[a]	na	Mostofa K et al. (unpublished data)
Kurose River (Izumi site), 34°N	unfiltered[b]	6 (dark)	118.5	433	na	−367	na	−85	Mostofa K et al. (unpublished data)
Kurose River (Hinotsume site), 34°N	0.20	10 (irradiated)	152.5	412	−90	na	−22	na	Mostofa K et al. (unpublished data)
Kurose River (Hinotsume site), 34°N	0.45	10 (dark)	152.5	412	na	−132	na	−32	Mostofa K et al. (unpublished data)
Kurose River (Hinotsume site), 34°N	unfiltered[b]	10 (irradiated)	152.5	441	−7	na	−2	na	Mostofa K et al. (unpublished data)
Kurose River (Hinotsume site), 34°N	unfiltered[b]	10 (dark)	152.5	441	na	−161	na	−37	Mostofa K et al. (unpublished data)
Kishon River, Israel	filtered (0.20)	10 (dark)	na	792–1100	na	−(167–350)	na	−(22–34)	Borisover et al. (2011)
Parker River, near caostal (USA)		15	nd	703	−27	na	−3.8	na	Pullin et al. (2004)

(continued)

Table 1 (continued)

Type of samples/Locations	Filtration size/type (µm)	Irradiation time (h, or, day)	Solar intensity (MJm^{-2})	DOC Before irradiation	Changes in DOC		Changes in DOC		References
					Photoirradiation	Microbial	Photoirradiation	Microbial	
				(µM C)	(µM C)		%		
Parker River, near caostal (USA) + ·OH		15	nd	703	−29	na	−4	na	Pullin et al. (2004)
Rio Negro water, Amazon River, 3°S	0.20	3 (irradiated)	nd	~835	−98	na	−15	na	Amon and Benner (1996)
Rio Negro water, Amazon River, 3°S	0.20	3 (dark)	nd	nm	nm	nm	nm	1.2–2.7	Amon and Benner (1996)
Rio Negro water, Amazon River, 3°S	0.20	10 h (1 day)	nd	800	−41	na	−5	na	Amon and Benner (1996)
Rio Negro water, Amazon River, 3°S	0.20	4 h	nd	~785	−18	na	−2	na	Amon and Benner (1996)
Pearl River, Mississippi: 30°N	0.22	21 (irradiated)	nd	~565	−113	NMD	−20	NMD	Shiller et al. (2006)
Laramie River, U.S.A (41°N)	0.20	72 h (irradiated)	nd	2681	−597	na	−22	NMD	Brooks et al. (2007)
Chimney Park Wetland, U.S.A (41°N)	0.20	72 h (irradiated)	nd	3610	−666	na	−19	NMD	Brooks et al. (2007)
Sturgeon River waters in Lake Superior (47°N)	0.20	5.5 h (0 m)	nd	2829	−32	na	−1	na	Ma and Green (2004)
Sturgeon River waters in Lake Superior (47°N)	0.20	15.5 (0 m)	nd	2829	−61	na	−2	na	Ma and Green (2004)
Sturgeon River waters in Lake Superior (47°N)	0.20	5.5 h (6.5 m)	nd	2829	−18	na	−0.6	na	Ma and Green (2004)

(continued)

Table 1 (continued)

Type of samples/Locations	Filtration size/type (μm)	Irradiation time (h, or, day)	Solar intensity (MJm^{-2})	DOC Before irradiation	Changes in DOC		Changes in DOC		References
					Photoirradiation	Microbial	Photoirradiation	Microbial	
				(μM C)	(μM C)		%		
Sturgeon River waters in Lake Superior (47°N)	0.20	15.5 (6.5 m)	nd	2829	−15	na	−0.5	na	Ma and Green (2004)
Sturgeon River waters in Lake Superior (47°N)	0.20	5.5 h (24 m)	nd	2829	−12	na	−0.4	na	Ma and Green (2004)
Sturgeon River waters in Lake Superior (47°N)	0.20	15.5 (24 m)	nd	2829	−25	na	−0.9	na	Ma and Green (2004)
Lakes									
Lake Biwa, 35°N: surface water (2.5 m)	0.10	12 (irradiated)	137	126	−7	na	−6	na	Mostofa KMG et al. (unpublished) (2004)
Lake Biwa, 35°N: surface water (2.5 m)	0.10	12 (dark)	na	126	na	−2	na	−2	Mostofa KMG et al. (unpublished) (2004)
Lake Biwa, 35°N: surface water (70 m)	0.10	12 (irradiated)	137	93	−8	na	−9	na	Mostofa KMG et al. (unpublished) (2004)
Lake Biwa, 35°N: surface water (70 m)	0.10	12 (dark)	na	93	na	−7	na	−8	Mostofa KMG et al. (unpublished) (2004)

(continued)

Table 1 (continued)

Type of samples/Locations	Filtration size/type (μm)	Irradiation time (h, or, day)	Solar intensity (MJm^{-2})	DOC Before irradiation	Changes in DOC		Changes in DOC		References
					Photoirradiation	Microbial	Photoirradiation	Microbial	
				(μM C)	(μM C)		%		
Lake Biwa, 35°N: surface water (2.5 m)	<5 kDa	12 (irradiated)	137	73	−5	na	−11	na	Mostofa KMG et al. (unpublished) (2004)
Lake Biwa, 35°N: surface water (2.5 m)	<5 kDa	12 (dark)	na	73	na	0	na	0	Mostofa KMG et al. (unpublished) (2004)
Lake Biwa, 35°N: surface water (70 m)	<5 kDa	12 (irradiated)	137	63	−10	na	−16	na	Mostofa KMG et al. (unpublished) (2004)
Lake Biwa, 35°N: surface water (70 m)	<5 kDa	12 (dark)	na	63	na	2[a]	na	3[a]	Mostofa KMG et al. (unpublished) (2004)
Lake Lacawac, 41°N: surface water	0.20	7 (irradiated)	nd	400	−80	NMD	−20	NMD	Morris and Hargreaves (1997)

(continued)

Table 1 (continued)

Type of samples/Locations	Filtration size/type (μm)	Irradiation time (h, or, day)	Solar intensity (MJm^{-2})	DOC Before irradiation (μM C)	Changes in DOC Photoirradiation (μM C)	Changes in DOC Microbial (μM C)	Changes in DOC Photoirradiation (%)	Changes in DOC Microbial (%)	References
Lake Waynewood, 41°N: surface water	0.20	7 (irradiated)	nd	440	−75	NMD	−17	NMD	Morris and Hargreaves (1997)
Lake Giles, 41°N: surface water	0.20	7 (irradiated)	nd	91	NPD	NMD	NPD	NMD	Morris and Hargreaves (1997)
Lake Tuscaloosa	0.22	70 (irradiated)	nd	~235	−97	nm	−41	nm	Vähätalo and Wetzel (2004)
Lake Bjän, southern Sweden (58°N)	0.20	75 h (irradiated)	nd	1217	−125	na	−10	na	Bertilsson and Allard (1996)
Lake Bjän, southern Sweden (58°N)	0.20	75 h (irradiated)	nd	1217	−9	na	na	−1	Bertilsson and Allard (1996)
Lake Savojärvi, southwestern Finland	0.45	4 h (irradiated)	nd	1769	−396	na	−22	na	Corin et al. (1996)
Extracted humic acid, Pond water, Norway	0.45	80 h (irradiated)	nd	5500	−1925	−35	−35	na	Corin et al. (1996)
Extracted fulvic acid, Pond water, Norway	0.45	80 h (irradiated)	nd	10417	−2500	−24	−24	na	Corin et al. (1996)
Lake Superior (47°N)	0.20	5.5 h (0 m)	nd	208	−47	na	−22	na	Ma and Green (2004)
Lake Superior (47°N)	0.20	15.5 (0 m)	nd	208	−48	na	−23	na	Ma and Green (2004)

(continued)

Table 1 (continued)

Type of samples/Locations	Filtration size/type (µm)	Irradiation time (h, or, day)	Solar intensity (MJm^{-2})	DOC Before irradiation	Changes in DOC		Changes in DOC		References
					Photoirradiation	Microbial	Photoirradiation	Microbial	
				(µM C)	(µM C)		%		
Lake Superior (47°N)	0.20	5.5 h (6.5 m)	nd	208	−8	na	−23	na	Ma and Green (2004)
Lake Superior (47°N)	0.20	15.5 (6.5 m)	nd	208	−48	na	−24	na	Ma and Green (2004)
Lake Superior (47°N)	0.20	5.5 h (24 m)	nd	208	−51	na	−4	na	Ma and Green (2004)
Lake Superior (47°N)	0.20	15.5 (24 m)	nd	208	−19	na	−9	na	Ma and Green (2004)
Lake Erie (42°N)	0.10	Surface water	nd	583, 625	525, 575	583, 667	(10–8)	0, (+)7	Winter et al. (2007)
Luther Marsh northeast (43°N)	0.10	Surface water	nd	817, 958	533, 792	817, 950	(35–17)	0, −1	Winter et al. (2007)
Mill Creek east of Cambridge (43°N)	0.10	Surface water	nd	825, 875	683, 583	833, 883	(17–30)	(+)1	Winter et al. (2007)
Bannister Lake southwest of Cambridge (43°N)	0.10	Surface water	nd	775, 867	617, 642	758, 867	(20–22)	(+)1, −2	Winter et al. (2007)
Sanctuary Pond (41°N)	0.10	Surface water	nd	917, 992	650, 750	892, 1008	(20–24)	(+)2, −3	Winter et al. (2007)
Aldrich humic acid in deionized water	0.10	Surface water	nd	908, 842	675, 600	808, 817	(26–29)	−(11–3)	Winter et al. (2007)
Seawaters									
Satilla Estuary	0.20	70 (irradiated)	nd	2046	−628	na	−31	na	Moran et al. (2000)
Satilla Estuary	0.20	70 (irradiated)	nd	1972	−706	na	−36	na	Moran et al. (2000)

(continued)

Table 1 (continued)

Type of samples/Locations	Filtration size/type (μm)	Irradiation time (h, or, day)	Solar intensity (MJm^{-2})	DOC Before irradiation	Changes in DOC		Changes in DOC		References
					Photoirradiation	Microbial	Photoirradiation	Microbial	
				(μM C)	(μM C)		%		
Satilla Estuary, control (nonincubated)	0.20	51 (dark incubated)	nd	1886–1946	na	(170–215)	na	(8–11)	Moran et al. (2000)
Satilla Estuary, control (incubated)	0.20	51 (dark incubated)	nd	1881–2004	na	(171–209)	na	(9–10)	Moran et al. (2000)
Satilla Estuary, 12–16 % photobleached	0.20	51 (dark incubated)	nd	1728–1874	na	(227–261)	na	(13–14)	Moran et al. (2000)
Satilla Estuary, 21–25 % photobleached	0.20	51 (dark incubated)	nd	1563–1739	na	(177–261)	na	(11–15)	Moran et al. (2000)
Satilla Estuary, 32–44 % photobleached	0.20	51 (dark incubated)	nd	1498–1548	na	(170–292)	na	(11–19)	Moran et al. (2000)
Satilla Estuary, 59–64 % photobleached	0.20	51 (dark incubated)	nd	1249–1386	na	(194–322)	na	(16–23)	Moran et al. (2000)
Seawater + humic substances (SH)	0.20	16 h (Artificial sunlight)	nd	695	−22	na	−3	na	Miller and Moran (1997)
Seawater + the control addition (SC)	0.20	16 h (Artificial sunlight)	nd	515	−21	na	−4	na	Miller and Moran (1997)
Artificial seawater + the control addition (AC)	0.20	16 h (Artificial sunlight)	nd	241	−7	na	−3	na	Miller and Moran (1997)
Seawater + humic substances (SH)	0.20	14 (Dark incubation)	nd	695	na	−67	na	−10	Miller and Moran (1997)

(continued)

Table 1 (continued)

Type of samples/Locations	Filtration size/type (μm)	Irradiation time (h, or, day)	Solar intensity (MJm^{-2})	DOC Before irradiation	Changes in DOC		Changes in DOC		References
					Photoirradiation	Microbial	Photoirradiation	Microbial	
				(μM C)	(μM C)		%		
Seawater + the control addition (SC)	0.20	14 (Dark incubation)	nd	515	na	−39	na	−8	Miller and Moran (1997)
Artificial seawater + the control addition (AC)	0.20	14 (Dark incubation)	nd	241	na	−13	na	−5	Miller and Moran (1997)
Baltic Sea, BY15, Gotland Deep (57°N)	0.20 (unfiltered)	6	nd	328	−10	nd	−3	nd	Skoog et al. (1996)
Baltic Sea, BY32, Norrköping Deep (58°N)	0.20 (unfiltered)	6	nd	347	−10	nd	−3	nd	Skoog et al. (1996)
Baltic Sea, F15, Sydostbrotten (63°N)	0.20 (unfiltered)	4	nd	331	−23	nd	−7	nd	Skoog et al. (1996)

[a] means an increase in DOC concentration occurred during incubation period
[b] means incubated unfiltered sample filtered before analytical measurements using Ekicrodisc 25 mm syringe filter with 0.45 mm HT-Tuffryn Membrane (PALL, Gelman Laboratory)
(-) means losses in DOC concentration,*na* not applicable; *nm* not exactly mentioned
NPD nosignificant photochemical degradation; *NMD* no significant microbial degradation

From the results of photoinduced degradation of DOM (Table 1), several key phenomena highlighted the DOM photoinduced degradation in natural waters: First, photoinduced degradation is greatly dependent on the initial concentration of DOC: high photo mineralization is observed in waters with low DOC concentration, while mineralization greatly decreases with increasing DOC concentration. Second, photo mineralization of DOM is typically high in source waters (stream waters) and then decreases in downstream rivers or lakes or seawater. Third, photoinduced degradation is a relatively rapid process for mineralization of DOM compared to microbial degradation in natural waters, except for rivers that contain sewage effluents. It is demonstrated that the photoinduced mineralization of DOM is relatively high during the first day of irradiation in experimental observations, and then almost gradually decreases during the irradiation period (Table 1).

The results of photoinduced degradation on molecular size fractions of DOM demonstrate that photo mineralization is approximately 6 % for large molecular fractions (<0.10 μm) in surface lake waters compared to 9 % in deep waters during 12 days irradiation (Table 1). The results also show that photo mineralization of molecular fractions <5 kDa is relatively higher in deeper waters (16 %) than in surface layers (11 %), and it is higher compared to the 0.1 μm fractions of DOM (Table 1). These results can highlight four important features about the photoinduced degradation of DOM. First, molecular fractions <0.1 μm in surface waters composed of approximately 35 % of autochthonously produced DOC during the summer stratification period may not be photolytically susceptible to mineralization (Mostofa et al. 2009a). Second, molecular fraction of <5 kDa are highly susceptible to photoinduced degradation. Third, DOM in deeper layers is highly susceptible to photo mineralization. It can be noted that the DOC in Lake Biwa is autochthounously produced (45 %) during the summer stratification period, as estimated from higher DOC in summer (135 μM C in August) than in winter (~93 μM C in January) samples during the vertical mixing period (Mostofa et al. 2005). The low photodegradative nature of surface DOM appears to be caused by autochthonous production (~45 % in Lake Biwa), thus autochthonous DOM might be resistant to photoinduced mineralization. This result is in agreement with earlier studies where phytoplankton-exudate DOM, which is the major DOM source from bacterial production (Azam and Cho 1987), is resistant to photoinduced degradation by natural sunlight (Thomas and Lara 1995)

The highly photo-reactive nature of DOM in deeper waters than in the surface layer appears to be caused by two phenomena. First, major fractions of DOM in deeper waters belong to low molecular weight substances (Yoshioka et al. 2007), which may be photolytically mineralized. Second, microbial assimilation or respiration of particulate organic matter (POM: ca. algae) can produce the autchthonous DOM, which are highly photosensitive and photodegradable (Mostofa K et al., unpublished; Zhang et al. 2009; Johannessen et al. 2007; Zhang et al. 2009). The experimental study suggests that the algal-derived DOM is photolytically decomposed by natural sunlight, which is a more efficient photoinduced substrate than is the allochthonous DOM (Mostofa K et al., unpublished; Johannessen et al. 2007; Hulatt et al. 2009). The autochthonous organic substances in deeper layers are typically distinguished by a red shift of the fluorescence peak compared to the upper surface layer (Yoshioka et al. 2007; Mostofa et al. 2005; Hayase and Shinozuka 1995). These autochthonous substances are also identical to the organic substances produced experimentally upon photoinduced and microbial assimilations of algae (Mostofa et al. 2009b; Fu et al. 2010). This hypothesis is supported by the features of autochthonous fulvic acid extracted from POM in sea waters or sediment pore waters, which typically show the fluorescence peak at longer wavelength regions (see also chapter "Fluorescent Dissolved Organic Matter in Natural Waters") (Li W et al., unpublished data; Komada et al. 2002; Burdige et al. 2004; Managaki and Takada 2005; Calace et al. 2006; Parlanti et al. 2000). Another possible explanation is that photolabile DOM in the surface water layers is probably quickly mineralized by sunlight, which leaves only the more photolytically refractory substances near the surface, while photolabile DOM in the deeper layers is more protected from mineralization by the lower sunlight intensity. Interestingly, groundwater DOM has been found to be significantly more susceptible to photo mineralization than surface lake water DOM (Vione et al. 2009).

2.2 Microbial Degradation of DOM in Natural Waters

Microbial actions can decompose the DOM, estimated as dissolved organic carbon (DOC) concentration, in natural waters. This has been verified in experiments conducted on waters under dark conditions. Microbial activity can decrease DOC concentrations either slowly or rapidly depending on the DOM sources during the incubation period (Fig. 1g–i). The initial DOC concentration, amount of DOC changes and its percentage (%), as well as other experimental conditions in microbial degradation experiments are presented in Table 1. It is demonstrated that the decrease of DOC concentration because of microbial activity for various natural waters is approximately 0–8 % in stream waters during 12–13 days, 1–85 % in downstream river waters during hours to 10 days, 0–8 % in lake waters during hours to 70 days, 8–23 % in estuarine waters during 51 days, 5–10 % in seawaters during 14 days of incubation period under dark conditions (Table 1) (Mostofa et al. 2007; Moran et al. 2000; Bertilsson and Allard 1996; Mostofa et al. 2005; Miller and Moran 1997; Mostofa and Sakugawa unpublished data ; Borisover et al. 2011; Winter et al. 2007). From the results of microbial degradation (Table 1) it is possible to generalize several key features commonly observed in natural waters: First, downstream DOM, particularly in sewerage-impacted rivers is significantly labile to microbial degradation. Second, upstream DOM is typically recalcitrant to microbial degradation (Fig. 1g). Third, microbial degradation is typically a slow process for the mineralization of DOM in natural waters, except for downstream rivers with sewage effluents. For example, DOC mineralization rapidly occurs in unfiltered and filtered dark samples (32–85 %) of downstream rivers where DOM is mostly fluxed from untreated sewerage effluents near urban areas (Table 1; Fig. 1i) (Mostofa et al. 2007; Mostofa and Sakugawa unpublished data).

The results of microbial degradation on molecular size fractions of DOM demonstrate that DOC mineralization is approximately 2 % for large molecular fractions (<0.10 μm) in surface layers and 8 % in deeper layers (Table 1). However, DOM with molecular fractions of <5 kDa is not altered at all under dark incubation (Table 1). These results can suggest four important features about microbial degradation of DOM in natural waters. First, molecular fractions of <0.1 μm in both lake surface and deeper waters are labile to microbial degradation, but deeper waters are much labile than surface waters. It is also suggested that autochthonous DOM in surface waters, typically produced during the summer stratification period, is less susceptible to microbial processes. It can be noted that the DOC concentration in Lake Biwa waters is autochthounously produced (45 %) during the summer stratification period, as estimated from the summer DOC values (135 μM C in August) that are higher than the winter ones (~93 μM C in January) during vertical mixing (Mostofa et al. 2005). Second, DOM with molecular weight <5 kDa in both surface and deeper layers is not susceptible to microbial degradation.

2.3 Mechanisms of the Photoinduced Degradation of DOM in Natural Waters

The $HO^\bullet$ radical plays a significant role in photoinduced degradation of DOM in natural waters (Zepp et al. 1992; Southworth and Voelker 2003; Zafiriou et al. 1984; Zika 1981; Voelker et al. 1997). Photoinduced degradation of DOM generally occurs upon direct absorption of UV and visible sunlight by functional groups in DOM, which are optically detected either as chromophores in CDOM or fluorophores in FDOM. Evidence has been provided that an electron transfer from functional groups on DOM can lead to the photoinduced formation of H_2O_2 in aqueous solution (Eqs. 3.13–3.18, see also chapter "Dissolved Organic Matter in Natural Waters") (Mostofa and Sakugawa 2009; Senesi 1990). H_2O_2 subsequently leads to the generation of $HO^\bullet$, by direct photochemistry or by Fenton/photo-Fenton/photo-ferrioxalate reaction systems. These processes can be involved into the photo transformation of DOM in natural waters. Therefore, a general mechanistic scheme for photoinduced degradation of DOM can be depicted as follows (Eqs. 2.1–2.4):

$$\mathrm{DOM} + O_2 + H_2O + H^+ \overset{h\upsilon}{\rightarrow} H_2O_2 + \mathrm{DOM}^{\bullet +} + O_2 + OH^- \tag{2.1}$$

$$H_2O_2 \overset{h\upsilon}{\rightarrow} 2HO^\bullet \tag{2.2}$$

$$\mathrm{DOM}^{\bullet +} + HO^\bullet \overset{h\upsilon}{\rightarrow} \left[\mathrm{DOM}^{\bullet +}HO^\bullet\right]^* \tag{2.3}$$

$$\left[\mathrm{DOM}^{\bullet +}HO^\bullet\right]^* \overset{h\upsilon}{\rightarrow} \mathrm{LMWDOM} + \mathrm{DIC} + CO_2 + \text{other byproducts} \tag{2.4}$$

First, H_2O_2 is formed photolytically through production of $O_2^{\bullet -}$ ion by the release of electrons from DOM chromophores or fluorophores, due to solar effects (Eq. 2.1) that have been discussed earlier in chapter "Photoinduced and Microbial Generation of Hydrogen Peroxide and Organic Peroxides in Natural Waters". Subsequent irradiation converts H_2O_2 into $HO^\bullet$ either photolytically (Eq. 2.2), or via Fenton and photo-Fenton reactions and the other processes that have been mentioned earlier in chapter "Photoinduced Generation of Hydroxyl Radical in Natural Waters". The generated $HO^{\bullet}$ can rapidly react with the $DOM^{\bullet +}$, initially formed during H_2O_2 production (Eq. 2.1), to form a complex $(DOM^{\bullet +}\ HO^{\bullet})^{*}$ (Eq. 2.3) that is subsequently transformed into low molecular weight DOM (LMWDOM), dissolved inorganic carbon (DIC), CO_2, and other byproducts (Eq. 2.4). It is noted that the sequential photoinduced degradation of functional groups in DOM will be elucidated in the next section.

The described mechanism might be applicable in merely DOM-rich natural waters. However, in iron-rich waters the degradation of DOM might be caused by the $HO^{\bullet}$ radicals mostly generated from Fenton reaction (Fenton 1894; Kang et al. 2000), photo-Fenton reaction (Zepp et al. 1992; Southworth and Voelker 2003; Voelker et al. 1997) and photo-ferrioxalate/H_2O_2 reaction (Safarzadeh-Amiri et al. 1997; Safarzadeh-Amiri et al. 1996; Jeong and Yoon 2005) depending on the concentrations of iron as well as oxalate ions in waters. The mechanisms for $HO^{\bullet}$ production from Fenton, photo-Fenton and photo-ferrioxalate/H_2O_2 reaction have been discussed earlier in chapter "Photoinduced Generation of Hydroxyl Radical in Natural Waters". Another most important pathway of $HO^{\bullet}$ production is the photolysis of NO_2^- and NO_3^-, thereby causing the indirect photodegradation of DOM by NO_2^- and NO_3^- in waters (Zafiriou and True 1979a, b; Takeda et al. 2004; Minella et al. 2011; Arakaki et al. 1999; Mack and Bolton 1999).

Therefore, the photoinduced transformation of DOM may undergo by two major pathways depending on the production of free radicals ($^{\bullet}OR$, R=H or alkyl group) in aqueous solution. First, direct photoinduced reactions of DOM, which take place by $HO^{\bullet}$ or other reactive species that may be photolytically generated from DOM components in natural waters. Second, indirect photoinduced reactions of DOM, which typically occur photolytically by $HO^{\bullet}$ that may be generated from Fenton reaction, photo-Fenton reaction, photo-ferrioxalate/H_2O_2 reaction, as well as NO_2^- and/or NO_3^- photolysis in natural waters. If the direct photoinduced processes dominate, the rates of photoinduced degradation as well as of product formation will be proportional to the amount of light absorbed by DOM components such as FDOM or CDOM (Cooper et al. 1989; Blough and Zepp 1995; Goldstone et al. 2002). The indirect photoinduced processes induce the homogeneous production of $HO^{\bullet}$, which subsequently leads to non-selective photoinduced degradation of all organic moieties in DOM in natural waters (Haag and Hoigné 1985; Zepp et al. 1987; Zhou and Mopper 1990; Nakatani et al. 2004). These results can suggest two important facts that occur in photoinduced reactions: (a) Photobleaching can typically proceed with photoproduction of LMW organic substances via direct mechanisms, especially in waters having high FDOM or CDOM content such as in river, lake and coastal waters, and (b) Photoinduced generation of LMW organic substances can typically proceed via indirect mechanisms. The photoinduced generation rate of $HO^{\bullet}$ in lake water by CDOM/FDOM and other sources was too low to account for the photoinduced mineralization of DOM. The latter process appears to be favoured in Fe-rich waters, and possibly involves the photochemistry of Fe(III)-DOC complexes (Vione et al. 2009).

2.4 Mechanisms for Photoinduced Degradation of DOM Functional Groups by HO•

It has been shown that the photoinduced generation of HO• from DOM occurs primarily through H_2O_2 that is produced via photoionization of the most electron-rich organic compounds. The process yields $DOM^{\bullet+}$, which initiates several other reactions (Eqs. 2.1, 2.2). The HO• radical can subsequently react with $DOM^{\bullet+}$ and initiate complex chain reactions (Eq. 2.3). The sequential reactions with HO• yields various end products such as LMWDOM, CO_2, DIC and CO (Eq. 2.4).

The group contribution method (GCM) by Minakata and his co-workers (Minakata et al. 2009) allows the prediction of the aqueous phase HO• rate constants for various functional groups of a given organic compound. It may pave the way to understand the mechanism for the degradation of organic compounds in Advanced Oxidation Processes (AOPs) such as O_3/H_2O_2, UV/H_2O_2 and UV/TiO_2, as well as in natural water photochemistry (Mostofa et al. 2007, 2011; Moran et al. 2000; Minakata et al. 2009; Huber et al. 2003; Rosenfeldt and Linden 2004; Westerhoff et al. 2005; Minakata et al. 2011). The rate constants are discussed in the following section. Photodecarboxylation ($RCOOH + 1/2O_2 + h\upsilon \rightarrow ROH + CO_2$) is one of the important reactions for generating CO_2 by degradation of LMWDOM such as RCOOH in aqueous media (Xie et al. 2004). Another possible pathway could involve phosgene, which is generated photolytically from photosensitive chloroform ($CHCl_3 + O_2 + h\upsilon \rightarrow COCl_2 + HCl$) (Shriner and Cox 1943). The phosgene ($COCl_2$) is highly photosensitive and highly reactive, and it could degrade fluorophores such as the amino groups ($RNH_2 + COCl_2 \rightarrow RN{=}CO + 2\,HCl$) or carboxylic functional groups ($RCO_2H + COCl_2 \rightarrow RC(O)Cl + HCl + CO_2$) (Mostofa et al. 2011; Shriner and Cox 1943). It is noted that macromolecules such as stream fulvic acid and humic acid of vascular plant origin are composed of various functional groups such as –COOH, methoxyl, alcoholic OH, phenolic OH, carbohydrate OH, –C=C–, C=O, aromatic carbon (17–30 %), aliphatic carbon (47–63 %) as well as N, S, and P-atom-containing functional groups (Malcolm 1985; Steelink 2002). Marine DOM of biological origin is composed of amino group in its molecular structure (Midorikawa and Tanoue 1998, 1996). The photoinduced reactions of natural organic matter can lead to the sequential degradation of various functional groups (Mostofa et al. 2011; Xie et al. 2004; Minakata et al. 2009, 2011; Shriner and Cox 1943; Li and Crittenden 2009), of organic molecules bound to fulvic and humic acids (Allard et al. 1994; Amador et al. 1989), and to the sequential decrease in fluorescence intensity of fulvic acid-like substances (peak C and A) with irradiation time (Mostofa et al. 2007, 2011; Moran et al. 2000). The general reaction mechanisms that HO radicals induce are a parent compound → aldehydes and ketones → carboxylic compounds → carbon dioxide and minerals (Bolton and Carter 1994). This reaction mechanism is mostly applicable to fulvic acid and humic acid of plant origin, autochthonous fulvic acid of algal origin, proteins and aromatic amino acids, and of all the compounds that can initiate the reaction through self-generation of HO• via H_2O_2 photo production.

2.4.1 Reaction rate Constants by Functional Group Contribution Method

Recently, it has been possible to determine the aqueous phase $HO^{\bullet}$ reaction rate constants by the functional GCM, which can be applied to the photoinduced degradation of a given organic compound in aqueous media (Minakata et al. 2009). The GCM is based on Benson's thermochemical group additivity (Benson 1976). Under the principle of group additivity, it is hypothesized that an observed experimental rate constant for a given organic compound is the combined rate of all elementary reactions involving $HO^{\bullet}$, which can be estimated using Arrhenius activation energy E_a and frequency factor A. Each reaction mechanism defines a base activation energy, E_a^{o}, and a functional group contribution of activation energy, E_a^{Ri}. The latter results from the neighboring (i.e., α-position) and/or the next-nearest neighboring (i.e., β-position) functional group (i.e., Ri). The GCM considers four reaction mechanisms that can be initiated by $HO^{\bullet}$ in the aqueous phase, which include (1) H-atom abstraction, (2) $HO^{\bullet}$ addition to C C doble bond(s) on alkenes, (3) $HO^{\bullet}$ addition to C=C doble bond(s) on aromatic compounds, and (4) $HO^{\bullet}$ interaction with sulfur (S)-, nitrogen (N)-, or phosphorus (P)-atom-containing compounds (Minakata et al. 2009). Accordingly, an overall reaction rate constant, k_{overall}, can be given by Eq. 2.5

$$k_{\text{overall}} = k_{\text{abs}} + k_{\text{add-alkene}} + k_{\text{add-aromatic}} + k_{\text{i}} \tag{2.5}$$

where, k_{abs}, $k_{\text{add-alkene}}$, $k_{\text{add-aromatic}}$, and k_{int} are the rate constants for the aforementioned reaction mechanisms (1)–(4), respectively.

Rate constant for hydrogen-atom abstraction (Minakata et al. 2009): For H-atom abstraction, the active bond is a C–H bond. In general, molecules are categorized based on the number of C–H bond(s) (i.e., CH_3R_1, $CH_2R_1R_2$, and $CHR_1R_2R_3$, where R_i is a functional group ($i = 1-3$)). Each of the fragments corresponds to a partial rate constant $k_{CH_3R_1}$, $k_{CH_2R_1R_2}$, and $k_{CHR_1R_2R_3}$, respectively. The C–H bond itself and adjacent functional group(s) contributes to the overall E_a as the base activation energy, E_a^0, and group contribution parameter, $E_{a,\text{abs}}$ Ri, due on the functional group R_i, respectively. For example, the base activation energy for H-atom abstraction from one of the primary C–H bonds is $E_{a,\text{prim}}0$. The $E_{a,\text{abs}}$ Ri indicates the electron-donating and—withdrawing ability of the functional group. An electron-donating functional group decreases the E_a and, hence, increases the overall reaction rate constant, and vice versa. Accordingly, the partial rate constant for the fragmented parts such as CH_3R_1, $CH_2R_1R_2$ and $CHR_1R_2R_3$ can be written as below (Eqs. 2.6–2.8)

$$k_{CH_3R_1} = 3A_{\text{prim}}e^{-\frac{E^0_{a,\text{prim}}+E^{R_1}_{a,\text{abs}}}{RT}} \tag{2.6}$$

$$k_{CH_2}R_1R_2 = 2A_{\text{sec}}e^{-\frac{E^0_{a,\text{sec}}+E^{R_1}_{a,\text{abs}}+E^{R_2}_{a,\text{abs}}}{RT}} \tag{2.7}$$

$$k_{CHR_1R_2R_3} = A_{\text{tert}}e^{-\frac{E^0_{a,\text{tert}}+E^{R_1}_{a,abs}+E^{R_2}_{a,abs}+E^{R_3}_{a,abs}}{RT}} \tag{2.8}$$

where 3 is the amount of primary C–H bonds, A_{prim} denotes the Arrhenius frequency factor for the reaction of HO• with CH_3R_1, R is the universal gas constant, and T denotes absolute temperature. However, for (Eqs. 2.6–2.8), the functional group contribution is ignored for cases where the neighboring functional groups have no effect on the H-atom abstraction (i.e.,$E^{-H}_{a,abs}$is zero, where a valence bond of the H-atom is expressed as a line before H).

In (Eqs. 2.6–2.8), the group rate constants, which represent H-atom abstraction from the primary, secondary and tertiary C–H bond are defined as $k^0{}_{prim}$, $k^0{}_{sec}$, and $k^0{}_{tert}$, respectively. They are expressed in (Eqs. 2.9–2.11).

$$k^0_{prim} = A^0_{prim} e^{-E^0_{a,prim}/RT} \tag{2.9}$$

$$k^0_{sec} = A^0_{sec} e^{-E^0_{a,sec}/RT} \tag{2.10}$$

$$k^0_{tert} = A^0_{tert} e^{-E^0_{a,tert}/RT} \tag{2.11}$$

In addition, the group rate constant k_{R4} is defined for the HO$^\bullet$ interaction with the functional group R_4 (e.g. –OH and –COOH). The group contribution factor, X_{Ri}, that represents the influence of functional group R_i can be denoted as (Eq. 2.12)

$$X_{R_i} = e^{-E^{Ri}_{a,abs}/RT} \tag{2.12}$$

The rate constant for H-atom abstraction, k_{abs}, can be written as the sum of the partial rate constants in (Eq. 2.13) because each reaction is independent from one another

$$k_{abs} = 3\sum_0^I k^0_{prim} X_{R_1} + 2\sum_0^J k^0_{sec} X_{R_1} X_{R_2} + \sum_0^k k^0_{tert} X_{R_1} X_{R_2} X_{R_3} + k_{R_4} \tag{2.13}$$

where, I, J, and K denote the number of the fragments CH_3R_1, CH_2R_2, and $CHR_1R_2R_3$, respectively.

As a typical example the rate constant calculation for 1,2-dichloro-3-bromopropane (CH_2Cl–CHCl–CH_2Br) can be written as below (Eq. 2.14)

$$k_{overall} = 2k^o_{sec} X_{-Cl} X_{-CHCl} + k^o_{tert} X_{-Cl} X_{-CH^2Cl} X_{-CH^2Br} + 2k^o_{sec} X_{-Br} X_{-CHCl-} \tag{2.14}$$

It is shown that group rate constants of $k^o{}_{prim}$, $k^o{}_{sec}$, and $k^o{}_{tert}$ are 1.18×10^8, 5.11×10^8, and 1.99×10^9 M^{-1} s^{-1}, respectively and follow the order $k^o{}_{tert} > k^o{}_{sec} > k^o{}_{prim}$ that is consistent with the radical stability of primary, secondary, and tertiary carbon-centered radicals due to the hyperconjugation. The term k_{R4} is accounted for by the group rate constants k_{-OH} and k_{-COOH}, respectively (Eq. 2.13). The k_{-OH} is 1.00×10^8 M^{-1} s^{-1}, representing 33, 8.5, and <5 % of the

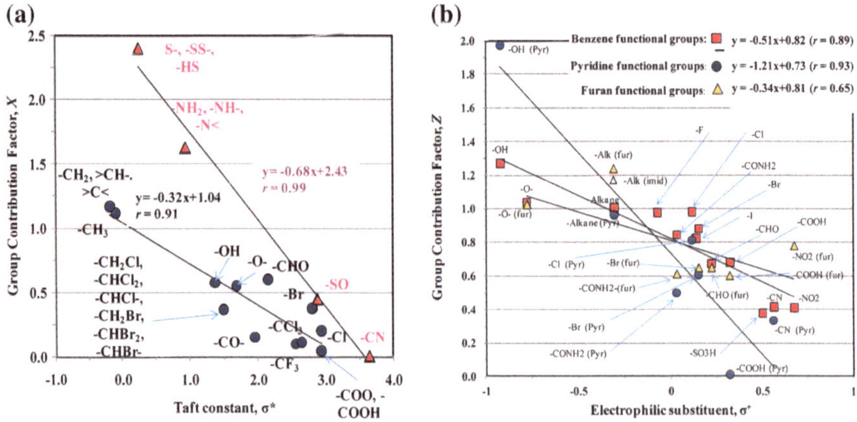

Fig. 2 Comparison of the group contribution factors for H-atom abstraction with the Taft constant, σ^* (a; Karelson 2000) and those for $HO^{\bullet}$ addition to aromatic compounds with electrophilic substituent parameter, σ^+ (Fig. b; Karelson 2000. Group contribution factors include ● alkyl, oxygenated, and halogenated functional groups and ▲ S–, N–, or P-atom-containing functional groups (Fig. a). Group contribution factors for benzene (■), pyridine (●), and furan (▲) compounds (Fig. b). The σ^* of [$-CHCl_2$], [–CO], [–COO, COOH], [–S–, –SS–, HS–], [$-NH_2$, –NH–, –N <] is an average of [CH_2Cl, CH_2Br, $CHCl_2$, $CHBr_2$], [$COCH_3$, COC_2H_5, $COC(CH_3)_3$, COC_6H_5, COF, COCl], [COOH, $COOC_2H_5$], [SCH_3, SC_2H_5, $SCH(CH_3)_2$], and [$NHCH_3$, $NH(CH_2)_3CH_3$, $N(C_2H_5)_2$], respectively. The σ^* of [–SO] and [–N–CO-] refer to [$S(O)CH_3$] and [$NHCOC_6H_5$], respectively. *Data source* Minakata et al. (2009)

H-atom abstraction from the O–H bond in methanol, ethanol, and other alcohol compounds, respectively, which is comparable with the experimental observations (Asmus et al. 1973). The k_{-COOH} is 7.0×10^5 M^{-1} s^{-1}, which is consistent with experimental data for oxalic acid (Getoff et al. 1971).

It is demonstrated that the group contribution factors for the H-atom abstraction linearly correlate with the Taft constant, σ^* (Karelson 2000) (Fig. 2). The alkyl functional groups may often weaken the C–H bond with release of the steric compression. The alkyl functional group moves apart to form a planar radical, thereby increasing the $HO^{\bullet}$ reactivity in the H-atom abstraction reactions. Therefore, X_{-CH3} and $X_{-CH2-} \approx X_{>CH-} \approx X_{>C<}$ values are greater than 1.0, which correspond to negative values of the Taft constant (Fig. 2). In contrast, low values of the group contribution factors for any functional groups indicate their electron-withdrawing ability ($\sigma^* > 0$).

Rate constant for $HO^{\bullet}$ addition to alkenes (Minakata et al. 2009): The detailed mechanisms of $HO^{\bullet}$ addition to alkenes in the aqueous phase are not well documented in earlier studies (Getoff 1991; Billamboz et al. 2010). It is generally considered that π-electrons in alkene compounds (>C=C<) absorb radiation to form an excited state, which then releases electron (e^-) to form H_2O_2 and $>C=C^+<$ (Eq. 2.1; chapter "Photoinduced and Microbial Generation of Hydrogen Peroxide and Organic Peroxides in Natural Waters", Eqs. 2.13–2.18). The $HO^{\bullet}$ then reacts with C^+ to form the reaction intermediates. The excitation of the π-electrons in >C=C< depends on the other functional groups bonded to the alkene.

Except ethylene, alkenes are categorized into six basic structures on the basis of the number of H atoms and their positions, including *cis* and *trans* conformations (i.e., >C=C< , H>C=C< , H_2C=C<, H>C=C<H (*cis*), H>C=C<H (*trans*), and H_2C=C<H). If the base structure is symmetrically associated with the number and position of hydrogen atom(s), the probability of $HO^\bullet$ addition to two unsaturated carbons is considered to be identical, whereas it is different for the asymmetrical base structure. This may reflect the differences in the *A* resulting from regioselectivity. Accordingly, the group rate constant, $k^o_{(structure)\text{-}h}$, and group contribution factor, Y_{Rl}, for $HO^\bullet$ addition to one of the base structures can be written using Arrhenius frequency factor, $A^o_{(structure)\text{-}h}$, and group contribution parameter, $E_{a,add\text{-}alkene}Rl$ of the functional group R_l (l denotes the number of functional groups, l = 1–4):

$$k^o{}_{(\text{structure})\text{-}h} = A^o{}_{(\text{structure})\text{-}h}e^{-[E^o{}_{a,(\text{structure})}]/RT} \tag{2.15}$$

$$Y_{R1} = e^{-[E_{a,(\text{structure})}R1]/RT} \tag{2.16}$$

where (structure) represents six base structures, $E^o_{a(structure)}$ denotes a base part of E_a for (structure), and h denotes a position for $HO^\bullet$ to add i.e., 1 and 2 for the addition to the left and right carbon, respectively. The rate constant for $HO^\bullet$ addition to alkene, $k_{add\text{-}alkene}$, can be written as below (Eq. 2.17)

$$k_{\text{add-alkene}} = \sum gk^o_{(\text{structure-}h)}Y_{R1} \tag{2.17}$$

where g indicates the 1 or 2 that represents asymmetrical and symmetrical addition, respectively. The rate constant for tetrachloroethylene ($Cl_2C=CCl_2$) as a typical example that is shown below (Eq. 2.18):

$$k = 2k^o_{>C=C<}Y_{-Cl}Y_{-Cl}Y_{-Cl}Y_{-Cl} \tag{2.18}$$

Few rate constants are reported for the conjugated and unconjugated dienes. It is shown that the group contribution factors do not linearly correlate with the Taft constant. Two reasons can be considered. First, the functional group contribution to the E_a does not follow the general inductive effect (i.e., Taft constant). Second, the experimental rate constants do not seem to follow the inductive effect (e.g., vinyl chloride > ethylene > vinyl alcohol) because of experimental errors or the existence of unknown reaction mechanisms. Considering new reaction mechanisms such as the excitation of alkenes by radiation may pave the way for future studies in that regard. Despite the observation of the nonlinear correlation between the group contribution factors and the Taft constant, 79 % of the calibrated rate constants were within the error goal, which might be acceptable for a rate constant estimator.

Rate constants for $HO^\bullet$ Addition to Aromatic Compounds (Minakata et al. 2009): The $HO^\bullet$ addition to the aromatic ring often occurs at rates close to diffusion-control. The electron-donating and -withdrawing functional groups on the aromatic ring can significantly affect the rate constants and the ratio of *ortho-*, *meta-*, *para-*, and *ipso*-addition. For the $HO^\bullet$ addition to aromatic compounds, the following points are considered: (i) Probability for the symmetrical HO$^\bullet$ addition to the benzene ring is identical (ii) Addition to the *ipso*-position is negligible for the aromatic compounds with single functional groups (e.g., <8 % for phenol and <1 % for chloro benzene) due to the significant steric effect (Raghavan and Steenken 1980; Merga et al. 1996; Mvula et al. 2001). Therefore, only when all positions on the aromatic ring are filled with functional groups, HO$^\bullet$ adds to the *ipso*-position with identical probability for all the available positions.

For the determination of the reaction rate constant, the E_a is a sum of two parts: (i) a base part, $E^o{}_a$, resulting from the HO$^\bullet$ addition to the aromatic ring depending on the number(s) and position(s) of the functional groups and (ii) group contribution parameter(s), $E_{a,\text{add-aromatic}}\text{Rm}$, due to the functional group(s), R_m (where m is the number of functional group(s), $m = 1$–6), on the aromatic ring. To reduce the number of group contribution factors to calibrate, it is assumed that A differs not by the type of the functional groups but by their number and position. Accordingly, the group rate constant, $k^o{}_{\text{(i-name)-j}}$, and the group contribution factor, Z_{Rm}, may be expressed as below (Eqs. 2.19, 2.20)

$$k^o_{\text{(i-name)-j}} = A^o_{\text{(i-name)-j}} e^{-[(E^o a\text{(i-name)})]/RT} \tag{2.19}$$

$$Z_{\text{Rm}} = e^{-\left(E_{a,\text{add-aromatic}}\text{Rm}\right)/RT} \tag{2.20}$$

where $A^o{}_{(i\text{-name})\text{-}j}$ denotes the Arrhenius frequency factor; $E^o{}_{a(i\text{-name})}$ denotes a base part of E_a; the name (benz, pyr, fur, imid, or triaz) denotes a compound that has a base structure of benzene, pyridine, furan, imidazole, or triazine, respectively; i denotes position(s) of the functional group, and j denotes position(s) for HO$^\bullet$ to add. The rate constant for the HO$^\bullet$ addition to aromatic compounds can be expressed by (Eq. 2.21)

$$k_{\text{add-aromatic}} = \sum n k^o{}_{\text{(i-name)-j}} Z_{\text{Rm}} \tag{2.21}$$

where n indicates the number of available position(s) to add. The rate constant for 1,4-*tert*-butylphenol [$(CH_3)_3C$–C_6H_4–OH] can be depicted as a typical example (Eq. 2.22)

$$\begin{aligned} k = &\left\{2k^o{}_{(1,4\text{-benz})-2,6} + 2k^o{}_{(1,4\text{-benz})-3,5}\right\} Z_{-\text{OH}} Z_{\text{-alkane}} \\ &+ 3 \times 3 \times k_{\text{prim}} X_{>\text{C}<} + k_{-\text{OH}} \end{aligned} \tag{2.22}$$

The group contribution factors for the HO$^\bullet$ addition to aromatic compounds, against electrophilic substituent constants σ^+ are depicted in Fig 2b for benzene ($r = 0.89$), pyridine ($r = 0.93$), and furan ($r = 0.65$) compounds (EPI 2007). The figure shows that the group contribution factors that are empirically derived from the experimental rate constants linearly correlate with the general electron-donating and -withdrawing property. It is shown that a total of 64 % of the rate constants for 64 compounds from the prediction is within the error goal.

Rate constants for HO$^\bullet$ interactions with S-, N-, or P-atom containing compounds (Minakata et al. 2009): The HO$^\bullet$ radical reacts with the S-, N-, or

P-atom-containing functional groups bonded to a given compound in the aqueous phase forming a $2\sigma/1\sigma^*$ two-center-three-electron (2c–3e) adduct (Bonifačič 1999). These functional groups also affect the H-atom abstraction reaction by donating or withdrawing electrons on the C–H bonds. The group rate constant, k_{R4} (Eq. 2.13) represents the reaction of HO$^\bullet$ with S-, N-, or P-atom-containing compounds. The influence of neighboring functional groups is considered as negligible. The rate constant, k, for HO$^\bullet$ addition to iminodiacetic acid (HOOC–CH_2–NH–CH_2–COOH) as a typical example is expressed below (Eq. 2.23):

$$k = 2 \times 2k^{o}_{sec} X_{-COOH} X_{-NH-} + k_{-NH-} + 2k_{-COOH} \quad (2.23)$$

It is shown that the group rate constants k_{-CN} and k_{-NH2} can be compared with the rate constants for compounds that react with HO$^\bullet$ via only interaction such as cyanogen and thiourea, respectively. The rate constant for thiourea (which has two –NH_2 groups) is approximately twice k_{-NH2}, because the electron positive –CS– functional group does not significantly affect the electron density of the N atom. The reaction of HO$^\bullet$ with urea is presumably different because the two amine functional groups of urea are bonded to the electron-negative functional group, –CO–. Thus, another group rate constant $k_{-N-CO-N-}$ is considered for methylurea, tetramethyl urea, and 1,3-dimethylurea. The magnitude of most group rate constants for the S-containing compounds is of the same order as for the amine-containing ones, but approximately 1 order of magnitude larger than for the amide-containing compounds. This might be caused by the electronegative –CO– functional group of the amide. The S-, N-, or P-atom-containing group contribution factors apparently play the same role as the functional groups for H-atom abstraction, i.e., $X_{Ri} = e^{-(Ea,\ absRi)/RT}$. However, it is anticipated that S-, N-, or P-atom-containing functional groups may have different effects on H-atom abstraction. The group contribution factors for –S, –S–S–, and –SH, and –NH_2, –NH–, and –N<, respectively, are assumed to be identical due to the following reasons: (1) limited data availability for single functional group compounds, (2) similar electron inductive ability, and (3) application for the gaseous phase. In addition, the same data sets for the S– and N-atom-containing compounds are used to calibrate the group rate constants, k_{-S-}, k_{-S-S-}, and k_{-SH}, and k_{-NH2}, k_{-NH-}, and $k_{-N<}$, respectively. These group rate constants are not assumed to be identical because the interaction of HO$^\bullet$ with each functional group might be more significant than the electron donating effects of the functional groups. For similar electron inductive ability, the Taft constant indicates similar values among the S– and N-atom-containing functional groups. For example, the Taft constants for SCH_3, SC_2H_5, and SH are 1.66, 1.44, and 1.52, respectively (Karelson 2000), and those for NH_2, $NHCH_3$, $N(CH_3)_2$, $NH(CH_2)_3CH_3$, and $N(C_2H_5)_2$ are 0.62, 0.94, 1.02, 1.08, and 1.00, respectively (Karelson 2000). These values are well distinguished from 3.61 of NH_3^+, 4.66 of NO_2, 4.16 of $N^+(CH_3)_3$, and 3.64 of CN. Finally, it is assumed that the group contributed factors for –S–,–S–S–, and –SH, and for –NH_2, –NH–, –N< , –NNO, and –NNO_2 are identical, which successfully predicted the gas-phase HO$^\bullet$ rate constants (Atkinson 1986, 1987; Kwok and Atkinson 1995). A linear correlation between

Table 2 Hydrogenotrophic homoacetogens and methanogens isolated from various environments

Microorganism	Growth temp (°C)	Optimum temp (°C)		References
Acetobacterium bakiia	1–30	20	Sediments of a polluted pond	Kotsyurbenko et al. (1995)
Acetobacterium paludosuma	1–30	20	Sediments of a fen	Kotsyurbenko et al. (1995)
Acetobacterium fumetariuma	1–35	30	Manure digested at low temp	Kotsyurbenko et al. (1995)
Acetobacterium tundraeb	1–30	20–25	Tundra wetlands	Simankova et al. (2000)
Methanogenic strain MSB	1–32	25–30	Sediments of a polluted pond	Kotsyurbenko et al. (2001)
Methanogenic strain MSP	4–35	25–30	Sediments of a polluted pond	Kotsyurbenko et al. (2001)
Methanobacterium strain MB4	5–30	25–30	Peat samples	Kotsyurbenko et al. (2007)

the group contribution factors of S-, N-, or P-atom-containing functional groups and the Taft constant, σ^*, is observed ($r = 0.99$) (Fig. 2) (Karelson 2000). The X_{Ri} values for S-, N-, or P-atom-containing functional groups are greater than those of the alkyl, oxygenated, and halogenated functional groups (Fig. 2). This suggests that S-, N-, or P-atom-containing functional groups donate more electrons toward the neighboring C–H bond(s), thereby enhancing the H-atom abstraction by $HO^{\bullet}$.

The GCM includes 66 group rate constants and 80 group contribution factors, which characterize each $HO^{\bullet}$ reaction mechanism with steric effects of the chemical structure groups and impacts of the neighboring functional groups, respectively (Minakata et al. 2009). The group contribution factors for H-atom abstraction and $HO^{\bullet}$ addition to the aromatic compounds linearly correlate with the Taft constants, σ^*, and the electrophilic substituent parameters, σ^+, respectively. The best calibrations for 83 % (257 rate constants) and predictions for 62 % (77 rate constants) of the rate constants are within 0.5–2 times the experimental values. Literature-reported experimental $HO^{\bullet}$ rate constants for 310 and 124 compounds are used for calibration and prediction, respectively.

Although there are a few tools available to determine aqueous phase hydroxyl radical reaction rate constants (Minakata et al. 2011; Herrmann 2003; Monod et al. 2005; Minakata and Crittenden 2011; Herrmann et al. 2010), the GCM is quoted as "The wide application range in combination with the user-friendliness makes it probably the best currently available estimation tool for HO radical reactions in aqueous solution. Overall, the method of Minakata et al. (2009) is currently the most broadly usable method for the prediction of HO radical reaction rates in aqueous solution (Herrmann et al. 2010). The GCM peer-reviewed paper provided both MS Excel spread sheet and compiled Fotran program as supportion information. Any users are able to access these programs and determine the aqueous phase HO· reaction rate constants with inputs of structural information of a compound of interest.

Various functional groups widely differ for their reaction rate constants with $HO^{\bullet}$ (Fig. 2). Similarly, the production rates of H_2O_2 and $HO^{\bullet}$ photolytically formed from different organic compounds are much varied (Table 1: chapter "Photoinduced and Microbial Generation of Hydrogen Peroxide and Organic Peroxides in Natural Waters"; Table 2: chapter "Photoinduced Generation of Hydroxyl Radical in Natural Waters"). Variations in the production rates depend on the chemical nature of the functional groups bonded to each organic compound. Therefore, it can be concluded that the functional groups have an important impact both on the photoinduced production of $HO^{\bullet}$ and on the $HO^{\bullet}$ reaction with organic compounds. Both issues are very significant for the photoinduced processes that involve dissolved organic matter in surface waters.

2.5 Mechanisms of Microbial Degradation of DOM in Natural Waters

The organic matter in wastes and biomass is diagenetically altered by complex microbial processes into various kinds of organic substances such as long-chain fatty acids, C_3 to C_5 organic acids, alcohols, aromatic compounds, humic substances (fulvic and humic acids) of terrestrial plant origin, autochthonous fulvic acid of algal origin, acetate, formate, methanol, CO_2, H_2, as well as minor products. These processes take place in waters, in soil environments or in sediment pore waters of lake and marine systems (Mostofa et al. 2009a; Conrad 1999; Lovley 2006; Li W et al., unpublished data; Burdige et al. 2004; Yang and Guo 1990; Leenheer and Croue 2003).

The functional groups of organic substances and the minor components may be subsequently converted into CO_2, methane and other products by fermentative microorganisms and Fe(III)-reducing microorganisms. These processes take place with simultaneous reduction of an array of electron acceptors, including oxygen, H_2, nitrate, manganese oxides, Fe(III) oxides, sulfate, H_2S, and humic substances in water (Lovley 2006; Lovley et al. 1996; Nagase and Matsuo 1982; Jetten et al. 1992; Coleman et al. 1993; Roden and Wetzel 1996; Pelmenschikov et al. 2002; Keppler et al. 2006; Itoh et al. 2008; Reguera et al. 2005).

Fe(III)-reducing microorganisms, commonly *Geobacter* species in temperate environments (Lovley et al. 2004), and Fe(III)-reducing archaea in warm environments (Kashefi et al. 2004) metabolize the fermentation products and the functional groups in organic substances. They are oxidized to CO_2, with Fe(III) oxides serving as the electron acceptor (Lovley 2006; Lovley et al. 1996). The mechanism for CO_2 formation from Fe(III) oxide in the presence of Geobactor spp. is depicted (Fig. 3) (Lovley et al. 1996):

The general reactions for microbial Fe(III) reduction coupled with the oxidation of fermentation products such as acetate (Eq. 2.24) and hydrogen (Eq. 2.25) are described below (Eqs. 2.24, 2.25) (Coleman et al. 1993; Lovley 1991).

$$4Fe_2O_3 + CH_3COO^- + 7H_2O \rightarrow 8Fe^{2+} + 2HCOL_3{}^- + 15OH^- \quad (2.24)$$

$$Fe_2O_3 + H_2 + H_2O \rightarrow 2Fe^{2+} + 4OH^- \quad (2.25)$$

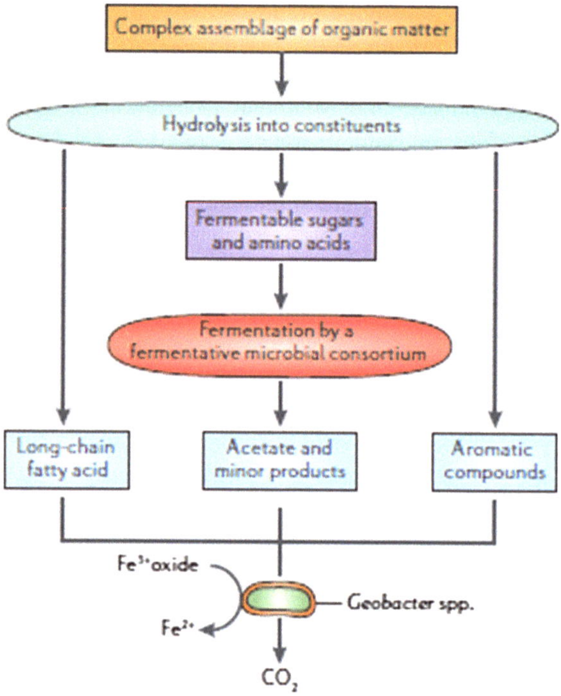

Fig. 3 Generalized pathway for the anaerobic oxidation of organic matter to carbon dioxide with Fe^{3+} oxide serving as an electron acceptor in temperate, freshwater and sedimentary environments. The process is mediated by a consortium of fermentative microorganisms and Geobacter species (ca. *Geobactter metallireducens*). *Data source* Lovley (2006)

$$Fe^{2+} + HCO_3^- + OH^- \rightarrow FeCO_3 + H_2O \quad (2.26)$$

The ferrous iron, hydroxyl ions and bicarbonate, produced together in (Eqs. 2.24, 2.25) can combine to form siderite in aquatic sediment porewaters (Eq. 2.26) (Coleman et al. 1993). Evidences from that research study show that two genera of sulphate-reducing bacteria, *Desulfobacter* and *Desulfovibrio*, can oxidize H_2 and acetate in aquatic sediment waters (Coleman et al. 1993). It is shown that H_2 is the most important electron donor for *Desulfovibrio* (Eq. 2.27), and acetate is the most environmentally significant electron donor for *Desulfobacter* (Eq. 2.28) sulphate reducing bacteria (Coleman et al. 1993):

$$4H_2 + SO_4{}^{2-} \rightarrow S^{2-} + 4H_2O \quad (2.27)$$

$$CH_3COO^- + SO_4{}^{2-} \rightarrow S^{2-} + 2HCO_3^- + H^+ \quad (2.28)$$

The study shows that *Desulfobulbus propionicus* can oxidize S to $SO_4{}^{2-}$ with an electrode serving as the electron acceptor (Lovley 2006). This is an important reaction at the anode surface in sediments, where high concentrations of sulphide can abiotically react with electrodes producing S° (Fig. 4) (Lovley 2006). This abiotic reaction merely yields two out of eight electrons potentially available from sulphide (S^{2-}) oxidation (Fig. 4). Oxidation of S° to $SO_4{}^{2-}$ extracts six electrons and regenerates $SO_4{}^{2-}$ as an electron acceptor for further microbial reduction by microorganisms in the family *Desulfobulbaceae* (Lovley 2006; Holmes et al. 2004).

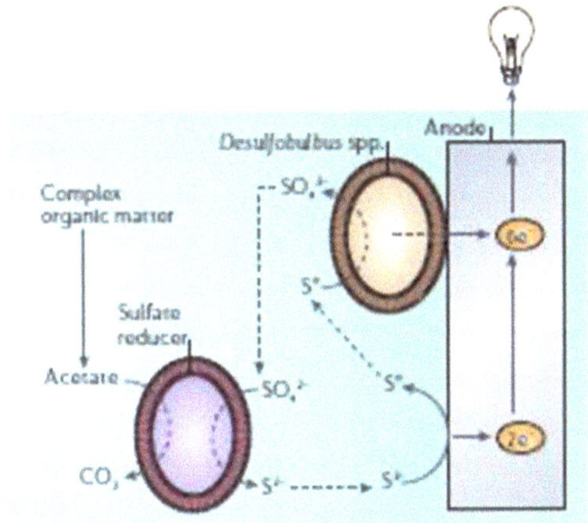

Fig. 4 Mechanism by which sulfate reducer can convert acetate to CO_2 and reduced sulphur compounds to electricity production in sediment microbial fuel cells in sulphide-rich sediments. *Data source* Lovley (2006)

A pure culture of *Desulfovibrio desulfuricans* can readily reduce Fe(III), but *Desulfobacter postgatei and Desulfobactercurvatus* cannot (Coleman et al. 1993). The experimental study showed the occurrence of the metabolism of Fe(III) and sulphate by *D. desulfuricans*; at low concentrations of H_2 in aquatic sediments, Fe(III) might be the predominant electron acceptor (Coleman et al. 1993). It has been evidenced that fermentation or methanogenesis do not metabolize the organics rapidly (Lovley 2006), but can produce a number of minor components such as acetate, formate, methanol, CO_2 and H_2 at the end of the metabolic process (Yang and Guo 1990; Roden and Wetzel 1996; Zeikus et al. 1975; Lovley and Klug 1986; Lovley and Phillips 1987). These products are subsequently used for methane formation.

Methanogenic bacteria are a diverse subgroup of *archaebacteria (Archaea)* that convert CO_2 into methane to provide energy (31 kcal/mol) for the cell (Eq. 2.29) (Thauer et al. 2008; Thauer 1998):

$$CO_2 + 4H_2 \rightarrow CH_4 + 2H_2O \tag{2.29}$$

The conversion of glucose to alcohols and fatty acids during the fermentation allows the utilization of the standard Gibbs free energy content (Conrad 1999; Thauer et al. 1977). The degradation of alcohols and fatty acids to acetate and H_2 caused by syntrophic bacteria is endoergonic under standard conditions (Conrad 1999; Thauer et al. 1977), but it can take place when it is combined with H_2-consuming methanogenesis (Conrad 1999). Hydrogenotrophic and acetotrophic methanogenesis may convert fermentation products or glucose to CH_4 and/or CO_2.

The mechanism for methane formation in methyl-coenzyme *M* reductase (MCR) has been evidenced using the B3LYP hybrid density functional method and chemical models consisting of 107 atoms (Pelmenschikov et al. 2002). In this mechanism, the reaction starts with CoB and methyl-CoM coenzymes and with the active Ni(I) state of the tetrapyrrole F_{430} prosthetic group, which then forms a

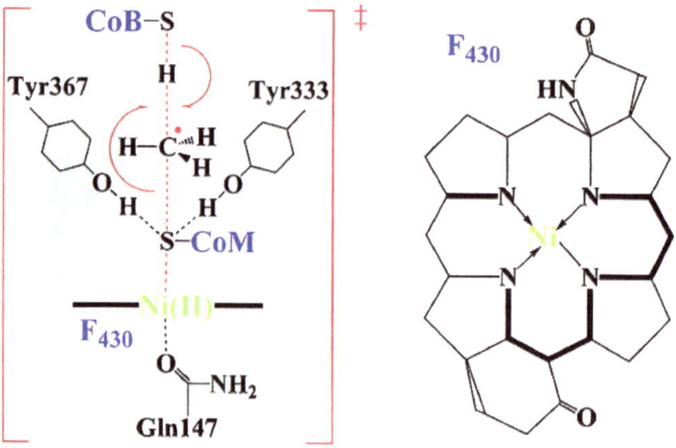

Fig. 5 The mechanism for methane formation in methyl-coenzyme *M* reductase (MCR) in methanogenesis. *Data source* Pelmenschikov et al. (2002)

free methyl radical at the transition state (Fig. 5). A methyl radical is then released from methyl-CoM, induced by the attack of Ni(I) on the methyl-CoM thioether sulfur, which oxidizes the metal center from Ni(I) to Ni(II). The latter forms a strong bond of 38.6 kcal/mol with the sulfur of CoM (Eq. 2.44):

$$\begin{aligned}&\text{CoB-S-H} + \text{CH}_3\text{-S-CoM} + \text{Ni(I)F}_{430} \rightarrow \\ &\text{CoB-S-H} + \text{CH}_3^{\bullet} + \text{CoM-S-Ni(II)F}_{430}\end{aligned} \tag{2.44}$$

The resulting methyl radical is rapidly quenched by hydrogen-atom transfer from the CoB thiol group, yielding the CH_4 and the CoB radical. The pathway has activation energy of approximately 20 kcal/mol, leading to stereoinversion at the reactive carbon (Eq. 2.45):

$$\begin{aligned}&\text{CoB-S-H} + \text{CH}_3^{\bullet} + \text{CoM-S-NI(II)F}_{430} \rightarrow \\ &\text{CH}_4 + \text{CoB-S}^{\bullet} + \text{CoM-S-NI(II)F}_{430}\end{aligned} \tag{2.45}$$

In the final step, formation of heterodisulfide CoB-S–S-CoM is proposed in which nickel is reduced back to Ni(I) (Eq. 2.46).

$$\text{CoB-S}^{\bullet} + \text{CoM-S-NI(II)F}_{430} \rightarrow \text{CoB-S-S-CoM} + \text{NI(I)F}_{430} \tag{2.46}$$

It can be noted that methyl-coenzyme M is 2-mercaptoethanesulfonic acid that is unique to the methanogens, and coenzyme B is 7-mercaptoheptanoylthreonine phosphate that includes an aliphatic linker of six methylene units between the phosphothreonine head group and the thiol group.

A recent study shows that MCR is the key enzyme in methane formation by methanogenic *Archaea* when it is incubated with the natural substrates

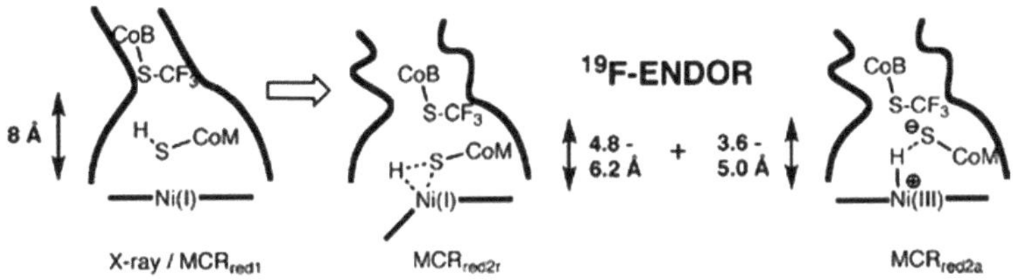

Fig. 6 The mechanism for methane formation in methyl-coenzyme *M* reductase (MCR) in methanogenesis in presence of natural substances. Data source Ebner et al. (2010)

(Ebner et al. 2010). In this mechanism, the enzyme converts the thioether methyl-coenzyme M, and the thiol coenzyme B, into methane and the heterodisulfide of coenzyme M and coenzyme B (Ebner et al. 2010). In the presence of the competitive inhibitor coenzyme M instead of methyl-coenzyme M, addition of coenzyme B to the active Ni(I) state of MCR_{red1} induces two new species called MCR_{red2a} and MCR_{red2r} (Fig. 6). The two MCR_{red2} signals can also be induced by the *S*-methyl- and the *S*-trifluoromethyl analogs of coenzyme B. It is thus suggested that the protein may undergo a conformational change upon formation of MCR_{red2} species in the transition from MCR_{red1}, which opens up the possibility that nickel coordination geometries other than square planar, tetragonal pyramidal, or elongated octahedral might occur in intermediates of the catalytic cycle (Ebner et al. 2010).

The degradation of specific aliphatic carbon or functional groups by microbial processes in natural waters may preferentially occur in macromolecules such as fulvic and humic acids of terrestrial plant origin, as well as autochthonous fulvic acid of algal origin. The microbial changes in the functional groups of organic substances are typical phenomena in sediment pore waters, where a decrease of the acidic functional groups as well as an increase of basic and neutral functional groups occurs with depth (Rosenfeld 1979; Burdige and Martens 1988; Wu and Tanoue 2001; Maita et al. 1982; Steinberg et al. 1987). Such changes in functional groups of autochthonous fulvic acid (C-like) can be understood from the vertical increase in fluorescence intensity with depth, identified with excitation and emission matrix (EEM) of pore water samples and their parallel factor (PARAFAC) modeling in the pore waters of lakes (Li W et al., unpublished data). The low values of fluorescence index for autochthonous fulvic acid (C-like) at deeper depth, compared with upper sediment pore water, confirm the changes with depth in the functional groups of that component (Li W et al., unpublished data). Such changes might be a useful indicator for complex microbial processing of the functional groups of autochthonous fulvic acids in the pore waters of lakes. Therefore, it is suggested that microbial degradation may diagenetically alter either the minor components (e.g. acetate) or the functional groups bound to macromolecules, such as fulvic and humic acid from terrestrial plants and autochthonous fulvic acid from algal biomass, generating CO_2, CH_4 and other products.

2.6 *Kinetics of Photoinduced Degradation of DOM*

Photoinduced degradation can decrease the concentration of dissolved organic carbon (DOC) as a function of the integrated solar intensity (Fig. 7) (Mostofa et al. 2005, 2007). The changes in the DOC concentration can be best fit with a first order reaction as reported below (Eq. 2.47):

$$Ln\,(DOC/DOC_o) = -k_1 S \tag{2.47}$$

where k_1 is the reaction rate constant for the photoinduced degradation of DOC, *DOC* is organic carbon concentration after irradiation and DOC_o the initial one, and *S* is the integrated solar intensity or photon energy (MJ m^{-2}) (Fig. 7a) (Mostofa et al. 2007).

Kinetics studies on the photoinduced degradation of DOM can explain several important phenomena in waters (Mostofa et al. 2007). First, stream DOM undergoes rapid photoinduced degradation (1.8–2.6 × 10^{-3} MJ^{-1} m^2 in waters of the Kago and Nishi-Mataya upstreams) (Mostofa et al. 2007). Second, microbial degradation under dark incubation is quite low or negligible for upstream DOM (0.7–4.6 × 10^{-4} MJ^{-1} m^2 for the same upstreams). Third, in rivers that include various sources of DOM the latter can be uniformly degraded both photolytically (9.5 × 10^{-4} MJ^{-1} m^2) and microbiologically (11 × 10^{-4} MJ^{-1} m^2) (Fig. 1c and d).

3 Factors Controlling the Photoinduced Degradation of DOM in Natural Waters

Photoinduced degradation of DOM depends on the sources of waters, concentration level and optical-chemical nature of DOM, time and space. Photoinduced degradation of DOM is an important phenomenon that plays a significant role in the biogeochemistry of the carbon cycle, biological activity and primary and secondary productions in natural waters (Mostofa et al. 2009a; Ma and Green

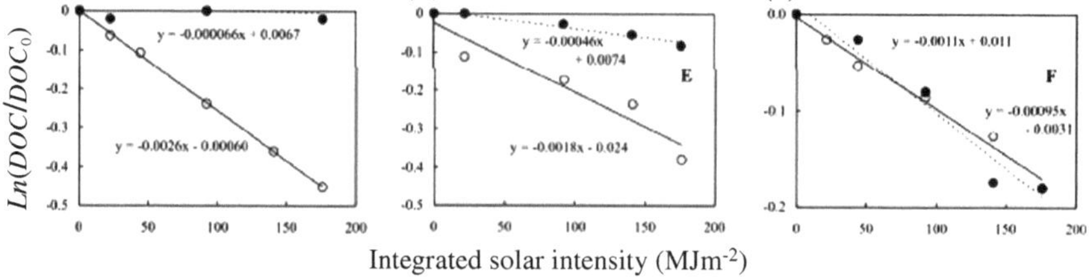

Fig. 7 Relationship between the $Ln(DOC/DOC_0)$ and the integrated solar intensity (MJm^{-2}) in the Kago upstream (**a**), Nishi-Mataya upstream (**b**), and in the downstream waters of Yasu River (**c**). *Data source* Mostofa et al. (2007)

2004; Vahatalo et al. 2000; Mostofa et al. 2000, 2007; Gao and Zepp 1998; Graneli et al. 1996, 1998; Wu et al. 2005; Mopper et al. 1991; Miller and Zepp 1995; Miller and Moran 1997; Rosenstock et al. 2005; Nieto-Cid et al. 2006). The Photoinduced degradation rate is mostly dependent on several major factors that are greatly related to variable local conditions and to the concentration levels and molecular nature of DOM. The key factors are: (1) Sunlight or solar radiation, (2) Water temperature, (3) Effects of total dissolved Fe and photo-Fenton reaction, (4) Occurrence and quantity of NO_2^- and NO_3^- ions, (5) Molecular nature of DOM, (6) pH and alkalinity of the waters, (7) Dissolved oxygen (O_2), (8) Depth of the water, (9) Physical mixing in the surface mixing zone, (10) Increasing UV-radiation during ozone hole events, (11) Global warming and (12) Salinity.

3.1 Sunlight or Solar Radiation

Solar radiation is the key factor for photoinduced degradation of DOM or organic contaminants in water (Morris and Hargreaves 1997; Reche et al. 1999; Mostofa and Sakugawa 2009; Vahatalo et al. 2000; Mostofa et al. 2007; Moran et al. 2000; Wu et al. 2005; Molot and Dillon 1997; Dobrović et al. 2007). Photoinduced degradation of DOM depends on the spectral wavelengths of solar radiation such as UV-A (315–400 nm), UV-B (280–315 nm), and visible light (400–700 nm) as well as their significant variations after penetration in the water column (Scully et al. 1996; Morris and Hargreaves 1997; Reche et al. 1999; Vahatalo et al. 2000; Graneli et al. 1996, 1998; Lindell et al. 1996; Kieber et al. 1990; Molot and Dillon 1997; de Haan 1993; Herndl et al. 1993; Valentine and Zepp 1993). DOM is typically susceptible to absorb UV radiation in sea and lake waters (Kirk 1994; Morris et al. 1995). The penetration of UV radiation in natural waters is greatly variable, with typical penetration depths in clear ocean water of ~20 m for UV-B and ~50 m for UV-A radiation, 5–10 m for UV-B radiation in oligotrophic oceans and 0.5–3 m in freshwaters (Kirk 1994; Smith and Baker 1981; Waiser and Robarts 2000). It can be expected that the photoinduced degradation of DOM is significantly dependent on the attenuation of downward irradiance in natural waters. It has been shown that the contribution of solar intensity to total photoinduced degradation of DOM in lakes is 39–69 % by UV-A, 9–17 % by UV-B, and 23–44 % by visible light radiation (Vahatalo et al. 2000). Photoinduced mineralization of natural DOC is increased <9 % when the UV-B radiation is doubled in humic lakes (Vahatalo et al. 2000). Control irradiation by wavelengths 254 nm (hereafter UV) and 185 nm (hereafter VUV) on DOM demonstrates that the DOM degradation rate at 185 nm increases approximately ten-fold compared to those at 254 nm. An increase in fluxes of the UV radiation can substantially increase the quantity of the reactive free radicals such as $HO^\bullet$ and H_2O_2 in waters (Qian et al. 2001; Rex et al. 1997; Yocis et al. 2000). Rex et al. 1997; Yocis et al. 2000; During an ozone hole event, the production rates of $HO^\bullet$ are greatly enhanced in Antarctic waters (Qian et al. 2001; Rex et al. 1997). The $HO^\bullet$ is the most powerful oxidizing agent that can be involved into the photoinduced degradation of DOM in waters, although there is evidence that the formation rate of $HO^{\bullet}$ may be insufficient to account for the DOM mineralization (Vione et al. 2009).

Variations in the spectral irradiance penetration among various waters and in the effect of radiation wavelength on DOM transformation might be caused by three factors. First, concentration levels and molecular nature of DOM can modify the absorption spectrum for a variety of waters. Second, contents of total iron, a major factor of $HO^{\bullet}$ production through photo-Fenton reaction and probably also of DOM mineralization through $HO^{\bullet}$-independent processes (Vione et al. 2009), are greatly varied for a variety of waters. Third, depletion of the stratospheric ozone layer may greatly increase the UV-B radiation, thereby enhancing the photoinduced mineralization of DOM by UV-B (Qian et al. 2001; Randall et al. 2005). An effort is still needed to accout for the different results obtained in different studies. On the one hand, a limited increase (~4 %) in DOM photoinduced mineralization has been observed in Brazil compared to Sweden, although the dose of UV-B was three-fold higher in Brazil than in Sweden (Graneli et al. 1998). On the other hand, half of the total photoinduced degradation of DOM was attributable to wavelengths shorter than 360 nm (Vahatalo et al. 2000).

3.2 Water Temperature

Air temperature is greatly varied from 0 to approximately 50 °C in different regions, which might control the water temperature (WT) and its variation in natural waters. A low WT can reduce the movement of the reactants in the aqueous solution, thereby causing a decrease in the reaction kinetics of DOM in waters. WT that is driven by solar intensity is directly related to the photoinduced generation of H_2O_2. The H_2O_2 in river shows a significantly higher production in summer and lower in winter (Mostofa and Sakugawa 2003, 2009). The lower production of H_2O_2 due to low WT and solar irradiance may subsequently decrease the production rate of $HO^{\bullet}$ (Qian et al. 2001) and, as a consequence, the photoinduced degradation rate of DOM in aqueous solution.

3.3 Effects of Total Dissolved Fe and Photo-Fenton Reaction

The concentration of total dissolved Fe is one of the most important factors for the photoinduced degradation of DOM in waters (Vione et al. 2009; Gao and Zepp 1998; Wu et al. 2005; Gennings et al. 2001), through the photo-Fenton reaction (Zepp et al. 1992; Southworth and Voelker 2003; McKnight et al. 1988; Arakaki and Faust 1998); or via $HO^{\bullet}$-independent processes (Vione et al. 2009). The generation rate of $HO^{\bullet}$ is much higher for elevated Fe levels in acidic waters (McKnight et al. 1988; Allen et al. 1996). The oxidation of Fe^{2+} by photo-generated H_2O_2 causes the production of $HO^{\bullet}$ and Fe^{3+}, but Fe^{2+} is regenerated from Fe^{3+} by several pathways via the process of $h\upsilon/H_2O_2/O_2^{\bullet-}$. The regeneration of Fe^{2+} greatly enhances the production of $HO^•$ (Vione et al. 2004) and leads to high photoinduced degradation of DOM in iron-rich waters. Addition to the water of fluoride ion or deferoxamine mesylate (DFOM) can form unreactive Fe^{3+} complexes, inhibiting iron photoreduction and slowing down the photoinduced degradation of DOM (Gao and Zepp 1998; Wu et al. 2005). The photoinduced formation of DIC, CO and NH_4^+ has been greatly affected by the addition of fluoride ion to the water of the River Satilla (Gao and Zepp 1998). Thus, the photo-Fenton reaction plays an important role in natural waters, especially in acidic waters. The photoinduced degradation rate constant of humic acid is significantly decreased by the addition of fluoride, but that of fulvic acid is not affected (Wu et al. 2005). Dissolved Fe is thus thought to play an important role in the photoinduced degradation of humic acid rather than fulvic acid. Due to the higher aromaticity of humic acid as compared to fulvic acid (30–51 % of aromatic carbon vs. 14–20 %) (Malcolm 1985; Gron et al. 1996), humic acid is more susceptible to react with $HO^•$ which is generated from the photo-Fenton reaction ($Fe^{2+} + H_2O_2 \rightarrow HO^• + {}^-OH + Fe^{3+}$) (Zepp et al. 1992; Senesi 1990; Minakata et al. 2009; Chen and Pignatello 1997). Therefore, it is likely that humic acid is the DOM component that undergoes the fastest photoinduced degradation in natural waters.

3.4 *Occurrence and Quantity of NO_2^- and NO_3^- Ions*

Photoinduced degradation of DOM can be affected by the occurrence and concentration levels of NO_2^- and NO_3^- ions, both of which are efficient in the production of $HO^•$ upon photolysis in waters (see also chapter "Photoinduced Generation of Hydroxyl Radical in Natural Waters") (Zafiriou and True 1979a, b; Takeda et al. 2004; Vione et al. 2006; Mack and Bolton 1999; Nakatani et al. 2004; Chin et al. 2004). Contribution of $HO^•$ production in sewage polluted rivers is 48–80 % from NO_2^- and 2–19 % from NO_3^-, but the contribution is 6–26 % and 1–49 %, respectively, in upstreams and clean rivers (Takeda et al. 2004; Nakatani et al. 2004). In anthropogenically polluted Rhône River Delta (S. France) and Lake Piccolo (NW Italy) the contribution of $HO^•$ production is accounted for by NO_2^- (62–63 %) and NO_3^- (27–38 %), while in the unpolluted and remote Lake Goose and Lake Divide (Wyoming, USA) the contribution of nitrate and nitrite is relatively lower, 0–11 % and <0.5 %, respectively (Minero et al. 2007). In seawater NO_2^- is the major source of $HO^•$ in Seto Inland Sea (7–75 %) and Yellow Sea (10–44 %) compared to NO_3^- (<1 % and 0.4–8 %, respectively) (Takeda et al. 2004). The two anions (NO_2^- and NO_3^-) collectively are dominant sources in both river and seawater, while their role in lake water is less important (Vione et al. 2006). Natural waters that include high concentration levels of NO_2^- and NO_3^- could induce degradation of DOM by photoinduced production of $HO^•$. However, it has been found that the rate of mineralization of DOM in acidified lake water far exceeds that rate of $HO^•$ generation by all the sources, which suggests that $HO^•$-independent processes (tentatively, photolysis of Fe(III)-DOM complexes) may also play an important role in DOM mineralization (Vione et al. 2009).

3.5 Molecular Characteristics of the DOM

Photoinduced degradation of DOM is significantly dependent on the molecular characteristics or the absorbing nature of the organic substances in the chromophoric dissolved organic matter (CDOM) or fluorescent dissolved organic matter (FDOM) in waters (Table 1). Chromophores in CDOM and fluorophores in FDOM are considered to be equivalent components with respect to photosensitization by sunlight. Photoinduced degradation of DOM takes place upon absorption of photons, which is predominantly dependent on the chemical nature of the organic substances present in DOM. The radiation absorption by organic matter increases along the spectrum from visible toward UV regions (Amador et al. 1989; Kieber et al. 1990). High molecular weight (HMW) DOM such as humic substances (fulvic and humic acids), and fluorescent whitening agents (FWAs) or components of detergents (DAS1 and DSBP) can absorb both visible and shorter wavelength regions (Kramer et al. 1996; Kieber et al. 1990; Sadtler 1968; Strome and Miller 1978). Many low molecular weight organic acids photo-generated from large CDOM or FDOM can absorb only in UV-C range, but they do not absorb radiation in the UV-B, UV-A or visible range (Sadtler 1968). For example, acetaldehyde absorbs light at 208–224 nm (Kieber et al. 1990; Mopper et al. 1991; Strome and Miller 1978), acetate at 204–270 nm (Dahlén et al. 1996; Wetzel et al. 1995), formaldehyde at 207–250 nm (Mopper et al. 1991; Mopper and Stahovec 1986), glyoxal at <240 nm (Mopper et al. 1991; Mopper and Stahovec 1986), malonate at 225–240 nm (Dahlén et al. 1996), pyruvate at 200–227 nm (Wetzel et al. 1995; Kieber et al. 1990; Mopper et al. 1991), and propanal at ~230 nm (Mopper and Stahovec 1986).

Fulvic acid absorbs radiation in both the visible and UV ranges (Fig. 1a) (Mostofa et al. 2005). The DOC concentrations varied between upstream (99 μM C, Kago) and downstream rivers (194 μM C, Yasu), but the absorption in the visible region is likely the same in both upstream (Fig. 1a) and downstream waters (Fig. 1b). Such an absorption is usually caused by HMW DOM. Humic acid is degraded more quickly than fulvic acid in water (Wu et al. 2005), probably because of the higher aromaticity (Malcolm 1985; Gron et al. 1996). The absorption of visible light by chromophores or fluorophores in HMW DOM causes decomposition, which is usually more marked in upstream (Fig. 1a) than in downstream waters (Fig. 1b). The following order for DOM photoinduced degradation can be proposed: upstream DOM > downstream rivers > lake > seawaters (Table 1; Fig. 1). Therefore, photoinduced degradation is greatly dependent on the molecular nature of DOM compositions in waters. Interestingly, the residence time of water in lakes and sea is much higher compared to the rivers. Possibly the lower photoinduced lability of the DOM found in lake or sea water is due to the fact that labile DOM in these environment has sufficient time to undergo photoinduced degradation.

3.6 pH and Alkalinity of Waters

Both pH and alkalinity, which can greatly vary among different waters, can influence the photoreaction rates of DOM, its chemical structure and speciation

(Vahatalo et al. 2000; Wu et al. 2005; Gennings et al. 2001; Molot et al. 2005). The photo-Fenton reaction is greatly influenced by pH. A decrease in pH greatly accelerated the photoinduced degradation of DOM in softwater stream (Molot et al. 2005), in Satilla river (Gao and Zepp 1998) as well as in lake water from NW Italy (Vione et al. 2009). The experimental study shows that after 69 hours of artificial irradiation without addition of KI, DOC loss is decreased as pH increases from pH 4 to 9 whilst addition of KI is significantly reduced loss of DOC at pH 4, 5 and 7 but not at pH 9 with the fraction of DOC lost by non-$HO^\bullet$ mechanisms gradually increasing from 58 % to 75 % between pH 4 and 7, and 100 % at pH 9 (Molot et al. 2005). Photoinduced degradation rates of DOC and fluorescence are greatly increased with a decrease in sample pH from 8 to 6 and then to 4 (Wu et al. 2005). Conversely, the production rates of $HO^\bullet$ in the Fenton or photo-Fenton reaction are greatly enhanced with a decrease in pH of natural waters (Zepp et al. 1992; Vione et al. 2009; Goldstone et al. 2002; Moffett and Zika 1987; Millero and Sotolongo 1989).

The apparent mechanism for enhanced photoinduced loss of DOC at low pH is oxidation to dissolved inorganic carbon (DIC) by reaction with $HO^\bullet$ produced via the iron-mediated photo-Fenton pathway (Zepp et al. 1992; Voelker et al. 1997). Therefore, high production rate of $HO^\bullet$ at low pH can accelerate the photoinduced degradation of DOM in waters. However, there is evidence that the production rate of $HO^\bullet$ in acidified lake water is unable to account for the rate of DOM mineralization, which suggests that additional mineralization processes would also be operational (Vione et al. 2009). The major terrestrial alkalinity-producing processes such as ionic exchange, weathering and biological assimilation of nitrate and other anions, mostly depend on the watershed geology, morphology, soil characteristics, and hydrological conditions (Psenner 1988). Watersheds of lakes exported more SO_4^{2-}, NO_3^- and H^+ than they received, and the lakes are the dominant acidity-consuming parts of the whole ecosystem, neutralizing 50–58 % of the H^+ input (Kopacek et al. 2003). Terrestrial fluxes of organic acid anions can also consume H^+ in natural lakes and are thought to be the third major internal alkalinity-producing mechanism after the biochemical reductions of NO_3^- and SO_4^{2-} (Kopacek et al. 2003; Cook et al. 1986; Schindler et al. 1986). An increase in alkalinity in waters can decrease the production of H_2O_2 by slowing the reaction of $O_2^{\bullet-}$ protonation ($2O_2^{\bullet-} + 2H^+ \rightarrow H_2O_2 + O_2$). A decrease in H_2O_2 production can reduce the photoinduced generation of $HO^\bullet$ through photo-Fenton reaction or direct photolysis, thereby decreasing the photoinduced degradation of DOM in natural waters.

3.7 Dissolved Oxygen

Dissolved oxygen (O_2) can enhance the photoinduced degradation of DOM in waters (Vahatalo et al. 2000; Amon and Benner 1996; Obernosterer et al. 2001; Laane et al. 1985; Lindell and Rai 1994; Reitner et al. 1997). Addition of O_2 to photoinduced reaction systems can greatly promote the photoinduced degradation rates (Gao and Zepp 1998; Stumm and Lee 1961; Miles and Brezonik 1981). Stumm and Lee 1961; Miles and Brezonik 1981; In iron-rich waters, the ferrous iron is often oxidized by dissolved O_2 with production of ferric oxide floc (Stumm and Lee 1961). O_2 is consumed at a rate of 0.02–0.09 mg L^{-1} h^{-1} in humic colored waters having pH 3–4 and total iron concentration of 0.1–2.0 mg L^{-1} under irradiation. The consumption rate is slightly lower (0.01–0.04 mg L^{-1} h^{-1}) under dark conditions (Miles and Brezonik 1981). Some standard organic compounds can consume O_2 at rates of 0.01–0.83 mg L^{-1} h^{-1} under irradiation and 0.01–0.70 mg $L^{-}1$ h^{-1} in the dark. These results have been obtained for a concentration of 100 mg L^{-1} of organic compounds in the presence of 6 mg L^{-1} of Fe(III) (Miles and Brezonik 1981). In photoexperiments conducted on Amazon river water samples, the O_2 consumption rate was 3.68 μM O_2 h^{-1} under irradiation and it was twelve times lower (0.30 μM O_2 h^{-1}) in the dark (Amon and Benner 1996). High rates of DOC loss and O_2 consumption are often observed in riverine DOM, with little or no additional production of biologically labile organic compounds. The photoinduced O_2 demand of surface water DOM in the Atlantic Ocean varied from 0.1 to 2.8 μmol O_2 L^{-1} d^{-1} in 12 h irradiation periods (Obernosterer et al. 2001). Rivers usually exhibit a higher O_2 consumption rate than seawaters. The O_2 consumption in waters is hypothesized to contribute to the generation of H_2O_2 through production of superoxide radical ion ($O_2^{\bullet-}$) as intermediate, upon monoelectronic reduction of O_2 by aquated electrons (e^-) produced by DOM (see chapter “Photoinduced and Microbial Generation of Hydrogen Peroxide and Organic Peroxides in Natural Waters”). Photolytically produced H_2O_2 can participate to the production of $HO^\bullet$, by either the photo-Fenton reaction or the direct photolysis, and such processes can contribute to the photoinduced degradation of DOM in waters.

3.8 Depth of the Water Column

The Photoinduced degradation of DOM is significantly dependent on the depth of the water column. Degradation is higher in the upper surface layer and gradually decreases with an increase in the water column depth (Ma and Green 2004; Vahatalo et al. 2000). Solar radiation can mineralize 19 mmol C m^{-3} d^{-1} at a depth of 1 cm, and the rate of mineralization decreases with increasing depth with an attenuation coefficient of 23 m^{-1} (Vahatalo et al. 2000). Most of the photo mineralization takes place in the top 10 cm in lakes (Vahatalo et al. 2000). The presence of low quantity of suspended solids or particulate matter allows for a deeper penetration of light in the water column, which can result into a greater potential for the photoinduced degradation of deeper DOM. Both river and lake DOM have a high potential to undergo photoinduced degradation in the surface layers (0 m), and photoinduced degradation gradually decreases in the deeper layers (6.5 and 24 m), as has been found in an in situ experimental study (Ma and Green 2004). Surface waters with a high level of DOC greatly inhibit the penetration of solar radiation into the deeper layers, whilst penetration of radiation at longer depth of the water column is usually observed in water bodies with low DOC (Morris et al. 1995). UV-B radiation penetrates at a depth of 0.1–5 m, while UV-A penetrates at 0.2–15 m (Farmer et al. 1993). Therefore, UV-A plays a more important role into the photoinduced degradation of DOM in deeper layers compared to UV-B (Piazena and Häder 1994; Blough et al. 1993). It can be concluded that the photoinduced degradation of DOM at any depth of the water column in freshwater systems and in oceans is greatly dependent on the penetration of light intensity.

3.9 Physical Mixing in the Surface Mixing Zone

Physical or turbulent mixing in the surface mixing zone of the water column might be an important factor to enhance the photoinduced degradation of DOM in waters. The mixing process allows the reactants of a chemical reaction to come more frequently in contact, thus accelerating the reaction rate. It has been shown that the production rates of H_2O_2 are higher for both Suwannee River Fulvic Acid (445 nM h^{-1}) and seawater (86 nM h^{-1}) in stirred samples compared to unstirred ones (211 and 51 nM h^{-1}, respectively) (Mostofa K, Sakugawa H unpublished data). These photoexperiments were conducted using a solar simulator . Simultaneously, the fluorescence intensity of fulvic acid is decreased to a higher extent in stirred samples compared to non-stirred ones. Therefore, physical mixing is an essential factor to increase the reaction rate or promote the photoinduced processes in natural surface waters. Moreover, production of H_2O_2 is merely observed in the surface mixing zone, where H_2O_2 is derived from the photoinduced degradation of DOM (Moore et al. 1993; Sikorski and Zika 1993a, b; Sakugawa et al. 2000; Johnson et al. 1989). Furthermore, the fluorescence intensity of fulvic acids is much lower in the surface mixing zone in lake or seawaters due to solar effects (Mostofa et al. 2005; Hayase and Shinozuka 1995). Therefore, physical mixing in the surface mixing zone is an important factor for promoting the photoinduced degradation of DOM in waters. Mixing processes are typically dependent on physical factors such as strong or weak wind, presence of artificial or natural dams, power-dam outfalls, stream riffles, waterfall, and finally water temperature which affects the stratification-stagnant regime of natural waters.

3.10 Increasing UV Radiation During Ozone Hole Events

The ozone hole due to stratospheric ozone depletion because of anthropogenic activities is a well-known phenomenon in the Antarctic (Qian et al. 2001; Jones and Shanklin 1995) and Arctic oceanic regions (Rex et al. 1997; Randall et al. 2005). Moreover, the incident UV-B radiation is increased at a rate of 10–20 % per decade at temperate latitudes (Kerr and McElroy 1993; Madronich 1992). An increase in UV-B radiation may greatly enhance the production of $HO^\bullet$ by inducing higher rates of photolysis of NO_2^- and NO_3^-, and of other redox reactions in natural waters (Qian et al. 2001; Randall et al. 2005). The $HO^•$ formation rates from nitrate, as well as DOM plus nitrite are significantly increased during ozone hole conditions, compared to non-ozone hole periods (Qian et al. 2001). Therefore, a higher production of $HO^•$ during ozone hole events can enhance the photoinduced degradation of DOM. UV transparency of the lake water column is also greatly enhanced during the summer season due to photoinduced degradation of DOM in the lake epilimnion (Morris and Hargreaves 1997). Diffuse attenuation coefficients are greatly varied (39–81 %) seasonally at the epilimnion, and minimum values occur during the summer season (Morris and Hargreaves 1997). Thus, an increase in incident UV radiation (280–400 nm) in response to stratospheric ozone depletion can increase the transformation of surface DOM and, by increasing the UV transparency of water, can also induce additional degradation of DOM in the deeper layers (Qian et al. 2001; Randall et al. 2005).

3.11 Global Warming

Global warming may expand the summer season (Huisman et al. 2006; Sarmiento et al. 2004; Schmittner 2005), which might accelerate the photoinduced degradation of DOM. For example, it might lead high production of $HO^•$ because of the increase in water temperature due to global warming. At the same time, there can be an increase in UV radiation during ozone hole events (Qian et al. 2001; Rex et al. 1997). Global warming may also affect (and possibly enhance) the water column transparency, which is modified on a variety of time scales, and the depth of the mixing layer, as well as lead to changes in climatologic factors such as cloud cover, particulate material and total content of column ozone. These factors may influence the incident UV radiation (Morris and Hargreaves 1997; Morris et al. 1995; Scully and Lean 1994). Global warming may affect the seasonal patterns of chlorophyll and nutrient concentrations in the deep chlorophyll maxima (DCM) in waters (Huisman et al. 2006; Mostofa et al. 2009b; Letelier et al. 2004). The combination of global warming and photoinduced degradation may significantly impact on primary production, species composition, global carbon cycle, biological activities, and finally the seasonal modifications of the water column in natural waters (Huisman et al. 2006; Häder et al. 2011). The effects of global warming on photoinduced degradation of DOM are extensively discussed in the global warming chapter (see chapter "Impacts of Global Warming on Biogeochemical Cycles in Natural Waters").

3.12 Salinity

DOM photoreactivity is significantly increased with salinity or addition of salts in natural waters (Osburn et al. 2009; Hernes and Benner 2003; Osburn and Morris 2003; Anastasio and Newberg 2007; Grebel et al. 2009). Controlled laboratory studies demonstrate that the presence of seawater concentrations of chloride and bromide ions can enhance absorbance photobleaching reaction rates by ~40 %, regardless of DOM source or the presence or absence of carbonate ions (Grebel et al. 2009). In another study, a decrease in CDOM photobleaching at 280 nm is detected when humic CDOM is added to an artificial salinity gradient used to mimic coastal mixing (Minor et al. 2006). Dissolved lignin phenols are significantly affected by salinity and at salinities >25 psu, photooxidation is a dominant factor influencing lignin compositions and concentrations (Hernes and Benner 2003).

The mechanism behind the high photoinduced degradation of DOM with salinity apparently involves two factors: first, irradiated CDOM can induce photoinduced production of hydrogen peroxide (H_2O_2) that is a $HO^{\bullet}$ source via photolysis or the Fenton reaction, and the photoinduced generation of H_2O_2 is enhanced by salinity. Trace metal ions (M) in salinity or sea waters can complex with DOM (M-DOM) forming a strong ϖ–electron bonding system between metal ions and the functional groups in DOM (see chapter "Complexation of Dissolved Organic Matter With Trace Metal Ions in Natural Waters" for in details explanation). This ϖ–electron in M-DOM complex is rapidly excited photolytically, which is responsible for high production of aqueous electrons (e_{aq}^-) and subsequently the high production of superoxide ion ($O_2^{\bullet-}$), H_2O_2 and $HO^{\bullet}$, respectively. Indeed, photogeneration of H_2O_2 from ultrafiltered river DOM is substantially increased with salinity, from 15 to 368 nM h^{-1} at circumneutral pH (Osburn et al. 2009). Salinity or NaCl salts can substantially increase the aqueous electrons (e_{aq}^-) from DOM components photolytically in aqueous media (Assel et al. 1998; Gopinathan et al. 1972). This effect subsequently can enhance the H_2O_2 production from DOM components in waters (Moore et al. 1993; Mostofa and Sakugawa 2009; Richard et al. 2007; Fujiwara et al. 1993). Recent studies observe that the sea-salt particulate matter extracted from coastal seawaters show substantially high $HO^{\bullet}$ production (rate: ~2778–27778 M s^{-1}), approximately 3–4 orders of magnitude greater than $HO^{\bullet}$ photoformation rates in surface seawater (Anastasio and Newberg 2007), which may support the above phenomena.

Second, the reaction of $HO^{\bullet}$ with halide ions (X^-) can form reactive halogen radicals ($BrX^{\bullet-}$) that can react with electron-rich functional groups within DOM more selectively than $HO^{\bullet}$ (Goldstone et al. 2002; Grebel et al. 2009; Salinity can significantly affect the CDOM or FDOM properties, which are responsible for their high photoinduced behavior, which are discussed in detail in other chapters (see chapters "Colored and Chromophoric Dissolved Organic Matter in Natural Waters", "Fluorescent Dissolved Organic Matter in Natural Waters" and "Complexation of Dissolved Organic Matter With Trace Metal Ions in Natural Waters").

4 Factors Controlling the Microbial Degradation of DOM in Waters

Microorganisms are generally responsible for catalyzing the oxidation of organic matter and for inducing changes in the functional groups of DOM, either in deeper waters or in soil and sediment pore waters (Mostofa et al. 2007; Moran et al.

2000; Lønborg and Søndergaard 2009; Lovley and Chapelle 1995; Coble 2007; Hopkinson et al. 2002; Koschorreck et al. 2008; Lønborg et al. 2009a, b). An increase in fulvic acid-like (or humic-like) fluorescence either in deeper waters of lakes and oceans or in dark incubated water samples are considered to be the effect of microbial degradation of organic matter and of the related transformation of the functional groups of DOM (Ma and Green 2004; Mostofa et al. 2007a; b; Moran et al. 2000; Hayase and Shinozuka 1995; Coble 2007; Coble 1996). Microbial degradation of DOM depends on several key factors that can be distinguished as: (1) Occurrence and nature of microbes in waters; (2) Sources of DOM and the quantity of its fermentation products; (3) Temperature; (4) pH; and (5) Sediment depths.

4.1 Occurrence and Nature of Microorganisms

Microbial degradation of organic matter and of functional groups of macromolecules depends on the occurrence and nature of microorganisms in waters (Lovley 2006; Uchida et al. 1998; Lovley et al. 1996; Kotsyurbenko et al. 2001; Coleman et al. 1993; Conrad et al. 1989; Conrad et al. 1989; Morvan et al. 1994). Recent studies demonstrated the presence of methanogens belonging to the *Methanomicrobiaceae*, *Methanobacteriaceae*, *Methanococcaceae*, *Methanosarcinaceae*, and *Methanosaetaceae*, as well as new archaeal lineages within the *Euryarchaeota* (Kotsyurbenko et al. 2007; Basiliko et al. 2003; Cadillo-Quiroz et al. 2006; Casper et al. 2003; Galand et al. 2002; Horn et al. 2003; Sizova et al. 2003; Upton et al. 2000; Utsumi et al. 2003). It is shown that methanogenic archaea and homoacetogenic bacteria are the main H_2 consumers in the absence of inorganic electron acceptors such as nitrate, ferric iron and sulfate, which compete for available H_2 in anoxic environments (Kotsyurbenko et al. 2001). Degradation of alcohols and fatty acids is usually enabled by syntrophy between H_2-producing syntrophic bacteria and H_2-consuming methanogenic archaea (Conrad 1999; Schink 1997). The most important reactions for hydrogenotrophic (Eq. 4.1) and acetotrophic methanogenesis (Eq. 4.1) for degradation of glucose can be expressed below (Eqs. 4.1, 4.2) (Conrad 1999; Thauer et al. 1977):

$$C_6H_{12}O_6 \rightarrow CH_3COOH + CO_2 + CH_4 \Delta G^{o\prime} = -346.8 \quad (4.1)$$

$$C_6H_{12}O_6 \rightarrow 3CO_2 + 3CH_4 \Delta G^{o\prime} = -418.1 \quad (4.2)$$

An experimental study shows that *D. desulfuricans* can reduce the Fe(III) and sulphate simultaneously at rates comparable to Fe(III) and sulphate reduction under non-limiting H_2 concentration, when only one of the electron acceptors is provided (Coleman et al. 1993). On the other hand, H_2 is metabolized by *D. desulfuricans* at lower concentrations with Fe(III) than with sulphate (Coleman et al. 1993). Interestingly, these bacteria do not metabolize H_2 below ~10^{-5} atm partial pressure (Cord-Ruwisch et al. 1988).

4.2 Sources of DOM and its Fermentation Products

The methanogenesis depends on the sources of organic matter such as vascular plants or algal biomass, the fermentation or degradation products of which are greatly varied in soil, peatland or sediment pore waters. For example, the disintegration or fermentation of vascular plant materials by aerobic and anaerobic bacteria can produce humic substances (fulvic and humic acid), structural polysaccharides, polyphenols, proteins, amino acids, carbohydrates and inorganic components in soil environments (Mostofa et al. 2009a; Malcolm 1985; Chefetz 2002; Cadillo-Quiroz et al. 2010; Hur 2011; Peña-Méndez et al. 2005). In contrast, among the DOM components that originated from algal or phytoplankton biomasses, one can find autochthonous fulvic acid, protein, amino sugars and labile polysaccharides (Mostofa et al. 2009a; Zhang et al. 2009; Li W et al., unpublished data; Parlanti et al. 2000; Benner and Kaiser 2003). The changes of the DOM by microbial processes significantly depend on its sources and composition and/or the mixing ratios of the individual original source materials in natural water (Hur 2011).

It is shown that fulvic and humic acids are composed of diverse functional groups such as –COOH, carboxyl, methoxyl, alcoholic OH, carbohydrate OH and phenolic OH. Low aromaticity is observed in fulvic acid (17 % of aromatic C and 63 % of aliphatic C) compared to humic acid (30 % and 47 %, respectively) (Malcolm 1985; Steelink 2002). Carbon distribution by solid-state CPMAS ^{13}C NMR shows about 24 % of C–O, 3 % of anomeric C, 12 % of C=C, 5 % of φ–O (φ = other elements except C), 16 % of COOH, 4 % of C=O. Elemental analysis showed 38 % of O, 0.87 % of N, 0.74 % of S and 0.62 % of P (Malcolm 1985). Although the chemical structure of autochthonous fulvic acid of algal origin is still unclear, the material is likely to be a macromolecule because of the similarity of its EEM spectra to standard Suwannee River Fulvic Acid (Fig. 1d,e, and f) (Mostofa et al. 2009b). This might be the reason of the effective degradation of the autochthonous fulvic acid of algal origin, which is observed in the EEM images in the lake sediment pore waters Li W et al. (unpublished data). Therefore, it is not surprising that functional groups bound to either fulvic and humic acids of terrestrial origin or autochthonous fulvic acid of algal origin are affected by microbial processes in the sediment waters.

Depending on the presence of either terrestrial plant material or algal or phytoplankton biomass, different fermentation products can be found in a variety of sediment waters. In a similar way, the contribution of H_2 to CH_4 production in different methanogenic sediments is quite variable: 32–46 % in Kichier Lake, 36–46 % in Lake Mendota, 15–39 % in Lake Washington, 17–31 % in anoxic paddy soil, 8 % in Colne Pt. Salt marsh, 4 % in Knaack Lake, 0 % in Lake Constance, 97 % in Kuznechika lake, 74–86 % in Octopus Spring mat, 76–82 % in Blelham Tarn, 71–80 % in Cape Lookout Bight, 100 % in Kings Lake Bog, 95–97 % in Bunger Hills (Antarctica), and 99–100 % in Lake Baikal deep sediments (Schulz and Conrad 1996; Ivanov et al. 1976; Winfrey and Zeikus 1979; Sandbeck and Ward 1981; Jones et al. 1982; Banat et al. 1983; Crill and Martens 1983; Phelps and Zeikus 1984; Kuivila et al. 1989; Lansdown et al. 1992; Rothfuss and Conrad 1992; Galchenko 1994; Namsaraev et al. 1995). It is shown that different types of plant material lead to different rates of acetate formation. There is also a stronger substrate-based coupling of root surface and methanogens in oligotrophic (bog) than in minerotrophic (fen) sites (Cadillo-Quiroz et al. 2010; Ström et al. 2003; Öquist and Svensson 2002). Seasonal algal or phytoplankton blooms might be responsible for formation of acetate and CH_4 in the sediments of deep lakes (Schulz and Conrad 1995). The acetate concentration profiles show maxima (~100 μM in 2 or 4 cm depth) in summer and minima (~5 μM over the entire depth) in winter, when the respective CH_4 concentrations are ~750 μM in summer and ~120 μM in winter (Schulz and Conrad 1995).

It is evidenced that gas bubbles contain about 60–70 % CH_4 with an average $\delta^{13}C$ of −56.2 % and δD of −354 %, and 2 % CO_2 with an average $\delta^{13}C$ of −14.1 % (Thebrath et al. 1993). These data indicate that CH_4 is produced from methyl carbon, *i.e.* mainly using acetate as fermentative substrate (Thebrath et al. 1993). In anoxic paddy soil, interspecies H_2 transfer within methanogenic bacterial associations (MBA) account for 95–97 % of the conversion of $^{14}CO_2$ to $^{14}CH_4$, and only 3–5 % of the $^{14}CH_4$ is produced from the turnover of dissolved H_2 (Conrad et al. 1989a, b). An experimental study demonstrates that the ratio of Fe(II) production to CO_2 production (3.9) is similar to that expected (4.0) for organic carbon oxidation coupled to Fe(III) oxide reduction (Fig. 8) (Roden and Wetzel 1996). The study also shows that the rates of CH_4 production are low during the Fe(III) reduction in oxidized sediments, but increase when the Fe(III) oxides are depleted to background levels (Fig. 8a). The rates of CO_2 and CH_4 production are about

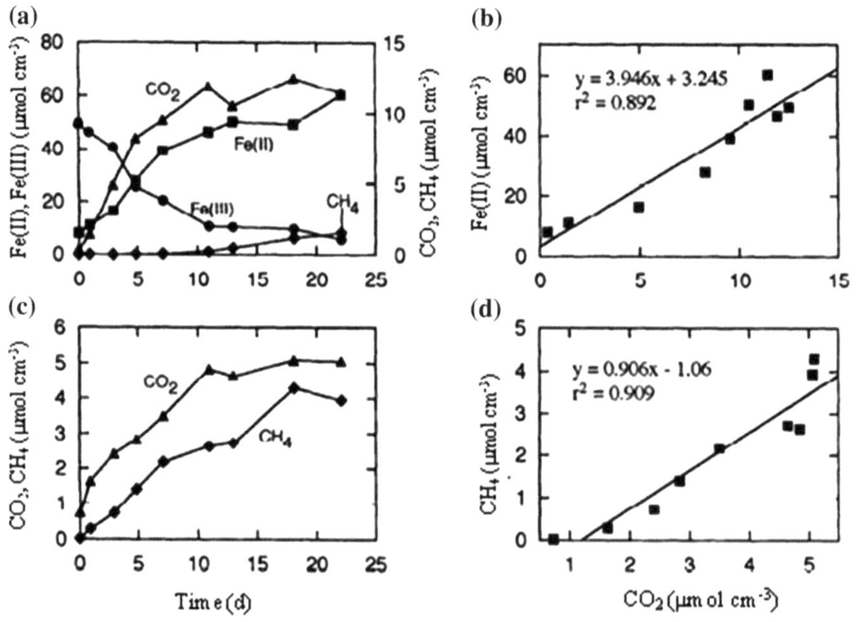

Fig. 8 Fe(III) reduction, CO_2 production, and CH_4 production in oxidized (**a**, **b**) and reduced (**c**, **d**) Talladega wetland sediment slurries. *Data source* Roden and Wetzel (1996)

equal during the incubation of reduced sediments (Fig. 8c,d) (Roden and Wetzel 1996). Therefore, the occurrence and the nature of organic matter and its fermentation or disintegration products are key factors for the production of CO_2, CH_4 and other end products in the aquatic environments.

4.3 Temperature

The growth of homoacetogenic bacteria and methanogenic archaea significantly depends on the ambient temperature (Table 2) (Kotsyurbenko et al. 2001; Kotsyurbenko et al. 2007; Thebrath et al. 1993; Westermann 1994; Kotsyurbenko et al. 1995; Simankova et al. 2000; Zinder 1990). It is shown that the microbial function is typically much higher at low temperature (5.0–7.0 °C), showing maximum bacterial abundance (3.9–7.9 $\times$ 10^5 cells ml^{-1}, mean = 6.4) and biomass (4.0–6.7 μg C L^{-1}, mean = 5.2). Lower values (1.3–2.5 $\times$ 10^5 cells ml^{-1}, mean = 1.8; and 1.3–2.4 μg C L^{-1}, mean = 1.7, respectively) have been found at higher temperature (7.5–11.1 °C) in open water in Lake La Caldera (Carrillo et al. 2002). Homoacetogenic bacteria and methanogenic archaea can consume H_2 over a temperature range of 1–35 °C, but their optimum temperature is often high (~20–30 °C, Table 2). Homoacetogenic *A. bakii, A. tundrae* and the methanogenic strain MSB have shown the largest temperature range for optimal H_2 consumption, which is extending at least from 4 to 30 °C (Kotsyurbenko et al. 2001). However, *A. fimetarium*, *A. paludosum* and strain MSP become less efficient toward H_2 consumption when the temperature decreases below 10 °C (Kotsyurbenko et al. 2001). Low temperatures are often favorable for acetogenesis, which becomes a quantitatively important process in anaerobic environments (Nozhevnikova et al. 1994; Kotsyurbenko et al. 1993).

At low temperature, homoacetogenic bacteria outcompetes methanogens for H_2 in laboratory experiments (Conrad et al. 1989; Kotsyurbenko et al. 1993). According to kinetic estimations, homoacetogens have a much higher growth rate at low temperature than methanogens (Kotsyurbenko et al. 1996). It is also shown that the contribution of methanogenic bacterial associations (MBA) to H_2-dependent methanogenesis is enhanced (it reaches 99 %) when the temperature is shifted from 30 to 17 °C, or when the soil is planted with rice (Conrad et al. 1989a, b). This enhancement is partially due to an increased utilization of dissolved H_2 by chloroform-insensitive non-methanogenic bacteria, most probably homoacetogens, so that CH_4 production is almost completely restricted to H_2-syntrophic MBA. Acetate is the precursor of approximately two-thirds of the methane produced in mesophilic (30–40 °C) and thermophilic (45–65 °C) anaerobic bioreactors (Zinder 1990). Increasing the incubation temperature of two swamp slurries from 2 to 37 °C resulted in a 8- to 18-fold increase in the H_2 partial pressure (Westermann 1994). The study also shows that the concentration of volatile fatty acids remained fairly constant except for butyrate, which decreased with increasing temperature.

Despite the constant low temperature (4 °C) during the summer and winter seasons in lake sediment pore waters, high variations in methane production in summer compared to winter are suggested to be caused by algal biomass blooms in surface waters (Schulz and Conrad 1995). In this case, the effect of temperature would not be significant in the microbial formation of methane from acetate in the sediment pore waters. The effect of temperature on methanogenesis mostly depends on the nature of organic sediments, presence of microorganisms and the fermentation or degradation products in water.

4.4 pH

The pH is an important factor that influences the rate of methanogenesis as well as the CH_4 production pathway and the methanogenic archaeal community in sediment waters. The methanogenesis is typically inhibited at low pH, where microbial turnover rates are slower, although significant methane production is still observed in acidic peat lands (Kotsyurbenko et al. 2007; Horn et al. 2003; Dunfield et al. 1993; Hornibrook et al. 2000; Goodwin and Zeikus 1987; Bräuer et al. 2004, 2006). Acetate as a major carbon source for methanogenesis may be unavailable to the methanogens at low pH because of the inhibitory effect of non-dissociated acetate toward methanogenesis (Fukuzaki et al. 1990). Low pH conditions may also reduce the microbial processes of H_2 production and consumption in anaerobic environments (Goodwin et al. 1988). An isolate from a landfill is able to grow at pH 5 (Lapado and Barlaz 1997) and an isolate from a peat land grows at pH 5.3 but generates some methane down to pH 3.1 (Williams and Crawford 1985). Acidotolerant hydrogenotrophic methanogenic consortia have been enriched from a peat bog at pH 4 (Sizova et al. 2003), and molecular analysis of an acidic peat bog reveals the presence of *Methanomicrobiaceae* and *Methanosarcinaceae* at pH 4.5 (Kotsyurbenko et al. 2004).

In acidic mining lakes, sulfate reduction often occurs when the pH in the sediment is almost neutral (Meier et al. 2004). An increase with depth of pH from 2.6 up to 6 enhanced the production of CH_4 and CO_2 in the sediment cores of Lake Caviahue (Koschorreck et al. 2008). In the most acidic surface layer of the sediment (pH < 4), methanogenesis is inhibited as suggested by a linear CH_4 concentration profile. In contrast, methanogenesis is highly active below 40 cm depth at high pH (>4). The carbon isotope composition of CH_4 is between –65 and –70 ‰, which is indicative of the biological origin of methane in Lake Caviahue. Therefore, it is suggested that the high biomass content of the sediment may induce high rates of sulfate reduction, which presumably raises the pH and creates favorable conditions for methanogens in deeper sediment layers (Koschorreck et al. 2008). On the other hand, the ratio of $\delta^{13}CO_2$ to $\delta^{13}CH_4$ increases from 1.053 at pH 6 up to 1.072 at pH 3.8, indicating a relative increase of hydrogenotrophic methanogenesis at low pH values (Hornibrook et al. 2000; Whiticar 1999). The genus *Methanobacterium* contains two alkaliphilic and one moderate acidophilic species, and collectively they have the widest growth ranges over a pH variation from 3.8 to 9.9 (Kotelnikova et al. 1998; Worakit et al. 1986; Patel et al. 1990). The pH is thus one of the important factors that can control the methanogenesis in sediment.

4.5 Sediment Depths

It is shown that the degradation of organic matter is mostly occurring in the upper sediment layer (1–10 cm depths) in lakes or in soil (Nakane et al. 1997; Li et al. unpublished data; Roden and Wetzel 1996; Schulz and Conrad 1995). However, the methanogenesis may also occur in deeper sediment layers under favorable conditions, either in peatlands or in sediment pore waters (Koschorreck et al. 2008; Nakagawa et al. 2002; Galand et al. 2005). In the sediments of Lake Caviahue the CH_4 concentration is steadily increased from 0 to 6.0 mM from 1 to 30 cm sediment depths (Fig. 9) (Koschorreck et al. 2008). The concentration of CH_4 typically reaches saturation (~1 mM) at 3–40 cm below the top few centimeters in unvegetated sediments. In vegetated sediments CH_4 concentrations are very low (0.0–0.1 mM) until 20 cm, after which they increase at ~1 mM level at 40 cm (Roden and Wetzel 1996). In humic bog lakes the deeper parts of the water column favor microdiversification of methanogens, whilst the periodically disturbed water column of shallower dimictic lakes promotes genetically more diverse methanogen communities (Milferstedt et al. 2010). In peatlands, hydrogenotrophic methanogenesis is the predominant pathway of CH_4 formation, accounting for 50 to 100 % of total CH_4 production, particularly in the deeper layers (Nakagawa et al. 2002; Galand et al. 2005). Therefore, sediment depths play an important role in the production of methane in sediment waters.

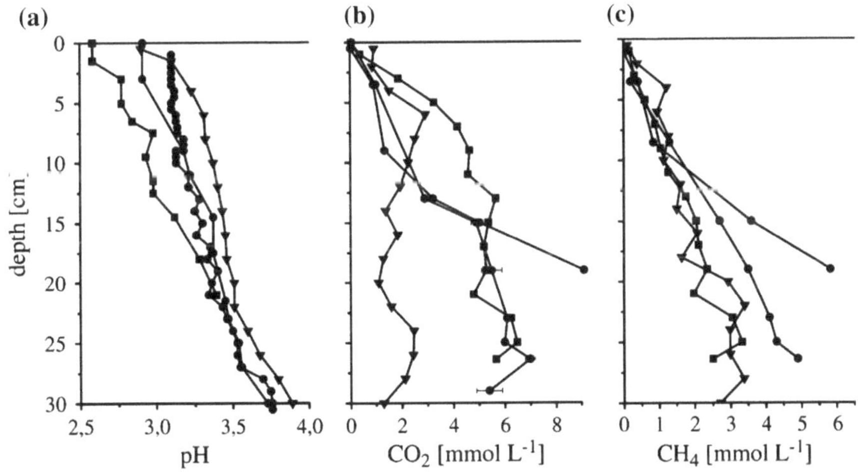

Fig. 9 Vertical profiles of pH (**a**), dissolved gases CO_2 (**b**) and CH_4 (**c**) in different sediment cores from Lake Caviahue. The pH is measured in KCl extracts during the field work in 2003 (■ 2001, ● 2003, ▼ 2004). *Data source* Koschorreck et al. (2008)

5 Photoproducts of DOM and Their Significance on Biogeochemical Cycles in Natural Waters

Photoinduced degradation of DOM in natural waters generally occurs by the sequential degradation of high molecular weight substances, producing low molecular weight compounds and ending up in mineralization yielding CO_2, CO, DIC, COS, and so on (Table 3) (Reche et al. 1999; Ma and Green 2004; Graneli et al. 1996, 1998; Xie et al. 2004; Mopper et al. 1991; Miller and Zepp 1995; Bertilsson and Tranvik 2000; Fujiwara et al. 1995; Bushaw et al. 1996; Miller and Moran 1997; Stiller and Nissenbaum 1999; White et al. 2010; Valentine and Zepp 1993; Mostofa K et al., unpublished data; Francko and Heath 1982; Fang 2004; Chen et al. 2001; Karl and Tien 1997; Jones 1991; Jones and Amador 1993). Photoinduced degradation on DOM can typically lead to a variety of photo products, which can be distinguished in: (1) Hydrogen peroxide and organic peroxides; (2) Low molecular weight organic substances; (3) Aromatic mono- and dibasic acids; (4) Microbiologically labile organic photoproducts; (5) Carbon-gas end photoproducts; (6) Nitrogenous compounds (e.g. NH_4^+); (7) Phosphate; and (8) Release of energy to the water ecosystem.

5.1 *Photoinduced Formation of Hydrogen Peroxide and Organic Peroxides*

The formation of hydrogen peroxide (H_2O_2) and organic peroxides (ROOH) is the primary step of the photoinduced processes involving DOM in waters. The concentration levels of H_2O_2 are significantly different for a variety of waters, ranging from 4 to 3200 nM in rivers, 10–800 nM in lakes, and 0–1700 nM in seawaters as mentioned in chapter "Photoinduced and Microbial Generation of Hydrogen Peroxide and Organic Peroxides in Natural Waters". The concentration levels of ROOH compounds are also highly variable in natural waters, showing low concentrations in rivers (0–200 nM) and relatively higher levels in seawater (1–389 nM). H_2O_2 and ROOH may form free radicals ($HO^\bullet$, $RO^\bullet$, R = H or alkyl group), either directly upon photolysis or indirectly by photo-Fenton reactions. The reactive radicals thus generated contribute to the degradation of the organic substances that make up DOM.

5.2 *Photoinduced Formation of Low Molecular Size Organic Substances*

Photoinduced degradation can convert the high molecular weight DOM into low molecular size organic substances in natural waters (Table 3) (Moran and Zepp 1997; Biddanda and Benner 1997; Kramer et al. 1996; Allard et al. 1994; Yoshioka et al. 2007;

Table 3 Concentration levels and photoproduction rates of various end products generated from photochemical degradation of dissolved organic matter in natural waters and in photoexperiments conducted conducted on natural waters as well as on standard organic subsatnces

Samples/Sources of water	Source of light/Natural level	H_2O_2 (nM)	ROOHs (nM)	LMW compounds (nM h^{-1})	DIC (nM h^{-1})	CO_2(nM or nM h^{-1})	CO (nM or nM h^{-1})	Hydrocarbons (nM)	Ammonium (nM or nM h^{-1})	COS (ng L^{-1})	PO_4^{3-} (μg L^{-1})	DOC (μM C)	DON (μM)	References
Rivers														
Suwannee River, FL (29°N)	Xe lamp	–	–	–	80–630	–	4.1–5.1	–	–	–	–	4100–5150	–	Miller and Zepp (1995)
Suwannee River, FL (29°N)	Xe lamp	–	–	–	–	–	41–1500	–	–	–	–	2250–3917	–	Valentine and Zepp (1993)
Sturgeon River (47°N)	Sunlight (0 m)	–	–	–	2424	–	–	–	–	–	–	2798	–	Ma and Green (2004)
Sturgeon River (47°N)	Sunlight (6.5 m)	–	–	–	606	–	–	–	–	–	–	2873	–	Ma and Green (2004)
Sturgeon River (47°N)	Sunlight (24 m)	–	–	–	2727	–	–	–	–	–	–	2818	–	Ma and Green (2004)
Sturgeon River (47°N)	Dark	–	–	–	99970	–	–	–	–	–	–	2798	–	Ma and Green (2004)
Houghton Marsh (47°N)	Xe lamp	–	–	–	–	–	120–1330	–	–	–	–	2000–3083	–	Valentine and Zepp (1993)
Okefenokee Swamp (31°N)	Xe lamp	–	–	–	–	–	1200	–	–	–	–	3083	–	Valentine and Zepp (1993)
Jordan River (31°N), East African Rift Valley	Natural level	–	–	–	–	–	–	–	20588	–	8.8–85	–	–	Stiller and Nissenbaum (1999)
En Feshkha (31°N), East African Rift Valley	Natural level	–	–	–	–	–	–	–	<588	–	0–6	–	–	Stiller and Nissenbaum (1999)

(continued)

Table 3 (contniued)

Samples/Sources of water	Source of light/Natural level	H_2O_2 (nM)	ROOHs (nM)	LMW compounds (nM h^{-1})	DIC (nM h^{-1})	CO_2(nM or nM h^{-1})	CO (nM or nM h^{-1})	Hydrocarbons (nM)	Ammonium (nM or nM h^{-1})	COS (ng L^{-1})	PO_4^{3-} (µg L^{-1})	DOC (µM C)	DON (µM)	References
Nahal David (31°N), East African Rift Valley	Natural level	–	–	–	–	–	–	–	1176	–	<5	–	–	Stiller and Nissenbaum (1999)
Ein Noit (31°N), East African Rift Valley	Natural level	–	–	–	–	–	–	–	50588	–	<5	–	–	Stiller and Nissenbaum (1999)
Suwannee River, whole	Xe lamp	–	–	–	–	–	–	–	360	–	–	ND	74	Bushaw et al. (1996)
Boreal Pond, whole (July)	Xe lamp	–	–	–	–	–	–	–	80–150	–	–	3000	55	Bushaw et al. (1996)
Okefenokee swamp, whole	Xe lamp	–	–	–	–	–	–	–	40–340	–	–	3840	86	Bushaw et al. (1996)
Rivers	Natural level	4 –3200	0–200	–	–	–	–	–	–	–	–	–	–	Table 1 (Photoinduced and Microbial Generation of Hydrogen Peroxide and Organic Peroxides in Natural Waters)
Satilla River: Air-saturation	Xe lamp	–	–	–	–	22950	–	–	–	–	–	1570	–	Xie et al. (2004)

(continued)

Table 3 (contniued)

Samples/Sources of water	Source of light/Natural level	H_2O_2 (nM)	ROOHs (nM)	LMW compounds (nM h^{-1})	DIC (nM h^{-1})	CO_2(nM or nM h^{-1})	CO (nM or nM h^{-1})	Hydrocarbons (nM)	Ammonium (nM or nM h^{-1})	COS (ng L^{-1})	PO_4^{3-} (μg L^{-1})	DOC (μM C)	DON (μM)	References
Satilla River: O_2-saturation	Xe lamp	–	–	–	–	26450	–	–	–	–	–	1620	–	Xie et al. (2004)
Satilla River: air-saturation + DFOM**	Xe lamp	–	–	–	–	5950	–	–	–	–	–	1588	–	Xie et al. (2004)
Satilla River: O_2-saturation + DFOM	Xe lamp	–	–	–	–	7150	–	–	–	–	–	1545	–	Xie et al. (2004)
Satilla River: N_2-saturation	Xe lamp	–	–	–	–	1913	–	–	–	–	–	1574	–	Xie et al. (2004)
Altamaha River: air-saturation	Xe lamp	–	–	–	–	13150	–	–	–	–	–	1442	–	Xie et al. (2004)
Altamaha River: O_2-saturation	Xe lamp	–	–	–	–	18325	–	–	–	–	–	1430	–	Xie et al. (2004)
Altamaha River: air-saturation + DFOM	Xe lamp	–	–	–	–	5428	–	–	–	–	–	1441	–	Xie et al. (2004)
Altamaha River: O_2-saturation + DFOM	Xe lamp	–	–	–	–	5622	–	–	–	–	–	1406	–	Xie et al. (2004)
Lakes														
Lake Superior (47°N)	Sunlight (0 m)	–	–	–	10606	–	–	–	–	–	–	161	–	Ma and Green (2004)
Lake Superior (47°N)	Sunlight (6.5 m)	–	–	–	10909	–	–	–	–	–	–	160	–	Ma and Green (2004)

(continued)

Table 3 (contniued)

Samples/Sources of water	Source of light/Natural level	H_2O_2 (nM)	ROOHs (nM)	LMW compounds (nM h^{-1})	DIC (nM h^{-1})	CO_2(nM or nM h^{-1})	CO (nM or nM h^{-1})	Hydrocarbons (nM)	Ammonium (nM or nM h^{-1})	COS (ng L^{-1})	PO_4^{3-} (μg L^{-1})	DOC (μM C)	DON (μM)	References
Lake Superior (47°N)	Sunlight (24 m)	–	–	–	3182	–	–	–	–	–	–	200	–	Ma and Green (2004)
Lake Superior (47°N)	Dark	–	–	–	175485	–	–	–	–	–	–	``	–	Ma and Green (2004)
38 Lakes (55 –71°N)	Artificial	–	–	2500–44200	1300–158500	–	–	–	–	–	–	167 –1833	–	Bertilsson and Tranvik (2000)
Kinoshe Lake (51°N)	Xe lamp	–	–	–	–	–	260	–	–	–	–	1250	–	Valentine and Zepp (1993)
5 Lakes (57°N)	Natural sunlight	–	–	–	800–3400	–	–	–	–	–	–	325–1617	–	Graneli et al. (1996)
10 Lakes (57°N in Sweden and 3–22°S in Brazil)	Natural sunlight	–	–	–	25833–1612833	–	–	–	–	–	–	242–3483	–	Graneli et al. (1998)
24 Lakes, Wisconsin and Michigan, U.S.A	Natural level	–	–	–	–	–	–	–	–	–	8.5–129.3	308–1792	–	Reche et al. (1999)
Crazy Eddie Bog, northeastern Ohio	Natural level	–	–	–	–	–	–	–	–	–	2.0–24.0	–	–	Francko and Heath (1982)
Lake Biwa, Japan, 35°N: 0.5 m	Natural level	–	–	–	–	–	–	–	–	–	0.003–0.011	–	–	Mostofa K et al. (unpublished)
Lake Biwa, Japan, 35°N: 10 m	Natural level	–	–	–	–	–	–	–	–	–	0.005–0.009	–	–	Mostofa K et al. (unpublished)

(continued)

Table 3 (contniued)

Samples/Sources of water	Source of light/Natural level	H_2O_2 (nM)	ROOHs (nM)	LMW compounds (nM h^{-1})	DIC (nM h^{-1})	CO_2(nM or nM h^{-1})	CO (nM or nM h^{-1})	Hydrocarbons (nM)	Ammonium (nM or nM h^{-1})	COS (ng L^{-1})	PO_4^{3-} (μg L^{-1})	DOC (μM C)	DON (μM)	References
Lake Biwa, Japan, 35°N: 20 m	Natural level	–	–	–	–	–	–	–	–	–	0.004–0.009	–	–	Mostofa K et al. (unpublished)
Lake Biwa, Japan, 35°N: 40 m	Natural level	–	–	–	–	–	–	–	–	–	0.003–0.008	–	–	Mostofa K et al. (unpublished)
Lake Biwa, Japan, 35°N: 80 m	Natural level	–	–	–	–	–	–	–	–	–	0.005–0.023	–	–	Mostofa K et al. (unpublished)
Lakes	Natural level	10–800	–	–	–	–	–	–	–	–	–	–	–	Table 1 (Photoinduced and Microbial Generation of Hydrogen Peroxide and Organic Peroxides in Natural Waters)
Estuaries and sea waters														
Ohta River Estuary (34°N), Japan	Natural (July)	–	–	–	–	–	–	–	–	54.4	–	~80–290	–	Fujiwara et al. (1995)
Ohta River Estuary (34°N), Japan	Natural (December)	–	–	–	–	–	–	–	–	23.9	–	“	–	Fujiwara et al. (1995)
Delaware estuary	Xe lamp	–	–	–	–	83–625	–	–	–	–	–	–	–	White et al. (2010)

(continued)

Table 3 (contniued)

Samples/Sources of water	Source of light/Natural level	H_2O_2 (nM)	ROOHs (nM)	LMW compounds (nM h^{-1})	DIC (nM h^{-1})	CO_2(nM or nM h^{-1})	CO (nM or nM h^{-1})	Hydrocarbons (nM)	Ammonium (nM or nM h^{-1})	COS (ng L^{-1})	PO_4^{3-} (μg L^{-1})	DOC (μM C)	DON (μM)	References
Seto Inland Sea (34°N), Japan (surface: 0–5 m)	Natural	–	–	–	–	–	–	–	–	~5–17	–	"	–	Fujiwara et al. (1995)
Seto Inland Sea (34°N), Japan (deeper: 20 m)	Natural	–	–	–	–	–	–	–	–	~3–5	–	"	–	Fujiwara et al. (1995)
Coastal waters (31°N), East African Rift Valley	Natural level	–	–	–	–	–	–	–	82353–2047059	–	5 –63	–	–	Stiller and Nissenbaum (1999)
East China Sea, 24–30°N, China	Natural level	–	–	–	–	–	–	–	–	–	1.6–96.3	–	–	Fang (2004)
East China Sea, 24–30°N, China	Natural level (summer)	–	–	–	–	–	–	–	–	–	2.6–22.7	–	–	Chen et al. (2001)
East China Sea, 24–30°N, China	Natural level (winter)	–	–	–	–	–	–	–	–	–	3.5–23.4	–	–	Chen et al. (2001)
North Pacific Ocean, Station ALOHA (22°N)	Natural level (0–150 m)	–	–	–	–	–	–	–	–	–	~0.3–9.6	–	–	Karl and Tien (1997)

(continued)

Table 3 (contniued)

Samples/Sources of water	Source of light/Natural level	H_2O_2 (nM)	RCOHs (nM)	LMW compounds (nM h^{-1})	DIC (nM h^{-1})	CO_2(nM or nM h^{-1})	CO (nM or nM h^{-1})	Hydrocarbons (nM)	Ammonium (nM or nM h^{-1})	COS (ng L^{-1})	PO_4^{3-} (μg L^{-1})	DOC (μM C)	DON (μM)	References
Intracoastal Waterway, caostal waters (30°N)	Xe lamp	–	–	–	–	–	110	–	–	–	–	267	–	Valentine and Zepp (1993)
Live Oak, Gulf coast of Florida (30°N)	Xe lamp	–	–	–	–	–	90	–	–	–	–	642	–	Valentine and Zepp (1993)
Sargasso Sea (photic zone)	Natural	–	–	–	–	–	5.8	–	–	–	–	–	–	Jones (1991)
Sargasso Sea (aphotic zone)	Natural	–	–	–	–	–	0.1	–	–	–	–	–	–	Jones (1991)
Sapelo Island Marsh, GA (31°N)	Natural	–	–	–	40–1280	–	3.3–4.9	–	–	–	–	500	–	Miller and Zepp (1995)
Gulf of Mexico (28°N)	Natural	–	–	–	120–400	–	15.8–24.8	–	–	–	–	170	–	Miller and Zepp (1995)
Seawater + humic substances (SH)	Artificial sunlight	–	–	–	2300	–	147	–	–	–	–	695	–	Miller and Moran (1997)
Seawater + the control addition (SC)	Artificial sunlight	–	–	–	1600	–	104	–	–	–	–	515	–	Miller and Moran (1997)
Artificial seawater + the control addition (AC)	Artificial sunlight	–	–	–	300	–	40	–	–	–	–	241	–	Miller and Moran (1997)

(continued)

Table 3 (contniued)

Samples/Sources of water	Source of light/Natural level	H_2O_2 (nM)	ROOHs (nM)	LMW compounds (nM h^{-1})	DIC (nM h^{-1})	CO_2(nM or nM h^{-1})	CO (nM or nM h^{-1})	Hydrocarbons (nM)	Ammonium (nM or nM h^{-1})	COS (ng L^{-1})	PO_4^{3-} (µg L^{-1})	DOC (µM C)	DON (µM)	References
Sargasso Sea waters (1–20 m)	Sunlight	–	–	4.1	–	–	4.1	–	–	–	–	–	–	Mopper et al. (1991)
Sargasso Sea waters (20–150 m)	Sunlight	–	–	3.3	–	–	9.6	–	–	–	–	–	–	Mopper et al. (1991)
Sargasso Sea waters (500–4000 m)	Sunlight	–	–	10.6	–	–	16.1	–	–	–	–	–	–	Mopper et al. (1991)
Southeastern Caribbean Sea & Gulf of Paria	Natural level (Spring)	–	–	–	–	–	0.9–6.3	–	–	–	–	–	–	Jones and Amador (1993)
Southeastern Caribbean Sea & Gulf of Paria	Natural level (Fall)	–	–	–	–	–	0.6–31.6	–	–	–	–	–	–	Jones and Amador (1993)
Seawater	Natural level	0–1700	1–389	–	–	–	–	–	–	–	–	–	–	Table 1 ("Photoinduced and Microbial Generation of Hydrogen Peroxide and Organic Peroxides in Natural Waters")

(continued)

Table 3 (contniued)

Samples/Sources of water	Source of light/Natural level	H_2O_2 (nM)	ROOHs (nM)	LMW compounds (nM h^{-1})	DIC (nM h^{-1})	CO_2(nM or nM h^{-1})	CO (nM or nM h^{-1})	Hydrocarbons (nM)	Ammonium (nM or nM h^{-1})	COS (ng L^{-1})	PO_4^{3-} (μg L^{-1})	DOC (μM C)	DON (μM)	References
Standard and extracted organic substances						–								
Fluka humic	Xe lamp	–	–	–	–	–	1500	–	–	–	–	2500	–	Valentine and Zepp (1993)
Contech fulvic	Xe lamp	–	–	–	–	–	1600	–	–	–	–	4333	–	Valentine and Zepp (1993)
Oyster River fulvic (43°N)	Xe lamp	–	–	–	–	–	3200	–	–	–	–	4667	–	Valentine and Zepp (1993)
Soil fulvic (43°N)	Xe lamp	–	–	–	–	–	3330	–	–	–	–	4667		Valentine and Zepp (1993)
Oyster River, fulvic acid	Xe lamp	–	–	–	–	–	–	–	320	–	–	2840	36	Bushaw et al. (1996)
Fluka, humic acid	Xe lamp	–	–	–	–	–	–	–	230	–	–	1942	51	Bushaw et al. (1996)
Boreal Pond, fulvic acid (August)	Xe lamp	–	–	–	–	–	–	–	65	–	–	1133	24	Bushaw et al. (1996)
Boreal Pond, inlet, fulvic acid (June)	Natural	–	–	–	–	–	–	–	370	–	–	895	–	Bushaw et al. (1996)
Satilla River estuary, fulvic acid	Natural	–	–	–	–	–	–	–	50	–	–	787	20	Bushaw et al. (1996)

* Rate estimated using the deduction of initial values (0 h) from the nearest detected values (2, 8, and 24 h) for Satilla River and (4 and 18 h) for Altamaha River waters

** DFOM indicates the deferoxamine mesylate that is a strong Fe(III)-complexing ligand that forms nearly photo-inert complexes

Amador et al. 1989; Malcolm 1990; Mopper et al. 1991; Bertilsson and Tranvik 2000; Chen et al. 1978; Corin et al. 1996; de Haan 1993; Sun et al. 1993; Hongve 1994; Peuravuori and Pihlaja 1997). The major low molecular size substances examined are polysaccharides, N-acetylamino sugars, polypeptides, lipids, proteins, n-C_{16} and n-C_{18} fatty acid methyl esters, etc. The conversion rate of DOM into identifiable organic photoproducts is 20 % of the bleaching rate of the DOM, leaving a vast unidentified pool of bleached organic matter in natural waters (Miller and Zepp 1995). The unidentified bleached DOM would account for a large proportion of the total biologically available photoproducts (Miller and Moran 1997).

Photoinduced degradation of DOM can produce a variety of low molecular weight (LMW) aliphatic organic compounds which are considered to be microbiologically labile in the aquatic environment (Moran and Zepp 1997; Dahlén et al. 1996; Wetzel et al. 1995; Corin et al. 1996). The most common labile LMW organic compounds include formaldehyde, formic acid, formate, acetaldehyde, acetate, acetic acid, hydroxyacetic acid, hydroxyacetate, acetone, propanal, oxalic acid, oxalate, citric acid, citrate, glyoxal, methylglyoxal, glyoxylic acid, glyoxylate, ketomalonic acid, malonic acid, malonate, levulinic acid, levulinate, succinic acid, succinate, pyruvic acid and pyruvate. Nine organic peroxides (ROOH) such as methyl hydroperoxide, hydroxymethyl hydroperoxide, ethyl hydroperoxide, 1-hydroxyethyl hydroperoxide, 2-hydroxyethyl hydroperoxide, 1-hydroxypropyl hydroperoxide, 2-hydroxypropyl hydroperoxide, 3-hydroxypropyl hydroperoxide, and bis(hydroxymethyl)peroxide are observed in air and rainwaters (Hellpointner and Gäb 1989; Hewitt and Kok 1991; Jackson and Hewitt 1996). Precipitation is a potential source of these peroxides into surface waters, where they are microbiologically labile in natural waters (Mostofa 2005). A rapid decrease of peracetic acid added to unfiltered river waters was detected in dark controlled samples, suggesting that the organic peroxides are microbiologically labile (Mostofa 2005).

Some long-chain aliphatic organic acids, such as 2-hydroxy propanoic acid, 3-oxobutanoic acid, 4-oxopentanoic acid, hexanoic acid, pentanedioic acid, octanoic acid, nonanoic acid, and decanoic acid, are produced by the Photoinduced degradation of humic substances extracted from lakes (Corin et al. 1996). The production of keto acids such as 3-oxobutanoic acid and 4-oxopentanoic acid is greatly enhanced by an increase of the UV-dose. They are mostly produced from fulvic acid rather than humic acid (Corin et al. 1996), probably because of the higher percentage of aliphatic carbon bound to fulvic acid (63 %) compared to humic acid (47 %) (Malcolm 1985). Carboxylic acids (oxalic, malonic, formic, acetic) are usually major products of the photoinduced degradation of DOM (25–34.4 %) (Ma and Green 2004; Bertilsson et al. 1999; Bertilsson and Tranvik 2000).

The total production rate of LMW organic substances is much higher in lakes (2500–44200 nM h^{-1}) than in seawater (3.3–10.6 nM h^{-1}) (Table 3). Variations of the photolytically produced LMW carboxylic acids in different lakes are linked to the presence of fulvic and humic acids in DOM (Bertilsson and Tranvik 1998). The LMW organic substances undergo a rather fast disappearance in natural waters, probably because of two major pathways. First of all, the LMW organic compounds can be rapidly assimilated by natural microorganisms or bacterial populations and thus participate to the food chains for the growth of microbes in natural waters. Moreover, the LMW organic compounds can be photolytically mineralized into final end photoproducts such as CO or CO_2, which can take part to the global carbon cycle (Miller and Zepp 1995; Kieber et al. 2001).

5.3 *Photoinduced Formation of Aromatic Mono- and Dibasic Acids*

Photoinduced degradation reactions can convert the high molecular weight DOM into a variety of aromatic mono- and dibasic acids, phenolic compounds, aromatic aldehydes and ketones in waters (Kramer et al. 1996; Wetzel et al. 1995; Chen et al. 1978; Corin et al. 1996; Leenheer and Croue 2003; Peuravuori and Pihlaja 1997; Haan et al. 1979; Choudhry 1981; Langvik et al. 1994; Schmitt-Kopplin et al. 1998). The aromatic compounds most commonly identified as photo products are 4-hydroxy- and 4-hydroxy-3-methoxybenzaldehyde, benzoic acid and its derivatives such as 2-hydroxy-, 3-hydroxy-, 4-hydroxy-, 2,4-dihydroxy-, 3,4-dihydroxy-, 3-methoxy- and 4-hydroxy-3-methoxy-benzoic acid, 1,2-, 1,3- and 1,4-benzenedicarboxylic acid, benzene-di, -tri-, tetra-, penta-, and -hexa-carboxylic acids, decanoic acid and its derivatives such as tetra-, hexa- and octa-decanoic acid, acetophenone and its derivatives, methoxybenzene and its derivatives, methoxytoluenes, methoxystyrene, phenols, methoxylated phenols and hydroxyfuran. Most of the LMW aromatic acids are formed from humic and fulvic acids during the first 2 h irradiation of the natural humic waters. Prolonged irradiation may lead to a decrease of their concentrations due to further mineralization to end products (Corin et al. 1996).

5.4 *Photoinduced Formation of Carbon-Gas End Photoproducts Including DIC*

The photo mineralization of DOM leads to the formation of carbon-gas end photoproducts, which include CO, CO_2, dissolved inorganic carbon (DIC, usually defined as the sum of an equilibrium mixture of dissolved CO_2, H_2CO_3, HCO_3^-, and CO_3^{2-}: (Eq. 5.1) and carbonyl sulfide (COS) in natural waters (Table 3) (Reche et al. 1999; Ma and Green 2004; Graneli et al. 1996, 1998; Clark et al. 2004; Xie et al. 2004; Borges et al. 2008; Kujawinski et al. 2009; Tranvik et al. 2009; Omar et al. 2010; Ballaré et al. 2011; Zepp et al. 2011; Mopper et al. 1991; Miller and Zepp 1995; Bertilsson and Tranvik 2000; Fujiwara et al. 1995; Bushaw et al. 1996; Miller and Moran 1997; White et al. 2010; Cai 2011; Johannessen et al. 2007; Valentine and Zepp 1993; Molot et al. 2005; Francko and Heath 1982; Fang 2004; Chen et al. 2001; Karl and Tien 1997; Jones 1991; Jones and Amador 1993; Cai and Wang 1998; Zepp et al. 1998; Cai et al. 1998, 1999; Fichot and

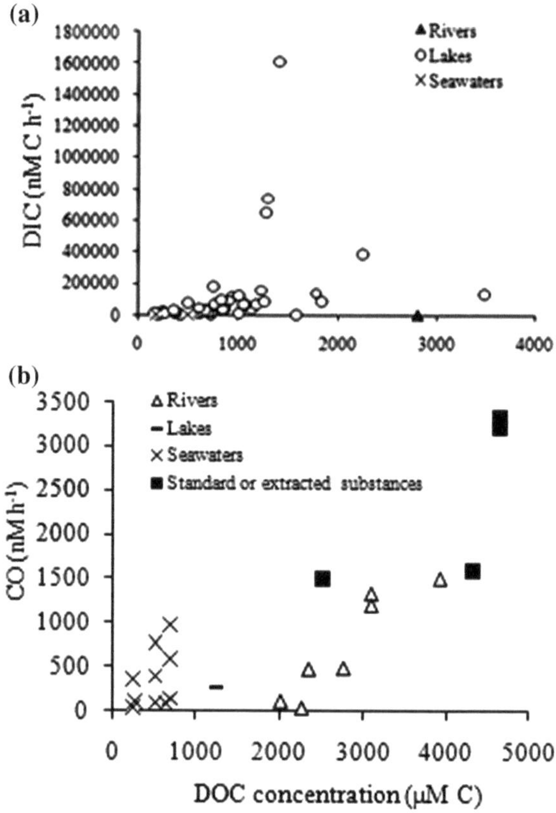

Fig. 10 Relationship between DOC concentration and DIC (Fig. a) or CO (Fig. b) photoproducts generated in photo experiments conducted on different sources of natural waters. *Data source* Table 4 and references therein

Miller 2010; Liu et al. 2010; Lohrenz et al. 2010). It is shown that gaseous CO_2 is rapidly dissolved in waters, which can be presented as (Eq. 5.1) (Liu et al. 2010):

$$CO_2 + H_2O \leftrightarrow H_2CO_2 \leftrightarrow H^+ + HCO_3{}^- \leftrightarrow 2H^+ + CO_3{}^{2-} \quad (5.1)$$

where the reaction (Eq. 5.1) is an equilibrium mixture of dissolved carbondioxide, carbonic acid, bicarbonate and carbonate ions. The proportion of each species depends on pH whereas at high pH the reaction shifts to the right hand side of (Eq. 5.1) and bicarbonate

($HCO_3{}^-$) at pH between 7 and 9 dominates, approximately 95 % of the carbon in the water. At high pH > 0.5, carbonate predominates (Dreybrodt 1988). DIC is also derived remarkably by carbonate dissolution with uptake of CO_2 in soil water (Eq. 5.2), and weathering as well as dissolution of silicate minerals (Eq. 5.3) (Liu et al. 2010; Dupré et al. 2003; Mortatti and Probst 2003).

$$CaCO_3 + CO_2 + H_2O = Ca^{2+} + 2HCO_3^- \quad (5.2)$$

$$2CO_2 + 3H_2O + CaSiO_3 + 2HCO_3^- + H_4SiO_4 \quad (5.3)$$

The concentration levels of carbon-gas photoproducts are highly variable among rivers, lakes and seawaters studied under light and dark conditions (Table 3;

Fig. 10). The production rate of DIC in irradiated samples is 80–2420 nM h^{-1} in rivers, 3180–1612800 nM h^{-1} in lakes and 40–13200 nM h^{-1} in seawaters. Under dark incubation the production rates are 99970 nM h^{-1} in rivers, 175500 nM h^{-1} in lakes, and 300–2300 nM h^{-1} in seawaters (Table 3).

The production rate of CO_2 in irradiated freshwater two coastal rivers is significantly high in air-saturation (13.2–23.0 μM h^{-1}) and O_2-saturation (18.3–26.5 μM h^{-1}), which are substantially decreased under air-saturation plus DFOM (5.4–6.0 μM h^{-1}), O_2-saturation plus DFOM (5.6–7.2 μM h^{-1}), and N_2-saturation (1.9 μM h^{-1}, measured for one river sample only) (Table 3) (Xie et al. 2004). Note that DFOM is the deferoxamine mesylate that is a strong Fe(III)-complexing ligand that forms nearly photo-inert complexes (Gao and Zepp 1998). The CO_2 production rate in the N_2-saturation river water is only ca. 10 % and 20 % of those in the O_2-saturation and air-saturation samples, respectively (Xie et al. 2004). This study observes that although CO_2 production in the O_2-saturation and DFOM samples is consistently higher than in the air-sat and DFOM samples, the difference between the two seldom exceeded 10 %, which indicates that in the presence of DFOM, iron rather than O_2 is the limiting factor for CO_2 production (Table 3) (Xie et al. 2004). These results suggest that although, O_2 and iron both can play very important roles in CO_2 production, photoinduced processes without the involvement of O_2 and iron (particularly the iron independent processes) can also contribute to CO_2 production in natural waters (Xie et al. 2004). CO_2 photoproduction rate is ~0.08–0.63 μM h^{-1} in estuarine water showing high production rate (~0.63 nM h^{-1}) in low salinity than in high salinity waters ~0.08 μM h^{-1} (White et al. 2010). Studies therefore observe that CO_2 is the largest carbon-containing product of DOM photodegradation in natural waters (Xie et al. 2004; Miller and Zepp 1995).

It is shown that the lakes produce DIC to a higher extent compared to rivers and seawaters (Fig. 10a). The DIC photoproducts in lakes are well correlated with DOC concentration, but no correlation was observed in rivers and seawaters (Fig. 10a). The high production of DIC in lakes could originate from the higher presence of low molecular weight organic substances (54–79 % in the <5 kDa range), which are mostly originated from fulvic and humic acids (Waiser and Robarts 2000; Yoshioka et al. 2007; Wu and Tanoue 2001). The photo production rate of DIC is higher in the upper surface layers (2.42 μM h^{-1} at 0 m), it then gradually decreases (0.60 μM h^{-1} at 6.5 m) and then increases again (2.73 μM h^{-1} at 24 m) in the deeper layers, but they are greatly lower than those of dark incubation samples (100.0 μM h^{-1}) in photoexperiments conducted on river water samples keeping in situ at different vertical depths in lake environment (Ma and Green 2004). For lake waters the production rates of DIC are gradually decreased from the surface (10.6–10.9 μM h^{-1} at 0–6 m) to deeper layers (3.2 μM h^{-1} at 24 m). However, under dark incubation the production rate is 175.5 μM h^{-1}, much higher than the rate of photoproduction. An increase in the DIC production rate of river waters kept in situ in the deeper layer of lakes (24 m depth) is likely caused by the much lower sunlight intensity that makes this setup equivalent to dark incubation. Therefore, high production of DIC under dark incubation allows the hypothesis that microbial degradation plays a significant and probably a major role in DOM mineralization in natural waters, especially in deeper lake or oceanic environments. The study shows that the global annual rate of photoinduced production of DIC (10^{14}–10^{15} mol DIC per year) (Johannessen 2000) might be on the same order of magnitude as that of sequestration of DIC by new production (~10^{15} mol DIC per year) (Liu et al. 2000).

CO production upon photoinduced degradation of DOM is highly variable for a variety of waters (Table 3). Production rates are 0.004–1.5 $\mu M\ h^{-1}$ in rivers, 0.26 $\mu M\ h^{-1}$ in lakes, 0.004–1.0 $\mu M\ h^{-1}$ in seawaters, and 1.5–3.3 $\mu M\ h^{-1}$ in standard dissolved fulvic and humic acids (Table 3). It is estimated that >95 % of the total water-column CO photoproduction occurs within the mixed layer on a global, yearly basis (Fichot and Miller 2010). It has been shown that the production rates of CO are almost linearly correlated with the concentration of DOC and of standard organic substances (Fig. 10b). The photoproduction of CO in the ocean is induced mainly by the UV component of solar radiation (Zepp et al. 1998; Atlas et al. 1994). Quantum yields (the quantum yield is the fraction of absorbed radiation that results in photoreaction) for CO production at wavelengths greater than 297 nm are highest in the UV-B region (Zepp et al. 1998). Turnover times for CO are in the order of hours, and they are generally lower (3–98 h) in fall and higher (2–108 h) in spring samples in the Caribbean Sea (Jones 1991; Jones and Amador 1993). The CO oxidation rate is lower in spring samples (20–340 pmol $L^{-1}\ h^{-1}$ except one sample, 980 pmol $L^{-1}\ h^{-1}$) than in fall samples (20–660 pmol $L^{-1}\ h^{-1}$ except one sample, 810 pmol $L^{-1}\ h^{-1}$). The concentration levels of CO are variable: 1–6 nM in spring and 0.6–32 nM in fall. The variations in the oxidation rates appear to be linked with two important phenomena. First, nitrifying and carboxydobacteria are both thought to have a role in oxidizing CO in the oceans (Conrad and Seiler 1980). Second, high concentrations of CO are able to inhibit marine nitrifying bacteria in natural waters (Jones and Amador 1993; Jones and Morita 1984).

In the estuary of River Ohta, the concentration of carbonyl sulfide (COS) was highest (54.4 ng L^{-1}) in the late afternoon (17:00) during the summer season (July), and lowest (23.9 ng L^{-1}) soon after noontime (14:00) during the winter season (December) (Table 3) (Fujiwara et al. 1995). In Seto Inland Sea the COS concentration was higher (~5–17 ng L^{-1}) in the surface layer (0–5 m), whilst it was lower (~3–5 ng L^{-1}) in the deeper layer (20 m) (Fujiwara et al. 1995). An increase in COS concentrations is often linked with an increase of solar radiation, lower concentrations being detected at night time and in the early morning. Moreover, higher concentrations are found in surface seawater than in the deeper layers, suggesting that COS is photolytically produced in natural waters.

Photo products such as H_2O_2, ROOH and CO_2, simultaneously generated during the photoinduced degradation of DOM, can be photosynthetically transformed into carbohydrates during the summer season in natural surface waters. The relevant processes can be depicted as follows (Eqs. 5.4, 5.5) (Mostofa et al. 2009a; Komissarov 1994, 1995, 2003):

$$\mathrm{DOM} + h\upsilon \rightarrow \mathrm{H_2O_2} + \mathrm{CO_2/CO/DIC} + \mathrm{LOWDOM} + \mathrm{E}\ (\pm) \quad (5.4)$$

$$\mathrm{x\ CO_{2(air)}} + \mathrm{y\ H_2O_{2(H_2O)}} + h\upsilon \rightarrow \mathrm{C_x\ (H_2O)_y} + (\mathrm{x} + \mathrm{y}/2)\ \mathrm{O_2} + \mathrm{E}\ (\pm) \quad (5.5)$$

This mechanism may help in the understanding of the production of autochthonous DOM, particularly carbohydrate compounds, during the summer season in natural waters. The new reaction mechanim for photosynthesis (Eq. 4.2) has been discussed in details in photosynthesis chapter (see chapter "Photosynthesis in Nature: A New Look").

5.5 Photoinduced Formation of Nitrogenous Compounds

Nitrogenous photoproducts include the ammonium (NH_4^+) and nitrate/nitrite (NO_3^-/NO_2^-) that are released by photoinduced degradation of humic substances and degradation of dissolved organic nitrogen (DON) in waters (Table 3) (Mostofa et al. 2011; Li et al. 2008; Bushaw et al. 1996; Mack and Bolton 1999; Carlsson et al. 1993; Stedmon et al. 2007). Ammonium is produced by transformation of aquatic dissolved organic matter in waters, and the production rates are 40–370 nM for 895–3840 μM C of DOC and 20–86 μM of dissolved organic nitrogen (Bushaw et al. 1996). DON concentrations are 7–26 μM in Lake Hongfeng and 14–47 μM in Lake Baihua (Li et al. 2008), approximately 8.35 μM in the epilimnion of Lake Biwa (Kim et al. 2006), and approximately 10 μM in coastal waters (Bronk 2002). The high concentrations of NO_2^- and NH_4^+ in Lakes Hongfeng and Baihua during the summer stratification period suggest the regeneration of inorganic nitrogen (NO_2^- and NH_4^+) in lakes (Li et al. 2008). Photoinduced respiration or assimilation of lake algae under natural sunlight can release NH_4^+ in waters, suggesting that autochthonous organic matter is a major source of NH_4^+ in natural waters Mostofa K et al. (unpublished). DOM in coastal waters derives from terrestrial humic substances (Carlsson et al. 1993). This leads to an increase of the nitrogen availability, which subsequently stimulates the rates of primary and secondary production. The uptake of inorganic nitrogen by bacteria during a phytoplankton bloom can be observed, particularly in lake or coastal waters where the inputs of terrestrial humic substances are much higher (Kirchman et al. 1991; Amon and Benner 1994). Photolytically produced ammonium can be assimilated by bacterial populations, which can lead to an increase in the production of autotrophic and heterotrophic biomass in planktonic environments. Photo release of inorganic nitrogen from DOM is an important source of nitrogen availability in several aquatic ecosystems, such as nitrogen-limited and high-latitude environments, and coastal waters where high primary and secondary production are usually occurring.

5.6 Photoinduced Formation of Phosphate

Photoinduced degradation of DOM can lead to the release of phosphate (PO_4^{3-}) in natural surface waters (Table 3) (Reche et al. 1999; Mostofa et al. 2011; Zhang et al. 2004; Fang 2004; Chen et al. 2001; Karl and Tien 1997; Suzumura and Ingall

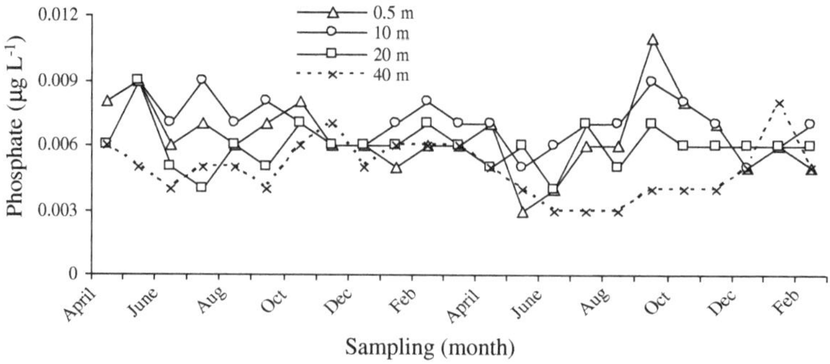

Fig. 11 Variation of phosphate concentration (month by month) during the period of April 1999 to February 2001 in the waters of Lake Biwa. *Data sources* Mostofa KMG et al. (unpublished) and Shiga prefecture office, Japan)

2004). Phosphate concentration in waters of Lake Biwa is often higher in the surface layer during the summer season, compared to other seasons or the deeper layer (Fig. 11) (Mostofa KMG et al., unpublished). This finding suggests that phosphate may be produced by photoinduced degradation processes in the surface waters of Lake Biwa. Indeed, the inorganic phosphorus is increased in the surface layer, whilst the organic phosphorus concentration is decreased and the surface concentration of organic phosphorus is lower compared to the deeper layers (Fang 2004). Therefore, the inorganic phosphorus may be produced from the decomposition of dissolved organic phosphorus in surface waters. The concentration levels of soluble reactive phosphorus were highest in the upper surface layers (0–30 m) in North Pacific Ocean (Karl and Tien 1997). Concentration levels of dissolved hydrophobic phosphorus were much higher in the surface layers (6.6–18 nM) and gradually decreased with depth (5–10 nM at below 400 m) (Suzumura and Ingall 2004). Total dissolved phosphorus (TDP) in coastal waters was much higher in surface waters (3.51–7.12 dpm m^{-3}) than in the deeper layers (0.83–2.0 dpm m^{-3}), and the surface concentrations at inshore stations (5.09–7.12 dpm m^{-3}) were significantly higher than at the offshore stations (3.51 dpm m^{-3}) (Zhang et al. 2004). From these data it can be inferred that inorganic phosphate might be an important photoproduct of DOM photoinduced degradation in surface waters.

5.7 Photoinduced and Microbial Processes for the Release of Energy and of End Products to the Water Ecosystem

Photoinduced degradation of DOM generally takes place through redox reactions in waters which can lead to energy changes (±) such as supply (+) or consumption (−) of energy in the aquatic environment. Energy changes also occur during photosynthesis in natural waters (Komissarov 1994, 1995, 2003). DOM with its content of organic

C and N is a thermodynamic anomaly that provides a major source of energy to drive aquatic and terrestrial ecosystems (Salonen et al. 1992; Wetzel 1984, 1992; Hedges et al. 2000; Tranvik 1992). Therefore, any changes in energy during the photoinduced degradation of DOM are thermodynamically vital for all the living organisms and for the natural water ecosystem. Photoinduced degradation of DOM in natural waters is thus interlinked with production of free radicals, microbial processes, photosynthesis, autochthonous DOM, nutrients, end photoproducts and their utilization as food for microorganisms. A conceptual schematic diagram for the photoinduced and microbial processes of DOM and POM, photoproducts and their importance in the aquatic environment is reported below (Fig. 12):

Finally, it can be concluded that most of the changes that take place in the natural ecosystem are closely interlinked.

6 Interactions Between Photoinduced and Microbial Processes in Natural Waters

The understanding of the interactions between photoinduced and microbial degradation of DOM has required a proper elucidation of the photoinduced processes. It is now considered that microbial degradation takes place at the same time as the photoinduced degradation process (Kopacek et al. 2003; Moran

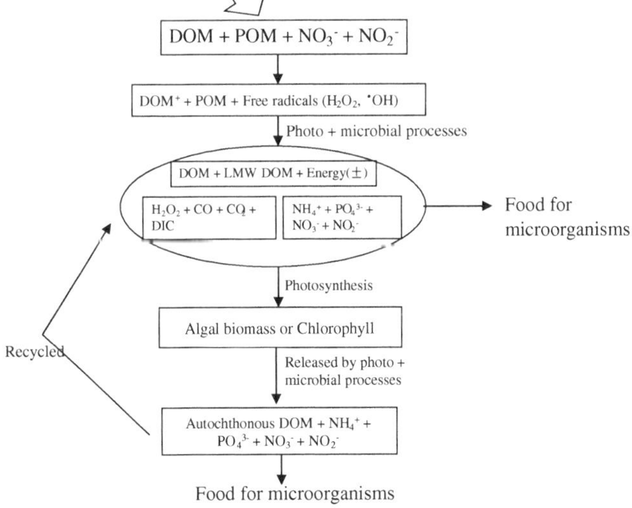

Fig. 12 Conceptual model on photoinduced degradation of DOM and its consecutive effects on key biogeochemical processes in natural waters. [*Data source* with modifications Mostofa et al. (2009b)

et al. 2000; Amon and Benner 1996; Vähätalo and Wetzel 2004; Miller and Moran 1997). In essence, during the photoinduced process various species among which are the superoxide radical ion ($O_2^{\bullet -}$), the $HO^{\bullet}$ and peroxides (H_2O_2 and ROOH) are generated either in surface waters or in aqueous solutions during laboratory irradiation experiments (Takeda et al. 2004; Moore et al. 1993; Mostofa and Sakugawa 2009; Southworth and Voelker 2003; Goldstone et al. 2002). The photogenerated reactive species are involved into the photoinduced degradation of DOM in waters. Simultaneously, these species can inhibit or deactivate the activity of catalase, peroxidase and superoxide dismutase associated with bacterial cells, particulate organic matter and DOM (Moffett and Zafiriou 1990; Tanaka et al. 1985; Serban and Nissenbaum 1986; Zepp et al. 1987). Bacterial cells can protect themselves from harmful oxidizing species such as H_2O_2, $O_2^{\bullet -}$ and $HO^{\bullet}$ by adjusting the level of their enzymes (Chance et al. 1979). Therefore, microbial degradation is expected to take place to a negligible extent during the photoinduced degradation of DOM in aqueous media.

The bacterial growth shows seasonal variations, reaching the maximum during spring to early summer and decreasing greatly during the summer season when the water temperature exceeds 25.5 °C in lakes (Zhao et al. 2003; Darakas 2002). Natural sunlight or UV-exposure can decrease the bacterial production by 15–80 %, which considerably inhibits the formation and biodegradation of DOM in natural surface waters (Amon and Benner 1996; Bertilsson and Tranvik 1998; Benner and Ziegler 1999; Naganuma et al. 1996; Tranvik and Kokalj 1998). However, other studies have found that solar exposure can enhance the bacterial growth by about 35–200 % (Lindell et al. 1996; Wetzel et al. 1995; Bushaw et al. 1996; Miller and Moran 1997; Herndl et al. 1993; Lindell and Rai 1994; Reitner et al. 1997; Jørgensen et al. 1998; Moran and Hodson 1994). Moreover, bacterial growth is usually observed in deeper waters (Benner and Ziegler 1999). The increase or the decrease of bacterial growth by sunlight depends on two key factors: (i) The production of reactive species (H_2O_2, ROOH, $HO^{\bullet}$) and of mineralization products such as CO_2, CO and DIC; (ii) The concentration level and molecular nature of DOM, the concentration of total dissolved iron for photo-Fenton reaction, water temperature, dissolved oxygen, physical mixing etc. Water temperature during the summer season is merely regulated by natural solar radiation, which can lead to high generation of free radicals and mineralization products in natural waters (Moore et al. 1993; Mostofa and Sakugawa 2009; Zafiriou et al. 1984; Zika 1981; Obernosterer et al. 2001; Fujiwara et al. 1993; Sakugawa et al. 2000). The free radicals, as strong oxidants and depending on their concentration, may have several deadly impacts on many stages of cell metabolism, including those involved into the induction of programmed cell death (Samuilov et al. 2001). This hypothesized effect is supported by the observation of an inhibition by UV radiation of the activity of microorganisms in water (Herndl et al. 1993; Lund and Hongve 1994; Karentz et al. 1994), and of the deadly impact of UV radiation on bacterial cells or microorganisms in the aquatic environment (Qian et al. 2001; Randall et al. 2005). Qian et al. 2001; Sunlight often affects bacterial growth merely in the upper surface waters, and UV-B radiation inhibited bacterial production by 39–82 % in a high mountain lake (Carrillo et al. 2002). Therefore, microbial activity may be less efficient during the photoinduced degradation of DOM in natural waters.

7 Scope of Future Challenges

Photoinduced and microbial degradation of DOM is an important research subject in the photochemistry of the aquatic environment. Till now, a few researches have been conducted on the degradation of bulk DOM in natural waters. Photoinduced and microbial degradation of organic substances such as standard fulvic acid, humic acid, fluorescent whitening agents (DAS1 and DSBP) and chlorophyll was seldom conducted to examine their end photoproducts in the aquatic environments. For a better elucidation of the degradation processes of DOM in natural waters, it is vital to conduct an extensive study on streams, rivers, lakes, coastal and seawaters. The photolytically produced low molecular weight (LMW) organic substances are microbiologically important, but in some cases they might be toxic. Till now only a few researches have been conducted to identify the LMW organic compounds in the aquatic environment. The effects of water temperature and pH on the coupled photoinduced and microbial degradation of DOM and of standard organic compounds have not been studied so far. Obviously, variation in water temperature and pH might have a significant role on degradation processes and on the concentration levels of the end photoproducts. Therefore, a number of important research needs for future challenges can be distinguished as: (i) Photoinduced and microbial degradation of various molecular size fractions of DOM for a variety of waters. (ii) An extensive study on the microbial degradation of DOM for a variety of waters, and the development of the mechanism for microbial degradation of DOM. (iii) Effect of temperature and pH on photoinduced and microbial degradation of DOM for a variety of waters and for standard organic substances. (iv) Interactions between photoinduced and microbial degradation of DOM, and its impact on microorganisms in the aquatic ecosystem. (v) Investigation on LMW organic substances produced from photoinduced and microbial degradation of DOM in natural waters and from standard HMW organic substances. (vi) Elucidation of the microbiological changes involving the macromolecules (fulvic acid, humic acid and autochthonous fulvic acid) under dark incubation. (vii) Understanding of the mechanisms of the photoinduced degradation of the macromolecules (fulvic acid, humic acid, and autochthonous fulvic acid) by natural sunlight in aqueous media. The mechanism depicted in this chapter may pave the way for future directions in the field. (viii) Refinement of the group contribution method (GCM) to predict $HO^{\bullet}$ reaction rate constants. Because of the limited data availability for the rate constants of various compounds in the aqueous phase, the GCM still has many gaps and its implementation should be the focus for future research (Minakata et al. 2009).

It is expected that new research works can focus on the combined photoinduced and microbial degradation of DOM and of the photoproducts of the process, to understand a whole feature of the biogeochemical process of DOM in the aquatic environment.

8 Nomenclature

DFOM	Deferoxamine mesylate
DOC	Dissolved organic carbon
DOM	Disscolved organic matter
DIC	Dissolved inorganic carbon ($CO_2 + H_2CO_3 + HCO_3^- + CO_3^{-2}$)
GCM	Group contribution method
H_2O_2	Hydrogen peroxide
LMW	Low molecular weight
$HO^{\bullet}$	Hydroxyl radical
SS PM	Sea-salt particulate matter

Problems

(1) Describe the various biogeochemical functions of DOM for photoinduced and microbial processes in natural waters
(2) Explain why the photoinduced degradation of DOM varies for a variety of natural waters.
(3) Explain the theoretical model for photoinduced degradation process in aqueous media.
(4) Explain the mechanism of the photoinduced degradation of DOM in natural waters.
(5) Explain the mechanism for the photoinduced degradation by $HO^{\bullet}$ of functional groups bonded to DOM and explain the group contribution method (GCM) to predict the $HO^{\bullet}$ reaction rate constants.
(6) What is the methanogenesis? Explain the chemical reactions that are involved in the formation of CH_4 from methyl-coenzyme *M* reductase.
(7) What are the controlling factors for the photoinduced degradation of DOM? Describe four important factors that can regulate the DOM Photoinduced degradation in aqueous solution.
(8) What are the controlling factors for the microbial degradation of DOM? Describe two of the important factors.
(9) List the various end photo products generated from the photoinduced degradation of DOM in natural waters. Explain what end products are vital for microbial processes in the aquatic environments.

(10) What is the new reaction for photosynthesis? Explain how photo processes can affect photosynthesis in waters.
(11) Explain the flow diagram on how photoinduced and microbial processes work together for the degradation of DOM and POM in natural waters.
(12) What are the key interactions between photoinduced and microbial processes in natural waters?

Acknowledgments We thank Mr. Cui Lifeng of Institute of Geochemistry, Chinese Academy of Sciences, China for his generous assistance. This work was financially supported jointly by the National Natural Science Foundation of China (Grant Nos.40525011, 1314765) and the Chinese Academy of Sciences (KZCX2-YW-102). This work was partly supported by Brook Byers Institute for Sustainable Systems, Georgia Institute of Technology, United States and National Science Foundation of the United States award 0854416; University Turin, Italy; Chinese Research Academy of Environmental Sciences, China; Northwest Missouri State University, USA; Kyoto University, Japan; and Hiroshima University, Japan. This study acknowledges the reprinted (adapted) with permission from Minakata D, Li K, Westerhoff P, Crittenden J, Development of a group contribution method to predict aqueous phase hydroxyl radical (HO•) reaction rate constants, Environ Sci Technol, 43(16), 6220–6227. Copyright (2009) American Chemical Society; reprinted by permission from Macmillan Publishers Ltd: [Nature Rev Microbiol] (Lovley DR, Bug juice: harvesting electricity with microorganisms, 4, 497–508), copyright (2006); reprinted (adapted) with permission from Pelmenschikov V, Blomberg MRA, Siegbahn PEM, Crabtree RH, A Mechanism from Quantum Chemical Studies for Methane Formation in Methanogenesis, JACS, 124, 4039–4049. Copyright (2002) American Chemical Society; reprinted (adapted) with permission from Ebner S, Jaun B, Goenrich M, Thauer RK, Harmer J, Binding of coenzyme B induces a major conformational change in the active site of methyl-coenzyme M reductase, JACS, 132, 567–575. Copyright (2010) American Chemical Society; reprinted from Journal of Volcanology and Geothermal Research, 178 (2), Koschorreck M, Wendt-Potthoff K, Scharf B, Richnow HH, Methanogenesis in the sediment of the acidic Lake Caviahue in Argentina, 197–204. Copyright (2008), with permission from Elsevier; Copyright (1996) by the Association for the Sciences of Limnol Oceanogr, Inc.; Copyright (2000) by the Association for the Sciences of Limnol Oceanogr, Inc.; Copyright (2011) by The Geochemical Society of Japan; and Shiga prefecture office, Japan.

References

Allard B, Boren H, Pettersson C, Zhang G (1994) Degradation of humic substances by UV irradiation. Environ Int 20:97–101

Allen JM, Lucas S, Allen SK (1996) Formation of hydroxyl radical (· OH) in illuminated surface waters contaminated with acidic mine drainage. Environ Toxicol Chem 15:107–113

Amador JA, Alexander M, Zika RG (1989) Sequential photochemical and microbial degradation of organic molecules bound to humic acid. Appl Environ Microbiol 55:2843–2849

Amon RMW, Benner R (1994) Rapid cycling of high-molecular-weight dissolved organic matter in the ocean. Nature 369:549–552

Amon R, Benner R (1996) Photochemical and microbial consumption of dissolved organic carbon and dissolved oxygen in the Amazon River system. Geochim Cosmochim Acta 60:1783–1792

Anastasio C, Newberg JT (2007) Sources and sinks of hydroxyl radical in sea-salt particles. J Geophys Res 112:D10306. doi:10.1029/2006JD008061

Arakaki T, Faust BC (1998) Sources, sinks, and mechanisms of hydroxyl radical (•OH) photoproduction and consumption in authentic acidic continental cloud waters from Whiteface mountain, New York: The role of the Fe (r)(r = II, III) photochemical cycle. J Geophys Res 103:3487–3504

Arakaki T, Miyake T, Shibata M, Sakugawa H (1999) Photochemical formation and scavenging of hydroxyl radical in rain and dew waters. Nippon Kagaku Kaishi 5:335–340 (In Japanese)
Arundhati Pal AKP (2008) Microbial extracellular polymeric substances: central elements in heavy metal bioremediation. Ind J Microbiol 48:49–64
Asmus K-D, Bonifačič M (1999) Sulfur-centered reactive intermediates as studied by radiation chemical and complementary techniques. In: Alfassi ZB (Ed), S-centered radicals, the chemistry of free radicals. John Wiley & Sons Ltd, Chichester. Chapter 5, pp 141–191
Asmus K, Möckel H, Henglein A (1973) Pulse radiolytic study of the site of hydroxyl radical attack on aliphatic alcohols in aqueous solution. J Phys Chem 77:1218–1221
Assel M, Laenen R, Laubereau A (1998) Ultrafast electron trapping in an aqueous NaCl-solution. Chem Phys Lett 289:267–274
Atkinson R (1986) Kinetics and mechanisms of the gas-phase reactions of the hydroxyl radical with organic compounds under atmospheric conditions. Chem Rev 86:69–201
Atkinson R (1987) A structure-activity relationship for the estimation of rate constants for the gas-phase reactions of OH radicals with organic compounds. Int J Chem Kinet 19:799–828
Atlas E, Lueb R, Madronich S, Prezelin B, Smith R (1994) Dissolved trace gas measurements and UV effects near the Antarctic Peninsula during ICECOLORS'93. EOS Trans AGU 75:377
Azam F, Cho B (1987) Bacterial utilization of organic matter in the sea. In: Fletcher M, Gray TRG, Jones JG (Eds), Ecology of microbial communities, Cambridge University Press, Cambridge, pp 261–281
Baker A, Inverarity R (2004) Protein-like fluorescence intensity as a possible tool for determining river water quality. Hydrol Process 18:2927–2945
Ballaré CL, Caldwell MM, Flint SD, Robinson S, Bornman JF (2011) Effects of solar ultraviolet radiation on terrestrial ecosystems. Patterns, mechanisms, and interactions with climate change. Photochem Photobiol Sci 10:226–241
Banat IM, Nedwell DB, Balba MT (1983) Stimulation of methanogenesis by slurries of saltmarsh sediment after the addition of molybdate to inhibit sulphate-reducing bacteria. J Gen Microbiol 129:123–129
Barbiero RP, Tuchman ML (2004) The deep chlorophyll maximum in lake superior. J Great Lakes Res 30:256–268
Basiliko N, Yavitt JB, Dees P, Merkel S (2003) Methane biogeochemistry and methanogen communities in two northern peatland ecosystems, New York State. Geomicrobiol J 20:563–577
Benner R, Biddanda B (1998) Photochemical transformations of surface and deep marine dissolved organic matter: effects on bacterial growth. Limnol Oceanogr 43:1373–1378
Benner R, Kaiser K (2003) Abundance of amino sugars and peptidoglycan in marine particulate and dissolved organic matter. Limnol Oceanogr 48:118–128
Benner R, Ziegler S (1999) Do photochemical transformations of dissolved organic matter produce biorefractory as well as bioreactive substrates? Microbial Biosystems: New Frontiers Procee 8th Int Sym On Microb Ecol In: Bell CR, Brylinsky M, Johnson-Green P (Ed) Atlantic Canada Soc for Micrb Ecol Halifax, Canada
Benson SW (1976) Thermochemical kinetics: methods for the estimation of thermochemical data and rate parameters. 2nd Edn,Wiley, Now York
Bertilsson S, Allard B (1996) Sequential photochemical and microbial degradation of refractory dissolved organic matter in a humic freshwater system. Ergebn Limnol 48:133–141
Bertilsson S, Tranvik LJ (1998) Photochemically produced carboxylic acids as substrates for freshwater bacterioplankton. Limnol Oceanogr 43:885–895
Bertilsson S, Tranvik LJ (2000) Photochemical transformation of dissolved organic matter in lakes. Limnol Oceanogr 45:753–762
Bertilsson S, Stepanauskas R, Cuadros-Hansson R, Granéli W, Wikner J, Tranvik L (1999) Photochemically induced changes in bioavailable carbon and nitrogen pools in a boreal watershed. Aquat Microb Ecol 19:47–56
Biddanda B, Benner R (1997) Carbon, nitrogen, and carbohydrate fluxes during the production of particulate and dissolved organic matter by marine phytoplankton. Limnol Oceanogr 42:506–518

Billamboz N, Grivet M, Foley S, Baldacchino G, Hubinois JC (2010) Radiolysis of the polyethylene/water system: studies on the role of hydroxyl radical. Radiat Phys Chem 79:36–40
Blokhina O, Virolainen E, Fagerstedt KV (2003) Antioxidants, oxidative damage and oxygen deprivation stress: a review. Ann Botany 91:179–194
Blough NV, del Vecchio R (2002) Chromophoric DOM in the coastal environment. In: Hansell DA, Carlson CA (eds) Biogeochemistry of marine dissolved organic matter. Elsevier Science, USA, p 774
Blough NV, Zepp RG (1995) Reactive oxygen species in natural waters. In: Valentine JS, Foote CS (eds) Active oxygen in chemistry. Blackie Academic and Professional, New York, pp 280–333
Blough N, Zafiriou O, Bonilla J (1993) Optical absorption spectra of waters from the Orinoco River outflow: terrestrial input of colored organic matter to the Caribbean. J Geophys Res 98:2271–2278
Bolton J, Carter S (1994) Homogeneous photodegradation of pollutants in contaminated water: An introduction. In: Helz G, Zepp R, Crosby D (Eds), Aquatic and Surface Photochemistry. CRC press, Boca Raton. Chap 33
Borges A, Ruddick K, Schiettecatte LS, Delille B (2008) Net ecosystem production and carbon dioxide fluxes in the Scheldt estuarine plume. BMC Ecol 8:15
Borisover M, Laor Y, Saadi I, Lado M, Bukhanovsky N (2011) Tracing organic footprints from industrial effluent discharge in recalcitrant riverine chromophoric dissolved organic matter. Water Air Soil Pollut 222:255–269
Bräuer S, Yavitt J, Zinder S (2004) Methanogenesis in McLean Bog, an acidic peat bog in upstate New York: stimulation by H_2/CO_2 in the presence of rifampicin, or by low concentrations of acetate. Geomicrob J 21:433–443
Bräuer SL, Cadillo-Quiroz H, Yashiro E, Yavitt JB, Zinder SH (2006) Isolation of a novel acidiphilic methanogen from an acidic peat bog. Nature 442:192–194
Brezonik PL, Fulkerson-Brekken J (1998) Nitrate-induced photolysis in natural waters: controls on concentrations of hydroxyl radical photo-intermediates by natural scavenging agents. Environ Sci Technol 32:3004–3010
Bronk DA (2002) Dynamics of DON. Biogeochemistry of marine dissolved organic matter. In: Hansell DA, Carlson CA (eds) Biogeochemistry of marine dissolved organic matter. Academic Press, San Diego, pp 153–249
Brooks ML, Meyer JS, McKnight DM (2007) Photooxidation of wetland and riverine dissolved organic matter: altered copper complexation and organic composition. Hydrobiology 579:95–113
Burdige DJ, Martens CS (1988) Biogeochemical cycling in an organic-rich coastal marine basin: 10. The role of amino acids in sedimentary carbon and nitrogen cycling. Geochim Cosmochim Acta 52:1571–1584
Burdige DJ, Kline SW, Chen W (2004) Fluorescent dissolved organic matter in marine sediment pore waters. Mar Chem 89:289–311
Bushaw KL, Zepp RG, Tarr MA, Schulz-Jander D, Bourbonniere RA, Hodson RE, Miller WL, Bronk DA, Moran MA (1996) Photochemical release of biologically available nitrogen from aquatic dissolved organic matter. Nature 381:404–407
Cadillo-Quiroz H, Bräuer S, Yashiro E, Sun C, Yavitt J, Zinder S (2006) Vertical profiles of methanogenesis and methanogens in two contrasting acidic peatlands in central New York State, USA. Environ Microbiol 8:1428–1440
Cadillo-Quiroz H, Yavitt JB, Zinder SH, Thies JE (2010) Diversity and community structure of Archaea inhabiting the rhizoplane of two contrasting plants from an acidic bog. Microb Ecol 59:757–767
Cai WJ (2011) Coastal ocean carbon paradox: CO_2 sinks or sites of terrestrial carbon incineration. Annu Rev Mar Sci 3:123–1453
Cai WJ, Wang Y (1998) The chemistry, fluxes, and sources of carbon dioxide in the estuarine waters of the Satilla and Altamaha Rivers, Georgia. Limnol Oceanogr 43:657–668
Cai WJ, Wang Y, Hodson RE (1998) Acid-base properties of dissolved organic matter in the estuarine waters of Georgia, USA. Geochim Cosmochim Acta 62:473–483

Cai WJ, Pomeroy LR, Moran MA, Wang Y (1999) Oxygen and carbon dioxide mass balance for the estuarine-intertidal marsh complex of five rivers in the southeastern US. Limnol Oceanogr 44:639–649
Calace N, Palmieri N, Mirante S, Petronio B, Pietroletti M (2006) Dissolved and particulate humic substances in water channels in the historic centre of Venice. Water Res 40:1109–1118
Call D, Logan BE (2008) Hydrogen production in a single chamber microbial electrolysis cell lacking a membrane. Environ Sci Technol 42:3401–3406
Carlsson P, Segatto AZ, Graneli E (1993) Nitrogen bound to humic matter of terrestrial origin-a nitrogen pool for coastal phytoplankton? Mar Ecol Prog Ser 97:105–116
Carpenter SR, Cole JJ, Kitchell JF, Pace ML (1998) Impact of dissolved organic carbon, phosphorus, and grazing on phytoplankton biomass and production in experimental lakes. Limnol Oceanogr 43:73–80
Carrillo P, Medina-Sánchez JM, Villar-Argaiz M (2002) The interaction of phytoplankton and bacteria in a high mountain lake: importance of the spectral composition of solar radiation. Limnol Oceanogr 47:1294–1306
Casper P, Chim Chan O, Furtado ALS, Adams DD (2003) Methane in an acidic bog lake: The influence of peat in the catchment on the biogeochemistry of methane. Aquat Sci 65:36–46
Chance B, Sies H, Boveris A (1979) Hydroperoxide metabolism in mammalian organs. Physiol Rev 59:527–605
Chaudhuri SK, Lovley DR (2003) Electricity generation by direct oxidation of glucose in mediatorless microbial fuel cells. Nature Biotechnol 21:1229–1232
Chefetz B (2002)
Chen RZ, Pignatello JJ (1997) Role of quinone intermediates as electron shuttles in Fenton and photo assisted Fenton oxidations of aromatic compounds. Environ Sci Technol 31:2399–2406
Chen Y, Khan S, Schnitzer M (1978) Ultraviolet irradiation of dilute fulvic acid solutions. Soil Sci Soc Am J 42:292–296
Chen JH, Lion LW, Ghiorse WC, Shuler ML (1995) Mobilization of adsorbed cadmium and lead in aquifer material by bacterial extracellular polymers. Water Res 29:421–430
Chen YL, Chen HY, Lee WH, Hung CC, Wong GTF, Kanda J (2001) New production in the East China Sea, comparison between well-mixed winter and stratified summer conditions. Contin Shelf Res 21:751–764
Cheng S, Logan BE (2007) Sustainable and efficient biohydrogen production via electrohydrogenesis. Proc Natl Aca Sci 104:18871
Chin YP, Miller PL, Zeng L, Cawley K, Weavers LK (2004) Photosensitized degradation of bisphenol A by dissolved organic matter. Environ Sci Technol 38:5888–5894
Choudhry GG (1981) Humic substances Part II†: photophysical, photochemical and free radical characteristics. Toxicol Environ Chem 4:261–295
Clark CD, Hiscocka WT, Millero FJ, Hitchcock G, Brand L, Miller WL, Ziolkowski L, Chen RF, Zika RG (2004) CDOM distribution and CO2 production on the Southwest Florida Shelf. Mar Chem 89:145–167
Coble PG (1996) Characterization of marine and terrestrial DOM in seawater using excitation-emission matrix spectroscopy. Mar Chem 51:325–346
Coble PG (2007) Marine optical biogeochemistry: the chemistry of ocean color. Chem Rev 107:402–418
Coleman ML, Hedrick DB, Lovley DR, White DC, Pye K (1993) Reduction of Fe(III) in sediments by sulphate-reducing bacteria. Nature 361:436–438
Conrad R (1999) Contribution of hydrogen to methane production and control of hydrogen concentrations in methanogenic soils and sediments. FEMS Microbiol Ecol 28:193–202
Conrad R, Seiler W (1980) Photooxidative production and microbial consumption of carbon monoxide in seawater. FEMS Microbiol Lett 9:61–64
Conrad R, Bak F, Seitz H, Thebrath B, Mayer H, Schütz H (1989a) Hydrogen turnover by psychrotrophic homoacetogenic and mesophilic methanogenic bacteria in anoxic paddy soil and lake sediment. FEMS Microbiol Lett 62:285–293

Conrad R, Mayer HP, Wüst M (1989b) Temporal change of gas metabolism by hydrogen-syntrophic methanogenic bacterial associations in anoxic paddy soil. FEMS Microbiol Lett 62:265–273
Cook R, Kelly C, Schindler D, Turner M (1986) Mechanisms of hydrogen ion neutralization in an experimentally acidified lake. Limnol Oceanogr 31:134–148
Cooper WJ, Zika RG, Petasne RG, Plane JMC (1988) Photochemical formation of hydrogen peroxide in natural waters exposed to sunlight. Environ Sci Technol 22:1156–1160
Cooper WJ, Zika RG, Petasne RG, Fischer AM (1989) Sunlight-induced photochemistry of humic substances in natural waters: major reactive species. Adv Chem Ser 219:333–362
Cord-Ruwisch R, Seitz HJ, Conrad R (1988) The capacity of hydrogenotrophic anaerobic bacteria to compete for traces of hydrogen depends on the redox potential of the terminal electron acceptor. Arch Microbiol 149:350–357
Corin N, Backlund P, Kulovaara M (1996) Degradation products formed during UV-irradiation of humic waters. Chemosphere 33:245–255
Crill PM, Martens CS (1983) Spatial and temporal fluctuations of methane production in anoxic coastal marine sediments. Limnol Oceanogr 28:1117–1130
Cuny P, Marty JC, Chiavérini J, Vescovali I, Raphel D, Rontani JF (2002) One-year seasonal survey of the chlorophyll photodegradation process in the northwestern Mediterranean Sea. Deep Sea Res Part II 49:1987–2005
Dahlén J, Bertilsson S, Pettersson C (1996) Effects of UV-A irradiation on dissolved organic matter in humic surface waters. Environ Int 22:501–506
Darakas E (2002) E. coli kinetics-effect of temperature on the maintenance and respectively the decay phase. Environ Monitor Assess 78:101–110
de Haan H (1993) Solar UV-light penetration and photodegradation of humic substances in peaty lake water. Limnol Oceanogr 38:1072–1076
del Vecchio R, Blough NV (2002) Photobleaching of chromophoric dissolved organic matter in natural waters: kinetics and modeling. Mar Chem 78:231–253
Deshmukh AP, Chen Y, Tarchitzky J, Chefetz B, Hatcher PG (2002) Structural characterization of soil organic matter and humic acids in particle-size fractions of an agricultural soil. Soil Sci Soc Am J 66:129–141
Dobrović S, Juretić H, Ružinski N (2007) Photodegradation of natural organic matter in water with UV irradiation at 185 and 254 nm: importance of hydrodynamic conditions on the decomposition rate. Sep Sci Technol 42:1421–1432
Dreybrodt W (1988) Processes in karst systems, physics, chemistry, and geology
Dunfield P, Dumont R, Moore TR (1993) Methane production and consumption in temperate and subarctic peat soils: response to temperature and pH. Soil Biol Biochem 25:321–326
Dupré B, Dessert C, Oliva P, Goddéris Y, Viers J, François L, Millot R, Gaillardet J (2003) Rivers, chemical weathering and Earth's climate. CR Geosci 335:1141–1160
Ebner S, Jaun B, Goenrich M, Thauer R, Harmer J (2010) Binding of coenzyme B induces a major conformational change in the active site of methyl-coenzyme M reductase. JACS 132:567–575
EPI (2007) Science advisory board review of the estimation programs interface (EPI) suite; office of the administrator science advisory board (SAB). U.S. Environmental Protection Agency, Washington
Fang TH (2004) Phosphorus speciation and budget of the East China Sea. Contin Shelf Res 24:1285–1299
Farmer CT, Moore CA, Zika RG, Sikorski RJ (1993) Effects of low and high orinoco river flow on the underwater light field of the Eastern Caribbean basin. J Geophys Res 98:2279–2288
Fenton HJ (1894) Oxidation of tartaric acid in presence of iron. J Chem Soc 65:899–910
Fichot CG, Miller WL (2010) An approach to quantify depth-resolved marine photochemical fluxes using remote sensing: Application to carbon monoxide (CO) photoproduction. Remote Sens Environ 114:1363–1377
Francko DA, Heath RT (1982) UV-sensitive complex phosphorus: association with dissolved humic material and iron in a bog lake. Limnol Oceanogr 27:564–569

Fu P, Mostofa KMG, Wu F, Liu CQ, Li W, Liao H, Wang L, Wang J, Mei Y (2010) Excitation-emission matrix characterization of dissolved organic matter sources in two eutrophic lakes (Southwestern China Plateau). Geochem J 44:99

Fujiwara K, Ushiroda T, Takeda K, Kumamoto YI, Tsubota H (1993) Diurnal and seasonal distribution of hydrogen peroxide in seawater of the Seto Inland Sea. Geochem J 27:103–115

Fujiwara K, Takeda K, Kumamoto Y (1995) Generations of carbonyl sulfide and hydrogen peroxide in the Seto Inland Sea—Photochemical reactions progressing in the coastal seawater. In: Sakai H, Nozaki Y (eds) Biogeochemical processes and Ocean flux in the Western Pacific. TERRAPUB, Tokyo, pp 101–127

Fukuzaki S, Nishio N, Nagai S (1990) Kinetics of the methanogenic fermentation of acetate. Appl Environ Microbiol 56:3158–3163

Galand PE, Saarnio S, Fritze H, Yrjälä K (2002) Depth related diversity of methanogen Archaea in Finnish oligotrophic fen. FEMS Microbiol Ecol 42:441–449

Galand P, Fritze H, Conrad R, Yrjälä K (2005) Pathways for methanogenesis and diversity of methanogenic archaea in three boreal peatland ecosystems. Appl Environ Microbiol 71:2195–2198

Galchenko V (1994) Sulfate Reduction, methane production, and methane oxidation in various water bodies of Banger-Hills Oasis of Antarctica. Microbiology 63:388–396

Gao H, Zepp RG (1998) Factors influencing photoreactions of dissolved organic matter in a coastal river of the southeastern United States. Environ Sci Technol 32:2940–2946

Gennings C, Molot LA, Dillon PJ (2001) Enhanced photochemical loss of organic carbon in acidic waters. Biogeochemistry 52:339–354

Getoff N (1991) Radiation-and photoinduced degradation of pollutants in water. A comparative study. Radiat Phys Chem 37:673–680

Getoff N, Schwoerer F, Markovic V, Sehested K, Nielsen SO (1971) Pulse radiolysis of oxalic acid and oxalates. J Phys Chem 75:749–755

Goldstone J, Pullin M, Bertilsson S, Voelker B (2002) Reactions of hydroxyl radical with humic substances: bleaching, mineralization, and production of bioavailable carbon substrates. Environ Sci Technol 36:364–372

Goodwin S, Zeikus JG (1987) Ecophysiological adaptations of anaerobic bacteria to low pH: analysis of anaerobic digestion in acidic bog sediments. Appl Environ Microbiol 53:57–64

Goodwin S, Conrad R, Zeikus J (1988) Influence of pH on microbial hydrogen metabolism in diverse sedimentary ecosystems. Appl Environ Microbiol 54:590–593

Gopinathan C, Damle P, Hart EJ (1972) Gamma-Ray irradiated sodium chloride as a source of hydrated electrons. J Phys Chem 76:3694–3698

Graneli W, Lindell M, Tranvik L (1996) Photo-oxidative production of dissolved inorganic carbon in lakes of different humic content. Limnol Oceanogr 41:698–706

Graneli W, Lindell M, de Faria B, de Assis EF (1998) Photoproduction of dissolved inorganic carbon in temperate and tropical lakes-dependence on wavelength band and dissolved organic carbon concentration. Biogeochemistry 43:175–195

Grebel JE, Pignatello JJ, Song W, Cooper WJ, Mitch WA (2009) Impact of halides on the photo bleaching of dissolved organic matter. Mar Chem 115:134–144

Gron C, Wassenaar L, Krog M (1996) Origin and structures of groundwater humic substances from three Danish aquifers. Environ Int 22:519–534

Guildford SJ, Hecky RE (2000) Total nitrogen, total phosphorus, and nutrient limitation in lakes and oceans: Is there a common relationship? Limnol Oceanogr 45:1213–1223

Haag WR, Hoigné J (1985) Photo-sensitized oxidation in natural water via OH radicals. Chemosphere 14:1659–1671

Haan H, Boer T, Halma G (1979) Curie point pyrolysis mass-spectrometry of fulvic acids from Tjeukemeer, The Netherlands. Freshwater Biol 9:315–317

Häder DP, Helbling E, Williamson C, Worrest R (2011) Effects of UV radiation on aquatic ecosystems and interactions with climate change. Photochem Photobiol Sci 10:242–260

Harvey HR, Tuttle JH, Bell JT (1995) Kinetics of phytoplankton decay during simulated sedimentation: Changes in biochemical composition and microbial activity under oxic and anoxic conditions. Geochim Cosmochim Acta 59:3367–3377

Hayase K, Shinozuka N (1995) Vertical distribution of fluorescent organic matter along with AOU and nutrients in the equatorial Central Pacific. Mar Chem 48:283–290
Hedges J, Eglinton G, Hatcher P, Kirchman D, Arnosti C, Derenne S, Evershed R, Kögel-Knabner I, De Leeuw J, Littke R (2000) The molecularly-uncharacterized component of non-living organic matter in natural environments. Org Geochem 31:945–958
Hellpointner E, Gäb S (1989) Detection of methyl, hydroxymethyl and hydroxyethyl hydroperoxides in air and precipitation. Nature 337:631–634
Henrichs SM, Doyle AP (1986) Decomposition of 14 C-labeled organic substances in marine sediments. Limnol Oceanogr 31:765–778
Herndl GJ, Müller-Niklas G, Frick J (1993) Major role of ultraviolet-B in controlling bacterio plankton growth in the surface layer of the ocean. Nature 361:717–719
Hernes PJ, Benner R (2003) Photochemical and microbial degradation of dissolved lignin phenols: Implications for the fate of terrigenous dissolved organic matter in marine environments. J Geophys Res 108:3291–3299
Herrmann H (2003) Kinetics of aqueous phase reactions relevant for atmospheric chemistry. Chem Rev 103:4691–4716
Herrmann H, Hoffmann D, Schaefer T, Bräuer P, Tilgner A (2010) Tropospheric aqueous-phase free-radical chemistry: radical Sources, spectra, reaction kinetics and prediction tools. Chem Phys Chem 11:3796–3822
Hewitt CN, Kok GL (1991) Formation and occurrence of organic hydroperoxides in the troposphere: laboratory and field observations. J Atmos Chem 12:181–194
Holmes D, Bond D, O'neil R, Reimers C, Tender L, Lovley D (2004) Microbial communities associated with electrodes harvesting electricity from a variety of aquatic sediments. Microb Ecol 48:178–190
Hongve D (1994) Sunlight degradation of aquatic humic substances. Acta Hydrochim Hydrobiol 22:117–120
Hopkinson CS, Vallino JJ, Nolin A (2002) Decomposition of dissolved organic matter from the continental margin. Deep Sea Res Part II 49:4461–4478
Horn MA, Matthies C, Küsel K, Schramm A, Drake HL (2003) Hydrogenotrophic methanogenesis by moderately acid-tolerant methanogens of a methane-emitting acidic peat. Appl Environ Microbiol 69:74–83
Hornibrook ERC, Longstaffe FJ, Fyfe WS (2000) Evolution of stable carbon isotope compositions for methane and carbon dioxide in freshwater wetlands and other anaerobic environments. Geochim Cosmochim Acta 64:1013–1027
Huber MM, Canonica S, Park GY, von Gunten U (2003) Oxidation of pharmaceuticals during ozonation and advanced oxidation processes. Environ Sci Technol 37:1016–1024
Huisman J, Thi NNP, Karl DM, Sommeijer B (2006) Reduced mixing generates oscillations and chaos in the oceanic deep chlorophyll maximum. Nature 439:322–325
Hulatt CJ, Thomas DN, Bowers DG, Norman L, Zhang C (2009) Exudation and decomposition of chromophoric dissolved organic matter (CDOM) from some temperate macroalgae. Estuar Coast Shelf Sci 84:147–153
Hur J (2011) Microbial changes in selected operational descriptors of dissolved organic matters from various sources in a watershed. Water Air Soil Pollut 215:465–476
Itoh M, Ohte N, Koba K, Sugimoto A, Tani M (2008) Analysis of methane production pathways in a riparian wetland of a temperate forest catchment, using $\delta^{13}C$ of pore water CH_4 and CO_2. J Geophys Res 113:G03005
Ivanov M, Belyaev S, Laurinavichus K (1976) Methods of quantitative investigation of microbiological production and utilization of methane. In: Schlegel HG, Gottschalk G, Pfennig N (eds) Microbial production and utilization of gases. E Goltze, Göttingen, pp 63–67
Jackson A, Hewitt CN (1996) Hydrogen peroxide and organic hydroperoxide concentrations in air in a eucalyptus forest in central Portugal. Atmos Environ 30:819–830
Jeong J, Yoon J (2005) pH effect on OH radical production in photo/ferrioxalate system. Water Res 39:2893–2900

Jetten MSM, Stams AJM, Zehnder AJB (1992) Methanogenesis from acetate: a comparison of the acetate metabolism in Methanothrix soehngenii and Methanosarcina spp. FEMS Microbiol Lett 88:181–197
Johannessen SC (2000) A photochemical sink for dissolved organic carbon in the ocean (176). PhD Thesis, Oceanography, Dalhousie University
Johannessen SC, Peña MA, Quenneville ML (2007) Photochemical production of carbon dioxide during a coastal phytoplankton bloom. Estuar Coast Shelf Sci 73:236–242
Johnson KS, Willason SW, Wiesenburg DA, Lohrenz SE, Arnone RA (1989) Hydrogen peroxide in the western Mediterranean Sea: A tracer for vertical advection. Deep Sea Res Part I 36:241–254
Jones RD (1991) Carbon monoxide and methane distribution and consumption in the photic zone of the Sargasso Sea. Deep Sea Res Part I 38:625–635
Jones RD, Amador JA (1993) Methane and carbon monoxide production, oxidation, and turnover times in the Caribbean Sea as influenced by the Orinoco River. J Geophys Res 98:2353–2359
Jones R, Morita R (1984) Effect of several nitrification inhibitors on carbon monoxide and methane oxidation by ammonium oxidizers. Can J Microbiol 30:1276–1279
Jones A, Shanklin J (1995) Continued decline of total ozone over Halley, Antarctica, since 1985. Nature 376:409–411
Jones JG, Simon BM, Gardener S (1982) Factors affecting methanogenesis and associated anaerobic processes in the sediments of a stratified eutrophic lake. J Gen Microbiol 128:1
Jørgensen NOG, Tranvik L, Edling H, Granéli W, Lindell M (1998) Effects of sunlight on occurrence and bacterial turnover of specific carbon and nitrogen compounds in lake water. FEMS Microbiol Ecol 25:217–227
Kaiser E, Sulzberger B (2004) Phototransformation of riverine dissolved organic matter (DOM) in the presence of abundant iron: effect on DOM bioavailability. Limnol Oceanogr 49:540–554
Kang SF, Liao CH, Po ST (2000) Decolorization of textile wastewater by photo-Fenton oxidation technology. Chemosphere 41:1287–1294
Karelson M (2000) Molecular descriptors in QSAR/QSPR. John Wiley & Sons, Inc: New York
Karentz D, Bothwell M, Coffin R, Hanson A, Herndl G, Kilham S, Lesser M, Lindell M, Moeller R, Morris D (1994) Impact of UV-B radiation on pelagic freshwater ecosystems: report of working group on bacteria and phytoplankton. Ergebn Limnol 43:31–31
Karl DM, Tien G (1997) Temporal variability in dissolved phosphorus concentrations in the subtropical North Pacific Ocean. Mar Chem 56:77–96
Kashefi KK, Holmes DE, Lovley DR, Tor JM (2004) In: (Wilcock WS, DeLong EF, Kelley DS, Baross JA, Cary SC (Ed), The subseafloor biosphere at mid-ocean ridges, American Geophysical Union, Washington pp 199–211
Keppler F, Hamilton JTG, Braß M, Röckmann T (2006) Methane emissions from terrestrial plants under aerobic conditions. Nature 439:187–191
Kerr J, McElroy C (1993) Evidence for large upward trends of ultraviolet-B radiation linked to ozone depletion. Science 262:1032–1034
Kieber DJ, McDaniel J, Mopper K (1989) Photochemical source of biological substrates in sea water: Implications for carbon cycling. Nature 341:637–639
Kieber RJ, Zhou X, Mopper K (1990) Formation of carbonyl compounds from UV-induced photodegradation of humic substances in natural waters: Fate of riverine carbon in the sea. Limnol Oceanogr 35:1503–1515
Kieber D, Mopper K, Qian J, Zafiriou O (2001) Photochemical formation of dissolved inorganic carbon in seawater and its impact on the marine carbon cycle. Paper presented at the Session 2: Photochemical reactions in surface waters ASLO 2001, Albuquerque, February 12–16
Kim DJ, Kim SH (2006) Effect of nitrite concentration on the distribution and competition of nitrite-oxidizing bacteria in nitratation reactor systems and their kinetic characteristics. Water Res 40:887–894
Kim C, Nishimura Y, Nagata T (2006) Role of dissolved organic matter in hypolimnetic mineralization of carbon and nitrogen in a large, monomictic lake. Limnol Oceanogr 51:70–78

Kirchman DL, Suzuki Y, Garside C, Ducklow HW (1991) High turnover rates of dissolved organic carbon during a spring phytoplankton bloom. Nature 352:612–614

Kirk J (1994) Optics of UV-B radiation in natural waters. Ergebn Limnol 43:1–16

Komada T, Schofield OME, Reimers CE (2002) Fluorescence characteristics of organic matter released from coastal sediments during resuspension. Mar Chem 79:81–97

Komissarov G (1994) Photosynthesis: a new look. Sci Russia 5:52–55

Komissarov G (1995) Photosynthesis as a Physicochemical Process. Khim Fiz 14:20–28

Komissarov G (2003) Photosynthesis: the physical-chemical approach. J Adv Chem Phys 2:28–61

Kopacek J, Hejzlar J, Kana J, Porcal P, Klementova S (2003) Photochemical, chemical, and biological transformations of dissolved organic carbon and its effect on alkalinity production in acidified lakes. Limnol Oceanogr 48:106–117

Koschorreck M, Wendt-Potthoff K, Scharf B, Richnow HH (2008) Methanogenesis in the sediment of the acidic Lake Caviahue in Argentina. J Volcanol Geothermal Res 178:197–204

Kotelnikova S, Macario AJL, Pedersen K (1998) Methanobacterium subterraneum sp. nov., a new alkaliphilic, eurythermic and halotolerant methanogen isolated from deep granitic groundwater. Int J Syst Bacteriol 48:357–367

Kotsyurbenko O, Nozhevnikova A, Zavarzin G (1993) Methanogenic degradation of organic matter by anaerobic bacteria at low temperature. Chemosphere 27:1745–1761

Kotsyurbenko O, Simankova M, Nozhevnikova A, Zhilina T, Bolotina N, Lysenko A, Osipov G (1995) New species of psychrophilic acetogens: *Acetobacterium bakii* sp. nov., *A. paludosum* sp. nov., *A. fimetarium* sp. nov. Arch Microbiol 163:29–34

Kotsyurbenko O, Nozhevnikova A, Soloviova T, Zavarzin G (1996) Methanogenesis at low temperatures by microflora of tundra wetland soil. Antonie Van Leeuwenhoek 69:75–86

Kotsyurbenko OR, Glagolev MV, Nozhevnikova AN, Conrad R (2001) Competition between homoacetogenic bacteria and methanogenic archaea for hydrogen at low temperature. FEMS Microbiol Ecol 38:153–159

Kotsyurbenko OR, Chin KJ, Glagolev MV, Stubner S, Simankova MV, Nozhevnikova AN, Conrad R (2004) Acetoclastic and hydrogenotrophic methane production and methanogenic populations in an acidic West-Siberian peat bog. Environ Microbiol 6:1159–1173

Kotsyurbenko O, Friedrich M, Simankova M, Nozhevnikova A, Golyshin P, Timmis K, Conrad R (2007) Shift from acetoclastic to H2-dependent methanogenesis in a West Siberian peat bog at low pH values and isolation of an acidophilic Methanobacterium strain. Appl Environ Microbiol 73:2344–2348

Kramer JB, Canonica S, Hoigné J, Kaschig J (1996) Degradation of fluorescent whitening agents in sunlit natural waters. Environ Sci Technol 30:2227–2234

Kuivila K, Murray J, Devol A, Novelli P (1989) Methane production, sulfate reduction and competition for substrates in the sediments of Lake Washington. Geochim Cosmochim Acta 53:409–416

Kujawinski EB, Longnecker K, Blough NV, Vecchio RD, Finlay L, Kitner JB, Giovannoni SJ (2009) Identification of possible source markers in marine dissolved organic matter using ultrahigh resolution mass spectrometry. Geochim Cosmochim Acta 73:4384–4399

Kulovaara M (1996) Light-induced degradation of aquatic humic substances by simulated sunlight. Int J Environ Anal Chem 62:85–95

Kwok ESC, Atkinson R (1995) Estimation of hydroxyl radical reaction rate constants for gas-phase organic compounds using a structure-reactivity relationship: an update. Atmos Environ 29:1685–1695

Laane R, Gieskes W, Kraay G, Eversdijk A (1985) Oxygen consumption from natural waters by photo-oxidizing processes. Neth J Sea Res 19:125–128

Langvik VA, Akerback N, Holmbom B (1994) Characterization of aromatic structures in humic and fulvic acids. Environ Int 20:61–65

Lansdown J, Quay P, King S (1992) CH_4 production via CO_2 reduction in a temperate bog: a source of $\delta^{13}C$-depleted CH_4. Geochim Cosmochim Acta 56:3493–3503

Lapado JA, Barlaz MA (1997) Isolation and characterization of refuse methanogens. J Appl Microbiol 82:751–758

Laurion I, Ventura M, Catalan J, Psenner R, Sommaruga R (2000) Attenuation of ultraviolet radiation in mountain lakes: Factors controlling the among-and within-lake variability. Limnol Oceanogr 45:1274–1288

Lee HS, Rittmann BE (2009) Evaluation of metabolism using stoichiometry in fermentative biohydrogen. Biotechnol Bioeng 102:749–758

Lee HS, Rittmann BE (2010) Characterization of energy losses in an upflow single-chamber microbial electrolysis cell. International J Hydrogen Energy 35:920–927

Lee HS, Salerno MB, Rittmann BE (2008) Thermodynamic evaluation on H_2 production in glucose fermentation. Environ Sci Technol 42:2401–2407

Leenheer J, Croue JP (2003) Characterizing aquatic dissolved organic matter. Environ Sci Technol 37:18–26

Lehmann MF, Bernasconi SM, Barbieri A, McKenzie JA (2002) Preservation of organic matter and alteration of its carbon and nitrogen isotope composition during simulated and in situ early sedimentary diagenesis. Geochim Cosmochim Acta 66:3573–3584

Letelier RM, Karl DM, Abbott MR, Bidigare RR (2004) Light driven seasonal patterns of chlorophyll and nitrate in the lower euphotic zone of the North Pacific Subtropical Gyre. Limnol Oceanogr 49:508–519

Li K, Crittenden J (2009) Computerized pathway elucidation for hydroxyl radical-induced chain reaction mechanisms in aqueous phase advanced oxidation processes. Environ Sci Technol 43:2831–2837

Li C, Fang HHP (2007) Fermentative hydrogen production from wastewater and solid wastes by mixed cultures. Crit Rev Environ Sci Technol 37:1–39

Li W, Wu F, Liu C, Fu P, Wang J, Mei Y, Wang L, Guo J (2008) Temporal and spatial distributions of dissolved organic carbon and nitrogen in two small lakes on the Southwestern China Plateau. Limnology 9:163–171

Lindell MJ, Rai H (1994) Photochemical oxygen consumption in humic waters. Ergebn Limnol 43:145–145

Lindell M, Granéli W, Tranvik L (1996) Impact of solar (UV)-radiation on bacterial growth in lakes. Aquat Microb Ecol 11:135–141

Liu K, Atkinson L, Chen C, Gao S, Hall J, Macdonald R, Talaue L (2000) Exploring continental margin carbon fluxes on a global scale. Eos Trans AGU 81(52):641. doi:10.1029/EO081i052p00641-01

Liu Z, Dreybrodt W, Wang H (2010) A new direction in effective accounting for the atmospheric CO_2 budget: Considering the combined action of carbonate dissolution, the global water cycle and photosynthetic uptake of DIC by aquatic organisms. Earth Sci Rev 99:162–172

Logan BE, Regan JM (2006) Electricity-producing bacterial communities in microbial fuel cells. TRENDS in Microbiol 14:512–518

Lohrenz SE, Cai WJ, Chen F, Chen X, Tuel M (2010) Seasonal variability in air-sea fluxes of CO_2 in a river-influenced coastal margin. J Geophys Res 115:C10034

Lønborg C, Søndergaard M (2009) Microbial availability and degradation of dissolved organic carbon and nitrogen in two coastal areas. Estuar Coast Shelf Sci 81:513–520

Lønborg C, Álvarez-Salgado XA, Davidson K, Miller AEJ (2009a) Production of bioavailable and refractory dissolved organic matter by coastal heterotrophic microbial populations. Estuar Coast Shelf Sci 82:682–688

Lønborg C, Davidson K, Álvarez-Salgado XA, Miller AEJ (2009b) Bioavailability and bacterial degradation rates of dissolved organic matter in a temperate coastal area during an annual cycle. Mar Chem 113:219–226

Lovley D (1991) Dissimilatory Fe(III) and Mn(IV) reduction. Microbiol Rev 55:259–287

Lovley DR (2006) Bug juice: harvesting electricity with microorganisms. Nature Rev Microbiol 4:497–508

Lovley DR, Chapelle FH (1995) Deep subsurface microbial processes. Rev Geophys 33:365–381

Lovley D, Klug M (1986) Model for the distribution of sulfate reduction and methanogenesis in freshwater sediments. Geochim Cosmochim Acta 50:11–18

Lovley DR, Phillips EJP (1987) Competitive mechanisms for inhibition of sulfate reduction and methane production in the zone of ferric iron reduction in sediments. Appl Environ Microbiol 53:2636–2641
Lovley DR, Coates JD, Blunt-Harris EL, Phillips EJP, Woodward JC (1996) Humic substances as electron acceptors for microbial respiration. Nature 382:445–448
Lovley DR, Holmes DE, Nevin KP (2004) Dissimilatory Fe(III) and Mn(IV) reduction. Adv Microb Physiol 49:219–286
Lund V, Hongve D (1994) Ultraviolet irradiated water containing humic substances inhibits bacterial metabolism. Water Res 28:1111–1116
Ma X, Green SA (2004) Photochemical transformation of dissolved organic carbon in lake superior–an in situ experiment. J Great Lakes Res 30:97–112
Mack J, Bolton JR (1999) Photochemistry of nitrite and nitrate in aqueous solution: a review. J Photochem Photobiol, A 128:1–13
Maddigapu P, Minella M, Vione D, Maurino V, Minero C (2011) Modeling phototransformation reactions in surface water bodies: 2, 4-dichloro-6-nitrophenol as a case study. Environ Sci Technol 45:209
Madronich S (1992) Implications of recent total atmospheric ozone measurements for biologically active ultraviolet radiation reaching the Earth's surface. Geophys Res Lett 19:37–40
Maita Y, Montani S, Ishii J (1982) Early diagenesis of amino acids in Okhotsk Sea sediments. Deep Sea Res Part A 29:485–498
Malcolm R (1985) Geochemistry of stream fulvic and humic substances
Malcolm RL (1990) The uniqueness of humic substances in each of soil, stream and marine environments. Analyt Chim Acta 232:19–30
Managaki S, Takada H (2005) Fluorescent whitening agents in Tokyo Bay sediments: molecular evidence of lateral transport of land-derived particulate matter. Mar Chem 95:113–127
McKnight D, Kimball B, Bencala K (1988) Iron photoreduction and oxidation in an acidic mountain stream. Science 240:637–640
Meier J, Babenzien H-D, Wendt-Potthoff K (2004) Microbial cycling of iron and sulfur in sediments of acidic and pH-neutral mining lakes in Lusatia (Brandenburg, Germany). Biogeochemistry 67:135–156
Merga G, Schuchmann HP, Rao BSM, von Sonntag C (1996) $^{\cdot}$OH radical-induced oxidation of chlorobenzene in aqueous solution in the absence and presence of oxygen. J Chem Soc, Perkin Trans 2:1097–1103
Midorikawa T, Tanoue E (1996) Extraction and characterization of organic ligands from oceanic water columns by immobilized metal ion affinity chromatography. Mar Chem 52:157–171
Midorikawa T, Tanoue E (1998) Molecular masses and chromophoric properties of dissolved organic ligands for copper (II) in oceanic water. Mar Chem 62:219–239
Miles CJ, Brezonik PL (1981) Oxygen consumption in humic-colored waters by a photochemical ferrous-ferric catalytic cycle. Environ Sci Technol 15:1089–1095
Milferstedt K, Youngblut ND, Whitaker RJ (2010) Spatial structure and persistence of methanogen populations in humic bog lakes. ISME J 4:764–776
Miller WL (1998) Effects of UV radiation on aquatic humus: photochemical principles and experimental considerations. In: Hessen DO, Tranvik LJ (eds) Aquatic humic substances. Springer, Berlin, pp 125–144
Miller WL, Moran MA (1997) Interaction of photochemical and microbial processes in the degradation of refractory dissolved organic matter from a coastal marine environment. Limnol Oceanogr 42:1317–1324
Miller WL, Zepp RG (1995) Photochemical production of dissolved inorganic carbon from terrestrial organic matter: significance to the oceanic organic. Geophys Res Lett 22:417–420
Millero FJ, Sotolongo S (1989) The oxidation of Fe(II) with H_2O_2 in seawater. Geochim Cosmochim Acta 53:1867–1873
Minakata D, Crittenden J (2011) Linear free energy relationships between aqueous phase Hydroxyl radical reaction rate constants and free energy of activation. Environ Sci Technol 45:3479–3486

Minakata D, Li K, Westerhoff P, Crittenden J (2009) Development of a group contribution method to predict aqueous phase hydroxyl radical ($HO^•$) reaction rate constants. Environ Sci Technol 43:6220–6227

Minakata D, Song W, Crittenden J (2011) Reactivity of aqueous phase hydroxyl radical with halogenated carboxylate anions: experimental and theoretical studies. Environ Sci Technol 45:6057–6065

Minella M, Rogora M, Vione D, Maurino V, Minero C (2011) A model approach to assess the long-term trends of indirect photochemistry in lake water. The case of Lake Maggiore (NW Italy). Sci Total Environ 409:3463–3471

Minero C, Chiron S, Falletti G, Maurino V, Pelizzetti E, Ajassa R, Carlotti ME, Vione D (2007) Photochemincal processes involving nitrite in surface water samples. Aquat Sci 69:71–85

Minor E, Pothen J, Dalzell B, Abdulla H, Mopper K (2006) Effects of salinity changes on the photodegradation and ultraviolet-visible absorbance of terrestrial dissolved organic matter. Limnol Oceanogr 51:2181–2186

Moffett J, Zafiriou O (1990) An investigation of hydrogen peroxide chemistry in surface waters of Vineyard Sound with H218O2 and 18O2. Limnol Oceanogr 35:1221–1229

Moffett JW, Zika RG (1987) Reaction kinetics of hydrogen peroxide with copper and iron in seawater. Environ Sci Technol 21:804–810

Molot LA, Dillon PJ (1997) Photolytic regulation of dissolved organic carbon in northern lakes. Global Biogeochem Cy 11:357–365

Molot LA, Hudson JJ, Dillon PJ, Miller SA (2005) Effect of pH on photo-oxidation of dissolved organic carbon by hydroxyl radicals in a coloured, softwater stream. Aquat Sci 67:189–195

Monod A, Poulain L, Grubert S, Voisin D, Wortham H (2005) Kinetics of OH-initiated oxidation of oxygenated organic compounds in the aqueous phase: new rate constants, structure–activity relationships and atmospheric implications. Atmos Environ 39:7667–7688

Moore CA, Farmer CT, Zika RG (1993) Influence of the Orinoco river on hydrogen peroxide distribution and production in the eastern Caribbean. J Geophys Res 98:2289–2298

Mopper K, Stahovec WL (1986) Sources and sinks of low molecular weight organic carbonyl compounds in seawater. Mar Chem 19:305–321

Mopper K, Zhou X (1990) Hydroxyl radical photoproduction in the sea and its potential impact on marine processes. Science 250:661–664

Mopper K, Zhou X, Kieber RJ, Kieber DJ, Sikorski RJ, Jones RD (1991) Photochemical degradation of dissolved organic carbon and its impact on the oceanic carbon cycle. Nature 353:60–62

Moran MA, Hodson RE (1994) Support of bacterioplankton production by dissolved humic substances from three marine environments. Mar Ecol Prog Ser 110:241–241

Moran MA, Zepp RG (1997) Role of photoreactions in the formation of biologically labile compounds from dissolved organic matter. Limnol Oceanogr 42:1307–1316

Moran MA, Sheldon WM Jr, Zepp RG (2000) Carbon loss and optical property changes during long-term photochemical and biological degradation of estuarine dissolved organic matter. Limnol Oceanogr 45:1254–1264

Morris DP, Hargreaves BR (1997) The role of photochemical degradation of dissolved organic carbon in regulating the UV transparency of three lakes on the Pocono Plateau. Limnol Oceanogr 42:239–249

Morris DP, Zagarese H, Williamson CE, Balseiro EG, Hargreaves BR, Modenutti B, Moeller R, Queimalinos C (1995) The attenuation of solar UV radiation in lakes and the role of dissolved organic carbon. Limnol Oceanogr 40:1381–1391

Mortatti J, Probst JL (2003) Silicate rock weathering and atmospheric/soil CO_2 uptake in the Amazon basin estimated from river water geochemistry: seasonal and spatial variations. Chem Geol 197:177–196

Morvan B, Dore J, Rieu-Lesme F, Foucat L, Fonty G, Gouet P (1994) Establishment of hydrogen-utilizing bacteria in the rumen of the newborn lamb. FEMS Microbiol Lett 117:249–256

Mostofa K (2005) Dynamics, characteristics and photochemical processes of fluorescent dissolved organic matter and peroxides in river water. Ph D Thesis September 2005, Hiroshima University, Japan
Mostofa KMG, Sakugawa H (2003) Spatial and temporal variation of hydrogen peroxide in stream and river waters: Effect of photo-bio-physio-chemical processes of aquatic matters. Abs 13th Ann VM Goldschmidt Conf, Kurashiki, Japan. Geochim Cosmochim Acta 67(18S):A309. In, 2003
Mostofa KMG, Sakugawa H (2009) Spatial and temporal variations and factors controlling the concentrations of hydrogen peroxide and organic peroxides in rivers. Environ Chem 6:524–534
Mostofa KMG, Honda Y, Sakugawa H (2005a) Dynamics and optical nature of fluorescent dissolved organic matter in river waters in Hiroshima Prefecture, Japan. Geochem J 39:257–271
Mostofa KMG, Yoshioka T, Konohira E, Tanoue E, Hayakawa K, Takahashi M (2005b) Three-dimensional fluorescence as a tool for investigating the dynamics of dissolved organic matter in the Lake Biwa watershed. Limnology 6:101–115
Mostofa KMG, Yoshioka T, Konohira E, Tanoue E (2007a) Dynamics and characteristics of fluorescent dissolved organic matter in the groundwater, river and lake water. Water Air Soil Pollut 184:157–176
Mostofa KMG, Yoshioka T, Konohira E, Tanoue E (2007b) Photodegradation of fluorescent dissolved organic matter in river waters. Geochem J 41:323–331
Mostofa K, Wu F, Yoshioka T, Sakugawa H, Tanoue E (2009a) Dissolved organic matter in the aquatic environments. In: Wu F, B X (eds) Natural organic matter and its significance in the environment. Science press, Beijing pp 3–66
Mostofa K, Liu C, Wu F, Fu P, Ying W, Yuan J (2009b) Overview of key biogeochemical functions in Lake Ecosystem: Impacts of organic matter pollution and global warming. Paper presented at the Proceedings of the 13th World Lake Conf Wuhan, Wuhan, China 1–5 Nov 2009
Mostofa KMG, Wu F, Liu CQ, Fang WL, Yuan J, Ying WL, Wen L, Yi M (2010) Characterization of Nanming river (Southwestern China) sewerage-impacted pollution using an excitation-emission matrix and PARAFAC. Limnology 11:217–231
Mostofa KMG, Wu F, Liu CQ, Vione D, Yoshioka T, Sakugawa H, Tanoue E (2011) Photochemical, microbial and metal complexation behavior of fluorescent dissolved organic matter in the aquatic environments. Geochem J 45:235–254
Mvula E, Schuchmann MN, von Sonntag C (2001) Reactions of phenol-OH-adduct radicals. Phenoxyl radical formation by water elimination vs. oxidation by dioxygen. J Chem Soc, Perkin Trans 2:264–268
Naganuma T, Konishil S, Nakane TIT (1996) Photodegradation or photoalteration? Microbial assay of the effect of UV-B on dissolved organic matter. Mar Ecol Prog Ser 135:309–310
Nagase M, Matsuo T (1982) Interactions between amino-acid-degrading bacteria and methanogenic bacteria in anaerobic digestion. Biotechnol Bioeng 24:2227–2239
Nakagawa F, Yoshida N, Nojiri Y, Makarov VN (2002) Production of methane from alasses in eastern Siberia: implications from its 14C and stable isotopic compositions. Global Biogeochem Cy 16:1041
Nakane K, Kohno T, Horikoshi T, Nakatsubo T (1997) Soil carbon cycling at a black spruce (Picea Mariana) forest stand in Saskatchewan, Canada. J Geophys Res 102:28785–28793
Nakatani N, Hashimoto N, Sakugawa H (2004) An evaluation of hydroxyl radical formation in river water and the potential for photodegradation of bisphenol A. In: Hill RJ, Leventhal J, Aizenshtat Z, Baedecker MJ, Claypool G, Eganhouse R, Goldhaber M, Peters K, (Eds), The Geochemical society special publication series 9, Geochemical investigations in earth and space Science: A Tribute to Isaac R Kaplan Elsevier, Amsterdam, pp 233–242
Namsaraev B, Dulov L, Sokolova E (1995) Bacterial methane production in the bottom sediments of Lake Baikal. Mikrobiologiya 64:411–412
Nieto-Cid M, Alvarez-Salgado X, Pérez F (2006) Microbial and photochemical reactivity of fluorescent dissolved organic matter in a coastal upwelling system. Limnol Oceanogr 51:1391–1400

Nozhevnikova A, Kotsyurbenko O, Simankova M (1994) Acetogenesis at low temperature. In: Drake H (ed) Acetogenesis. Chapman and Hall, New York, pp 416–431
Obernosterer I, Ruardij P, Herndl GJ (2001) Spatial and diurnal dynamics of dissolved organic matter (DOM) fluorescence and H_2O_2 and the photochemical oxygen demand of surface water DOM across the subtropical Atlantic Ocean. Limnol Oceanogr 46:632–643
Omar A, Olsen A, Johannessen T, Hoppema M, Thomas H, Borges A (2010) Spatiotemporal variations of fCO_2 in the North Sea. Ocean Sci 6:77–89
Öquist M, Svensson B (2002) Vascular plants as regulators of methane emissions from a subarctic mire ecosystem. J Geophys Res 107:4580
Osburn CL, Morris DP (2003) Photochemistry of chromophoric dissolved organic matter in natural waters. UV Eff Aquat Org Ecosyst 1:185–217
Osburn CL, O'Sullivan DW, Boyd TJ (2009) Increases in the longwave photobleaching of chromophoric dissolved organic matter in coastal waters. Limnol Oceanogr 54:145–159
O'Sullivan DW, Neale PJ, Coffin RB, Boyd TJ, Osburn CL (2005) Photochemical production of hydrogen peroxide and methylhydroperoxide in coastal waters. Mar Chem 97:14–33
Palenik B, Morel F (1988) Dark production of H_2O_2 in the Sargasso Sea. Limnol Oceanogr 33:1606–1611
Parlanti E, Wörz K, Geoffroy L, Lamotte M (2000) Dissolved organic matter fluorescence spectroscopy as a tool to estimate biological activity in a coastal zone submitted to anthropogenic inputs. Org Geochem 31:1765–1781
Patel G, Sprott G, Fein J (1990) Isolation and characterization of Methanobacterium espanolae sp. nov., a mesophilic, moderately acidiphilic methanogen. Int J Syst Bacteriol 40:12–18
Pelmenschikov V, Blomberg MRA, Siegbahn PEM, Crabtree RH (2002) A mechanism from quantum chemical studies for methane formation in methanogenesis. JACS 124:4039–4049
Peña-Méndez EM, Havel J, Patočka J (2005) Humic substances–compounds of still unknown structure: applications in agriculture, industry, environment, and biomedicine. J Appl Biomed 3:13–24
Peuravuori J, Pihlaja K (1997) Pyrolysis electron impact mass spectrometry in studying aquatic humic substances. Analyt Chim Acta 350:241–247
Phelps T, Zeikus J (1984) Influence of pH on terminal carbon metabolism in anoxic sediments from a mildly acidic lake. Appl Environ Microbiol 48:1088–1095
Piazena H, Häder DP (1994) Penetration of solar UV irradiation in coastal lagoons of the southern Baltic Sea and its effect on phytoplankton communities. Photochem Photobiol 60:463–469
Premier G, Kim J, Massanet-Nicolau J, Kyazze G, Esteves S, Penumathsa B, Rodríguez J, Maddy J, Dinsdale R, Guwy A (2012) Integration of biohydrogen, biomethane and bioelectrochemical systems. Renewable Energy (in press)
Psenner R (1988) Alkalinity generation in a soft-water lake: watershed and in-lake processes. Limnol Oceanogr 33:1463–1475
Pullin MJ, Bertilsson S, Goldstone JV, Voelker BM (2004) Effects of sunlight and hydroxyl radical on dissolved organic matter: Bacterial growth efficiency and production of carboxylic acids and other substrates. Limnol Oceanogr 49:2011–2022
Qian J, Mopper K, Kieber DJ (2001) Photochemical production of the hydroxyl radical in Antarctic waters. Deep Sea Res Part I 48:741–759
Raghavan N, Steenken S (1980) Electrophilic reaction of the hydroxyl radical with phenol. Determination of the distribution of isomeric dihydroxycyclohexadienyl radicals. JACS 102:3495–3499
Randall C, Harvey V, Manney G, Orsolini Y, Codrescu M, Sioris C, Brohede S, Haley C, Gordley L, Zawodny J (2005) Stratospheric effects of energetic particle precipitation in 2003–2004. Geophys Res Lett 32:L05802
Ratte M, Bujok O, Spitzy A, Rudolph J (1998) Photochemical alkene formation in seawater from dissolved organic carbon: Results from laboratory experiments. J Geophys Res 103:5707–5717
Reche I, Pace ML, Cole JJ (1999) Relationship of trophic and chemical conditions to photobleaching of dissolved organic matter in lake ecosystems. Biogeochemistry 44:259–280
Reguera G, McCarthy KD, Mehta T, Nicoll JS, Tuominen MT, Lovley DR (2005) Extracellular electron transfer via microbial nanowires. Nature 435:1098–1101

Reitner B, Herndl GJ, Herzig A (1997) Role of ultraviolet-B radiation on photochemical and microbial oxygen consumption in a humic-rich shallow lake. Limnol Oceanogr 42:950–960
Rex M, Harris NRP, von der Gathen P, Lehmann R, Braathen GO, Reimer E, Beck A, Chipperfield MP, Alfier R, Allaart M (1997) Prolonged stratospheric ozone loss in the 1995–96 Arctic winter. Nature 389:835–838
Richard LE, Peake BM, Rusak SA, Cooper WJ, Burritt DJ (2007) Production and decomposition dynamics of hydrogen peroxide in freshwater. Environ Chem 4:49–54
Roberts EC, Priscu JC, Wolf C, Lyons WB, Laybourn-Parry J (2004) The distribution of microplankton in the McMurdo Dry Valley Lakes, Antarctica: response to ecosystem legacy or present-day climatic controls? Polar Biol 27:238–249
Rochelle-Newall E, Fisher T (2002) Production of chromophoric dissolved organic matter fluorescence in marine and estuarine environments: an investigation into the role of phytoplankton. Mar Chem 77:7–21
Roden EE, Wetzel RG (1996) Organic carbon oxidation and suppression of methane production by microbial Fe(III) oxide reduction in vegetated and unvegetated freshwater wetland sediments. Limnol Oceanogr 41:1733–1748
Rontani JF (2001) Visible light-dependent degradation of lipidic phytoplanktonic components during senescence: a review. Phytochemistry 58:187–202
Rosenfeld JK (1979) Amino acid diagenesis and adsorption in nearshore anoxic sediments. Limnol Oceanogr 24:1014–1021
Rosenfeldt EJ, Linden KG (2004) Degradation of endocrine disrupting chemicals bisphenol A, ethinyl estradiol, and estradiol during UV photolysis and advanced oxidation processes. Environ Sci Technol 38:5476–5483
Rosenstock B, Zwisler W, Simon M (2005) Bacterial consumption of humic and non-humic low and high molecular weight DOM and the effect of solar irradiation on the turnover of labile DOM in the Southern Ocean. Microb Ecol 50:90–101
Rothfuss F, Conrad R (1992) Vertical profiles of CH_4 concentrations, dissolved substrates and processes involved in CH_4 production in a flooded Italian rice field. Biogeochemistry 18:137–152
Rozendal RA, Hamelers HVM, Molenkamp RJ, Buisman CJN (2007) Performance of single chamber biocatalyzed electrolysis with different types of ion exchange membranes. Water Res 41:1984–1994
Sadtler (1968) The Sadtler standard spectra 1968 edition. Sadtler Research Lab, Inc, Philadelphia
Safarzadeh-Amiri A, Bolton JR, Cater SR (1996) Ferrioxalate-mediated solar degradation of organic contaminants in water. Sol Energy 56:439–443
Safarzadeh-Amiri A, Bolton JR, Cater SR (1997) Ferrioxalate-mediated photodegradation of organic pollutants in contaminated water. Water Res 31:787–798
Sakugawa H, Takami A, Kawai H, Takeda K, Fujiwara K, Hirata S (2000a) The occurrence of organic peroxide in seawater. In: Handa N, Tanoue E, Hama T (eds) Dynamics and characterization of marine organic matter. TERRAPUB, Kluwer, pp 231–240
Salonen K, Vähätalo A (1994) Photochemical mineralisation of dissolved organic matter in Lake Skjervatjern. Environ Int 20:307–312
Salonen K, Kairesalo T, Jones R (1992) Dissolved organic matter in lacustrine ecosystems: Energy-source and system regulator. Hydrobiology 229:1–291
Samuilov V, Bezryadnov D, Gusev M, Kitashov A, Fedorenko T (2001) Hydrogen peroxide inhibits photosynthetic electron transport in cells of cyanobacteria. Biochemistry (Moscow) 66:640–645
Sandbeck KA, Ward DM (1981) Fate of immediate methane precursors in low-sulfate, hot-spring algal-bacterial mats. Appl Environ Microbiol 41:775–782
Sarmiento JL, Slater R, Barber R, Bopp L, Doney SC, Hirst A, Kleypas J, Matear R, Mikolajewicz U, Monfray P (2004) Response of ocean ecosystems to climate warming. Global Biogeochem Cy 18:3001–3023
Schindler D, Turner M, Stainton M, Linsey G (1986) Natural sources of acid neutralizing capacity in low alkalinity lakes of the Precambrian Shield. Science 232:844–847
Schink B (1997) Energetics of syntrophic cooperation in methanogenic degradation. Microbiol Mol Biol Rev 61:262–280

Schmitt-Kopplin P, Hertkorn N, Schulten HR, Kettrup A (1998) Structural changes in a dissolved soil humic acid during photochemical degradation processes under O_2 and N_2 atmosphere. Environ Sci Technol 32:2531–2541

Schmittner A (2005) Decline of the marine ecosystem caused by a reduction in the Atlantic overturning circulation. Nature 434:628–633

Schulz S, Conrad R (1995) Effect of algal deposition on acetate and methane concentrations in the profundal sediment of a deep lake (Lake Constance). FEMS Microbiol Ecol 16:251–260

Schulz S, Conrad R (1996) Influence of temperature on pathways to methane production in the permanently cold profundal sediment of Lake Constance. FEMS Microbiol Ecol 20:1–14

Schwarzenbach RPGP, Imboden DM (1993) Environmental organic chemistry. Wiley, New York, pp 436–484

Scully N, Lean D (1994) The attenuation of ultraviolet radiation in temperate lakes (With 8 figures and 5 tables). Arch Hydrobiol Beih Ergebn Limnol 43:135–144

Scully N, McQueen D, Lean D, Cooper W (1996) Hydrogen peroxide formation: the interaction of ultraviolet radiation and dissolved organic carbon in lake waters along a 43–75? N Gradient. Limnol Oceanogr 41:540–548

Senesi N (1990) Molecular and quantitative aspects of the chemistry of fulvic acid and its interactions with metal ions and organic chemicals: Part II. The fluorescence spectroscopy approach. Analyt Chim Acta 232:77–106

Serban A, Nissenbaum A (1986) Humic acid association with peroxidase and catalase. Soil Biol Biochem 18:41–44

Shiller AM, Duan S, van Erp P, Bianchi TS (2006) Photo-oxidation of dissolved organic matter in river water and its effect on trace element speciation. Limnol Oceanogr 51(4):1716–1728

Shriner RLHW, Cox RFB (1943) p-Nitrophenyl Isocyanate. Org Synth Coll 2:453–455

Siegel DA, Michaels AF (1996) Quantification of non-algal light attenuation in the Sargasso sea: implications for biogeochemistry and remote sensing. Deep Sea Res Part II 43:321–345

Sikorski RJ, Zika R (1993a) Modeling mixed-layer photochemistry of H_2O_2: physical and chemical modeling of distribution. J Geophys Res 98:2329–2340

Sikorski RJ, Zika R (1993b) Modeling mixed-layer photochemistry of H_2O_2: optical and chemical modeling of production. J Geophys Res 98:2315–2328

Simankova MV, Kotsyurbenko OR, Stackebrandt E, Kostrikina NA, Lysenko AM, Osipov GA, Nozhevnikova AN (2000) *Acetobacterium tundrae* sp. nov., a new psychrophilic acetogenic bacterium from tundra soil. Arch Microbiol 174:440–447

Sizova MV, Panikov NS, Tourova TP, Flanagan PW (2003) Isolation and characterization of oligotrophic acido-tolerant methanogenic consortia from a Sphagnum peat bog. FEMS Microbiol Ecol 45:301–315

Skoog A, Wedborg M, Fogelqvist E (1996) Photobleaching of fluorescence and the organic carbon concentration in a coastal environment. Mar Chem 55:333–345

Smith RC, Baker KS (1981) Optical properties of the clearest natural waters (200–800 nm). Appl Opt 20:177–184

Song R, Westerhoff P, Minear RA, Amy GL (1996) Interaction between bromine and natural organic matter. (Minear RA, Amy GL (eds), Water disinfection and natural organic matter, Americal Chemical Society, Washington, pp 298–321

Southworth BA, Voelker BM (2003) Hydroxyl radical production via the photo-Fenton reaction in the presence of fulvic acid. Environ Sci Technol 37:1130–1136

Stedmon CA, Markager S (2005) Tracing the production and degradation of autochthonous fractions of dissolved organic matter by fluorescence analysis. Limnol Oceanogr 50:1415–1426

Stedmon CA, Markager S, Tranvik L, Kronberg L, Slätis T, Martinsen W (2007) Photochemical production of ammonium and transformation of dissolved organic matter in the Baltic Sea. Mar Chem 104:227–240

Steelink C (2002) Peer Reviewed: investigating humic acids in soils. Anal Chem 74:326–333

Steinberg SM, Venkatesan MI, Kaplan IR (1987) Org Geochem of sediments from the continental margin off southern New England, U.S.A.—part I. Amino acids, carbohydrates and lignin. Mar Chem 21:249–265

Stiller M, Nissenbaum A (1999) Geochemical investigation of phosphorus and nitrogen in the hypersaline Dead Sea. Geochim Cosmochim Acta 63:3467–3475

Ström L, Ekberg A, Mastepanov M, Røjle Christensen T (2003) The effect of vascular plants on carbon turnover and methane emissions from a tundra wetland. Glob Change Biol 9:1185–1192

Strome D, Miller MC (1978) Photolytic changes in dissolved humic substances. Int Ver Theor Angew Limnol 20:1248–1254

Stumm W, Lee GF (1961) Oxygenation of ferrous iron. Ind Eng Chem 53:143–146

Sun MY, Lee C, Aller RC (1993) Anoxic and Oxic Degradation of ^14C-labeled Chloropigments and a 14C-labeled Diatom in Long Island Sound sediments. Limnol Oceanogr 38:1438–1451

Suzumura M, Ingall ED (2004) Distribution and dynamics of various forms of phosphorus in seawater: insights from field observations in the Pacific Ocean and a laboratory experiment. Deep Sea Res Part I 51:1113–1130

Takeda K, Takedoi H, Yamaji S, Ohta K, Sakugawa H (2004) Determination of hydroxyl radical photoproduction rates in natural waters. Anal Sci 20:153–158

Tanaka K, Suda Y, Kondo N, Sugahara K (1985) O_3 tolerance and the ascorbate-dependent H_2O_2 decomposing system in chloroplasts. Plant Cell Physiol 26:1425–1431

Thauer RK (1998) Biochemistry of methanogenesis: a tribute to Marjory Stephenson. Microbiology 144:2377–2406

Thauer RK, Jungermann K, Decker K (1977) Energy conservation in chemotrophic anaerobic bacteria. Bacteriolog Rev 41:100

Thauer RK, Kaster AK, Seedorf H, Buckel W, Hedderich R (2008) Methanogenic archaea: ecologically relevant differences in energy conservation. Nature Rev Microbiol 6:579–591

Thebrath B, Rothfuss F, Whiticar M, Conrad R (1993) Methane production in littoral sediment of Lake Constance. FEMS Microbiol Lett 102:279–289

Thomas DN, Lara RJ (1995) Photodegradation of algal derived dissolved organic carbon. Mar Ecol Prog Ser Oldendorf 116:309–310

Tranvik LJ (1992) Allochthonous dissolved organic matter as an energy source for pelagic bacteria and the concept of the microbial loop. Hydrobiology 229:107–114

Tranvik L, Kokalj S (1998) Decreased biodegradability of algal DOC due to interactive effects of UV radiation and humic matter. Aquat Microb Ecol 14:301–307

Tranvik LJ, Downing JA, Cotner JB, Loiselle SA, Striegl RG, Ballatore TJ, Dillon P, Knoll L, Kutser T, Larsen S (2009) Lakes and reservoirs as regulators of carbon cycling and climate. Limnol Oceanogr 54:2298–2314

Twardowski MS, Donaghay PL (2002) Photobleaching of aquatic dissolved materials: Absorption removal, spectral alteration, and their interrelationship. J Geophys Res 107:3091

Uchida M, Nakatsubo T, Horikoshi T, Nakane K (1998) Contribution of micro-organisms to the carbon dynamics in black spruce (Picea Mariana) forest soil in Canada. Ecol Res 13:17–26

Uchida M, Nakatsubo T, Kasai Y, Nakane K, Horikoshi T (2000) Altitudinal differences in organic matter mass loss and fungal biomass in a subalpine coniferous forest, Mt. Fuji, Japan. Arct Antarct Alp Res 32:262–269

Upton M, Hill B, Edwards C, Saunders J, Ritchie D, Lloyd D (2000) Combined molecular ecological and confocal laser scanning microscopic analysis of peat bog methanogen populations. FEMS Microbiol Lett 193:275–281

Utsumi M, Belova SE, King GM, Uchiyama H (2003) Phylogenetic comparison of methanogen diversity in different wetland soils. J Gen Appl Microbiol 49:75–83

Vähätalo AV, Järvinen M (2007) Photochemically produced bioavailable nitrogen from biologically recalcitrant dissolved organic matter stimulates production of a nitrogen-limited microbial food web in the Baltic Sea. Limnol Oceanogr 52:132–143

Vähätalo AV, Wetzel RG (2004) Photochemical and microbial decomposition of chromophoric dissolved organic matter during long (months–years) exposures. Mar Chem 89:313–326

Vahatalo AV, Salkinoja-Salonen M, Taalas P, Salonen K (2000) Spectrum of the quantum yield for photochemical mineralization of dissolved organic carbon in a humic lake. Limnol Oceanogr 45:664–676

Valentine RL, Zepp RG (1993) Formation of carbon monoxide from the photodegradation of terrestrial dissolved organic carbon in natural waters. Environ Sci Technol 27:409–412

Vione D, Merlo F, Maurino V, Minero C (2004) Effect of humic acids on the Fenton degradation of phenol. Environ Chem Lett 2:129–133

Vione D, Falletti G, Maurino V, Minero C, Pelizzetti E, Malandrino M, Ajassa R, Olariu RI, Arsene C (2006) Sources and sinks of hydroxyl radicals upon irradiation of natural water samples. Environ Sci Technol 40:3775–3781

Vione D, Minero C, Maurino V, Olariu R-L, Arsene C, KMG M (2011) Photochemical transformation processes in surface waters. In: Maes KJ, Willems JM (Eds), Photochemistry: UV/VIS spectroscopy, photochemical reactions and photosynthesis, Nova Science Publishers, Chapter 9

Vione D, Lauri V, Minero C, Maurino V, Malandrino M, Carlotti ME, Olariu RI, Arsene C (2009) Photostability and photolability of dissolved organic matter upon irradiation of natural water samples under simulated sunlight. Aquat Sci 71:34–45

Voelker BM, Morel FMM, Sulzberger B (1997) Iron redox cycling in surface waters: effects of humic substances and light. Environ Sci Technol 31:1004–1011

von Gunten U, Oliveras Y (1997) Kinetics of the reaction between hydrogen peroxide and hypobromous acid: implication on water treatment and natural systems. Water Res 31:900–906

Waiser MJ, Robarts RD (2000) Changes in composition and reactivity of allochthonous DOM in a prairie saline lake. Limnol Oceanogr 45:763–774

Weiss R, Carmack ECC, Koropalov V (1991) Deep-water renewal and biological production in Lake Baikal. Nature 349:665–669

Westerhoff P, Yoon Y, Snyder S, Wert E (2005) Fate of endocrine-disruptor, pharmaceutical, and personal care product chemicals during simulated drinking water treatment processes. Environ Sci Technol 39:6649–6663

Westermann P (1994) The effect of incubation temperature on steady-state concentrations of hydrogen and volatile fatty acids during anaerobic degradation in slurries from wetland sediments. FEMS Microbiol Ecol 13:295–302

Wetzel RG (1984) Detrital dissolved and particulate organic carbon functions in aquatic ecosystems. Bull Mar Sci 35:503–509

Wetzel RG (1992) Gradient-dominated ecosystems: sources and regulatory functions of dissolved organic matter in freshwater ecosystems. Hydrobiology 229:181–198

Wetzel RG, Hatcher PG, Bianchi TS (1995) Natural photolysis by ultraviolet irradiance of recalcitrant dissolved organic matter to simple substrates for rapid bacterial metabolism. Limnol Oceanogr 40:1369–1380

White EM, Vaughan PP, Zepp RG (2003) Role of the photo-Fenton reaction in the production of hydroxyl radicals and photobleaching of colored dissolved organic matter in a coastal river of the southeastern United States. Aquat Sci 65:402–414

White EM, Kieber DJ, Sherrard J, Miller WL, Mopper K (2010) Carbon dioxide and carbon monoxide photoproduction quantum yields in the Delaware Estuary. Mar Chem 118:11–21

Whiticar MJ (1999) Carbon and hydrogen isotope systematics of bacterial formation and oxidation of methane. Chem Geol 161:291–314

Williams RT, Crawford RL (1985) Methanogenic bacteria, including an acid-tolerant strain, from peatlands. Appl Environ Microbiol 50:1542–1544

Winfrey M, Zeikus J (1979) Anaerobic metabolism of immediate methane precursors in Lake Mendota. Appl Environ Microbiol 37:244–253

Winter AR, Fish TAE, Playle RC, Smith DS, Curtis PJ (2007) Photodegradation of natural organic matter from diverse freshwater sources. Aquat Toxicol 84:215–222

Worakit S, Boone D, Mah R, Abdel-Samie M, El-Halwagi M (1986) *Methanobacterium alcaliphilum* sp. nov., an H_2-utilizing methanogen that grows at high pH values. Int J Syst Bacteriol 36:380–382

Wu F, Tanoue E (2001) Molecular mass distribution and fluorescence characteristics of dissolved organic ligands for copper (II) in Lake Biwa, Japan. Org Geochem 32:11–20
Wu FC, Mills RB, Cai YR, Evans RD, Dillon PJ (2005) Photodegradation-induced changes in dissolved organic matter in acidic waters. Can J Fish Aquat Sci 62:1019–1027
Xie H, Zafiriou OC, Cai WJ, Zepp RG, Wang Y (2004) Photooxidation and its effects on the carboxyl content of dissolved organic matter in two coastal rivers in the southeastern United States. Environ Sci Technol 38:4113–4119
Yamashita Y, Tanoue E (2004) In situ production of chromophoric dissolved organic matter in coastal environments. Geophys Res Lett 31 [doi: 101029/2004GL019734]
Yamashita Y, Tanoue E (2008) Production of bio-refractory fluorescent dissolved organic matter in the ocean interior. Nature Geosci 1:579–582
Yang ST, Guo M (1990) Kinetics of methanogenesis from whey permeate in packed bed immobilized cells bioreactor. Biotechnol Bioeng 36:427–436
Yocis BH, Kieber DJ, Mopper K (2000) Photochemical production of hydrogen peroxide in Antarctic waters. Deep Sea Res Part I 47:1077–1099
Yoshioka T, Mostofa KMG, Konohira E, Tanoue E, Hayakawa K, Takahashi M, Ueda S, Katsuyama M, Khodzher T, Bashenkhaeva N (2007) Distribution and characteristics of molecular size fractions of freshwater-dissolved organic matter in watershed environments: its implication to degradation. Limnology 8:29–44
Zafiriou OC, True MB (1979a) Nitrate photolysis in seawater by sunlight. Mar Chem 8:33–42
Zafiriou OC, True MB (1979b) Nitrite photolysis in seawater by sunlight. Mar Chem 8:9–32
Zafiriou OC, Joussot-Dubien J, Zepp RG, Zika RG (1984) Photochemistry of natural waters. Environ Sci Technol 18:356A–371A
Zafiriou OC, True MB, Hayon E (1987) Consequences of OH radical reaction in sea water: formation and decay of Br2− ion radical. In: Photochemistry of environmental aquatic systems Am Chem Soc, Washington, pp 89–105
Zeikus J, Weimer P, Nelson D, Daniels L (1975) Bacterial methanogenesis: acetate as a methane precursor in pure culture. Arch Microbiol 104:129–134
Zepp RG, Hoigne J, Bader H (1987a) Nitrate-induced photooxidation of trace organic chemicals in water. Environ Sci Technol 21:443–450
Zepp RG, Skurlatov Y, Pierce J (1987b) Algal-induced decay and formation of hydrogen peroxide in water: its possible role in oxidation of anilines by algae. In: Zika RG, Cooper WJ (eds) Photochemistry of environmental aquatic systems, ACS symp Ser 327. Am Chem Soc, Washington, pp 213–224
Zepp RG, Faust BC, Hoigne J (1992) Hydroxyl radical formation in aqueous reactions (pH 3–8) of iron (II) with hydrogen peroxide: the photo-Fenton reaction. Environ Sci Technol 26:313–319
Zepp RG, Callaghan T, Erickson D (1998) Effects of enhanced solar ultraviolet radiation on biogeochemical cycles. J Photochem Photobiol, B 46:69–82
Zepp R, Erickson D III, Paul N, Sulzberger B (2011) Effects of solar UV radiation and climate change on biogeochemical cycling: interactions and feedbacks. Photochem Photobiol Sci 10:261–279
Zhang Y, Zhu L, Zeng X, Lin Y (2004) The biogeochemical cycling of phosphorus in the upper ocean of the East China Sea. Estuar Coast Shelf Sci 60:369–379
Zhang Y, van Dijk MA, Liu M, Zhu G, Qin B (2009a) The contribution of phytoplankton degradation to chromophoric dissolved organic matter (CDOM) in eutrophic shallow lakes: field and experimental evidence. Water Res 43:4685–4697
Zhang Y, Liu M, Qin B, Feng S (2009b) Photochemical degradation of chromophoric-dissolved organic matter exposed to simulated UV-B and natural solar radiation. Hydrobiology 627:159–168
Zhang D, Pan X, Mostofa KMG, Chen X, Mu G, Wu F, Liu J, Song W, Yang J, Liu Y (2010) Complexation between Hg(II) and biofilm extracellular polymeric substances: an application of fluorescence spectroscopy. J Hazard Mat 175:359–365

Zhao Y, Yu Y, Feng W, Shen Y (2003) Growth and production of free-living heterotrophic nanoflagellates in a eutrophic lake–Lake Donghu, Wuhan, China. Hydrobiology 498:85–95
Zhou X, Mopper K (1990) Determination of photochemically produced hydroxyl radicals in seawater and freshwater. Mar Chem 30:71–88
Zika RG (1981) Marine organic photochemistry. In: Elsevier Oceanography Series, vol 31, pp 299–325
Zinder SH (1990) Conversion of acetic acid to methane by thermophiles. FEMS Microbiol Lett 75:125–137
Zinder SH (1993) Physiological ecology of methanogens. In: Ferry JG (Ed) Methanogenesis: Ecology, Physiology, Biochemistry and Genetics, Chapman and Hall, New York pp 128–206

Colored and Chromophoric Dissolved Organic Matter in Natural Waters

Khan M. G. Mostofa, Cong-qiang Liu, Davide Vione, M. Abdul Mottaleb, Hiroshi Ogawa, Shafi M. Tareq and Takahito Yoshioka

1 Introduction

Chromophoric and colored dissolved organic matter (CDOM) is the opitically active components of bulk dissolved organic matter (DOM) composing of a complex mixture of organic compounds of both allochthonous and autochthonous origin, which absorb light at both ultraviolet (UV) and visible wavelengths (Bricaud et al. 1981; Arrigo and Brown 1996; Nelson et al. 1998; Nelson et al. 2004; Warnock et al. 1999; Nelson and Siegel 2002; Vähätalo and Wetzel 2004; Coble 2007; Zhang et al. 2009; Mostofa et al. 2009). Allochthonous organic substances

K. M. G. Mostofa (✉) · C. Q. Liu
State Key Laboratory of Environmental Geochemistry, Chinese Academy of Sciences, Institute of Geochemistry, Guiyang 550002, PR, China
e-mail: mostofa@vip.gyig.ac.cn

D. Vione
Dipartimento di Chimica Analitica, University of Turin, I-10125, Turin, Italy
Centro Interdipartimentale NatRisk, I-10095 Grugliasco, (TO), Italy

M. A. Mottaleb
Department of Chemistry/Physics, Center for Innovation and Entrepreneurship (CIE), Northwest Missouri State University, 800 University Drive, Maryville, MO 64468, USA

H. Ogawa
Atmospheric and Ocean Research Institute, The University of Tokyo, 1-15-1, Minamidai, Nakano, Tokyo 164-8639, Japan

S. M. Tareq
Department of Environmental Sciences, Jahangirnagar University, Dhaka 1342, Bangladesh

T. Yoshioka
Field Science Education and Research Center, Kyoto University, Kitashirakawa Oiwake-cho, Sakyo-ku, Kyoto 606-8502, Japan

K. M. G. Mostofa et al. (eds.), *Photobiogeochemistry of Organic Matter*, Environmental Science and Engineering, DOI: 10.1007/978-3-642-32223-5_5,

are generally derived from terrestrial plant materials in soil, while autochthonous organic substances are produced mostly from algae, phytoplankton and bacteria within the water column. Allochthonous CDOM (mostly fulvic and humic acids) originating in terrestrial environments flows through rivers and estuaries onto coastal shelves and then reaches the open ocean. During such transport it experiences large changes in ionic composition and physicochemical environment (Mostofa et al. 2009; Malcolm 1985; Malcolm 1990; Wetzel 1992; Nakane et al. 1997; Uchida et al. 2000; Vodacek et al. 1997; Mitra et al. 2000; Fahey et al. 2005; Murphy et al. 2008). Autochthonous CDOM of algal, phytoplankton and bacterial origin is generally composed of autochthonous fulvic acids, carbohydrates, amino acids, proteins, lipids, organic acids and so on (Nelson et al. 2004; Coble 2007; Zhang et al. 2009; Coble 1996; Tanoue 2000; Jennings and Steinberg 1994; Rochelle-Newall and Fisher 2002; Yamashita and Tanoue 2003; Yamashita and Tanoue 2004; Yamashita and Tanoue 2008; Stedmon and Markager 2005; Stedmon et al. 2007; Stedmon et al. 2007; Wada et al. 2007; Helms et al. 2008; Hulatt et al. 2009; Ortega-Retuerta et al. 2009; Mostofa et al. 2009). Phytoplankton is capable of releasing 10–60 % of the carbon and 15–50 % of the nitrogen assimilated during photosynthesis (Sundh 1992; Bronk et al. 1994; Braven et al. 1995; Malinsky-Rushansky and Legrand 1996; Slawyk et al. 1998; Slawyk et al. 2000). As a consequence, CDOM levels are significantly increased after phytoplankton blooms in natural waters (Billen and Fontigny 1987; Ittekkot 1982). The contributions of allocthonous fulvic and humic acids, autochthonous fulvic acids, carbohydrates, amino acids, proteins, lipids and organic acids are significantly different among rivers, lakes and oceans. Such differences are discussed in detail in the DOM chapter.

The absorption and scattering coefficients of water were determined firstly by the aquatic scientist Gamburtsev in 1924 (cited in Kozlyaninov (MV 1972)) and then in other studies (Clarke and James 1939; Duntley 1942; Yentsch 1960; Preisendorfer 1961; Latimer 1963; Sullivan 1963; Kozlyaninov and Pelevin 1966; Jerlov 1968). Jerlov (1968) has been the first to hypothesize that CDOM in natural waters absorbs maximally in the blue region of the spectrum and that its absorption decreases exponentially with increasing wavelength, up to the photosynthetically available radiation (PAR) waveband. Since then a lot of research studies have been conducted on the absorption and scattering properties of CDOM, chlorophyll and particulate material in natural waters (Kozlyaninov and Pelevin 1966; Clarke et al. 1970; Tyler and Smith 1970; Lorenzen 1972; Petzold 1972; Morel 1973; Arvesen et al. 1973; Clarke and Ewing 1974; Duntley et al. 1974; Gordon et al. 1975; Maul and Gordon 1976; Prieur 1976; Morel and Prieur 1977; Bukata et al. 1979; Kirk 1981; Gordon and Morel 1983; Kirk 1988).

High molecular weight DOM, and particularly allochthonous fulvic and humic acids and autochthonous fulvic acids, absorb radiation along a broad spectrum from 250 to 800 nm (Warnock et al. 1999; Zhang et al. 2009; Jennings and Steinberg 1994; Rochelle-Newall and Fisher 2002; Wada et al. 2007; Helms et al. 2008; Hulatt et al. 2009; Ishiwatari 1973; Lawrence 1980; Zepp and Schlotzhauer 1981; Hayase and Tsubota 1985; Davies-Colley and Vant 1987; Zhang et al.

2009; Du et al. 2010). The humic substances (fulvic and humic acids) of allochthonous origin and autochthonous fulvic acids of algal origin are colored as they absorb visible light. They can thus be denoted both as colored and as chromophoric DOM (CDOM). Other terminology includes *yellow substances* or *Gelbstoff* in natural waters. On the other hand, low molecular weight CDOM also shows absorption at shorter wavelengths, such as formaldehyde at 207–250 nm, acetaldehyde at 208–224 nm, acetate at 204–270 nm, glyoxal at <240 nm, malonate at 225–240 nm, and pyruvate at 200–227 nm (Strome and Miller 1978; Mopper and Stahovec 1986; Kieber et al. 1990; Mopper et al. 1991; Wetzel et al. 1995; Dahlén et al. 1996). These LMW organic substances do not show any color as they do not absorb in the visible, and they can thus be denoted as CDOM but not as colored DOM.

Different variables such as the occurrence of allochthonous fulvic and humic acids and/or autochthonous fulvic acids, particulate materials (e.g. algae), chlorophyll *a* or phytoplankton, as well as incident light wavelengths and solar zenith angle, snow, ice, and water itself are responsible for the absorption and scattering of UV and PAR radiation, which may affect deeper waters in lakes and oceans (Smith and Baker 1981; Kirk 1984, 1991, 1994; Carder et al. 1989; McKnight et al. 1994; Scully and Lean 1994; Morris et al. 1995; Vernet and Whitehead 1996; Laurion et al. 1997; Sommaruga and Psenner 1997; Kahru and Mitchell 1998; Laurion et al. 2000; Markager and Vincent 2000; Belzile et al. 2000; Shank et al. 2005; Warren et al. 2006; Green et al. 2008; Hayakawa and Sugiyama 2008; Zhao et al. 2009; Shank et al. 2010).

CDOM absorption properties are significantly affected by several factors: DOM content and its chemical nature (Ishiwatari 1973; Lawrence 1980; Zepp and Schlotzhauer 1981; Hayase and Tsubota 1985; McKnight et al. 1994; Dubach et al. 1964; Zanardi-Lamardo et al. 2004; Singh et al. 2010; Singh et al. 2010; Minero et al. 2007; Minero et al. 2007; Vione et al. 2010), occurrence and types of suspended particulate matter (Zhang et al. 2009; Dupouy et al. 2010; Uusikivi et al. 2010; Gregg and Casey 2009; O'Donnell et al. 2010), photoinduced and microbial processes in freshwater and marine environments (Shank et al. 2010; Boehme and Coble 2000; Moran et al. 2000; Winch et al. 2002; Hernes and Benner 2003; Pullin et al. 2004; Scully et al. 2004; Winter et al. 2007; Zhang et al. 2007; Zhang and Qin 2007; Ma and Green 2004), and salinity (Singh et al. 2010; Singh et al. 2010; Hernes and Benner 2003; del Vecchio and Blough 2004; Blough and del Vecchio 2002; Fournier 2007). The CDOM plays several key roles in biogeochemical processes such as photoinduced reactions and biological processes, and it minimizes the deleterious consequences of UV radiation exposure for phytoplankton and other organisms in natural waters (Wetzel et al. 1995; Schindler et al. 1996; Williamson et al. 1996; Yan et al. 1996; Moran and Zepp 1997; Thomas-Smith and Blough 2001; Mostofa and Sakugawa 2009). Several reviews have been carried out on CDOM including its optical and chemical properties, sinks and distribution, relationship of CDOM with salinity, DOC and fluorescence, photochemistry, ocean color, and finally remote sensing applications in natural waters (Coble 2007; del Vecchio and Blough 2004; Blough and del Vecchio 2002; del Castillo 2005).

This chapter will give a general overview on the sources and nature of colored and CDOM, biogeochemical functions of CDOM, optical variables, chromophores in the CDOM, and the theory of CDOM absorbance in natural waters. Moreover, the controlling factors of CDOM that are most important in natural waters will be discussed. Two key factors such as photoinduced and microbial changes in CDOM are also discussed. Finally, a discussion is devoted to the importance of CDOM in natural waters.

1.1 Key Biogeochemical Functions of CDOM

The biogeochemical functions of DOM (discussed in chapter "Dissolved Organic Matter in Natural Waters") have many similarities with those of CDOM, because DOM contains both colored and chromophoric DOM. The key biogeochemical functions of CDOM associated with natural waters are reported as follows. (i) Distribution of colored DOM in the ocean could be a useful indicator for better understanding DOC contents and distribution (del Vecchio and Blough 2004; Hayase et al. 1987; Vodacek et al. 1995; Ferrari et al. 1996; Stabenau and Zika 2004; Prahl and Coble 1994), salinity distribution (Laane and Kramer 1990; Dorsch and Bidleman 1982; Willey and Atkinson 1982; Carder et al. 1993), water mass mixing (Hujerslev et al. 1996; Nieke et al. 1997) and pollutant dispersion (Laane and Kramer 1990). (ii) Colored DOM (mostly allochthonous fulvic and humic acids, and autocthonous fulvic acids) is responsible for the photoinduced generation of hydrogen peroxide (H_2O_2) and organic peroxides (ROOH) and for initiation of photo-oxidation processes in natural waters upon sunlight absorption (Mostofa and Sakugawa 2009; Moore et al. 1993; Vione et al. 2006; Richard et al. 2007; al Housari et al. 2010; Clark et al. 2009; Minella et al. 2011). (iii) Colored DOM is partially responsible for the photoinduced generation of hydroxyl radicals ($HO^{\bullet}$), strong oxidizing agents that may decompose the colored CDOM itself and other organic substances in aqueous media (Minero et al. 2007; Vione et al. 2006). The production of $HO^{\bullet}$ by CDOM could take place upon water oxidation by the triplet states $^3CDOM^*$ or through formation of H_2O_2, either upon direct photoinduced dissociation ($H_2O_2 + h\upsilon \rightarrow 2HO^{\bullet}$) or through Fenton or photo-Fenton decomposition (Zepp et al. 1992; Legrini et al. 1993; Von Sonntag et al. 1993; Takeda et al. 2004; Nakatani et al. 2007). (iv) The absorption of light by chromophores in CDOM initiates several important processes in natural waters such as the release of heat (Kirk 1988), the production of luminescence (Coble 1996; Vodacek et al. 1995; del Castillo and Miller 2008; Blough and Green 1995) and the formation of numerous photo products (Ma and Green 2004; Moran and Zepp 1997; Corin et al. 1996; Miller 1998; Blough and Zepp 1995). (v) Photodegradation of CDOM in surface waters results in the direct loss of absorption and fluorescence that lead to a decrease in the number of the chromophores of CDOM, thereby causing a decline of CDOM's absorption properties and photoinduced nature (Moran et al. 2000; Skoog et al. 1996; Reche et al. 1999; Whitehead and Vernet 2000; Twardowski and Donaghay 2001; del Vecchio and Blough 2002; Mostofa et al.

2007; Mostofa et al. 2007; Patsayeva et al. 1991; Kouassi and Zika 1990; Kouassi et al. 1990). (vi) CDOM absorbs light of both ultraviolet and visible wavelengths, which reduces the amount of PAR for photosynthesis (Bricaud et al. 1981; Arrigo and Brown 1996; Kirk 1994; Kalle 1966) and influences ocean color as determined by remote sensing (del Castillo and Miller 2008; Carder et al. 1991; Carder et al. 1999; Hoge et al. 1995; Hoge et al. 2001). (vii) Phytoplankton usually absorbs strongly in the blue and weakly in the green part of the spectrum. This is applied for the estimation of the chlorophyll *a* concentration within the upper layer of the water column, followed by empirical algorithms of ocean color based on blue-to-green ratios (McClain et al. 2004; Coble et al. 1998; O'Reilly et al. 1998). (viii) CDOM absorbance could be used for determining the chemical composition and structure of CDOM (González-Vila et al. 2001; Hedges et al. 2000; Osburn et al. 2001; Stabenau et al. 2004; Bracchini et al. 2010). (ix) The spectral slope (S) is often used as a proxy for CDOM composition, including the ratio of fulvic to humic acids and molecular weight (Twardowski et al. 2004). Salinity is negatively related to the spectral slopes from 275 to 295 nm, which might be considered a good proxy for DOM molecular weight (Ortega-Retuerta et al. 2010). (x) Finally, CDOM is responsible for the global carbon cycle and for biogeochemical processes through production, distribution, transport and decomposition of organic compounds in natural waters (Moran et al. 2000; Ma and Green 2004; Mostofa et al. 2007; Hedges et al. 2000; Hedges 1992; Amon and Benner 1994; Takahashi et al. 1995; Vähätalo et al. 2000; Ogawa and Tanoue 2003; Medina-Sánchez et al. 2006).

2 CDOM in Natural Waters

Colored and chromophoric dissolved organic matter (CDOM) is defined as a optically active component of bulk DOM of both allochthonous and autochthonous origin that absorbs light over a broad range of ultra-violet (UV) and visible wavelengths (200–800 nm), exhibiting absorbance spectra that generally decrease with increasing wavelength for freshwater and marine water (Fig. 1) (Bricaud et al. 1981; Arrigo and Brown 1996; Nelson et al. 1998; Warnock et al. 1999; Nelson and Siegel 2002; Vähätalo and Wetzel 2004; Coble 2007; Zhang et al. 2009). The absorbance of CDOM is useful for the knowledge of the contents of the materials present and for the identification of absorbance changes due to physical, photoinduced and biological processes in a variety of waters (Vähätalo and Wetzel 2004; Coble 2007; Vodacek et al. 1997; del Vecchio and Blough 2004; del Vecchio and Blough 2002; Vähätalo et al. 2000). It has been reported that there are differences in levels and optical properties between freshwater and marine water CDOM. Extreme enrichment in CDOM is usually observed in freshwater environments (Fig. 1) (Vähätalo and Wetzel 2004; del Vecchio and Blough 2004; del Vecchio and Blough 2002; Conmy et al. 2004; Kowalczuk et al. 2003). CDOM shows absorption at wavelengths 450–800 nm in freshwater (Kowalczuk et al. 2003), but usually not in marine waters. Carbon-specific absorption coefficients for riverine

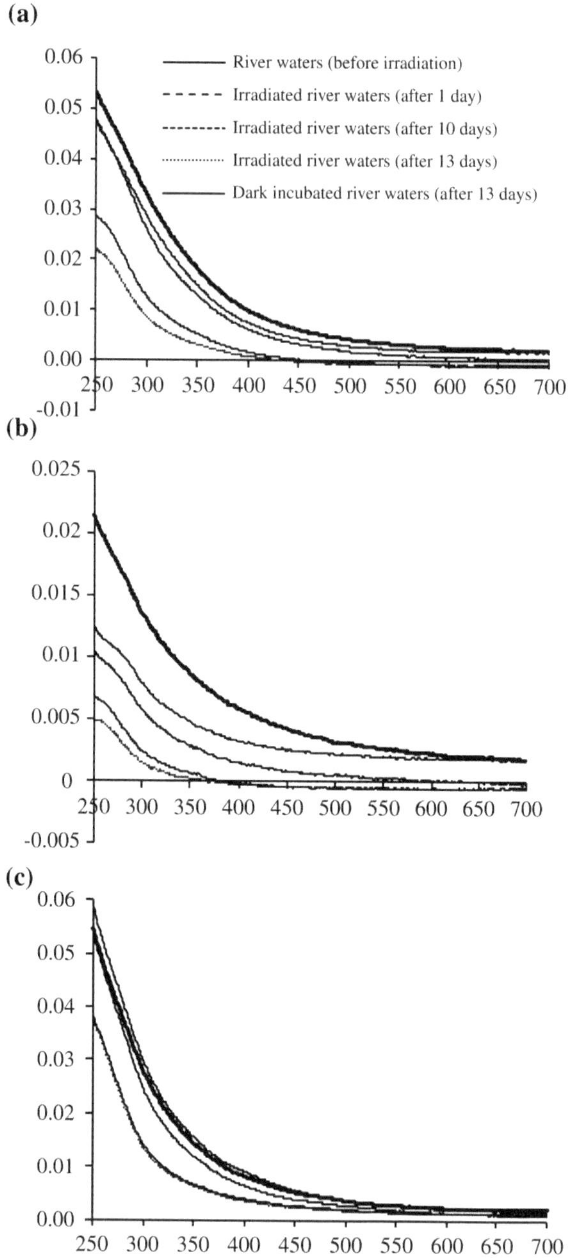

Fig. 1 Absorption spectra of upstream CDOM (**a**, **b** Nishi-Mattaya and Higashi-Mataya, Lake Biwa watershed, Japan), and downstream CDOM (**c** Yasu River, Lake Biwa watershed, Japan). *Data source* Mostofa KMG et al. (unpublished data)

sources of CDOM are 10–150 times higher than those of marine CDOM, whilst coastal waters that receive riverine inputs typically possess higher CDOM absorption coefficients than open ocean waters (Arrigo and Brown 1996; Vähätalo and Wetzel 2004; Carder et al. 1989; del Vecchio and Blough 2004; del Vecchio and Blough 2002; Kowalczuk et al. 2003; Green and Blough 1994). This might be due to the large quantities of humic substances in freshwaters, which usually absorb radiation at $\lambda > 450$ nm. However, riverine input of chromophores contained in freshwater CDOM to the coastal marine environment are decomposed by photodegradation in the coastal area (Vähätalo and Wetzel 2004; del Vecchio and Blough 2002; Vähätalo et al. 2000). The combination of photodegradation and dilution prevents large amount of freshwater CDOM to reach the open ocean. Autochthonous CDOM usually increases relative to chlorophyll *a* concentration in the surface waters of lakes and oceans (Zhang et al. 2009; Sasaki et al. 2005; Mostofa et al. 2011). It is therefore suggested that phytoplankton degradation begins after the spring bloom in surface waters.

2.1 Sources and Nature of CDOM in Natural Waters

CDOM is generally originated from two key sources, allochthonous and autochthonous. The key sources of allochthonous DOM that includes fulvic and humic acids (humic substances) are vascular plant material and particulate detrital pools in terrestrial soil ecosystems. On the other hand, the key contributors for autochthonous DOM in natural waters and in sediment pore waters are phytoplankton or algal biomass, bacteria, coral, coral reef, submerged aquatic vegetation, seagrass and marsh- and mangrove forest. The sources and nature of CDOM are very similar to those of DOM, which have been explained in detail in chapter "Dissolved Organic Matter in Natural Waters".

2.2 Chromophores in CDOM

A chromophore is defined as a part of an organic molecule (functional group) with or without electron-donating heteroatoms such as N, O, and S, as a functional group with a highly unsaturated aliphatic carbon chain, or as a molecule with a structure that can hold up an electron or has extensive π-electron systems, which can absorb photons (light energy) with significant efficiency, causing the promotion from the ground state to an excited one. The key chromophores in a molecule or DOM in natural waters are –OH, –CH=O, –C=O, –COOH, $-COOCH_3$, $-OCH_3$, $-NH_2$, –NH–, –CH=CH–, –CH=CH–COOH, $-OCH_3$, $-(NH_2)CH-COOH$, S-, O- or N-containing aromatic compounds, Schiff-base derivatives (–N=C–C=C–N–) and so on (Mostofa et al. 2009a; Malcolm 1985; Strome and Miller 1978; Mopper and Stahovec 1986; Kieber et al. 1990; Mopper et al. 1991; Wetzel et al. 1995; Mostofa and Sakugawa 2009; Corin et al. 1996; Senesi 1990; Leenheer and Croue 2003; Peña-Méndez et al. 2005; Seitzinger et al. 2005; Zhang et al. 2005). Fulvic and humic acids of vascular plant origin have a molecular structure containing carboxy- and methoxy- benzenes and phenolic groups, carboxyl groups, alcoholic OH, carbohydrate OH, –C=C–, hydroxycoumarin-like structures, chromone, xanthone, quinoline, and O-, N-, S- and P-containing functional groups. They include aromatic carbon (17–30) and aliphatic carbon (47–63 %) (Malcolm 1985; Senesi 1990; Leenheer and Croue 2003; Steelink 2002). All of these functional groups can be considered as key chromophores in fulvic and humic acids in natural waters.

2.3 Theory of CDOM Absorbance

Photon absorption by a CDOM chromophore in aqueous solution firstly induces the excitation of an electron from its ground state to an excited one (Fig. 2), (Mostofa et al. 2009; Senesi 1990). Three types of electronic transitions occur with the CDOM chromophores in natural waters due to the absorption of UV or visible radiation: (i) transitions involving π, σ, and *n* electrons; (ii) transitions involving charge-transfer electrons, and (iii) transitions involving *d*- and *f*- orbital electrons in metals. The CDOM chromophores (e.g. fulvic acid, humic acid and tryptophan) mostly undergo transitions involving *n* or π electrons to the π^* excited state or charge-transfer electrons, and excitation of unpaired electrons in *d*- and *f*- orbitals (e.g. fulvic acid complexes with transition metals such as Cu(II) and Fe(III), having unpaired electrons) (Senesi 1990; Schulman 1985; Voelker and Sulzberger 1996; Senesi 1990; Fox 1990; Morales et al. 1997; Grabowski et al. 2003). The chromophores of CDOM absorb radiation in the wavelength range 200–700 nm and control the penetration along the water column of UV-B (280–320 nm), UV-A (320–400 nm) and total UV radiation (280–400 nm), as

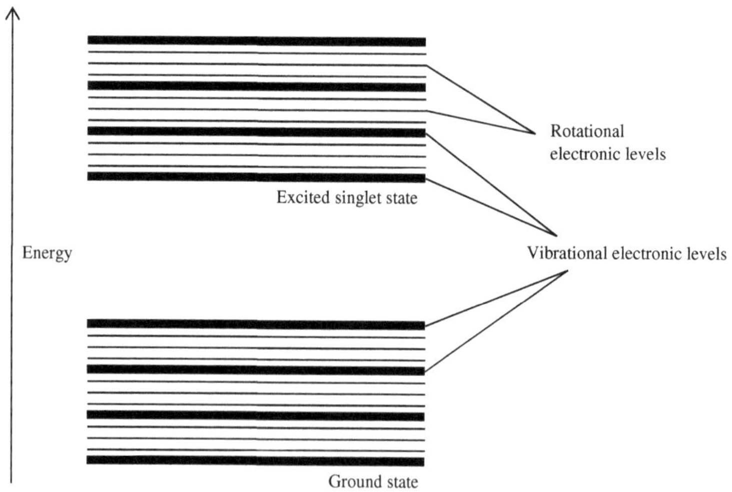

Fig. 2 A schematic energy level diagram for an organic molecule showing their rotational and vibrational electronic levels

well as of photosynthetically active radiation (PAR, 400–700 nm). CDOM absorbance in aqueous solution is directly proportional to the path length (l) and the concentration of the absorbing CDOM species (c). The *Beer's Law* states that $A = \boldsymbol{\varepsilon l c}$, where $\boldsymbol{\varepsilon}$ is a constant of proportionality called the molar absorption coefficient. The attenuation of radiation in solution increases with increasing CDOM absorbance.

As far as PAR radiation is concerned, CDOM in natural waters absorbs maximally in the blue region of the spectrum and its absorption decreases exponentially with increasing wavelength. CDOM spectra have been fitted to an exponential function by Bricaud and his co-workers (Bricaud et al. 1981) as described below (Eq. 2.1):

$$a_{\mathrm{CDOM}}(\lambda) = a_{\mathrm{CDOM}}(\lambda_{\mathrm{o}})\mathrm{e}^{-S(\lambda-\lambda\mathrm{o})} \quad (2.1)$$

where $a_{\mathrm{CDOM}}(\lambda)$ and $a_{\mathrm{CDOM}}(\lambda_{\mathrm{O}})$ are the absorption coefficients at a specific wavelength (λ) and at a reference wavelength (λ_{o}), respectively. S is the slope of the absorption spectrum, determined using either a nonlinear least-squares fitting routine over the wavelength range 290–700 nm or a linear least-squares fit of the log-transformed absorption data, over the range from 290 nm to the instrument detection limit (Vodacek et al. 1997). It is shown that the S value is significantly affected by the presence of different organic substances. Humic acids have a lower S value than fulvic acids (Zepp and Schlotzhauer 1981; Carder et al. 1989; Markager and Vincent 2000. Jerlov (1968) suggested an S value of 15 μm^{-1} whilst Bricaud et al. (Bricaud et al. 1981) showed S to vary between 10 and 20 μm^{-1} with a mean value of 14 μm^{-1}. The absorption coefficient is determined using the following equation

$$a_{\mathrm{CDOM}}(\lambda) = 2.303 \times A_{\mathrm{CDOM}}(\lambda)/l \quad (2.2)$$

where $A_{\mathrm{CDOM}}(\lambda)$ is the absorbance ($\log I_{\mathrm{o}}/I$), and l is the path length (in meters). The spectral slope of the absorption spectrum is widely used in remote sensing in coastal and marine environments (Vodacek et al. 1995; Hoge et al. 1995).

2.4 Absorption Coefficients of CDOM and Particulate Matter in Natural Waters

CDOM of both autochthonous and allochthonous origin generally exhibits UV and visible absorption spectra with low absorbance at longer wavelengths, and the absorbance increases with decreasing wavelength from 700 to 200 nm (Fig. 1) (Rochelle-Newall and Fisher 2002; Wada et al. 2007; Helms et al. 2008; Hulatt et al. 2009; Hayase and Tsubota 1985; Davies-Colley and Vant 1987; Zhang et al. 2009; Du et al. 2010; Odriozola et al. 2007; Bricaud et al. 2010). Allochthonous fulvic acids generally exhibit monotonous spectra whilst allochthonous humic acids have a shoulder around 400 nm in aqueous media (Fig. 3) (Ishiwatari 1973; Lawrence 1980; Zepp and Schlotzhauer 1981; Hayase and Tsubota 1985). Dissolved mycosporine based amino acids (MAAs), released by dinoflagellates,

Fig. 3 Relationship between absorption coefficients examined at 320 and 480 nm and molecular weight fractionations of extracted humic acid (HA) and fulvic acid (FA) in sediment samples collected from Tokyo Bay using Amicon Diaflo Ultrafilters. Figure produced from *Data source* Hayase and Tsubota (1985)

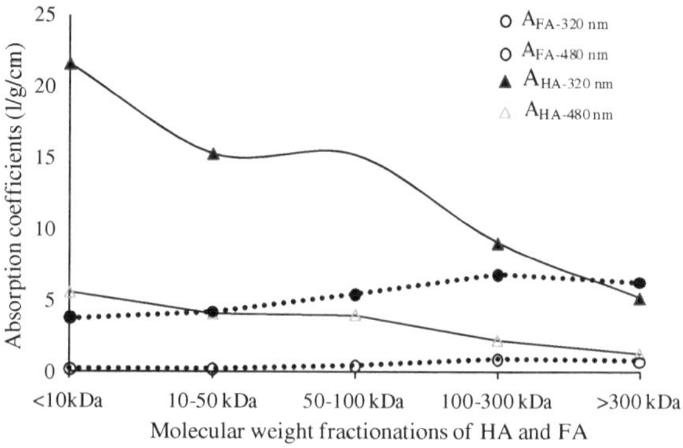

may have a large influence on the absorption of UV light through the water column (Vernet and Whitehead 1996; Kahru and Mitchell 1998; Whitehead and Vernet 2000; Vernet et al. 1989). The MAAs are frequently identified in a wide phyletic assortment of marine organisms and can provide partial photoprotection from UV radiation (Karentz et al. 1991; Bandaranayake 1998; Jeffrey et al. 1999). It is reported that *Lingulodinium polyedrum* (formerly known as *Gonyaulax polyedra* (Dodge 1989)) produces seven types of MAAs with absorption maxima ranging between 310 and 364 nm (Vernet and Whitehead 1996). It is shown that radiation absorption by dissolved MAAs, assessed from the measured MAAs concentrations, significantly correlates with DOM UV absorption ($r^2 = 0.77$) and accounts for up to 10 % of the total DOM absorption at 330 nm (Whitehead and Vernet 2000). The chlorophyll-specific absorption coefficient for phytoplankton is high during the summer season and low during the spring bloom season. This might be the consequence of the occurrence of large diatoms during the spring bloom, while small phytoplankton dominates during the summer period (Sasaki et al. 2005). In addition, the autochthonous CDOM originated from algal or phytoplankton biomass shows strong absorption at 250–700 nm (Zhang et al. 2009; Jennings and Steinberg 1994; Rochelle-Newall and Fisher 2002; Wada et al. 2007; Helms et al. 2008; Hulatt et al. 2009). A study of Bay waters has shown that the relative contribution of CDOM absorption to the total absorption coefficient is >50 % (Sasaki et al. 2005).

Particulate material (phytoplankton or algae and detritus) have absorption spectra in the UV and visible region with several shoulders (Fig. 4a) (Zhang et al. 2009; Odriozola et al. 2007; Smith et al. 2004; Vantrepotte et al. 2007). Phytoplankton shows a strong absorption around 440 and a relatively weak absorption around 680 nm in aqueous media (Fig. 4a), (Zhang et al. 2009). The absorption coefficients of phytoplankton at 440 nm are significantly higher (63–88 %) than those of detritus studied during dry and wet seasons in seawater (Odriozola et al. 2007).

Blue light attenuation in the water column is often dominated by CDOM rather than by phytoplankton absorption. For instance, average CDOM absorption in

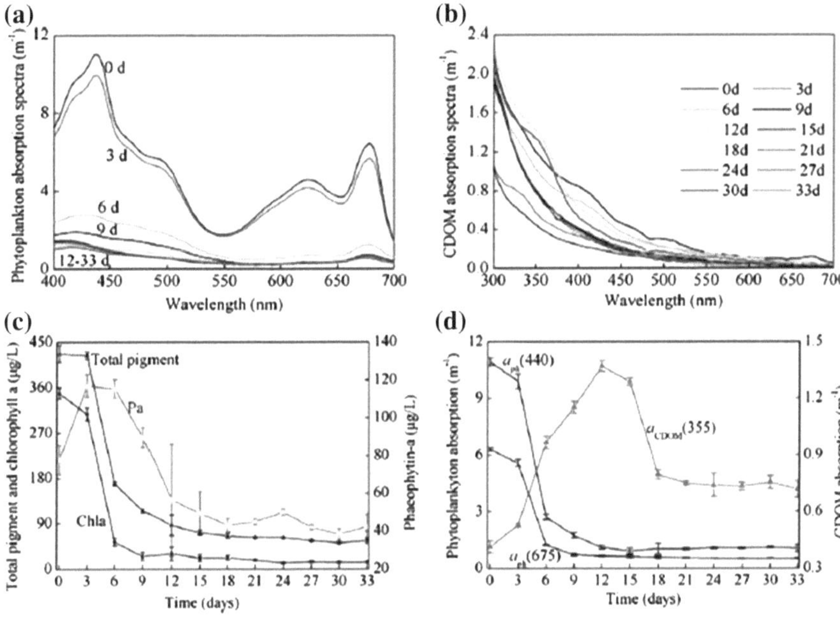

Fig. 4 Changes in the mean absorption spectra (n = 3) of (**a**) phytoplankton pigment absorption [$a_{ph}(\lambda)$], (**b**) CDOM absorption [$a_{CDOM}(\lambda)$], (**c**) The concentrations of total pigment, chlorophyll *a* and Phaeophytin-*a* (P_a, and (**d**) phytoplankton pigment absorption at the Chl *a* absorption maxima at 440 and 675 nm, and CDOM absorption at 355 nm during the degradation experiment period (0–33 days). Error bar indicates the means and standard deviations (n = 3). *Data source* Zhang et al. (2009)

Tampa Bay at 443 nm is about five times higher in June and about ten times higher in October than phytoplankton pigment absorption at 443 nm (Chen et al. 2007). It has been shown that the impact of the suspended and dissolved matter on lake-water optical quality is influenced by wind resuspension of particulate matter: the relative role of dissolved matter in the absorption of UV and visible radiation prevails for low wind speed conditions (<2.2 m s^{-1}), while at high wind speed (3.9 m s^{-1}) particulate matter resuspension can strongly influence the attenuation and the extinction measurements (Bracchini et al. 2005).

3 Apparent and Inherent Optical Properties of Natural Waters on Solar Altitude

The nature of the light field in a water body under a given incident solar irradiance is a function of inherent optical properties such as the absorption coefficient, the scattering coefficient and the normalized volume scattering function (which characterizes the angular distribution of single-event scattering). Sometimes the volume scattering function is referred to as the scattering phase function of the aquatic medium (Preisendorfer 1961; Kirk 1991). Optical properties of water at any given point in the medium are dependent on the irradiance distribution at that point (Preisendorfer 1961). The apparent optical properties of water bodies (e.g. vertical attenuation coefficient, irradiance and reflectance) are significantly dependent on the inherent optical properties (e.g. absorption, scattering and backscattering coefficients), which depend on the shape of the volume scattering function (Kirk 1984, 1991, 1994; Belzile et al. 2002). Kirk (1984, 1991, 1994) derived an empirical relationship between the apparent and inherent optical properties of natural waters, depending on the angle of incidence of the photons on the surface, based on field observation data followed by Monte Carlo modeling of the underwater light field. He stated that the vertical attenuation coefficient for downward irradiance at the midpoint of the euphotic zone, K_d, can be expressed as a function of the absorption coefficient, the scattering coefficient, and the solar altitude in accordance with (Eq. 2.3):

$$K_d = \frac{1}{\mu_0}\left[a^2 + G(\mu_0)\,ab\right]^{1/2} \tag{2.3}$$

where a is the absorption coefficient, b is the scattering coefficient, μ_0 is the cosine of the zenith angle of refracted solar photons (direct beam) just beneath the surface (calculated from the incident zenith angle using Snell's Law). $G(\mu_0)$ is a coefficient function that specifies the relative contribution of scattering to the vertical attenuation of irradiance. Its value is determined by the shape of the volume scattering function $\beta(\theta)$ and by μ_0. $G(\mu_0)$ is a linear function of μ_0 that can be expressed as (Eq. 2.4):

$$G(\mu_0) = g_1\mu_0 - g_2 \tag{2.4}$$

where g_1 and g_2 are numerical constants that depend on the volume scattering function used in the calculations.

In case of irradiance reflectance, Gordon et al. (1975) fit their data to another power series that has been simplified by Jerlov (1976) to (Eq. 2.5):

$$R(0) = \text{Constant}\,\frac{b_b}{a + b_b} \tag{2.5}$$

where $R(0)$ is the irradiance reflectance just below the surface, b_b is the backscattering coefficient, and the constant has the value 0.32 for zenith sun and 0.37 for an overcast sky. The irradiance reflectance can be expressed as (Eq. 2.6) (Kirk 1981)

$$R(0) = 0.328 b_b/a \tag{2.6}$$

This equation is in good agreement with the result obtained by Prieur (1976) using a different calculation procedure (Eq. 2.7):

$$R(0) = 0.33 b_b/a \tag{2.7}$$

These results show that the irradiance reflectance just below the surface is proportional to the backscattering coefficient and inversely proportional to the absorption coefficient in accordance with (Eq. 2.8)

$$R(0) = C(\mu_0)\, b_b/a \tag{2.8}$$

where $C(\mu_0)$ is a function of the angular distribution of the incident light flux that is constant for a given angular distribution. Other studies also show similar results where the reflectance increases as solar altitude decreases, indicating that $C(\mu_0)$ is a function of μ_0 (Gordon et al. 1975; Kirk 1981). Kirk's equation has been widely used to determine *a* and *b* in earlier studies (Belzile et al. 2002; Weidemann and Bannister 1986; Shooter et al. 1998).

3.1 Optical Variables for the Attenuation of UV and Photosynthetically Available Radiation

The absorption and scattering coefficients of different optical variables to UV and PAR are identified in natural waters (Smith and Baker 1981, Kirk 1984, 1991, 1994; Carder et al. 1989; McKnight et al. 1994; Scully and Lean 1994; Morris et al. 1995; Vernet and Whitehead 1996; Laurion et al. 1997; Sommaruga and Psenner 1997; Kahru and Mitchell 1998; Laurion et al. 2000; Markager and Vincent 2000; Belzile et al. 2000; O'Donnell et al. 2010; Belzile et al. 2002; Pierson et al. 2008; Pérez et al. 2010; Morris 2009). Reflectance, scattering and absorption of light occur in any surface as a function of latitude. Once light penetrates the air–water interface, it can either be scattered or absorbed by the constituents present in natural waters. The various optical variables in natural waters can be discriminated as below: (1) Content of CDOM, (2) Nature and molecular weight fractions of DOM, (3) Absorptivity (a_{CDOM}) and fluorescence (F_{CDOM}) of CDOM at specific wavelengths, (4) Effect of variation in incident light wavelengths and solar zenith angle, (5) Particulate materials, (6) Chlorophyll *a* concentration, (7) Water, and (8) Ice in Arctic and Antarctic regions.

3.1.1 Contents of CDOM

CDOM is one of the key factors explaining the role of absorption and scattering to the attenuation of UV and photosynthetically available radiation in natural waters (Vodacek et al. 1997; Scully and Lean 1994; Morris et al. 1995; Laurion et al. 1997; Hayakawa and Sugiyama 2008; Smith et al. 2004; Belzile et al. 2002; Pierson et al. 2008; Smith and Baker 1979; Morris and Hargreaves 1997; Vincent et al. 1998; Pienitz and Vincent 2000; Kratzer et al. 2008; Devlin et al. 2009). DOM controls the downward irradiance flux through the water column of UV-B (280–320 nm), UV-A (320–400 nm), total UV (280–400 nm) and photosynthetically active radiation (PAR, 400–700 nm) (Markager and Vincent 2000; Morris and Hargreaves 1997; Tranvik and Kokalj 1998; Bertilsson and Tranvik 2000). Post bloom increases in DOM concentration from phytoplankton biomass cause increased DOM absorption, which in turn decreases the UV transmission through the water column (Whitehead and Vernet 2000). Mineralization of DOM by photoinduced and microbial processes can decompose the chromophores in CDOM, thereby reducing the absorption of UV and visible radiation. The consequence is an increase of UV transparency of surface waters (Nelson et al. 1998; Vodacek et al. 1997; Kieber et al. 1990; Moran et al. 2000; Skoog et al. 1996; Reche et al. 1999; Whitehead and Vernet 2000; Twardowski and Donaghay 2001; del Vecchio and Blough 2002; Mostofa et al. 2007; Mostofa et al. 2007; Patsayeva et al. 1991; Kouassi and Zika 1990; Kouassi et al. 1990; Morris and Hargreaves 1997; Allard et al. 1994; Zepp et al. 1995). CDOM accounts on average for 17–98 % of the total light attenuation coefficient in lake, estuarine and coastal seawaters (Odriozola et al. 2007; Lund-Hansen 2004; Obrador and Pretus 2008; Effler et al. 2010). CDOM shows variable and important contributions in summer (10–90 %) along the Patagonia shelf-break front (Ferreira et al. 2009). Absorption by CDOM and water together contributes to 88–94 % of UV radiation attenuation in tributaries, but only 37–77 % in lakes (Smith et al. 2004). These studies show that light attenuation by CDOM significantly depends on the occurrence of suspended particulate matter and phytoplankton with high contents of Chl *a* and water. The total content of DOC could be a useful estimate of UV transparency in natural surface waters.

3.1.2 Nature and Molecular Weight Fractions of CDOM

The absorption and scattering coefficients for UV and PAR are significantly dependent on the chemical nature and variability of organic compounds (Ishiwatari 1973; Lawrence 1980; Zepp and Schlotzhauer 1981; Hayase and Tsubota 1985; McKnight et al. 1994; Dubach et al. 1964; Zanardi-Lamardo et al. 2004; Singh et al. 2010). Fulvic and humic acids are typically a mixture of organic compounds with molecular weights ranging from <100 to >300,000 Daltons in natural waters (Hayase and Tsubota 1985; Ghassemi and Christman 1968; Thurman 1985; Ma and Ali 2009). Absorption coefficients of allochthonous fulvic and humic acids are significantly dependent on the molecular weight fractions (Fig. 3) (Ishiwatari 1973; Lawrence 1980; Zepp and Schlotzhauer 1981; Hayase and Tsubota 1985; Dubach et al. 1964; Ghassemi and Christman 1968). In addition, autochthonous CDOM of algal origin showed efficient absorption over the entire spectrum at 250–700 nm during 33 days of dark incubation (Fig. 4) (Zhang et al. 2009). This suggests that autochthonous CDOM has a similar absorption pattern as allochthonous fulvic acids. The absorption coefficients (l/g/cm) of fulvic acid extracted from sediment pore waters are much higher at 320 nm than at 480 nm in all the molecular weight fractions. However, absorption is low in low molecular weight fractions (<10 kDa) and generally increases with increasing molecular weight up to >300 kDa (Fig. 3) (Hayase and Tsubota 1985).

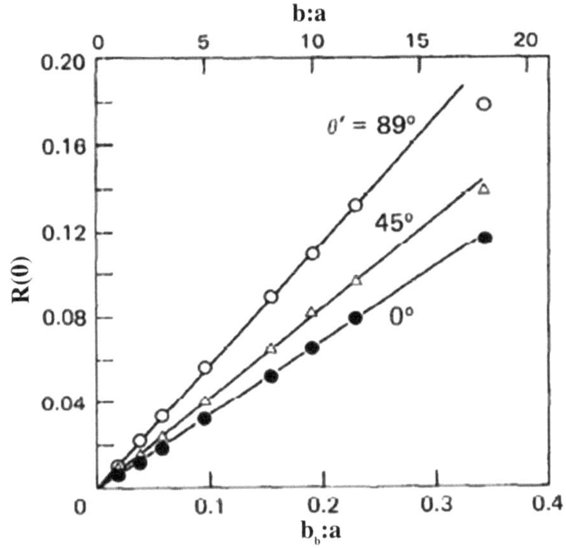

Fig. 5 Linear relationship between irradiance reflectance and the ratio of the backscattering coefficient (b_b) to the absorption coefficient at three different angles of incidence. *Data source* Kirk (1984)

On the other hand, the absorption coefficients (l/g/cm) of whole humic acid extracted from sediment pore waters are much higher at 320 than at 480 nm, but absorption is much higher in low molecular weight fractions (<10 kDa) and typically decreases with increasing molecular weight up to >300 kDa (Fig. 3) (Hayase and Tsubota 1985). A small peak shoulder is observed for molecular weight fractions of humic acid of 50–100 kDa (Fig. 3), which might account for the shoulder detected in whole humic acid absorbance spectra (Ishiwatari 1973; Hayase and Tsubota 1985). Therefore, the absorption coefficient (l/g/cm) of humic acid decreases whilst that of fulvic acid increases with increasing molecular weight. Similarly, soil fulvic acid shows a parallel increase in absorption coefficients with molecular weight (Dubach et al. 1964). Such opposite trends of the absorption coefficients between fulvic and humic acids may result from the difference in their molecular structures (Figs. 5 and 6).

3.1.3 Absorption (a_{CDOM}) and Fluorescence (F_{CDOM}) of CDOM at Specific Wavelengths

The variations of the absorption (a_{CDOM}) and fluorescence (F_{CDOM}) of CDOM at specific wavelengths are significant depending on the nature and sources of CDOM in waters (Zhang et al. 2009; Coble 1996; Hayase and Tsubota 1985; Moran et al. 2000; Winter et al. 2007; del Vecchio and Blough 2004; del Vecchio and Blough 2002; Belzile et al. 2002; Mostofa KMG et al. unpublished data). Fluorescence spectroscopy has been applied to identify the fulvic and humic acids of allochthonous origin, autochthonous fulvic acids (also denoted as marine humic-like) of algal origin, aromatic amino acids (tryptophan, tyrosine and phenylalanine),

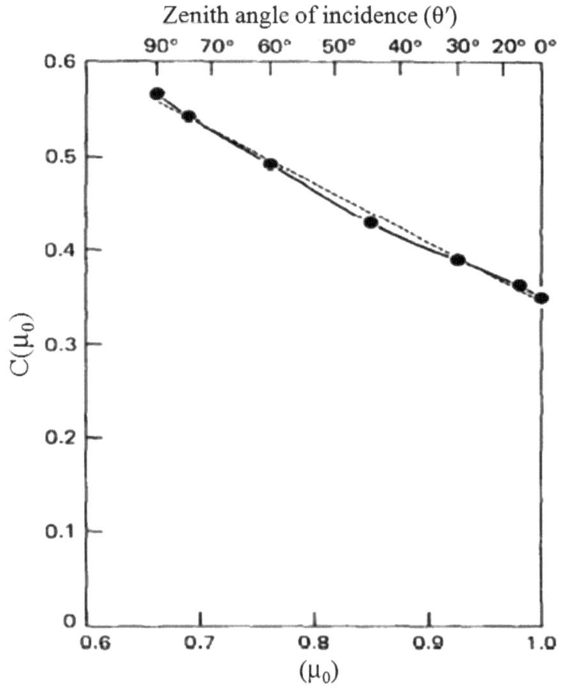

Fig. 6 Relationship between $C(\mu_0)$ with μ_0 with zenith angle of incidence (θ'). The continuous curve is drawn to join all the points where the broken line is the best straight line that can be drawn through the points. *Data source* Kirk (1984)

proteins and some anthropogenic compounds (such as fluorescent whitening agents and detergents) in natural waters (Coble 2007; Coble 1996; Yamashita and Tanoue 2003; Mostofa and Sakugawa 2009; Mostofa et al. 2010). These substances have fluorescence peaks at specific excitation-emission maxima, namely in the C-, A-, T- and T_{UV}-regions. Allochthounous fulvic and humic acids show absorption over the entire spectrum (250–700 nm), with a small peak shoulder for humic acid of 50–100 kDa (Fig. 3) (Ishiwatari 1973; Hayase and Tsubota 1985). The absorption coefficient (l/g/cm) of extracted humic acid typically decreases whilst that of fulvic acid increases with increasing molecular weight, as already reported. In addition, allochthonous fulvic acids of upstream and downstream waters generally show relatively high absorption at longer wavelengths (500–700 nm, see Fig. 1). Such absorption is either greatly decreased or not detectable in lake and marine waters (Moran et al. 2000; Winter et al. 2007; del Vecchio and Blough 2004; del Vecchio and Blough 2002; Belzile et al. 2002). Autochthonous CDOM of algal origin shows strong absorption over the entire 250–700 nm spectrum, and the fluorescence excitation-emission maxima resemble those of allochthonous fulvic acids in water (Fig. 4) (Zhang et al. 2009; Mostofa et al. 2009; Zhang et al. 2009). The absorption of both allochthonous and autochthonous fluorescent substances is very variable at specific wavelengths because these components can be significantly degraded by both photoinduced and microbial processes (Zhang et al. 2009; Moran et al. 2000; Winter et al. 2007; del Vecchio and Blough 2004; del Vecchio and Blough 2002; Mostofa et al. 2007;

Mostofa et al. 2010). As a consequence, the absorption of CDOM at specific wavelengths spectrum is found to considerably change for an array of waters. CDOM absorption contributed in average to 66 and 40 % of the UV diffuse attenuation coefficients in clear and turbid water, respectively (Belzile et al. 2002).

3.1.4 Effect of Variation in Incident Light Wavelengths and Solar Zenith Angle

The inherent and the apparent optical properties of natural waters are significantly dependent on the angle of the light flux incident on the water surface. They vary with solar altitude and with the proportion of diffuse and direct solar radiation (Kirk 1984, 1991, 1994; Morris 2009; Morel and Bélanger 2006). Remembering the a and b coefficients of Eqs. (2.5–2.8), when the solar zenith angle increases from 0° to 45° and finally to 89°, the ratio of the vertical attenuation coefficient to the absorption coefficient (K_d,(z_m)/a) increases by 15 and 41 %, respectively, when b:a = 1, by 8 and 22 % when b:a = 5, and by 5 and 12 % when b:a = 10 (Kirk 1984). K_d is thus rather insensitive to solar altitude in highly scattering waters (high b), but a considerable effect of solar altitude is observed in clear oceanic waters with low values of b:a. For all natural waters the shape of the volume scattering function is such that there is much more scattering in a forward than in a backward direction (Kirk 1984). When the incident beam moves away from the vertical, an increasing proportion of the more intense forward scattering becomes upward rather than downward scattering, thereby increasing the irradiance reflectance with decreasing solar altitude (increasing zenith angle) (Kirk 1984).

3.1.5 Suspended Particulate Matter

Absorption and scattering play a major role in UV and PAR attenuation by suspended particulate matter (SPM) such as phytoplankton pigments, algae, living heterotrophs, mineral sediments and detritus (organic, inorganic and mineral constituents) (Zhang et al. 2009, Kirk 1984, 1991, 1994; Laurion et al. 2000; Hayakawa and Sugiyama 2008; Dupouy et al. 2010; Uusikivi et al. 2010; Odriozola et al. 2007; Vantrepotte et al. 2007; Belzile et al. 2002; Pierson et al. 2008; Kratzer et al. 2008; Devlin et al. 2009; Hodoki and Watanabe 1998; Smith et al. 1999; Stambler 2005; Bowers and Binding 2006; Binding et al. 2008; Devlin et al. 2008; Foden et al. 2008). The spectral absorption coefficients of particulate matter (PM) are about twice higher in UV than in PAR wavelengths in the Baltic Sea ice (Uusikivi et al. 2010). Particulate absorption coefficients are appreciable in magnitude, averaging up to 103 % of a_{CDOM} at 380 nm and reflecting significant influence of both algal and detrital particles in lake and tributary waters (Smith et al. 2004). PM absorption spectra can include significant contribution by mycosporine-like amino acids between 320 and 345 nm (Uusikivi et al. 2010).

Experimental studies on phytoplankton biomass showed that phytoplankton has a strong absorption capacity at 400–700 nm and that its absorption decreases during 33 days of dark incubation (Fig. 4) (Zhang et al. 2009). An increase in UV absorption by PM may result from additional algal biomass during the phytoplankton bloom (Whitehead and Vernet 2000). The algal community composition in term of dominant cell size and, therefore, of pigment packaging is the key factor driving the phytoplankton specific absorption in the water column (Vantrepotte et al. 2007). The scattering coefficient of particulate materials increases approximately linearly with decreasing wavelength where suspended sediments dominate the optical signal in natural waters (Belzile et al. 2002; Morel 1988; Ahn et al. 1992; Roesler and Zaneveld 2258; Haltrin 1999; Pegau et al. 1999; Roesler 1998; Morel and Loisel 1998). The experimental and theoretical efficiency factor for scattering by the picocyanobacteria *Synechococcus* sp., *Synechocystis* sp. and *Anacystis marina* increases with decreasing wavelength (Ahn et al. 1992). However, an increase of the scattering efficiency with decreasing wavelength is not systematically detected in larger algal species (Ahn et al. 1992). The spectral slope coefficients (300–700 nm) of CDOM samples increase by as much as 20 % after mixing with 10 g L^{-1} sediment and by 5 % after mixing with 1 g L^{-1} sediment (Shank et al. 2005). This suggests that sorption to particles has the potential to significantly alter the optical properties of CDOM in the water column of turbid shallow environments or in areas of high benthic exchange (Shank et al. 2005).

Phytoplankton is the key driver of the spatial–temporal variations of the light attenuation coefficient, which accounts on average for 44 % of the total light attenuation (Obrador and Pretus 2008). The backscattering coefficient is highly correlated with turbidity and suspended matter ($R^2 = 0.98$), but it is poorly correlated to chlorophyll ($R^2 = 0.42$) (Dupouy et al. 2010), suggesting the importance of the inherent optical properties of PM in waters. The concentration-specific absorption coefficient of mineral particles is generally found to decrease exponentially with wavelength towards a constant non-zero value in the red (Bowers and Binding 2006). Specific scattering coefficients of mineral particles show a tendency to decrease from the open ocean into energetic shelf seas and estuaries, but then to increase again within shelf seas as turbulent energy increases (Bowers and Binding 2006). Light attenuation and scattering by particles can account for 11–52 % of the total attenuation/scattering in a variety of waters (Smith et al. 2004; Belzile et al. 2002; Lund-Hansen 2004; Smith et al. 1999). PM can contribute an estimated 25–90 % of the attenuation coefficients for the first-year sea ice at wavelengths <500 nm (Fritsen et al. 2011). The total particulate absorption coefficients at 300 nm are 0.1–0.3 m^{-1} in Southern Ocean waters (Holm-Hansen et al. 1993). Specific absorption coefficients for Antarctic phytoplankton is 0.1 m^2 (mg chl $a)^{-1}$ within the UV range (Mitchell et al. 1989; Arrigo 1994). Bacterial attenuation at 390 nm ranges from 0.002 m^{-1} for *Micrococcus* sp. to 2.80 m^{-1} for *Moraxella* sp. at concentrations of 1012 cells m^{-3}, and increases markedly at shorter wavelength (Kopelevich et al. 1987). These studies show that light attenuation by suspended particles is very variable depending on the water (clear or turbid) and the particle loading.

3.1.6 Total Content of Chlorophyll *a*

Chloropigments (chlorophyll *a* and carotenoids) could be important determinants of UV attenuation in natural waters (Zhang et al. 2009; Dupouy et al. 2010; Williamson et al. 1996; Belzile et al. 2002; Kratzer et al. 2008; Devlin et al. 2009; Lund-Hansen 2004; Morel and Bélanger 2006; Smith et al. 1999; Stambler 2005; Baker and Smith 1982; Gallegos and Bergstrom 2005). Chlorophyll *a* (Chl *a*) or phytoplankton biomass shows an absorption maximum around 440 nm (Fig. 4a, c, d) (Zhang et al. 2009; Bowers et al. 2000). Phytoplankton absorption is maximal when Chl *a* and phaeophytin-*a* are detected at the highest levels (Fig. 4a, c, d) (Zhang et al. 2009). Extraordinary spring blooms of the dinoflagellate *Prorocentrum minimum* can produce very high concentrations of chlorophyll, which increase for instance light attenuation in the upper Chesapeake Bay (Gallegos and Bergstrom 2005). Chlorophyll *a* specific absorption coefficients for both UV and PAR domains are representative of the dominant picophytoplankton in the Red Sea (Stambler 2005).

The study showed a deep chlorophyll maximum at about 50–60 m, with ~1 × 10^8 cells L^{-1} dominated by high concentrations of *Prochlorococcus* (~75 %), whereas in the Gulf of Eilat (Aqaba) ~4 × 10^7 cells L^{-1} have been reported. Eukaryotic algae (~20 %), cyanobacteria (*Synechococcus*) (~50 %) and *Prochlorococcus* (~25 %), are distributed throughout the water column in the Red Sea (Stambler 2005). Microbial degradation of phytoplankton or chlorophyll a are responsible for the decrease in Chl *a* or phytoplankton absorption in waters (Zhang et al. 2009). The share of light attenuation by phytoplankton is on average 32 % and reaches up to 74 % at high Chl *a* concentrations in estuarine-coastal waters (Lund-Hansen 2004). Phytoplankton absorption is the dominant optical component of light absorption (60–85 %) in spring along the Patagonia shelf-break front (Ferreira et al. 2009). Therefore the CDOM absorption is significantly dependent on the contents of phytoplankton or total contents of Chl *a* in natural waters.

3.1.7 Water

Light absorption by natural waters depends on the water quality (clear or turbid, presence of particulate matter, and CDOM content) (Kirk 1984; Hayakawa and Sugiyama 2008; Gregg and Casey 2009; Fournier 2007; Belzile et al. 2002; Pérez et al. 2010; Morris 2009; Lund-Hansen 2004; Effler et al. 2010). Recent studies show that water even in its purest form exhibits a complex absorption spectrum and a significant amount of scattering caused by refractive index fluctuations (Fournier 2007). The optical properties of natural waters are typically function of the underwater irradiance and of either the vertical attenuation coefficient for downward irradiance (K_d) or the irradiance reflectance. It is $R = E_u/E_d$, where E_u and E_d are the upward and downward irradiance at a given depth (Kirk 1984). These optical properties significantly depend on the nature of the light field within the water body and vary with depth and solar altitude (Kirk 1984; Belzile et al. 2002). The vertical attenuation coefficient (K_d PAR) has been found to vary from 0.40 to 47 m^{-1} in sixteen Argentinean shallow lakes. High K_d PAR values (>13 m^{-1}) have been detected in highly turbid lakes, medium K_d PAR values (<10 m^{-1}) in clear-vegetated lakes, and very low K_d PAR values in Patagonian lakes (<2.5 m^{-1}) (Pérez et al. 2010). Depending on the occurrence of key absorbance variables such as high CDOM, particulate material and chlorophyll, the absorption of water can vary considerably. Light attenuation by water contributes on average 0.3–9 % in UV and PAR, although it is highly variable between clear and turbid waters (Belzile et al. 2002; Lund-Hansen 2004).

3.1.8 Snow and Ice in Arctic and Antarctic Regions

Absorption and scattering by snow and ice significantly affect the UV and PAR attenuation, particularly in the Arctic and Antarctic region (Belzile et al. 2000; Warren et al. 2006; Grenfell and Perovich 1984; Buckley and Trodahl 1987; Perovich 1993; Trodahl and Buckley 1990; Arrigo et al. 1991; Perovich et al. 1998; Norman et al. 2011). Snow is a scattering-dominated medium, the scattering of which is independent of wavelength between 350 and 600 nm. The attenuation of solar radiation in snow can be used to infer the spectral absorption coefficient of pure ice, by reference to a known value at 600 nm (Warren et al. 2006). The spectral downwelling diffuse attenuation coefficient is caused by both scattering and absorption within the medium. Scattering by snow depends on grain size, snow density and water content, whilst scattering by ice depends on ice structure and particle back-scattering (Buckley and Trodahl 1987; Trodahl and Buckley 1990; Arrigo et al. 1991; Perovich et al. 1998).

It has been shown that the UV-B transmittance through 1.7 m–thick first-year ice decreases from 2–1 to 0.2–0.1 % from the end of October to mid-November in McMurdo Sound (Trodahl and Buckley 1990). The decrease in transmittance is the effect of the formation of a highly scattering layer, subsequent to ice-surface drainage. UV-B transmittance at 320 nm for 1.6 m of snow-covered first-year ice also decreases by an order of magnitude from 0.3 % in April to 0.03 % in June in the Chukchi Sea (Perovich et al. 1998). A bloom of ice algae at the bottom of the ice can also reduce the UV radiation transmittance (Perovich et al. 1998). Belzile et al. (2000) report that about 2–13 % of incident UV-B irradiance is transmitted through snow, ice and ice algae biomass, whilst transmittance increases to 5–19 % for UV-A and to 5–12 % for PAR. An influence of ice algae on PAR transmission is also observed (Belzile et al. 2000; Arrigo et al. 1991; Palmisano et al. 1987). The absorption of irradiance depends on the absoprtion by pure ice and brine, CDOM and particulate organic matter (POM) (Belzile et al. 2000; Uusikivi et al. 2010; Fritsen et al. 2011; Grenfell and Perovich 1984; Perovich et al. 1998; Norman et al. 2011; Warren et al. 1993). In Baltic Sea ice organic matter, both particulate and dissolved, influences the optical properties of sea ice and strongly modifies the UV radiation exposure of biological communities in and under snow-free sea ice (Uusikivi et al. 2010). Note that transmittance, $T(\lambda)$, is the ratio of the downwelling irradiance at the lower surface of the ice, $E_d(z_{ice}, \lambda)$, to the incident irradiance, $E_d(0, \lambda)$ (Perovich 1993). The transmittance depends on the spectral reflection coefficient, $\alpha(\lambda)$—that is, the fraction of $E_d(0, \lambda)$ that is reflected—and on the attenuation of irradiance by snow and ice, according to Beer's Law.

4 Factors Affecting Absorption of Radiation by CDOM

CDOM absorption differs considerably in a variety of natural waters and depends on several factors, which can be distinguished as follows: (i) Contents and molecular nature of CDOM; (ii) Occurrence and type of sediments; (iii) Photoinduced degradation; (iv) Microbial degradation; and (v) Salinity.

4.1.1 Contents and Molecular Nature of DOM

The CDOM absorption depends on total DOM contents and on its molecular nature (Fig. 7) (Vodacek et al. 1997; Ishiwatari 1973; Lawrence 1980; Zepp and Schlotzhauer 1981; Hayase and Tsubota 1985; McKnight et al. 1994; Dubach et al. 1964; del Vecchio and Blough 2004; Belzile et al. 2002; Vincent et al. 1998; Pienitz and Vincent 2000). DOC contents are very much correlated with CDOM absorption (a_{CDOM}) in natural waters, with the exception of surface waters during the summer stratification period (Rochelle-Newall and Fisher 2002; del Vecchio and Blough 2004; Vodacek et al. 1995; Ferrari et al. 1996; Ferrari 2000; Klinkhammer et al. 2000; Chen et al. 2002; Kowalczuk et al. 2010). The results typically suggest that the CDOM fraction often increases linearly with the DOC content, whilst the non-CDOM fraction of DOC remains relatively constant at approximately 50–100 μM C (Fig. 7). It has been shown that humic-like CDOM components with excitation maxima at longer wavelengths have significantly higher non-absorbing DOC compared to humic-like CDOM components with excitation maxima at shorter wavelengths (Kowalczuk et al. 2010). The relationship between the DOC concentration and the intensity of one of the protein-like components can result in significantly reduced non-absorbing DOC fraction (Kowalczuk et al. 2010). This study suggests that the relative proportion of humic-like CDOM components (characterized by excitation maximum at longer wavelengths) and the main protein-like component have the highest impact on the absorption at 350 nm (Kowalczuk et al. 2010). Moreover, two phenomena are responsible for the observed differences in CDOM absorption in surface waters. First, CDOM properties (chromophores in CDOM) are significantly altered by exposure to natural sunlight in surface waters, which reduces the CDOM absorption. Second, new CDOM is produced from algal or phytoplankton biomass under

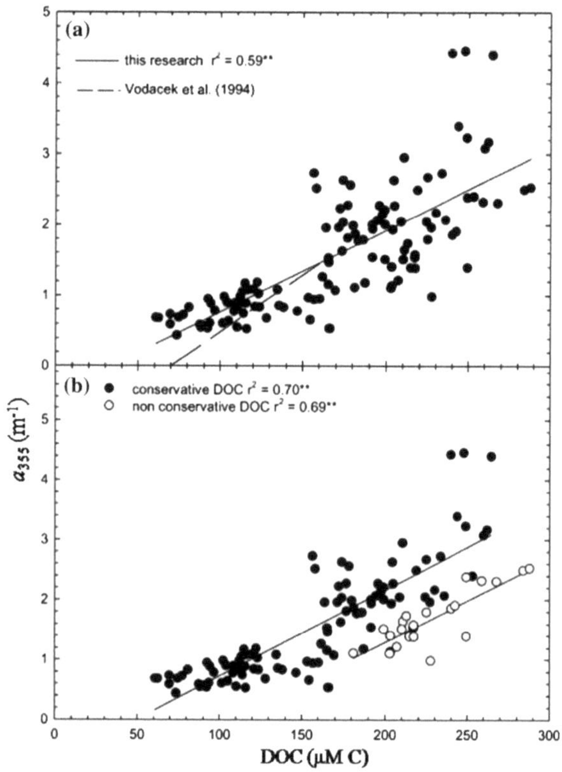

Fig. 7 Relationship between CDOM absorption (a_{355}, m^{-1}) and DOC concentrations in Chesapeake Bay. *Panel* (**a**) shows DOC and CDOM absorption for all cruises, and in *Panel* (**b**) DOC concentrations are separated into conservative and non-conservative groups. *Dashed line* on *Panel* (**a**) represents data of (Vodacek et al. 1995). *Data source* Rochelle-Newall and Fisher (2002)

both photoinduced and microbial assimilations during the summer stratification period in surface waters (Zhang et al. 2009; Rochelle-Newall and Fisher 2002; Yamashita and Tanoue 2004; Yamashita and Tanoue 2008; Mostofa et al. 2009; Rochelle-Newall 1999; Mostofa et al. 2005; Parlanti et al. 2000). This material does not show strong CDOM absorption. CDOM absorption is strongly dependent on the molecular nature of DOM (Lawrence 1980; Zepp and Schlotzhauer 1981; Hayase and Tsubota 1985; Ghassemi and Christman 1968; Thurman 1985; Ma and Ali 2009), fulvic and humic acids depending on their sources and molecular weight are significantly different as far as radiation absorption is concerned.

4.2 Occurrence and Nature of Sediments

DOM composition depends significantly on the sources and nature of sediments, such as vascular plant material and algal biomass in soil or sediment pore waters (Zhang et al. 2009; Mostofa et al. 2009; Malcolm 1985; Zhao et al. 2009; Parlanti et al. 2000; Fu et al. 2010; Li W et al. unpublished data). Vascular plant material of terrestrial origin in waters or pore waters is mostly responsible for microbial production of allochthonous humic substances (fulvic and humic acids), carbohydrates, amino acids and so on (Mostofa et al. 2009; Malcolm 1985; Ittekkot et al. 1985; Guéguen et al. 2006). The terrestrial run-off through rivers may bring the plant material to many surface waters, such transport depending on precipitation and on the type and density of terrestrial plants. On the other hand, algal biomass and microbes which develop in the photic zone may release autochthonous fulvic acids, amino acids, carbohydrates, proteins, fatty acids, peptides, organic acids and other compounds (Zhang et al. 2009; McCarthy et al. 1996; Biddanda and Benner 1997; Wakeham et al. 1997; Rosenstock et al. 2005; Hama and Handa 1992). Sometimes such a process is photolytically enhanced. Major seasonal differences in the spectral slope values show that phytoplankton degradation is one of the important sources of CDOM in summer, whereas in other seasons CDOM mainly reaches lake water from river input (Zhang et al. 2009; Zhang and Qin 2007). At the same time, sediment-trap studies demonstrate that only 1–35 % of the organic carbon (viz. algae) synthesized in the photic zone reaches the sediment surface in marine and lacustrine waters (Bernasconi et al. 1997; Hernes et al. 2001; Lehmann et al. 2002). Such algal material releases the same autochthonous organic substances in pore waters as it does in the upper parts of the water column (Burdige et al. 2004; Li W et al., unpublished data). Some DOM components from the pore-water sediment surface can mix up with surface waters during the vertical mixing (overturn) period in lakes or oceans. Allochthonous and autochthous DOM has very variable absorption properties (Zhang et al. 2009; Vodacek et al. 1997; Zepp and Schlotzhauer 1981; Hayase and Tsubota 1985; McKnight et al. 1994; del Vecchio and Blough 2004; Vodacek et al. 1995; Belzile et al. 2002; Vincent et al. 1998; Pienitz and Vincent 2000). Therefore, DOM composition also depends on the occurrence and nature of sediments in natural waters.

4.3 Photoinduced Degradation of CDOM in Natural Waters

Photoinduced processes can decompose the chromophores in CDOM and thus decrease the CDOM absorption (Vähätalo and Wetzel 2004; Zhang et al. 2009; Shank et al. 2010; Moran et al. 2000; Winter et al. 2007; del Vecchio and Blough 2002; Norman et al. 2011; Zagarese et al. 2001). A theoretical model for the photoinduced degradation of CDOM and its effects on absorption properties are discussed in this section.

4.3.1 Theoretical Model for Photoinduced Degradation of CDOM

The CDOM chromophores absorb photons and sometimes undergo degradation, while DOM undergoes partial mineralization to hydrogen peroxide (H_2O_2), CO_2, DIC (sum of dissolved CO_2, H_2CO_3, HCO_3^-, and CO_3^{2-}), COS, CO, ammonium, gaseous hydrocarbons, organic peroxides (ROOH), low molecular weight (LMW) DOM, and so on in upper surface waters (Fig. 1) (Vähätalo and Wetzel 2004; Ma and Green 2004; Moran and Zepp 1997; Mostofa and Sakugawa 2009;

del Vecchio and Blough 2002; Vähätalo et al. 2000; Bertilsson and Tranvik 2000; Allard et al. 1994; Amador et al. 1989; Fujiwara et al. 1995; Bertilsson and Allard 1996; Granéli et al. 1996; Granéli et al. 1998; Miller and Moran 1997; Clark et al. 2004; Xie et al. 2004; Johannessen et al. 2007; Fichot and Miller 2010). In surface waters, the rate of photoinduced mineralization of CDOM (pm_z, mol C m^{-3} d^{-1}), modified by Vähätalo et al. (2000) from Schwarzenbach et al. (1993) and Miller (1998), can be expressed as follows (Eq. 4.1):

$$pm_z = \int_{\lambda\min}^{\lambda\max} \varphi_\lambda Q_{s,z,\lambda} a_{CDOM,\lambda} d\lambda \tag{4.1}$$

where φ_λ is the spectrum of the apparent quantum yield for photoinduced mineralization (mol produced DIC/mol absorbed photons), $Q_{s,z,\lambda}$ is the scalar photon flux density spectrum at the depth z (also referred to as actinic flux, mol photons m^{-2} d^{-1}), and $_\lambda a_{CDOM,\lambda}$ is the absorption spectrum of CDOM (m^{-1}). The parameters λ_{max} and λ_{min} are the minimum and maximum wavelengths contributing to photoinduced mineralization.

In the whole water column the rate of photoinduced mineralization, modified by Vähätalo et al. (2000) from Miller (1998), can be expressed as follows (Eq. 4.2):

$$pm = \int_{\lambda\min}^{\lambda\max} \varphi_\lambda Q_{a,\lambda} (a_{CDOM,\lambda}/a_{tot,\lambda}) d\lambda \tag{4.2}$$

where $Q_{a,\lambda}$ represents the photons absorbed by the water column (mol photons m^{-2} d^{-1}) and the $a_{CDOM,\lambda}/a_{tot,\lambda}$ ratio expresses how much CDOM contributes to the total absorption. In infinitely deep waters, $Q_{a,\lambda}$ roughly equals the downward vector photon flux density just below the surface ($Q_{d,v,0\text{-}\lambda}$) (Sikorski and Zika 1993; Sikorski and Zika 1993).

The quantum yields related to CDOM decrease exponentially with increasing wavelength (Moran and Zepp 1997; Vähätalo et al. 2000; Sikorski and Zika 1993; Ratte et al. 1998; Gao and Zepp 1998. A generalized equation linking quantum yield and wavelength (Vähätalo et al. 2000) can be expressed as below (Eq. 4.3):

$$\varphi_\lambda = c \times 10^{-d\lambda} \tag{4.3}$$

where c (dimensionless) and d (nm^{-1}) are positive constants and λ is wavelength (nm). Different combinations of c and d can cover a wide range of exponential relationships between quantum yield and wavelength.

4.3.2 CDOM Absorption Loss in Long- and Short-Wavelengths Due to Photoinduced Degradation

Photoinduced degradation rapidly lowers the CDOM absorption coefficients across the entire spectrum, both in natural waters and in standard organic substances (Fig. 1) (Vodacek et al. 1997; Zhang et al. 2009; Shank et al. 2010; Moran et al. 2000; Hernes and Benner 2003; Winter et al. 2007; del Vecchio and Blough

2002; Ortega-Retuerta et al. 2010; Norman et al. 2011; Kowalczuk 1999; Kitidis et al. 2006). Absorption losses are likely different for a variety of natural waters (Table 1) (Zhang et al. 2009; Moran et al. 2000; Winter et al. 2007; Mostofa KMG et al., unpublished data; Norman et al. 2011). They are of order 77–97 % at 340–350 nm for upstream CDOM (Kago and Nishi-Mataya upstream, Japan) and 58–59 % at 340–350 nm for downstream CDOM (Yasu River, Japan) after 13 days of irradiation (Table 1) (Mostofa KMG et al. unpublished data). The CDOM absorption is entirely quenched at 700–444 nm for Kago upstream and 700–366 nm for Nishi-Mataya upstream, but downstream CDOM is little decomposed (14–45 %) at 600–700 nm (Fig. 1). Note that the upstream CDOM is mostly made up of fulvic acids having low DOC concentrations (99 and 38 μM C for NM upstream), whilst downstream CDOM has different origin such as autochthonous (protein-like or tryptophan-like), allochthonous (fulvic acids), and agricultural sources with relatively high levels of DOC (e.g., 194 μM C for Yasu River) (Mostofa et al. 2007; Mostofa et al. 2005). This suggests that autochthonous CDOM may originate in the river bed during the summer season and agricultural CDOM may be released from nearby agricultural fields.

The differences in CDOM absorption losses between upstream and downstream river waters suggest three issues. First, absorption losses depend on CDOM source and composition. Second, fulvic acids in upstream river waters are highly decomposed, as demonstrated by the complete loss of absorption in the longer wavelength region (from 366 to 700 nm). Such absorption losses are accompanied by high losses (72–84 % at peak C-region) in the fluorescence intensity of fulvic acids (Mostofa et al. 2007). Third, CDOM absorption losses are relatively limited in downstream river waters. A possible reason is that this CDOM may be a mixture os compounds originating from several sources such as autochthonous, allochthonous and agricultural. Because of the high decrease of the fluorescence intensity of allochthonous fulvic acids (80 % in downstream waters) and tryptophan (59 % in downstream waters) detected in earlier studies (Mostofa et al. 2007), it is suggested that the remaining autochthonous and agricultural CDOM might be recalcitrant or refractory to photoinduced degradation.

In addition, CDOM absorption losses are 55–76 % at 340 nm in the water of various lakes and ponds after 13 days irradiation (Table 1) (Winter et al. 2007). However, much lower absorption losses have been observed in the case of Lake Taihu after 12 days irradiation: 30 % at 355 nm and 21 % at 280 nm (Table 1) (Zhang et al. 2009). CDOM absorption losses are in the range of 50–64 % at 350 nm for estuarine CDOM after 70 days irradiation period (Table 1) (Moran et al. 2000). It is estimated that approximately 70 % of terrestrial CDOM is lost by photo-oxidation on the Middle Atlantic Bight shelf (Vodacek et al. 1997). In Antarctic surface waters, sea ice CDOM susceptibility to photo-bleaching in an in situ 120 h exposure showed a loss in CDOM absorption of 53 % at 280 nm, 58 % at 330 nm, and 30 % at 375 nm (Norman et al. 2011). This result suggests that Antarctic CDOM is more photosensitive than average lake or seawater CDOM. Absorption losses for standard Aldrich humic acid are 42–47 % at 340 nm in deionized water after 13 days irradiation (Table 1) (Winter et al. 2007).

Table 1 Changes in the specific absorbance coefficients of CDOM at different wavelength ranges by photoinduced degrdation on natural waters

Samples	Incubation time	Changes in the specific absorbance coefficients (SAC) at different wavelenths (%)		References
	(h or days)	Photoinduced degradation	Microbial degradation	
Rivers				(Mostofa KMG et al. unpublished data)
Upstream waters (Kago, Japan) at 340–350 nm	13	−(77–79)	+(20–22)	(Mostofa KMG et al. unpublished data)
Upstream waters (Kago, Japan) at 444–700 nm	13	−100	+(19–38)	(Mostofa KMG et al. unpublished data)
Upstream waters (Nishi-Mataya, Japan) at 340–350 nm	13	−(93–97)	+(78–81)	(Mostofa KMG et al. unpublished data)
Upstream waters (Nishi-Mataya, Japan) at 366–700 nm	13	−100	+(4–25)	(Mostofa KMG et al. unpublished data).
Downstream waters (Yasu River, Japan) at 340–350 nm	13	−(58–59)	−(6–7)	(Mostofa KMG et al. unpublished data)
Downstream waters (Yasu River, Japan) at 600–700 nm	13	−(14–45)	+(8–49)	(Mostofa KMG et al. unpublished data).
Lakes				
Lake Taihu, China at 355 nm	12	−30	nd	Zhang et al. (2009b)
Lake Taihu, China at 280 nm	12	−21	nd	Zhang et al. (2009b)
Lake Taihu, China at 280 nm	2–12 h (UV)	−(8–19)	nd	Zhang et al. (2009b)
Lake Erie water (42°N) at 340 nm	13	−(75–76)	−(12–32)	Winter et al. (2007)
Luther Marsh northeast water (43°N) at 340 nm	13	−(61–63)	+ (2–3)	Winter et al. (2007)
Mill Creek east of Cambridge water (43°N) at 340 nm	13	−(56–72)	−1, + 2	Winter et al. (2007)
Bannister Lake southwest of Cambridge water (43°N) at 340 nm	13	−(55–62)	+3, −3	Winter et al. (2007)
Sanctuary Pond water (41°N) at 340 nm	13	−(64–74)	+3, −4	Winter et al. (2007)
Aldrich humic acid in deionized water at 340 nm	13	−(42–47)	+(2–12)	Winter et al. (2007)
Estuaries and oceans				
Satilla Estuary (Georgia, USA) at 350 nm	70	−(59–64)	−(2–4)	Moran et al. (2000)
Satilla Estuary (Georgia, USA) at 350 nm	70	−(50–53)	+0.6, −7	Moran et al. (2000)
Satilla Estuary (Georgia, USA) at 250–500 nm	51	nd	−(4–11)	Moran et al. (2000)
Antarctic sea ice and oceanic water at 375 nm	120 h	−30	nd	Norman et al. (2011)
Antarctic sea ice and oceanic water at 330 nm	121 h	−58	nd	Norman et al. (2011)
Antarctic sea ice and oceanic water at 280 nm	122 h	−53	nd	Norman et al. (2011)

Irradiation time-hours (h) mentioned with each time as ‘h’ and ‘days’ mentioned as a whole digit only
nd Not detected

The decrease in CDOM absorption is relatively lower in the waters of Bay and oceans (Blough and del Vecchio 2002; del Vecchio and Blough 2002.

Generally speaking, CDOM absorption losses in lakes, estuaries, Bay and oceans are significantly lower than for upstream CDOM. The main reason might be the autochthonous sources of most of the CDOM in these waters, which make the corresponding CDOM less susceptible to photoinduced degradation. Indeed, the fraction of autochthonous CDOM is maximum (25–98 %) in lakes and oceans, whilst allochthonous humic substances (mostly fulvic acids) are 2–75 % (see also chapter "Dissolved Organic Matter in Natural Waters") (Mostofa et al. 2009; McCarthy et al. 1996; Biddanda and Benner 1997; Moran et al. 1991; Moran and Hodson 1994; Benner and Kaiser 2003). In addition, the amount of allochthonous CDOM (mostly fulvic and humic acids) is considerably decreased in the transport from rivers to lakes and oceans because of photoinduced decomposition by natural sunlight (Vähätalo and Wetzel 2004; Vodacek et al. 1997; Mopper et al. 1991; Wetzel et al. 1995; Moran et al. 2000; Skoog et al. 1996; Mostofa et al. 2007; Bertilsson and Tranvik 2000; Amon and Benner 1996; Twardowski and Donaghay 2002; Waiser and Robarts 2004; Wu et al. 2005; Brooks et al. 2007). The experimental results demonstrate the photoreactive nature of CDOM, with half-lives from 2.1 to 5.1 days due to photobleaching in the upper layer and duplication times from 4.9 to 15.7 days due to photohumification. Such results highlight the highly dynamic nature of CDOM in the Southern Ocean (Ortega-Retuerta et al. 2010). In addition, the high susceptibility to photobleaching of CDOM in Antarctic ice waters might be the effect of the presence of fresh CDOM in bulk ice samples, due to elevated in situ production (Norman et al. 2011). The fresh CDOM in Antarctic ice waters is characterized by low S and high a_{375}. In contrast, aged material present in brine and seawater samples is characterised by high S values and low a_{375} (Norman et al. 2011). Therefore, photoinduced degradation is one of the key factors that can regulate the CDOM absorption depending on its composition for a variety of natural waters.

4.3.3 Changes in Spectral Slope Due to Photoinduced Degradation

Photoinduced degradation can alter the spectral slope of CDOM (S) either in natural surface waters or in experimental observations under solar irradiation (Fig. 8) (Vodacek et al. 1997; Helms et al. 2008; Zhang et al. 2009; Shank et al. 2010; Moran et al. 2000; Zhang and Qin 2007; del Vecchio and Blough 2002; Mostofa KMG et al., unpublished data; Xie et al. 2004; Twardowski and Donaghay 2002; Tzortziou et al. 2007; Whitehead et al. 2000). Two key phenomena are generally observed: the first one is an increase of S because of CDOM photobleaching by solar radiation (Fig. 8) (Helms et al. 2008; Zhang et al. 2009; Shank et al. 2010; Moran et al. 2000; del Vecchio and Blough 2002; Xie et al. 2004; Twardowski and Donaghay 2002; Whitehead et al. 2000). It is suggested that photobleaching can be caused by the transformation of high-molecular weight CDOM complexes that absorb at longer wavelengths into smaller complexes that absorb at shorter wavelengths. The opposite effect can also be observed: in some

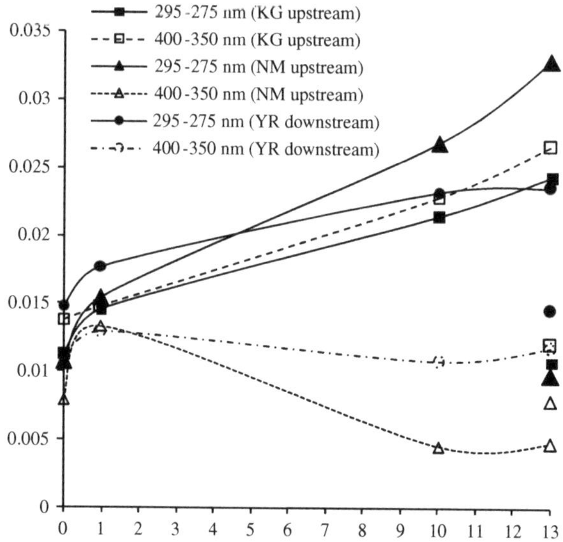

Fig. 8 Changes in the spectral slope values (*S*) in the upstream (Kago, KG and Nishi-Mataya, NM) and downstream waters (Yasu River) due to photochemical and microbial degradation of CDOM during the 13 days of irradiation and dark incubation period. Microbial degradation is presented for 13 days using the similar symbols for the respective samples. *Data source* Mostofa KMG et al. (unpublished data)

cases, CDOM photobleaching by solar radiation decreases *S* (Fig. 8) (Helms et al. 2008; Stabenau et al. 2004; Morris and Hargreaves 1997; Gao and Zepp 1998; Tzortziou et al. 2007; Miller 1994; Zepp et al. 1998; del Castillo et al. 1999). It is possible that *S* variation is caused by the different protocols employed in its calculation (linear function, LF versus non-linear function, NLF) or by the different spectral ranges adopted in the irradiation experiments (Zhang et al. 2009; del Vecchio and Blough 2002). Spectral wavelength ranges used to calculate *S* are most often 275–295 or 350–400 nm. Deviation of *S* for different spectral ranges is mainly caused by the occurrence of different chromophores in CDOM, which show variable reactivity toward photoinduced degradation.

However, variation in *S* is also observed when the same spectral range is considered and the same calculation method is adopted, in a variety of natural waters and in their photobleached samples (Fig. 8). Photoinduced degradation increases *S* at 275–295 nm ($S_{275-295}$) for upstream (115 % for Kago, KG and 207 % for Nishi-Mataya, NM) and downstream DOM (59 % for Yasu River, YR). In contrast, *S* at 350–400 nm ($S_{350-400}$) is increased for upstream (92 % for KG) and downstream DOM (6 % for YR), but is decreased for upstream DOM (41 % for NM) during 13 days irradiation (Fig. 8). Earlier studies have shown that upstream DOM is mainly composed of fulvic acids whist downstream DOM is contributed by several sources including autochthonous, allochthonous and agricultural DOM (Mostofa et al. 2007; Mostofa et al. 2005). *S* also increases at 290-500 nm in mangrove and Sargassum CDOM after 48 h irradiation (Shank et al. 2010). The highest increase of *S* has been observed upon irradiation at shorter wavelengths, while irradiation at $\lambda > 400$ nm produced small losses in absorption and little changes in *S* in Bay waters (del Vecchio and Blough 2002). Maxima of $S_{290-350}$ and $S_{250-650}$ and minima of a_{300} have been attributed to CDOM photo-oxidation in the surface waters of the Atlantic Ocean (Kitidis et al. 2006).

The variation of S suggests two important characteristics of natural waters. The first is that CDOM chromophores are decomposed photolytically and their decomposition rates are variable depending on the chemical nature (allochthonous or autochthonous) of CDOM and on its molecular structure. The second issue is that the decomposition rates of the different CDOM chromophores affect in different ways the values of S in different spectral ranges. The overall effect depends also in this case on the CDOM origin and composition. It has been shown for instance that decomposition of CDOM fractions with higher-than-average concentrations of carboxyl-, hydroxyl- and ester-substituted aromatic rings, upon either photoinduced oxidation or chlorine addition, decreases the intensity and width of the electron-transfer and benzenoid bands (Korshin et al. 1997). Such a finding suggests a correlation between the spectral slope (S) alterations by photoinduced degradation and the modification of mean molecular size and molecular structure of CDOM. The latter have generally been found to decrease from rivers to lakes and oceans (Moran et al. 2000; Moran and Zepp 1997; Corin et al. 1996; Allard et al. 1994; Amador et al. 1989; Wu et al. 2005; Yoshioka et al. 2007; Clark et al. 2008). Moreover, the spectral slope S (Jerlov 1968) as well as the carbon-specific absorptivity could be useful indicators to examine photodegradation processes in natural waters (Vodacek et al. 1997; Twardowski and Donaghay 2001; Morris and Hargreaves 1997; Whitehead et al. 2000).

4.3.4 Effect of Monochromatic and Polychromatic Irradiation on CDOM Absorption

Monochromatic irradiation of Suwannee River Fulvic Acid (SRFA) and natural waters can result in the loss of absorption across the entire spectrum and the largest absorption losses are often observed at the irradiation wavelength, λ_{irr} (Fig. 9) (del Vecchio and Blough 2002). Outside the band of direct bleaching, the loss of absorption appears to be fairly uniform across the examined spectral range. Major secondary bands of absorption loss outside λ_{irr} are not evident in the difference spectra (Fig. 9b, d, f). The high losses of absorption at λ_{irr} can mostly be attributed to the direct photoinduced destruction of the chromophore(s) absorbing at this wavelength. The kinetics of absorption loss at both λ_{irr} and at the wavelengths outside of this band exhibit excellent fits to either a single exponential or a sum of two exponentials functions. The overall rate of absorption loss is always higher at the λ_{irr}. The smaller, indirect absorption losses observed outside λ_{irr} could be produced by two effects (del Vecchio and Blough 2002): (i) the direct photoinduced destruction of chromophore(s) having absorption bands both at λ_{irr} and outside λ_{irr}; (ii) the production of reactive oxygen species from primary photochemistry at λ_{irr} that react secondarily to destroy chromophores absorbing at wavelengths outside λ_{irr}. The uniform loss of absorption outside the primary bleaching band (e.g., away from λ_{irr}) suggests that indirect photobleaching could result from the indiscriminate destruction of chromophores by reactive oxygen species produced by the primary photochemistry. It is shown that the reactive oxygen species

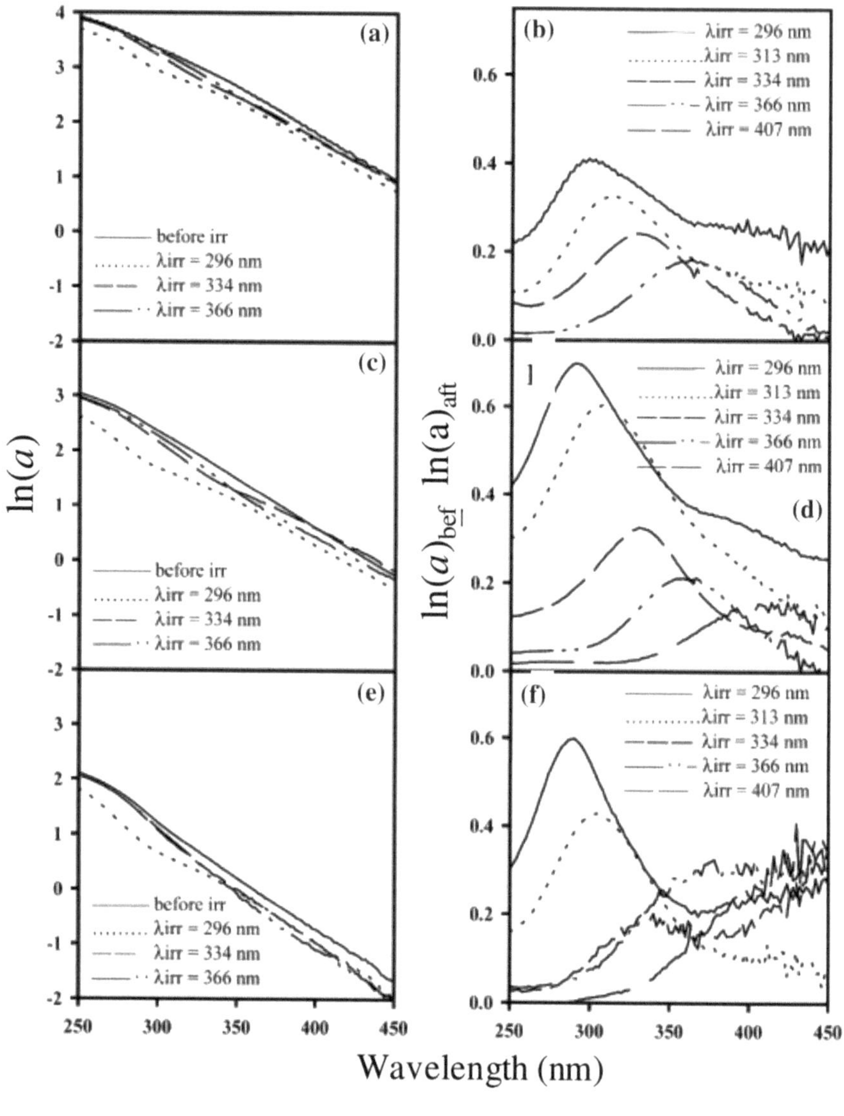

Fig. 9 Log-linearized absorption spectra obtained before and after monochromatic irradiation (10 nm band pass) (*left panels*) and difference spectra obtained by subtracting the ln (**a**) after irradiation from the original spectrum (*right panels*) for SRFA (**a** and **b**; 1 cm optical path length), Delaware Bay water (**c** and **d**; 5-cm optical path length), and Chesapeake Bay water (**e** and **f**, 5- and 10-cm optical path lengths). All spectra reported on the *left panels* are the result of independent experiments. *Data source* del Vecchio and Blough (2002)

(e.g. hydrogen peroxide, the hydroxyl radical, singlet oxygen, superoxide) are photolytically generated from CDOM in waters (Thomas-Smith and Blough 2001; Mostofa and Sakugawa 2009; al Housari et al. 2010; Minella et al. 2011; Zepp et al. 1998; Vaughan and Blough 1998; Zafiriou et al. 1998). The wavelength dependence observed for reactive intermediate production is consistent with the decrease in the efficiency of the primary photobleaching: because the efficiency of reactive intermediate production decreases with increasing wavelength, any indirect photobleaching caused by reactions with these reactive intermediates would be expected to follow the same trend (del Vecchio and Blough 2002).

Polychromatic irradiation of SRFA and natural waters can result in the loss of absorption across the entire spectrum and the bleaching is often more pronounced in the spectral region that is transmitted by the cut-off filter (Fig. 10) (del Vecchio and Blough 2002), in analogy with the results obtained for monochromatic irradiation. Coherently, absorption losses increase with decreasing λ of the cut-off filters (Fig. 9) (del Vecchio and Blough 2002). The results show that the relative loss of absorption is higher at longer wavelengths, although the efficiency of direct photobleaching decreases significantly with increasing wavelength (del Vecchio and Blough 2002). This result can be attributed to two factors (del Vecchio and Blough 2002): (i) the higher number of longer-wavelength photons produced by the light source; (ii) the higher rates of indirect absorption loss produced at longer wavelengths by the absorption of short wavelength photons. The losses of absorption at longer wavelengths lead to an increase of S when the spectral data are fit to a single exponential function using either linear or non-linear least squares methods (del Vecchio and Blough 2002).

Using a 320-nm filter, the spectral dependence of a solar simulator is similar to that of ground-level solar spectrum. The changes in S for $\lambda_{irr} > 320$ nm (Fig. 10) should thus reflect the changes in an optically thin section of surface water. The results indicate that in waters where the penetration depths of the photolytically active UV-B and UV-A wavelengths are comparable to the mixed layer depth, the loss of CDOM absorption and the increase in S in the mixed layer will be relatively rapid. If the penetration depths are much shallower than the mixed layer depth, absorption losses and changes in S in the mixed layer will be very small even over extended periods of time (del Vecchio and Blough 2002).

4.3.5 Factors Controlling the Photoinduced Degradation of CDOM Absorption

Photodegradation of CDOM depends on the sources of water, CDOM concentration, optical-chemical CDOM nature, time, space, sunlight irradiance, water chemical conditions, DOM contents, mixing regime, rain or precipitation and so on (Ma and Green 2004; Reche et al. 1999; Whitehead and Vernet 2000; Gonsior et al. 2008). It has been shown that photobleaching varies significantly depending on the lamp distance from the samples. The decrease of CDOM absorption is 8–19 % at 5 cm lamp distance, but only 2–5 % when the lamp is positioned at 45 cm, during a 2–12 h irradiation period using a UV-B lamp (Zhang et al. 2009). Moreover, the key factors that affect CDOM photobleaching are: (1) Solar radiation; (2) Water temperature; (3) Effects of total dissolved Fe and photo-Fenton reaction; (4) Occurrence and quantity of NO_2^- and NO_3^- ions; (5) Molecular nature of DOM; (6) pH and alkalinity of the water; (7) Dissolved oxygen (O_2; (8) Depth

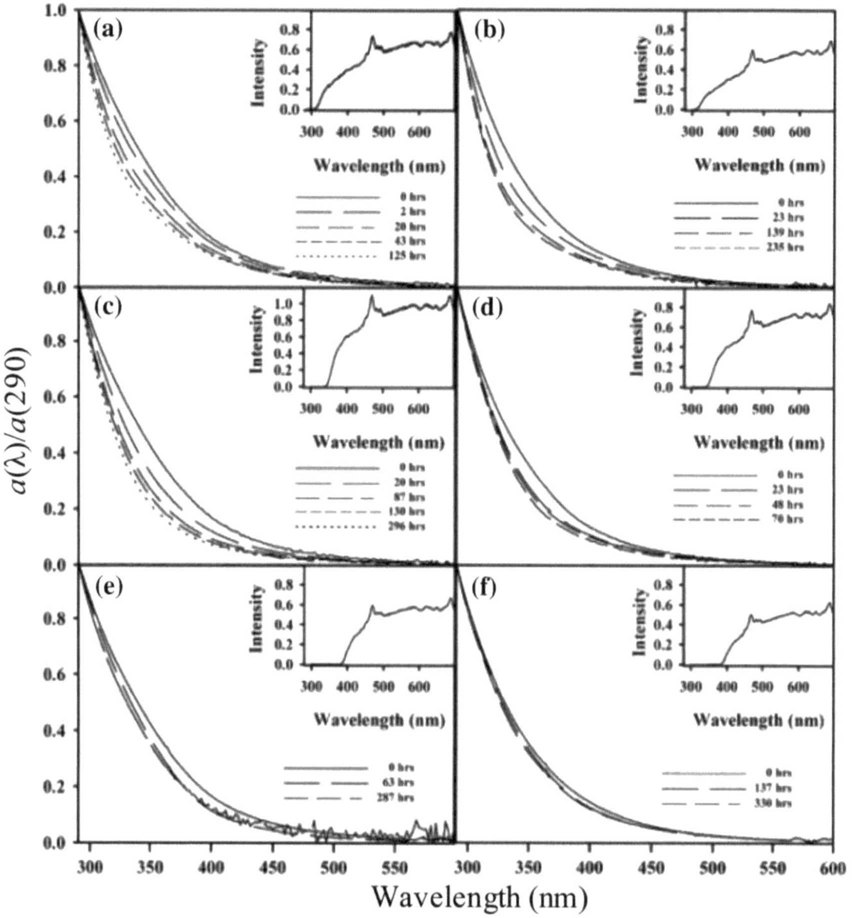

Fig. 10 Absorption spectra normalized to *a*(290) acquired during polychromatic irradiations for SRFA (*left panels*) and Delaware Bay water (*right panels*) using a cut-off filter at 320 (**a**, **b**), 360 (**c**, **d**), and 400 nm (**e**, **f**). Insets: spectral irradiance of the source in units of 10^{15} photons cm^{-2} s^{-1} nm^{-1}. Note that in the absence of an increase in the S, these spectra would be superimposable regardless of the fitting procedure. *Data source* del Vecchio and Blough (2002)

of the water; (9) Physical mixing in the surface mixing zone; (10) Increasing UV-radiation during ozone hole events; (11) Global warming and (12) Salinity. These factors are discussed in chapter "Photoinduced and Microbial Degradation of Dissolved Organic Matter in Natural Waters".

4.4 Microbial Degradation of CDOM

Microbial degradation significantly changes the CDOM absorption and spectral slope properties in natural waters (Fig. 1; Table 1) (Helms et al. 2008; Moran et al. 2000; Winter et al. 2007; Mostofa KMG et al., unpublished data; Brown 1977). After 13 days incubation, CDOM absorption has been found to increase over the entire spectrum in upstream waters. Absorption increase was 20–81 % at 340–350 nm and 4–38 % at 600–700 nm, but it was highest at 312–410 nm (77–88 %) for Nishi-Mataya upstream and at 440–570 nm (39–49 %) for Kago upstream (Fig. 1; Table 1). In downstream waters, CDOM absorption was decreased (6–7 %) at 340–350 nm but increased (8–49 %) at 600–700 nm. The maximum decrease occurred at 390–415 nm (9–11 %) (Fig. 1; Table 1). CDOM absorption was also decreased in pond (4 % at 340 nm), lakes (1–32 % at 340 nm), estuaries (2–4 % at 350 nm and 4–11 % at 250–500 nm). In other cases, very small increases have been observed in pond (3 % at 340 nm), marsh (2–3 % at 340 nm), lakes (2–3 % at 340 nm) and estuaries (1 %) (Table 1) (Moran et al. 2000; Winter et al. 2007). These results show a significant microbial effect on CDOM in natural waters (Table 1; Fig. 1).

Such an effect shows several characteristic phenomena. First, an increase in CDOM absorption over the entire spectrum in upstream waters might be due to a microbial alteration of the composition of fulvic acids. In fact, it has been known that upstream CDOM is mainly composed of fulvic acids (Mostofa et al. 2007; Mostofa et al. 2005). Second, an increase in absorption at longer wavelength and a decrease at shorter wavelength in the waters of downstream river might be due to the presence of various CDOM sources. Note that upstream water is one of the sources of the downstream one, and downstream fulvic acids (FA) might derive from upstream FA upon transformation induced by photochemistry and/or by microorganisms. The latter process is rather slow but could account for the increase of downstream CDOM absorption in the longer wavelength region, because a similar phenomenon is also observed in upstream waters. Conversely, the autochthonous and agricultural CDOM in downstream waters are likely to undergo rapid microbial degradation that, given the different nature of this kind of CDOM, might result in a decrease of CDOM absorption in the shorter wavelength region. Downstream DOM is in fact derived from several sources including autochthonous (protein-like or tryptophan-like), allochthonous (mostly fulvic acids of upstream origin) and agricultural DOM that is released from nearby agricultural fields (Mostofa et al. 2007; Mostofa et al. 2005).

CDOM absorption is also found either to decrease or to increase in ponds, lakes, marshes and estuaries that are relatively similar to downstream river environments. The CDOM in these natural waters generally consists of both allochthonous (mostly fulvic acids) and autochthonous material. Autochthonous organic substances are microbially labile and their absorption is decreased by microbial degradation, differently from allochthonous fulvic substances. This can explain the rather complex effects of microbial processing on CDOM absorption in the different wavelength ranges.

Experimental studies also show that a large amount of high molecular weight CDOM is produced during phytoplankton lysis. $S_{300-500}$ is decreased in the first 9 days when CDOM composition is changed due to increasing microbial activity, which is expected to decrease the molecular weight of organic substances (Fig. 4b) (Zhang et al. 2009; Mostofa KMG et al., unpublished data).

Alteration of autochthonous fulvic acids (C-like) changes their fluorescence intensity during a long incubation period in the dark (120 days) at room temperature, possibly because of microbial assimilation of lake algae (Mostofa KMG et al., unpublished data). The autochthonous fulvic acid (C-like) of algal origin is identified using the PARAFAC model on the EEM spectra of the samples. The aerobic microbial incubation in the dark results in a statistically significant decrease of *S* over timescales of days to weeks, due to microbial production or to selective preservation of long-wavelength absorbing substances (Helms et al. 2008).

Microbial degradation typically alters *S* in natural waters (Fig. 8) (Helms et al. 2008; Mostofa KMG et al., unpublished data; Brown 1977). It has been observed a decrease in *S* of 1–13 % at both 275–295 nm ($S_{275-295}$) and 350–400 nm ($S_{350-400}$) for upstream and downstream DOM, with the esception of 350–400 nm ($S_{350-400}$) in downstream DOM where *S* has been found to increase (~5 %) after 13 days of dark incubation (Fig. 8). Because of significant variations in CDOM composition between upstream and downstream samples, it is suggested that microbial degradation of CDOM depends on its chemical nature and on its sources.

4.4.1 Factors Controlling the Microbial Degradation of CDOM Absorbance

Microbial degradation of DOM depends on several key factors that can be listed as: (1) Occurrence and nature of microbes in waters; (2) Sources of DOM and amount of bacterial fermentation products; (3) Temperature; (4) pH; and (5) Sediment depths in pore waters. These factors are discussed in detail in chapter "Photoinduced and Microbial Degradation of Dissolved Organic Matter in Natural Waters".

4.5 Salinity

The absorption properties of CDOM are modified when the terrestrial riverine input of CDOM is mixed with costal seawaters (Singh et al. 2010; Hernes and Benner 2003; del Vecchio and Blough 2004; Blough and del Vecchio 2002; Fournier 2007; Nieke et al. 1997; Twardowski and Donaghay 2001; del Castillo et al. 1999; Gonsior et al. 2008; Uher et al. 2001; Blough et al. 1993; Sholkovitz 1976). A recent study shows that the addition of various salts, which are present in seawater, gives rise to an extra absorption in the far UV and to an increase of the amount of scattering because small variation in salt concentration can cause refractive index fluctuations (Fournier 2007). A linear inverse relationship of CDOM absorption to salinity could be a useful indicator of salinity in coastal waters strongly affected by river input (Singh et al. 2010; del Vecchio and Blough 2004; Nieke et al. 1997; Blough et al. 1993). However, at lower salinity minor deviations from linearity may result from consumption or production of CDOM in coastal waters (del Vecchio and Blough 2004; Twardowski and Donaghay 2001; del Castillo et al. 1999; Uher et al. 2001; Blough et al. 1993; Sholkovitz 1976). In addition, a nonlinear dependence may result from the conservative mixing of multiple water masses containing variable CDOM (del Vecchio and Blough 2004; Blough and del Vecchio 2002; Hujerslev et al. 1996; Chen et al. 2007; Blough et al. 1993). Photoinduced degradation can greatly decrease the CDOM absorbance in intermediate- to high-salinity surface waters under stratified conditions during the summer period (Vodacek et al. 1997; del Vecchio and Blough 2004; Chen et al. 2007; Osburn and Morris 2003; Osburn et al. 2009). Dissolved lignin phenols are significantly affected by salinity and two key phenomena are generally detected: First, a nonconservative decrease in dissolved high molecular weight (HMW) lignin phenols at salinity <25 psu is likely due to flocculation and microbial degradation. In contrast, LMW dissolved lignin phenols mix conservatively (Hernes and Benner 2003). Second, at salinity >25 psu photooxidation is a dominant factor influencing lignin composition and concentration (Hernes and Benner 2003).

CDOM photoreactivity can increase with salinity across an estuarine gradient. Shortwave CDOM absorption loss (e.g. at 280 nm) does not change with salinity, but longwave CDOM absorption loss (e.g. at 440 nm) is often decreased by 10–40 % with increasing salinity (Osburn and Morris 2003; Osburn et al. 2009). In another study, a decrease in CDOM photobleaching at 280 nm is detected when humic CDOM is added to an artificial salinity gradient used to mimic coastal mixing (Minor et al. 2006). The decrease of the absorption properties of CDOM with salinity can be accounted for by several factors: (i) Mixing of CDOM-rich riverine water with CDOM-poor coastal water (del Vecchio and Blough 2004; Gonsior et al. 2008; Blough et al. 1993); (ii) Photodegradation of chromophores present in riverine CDOM after they reach the coastal regions during the summer stratification period (Vodacek et al. 1997; Moran et al. 2000; del Vecchio and Blough 2004; Blough and del Vecchio 2002; Whitehead and Vernet 2000; del Vecchio and Blough 2002; Osburn et al. 2009); (iii) Microbial degradation, in particular of the autochthonous fraction that is the major part of CDOM in marine waters (Table 1) (Moran et al. 2000; Winter et al. 2007; Moran and Hodson 1994; Brown 1977; Opsahl and Benner 1998); (iv) Flocculation and precipitation of riverine CDOM because of increased salinity (Blough et al. 1993; Sholkovitz 1976; Sieburth and Jensen 1968; Fox 1991); and possibly (v) Enhanced CDOM photodegradation in saline waters.

The mechanism behind the latter process apparently involves two factors: first, irradiated CDOM can induce photoinduced production of hydrogen peroxide (H_2O_2) that is a $HO^{\bullet}$ source via photolysis or the photo-Fenton reaction, and the photoinduced generation of H_2O_2 is enhanced by salinity. Trace metal ions (M) in salinity or sea waters can complex with DOM (M-DOM) forming a strong π–electron bonding system between metal ions and the functional groups in DOM (see chapter "Complexation of Dissolved Organic Matter with Trace Metal Ions in Natural Waters" for in details explanation). This π–electron in M-DOM complex is rapidly excited photolytically, which is responsible for high production of aqueous electrons (e_{aq}-) and subsequently the high production of superoxide ion

($O_2^{\bullet -}$), H_2O_2 and $HO^\bullet$, respectively. Indeed, photogeneration of H_2O_2 from ultrafiltered river DOM is substantially increased with salinity, from 15 to 368 nM h^{-1} at circumneutral pH (Osburn et al. 2009). Salinity or NaCl salts are responsible for generating high production of aqueous electrons (e_{aq}^-) photolytically in aqueous media (Gopinathan et al. 1972; Assel et al. 1998) that may subsequently enhance the H_2O_2 production in waters (Mostofa and Sakugawa 2009; Moore et al. 1993; Richard et al. 2007; Fujiwara et al. 1993).

Recent studies observe that the sea-salt particulate matter extracted from coastal seawaters show substantially high $HO^\bullet$ production (rate: ~2778–27778 M s^{-1}), approximately 3–4 orders of magnitude greater than $HO^\bullet$ photoformation rates in surface seawater (Anastasio and Newberg 2007). Note that comparison of river and salinity of sea waters shows that Na^+, Ca^{2+}, Mg^{2+}, K^+, HCO_3^-, Cl^- and SO_4^{2-} are typically 1,670 times, 27 times, 330 times, 170 times, 2.4 times, 2,400 times and 245 times, respectively, higher than those in rivers (Livingstone 1963; Hem 1985). The order of the other cations is $Mg^{2+} > Ca^{2+} > K^+ > Sr^{2+}$ and the anion Cl^- is approximately equal to the sum of the cations and the other anions are SO_4^{2-}, HCO_3^-, Br^-, and F^- (Livingstone 1963; Hem 1985; Carpenter and Manella 1973). Second, the reaction of $HO^\bullet$ with halide ions (X^-) can form reactive halogen radicals ($X_2^{\bullet -}$) that can react with electron-rich functional groups within DOM more selectively than $HO^\bullet$ (Zafiriou et al. 1987; Song et al. 1996; Von Gunten and Oliveras 1997; Goldstone et al. 2002; Grebel et al. 2009). The absorption of radiation by CDOM is usually increased in ionic solutions of NaCl, which might suggest an increase of CDOM absorption in coastal waters. However, such an effect is more than compensate for by the efficient photodegradation of CDOM chromophores in saline waters.

5 All Colored DOM is Chromophoric DOM, But Not All Chromophoric DOM is Colored

The main chromophores in colored or chromophoric DOM are Schiff-base derivatives (–N=C–C=C–N–) and groups such as –COOH, –$COOCH_3$, –OH, –OCH_3, –CH=CH–, –CH=O, –C=O, –NH_2, –NH–, –CH=CH–COOH, –OCH_3, –CH_2–(NH_2)CH–COOH, and S-, O- or N-containing aromatic compounds or functional groups (Mostofa et al. 2009; Malcolm 1985; Corin et al. 1996; Senesi 1990; Leenheer and Croue 2003; Peña-Méndez et al. 2005; Zhang et al. 2005; Steelink 2002; Seitzinger et al. 2005). The allochthonous fulvic and humic acids (humic substances) of vascular plant origin and the autochthonous fulvic acids of algal (or phytoplankton) origin show absorbance in a wide wavelength interval, 200–800 nm (Figs. 2, 3, 4) (Zhang et al. 2009; Ishiwatari 1973; Lawrence 1980; Zepp and Schlotzhauer 1981; Hayase and Tsubota 1985; Dubach et al. 1964), and contain the above mentioned chromophores (or functional groups) in their molecular structure. These macromolecular substances are both colored DOM (they absorb in the visible) and chromophoric DOM. On the other hand, there are a lot of organic substances including most notably low molecular weight (LMW) CDOM that are not colored, not being able to absorb radiation in the visible range. For example, acetaldehyde absorbs light at 208–224 nm (Strome and Miller 1978; Kieber et al. 1990; Mopper et al. 1991), acetate at 204–270 nm (Wetzel et al. 1995; Dahlén et al. 1996), formaldehyde at 207–250 nm (Mopper and Stahovec 1986; Kieber et al. 1990; Mopper et al. 1991), glyoxal at <240 nm (Mopper and Stahovec 1986; Mopper et al. 1991, malonate at 225–240 nm (Dahlén et al. 1996), pyruvate at 200–227 nm (Kieber et al. 1990; Mopper et al. 1991; Wetzel et al. 1995) and propanal at ~230 nm (Mopper and Stahovec 1986). These organic compounds are definitely not colored, but they contribute to the absorption of radiation by water in the relevant wavelength ranges. It is thus shown that, while all colored DOM is also chromophoric DOM, not all the chromophoric DOM is also colored.

6 Importance of CDOM Studies in Natural Waters

CDOM is a major bio-optical parameter because of its strong light-absorbing properties. These properties are involved in some very important biogeochemical processes and are very useful for detection techniques. Implications are: (i) remote sensing of CDOM in natural waters; (ii) DOM dynamics in natural waters; (iii) Photoinduced degradation of CDOM and its impact in natural waters; and (iv) Protection of microorganisms from UV radiation by CDOM.

6.1 Remote Sensing of CDOM in Natural Waters

Remote sensing is widely used to estimate the ocean color constituents such as chlorophyll *a* and algae, and also to assess primary productivity, occurrence of toxic dinoflagellate, total suspended solids (TSS), tripton (inorganic suspended particulate matter), inherent optical properties, CDOM contents, diffuse attenuation coefficients (K_d), DOC concentration and transport from rivers to lakes and oceans (Coble 2007; Del Castillo and Miller 2008; Carder et al. 1991; Carder et al. 1999; McClain et al. 2004; O'Reilly et al. 1998; Ferreira et al. 2009; Tzortziou et al. 2007; Son et al. 2011; Sathyendranath et al. 1989; Woodruff et al. 1999; Stramski et al. 2001; Volpe et al. 2011; van der Woerd et al. 2011; Le et al. 2011; Carvalho et al. 2011; Santini et al. 2010; Matthews et al. 2010; Doxaran et al. 2002; Cui et al. 2010; Zibordi et al. 2009; Werdell et al. 2009; Tomlinson et al. 2009; Friedrichs et al. 2009; Van Der Woerd and Pasterkamp 2008; Hunter et al. 2008; Brown et al. 2008; Zawada et al. 2007; Mélin et al. 2007; Tzortziou et al. 2006; Koponen et al. 2007; Kishino et al. 2005; Phinn et al. 2005; Vahtmäe et al. 2006; Zimba and Gitelson 2006). Satellite remote sensing is also used to monitor cyanobacterial blooms in natural waters, which can be detected from a small peak in reflectance spectra near 650 nm that is specific of cyanobacteria (Matthews et al. 2010; Kutser et al. 2006;

Becker et al. 2009; Hunter et al. 2010). Remote sensing is currently used to monitor broad changes in phytoplankton communities, exploiting the spectral dissimilarities of brown, green, blue-green and red algae in inland waters. This technique is an extremely useful tool for limnological research and water resource management (Hunter et al. 2008). Water quality (Secchi depth, K_d in PAR, tripton, CDOM) and substrate cover type (seagrass, algae, sand) parameters, which vary in sub-tropical and tropical coastal environments may also affect the satellite image data (Phinn et al. 2005) and can thus influence the remote sensing information. A combination of a chlorophyll anomaly (spectral shape at 490 nm) and a backscatter ratio can provide an improvement in satellite detection of the toxic dinoflagellate *Karenia brevis*. It is possible to increase the detection accuracy by 30–50 % in seawaters (Tomlinson et al. 2009; Cannizzaro et al. 2008). The remote sensing application has also been used to characterize high concentrations of suspended sediment and to map chlorophyll *a* (Chl *a*) or phytoplankton and non-phytoplankton suspended matter distribution in lakes and oceans (Ferreira et al. 2009; Cannizzaro and Carder 2006; Gons et al. 2008; Oyama et al. 2009; González Vilas et al. 2011).

The ocean color depends on the optical variables (Coble 2007; Del Castillo and Miller 2008; Carder et al. 1991; Hoge et al. 1995; Hoge et al. 2001; Sathyendranath et al. 1989; Stramski et al. 2001; Brown et al. 2008; Mélin et al. 2007; Lee et al. 1994; Kahru and Mitchell 2001; Siegel et al. 2002; Nair et al. 2008; Siegel et al. 2005). The key factors are (i) phytoplankton species (or algae) and their variability; (ii) the amount of colored DOM; (iii) the amount and size of organic particles; (iv) the contents of inorganic particles (tripton); and (v) water itself. CDOM can be estimated by using ocean color with various levels of success (del Castillo and Miller 2008; Carder et al. 1999; Hoge et al. 1995; Hoge et al. 2001; Lee et al. 1994; Kahru and Mitchell 2001; Siegel et al. 2002; Siegel et al. 2005). In open ocean waters, far from the influence of terrestrial runoff that mainly affects coastal waters (case 1 waters), the spectral quality and intensity of light leaving the ocean depends first of all on the concentration of phytoplankton (Morel and Prieur 1977; Morel 1980). Empirical algorithms of ocean color based on blue-to-green ratios are used to estimate the chlorophyll *a* concentration within the upper layer of the water column (McClain et al. 2004; O'Reilly et al. 1998; Harding et al. 2005). The second-order variability for given chlorophyll levels depends on two main sources: (i) the amount of non-algal absorption, especially due to colored dissolved organic matter; and (ii) the amplitude of the backscattering coefficient of particles (Brown et al. 2008). Remote sensing of surface waters in the open ocean (case 2 waters) could be used in conjunction with the inversion of UV-blue wavelengths, to separate the contribution of non-algal particles and of colored dissolved organic matter to the total light absorption, and to monitor non-algal suspended particle concentration and distribution (Tzortziou et al. 2007). It is shown that phytoplankton typically absorbs strongly in the blue and weakly in the green. CDOM absorption thus overlaps to that of phytoplankton and non-algal particulate matter in the blue part of the visible spectrum. This issue might affect the primary productivity and the remote sensing estimation of phytoplankton biomass and of total suspended matter concentration (Zhang et al. 2009; Carder et al. 1991; Doxaran et al. 2002).

Ocean color remote sensing could be applied to estimate DOC transport in river-ocean transects, depending on the robust empirical relationships between DOC and CDOM, CDOM and salinity, and salinity and river flow (del Castillo and Miller 2008). A flow chart is depicting the method and rationale for the estimate of DOC transport (Fig. 11). The success of this approach depends on four conditions (del Castillo and Miller 2008): (1) DOC and CDOM must behave conservatively at the study site; (2) The relationship between CDOM and DOC in the river end member must remain constant; (3) One should be able to derive CDOM from satellite ocean color measurements; (4) Salinity in the study area should

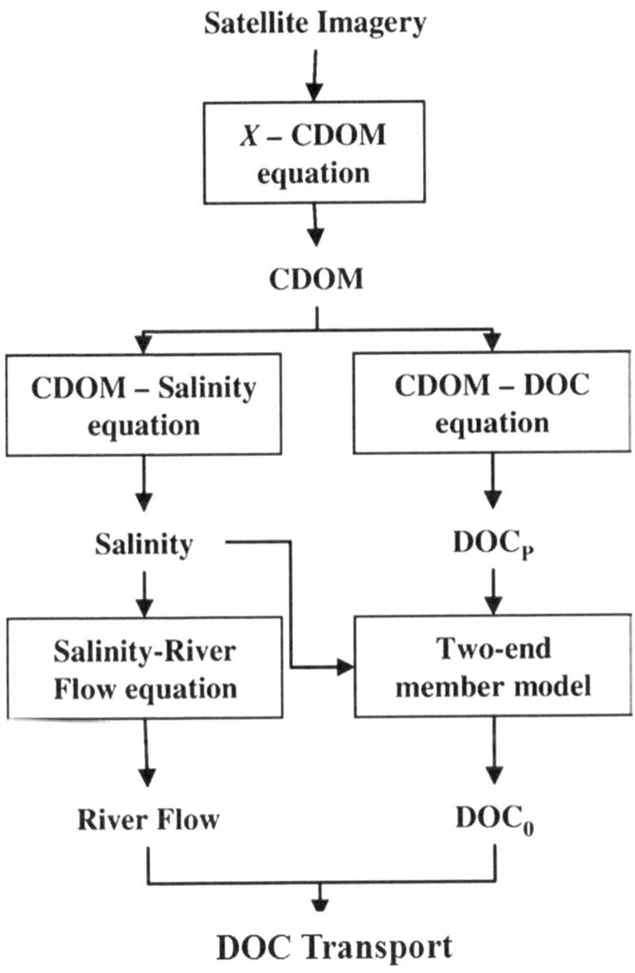

Fig. 11 Flow chart showing the method and rational of DOC transport. Boxes represent empirical relationship, plain text represent the outputs of these relationships. *X* is the remote sensing reflectance ratio used in this study, and DOC_p and DOC_0 are the concentrations of DOC in the river plume and river end-member respectively. *Data source* del Castillo and Miller (2008)

correlate with river flow. Note that the remote sensing device generally detects radiation that is reflected or backscattered from the target that initially absorbs the radiation.

6.2 DOM Dynamics in Natural Waters

Good correlations between CDOM absorption and DOM (DOC concentration) are usually characterized by a positive intercept on the DOC axis at 50–100 μM, because offshore waters contain only very small amounts of CDOM (Vodacek et al. 1997; Rochelle-Newall and Fisher 2002; del Vecchio and Blough 2004; Vodacek et al. 1995; Klinkhammer et al. 2000; Chen et al. 2002; Gueguen et al. 2011). The non-absorbing DOC fraction varies with the qualitative composition of the CDOM (Kowalczuk et al. 2010). The CDOM/DOC dependence is typically changed in surface waters during the summer stratification period, when the water samples show lower absorption to DOC ratios compared to waters from below the mixed layer or collected in other seasons (Vodacek et al. 1997). This might be the effect of high autochthonous production of non-absorbing DOM in surface waters during the summer stratification period (Mitra et al. 2000; Ogawa and Tanoue 2003; Mostofa et al. 2005; Fu et al. 2010; Yoshioka et al. 2002; Hayakawa et al. 2003; Hayakawa 2004; Ogawa and Ogura 1992), and/or of photoinduced degradation of CDOM with production of non-absorbing compounds (Coble 2007; Vodacek et al. 1997; Mostofa et al. 2005). Note that a rough estimate shows that the increase in autochthonous DOC contents during the summer stratification period is 0–88 in lakes and 0–194 % in oceans, determined by comparing the epilimnionetic DOM with that of the hypolimnion (Mostofa et al. 2009).

The predominant presence of colored DOM such as allochthonous fulvic and humic acids is responsible for the good correlation that is usually observed between CDOM absorbance and DOM contents in a variety of waters (Vione et al. 2010). The contributions of allochthonous humic substances (fulvic and humic acids) in rivers are 30–85 % (the ratio of fulvic acid to humic acid is 9:1 for lower stream DOC and it decreases to 4:1 or less for higher DOC stream), 15–60 % in lakes (30–60 % during winter and 15–40 % during the summer period), 1–75 % in shelf seawater (38 % of marsh origin and or 62 % of river origin), and 2–38 % in ocean (see chapter "Dissolved Organic Matter in Natural Waters" for detailed description) (Mostofa et al. 2009; Moran et al. 1991; Moran and Hodson 1994). The seasonal and spatial–temporal variations of the CDOM absorbance to DOC ratios are dependent on the presence of colored DOM compounds in natural waters.

A recent study demonstrates that DOC transport can be determined using ocean color remote sensing if the empirical relationships between DOC and CDOM, CDOM and salinity, and salinity and river flow are known (del Castillo and Miller 2008). It has been shown that remote sensing estimates of river flow and DOC transport are correlated well (r^2 = ~ 0.70) with field observation data, showing low variability in DOC concentrations in the river end-member (7–11 %), and high seasonal variability in river flow (~50 %) in the Mississippi River Plume (del Castillo and Miller 2008). This result can be influenced by several biogeochemical processes such as high DOM photodegradation, biodegradation, production and flocculation, as well as extreme precipitation caused by natural disaster (del Castillo and Miller 2008; Wright 2005). These biogeochemical processes have little or negligible effects in low salinity waters of river plumes due to the predominance of riverine CDOM (Blough and del Vecchio 2002; del Castillo and Miller 2008; del Vecchio and Blough 2002; del Castillo et al. 1999; del Castillo et al. 2000; del Castillo et al. 2001; Mantoura and Woodward 1983).

6.3 Photoinduced Degradation of CDOM and Its Impact in Natural Waters

Photoinduced degradation of CDOM by sunlight can affect its optical and chemical properties, by inducing decomposition of the CDOM chromophores and thus reducing CDOM absorptivity of UV and visible radiation (Kieber et al. 1990; Moran et al. 2000; Skoog et al. 1996; Reche et al. 1999; Whitehead and Vernet 2000; Twardowski and Donaghay 2001; del Vecchio and Blough 2002; Mostofa et al. 2007; Patsayeva et al. 1991; Kouassi and Zika 1990; Kouassi et al. 1990; Morris and Hargreaves 1997; Allard et al. 1994; Fichot and Miller 2010). The effect of the photoinduced degradation of CDOM is an increase of UV transparency in surface waters (Nelson et al. 1998; Vodacek et al. 1997; Kieber et al. 1990; Morris and Hargreaves 1997; Zepp et al. 1995). However, CDOM absorption losses by photoinduced degradation can also result in a variety of changes in CDOM composition, which can be listed as follows: (i) Formation of strong oxidants such as singlet oxygen, superoxide, hydroxyl radical, hydrogen peroxide, organic peroxides during the photodegradation of CDOM may have severe and chronic toxic effects on aquatic organisms and important ecological consequences in aquatic environments (Williamson et al. 1996; Thomas-Smith and Blough 2001; Mostofa and Sakugawa 2009; Moore et al. 1993; Zepp et al. 1998; Vaughan and Blough 1998; Zafiriou et al. 1998; Xenopoulos and Bird 1997; Palenik et al. 1991); (ii) Formation of low molecular weight organic substances, which is generally more important in lakes and oceans than in rivers (Moran and Zepp 1997; Corin et al. 1996; Biddanda and Benner 1997; Yoshioka et al. 2007); (iii) Formation of biologically labile compounds that enhance biodegradation (Wetzel et al. 1995; Moran and Zepp 1997); (iv) Photo formation of carbon-gas end photoproducts (CO_2, CO), DIC, COS and so on in natural waters (Ma and Green 2004; Bertilsson and Tranvik 2000; Miller and Moran 1997; Fichot and Miller 2010; Weiss et al. 1995); (v) Release of nitrogen compounds (e.g. NH_4^+) and phosphate, which may typically be produced by degradation of dissolved organic nitrogen (DON) and dissolved organic phosphorus (DOP) in the epilimnion of natural waters (Mostofa et al. 2011; Zagarese et al. 2001; Kim et al. 2006; Vähätalo and Järvinen 2007; Li et al. 2008; Zhang et al. 2004). These nutrients are used by algae and bacteria. (viii) Finally, releases of energy to the water ecosystem (Wetzel 1992; Hedges et al. 2000; Tranvik 1992).

The decrease of CDOM absorption over the entire spectrum, induced by photoinduced processes is accompanied by the decrease of total fluorescence intensity at peak C-, A-, T- and T_{UV}-regions of various fluorescent substances (Zhang et al. 2009; Coble 1996; Moran et al. 2000; Mostofa et al. 2007; Mostofa et al. 2010). Photoinduced DOC mineralization (decrease of DOC concentration) is also observed (Moran et al. 2000; del Vecchio and Blough 2002; Mostofa et al. 2007; Mostofa et al. 2005; Frimmel and Bauer 1987; Vione et al. 2009; de Haan 1993). As a matter of fact, the impacts of photoinduced degradation on CDOM absorption are similar to those on DOC degradation that have been explained in details in earlier chapter (see chapter "Photoinduced and Microbial Degradation of Dissolved Organic Matter in Natural Waters").

6.4 Protection of Microorganisms from UV Radiation by CDOM

Natural UV radiation (280–400 nm) is a selective and strong environmental factor that damages the cell structures including proteins, lipids, membranes, pigments and DNA. It affects the productivity of freshwater and marine organisms (Marchant et al. 1991; Vincent and Roy 1993; Bothwell et al. 1994; Banaszak and Trench 1995; Leavitt et al. 1997; Poli et al. 2004; Lesser 2006; Valko et al. 2006; Xiong et al. 1997; Teai et al. 1998). Such impacts are caused by the UV-radiation induced production of strong reactive oxygen species such as superoxide radicals, singlet oxygen, hydrogen peroxide and hydroxyl radicals. The main processes involved are the photo-Fenton reaction and the photolysis of NO_2^-, NO_3^- and DOM (see chapter "Photoinduced Generation of Hydroxyl Radical in Natural Waters"). UV attenuation by CDOM is important in minimizing the deleterious consequences of UV radiation on phytoplankton and other organisms in natural waters (Wetzel 1992; Morris et al. 1995; Schindler et al. 1996; Williamson et al. 1996; Yan et al. 1996; Smith and Baker 1979). It is shown that postbloom increases in DOM concentration induced by grazing and decomposition of phytoplankton biomass cause an increase of DOM absorption and a related decrease of UV transmission through the water column (Whitehead and Vernet 2000). The partial UV protection by autochthonous DOM has the consequence that shallow blooming phytoplankton may assist the development of a subsequent bloom in deep waters (Whitehead and Vernet 2000). On the other hand, any changes in hydrology (e.g., high temperature, drier climate) or geochemistry (e.g., increased atmospheric deposition of strong acids) due to global warming can reduce DOM concentrations in surface waters. The consequence might be an increased exposure of aquatic organisms to both UV-A and UV-B radiation (Schindler et al. 1996; Yan et al. 1996; Morris and Hargreaves 1997).

7 Scope of the Future Challenges

CDOM is generally produced from two major sources: allochthonous and autochthonous. Absorption coefficients of CDOM have often been determined jointly in earlier studies. Recent studies show that autochthonous DOM is significantly produced from photo- and microbial respiration or assimilation of algae or phytoplankton biomass, yielding compounds that are very similar to allochthonous fulvic and humic acids (Zhang et al. 2009; Mostofa et al. 2009; Yamashita and Tanoue 2004; Yamashita and Tanoue 2008; Mostofa et al. 2009). The specific absorption properties of allochthonous and autochthonous CDOM could be important to understand the physical, photoinduced and microbial changes of those two CDOM pools in natural waters. Remote sensing is widely applied to characterize various parameters as explained earlier, but a key application could be the detection of algal blooms in natural waters. The CDOM research for future challenges can be distinguished as: (i) Investigation on CDOM absorption of allochthothous fulvic and humic acids compared to autochthonous DOM originated from algae or phytoplankton biomass in natural waters. (ii) Investigation on the absorption of radiation by various amino acids or proteins originated from phytoplankton with respect to their standard substances. (iii) Study of the photoinduced and microbial changes of CDOM absorption for both allochthonous and autochthonous DOM. (iv) Investigation on the differences of photoinduced reactivity between allochthonous (fulvic and humic acids) and autochthonous DOM of algal or phytoplankton origin (termed autochthonous fulvic acid, see chapter "Fluorescent Dissolved Organic Matter in Natural Waters"). (v) Application of remote sensing in investigation of algal blooms in natural waters.

Problems

(1) Define the chromophoric dissolved organiu matter (CDOM)?
(2) Define the chromophores in CDOM. Mention the key chromophores in CDOM found in natural waters.
(3) Explain the CDOM absorbance theorem.
(4) What are the optical variables for the attenuation of UV and photosynthetically available radiation in waters? Mention the key optical variable and how it affects the absorption properties of waters.
(5) Explain the effects of suspended particulate matter and chlorophyll on absorption coefficients in waters.
(6) What are the controlling factors that affect CDOM absorption in waters? Explain the role of CDOM contents on absorption properties.

(7) Explain the effects of photoinduced and microbial processes on CDOM absorption properties in waters.
(8) Explain the effects of photoinduced and microbial degradation of CDOM on spectral slopes.
(9) What are the effects of monochromatic and polychromatic irradiation on CDOM absorption spectra?
(10) Why are the CDOM absorption losses by irradiation significantly different for a variety of natural waters?
(11) How does salinity affect CDOM absorption in oceans?
(12) 'All colored DOM is chromophoric DOM, but not all chromophoric DOM is colored'- Explain this concept.
(13) Explain shortly the applications of remote sensing to surface waters.
(14) Explain the impacts of photoinduced degradation of CDOM in waters.

Acknowledgments We thank Dr. Tu Chenglong of Institute of Geochemistry, Chinese Academy of Sciences, China for his generous assistance during the manuscript preparation. This work was financially supported by the Institute of Geochemistry, the Chinese Academy of Sciences, China. This work was partly supported by University of Turin, Italy; Atmospheric and Ocean Research Institute, The University of Tokyo, Japan; Center for Innovation and Entrepreneurship, Northwest Missouri State University, USA; and Jahangirnagar University, Bangladesh. This study acknowledges the reprinted from Geochim Cosmochim Acta, Hayase K, Tsubota H, Sedimentary humic and fulvic acids as fluorescent organic materials, 159–163, Copyright (1985), with permission from Elsevier; reprinted from Water Res, 43(18), Zhang Y, van Dijk MA, Liu M, Zhu G, Qin B, The contribution of phytoplankton degradation to CDOM in eutrophic shallow lakes: Field and experimental evidence, 4685–4697, Copyright (2009), with permission from Elsevier; reprinted from Mar Chem, 77(1), Rochelle-Newall EJ, Fisher TR, Chromophoric dissolved organic matter and dissolved organic carbon in Chesapeake Bay, 23– 41, Copyright (2002), with permission from Elsevier; reprinted from Mar Chem, 78(4), del Vecchio R, Blough NV, Photobleaching of chromophoric dissolved organic matter in natural waters: kinetics and modeling, 231–253, Copyright (2002), with permission from Elsevier; and reprinted from Remote Sens Environ, 112, del Castillo CE, Miller RL, On the use of ocean color remote sensing to measure the transport of dissolved organic carbon by the Mississippi River Plume, 836–844, Copyright (2008), with permission from Elsevier; Copyright (1981) by the American Society of Limnology and Oceanography, Inc; Copyright (1984) by the Association for the Sciences of Limnology and Oceanography, Inc.; Copyright (1994) by the American Society of Limnology and Oceanography, Inc; and Copyright (2000) by the Association for the Sciences of Limnology and Oceanography, Inc.

References

Ahn YH, Bricaud A, Morel A (1992) Light backscattering efficiency and related properties of some phytoplankters. Deep Sea Res Part I 39:1835–1855
al Housari F, Vione D, Chiron S, Barbati S (2010) Reactive photoinduced species in estuarine waters. Characterization of hydroxyl radical, singlet oxygen and dissolved organic matter triplet state in natural oxidation processes. Photochem Photobiol Sci 9:78–86
Allard B, Boren H, Pettersson C, Zhang G (1994) Degradation of humic substances by UV irradiation. Environ Int 20:97–101
Amador JA, Alexander M, Zika RG (1989) Sequential photochemical and microbial degradation of organic molecules bound to humic acid. Appl Environ Microbiol 55:2843–2849

Amon RMW, Benner R (1994) Rapid cycling of high-molecular-weight dissolved organic matter in the ocean. Nature 369:549–552

Amon R, Benner R (1996) Photochemical and microbial consumption of dissolved organic carbon and dissolved oxygen in the Amazon River system. Geochim Cosmochim Acta 60:1783–1792

Anastasio C, Newberg JT (2007) Sources and sinks of hydroxyl radical in sea-salt particles. J Gepphys Res 112:D10306. doi:10.1029/2006JD008061

Arrigo KR (1994) Impact of ozone depletion on phytoplankton growth in the Southern Ocean: large-scale spatial and temporal variability. Mar Ecol Prog Ser 114:1–1

Arrigo K, Brown C (1996) Impact of chromophoric dissolved organic matter on UV inhibition of primary productivity in the sea. Mar Ecol Prog Ser 140:207–216

Arrigo KR, Sullivan CW, Kremer JN (1991) A bio-optical model of Antarctic sea ice. J Geophys Res 96:10581–10592

Arvesen JC, Millard J, Weaver E (1973) Remote sensing of chlorophyll and temperature in marine and fresh waters. Astronautica Acta 18:229–239

Assel M, Laenen R, Laubereau A (1998) Ultrafast electron trapping in an aqueous NaCl-solution. Chem Phys Lett 289:267–274

Baker KS, Smith RC (1982) Spectral irradiance penetration in natural waters. The role of solar ultraviolet radiation in marine ecosystems In: Calkins J (ed) Role of solar ultraviolet radiation in marine ecosystems, Plenum Press, New York, pp 233–246

Banaszak AT, Trench RK (1995) Effects of ultraviolet (UV) radiation on marine microalgal-invertebrate symbioses. I. Response of the algal symbionts in culture an <i> in hospite </i>. J Exp Mar Biol Ecol 194:213–232

Bandaranayake WM (1998) Mycosporines: are they nature's sunscreens? Nat Prod Rep 15:159–172

Becker RH, Sultan MI, Boyer GL, Twiss MR, Konopko E (2009) Mapping cyanobacterial blooms in the Great Lakes using MODIS. J Great Lakes Res 35:447–453

Belzile C, Johannessen SC, Gosselin M, Demers S, Miller WL (2000) Ultraviolet attenuation by dissolved and particulate constituents of first-year ice during late spring in an Arctic polynya. Limnol Oceanogr 45:1265–1273

Belzile C, Vincent WF, Kumagai M (2002) Contribution of absorption and scattering to the attenuation of UV and photosynthetically available radiation in Lake Biwa. Limnol Oceanogr 47:95–107

Benner R, Kaiser K (2003) Abundance of amino sugars and peptidoglycan in marine particulate and dissolved organic matter. Limnol Oceanogr 48:118–128

Bernasconi SM, Barbieri A, Simona M (1997) Carbon and nitrogen isotope variations in sedimenting organic matter in Lake Lugano. Limnol Oceanogr 42:1755–1765

Bertilsson S, Allard B (1996) Sequential photochemical and microbial degradation of refractory dissolved organic matter in a humic freshwater system. Arch Hydrobiol Beih Ergebn Limnol 48:133–141

Bertilsson S, Tranvik LJ (2000) Photochemical transformation of dissolved organic matter in lakes. Limnol Oceanogr 45:753–762

Biddanda B, Benner R (1997) Carbon, nitrogen, and carbohydrate fluxes during the production of particulate and dissolved organic matter by marine phytoplankton. Limnol Oceanogr 42:506–518

Billen G, Fontigny A (1987) Dynamics of a Phaeosystis-dominated spring bloom in Belgian coastal waters. 11. Bacterioplankton dynamics. Mar Ecol Prog Ser 37:249–257

Binding CE, Jerome JH, Bukata RP, Booty WG (2008) Spectral absorption properties of dissolved and particulate matter in Lake Erie. Remote Sens Environ 112:1702–1711

Blough N, del Vecchio R (2002) Chromophoric DOM in the coastal environment. In: Hansell DA, Carlson CA (eds) Biogeochemistry of marine dissolved organic matter. Elsevier Science, USA, p 774

Blough N, Green S (1995) Spectroscopic characterization and remote sensing of non-living organic matter. In: Zepp RG, Sonntag C (eds) The Dahlem conference report—the role of nonliving organic matter in the Earth's Carbon Cycle. Wiley, pp 23–45

Blough N, Zepp S (1995) Reactive oxygen species in natural waters. In: Foote CS et al (eds) Active Oxygen in Chemistry. Chapman & Hall, Glasgow, UK, pp 280–333
Blough N, Zafiriou O, Bonilla J (1993) Optical absorption spectra of waters from the Orinoco River outflow: terrestrial input of colored organic matter to the Caribbean. J Geophys Res 98:2271–2278
Boehme JR, Coble PG (2000) Characterization of colored dissolved organic matter using high-energy laser fragmentation. Environ Sci Technol 34:3283–3290
Bothwell ML, Sherbot DMJ, Pollock CM (1994) Ecosystem response to solar ultraviolet-B radiation: influence of trophic-level interactions. Science 265:97–100
Bowers D, Binding C (2006) The optical properties of mineral suspended particles: a review and synthesis. Estuar Coast Shelf Sci 67:219–230
Bowers D, Harker G, Smith P, Tett P (2000) Optical properties of a region of freshwater influence (the Clyde Sea). Estuar Coast Shelf Sci 50:717–726
Bracchini L, Dattilo AM, Falcucci M, Loiselle SA, Hull V, Arena C, Rossi C (2005) Spatial and temporal variations of the inherent and apparent optical properties in a shallow coastal lake. J Photochem Photobiol B Biol 80:161–177
Bracchini L, Tognazzi A, Dattilo AM, Decembrini F, Rossi C, Loiselle SA (2010) Sensitivity analysis of CDOM spectral slope in artificial and natural samples: an application in the central eastern Mediterranean Basin. Aquat Sci 72:485–498
Braven J, Butler EI, Chapman J, Evens R (1995) Changes in dissolved free amino acid composition in sea water associated with phytoplankton populations. Sci Total Environ 172:145–150
Bricaud A, Morel A, Prieur L (1981) Absorption by dissolved organic matter of the sea (yellow substance) in the UV and visible domains. Limnol Oceanogr 26:43–53
Bricaud A, Babin M, Claustre H, Ras J, Tièche F (2010) Light absorption properties and absorption budget of Southeast Pacific waters. J Geophys Res 115:C08009
Bronk DA, Glibert PM, Ward BB (1994) Nitrogen uptake, dissolved organic nitrogen release, and new production. Science 265:1843–1846
Brooks ML, Meyer JS, McKnight DM (2007) Photooxidation of wetland and riverine dissolved organic matter: altered copper complexation and organic composition. Hydrobiologia 579:95–113
Brown M (1977) Transmission spectroscopy examinations of natural waters: C. Ultraviolet spectral characteristics of the transition from terrestrial humus to marine yellow substance. Estuar Coast Mar Sci 5:309–317
Brown CA, Huot Y, Werdell PJ, Gentili B, Claustre H (2008) The origin and global distribution of second order variability in satellite ocean color and its potential applications to algorithm development. Remote Sens Environ 112:4186–4203
Buckley R, Trodahl H (1987) Scattering and absorption of visible light by sea ice
Bukata R, Jerome J, Bruton J, Jain S (1979) Determination of inherent optical properties of Lake Ontario coastal waters. Appl Optics 18:3926–3932
Burdige DJ, Kline SW, Chen W (2004) Fluorescent dissolved organic matter in marine sediment pore waters. Mar Chem 89:289–311
Cannizzaro JP, Carder KL (2006) Estimating chlorophyll <i> a </i> concentrations from remote-sensing reflectance in optically shallow waters. Remote Sens Environ 101:13–24
Cannizzaro JP, Carder KL, Chen FR, Heil CA, Vargo GA (2008) A novel technique for detection of the toxic dinoflagellate, Karenia brevis, in the Gulf of Mexico from remotely sensed ocean color data. Continent Shelf Res 28:137–158
Carder KL, Steward RG, Harvey GR, Ortner PB (1989) Marine humic and fulvic acids: Their effects on remote sensing of ocean chlorophyll. Limnol Oceanogr 34:68–81
Carder K, Hawes S, Baker K, Smith R (1991) Reflectance model for quantifying chlorophyll a in the presence oi productivity degradation products. J Geophys Res 96:20599–20611
Carder KL, Steward RG, Chen RF, Hawes S, Lee Z, Davis CO (1993) AVIRIS calibration and application in Coastal Ocean environments: tracers of soluble and particulate constituents of the Tampa Bay coastal plume. Photogramm Eng Remote Sens 59:339–344

Carder KL, Chen F, Lee Z, Hawes S, Kamykowski D (1999) Semianalytic moderate-resolution imaging spectrometer algorithms for chlorophyll a and absorption with bio-optical domains based on nitrate-depletion temperatures. J Geophys Res 104:5403
Carpenter J, Manella M (1973) Magnesium to chlorinity ratios in seawater. J Geophys Res 78:3621–3626
Carvalho GA, Minnett PJ, Banzon VF, Baringer W, Heil CA (2011) Long-term evaluation of three satellite ocean color algorithms for identifying harmful algal blooms (Karenia brevis) along the west coast of Florida: a matchup assessment. Remote Sens Environ 115:1–18
Chen R, Zhang Y, Vlahos P, Rudnick S (2002) The fluorescence of dissolved organic matter in the Mid-Atlantic Bight. Deep Sea Res Part II 49:4439–4459
Chen Z, Hu C, Conmy RN, Muller-Karger F, Swarzenski P (2007) Colored dissolved organic matter in Tampa Bay, Florida. Mar Chem 104:98–109
Clark CD, Hiscock WT, Millero FJ, Hitchcock G, Brand L, Miller WL, Ziolkowski L, Chen RF, Zika RG (2004) CDOM distribution and CO_2 production on the Southwest Florida Shelf. Mar Chem 89:145–167
Clark CD, Litz LP, Grant SB (2008) Salt marshes as a source of chromophoric dissolved organic matter (CDOM) to Southern California coastal waters. Limnol Oceanogr 53:1923–1933
Clark CD, de Bruyn WJ, Jones JG (2009) Photochemical production of hydrogen peroxide in size-fractionated Southern California coastal waters. Chemosphere 76:141–146
Clarke GL, Ewing GC (1974) Remote spectroscopy of the sea for biological production studies. In: Jerlov NG, Steemann Nielsen E (eds) Optical aspects of oceanography, Academic, Largo, pp 389–414
Clarke G, James HR (1939) Laboratory analysis of the selective absorption of light by sea water. J Opt Sot Am 29:43–55
Clarke GL, Ewing GC, Lorenzen CJ (1970) Spectra of backscattered light from the sea obtained from aircraft as a measure of chlorophyll concentration. Science 167:1119–1121
Coble PG (1996) Characterization of marine and terrestrial DOM in seawater using excitation-emission matrix spectroscopy. Mar Chem 51:325–346
Coble PG (2007) Marine optical biogeochemistry: the chemistry of ocean color. Chem Rev 107:402–418
Coble PG, del Castillo CE, Avril B (1998) Distribution and optical properties of CDOM in the Arabian Sea during the 1995 Southwest Monsoon. Deep-Sea Res Part II 45:2195–2223
Conmy RN, Coble PG, Chen RF, Gardner GB (2004) Optical properties of colored dissolved organic matter in the Northern Gulf of Mexico. Mar Chem 89:127–144
Corin N, Backlund P, Kulovaara M (1996) Degradation products formed during UV-irradiation of humic waters. Chemosphere 33:245–255
Cui T, Zhang J, Groom S, Sun L, Smyth T, Sathyendranath S (2010) Validation of MERIS ocean-color products in the Bohai Sea: a case study for turbid coastal waters. Remote Sens Environ 114:2326–2336
Dahlén J, Bertilsson S, Pettersson C (1996) Effects of UV-A irradiation on dissolved organic matter in humic surface waters. Environ Int 22:501–506
Davies-Colley R, Vant W (1987) Absorption of light by yellow substance in freshwater lakes. Limnol Oceanogr 32:416–425
de Haan H (1993) Solar UV-light penetration and photodegradation of humic substances in peaty lake water. Limnol Oceanogr 38:1072–1076
del Castillo CE (2005) Remote sensing of colored dissolved organic matter in coastal environments. In: Miller RM, del Castillo CE, McKee B (Eds) Remote sensing of aquatic coastal environments. Springer, New York, p 345
del Castillo CE, Miller RL (2008) On the use of ocean color remote sensing to measure the transport of dissolved organic carbon by the Mississippi River Plume. Remote Sens Environ 112:836–844
del Castillo CE, Coble PG, Morell JM, López JM, Corredor JE (1999) Analysis of the optical properties of the Orinoco River plume by absorption and fluorescence spectroscopy. Mar Chem 66:35–51

del Castillo CE, Gilbes F, Coble PG, Muller-Karger FE (2000) On the dispersal of riverine colored dissolved organic matter over the West Florida Shelf. Limnol Oceanogr 45:1425–1432

del Castillo CE, Coble PG, Conmy RN, Müller-Karger FE, Vanderbloemen L, Vargo GA (2001) Multispectral in situ measurements of organic matter and chlorophyll fluorescence in seawater: documenting the intrusion of the Mississippi River plume in the West Florida Shelf. Limnol Oceanogr 46:1836–1843

del Vecchio R, Blough NV (2002) Photobleaching of chromophoric dissolved organic matter in natural waters: kinetics and modeling. Mar Chem 78:231–253

del Vecchio R, Blough NV (2004) Spatial and seasonal distribution of chromophoric dissolved organic matter and dissolved organic carbon in the Middle Atlantic Bight. Mar Chem 89:169–187

Devlin M, Barry J, Mills D, Gowen R, Foden J, Sivyer D, Tett P (2008) Relationships between suspended particulate material, light attenuation and Secchi depth in UK marine waters. Estuar Coast Shelf Sci 79:429–439

Devlin M, Barry J, Mills D, Gowen R, Foden J, Sivyer D, Greenwood N, Pearce D, Tett P (2009) Estimating the diffuse attenuation coefficient from optically active constituents in UK marine waters. Estuar Coast Shelf Sci 82:73–83

Dodge JD (1989) Some revisions of the family Gonyaulacaceae (Dinophyceae) based on scanning electron microscope study. Bot Mar 32:275–298

Dorsch JE, Bidleman TF (1982) Natural organics as fluorescent tracers of river-mixing. Estuar Coastal Shelf Sci 15:701–707

Doxaran D, Froidefond JM, Lavender S, Castaing P (2002) Spectral signature of highly turbid waters: application with SPOT data to quantify suspended particulate matter concentrations. Remote Sens Environ 81:149–161

Du C, Shang S, Dong Q, Hu C, Wu J (2010) Characteristics of chromophoric dissolved organic matter in the nearshore waters of the western Taiwan Strait. Estuar Coast Shelf Sci 88:350–356

Dubach P, Mehta N, Jakab T, Martin F, Roulet N (1964) Chemical investigations on soil humic substances. Geochim Cosmochim Acta 28:1567–1578

Duntley SQ (1942) The optical properties of diffusing materials. JOSA 32:61–61

Duntley S, Wilson WH, Edgerton CF (1974) Ocean color analysis, part 1. Scripps Inst Oceanogr Ref 74–10:l–37

Dupouy C, Neveux J, Ouillon S, Frouin R, Murakami H, Hochard S, Dirberg G (2010) Inherent optical properties and satellite retrieval of chlorophyll concentration in the lagoon and open ocean waters of New Caledonia. Mar Pollut Bull 61:503–518

Effler SW, Perkins MG, Peng F, Strait C, Weidemann AD, Auer MT (2010) Light-absorbing components in Lake Superior. J Great Lakes Res 36:656–665

Fahey T, Siccama T, Driscoll C, Likens G, Campbell J, Johnson C, Battles J, Aber J, Cole J, Fisk M (2005) The biogeochemistry of carbon at Hubbard Brook. Biogeochemistry 75:109–176

Ferrari GM (2000) The relationship between chromophoric dissolved organic matter and dissolved organic carbon in the European Atlantic coastal area and in the West Mediterranean Sea (Gulf of Lions). Mar Chem 70:339–357

Ferrari GM, Dowell MD, Grossi S, Targa C (1996) Relationship between the optical properties of chromophoric dissolved organic matter and total concentration of dissolved organic carbon in the southern Baltic Sea region. Mar Chem 55:299–316

Ferreira A, Garcia VMT, Garcia CAE (2009) Light absorption by phytoplankton, non-algal particles and dissolved organic matter at the Patagonia shelf-break in spring and summer. Deep Sea Res Part I 56:2162–2174

Fichot CG, Miller WL (2010) An approach to quantify depth-resolved marine photochemical fluxes using remote sensing: Application to carbon monoxide (CO) photoproduction. Remote Sens Environ 114:1363–1377

Foden J, Sivyer D, Mills D, Devlin M (2008) Spatial and temporal distribution of chromophoric dissolved organic matter (CDOM) fluorescence and its contribution to light attenuation in UK waterbodies. Estuar Coast Shelf Sci 79:707–717

Jonasz M, Fournier, GR (2007) Optical properties of pure water, seawater, and natural waters. Light scattering by particles in water, Academic Press, Amsterdam, pp 33–85

Fox MA (1990) Photochemical electron transfer. Photochem Photobiol 52:617–627

Fox LE (1991) The transport and composition of humic substances in estuaries. In: Baker RA (ed) Organic substances and sediments in water, vol 1., Humics and SoilsLewis Publisher, Chelsea, pp 129–162

Friedrichs MAM, Carr ME, Barber RT, Scardi M, Antoine D, Armstrong RA, Asanuma I, Behrenfeld MJ, Buitenhuis ET, Chai F (2009) Assessing the uncertainties of model estimates of primary productivity in the tropical Pacific Ocean. J Mar Syst 76:113–133

Frimmel F, Bauer H (1987) Influence of photochemical reactions on the optical properties of aquatic humic substances gained from fall leaves. Sci Total Environ 62:139–148

Fritsen CH, Wirthlin ED, Momberg DK, Lewis MJ, Ackley SF (2011) Bio-optical properties of Antarctic pack ice in the early austral spring. Deep Sea Res Part II 58:1052–1061

Fu P, Mostofa KMG, Wu F, Liu CQ, Li W, Liao H, Wang L, Wang J, Mei Y (2010) Excitation-emission matrix characterization of dissolved organic matter sources in two eutrophic lakes (Southwestern China Plateau). Geochem J 44:99–112

Fujiwara K, Ushiroda T, Takeda K, Kumamoto YI, Tsubota H (1993) Diurnal and seasonal distribution of hydrogen peroxide in seawater of the Seto Inland Sea. Geochem J 27:103–115

Fujiwara K, Takeda K, Kumamoto Y (1995) Generations of carbonyl sulfide and hydrogen peroxide in the Seto Inland Sea—Photochemical reactions progressing in the coastal seawater. In: Sakai H, Nozaki Y (eds) Biogeochemical processes and Ocean Flux in the Western Pacific. Terrapub, Tokyo, pp 101–127

Gallegos CL, Bergstrom PW (2005) Effects of a <i> Prorocentrum minimum </i> bloom on light availability for and potential impacts on submersed aquatic vegetation in upper Chesapeake Bay. Harmful algae 4:553–574

Gao H, Zepp RG (1998) Factors influencing photoreactions of dissolved organic matter in a coastal river of the Southeastern United States. Environ Sci Technol 32:2940–2946

Ghassemi M, Christman R (1968) Properties of the yellow organic acids of natural waters. Limnol Oceanogr 13:583–597

Goldstone J, Pullin M, Bertilsson S, Voelker B (2002) Reactions of hydroxyl radical with humic substances: Bleaching, mineralization, and production of bioavailable carbon substrates. Environ Sci Technol 36:364–372

Gons HJ, Auer MT, Effler SW (2008) MERIS satellite chlorophyll mapping of oligotrophic and eutrophic waters in the Laurentian Great Lakes. Remote Sens Environ 112:4098–4106

Gonsior M, Peake BM, Cooper WJ, Jaffé R, Young H, Kahn AE, Kowalczuk P (2008) Spectral characterization of chromophoric dissolved organic matter (CDOM) in a fjord (Doubtful Sound, New Zealand). Aquat Sci 70:397–409

González Vilas L, Spyrakos E, Torres Palenzuela JM (2011) Neural network estimation of chlorophyll a from MERIS full resolution data for the coastal waters of Galician rias (NW Spain). Remote Sens Environ 115:524–535

González-Vila F, Lankes U, Lüdemann HD (2001) Comparison of the information gained by pyrolytic techniques and NMR spectroscopy on the structural features of aquatic humic substances. J Anal Appl Pyrolysis 58:349–359

Gopinathan C, Damle P, Hart EJ (1972) gamma.-Ray irradiated sodium chloride as a source of hydrated electrons. J Phys Chem 76:3694–3698

Gordon HR, Morel AY (1983) Remote assessment of ocean color for interpretation of satellite visible imagery: a review. SpringerVerlag, New York

Gordon HR, Brown OB, Jacobs MM (1975) Computed relationships between the inherent and apparent optical properties of a flat homogeneous ocean. Appl Optics 14:417–427

Grabowski ZR, Rotkiewicz K, Rettig W (2003) Structural changes accompanying intramolecular electron transfer: focus on twisted intramolecular charge-transfer states and structures. Chem Rev 103:3899–4032
Granéli W, Lindell M, Tranvik L (1996) Photo-oxidative production of dissolved inorganic carbon in lakes of different humic content. Limnol Oceanogr 41:698–706
Granéli W, Lindell M, de Faria BM, de Assis Esteves F (1998) Photoproduction of dissolved inorganic carbon in temperate and tropical lakes–dependence on wavelength band and dissolved organic carbon concentration. Biogeochemistry 43:175–195
Grebel JE, Pignatello JJ, Song W, Cooper WJ, Mitch WA (2009) Impact of halides on the photobleaching of dissolved organic matter. Mar Chem 115:134–144
Green SA, Blough NV (1994) Optical absorption and fluorescence properties of chromophoric dissolved organic matter in natural waters. Limnol Oceanogr 39:1903–1916
Green RE, Gould RW, Ko DS (2008) Statistical models for sediment/detritus and dissolved absorption coefficients in coastal waters of the northern Gulf of Mexico. Continent Shelf Res 28:1273–1285
Gregg WW, Casey NW (2009) Skill assessment of a spectral ocean–atmosphere radiative model. J Mar Systems 76:49–63
Grenfell TC, Perovich DK (1984) Spectral albedos of sea ice and incident solar irradiance in the southern Beaufort Sea. J Geophys Res 89:3573–3580
Gueguen C, Granskog MA, McCullough G, Barber DG (2011) Characterisation of colored dissolved organic matter in Hudson Bay and Hudson Strait using parallel factor analysis. J Mar Syst 88:423–433
Guéguen C, Guo L, Wang D, Tanaka N, Hung CC (2006) Chemical characteristics and origin of dissolved organic matter in the Yukon River. Biogeochemistry 77:139–155
Haltrin VI (1999) Chlorophyll-based model of seawater optical properties. Appl Optics 38:6826–6832
Hama J, Handa N (1992) Diel variation of water-extractable carbohydrate composition of natural phytoplankton populations in Kinu-ura Bay. J Exp Mar Biol Ecol 162:159–176
Harding LW, Magnuson A, Mallonee ME (2005) SeaWiFS retrievals of chlorophyll in Chesapeake Bay and the mid-Atlantic bight. Estuar Coast Shelf Sci 62:75–94
Hayakawa K (2004) Seasonal variations and dynamics of dissolved carbohydrates in Lake Biwa. Org Geochem 35:169–179
Hayakawa K, Sugiyama Y (2008) Spatial and seasonal variations in attenuation of solar ultraviolet radiation in Lake Biwa, Japan. J Photochem Photobiol B Biol 90:121–133
Hayakawa K, Sekino T, Yoshioka T, Maruo M, Kumagai M (2003) Dissolved organic carbon and fluorescence in Lake Hovsgol: factors reducing humic content of the lake water. Limnology 4:25–33
Hayase K, Tsubota H (1985) Sedimentary humic acid and fulvic acid as fluorescent organic materials. Geochim Cosmochim Acta 49:159–163
Hayase K, Yamamoto M, Nakazawa I, Tsubota H (1987) Behavior of natural fluorescence in Sagami Bay and Tokyo Bay, Japan—vertical and lateral distributions. Mar Chem 20:265–276
Hedges JI (1992) Global biogeochemical cycles: progress and problems. Mar Chem 39:67–93
Hedges J, Eglinton G, Hatcher P, Kirchman D, Arnosti C, Derenne S, Evershed R, Kögel-Knabner I, de Leeuw J, Littke R (2000) The molecularly-uncharacterized component of nonliving organic matter in natural environments. Org Geochem 31:945–958
Helms JR, Stubbins A, Ritchie JD, Minor EC, Kieber DJ, Mopper K (2008) Absorption spectral slopes and slope ratios as indicators of molecular weight, source, and photobleaching of chromophoric dissolved organic matter. Limnol Oceanogr 53:955–969
Hem JD (1985) Study and interpretation of the chemical characteristics of natural water. vol 2254, Department of the Interior, US Geological Survey
Hernes PJ, Benner R (2003) Photochemical and microbial degradation of dissolved lignin phenols: implications for the fate of terrigenous dissolved organic matter in marine environments. J Geophys Res 108:3291

Hernes PJ, Peterson ML, Murray JW, Wakeham SG, Lee C, Hedges JI (2001) Particulate carbon and nitrogen fluxes and compositions in the central equatorial Pacific. Deep Sea Res Part I 48:1999–2023

Hodoki Y, Watanabe Y (1998) Attenuation of solar ultraviolet radiation in eutrophic freshwater lakes and ponds. Japanese J Limnol 59:27–37

Hoge FE, Williams ME, Swift RN, Yungel JK, Vodacek A (1995) Satellite retrieval of the absorption coefficient of chromophoric dissolved organic matter in continental margins. J Geophys Res 100:28,847–28,854

Hoge FE, Wright CW, Lyon PE, Swift RN, Yungel JK (2001) Inherent optical properties imagery of the western North Atlantic Ocean—Horizontal spatial variability of the upper mixed layer. J Geophys Res 106:31,129–31,140

Holm-Hansen O, Lubin D, Helbling EW (1993) Ultraviolet radiation and its effects on organisms in aquatic environments. In: Young A (ed) Environmental UV photobiology. Plenum Press, New York, pp 379–425

Hujerslev N, Holt N, Aarup T (1996) Optical measurements in the North Sea-Baltic Sea transition zone. I. On the origin of the deep water in the Kattegat. Continent Shelf Res 16:1329–1342

Hulatt CJ, Thomas DN, Bowers DG, Norman L, Zhang C (2009) Exudation and decomposition of chromophoric dissolved organic matter (CDOM) from some temperate macroalgae. Estuar Coast Shelf Sci 84:147–153

Hunter PD, Tyler AN, Présing M, Kovács AW, Preston T (2008) Spectral discrimination of phytoplankton colour groups: the effect of suspended particulate matter and sensor spectral resolution. Remote Sens Environ 112:1527–1544

Hunter PD, Tyler AN, Carvalho L, Codd GA, Maberly SC (2010) Hyperspectral remote sensing of cyanobacterial pigments as indicators for cell populations and toxins in eutrophic lakes. Remote Sens Environ 114:2705–2718

Ishiwatari R (1973) Chemical characterization of fractionated humic acids from lake and marine sediments. Chem Geol 12:113–126

Ittekkot V (1982) Variations of dissolved organic matter during a plankton bloom: qualitative aspects, based on sugar and amino acid analyses. Mar Chem 11:143–158

Ittekkot V, Safiullah S, Mycke B, Seifert R (1985) Seasonal variability and geochemical significance of organic matter in the River Ganges, Bangladesh. Nature 317:800–802

Jeffrey S, MacTavish H, Dunlap W, Vesk M, Groenewoud K (1999) Occurrence of UVA-and UVB-absorbing compounds in 152 species (206 strains) of marine microalgae. Mar Ecol Prog Ser 189:35–51

Jennings J, Steinberg P (1994) In situ exudation of phlorotannins by the sublittoral kelp Ecklonia radiata. Mar Biol 121:349–354

Jerlov NG (1968) Optical oceanography, vol 5. Elsevier, Amsterdam

Jerlov N (1976) Marine optics, Elsevier, New York, 231 p

Johannessen SC, Peña MA, Quenneville ML (2007) Photochemical production of carbon dioxide during a coastal phytoplankton bloom. Estuar Coast Shelf Sci 73:236–242

Kahru M, Mitchell BG (1998) Spectral reflectance and absorption of a massive red tide. J Geophys Res 103:21,601–621,609

Kahru M, Mitchell BG (2001) Seasonal and nonseasonal variability of satellite-derived chlorophyll and colored dissolved organic matter concentration in the California current. J Geophys Res 106:2517–2529

Kalle K (1966) The problem of the gelbstoff in the sea. Oceanogr Mar Biol Annu Rev 4:91–104

Karentz D, McEuen F, Land M, Dunlap W (1991) Survey of mycosporine-like amino acid compounds in Antarctic marine organisms: potential protection from ultraviolet exposure. Mar Biol 108:157–166

Kieber RJ, Zhou X, Mopper K (1990) Formation of carbonyl compounds from UV-induced photodegradation of humic substances in natural waters: fate of riverine carbon in the sea. Limnol Oceanogr 35:1503–1515

Kim C, Nishimura Y, Nagata T (2006) Role of dissolved organic matter in hypolimnetic mineralization of carbon and nitrogen in a large, monomictic lake. Limnol Oceanogr 51:70–78
Kirk J (1981a) Monte Carlo study of the nature of the underwater light held in, and the relationships between optical properties of, turbid yellow waters. Aust J Mar Freshwater Res 32:533–539
Kirk J (1981b) Estimation of the scattering coefficient of natural waters using underwater irradiance measurements. Aust J Mar Freshwater Res 32:533–539
Kirk J (1984) Dependence of relationship between inherent and apparent optical properties of water on solar altitude. Limnol Oceanogr 29:350–356
Kirk J (1988) Optical water quality, what does it mean and how should we measure it? J Water Pollut Control Federation 60:194–197
Kirk JTO (1991) Volume scattering function, average cosines, and the underwater light field. Limnol Oceanogr 36:455–467
Kirk JTO (1994) Characteristics of the light field in highly turbid waters: a monte carlo study. Limnol Oceanogr 39:702–706
Kishino M, Tanaka A, Ishizaka J (2005) Retrieval of Chlorophyll *a*, suspended solids, and colored dissolved organic matter in Tokyo Bay using ASTER data. Remote Sens Environ 99:66–74
Kitidis V, Stubbins AP, Uher G, Upstill Goddard RC, Law CS, Woodward EMS (2006) Variability of chromophoric organic matter in surface waters of the Atlantic Ocean. Deep Sea Res Part II 53:1666–1684
Klinkhammer G, McManus J, Colbert D, Rudnicki M (2000) Behavior of terrestrial dissolved organic matter at the continent-ocean boundary from high-resolution distributions. Geochim Cosmochim Acta 64:2765–2774
Kopelevich O, Rodionov V, Stupakova T (1987) Effect of bacteria on optical characteristics of ocean water. Oceanology 27:696–700
Koponen S, Attila J, Pulliainen J, Kallio K, Pyhälahti T, Lindfors A, Rasmus K, Hallikainen M (2007) A case study of airborne and satellite remote sensing of a spring bloom event in the Gulf of Finland. Continent Shelf Res 27:228–244
Korshin GV, Li CW, Benjamin MM (1997) Monitoring the properties of natural organic matter through UV spectroscopy: a consistent theory. Water Res 31:1787–1795
Kouassi AM, Zika RG (1990) Light-induced alteration of the photophysical properties of dissolved organic matter in seawater Part I. Photoreversible properties of natural water fluorescence. Netherlands J Sea Res 27:25–32
Kouassi AM, Zika RG, Plane J (1990) Light-induced alteration of the photophysical properties of dissolved organic matter in seawater: Part II. Estimates of the environmental rates of the natural water fluorescence Part II. Estimates of the environmental rates of the natural water fluorescence. Netherlands J Sea Res 27:33–41
Kowalczuk P (1999) Seasonal variability of yellow substance absorption in the surface layer of the Baltic Sea. J Geophys Res 104:30047–30058
Kowalczuk P, Cooper WJ, Whitehead RF, Durako MJ, Sheldon W (2003) Characterization of CDOM in an organic-rich river and surrounding coastal ocean in the South Atlantic Bight. Aquat Sci 65:384–401
Kowalczuk P, Cooper WJ, Durako MJ, Kahn AE, Gonsior M, Young H (2010) Characterization of dissolved organic matter fluorescence in the South Atlantic Bight with use of PARAFAC model: Relationships between fluorescence and its components, absorption coefficients and organic carbon concentrations. Mar Chem 118:22–36
Kratzer S, Brockmann C, Moore G (2008) Using MERIS full resolution data to monitor coastal waters–A case study from Himmerfjärden, a fjord-like bay in the northwestern Baltic Sea. Remote Sens Environ 112:2284–2300
Kutser T, Metsamaa L, Strömbeck N, Vahtmäe E (2006) Monitoring cyanobacterial blooms by satellite remote sensing. Estuar Coast Shelf Sci 67:303–312
Laane R, Kramer K (1990) Natural fluorescence in the North Sea and its major estuaries. Netherlands J Sea Res 26:1–9

Laurion I, Vincent WF, Lean DRS (1997) Underwater ultraviolet radiation: development of spectral models for northern high latitude lakes. Photochem Photobiol 65:107–114

Laurion I, Ventura M, Catalan J, Psenner R, Sommaruga R (2000) Attenuation of ultraviolet radiation in mountain lakes: factors controlling the among-and within-lake variability. Limnol Oceanogr 45:1274–1288

Lawrence J (1980) Semi-quantitative determination of fulvic acid, tannin and lignin in natural waters. Water Res 14:373–377

Le C, Li Y, Zha Y, Sun D, Huang C, Zhang H (2011) Remote estimation of chlorophyll a in optically complex waters based on optical classification. Remote Sens Environ 115:725–737

Leavitt PR, Vinebrooke RD, Donald DB, Smol JP, Schindler DW (1997) Past ultraviolet radiation environments in lakes derived from fossil pigments. Nature 388:457–459

Lee Z, Carder KL, Hawes SK, Steward RG, Peacock TG, Davis CO (1994) Model for the interpretation of hyperspectral remote-sensing reflectance. Appl Optics 33:5721–5732

Leenheer JA, Croue JP (2003) Peer reviewed: characterizing aquatic dissolved organic matter. Environ Sci Technol 37:18–26

Legrini O, Oliveros E, Braun A (1993) Photochemical processes for water treatment. Chem Rev 93:671–698

Lehmann MF, Bernasconi SM, Barbieri A, McKenzie JA (2002) Preservation of organic matter and alteration of its carbon and nitrogen isotope composition during simulated and in situ early sedimentary diagenesis. Geochim Cosmochim Acta 66:3573–3584

Lesser MP (2006) Oxidative stress in marine environments: biochemistry and physiological ecology. Annu Rev Physiol 68:253–278

Li W, Wu F, Liu C, Fu P, Wang J, Mei Y, Wang L, Guo J (2008) Temporal and spatial distributions of dissolved organic carbon and nitrogen in two small lakes on the Southwestern China Plateau. Limnology 9:163–171

Livingstone DA (1963) Chemical composition of rivers and lakes. In: Fleischer M (ed) Data of geochemistry, US geological survey, US Govt Printing office, Washington DC, pp 58–661

Lorenzen CJ (1972) Extinction of light in the ocean by phytoplankton. J Conseil 34:262–267

Lund-Hansen LC (2004) Diffuse attenuation coefficients Kd (PAR) at the estuarine North Sea-Baltic Sea transition: time-series, partitioning, absorption, and scattering. Estuar Coast Shelf Sci 61:251–259

Ma X, Ali N (2009) Dissolved organic matter in the aquatic environments. In: Wu FC, Xing B (eds) Natural organic matter and its significance in the environment, Ch 2. Science Press, Beijing, pp 66–89

Ma X, Green SA (2004) Photochemical transformation of dissolved organic carbon in lake superior—an in situ experiment. J Great Lakes Res 30:97–112

Malcolm RL (1985) Geochemistry of stream fulvic and humic substances. In: Aiken GR, McKnight DM, Wershaw RL, MacCarthy P (eds) Humic substances in soil, sediment, and water: geochemistry, Isolation and Characterization, Wiley, pp 181–209

Malcolm RL (1990) The uniqueness of humic substances in each of soil, stream and marine environments. Anal Chim Acta 232:19–30

Malinsky-Rushansky N, Legrand C (1996) Excretion of dissolved organic carbon by phytoplankton of different sizes and subsequent bacterial uptake. Mar Ecol Prog Ser 132:249–255

Mantoura R, Woodward E (1983) Conservative behaviour of riverine dissolved organic carbon in the Severn Estuary: chemical and geochemical implications. Geochim Cosmochim Acta 47:1293–1309

Marchant H, Davidson A, Kelly G (1991) UV-B protecting compounds in the marine algaPhaeocystis pouchetii from Antarctica. Mar Biol 109:391–395

Markager S, Vincent WF (2000) Spectral light attenuation and the absorption of UV and blue light in natural waters. Limnol Oceanogr 45:642–650

Matthews MW, Bernard S, Winter K (2010) Remote sensing of cyanobacteria-dominant algal blooms and water quality parameters in Zeekoevlei, a small hypertrophic lake, using MERIS. Remote Sens Environ 114:2070–2087

Maul GA, Gordon HR (1976) On the use of the earth resources technology satellite (LANDSAT-1) in optical oceanography. Remote Scns Environ 4:95–128

McCarthy M, Hedges J, Benner R (1996) Major biochemical composition of dissolved high molecular weight organic matter in seawater. Mar Chem 55:281–297

McClain CR, Signorini SR, Christian JR (2004a) Subtropical gyre variability observed by ocean-color satellites. Deep-Sea Res Part II 51:281–301

McClain CR, Feldman GC, Hooker SB (2004b) An overview of the SeaWiFS project and strategies for producing a climate research quality global ocean bio-optical time series. Deep Sea Res Part II 51:5–42

McKnight DM, Andrews ED, Spulding SA, Aiken GR (1994) Aquatic fulvic acids in algal-rich Antarctic ponds. Limnol Oceanogr 39:1972–1979

Medina-Sánchez JM, Villar-Argaiz M, Carrillo P (2006) Solar radiation-nutrient interaction enhances the resource and predation algal control on bacterioplankton: a short-term experimental study. Limnol Oceanogr 51:913–924

Mélin F, Zibordi G, Berthon JF (2007) Assessment of satellite ocean color products at a coastal site. Remote Sens Environ 110:192–215

Miller W (1994) Recent advances in the photochemistry of natural dissolved organic matter. In: Helz GR, Zepp RG, Crosby DG (eds) Aquatic and surface photochemistry CRC Press. USA, Boca Raton, pp 111–128

Miller WL (1998) Effects of UV radiation on aquatic humus: photochemical principles and experimental considerations. In: Hessen DO, Tranvik LJ (ed) Aquatic humic substances, Springer, New York, pp 125–143

Miller WL, Moran MA (1997) Interaction of photochemical and microbial processes in the degradation of refractory dissolved organic matter from a coastal marine environment. Limnol Oceanogr 42:1317–1324

Minella M, Romeo F, Vione D, Maurino V, Minero C (2011) Low to negligible photoactivity of lake-water matter in the size range from 0.1 to 5 mu m. Chemosphere 83:1480–1485

Minero C, Lauri V, Falletti G, Maurino V, Pelizzetti E, Vione D (2007a) Spectrophotometric characterisation of surface lakewater samples: implications for the quantification of nitrate and the properties of dissolved organic matter. Anal Chim 97:1107–1116

Minero C, Chiron S, Falletti G, Maurino V, Pelizzetti E, Ajassa R, Carlotti ME, Vione D (2007b) Photochemincal processes involving nitrite in surface water samples. Aquat Sci 69:71–85

Minor E, Pothen J, Dalzell B, Abdulla H, Mopper K (2006) Effects of salinity changes on the photodegradation and ultraviolet-visible absorbance of terrestrial dissolved organic matter. Limnol Oceanogr 51:2181–2186

Mitchell B, Vernet M, Holm-Hansen O (1989) Ultraviolet light attenuation in Antarctic waters in relation to particulate absorption and photosynthesis. Antarct J US 24:179–180

Mitra S, Bianchi TS, Guo L, Santschi PH (2000) Terrestrially derived dissolved organic matter in the Chesapeake Bay and the Middle Atlantic Bight. Geochim Cosmochim Acta 64:3547–3557

Moore CA, Farmer CT, Zika RG (1993) Influence of the Orinoco River on hydrogen peroxide distribution and production in the eastern Caribbean. J Geophys Res 98:2289–2298

Mopper K, Stahovec WL (1986) Sources and sinks of low molecular weight organic carbonyl compounds in seawater. Mar Chem 19:305–321

Mopper K, Zhou X, Kieber RJ, Kieber DJ, Sikorski RJ, Jones RD (1991) Photochemical degradation of dissolved organic carbon and its impact on the oceanic carbon cycle. Nature 353:60–62

Morales SR, Vargas CV, Hernandez TC (1997) Bridge effect in charge transfer absorption bands of para-substituted beta-nitrostyrenes. Spectroscopy 13:201–206

Moran MA, Hodson RE (1994) Dissolved humic substances of vascular plant origin in a coastal marine environment. Limnol Oceanogr 39:762–771

Moran MA, Zepp RG (1997) Role of photoreactions in the formation of biologically labile compounds from dissolved organic matter. Limnol Oceanogr 42:1307–1316

Moran MA, Pomeroy LR, Sheppard ES, Atkinson LP, Hodson RE (1991) Distribution of terrestrially derived dissolved organic matter on the southeastern US continental shelf. Limnol Oceanogr 36:1134–1149

Moran MA, Sheldon WM Jr, Zepp RG (2000) Carbon loss and optical property changes during long-term photochemical and biological degradation of estuarine dissolved organic matter. Limnol Oceanogr 45:1254–1264

Morel A (1973) Measurements of spectral and total radiant flux. In: SCOR data rep discoverer expedition, Scripps Inst Oceanogr Ref 73–16(1):Fl–F341

Morel A (1980) In-water and remote measurements of ocean color. Bound-Layer Meteorol 18:177–201

Morel A (1988) Optical modeling of the upper ocean in relation to its biogenous matter content(case I waters). J Geophys Res 93:749–710

Morel A, Bélanger S (2006) Improved detection of turbid waters from ocean color sensors information. Remote Sens Environ 102:237–249

Morel A, Loisel H (1998) Apparent optical properties of oceanic water: dependence on the molecular scattering contribution. Appl Optics 37:4765–4776

Morel A, Prieur L (1977) Analysis of variations in ocean color. Limnol Oceanogr 22:709–722

Morris DP (2009) Optical properties of water. In: Gene EL (ed) Encyclopedia of Inland waters Oxford, Academic Press, Massachusetts, pp 682–689

Morris DP, Hargreaves BR (1997) The role of photochemical degradation of dissolved organic carbon in regulating the UV transparency of three lakes on the Pocono Plateau. Limnol Oceanogr 42:239–249

Morris DP, Zagarese H, Williamson CE, Balseiro EG, Hargreaves BR, Modenutti B, Moeller R, Queimalinos C (1995) The attenuation of solar UV radiation in lakes and the role of dissolved organic carbon. Limnol Oceanogr 40:1381–1391

Mostofa KMG, Sakugawa H (2009) Spatial and temporal variations and factors controlling the concentrations of hydrogen peroxide and organic peroxides in rivers. Environ Chem 6:524–534

Mostofa KMG, Yoshioka T, Konohira E, Tanoue E, Hayakawa K, Takahashi M (2005) Three-dimensional fluorescence as a tool for investigating the dynamics of dissolved organic matter in the Lake Biwa watershed. Limnology 6:101–115

Mostofa KMG, Yoshioka T, Konohira E, Tanoue E (2007a) Dynamics and characteristics of fluorescent dissolved organic matter in the groundwater, river and lake water. Water Air Soil Pollut 184:157–176

Mostofa KMG, Yoshioka T, Konohira E, Tanoue E (2007b) Photodegradation of fluorescent dissolved organic matter in river waters. Geochem J 41:323–331

Mostofa K, Wu FC, Yoshioka T, Sakugawa H, Tanoue E (2009a) Dissolved organic matter in the aquatic environments. In: Wu FC, Xing B (eds) Natural organic matter and its significance in the environment. Science Press, Beijing, pp 3–65

Mostofa KMG, Liu C, Wu FC, Fu PQ, Ying WL, Yuan J (2009) Overview of key biogeochemical functions in lake ecosystem: impacts of organic matter pollution and global warming, Keynote Speech, proceedings of the 13 th world lake conference Wuhan, China, 1–5 Nov 2009, pp 59–60

Mostofa KMG, Wu F, Liu CQ, Fang WL, Yuan J, Ying WL, Wen L, Yi M (2010) Characterization of Nanming River (southwestern China) sewerage-impacted pollution using an excitation-emission matrix and PARAFAC. Limnology 11:217–231

Mostofa KMG, Wu F, Liu CQ, Vione D, Yoshioka T, Sakugawa H, Tanoue E (2011) Photochemical, microbial and metal complexation behavior of fluorescent dissolved organic matter in the aquatic environments. Geochem J 45:235–254

Murphy KR, Stedmon CA, Waite TD, Ruiz GM (2008) Distinguishing between terrestrial and autochthonous organic matter sources in marine environments using fluorescence spectroscopy. Mar Chem 108:40–58

MV K (1972) The basic relationships between the hydra-optical parameters. In: Shifrin KS (ed) Optics of the ocean and the atmosphere, Nauka, pp 5–24 [in Russian]

Nair A, Sathyendranath S, Platt T, Morales J, Stuart V, Forget MH, Devred E, Bouman H (2008) Remote sensing of phytoplankton functional types. Remote Sens Environ 112:3366–3375
Nakane K, Kohno T, Horikoshi T, Nakatsubo T (1997) Soil carbon cycling at a black spruce (Picea mariana) forest stand in Saskatchewan, Canada. J Geophys Res-All Series- 103:28–28
Nakatani N, Ueda M, Shindo H, Takeda K, Sakugawa H (2007) Contribution of the photo-Fenton reaction to hydroxyl radical formation rates in river and rain water samples. Anal Sci 23:1137–1142
Nelson NB, Siegel DA (2002) Chromophoric DOM in the open ocean. In: Carlson CA (ed) Hansell DA. Biogeochemistry of Marine Dissolved Organic Matter Academic Press, San Diego, pp 547–578
Nelson N, Siegel D, Michaels A (1998) Seasonal dynamics of colored dissolved material in the Sargasso Sea. Deep Sea Res Part I 45:931–957
Nelson NB, Carlson CA, Steinberg DK (2004) Production of chromophoric dissolved organic matter by Sargasso Sea microbes. Mar Chem 89:273–287
Nieke B, Reuter R, Heuermann R, Wang H, Babin M, Therriault J (1997) Light absorption and fluorescence properties of chromophoric dissolved organic matter (CDOM), in the St. Lawrence Estuary (Case 2 waters). Continent Shelf Res 17:235–252
Norman L, Thomas DN, Stedmon CA, Granskog MA, Papadimitriou S, Krapp RH, Meiners KM, Lannuzel D, van der Merwe P, Dieckmann GS (2011) The characteristics of dissolved organic matter (DOM) and chromophoric dissolved organic matter (CDOM) in Antarctic sea ice. Deep Sea Res Part II 58:1075–1091
Obrador B, Pretus JL (2008) Light regime and components of turbidity in a Mediterranean coastal lagoon. Estuar Coast Shelf Sci 77:123–133
O'Donnell DM, Effler SW, Strait CM, Leshkevich GA (2010) Optical characterizations and pursuit of optical closure for the western basin of Lake Erie through in situ measurements. J Great Lakes Res 36:736–746
Odriozola AL, Varela R, Hu C, Astor Y, Lorenzoni L, Müller-Karger FE (2007) On the absorption of light in the Orinoco River plume. Continent Shelf Res 27:1447–1464
Ogawa H, Ogura N (1992) Comparison of two methods for measuring dissolved organic carbon in sea water. Nature 356:696–698
Ogawa H, Tanoue E (2003) Dissolved organic matter in oceanic waters. J Oceanogr 59:129–147
Opsahl S, Benner R (1998) Photochemical reactivity of dissolved lignin in river and ocean waters. Limnol Oceanogr 43:1297–1304
O'Reilly JE, Maritorena S, Mitchell BG, Siegel DA, Carder KL, Garver SA, Kahru M, McClain C (1998) Ocean color chlorophyll algorithms for SeaWiFS. J Geophys Res 103:24937–24953
Ortega-Retuerta E, Frazer TK, Duarte CM, Ruiz-Halpern S, Tovar-Sánchez A, Arrieta JM, Reche I (2009) Biogeneration of chromophoric dissolved organic matter by bacteria and krill in the Southern Ocean. Limnol Oceanogr 54:1941–1950
Ortega-Retuerta E, Reche I, Pulido-Villena E, Agustí S, Duarte C (2010) Distribution and photoreactivity of chromophoric dissolved organic matter in the Antarctic Peninsula (Southern Ocean). Mar Chem 118:129–139
Osburn CL, Morris DP (2003) Photochemistry of chromophoric dissolved organic matter in natural waters. In: Hebling EW, Zagarese H (eds) UV effects in aquatic organisms and ecosystems. Royal Society of Chemistry, , pp 187–217
Osburn CL, Morris DP, Thorn KA, Moeller RE (2001) Chemical and optical changes in freshwater dissolved organic matter exposed to solar radiation. Biogeochemistry 54:251–278
Osburn CL, O'Sullivan DW, Boyd TJ (2009) Increases in the longwave photobleaching of chromophoric dissolved organic matter in coastal waters. Limnol Oceanogr 54:145–159
Oyama Y, Matsushita B, Fukushima T, Matsushige K, Imai A (2009) Application of spectral decomposition algorithm for mapping water quality in a turbid lake (Lake Kasumigaura, Japan) from Landsat TM data. ISPRS J Photogram Rem Sens 64:73–85
Palenik B, Price NM, Morel FMM (1991) Potential effects of UV-B on the chemical environment of marine organisms: a review. Environ Pollut 70:117–130

Palmisano AC, SooHoo JB, Moe RL, Sullivan CW (1987) Sea ice microbial communities. VII. Changes in underice spectral irradiance during the development of Antarctic sea ice microbial communities. Mar Ecol Prog Ser 35:165–173
Parlanti E, Wörz K, Geoffroy L, Lamotte M (2000) Dissolved organic matter fluorescence spectroscopy as a tool to estimate biological activity in a coastal zone submitted to anthropogenic inputs. Org Geochem 31:1765–1781
Patsayeva SV, Fadeev VV, Filippova EM, Chubarov VV, Yuzhakov VI (1991) The effect of temperature and ultraviolet radiation influence on the luminescence spectrum characteristics of dissolved organic substance. Moscow Univ Phys Bull 46:66–69
Pegau WS, Zanevcld JRV, Barnard AH, Maske H, Borrego SÁ, Lara RL, Duarte RC (1999) Inherent optical properties in the Gulf of California. Cienc Mar 25:469–485
Peña-Méndez EM, Havel J, Patocka J (2005) Humic substances–compounds of still unknown structure: applications in agriculture, industry, environment, and biomedicine. J Appl Biomed 3:13–24
Pérez GL, Torremorell A, Bustingorry J, Escaray R, Pérez P, Diéguez M, Zagarese H (2010) Optical characteristics of shallow lakes from the Pampa and Patagonia regions of Argentina. Limnologica 40:30–39
Perovich D (1993) A theoretical model of ultraviolet light transmission through Antarctic sea ice. J Geophys Res 98:22579–22587
Perovich DK, Roesler CS, Pegau WS (1998) Variability in Arctic sea ice optical properties. J Geophys Res 103:1193–1208
Petzold TJ (1972) Volume scattering functions for selected ocean waters. Scripps Inst Oceanogr Ref 72–78:79
Phinn S, Dekker A, Brando V, Roelfsema C (2005) Mapping water quality and substrate cover in optically complex coastal and reef waters: an integrated approach. Mar Pollut Bull 51:459–469
Pienitz R, Vincent WF (2000) Effect of climate change relative to ozone depletion on UV exposure in subarctic lakes. Nature 404:484–487
Pierson DC, Kratzer S, Strömbeck N, Håkansson B (2008) Relationship between the attenuation of downwelling irradiance at 490 nm with the attenuation of PAR (400–700 nm) in the Baltic Sea. Remote Sens Environ 112:668–680
Poli G, Leonarduzzi G, Biasi F, Chiarpotto E (2004) Oxidative stress and cell signalling. Curr Med Chem 11:1163–1182
Prahl FG, Coble PG (1994) Input and behavior of dissolved organic carbon in the Columbia River estuary. Olsen & Olsen, Fredensborg, Denmark, pp 451–457
Preisendorfer RW (1961) Application of radiative transfer theory to light measurements in the sea. IUGG Monogr 10:1 1–29
Prieur L (1976) Transfer radiatif dans les eaux demer Application a la determination de parameters optiques caractkrisant leur teneur en substances dissoutcs et leur contenu en particulcs. Univ Pierre et Marie Curie, DS thesis
Pullin MJ, Bertilsson S, Goldstone JV, Voelker BM (2004) Effects of sunlight and hydroxyl radical on dissolved organic matter: bacterial growth efficiency and production of carboxylic acids and other substrates. Limnol Oceanogr 49:2011–2022
Ratte M, Bujok O, Spitzy A, Rudolph J (1998) Photochemical alkene formation in sea water from dissolved organic carbon: results from laboratory experiments. J Geophys Res 103:5707–5717
Reche I, Pace ML, Cole JJ (1999) Relationship of trophic and chemical conditions to photobleaching of dissolved organic matter in lake ecosystems. Biogeochemistry 44:259–280
Richard LE, Peake BM, Rusak SA, Cooper WJ, Burritt DJ (2007) Production and decomposition dynamics of hydrogen peroxide in freshwater. Environ Chem 4:49–54
Rochelle-Newall E (1999) Dynamics of chromophoric dissolved organic matter and dissolved organic carbon in experimental mesocosms. Int J Remote Sens 20:627–641
Rochelle-Newall E, Fisher T (2002) Chromophoric dissolved organic matter and dissolved organic carbon in Chesapeake Bay. Mar Chem 77:23–41

Roesler CS (1998) Theoretical and experimental approaches to improve the accuracy of particulate absorption coefficients derived from the quantitative filter technique. Limnol Oceanogr 43(7):1649–1660

Roesler CS, Zaneveld JRV High-resolution vertical profiles of spectral absorption, attenuation, and scattering coefficients in highly stratified waters. In: Ocean Optics XII Proc SPIE 2258, pp 309–319

Rosenstock B, Zwisler W, Simon M (2005) Bacterial consumption of humic and non-humic low and high molecular weight DOM and the effect of solar irradiation on the turnover of labile DOM in the Southern Ocean. Microb Ecol 50:90–101

Santini F, Alberotanza L, Cavalli RM, Pignatti S (2010) A two-step optimization procedure for assessing water constituent concentrations by hyperspectral remote sensing techniques: an application to the highly turbid Venice lagoon waters. Remote Sens Environ 114:887–898

Sasaki H, Miyamura T, Saitoh S, Ishizaka J (2005) Seasonal variation of absorption by particles and colored dissolved organic matter (CDOM) in Funka Bay, southwestern Hokkaido, Japan. Estuar Coast Shelf Sci 64:447–458

Sathyendranath S, Prieur L, Morel A (1989) A three-component model of ocean colour and its application to remote sensing of phytoplankton pigments in coastal waters. Int J Remote Sens 10:1373–1394

Schindler DW, Curtis PJ, Parker BR, Stainton MP (1996) Consequences of climate warming and lake acidification for UV-B penetration in North American boreal lakes. Nature 379:705–708

Schulman SG (1985) Molecular Luminescence Spectroscopy: Methods and Applications, vol 1. Wiley-Interscience, New York

Schwarzenbach RP, Gschwend PM, Imboden DM (1993) Environmental organic chemistry. Wiley, New York, pp 436–484

Scully N, Lean D (1994) The attenuation of ultraviolet radiation in temperate lakes. (With 8 figures and 5 tables). Arch Hydrobiol Beih Ergebn Limnol 43:135–144

Scully NM, Maie N, Dailey SK, Boyer JN, Jones RD, Jaffé R (2004) Early diagenesis of plant-derived dissolved organic matter along a wetland, mangrove, estuary ecotone. Limnol Oceanogr 49:1667–1678

Seitzinger S, Hartnett H, Lauck R, Mazurek M, Minegishi T, Spyres G, Styles R (2005a) Molecular-level chemical characterization and bioavailability of dissolved organic matter in stream water using electrospray-ionization mass spectrometry. Limnol Oceanogr 50:1–12

Seitzinger S, Hartnett H, Lauck R, Mazurek M, Minegishi T, Spyres G, Styles R (2005b) Molecular-level chemical characterization and bioavailability of dissolved organic matter in stream water using electrospray-ionization mass spectrometry. Limnol Oceanogr 50:1–12

Senesi N (1990a) Molecular and quantitative aspects of the chemistry of fulvic acid and its interactions with metal ions and organic chemicals: Part II. The fluorescence spectroscopy approach. Anal Chim Acta 232:77–106

Senesi N (1990b) Molecular and quantitative aspects of the chemistry of fulvic acid and its interactions with metal ions and organic chemicals : Part I. The electron spin resonance approach. Anal Chim Acta 232:51–75

Shank GC, Zepp RG, Whitehead RF, Moran MA (2005) Variations in the spectral properties of freshwater and estuarine CDOM caused by partitioning onto river and estuarine sediments. Estuar Coast Shelf Sci 65:289–301

Shank GC, Zepp RG, Vähätalo A, Lee R, Bartels E (2010) Photobleaching kinetics of chromophoric dissolved organic matter derived from mangrove leaf litter and floating Sargassum colonies. Mar Chem 119:162–171

Sholkovitz E (1976) Flocculation of dissolved organic and inorganic matter during the mixing of river water and seawater. Geochim Cosmochim Acta 40:831–845

Shooter D, Davies-Colley RJ, Kirk JTO (1998) Light absorption and scattering by ocean waters in the vicinity of the Chatham Rise, South Pacific Ocean. Marine and fresh. Water Res 49:455–461

Sieburth JMN, Jensen A (1968) Studies on algal substances in the sea. I. Gelbstoff (humic material) in terrestrial and marine waters. J Exp Mar Biol Ecol 2:174–189

Siegel D, Maritorena S, Nelson N, Hansell D, Lorenzi-Kayser M (2002) Global distribution and dynamics of colored dissolved and detrital organic materials. J Geophys Res 107:14

Siegel D, Maritorena S, Nelson N, Behrenfeld M, McClain C (2005) Colored dissolved organic matter and its influence on the satellite-based characterization of the ocean biosphere. Geophys Res Lett 32:L20605

Sikorski RJ, Zika R (1993a) Modeling mixed-layer photochemistry of H2O2: Physical and chemical modeling of distribution. J Geophys Res 98:2329–2340

Sikorski RJ, Zika R (1993b) Modeling mixed-layer photochemistry of H2O2: Optical and chemical modeling of production. J Geophys Res 98:2315–2328

Singh S, D'Sa E, Swenson E (2010a) Seasonal variability in CDOM absorption and fluorescence properties in the Barataria Basin, Louisiana, USA. J Environ Sci 22:1481–1490

Singh S, D'Sa EJ, Swenson EM (2010b) Chromophoric dissolved organic matter (CDOM) variability in Barataria Basin using excitation–emission matrix (EEM) fluorescence and parallel factor analysis (PARAFAC). Sci Total Environ 408:3211–3222

Skoog A, Wedborg M, Fogelqvist E (1996) Photobleaching of fluorescence and the organic carbon concentration in a coastal environment. Mar Chem 55:333–345

Slawyk G, Raimbault P, Garcia N (1998) Measuring gross uptake of 15 N-labeled nitrogen by marine phytoplankton without particulate matter collection: evidence of low 15 N losses to the dissolved organic nitrogen pool. Limnol Oceanogr 43:1734–1739

Slawyk G, Raimbault P, Garcia N (2000) Use of 15 N to measure dissolved Organic Nitrogen release by Marine Phytoplankton (reply to comment by Bronk and Ward). Limnol Oceanogr 45:1884–1886

Smith RC, Baker KS (1979) Penetration of UV-B and biologically effective dose-rates in natural waters. Photochem Photobiol 29:311–323

Smith RC, Baker KS (1981) Optical properties of the clearest natural waters (200–800 nm). Appl Optics 20:177–184

Smith R, Furgal J, Charlton M, Greenberg B, Hiriart V, Marwood C (1999) Attenuation of ultraviolet radiation in a large lake with low dissolved organic matter concentrations. Can J Fish Aquat Sci 56:1351–1361

Smith REH, Allen CD, Charlton MN (2004) Dissolved organic matter and ultraviolet radiation penetration in the Laurentian Great Lakes and tributary waters. J Great Lakes Res 30:367–380

Sommaruga R, Psenner R (1997) Ultraviolet radiation in a high mountain lake of the Austrian Alps: air and underwater measurements. Photochem Photobiol 65:957–963

Son SH, Wang M, Shon JK (2011) Satellite observations of optical and biological properties in the Korean dump site of the Yellow Sea. Remote Sens Environ 115:562–572

Song R, Westerhoff P, Minear RA, Amy GL (1996) Interactions between bromine and natural organic matter. In: Minear RA, Amy GL (eds) Water disinfection and natural organic matter. Americal Chemical Society, Washington, DC, pp 298–321

Stabenau ER, Zika RG (2004) Correlation of the absorption coefficient with a reduction in mean mass for dissolved organic matter in southwest Florida river plumes. Mar Chem 89:55–67

Stabenau ER, Zepp RG, Bartels E, Zika RG (2004) Role of the seagrass Thalassia testudinum as a source of chromophoric dissolvedorganic matter in coastal south Florida. Mar Ecol Prog Ser 282:59–72

Stambler N (2005) Bio-optical properties of the northern Red Sea and the Gulf of Eilat (Aqaba) during winter 1999. J Sea Res 54:186–203

Stedmon CA, Markager S (2005) Resolving the variability in dissolved organic matter fluorescence in a temperate estuary and its catchment using PARAFAC analysis. Limnol Oceanogr 50:686–697

Stedmon CA, Markager S, Tranvik L, Kronberg L, Slätis T, Martinsen W (2007a) Photochemical production of ammonium and transformation of dissolved organic matter in the Baltic Sea. Mar Chem 104:227–240
Stedmon CA, Thomas DN, Granskog M, Kaartokallio H, Papadimitriou S, Kuosa H (2007b) Characteristics of dissolved organic matter in Baltic coastal sea ice: Allochthonous or autochthonous origins? Environ Sci Technol 41:7273–7279
Steelink C (2002) Peer reviewed: investigating humic acids in soils. Anal Chem 74:326–333
Stramski D, Bricaud A, Morel A (2001) Modeling the inherent optical properties of the ocean based on the detailed composition of the planktonic community. Appl Optics 40:2929–2945
Strome DJ, Miller MC (1978) Photolytic changes in dissolved humic substances. Int Ver Theor Angew Limnol 20:1248–1254
Sullivan SA (1963) Experimental study of the absorption in distilled water, artificial sea water, and heavy water in the visible region of the spectrum. J Opt Sot Am 53:962–967
Sundh I (1992) Biochemical composition of dissolved organic carbon released from natural communities of lake phytoplankton. Arch Hydrobiol 125:347–369
Takahashi M, Hama T, Matsunaga K, Handa N (1995) Photosynthetic organic carbon production and respiratory organic carbon consumption in the trophogenic layer of Lake Biwa. J Plankton Res 17:1017–1025
Takeda K, Takedoi H, Yamaji S, Ohta K, Sakugawa H (2004) Determination of hydroxyl radical photoproduction rates in natural waters. Anal Sci 20:153–158
Tanoue E (2000) Proteins in the sea—synthesis. In: Handa N, Tanoue E, Hama T (eds) Dynamics and characterization of marine organic matter. TERRAPUB/Kluwer, Tokyo, pp 383–463
Teai T, Drollet JH, Bianchini JP, Cambon A, Martin PMV (1998) Occurrence of ultraviolet radiation-absorbing mycosporine-like amino acids in coral mucus and whole corals of French Polynesia. Mar Freshwater Res 49:127–132
Thomas-Smith TE, Blough NV (2001) Photoproduction of hydrated electron from constituents of natural waters. Environ Sci Technol 35:2721–2726
Thurman E (1985) Humic substances in groundwater. In: Aiken GR, McKnight DM, Wershaw RL MacCarthy P (eds) Humic substances in soil, sediment, and water: geochemistry, isolation, and characterization, Wiley, Hoboken, pp 87–103
Tomlinson M, Wynne T, Stumpf R (2009) An evaluation of remote sensing techniques for enhanced detection of the toxic dinoflagellate, Karenia brevis. Remote Sens Environ 113:598–609
Tranvik LJ (1992) Allochthonous dissolved organic matter as an energy source for pelagic bacteria and the concept of the microbial loop. Hydrobiologia 229:107–114
Tranvik L, Kokalj S (1998) Decreased biodegradability of algal DOC due to interactive effects of UV radiation and humic matter. Aquat Microb Ecol 14:301–307
Trodahl H, Buckley R (1990) Enhanced ultraviolet transmission of Antarctic sea ice during the austral spring. Geophys Res Lett 17:2177–2179
Twardowski MS, Donaghay PL (2001) Separating in situ and terrigenous sources of absorption by dissolved materials in coastal waters. J Geophys Res 106:2545–2560
Twardowski MS, Donaghay PL (2002) Photobleaching of aquatic dissolved materials: absorption removal, spectral alteration, and their interrelationship. J Geophys Res 107:3091
Twardowski MS, Boss E, Sullivan JM, Donaghay PL (2004) Modeling the spectral shape of absorption by chromophoric dissolved organic matter. Mar Chem 89:69–88
Tyler JE, Smith RC (1970) Measurements of spectral irradiance underwater, vol 1. Gordon & Breach Publishing Group, Newark
Tzortziou M, Herman JR, Gallegos CL, Neale PJ, Subramaniam A, Harding LW Jr, Ahmad Z (2006) Bio-optics of the Chesapeake Bay from measurements and radiative transfer closure. Estuar Coast Shelf Sci 68:348–362
Tzortziou M, Osburn CL, Neale PJ (2007a) Photobleaching of dissolved organic material from a Tidal Marsh-Estuarine system of the Chesapeake Bay. Photochem Photobiol 83:782–792

Tzortziou M, Subramaniam A, Herman JR, Gallegos CL, Neale PJ, Harding LW (2007b) Remote sensing reflectance and inherent optical properties in the mid Chesapeake Bay. Estuar Coast Shelf Sci 72:16–32
Uchida M, Nakatsubo T, Kasai Y, Nakane K, Horikoshi T (2000) Altitude differences in organic matter mass loss and fungal biomass in a subalpine coniferous forest, Mr Fuji, Japan. Arctic Antarctic Alp Res 32:262–269
Uher G, Hughes C, Henry G, Upstill-Goddard RC (2001) Non-conservative mixing behavior of colored dissolved organic matter in a humic-rich, turbid estuary. Geophys Res Lett 28:3309–3312
Uusikivi J, Vähätalo AV, Granskog MA, Sommaruga R (2010) Contribution of mycosporine-like amino acids and colored dissolved and particulate matter to sea ice optical properties and ultraviolet attenuation. Limnol Oceanogr 55:703
Vähätalo AV, Järvinen M (2007) Photochemically produced bioavailable nitrogen from biologically recalcitrant dissolved organic matter stimulates production of a nitrogen-limited microbial food web in the Baltic Sea. Limnol Oceanogr 52:132–143
Vähätalo AV, Wetzel RG (2004) Photochemical and microbial decomposition of chromophoric dissolved organic matter during long (months–years) exposures. Mar Chem 89:313–326
Vähätalo AV, Salkinoja-Salonen M, Taalas P, Salonen K (2000) Spectrum of the quantum yield for photochemical mineralization of dissolved organic carbon in a humic lake. Limnol Oceanogr 45:664–676
Vahtmäe E, Kutser T, Martin G, Kotta J (2006) Feasibility of hyperspectral remote sensing for mapping benthic macroalgal cover in turbid coastal waters—a Baltic Sea case study. Remote Sens Environ 101:342–351
Valko M, Rhodes C, Moncol J, Izakovic M, Mazur M (2006) Free radicals, metals and antioxidants in oxidative stress-induced cancer. Chem Biol Interact 160:1–40
Van Der Woerd HJ, Pasterkamp R (2008) HYDROPT: a fast and flexible method to retrieve chlorophyll-a from multispectral satellite observations of optically complex coastal waters. Remote Sens Environ 112:1795–1807
van der Woerd HJ, Blauw A, Peperzak L, Pasterkamp R, Peters S (2011) Analysis of the spatial evolution of the 2003 algal bloom in the Voordelta (North Sea). J Sea Res 65:195–204
Vantrepotte V, Brunet C, Mériaux X, Lécuyer E, Vellucci V, Santer R (2007) Bio-optical properties of coastal waters in the Eastern English Channel. Estuar Coast Shelf Sci 72:201–212
Vaughan PP, Blough NV (1998) Photochemical formation of hydroxyl radical by constituents of natural waters. Environ Sci Technol 32:2947–2953
Vernet M, Whitehead K (1996) Release of ultraviolet-absorbing compounds by the red-tide dinoflagellate *Lingulodinium polyedra*. Mar Biol 127:35–44
Vernet M, Neori A, Haxo F (1989) Spectral properties and photosynthetic action in red-tide populations of Prorocentrum micans and Gonyaulax polyedra. Mar Biol 103:365–371
Vincent WF, Roy S (1993) Solar ultraviolet-B radiation and aquatic primary production: damage, protection, and recovery. Environ Rev 1:1–12
Vincent WF, Rae R, Laurion I, Howard-Williams C, Priscu JC (1998) Transparency of Antarctic ice-covered lakes to solar UV radiation. Limnol Oceanogr 43:618–624
Vione D, Falletti G, Maurino V, Minero C, Pelizzetti E, Malandrino M, Ajassa R, Olariu RI, Arsene C (2006) Sources and sinks of hydroxyl radicals upon irradiation of natural water samples. Environ Sci Technol 40:3775–3781
Vione D, Lauri V, Minero C, Maurino V, Malandrino M, Carlotti ME, Olariu RI, Arsene C (2009) Photostability and photolability of dissolved organic matter upon irradiation of natural water samples under simulated sunlight. Aquat Sci 71:34–45
Vione D, Das R, Rubertelli F, Maurino V, Minero C, Barbati S, Chiron S (2010) Modelling the occurrence and reactivity of hydroxyl radicals in surface waters: implications for the fate of selected pesticides. Int J Environ Anal Chem 90:260–275
Vodacek A, Hoge FE, Swift RN, Yungel JK, Peltzer ET, Blough NV (1995) The use of in situ and airborne fluorescence measurements to determine UV absorption coefficients and DOC concentrations in surface waters. Limnol Oceanogr 40:411–415

Vodacek A, Blough NV, DeGrandpre MD, Peltzer ET, Nelson RK (1997) Seasonal variation of CDOM and DOC in the Middle Atlantic Bight: terrestrial inputs and photooxidation. Limnol Oceanogr 42:674–686

Voelker BM, Sulzberger B (1996) Effects of fulvic acid on Fe(II) oxidation by hydrogen peroxide. Environ Sci Technol 30:1106–1114

Volpe V, Silvestri S, Marani M (2011) Remote sensing retrieval of suspended sediment concentration in shallow waters. Remote Sens Environ 115:44–54

Von Gunten U, Oliveras Y (1997) Kinetics of the reaction between hydrogen peroxide and hypobromous acid: implication on water treatment and natural systems. Water Res 31:900–906

Von Sonntag C, Mark G, Mertens R, Schuchmann M, Schuchmann H (1993) UV radiation and/or oxidants in water pollution control. J Water Supply Res Technol—Aqua 42:201–211

Wada S, Aoki MN, Tsuchiya Y, Sato T, Shinagawa H, Hama T (2007) Quantitative and qualitative analyses of dissolved organic matter released from <i> Ecklonia cava </i> Kjellman, in Oura Bay, Shimoda, Izu Peninsula, Japan. Journal of experimental Mar Biol Ecol 349:344–358

Waiser MJ, Robarts RD (2004) Photodegradation of DOC in a shallow prairie wetland: evidence from seasonal changes in DOC optical properties and chemical characteristics. Biogeochemistry 69:263–284

Wakeham SG, Lee C, Hedges JI, Hernes PJ, Peterson MJ (1997) Molecular indicators of diagenetic status in marine organic matter. Geochim Cosmochim Acta 61:5363–5369

Warnock RE, Gieskes WWC, van Laar S (1999) Regional and seasonal differences in light absorption by yellow substance in the Southern Bight of the North Sea. J Sea Res 42:169–178

Warren SG, Roesler CS, Morgan VI, Brandt RE, Goodwin ID, Allison I (1993) Green icebergs formed by freezing of organic-rich seawater to the base of Antarctic ice shelves. J Geophys Res 98:6921–6928

Warren SG, Brandt RE, Grenfell TC (2006) Visible and near-ultraviolet absorption spectrum of ice from transmission of solar radiation into snow. Appl Optics 45:5320–5334

Weidemann A, Bannister T (1986) Absorption and scattering coefficients in Irondequoit Bay. Limnol Oceanogr 31:567–583

Weiss PS, Andrews SS, Johnson JE, Zafiriou OC (1995) Photoproduction of carbonyl sulfide in South Pacific Ocean waters as a function of irradiation wavelength. Geophys Res Lett 22:215–218

Werdell PJ, Bailey SW, Franz BA, Harding LW Jr, Feldman GC, McClain CR (2009) Regional and seasonal variability of chlorophyll-a in Chesapeake Bay as observed by SeaWiFS and MODIS-Aqua. Remote Sens Environ 113:1319–1330

Wetzel RG (1992) Gradient-dominated ecosystems: sources and regulatory functions of dissolved organic matter in freshwater ecosystems. Hydrobiologia 229:181–198

Wetzel RG, Hatcher PG, Bianchi TS (1995) Natural photolysis by ultraviolet irradiance of recalcitrant dissolved organic matter to simple substrates for rapid bacterial metabolism. Limnol Oceanogr 40:1369–1380

Whitehead K, Vernet M (2000) Influence of mycosporine-like amino acids (MAAs) on UV absorption by particulate and dissolved organic matter in La Jolla Bay. Limnol Oceanogr 45:1788–1796

Whitehead RF, de Mora S, Demers S, Gosselin M, Monfort P, Mostajir B (2000) Interactions of ultraviolet-B radiation, mixing, and biological activity on photobleaching of natural chromophoric dissolved organic matter: a mesocosm study. Limnol Oceanogr 45:278–291

Willey JD, Atkinson LP (1982) The effects of seawater magnesium on natural fluorescence during estuarine mixing, and implications for tracer applications. Estuar Coast Shelf Sci 14:49–59

Williamson CE, Stemberger RS, Morris DP, Frost TM, Paulsen SG (1996) Ultraviolet radiation in North American lakes: attenuation estimates from DOC measurements and implications for plankton communities. Limnol Oceanogr 41:1024–1034

Winch S, Ridal J, Lean D (2002) Increased metal bioavailability following alteration of freshwater dissolved organic carbon by ultraviolet B radiation exposure. Environ Toxicol 17:267–274

Winter AR, Fish TAE, Playle RC, Smith DS, Curtis PJ (2007) Photodegradation of natural organic matter from diverse freshwater sources. Aquat Toxicol 84:215–222
Woodruff DL, Stumpf RP, Paerl HW (1999) Remote estimation of water clarity in optically complex estuarine waters. Remote Sens Environ 68:41–52
Wright VM (2005) The seasonal dynamics of colored dissolved organic matter in the mississippi river plume and Northern Gulf of Mexico. University of Southern Mississippi, MS Thesis, p 68
Wu FC, Mills RB, Cai YR, Evans RD, Dillon PJ (2005) Photodegradation-induced changes in dissolved organic matter in acidic waters. Can J Fish Aquat Sci 62:1019–1027
Xenopoulos MA, Bird DF (1997) Effect of acute exposure to hydrogen peroxide on the production of phytoplankton and bacterioplankton in a mesohumic lake. Photochem Photobiol 66:471–478
Xie H, Zafiriou OC, Cai WJ, Zepp RG, Wang Y (2004) Photooxidation and its effects on the carboxyl content of dissolved organic matter in two coastal rivers in the southeastern United States. Environ Sci Technol 38:4113–4119
Xiong F, Komenda J, Kopecký J, Nedbal L (1997) Strategies of ultraviolet-B protection in microscopic algae. Physiol Plant 100:378–388
Yentsch CS (1960) The influence of phytoplankton pigments on the colour of sea water. Deep Sea Res 7:1−9
Latimer P (1963) Is selective scattering a universal phenomenon. In: Studies on microalgae and photosynthetic bacteria. Jap Sot Plant Physiol, Univ Tokyo, pp 213−225
Kozlyaninov M, Pelevin V (1966) On the application of a one-dimensional approximation in the investigation of the propagation of optical radiation in the sea. [in Russian] Tr Inst Okeanol Akad Nauk SSSR 77, pp 73−79 Also 1966 US Dep Comm, Jt Publ Res Ser Rep 36:54−63
Yamashita Y, Tanoue E (2003) Chemical characterization of protein-like fluorophores in DOM in relation to aromatic amino acids. Mar Chem 82:255–271
Yamashita Y, Tanoue E (2004) In situ production of chromophoric dissolved organic matter in coastal environments. Geophys Res Lett 31:L14302
Yamashita Y, Tanoue E (2008) Production of bio-refractory fluorescent dissolved organic matter in the ocean interior. Nature Geosci 1:579–582
Yan ND, Keller W, Scully NM, Lean DRS, Dillon PJ (1996) Increased UV-B penetration in a lake owing to drought-induced acidification. Nature 381:141–143
Yoshioka T, Ueda S, Khodzher T, Bashenkhaeva N, Korovyakova I, Sorokovikova L, Gorbunova L (2002) Distribution of dissolved organic carbon in Lake Baikal and its watershed. Limnology 3:159–168
Yoshioka T, Mostofa KMG, Konohira E, Tanoue E, Hayakawa K, Takahashi M, Ueda S, Katsuyama M, Khodzher T, Bashenkhaeva N (2007) Distribution and characteristics of molecular size fractions of freshwater-dissolved organic matter in watershed environments: its implication to degradation. Limnology 8:29–44
Zafiriou OC, True MB, Hayon E (1987) Consequences of OH radical reaction in sea water: Formation and decay of Br2-ion radical. ACS Publications
Zafiriou OC, Voelker BM, Sedlak DL (1998) Chemistry of the superoxide radical (O2-) in seawater: Reactions with inorganic copper complexes. The J Phys Chem A 102:5693–5700
Zagarese H, Diaz M, Pedrozo F, Ferraro M, Cravero W, Tartarotti B (2001) Photodegradation of natural organic matter exposed to fluctuating levels of solar radiation. J Photochem Photobiol B: Biol 61:35–45
Zanardi-Lamardo E, Moore CA, Zika RG (2004) Seasonal variation in molecular mass and optical properties of chromophoric dissolved organic material in coastal waters of southwest Florida. Mar Chem 89:37–54
Zawada DG, Hu C, Clayton T, Chen Z, Brock JC, Muller-Karger FE (2007) Remote sensing of particle backscattering in Chesapeake Bay: a 6-year SeaWiFS retrospective view. Estuar Coast Shelf Sci 73:792–806
Zepp RG, Schlotzhauer PF (1981) Comparison of photochemical behavior of various humic substances in water: III. Spectroscopic properties of humic substances. Chemosphere 10:479–486

Zepp RG, Faust BC, Hoigne J (1992) Hydroxyl radical formation in aqueous reactions (pH 3–8) of iron (II) with hydrogen peroxide: the photo-Fenton reaction. Environ Sci Technol 26:313–319

Zepp RG, Callaghan TV, Erickson DJ (1995) Effects of increased solar ultraviolet radiation on biogeochemical cycles. Ambio 24:181–187

Zepp RG, Callaghan T, Erickson D (1998) Effects of enhanced solar ultraviolet radiation on biogeochemical cycles. J Photochem Photobiol B: Biol 46:69–82

Zhang YL, Qin BQ (2007) Variations in spectral slope in Lake Taihu, a large subtropical shallow lake in China. J Great Lakes Res 33:483–496

Zhang Y, Zhu L, Zeng X, Lin Y (2004) The biogeochemical cycling of phosphorus in the upper ocean of the East China Sea. Estuar Coast Shelf Sci 60:369–379

Zhang X, Minear RA, Barrett SE (2005) Characterization of high molecular weight disinfection byproducts from chlorination of humic substances with/without coagulation pretreatment using UF-SEC-ESI-MS/MS. Environ Sci Technol 39:963–972

Zhang Y, Zhang B, Wang X, Li J, Feng S, Zhao Q, Liu M, Qin B (2007) A study of absorption characteristics of chromophoric dissolved organic matter and particles in Lake Taihu, China. Hydrobiologia 592:105–120

Zhang Y, Van Dijk MA, Liu M, Zhu G, Qin B (2009a) The contribution of phytoplankton degradation to chromophoric dissolved organic matter (CDOM) in eutrophic shallow lakes: Field and experimental evidence. Water Res 43:4685–4697

Zhang Y, Liu M, Qin B, Feng S (2009b) Photochemical degradation of chromophoric-dissolved organic matter exposed to simulated UV-B and natural solar radiation. Hydrobiologia 627:159–168

Zhao J, Cao W, Wang G, Yang D, Yang Y, Sun Z, Zhou W, Liang S (2009) The variations in optical properties of CDOM throughout an algal bloom event. Estuar Coast Shelf Sci 82:225–232

Zibordi G, Berthon JF, Mélin F, D'Alimonte D, Kaitala S (2009) Validation of satellite ocean color primary products at optically complex coastal sites: Northern Adriatic Sea, Northern Baltic Proper and Gulf of Finland. Remote Sens Environ 113:2574–2591

Zimba PV, Gitelson A (2006) Remote estimation of chlorophyll concentration in hyper-eutrophic aquatic systems: model tuning and accuracy optimization. Aquaculture 256:272–286

Fluorescent Dissolved Organic Matter in Natural Waters

Khan M. G. Mostofa, Cong-qiang Liu, Takahito Yoshioka, Davide Vione, Yunlin Zhang and Hiroshi Sakugawa

1 Introduction

Dissolved organic matter (DOM), of allochthonous and autochthonous origin, is a heterogeneous mixture of organic compounds, with molecular weights ranging from less than 100 to over 300,000 Da in natural waters (Hayase and Tsubota 1985; Thurman 1985; Ma and Ali 2009; Mostofa et al. 2009a). The DOM components are involved into key biogeochemical processes such as global carbon cycle, nutrient dynamics, photosynthesis, biological activity and finally as energy sources in the aquatic environments (Mostofa et al. 2009a, b; Hedges 1992). Among the

K. M. G. Mostofa (✉) · C. Q. Liu
State Key Laboratory of Environmental Geochemistry, Institute of Geochemistry, Chinese Academy of Sciences, Guiyang 550002, People's Republic of China
e-mail: mostofa@vip.gyig.ac.cn

T. Yoshioka
Field Science Education and Research Center, Kyoto University, Kitashirakawa Oiwake-cho, Sakyo-ku, Kyoto 606-8502, Japan

D. Vione
Dipartimento Chimica Analitica, University of Torino, I-10125 Turin, Italy

Centro Interdipartimentale NatRisk, I-10095 Grugliasco, (TO), Italy

Y. Zhang
Taihu Lake Laboratory Ecosystem Research Station, State Key Laboratory of Lake Science and Environment, Nanjing Institute of Geography and Limnology, Chinese Academy of Sciences, 73 East Beijing Road, Nanjing 210008, People's Republic of China

H. Sakugawa
Department of Environmental Dynamics and Management, Graduate School of Biosphere Science, Hiroshima University, 1-7-1, Kagamiyama, Higashi-Hiroshima 739-8521, Japan

K. M. G. Mostofa et al. (eds.), *Photobiogeochemistry of Organic Matter*, Environmental Science and Engineering, DOI: 10.1007/978-3-642-32223-5_6,

DOM components, only a limited fraction of organic compounds show fluorescence properties. These compounds are termed the fluorescent DOM (FDOM). In the pioneering works conducted by Kalle (1949, 1963) and Duursma (1974), the fluorescence of terrestrial humic substances served as a tracer of soil organic matter in freshwater and seawater environments. The fluorescence of humic substances has then been used to distinguish the mixing of river water with seawater as well as their sources (Otto 1967; Zimmerman and Rommets 1974; Dorsch and Bidleman 1982; Willey 1984; Hayase et al. 1987).

A number of excitation–emission (Ex/Em) maxima are detected for humic substances depending on their sources and nature (either fulvic acid or humic acid) in soil and natural waters (Christman and Ghassemi 1966; Ghassemi and Christman 1968; Levesque 1972; Almgren et al. 1975; Brun and Milburn 1977; Gosh and Schnitzer 1980; Momzikoff et al. 1992). The effects of the molecular weight of fulvic and humic acids (<10,000 to >300,000 Da) on fluorescence properties have been examined in earlier studies, particularly in lignin sulfonates (Christman and Minear 1967), soil fulvic and humic acids (Levesque 1972; McCreary and Snoeyink 1980), natural waters (Hall and Lee 1974; Stewart and Wetzel 1980; Visser 1984) and sediment pore waters (Hayase and Tsubota 1985, 1983). The results suggest that the Ex/Em wavelength maxima for humic acid are often present at longer wavelengths than those of fulvic acid. The position of these maxima is independent of the molecular weight, whilst smaller molecules in terrestrial fulvic and humic acids typically exhibit higher fluorescence intensity than the larger ones. It is also suggested that the fluorophores in humic acid are aromatic compounds with higher molecular weight compared to those in fulvic acid (Hayase and Tsubota 1985). The fluorescence quantum yields of commercial, soil and aquatic humic substances excited at 350 nm have been determined by Zepp and Scholtzhauer (1981). Linear correlations are observed between pH and the fluorescence intensity at the Ex/Em peaks of fulvic and humic acids in natural waters and in microbial cultures (Visser 1984). The fluorescence characteristics are different between coastal marine sedimentary humic and fulvic acids (Hayase and Tsubota 1985). The vertical distribution of humic-like fluorescent substances has been examined in marine waters (Hayase et al. 1987, 1988; Chen and Bada 1989, 1990, 1992; Hayase and Shinozuka 1995). The fluorescence intensity of humic-like substances is correlated with phosphate and nitrate in the deeper marine waters, suggesting that the production of humic-like substances and nutrients (phosphate and nitrate) results from the decomposition of settling particles in the water column (Hayase and Shinozuka 1995). Senesi (1990a) summarized the fluorescence properties of fulvic acid-like components in freshwater and seawater. All of these studies are two-dimensional and do not distinguish well the fluorescence excitation–emission (Ex/Em) peak positions that can be used for the characterization of the fluorophores of humic substances.

Coble et al. (1990) have firstly applied the three-dimensional fluorescence (excitation–emission matrix) spectroscopy (EEMS) to marine FDOM to distinguish between the humic-like and protein-like fluorescence peaks in seawater. Coble (1996) summarized the various fluorescence peaks that can be distinguished in river, lake and marine FDOM. They identified the Ex/Em wavelengths for humic-like peaks at Ex/Em = 260/380–460 nm (Peak A) and 350/420–480 nm (Peak C), marine humic-like (recently called photobleached fulvic-like) at 312/380–420 nm (Peak M) and two protein-like peaks, i.e., tryptophan-like (275/340 nm; Peak T) and tyrosine-like (275/310 nm; Peak B) peaks. Since then a number of studies have identified humic-like and protein-like substances in natural waters (Mopper and Schultz 1993; de Souza-Sierra et al. 1994; Determann et al. 1994; 1996; Mayer et al. 1999; Parlanti et al. 2000; Yamashita and Tanoue 2003a). Mostofa et al. (2005a) have characterized the fluorescent whitening agents (FWAs) or components of household detergents (DAS1 and DSBP) in terms of their fluorescence characteristics at Ex/Em = 330–350/430–449 nm (Peak W) in sewerage-impacted rivers.

To find out more useful information in EEM spectra, Principal Component Analysis (PCA), a multivariate data analysis method, has been applied to the study of EEMs in marine science. PCA is a more comprehensive data analysis method than the traditional 'peak picking' techniques (Persson and Wedborg 2001). However, the two-way PCA models are insufficient for the modeling of the essentially three-way character of EEMs (Bro 1997). Recently, parallel factor analysis (PARAFAC), a statistical modeling approach, has been successfully applied to decompose EEMs of complex mixtures in aqueous solution into their individual fluorescent components (Bro 1997, 1998, 1999; Ross et al. 1991; Jiji et al. 1999; Baunsgaard et al. 2000; 2001; da Silva et al. 2002; Stedmon et al. 2003). The combination of EEM and PARAFAC is widely applied to isolate and distinguish the fluorescent components in terrestrial soil, pore waters and natural waters (Fulton et al. 2004; Cory and McKnight 2005; Hall et al. 2005; Stedmon and Markager 2005a, 2005b; Ohno and Bro 2006; Muller-Karger et al. 2005; Stedmon et al. 2007a, 2007b; Mostofa et al. 2010).

FDOM components can undergo photoinduced decomposition by natural sunlight in surface waters or in laboratory conditions. Photoinduced decomposition has been observed for FDOM in rivers (Mostofa et al. 2005a, 2007a, 2010; Gao and Zepp 1998; White et al. 2003; Patel-Sorrentino et al. 2004; Brooks et al. 2007), lakes (Ma and Green 2004; Garcia et al. 2005; Winter et al. 2007; Mostofa KMG et al., unpublished data), estuaries (Skoog et al. 1996; Moran et al. 2000; Osburn et al. 2009), wetlands (Brooks et al. 2007; Waiser and Robarts 2004), marine waters (Stedmon et al. 2007a, 2007b; Skoog et al. 1996; Ferrari et al. 1996a; Kieber et al. 1997; Miller et al. 2002; Lepane et al. 2003; Bertilsson et al. 2004; Boehme et al. 2004; del Vecchio and Blough 2004; Zanardi-Lamardo et al. 2004; Abboudi et al. 2008), extracted or standard fulvic acid and humic acid (Mostofa et al. 2005a; Winter et al. 2007; Lepane et al. 2003; Fukushima et al. 2001; del Vecchio and Blough 2002; Uyguner and Bekbolet 2005; Mostofa and Sakugawa 2009), and fluorescent whitening agents, standard or dissolved in natural waters (Mostofa et al. 2005a, 2010; Poiger et al. 1999). FDOM components are also decomposed microbiologically, in deep natural waters or upon dark incubation under laboratory conditions. A similar behavior has been observed for extracted or standard fulvic and humic acids (Mostofa et al. 2010, 2007a; Garcia et al. 2005; Moran et al. 2000; Lepane et al. 2003; Abboudi et al. 2008). Photochemistry is highly susceptible to degrade flurophores at both peak C- and A-regions, whilst microbial degradation is more susceptible to decompose flurophores at peak A- and T_{UV}-regions. The latter fluorescent compounds are occasionally unable to undergo photoinduced decomposition, particularly the fluorophores at the T_{UV}-region (Mostofa et al. 2010).

A review by Leenheer and Croué (2003) includes the comparison between the fluorescence properties of various fluorophores within natural organic matter (NOM). Another review by Hudson et al. (2007) focuses on the effect of metal ions on DOM fluorescence and photodegradation, and on the application of DOM fluorescence in natural waters. A recent review by Coble (2007) covers the topic of marine optical biogeochemistry and discusses the chemical properties and the sinks of chromophoric or colored dissolved organic matter (CDOM), as well as its fluorescence characteristics. Another review covers the application of fluorescence to the identification and monitoring of sewerage-derived DOM and to the impact of treatment processes on fluorescence. It provides useful information on how fluorescence could be a potential tool for recycled water systems (Henderson et al. 2009). Finally, another review summarizes the fluorescence properties of various organic components. From field observations in natural waters and the use of standards, it can be derived that all the fluorescent components in DOM can be grouped in four regions: peak C-region (280–400/380–550 nm), peak A-region (220–280/380–550 nm), peak T-region (260–285/290–380 nm), and peak T_{UV}-region (215–260/280–380 nm) (Mostofa et al. 2009a).

This chapter will provide a general overview on the fluorophores, on the fluorescence properties of key organic substances in combination with their molecular characteristics, and on PARAFAC modeling to identify the fluorescent components. This paper will deal with the identification of autochthonous DOM, and in particular of autochthonous fulvic acids of algal or phytoplankton origin. It will be discussed how these autochthonous fluorescent components differ from allochthonous fulvic and humic acids. This review will extensively discuss the key factors that significantly affect the fluorescence properties of FDOM. It will also address the photoinduced and microbial FDOM degradation as well as the mechanisms, the controlling factors and their significance to understand the biogeochemical FDOM activity in freshwater and marine environments. Finally, a comparison will be provided of the relative importance of studying FDOM versus CDOM absorbance, as well as how fluorophores in FDOM differ from chromophores in CDOM.

2 Principle of Fluorescence (Excitation–Emission Matrix) Spectroscopy

Fluorescence (excitation–emission matrix, EEM) spectroscopy (EEMS) gives a three-dimensional image of an aqueous solution that is measured for the fluorescence intensity of the fluorophores as a function of the excitation and emission wavelengths. EEM spectra are a combination of multiple emission spectra at a range of excitations. EEMS finds wide applications due to its precise, quick and relatively simple characterization of DOM fractions in natural waters. The principles of fluorescence spectroscopy have been summarized in earlier studies (Senesi 1990a; Hudson et al. 2007; Guilbault 1990; Grabowski et al. 2003; Oheim et al. 2006). An organic molecule has a series of closely spaced energy levels, and one of its electrons can be excited from a lower to a higher level upon absorption of a discrete quantum of light that is equal in energy to the difference between the two energy states (Fig. 1) (Senesi 1990a). Fluorescence can be simply defined as the emission of a photon at a longer wavelength (lower energy, $h\upsilon_F$) that occurs when the electron returns to the ground state. Radiation absorption occurs at a timescale of approximately 10^{-15} s, emission of fluorescence photons on a timescale of about 10^{-8} s, and internal conversion typically on a time scale of about 10^{-12} s or less (Fig. 1). Fluorescence is basically the reverse of absorption (Senesi 1990a).

When a fluorophore or fluorescent molecule absorbs a photon with a frequency υ, which corresponds to a photon energy $h\upsilon_{ex}$ (h = Planck's constant), its fluorescence emission can simply be depicted by the wave function ψ as below (Eqs. 2.1, 2.2):

$$\text{Excitation (absorption)}: \psi_0 + h\upsilon_{ex} \rightarrow \psi_e \tag{2.1}$$

$$\text{Fluorescence (emission)}: \psi_e \rightarrow \psi_0 + h\upsilon_F + \text{heat} \tag{2.2}$$

where ψ_0 is termed the ground state of the fluorophore and ψ_e is its first electronically excited state. The fluorescence emission energy, $h\upsilon_F$, varies depending on the return of the photon to the ground state level (ψ_0). A fluorophore in its excited state, ψ_1, can lose its energy by internal conversion such as 'non-radiative relaxation', where the excitation energy is dissipated as heat (vibrational relaxation) to the solvent.

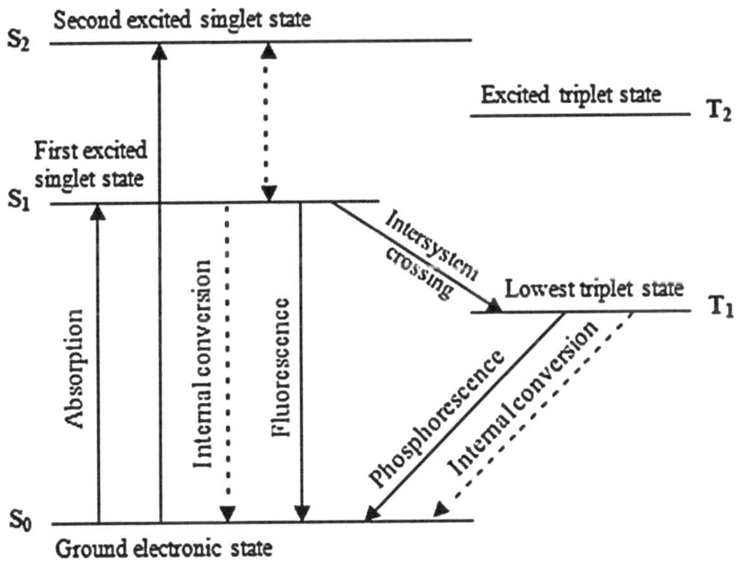

Fig. 1 Schematic energy level diagram for a diatomic molecule illustrating principal excited-state processes. *Data source* Senesi (1990a)

The key processes that compete with fluorescence emission from the lowest excited singlet to the ground state are internal conversion, intersystem crossing and photodegradation (Fig. 1) (Senesi 1990a). The internal conversion depends on several factors such as increasing solvation, increasing temperature and molecular flexibility. Such factors increase the interaction of a molecule with its medium and accelerate the rate of internal conversion by collisional deactivation. Intersystem crossing involves the transition from the lowest vibrational level of an excited singlet state to an upper vibrational level of a triplet state, or vice versa. A change of spin occurs in intersystem crossing, but its rate is usually slower compared to that of internal conversion. Photodegradation causes a decrease in fluorescence intensity, and its effect might be higher for light-sensitive organic substances. The probability of a photodegradation process depends primarily on the energy difference between the ground state and the first excited singlet, i.e. it increases when the energy content of the excited state is increased.

The fluorescence quantum yield (or efficiency) (Φ_f) is defined as the ratio of the number of emitted fluorescence photons to the number of photons absorbed (Senesi 1990a).

$$\Phi_f = \frac{\text{number of photons as fluorescence}}{\text{number of photons absorbed}}$$

The fluorescence efficiency determines the effectiveness with which the absorbed energy is re-emitted. It depends on several factors such as the molecular structure of the fluorescent molecule and its absorption nature, the non-radiative processes, the temperature and the wavelength used for excitation (Senesi 1990a; Wehry 1973).

The probability of finding a molecule in the excited state at a time t after the excitation source is turned off can be expressed as $\exp(-t/\tau)$ where τ is the fluorescence lifetime (Senesi 1990a). The fluorescence lifetime of a molecule is defined as the mean lifetime of the excited state before photon emission. The fluorescence intensity (F) typically follows a first-order kinetics and can be written as follows (Senesi 1990a):

$$F(t) = F_0 \exp(-t/\tau) \tag{2.3}$$

where $F(t)$ is the fluorescence intensity at the time t, F_0 the initial maximum intensity with the excitation source on, i.e. during excitation, t is the time elapsed after the excitation source is turned off, and τ is the fluorescence lifetime or the decay rate of fluorescence.

Fluorescence lifetimes or decay rates (τ) depend on the overall rates at which the excited state is deactivated with both radiative and non-radiative processes. It is $\tau_{tot}^{-1} = \tau_{rad}^{-1} + \tau_{nrad}^{-1}$. Fluorescence lifetimes for commonly used fluorescent molecules are typically of the order of nanoseconds. The intrinsic or natural lifetime (τ_0) corresponds to an absolute quantum efficiency (Φ_0) equal to 1. It happens when fluorescence is the only mechanism by which the excited state returns to the ground state. When non-radiative processes are deactivating the excited state, the measured lifetime (τ) can be expressed as (Eq. 2.4) (Senesi 1990a):

$$\tau = \Phi_0 \tau_0 \tag{2.4}$$

where $\Phi < \Phi_0$ is the fluorescence quantum efficiency. Fluorescence intensity (F) is proportional to the number of excited states which, in turn, depends on the concentration of the absorbing molecules in solution. According to a basic equation, F can be expressed as follows (Senesi 1990a):

$$F = \Phi I_0 \left[1 - \exp\left(-\varepsilon bc\right)\right] \tag{2.5}$$

where Φ is the quantum efficiency, I_0 is the intensity of the incident radiation that is directly proportional to the fluorescence intensity (F), ε is the (neperian) molar absorptivity of the molecule at the excitation wavelength (higher ε values produce higher fluorescence intensities), b is the path length of the cell and c is the molar concentration of the molecule.

For very diluted solutions, where εbc is sufficiently small, Eq. 2.5 can be expressed as follows (Senesi 1990a):

$$F = \Phi I_0 \varepsilon bc \tag{2.6}$$

Equation 2.6 predicts a linear relationship between fluorescence intensity and concentration of the molecule when εbc is small. If the concentration of the molecule and, as a consequence, εbc increases, a non-linear relationship is followed by Eq. (2.5). Small εbc values indicate that the fluorescence intensity of the molecule is essentially homogeneous throughout the sample. On the other hand, larger εbc values result in a fluorescence intensity of the molecule that is no longer homogeneous within the cell, but is increasingly localized at its front surface (Senesi 1990a).

2.1 *Fluorophores in the Fluorescent Molecule and their Controlling Factors*

A fluorophore is defined as a part of an organic molecule, with or without electron-donating heteroatoms such as N, O, and S, or as a functional group of a highly unsaturated aliphatic molecule with a structure that can hold up an excited electron, or having extensive π-electron systems, which exhibits fluoresce with significant efficiency (Mostofa et al. 2009a). The major fluorophores in various fluorescent organic molecules in natural waters are composed of Schiff-base derivatives (–N=C–C=C–N–), –COOH, –$COOCH_3$, –OH, –OCH_3, –CH=O, –C=O, –NH_2, –NH–, –CH=CH–COOH, –OCH_3, –CH_2–(NH_2)CH–COOH, S-, O- or N-containing aromatic compounds (Mostofa et al. 2009a; Senesi 1990a; Leenheer and Croué 2003; Malcolm 1985; Corin et al. 1996; Peña-Méndez et al. 2005; Seitzinger et al. 2005; Zhang et al. 2005).

According to the basic principle of fluorescence, the elements O, N, S, and P as well as their related functional groups (C–O, C=C, φ–O, COOH, and C=O) in fulvic acid can show fluorescence properties. Each of the functional groups in fulvic acid is referred to as fluorophore. All fluorophores in a mother molecule can exhibit fluorescence properties and any change in the molecule can have an effect on the overall fluorescence properties (Senesi 1990a). Fluorophores present in allochthonous fulvic acid (or allochthonous humic acid or autochthonous fulvic acid) either at peak C-region (Ex/Em = 280–400/380–550 nm) or A-region (Ex/Em = 215–280/380–550 nm) in EEM spectra can be denoted as fluorochrome. The molecular structure of fulvic acid is not yet known because of the complicated chemical composition and relatively large molecular size. However, fulvic and humic acids of vascular plant origin have allowed a partial identification of their molecular structure as benzene-containing carboxyl, methoxylate and phenolic groups, carboxyl, alcoholic OH, carbohydrate OH, –C=C–, hydroxycoumarin-like structures, fluorophores containing Schiff-base derivatives, chromone, xanthone, quinoline ones, as well as functional groups containing O, N, S, and P atoms. Such functional groups include aromatic carbon (17–30 %) and aliphatic carbon (47–63 %) (Leenheer and Croué 2003; Malcolm 1985; Senesi 1990b; Steelink 2002). All of the cited functional groups can be considered as major fluorophores in fulvic and humic acids in natural waters. They can display two fluorescence peaks: peak C at longer wavelength (or peak C-region) and peak A at shorter ultraviolet (UV) wavelengths (or peak A-region) (Mostofa et al. 2009a, 2005a, 2010, 2007a; Senesi 1990a; Coble et al. 1990; Coble 1996, 2007; Mostofa KMG et al., unpublished data; Komaki and Yabe 1982; Schwede-Thomas et al. 2005; Nakajima 2006). The electronic transition of the lowest energy that involves a fluorophore in a molecule exhibits fluorescence (Ex/Em) with the highest intensity at peak C and A-regions. When a fluorophore is degraded by photolytic processes, another lowest energy fluorophore will subsequently produce the fluorescence peak in the respective regions. Therefore, a particular peak (e.g., peak C, peak A, peak T or T_{UV}) of a fluorescent molecule is the outcome of the contribution of all fluorophores present in the molecule itself.

The fluorescence properties of an organic molecule containing fluorophores depend on several inner (or internal) and external (local physical conditions in the fluorophore's microenvironment) factors associated with chemical structure (Mostofa et al. 2009a; Senesi 1990a, 1990b; Lakowicz 1999; Tadrous 2000; Wu et al. 2002, 2004a; Baker 2005). The inner or internal factors are: (i) the probability of absorbing a photon; (ii) the number of fluorophores or functional groups present in the molecule; and (iii) the quantum yield that measures the probability of radiative decay from the excited state; (iv) the extension of the π-electron system, which reduces the excitation energy and shifts the emission wavelengths toward higher values; (v) heteroatom substitution on aromatic compounds; (vi) electron withdrawing (meta-directing) functional groups in aromatic compounds, which reduce the fluorescence intensity; (vii) electron-donating (ortho–para directing) functional groups in aromatic compounds, which increase the fluorescence efficiency; (viii) functional groups such as carbonyl, hydroxide, alkoxide and amino ones, which shift fluorescence toward longer wavelengths; (ix) an increase in structural rigidity that inhibits the internal conversion, thereby leading to an increase in fluorescence; (x) an increase in the solution redox potential, which enhances fluorescence; and (xi) the concentrations of solutes in aqueous media that would normally cause the fluorescence intensity to decrease when concentration is high.

External factors are: (xii) pH, considering that the fluorescence intensity markedly increases with increasing pH; (xiii) exposure of the fluorophores to an heat source, where increasing temperature causes fluorescence quenching; (xiv) complexation of metal ions with florophores of an organic molecule which can change the fluorescence intensity of that flurophore either enhanced or quenched compared to the original fluorescent DOM (see also chapter "Complexation of Dissolved Organic Matter with Trace Metal ions in Natural Waters") (Wu et al. 2004a, 2004b, 2004c; Smith and Kramer 1999; Ohno et al. 2007; Fu et al. 2007; Zhang et al. 2010; Manciulea et al. 2009; Mounier et al. 2011); (xv) inter- and intramolecular fluorescence quenching of the florophores in organic substance in the presence of other organic components (Marmé et al. 2003; Sun et al. 2012); (xvi) any changes in the molecule upon photoinduced or microbial degradation processes can alter its fluorescence properties, such as peak position and fluorescence intensity (Mostofa et al. 2009a; Senesi 1990a; Wu et al. 2004a). Light exposure can induce in a substantial decrease in the loss of fluorophore binding sites and the stability constant (Wu et al. 2004a; Kulovaara et al. 1995; Bertilsson and Tranvik 2000); and (xvii) any changes in the molecule upon microbial degradation processes can alter its fluorescence properties, such as peak position and fluorescence intensity (Mostofa et al. 2009a, 2007a, 2005b; Senesi 1990a; Moran et al. 2000).

2.2 Fluorescent Dissolved Organic Matter (FDOM) and its Characterization Using EEM Spectroscopy

Fluorescent dissolved organic matter (FDOM) is operationally defined as the dissolved organic matter (DOM) fraction that shows significant fluorescence efficiency or intensity at a particular excitation–emission wavelength (Mostofa et al. 2009a). The FDOM species that are commonly detected in aqueous solution are summarized in Table 1 (Senesi 1990a; Coble et al. 1990, 1998; Coble 1996, 2007; Mayer et al. 1999; Parlanti et al. 2000; Yamashita and Tanoue 2003a; Mostofa et al. 2005a, 2010; Mostofa KMG et al., unpublished data; Mostofa and Sakugawa 2009; Komaki and Yabe 1982; Schwede-Thomas et al. 2005; Nakajima 2006; Baker 2005; Zhang et al. 2010; Sugiyama et al. 2005; Liu and Fang 2002; Provenzano et al. 2004; Lu and Allen 2002; Moberg et al. 2001; Determann et al. 1998; Matthews et al. 1996; Baker and Curry 2004). They are humic substances (fulvic and humic acids) of vascular plant origin, marine humic-like compounds, autochthonous fulvic acids of algal origin, derived from photoinduced and microbial assimilation of lake algae, aromatic amino acids (tryptophan, tyrosine and phenylalanine), proteins, diaminostilbene-type (DAS1) and distyryl biphenyl (DSBP) fluorescent whitening agents (FWAs), other components of household detergents, and chlorophyll (Table 1).

Table 1 Fluorescence excitation/emission (Ex/Em) wavelengths of standard substances and the subsequent characteristic peaks of the reference components in natural waters

Samples	Fluorescence properties			Peak A-region	Peak T-region	Peak T_{UV}-region	References
Standard substances/Component type	Peak C-region						
	Peak C	Peak M_p	Peak W				
	Ex/Em (nm)						
Standard Suwannee River Fulvic Acid dissolved in Milli-Q waters (n = 5)[b]	330/462	–	–	250/462	–	–	Mostofa et al. (2005a), Mostofa KMG et al., (unpublished data)
Standard Suwannee River Fulvic Acid dissolved in Milli-Q waters (n = 5)[b]	–	–	–	230/441	–	–	Mostofa et al. (2005a), Mostofa KMG et al., (unpublished data)
Standard Suwannee River Fulvic Acid dissolved in Milli-Q waters	325/442	–	–	255/450	–	–	Nakajima (2006)
Standard Suwannee River Fulvic Acid dissolved in sea waters	345/452	–	–	255/451	–	–	Nakajima (2006)
Fulvic acids (IHSS standard, n = 5)	333 ± 4/452 ± 12	–	–	–	–	–	Baker (2005)
Suwannee River Fulvic Acid	325/450	–	–	260/460	–	–	Coble et al. (1990)
Fulvic acid, extracted from River	–	300–310/420–430	–	260–270/430–440	–	–	(Schwede-Thomas et al. (2005)
Fulvic acid, extracted from Lake	–	305/448	–	240/440	–	–	Mostofa KMG et al., (unpublished data)
Suwannee River and Pine Barrens (pH: 3–6)	330–340/451–467	–	–	240–260/427–468	–	–	Schwede-Thomas et al. (2005)
Fulvic acid (SJF)	–	310/419	–	–	–	–	Coble (1996), (2007)

(contniued)

Table 1 (contniued)

Samples	Fluorescence properties			Peak A-region	Peak T-region	Peak T_{UV}-region	References
Standard substances/Component type	Peak C-region						
	Peak C	Peak M_p	Peak W				
	Ex/Em (nm)						
Soil fulvic acid (Standard)	–	320/440	–	270–280/430–440	–	–	Sugiyama et al. (2005)
Standard Suwannee River Humic Acid[b]	350/461	300/461	–	255/461	–	–	Mostofa et al. (2005a)
Standard Suwannee River Humic Acid[b]	–	–	–	265, 230/436	–	–	Mostofa et al. (2005a)
Standard Suwannee River Humic Acid (n = 3)	320–345/478–498	–	–	–	–	–	Baker (2005)
Humic acid (EEP2)	–	310/428	–	–	–	–	Coble (1996), (2007)
Humic acid (EEP2)	–	310/423	–	–	–	–	Coble (1996), (2007)
Humic acid, extracted from Lake	–	295/464	–	255/464	–	–	Mostofa KMG et al., (unpublished data)
Protein-like, detected in sewerage samples[a]	–	–	–	–	280/339–346	230/339–346	Mostofa et al. (2010)
Protein-like, extracted from EPS	–	–	–	–	280–285/340–350	225/340–350	Liu and Fang (2002)
Aromatic protein or soluble microbial by-products-like	–	–	–	–	270–280/320–350	220–230/340–350	Mayer et al. (1999)
Tryptophan standard dissolved in Milli-Q waters (n = 3)[b]	–	–	–	–	275–280/352–356	225/343–358	Mostofa et al. (2005a), (2010)
Tryptophan standard dissolved in Milli-Q waters	–	–	–	–	275/357	Peak	Nakajima (2006)

(contniued)

Table 1 (contniued)

Samples	Fluorescence properties			Peak A-region	Peak T-region	Peak T_{UV}-region	References
Standard substances/Component type	Peak C-region						
	Peak C	Peak M_P	Peak W				
	Ex/Em (nm)						
Tryptophan standard dissolved in sea waters	–	–	–	–	275/355	Peak[c]	Nakajima (2006)
Tryptophan standard (n = 5)	–	–	–	–	283 ± 3/351 ± 2	231 ± 3/353 ± 4	Baker (2005)
Tryptophan	–	–	–	–	280/357	227/351	Baker (2005)
Tryptophan[b]	–	–	–	–	280/342–346	Peak[c]	Yamashita and Tanoue (2003a)[b]
Tryptophan-like, extracted from EPS	–	–	–	–	275–280/ 322–336	220–230/ 328–334	Zhang et al. (2010)
Tyrosine standard dissolved in Milli-Q waters	–	–	–	–	275/303	230/304	Nakajima (2006)
Tyrosine standard dissolved in sea waters	–	–	–	–	275/304	230/307	Nakajima (2006)
Tyrosine standard dissolved in Milli-Q waters	–	–	–	–	270/314	Peak[c]	Mostofa and Sakugawa (2009)
Tyrosine[b]	–	–	–	–	270–275/ 300–302	Peak[c]	Yamashita and Tanoue (2003a)[b]
Tyrosine-like	–	–	–	–	275/310	Peak[c]	Coble (1996)
Tyrosine	–	–	–	–	275/303	Peak[c]	Parlanti et al. (2000)
Tyrosine-like, protein-like	–	–	–	–	275/310	Peak[c]	Provenzano et al. (2004)
Tyrosine-like, protein-like	–	–	–	–	265–280/ 293–313	Peak[c]	Lu and Allen (2002)
Phenylalanine standard dissolved in Milli-Q waters[b]	–	–	–	–	255–265/ 284–285	Peak[c]	Yamashita and Tanoue (2003a)[b]

(contniued)

Table 1 (contniued)

Samples	Fluorescence properties			Peak A-region	Peak T-region	Peak T_{UV}-region	References
Standard substances/Component type	Peak C-region						
	Peak C	Peak M_p	Peak W				
	Ex/Em (nm)						
Phenylalanine standard dissolved in Milli-Q waters	–	–	–	–	260/286	Peak	Nakajima (2006)
Phenylalanine standard dissolved in sea waters	–	–	–	–	260/284	Peak	Nakajima (2006)
Distyryl biphenyl (DSBP), FWAs dissolved in Milli-Q waters (n = 3)	–	–	350/436	235–265/435–445	–	–	Mostofa et al. (2010)
DSBP, FWAs dissolved in Milli-Q waters	–	–	350/433	245/431	–	–	Nakajima (2006)
DSBP, FWAs dissolved in sea waters	–	–	345/435	245/437	–	–	Nakajima (2006)
DSBP, FWAs (n = 2)	–	–	355/430–432	Peak	–	–	Komaki and Yabe (1982)
Diaminostilbene-type (DAS1), FWAs dissolved in Milli-Q waters (n = 4)	–	–	335–355/438–449	240–245/434–446	–	–	Mostofa et al. (2010)
DAS1, FWAs dissolved in Milli-Q waters	–	–	335/435	250/439	–	–	Nakajima (2006)
DAS1, FWAs (n = 2)	–	–	340–343/430–432	Peak	–	–	Komaki and Yabe (1982)
DAS1, FWAs dissolved in sea waters	–	–	345/436	250/433	–	–	Nakajima (2006)
Chlorophyll-like, pigment-like	398/660	–	–	–	–	–	Coble et al. (1998)
Chlorophyll *a*	431/670	–	–	–	–	–	Moberg et al. (2001)

(contniued)

Table 1 (contniued)

Samples	Fluorescence properties			Peak A-region	Peak T-region	Peak T_{UV}-region	References
Standard substances/Component type	Peak C-region						
	Peak C	Peak M_p	Peak W				
	Ex/Em (nm)						
Chlorophyll *b*	435/659	–	–	–	–	–	Moberg et al. (2001)
Algae or phytoplankton, resuspensions in Milli-Q waters, collected from lake waters[a]	–	–	–	–	280/346, 270/327	230/346, 230/327	Mostofa KMG et al., (unpublished data)
Algae or phytoplankton, resuspensions in River waters, collected from lake waters[a]	–	–	–	–	285/340, 270/336	230/336	Mostofa KMG et al., (unpublished data)
Algae and bacteria, collected from marine waters	–	–	–	–	280/340	230/340, 230/305	Determann et al. (1998)
Corals	310–390/430–490	–	–	–	280/320–350	–	Matthews et al. (1996)
1,4-Dichlorobenzene dissolved in Milli-Q waters	–	–	–	–	–	225/294–299	Mostofa KMG et al., (unpublished data)
Anthracene dissolved in Milli-Q waters	340/401–405	–	–	250/401–404	–	–	Mostofa KMG et al., (unpublished data)
Phenanthrene dissolved in Milli-Q waters	350/367, 348	–	–	–	290/349, 366	–	Mostofa KMG et al., (unpublished data)
Perylene dissolved in Milli-Q waters	–	–	–	–	250/366, 348	–	Mostofa KMG et al., (unpublished data)

(contniued)

Table 1 (contniued)

Samples	Fluorescence properties			Peak A-region	Peak T-region	Peak T_{UV}-region	References
Standard substances/Component type	Peak C-region						
	Peak C	Peak M_p	Peak W				
	Ex/Em (nm)						
Naphthalene in landfill leachate	–	–	–	–	–	220–230/ 340–370	Baker and Curry (2004)
Melanoidin	–	363/458	–	–	–	–	Coble (1996), (2007)
Phenol	–	–	–	–	270/297	–	Coble (1996), (2007)
Phenol dissolved in Milli-Q water	–	–	–	–	270/299	–	Nakajima (2006)
Phenol dissolved in sea water	–	–	–	–	270/298	–	Nakajima (2006)
4-Biphenyl carboxaldehyde	–	305/410	–	–	–	255/315	Mostofa et al. (2010)
o-Cresol	–	–	–	–	275/303	215/304	Mostofa et al. (2010)
p-Cresol	–	–	–	–	280/309	225/309	Mostofa et al. (2010)
p-Hydroxyphenyl acetic acid	–	–	–	–	280/305	230/304	Mostofa et al. (2010)
Benzoic acid dissolved in Milli-Q waters	–	–	–	–	280/311	–	Nakajima (2006)
Benzoic acid dissolved in sea waters	–	300/396	–	–	–	–	Nakajima (2006)
o-Hydroxy benzoic acid or salicylic acid	–	300/407	–	235/410	–	–	Coble (1996), (2007)
o-Hydroxy benzoic acid or salicylic acid	–	314/410	–	–	–	–	Senesi (1990a)
3-Hydroxybenzoic acid	–	314/423	–	–	–	–	Senesi (1990a)
p-Hydroxybenzoic acid dissolved in Milli-Q water	–	–	–	–	–	255/318	Nakajima (2006)

(contniued)

Table 1 (contniued)

Samples	Fluorescence properties			Peak A-region	Peak T-region	Peak T_{UV}-region	References
Standard substances/Component type	Peak C-region						
	Peak C	Peak M_p	Peak W				
	Ex/Em (nm)						
p-Hydroxybenzoic acid dissolved in sea water	–	–	–	–	280/388	–	Nakajima (2006)
Methyl salicylate	366/448	302/448	–	–	–	–	Senesi (1990a)
p-Hydroxy benzaldehyde dissolved in Milli-Q waters	–	325/365	–	–	–	235/353	Nakajima (2006)
p-Hydroxy benzaldehyde dissolved in sea waters	330/372	–	–	240/389	–	–	Nakajima (2006)
p-Hydroxy acetophenone dissolved in Milli-Q waters	–	310/347	–	–	–	225/353	Nakajima (2006)
p-Hydroxy acetophenone dissolved in sea waters	340/382	–	–	240/386	–	–	Nakajima (2006)
Protocatechuic acid (ionized)	340–370/455	–	–	–	–	–	Senesi (1990a)
3-Hydroxycinnamic acid	–	310/407	–	–	–	–	Senesi (1990a)
Caffeic acid	365/450	–	–	–	–	–	Senesi (1990a)
Ferulic acid	350/440	–	–	–	–	–	Senesi (1990a)
ß-Naphthols (ionized)	350/460	–	–	–	–	–	Senesi (1990a)
Xanthone	–	410/456	–	–	–	–	Senesi (1990a)
3-Hydroxyxanthone	343, 365/465	–	–	–	–	–	Senesi (1990a)
3-Hydroxy quinoline	350/450	–	–	–	–	–	Senesi (1990a)

M_p means fluorescence Ex/Em maxima of fulvic acid which is photobleached by photochemical processes or by any other natural processes
[a]Indicates the organic components identified using PARAFAC modeling on sample EEM data and river EEM data is deducted from samples in case of algae
[b]Ranges expresses the authentic standard at various concentrations (1–5 mg L^{-1}) and mechanical reproducibility
peak̃ indicates the occurrence of a peak that do not indentify
EPS extracellular polymeric substances

Coble (1996) firstly designated the fluorescence Ex/Em wavelength peaks of humic-like substances (peak C, peak M, and peak A) as well as of protein-like, tryptophan-like and tyrosine-like ones (peak T and peak B). Parlanti et al. (2000) slightly modified the wavelength ranges and designated them with new letters (α, β and ά, and γ and δ, respectively). Chen et al. (2003) selected the wavelength boundaries at various regions (Region I, Region II, Region III, Region IV, and Region V) to define DOM in natural waters. To account for all the FDOM components of allochthonous, autochthonous and anthropogenic sources, it is desirable to generalize the peak positions by combining the peak regions (Chen et al. 2003) and specifying the letters as made by Coble (Coble 1996). All aquatic scientists have accepted Coble`s specification, and further speciation makes it more complicated. Overviewing and justifying the wavelength ranges of the fluorescence Ex/Em peaks of Coble (1996) and Parlanti et al. (2000) with those of the field observations, Mostofa et al. (2009a) summarize and then specify the fluorescence peaks of various FDOM components at four regions: peak C-region (280–400/380–550 nm), peak A-region (215–280/380–550 nm), peak T-region (260–285/290–380 nm), and peak T_{UV}-region (215–260/280–380 nm) (Table 1) (Mostofa et al. 2009a).

Peak C-region accounts for the broader excitation–emission wavelength ranges at Ex/Em = 280–400/380–550 nm, which include the humic substances (fulvic acid and humic acid) of terrestrial vascular plant origin (C-like and M-like), autochthonous fulvic acids (C-like and M-like) of algal or phytoplankton origin, photo-bleached allochthonous fulvic acid, fluorescent whitening agents (FWAs) such as DAS1 and DSBP as well as few standard organic substances (Tables 1, 2) (Coble et al. 1990, 1998; Coble 1996, 2007; Mopper and Schultz 1993; Parlanti et al. 2000; Yamashita and Tanoue 2003a; Mostofa et al. 2005a; Stedmon et al. 2003, 2007a, 2007b; Fulton et al. 2004; Mostofa et al. 2010, 2007a; 2005b; Mostofa KMG et al., unpublished data; Moran et al. 2000; Komaki and Yabe 1982; Schwede-Thomas et al. 2005; Baker 2005; Sugiyama et al. 2005; Moberg et al. 2001; Chen et al. 2003; Klapper et al. 2002; Komada et al. 2002; Nagao et al. 2003; Boyd and Osburn 2004; Burdige et al. 2004; Conmy et al. 2004; Fu et al. 2006, 2010; Mostofa et al. 2007b; Gao et al. 2010).

Peak A-region accounts for the shorter wavelength region at Ex/Em = 215–280/380–550 nm, and all FDOM components showing fluorescence at the peak C-region can display their secondary fluorescence peaks at the A-region (Tables 1, 2) (Mostofa et al. 2009a, 2005a, 2010, 2007b; Coble et al. 1990; Coble 1996; Mopper and Schultz 1993; Parlanti et al. 2000; Fulton et al. 2004; Komaki and Yabe 1982; Schwede-Thomas et al. 2005; Baker and Curry 2004; Chen et al. 2003; Klapper et al. 2002; Boyd and Osburn 2004; Burdige et al. 2004; Fu et al. 2006, 2010; Suzuki et al. 1997; Zhang et al. 2009a). Peak T-region accounts for the fluorescence peaks at Ex/Em = 260–285/290–380 nm, which includes the primary and secondary fluorescence peaks of various organic substances such as protein-like, aromatic amino acids (tryptophan-like, tyrosine-like and phenylalanine-like), phenol-like compounds, algae, corals, benzoic acid, *p*-hydroxy benzoic acid, perylene, phenanthrene, *o*-cresol and *p*-cresol (Table 1) (Mostofa et al.

Table 2 The fluorescent components identified using PARAFAC modeling on the sample's EEM and their characteristic peaks as excitation (Ex)–Emission (Em) maxima

Fluorescent components	Components identified and sources	Fluorescence properties								References
		Peak C-region				Peak A-region	Peak T-region	Peak T_{UV}-region		
		Peak C	Peak M_p	Peak M	Peak W			Major	Minor	
		EX/Em (nm)								
Allochthonous fulvic acids (C-like, A-like, and M-like)										
Suwannee River Fulvic Acid (C-like), standard, dissolved in Milli-Q waters (n = 5)	Component 1, allochthonous	330/462	–	–	–	250/462	–	–	–	Mostofa et al. (2005a)[c], Mostofa KMG et al., (unpublished data)
Fulvic acid (C-like), Yellow River, upstream waters, China	Component 1	335/449	–	–	–	250/449	–	–	–	Mostofa KMG et al., (unpublished data)
Fulvic acid (C-like), Yellow River, mainstream waters, China	Component 1	340/440	–	295/440	–	250/440	–	–	–	Mostofa KMG et al., (unpublished data)
Fulvic acid (C-like), Nanming River, China	Component 1	300–310/ 423–448	–	–	–	235–255/ 425–447	–	–	–	Mostofa et al. (2010)
Fulvic acid (C-like)?, urban sewerage samples[RU]	Component 5	380/467	–	–	–	260/467	–	–	–	Guo et al. (2010)
Fulvic acid (C-like)?, drinking water treatment plant[RU]	Component 3	330/420	–	–	–	<250/420	–	–	–	Baghoth et al. (2010)
Fulvic acid (C-like), The Second Song Hua Jiang River, North-East China	Component 1	325–340/ 449–458	–	–	–	255–260/ 449–458	–	–	–	Mostofa KMG et al., (unpublished data)
Fulvic acid (C-like), LiaoHe River, North-East China	Component 1	330/449	–	–	–	260/449	–	–	–	Mostofa KMG et al., (unpublished data)
Fulvic acid (C-like), Yellow River, China (12 days dark incubation)	Component 1	300/449	–	–	–	250/449	–	–	–	Mostofa KMG et al., (unpublished data)
Fulvic acid (C-like), streams, springs and thermokarsts, CPCRW, Alaska	Component 1[c], allochthonous	–		Peak′	–	Peak′	–	–	–	Balcarczyk et al. (2009)

(contniued)

Table 2 (contniued)

Fluorescent components	Components identi-fied and sources	Fluorescence properties								References
		Peak C-region				Peak A-region	Peak T-region	Peak T_{UV}-region		
		Peak C	Peak M_P	Peak M	Peak W			Major	Minor	
		EX/Em (nm)								
Fulvic acid (C-like), Occoquan Watershed (Northern Virginia, US)[RU]	Component 1, allochthonous	350/456	–	–	–	240/456	–	–	–	Holbrook et al. (2006)
Fulvic acid (C-like), Nishi-Mataya and Higashi-Mataya upstreams, Japan	Component 1	330–335/460–463	–	–	–	250–255/460–463	–	–	–	Mostofa et al. (2005b)[c]
Fuvic Acid (C-like), Yasu River, Lake Biwa watershed, Japan	Component 1	–	310/464	–	–	250/464	–	–	–	Mostofa et al. (2005b)[c]
Fuvic Acid (C-like), 3 Rivers (Ane, Echi and Amano), Lake Biwa watershed, Japan	Component 1	330–335/455–462	–	–	–	255/455–462	–	–	–	Mostofa et al. (2005b)[c]
Fulvic acid (C-like), Lake Biwa, epilimnion (0–20 m), during summer period	Component 1, allochthonous	–	295–310/449–450	–	–	250–255/449–450	–	–	–	Mostofa et al. (2005b)[c]
Fuvic acid (C-like), Lake Biwa, epilimnion (0–20 m), during winter period	Component 1, allochthonous	–	310/443	–	–	255–260/443	–	–	–	Mostofa et al. (2005b)[c]
Fulvic Acid (C-like), Lake Biwa, hypolimnion (40–80 m), during summer period	Component 1, allochthonous	–	300–305/444–461	–	–	255–260/444–461	–	–	–	Mostofa et al. (2005b)[c]
Fulvic acid (C-like), Lake Biwa, epilimnion (40–80 m), during winter period	Component 1 allochthonous	–	305–310/450–464	–	–	255–260/450–464	–	–	–	Mostofa et al. (2005b)[c]
Fulvic acid (C-like)?, Urdaibai and Foz Estuaries, Iberian Peninsula	Component 1, allochthonous	305/439	–	–	–	260/439	–	–	–	Santín et al. (2009)

(contniued)

Table 2 (contniued)

Fluorescent components	Components identified and sources	Fluorescence properties								References
		Peak C-region				Peak A-region	Peak T-region	Peak T_{UV}-region		
		Peak C	Peak M_p	Peak M	Peak W			Major	Minor	
		EX/Em (nm)								
Fulvic acid (C-like)?, Horsens Estuary, Jutland Peninsula, Denmark[RU]	Component 2, allochthonous	385/504	–	–	–	250/504	–	–	–	Stedmon and Markager (2005b)
Fulvic acid (C-like)?, Estuary of Horsens Fjord, Denmark[RU]	Component 3, allochthonous	360/478	–	–	–	270/478	–	–	–	Stedmon et al. (2003)
Fulvic acid (C-like), coastal shelf, South Atlantic Bight	Component 1, allochthonous	Peak′	–	–	–	250/452	–	–	–	Kowalczuk et al. (2009)
Fulvic acid (C-like)?, Bay waters, Barataria Basin (Louisiana, USA)[RUc]	Allochthonous	410/520	–	–	–	250/520	–	–	–	Singh et al. (2010)
Fulvic acid (C-like), Bay waters, Barataria Basin (Louisiana, USA)[RUc]	Component 1, allochthonous	340/440	–	–	–	250/440	–	–	–	Singh et al. (2010)
Fulvic acid (C-like)?, ground water, fresh and Florida Bay waters, Florida coastal Everglades	Component 6, allochthonous	Peak′	–	–	–	Peak′	–	–	–	Chen et al. (2010)
Fulvic acid-like?, glacial ice samples, Antarctic and Arctic Ocean	Component 3, autochthonous	Peak′	–	–	–	<250/446	–	–	–	Dubnick et al. (2010)
Fulvic acid-like (C-like)?, water extractable from sugar maple leaves	Component 5	315/429	–	–	–	Peak′	–	–	–	Hunt et al. (2008)
Photobleached fulvic acid (C-like), Yellow River, China (3 h sunlight irradiation)	Component 1	–	300–305/449	–	–	250/449	–	–	–	Mostofa KMG et al., (unpublished data)

(contniued)

Table 2 (contniued)

Fluorescent components	Components identified and sources	Fluorescence properties								References
		Peak C-region				Peak A-region	Peak T-region	Peak T_{UV}-region		
		Peak C	Peak M_p	Peak M	Peak W			Major	Minor	
		EX/Em (nm)								
Photobleached fulvic acid (C-like), Seto Inland Sea, Japan	Component 2, allochthonous	–	320–325/449–454	–	–	255–260/449–454	–	–	–	Mostofa KMG et al., (unpublished data)
Photobleached fulvic acid (C-like)?, Liverpool Bay, Irish Sea	Component 1	–	320/422	–	–	<250/422	–	–	–	Yamashita et al. (2011)
Fulvic acid (C-like), plant biomass, animal manure, and soils	Components 3, allochthonous	315/447	–	–	–	~250/447	–	–	–	Ohno and Bro (2006)
Suwannee River Fulvic Acid (A-like), standard, dissolved in Milli-Q waters (n = 5)	Component 2, allochthonous	–	–	–	–	230/441	–	–	–	Mostofa et al. (2005a)[c], Mostofa KMG et al., (unpublished data)
Fulvic acid (A-like), streams, springs and thermokarsts, CPCRW, Alaska	Component 2[c], allochthonous	–	–	–	–	Peak′	–	–	–	Balcarczyk et al. (2009)
Fulvic acid (A-like), Nishi-Mataya and Higashi-Mataya upstreams, Japan	Component 2	–	–	290–295/414	–	225/414	–	–	–	Mostofa et al. (2005b)[c]
Fuvic Acid (A-like), 3 Rivers (Ane, Echi and Amano), Lake Biwa watershed, Japan	Component 2	–	–	285–290/442	–	225/442	–	–	–	Mostofa et al. (2005b)[c]
Fulvic acid (A-like), Lake Biwa, epilimnion (0–20 m), during summer period	Component 2, allochthonous	–	–	–	–	225/442	–	–	–	Mostofa et al. (2005b)[c]

(contniued)

Table 2 (contniued)

Fluorescent components	Components identified and sources	Fluorescence properties								References
		Peak C-region				Peak A-region	Peak T-region	Peak T_{UV}-region		
		Peak C	Peak M_p	Peak M	Peak W			Major	Minor	
		EX/Em (nm)								
Fuvic acid (A-like), Lake Biwa, epilimnion (0–20 m), during winter period	Component 2, allochthonous	–	–	–	–	225/432–439	–	–	–	Mostofa et al. (2005b)[c]
Fulvic Acid (A-like), Lake Biwa, hypolimnion (40–80 m), during summer period	Component 2, allochthonous	–	–	–	–	225/432–442	–	–	–	Mostofa et al. (2005b)[c]
Fulvic acid (A-like), Lake Biwa, epilimnion (40–80 m), during winter period	Component 2, allochthonous	–	–	–	–	225/433–434	–	–	–	Mostofa et al. (2005b)[c]
Fulvic acid (A-like)?, Horsens Estuary, Jutland Peninsula, Denmark[RU]	Component 1, allochthonous	–	–	–	–	250/448	–	–	–	Stedmon and Markager (2005b)
Fulvic acid (A-like)?, Estuary of Horsens Fjord, Denmark[RU]	Component 1, allochthonous	–	–	–	–	240/436	–	–	–	Stedmon et al. (2003)
Fulvic acid (A-like)?, Estuary of Horsens Fjord, Denmark[RU]	Component 2, allochthonous	–	–	–	–	240/416	–	–	–	Stedmon et al. (2003)
Fulvic acid (A-like)?, deep waters of the Okhotsk Sea and North Pacific Ocean	Component 3, allochthonous	–	–	–	–	<260/460–480?	–	–	–	Yamashita et al. (2010)
Fulvic acid (M-like), Yellow River, mainstream waters, China	Component 2	–	–	290/429	–	235/429	–	–	–	Mostofa KMG et al., (unpublished data)

(contniued)

Table 2 (contniued)

Fluorescent components	Components identified and sources	Fluorescence properties								References
		Peak C-region				Peak A-region	Peak T-region	Peak T_{UV}-region		
		Peak C	Peak M_p	Peak M	Peak W			Major	Minor	
		EX/Em (nm)								
Fulvic acid (M-like), The Second Song Hua Jiang River, North-East China	Component 2	–	–	300/392–411	–	230–235/392–411	–	–	–	Mostofa KMG et al., (unpublished data)
Fulvic acid (M-like), LiaoHe River, North-East China	Component 2	–	–	285/387	–	230/387	–	–	–	Mostofa KMG et al., (unpublished data)
Fulvic acid (M-like), main-stream of NenJiang River, North-East China	Component 2	–	–	290/417	–	235/417	–	–	–	Mostofa KMG et al., (unpublished data)
Fulvic acid (M-like), tributaries of NenJiang River, North-East China	Component 2	–	–	310/417	–	235/417	–	–	–	Mostofa KMG et al., (unpublished data)
Fulvic acid (M-like)?, drinking water treatment plant[RU]	Component 2	–	–	320/410	–	250/410	–	–	–	Baghoth et al. (2010)
Fulvic acid (M-like), Occoquan Watershed (Northern Virginia, US)[RU]	Component 2, allochthonous	–	–	305/396	–	240/396	–	–	–	Holbrook et al. (2006)
Fulvic acid (M-like)?, river and coastal waters	Component 1. allochochthonous	–		305/428	–	<260/428	–	–	–	Yamashita and Jaffé (2008)
Fulvic acid (M-like)?, ground water, fresh and Florida Bay waters, Florida coastal Everglades	Component 3 allochthonous	–	–	Peak′	–	Peak′	–	–	–	Chen et al. (2010)
Fulvic acid (M-like)?, river, esturine and coastal marine waters	Component 5	–	–	Peak′?	–	240/414	–	–	–	Fellman et al. (2010)

(continued)

Table 2 (contniued)

Fluorescent components	Components identified and sources	Fluorescence properties								References
		Peak C-region				Peak A-region	Peak T-region	Peak T_{UV}-region		
		Peak C	Peak M_p	Peak M	Peak W			Major	Minor	
		EX/Em (nm)								
Fulvic acid (M-like)?, water extractable from sugar maple leaves	Component 1	–	–	312/417	–	240/417	–	–	–	Hunt et al. (2008)
Allochthonous humic acids (*C-like, A-like, and M-like*)										
Suwannee River Humic Acid (C-like), standard, dissolved in Milli-Q waters (n = 4)	Component 1, allochthonous	350/461	300/461	–	–	255/461	–	–	–	Mostofa et al. (2005a)
Humic acid (C-like), streams, springs and thermokarsts, CPCRW, Alaska	Component 4, allochthonous	Peak′	–	–	–	Peak′	–	–	–	Balcarczyk et al. (2009)
Humic acid (C-like)?, soil and stream waters, temperate rainforest watersheds	Allochthonous	330/460–480	–	–	–	<250/450–460	–	–	–	Fellman et al. (2009)
Humic acid (C-like)?, soil solution samples, forest and wetland soils, rainforest watersheds	Allochthonous	370/440	–	–	–	<250/440	–	–	–	Fellman et al. (2008)
Humic acid (C-like)?, ground water, fresh and Florida Bay waters, Florida coastal Everglades	Component 5	Peak′	–	–	–	Peak′	–	–	–	Chen et al. (2010)
Humic acid (C-like), river and coastal waters	Component 2, allochthonous	340, 405/>500	–	–	–	<260/>500	–	–	–	Yamashita and Jaffé (2008)
Humic acid (C-like)?, river, esturine and coastal marine waters	Component 2	330/456–480	–	–	–	Peak′?	–	–	–	Fellman et al. (2010)

(contniued)

Table 2 (contniued)

Fluorescent components	Components identi-fied and sources	Fluorescence properties								References
		Peak C-region				Peak A-region	Peak T-region	Peak T_{UV}-region		
		Peak C	Peak M_p	Peak M	Peak W			Major	Minor	
		EX/Em (nm)								
Humic acid (C-like)?, river, esturine and coastal marine waters	Component 3	290/510	–	–	–	–	–	–	–	Fellman et al. (2010)
Humic acid (C-like), Urdaibai and Foz Estuaries, Iberian Peninsula	Component 2, allochthonous	385/>500	–	–	–	256/>500	–	–	–	Santín et al. (2009)
Humic acid (C-like)?, Liverpool Bay, Irish Sea	Component 3	300, 410/510	–	–	–	Peak′	–	–	–	Yamashita et al. (2011)
Humic acid (C-like), coastal shelf, South Atlantic Bight	Component 4, allochthonous	390/508	–	–	–	270/508	–	–	–	Kowalczuk et al. (2009)
Humic acid (C-like), north Pacific and Atlantic oceans (BWE7 model)	Component 3^c, allochthonous	370/490	–	–	–	260/490	–	–	–	Murphy et al. (2008)
Humic acid (C-like), north Pacific and Atlantic oceans (Kauai model)	Component 1 (P3), autochthonous	380/498	–	–	–	<260/498	–	–	–	Murphy et al. (2008)
Humic acid (C-like)?, water extractable from sugar maple leaves	Component 2	351/459	–	–	–	240/459	–	–	–	Hunt et al. (2008)
Humic acid (C-like)?, compost products solution	Component 2	350/450	–	–	–	250/450	–	–	–	Yu et al. (2010)
Humic acid (C-like)?, municipal leachate samples	Component 2	360/458	–	300/458	–	250/458	–	–	–	Wu et al. (2011)
Humic acid (C-like)?, drinking water treatment plant^{RU}	Component 1	360/480	–	–	–	260/480	–	–	–	Baghoth et al. (2010)
Humic acid (C-like), plant biomass, animal manure, and soils	Component 1, allochthonous	350–360/460–480	–	–	–	Peak′	–	–	–	Ohno and Bro (2006)

(continued)

Table 2 (contniued)

Fluorescent components	Components identified and sources	Fluorescence properties								References
		Peak C-region				Peak A-region	Peak T-region	Peak T_{UV}-region		
		Peak C	Peak M_p	Peak M	Peak W			Major	Minor	
		EX/Em (nm)								
Suwannee River Humic Acid (A-like), standard, dissolved in Milli-Q waters (n = 4)	Component 2, allochthonous	–	–	–	–	265, 230/436	–	–	–	Mostofa et al. (2005a)[c]
Humic acid (A-like)?, soil and stream waters, temperate rainforest watersheds	Allochthonous	–	–	–	–	<250/400	–	–	–	Fellman et al. (2009)
Humic acid (A-like)?, soil and stream waters, temperate rainforest watersheds	Allochthonous	–	–	–	–	<250/400	–	–	–	Fellman et al. (2009)
Humic acid (A-like)?, ground water, fresh and Florida Bay waters, Florida coastal Everglades	Component 2, allochthonous	–	–	–	–	Peak′	–	–	–	Chen et al. (2010)
Humic acid (A-like)?, river, esturine and coastal marine waters	Component 1	–	–	–	–	<250/ 450–470	–	–	–	Fellman et al. (2010)
Humic acid (A-like)?, water extractable from sugar maple leaves	Component 3	–	–	–	–	240/483	–	–	–	Hunt et al. (2008)
Humic acid (A-like)?, municipal leachate samples	Component 4	–	–	–	–	220/432	–	–	–	Wu et al. (2011)
Humic acid (M-like)?, soil solution samples, forest and wetland soils, rainforest watersheds	Allochthonous	–	–	300/416	–	240/416	–	–	–	Fellman et al. (2008)

(contniued)

Table 2 (contniued)

Fluorescent components	Components identified and sources	Fluorescence properties								References
		Peak C-region				Peak A-region	Peak T-region	Peak T_{UV}-region		
		Peak C	Peak M_p	Peak M	Peak W			Major	Minor	
		EX/Em (nm)								
Humic acid (M-like)?, soil and stream waters, temperate rainforest watersheds	Allochthonous	–	–	295/414	–	<250/414	–	–	–	Fellman et al. (2009)
Humic acid (M-like)?, municipal leachate samples	Component 1	–	–	330/412	–	240/412	–	–	–	Wu et al. (2011)
Humic acid (M-like), plant biomass, animal manure, and soils	Components 2, allochthonous	–		290–300/465	–	>240/465	–	–	–	Ohno and Bro (2006)
Autochthonous fulvic acid (C-like): biologically or photochemically produced										
–										
Autochthonous fulvic acid (C-like), mainstream of NenJiang River, North-East China	Component 1	350/460	–	–	–	260/460	–	–	–	Mostofa KMG et al., (unpublished data)
Autochthonous fulvic acid (C-like), tributaries of NenJiang River, North-East China	Component 1	340/460	–	–	–	260/460	–	–	–	Mostofa KMG et al., (unpublished data)
Autochthonous fulvic acid (C-like), surface waters (0–25 m), Lake Hongfeng, China	Component 1, algae or phytoplankton	335–340/442–464	–	–	–	260/442–464	–	–	–	Fu et al. (2010)[c]
Autochthonous fulvic acid (C-like), algae origin, microbial assimilations in Milli-Q waters	Component 1, algae or phytoplankton	340/442	–	290/442	–	260/442	–	–	–	Mostofa KMG et al., (unpublished data)

(continued)

Table 2 (contniued)

Fluorescent components	Components identified and sources	Fluorescence properties								References
		Peak C-region				Peak A-region	Peak T-region	Peak T_{UV}-region		
		Peak C	Peak M_p	Peak M	Peak W			Major	Minor	
		EX/Em (nm)								
Autochthonous fulvic acid (C-like), algae origin, microbial assimilations in River waters	Component 1, algae or phytoplankton	340–455	–	295–300/ 430–448	–	260–265/ 436–448	–	–	–	Mostofa KMG et al., (unpublished data)
Autochthonous fulvic acid (C-like), algae origin, photo assimilations in Milli-Q waters	Component 1, algae or phytoplankton	340/448	–	–	–	260/448	–	–	–	Mostofa KMG et al., (unpublished data)
Autochthonous fulvic acid (C-like), algae origin, photo assimilations in River waters	Component 1, algae or phytoplankton	340/454	–	–	–	270/454	–	–	–	Mostofa KMG et al., (unpublished data)
Autochthonous fulvic acid (C-like), algae origin, under a 12:12 h light/dark cycle	Autochthonous	355/445	–	–	–	280/435	–	–	–	Zhao et al. (2009)
Autochthonous fulvic acid (C-like) or hydrophilic DOM, three phytoplankton: 12:12 h light/dark cycle	Autochthonous	340–350/420–440	–	–	–	260–280/ 425–445	–	–	–	Aoki et al. (2008)
Autochthonous fulvic acid (C-like) or hydrophobic acid, three phytoplankton: 12:12 h light/dark cycle	Autochthonous	330–350/ 435–440	–	–	–	250–290 /430–455	–	–	–	Aoki et al. (2008)
Autochthonous fulvic acid (C-like), streams, springs and thermokarsts, CPCRW, Alaska	Component 6, algae	Peak′	–	–	–	Peak′	–	–	–	Balcarczyk et al. (2009)

(contniued)

Table 2 (contniued)

Fluorescent components	Components identified and sources	Fluorescence properties								References
		Peak C-region				Peak A-region	Peak T-region	Peak T_{UV}-region		
		Peak C	Peak M_p	Peak M	Peak W			Major	Minor	
		EX/Em (nm)								
Autochthonous fulvic acid (C-like)?, fresh and Florida Bay waters, Florida coastal Everglades	Component 1, autochthonous	Peak′	–	–	–	Peak′	–	–	–	Chen et al. (2010)
Autochthonous fulvic acid (C-like), Lake Taihu	Component 1, phytoplankton	365/453	–	–	–	270/453	–	–	–	Zhang et al. (2009a)
Autochthonous fulvic acid (C-like)?, Sepetiba Bay, Brazil	Component 1, phytoplankton	350/400–450	–	–	–	275/400–450	–	–	–	Luciani et al. (2008)
Autochthonous fulvic acid (C-like), Sea ice, Baltic Sea coastal regions[RU]	Component 1	Peak′	–	–	–	Peak′	–	–	–	Stedmon et al. (2007a)
Autochthonous fulvic acid (C-like), deep waters of the Okhotsk Sea and North Pacific Ocean	Component 1	370/466	–	–	–	<260/466	–	–	–	Yamashita et al. (2010)
Autochthonous fulvic acid (C-like)?, Southern Ocean	Component 3	340/420	–	–	–	260/420	–	–	–	Wedborg et al. (2007)
Autochthonous fulvic acid (C-like), north Pacific and Atlantic oceans (Kauai model)	Component 8 (P8)[c], autochthonous	355/434	–	–	–	260/434	–	–	–	Murphy et al. (2008)
Humic-like, marine waters	Algae or phytoplankton	330–350/420–480	–	–	–	250–260/420–480	–	–	–	Coble (1996), Parlanti et al. (2000)
Autochthonous fulvic acid (M-like): biologically produced										
Marine humic-like, marine waters	Algae or phytoplankton	–	–	310–320/380–420	–	Peak′	–	–	–	Coble (1996), Parlanti et al. (2000)

(continued)

Table 2 (contniued)

Fluorescent components	Components identified and sources	Fluorescence properties								References
		Peak C-region				Peak A-region	Peak T-region	Peak T_{UV}-region		
		Peak C	Peak M_P	Peak M	Peak W			Major	Minor	
		EX/Em (nm)								
Autochthonous fulvic acid (M-like), surface waters (0–25 m), Lake Hongfeng, China	Component 2, algae or phytoplankton	–	–	295–300/396–422	–	235–240/296–422	–	–	–	Fu et al. (2010)[c]
Autochthonous fulvic acid (M-like), microbial assimilations of lake algae in river waters	Component 2, algae or phytoplankton	–	–	300/405	–	240/405	–	–	–	Mostofa KMG et al., (unpublished data)
Autochthonous fulvic acid (M-like), Lake Taihue	Component 4, phytoplankton	–	–	315/372	–	–	–	–	–	Zhang et al. (2009a)
Autochthonous fulvic acid (M-like)?, Lake Taihue	Component 3, phytoplankton	–	–	330/412	–	255/412	–	–	–	Zhang et al. (2009a)
Autochthonous fulvic acid (M-like), Lake Taihue[c]	Component 1, algae or phytoplankton	–	–	322/407	–	Peak′	–	–	–	Wang et al. (2007)
Autochthonous fulvic acid (M-like) or hydrophilic DOM, three phytoplankton: 12:12 h light/dark cycle	Autochthonous	–	–	320/385	–	–	–	–	–	Aoki et al. (2008)
Autochthonous fulvic acid (M-like) or hydrophobic acid, three phytoplankton: 12:12 h light/dark cycle	Autochthonous	–	–	330/385	–	–	–	–	–	Aoki et al. (2008)
Autochthonous fulvic acid (M-like), microbial assimilations of marine algae in Milli-Q waters	Component 2, algae or phytoplankton	–	–	290/400–410	–	Peak′	–	–	–	Parlanti et al. (2000)

(contniued)

Table 2 (contniued)

Fluorescent components	Components identi-fied and sources	Fluorescence properties								References
		Peak C-region				Peak A-region	Peak T-region	Peak T_{UV}-region		
		Peak C	Peak M_p	Peak M	Peak W			Major	Minor	
		EX/Em (nm)								
Autochthonous fulvic acid (M-like), microbial assimilations of marine algae in sea waters	Component 2, algae or phytoplankton	–	–	300–310/ 400–410	–	Peak′	–	–	–	Parlanti et al. (2000)
Autochthonous fulvic acid (M-like)?, streams, springs and thermokarsts, CPCRW, Alaska	Component 8, algae	–	–	Peak′	–	Peak′	–	–	–	Balcarczyk et al. (2009)
Autochthonous fulvic acid (M-like), microbially produced in mesocosm experiment[RU]	Component 3, algae or phytoplankton	–	–	295/398	–	Peak′	–	–	–	Stedmon and Markager (2005a)
Autochthonous fulvic acid (M-like)?, microbially produced in mesocosm experiment[RU]	Component 5, phytoplankton	–	–	345/434	–	Peak′	–	–	–	Stedmon and Markager (2005a)
Autochthonous fulvic acid (M-like), river and coastal waters	Component 4 autochthonous	–	–	305/378	–	<260/378	–	–	–	Yamashita and Jaffé (2008)
Autochthonous fuvic acid (M-like)?, Florida Bay waters, Florida coastal Everglades	Component 4	–	–	Peak′	–	Peak′	–	–	–	Chen et al. (2010)
Photobleached autochthonous fulvic acid (M-like)?, Liverpool Bay, Irish Sea	Component 4	–		295/358	–	<250/358	–	–	–	Yamashita et al. (2011)

(continued)

Table 2 (contniued)

Fluorescent components	Components identified and sources	Fluorescence properties								References
		Peak C-region				Peak A-region	Peak T-region	Peak T_{UV}-region		
		Peak C	Peak M_p	Peak M	Peak W			Major	Minor	
		EX/Em (nm)								
Autochthonous fulvic acid (M-like)?, Sepetiba Bay, Brazil	Component 3, phytoplankton	–	–	320/380–420	–	Peak′	–	–	–	Luciani et al. (2008)
Autochthonous fulvic acid (M-like)?, river, esturine and coastal marine waters	Component 4	–	–	Peak′?	–	240/384	–	–	–	Fellman et al. (2010)
Autochthonous fulvic acid (M-like), Urdaibai and Foz Estuaries, Iberian Peninsula	Component 3, autochthonous	–	–	320/388	–	Peak′	–	–	–	Santín et al. (2009)
Autochthonous fulvic acid (M-like), Horsens Estuary, Jutland Peninsula, Denmark[RU]	Component 6[c], autochthonous	–	–	320/400	–	250/400	–	–	–	Stedmon and Markager (2005b)
Autochthonous fulvic acid (M-like), Estuary of Horsens Fjord, Denmark[RUc]	Component 4[c], autochthonous	–	–	325/416	–	250/416	–	–	–	Stedmon et al. (2003)
Autochthonous fulvic acid (M-like)?, Horsens Estuary, Jutland Peninsula, Denmark[RU]	Component 5, allochthonous/agriculture		–	325/428	–	–	–	–	–	Stedmon and Markager (2005b)
Autochthonous fulvic acid (M-like), coastal waters, Isa Bay	Component 6, autochthonous	–	–	325/385	–	260/385	–	–	–	Yamashita et al. (2008)
Autochthonous fulvic acid (M-like), coastal shelf, South Atlantic Bight	Component 3, autochthonous	–	–	310/400	–	250/400	–	–	–	Kowalczuk et al. (2009)

(contniued)

Table 2 (contniued)

Fluorescent components	Components identi-fied and sources	Fluorescence properties								References
		Peak C-region				Peak A-region	Peak T-region	Peak T_{UV}-region		
		Peak C	Peak M_p	Peak M	Peak W			Major	Minor	
		EX/Em (nm)								
Autochthonous fulvic acid (M-like), deep waters of the Okhotsk Sea and North Pacific Ocean	Component 2	–	–	325/385	–	<260/385	–	–	–	Yamashita et al. (2010)
Autochthonous fulvic acid (M-like), north Pacific and Atlantic oceans (BWE7 model)	Component 2, autochthonous	–	–	315/418	–	Peak′	–	–	–	Murphy et al. (2008)
Autochthonous fulvic acid (M-like), north Pacific and Atlantic oceans (Kauai model)	Component 1 (P1), autochthonous	–	–	310/414	–	260/414	–	–	–	Murphy et al. (2008)
Autochthonous fulvic acid (M-like?, drinking water treatment plant[RU]	Component 6	–	–	300/406	–	<250/406	–	–	–	Baghoth et al. (2010)
Autochthonous fulvic acid (M-like)?, compost products solution	Component 1	–	–	330/410	–	230/410	–	–	–	Yu et al. (2010)
Autochthonous fulvic acid (M-like), pore waters, four lakes, China	Component 2, algae or phytoplankton	–	–	300–310/ 396–416	–	225–240/ 396–416	–	–	–	Li et al., Characteristics of sediment pore water dissolved organic matter in four Chinese lakes using EEM spectroscopy and PARAFAC mod-eling, (unpublished data)

(continued)

Table 2 (contniued)

Fluorescent components	Components identified and sources	Fluorescence properties								References
		Peak C-region				Peak A-region	Peak T-region	Peak T_{UV}-region		
		Peak C	Peak M_p	Peak M	Peak W			Major	Minor	
		EX/Em (nm)								
Protein-like substance										
Protein-like, streams, springs and thermokarsts, CPCRW, Alaska	Component 5, autochthonous	–	–	–	–	–	Peak′	Peak′	–	Balcarczyk et al. (2009)
Protein-like, sewerage drainage samples, Nanming River, China	Component 1	–	–	–	–	–	280/339–346	230/338–351	–	Mostofa et al. (2010)
Protein-like, washing samples collected after washing cloths	Component 1	–	–	–	–	–	280/344	235/348	–	Mostofa et al. (2010)
Protein-like, hydrophilic DOM fraction, three phytoplankton: 12:12 h light/dark cycle	Autochthonous	–	–	–	–	–	270–290/335–375	Peak′	–	Aoki et al. (2008)
Protein-like, hydrophobic acid fraction, three phytoplankton: 12:12 h light/dark cycle	Autochthonous	–	–	–	–	–	270–290/250–365	Peak′	–	Aoki et al. (2008)
Protein-like?, ground water, fresh and Florida Bay waters, Florida coastal Everglades	Component 8	–	–	–	–	–	Peak′	–	–	Chen et al. (2010)
Protein-like, coastal shelf, South Atlantic Bight	Component 6, autochthonous	–	–	–	–	–	290/356	250/356	–	Kowalczuk et al. (2009)
Protein-like?, Baltic Sea	Component 4, autochthonous	–	–	–	–	–	Peak′	Peak′	–	Stedmon et al. (2007a)

(contniued)

Table 2 (contniued)

Fluorescent components	Components identified and sources	Fluorescence properties								References
		Peak C-region				Peak A-region	Peak T-region	Peak T_{UV}-region		
		Peak C	Peak M_p	Peak M	Peak W			Major	Minor	
		EX/Em (nm)								
Protein-like, north Pacific and Atlantic oceans (BWE7 model)	Component 6, autochthonous	–	–	–	–	–	280/328	Peak′	–	Murphy et al. (2008)
Protein-like, north Pacific and Atlantic oceans (BWE7 model)	Component 7, autochthonous	–	–	–	–	–	300/338	240/338	–	Murphy et al. (2008)
Protein-like?, glacial ice samples, Antarctic and Arctic Ocean	Component 5, autochthonous	–	–	–	–	–	275/320	–	–	Dubnick et al. (2010)
Aromatic amino acids										
Tryptophan-like, soil and stream waters, temperate rainforest watersheds	Allochthonous	–	–	–	–	–	280/330–340	Peak	–	Fellman et al. (2009)
Tryptophan-like, Yasu River, Lake Biwa watershed, Japan	Component 2, autochthonous	–	–	–	–	–	280/344	230/344	–	Mostofa et al. (2005a)
Tryptophan-like, Occoquan Watershed (Northern Virginia, US)	Component 3, autochthonous	–	–	–	–	–	280/340	230/340	–	Holbrook et al. (2006)
Tryptophan-like, soil solution samples, forest and wetland soils, rainforest watersheds	Allochthonous	–	–	–	–	–	280/330–340	Peak	–	Fellman et al. (2008)
Tryptophan-like, Nanming River, China	Anthropogenic sources	–	–	–	–	–	275–280/333–351	225–235/338–351	–	Mostofa et al. (2010)
Tryptophan-like, urban sewerage samples[RU]	Component [illegible]	–	–	–	–	–	275/339	220/339	–	Guo et al. (2010)

(continued)

Table 2 (contniued)

Fluorescent components	Components identified and sources	Fluorescence properties								References
		Peak C-region				Peak A-region	Peak T-region	Peak T_{UV}-region		
		Peak C	Peak M_p	Peak M	Peak W			Major	Minor	
		EX/Em (nm)								
Tryptophan-like, drinking water treatment plant[RU]	Component 4, autochthonous	–	–	–	–	–	290/360	<250/360	–	Baghoth et al. (2010)
Tryptophan-like, municipal leachate samples	Component 3	–	–	–	–	–	280/340	230/340	–	Wu et al. (2011)
Tryptophan-like, Estuary of Horsens Fjord, Denmark[RU]	Component 5, autochthonous	–	–	–	–	–	280/368	240/368	–	Stedmon et al. (2003)
Tryptophan-like, microbially produced in mesocosm experiment[RU]	Component 6, algae- or phytoplankton	–	–	–	–	–	280/338	Peak′	–	Stedmon and Markager (2005a)
Tryptophan-like, Horsens Estuary, Jutland Peninsula, Denmark[RU]	Component 7, autochthonous	–	–	–	–	–	280/344	Peak′	–	Stedmon and Markager (2005b)
Tryptophan-like, river and coastal waters	Component 7, autochthonous	–	–	–	–	–	295/340	Peak′	–	Yamashita and Jaffé (2008)
Tryptophan-like, river, esturine and coastal marine waters	Component 7	–	–	–	–	–	280/330–340	Peak′	–	Fellman et al. (2010)
Tryptophan-like, Liverpool Bay, Irish Sea	Component 5, autochthonous	–	–	–	–	–	280/334	Peak′?	–	Yamashita et al. (2011)
Tryptophan-like, marine waters	Autochthonous	–	–	–	–	–	270–280/320–350	Peak′	–	Coble (1996), Parlanti et al. (2000)
Tryptophan-like, glacial ice samples, Antarctic and Arctic Ocean	Component 2, autochthonous	–	–	–	–	–	280/348	<250/348	–	Dubnick et al. (2010)
Tryptophan-like, compost products solution	Component 3	–	–	–	–	–	280/340	220/340	–	Yu et al. (2010)
Tryptophan-like, plant biomass, animal manure, and soils	Components 4, allochthonous	–	–	–	–	–	270/354	Peak′	–	Ohno and Bro (2006)

(contniued)

Table 2 (contniued)

Fluorescent components	Components identified and sources	Fluorescence properties								References
		Peak C-region				Peak A-region	Peak T-region	Peak T_{UV}-region		
		Peak C	Peak M_P	Peak M	Peak W			Major	Minor	
		EX/Em (nm)								
Tyrosine-like, streams, springs and thermokarsts, CPCRW, Alaska	Component 9, autochthonous	–	–	–	–	–	Peak′	Peak′	–	Balcarczyk et al. (2009)
Tyrosine-like, soil and stream waters, temperate rainforest watersheds	Allochthonous	–	–	–	–	–	275/304–306	Peak′	–	Fellman et al. (2009)
Tyrosine-like, soil solution samples, forest and wetland soils, rainforest watersheds	Allochthonous	–	–	–	–	–	275/304–306	Peak′	–	Fellman et al. (2008)
Tyrosine-like, Urdaibai and Foz Estuaries, Iberian Peninsula	Component 4, autochthonous	–	–	–	–	–	275/304	Peak′	–	Santín et al. (2009)
Tyrosine-like, microbially produced in mesocosm experiment[RU]	Component 4	–	–	–	–	–	275/306	Peak′	–	Stedmon and Markager (2005a)
Tyrosine-like, Horsens Estuary, Jutland Peninsula, Denmark[RU]	Component 8, autochthonous	–	–	–	–	–	275/304	Peak′	–	Stedmon and Markager (2005b)
Tyrosine-like, ground water, Fresh and Florida Bay waters, Florida coastal Everglades	Component 7	–	–	–	–	–	Peak′	Peak′	–	Chen et al. (2010)
Tyrosine-like, river, esturine and coastal marine waters	Component 8	–	–	–	–	–	275/304–306	Peak′	–	Fellman et al. (2010)
Tyrosine-like, coastal waters	Component 8 autochthonous	–	–	–	–	–	275/324	Peak′	–	Yamashita and Jaffé (2008)
Tyrosine-like, Liverpool Bay, Irish Sea	Component 6, autochthonous	–	–	–	–	–	275/302	Peak′?	–	Yamashita et al. (2011)

(continued)

Table 2 (contniued)

Fluorescent components	Components identified and sources	Fluorescence properties								References
		Peak C-region				Peak A-region	Peak T-region	Peak T_{UV}-region		
		Peak C	Peak M_p	Peak M	Peak W			Major	Minor	
		EX/Em (nm)								
Tyrosine-like, coastal waters, Isa Bay	Component 7	–	–	–	–	–	270/299	Peak′	–	Yamashita et al. (2008)
Tyrosine-like, coastal shelf, South Atlantic Bight	Component 5, autochthonous	–	–	–	–	–	270/332	Peak′	–	Kowalczuk et al. (2009)
Tyrosine-like, deep waters of the Okhotsk Sea and North Pacific Ocean	Component 4	–	–	–	–	–	275/306	Peak′?	–	Yamashita et al. (2010)
Tyrosine-like, north Pacific and Atlantic oceans (BWE7 model)	Component 1, autochthonous	–	–	–	–	–	275/300	Peak′	–	Murphy et al. (2008)
Tyrosine-like, marine waters	Autochthonous	–	–	–	–	–	275/310	Peak′	–	Coble (1996)
Tyrosine-like, water extractable from sugar maple leaves	Component 4	–	–	–	–	–	270/310	Peak′	–	Hunt et al. (2008)
Tyrosine-like, drinking water treatment plant[RU]	Component 7	–	–	–	–	–	270/306	Peak′?	–	Baghoth et al. (2010)
Tyrosine-like: plant biomass, animal manure, and soils	Components 5, allochthonous	–	–	–	–	–	273/309	Peak′	–	Ohno and Bro (2006)
Phenylalanine	Autochthonous	–	–	–	–	–	255–265/284–285	–	–	Yamashita and Tanoue (2003a)[b]
Phenylalanine-like?, glacial ice samples, Antarctic and Arctic Ocean	Component 1	–	–	–	–	–	265/306	Peak′	–	Dubnick et al. (2010)
Detergent-like or fluorescent whitening agents (FWAs)-like										
Detergents dissolved in Milli-Q waters, Nafine Chem Ind Ltd, China	Component 1	–	–	–	345/430–435	240/429–433	–	–	–	Mostofa et al. (2010), Mostofa KMG et al., (unpublished data)

(contniued)

Table 2 (contniued)

Fluorescent components	Components identified and sources	Fluorescence properties								References
		Peak C-region				Peak A-region	Peak T-region	Peak T_{UV}-region		
		Peak C	Peak M_p	Peak M	Peak W			Major	Minor	
		EX/Em (nm)								
Detergents dissolved in Milli-Q waters, Nafine Chem Ind Ltd, China	Component 2	–	–	–	–	–	–	–	225/287–289	Mostofa et al. (2010), Mostofa KMG et al., (unpublished data)
Detergents dissolved in Milli-Q waters, Nice Group Co Ltd, China	Component 1	–	–	–	345/430	240/427	–	–	–	Mostofa et al. (2010), Mostofa KMG et al., (unpublished data)
Detergents dissolved in Milli-Q waters, Nice Group Co Ltd, China	Component 2	–	–	–	–	–	–	–	225/287	Mostofa et al. (2010), Mostofa KMG et al., (unpublished data)
Detergent-like, detected in river water samples	Component 1	–	–	–	335–345/ 432–437	255/425–447	–	–	–	Mostofa et al. (2010), Mostofa KMG et al., (unpublished data)
Detergent-like, detected in river water samples	Component 2	–	–	–	–	–	–	–	225/289 –294	Mostofa et al. (2010), Mostofa KMG et al., (unpublished data)
Detergent-like, Izumi and Hinotsume, Kurose River, Japan	Component 1	–	–	–	350/437	250/437	–	–	–	Mostofa et al. (2005a)
Detergent-like, detected in sewerage samples	Component 1	–	–	–	335–345/ 432–437	240–250/ 425–443	–	–	–	Mostofa et al. (2010), Mostofa KMG et al., (unpublished data)

(continued)

Table 2 (contniued)

Fluorescent components	Components identified and sources	Fluorescence properties								References
		Peak C-region				Peak A-region	Peak T-region	Peak T_{UV}-region		
		Peak C	Peak M_p	Peak M	Peak W			Major	Minor	
		EX/Em (nm)								
Detergent-like?, urban sewerage samples[RU]	Component 4	–	–	–	340/422	230/422	–	–	–	Guo et al. (2010)
Detergent-like, detected in sewerage samples	Component 2	–	–	–	–	–	–	–	225–230/ 291–296	Mostofa et al. (2010), Mostofa KMG et al., (unpublished data)
Detergent-like?, drinking water treatment plant[RU]	Component 5	–	–	–	340/440	250/440	–	–	–	Baghoth et al. (2010)
Algae or phytoplankton										
Algae (green) or bacteria in Milli-Q or river water samples, collected from lake surface waters	Component 1	–	–	–	–	–	280–285/ 340–346	230/346	–	Mostofa KMG et al., (unpublished data)[a]
Algae (green) or bacteria in Milli-Q or river water samples, collected from lake surface waters	Component 2	–	–	–	–	–	270/327–336	230/327–336	–	Mostofa KMG et al., (unpublished data)[a]

RU indicates the Raman Unit (nm-1) calibration of fluorescence intensity of the samples studied whereas excitation and emission wavelength is much differed from standard quinine sulfate unit (QSU) calibration (Mostofa et al. 2005b)

Peak M_p means fluorescence Ex/Em maxima of fulvic acid which is photobleached by photochemical processes or by any other natural processes

Peak M indicates the allochthonous and autochthonous fulvic acid (M-like)

[a]Indicates river EEM data is deducted from sample before PARAFAC modeling on the matrix data

[b]Ranges expresses the authentic standard at various concentrations (1–5 mg L^{-1}) and mechanical reproducibility

[c]Indicates that the components are identified using the PARAFAC model on the original EEM of that published paper

Peak′ means the peak is at this region, but not specified in the paper; *UD* unpublished data

? indicates the values mentioned here are different than in the paper

2009a, 2005a, 2010, 2007b; Mopper and Schultz 1993; Yamashita and Tanoue 2003a; Komaki and Yabe 1982; Nakajima 2006; Baker 2005; Chen et al. 2003; Burdige et al. 2004; Fu et al. 2006). Peak T_{UV}-region depicts the shorter (UV) wavelength ranges at Ex/Em = 215–260/280–380 nm, which includes mostly the secondary fluorescence peaks of various fluorescent organic substances such as proteins, aromatic amino acids (tryptophan-like, tyrosine-like and phenylalanine-like), algae, detergent component, phenol-like compounds, naphthalene, *o*-cresol, *p*-cresol, *p*-hydroxy benzaldehyde, *p*-hydroxy acetophenone, 1,4-dichlorobenzene, and 4-biphenyl carboxaldehyde (Table 1) (Mostofa et al. 2009a, 2010; Mopper and Schultz 1993; Yamashita and Tanoue 2003a; Nakajima 2006; Baker 2005; Baker and Curry 2004; Chen et al. 2003; Burdige et al. 2004; Fu et al. 2006). Note that the fluorescence intensity expressed as QSU (quinine sulphate unit) is considered to identify the authentic or valid excitation–emission wavelength maxima. In contrast, the maxima of the calibrated fluorescence intensity obtained using the Raman Unit (RU: nm^{-1}) method can be significantly shifted, particularly at the peak C-region (Mostofa et al. 2005b). Such issues are discussed in details later, in the fluorescence intensity normalization section.

2.3 PARAFAC Modeling in FDOM Study

Parallel factor (PARAFAC) modeling is a three-way multivariate analysis that can be applied on an additive mixture of fluorescence signals obtained from excitation–emission matrix spectra. PARAFAC is capable of isolating and quantifying the individual fluorescence component signals in terms of fluorescence intensity of FDOM in natural waters or in mixtures.

From the PARAFAC model (Harshman 1970) it can be implied that for any fluorophore, the emission intensity at a specific wavelength j that corresponds to excitation at the wavelength k can be expressed as follows (Eq. 2.7):

$$x_{jk} = ab_jc_k \tag{2.7}$$

where x_{jk} is the intensity of the light at the emission wavelength j and excitation wavelength k, a is the concentration (in arbitrary units) of the analyte, b_j is the relative emission at the wavelength j, and c_k is the relative amount of light absorbed at the excitation wavelength k. For any number of analytes and samples, the PARAFAC model can be developed into a set of trilinear terms and a residual array as (Eq. 2.8) (Stedmon et al. 2003)

$$x_{ijk} = \sum_{f=1}^{F} a_{ij}b_{jf}c_{kf} + \varepsilon_{ijk},\ i = 1,\ldots,I;\ j = 1,\ldots,J;\ k = 1,\ldots,K \tag{2.8}$$

where x_{ijk} is the fluorescence intensity of the ith sample at the emission wavelength j and excitation wavelength k. a_{if} is directly proportional to the concentration (in arbitrary units) of the analyte f in sample i. b_{jf} is directly proportional to the quantum efficiency of fluorescence of the analyte f at the emission wavelength j. Similarly, c_{kf} is linearly related to the specific absorption coefficient at excitation wavelength k. F is the number of components in the model and ε_{ijk} is the residual matrix that indicates the variability not accounted for by the model.

Three steps are followed before running EEM data in the PARAFAC model (Bro 1997; Stedmon et al. 2003). First, the Milli-Q water blank is subtracted from every sample. Second, all values of the Raleigh light scattering are properly eliminated from the data of sample's EEMs to avoid any effect on the component numbers. Third, non-negative constraints are applied in the PARAFAC modeling to avoid negative values of excitation, emission and concentration in any model component.

By applying bilinear models to the EEM data (Bro 1999), it is possible to judge the residuals of the fit. If systematic variation is left in, the residuals that indicate more components can be extracted. If a plot of the residual sum of squares versus the number of components sharply flattens out for a certain number of components, this indicates the true number of components. To calculate variance-like estimators, the degrees of freedom are expressed as follows (Eq. 2.9):

$$\mathrm{Dof}\,(F) = IJK - F\,(I + J + K - 2) \tag{2.9}$$

for a trilinear PARAFAC model where I, J and K are the dimensions of the first, second and third mode, respectively, and F is the number of components in the model (Bro 1997). In the PARAFAC model, it is important to select the true number of components, ranging from 1 until the proper components are identified. The true number (Bro 1997) of components is determined on the basis of the residuals, the core consistency (that must be 100 %), the number of iterations (which should be near zero) and the findings of the EEM spectra for the respective samples, with reference to the various standard substances. The loadings of the emission and excitation wavelengths are often used to check the variability of the selected components (Stedmon et al. 2003). The three key ways of determining the true number of components are (Bro 1997): (i) Spilt-half experiments, (ii) judging residuals, and (iii) comparison with the external knowledge of the original EEM images and data being modeled. The other most important factors that one needs to know before PARAFAC modeling are: (i) Selection of the proper excitation–emission wavelength ranges for the measured samples. Such ranges significantly affect the shape and images of the isolated components, as well as the reproducibility of fluorescence intensity. Depending on the nature of FDOM components in natural waters, the most chosen wavelength ranges could be 220–400 nm for excitation and 280–550 nm for emission. (ii) Similar types of samples must be modeled individually. For example, individual modeling should be made of upstream rivers with merely natural sources of DOM, downstream rivers with a variety of DOM sources, and lake waters with both autochthonous and allochthonous sources of DOM. PARAFAC can identify the key fluorescent components in DOM, but it cannot isolate the minor ones that remain as a residue (Stedmon et al. 2003).

2.3.1 Characterization of FDOM Using EEM in Combination with PARAFAC

PARAFAC modeling, a three-way method with its origin in psychometrics (Harshman 1970; Caroll and Chang 1970), can be effectively applied to isolate the EEMs of either an aqueous mixture of organic components or of natural DOM into their individual fluorescence components (Bro 1997, 1998, 1999; Ross et al. 1991; Jiji et al. 1999; Baunsgaard et al. 2000, 2001; da Silva et al. 2002; Stedmon et al. 2003, 2007a, 2007b; Cory and McKnight 2005; Mostofa et al. 2010; Wedborg et al. 2007; Hiriart-Baer et al. 2008; Hunt et al. 2008; Luciani et al. 2008; Kowalczuk et al. 2009; Ohno et al. 2009; Zhao et al. 2009; Baghoth et al. 2010; Chen et al. 2010; Dubnick et al. 2010; Fellman et al. 2008, 2009, 2010; Guo et al. 2010; Singh et al. 2010; Yu et al. 2010; Wu et al. 2011; Yamashita et al. 2010, 2011). The EEMs often involve various types of overlapping peaks because of the natural DOM composition, which makes it difficult to identify the fluorescent component peaks and their intensities. PARAFAC, a statistical modeling approach, can isolate the fluorescent components from EEMs and then determine the concentration of the fluorescing compounds.

The EEM in combination with PARAFAC analysis has been applied to separate and identify the various DOM components and their concentrations in rivers and freshwaters (Mostofa et al. 2010; Chen et al. 2010; Fellman et al. 2009, 2010; Holbrook et al. 2006; Hua et al. 2007; Balcarczyk et al. 2009), lakes (Cory and McKnight 2005; Chin et al. 1994; Gron et al. 1996), wetlands (Fellman et al. 2008; Holbrook et al. 2006), estuaries (Stedmon et al. 2003; Hall et al. 2005; Stedmon and Markager 2005a, 2005b; Fellman et al. 2010; Santín et al. 2009), bays and marine waters (Luciani et al. 2008; Fellman et al. 2010; Singh et al. 2010; Yamashita et al. 2010, 2011, 2008; Murphy et al. 2008), lake sediment pore waters (Li et al., Characteristics of sediment pore water dissolved organic matter in four Chinese lakes using EEM spectroscopy and PARAFAC modeling, unpublished data), wastewaters (Baghoth et al. 2010; Guo et al. 2010; Wu et al. 2011), landfill leachate (Baker and Curry 2004; Lu et al. 2009) and soil (Ohno and Bro 2006; Fellman et al. 2008, 2009). PARAFAC is also adopted in a wide range of different applications, such as the identification of autochthonous fulvic acids of algal origin that can be distinguished from allochthonous fulvic acid (Mostofa et al. 2009b; Stedmon et al. 2007b; Mostofa KMG et al., unpublished data; Zhang et al. 2009a; Li et al., Characteristics of sediment pore water dissolved organic matter in four Chinese lakes using EEM spectroscopy and PARAFAC modeling, unpublished data). PARAFAC can also be used to trace photoinduced and microbial changes in DOM components and their intensities (Cory and McKnight 2005; Stedmon and Markager 2005a; Stedmon et al. 2007a; Mostofa et al. 2010), in water source categorization (Hua et al. 2007) and in correlation with water quality parameters (Hayase and Tsubota 1985), in the identification of changes in the fulvic acid redox state (Fulton et al. 2004; Cory and McKnight 2005), and finally in studying interactions between trace metals and DOM (Yamashita and Jaffé 2008).

The most common fluorescent DOM components isolated from DOM in natural waters using PARAFAC modeling are allochthonous fulvic acids and humic acids of vascular plant origin, autochthonous fulvic acids (termed C-like and M-like based on the component's peak positions) of algal (or phytoplankton) origin, aromatic amino acids (tryptophan, tyrosine and phenylalanine), FWAs (DAS1 and DSBP), green algae, chlorophyll *a* (Chl *a*) and chlorophyll *b* (Table 2) (Coble 1996; Parlanti et al. 2000; Yamashita and Tanoue 2003a; Mostofa et al. 2005a, Mostofa et al. 2005b, 2010; Stedmon et al. 2003, 2007a; Stedmon and Markager 2005a, 2005b; Ohno and Bro 2006; Mostofa KMG et al., unpublished data; Fu et al. 2010; Zhang et al. 2009a; Wedborg et al. 2007; Hunt et al. 2008; Luciani et al. 2008; Kowalczuk et al. 2009; Zhao et al. 2009; Baghoth et al. 2010; Chen et al. 2010; Dubnick et al. 2010; Fellman et al. 2008, 2009, 2010; Guo et al. 2010; Singh et al. 2010; Yu et al. 2010; Wu et al. 2011; Yamashita et al. 2010, 2011; Holbrook et al. 2006; Balcarczyk et al. 2009; Santín et al. 2009; Murphy et al. 2008; Yamashita et al. 2008; Li et al., Characteristics of sediment pore water dissolved organic matter in four Chinese lakes using EEM spectroscopy and PARAFAC modeling, unpublished data; Yamashita and Jaffé 2008; Aoki et al. 2008; Wang et al. 2007). These fluorescent components generally exhibit fluorescence at peak C-, peak A-, peak T- and T_{UV}-regions (Table 2). Their fluorescence properties are depicted in Sect. 2.5, as a function of the molecular structure of the respective components.

2.4 Fluorescence Intensity of a Molecule and its Normalization

A fluorescent molecule has one, two or several fluorescence peaks, and each peak can be denoted as fluorophore or fluorochrome. For example, fulvic acid has two fluorescence peaks in EEM spectra, namely peak C at longer wavelength and peak A in the shorter-wavelength UV region. The fluoresce intensity of fulvic acid is higher at peak A-region compared to the peak C-region (Mostofa et al. 2009a, 2005a, 2010; Coble 1996; Yamashita and Jaffé 2008). On the other hand, humic acid has several fluorescence peaks such as peak M, peak C and peak A. The peak A-region has usually the highest fluorescence intensity (Mostofa et al. 2009a, 2005a, 2010; Coble 1996; Yamashita and Jaffé 2008). Tryptophan amino acid has two fluorescence peaks: peak T at relatively longer wavelengths and peak T_{UV} in the shorter-wavelength UV region (Mostofa et al. 2009a, 2005a, 2010; Coble 1996; Yamashita and Jaffé 2008). The latter has usually the highest fluorescence intensity. The fluorophore or flurochrome at peak C or peak A of humic and fulvic acids are the result of the contribution of several fluorophores, because the macromolecular structures of these compounds generally include a number of functional groups (Mostofa et al. 2009a; Senesi 1990a; Malcolm 1985). Tryptophan amino acid has only one florophore at each peak position (T and T_{UV}) that can be denoted as single fluorophore.

The fluorescence intensity of a sample is normalized to monitor the stability of the light (or energy) emitted by the xenon lamp in the fluorometer, to compare with former published results and is vital for theoretical and technical fluorescence measurement, which has been expressed by means of three methods in earlier studies. (i) The typically used standard quinine sulfate method is the fluorescence intensity normalized to that of aqueous solutions of quinine sulfate monohydrate (at ppm or ppb levels) in either 0.05–0.1 M or 0.1 N solution of sulfuric acid (H_2SO_4). However, aquatic scientists use different scaling units to express the fluorescence intensity, such as millifluorescence (mFI) (Dorsch and Bidleman 1982; Hayase et al. 1987), the fluorescence unit (flu) (Chen and Bada 1992), Raman-normalized quinine sulfate equivalents (QSE) (Coble et al. 1998; Kowalczuk et al. 2009), and the quinine sulfate unit (QSU) (Coble 1996; Mopper and Schultz 1993; Yamashita and Tanoue 2003a; Mostofa et al. 2005a; Nagao et al. 2003; Burdige et al. 2004; Zhang et al. 2009a, 2009b; Coble et al. 1993; Obernosterer and Herndl 2000). (ii) The arbitrary unit method is the direct fluorescence intensity that is primarily detected by the fluorescence spectrophotometer (Mayer et al. 1999; Cory and McKnight 2005; Fu et al. 2007, 2006; Baker and Curry 2004; Chen et al. 2003; Klapper et al. 2002; Yue et al. 2006). The Raman peak intensity of Milli-Q water at Ex/Em = 348 or 350 or 275/303 nm over the analysis period (or the QS solution as mentioned before) is used to monitor the stability of light emitted by the xenon lamp in the fluorometer. (iii) The Raman Unit (RU) method is the corrected fluorescence intensity, where the Raman signals are corrected by the baseline and integrated over the entire Raman peak for each excitation wavelength. Then, the fluorescence intensities are divided by the Raman area for the corresponding excitation wavelength to obtain a RU (nm^{-1}) (Determann et al. 1994, 1996; Stedmon et al. 2003; Fulton et al. 2004; Mostofa et al. 2005b; Matthews et al. 1996; Nieke et al. 1997; Hayakawa et al. 2003; Yoshioka et al. 2007; Huguet et al. 2009). The RU calibration processes are difficult at the shorter wavelength regions due to noise, whilst the fluorescence intensities at longer excitation wavelengths tend to be enhanced (Mostofa et al. 2005b). Because of RU calibration on fluorescence intensities over all the excitation wavelengths, it is possible to have artifacts that produce unusual fluorescent components that are not shown in the original EEM spectra.

For example, the results of PARAFAC modeling on QSU and RU EEM data of upstream (Nishi-Mataya) and lake waters (Lake Biwa) show the presence of fulvic acid-like substance (components 1 and 2) with two fluorescence peaks for component 1, at Ex/Em = 320/442 (peak C) and 255/442 nm (peak A) (Fig. 2a). There is no specific peak for component 2, which only has strong fluorescence intensity at Ex/Em = 225–230/428 nm (Fig. 2b) in QSU of upstream waters. Photobleached fulvic acid-like compounds have two fluorescence peaks at Ex/Em = 310/450 nm (peak C) and 250/450 nm (peak A) (component 1, Fig. 2e), while component 2 is associated to photobleached autochthonous fulvic acid-like material with a weak peak at Ex/Em = 280/442 nm (component 2, Fig. 2f) in QSU of lake water. On the other hand, in RU units the respective fluorescent components are composed of the two fluorescence peaks at Ex/Em = 370/474 nm (peak C) and 270/474 nm (peak A) for component 1 (Fig. 2c). Component 2 has a peak at Ex/Em = 330/427 nm (peak C) in upstream waters. In lake waters, the fluorescence peaks for component 1 are 350/461 nm and 260/461 nm (Fig. 2g), and the peaks for component 2 are

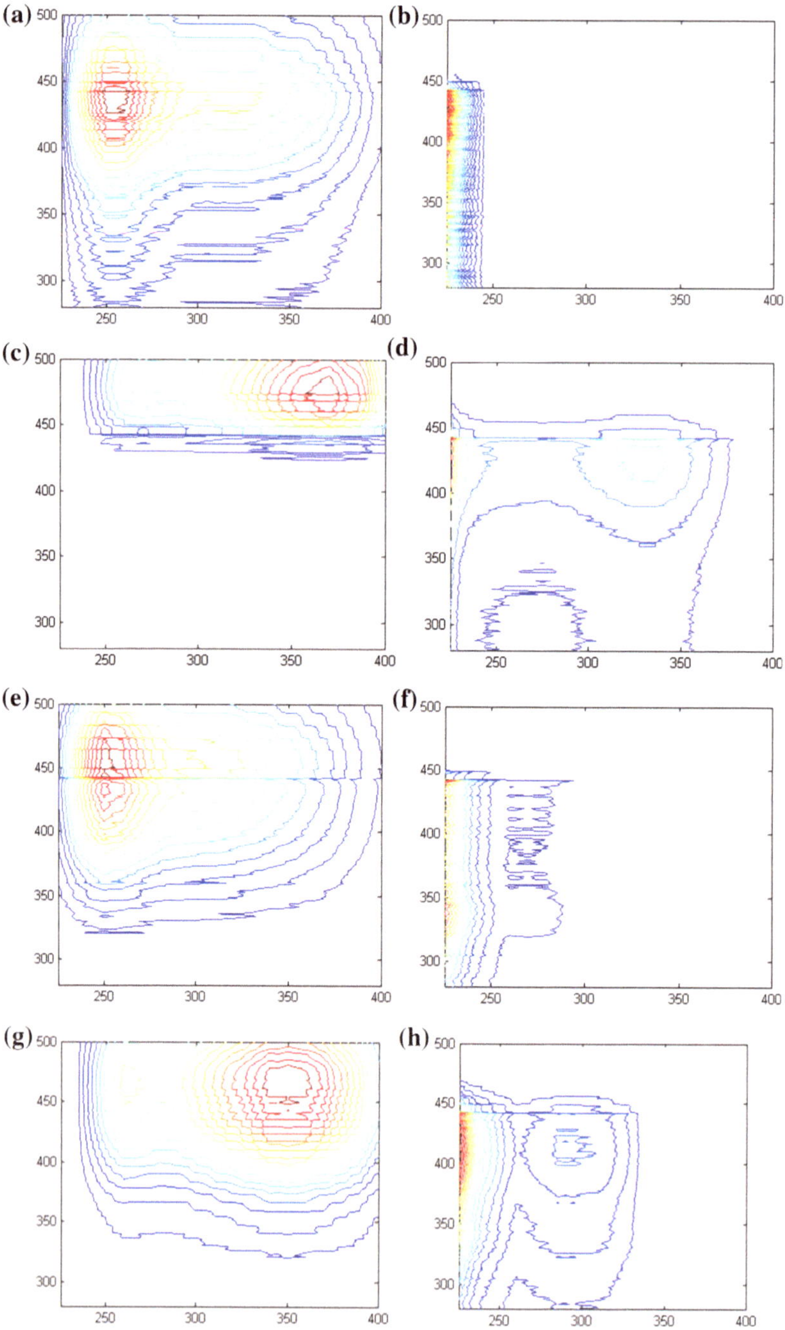

Fig. 2 Differences in the EEM images of fluorescent components identified using PARAFAC modeling on a.u. (or QSU) calibration (**a**, **b** Nishi-Mataya upstream and **e**, **f**: Lake Biwa surface waters) and Raman Unit calibration (**c**, **d** Nishi-Mataya upstream and **g**, **h**: Lake Biwa surface waters). PARAFAC analysis is conducted on earlier published and their respective a.u. *Data source* Mostofa et al. (2005b)

295/409 nm and 230/409 nm (Fig. 3h). The results show that the excitation–emission (Ex/Em) wavelengths of both peaks C and A in fulvic acid-like substances are shifted in the RU calibration compared to QSU. The fluorescence intensities of peak C in RU calibration are higher than in the peak A-region, which is entirely opposite to QSU calibration. The fulvic acid-like components in QSU of upstream waters (Fig. 2a, b) are characteristically similar to standard Suwannee River Fulvic Acid (Fig. 2a, b). The fulvic acid-like component (component 1) and autochthonous fulvic acid-like component (component 2) in QSU data of lake waters are similar to their respective photo-bleaching components, which have been detected in lake surface waters during the summer stratification period (Mostofa et al. 2005b). It can be noted that the PARAFAC modeling has been carried out on published data of monthly samples collected at Nishi-Mayata upstream (April, May, July, and August) and Lake Biwa surface waters (2.5, 10 and 20 m depth for April, July, August and September) (Mostofa et al. 2005b). The results indicate once more that both the excitation–emission wavelength peaks and their respective fluorescence intensities are significantly changed when using RU calibration. Therefore, it is strongly recommended that the calibration of the fluorescence intensity data of EEM spectra is carried out using either the QSU method or arbitrary units, avoiding the RU calibration method. It is also recommended that before measurement of the samples, Xe-light in the fluorescence spectrophotometer is corrected according to the instrument's guidelines using a Rhodamine B solution.

2.5 EEM Properties and Molecular Characteristics of Key FDOM Components Identified by PARAFAC

The molecular, chemical and EEM properties of the various FDOM components in natural waters are discussed below.

Allochthonous Fulvic Acids (C-like, A-like and M-like)

Standard Suwannee River Fulvic Acid (SRFA) is composed of two fluorescent components identified by PARAFAC modeling on its EEM. The first component is denoted as allochthonous fulvic acid (C-like), which includes two fluorescence peaks at Ex/Em = 295–410/439–520 nm (peak C-region) and at Ex/Em = 240–270/439–520 nm (peak A-region) (Figs. 2a, 3a; Table 2). The peak A of the allochthonous fulvic acid (C-like) often has higher fluorescence intensity (by ~1.30–3.0 times) compared to the peak C-region. Note that the Raman Unit (RU) calibration often returns longer excitation wavelengths, particularly at the peak C-region, compared to standard quinine sulfate (or other) calibration. The allochthonous fulvic acid (C-like) is identified at Ex/Em = 325–340/442–462 nm and 250–260/450–451 nm for standard SRFA dissolved in Milli-Q waters. The corresponding wavelengths are 345/452 and 255/451 nm for SRFA dissolved in seawater; 300–340/419–467 and 240–280/427–468 nm for fulvic acid extracted from rivers and lakes; 300–340/423–464 and 235–260/425–464 nm for various upstream and rivers, but not in the water Raman Unit (RU: nm^{-1}) calibration (330–380/420–467 and 240–260/4420–467 nm, respectively); 295–310/443–464 and 250–260/443–464 nm in lakes; 305/439 and 260/439 nm in estuaries, which shifts to 360–385/478–504 and 250–270/478–504 nm in RU calibration; 320–325/422–454 and <250/422–454 nm in bay and marine waters (but one finds 340–410/440–520 and 250/440–520 nm upon RU calibration in bay waters from the Barataria Basin); 300–305/449 and 250/449 nm in irradiated river waters; 315/429 nm in water extracted from sugar maple leaves; 315/447 and ~250/447 nm in plant biomass, manure and soil (Tables 1, 2) (Stedmon et al. 2003; Stedmon and Markager 2005b; Ohno and Bro 2006; Mostofa et al. 2010; Mostofa KMG et al., unpublished data; Nakajima 2006; Hunt et al. 2008; Kowalczuk et al. 2009; Baghoth et al. 2010; Chen et al. 2010; Dubnick et al. 2010; Fellman et al. 2010; Guo et al. 2010; Singh et al. 2010; Yamashita et al. 2010, 2011; Balcarczyk et al. 2009; Santín et al. 2009). Note that longer excitation–emission maxima have been observed at the C-region for SRFA dissolved in sea water compared to Milli-Q water. This is presumably linked to the formation of complexes of trace elements in seawater with the functional groups (or fluorophores) bound in SRFA. Complex formation can significantly enhance electron promotion at peak C from the ground to the excited state by longer wavelength energy. The lower excitation energy can shift the Ex/Em maxima to longer wavelengths. This will be explained in detail in the section that deals with the effect of salinity.

The second component is denoted as allochthonous fulvic acid (A-like) and is typically composed of a strong fluorescence shoulder (or peak) at Ex/Em = 225–250/413–448 nm (peak A-region) and of a minor peak at Ex/Em = 280–295/414–442 nm (peak C-region) (Fig. 3b). The allochthonous fulvic acid (A-like) is identified at 230/441 nm (peak A-region) in SRFA dissolved in Milli-Q waters; 225–230/414–442 nm (peak C-region) and 285–295/414–442 nm (minor peak at peak C-region) in upstream waters (Figs. 2b, 3b); 225/432–442 nm in lakes; 240–250/416–448 nm in estuaries, and <260/(not mentioned) nm in marine waters (Table 2) (Stedmon et al. 2003; Stedmon and Markager 2005b; Mostofa KMG et al., unpublished data; Yamashita et al. 2010; Balcarczyk et al. 2009).

The third component of allochthonous fulvic acid exhibits fluorescence excitation–emission maxima at Ex/Em = 285–310/387–429 nm (peak C-region) and Ex/Em = 230–260/387–429 nm (peak A-region). The fluorescence intensity at peak A-region is >1.50 times higher than at C-region (Fig. 3c; Table 2). The Ex/Em maxima of this component at peak C-region occur at relatively shorter wavelengths compared to those of the allochthonous fulvic acid (C-like). Therefore, the third component can be denoted as allochthonous fulvic acid (M-like). Allochthonous fulvic acid (M-like) has been detected at Ex/Em = 285–310/387–429 nm at peak C-region and 230–240/387–429 nm at peak A-region in the waters of the Yellow River and Heilongjiang (Amur) River watershed (China); 305/396 and 240/396 nm in Occoquan Watershed (USA); 305/428 and <240–260/414–428 nm in marine waters; 320/410 and 250/410 nm in drinking water treatment plants, and 312/417

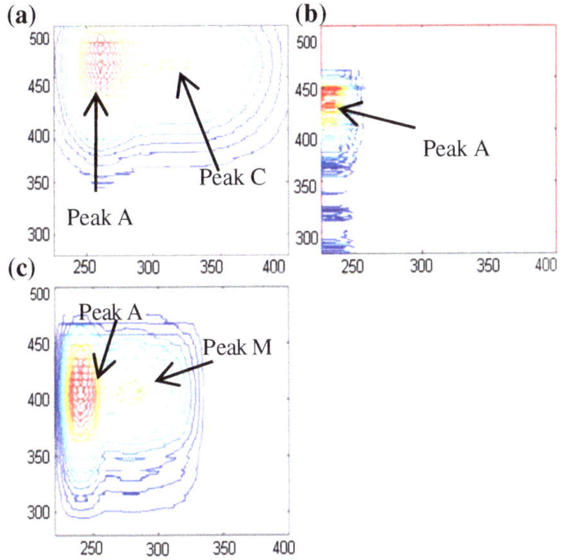

Fig. 3 The fluorescent components of allochthonous fulvic acid (C-like) (**a**) and allochthonous fulvic acid (A-like) (**b**) of standard Suwannee River Fulvic Acid's aqueous samples and of allochthonous fulvic acid (M-like) (**c**) of upstream waters (Yellow River, China) identified using PARAFAC modeling on their respective EEM spectra. Allochthonous fulvic acids (C-like and A-like) are similar to those of stream allocthonous fulvic acid (C-like and A-like) (Fig. 2a, b) of upstream water samples (Nishi-Mataya upstream). PARAFAC analysis is conducted on earlier published and their respective a.u. data (*Data source* Mostofa et al. 2005a; Mostofa KMG et al., unpublished data).

and 240/417 nm in water extracted from sugar maple leaves (Fig. 3c; Table 2) (Mostofa KMG et al., unpublished data; Hunt et al. 2008; Baghoth et al. 2010; Chen et al. 2010; Fellman et al. 2010; Holbrook et al. 2006; Yamashita and Jaffé 2008). The origin of this component is presumably located in the terrestrial soil ecosystem. Experimental results show that allochthonous fulvic acid (M-like) is entirely decomposed photolytically in the upstream waters and in the main channel waters of the Yellow River, because this component is minor (~5 %) compared to the major component (~89 %) of allochthonous fulvic acid (C-like) in the total Yellow River DOM (Mostofa KMG et al., unpublished data). On the other hand, the allochthonous fulvic acid (M-like) undergoes complete microbial degradation after 12 days of dark incubation in both upstream and main channel filtered waters of the Yellow River at room temperature. This indicates that the allochthonous fulvic acid (M-like) is both photolytically and microbially labile in natural waters.

The molecular structure of fulvic acid is not yet known because of the complicated chemical composition and relatively large molecular size. The molecular weight of fulvic acid is approximately 2310 Da (Chin et al. 1994); it is generally composed of relatively few aromatic groups, which yield 14–20 % of aromatic carbon compared to 30–51 % for humic acid (Malcolm 1985; Steelink 2002; Gron et al. 1996). The fluorophores associated with the low molecular weight fraction of fulvic acid (<10 kDa) exhibit relatively high fluorescence intensity, which gradually decreases with an increase in molecular weight (Hayase and Tsubota 1985; Levesque 1972; Gosh and Schnitzer 1980; McCreary and Snoeyink 1980; Visser 1984). The fulvic acid fluorophores have relatively smaller functional groups and show higher fluorescence intensity at peak C- and A-regions compared to humic acids (Hayase and Tsubota 1985; Mostofa et al. 2009a, 2005a).

Photobleached Allochthonous Fulvic Acids

Photobleached fulvic acid generally arises from the degradation of fluorophores bound to the allochthonous fulvic acid (generally C-like), which thus results in the excitation–emission maxima of peak C having relatively shorter wavelengths compared to those of the initial fluorescence maxima (Fig. 3d, e). Photoinduced degradation causes decomposition of allochthonous fulvic acids (C-like, A-like and M-like) in natural waters (Fig. 3d, e) (Mostofa et al. 2005a; Mostofa et al. 2005b, 2010, 2007a; Moran et al. 2000). In field observations of the waters of Lake Biwa, the fluorescence excitation–emission maxima at the peak C-region are detected at

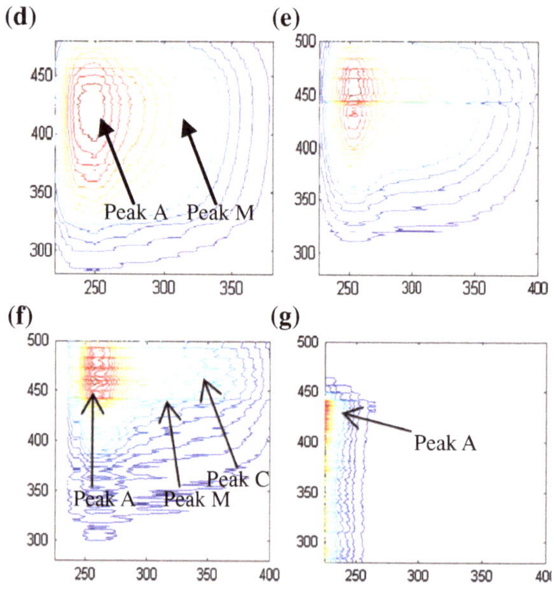

d, **e** The fluorescent components of standard Suwannee River Humic Acids in EEM data of its aqueous samples identified using PARAFAC modeling. PARAFAC analysis is conducted on earlier published and their respective a.u. data. *Data source* Mostofa et al. (2005a). **f**, **g** The fluorescent component of photobleached fulvic acids of irradiated river waters (3 h irradiation by midday sunlight, Nanming River, China) and lake surface waters during the summer stratification period (2.5–20 m, Lake Biwa, Japan) identified using PARAFAC modeling. *Data source* Mostofa et al. (2010, 2005b).

shorter wavelengths (Ex/Em = 295–310/443–450 nm) in surface waters (0–20 m) compared to the deep water layers (40–80 m) (Ex/Em = 300–310/444–464 nm). Furthermore, maxima are 320–325/422–454 nm in sea waters and 300–305/449 and 250/449 nm in irradiated river waters (Table 2) (Mostofa KMG et al., unpublished data; Mostofa et al. 2005b; Yamashita et al. 2011). The excitation–emission maxima of the fluorescence peak C are shifted in irradiated samples from longer to shorter wavelength regions, which gives a blue-shift of the peak position (Mostofa et al. 2005a, 2007a; Moran et al. 2000). Such blue-shift phenomenon is the photoinduced result of the mineralization of fluorophores that are present in the molecular structure of fulvic acids. Irradiation decomposes the allochthonous fulvic acid (A-like and M-like) entirely (Table 2) (Mostofa KMG et al., unpublished data). Photobleached fulvic acid is generally detected in the surface waters of rivers, lakes, estuaries and oceans (Table 2) (Brooks et al. 2007; Mostofa et al. 2007a, 2005b; Garcia et al. 2005; Skoog et al. 1996; Moran et al. 2000; Osburn et al. 2009; Lepane et al. 2003; Abboudi et al. 2008; Poiger et al. 1999; Zhang et al. 2009b).

Allochthonous Humic Acids (C-like, A-like and M-like)

The standard Suwannee River Humic Acid (SRHA) has three fluorescent components, namely allochthonous humic acid (C-like), humic acid (A-like) and humic acid (M-like). They can be identified using PARAFAC modeling of EEM spectra in a variety of waters (Fig. 3f, g; Table 2). The allochthonous humic acid (C-like) shows three fluorescence peaks, of which two are at Ex/Em = 285–340/460–480 nm (shorter wavelength) and at 350–405/480–508 nm (longer wavelength, peak C-region). The third peak is located at Ex/Em = 240–270/460–508 nm in the peak A-region (Fig. 3f; Table 2). Allochthonous humic acid (C-like) has been detected at Ex/Em = 320–350/461–498 and 300/461 nm (peak C-region) and at 255/461 nm (peak A-region) in standard SRFA dissolved in Milli-Q water; at 295–310/423–464 and 255/464 nm in extracted humic acid; at 330/456–480 and <250/450–460 nm in river and soil waters; at 385/>500 and 256/>500 nm in estuaries; at 300–340/>500–510, 290–405/>500–510 and <260–270/>508 nm in bay and marine waters; at 370–380/490–498 and <260/490–498 nm in the north Pacific and Atlantic ocean; at 360/458–480, 300/458 and 250–260/458–480 nm in drinking water treatment plants and municipal wastes; at 370/440 and <250/440 nm in soil; at 351/459 and 240/459 nm in water extracted from sugar maple leaves; finally, at 350–360/460–480 nm in plant biomass, manure and soil (Fig. 3f; Table 2) (Coble 1996; Mostofa et al. 2005a; Ohno and Bro 2006; Baker 2005; Hunt et al. 2008; Kowalczuk et al. 2009; Baghoth et al. 2010; Chen et al. 2010; Fellman et al. 2008, 2010; Guo et al. 2010; Yu et al. 2010; Wu et al. 2011; Yamashita et al. 2011; Balcarczyk et al. 2009; Santín et al. 2009; Murphy et al. 2008; Yamashita and Jaffé 2008).

The allochthonous humic acid (A-like) can exhibit a strong shoulder (not often a clear peak) at peak A-region in a wide range of emission wavelengths.

The C-region peak does not appear because of relatively strong fluorescence intensity at the peak A-region, which explains why this component can be denoted as humic acid (A-like) (Fig. 3g; Table 2). The allochthonous humic acid (A-like) has been identified at Ex/Em = ~ 230/436 nm, with a minor peak at 265/436 nm, in the peak A-region of SRHA dissolved in Milli-Q water; at <250/400 nm in stream waters; at <250/450–470 in estuaries and coastal marine waters; at 240/483 nm in water extracted from sugar maple leaves; and at 220/432 nm in municipal leachate samples (Fig. 3g; Table 2) (Mostofa et al. 2005a; Hunt et al. 2008; Chen et al. 2010; Fellman et al. 2009, 2010; Wu et al. 2011; Murphy et al. 2008). The EEM images and the fluorescence properties of allochthonous humic acid (A-like) are apparently similar to those of allochthonous fulvic acid (A-like). Another component of allochthonous humic acid (M-like) is presumably occurring in soil samples together with allochthonous humic acid (C-like). The allochthonous humic acid (M-like) is often detected at Ex/Em = 300/416 nm in the peak C-region and at 240/416 nm in the peak A-region in soil; at 295/414 and <250/414 nm in stream waters and soil; at 330/412 and 240/412 nm in municipal leachate samples; at 295–300/465 and >240/465 nm in plant biomass, manure and soil (Ohno and Bro 2006; Fellman et al. 2008, 2009; Wu et al. 2011).

The allochthonous humic acid (C-like) can show a fluorescence peak at longer Ex/Em wavelengths in the peak C-region, whereas allochthonous fulvic acid (C-like) often shows fluorescence at shorter Ex/Em wavelengths. Allochthonous humic acids (C-like, A-like and M-like) are not often detected in natural waters, particularly in waters with low DOC concentration. Allochthonous fulvic acid is generally more concentrated than humic acid in natural water, with a ratio of fulvic to humic acid that is typically 9:1. In some cases, especially for high-DOC waters, the ratio decreases to 4:1 or less (Malcolm 1985; Peuravuori and Pihlaja 1999). For this reason, allochthonous humic acids are not often observed. The molecular structure of humic acid is not yet defined due to its complex chemical composition. It is generally highly aromatic in nature, with 30–51 % of aromatic carbon compared to 14–20 % of aromatic carbon in fulvic acid (Malcolm 1985; Steelink 2002; Gron et al. 1996). The high aromaticity and the functional groups of humic acid are responsible for the appearance of several fluorescence peaks at peak C-regions and A-regions (Tables 1, 2). The fluorophores associated with the low molecular weight fraction of humic acid (<10 kDa) exhibit relatively high fluorescence intensity. Fluorescence sharply decreases in the molecular weight fractions of 100–300 kDa and further increases for humic acid fractions >300 kDa (Hayase and Tsubota 1985; Levesque 1972; Gosh and Schnitzer 1980; McCreary and Snoeyink 1980; Visser 1984). The fluorophores bound to humic acid are functional groups with relatively high molecular weight. They show fluorescence in the peak C-region as well as peak A-region (Hayase and Tsubota 1985; Mostofa et al. 2009a, 2005a; Kowalczuk et al. 2009; Santín et al. 2009; Yamashita and Jaffé 2008). Humic acid undergoes photoinduced decomposition by sunlight in natural waters.

Autochthonous Fulvic Acids (C-like and M-like)

Two fluorescent components can be autochthonously produced from algae or phytoplankton biomass under photorespiration or microbial respiration (or assimilation) in natural waters (Fig. 3h–l; Table 2) (Mostofa et al. 2009a, b; Coble 1996, 2007; Parlanti et al. 2000; Stedmon et al. 2007a; Zhang et al. 2009a; Balcarczyk et al. 2009; Murphy et al. 2008; Aoki et al. 2008). The first component can exhibit either two or three fluorescence peaks, at Ex/Em = 330–370/434–480 nm and 290–300/430–448 nm in the peak C-region as well as 250–270/434–480 nm in the peak A-region (Fig. 3h–j; Tables 1, 2). The EEM images of the first autochthonous fluorescent component are similar to those of the allochthonous fulvic acid (C-like). Therefore, this component can be denoted as autochthonous fulvic acid (C-like). The early stage of autochthonous fulvic acid (C-like) when it originates from algae can exhibit three fluorescence peaks, which are subsequently altered into two peaks (Mostofa KMG et al., unpublished data). The EEM spectra shows that the fluorescence intensity of autochthonous fulvic acid

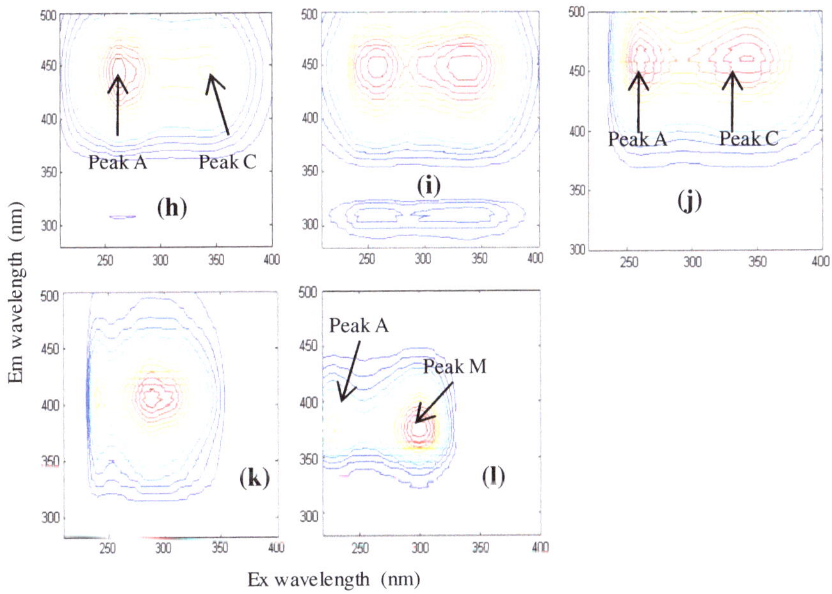

Fig. 3 Continued

The fluorescent components of autochthonous fulvic acid (C-like) under microbial respiration or assimilation of lake algae (**h**), autochthonous fulvic acid (C-like) under photorespiration or assimilation of lake algae (**i**) in natural waters, Tributary of NenJiang River, China (**j**) as well as autochthonous fulvic acid (M-like) in microbial respiration of lake algae (**k**) and in surface waters of Lake Hongfeng, China (**l**) identified using PARAFAC modeling on the EEM spectra of their respective samples. *Data source* Mostofa KMG et al., (unpublished data).

(C-like) at the peak C-region is relatively high, or similar to the intensity at the peak A-region (Fig. 3h–j; Table 2) (Mostofa et al. 2009b; Stedmon and Markager 2005a; Balcarczyk et al. 2009). In contrast, the fluorescence intensity of allochthonous fulvic acid (C-like) is higher at peak A-region than at peak C-region (Figs. 2a, 3a). The autochthonous fulvic acid (C-like) showed fluorescence peaks at Ex/Em = 340–350/460 nm in the peak C-region and at 260/460 nm in the peak A-region in mainstream and tributaries of NenJiang River, China; at 335–340/442–464 nm and 260/442–464 nm in surface waters of Lake Hongfeng, China; at ~360/~460 nm (peak A-region not mentioned) in streams; at 340–350/460, 340/448–454 and 260–270/448–454 nm upon photo-assimilation of algae in Milli-Q and river water; at 340/448–455, 290–300/430–448, and 260–448 nm upon microbial respiration (or assimilation) in Milli-Q and river water; at 365/453 and 270/453 nm upon microbial respiration (or assimilation) of phytoplankton in isotonic water (0.5 ‰ salinity); at 355/445 and 280/435 nm when originateng from algae under both light and dark incubation (12 h each); at 350/400–450 and 275/400–450 nm in bay waters (Brazil); at 370/466 and <260/466 nm in the deep waters of Okhotsk Sea and the North Pacific Ocean; at 340/420 and 260/420 nm in the Southern ocean; at 355/434 and 260/434 nm in north Pacific and Atlantic oceans; and at 330–350/420–480 and 250–260/420–480 nm in marine waters (Table 2) (Mostofa et al. 2009b; Coble 1996, 2007; Parlanti et al. 2000; Stedmon et al. 2007a; Zhang et al. 2009a; Wedborg et al. 2007; Luciani et al. 2008; Zhao et al. 2009; Chen et al. 2010; Yamashita et al. 2010; Balcarczyk et al. 2009; Murphy et al. 2008).

On the other hand, the second fluorescent component of algal or phytoplankton origin can exhibit one strong fluorescence peak in the peak C-region and one minor peak (not often shown) in the peak A-region. The peak in the C-region is located at shorter wavelengths compared to the first component of autochthonous fulvic acid (C-like) (Fig. 3k, l) (Mostofa et al. 2009b; Coble 1996, 2007; Parlanti et al. 2000; Stedmon et al. 2003, 2007a; Stedmon and Markager 2005a, 2005b; Zhang et al. 2009a; Balcarczyk et al. 2009; Murphy et al. 2008; Yamashita et al. 2008; Yamashita and Jaffé 2008; Wang et al. 2007). Considering the EEM images of the second fluorescent component, which are similar to those of allochthonous fulvic acid (M-like) (Fig. 3c) and of marine humic acid denoted as M (Coble 1996), such component can be denoted as autochthonous fulvic acid (M-like). The autochthonous fulvic acid (M-like) of algae or phytoplankton origin usually exhibits two fluorescence peaks at Ex/Em = 290–330/358–434 nm in the peak C-region and at 225–360/358–416 nm in the peak A-region (Table 2). The autochthonous fulvic acid (M-like) is often detected in several studies in field and experimental observations, at Ex/Em = 295–300/396–422 nm in the peak C-region and at 235–240/396–422 nm in the peak A-region in surface waters (0–25 m) of Lake Hongfeng, China; at 315/372 nm in the waters of Lake Taihu; at 322/407 nm in the waters of Lake Taihue; at 300/406 and 240/405 nm upon microbial respiration (or assimilation) of algae in river water; at 290–310/400–410 nm upon microbial assimilation of marine algae in Milli-Q water; at 295/398 nm upon microbial production in a mesocosm experiment; at 320–325/388–428 nm in estuaries;

at 305/378 and <260/378 nm in river and coastal waters; at 295–325/358–420 and 260/385 nm in bay waters; at 310/380–420 and 240–250/384–400 nm in the coastal waters of South Atlantic Bight; at 325/385 and <260/385 nm in the deep waters of the Okhotsk Sea and North Pacific Ocean; at 310–315/414–418 and 260/414 nm in the waters of the north Pacific and Atlantic oceans; at 310–320/380–420 nm in marine waters; at 300/406 and <250/410 nm in drinking water treatment plants; and at 330/410 and 230/410 nm in compost products solutions (Fig. 3k, l; Tables 1, 2) (Mostofa et al. 2009b; Coble 1996, 2007; Parlanti et al. 2000; Stedmon et al. 2003, 2007a; Stedmon and Markager 2005a, 2005b; Zhang et al. 2009a; Luciani et al. 2008; Baghoth et al. 2010; Chen et al. 2010; Fellman et al. 2010; Yu et al. 2010; Yamashita et al. 2010, 2011, 2008; Balcarczyk et al. 2009; Murphy et al. 2008; Li et al., Characteristics of sediment pore water dissolved organic matter in four Chinese lakes using EEM spectroscopy and PARAFAC modeling, unpublished data; Yamashita and Jaffé 2008; Wang et al. 2007). The autochthonous fulvic acid (M-like) shows much higher fluorescence intensity at peak C-region than at peak A-region (Fig. 3k, l). In contrast, allochthonous fulvic acid (M-like) shows more intense fluorescence (often twofold) at peak A-region than at peak C-region (Fig. 3c). The variation in the Ex/Em wavelengths of the peaks for autochthonous fulvic acid (C-like and M-like) can be caused by pH, ionic strength, the presence of the trace elements that form complexes with DOM, water origin (Milli-Q, river, lake and seawater) and solvents, content of DOM as well as presence and nature of its components, and finally instrumentation (Mostofa et al. 2009a; Senesi 1990a; Coble et al. 1990; Wu and Tanoue 2001a; Wu et al. 2005; Lochmuller and Saavedra 1986). The molecular weight of autochthonous fulvic acids has been determined as ~<1900 for those originating from photoinduced assimilation of algae under natural sunlight, and ~<1700 upon microbial assimilation of algae in Milli-Q water (Mostofa KMG et al., unpublished data). The functional groups of autochthonous fulvic acid (C-like and M-like) are entirely unknown, which could be the focus for future research challenges. Because of higher fluorescence intensity at the peak C-region compared to allochthonous fulvic and humic acids, autochthonous fulvic acids of algal origin are expected to be quite rich in fluorophores. Autochthonous fulvic acid (C-like) of microbial/algal origin is rapidly decomposed by natural sunlight. This might be an effect of high fluorescence intensity at the peak C-region, which may be linked to high reactivity in natural waters (Mostofa KMG et al., unpublished data). It has recently been found that algal-derived DOM is a more efficient photoinduced substrate than allochthonous DOM (Johannessen et al. 2007; Hulatt et al. 2009).

Protein-Like Component

The protein-like component shows two fluorescence peaks at Ex/Em = 280–300/328–356 nm (peak T) in the peak T-region and a peak at Ex/Em = 235–250/338–356 nm (peak T_{UV}) in the T_{UV}-region (Fig. 3m, n; Tables 1, 2). The EEM images of the protein-like component show that the fluorescence intensity at

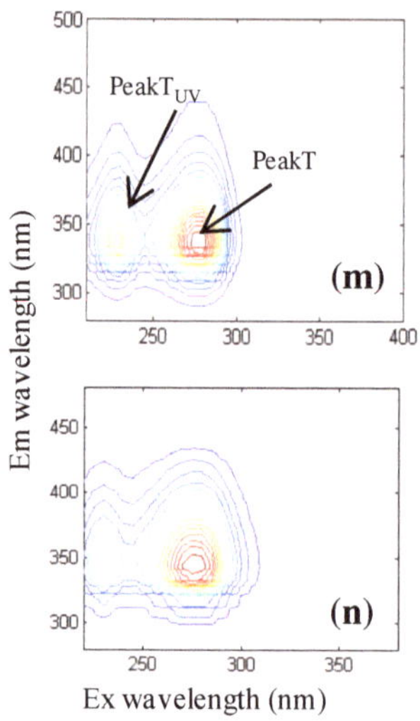

The fluorescent components of protein-like component of algae origin in aqueous samples (**m**), and in sewerage drainage waters (**n** Nanming River, China) identified using PARAFAC modeling on the EEM spectra of their respective samples. (*Data source* Mostofa et al. 2010; Mostofa KMG et al., unpublished data). The fluorescent components of standard tryptophan amino acid

the peak T-region is much higher than at the peak T_{UV}-region (Fig. 3m, n). The protein-like component has been identified at Ex/Em = 280/339–346 nm at the peak T-region and at 230–235/339–346 nm at the peak T_{UV}-region in sewerage waters and washing waters; at 280–285/340–350 and 225/340–350 nm in extracted protein from extracellular polymeric substances (EPS); at 270–280/320–350 and 225/343–358 nm in aromatic proteins or soluble microbial by-products; at 290/356 and 250/356 nm in coastal shelf waters; at 280–300/328–338 and 240/338 nm in north Pacific and Atlantic ocean; and at 275/320 nm in glacial ice samples from the Antarctic and Arctic ocean (Fig. 3m, n; Tables 1, 2) (Mayer et al. 1999; Mostofa et al. 2010; Liu and Fang 2002; Kowalczuk et al. 2009; Chen et al. 2010; Dubnick et al. 2010; Balcarczyk et al. 2009; Murphy et al. 2008). The protein-like component belongs to a molecular size range of >0.1 μm in natural waters (Wu and Tanoue 2001a).

Aromatic Amino Acids (Tryptophan, Tyrosine and Phenylalanine)

The tryptophan-like, tyrosine-like and phenylalanine-like components are often detected in natural waters (Tables 1, 2). The tryptophan-like component has two fluorescence peaks; the first one is located at Ex/Em = 275–285/321–360 nm in the longer wavelength region (peak T), the second one at Ex/Em = 225/340–360 nm in the shorter wavelength region (peak T_{uv}) (Fig. 3o, p; Tables 1, 2). Tryptophan-like component is has been detected at Ex/Em = 275–285/342–357 nm in tryptophan standard dissolved in Milli-Q water and at 275/355 when dissolved in seawater

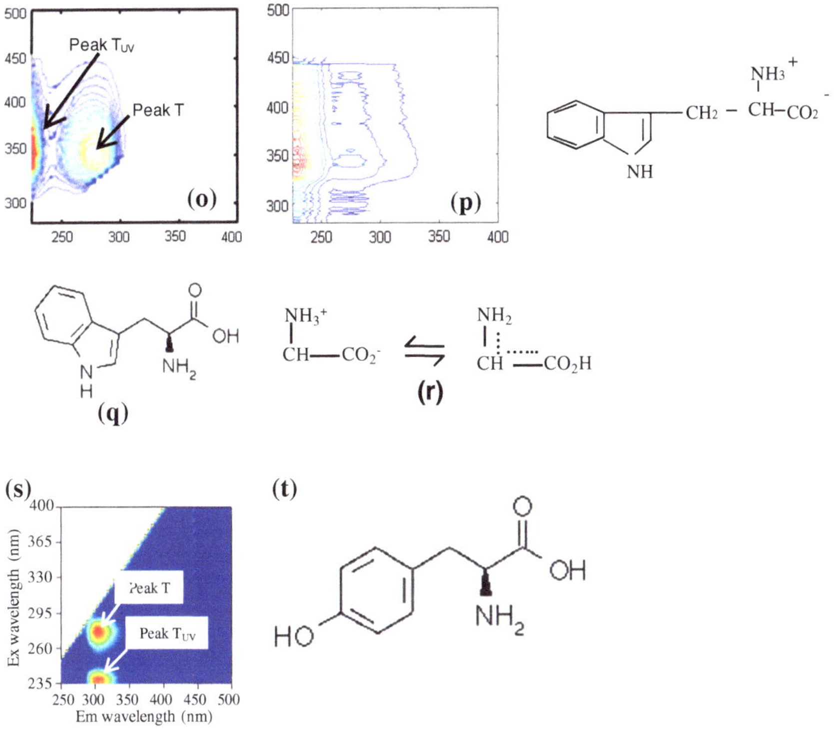

Fig. 3 Continued
(**o**) in EEM data of its aqueous samples and in river waters (Yasu River, Japan) identified using PARAFAC modeling. PARAFAC analysis is conducted on earlier published and their respective a.u. data (*Data source* Mostofa et al. 2005a, 2005b). The molecular structure of tryptophan (**q**) and its resonance configuration (**r**). The fluorescent EEM spectra of standard tyrosine amino acid (**s**) dissolved in Milli-Q waters (*Data source* Nakajima 2006).

(275/357 nm in Milli-Q water); at 275–280/322–336 and 220–230/328–334 nm in tryptophan extracted from EPS; at 275–285/330–351 and 225–235/338–351 nm in various freshwaters; at 275–290/339–360 and 230/340 nm in sewerage samples and drainking water treatment plants; at 280/338–368 and 240/368 nm in estuaries; at 280–295/330–340 nm in bay and coastal waters; at 270–280/320–350 nm in marine waters; at 280/348 and <250/348 nm in ice samples from the Antarctic and Arctic ocean; at 280/330–340 nm in soil; at 280/340 and 220/340 nm in compost products solutions; and at 270/354 nm in plant biomass, animal manure and soil (Coble 1996; Parlanti et al. 2000; Yamashita and Tanoue 2003a; Mostofa et al. 2005a, 2010; Stedmon et al. 2003; Stedmon and Markager 2005a, 2005b; Ohno and Bro 2006; Nakajima 2006; Baker 2005; Zhang et al. 2010; Provenzano et al. 2004; Lu and Allen 2002; Dubnick et al. 2010; Fellman et al. 2008, 2009, 2010; Guo et al. 2010; Yu et al. 2010; Wu et al. 2011; Yamashita et al. 2011; Holbrook et al. 2006;

Yamashita and Jaffé 2008). The molecular formula of tryptophan is $C_{11}H_{12}N_2O_2$ and its molecular weight is 204.23. The chemical structure of tryptophan, $C_8H_5(NH)–CH_2(NH_3^+)CHCOO^-$, is relatively simple (Fig. 3q). The fluorophore at peak T is probably linked to the functional group, $–CH_2–(NH_3^+)–CH–COO^-$, while the fluorophore at peak T_{uv} is probably connected to the $C_8H_5(NH)–$ group that contains the aromatic ring (Mostofa et al. 2009a). The resonance configuration of the functional group, $–CH_2–(NH_3^+)–CH–COO^-$ (Fig. 3r), may confirm the peak position of the T-region at longer wavelengths than the T_{UV}-region. The fluorescence intensity of tryptophan at peak T_{uv} is much stronger (two to threefold higher) than that of peak T. It might be a useful indicator to differentiate the tryptophan-like component from the protein-like component, which typically shows higher fluorescence at the peak T-region than at the peak T_{UV}-region (Fig. 3m, n). Tryptophan is derived microbially from algae, phytoplankton and bacteria in freshwater, marine and sediment pore waters (Chen and Bada 1989; Coble 1996; Yamashita and Tanoue 2003a; Mostofa et al. 2005b; Determann et al. 1998; Baek et al. 1988; Petersen 1989; Wu and Tanoue 2001b; Wu et al. 2001; Cammack et al. 2004). Its very intense fluorescence in aqueous solution could be connected to the functional group $–CH_2–(NH_3^+)–CH–COO^-$ (Fig. 3r). Tryptophan is decomposed both photolytically and microbially in natural waters (Mostofa et al. 2010, 2007a; Winter et al. 2007; Moran et al. 2000). The fluorescence Ex/Em wavelength maxima of tryptophan mostly depend on the polarity of the solvent and the type of the protein. The fluorescence of protein-bound tryptophan is in fact shifted to shorter wavelengths due to shielding from water (Lakowicz 1983; Wolfbeis 1985).

The tyrosine-like component can exhibit two fluorescence peaks at Ex/Em = 270–280/293–314 nm in the peak T-region and at 230/304–307 nm in the peak T_{UV}-region (Fig. 3s; Tables 1, 2). The tyrosine-like component has been detected at Ex/Em = 270–275/303–314 nm (peak T-region) and at 230/304 nm (peak T_{UV}-region) in a tyrosine standard dissolved in Milli-Q water, and at 275/304 and 230/307 nm when the standard was dissolved in seawater; at 265–280/293–313 nm in rivers and other freshwater systems; at 275/<300 nm in lakes; at 275/304–306 nm in estuaries; at 270–275/299–332 nm in sea water; at 275/304–306 in soil; at 270/310 nm in water extracted from sugar maple leaves; at 270/306 nm in drinking water treatment plants; and at 273/309 nm in plant biomass, animal manure and soil (Fig. 3s; Tables 1, 2) (Coble 1996; Parlanti et al. 2000; Yamashita and Tanoue 2003a; Mostofa and Sakugawa 2009; Nakajima 2006; Provenzano et al. 2004; Lu and Allen 2002; Zhang et al. 2009a; Hunt et al. 2008; Kowalczuk et al. 2009; Baghoth et al. 2010; Chen et al. 2010; Fellman et al. 2008, 2009, 2010; Yamashita et al. 2010, 2011; Murphy et al. 2008; Yamashita and Jaffé 2008). Tryptophan and tyrosine when present together in peptides (component 4) are detected from two peaks at Ex/Em = 275/306, 338 nm (T-region) during algal blooming periods, as derived from a mesocom experiment (Table 2) (Stedmon and Markager 2005a). Tyrosine is derived microbially from algal biomass in freshwater and marine environments (Coble 1996; Yamashita and Tanoue 2003a; Stedmon and Markager 2005a; Determann et al. 1998). Comparison of tyrosine and trytophan concentrations with their respective fluorescence intensities in seawater samples as well as standard samples suggests that the fluorescence of tryptophan is approximately four times higher than that of tyrosine (Yamashita and Tanoue 2003a; Mostofa KMG and Sakugawa LH, unpublished data). The molecular formula of tyrosine is $C_9H_{11}NO_3$ and its molecular weight is 181.19. The different chemical structure of tyrosine compared to tryptophan (Fig. 3t) could account for its much lower fluorescence intensity (Yamashita and Tanoue 2003a; Mostofa KMG et al., unpublished data).

Phenylalanine shows two fluorescence peaks at Ex/Em = 255–265/284–286 nm (peak T-region) and at Ex/Em = ~ 220/284–286 nm (peak T_{UV}-region) in marine waters and in standards dissolved in Milli-Q or seawater (Fig. 3u; Tables 1, 2). The phenylalanine-like component has been identified at Ex/Em = 260/286 nm for a phenylalanine standard dissolved in Milli-Q water, and at 260/284 nm for a phenylalanine standard dissolved in sea water; at 255–265/284–285 nm in marine waters; and at 265/306 nm in ice samples from the Antarctic and Arctic Ocean (Yamashita and Tanoue 2003a; Nakajima 2006; Dubnick et al. 2010). The molecular formula of phenylalanine is $C_9H_{11}NO_2$ and its molecular weight is 165.19. The absence of the OH group in the benzene ring of phenylalanine (Fig. 3v) can account for the reduced fluorescence intensity when compared to tyrosine, and for the presence of fluorescence excitation–emission maxima at shorter wavelength regions than for tyrosine or tryptophan.

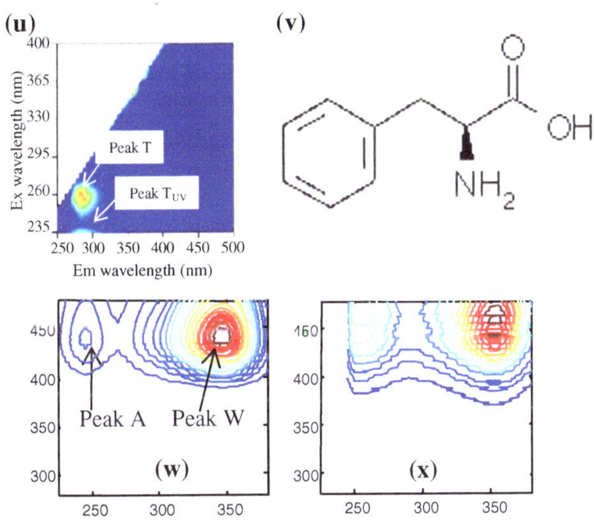

The molecular structure of tyrosine (**t**). **u**, **v** The fluorescent EEM spectra of standard phenylalanine amino acid (**u**) dissolved in Milli-Q waters (*Data source* Nakajima 2006). The molecular structure of phenylalanine (**v**). The fluorescent components of standard DSBP (**w**) in its aqueous samples and in downstream waters (**x** Kurose River, Japan) identified using PARAFAC modeling on their respective EEM data (*Data source* Mostofa and Sakugawa 2009).

Distyryl biphenyl

The aqueous solutions of distyryl biphenyl (DSBP) show two fluorescence peaks at Ex/Em = 350/436 nm (C-region) and at 235–265/435–445 nm (A-region) (Fig. 3w,x; Table 1). The EEM images and the fluorescence peaks of DSBP (Fig. 3w, x) are similar to those of autochthonous fulvic acid (C-like), showing strong fluorescence intensity at the peak C-region than at the peak A-region (Fig. 3h–j). The DSBP component (C-like) has been detected at Ex/Em = 350–355/430–436 nm in the peak C-region and at 235–265/431–446 nm in the peak A-region for a DSBP standard dissolved in Milli-Q water, and at 345/435 and 245/437 nm for DSBP dissolved in seawater (Table 1) (Mostofa et al. 2010; Komaki and Yabe 1982; Nakajima 2006). DSBP is one of the key components of fluorescent whitening agents in natural waters (Managaki and Takada 2005). DSBP is rapidly decomposed by natural sunlight, at a much faster rate compared to DAS1 in natural waters (Mostofa et al. 2005a; Poiger et al. 1999). The molecular structure of (DSBP), 4,4′-bis[(2-sulfostyryl) biphenyl, is shown in (Fig. 3y), with a molecular weight of 562 Da. The DSBP photoinduced decomposition is considered to be caused by an oxidative cleavage of the double bond, followed by the production of various aldehydes such as 2-sulfonic acid benzaldehyde, 4-aldehyde-4′-(2-sulfostyryl)biphenyl (4-benzaldehyde-2′-sulfonic acid-stilbene) and 4,4′-bisaldehyde biphenyl (Guglielmetti 1975; Kramer et al. 1996).

Diaminostilbene-type

The Diaminostilbene-type (DAS1) shows several fluorescence excitation–emission maxima at Ex/Em = 335–355/430–449 nm (peak C-region) and at 240–250/433–446 nm (peak A-region) in aqueous solution (Fig. 3z; Table 1). DAS1 fluorescence peaks have been identified at Ex/Em = 335–355/435–449 and 240–250/434–446 nm for standard DAS1 in Milli-Q water and at 345/436 and 250/433 nm for standard DAS1 dissolved in sea water; and at 340–343/430–432 nm for standard DAS1 in aqueous solutions (Fig. 3z) (Mostofa et al. 2010; Komaki and Yabe 1982; Nakajima 2006). DAS1 is rapidly decomposed by natural sunlight, although it undergoes relatively slower photodegradation compared to DSBP (Mostofa et al. 2005a; Poiger et al. 1999). The molecular structure of DAS1, 4,4′-bis[(4-anilino-6-morphilino-s-triazine-2-yl)amino] 2,2′-stilbenedisulfonate is shown in (Fig. 3a′). Itsphotodegradation yields alcohols, aldehydes and some unidentified products. It is reported that the degradation of DSBP decreases in the presence of DOM, but it is not hindered by DOM in natural waters (Kramer et al. 1996). Its molecular weight is 924 Da.

Detergents (Commercial or Household)

The household detergents are composed of two fluorescent components identified using PARAFAC modeling on the EEM spectra of the detergents solutions (Fig. 3b′, c′). The first component is denoted as detergent component (C-like),

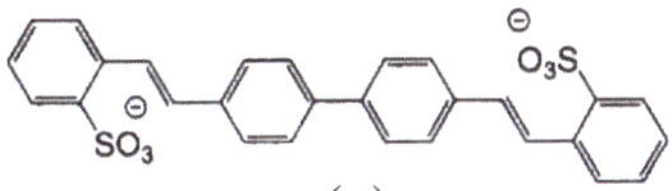

(y)

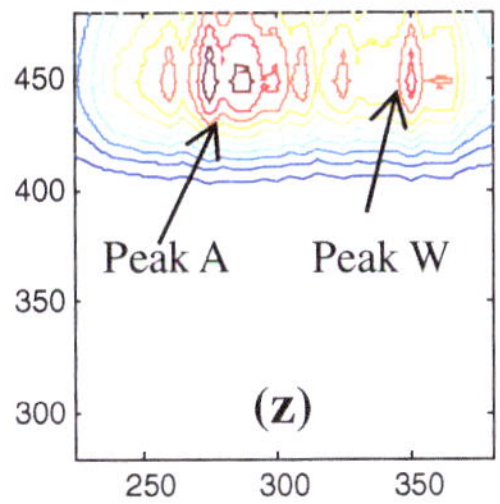

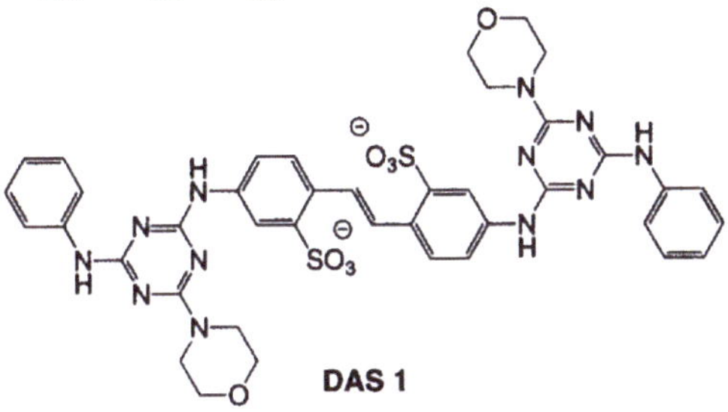

(a′)

The molecular structure of DSBP (**y**). **z**, **a′** The fluorescent components of standard DAS1 (**z**) identified using PARAFAC modeling on EEM data of its aqueous samples (*Data source* Mostofa and Sakugawa 2009). The molecular structure of DAS1 (y).

with two fluorescence peaks at Ex/Em = 335–345/430–437 nm (peak C-region) and at 240–255/425–447 nm (peak A-region), respectively (Fig. 3b′). The second fluorescent component is denoted as detergent component (T_{UV}-like), with a fluorescence peak at Ex/Em = 225–230/287–296 nm (peak T_{UV}-region) (Fig. 3c′; Table 2). Detergent component (C-like) has been detected at Ex/Em = 345/430–435 and 240/427–433 nm for household detergents (Nafine Chem Ind Ltd and Nice group Co Ltd, China) dissolved in Milli-Q water; at 335–350/432–437 and at 250–255/425–447 nm in river waters; at 335–345/422–437 and 240–250/422–443 nm in sewerage waters; and at 340/440 and 250/440 nm in drinking water treatment plants (Tables 1, 2) (Mostofa et al. 2010; Baghoth et al. 2010; Guo et al. 2010). The detergent component (T_{UV}-like) has been detected at Ex/Em = 225–230/291–296 nm in sewerage waters (Table 2) (Mostofa et al. 2010). Household detergents are generally detected by EEM spectroscopy (Mostofa et al. 2005a, 2010; Komaki and Yabe 1982) and other spectroscopic methods (Kramer et al. 1996) in effluent discharged by households located in towns. The detergent component (C-like) is rapidly decomposed by natural sunlight whist it is refractory to microbial degradation (Mostofa et al. 2010). On the other hand, the detergent

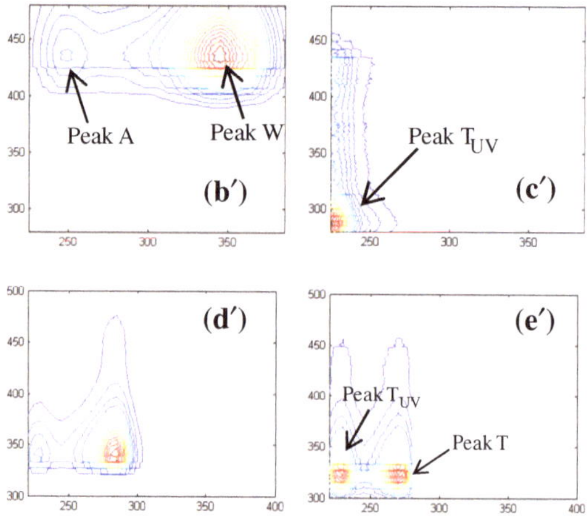

The fluorescent components of commercial or household detergents (**b′**, **c′**) indentified using PARAFAC modeling on EEM data of its aqueous samples (*Data source* Mostofa et al. 2010). The fluorescent components of lake green algae (**d′**, **e′**) isolated and resuspensions in aqueous media (Milli-Q and river waters) identified using PARAFAC modeling on respective EEM data (*Data source* Mostofa KMG et al., unpublished data).

component (T_{UV}-like) is refractory to photoinduced degradation but it is labile to microbial degradation (Mostofa et al. 2010). The form and composition of detergents is variable and recently includes synthetic or naturally occurring polymers with molecular weight ranging from <1000 to >1,000,000 Da (McCullen 1996). Surfactants can have different functionalities (anionic, nonionic and cationic) and commonly used commercial ones are linear alkylbenzene sulphonates, alkyl ethoxy sulphates, alkyl sulphates, alkylphenol ethoxylates, akyl ethoxylates, and quaternary ammonium compounds (McCullen 1996; Ying 2006).

Algae or Phytoplankton

Algae or phytoplankton show several peaks such as 280–285/340–346 nm at peak T-region and 230/327–346 or 230/305 nm at peak T_{UV}-region when resuspended in water (Tables 1, 2) (Mostofa KMG et al., unpublished data; Determann et al. 1998). PARAFAC modeling on the EEM spectra of algae in Milli-Q water show the two fluorescent components of algae (Fig. 3d′, e′; Table 2). The first algae component has a strong fluorescence peak at Ex/Em = 280–285/340–346 nm (peak T-region) and at 230/346 nm (peak T_{UV}-region) (Fig. 3d′), and is similar to the protein-like component that has a much more intense fluorescence at peak T-region than at peak T_{UV}-region (Fig. 3m, n). The second fluorescent component of algae shows a fluorescence peak at Ex/Em = 270/327–336 nm (peak T-region) and at 230/327–336 nm (peak T_{UV}-region) (Fig. 3e′; Table 2). The algae or bacteria collected from marine waters can exhibit fluorescence at Ex/Em = 280/340 nm (peak T-region)

(f′)

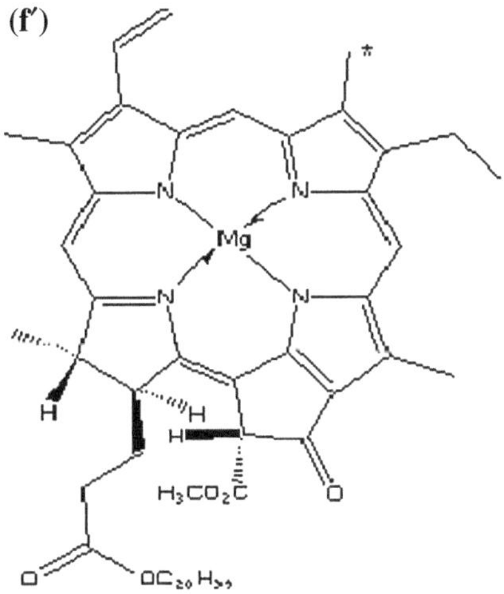

The molecular structure of chlorophyll *a* (**f′**) (*Data source* Clarke et al. 1976)

and two peaks at Ex/Em = 230/340 and 230/305 nm (peak T_{UV}-region) (Determann et al. 1998). Note that green algae have been collected from surface waters of Lake Hongfeng (China) during the summer season using GF/F filters, and their EEM properties have been determined after re-suspension in Milli-Q and river waters.

Chlorophyll a and Chlorophyll b

Chlorophyll *a* (Chl *a*) shows fluorescence at Ex/Em = 431/670 nm and chlorophyll *b* at Ex/Em = 435/659 nm (Moberg et al. 2001). The molecular formula of Chl *a* is $C_{55}H_{72}MgN_4O_5$ and it molecular weight is 893.49; its chemical structure is depicted in Fig. 3f′. The molecular formula of chlorophyll *b* is $C_{55}H_{70}MgN_4O_6$ and its molecular weight is 906.51. The chemical structure of chlorophyll *b* is similar to that of Chl *a*, just with the replacement of a methyl group [$-CH_3$, marked with an asterisk (*)] with an aldehyde one (–CHO). Photoexperiments conducted on sedimentary chloropigments using ^{14}C-labeled algal cells in combination with field observations, demonstrate that a major fraction of Chl *a* is rapidly degraded to soluble colorless compounds (Mostofa et al. 2009a; Klein et al. 1986; Bianchi et al. 1988; Sun et al. 1993). Only a minor fraction of Chl *a* (~30–40 %) is degraded to pheophytin *a* (Klein et al. 1986; Bianchi et al. 1988; Sun et al. 1993).

Identification of Allochthonous Fulvic and Humic Acids from Autochthonous Fulvic Acid (C-like and M-like) Using Fluorescence Index

The key component of autochthonous DOM is termed as marine humic-like substances (Coble 1996), sedimentary fulvic acid (Hayase et al. 1987, 1988) or marine fulvic acids (Malcolm 1990), without a coherent terminology. Recent studies demonstrate that the two fluorescent components, termed as autochthonous fulvic acid (C-like) (Fig. 3h–j) and as autochthonous fulvic acid (M-like) (Fig. 3k, l), are primarily produced under photoinduced or microbial respiration (or assimilation) of algae or phytoplankton biomass (Mostofa et al. 2009b; Stedmon and Markager 2005a; Zhang et al. 2009a). PARAFAC modeling on EEM spectra of algal-originated DOM suggests that the fluorescence peaks and the images of the first fluorescent component are similar to allochthonous fulvic acid (C-like) (Figs. 2a, 3a). Therefore, this component is indicated as autochthonous fulvic acid (C-like) of algal origin (Fig. 3h–j). On the other hand, the fluorescence peaks and the images of the second fluorescent component (Fig. 3k, l) are similar to allochthonous fulvic acid (M-like) (Fig. 3c) and to the marine humic-like substances (Coble 1996). Therefore, this component is denoted as autochthonous fulvic acid (M-like) of algal origin. The fluorescence intensities and the excitation–emission maxima of these two fluorescent components are significantly different depending on the respective peak positions. Considering the similarities between EEM images of the algal originated fluorescent component and allochthonous fulvic acid (C-like), it is suggested to denote the first and the second fluorescent components as 'autochthonous fulvic acid (C-like)' and 'autochthonous fulvic acid (M-like)', respectively (Fig. 3h–l). Similarly, allochthonous fulvic acids can be denoted as 'allochthonous fulvic acid (C-like)' and 'allochthonous fulvic acid (A-like)', respectively (Fig. 3a, b). The allochthonous fulvic acid (A-like) shows only one shoulder or strong fluorescence intensity at peak A-region, which may not be classified as a peak in standard SRFA and SRHA as well as in field observations (Tables 1, 2; Fig. 3b). Apparently, autochthonous fulvic acids often show higher fluorescence intensities at peak C-region than at peak A-region (Fig. 3h–l), whilst allochthonous fulvic acids (C-like) often show opposite behavior (Fig. 3a–c). The differences in fluorescence intensities at peak C- and A-regions could be useful to distinguish between allochthonous and autochthonous fulvic acids using the fluorescence index (Mostofa et al. 2009b; Huguet et al. 2009; Battin 1998; Zsolnay et al. 1999; McKnight et al. 2001). The fluorescence index ($f_{450/500}$) is defined as the ratio of fluorescence intensity at Ex/Em 370/450 nm to that at Ex/Em = 370/500 nm, which can provide a basis for estimating the degree of aromaticity—and potentially for discriminating the sources—of DOM (Battin 1998; McKnight et al. 2001). However, the index $f_{450/500}$ does not distinguish the autochthonous fulvic acid (C-like) of algal origin (1.75–2.59) from allochthonous fulvic acid (1.30–3.22) and allochthonous humic acid (1.28–1.51), which can be identified using PARAFAC modeling of a variety of DOM sources in natural waters (Fig. 4; Table 3) (Mostofa et al. 2005a, 2007a, 2005b; Mostofa KMG et al., unpublished data; Fu et al. 2010; Li et al., Characteristics of sediment pore water dissolved organic matter in four Chinese lakes using EEM spectroscopy and PARAFAC modeling, unpublished data). Further, this index can also fail when applied to a variety of natural waters (Huguet et al. 2009; Jaffé et al. 2004). Another fluorescence index (HIX) has been developed to estimate the degree of maturation of DOM in soil (Zsolnay et al. 1999). HIX is defined as the ratio H/L

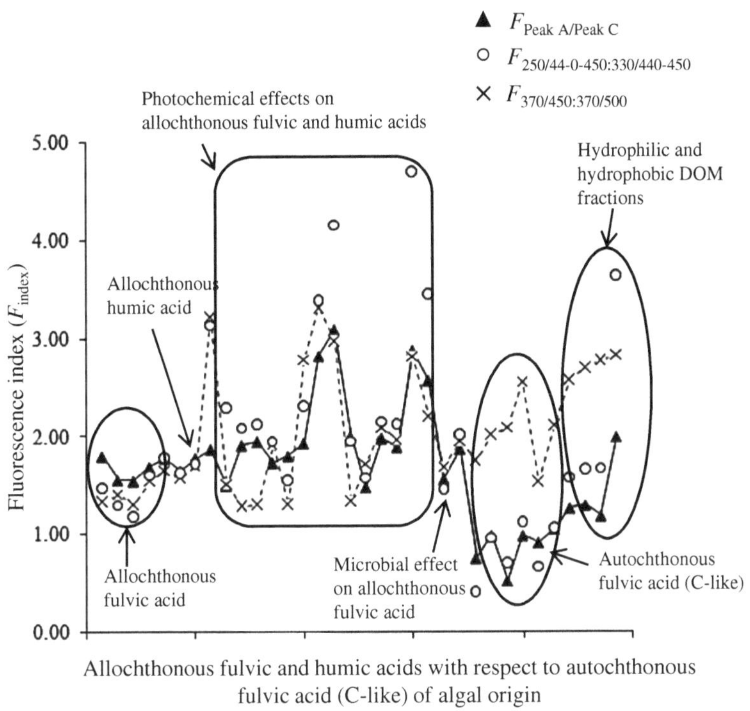

Fig. 4 The fluorescence index (F_{index}) values of allochthonous fulvic and humic acids as well as their photochemical and microbial changes with respect to autochthonous fulvic acid (C-like) of algal origin in waters

of two spectral region areas from the emission spectrum scanned for an excitation at 254 nm. The two areas are calculated between emission wavelengths 300 and 345 nm for L and between 435 and 480 nm for H. The HIX index has recently been applied to a variety of aquatic samples (Huguet et al. 2009; Vacher 2004). High HIX values (between 10 and 16) are a sign of strongly humified OM, mainly of terrestrial origin, whereas low values (<4) are associated with autochthonous OM (Huguet et al. 2009). However, HIX does not distinguish the allochthonous fulvic acid from autochthonous fulvic acid (C-like), identified using PARAFAC modeling on EEM samples in natural waters (Table 3). Indeed, HIX often shows negative values for autochthonous fulvic acid (C-like), and is thus unable to distinguish them from allochthonous fulvic acid.

To identify the PARAFAC fluorescent components, a new fluorescence index ($F_{250:330/440–450}$) has been used and is defined as the ratio of the fluorescence intensity at Ex/Em = 250/440–450 nm and at Ex/Em = 330/440–450 nm. In the case of the emission wavelengths (400–450 nm), average fluorescence intensities are used (Mostofa et al. 2009b). In the case of standard fulvic acid, the $F_{250:330/440–450}$ values vary significantly depending on the number of samples identified using PARAFAC modeling (Table 1; Fig. 4). A very useful fluorescence index

Table 3 The fluorescence index (F_{index}) of allochthonous fulvic and humic acids in contrast to autochthonous fulvic acid (C-like) of algal origin and their various sources identified using PARAFAC modelling on EEM spectra in water samples

DOM sources	Fluorescence index (F_{index})				References
	$F_{Peak\ A/Peak\ C}$	$F_{250:330/440–450}$	$f_{450/500}$	HIX**	
Fulvic acid of forest origin					
Standard Suwannee River Fulvic Acid (n = 3; 1–2 mg L^{-1})	1.80	1.47	1.33	23	Mostofa et al. (2005a)[a]
Standard Suwannee River Fulvic Acid (n = 4; 1–5 mg L^{-1})	1.56	1.29	1.40	25	Mostofa et al. (2005a)[a]
Standard Suwannee River Fulvic Acid (n = 5; 0.5–5 mg L^{-1})	1.55	1.18	1.30	−59	Mostofa et al. (2005a)[a]
Standard Suwannee River Humic Acid (n = 3; 1–5 mg L^{-1})	1.91	2.08	1.28	14	Mostofa et al. (2005a)[a]
Standard Suwannee River Humic Acid (n = 4; 1–5 mg L^{-1})	1.95	2.12	1.30	12	Mostofa et al. (2005a)[a]
Fulvic acid, upstream: Nishi-Mataya and Higashi-Mataya (n = 6)	1.69–1.78	1.60–1.78	1.55–1.65	12, 35	Mostofa et al. (2005b)[a]
Fulvic acid-like, upper branch waters (Z), Yellow River, China (n = 7)	1.30	1.36	2.02	25	Mostofa KMG et al., (unpublished data)
Fulvic acid, upper main channel waters (G), Yellow River, China (n = 18)	1.87	3.13	3.22	8	Mostofa KMG et al., (unpublished data)
Fulvic acid, downstream: Amano, Echi and Ane Rivers (n = 6)	1.65–1.78	1.62–1.72	1.58–1.70	4, 12, 17	Mostofa et al. (2005b)[a]
Photochemical effects on standards and river waters					
Irradiated standard Suwannee River Fulvic Acid (n = 2; 10 h and 20 h)	1.73	1.94	1.91	5	Mostofa KMG et al., (unpublished data)
Irradiated standard Suwannee River Humic Acid (10 h)	1.94	2.00	1.33	7	Mostofa KMG et al., (unpublished data)
Irradiated upstream river waters, Kago upstream, Lake Biwa watershed, Japan	1.8	1.55	1.30	2	Mostofa et al. (2007a)[a]
Irradiated fulvic acid, upper branch waters (Z), Yellow River, China (3 h, n = 2)	1.93	2.3	2.78	−8	Mostofa KMG et al., (unpublished data)

(continued)

Table 3 (continued)

DOM sources	Fluorescence index (F_{index})				References
	$F_{Peak A/Peak C}$	$F_{250:330/440–450}$	$f_{450/500}$	HIX**	
Irradiated fulvic acid, upper main channel waters (G), Yellow River, China (3 h, n = 2)	2.82	3.38	3.32	−14	Mostofa KMG et al., (unpublished data)
Irradiated downstream river waters, Yasu River, Lake Biwa watershed, Japan	3.09	4.15	2.98	−31	Mostofa et al. (2007a)[a]
Microbial effects on River fulvic acid					
Fulvic acid, upper branch waters (Z), Yellow River, China (12 days)	1.77	2.78	3.55	3	Mostofa KMG et al., (unpublished data)
Fulvic acid, upper main channel waters (G), Yellow River, China (12 days)	2.04	3.68	3.98	3	Mostofa KMG et al., (unpublished data)
Fulvic acid affected by agricultural and sewerage activities					
Fulvic acid, Yasu River, Lake Biwa watershed (n = 5)	2.18	2.36	1.8	−21	Mostofa et al. (2005b)
Extracted DOM fractions from Lake Hongfeng, China					
Fulvic acid, extracted from lake surface waters	1.30	1.90	2.20	–	Mostofa KMG et al., (unpublished data)
Humic acid, extracted from lake surface waters	1.50	2.29	1.51	–	Mostofa KMG et al., (unpublished data)
Hydrophilic acids (HIA), extracted from lake surface waters	1.26	1.57	2.57	–	Mostofa KMG et al., (unpublished data)
Hydrophilic bases (HIB), extracted from lake surface waters	1.29	1.65	2.70	–	Mostofa KMG et al., (unpublished data)
Hydrophilic neutrals (HIN), extracted from lake surface waters	1.18	1.66	2.77	–	Mostofa KMG et al., (unpublished data)
Hydrophobic neutrals (HON), extracted from lake surface waters	1.99	3.63	2.83	–	Mostofa KMG et al., (unpublished data)
Photobleaching allochthonous fulvic acid in lake					
Fulvic acid: 0–20 m, summer period, Lake Biwa	1.67–1.82	1.84–2.07	1.89–2.08	4, 6, 9	Mostofa et al. (2005b)[a]

(continued)

Table 3 (continued)

DOM sources	Fluorescence index (F_{index})				References
	$F_{Peak A/Peak C}$	$F_{250:330/440–450}$	$f_{450/500}$	HIX**	
Fulvic acid: 40–80 m, summer period, Lake Biwa	1.58–1.80	1.69–1.95	1.78–1.94	4, 6, 10	Mostofa et al. (2005b)[a]
Fulvic acid: 0–20 m, winter and vertical mixing period, Lake Biwa	1.48–1.98	1.57–2.14	1.71–2.02	4, 5	Mostofa et al. (2005b)[a]
Fulvic acid: 40–80 m, winter and vertical mixing period, 40–70 m, Lake Biwa	1.56–1.87	1.45–2.01	1.68–1.87	6, 7, 9	Mostofa et al. (2005b)[a]
Autochthonous fulvic acid (C-like) during photo- and microbial assimilations of algae					
Algal biomass + Milli-Q water, photo-assimilations: 6-h	0.98	0.95	2.01	11	Mostofa KMG et al., (unpublished data)
Algal biomass + River waters, photo-assimilations: 6-h	0.75	0.40	1.75	−96	Mostofa KMG et al., (unpublished data)
Algal biomass + Milli-Q waters, microbial-assimilations: 1–10 days (n = 5)	0.52	0.96	2.55	−3	Mostofa KMG et al., (unpublished data)
Algal biomass + Milli-Q waters, microbial-assimilations: 20–70 days (n = 6)	0.52	0.7	2.51	−54	Mostofa KMG et al., (unpublished data)
Algal biomass + River waters, microbial-assimilations: 1–10 days (n = 5)	1.53	1.23	2.59	−32	Mostofa KMG et al., (unpublished data)
Algal biomass + River waters, microbial-assimilations: 20–70 days (n = 6)	0.98	0.94	2.47	−37	Mostofa KMG et al., (unpublished data)
Algal biomass + River waters, microbial-assimilations: 80–180 days (n = 7)	0.87	1.11	2.08	−31	Mostofa KMG et al., (unpublished data)
Autochthonous fulvic acid (C-like) in river waters					
Autochthonous fulvic acid (C-like), Main channel of NenJiang (MNJ) River, China	1.04	0.8	0.35	−57	Mostofa KMG et al., (unpublished data)

(continued)

Table 3 (continued)

DOM sources	Fluorescence index (F_{index})				References
	$F_{Peak\ A/Peak\ C}$	$F_{250:330/440–450}$	$f_{450/500}$	HIX**	
Autochthonous fulvic acid (C-like), Tributaries of NenJiang (TNJ) River, China	0.85	1.05	0.37	−49	Mostofa KMG et al., (unpublished data)
Autochthonous fulvic acid (C-like), The Second SongHua (SH) River, China	1.14–1.31	0.98–1.21	1.54–1.71	−178	Mostofa KMG et al., (unpublished data)
Autochthonous fulvic acid (C-like), LiaoHe (LH) River, China	1.24	1.12	0.46	29	Mostofa KMG et al., (unpublished data)
Autochthonous and allochthonous lake DOM					
Epilimnion, 0–8 m (Summer: May–Sept), Lake Hongfeng	1.27–1.59	0.97–1.21	1.58–1.86	–	Fu et al. (2010)[a]
Epilimnion, 0–8 m (Winter: Nov–Feb.), Lake Hongfeng	1.57–1.61	1.20–1.50	1.83–2.25	–	Fu et al. (2010)[a]
Hypolimnion, 10–25 m (Summer: May–Sept), Lake Hongfeng	1.28–1.52	0.95–1.20	1.53–1.80	–	Fu et al. (2010)[a]
Hypolimnion, 10–25 m (Winter: Nov–Feb), Lake Hongfeng	1.52–1.64	1.20–1.25	1.80–2.22	–	Fu et al. (2010)[a]
Epilimnion, 0–8 m (Summer: May–Sept), Lake Baihua	1.37–1.55	1.10–1.26	1.82–2.14	–	Fu et al. (2010)[a]
Epilimnion, 0–8 m (Winter: Nov–Feb), Lake Baihua	1.25–1.33	1.25–1.30	2.19–2.23	–	Fu et al. (2010)[a]
Hypolimnion, 10–25 m (Summer: May–Sept), Lake Baihua	1.38–1.39	0.91–1.57	1.77–1.82	–	Fu et al. (2010)[a]
Hypolimnion, 10–25 m (Winter: Nov–Feb), Lake Baihua	1.39	1.53	1.73	–	Fu et al. (2010)[a]
Photobleaching allochthonous fulvic acid (C-like) in seawaters					
Fulvic acid-like: 0–15 m, Station 0, Seto Inland Sea, Japan (n = 11)	1.90	2.12	1.96	–	Mostofa KMG et al., (unpublished data)
Fulvic acid-like: 0–30 m, Stations 2 and 4, Seto Inland Sea, Japan (n = 12)	1.89	2.29	2.00	–	Mostofa KMG et al., (unpublished data)

(continued)

Table 3 (continued)

DOM sources	Fluorescence index (F_{index})				References
	$F_{Peak A/Peak C}$	$F_{250:330/440–450}$	$f_{450/500}$	HIX**	
Fulvic acid-like: 0–50 m, Stations 11 and 21, Seto Inland Sea, Japan (n = 13)	2.03	2.77	1.99	–	Mostofa KMG et al., (unpublished data)
Fulvic acid-like: 0–40 m, Stations 22 and 23, Seto Inland Sea, Japan (n = 12)	2.88	4.70	2.81	–	Mostofa KMG et al., (unpublished data)
Fulvic acid-like: 60–300 m, Stations 22 and 23, Seto Inland Sea, Japan (n = 9)	2.57	3.45	2.20	–	Mostofa KMG et al., (unpublished data)
Autochthonous and allochthonous fulvic acid (C-like): lake sediment pore waters					
Sediment pore waters, Lake Dianchi, China: 1–20 cm depths	2.07	1.93	2.73	–	Li et al., Characteristics of sediment pore water dissolved organic matter in four Chinese lakes using EEM spectroscopy and PARAFAC modeling, (unpublished data)
Sediment pore waters, Lake Dianchi, China: 22–40 cm depths	1.68	1.52	2.73	–	Li et al., Characteristics of sediment pore water dissolved organic matter in four Chinese lakes using EEM spectroscopy and PARAFAC modeling, (unpublished data)
Sediment pore waters, Lake Dianchi, China: 42–55 cm depths	1.21	1.25	2.05	–	Li et al., Characteristics of sediment pore water dissolved organic matter in four Chinese lakes using EEM spectroscopy and PARAFAC modeling, (unpublished data)
Sediment pore waters, Lake Chenghai, China: 1–20 cm depths	1.24	1.32	2.13	–	Li et al., Characteristics of sediment pore water dissolved organic matter in four Chinese lakes using EEM spectroscopy and PARAFAC modeling, (unpublished data)
Sediment pore waters, Lake Chenghai, China: 22–40 cm depths	1.05	0.66	1.53	–	Li et al., Characteristics of sediment pore water dissolved organic matter in four Chinese lakes using EEM spectroscopy and PARAFAC modeling, (unpublished data)
Sediment pore waters, Lake Chenghai, China: 42–87 cm depths	1.06–1.17	0.86–1.12	1.76–2.12	–	Li et al., Characteristics of sediment pore water dissolved organic matter in four Chinese lakes using EEM spectroscopy and PARA-FAC modeling, (unpublished data)

Table 3 (continued)

DOM sources	Fluorescence index (F_{index})				References
	$F_{Peak A/Peak C}$	$F_{250:330/440–450}$	$f_{450/500}$	HIX**	
Sediment pore waters, Lake Qinghai, China: 1–20 cm depths	0.91	0.80	1.67	–	Li et al., Characteristics of sediment pore water dissolved organic matter in four Chinese lakes using EEM spectroscopy and PARAFAC modeling, (unpublished data)
Sediment pore waters, Lake Qinghai, China: 22–40 cm depths	0.91	0.83	2.05	–	Li et al., Characteristics of sediment pore water dissolved organic matter in four Chinese lakes using EEM spectroscopy and PARAFAC modeling, (unpublished data)
Sediment pore waters, Lake Qinghai, China: 41–50 cm depths	0.92	0.77	2.00	–	Li et al., Characteristics of sediment pore water dissolved organic matter in four Chinese lakes using EEM spectroscopy and PARAFAC modeling, (unpublished data)
Sediment pore waters, Lake Bosten, China: 1–20 cm depths	1.22	1.25	2.13	–	Li et al., Characteristics of sediment pore water dissolved organic matter in four Chinese lakes using EEM spectroscopy and PARAFAC modeling, (unpublished data)
Sediment pore waters, Lake Bosten, China: 22–40 cm depths	0.97	0.76	1.62	–	Li et al., Characteristics of sediment pore water dissolved organic matter in four Chinese lakes using EEM spectroscopy and PARAFAC modeling, (unpublished data)
Sediment pore waters, Lake Bosten, China: 41–50 cm depths	1.03	1.05	2.11	–	Li et al., Characteristics of sediment pore water dissolved organic matter in four Chinese lakes using EEM spectroscopy and PARAFAC modeling, (unpublished data)

$F_{250:330/340–450}$ is defined by Mostofa et al. (2009b); $F_{370/350}$ is defined by McKnight et al. (2001); HIX is defined by Zsolnay et al. (1999)
$F_{peak A/Peak C}$ indicates the ratios of the maximum fluorescence intensity at peak A-region and peak C-region
HIX** Indicates that values are determined at excitation wavelength 255 nm instead of 254 nm
[a]Results obtained after applying the FARAFAC modeling on the publish sample's EEM spectra

is $F_{\text{Peak A/Peak C}}$, which is simple and can be easily applied to fluorescent components identified using PARAFAC modeling (Table 3; Fig. 4). The $F_{\text{Peak A/Peak C}}$ is defined as the ratio of the maximum fluorescence intensity at peak A (peak A-region) to that at peak C (peak C-region). For example, the maximum fluorescence intensity of peak A for standard Suwannee River Fulvic Acid (n = 5: 0.5–5 mg L^{-1}) identified using PARAFAC modeling (component 1) is 60 a.u. at Ex/Em = 260/463 nm and that for peak C is 39 a.u. at Ex/Em = 330/463 nm for the same component, which leads to $F_{\text{Peak A/Peak C}} = 1.55$ (Table 1). The comparisons of the $F_{\text{Peak A/Peak C}}$ values with other indices demonstrate that this new index is very useful to identify the allochthonous fulvic acid (C-like) (1.30–1.80) and allochthonous humic acid (1.50–1.95) from autochthonous fulvic acid of algal origin (0.52–0.98) (Table 3, Fig. 4). Note that the $F_{\text{Peak A/Peak C}}$ of autochthonous fulvic acid (C-like) is relatively high (1.53) in the initial 10 days of microbial assimilation of algae mixed with river waters (Table 3). Photoinduced degradation of fulvic acid in surface waters mostly causes a decrease in fluorescence intensity, which is more marked at peak C than at peak A and ultimately causes an increase of the $F_{\text{Peak A/Peak C}}$ values. This has also been found in irradiated SRFA (1.73), irradiated SRHA (1.94), irradiated fulvic acid in Kago upstream (1.80), Yellow River upper waters (1.93), Yellow River downstream waters (2.82), Yasu River waters (3.09) and lake surface waters (0–20 m) during the summer stratification period in Lake Biwa (1.67–1.82). Therefore, high values (>1.30) of $F_{\text{Peak A/Peak C}}$ can indicate the presence of photobleached fulvic or humic acid whilst low values (<1) are associated with autochthonous fulvic acid (C-like) of algal origin.

Low values of $F_{\text{Peak A/Peak C}}$ for autochthonous fulvic acid (C-like) of algal origin also suggests the presence of high aromaticity with more functional groups or fluorophores at peak C-region than at peak A region. Indeed, the highest fluorescence intensity of algal-origin autochthonous DOM is often detected at peak C-region than at peak A-region (Fig. 3h–j). On the other hand, fulvic acid (terrestrial in origin) shows lower fluorescence intensity at peak C-region than at peak A-region (Fig. 3a), which indicates the presence of relatively low aromaticity with a low number of functional groups or fluorophores and higher content of aliphatic carbon (Mostofa et al. 2009a, b; Malcolm 1985). Such differences in fluorescence intensities or fluorophores (functional groups) are attributed to differences in the f_{index} values, which would be useful in characterizing allochthonous fulvic acid and autochthonous fulvic acid. However, there is no study conducted on the chemical composition of algal-origin autochthonous DOM, which would be the focus for future research. Some researchers believe that autochthonously produced fulvic acid is similar to terrestrial fulvic acid, because autochthonous DOM shows a yellow color as well as similar fluorescence properties. Therefore, the f_{index} values can be useful to distinguish between the two materials. On the other hand, allochthonous fulvic acid (M-like) can merely be distinguished from the maximum fluorescence intensity of peak A (peak A-region) and low fluorescence intensity of peak C (peak C-region) (Fig. 3c) whilst autochthonous fulvic acid (M-like) of algal (or phytoplankton) origin shows an entirely opposite behavior (maximum fluorescence at peak C and low fluorescence at peak A) (Fig. 3k, l).

2.6 Relationship Between FDOM and DOM

Fulvic acid-like fluorescence intensity (FI) is significantly correlated with DOC concentration both linearly and non-linearly as a consequence of the effects of sunlight on river water. However, non-linear correlation is more significant than the linear one (Fig. 5). The extrapolation of the linear relationship suggests that fulvic acid contributes approximately 45–63 % of the total DOM in Kago (KG) upstream, approximately 53 % in Nishi-Mataya (NM) upstream and approximately 73 % in the downstream waters of the Yasu River (YR) (Mostofa et al. 2007a). The fulvic acid contribution in Japanese rivers estimated from DOC and fulvic acid-like FI is similar to that reported in other studies of river water (40–80 %) (Mostofa et al. 2009a; Malcolm 1985; Peuravuori and Pihlaja 1999; Mostofa 2005). The higher presence of allochthonous fulvic acid and humic acids in river water is responsible for the good correlations with DOM found in natural waters (Mostofa et al. 2005a; Fu et al. 2007, 2010; Westerhoff et al. 2001). A gradual decrease in fulvic acid-like FI with decreasing DOC concentration (Fig. 5) can be explained if losses in FI and DOC are mostly involving the fluorophores in fulvic acid. The latter consist of some repeating functional groups with highly variable composition, having aromatic rings and highly unsaturated aliphatics with extensive π-electron systems (Mostofa et al. 2009b, 2010; Senesi 1990a; Malcolm 1985; Corin et al. 1996; Wu et al. 2005). The fluorophores present in

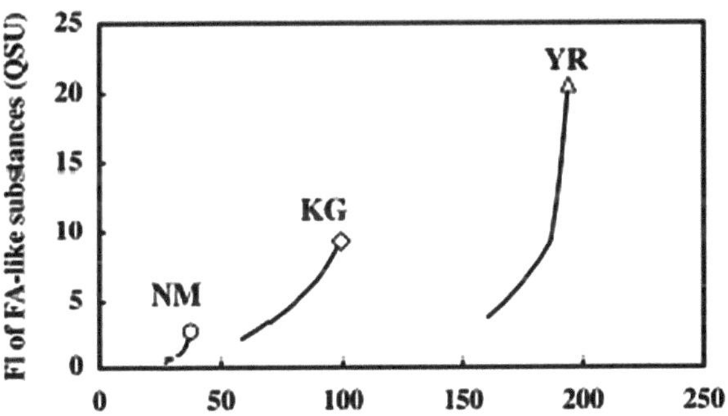

Fig. 5 Changes in DOC concentration and fulvic acid (FA)-like fluorescence intensity (FI) during photodegradation of DOM modeled by the first-order kinetics. Initial DOC concentration and FI are presented by the *open circle* (the Nishi-Mataya, NM upstream); *open diamond* (the Kago, KG upstream); and *open triangle* (Yasu River, YR downstream) for their respective samples collected from Lake Biwa watershed, Japan. *Data source* Mostofa et al. (2007a)

any macromolecule (e.g., fulvic acids and humic acids) at any peak C- or peak A-regions can be referred to as 'fluorochrome'. Any changes of the fluorochrome by photoinduced or microbial degradation can cause changes in the fluorescence properties of that molecule. Conversely, any DOM fraction that does not show fluorescence and is mostly composed of aliphatic C chain, which is photolytically inactive, is termed as 'non-fluorochrome'. On the basis of fluorescence characteristics, DOM can be separated into two major parts: fluorochrome and non-fluorochrome. They allow the photolytically sensitive fractions of DOM in waters to be distinguished. Therefore, any changes in the chemical composition of DOM by photoinduced processes can be examined by the determination of fluorescence characteristics in the asquatic environments (Mostofa et al. 2007a; Ma and Green 2004; Moran et al. 2000; Wu et al. 2007).

3 Factors Affecting the Fluorescence Properties of FDOM in Natural Waters

The fluorescence properties of FDOM are significantly affected by several factors in natural waters. They are: (i) Autochthonous origin of FDOM; (ii) Photodegradation of FDOM; (iii) Microbial degradation of FDOM; (iv) Complex formation between trace elements and FDOM; (v) Salinity; (vi) pH, and (vii) Temperature.

3.1 Autochthonous Origin of FDOM in Natural Waters

Photorespiration or photoinduced assimilation of organic matter (e.g. algal or phytoplankton biomass) can produce new DOM or FDOM in the aquatic environments (Fig. 3i, j) (Mostofa et al. 2009b; Fu et al. 2010; Aoki et al. 2008; Thomas and Lara 1995; Rochelle-Newall and Fisher 2002; Hiriart-Baer and Smiith 2005). It has been shown that the fluorescence intensity of FDOM is gradually increased upon 6 h sunlight irradiation in the presence of re-suspended algal biomass, collected by filtration of water (~0 m depth) from Lake Hongfeng (China) using GF/F filters during the summer season (Mostofa et al. 2009b). These results imply that photoinduced processes play an important role both in the decomposition of FDOM and in its production. They also play a key role in the biogeochemical cycles in the aquatic environments. Also the microbial degradation or assimilation of organic matter (e.g. algal or phytoplankton biomass) in in vitro experiments or under dark incubation may produce new DOM or FDOM in natural waters (Fig. 3h, j–l) (Mostofa et al. 2009b; Stedmon and Markager 2005a; Fu et al. 2010; Zhao et al. 2009; Aoki et al. 2008; Zhang et al. 2009b; Rochelle-Newall and Fisher 2002; Yamashita and Tanoue 2004, 2008; Miller et al. 2009). The fluorescence intensity of microbiologically produced FDOM was gradually increased after 20 days dark incubation at room temperature or upon resuspension of algal biomass, which was collected through filtration of surface lake waters (~0 m) of Lake Hongfeng (China) using GF/F filters during the summer season (Mostofa et al. 2009b). One generally observes high fluorescence intensity in deeper lake or oceanic environments (Hayase et al. 1987, 1988; Hayase and Shinozuka 1995; Mostofa et al. 2005b), which might be the effect of microbial release of FDOM.

Studies on phytoplankton shows that fulvic-like and protein-like fluorescent components are released during the cultivation of three kinds of phytoplankton (*Microcystis aeruginosa* and *Staurastrum dorcidentiferum* of green algae and *Cryptomonas ovata* of dark-brown whip-hair algae collected from lake waters) under a 12:12 h light/dark cycle in an MA medium and an improved VT medium at 20 °C (Aoki et al. 2008). The results demonstrate that produced new DOM from three phytoplanktons can exhibit the different Ex/Em fluorescence properties whereas *Microcystis* can produce the hydrophilic DOM fraction with fluorescence peak at Ex/Em = 340/430 nm (peak C) and Ex/Em = 260/445 nm (peak A) whilst hydrophobic acid (or autochthonous fulvic acids) fraction at Ex/Em = 330/440 nm and 250/455 nm. These two fractions also show the protein-like peak at Ex/Em = 290/335 nm and 280/350 nm (peak T) and the autochthonous fulvic acid (M-like) at 320/385 nm and 330/385 nm (peak C-region), respectively (Table 2) (Aoki et al. 2008). Correspondingly, *Staurastrum* can produce hydrophilic DOM fraction at Ex/Em = 340/420 nm and 280/425 nm and hydrophobic acid fraction at Ex/Em = 340/435 nm and 290/430 nm whereas these two fractions display merely the protein-like peak at Ex/Em = 270/375 nm and 290/365 nm, respectively and do not show any autochthonous fulvic acid (M-like) fluorescence (Table 2) (Aoki et al. 2008). Finally, *Cryptomonas* can produce the hydrophilic DOM fraction at Ex/Em = 350/440 nm and 280/440 nm and the hydrophobic acid fraction at Ex/Em = 350/440 nm and 290/450 nm whereas these two fraction also exhibit the protein-like peak at Ex/Em = 270/355 nm and 270/350 nm, respectively and do not exhibit the autochthonous fulvic acid (M-like) fluorescence (Table 2) (Cammack et al. 2004). Similarly, cultivation of three kinds of phytoplankton (*Prorocentrum donghaiense, Heterosigma akashiwo* and *Skeletonema costatum* collected from sea water) can produce the visible humic-like (C-like and M-like) and the protein-like or the tyrosine-like components in waters (Zhao et al. 2009, 2006). Therefore, production of the autochthonous DOM is largely dependent on the phytoplankton communities in natural waters.

3.2 Photodegradation of FDOM in Natural Waters

Photodegradation can sequentially change the optical properties (fluorescence peak and fluorescence intensity) of FDOM in waters. Photodegradation can change the fluorescence peak position (Ex/Em) of various FDOM components in waters (Mostofa et al. 2005a, 2010, 2007b, 2011; Moran et al. 2000; Miller et al. 2009). For instance, photodegradation can alter the terrestrial fulvic acid (C-like) fluorescence of peak C into the photo-bleached fluorescence peak (peak M_P), which can show the highest fluorescence intensity in some natural waters (Mostofa et al. 2007a, 2005b; Moran et al. 2000; Komada et al. 2002; Burdige et al. 2004). Due to photodegradation, the new photo-bleached peak M_P is shifted at shorter excitation–emission wavelengths (Fig. 3a, d, e). Such a change in the fluorescence peak caused by photodegradation is termed as `blue-shift`. Blue-shift is commonly observed in surface lake or seawaters where photodegradation is important due to exposure to natural sunlight (Fig. 3d, e). Photoinduced effects can decrease the fluorescence intensity (FI) of fulvic acid-like (peak C), FWAs-like (peak W), and tryptophan-like (peak T) compounds, which are commonly observed in natural waters in field and experimental observations (Fig. 5; Table 4) (Hayase and Shinozuka 1995; Mostofa et al. 2005a; 2005b, 2010, 2007a, 2011; Stedmon et al. 2007a; Brooks et al. 2007; Garcia et al. 2005; Winter et al. 2007; Mostofa KMG et al., unpublished data; Skoog et al. 1996; Moran et al. 2000; Osburn et al. 2009; Lepane et al. 2003; Abboudi et al. 2008; Poiger et al. 1999; Fu et al. 2010; Borisover et al. 2009; Yamashita and Tanoue 2008; Vodacek et al. 1997; Yamashita et al. 2007; Shank et al. 2010).

Fulvic Acid-like Components in Natural Waters

Fluorescence intensity losses of fulvic acid-like substances by photoinduced degradation are 1–84 % in rivers, 16–83 % in lakes, 19–67 % in estuaries, and 9–84 % in sea waters studied experimentally in the course of short (hours) to long-term (days or months) irradiation (Table 4). In lake water after 12 days irradiation, the losses of fulvic acid-like fluorescence intensity have been 36 % at the surface (2.5 m) and 48 % in deeper waters (70 m) for DOM fractions of <0.1 μm. In the case of DOM molecular-weight fractions below 5 kDa, the corresponding losses have been 16 % in surface waters (2.5 m) and 50 % in deeper waters (70 m). The low fluorescence intensity decrease in the case of surface-water DOM with molecular weight below 5 kDa may be explained by the fact that the corresponding samples have been collected during an ongoing summer stratification period (September). Therefore, the photosensitive DOM fractions had probably already undergone photoinduced decomposition before sample collection. The higher fluorescence intensity decrease observed for deep-water DOM may be accounted for by the fact that deep waters undergo photoinduced degradation processes to a lesser extent because of the reduced sunlight irradiance compared to surface waters (Laurion et al. 2000). As a consequence, deep-water samples may contain significant amounts of photosensitive DOM components, which have not been degraded in the natural environment and can undergo photoinduced decomposition when irradiated in the laboratory (Table 4). For similar reasons, photoinduced DOM mineralization is very difficult to be observed in surface lake water samples and is much easier to be detected upon irradiation of groundwater (Vione et al. 2009). In estuarine water it has been observed a fluorescence intensity decrease in fulvic acid-like

Table 4 Photochemical and microbial changes of fluorescence intensity (FI) of fulvic acid-like (peak C), fluorescent whitening agents (FWAs)-like (peak W) tryptophan-like, tyrosine-like or protein-like (peak T) substances as a result of photoirradiation experiments conducted on standard substance and natural waters

Type of samples/locations	Origin of DOM	Filtration size/type (μm)	Irradiation time (h/days)	Solar intensity (MJm^{-2})	Changes in fluorescence intensities: Photoirradiation	Microbial	Photoirradiation	Microbial	References
					Peak C-region		Peak T- and T_{UV}-regions		
					%		%		
Suwannee River Fulvic Acid (SRFA): 1 mg L^{-1}	Allochthonous/plant material	MQ water	10 h (Xe lamp)	nd	−42	(+) 0.1	na	na	Mostofa et al. (2011)
SRFA:1 mg L^{-1}	Allochthonous/plant material	MQ water	3 h (Xe lamp)	nd	−19	nd	na	na	Mostofa et al. (2011)
SRFA:1 mg L^{-1} + 50 μM NO_2^-	Allochthonous/plant material	MQ water	3 h (Xe lamp)	nd	−22	nd	na	na	Mostofa et al. (2011)
SRFA: 3 mg L^{-1}	Allochthonous/plant material	MQ water	3 h (Xe lamp)	nd	−23	nd	na	na	Mostofa et al. (2011)
SRFA: 5 mg L^{-1}	Allochthonous/plant material	MQ water	3 h (Xe lamp)	nd	−20	nd	na	na	Mostofa et al. (2011)
Suwannee River Humic Acid (SRHA): 1 mg L^{-1}	Allochthonous/plant material	MQ water	10 h (Xe lamp)	nd	(+) 70	nd	na	na	Mostofa et al. (2011)
SRHA: 3 mg L^{-1}	Allochthonous/plant material	MQ water	3 h (Xe lamp)	nd	−17	nd	na	na	Mostofa et al. (2011)
SRHA: 5 mg L^{-1}	Allochthonous/plant material	MQ water	3 h (Xe lamp)	nd	(+) 5	nd	na	na	Mostofa et al. (2011)
Tryptophan standard: 1 mg L^{-1}	–	MQ water	10 h (Xe lamp)	nd	na	nd	−63	−0.1	Mostofa et al. (2011)
Tryptophan standard: 3 mg L^{-1}	–	MQ water	3 h (Xe lamp)	nd	na	nd	−23	nd	Mostofa et al. (2011)
Tryptophan standard: 5 mg L^{-1}	–	MQ water	3 h (Xe lamp)	nd	na	nd	−20	nd	Mostofa et al. (2011)
Tyrosine standard: 1 mg L^{-1}	–	MQ water	10 h (Xe lamp)	nd	na	nd	−4	(+) 5	Mostofa KMG et al., (unpublished data)
Tyrosine standard: 3 mg L^{-1}	–	MQ water	3 h (Xe lamp)	nd	na	nd	−36	nd	Mostofa KMG et al., (unpublished data)
DSBP standard: 1 mg L^{-1}	–	MQ water	10 h (Xe lamp)	nd	−94	(+) 0.1	na	na	Mostofa et al. (2011)
DSBP standard: 1 mg L^{-1}	–	MQ water	0.5 h (Xe lamp)	nd	−40	nd	nd	nd	Mostofa et al. (2005a)
DSBP standard: 3 mg L^{-1}	–	MQ water	3 h (Xe lamp)	nd	−73	nd	nd	nd	Mostofa et al. (2011)
DSBP standard: 5 mg L^{-1}	–	MQ water	3 h (Xe lamp)	nd	−60	nd	nd	nd	Mostofa et al. (2011)

(continued)

Table 4 (continued)

Type of samples/locations	Origin of DOM	Filtration size/type (μm)	Irradiation time (h/days)	Solar intensity (MJm^{-2})	Changes in fluorescence intensities				References
					Photoirradiation	Microbial	Photoirradiation	Microbial	
					Peak C-region		Peak T- and T_{UV}-regions		
					%		%		
DAS1 standard: 1 mg L^{-1}	–	MQ water	10 h (Xe lamp)	nd	−93	nd	nd	nd	Mostofa et al. (2011)
DAS1 standard: 1 mg L^{-1}	–	MQ water	0.5 h (Xe lamp)	nd	−45	nd	nd	nd	Mostofa et al. (2011)
Fulvic acid, extracted from Göta River: 6.3 mg L^{-1}	Allochthonous/plant material	NaOH + MQ water	13 (UV-B lamp)	nd	−32	na	nd	nd	Lepane et al. (2003)
Humic acid, extracted from Göta River: 6.5 mg L^{-1}	Allochthonous/plant material	NaOH + MQ water	13 (UV-B lamp)	nd	(+) 4	na	nd	nd	Lepane et al. (2003)
Fulvic acid, extracted from Göta River: 6.3 mg L^{-1}	Allochthonous/plant material	NaOH + MQ water	9 (dark)	nd	na	(+) 8	nd	nd	Lepane et al. (2003)
Humic acid, extracted from Göta River: 6.5 mg L^{-1}	Allochthonous/plant material	NaOH + MQ water	9 (dark)	nd	na	(+) 6	nd	nd	Lepane et al. (2003)
Aldrich fulvic acid	Allochthonous/plant material	Deionized water	13 (irradiated)	nd	−(26–41)	na	nd	nd	Winter et al. (2007)
Aldrich fulvic acid	Allochthonous/plant material	Deionized water	13 (irradiated)	nd	−(21–45)	na	nd	nd	Winter et al. (2007)
Kago upstream, Japan (35°N)	Allochthonous/plant material	0.45	12 (irradiated)	192	−74	na	np	np	Mostofa et al. (2005b)
Kago upstream, Japan (35°N)	Allochthonous/plant material	0.45	12 (dark)	192	na	(+) 3	np	np	Mostofa et al. (2005b)
Kago upstream, Japan (35°N)	Allochthonous/plant material	0.45	13 (irradiated)	176	−72	na	np	np	Mostofa et al. (2007a)
Kago upstream, Japan (35°N)	Allochthonous/plant material	0.45	13 (dark)	176	na	(+) 15	np	np	Mostofa et al. (2007a)
Nishi-Mataya upstream, Japan (35°N)	Allochthonous/plant material	0.45	13 (irradiated)	176	−84	na	np	np	Mostofa et al. (2007a)
Nishi-Mataya upstream, Japan (35°N)	Allochthonous/plant material	0.45	13 (dark)	176	na	(+) 6	np	np	Mostofa et al. (2007a)
Nishi-Mataya upstream, Japan (35°N)	Allochthonous/plant material	0.45	12 (irradiated)	192	−78	na	−40	na	Mostofa et al. (2007a)

(continued)

Table 4 (continued)

Type of samples/locations	Origin of DOM	Filtration size/type (μm)	Irradiation time (h/days)	Solar intensity (MJm^{-2})	Changes in fluorescence intensities: Photoirradiation, Peak C-region (%)	Microbial, Peak C-region (%)	Photoirradiation, Peak T- and T_{UV}-regions (%)	Microbial, Peak T- and T_{UV}-regions (%)	References
Nishi-Mataya upstream, Japan (35°N)	Allochthonous/plant material	0.45	12 (dark)	192	na	(+) 5	na	np	Mostofa et al. (2007a)
Yasu River, Japan (35°N)	Allochthonous/autochthonous	0.45	13 (irradiated)	176	−80	na	−59	na	Mostofa et al. (2007a)
Yasu River, Japan (35°N)	Allochthonous/autochthonous	0.45	13 (dark)	176	na	(+) 14	na	(+) 6	Mostofa et al. (2007a)
Kurose River (Izumi site), 34°N	Allochthonous/anthropogenic	0.20	6 (irradiated)	118.5	−77	na	−10	na	Mostofa et al. (2011)
Kurose River (Izumi site), 34°N	Allochthonous/anthropogenic	0.45	6 (dark)	118.5	na	(+) 6	na	(+) 4	Mostofa et al. (2011)
Kurose River (Izumi site), 34°N	Allochthonous/anthropogenic	Unfiltered	6 (irradiated)	118.5	−77	na	−5	na	Mostofa et al. (2011)
Kurose River (Izumi site), 34°N	Allochthonous/anthropogenic	Unfiltered	6 (dark)	118.5	na	nd	na	−17	Mostofa et al. (2011)
Kurose River (Hinotsume site), 34°N	Allochthonous/anthropogenic	0.20	10 (irradiated)	152.5	−76	na	−21	na	Mostofa et al. (2011)
Kurose River (Hinotsume site), 34°N	Allochthonous/anthropogenic	0.45	10 (dark)	152.5	na	nd	na	−11	Mostofa et al. (2011)
Kurose River (Hinotsume site), 34°N	Allochthonous/anthropogenic	Unfiltered	10 (irradiated)	152.5	−81	na	−19	na	Mostofa et al. (2011)
Kurose River (Hinotsume site), 34°N	Allochthonous/anthropogenic	Unfiltered	10 (dark)	152.5	na	nd	na	−13	Mostofa et al. (2011)
Nanming River, (Near Institute), 26°N	Allochthonous/anthropogenic	Unfiltered	3 h (irradiated)	nd	−27	na	−32	na	Mostofa et al. (2010)
Detergent component (C-like), drain samples, (near institute), 26°N	–	Unfiltered	3 h (irradiated)	nd	−34	na	−50	na	Mostofa et al. (2010)

(continued)

Table 4 (continued)

Type of samples/locations	Origin of DOM	Filtration size/type (μm)	Irradiation time (h/days)	Solar intensity (MJm^{-2})	Changes in fluorescence intensities				References
					Photoirradiation	Microbial	Photoirradiation	Microbial	
					Peak C-region		Peak T- and T_{UV}-regions		
					%		%		
Detergent component (T_{UV}-like), drain samples, (near institute), 26°N	–	Unfiltered	3 h (irradiated)	nd	na	na	na	(+) 4	Mostofa et al. (2010)
Detergent component (C-like), component 2, commercial detergents	–	MQ water	3 h (irradiated)	nd	−88	na	np	na	Mostofa et al. (2010)
Detergent component (T_{UV}-like), component 2, commercial detergents	–	MQ water	3 h (irradiated)	nd	na	na	na	(+) 9	Mostofa et al. (2010)
Detergent component (C-like), river water + commercial detergent	–	Unfiltered	10 (dark)	nd	na	(+) 21	na	na	Mostofa et al. (2010)
Tryptophan-like (T-like): River water + commercial detergent	–	Unfiltered	10 (dark)	nd	na	na	na	−24	Mostofa et al. (2010)
Detergent component (T_{UV}-like): river water + commercial detergent	–	Unfiltered	10 (dark)	nd	na	na	na	−84	Mostofa et al. (2010)
Detergent component (C-like), drain samples, (near institute), 26°N	–	Unfiltered	10 (dark)	nd	na	(+) 8	na	na	Mostofa et al. (2010)
Tryptophan-like or protein-like (T-like), drain samples, (near institute), 26°N	Sewerage	Unfiltered	10 (dark)	nd	na	na	na	−67	Mostofa et al. (2010)
Detergent component (T_{UV}-like), drain samples, (near institute), 26°N	Sewerage	Unfiltered	10 (dark)	nd	na	na	na	−90	Mostofa et al. (2010)
Commercial detergent (C-like), component 1	–	MQ water	10 (dark)	nd	na	(+) 14	na	na	Mostofa et al. (2010)

(continued)

Table 4 (continued)

Type of samples/locations	Origin of DOM	Filtration size/type (μm)	Irradiation time (h/days)	Solar intensity (MJm^{-2})	Changes in fluorescence intensities				References
					Photoirradiation	Microbial	Photoirradiation	Microbial	
					Peak C-region		Peak T- and T_{UV}-regions		
					%		%		
Commercial detergent (T_{UV}-like), component 2	–	MQ water	10 (dark)	nd	na	na	na	−15	Mostofa et al. (2010)
River water + DSBP	–	River water	12 h (summer)	nd	−31	na	nd	nd	Poiger et al. (1999)
River water + DAS1	–	River water	12 h (summer)	nd	−12	na	nd	nd	Poiger et al. (1999)
Fulvic acid (C-like: comp 1), Yellow River, China	Allochthonous/plant material	River water	3 h (summer)	nd	−(43–46)	na	nd	nd	Mostofa KMG et al., (unpublished data)
Fulvic acid (M-like: comp 2), Yellow River, China	Allochthonous/plant material	River water	3 h (summer)	nd	−(5–68)	na	nd	nd	Mostofa KMG et al., (unpublished data)
Fulvic acid (C-like: comp 1), Yellow River, China	Allochthonous/plant material	River water	12 (dark)	nd	na	(+) (52–81)	nd	nd	Mostofa KMG et al., (unpublished data)
Fulvic acid (M-like: comp 2), Yellow River, China	Allochthonous/plant material	River water	12 (dark)	nd	na	−100	nd	(+) 8; (+) 33	Mostofa KMG et al., (unpublished data)
Mackenzie River, 68°N	Allochthonous/plant material	0.20	72 h (summer)	nd	−(13–45)	nd	nd	nd	Osburn et al. (2009)
Mackenzie River, 68°N	Allochthonous/plant material	0.20	72 h (spring)	nd	−33	nd	nd	nd	Osburn et al. (2009)
Mackenzie River, 68°N	Allochthonous/plant material	0.20	72 h (autumn)	nd	−(1–10)	nd	nd	nd	Osburn et al. (2009)
Laramie River-DOM, 41°N	Allochthonous/plant material	River water	72 h (sunlight)	nd	−23	na	nd	nd	Brooks et al. (2007)
Chimney Park Wetland-DOM, 41°N	Allochthonous/plant material	River water	72 h (sunlight)	nd	−7	na	nd	nd	Brooks et al. (2007)
Lake Biwa, 35°N: surface water (2.5 m)	Allochthonous/ autochthonous	0.10	12 (irradiated)	137	−36	na	−18	na	Mostofa et al. (2011)
Lake Biwa, 35°N: surface water (2.5 m)	Allochthonous/ autochthonous	0.10	12 (dark)	nd	na	(+) 31	na	(+) 68	Mostofa et al. (2011)
Lake Biwa, 35°N: deeper water (70 m)	Allochthonous/ autochthonous	0.10	12 (irradiated)	nd	−48	na	−7	na	Mostofa et al. (2011)

(continued)

Table 4 (continued)

Type of samples/locations	Origin of DOM	Filtration size/type (μm)	Irradiation time (h/days)	Solar intensity (MJm^{-2})	Changes in fluorescence intensities				References
					Photoirradiation	Microbial	Photoirradiation	Microbial	
					Peak C-region		Peak T- and T_{UV}-regions		
					%		%		
Lake Biwa, 35°N: deeper water (70 m)	Allochthonous/autochthonous	0.10	12 (dark)	nd	na	0	na	(+) 5	Mostofa et al. (2011)
Lake Biwa, 35°N: surface water (2.5 m)	Allochthonous/autochthonous	<5 kDa	12 (irradiated)	nd	−16	na	(+) 4	na	Mostofa et al. (2011)
Lake Biwa, 35°N: surface water (2.5 m)	Allochthonous/autochthonous	<5 kDa	12 (dark)	nd	na	(+) 102	na	(+) 51	Mostofa et al. (2011)
Lake Biwa, 35°N: deeper water (70 m)	Allochthonous/autochthonous	<5 kDa	12 (irradiated)	nd	−50	na	−19	na	Mostofa et al. (2011)
Lake Biwa, 35°N: deeper water (70 m)	Allochthonous/autochthonous	<5 kDa	12 (dark)	nd	na	(+) 20	na	(+) 28	Mostofa et al. (2011)
Four Lakes (45°N)	Allochthonous/autochthonous	0.45	13	nd	−(22–31)	−(0–5)	nd	nd	Garcia et al. (2005)
Lake Taihu, China	Allochthonous/autochthonous	0.22	12	nd	−70	nd	−60	nd	Zhang et al. (2009b)
Fulvic acid, Mill Creek: surface water, 43°N	Allochthonous/autochthonous	0.45	12	nd	−(48–79)	na	na	na	Winter et al. (2007)
Fulvic acid, Bannister Lake: surface water, 43°N	Allochthonous/autochthonous	0.45	12	nd	−(75–83)	na	na	na	Winter et al. (2007)
Fulvic acid and tryptophan, Lake Erie: surface water, 42°N	Allochthonous/autochthonous	0.45	12	nd	−(74–77)	na	na	na	Winter et al. (2007)
Fulvic acid, Sanctuary Pond: surface water, 41°N	Allochthonous/autochthonous	0.45	12	nd	−(71–79)	na	na	na	Winter et al. (2007)
Humic acid, Luther Marsh: surface water, 43°N	Allochthonous/autochthonous	0.45	12	nd	−(64–65)	na	na	na	Winter et al. (2007)
Humic acid, Mill Creek: surface water, 43°N	Allochthonous/autochthonous	0.45	12	nd	−(91–100)	na	na	na	Winter et al. (2007)
Humic acid, Bannister Lake: surface water, 43°N	Allochthonous/autochthonous	0.45	12	nd	−71	na	na	na	Winter et al. (2007)

(continued)

Table 4 (continued)

Type of samples/locations	Origin of DOM	Filtration size/type (μm)	Irradiation time (h/days)	Solar intensity (MJm^{-2})	Changes in fluorescence intensities: Photoirradiation, Peak C-region, %	Microbial, Peak C-region, %	Photoirradiation, Peak T- and T_{UV}-regions, %	Microbial, Peak T- and T_{UV}-regions, %	References
Humic acid, Lake Erie: surface water, 42°N	Allochthonous/ autochthonous	0.45	12	nd	np	na	na	na	Winter et al. (2007)
Humic acid, Sanctuary Pond: surface water, 41°N	Allochthonous/ autochthonous	0.45	12	nd	−(63–81)	na	na	na	Winter et al. (2007)
Tryptophan, Luther Marsh: surface water, 43°N	Allochthonous/ autochthonous	0.45	12	nd	na	na	(+) 62	0	Winter et al. (2007)
Tryptophan, Mill Creek: surface water, 43°N	Allochthonous/ autochthonous	0.45	12	nd	na	na	−(81–88)	0	Winter et al. (2007)
Tryptophan, Lake Erie: surface water, 42°N	Allochthonous/ autochthonous	0.45	12	nd	na	na	−(72–82)	0	Winter et al. (2007)
Tryptophan, Fish food	Allochthonous/ autochthonous	0.45	12	nd	na	na	−(54–66)	(+) (20–40)	Winter et al. (2007)
Tyrosine, Bannister Lake: surface water, 43°N	Allochthonous/ autochthonous	0.45	12	nd	na	na	(+) 68, −34	0	Winter et al. (2007)
Tyrosine, Lake Erie: surface water, 42°N	Allochthonous/ autochthonous	0.45	12	nd	na	na	(+) 1, −19	0	Winter et al. (2007)
Tyrosine, Sanctuary Pond: surface water, 41°N	Allochthonous/ autochthonous	0.45	12	nd	na	na	−(73–100)	0	Winter et al. (2007)
Satilla Estuary	Allochthonous/ autochthonous	0.20	70 (irradiated)	nd	−61	na	−45	na	Moran et al. (2000)
Satilla Estuary	Allochthonous/ autochthonous	0.20	70 (irradiated)	nd	−67	"	−37	"	Moran et al. (2000)
Satilla Estuary	Allochthonous/ autochthonous	0.20	51 (dark)	nd	na	−12	na	(+) 112	Moran et al. (2000)
Satilla Estuary	Allochthonous/ autochthonous	0.20	51 (dark)	nd	na	−1	na	(+) 23	Moran et al. (2000)
Estuary, Beaufort Sea, 69°N	Allochthonous/ autochthonous	0.20	72 h (summer)	nd	−(47–60)	nd	nd	nd	Osburn et al. (2009)

(continued)

Table 4 (continued)

Type of samples/locations	Origin of DOM	Filtration size/type (μm)	Irradiation time (h/days)	Solar intensity (MJm^{-2})	Changes in fluorescence intensities: Photoirradiation, Peak C-region (%)	Microbial, Peak C-region	Photoirradiation, Peak T- and T_{UV}-regions (%)	Microbial, Peak T- and T_{UV}-regions	References
Estuary, Beaufort Sea, 69ºN	Allochthonous/ autochthonous	0.20	72 h (spring)	nd	−33	nd	nd	nd	Osburn et al. (2009)
Estuary, Beaufort Sea, 69ºN	Allochthonous/ autochthonous	0.20	72 h (autumn)	nd	−19	nd	nd	nd	Osburn et al. (2009)
Shelf, Beaufort Sea, 69–70ºN	Allochthonous/ autochthonous	0.20	72 h (summer)	nd	−(67–75)	nd	nd	nd	Osburn et al. (2009)
Shelf, Beaufort Sea, 70ºN	Allochthonous/ autochthonous	0.20	72 h (spring)	nd	−(46–61)	nd	nd	nd	Osburn et al. (2009)
Shelf, Beaufort Sea, 70–71ºN	Allochthonous/ autochthonous	0.20	72 h (autumn)	nd	−(29–84)	nd	nd	nd	Osburn et al. (2009)
Gulf, Beaufort Sea, 70ºN	Allochthonous/ autochthonous	0.20	72 h (spring)	nd	−66	nd	nd	nd	Osburn et al. (2009)
Gulf, Beaufort Sea, 70–71ºN	Allochthonous/ autochthonous	0.20	72 h (autumn)	nd	−(50–61)	nd	nd	nd	Osburn et al. (2009)
Gulf, Beaufort Sea, 70–71ºN	Allochthonous/ autochthonous	0.20	72 h (winter)	nd	−(21–67)	nd	nd	nd	Osburn et al. (2009)
Fulvic acid (C-like)?, component 1, Baltic Sea: 55–65ºN	Allochthonous/ autochthonous	0.20	48 h (UV−A lamp)	nd	−65	na		na	Stedmon et al. (2007a)
Autochthonous fulvic acid (M-like)?, component 2	Allochthonous/ autochthonous	0.20	48 h (UV-A lamp)	nd	−69	na	na	na	Stedmon et al. (2007a)
Protein-like?, component 4	Allochthonous/ autochthonous	0.20	48 h (UV-A lamp)	nd	−54	na	na	na	Stedmon et al. (2007a)
Tyrosine-like?, component 5	Allochthonous/ autochthonous	0.20	48 h (UV-A lamp)	nd	na	na	−26	na	Stedmon et al. (2007a)
Fulvic acid (A-like)?, component 6	Allochthonous/ autochthonous	0.20	48 h (UV-A lamp)	nd	(+) (252–2740)	na	na	na	Stedmon et al. (2007a)
Baltic Sea, BY15: 0–30 m depth	Allochthonous/ autochthonous	0.20 (unfiltered)	5	nd	−(44–52)	nd	nd	nd	Skoog et al. (1996)
Baltic Sea, BY15: 100–240 m depth	Allochthonous/ autochthonous	0.20 (unfiltered)	5	nd	−(58–65)	nd	nd	nd	Skoog et al. (1996)

(continued)

Table 4 (continued)

Type of samples/locations	Origin of DOM	Filtration size/type (μm)	Irradiation time (h/days)	Solar intensity (MJm^{-2})	Changes in fluorescence intensities				References
					Photoirradiation	Microbial	Photoirradiation	Microbial	
					Peak C-region		Peak T- and T_{UV}-regions		
					%		%		
Baltic Sea, BY32: 0–50 m depth	Allochthonous/ autochthonous	0.20 (unfiltered)	12	nd	−(61–70)	nd	nd	nd	Skoog et al. (1996)
Baltic Sea, BY32: 0–50 m + chloroform	Allochthonous/ autochthonous	0.20 (unfiltered)	12	nd	−(59–81)	nd	nd	nd	Skoog et al. (1996)
Baltic Sea, BY32: 100–190 m depth	Allochthonous/ autochthonous	0.20 (unfiltered)	12	nd	−(73–75)	nd	nd	nd	Skoog et al. (1996)
Baltic Sea, BY32: 100–190 m + Chloroform	Allochthonous/ autochthonous	0.20 (unfiltered)	12	nd	−(83–84)	nd	nd	nd	Skoog et al. (1996)
Baltic Sea, F15: 0–50 m depth	Allochthonous/ autochthonous	0.20 (unfiltered)	4	nd	−(48–49)	nd	nd	nd	Skoog et al. (1996)
Baltic Sea, F15: 0–50 m depth	Allochthonous/ autochthonous	0.20 (unfiltered)	4	nd	−50	nd	nd	nd	Skoog et al. (1996)
Baltic Sea, F15: 0–50 m depth	Allochthonous/ autochthonous	0.20 (filtered)	4	nd	−(50–51)	nd	nd	nd	Skoog et al. (1996)
Baltic Sea, F15: 100 m depth	Allochthonous/ autochthonous	0.20 (filtered)	4	nd	−56	nd	nd	nd	Skoog et al. (1996)
Mediterranean Sea, 42°N: Canet lagons	Allochthonous/ autochthonous	0.20 (filtered)	8 h (summer)	nd	−22	(+) 8	nd	nd	Abboudi et al. (2008)
Mediterranean Sea, 42°N: Leucate lagons	Allochthonous/ autochthonous	0.20 (filtered)	8 h (summer)	nd	−9	(+) 0.4	nd	nd	Abboudi et al. (2008)
Mediterranean Sea, 42°N: coastal waters (SOLA)	Allochthonous/ autochthonous	0.20 (filtered)	8 h (summer)	nd	−34	−2	nd	nd	Abboudi et al. (2008)
Seawater, Gotland Deep: 40 m, 57°N	Allochthonous/ autochthonous	0.45	13 (UV-B lamp)	nd	−32	nd	nd	nd	Lepane et al. (2003)

nd not detected, *na* not applicable, *np* no significant fluorescence peak observed in samples
(−) and (+) means a decrease in fluorescence and an increase in fluorescence, respectively, of the respective peaks
Irradiation time-hours (h) mentioned with each time as 'h' and 'days' mentioned as a whole digit only

substances of 19–67 % during 72 h to 70 days irradiation (Table 4) (Moran et al. 2000; Osburn et al. 2009). In unfiltered seawater samples from the Baltic Sea irradiated for 4–5 days, the corresponding fluorescence intensity decrease has been of 44–52 % at the surface (0–50 m) and 56–65 % in the deeper layers (100–240 m). In some cases the decrease has been more marked, i.e. 61–70 % at the surface (0–50 m) and 73-75 % in the deeper layer (100–190 m). Interestingly the addition of chloroform significantly enhanced photodegradation, yielding a fulvic acid-like fluorescence decrease of 59–81 % in surface samples (0–50 m) and of 83–84 % in deep-water ones (Table 4) (Skoog et al. 1996). The mechanism behind the increased FDOM photodegradation upon addition of chloroform may be the production of phosgene in the presence of O_2 ($CHCl_3 + O_2 + h\upsilon \rightarrow COCl_2 + HCl$). Phosgene is highly reactive toward the degradation of the fluorophores, such as the amino groups ($RNH_2 + COCl_2 \rightarrow RN = CO + 2HCl$) or carboxylic acids ($RCO_2H + COCl_2 \rightarrow RC(O)Cl + HCl + CO_2$) (Mostofa et al. 2011; Shriner et al. 1943). Such processes would contribute to the decrease of DOM fluorescence in natural waters (Mostofa et al. 2009a, 2011).

Photoinduced degradation of Mediterranean Sea samples (8 h sunlight exposure) showed a decrease in the fluorescence of fulvic acid-like or humic-like fluorophores (peak C), in the range of 9–22 % for lagoon water and approaching 34 % for coastal water (Table 4) (Abboudi et al. 2008). Similarly, photoinduced degradation of waters collected from Mackenzie River and Beaufort Sea (Estuary, Shelf and Gulf) demonstrates that the degradation of fulvic acid-like fluorophores (peak C) is usually higher during summer irradiation than in spring, autumn and winter (Table 4). The photodegradation of fulvic acid-like fluorophore (peak C) is relatively higher in Beaufort Sea samples (47–60 % in Estuary during summer; 67–75 % in Shelf during summer; 66 % in Gulf during spring; 72 h irradiation) than in Satilla Estuary (61–67 %, 70 days), Baltic Sea (44–52 % in surface waters, 4–5 days), and Gotland Deep seawater (32 %, 13 days) (Table 4) (Stedmon et al. 2007a; Skoog et al. 1996; Moran et al. 2000; Osburn et al. 2009; Lepane et al. 2003). The high photodegradation of fulvic acid-like substances in Beaufort Sea samples has been explained by the occurrence of two phenomena. Firstly, in many cases a significant fraction of the fulvic acid-like substances are of autochthonous origin, which makes them highly susceptible to photodegradation (Mostofa KMG et al., unpublished data; Johannessen et al. 2007). Secondly, in the case of the Beaufort Sea the fulvic acid-like substances have allochthonous origin as they mainly derive from riverine input. Photoinduced degradation of these compounds is poorly effective due to low water temperature in the Beaufort Sea (−0.54 to 21.81 °C in Estuary, −1.36 to 9.23 °C in Shelf, and −1.68 to 0.12 °C in Gulf samples) (Osburn et al. 2009). Therefore, unaffected allochthonous fulvic acid-like substances are highly susceptible to degradation upon laboratory irradiation. The case of the Beaufort Sea may be a particular one, however, because it has been reported that DOM (or FDOM) components are produced from microbial assimilation of phytoplankton biomass or organic matter in natural waters (Mostofa et al. 2009a; Parlanti et al. 2000; Stedmon et al. 2007a, 2007b; Fu et al. 2010; Rochelle-Newall and Fisher 2002; Yamashita and Tanoue 2004; Rochelle-Newall et al. 1999). Photoinduced degradation thus leads to hypotheses about several characteristic chemical and optical features of FDOM, which can be listed as follows (Mostofa et al. 2011): (i) In upstream and downstream rivers the fluorescence is predominantly caused by fulvic and humic acids. In contrast, in the surface layer of lakes and oceans the fluorescence of various FDOM components is rapidly depleted by exposure to natural sunlight (Hayase and Shinozuka 1995; Mostofa et al. 2005b; Fu et al. 2010; Yamashita and Tanoue 2008). As a consequence, the FDOM sampled from these environments is relatively less susceptible to undergo further photoinduced degradation in the laboratory, as was for instance the case for Lake Biwa (Table 4). (ii) High losses of fulvic acid-like fluorescence have been observed upon irradiation of water samples from the deeper layers of lakes and seas. They are more pronounced compared to surface water. A reasonable explanation for this phenomena can include two facts. First, the higher occurrence of fulvic or humic acids in the deeper layers may result directly from terrestrial sources through riverine input without degradation in surface waters (Table 4) (Mopper et al. 1991). Second, the releases of autochthonous fulvic acid (C-like) can occur microbially from algal biomass or phytoplankton in deeper waters (Mostofa et al. 2009a, b; Zhang et al. 2009a; Yamashita and Tanoue 2004, 2008). Autochthonous material is highly susceptible to undergo photoinduced decomposition. It has recently been shown that algal-derived CDOM is a more efficient photoinduced substrate than allochthonous fulvic acid (Mostofa KMG et al., unpublished data; Johannessen et al. 2007; Hulatt et al. 2009). (iii) By comparison of the initial and final photo-bleached components of fulvic acid (C-like) using PARAFAC analysis, it is estimated that the decrease in fluorescence was highest (28–30 %) in the longer wavelength regions (Ex/Em = 335–350/430–450 nm) than at peak M (17 % at 310/450 nm) and peak A (20 % at 250/440 nm) in downstream river (Mostofa et al. 2010). This suggests that the fluorophore at the longer Ex/Em wavelength in fulvic acid is susceptible to undergo rapid photoinduced degradation in aqueous media. Thus, photodegradation would be useful in the removal of major anthropogenic fluorescent organic contaminants, particularly the fluorophores at the longer Ex/Em wavelengths in rivers (Mostofa et al. 2010). (iv) Photo-induced losses of fulvic acid-like fluorescence intensity are gradually reduced in the transition from river to lake, estuary and sea water (Yamashita and Tanoue 2003a; Mostofa et al. 2007a, 2005b; Vodacek et al. 1997; Cory et al. 2007). The cause might be linked to the prior losses of fluorescence intensity in stagnant lake or seawaters by photodegradation. In contrast, photodegradation in rivers is less effective due to continuous transport of water. Photodegradation changes the excitation–emission spectra by introducing a shift to shorter wavelengths. This might constitute evidence of the alteration of existing fluorophores or of the appearance of new fluorescent organic substances (Mostofa et al. 2009a). Examples of fluorescent substances arising from FDOM photodegradation could be salicylic acid (Ex/Em = 314/410 nm), 3-hydroxybenzoic acid (Ex/Em = 314/423 nm), and 3-hydroxycinnamic acid (Ex/Em = 310/407 nm). These molecules are characterized by fluorescence at relatively short wavelengths (Mostofa et al. 2009a).

Photodegradation of Fulvic Acid and Humic Acid

The fluorescence of standard or extracted fulvic and humic acids is typically decreased by photoinduced degradation under sunlight. The fluorescence of SRFA dissolved in Milli-Q water is photolytically decreased under simulated sunlight (by 42 % in 1 mg L^{-1} SRFA for 10 h, 23 % for 3 mg L^{-1} and 3 h, 20 % for 5 mg L^{-1} and 3 h, and by 22 % with 1 mg L^{-1} SRFA + 50 μM NO_2^- for 3 h) (Table 4). Extracted fulvic acid from Göta River shows a decrease in fluorescence (32 %) in alkaline samples (6.3 mg L^{-1} fulvic acid in 0.5 M NaOH solution) after 13 days UV-B irradiation (Table 4). A 6-h summer sunlight exposure of fulvic and humic acids extracted from lake, pond and marsh showed that the decrease of humic acid fluorescence was relatively higher (64–100 %) compared to fulvic acid (48–83 %) (Table 4). It is reported that the fluorescence of humic acid is highly depleted in acidic samples, and undergoes a more pronounced decrease compared to fulvic acid even at higher pH (Wu et al. 2005). Correspondingly, photoirradiation can decompose 35 % of extracted Nordic Reference humic acid (NoHA) and 24 % of extracted Nordic Reference fulvic acid (NoFA) from humus-rich pond water in photoexperiments conducted using a solar simulator (Corin et al. 1996). The reported results suggest that the photoinduced degradation of humic acid is pH and concentration dependent, but the reason behind this phenomenon is still unknown. However, the relatively high photolability of humic acid can be in agreement with the high level of aromaticity (30–51 %), in particular when compared to fulvic acid (14–21 %) (Malcolm 1985; Gron et al. 1996; Wu et al. 2005).

The rate constants for the decrease in fluorescence and for DOC loss are significantly higher for humic than for fulvic acid, as obtained by photoexperiments carried out at different pH levels on extracted humic and fulvic acid from upstreams (Wu et al. 2005). Thus, photodegradation of the humic acid fraction is significantly higher than the fulvic acid fraction and is more sensitive to pH. Interestingly, the higher photolability of humic compared to fulvic acid correlates well with the higher production rate of H_2O_2 upon irradiation of Suwannee River humic Acid (179×10^{-2} M s^{-1}) than for Suwannee River fulvic Acid (69×10^{-2} M s^{-1}) (Mostofa and Sakugawa 2009). This might imply that the production of H_2O_2 is a primary step for the photoinduced degradation of DOM in aqueous solution. Humic acid could thus be the primary target of DOM photodegradation in natural waters (Wu et al. 2005). In contrast, fulvic acid is photolytically more stable than humic acid in aqueous media and may play a vital role in biogeochemical processes due to its longer lifetime in natural waters.

It can be noted that the fluorescence intensity of humic acid is increased by irradiation in some particular cases. Thus, increases have been observed of ca. 70 % for Suwannee River Humic Acid (SRHA) (1 mg L^{-1}, 10 h), 5 % for SRHA (5 mg L^{-1}, 3 h), and 4 % in alkaline samples (6.5 mg L^{-1} Göta River humic acid in 0.5 M NaOH) (Table 4). The reason behind such a phenomenon may be the generation of aromatic photoproducts upon irradiation of humic acid (Corin et al. 1996). Some of these photoproducts may show fluorescence at peak C-region, for example 3-hydroxybenzoic acid at Ex/Em = 314/423 nm, 3-hydroxycinnamic acid at Ex/Em = 310/407 nm, and methyl salicylate at Ex/Em = 366/448 nm (Mostofa et al. 2010). This may produce an increase of humic acid fluorescence. Weak light intensity may prolong the lifetime of humic acid and its aromatic photoproducts, which may result in a fluorescence increase. On the other hand, intense and prolonged irradiation may decompose humic acid and its photoproducts rapidly (Corin et al. 1996), which causes a decrease in fluorescence.

Photodegradation of Aromatic Amino Acids and Protein

Photodegradation experiments have shown a decrease in tryptophan-like fluorescence (peak T) of 5–59 % in river waters, 7–88 % in lake waters, 37–45 % in estuarine waters, and 54 ± 6 % in sea waters (Table 4) (Stedmon et al. 2007a; Mostofa et al. 2010, 2011; Winter et al. 2007; Moran et al. 2000). Similarly, the decrease in tyrosine-like fluorescence intensity upon irradiation was 19–100 % in lakes, 26 ± 9 % in sea waters and 4–36 % for standard tyrosine dissolved in Milli-Q water. A photolytically-induced increase of tyrosine like fluorescence (1–68 %) was also observed in some lake waters (Table 4) (Stedmon et al. 2007a; Winter et al. 2007). A decrease of fluorescence upon irradiation has also been observed as 63 % for standard tryptophan at 1 mg L^{-1} in milli-Q water and 10 h, 23 % for 3 mg L^{-1} and 3 h, and 20 % for 5 mg L^{-1} and 3 h (Table 4) (Mostofa et al. 2011). Some amino acids including tryptophan are degradable by photoinduced processes due to their high chemical reactivity (Yamashita and Tanoue 2003a; Rosenstock et al. 2005). The decrease in tryptophan-like fluorescence intensity was lower (59 %) compared than that of fulvic acid (80 %) in rivers (Mostofa et al. 2007b). Therefore, protein- or tryptophan-like components are photolytically degradable but they are less susceptible to photoinduced degradation compared to fulvic acid in natural waters.

Photodegradation of FWAs and Other Substances in Aqueous Media

The FWAs (DAS1 and DSBP) and household detergents are present in significant amount in some polluted rivers, lakes, coastal sea waters and sediments (Mostofa et al. 2005a, 2010, 2011; Poiger et al. 1999, 1996; Komaki and Yabe 1982; Managaki and Takada 2005; Kramer et al. 1996; Stoll et al. 1998; Stoll and Giger 1998; Baker 2002; Yamaji et al. 2010). The observed fluorescence intensity losses of FWAs or detergent components after up to 10 days irradiation have been 12–81 % in rivers, 34 % in drain samples, and 60–94 % in Milli-Q water (Table 4) (Mostofa et al. 2005a, 2010, 2011). It has been shown that losses in fluorescence intensity of peak W are 76–81 % during 6–10 days irradiation in river water (Table 4). In the case of distyryl biphenyl (DSBP), the decrease in fluorescence intensity is 94 % for 1 mg L^{-1} and 10 h, 73 % for 3 mg L^{-1} and 3 h, 60 % for 5 mg L^{-1} and 3 h; and 31 %

in river water mixed with DSBP during 12 h summer irradiation (Poiger et al. 1999; Mostofa et al. 2011). For diaminostilbene (DAS1) it has been observed a 93 % decrease at 1 mg L^{-1} initial concentration and 10 h irradiation under a solar simulator; and a 12 % decrease in river water mixed with DSBP during 12 h summer irradiation (Table 4) (Poiger et al. 1999; Mostofa et al. 2011). Photoinduced degradation can also decrease 53 % of DAS1 and 81 % of DSBP contents in lake surface waters (Stoll et al. 1998). Irradiation experiments for the standard DSBP and DAS1 using a solar simulator indicate that the fluorescence intensities of FWAs are rapidly decreased, by 40 % for DSBP and 45 % for DAS1 after of 30 min irradiation, but with no shift of their Ex/Em wavelengths (Mostofa et al. 2005a). It is estimated in field observations that the observed removal of FWAs during transport (12 h residence time) is 31 and 12 % for DSBP and DAS 1, respectively, corresponding to half-lives of 0.9 and 2.7 days, respectively, under cloudless summer skies (Poiger et al. 1999). In another study, a mass balance calculation and DSBP/DAS1 ratio shows that ~95 % of DSBP and ~55 % of DAS1 supplied in sewage were decomposed photolytically by natural sunlight in inflowing rivers and in lake, while sedimentation to the lake bottom was insignificant for DSBP and reached ~35 % for DAS1 (Yamaji et al. 2010). More intense photodegradation of FWAs, especially the more photodegradable DSBP, has been observed in Lake Biwa, Japan, than in Lake Greifensee, Switzerland, possibly because of the longer residence time of water in the larger Lake Biwa (Yamaji et al. 2010). A FWAs-salinity diagram in the Tamagawa Estuary shows fairly conservative behavior of the FWAs with ~20 % and ~10 % removal of DSBP and DAS1, respectively, which is thought to be caused by photodegradation (Hayashi et al. 2002). The DSBP/DAS1 ratio also shows a decreasing trend from sewage effluents to rivers and to the Tokyo Bay, indicating selective photodegradation of DSBP (Hayashi et al. 2002). These results suggest that DSBP is more susceptible to photoinduced degradation than DAS1 in natural waters.

For commercial household detergent, the decrease in fluorescence intensity of detergent component (C-like, component 1) is 88 % at peak C-region and 70 % at peak A-region in Milli-Q water during the 3 h of direct sunlight irradiation, under noon summer clear sky conditions (Table 4) (Mostofa et al. 2010). The detergent component (T_{UV}-like) does not decompose photolytically, rather an increase in fluorescence is detected such as 4 % in sewerage drainage samples and 9 % in commercial detergents samples dissolved in Milli-Q water (Table 4) (Mostofa et al. 2010). In sewerage-impacted rivers, the fluorescence intensity of the detergent-like component (peak W) was significantly lower (28 %) at noon time (12:00–13:00 p.m.) than before sunrise (Mostofa et al. 2005a). This indicates that detergent-like compounds may have been decomposed photolytically by natural sunlight during the water transport (Mostofa et al. 2005a). In summary, FWAs and household detergents are highly susceptible to photoinduced degradation upon irradiation in the laboratory as well as in field observations (Table 4) (Mostofa et al. 2005a; Baker 2002).

Mechanism for Photodegradation of Fluorophores in FDOM in Aqueous Media

Sequential photodegradation is observed for FDOM fluorophores or functional groups in FDOM macromolecules, particularly fulvic and humic acids. The sequential degradation of fluorophores is generally caused by the presence of diverse functional groups in their molecular structure (Mostofa et al. 2009a; Senesi 1990a; Leenheer and Croué 2003; Malcolm 1985; Corin et al. 1996; Peña-Méndez et al. 2005; Seitzinger et al. 2005; Zhang et al. 2005). This phenomenon can be understood from the sequential decrease in the fluorescence intensity of fluvic acid-like components (peak C- and A-regions) with irradiation time (Mostofa et al. 2005a, 2007a; Moran et al. 2000). The sequential degradation of various functional groups bound to fulvic and humic acids has also been observed in natural waters (Shriner et al. 1943; Amador et al. 1989; Allard et al. 1994; Xie et al. 2004; Li and Crittenden 2009; Minakata et al. 2009).

Absorption of photon or light by a fluorophore (or functional group) is generally caused by its lowest energy excitation, then by the next lowest energy excitation caused by another fluorophore in the molecule and so on (Mostofa et al. 2009a; Senesi 1990a). Fluorophore excitation is the first step for the generation of H_2O_2 in aqueous media according to (Eq. 3.1). Photoirradiation converts H_2O_2 into $HO^{\bullet}$ (Eq. 3.2), photolytically or by Fenton and photo-Fenton reactions (see chapter "Photoinduced Generation of Hydroxyl Radical in Natural Waters"). The $HO^{\bullet}$ radical can then react with the initial excited fluorophore and decompose it (Eq. 3.3). Therefore, a scheme for the photoinduced degradation of fluorophores in macromolecules can be depicted as below (Eqs. 3.1–3.4; see chapters "Photoinduced and Microbial Generation of Hydrogen Peroxide and Organic Peroxides in Natural Waters", "Photoinduced Generation of Hydroxyl Radical in Natural Waters" and "Photoinduced and Microbial Degradation of Dissolved Organic Matter in Natural Waters"):

$$\mathrm{FDOM} + \mathrm{O_2} + \mathrm{HO_2} + \mathrm{H^+} \overset{h\nu}{\rightarrow} \mathrm{H_2O_2} + \mathrm{FDOM^{\bullet +}} + \mathrm{O_2} + \mathrm{OH^-} \tag{3.1}$$

$$\mathrm{H_2O_2} \overset{h\nu}{\rightarrow} 2\mathrm{HO^{\bullet}} \tag{3.2}$$

$$\mathrm{FDOM^{\bullet +}} + \mathrm{HO^{\bullet}} \overset{h\nu}{\rightarrow} \left[\mathrm{FDOM^{\bullet +}HO^{\bullet}}\right]^* \tag{3.3}$$

$$[\mathrm{FDOM^{\bullet +}HO^{\bullet}}]^* \overset{h\nu}{\rightarrow} \mathrm{LMWDOM} + \mathrm{DIC} + \mathrm{CO_2} + \text{other end products} \tag{3.4}$$

One of the pathways that can lead to H_2O_2 formation is the production of $O_2^{\bullet\,-}$ from O_2 upon release of electrons from irradiated FDOM fluorophores or chromophores (Eq. 3.1, see chapter "Photoinduced and Microbial Generation of Hydrogen Peroxide and Organic Peroxides in Natural Waters"). Reaction (3.4) produces low molecular weight DOM (LMWDOM), dissolved inorganic carbon (DIC), CO_2, and other end products (see chapter "Photoinduced and Microbial Degradation of Dissolved Organic Matter in Natural Waters").

Radiation absorption by the next lowest energy fluorophore would then produce further reactive species and cause sequential degradation of the fluorophores, and so on till the entire degradation of the parent molecule. Photoinduced degradation of organic substances can also occur by other processes. For instance, phosgene ($COCl_2$) is highly photosensitive and highly reactive (Shriner et al. 1943) as explained previously. The sequential photodegradation mechanism is applicable to various FDOM such as fulvic acid and humic acid of plant origin, autochthonous fulvic acid of algal origin, proteins and aromatic amino acids.

Controlling Factors for Photodegradation of FDOM in Natural Waters

Photodegradation of FDOM depends on several key factors that are similar to the photodegradation of DOM (Mostofa et al. 2011). The photoinduced degradation of FDOM depends on the several factors in the aquatic environments: (i) The nature or the quality of the organic components of DOM; (ii) The concentration or the quantity of the organic DOM components; (iii) The pH of the sample solution that may affect the photo-induced generation of $HO^•$, a strong oxidizing agent that is involved in the photodegradation of DOM (Bertilsson and Tranvik 2000; Wu et al. 2005; Kwan and Voelker 2002). pH also influences the photoactivity of Fe species that take part to DOM photomineralization (Vione et al. 2009); (iv) The presence and quantity of Fe in the water samples that may provide $HO^•$ through photo-Fenton reaction ($H_2O_2 + Fe^{2+} \rightarrow Fe^{3+} + HO^• + OH^-$) or induce DOM transformation though irradiated Fe-DOM complexes (Wu et al. 2005; Miles and Brezonik 1981; Zepp et al. 1992; Southworth and Voelker 2003); *(v)* The concentration of O_2 that can assist in the production of $HO^•$ or H_2O_2 (Miles and Brezonik 1981); (vi) The occurrence of NO_2^- and NO_3^-, further sources of $HO^•$ that may enhance the photoinduced decrease of DOM fluorescence (Table 4) (Zinder 1993). For example, irradiation experiments using a solar simulator have shown that addition of NO_2^- to standard SRFA can slightly enhance the decrease of fluorescence, which reaches approximately 22 % with 1 mg L^{-1} SRFA + 50 μM NO_2^- upon 3 h irradiation compared to 19 % with 1 mg L^{-1} SRFA after 3 h (Table 4). (vii) The light intensity (UV-B, UV-A and PAR: photosynthetically active radiation) is a key factor in the photoinduced reactions and controls the production of reactive transients that correspondingly enhance the photodegradation processes (Garcia et al. 2005; Bertilsson and Tranvik 2000; Granéli et al. 1998; Qian et al. 2001; Randall et al. 2005). Interestingly, the decrease of fluorescence upon addition of NO_2^- that is a major $HO^•$ source (Mack and Bolton 1999) was relatively limited (3 %). This finding would be compatible with SRFA photooxidation primarily occurring because of the photo-induced generation of $HO^•$ produced photolytically by SRFA itself, or through other processes. It can be noted that the production rate of H_2O_2 from SRFA is 69×10^{-12} M s^{-1} (Mostofa and Sakugawa 2009) and a relatively low level of H_2O_2 can accelerate the photoinduced degradation of humic acid in aqueous media (Wang et al. 2001). This hypothesis is in agreement with the assumption that part of the production of $HO^•$ by DOM under irradiation derives from H_2O_2 photogeneration. It is also in agreement with the results of field observations that the DOM fluorescence decreases with an increase of H_2O_2 concentration over the course of the day in marine surface waters (Obernosterer et al. 2001). An alternative possibility is that also other species different from $HO^•$ may induce the transformation of DOM. Interestingly, the generation rate of $HO^•$ was largely unable to account for the photoinduced mineralization of acidified lake-water or filtered groundwater DOM (Vione et al. 2009).

In addition, photodegradation of FDOM depends on several key factors such as sunlight incident doses, water chemistry, DOM contents, mixing regime and so on (White et al. 2003; Ma and Green 2004; Mostofa et al. 2011; Reche et al. 1999). Moreover, the key factors that affect the FDOM photobleaching are: (1) Solar radiation, (2) Water temperature, (3) Effects of total dissolved Fe and photo-Fenton reaction, (4) Occurrence and quantity of NO_2^- and NO_3^- ions, (5) Molecular nature of DOM, (6) pH and alkalinity of the waters, (7) Dissolved oxygen (O_2), (8) Depth of the water, (9) Physical mixing in the surface mixing zone, (10) Increasing UV-radiation during ozone hole event, and (11) Global warming. These factors are similar to those affecting the photoinduced degradation of DOM (see the chapter Photoinduced and Microbial Degradation of Dissolved Organic Matter in Natural Waters).

3.3 Microbial Degradation of FDOM in Natural Waters

Microbial degradation by autotrophs (plants, algae and bacteria) and heterotrophs (animals, fungi and bacteria) induces changes in FDOM in the deeper layers of natural waters (rivers, lakes, and oceans), pore waters and soil waters. The effects have been highlighted in field observations and experimentally under dark incubation. Microbial processes thus alter the fluorescence properties of various FDOM such as fulvic acid, humic acid, aromatic amino acids and FWAs (DAS1 and DSBP) (Fig. 5) (Mostofa et al. 2010, 2007a, 2005b, 2007b, 2011; Ma and Green 2004; Moran et al. 2000).

Fulvic Acid and Humic Acid of Terrestrial Plant Material Origin

The microbial degradation can alter the fulvic acid-like fluorescence intensities at peaks A-, C- and M-regions and their excitation–emission (Ex/Em) peak positions in natural waters (Ma and Green 2004; Moran et al. 2000; Mostofa et al. 2007b, 2011; Yamashita and Tanoue 2008). Allochthonous fulvic acid (C-like) fluorescence is increased by approximately 3–81 % in rivers due to microbial degradation, for an incubation period ranging from hours to 13 days (Table 4). Allochthonous fulvic acid (M-like) fluorescence is entirely decomposed microbially after 12 days of dark incubation at room temperature (Table 4). In lakes, the fulvic acid-like fluorescence is increased by up to 31 % in molecular fractions <0.1 μm and 102 % in <5 kDa fractions in surface waters. In deeper DOM fractions the corresponding increases are 0 and 20 % under dark incubation for 12 days (Table 4) (Mostofa et al. 2011). In contrast, fulvic acid-like fluorescence is decreased microbially (Table 4) (Garcia et al. 2005). In estuarine water, the decrease in fulvic acid-like fluorescence is relatively high: 11 % at peak A, 1–12 % at peak C, and 1–12 % at peak M in replicate samples during a 51 days incubation period (Table 4) (Moran et al. 2000). In Mediterranean Sea samples, DOM fluorescence is either increased (0.4–8 %) after 8 h dark incubation in lagoon water, or decreased (2 %) in coastal water (Table 4) (Abboudi et al. 2008). Upon microbial processing, the fluorescence of standard SRFA does not change significantly after 10 h incubation (Table 4). A fluorescence increase after a 9-day incubation period has been detected in fulvic (8 %) and humic acid (6 %) extracted from Göta River (Table 4). These results may lead to the hypothesis that there are several characteristic chemical and optical features of fulvic acid and FWAs in natural waters, which can be classified as: (i) Fluorescent compounds, particularly fulvic and humic acids (C-like) in stream, are typically recalcitrant to microbial degradation. Microbes are not capable of decomposing the fluorophores in the longer wavelength region, particularly the peak C-region in allochthonous fulvic and humic acids. Fluorophores of humic substances at peak C-region are mostly composed of aromatic molecules associated with functional groups having extensive π-electron systems, or with specific repeating functional groups in the carbon matrix of peak C-region, which are highly recalcitrant to microbial degradation (Mostofa et al. 2009a, 2010; Malcolm 1985; Geller 1986; Münster 1991). (ii) Allochthonous fulvic acid (M-like) is highly labile to biological degradation, thus it has opposite behavior than allochthonous fulvic acid (C-like). (iii) Under dark incubation, the increase of fulvic acid-like fluorescence is typically higher in surface lake water compared to the deeper layers. This finding allows the hypothesis that surface photo-bleached fulvic acid is highly labile to microbial changes, which can lead to a significant increase in fulvic acid-like fluorescence. (iv) Finally, the decrease in fulvic acid-like fluorescence is insignificant (0–5 %). This fluorescence may presumably result from the decomposition of other fluorescent components present in the same peak position.

Autochthonous Fulvic Acid (C-like) of Algal Origin

The fluorescence intensities of peaks A and C of autochthonous fulvic acid (C-like), produced during the microbial assimilation of algal biomass (Algae + river waters) during long-term dark incubation (180 days), were maximal at the 4th day of incubation. The intensities then became lowest at the 20th day, gradually increased until the 70th day and increased again after the 80th day till the 180th (Fig. 6). This result suggests that the microbial processes can alter rapidly the fluorophores bound at peaks A- and C-regions of autochthonous fulvic acid (C-like) during the initial 20 days of incubation. It follows a slower microbial alteration of fluorophores from 20 to 70 days, and an even slower alteration from 80 to 180 days. From 20 days onward, fluorescence intensities of both peaks gradually increase. This result can be interpreted by the ratios of fluorescence intensities of peak A divided by peak C of autochthonous fulvic acid (C-like). Such a

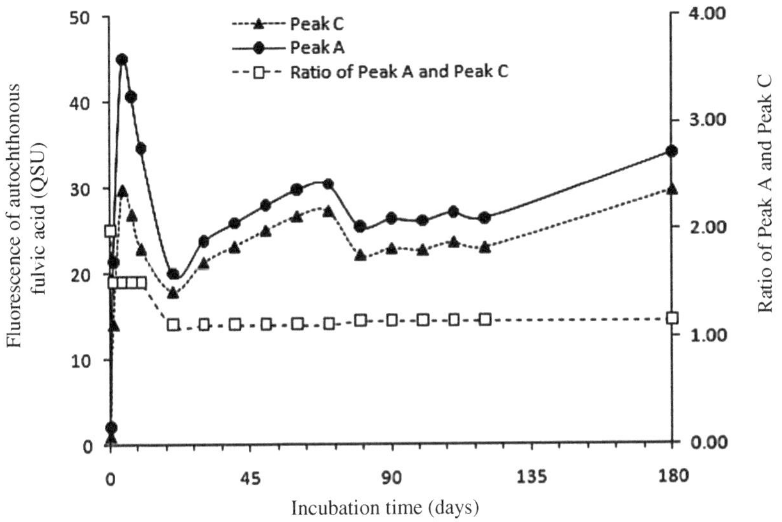

Fig. 6 The changes in the fluorescence intensities of peaks A and C and their fluorescence intensities ratios of autochthonous fulvic acid (C-like) originated under microbial assimilations of lake algal biomass during the long-term dark incubation period (1–180 days). The fluorescence intensities of both peaks A and C are the average of duplicate samples. *Data source* Mostofa KMG et al. (unpublished data)

ratio becomes high (peak A/peak C = 1.53) from 1 to 10 days, then it decreases to 0.98 from 20 to 70 days, and decreases again to 0.87 between 80 and 180 days of incubation (Table 3; Fig. 6). It is in agreement with earlier results concerning the production of CDOM in resuspension of particulate matter in an isotonic solution (0.5 ‰ salinity, no N and P). The measured spectral slope $S_{300-500}$ was highest at the 6th day, reached the lowest level at the 9th day, increased during days 12–18 and after 21 days it remained approximately constant (Zhang et al. 2009a). Timescale variation in the release of new DOM depends on the several key factors such as contents of nutrients, occurrence of microorganisms, salinity, and nature of particulate materials. Microbial degradation can also increase the spectral slope S_{350}–S_{400} (nm^{-1}), which is an index of high molecular weight substances and can decrease S_{275}–S_{295} (nm^{-1}), which is related to low molecular weight DOM (Helms et al. 2008).

The reported findings allow three characteristic phenomena to be hypothesized for the microbial degradation of autochthonous fulvic acid (C-like) of algal origin. Firstly, autochthonous fulvic acid (C-like) is primarily released in the initial phase (4–6 days), and afterward the fluorophores bound at peaks A- and C-regions may be rapidly decomposed by microbial processes (4–20 days). The result is the decrease either in the fluorescence intensities of both peaks A- and C-regions of autochthonous fulvic acid (C-like) or in the overall absorbance properties of CDOM. Secondly, the fluorophores connected to the peak A-region of autochthonous fulvic acid (C-like) may undergo faster microbial decomposition compared to those at peak C-region during the second phase (10–20 days). The result is a decrease in both the fluorescence intensities and the ratios of fluorescence intensities of peak A- and peak C-regions. Thirdly, microbial processes may slowly alter the fluorophores bound at both peak A- and C-regions during the third phase (20–70 days), and much more slowly during the fourth phase (80–180 days). The gradual increase in the fluorescence intensities of both peaks A and C suggest that a gradual conversion could be operational of autochthonous fulvic acid (C-like) into compounds highly recalcitrant or refractory to microbial degradation. These modifications are in agreement with earlier studies, which suggest that microbial processes induce rapid decomposition of the 'labile' fraction such as monosaccharides (e.g. glucose), amino acids and fatty acids, which are the monometric molecules that make up carbohydrates, proteins and lipids. In contrast, the 'refractory' fraction is decomposed more slowly (Zhang et al. 2009a; Hama 1991; Hama et al. 2004; Wakeham and Lee 1993; Wakeham et al. 1997; Harvey and Macko 1997; Hanamachi et al. 2008).

Microbial Degradation of FWAs (DAS1 and DSBP)

FWAs-like fluorescence (peak W) is often increased, by 6–14 % in rivers, by 8 % in drain samples, by 14 % for commercial detergents in Milli-Q water, and by 21 % in river waters plus commercial detergents after 6–10 days dark incubation (Table 4) (Mostofa et al. 2011). In rivers, commercial detergents or FWAs-like components typically undergo an increase in fluorescence after microbial degradation. Such a behavior is similar to that of fulvic acid under dark incubation in natural waters. Upon microbial processing, the fluorescence of standard DSBP does not change significantly upon 10 h of dark incubation (Table 4). Commercial detergent (component 2) that shows fluorescence peak at Ex/Em = 225–230/287–296 nm (peak T_{UV}-region), can be microbially decomposed by approximately 84 % in river plus detergent samples, 90 % in sewerage drain samples, and 15 % in commercial detergents samples dissolved in Milli-Q waters, within 10 days of dark incubation (Table 4) (Mostofa et al. 2010). These results suggest that highly polluted waters can rapidly decompose fluorophores at peak T_{UV}-region, i.e., partly the commercial detergents. In contrast, detergent components or FWAs (C-like) are unaltered microbially in natural waters. Microbes are primarily unable to decompose the FWAs (DSBP and DAS1) because of their complex molecular structure composed of a number of aromatic rings with several functional groups (Fig. 3y, a′) (Mostofa et al. 2010)).

Microbial Degradation of Aromatic Amino Acids

The fluorescence of tryptophan-like components under dark incubation is typically decreased, by approximately 13–24 % in unfiltered river waters, by 67 % in unfiltered sewerage drain samples, and by 11 % in filtered river samples (Table 4) (Mostofa et al. 2010). On the other hand, an increase in tryptophan-like fluorescence is often observed in filtered river waters (4–6 %), in lake water in the molecular fractions <0.1 μm (68 % in surface water and 5 % in deep water) and <5 kDa (51 % in surface water and 28 % in deep water), and in estuaries (23–112 %) (Table 4). Upon microbial processing, the standard tryptophan fluorescence does not change significantly after 10 h incubation (Table 4). From these results it can be concluded that there are several characteristic phenomena concerning microbial degradation of tryptophan-like components in natural waters. First, tryptophan-like components are microbiologically labile but microbial degradation is a relatively slow process whilst photodegradation is rapid (Moran et al. 2000; Mostofa et al. 2007b; Baker and Inverarity 2004). Second, an increase in tryptophan-like fluorescence in filtered samples and a decrease in unfiltered samples can be rationalized considering that the filtration processes may deactivate or hinder the bacterial activity. Therefore, if the fluorescence intensity decrease in unfiltered samples may be due to the microbial degradation of tryptophan, the increase in filtered samples might be the result of the binding of tryptophan-like components to humic substances (Volk et al. 1997). Interestingly, an increase of fulvic acid-like FI is typically observed under dark incubation (Mostofa et al. 2007b) and in deep lake or seawaters (Hayase and Shinozuka 1995; Mostofa et al. 2005b). Finally, photo-bleached tryptophan-like DOM is resistant to microbial processes in natural waters. It has been shown that 70 % of the dissolved amino acids (DAA) and dissolved carbohydrates (DCHO) associated with the humic fraction are consumed by microbial degradation in natural waters (Rosenstock and Simon 2003).

3.3.1 Mechanism for Microbial Degradation of Fluorophores in FDOM

The microbial degradation of high molecular weight (HMW) DOM such as fulvic and humic acids (humic substances) of vascular plant origin and autochthonous fulvic acid of algal origin can increase the fluorescence intensities at both peak A- and C-regions. It is generally considered that the peak A-region is linked with aliphatic moieties and functional groups with less aromaticity, whilst the peak C-region is characterized by high aromaticity and functional groups with repeated structural units. Therefore, microbial degradation can effectively modify the aliphatic part of HMW DOM, which can enhance the fluorescence intensity mostly at peak A-region. The microbial increase of fluorescence intensity of HMW DOM is the result of changes in the molecular structure by several pathways.

Firstly, microbes can degrade aliphatic carbon (e.g. carbohydrates) or the functional groups of macromolecules such as fulvic and humic acids of vascular plant origin, as well as autochthonous fulvic acids of algal or phytoplankton origin, with subsequent release of a variety of end products such as CH_4, CO_2, DIC, PO_4^{3-}, NH_4^+, H_2O_2 and organic peroxides (see also chapters "Photoinduced and Microbial Generation of Hydrogen Peroxide and Organic Peroxides in Natural Waters", "Photoinduced and Microbial Degradation of Dissolved Organic Matter in Natural Waters" and "Impacts of Global Warming on Biogeochemical Cycles in Natural Waters") (Ma and Green 2004; Mostofa and Sakugawa 2009; Palenik and Morel 1988; Conrad 1999; Lovley et al. 1996). Secondly, methanogenesis caused by microorganisms (methanogens and acetogens) is an important anaerobic process that can produce CH_4 and CO_2 by converting either acetate (and formate) or H_2/CO_2 in anaerobic environments (Conrad 1999; Lovley et al. 1996; Zinder 1993; Kotsyurbenko et al. 2001). It is presumably considered that the carbohydrate fraction or aliphatic carbon bound in macromolecules (allochthonous fulvic and humic acids) (Malcolm 1985; Peuravuori and Pihlaja 1999) may alter by the methanogenesis. This process can change the molecular structure either by modifying the existing functional groups in the macromolecules or by creating a new π-electron bonding system in the molecule.

3.3.2 Factors Affecting the Microbial Degradation of FDOM in Waters

An increase in fulvic acid-like or humic-like fluorescence at peak C- and A-regions as well as a decrease in fluorescence of aromatic amino acids either in deeper waters of lakes and ocean or in dark incubated water samples is an effect of the microbial degradation of organic matter and the related functional groups (Hayase and Shinozuka 1995; Coble 1996, 2007; Mostofa et al. 2010, 2007a, 2007b, 2011; Ma and Green 2004; Moran et al. 2000). Microbial degradation of DOM thus depends on several key factors that can be distinguished as: (1) Occurrence and nature of microbes in waters; (2) Sources of DOM and the quantity of their fermentation products; (3) Temperature; (4) pH; and (5) Sediment depths in pore waters.

3.4 Complex Formation of Trace Elements with FDOM

Trace elements can significantly affect the fluorescence properties of FDOM in natural waters (Mostofa et al. 2009a; Wu et al. 2004a, 2004b; Fu et al. 2007; Lu and Jaffé 2001). The trace elements or metals (M) can generally form complexes with the fluorophores or functional groups in fluorescent dissolved organic matter (FDOM), which are termed as M-DOM or M-FDOM. The relevant trace elements are transition metals such as Fe, V, Ce, Th, U, Mo, Cu, Mn, Ni, Co, Cr, Zn, Pb, Cd, Hg and UO_2(II), metal/metalloid such as Sb(III) and Al, as well as the alkaline earth elements (see also chapter "Complexation of Dissolved Organic Matter with Trace Metal Ions in Natural Waters") (Mostofa et al. 2009a, 2011; Wu et al. 2004a, 2004b, 2004c; Zhang et al. 2010; Yamashita and Jaffé 2008; Lu and Jaffé 2001). The relevant organic substances in M-DOM complexation are fulvic acid, humic acid, tryptophan, cysteine, selenoprotein P, extracellular polymeric substances (EPS), the Schiff base 2-[4-dimethylaminocinnamalamino]—benzoic acid, phenols and polyphenols (Mostofa et al. 2009a, 2011; Wu et al. 2004a, 2004b; Lu and Jaffé 2001). The fluorophores in FDOM or functional groups in DOM are responsible for the formation of complex with trace elements. Therefore, the fluorescence intensity is either enhanced or quenched due to the complexation of FDOM fluorophores with trace elements (Wu et al. 2004a, 2004b; Fu et al. 2007; Lu and Jaffé 2001; Cabaniss and Shuman 1988; Cabaniss 1992).

The complexation of trace elements with fluorescent substances does not only affect the fluorescence intensity, but also the fluorescence peak position of the respective fluorophore (Wu et al. 2004c). Both excitation and emission wavelengths of the respective fluorophore peak in fulvic acid gradually increase with increasing reaction time (Wu et al. 2004c). It has been hypothesized that donation of electrons occurs from functional groups or fluorophores in DOM to empty *d*-orbitals in transition metals or in partially empty *p*-orbitals in metal/metalloid ($F{:} + M^{n+} \rightarrow F{:}M^{n+}$), thereby causing a strong π-electron bonding system between DOM and metals (Mostofa et al. 2009a, 2011). Donation of electrons from functional groups or fluorophores in DOM causes the *d*-orbitals to be either stabilized or destabilized in the complex compound, thereby causing the fluorescence to either decrease or increase, respectively, in the M-DOM complexes (Mostofa et al. 2009a, 2011).

3.5 Salinity

Salinity can significantly affect the fluorescence properties of FDOM in natural waters (Dorsch and Bidleman 1982; Hayase et al. 1987; Coble 1996; Determann et al. 1996; Parlanti et al. 2000; Nakajima 2006; Laane 1980; Willey and Atkinson 1982; Berger et al. 1984; Laane and Kramer 1990; de Souza Sierra et al. 1997; Boyd et al. 2010). The fluorescence intensity decreases linearly with salinity (Dorsch and Bidleman 1982; Hayase et al. 1987; Laane 1980; Willey and Atkinson 1982; Berger et al. 1984; Laane and Kramer 1990). It is shown that the fluorescence intensities of fulvic acid are quenched significantly with modest saline mixing (Boyd et al. 2010). Two types of result are detected during the mixing of freshwater and seawater (Determann et al. 1996; de Souza Sierra et al. 1997). First, a slow blue-shift of the fluorescence at peak C-region is detected for humic (fulvic)-like fluorophores during the initial mixing of freshwater to seawater between salinity 0 and 32 (de Souza Sierra et al. 1997). Secondly, for higher salinity (>32) a rapid wavelength shift is detected until the salinity reaches the maximum seawater value (de Souza Sierra et al. 1997).

On the other hand, salinity is presumably considered to shift the excitation–emission wavelengths of freshwater fulvic acid (peak C) from the shorter wavelengths found in freshwater rivers (325–340/450–475 nm) and lakes (310–350/410–464 nm) to longer wavelength regions (350–365/446–465 nm) in marine environments (Mostofa et al. 2009a, 2005a; Coble 1996; Parlanti et al. 2000; Yamashita and Tanoue 2003a; Nakajima 2006). The mixing of standard organic substances with Milli-Q and seawater shows that the excitation–emission wavelength maxima of SRFA, DAS1, tyrosine, benzoic acid, *p*-hydroxybenzoic acid, *p*-hydroxybenzaldehyde and *p*-hydroxyacetophenone are significantly shifted from shorter to longer wavelength regions in seawater (Table 1) (Nakajima 2006). For example, the fluorescence peak C of SRFA dissolved in seawater is detected at Ex/Em = 345/452 nm, whilst the same peak in Milli-Q water is detected at Ex/Em = 325/442 nm. Peak A remains almost the same in both aqueous media (Table 1) (Nakajima 2006). The fluorescence peak C of autochthonous fulvic acid (C-like) of algal origin is detected at Ex/Em = 340/442–448 nm in Milli-Q water, and at Ex/Em = 340/454–455 nm in river waters during the photo- and microbial assimilations of algae (Table 2) (Mostofa KMG et al., unpublished data). In another study, the same fluorescence peak C of autochthonous fulvic acid (C-like) of algal origin has been detected at Ex/Em = 365/453 nm and 270/453 nm in an isotonic solution during the microbial assimilation of lake phytoplankton (0.5 ‰ salinity) (Table 2) (Zhang et al. 2009a). The autochthonous fulvic acid or marine humic-like of algal origin (peak M) at peak C-region has been found to shift from 290/400–410 nm in Milli-Q water to 300–310/400–410 nm in seawater (Table 2) (Parlanti et al. 2000)). Such a shift in excitation and emission wavelength maxima is presumably caused by the anions and cations present in sea water and is termed the red shift of fulvic acid-like fluorescence. The mechanism behind the red shift in sea water is attributed to complex formation between the functional groups (or flurophore at peak C- and A-regions) in fulvic acid and trace elements or ions. The complexation of trace elements with the functional groups (or fluorophores) bound at peak C or peak A in SRFA can significantly enhance the electron transfer from the ground state to the excited state by longer wavelength energy. This effect shifts the excitation–emission maxima of the peak C or peak A to longer wavelength regions. Such a shift in both excitation–emission wavelengths takes place during the initial complexation processes and increases with time (Wu et al. 2004a, 2004c). This is evidenced by the photoinduced formation of aqueous electrons (e_{aq}^-) from organic substances and by their high production in NaCl-mixed solutions compared to Milli-Q water (Gopinathan et al. 1972; Zepp et al. 1987; Fujiwara et al. 1993; Assel et al. 1998; Richard and Canonica 2005).

On the other hand, the mixing of some standard FDOM (e.g. DSBP, phenol, and tryptophan) with seawater shows that the fluorescence excitation–emission wavelength maxima (peak C-region and peak T-region) are shifted to shorter wavelengths compared to Milli-Q waters (Nakajima 2006). Such changes in fluorescence excitation–emission maxima are termed as blue-shift of the fluorophores in FDOM. In some cases the blue-shift of the fluorescence peaks could be caused by the loss of high molecular weight fluorescent components by physicochemical modifications such as flocculation, aggregation or precipitation during the initial mixing (de Souza Sierra et al. 1997; Sholkovitz 1976; Carlson and Mayer 1983; McCarthy et al. 1996; Van Heemst et al. 2000; Benner and Opsahl 2001). In the case of smaller molecules, the blue-shift phenomenon is presumably caused by complex formation between anions or cations and the fluorophores (or functional groups) of few fluorescent organic components. This may increase the excitation energy of the fluorophores bound to peak C or peak A-region and change the excitation–emission wavelengths from longer to shorter wavelength regions.

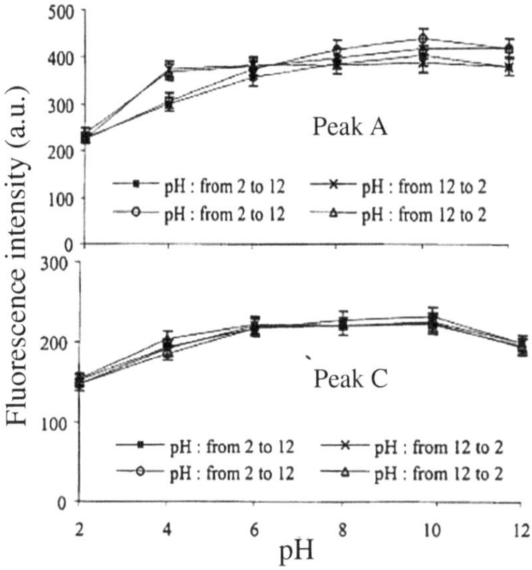

Fig. 7 Changes in the fluorescence intensities of the fluorophores regarding peak A- (**a**) and C-(**b**) regions with solution pH for bulk sample collected from Rio Solimoes in Amazon basin rivers. The *error bar* indicates their standard deviation. *Data source* Patel-Sorrentino et al. (2002)

3.6 pH

The fluorescence properties of various FDOM components are significantly changed by pH variations (2–12) in aqueous solution (Figs. 6, 7, 8) (Gosh and Schnitzer 1980; Henderson et al. 2009; Zhang et al. 2010; Laane 1982; Vodacek and Philpot 1987; Pullin and Cabaniss 1995; Mobed et al. 1996; Patel-Sorrentino et al. 2002; Baker et al. 2007; Spencer et al. 2007). The fluorescence intensities at peak C- and peak A-regions for fulvic acid in Amazon basin rivers are significantly increased up to pH 11, and then decrease in the pH interval 11–12 (Fig. 7) (Patel-Sorrentino et al. 2002). The ratios of fluorescence intensities of peak A and peak C are independent of the molecular fractions of particulate (>0.22 μm), colloidal and dissolved (<5 kDa) organic matter in natural waters (Patel-Sorrentino et al. 2002). The fluorescence intensity of peak C (Ex/Em = 320–340/410–430 nm, presumably caused by fulvic acid) is increased markedly between pH 2 and 6 and then decreases at pH 8–10. In contrast, the fluorescence of peak C (Ex/Em = 370–390/460–480 nm, possibly caused by humic acid) is unaltered at higher pH (Henderson et al. 2009; Spencer et al. 2007).

In the case of bulk lake DOM, the autochthonous fulvic acids (C-like and M-like, respectively, of algal origin) identified by PARAFAC modeling are detected at pH 8–10, but the C-like component is absent at pH 2–4 (Fig. 8) (Mostofa KMG et al., unpublished data). The fluorescence intensities of the M-like component are significantly influenced by pH in the peak A-region: compared to the initial lake-water pH (7.5), a 79 % increase is observed at pH 2, it decreases to 59 % at pH 4 and then gradually increases to 88 % at pH 10. The

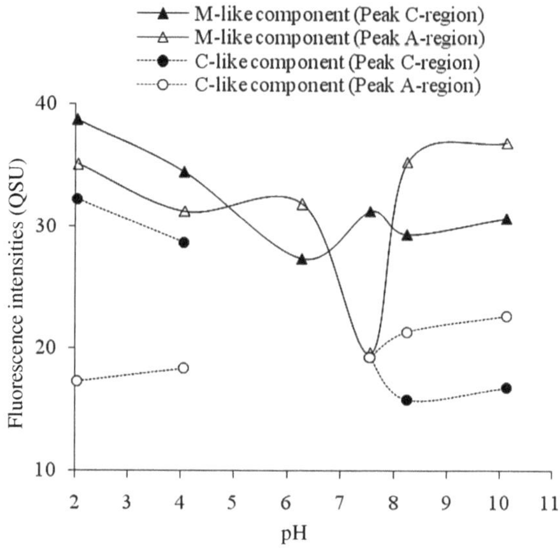

Fig. 8 Changes in the fluorescence intensities of algal-originated autochthonous fulvic acids (C-like and M-like component) identified in lake DOM (Lake Hongfeng, China) and its pH effect. The fluorescent components are identified using PARAFAC model on sample's EEM spectra. *Data source* Mostofa KMG et al. (unpublished data)

C-region is less affected by pH: there is a 24 % increase at pH 2, a decrease at pH 6 and a further increase up to pH 10. In contrast, the fluorescence intensity of the C-like component is significantly affected at the peak C-region: compared to the initial pH of 7.5 there is a 67 % increase at pH 2, then a gradual decrease up to the lowest intensity observed at pH 10. In the case of the peak A-region, one sees a 10 % decrease followed by a gradual increase up to a value that is 18 % higher compared to the initial one (Mostofa KMG et al., unpublished data). Therefore, the effect of pH on algal-originated autochthonous fulvic acids (M- and C-like) is quite different compared to allochthonous fulvic acids.

An increasing fluorescence intensity of humic substances has been detected as pH increases from 4 to 5.5, above which the increase is less important (Vodacek and Philpot 1987). In fulvic acid standards and wastewater treatment plant samples, when lowering the pH from 7 to 3 the decrease of the fluorescence intensity is 30–40 % at peaks C- and A-regions as well as over most of the EEM range, including the peak T- and T_{UV}-region (Westerhoff et al. 2001). For the fulvic acid-like component, the excitation–emission wavelengths for peak C undergo a red shift with increasing pH (Mostofa KMG et al., unpublished data; Westerhoff et al. 2001; Spencer et al. 2007). The pH effect on the complex formation of trace elements with DOM shows that, for DOM, the fluorescence index ($f_{450/500 \text{ at Ex}370 \text{ nm}}$) has a decreasing trend with increasing pH. For DOM + Hg(II) complexation, one sees an increase of the fluorescence index till pH 8 followed by a decrease up to pH 10 (Fu et al. 2007). This suggests that the fluorescence properties of DOM might be affected by several factors such as pH, coexisting metal ions and other organic substances. In addition, the fluorescence properties of three fluorescent whitening agents (FWAs) are modified by pH in the 3–7 range, and the largest pH effect has been detected for the distyrylbiphenyl (DSBP) (Westerhoff et al. 2001).

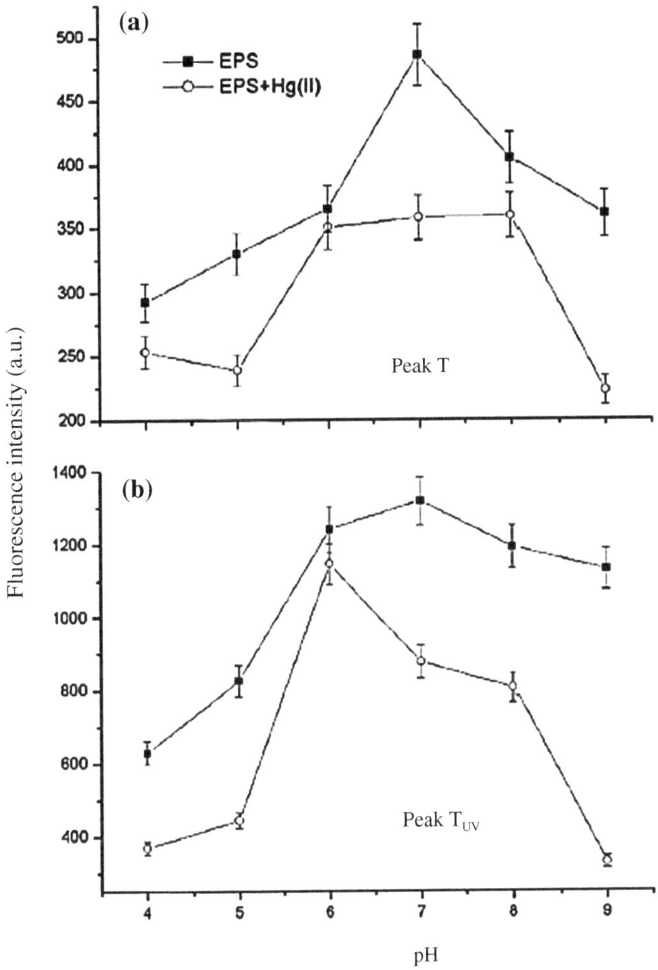

Fig. 9 Changes in the fluorescence intensities of peak T (**a**) and peak T_{UV} (**b**) for extracellular polymeric substances (EPS) with solution pH in the absence and presence of 3.0 mg L^{-1} Hg(II). The *error bar* indicates the standard deviation of three independent measurements. *Data source* Zhang et al. (2010)

In case of tryptophan-like substance or extracellular polymeric substances (EPS), the fluorescence intensities at peak T- and peak T_{UV}-regions are the highest at neutral pH (7.0) and often decrease when the solution pH increases (9.0) or decreases (4.0) (Fig. 9) (Zhang et al. 2010). The EPS is mostly composed of tryptophan-like substances that show a twice higher fluorescence intensity at peak T_{UV} than at peak T (Fig. 9). The pH effect on tryptophan-metal complexation has some characteristic features. The fluorescence intensity of peak T_{UV} is highest at pH 6 and peak T is similar at pH 6–8, differently from the trend of tryptophan fluorescence (Fig. 9) (Zhang et al. 2010). The fluorescence intensities of tryptophan standards decrease by up to 15 % at pH <4.5, there is little effect at pH 5–8, and fluorescence is enhanced by up to 30 % at pH > 8. The peak B (tyrosine-like) is more sensitive to pH changes than the other peaks (Hudson et al. 2007; Reynolds 2003).

Therefore, the fluorescence properties of FDOM are significantly affected by pH, at a different extent for a variety of waters. The pH effect depends on several factors such as the sources and chemical nature of DOM, the occurrence of different functional groups, the presence of other organic substances and the contents of trace elements. Four possible mechanisms are proposed from earlier studies for the pH effect (Gosh and Schnitzer 1980; Henderson et al. 2009; Westerhoff et al. 2001; Laane 1982; Patel-Sorrentino et al. 2002; Myneni et al. 1999): (i) the alteration of the molecular orbitals of excitable electrons; (ii) physical changes in the molecular shape caused by changes in charge density (for instance, humic substances have a linear structure at high pH and coil when pH decreases); (iii) competition between H^+ and metal ions to form complexes with the fluorescent substances; and (iv) conformational changes in the molecules that can expose or hide their fluorescent parts. These mechanisms for the pH effect are not well understood. However, because changes in the fluorescence properties due to pH variation from pH 2 to 12 are reversible, it is excluded that irreversible structural changes may occur (Vodacek and Philpot 1987; Patel-Sorrentino et al. 2002).

Reversible pH-induced changes in the fluorescence properties may be caused by two phenomena. First, the H^+ or OH^- ions can alter the availability of electrons to be excited in a specific functional group or fluorophore in a fluorescent molecule, which can significantly modify the fluorescence intensity. Usually, the intensity is increased by an enhancement of electron excitation and decreased by an inhibition. For example, under neutral conditions the functional group ($-CH_2-(NH_3^+)-CH-COO^-$) bound to peak T-like flurophore in tryptophan can show a resonance configuration that favors electron excitation and results into the highest fluorescence intensity (Fig. 9). The availability of electrons to be excited, and the fluorescence intensity as a consequence, decreases both under acidic conditions ($-CH_2-(NH_2)-CH-COOH$) and under basic ones ($-CH_2-(NH_2)-CH-COO^-$). In addition, the availability of non-bonding electrons (:NH–) in another functional group of tryptophan ($C_8H_5(NH)-$) bound to the peak T_{UV}-like flurophore would be highest under neutral conditions, because there is no solvent effect on :NH–. In contrast, the non-bonding electrons of :NH– can react either with H^+ or with OH^-. In both cases the reaction can significantly reduce the availability of non-bonding electrons and, as a consequence, the fluorescence intensity in acidic and in basic solutions.

The main functional groups bound to fulvic and humic acids are –COOH, $-COOCH_3$, –OH, $-OCH_3$, –CH=O, –C=O, $-NH_2$, –NH–, –CH=CH–COOH, $-OCH_3$, S-, O- or N-containing aromatic compounds, although their parent molecular structures are unknown (Mostofa et al. 2009a; Senesi 1990a; Leenheer and Croué 2003; Malcolm 1985; Corin et al. 1996; Peña-Méndez et al. 2005; Seitzinger et al. 2005; Zhang et al. 2005). Such diverse functional groups have strong affinity for complex formation with metal ions. Therefore, the pH effect on fulvic and humic acids shows a different pattern compared to the tryptophan molecule.

3.7 Temperature

The fluorescence intensity of FDOM is inversely related to temperature because of an increased collisional quenching of fluorescence at higher temperatures (Wehry 1973; Vodacek and Philpot 1987). Within the range 10–45 °C, the fluorescence intensity can increase by approximately 1 % with a 1 °C decrease in temperature in the case of tryptophan-like, humic-like and fulvic-like substances, depending on colloid size and fluorophore (Henderson et al. 2009; Baker 2005; Vodacek and Philpot 1987; Elliott et al. 2006; Seredyńska-Sobecka et al. 2007). In contrast, the fluorescence intensities of tryptophan standards are almost unaffected by a temperature variation of ±8 °C (Reynolds 2003). The thermal quenching of fluorescence can be significant because of variation in water temperature between the summer and winter seasons as well as the high variation between boreal, tropical and Antarctic-Arctic regions. The effect of temperature on fluorescence quenching is linear and reversible, and it can be prevented in the laboratory by measuring the fluorescence of samples at a constant temperature (Vodacek and Philpot 1987). The thermal quenching effects can also be overcome by applying simple correction factors, but such factors may be different for fluorophores of different size fractions (Seredyńska-Sobecka et al. 2007). The mechanism of the temperature effect on fluorescence is that a rise in water temperature increases the likelihood that an excited electron will return to its ground state by radiationless decay, leading to reduced fluorescence intensity (Henderson et al. 2009). It is suggested that a variation of water temperature across a range of 20 °C or more between summer and winter would lead to a corresponding decrease by 20 % of the fluorescence intensity during summer (Henderson et al. 2009). Fluorescence changes caused by temperature may have no effect on the structure of the DOM. However, it has been shown that non-reversible changes may occur, possibly as a result of the application of a light-source that may cause photodegradation or thermal decomposition (Vodacek and Philpot 1987).

4 Kinetics of Photodegradation of the Fluorescence Intensity of Fulvic Acid and Tryptophan

Fulvic acid and tryptophan-like fluorescence intensity (FI) decreases monotonically with the number of absorbed UV photons or with integrated solar intensity, as a result of solar effects on water (Fig. 10) (Mostofa et al. 2007a). Photodegradation of fulvic acid often follows a two-step kinetics (Mostofa et al. 2007a; Ma and Green 2004), while tryptophan is photodegraded in a single step (Mostofa et al. 2007a). A decrease of FI can be best fit to a first-order kinetics as follows (Eq. 4.1):

$$\mathrm{Ln}\,(FI/FI_o) = -k_2 S \tag{4.1}$$

where k_2 is the reaction rate constant for photodegradation of FI in waters, FI is the fluorescence intensity obtained after illumination, FI_o is the initial fluorescence

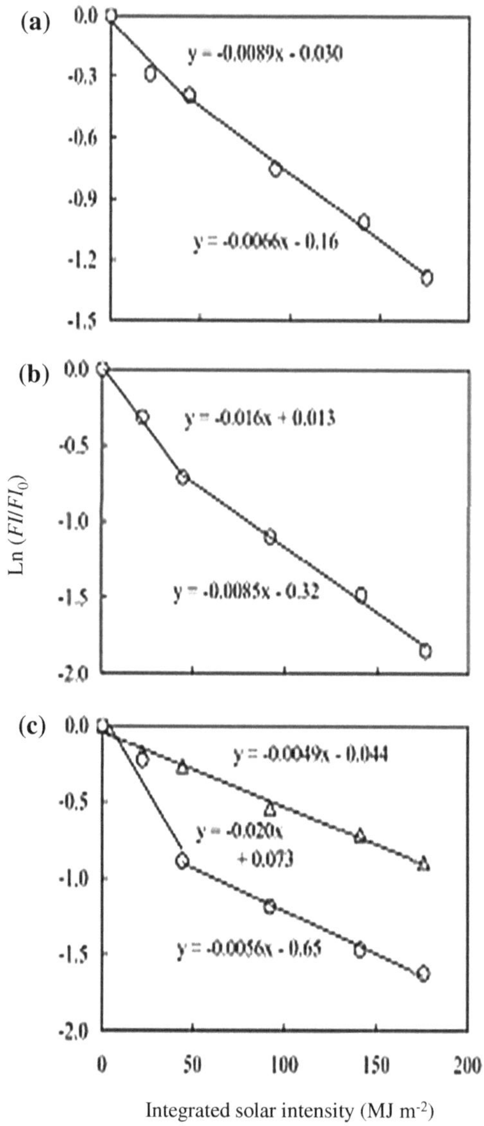

Fig. 10 Relationships between the Ln(FI/FI_0) and the integrated solar intensity and the integrated solar intensity for upstream waters (**a** and **b** Kago upstream and Nishi-Mataya upstream, respectively) and downstream waters (**c** Yasu River). *Open circle* indicates the changes in fulvic acid-like fluorescence intensity (FI) and *open triangle* in **c** shows the change in the protein-like FI under the irradiated condition. *Data source* Mostofa et al. (2007a)

intensity, and S is the integrated solar intensity (MJ m^{-2}). Photodegradation of FI can be clearly understood from the relationship between the S and Ln(FI/FI_o) (Mostofa et al. 2007a).

In river water, the reaction rate constant of fulvic acid-like FI is significantly higher in the first step of photodegradation (8.9–20 $\times$ 10^{-3} MJ^{-1} m^2) than in the second step (5.6–8.5 $\times$ 10^{-3} MJ^{-1} m^2) (Mostofa et al. 2007a). The photodegradation rate constant of protein-like FI often follows one-step kinetics with respect to integrated solar radiation. The photodegradation rate constant of tryptophan-like

FI (4.9×10^{-3} MJ^{-1} m^2) is quite similar to that observed in the second step of fulvic acid (5.6–8.5 $\times 10^{-3}$ MJ^{-1} m^2). From the results one can hypothesize several important photoinduced characteristics of fulvic acid and tryptophan components in waters. The fast first-step photodegradation of fulvic acid suggests the existence of highly sensitive fluorophores (and probably of a discrete class of them) that might undergo quick photoinduced decomposition (del Vecchio and Blough 2002). Schiff-base derivatives (–N=C–C=C–N–) are highly sensitive fluorophores that are commonly detected in DOM humic substances (fulvic and humic acids). They show a fluorescence peak at Ex/Em = 360–390/450–470 nm (Laane 1984) and could well be involved in the initial high losses of FI by fulvic acid in DOM in the aquatic environment. The almost linear reaction rate observed in the second step of fulvic acid photodegradation suggests the presence of homogeneous fluorophores which might be photolytically decomposed, although less quickly than the former ones. The linear photodegradation of tryptophan-like FI could be due to the presence of merely one type of fluorophore (–CH_2–$CH(NH_2)$–COOH), which would undergo gradual photoinduced decomposition. It is finally possible (unless it is a mere coincidence) that the similarity of the reaction rate constants of tryptophan-like FI and of fulvic acid in the second step are due to the presence of similar fluorophores (Mostofa et al. 2007a).

5 Ecological Significance of Photoinduced and Microbial Degradation of FDOM in Natural Waters

5.1 Ecological Significance of Photoinduced Degradation of FDOM

The decrease in fluorescence intensity of various FDOM samples reflects the sequential degradation and mineralization of the corresponding fluorophores or functional groups that are present in the chemical structure of FDOM (Corin et al. 1996; Mostofa et al. 2011; Amador et al. 1989; Bertilsson and Tranvik 1998). Photoinduced degradation modifies the fluorescence properties, and in particular the excitation–emission wavelengths (peaks A and C) of fulvic acid in the aquatic environments (Mostofa et al. 2007a, 2007b; Moran et al. 2000). From the photoinduced degradation of bog DOM and of International Humic Substances Society Nordic fulvic acid, it has been highlighted the losses of carbohydrates, secondary alcohols, protonated and substituted aromatic compounds, carboxyl, amide, ester, ketone and quinones, with no changes in aliphatic carbon (Osburn et al. 2001). Photoinduced changes in FDOM correspond to a decrease in the dissolved organic carbon (DOC) concentration (Brooks et al. 2007; Garcia et al. 2005; Moran et al. 2000; Osburn et al. 2009; Mostofa et al. 2007b; Vähätalo and Wetzel 2004) and to the generation of photoproducts. These processes can be summarized as: (i) Conversion of high-molecular weight into low-molecular weight DOM, which is generally observed in experimental and field observations of natural waters (Corin et al. 1996; Yoshioka et al. 2007; Wu et al. 2005; Morris and Hargreaves 1997). (ii) Formation of microbiologically labile organic substances, which is commonly observed in the epilimnion of natural waters (Bertilsson and Tranvik 2000, 1998; Moran and Zepp 1997). (iii) Formation of CO, CO_2 and dissolved inorganic carbon (DIC, which is usually defined as the sum of dissolved CO_2, H_2CO_3, HCO_3^-, and CO_3^{2-}), which is generally observed upon photodegradation of DOM (Ma and Green 2004; Bertilsson and Tranvik 2000; Granéli et al. 1998; Valentine and Zepp 1993; Miller and Moran 1997). (iv) Formation of N-containing (NH_4^+ or NO_2^-) and P-containing inorganic compounds, which may typically be produced by degradation of dissolved organic nitrogen (DON) and dissolved organic phosphorus (DOP) in the epilimnion of natural waters (Bronk 2002; Zhang et al. 2004; Kim et al. 2006; Vähätalo and Järvinen 2007; Li et al. 2008). (v) Energy changes ($\pm$), such as supply (+) or consumption ($-$) of energy because of the photodegradation of DOM, with (+) representing the photoinduced formation of biologically labile compounds and ($-$) the abiotic mineralization of DOM (Wetzel 1992; Tranvik 1992; Hedges et al. 2000).

5.2 Ecological Significance of Microbial Degradation of FDOM

The major changes in FDOM components and organic matter by microbial degradation can be discriminated as (Mostofa et al. 2009b): (i) Microbial assimilation of organic matter (e.g. algae or phytoplankton) can produce autochthonous DOM or FDOM at different rates in lake waters, and it can simultaneously produce nutrients (PO_4^{3-}, NH_4^+), H_2O_2, organic peroxides, and DIC (Ma and Green 2004; Mostofa and Sakugawa 2009; Zhang et al. 2009a; Yamashita and Tanoue 2008; Palenik and Morel 1988; Weiss et al. 1991; Harvey et al. 1995; Lehmann et al. 2002). (ii) The fluorescence intensities of fulvic and humic acids are usually increased under dark incubation. It is suggested that these compounds are usually recalcitrant to microbial degradation, and microbial effects typically cause a change in the chemical compositions of aliphatic carbon, which may enhance the fluorescence intensity (Mostofa et al. 2009a; Ma and Green 2004; Moran et al. 2000). In contrast the fluorescence of tryptophan-like or protein-like components is often decreased under dark incubation, suggesting that tryptophan-like or protein-like components are labile to microbial degradation (Mostofa et al. 2010; Baker and Inverarity 2004). (iii) Changes in the FDOM components by microbial degradation under dark incubation induce the release of a variety of microbial products such as DIC, PO_4^{3-}, NH_4^+, H_2O_2, organic peroxides and so on (Ma and Green 2004; Moran et al. 2000; Mostofa and Sakugawa 2009; Palenik and Morel 1988). It is thus suggested that microbial processes can induce important changes in DOM composition in natural waters. (iv) Extracellular polymeric substances (EPSs), biologically produced by most bacteria, are composed of a mixture of polysaccharides, mucopolysaccharide and proteins. EPSs mostly show the Ex/

Em peak which is similar to those of protein-like or tryptophan-like fluorescence (Table 2) (Zhang et al. 2010). EPSs produced by anaerobic sludge under sulfate-reducing conditions are capable of biosorption of heavy metals to remove from the waste water treatment plant (Zhang et al. 2010).

6 FDOM Study: A Useful Indicator of DOM Dynamics in Natural Waters

The EEMS and its combination with PARAFAC modeling could be useful to identify the fluorescent organic substances, their sources and their physical, photoinduced and microbial alterations in water (Mostofa et al. 2009a; Hudson et al. 2007; Coble 2007). The main applications are as follows: (i) Identification of the allochthonous fulvic and humic acid of vascular plant origin and of their terrestrial sources (Mostofa et al. 2005a; Stedmon et al. 2003; Ohno and Bro 2006; Singh et al. 2010; Holbrook et al. 2006; Balcarczyk et al. 2009; Santín et al. 2009; Yamashita and Jaffé 2008); (ii) Identification of autochthonous fulvic acids (C-like and M-like) of algal origin and of their sources in water (Stedmon and Markager 2005a, 2005b; Stedmon et al. 2007a; Mostofa et al. 2005b; Zhang et al. 2009a; Kowalczuk et al. 2009; Balcarczyk et al. 2009; Santín et al. 2009; Murphy et al. 2008; Yamashita and Jaffé 2008; Cammack et al. 2004; Nieto-Cid et al. 2005; Boehme and Wells 2006); (iii) Identification of proteins, of aromatic amino acids (tryptophan-like, tyrosine-like and phenylalanine-like) and of their autochthonous sources in water (Yamashita and Tanoue 2003a; Stedmon and Markager 2005a, 2005b; Mostofa et al. 2010; Zhang et al. 2009a; Kowalczuk et al. 2009; Balcarczyk et al. 2009; Santín et al. 2009; Murphy et al. 2008; Yamashita and Jaffé 2008; Boehme and Wells 2006); (iv) Detection of fluorescent whitening agents, of components of detergents and of their anthropogenic sources (Mostofa et al. 2005a, 2010; Komaki and Yabe 1982; Westerhoff et al. 2001; Baker 2002, 2001); (v) Identification of various DOM components of terrestrial, algal and anthropogenic origin in sediment pore waters (Burdige et al. 2004; Fu et al. 2006; Li et al., Characteristics of sediment pore water dissolved organic matter in four Chinese lakes using EEM spectroscopy and PARAFAC modeling, unpublished data); (vi) Characterization of chemical properties of humic substances from soil and compost (Fuentes et al. 2006); (vii) Detection of photoinduced alterations in DOM and in its optical-chemical properties in water (Mostofa et al. 2005a, 2005b; Skoog et al. 1996; Moran et al. 2000; del Vecchio and Blough 2002; Zhang et al. 2009b; Wu et al. 2005), and (viii) Identification of changes in the redox state of fulvic acid (reduced and oxidized) caused by microbial processes in water (Fulton et al. 2004).

EEMS could be useful as a potential monitoring tool to control organic matter pollution (Mostofa et al. 2009a; Hudson et al. 2007; Henderson et al. 2009). It could be applied: (i) To analyze drinking water and sewerage-impacted wastewater (Mostofa et al. 2010; Chen et al. 2003; Baker et al. 2004; Holbrook et al. 2005; Hudson et al. 2008); (ii) To detect pollution levels of anthropogenic DOM in freshwater (Mostofa et al. 2005a; Mostofa et al. 2010; Westerhoff et al. 2001; Baker 2002; Baker et al. 2004); (iii) To monitor the microbial decomposition of DOM (Hayase et al. 1988; Moran et al. 2000; Nieto-Cid et al. 2006); and (iv) to monitor the molecular weight distribution of DOM (Fu et al. 2006; Yoshioka et al. 2007; Wu et al. 2003a; Belzile and Guo 2006; Huguet et al. 2010) and changes in its composition at the watershed level (Mostofa et al. 2005b; Chen et al. 2003; Baker et al. 2004), or from coastal waters to open oceans (Yamashita and Tanoue 2003b).

Finally, the EEMS could be useful in technology development or fundamental research, in the field or at the level of molecular science (Mostofa et al. 2009a). The EEMS has already been applied: (i) In biomedicine or biotechnology to control fermentation in bioreactors (Li and Humphrey 1990) and to detect bacterial biofilms (Angell et al. 1993); (ii) In the detection of natural substances in water such as peroxides (hydrogen peroxide and organic peroxides), allochthonous fulvic and humic acids, proteins, amino acids and so on (Yamashita and Tanoue 2003a, 2003b; Nagao et al. 2003; Fujiwara et al. 1993; Wu et al. 2003a, 2003b); (iii) In the examination of the biological activity in cultures of marine bacteria, algae and coral extracts (Determann et al. 1998; Matthews et al. 1996; Cammack et al. 2004; Elliott et al. 2006); (iv) In the identification of the chemical properties at the molecular level, which arise by interaction between DOM and trace elements (Senesi 1990a, 1990b; Wu et al. 2004a, 2004b, 2007; Yamashita and Jaffé 2008), and (v) in the study of the interaction of DNA with fluorescent substances, e.g. in the framework of DNA–protein interaction (Taylor et al. 2000).

6.1 Are FDOM Studies Superior to CDOM?

The absorption spectra of chromophoric or colored dissolved organic matter (CDOM) usually do not show any specific identifiable peak for freshwater and marine CDOM. CDOM absorption and fluorescence (fulvic or humic acid-like) are significantly correlated with each other in a variety of waters (Ferrari et al. 1996a; del Vecchio and Blough 2004, 2002; Nieke et al. 1997; Vodacek et al. 1995; Ferrari et al. 1996b; Ferrari 2000; Green and Blough 1994; Seritti et al. 1998; Blough and del Vecchio 2002; Stabenau and Zika 2004). The absorbance of CDOM is useful for one to know the contents of the materials present as well as to identify changes in absorbance of total DOM due to physical, photoinduced and biological processes (del Vecchio and Blough 2004, 2002; Coble 2007; Vodacek et al. 1997; Vähätalo and Wetzel 2004; Vähätalo et al. 2000). The slope of the absorption spectrum is widely used in remote sensing in coastal and marine environments (Vodacek et al. 1995; Hoge et al. 1995). It has been reported that there are differences in levels and optical properties between freshwater and marine CDOM. Extreme enrichment in CDOM is usually observed in freshwater environments (Del Vecchio and Blough 2004, 2002; Conmy et al. 2004; Vähätalo and Wetzel 2004; Kowalczuk et al. 2003). Freshwater CDOM absorbs radiation at wavelengths 450–800 nm (Kowalczuk et al. 2003), which is usually not observed in marine waters. This might be due to the large amount of humic substances in freshwater, which absorb radiation at >450 nm. The riverine input of chromophores contained in freshwater CDOM to the coastal marine environment usually meets photodegradation in the coastal areas, which significantly reduces the CDOM content of seawater (del Vecchio and Blough 2002; Vähätalo and Wetzel 2004; Vähätalo et al. 2000).

Excitation of electrons is a typical phenomenon in both CDOM chromophores and FDOM fluorophores (Senesi 1990a; Wu et al. 2005). Therefore, the DOM components contributing to CDOM and FDOM would be partially the same. On the other hand, the fluorescent components in DOM are identified and distinguished on the basis of specific excitation–emission (Ex/Em) wavelength maxima in EEM spectra, upon PARAFAC modeling (Coble 1996; Fulton et al. 2004; Cory and McKnight 2005; Hall et al. 2005; Stedmon and Markager 2005a, 2005b; Ohno and Bro 2006; Stedmon et al. 2007a, 2007b; Mostofa et al. 2010; Wu et al. 2003a). In contrast, it is not possible to identify and distinguish the specific CDOM components due to the absence of peaks in the CDOM absorption spectra (del Vecchio and Blough 2002; Vähätalo and Wetzel 2004; Vähätalo et al. 2000).

The fluorescent organic substances that are usually identified in natural waters using EEM spectra in combination with PARAFAC modeling are fulvic acid-like, humic acid-like, autochthonous fulvic acids (C-like and M-like), protein-like, tryptophan-like, tyrosine-like components, and fluorescent whitening agents (FWAs)-like (Tables 1, 2). On the other hand, absorption spectra at 350, 355 or 375 nm have been used to monitor the CDOM absorption properties (del Vecchio and Blough 2004, 2002; Kowalczuk et al. 2003, 2005), and the specific UV absorbance (SUVA) at 254 or 280 nm has been adopted to estimate the aromatic carbon contents and to understand the chemical characteristics of DOM (Chin et al. 1994; Croué et al. 2003; Weishaar et al. 2003; Świetlik and Sikorska 2004).

6.2 How Do Fluorophores in FDOM Differ from Chromophores in CDOM?

The fluorophores in FDOM are expected to be fundamentally similar to the chromophores in CDOM. For example, tryptophan amino acid ($C_8H_5(NH)$–$CH_2(NH_3^+)CHCOO^-$) has two fluorophores such as $–CH_2–(NH_3^+)–CH–COO^-$ (peak T) and $C_8H_5(NH)–$ (peak T_{UV}). The two fluorophores absorb photons and are thus responsible for tryptophan absorption properties as well. In addition, macromolecules such as allochthonous fulvic acid or humic acid are composed of a number of fluorophores such as Schiff-base derivatives (–N=C–C=C–N–), –COOH, $–COOCH_3$, –OH, $–OCH_3$, –CH=O, –C=O, $–NH_2$, –NH–, –CH=CH–COOH, $–OCH_3$, S-, O- or N-containing aromatic compounds, and so on (Mostofa et al. 2009a; Senesi 1990a; Leenheer and Croué 2003; Malcolm 1985; Corin et al. 1996; Peña-Méndez et al. 2005; Seitzinger et al. 2005; Zhang et al. 2005). These fluorophores need photon absorption for their initial excitation, thus they are fundamentally the chromophores in the respective organic molecule. The fluorophores present in allochthonous fulvic acid can have two fluorescence peaks at peak C-region and peak A-region, but allochthonous humic acid shows several peaks at peak C-region.

On the other hand, allochthonous fulvic acid generally exhibits monotonous absorption spectra whilst allochthonous humic acid has a shoulder at around 400 nm in aqueous media (Hayase and Tsubota 1985; Zepp and Scholtzhauer 1981; Ishiwatari 1973; Lawrence 1980). In addition, CDOM generally exhibits low absorbance at longer wavelengths and the absorbance increases with decreasing wavelength from 700 to 200 nm (Hayase and Tsubota 1985).

Allochthonous fulvic and humic acids, as well as autochthonous fulvic acids are part of the colored DOM (CDOM) and absorb radiation at 200–800 nm. Along with them, also FDOM, protein-like, tryptophan-like, tyrosine-like, FWAs-like and other fluorescent components absorb radiation at 200–800 nm. In addition, there is a vast number of allochthonous and autochthonous non-fluorescent organic substances. They do not display fluorescence properties, but absorb radiation at specific wavelength ranges. For example, acetaldehyde absorb light at 208–224 nm (Mopper et al. 1991; Kieber et al. 1990), acetate at 204–270 nm (Wetzel et al. 1995; Dahlén et al. 1996), formaldehyde at 207–250 nm (Mopper et al. 1991; Kieber et al. 1990), glyoxal at <240 nm (Mopper et al. 1991), malonate at 225–240 nm (Dahlén et al. 1996) and so on. All these organic molecules are termed as CDOM but they do not belong to FDOM because as they do not show fluorescence properties. Therefore 'all fluorescent DOM (FDOM) is colored or chromophoric DOM (CDOM), but not all CDOM is also FDOM'.

7 Scope of the Future Challenges

Autochthonous fulvic acids (C-like and M-like) of algal origin show fluorescence properties at peak C- and A- regions, for which they show a similar behavior as allochthonous fulvic and humic acids. Researchers did not distinguish between the photoinduced and microbial degradation of the autochthonous DOM (fulvic acids) nor its differentiation with terrestrial fulvic acid, and this should be a key focus for future research. Two types of autochthonous fulvic acids (C-like and M-like) can be distinguished based on the presence of fluorophores. This material is originated from algal biomass or phytoplankton biomass. Among these two fulvic acids, the C-like fulvic acid is produced photolytically and microbially and undergoes rapid photoinduced degradation, therefore it does not appear as a key component in natural waters. However, it is important to extract the autochthonous fulvic acids from water, to identify them using other spectroscopic methods and to make relationship with fluorescence properties. The extraction of autochthonous fulvic acids, which represent key DOM sources in lake and marine waters, and the study of their photoinduced and biological changes would gain useful information by using EEM-PARAFAC.

A few studies have been conducted on the photoinduced and microbial changes of FDOM in natural waters. It should be important to conduct experiments in Asia, Africa and Latin America, because in these continents there are regions where freshwaters are highly contaminated with untreated sewerage and industrial effluents and few studies are presently available. Emerging contaminants such as pharmaceuticals, hormones, endocrine disrupting compounds and so on are widely detected in natural waters (Richardson 2007). The application of EEM-PARAFAC to identify emerging contaminants would constitute a new dimension to control and detect these harmful organic substances in aquatic environments. Finally, PARAFAC modeling of sample EEM spectra could be a useful parameter for the identification of DOM components and for the elucidation of photoinduced, biological and any other changes in the molecular properties of the fluorescent components. Some researchers use the Raman Unit fluorescence (nm^{-1}) EEM data for PARAFAC modeling, which causes significant changes in the component's fluorescence excitation–emission maxima and in the fluorescence intensity compared to QSU or a.u. calibration. It is strongly suggested to use either QSU or a.u. calibration, which allows a better comparison of the fluorescence properties and their application to DOM dynamics in natural waters.

8 Nomenclature

a.u.	Arbitrary unit
CDOM	Colored or chromophoric dissolved organic matter
DAS1	Diaminostilbene-type
DOC	Dissolved organic carbon
DOM	Dissolved organic matter
DSBP	Distyryl biphenyl
EEM	Excitation–emission matrix
EEMS	Excitation–emission matrix spectroscopy
FDOM	Fluorescent dissolved organic matter
FI	Fluorescence intensity
F_{index}	Fluorescence index
Fulvic acids	Two components are identified in standard SRFA
FWAs	Fluorescent whitening agents
Humic acids	Two components are identified in standard SRHA
NoHA	Nordic Reference-humic acid
NoFA	Nordic Reference-fulvic acid
PARAFAC	Parallel factor
QSU	Quinine sulfate unit
RU	Raman Unit

SRFA Suwannee River Fulvic Acid
SRHA Suwannee River Humic Acid

Problems

(1) Explain the principles of excitation–emission matrix spectroscopy.
(2) What is the fluorophore in a fluorescent molecule? What are the controlling factors that affect the fluorophores?
(3) Mention the key fluorescent substances detected in natural waters and explain how the fluorescence properties of allochthonous fulvic acids differ from those of humic acids.
(4) How is it possible to distinguish the fluorescence properties of fulvic acids of vascular plant origin from those of autochthonous fulvic acids of algal origin?
(5) What is PARAFAC modeling? Explain the importance of PARAFAC modeling in the separation and identification of the fluorescent components in EEM spectroscopy.
(6) Mention the key factors affecting the fluorescence properties of FDOM in natural waters, and explain how pH variation affects the fluorescence properties.
(7) Explain the effect of photoinduced degradation on the fluorescence properties of humic substances (fulvic and humic acids) in water, and provide the possible mechanism.
(8) What are the importance and impact of photoinduced degradation of FDOM in natural waters?
(9) Mention which FDOM components are significantly affected by microbial degradation in waters.
(10) Explain the changes of the fluorescent properties of various FDOM components due to microbial degradation in aqueous media.
(11) Mention the possible mechanisms for microbial degradation of fluorescent substances in aqueous media?
(12) What are the importance and impacts of microbial degradation of fluorescent substances in aqueous media?
(13) Explain the sentence 'All fluorescent DOM (FDOM) is colored or chromophoric DOM (CDOM), but not all CDOM is also FDOM'.
(14) in what respect are FDOM studies superior to CDOM one?
(15) Explain how the fluorophores in FDOM differ from chromophores in CDOM.

Acknowledgments We thank Dr. Kazuhide Hayakawa of Lake Biwa Environmental Research Institute, Japan for his valuable comments; Dr. Jie Yuan and Dr. Ling Li of Institute of Geochemistry, Chinese Academy of Sciences for their generous help. This work was financially supported by the Institute of Geochemistry, the Chinese Academy of Sciences, Guiyang, China. This work was partly supported by Nanjing Institute of Geography and Limnology, Chinese Academy of Sciences, China; Kyoto University, Japan; University Turin, Italy; and Hiroshima University, Japan. This chapter acknowledges the reprinted from Senesi (1990a), Copyright (1990), with permission from Elsevier; reprinted from Zhang et al. (2010), Copyright (2010), with permission from Elsevier; Copyright (2005, 2007) by The Geochemical Society of Japan; reprinted from Stedmon et al. (2003), Copyright (2003), with permission from Elsevier; reprinted from Bro (1997), Copyright (1997), with permission from Elsevier; copyright (2005, 2010) by the Japanese Society of Limnology; Copyright (2009) CSIRO; reprinted from Patel-Sorrentino et al. (2002), Copyright (2002), with permission from Elsevier; and reprinted (adapted) with permission from Clarke et al. (1976), Copyright (1976) American Chemical Society.

References

Abboudi M, Jeffrey W, Ghiglione JF, Pujo-Pay M, Oriol L, Sempéré R, Charrière B, Joux F (2008) Effects of photochemical transformations of dissolved organic matter on bacterial metabolism and diversity in three contrasting coastal sites in the Northwestern Mediterranean Sea during summer. Microb Ecol 55:344–357

Allard B, Borén H, Pattersson C, Zhang G (1994) Degradation of humic substances by UV irradiation. Environ Int 20:97–101

Almgren T, Josefsson B, Nyquist G (1975) A fluorescence method for studies of spent sulfite liquor and humic substances in sea water. Anal Chem Acta 78:411–422

Amador JA, Alexander M, Zika RG (1989) Sequential photochemical and microbial degradation of organic molecules bound to humic acid. App Environ Microb 55:2843–2849

Angell P, Arrage AA, Mittelman MW, White DC (1993) On line, non-destructive biomass determination of bacterial biofilms by fluorometry. J Microbiol Meth 18:317–327

Aoki S, Ohara S, Kimura K, Mizuguchi H, Fuse Y, Yamada E (2008) Characterization of fluorophores released from three kinds of lake phytoplankton using gel chromatography and fluorescence spectrophotometry. Anal Sci 24:1461–1467

Assel M, Laenen R, Laubereau A (1998) Ultrafast electron trapping in an aqueous NaCl-solution. Chem Phys Lett 289:267–274

Baek M, Nelson WH, Hargraves PE, Tanguay JF, Suib SL (1988) The steady state and decay characteristics of protein tryptophan fluorescence from algae. Appl Spectrosc 42:1405–1412

Baghoth SA, Sharma SK, Amy GL (2010) Tracking natural organic matter (NOM) in a drinking water treatment plant using fluorescence excitation emission matrices and PARAFAC. Water Res 45:797–809

Baker A (2001) Fluorescence excitation-emission matrix characterization of some sewage impacted rivers. Environ Sci Technol 35:948–953

Baker A (2002) Fluorescence excitation-emission matrix characterization of some farm wastes: Implications for water quality monitoring. Water Res 36:189–194

Baker A (2005) Thermal fluorescence quenching properties of dissolved organic matter. Water Res 39:4405–4412

Baker A, Curry M (2004) Fluorescence of leachates from three contrasting landfills. Water Res 38:2605–2613

Baker A, Inverarity R (2004) Protein-like fluorescence intensity as a possible tool for determining river water quality. Hydrol Process 18:2927–2945

Baker A, Ward D, Lieten SH, Periera R, Simpson EC, Slater M (2004) Measurement of protein-like fluorescence in river and waste water using a handheld spectrophotometer. Water Res 38:2934–2938

Baker A, Elliott S, Lead JR (2007) Effects of filtration and pH perturbation on freshwater organic matter fluorescence. Chemosphere 67:2035–2043

Balcarczyk KL, Jones Jr JB, Rudolf Jaffe' R, Maie N (2009) Stream dissolved organic matter bioavailability and composition in watersheds underlain with discontinuous permafrost. Biogeochemistry 94:255–270

Battin TJ (1998) Dissolved organic matter in a blackwater tributary of the upper Orinoco River, Venezuela. Org Geochem 28:561–569

Baunsgaard D, Munck L, Nørgaard L (2000) Evaluation of the quality of solid sugar samples by fluorescence spectroscopy and chemometrics. Appl Spectrosc 54:438–444

Baunsgaard D, Nørgaard L, Godshall MA (2001) Specific screening for color precursors and colorants in beet and cane sugar liquors in relation to model colorants using spectrofluorometry evaluated by HPLC and multiway data analysis. J Agric Food Chem 49:1687–1694

Belzile C, Guo L (2006) Optical properties of low molecular weight and colloidal organic matter: application of the ultrafiltration permeation model to DOM absorption and fluorescence. Mar Chem 98:183–196

Benner R, Opsahl S (2001) Molecular indicators of the sources and transformations of dissolved organic matter in the Mississippi river plume. Org Geochem 32:597–611

Berger P, Laane RPWM, Ilahude AG, Ewald M, Courtot P (1984) Comparative study of dissolved fluorescent matter in four west-european estuaries. Oceanol Acta 7:309–313

Bertilsson S, Tranvik LJ (1998) Photochemically produced carboxylic acids as substrates for freshwater bacterioplankton. Limnol Oceanogr 43:885–895

Bertilsson S, Tranvik LJ (2000) Photochemical transformation of dissolved organic matter in lakes. Limnol Oceanogr 45:753–762

Bertilsson S, Carlsson P, Granéli W (2004) Influence of solar radiation on the availability of dissolved organic matter to bacteria in the Southern Ocean. Deep Sea Res II 51:2557–2568

Bianchi TS, Dawson R, Sawangwong P (1988) The effects of macrobenthic deposit-feeding on the degradation of chloropigments in sandy sediments. J Exp Mar Biol Ecol 122:243–255

Blough NV, Del Vecchio R (2002) Chromophoric DOM in the coastal environment. In: Hansell DA, Carlson CA (eds) Biogeochemistry of marine dissolved organic matter. Elsevier Science, New York, p 774

Boehme J, Wells M (2006) Fluorescence variability of marine and terrestrial colloids: examining size fractions of chromophoric dissolved organic matter in the Damariscotta River estuary. Mar Chem 101:95–103

Boehme J, Coble PG, Conmy RN, Stovall-Leonard A (2004) Examining CDOM fluorescence variability using principal component analysis: seasonal and regional modelling of three-dimensional fluorescence in the Gulf of Mexico. Mar Chem 89:3–14

Borisover M, Laor Y, Parparov A, Bukhanovsky N, Lado M (2009) Spatial and seasonal patterns of fluorescent organic matter in Lake Kinneret (Sea of Galilee) and its catchment basin. Water Res 43:3104–3116

Boyd TJ, Osburn CL (2004) Changes in CDOM fluorescence from allochthonous and autochthonous sources during tidal mixing and bacterial degradation in two coastal estuaries. Mar Chem 89:189–210

Boyd TJ, Barham BP, Hall GJ, Schumann BS, Paerl RW, Osburn CL (2010) Variation in ultrafiltered and LMW organic matter fluorescence properties under simulated estuarine mixing transects: 2 Mixing with photoexposure. J Geophys Res 115:G00F14. doi:101029/2009JG000994

Bro R (1997) PARAFAC tutorial and applications. Chemom Intell Lab Syst 38(2):149–171

Bro R (1998) Multi-way analysis in the food industry Models, algorithms, and applications. PhD thesis, University of Amsterdam, Netherlands

Bro R (1999) Exploratory study of sugar production using fluorescence spectroscopy and multiway analysis. Chemom Intell Lab Syst 46:133–147

Bronk DA (2002) Dynamics of DON. In: Carlson CA, Hansell DA (eds) Biogeochemistry of marine dissolved organic matter. Academic Press, San Diego, pp 153–249

Brooks ML, Meyer JS, McKnight DM (2007) Photooxidation of wetland and riverine dissolved organic matter: altered copper complexation and organic composition. Hydrobiologia 579:95–113

Brun GL, Milburn DLD (1977) Automated fluorescence determination of humic substances in natural water. Anal Lett 10:1209–1219

Burdige DJ, Kline SW, Chen W (2004) Fluorescent dissolved organic matter in marine sediment pore waters. Mar Chem 89:289–311
Cabaniss SE (1992) Synchronous fluorescence spectra of metal-fulvic acid complexes. Environ Sci Technol 26:1133–1139
Cabaniss SE, Shuman MS (1988) Fluorescence quenching measurements of copper–fulvic acid binding. Anal Chem 60:2418–2421
Cammack W, Kalff J, Prarie Y, Smith EM (2004) Fluorescent dissolved organic matter in lakes: relationship with heterotrophic metabolism. Limnol Oceanogr 49:2034–2045
Carlson DJ, Mayer LM (1983) Relative influence of riverine and macroalgal phenolic materials on UV absorbance in temperate coastal waters. Can J Fishering Aquatic Sci 40:1258–1263
Caroll JD, Chang J (1970) Analysis of individual differences in multidimensional scaling via an N-way generalisation of and Eckart–Young decomposition. Psychometrika 35:283
Chen RF, Bada JL (1989) Seawater and porewater fluorescence in the Santa Barbara Basin. Geophys Res Lett 16:687–690
Chen RF, Bada JL (1990) A laser-based fluorometry system for investigations of seawater and pore water fluorescence. Mar Chem 31:219–230
Chen RF, Bada JL (1992) The fluorescence of dissolved organic matter in seawater. Mar Chem 37:191–221
Chen W, Westerhoff P, Leenheer JA, Booksh K (2003) Fluorescence excitation-emission matrix regional integration to quantify spectra for dissolved organic matter. Environ Sci Tech 37:5701–5710
Chen M, Price RM, Yamashita Y, Jaffé R (2010) Comparative study of dissolved organic matter from groundwater and surface water in the Florida coastal Everglades using multidimensional spectrofluorometry combined with multivariate statistics. Appl Geochem 25:872–880
Chin YP, Aiken GR, ØLoughlin E (1994) Molecular weight, polydispersity, and spectroscopic properties of aquatic humic substances. Environ Sci Technol 28:1853–1858
Christman RF, Ghassemi M (1966) Chemical nature of organic color in water. J Am Water Works Assoc 58:723–741
Christman RF, Minear RA (1967) Fluorometric detection of lignin sulfonates. Trend Eng 19:3–7 (University of Washington, Seattle, College of Engineering)
Clarke RH, Connors RE, Schaafsma TJ, Kleibeuker JF, Platenkamp RJ (1976) The triplet state of chlorophylls. J Am Chem Soc 98(12):3674–3677
Coble PG (1996) Characterization of marine and terrestrial DOM in sea water using excitation-emission matrix spectroscopy. Mar Chem 52:325–336
Coble PG (2007) Marine optical biogeochemistry: the chemistry of ocean color. Chem Rev 107:402–418
Coble PG, Green SA, Blough NV, Gagosian RB (1990) Characterization of dissolved organic matter in the Black Sea by fluorescence spectroscopy. Nature 348:432–435
Coble PG, Schultz CA, Mopper K (1993) Fluorescence contouring analysis of DOC inter calibration samples: a comparison of techniques. Mar Chem 41:173–178
Coble PG, Del Castillo CE, Avril B (1998) Distribution and optical properties of CDOM in the Arabian Sea during the 1995 Southwest Monsoon. Deep-Sea Res II 45:2195–2223
Conmy RN, Coble PG, Chen RF, Gardner GB (2004) Optical properties of colored dissolved organic matter in the Northern Gulf of Mexico. Mar Chem 89:127–144
Conrad R (1999) Contribution of hydrogen to methane production and control of hydrogen concentrations in methanogenic soils and sediments. FEMS Microbiol Ecol 28:193–202
Corin N, Backlund P, Kulovaara M (1996) Degradation products formed during UV-irradiation of humic waters. Chemosphere 33:245–255
Cory RM, McKnight DM (2005) Fluorescence spectroscopy reveals ubiquitous presence of oxidized and reduced quinines in dissolved organic matter. Environ Sci Technol 39:8142–8149
Cory RM, McKnight DM, Chin Y-P, Miller P, Jaros CL (2007) Chemical characteristics of fulvic acids from Arctic surface waters: microbial contributions and photochemical transformations. J Geophys Res 112:G04S51. doi:101029/2006JG000343

Croué JP, Benfdetti MF, Violleauand D, Leenheer JA (2003) Characterization and copper binding of humic and nonhumic organic matter isolated from the south platte river: evidence for the presence of nitrogenous binding site. Environ Sci Technol 37:328–336
da Silva JCGE, Leitao JMM, Costa FS, Ribeiro JLA (2002) Detection of verapamil drug by fluorescence and trilinear decomposition techniques. Anal Chim Acta 453:105–115
Dahlén J, Bertilsson S, Pettersson C (1996) Effects of UV-A irradiation on dissolved organic matter in humic surface waters. Environ Int 22:501–506
de Souza Sierra MM, Donard OFX, Lamotte M (1997) Spectral identification and behaviour of dissolved organic fluorescent materials during estuarine mixing processes. Mar Chem 58:51–58
de Souza-Sierra MM, Donard OFX, Lamotte M, Belin C, Ewald M (1994) Fluorescence spectroscopy of coastal and marine waters. Mar Chem 47:127–144
del Vecchio R, Blough NV (2002) Photobleaching of chromophoric dissolved organic matter in natural waters: kinetics and modeling. Mar Chem 78:231–253
del Vecchio R, Blough NV (2004) Spatial and seasonal distribution of chromophoric dissolved organic matter and dissolved organic carbon in the Middle Atlantic Bight. Mar Chem 89:169–187
Determann S, Reuter R, Wanger P, Willkomm R (1994) Fluorescence matter in the eastern Atlantic Ocean. Part 1: Method of measurement and near-surface distribution. Deep-Sea Res 41:659–675
Determann S, Reuter R, Willkomm R (1996) Fluorescence matter in the eastern Atlantic Ocean. Part 2: Vertical profiles and relation to water masses. Deep-Sea Res 43:345–360
Determann S, Lobbes JM, Reuter R, Rullkotter J (1998) Ultraviolet fluorescence excitation and emission spectroscopy of marine algae and bacteria. Mar Chem 62:137–156
Dorsch JE, Bidleman TF (1982) Natural organics as fluorescent tracers of river-sea mixing. Estuar Coast Shelf Sci 15:701–707
Dubnick A, Barker J, Sharp M, Wadham J, Lis G, Telling J, Fitzsimons S, Jackson M (2010) Characterization of dissolved organic matter (DOM) from glacial environments using total fluorescence spectroscopy and parallel factor analysis. Ann Glaciol 51:111–122
Duursma EK (1974) The fluorescence of dissolved organic matter in the sea. In: Jerlov NG, Nielsen ES (eds) Optical aspects of oceanography. Academic Press, New York, pp 237–256
Elliott S, Lead JR, Baker A (2006) Thermal quenching of fluorescence of freshwater, planktonic bacteria. Anal Chim Acta 564:219–225
Fellman JB, D'Amore DV, Hood E, Boone RD (2008) Fluorescence characteristics and biodegradability of dissolved organic matter in forest and wetland soils from coastal temperate watersheds in southeast Alaska. Biogeochemistry 88:169–184
Fellman JB, Hood E, D'Amore DV, Edwards RT, White D (2009) Seasonal changes in the chemical quality and biodegradability of dissolved organic matter exported from soils to streams in coastal temperate rainforest watersheds. Biogeochemistry 95:277–293. doi:10.1007/s10533-009-9336-6
Fellman JB, Spencer RGM, Hernes PJ, Edwards RT, D'Amore DV, Hood E (2010) The impact of glacier runoff on the biodegradability and biochemical composition of terrigenous dissolved organic matter in near-shore marine ecosystems. Mar Chem 121:112–122
Ferrari GM (2000) The relationship between chromophoric dissolved organic matter and dissolved organic carbon in the European Atlantic coastal area and in the West Mediterranean Sea (Gulf of Lions). Mar Chem 70:339–357
Ferrari GM, Dowell MD, Grossi S, Targa C (1996a) Relationship between the optical properties of chromophoric dissolved organic matter and total concentration of dissolved organic carbon in the southern Baltic Sea region. Mar Chem 55:299–316
Ferrari GM, Hoepffner N, Mingazzini M (1996b) Optical properties of the water in a deltaic environment: prospective tool to analyze satellite data in turbid waters. Remote Sens Environ 56:69–80
Fu P, Wu FC, Liu C, Wei Z, Bai Y, Liao H (2006) Spectroscopic characterization and molecular weight distribution of dissolved organic matter in sediment porewaters from Lake Erhai, Southwest China. Biogeochemistry 81:179–189

Fu PQ, Wu FC, Liu CQ, Wang F, Li W, Yue L, Guo QJ (2007) Fluorescence characterization of dissolved organic matter in an urban river and its complexation with Hg(II). Appl Geochem 22:1668–1679

Fu P, Mostofa KMG, Wu FC, Liu CQ, Li W, Liao H, Wang L, Wang J, Mei Y (2010) Excitation-emission matrix characterization of dissolved organic matter sources in two eutrophic lakes (Southwestern China Plateau). Geochem J 44:99–112

Fuentes M, González-Gaitano G, José M, García-Mina JM (2006) The usefulness of UV-vissible and fluorescence spectroscopies to study the chemical nature of humic substances from soils and composts. Org Geochem 37:1949–1959

Fujiwara K, Ushiroda T, Takeda K, Kumamoto Y, Tsubota H (1993) Diurnal and seasonal distribution of hydrogen peroxide in seawater of Seto Inland Sea. Geochem J 27:103–115

Fukushima M, Tatsumi K, Nagao S (2001) Degradation characteristics of humic acid during photo-fenton processes. Environ Sci Technol 35:3683–3690

Fulton JR, McKnight DM, Foreman CM, Cory RM, Stedmon C, Blunt E (2004) Changes in fulvic acid redox state through the oxycline of a permanently ice-covered Antarctic lake. Aquat Sci 66:27–46

Gao H, Zepp RG (1998) Factors influencing photoreactions of dissolved organic matter in a coastal river of the Southeastern United States. Environ Sci Technol 32:2940–2946

Gao L, Fan D, Li D, Cai J (2010) Fluorescence characteristics of chromophoric dissolved organic matter in shallow water along the Zhejiang coasts, Southeast China. Mar Environ Res 69:187–197

Garcia E, Amyot M, Ariya PA (2005) Relationship between DOC photochemistry and mercury redox transformations in temperate lakes and wetlands. Geochim Cosmochim Acta 69:1917–1924

Geller A (1986) Comparison of mechanisms enhancing biodegradability of refractory lake water constituents. Limnol Oceanogr 31:755–764

Ghassemi M, Christman RF (1968) Properties of the yellow organic acids of natural waters. Limnol Oceanogr 13:583–597

Gopinathan C, Damle PS, Hart EJ (1972) Gamma-ray irradiated sodium chloride as a source of hydrated electrons. J Phys Chem 76:3694–3698

Gosh K, Schnitzer M (1980) Fluorescence excitation spectra of humic substances. Can J Soil Sci 60:373–379

Grabowski ZR, Rotkiewicz K, Rettig W (2003) Structural changes accompanying intramolecular electron transfer: focus on twisted intramolecular charge-transfer states and structures. Chem Rev 103:3899–4032

Granéli W, Lindell M, Marcal De Farria B, De Assis Esteves F (1998) Photoproduction of dissolved inorganic carbon in temperate and tropical lakes-dependence on wavelength band and dissolved organic carbon concentration. Biogeochemistry 43:175–195

Green SA, Blough NV (1994) Optical absorption and fluorescence properties of chromophoric dissolved organic matter in natural waters. Limnol Oceanogr 39:1903–1916

Gron C, Wassenaar L, Krog M (1996) Origin and structures of groundwater humic substances from three Danish aquifers. Environ Int 22:519–534

Guglielmetti L (1975) Photochemical and biological degradation of water-soluble FWAs. In: Anliker R, Müller G (eds) Fluorescent whitening agents. Georg Thieme Publishers, Stuttgart, pp 180–190

Guilbault GG (1990) Practical fluorescence: theory, methods and techniques, 2nd edn. Dekker, New York

Guo W, Xu J, Wang J, Wen Y, Zhuo J, Yan Y (2010) Characterization of dissolved organic matter in urban sewage using excitation emission matrix fluorescence spectroscopy and parallel factor analysis. J Environ Sci 22:1728–1734

Hall KJ, Lee GF (1974) Molecular size and spectral characterization of organic matter in a meromictic lake. Water Res 8:239–251

Hall GJ, Clow KE, Kenny JE (2005) Estuarial fingerprinting thorough multidimensional fluorescence and multivariate analysis. Environ Sci Technol 39:7560–7567

Hama T (1991) Production and turnover rates of fatty acids in marine particulate matter through phytoplankton photosynthesis. Mar Chem 33:213–227

Hama T, Yanagi K, Hama J (2004) Decrease in molecular weight of photosynthetic products of marine phytoplankton during early diagencsis. Limnol Oceanogr 49:471–481
Hanamachi Y, Hama T, Yanai T (2008) Decomposition process of organic matter derived from freshwater phytoplankton. Limnology 9:57–69
Harshman RA (1970) Foundations of the PARAFAC procedure: model and conditions for an explanatory multi-mode factor analysis. UCLA Work Pap Phonetics 16:1–84
Harvey HR, Macko SA (1997) Kinetics of phytoplankton decay during simulated sedimentation: changes in lipids under oxic and anoxic conditions. Org Geochem 27:129–140
Harvey HR, Tuttle JH, Bell JT (1995) Kinetics of phytoplankton decay during simulated sedimentation: changes in biochemical composition and microbial activity under oxic and anoxic conditions. Geochim Cosmochim Acta 59:3367–3377
Hayakawa K, Sekino T, Yoshioka T, Maruo M, Kumagai M (2003) Dissolved organic carbon and fluorescence in Lake Hovsgol: factors reducing humic content of the lake water. Limnology 4:25–33
Hayase K, Shinozuka N (1995) Vertical distribution of fluorescent organic matter along with AOU and nutrients in the Equatorial Pacific. Mar Chem 48:282–290
Hayase K, Tsubota H (1983) Sedimentary humic acid and fulvic acid as surface active substances. Geochim Cosmochim Acta 47:947–952
Hayase K, Tsubota H (1985) Sedimentary humic and fulvic acids as fluorescent organic materials. Geochim Cosmochim Acta 49:159–163
Hayase K, Yamamoto M, Nakazawa I, Tsubota H (1987) Behavior of natural fluorescence in Sagami Bay and Tokyo Bay, Japan—vertical and lateral distributions. Mar Chem 20:265–276
Hayase K, Tsubota H, Sunada I, Goda S, Yamazaki H (1988) Vertical distribution of fluorescent organic matter in the North Pacific. Mar Chem 25:373–381
Hayashi Y, Managaki S, Takada H (2002) Fluorescent whitening agents in Tokyo Bay and adjacent rivers: their application as anthropogenic molecular markers in coastal environments. Environ Sci Technol 36:3556–3563
Hedges JI (1992) Global biogeochemical cycles: progress and problems. Mar Chem 39:67–93
Hedges JI, Eglinton G, Hatcher PG, Kirchman DL, Arnosti C, Dereenne S, Evershed RP, Kögel-Knabner I, de Leeuw JW, Littke R, Michaelis W, Rullkötter J (2000) The molecularly-uncharacterized component of nonliving organic matter in natural environments. Org Geochem 31:945–958
Helms JR, Stubbins A, Ritchie JD, Minor EC, Kieber DJ, Mopper K (2008) Absorption spectral slopes and slope ratios as indicators of molecular weight, source, and photobleaching of chromophoric dissolved organic matter. Limnol Oceanogr 53:955–969
Henderson RK, Baker A, Murphy KR, Hambly A, Stuetz RM, Khan SJ (2009) Fluorescence as a potential monitoring tool for recycled water systems: a review. Water Res 43:863–881
Hiriart-Baer VP, Smiith REH (2005) The effect of ultraviolet radiation on freshwater planktonic primary production: the role of recovery and mixing processes. Limnol Oceanogr 50:1352–1361
Hiriart-Baer VP, Diep N, Smith REH (2008) Dissolved organic matter in the great lakes: role and nature of allochthonous material. J Great Lakes Res 34:383–394
Hoge FE, Williams ME, Swift RN, Yungel JK, Vodacek A (1995) Satellite retrieval of the absorption coefficient of chromophoric dissolved organic matter in continental margins. J Geophy Res 100(24847):24854
Holbrook RD, Breidenich J, DeRose PC (2005) Impact of reclaimed water on select organic matter properties of a receiving stream-fluorescence and perylene sorption behavior. Environ Sci Technol 39:6453–6460
Holbrook RD, Yen JH, Grizzard TJ (2006) Characterizing natural organic material from the Occoquan Watershed (Northern Virginia, US) using fluorescence spectroscopy and PARAFAC. Sci Total Environ 361:249–266
Hua B, Dolan F, McGhee C, Clevenger TE, Deng B (2007) Water-source characterization with fluorescence EEM spectroscopy: PARAFAC analysis. Int J Environ Anal Chem 87:135–147
Hudson N, Baker A, Renolds D (2007) Fluorescence analysis of dissolved organic matter in natural, waste and polluted waters—a review. River Res Appl 23:631–649

Hudson N, Baker A, Ward D, Reynolds DM, Brunsdon C, Carliell-Marquet C, Browning S (2008) Can fluorescence spectrometry be used as a surrogate for the biochemical oxygen demand (BOD) test in water quality assessment? An example from South West England. Sci Total Environ 391:149–158
Huguet A, Vacher L, Relexans S, Saubusse S, Froidefond J-M, Parlanti E (2009) Properties of fluorescent dissolved organic matter in the Gironde Estuary. Org Geochem 40:706–719
Huguet A, Vacher L, Saubusse S, Etcheber H, Abril G, Relexans S, Ibalot F, Parlanti E (2010) New insights into the size distribution of fluorescent dissolved organic matter in estuarine waters. Org Geochem 41:595–610
Hulatt CJ, Thomas DN, Bowers DG, Norman L, Zhang C (2009) Exudation and decomposition of chromophoric dissolved organic matter (CDOM) from some temperate macroalgae. Estuar Coast Shelf Sci 84:147–153
Hunt JF, Ohno T, Fernandez IJ (2008) Influence of foliar phosphorus and nitrogen contents on chemical properties of water extractable organic matter derived from fresh and decomposed sugar maple leaves. Soil Biol Biochem 40:1931–1939
Ishiwatari R (1973) Chemical characterization of fractionated humic acids from lake and marine sediments. Chem Geol 12:113–126
Jaffé R, Boyer JN, Lu X, Maie N, Yange C, Scully NM, Mock S (2004) Source characterization of dissolved organic matter in a subtropical mangrovedominated estuary by fluorescence analysis. Mar Chem 84:195–210
Jiji RD, Cooper GA, Booksh KS (1999) Excitation–emission matrix fluorescence based determination of carbamate pesticides and polycyclic aromatic hydrocarbons. Anal Chim Acta 397:61–72
Johannessen SC, Peña MA, Quenneville ML (2007) Photochemical production of carbon dioxide during a coastal phytoplankton bloom. Estuar Coast Shelf Sci 73:236–242
Kalle K (1949) Fluoreszenz und Gelbstoff im Bottnischem und Finnishchen Meerbusen. Dtsch Hydrogr Z 2:9–124
Kalle K (1963) Uber das Verhalten und die Herkunft der in den Gewassern und in der Atmospheare vorhandenen himmelblauen Fluoreszenz. Dtsch Hydrogr Z 16:153–166
Kieber RJ, Zhou X, Mopper K (1990) Formation of carbonyl compounds from UV-induced photodegradation of humic substances in natural waters: fate of riverine carbon in the sea. Limnol Oceanogr 35:1503–1515
Kieber RJ, Hydro LH, Seaton PJ (1997) Photooxidation of triglycerides and fatty acids in seawater: implication toward the formation of marine humic substances. Limnol Oceanogr 42:1454–1462
Kim C, Nishimura Y, Nagata T (2006) Role of dissolved organic matter in hypolimnetic mineralization of carbon and nitrogen in a large, monomictic lake. Limnol Oceanogr 51:70–78
Klapper L, McKnight DM, Fulton JR, Blunt-Harris EL, Nevin KP, Lovley DR, Hatcher PG (2002) Fulvic acid oxidation state detection using fluorescence spectroscopy. Environ Sci Technol 36:3170–3175
Klein B, Gieskes WWC, Kraay GG (1986) Digestion of chlorophylls and carotenoids by the marine protozoan Oxyrrhis marina studied by h.p.l.c analysis of algal pigments. J Plankton Res 8:827–836
Komada T, Schofield OME, Reimers CE (2002) Fluorescence characteristics of organic matter released from coastal sediments during resuspension. Mar Chem 79:81–97
Komaki M, Yabe A (1982) Fluorometric analysis of fluorescent brightening agents in natural waters. Chem Soc J 5:859–867 (in Japanese)
Kotsyurbenko OR, Glagolev MV, Nozhevnikova AN, Conrad R (2001) Competition between homoacetogenic bacteria and methanogenic archaea for hydrogen at low temperature. FEMS Microbiol Ecol 38:153–159
Kowalczuk P, Cooper WJ, Whitehead RF, Durako MJ, Sheldon W (2003) Characterization of CDOM in an organic-rich river and surrounding coastal ocean in the South Atlantic Bight. Aquat Sci 65:384–401

Kowalczuk P, Stoń-Egiert J, Cooper WJ, Whitehead RF, Durako MJ (2005) Characterization of chromophoric dissolved organic matter (CDOM) in the Baltic Sea by excitation emission matrix fluorescence spectroscopy. Mar Chem 96:273–292

Kowalczuk P, Durako MJ, Young H, Kahn AE, Cooper WJ, Gonsior M (2009) Characterization of dissolved organic matter fluorescence in the South Atlantic Bight with use of PARAFAC model: Interannual variability. Mar Chem 113:182–196

Kramer JB, Canonica S, Hoigne J, Kaschig J (1996) Degradation of fluorescent whitening agents in sunlit natural waters. Environ Sci Technol 30:2227–2234

Kulovaara M, Corin N, Backlund P, Tervo J (1995) Impact of UV254-radiation on aquatic humic substances. Chemosphere 33:783–790

Kwan WP, Voelker BM (2002) Decomposition of hydrogen peroxide and organic compounds in the presence of dissolved iron and ferrihydrite. Environ Sci Technol 36:1467–1476

Laane RWPM (1980) Conservative behavior of dissolved organic carbon in the Ems-Dollart Estuary and the Western Sea. Neth J Sea Res 14:192–199

Laane RWPM (1982) Influence of pH on the fluorescence of dissolved organic matter. Mar Chem 11:395–401

Laane RWPM (1984) Comment on the structure of marine fulvic and humic acids. Mar Chem 15:85–87

Laane RWPM, Kramer KJM (1990) Natural fluorescence in the North Sea and its major estuaries. Neth J Sea Res 26:1–9

Lakowicz JR (1983) Principles of fluorescence spectroscopy. Plenum Press, New York

Lakowicz JR (1999) Principles of fluorescence spectroscopy, 2nd edn. Kluwer Academic, New York

Laurion I, Ventura M, Catalan J, Psenner R, Sommarruga R (2000) Attenuation of ultraviolet radiation in mountain lakes: factors controlling the among- and within-lake variability. Limnol Oceanogr 45:1274–1288

Lawrence J (1980) Semi-quantitative determination of fulvic acid, tannin and lignin in natural waters. Water Res 14:373–377

Leenheer JA, Croué JP (2003) Characterizing aquatic dissolved organic matter. Environ Sci Technol 37:18–26

Lehmann MF, Bernasconi SM, Peichert P, McKenzie JA (2002) Preservation of organic matter and alteration of its carbon and nitrogen isotope composition during simulated and in situ early sedimentary diagenesis. Geochim Cosmochim Acta 66:3573–3584

Lepane V, Persson T, Wedborg M (2003) Effects of UV-B radiation on molecular weight distribution and fluorescence from humic substances in riverine and low-salinity water. Estuar Coastal Shelf Sci 56:161–173

Levesque M (1972) Fluorescence and gel filtration of humic compounds. Soil Sci 113:346–353

Li K, Crittenden J (2009) Computerized pathway elucidation for hydroxyl radical-induced chain reaction mechanisms in aqueous phase advanced oxidation processes. Environ Sci Technol 43:2831–2837

Li JK, Humphrey AE (1990) Use of fluorometry for monitoring and control of a bioreactor. Biotechnol Bioeng 37:1043–1049

Li W, Wu FC, Liu CQ, Fu PQ, Wang J, Mei Y, Wang L, Guo J (2008) Temporal and spatial distributions of dissolved organic carbon and nitrogen in two small lakes on the Southwestern China Plateau. Limnology 9:163–171

Liu H, Fang HHP (2002) Characterization of electrostatic binding sites of extracellular polymers by linear programming analysis of titration data. Biotechnol Bioeng 80:806–811

Lochmuller CH, Saavedra SS (1986) Conformational changes in a soil fulvic acid measured by time dependent fluorescence depolarization. Anal Chem 38:1978–1981

Lovley DR, Coates JD, Blunt-Harris EL, Phillips EJP, Woodward JC (1996) Humic substances as electron acceptors for microbial respiration. Nature 382:445–448

Lu YF, Allen HE (2002) Characterization of copper complexation with natural dissolved organic matter (DOM)—link to acidic moieties of DOM and competition by Ca and Mg. Water Res 36:5083–5101

Lu XQ, Jaffé R (2001) Interaction between Hg(II) and natural dissolved organic matter: a fluorescence spectroscopy based study. Water Res 35:1793–1803

Lu F, Chang C-H, Lee D-J, He P-J, Shao L-M, Su A (2009) Dissolved organic matter with multi-peak fluorophores in landfill leachate. Chemosphere 74:575–582

Luciani X, Mounier S, Paraquetti HHM, Redon R, Lucas Y, Bois A, Lacerda LD, Raynaud M, Ripert M (2008) Tracing of dissolved organic matter from the SEPETIBA Bay (Brazil) by PARAFAC analysis of total luminescence matrices. Mar Environ Res 65:148–157

Ma XD, Ali N (2009) Detection of a DNA-like materials in Suwannee River Fulvic Acid. In: Wu FC, Xing B (eds) Natural organic matter and its significance in the environment. Science Press, Beijing, pp 66–89

Ma X, Green SA (2004) Photochemical transformation of dissolved organic carbon in Lake Superior—an in situ experiment. J Great Lakes Res 30(suppl 1):97–112

Mack J, Bolton JR (1999) Photochemistry of nitrite and nitrate in aqueous solution: a review. J Photochem Photobiol A Chem 128:1–13

Malcolm RL (1985) Geochemistry of stream fulvic and humic substances. In: Aiken GR, McKnight DM, Wershaw RL, MacCarthy P (eds) Humic substances in soil, sediment, and water: geochemistry, isolation and characterization. Wiley, New York, pp 181–209

Malcolm RL (1990) The uniqueness of humic substances in each of soil, stream and marine environments. Anal Chim Acta 232:19–30

Managaki S, Takada H (2005) Fluorescent whitening agents in Tokyo Bay sediments: molecular evidence of lateral transport of land-derived particulate matter. Mar Chem 95:113–127

Manciulea A, Baker A, Lead JR (2009) A fluorescence quenching study of the interaction of Suwannee River fulvic acid with iron oxide nanoparticles. Chemosphere 76:1023–1027

Marmé N, Knemeyer J-P, Sauer M, Wolfrum J (2003) Inter- and intramolecular fluorescence quenching of organic dyes by tryptophan. Bioconjugate Chem 14:1133–1139

Matthews BJH, Jones AC, Theodorou NK, Tudhope AW (1996) Excitation-emission matrix fluorescence spectroscopy applied to humic acid in coral reefs. Mar Chem 55:312–317

Mayer LM, Schick LL, Loder TC (1999) Dissolved protein fluorescence in two Maine estuaries. Mar Chem 64:171–179

McCarthy MD, Hedges JI, Benner R (1996) Major biochemical composition of dissolved high molecular weight organic matter in seawater. Mar Chem 55:281–297

McCreary JJ, Snoeyink VL (1980) Characterization and activated carbon adsorption of several humic substances. Water Res 14:151–160

McCullen WL (1996) Polymers for detergents: current technology and future trends. In: Coffey RT (ed) New horizons: an AOCS/CSMA detergent industry conference. The American Oil Chemists Society—Science, Washington, DC, pp 42–55

McKnight DM, Boyer EW, Westerhoff PK, Doran PT, Kulbe T, Andersen DT (2001) Spectrofluorometric characterization of dissolved organic matter for indication of precursor organic material and aromaticity. Limnol Oceanogr 46:38–48

Miles CJ, Brezonik L (1981) Oxygen consumption in humic-colored waters by a photochemical ferrous-ferric catalytic cycle. Environ Sci Technol 15:1089–1095

Miller WL, Moran MA (1997) Interaction of photochemical and microbial processes in the degradation of refractory dissolved organic matter from a coastal marine environment. Limnol Oceanogr 42:1317–1324

Miller WL, Moran MA, Sheldon WM, Zepp RG, Opsahl S (2002) Determination of apparent quantum yield spectra for the formation of biologically labile photoproducts. Limnol Oceanogr 47:343–352

Miller MP, McKnight DM, Chapra SC (2009) Production of microbially-derived fulvic acid from photolysis of quinone-containing extracellular products of phytoplankton. Aquat Sci 71:170–178

Minakata D, Li K, Westerhoff P, Crittenden J (2009) Development of a group contribution method to predict aqueous phase hydroxyl radical (HO•) reaction rate constants. Environ Sci Technol 43:6220–6227

Mobed JJ, Hemmingsen SL, Autry JL, McGown LB (1996) Fluorescence characterization of IHSS humic substances: total luminescence spectra with absorbance correction. Environ Sci Technol 30:3061–3065

Moberg L, Robertsson G, Karlberg B (2001) Spectrofluorimetric determination of chlorophylls and pheopigments using parallel factor analysis. Talanta 54:161–170

Momzikoff A, Dallot S, Pizay M-D (1992) Blue and yellow fluorescence of filtered seawater in a frontal zone (Ligurian Sea, northwest Mediterranean Sea). Deep-Sea Res 39:1481–1498

Mopper K, Schultz CA (1993) Fluorescence as a possible tool for studying the nature and water column distribution of DOC components. Mar Chem 41:229–238

Mopper K, Zhou X, Kieber RJ, Kieber DJ, Sikorski RJ, Jones RD (1991) Photochemical degradation of dissolved organic carbon and its impact on the oceanic carbon cycle. Nature 353:60–62

Moran MA, Zepp RG (1997) Role of photoreactions in the formation of biologically labile compounds from dissolved organic matter. Limnol Oceanogr 42:1307–1316

Moran MA Jr, Sheldon WM, Zepp RG (2000) Carbon loss and optical property changes during long-term photochemical and biological degradation of estuarine dissolved organic matter. Limnol Oceanogr 45:1254–1264

Morris DP, Hargreaves BR (1997) The role of photochemical degradation of dissolved organic carbon in regulating the UV transparency of three lakes on the Pocono Plateau. Limnol Oceanogr 42:239–249

Mostofa KMG (2005) Dynamics, characteristics and photochemical processes of fluorescent dissolved organic matter and peroxides in river water. Ph D Thesis, Sept 2005, Hiroshima University, Japan

Mostofa KMG, Sakugawa H (2009) Spatial and temporal variations and factors controlling the concentrations of hydrogen peroxide and organic peroxides in rivers. Environ Chem 6:524–534

Mostofa KMG, Honda Y, Sakugawa H (2005a) Dynamics and optical nature of fluorescent dissolved organic matter in river waters in Hiroshima prefecture, Japan. Geochem J 39:257–271

Mostofa KMG, Yoshioka T, Konohira E, Tanoue E, Hayakawa K, Takahashi M (2005b) Three-dimensional fluorescence as a tool for investigating the dynamics of dissolved organic matter in the Lake Biwa watershed. Limnology 6:101–115

Mostofa KMG, Yoshioka T, Konohira E, Tanoue E (2007a) Photodegradation of fluorescent dissolved organic matters in river waters. Geochem J 41:323–331

Mostofa KMG, Yoshioka T, Konohira E, Tanoue E (2007b) Dynamics and characteristics of fluorescent dissolved organic matter in the groundwater, river and lake water. Water Air Soil Pollut 184:157–176

Mostofa KMG, Wu FC, Yoshioka T, Sakugawa H, Tanoue E (2009a) Dissolved organic matter in the aquatic environments. In: Wu FC, Xing B (eds) Natural organic matter and its significance in the environment. Science Press, Beijing, pp 3–65

Mostofa KMG, Liu CQ, Wu FC, Fu PQ, Ying WL, Yuan J (2009b) Overview of key biogeochemical functions in lake ecosystem: impacts of organic matter pollution and global warming. Proceedings of 13th World Lake Conference, Wuhan, China, 1–5 Nov 2009, Keynote Speech, pp 59–60

Mostofa KMG, Wu FC, Liu CQ, Fang WL, Yuan J, Ying WL, Wen L, Yi M (2010) Characterization of Nanming River (Southwestern China) impacted by sewerage pollution using excitation-emission matrix and PARAFAC. Limnology 11:217–231. doi:101007/s10201-009-0306-4

Mostofa KMG, Wu FC, Liu CQ, Yoshioka T, Sakugawa H, Tanoue E (2011) Photochemical, microbial and metal complexation behavior of fluorescent dissolved organic matter in the aquatic environments. Invited review. Geochem J 45:235–254

Mounier S, Zhao H, Garnier C, Redon R (2011) Copper complexing properties of dissolved organic matter: PARAFAC treatment of fluorescence quenching. Biogeochemistry. doi:10.1007/s10533-010-9486-6

Muller-Karger FE, Hu C, Andrefouet S, Varela R, Thunell R (2005) The color of the coastal ocean and applications in the solution of research and management problems. Springer, Dordrecht

Münster U (1991) Extracellular enzyme activity in eutrophic and polyhumic lakes. In: Chrost RJ (ed) Microbial enzymes in aquatic environments. Springer, New York, pp 96–122

Murphy KR, Stedmon CA, Waite TD, Ruiz GM (2008) Distinguishing between terrestrial and autochthonous organic matter sources in marine environments using fluorescence spectroscopy. Mar Chem 108:40–58

Myneni SCB, Brown JT, Martinez GA, Meyer-Ilse W (1999) Imaging of humic substance macromolecular structures in water and soils. Science 286:1335–1337

Nagao S, Matsunaga T, Suzuki Y, Ueno T, Amano H (2003) Characteristics of humic substances in the Kuji River as determined by high-performance size exclusion chromatography with fluorescence detection. Water Res 37:4159–4170

Nakajima H (2006) Studies on photochemical degradation processes of dissolved organic matter in seawater. MS Thesis, Hiroshima University, pp 1–173

Nieke B, Reuter R, Heuermann R, Wang H, Babin M, Therriault JC (1997) Light absorption and fluorescence properties of chromophoric dissolved organic matter (CDOM), in the St Lawrence estuary (Case 2 waters). Cont Shelf Res 17:235–252

Nieto-Cid M, Álvarez-Salgado A, Gago J, Pérez FF (2005) DOM fluorescence, a tracer for biogeochemical processes in a coastal upwelling system (NW Iberian Peninsula). Mar Ecol Prog Ser 297:33–50

Nieto-Cid M, Álvarez-Salgado A, Pérez FF (2006) Microbial and photochemical reactivity of fluorescent dissolved organic matter in a coastal upwelling system. Limnol Oceanogr 51:1391–1400

Obernosterer I, Herndl GJ (2000) Differences in the optical and biological reactivity of the humic and nonhumic dissolved organic carbon component in two contrasting coastal marine environments. Limonl Oceanogr 45:1120–1129

Obernosterer I, Ruardij P, Herndl GJ (2001) Spatial and diurnal dynamics of dissolved organic matter (DOM) fluorescence and H2O2 and the photochemical oxygen demand of surface water DOM across the subtropical Atlantic Ocean. Limnol Oceanogr 46:632–643

Oheim M, Michael DJ, Geisbauer M, Madsen D, Chow RH (2006) Principles of two-photon excitation fluorescence microscopy and other nonlinear imaging approaches. Adv Drug Deliv Rev 58:788–808

Ohno T, Bro R (2006) Dissolved organic matter characterization using multiway spectral decomposition of fluorescence landscapes. Soil Sci Soc Am J 70:2028–2037

Ohno T, Amirbahman A, Bro R (2007) Parallel factor analysis of excitation–emission matrix fluorescence spectra of water soluble soil organic matter as basis for the determination of conditional metal binding parameters. Environ Sci Technol 42:186–192

Ohno T, Hunt JF, Gray KA (2009) Multi-dimensional fluorescence spectroscopy and PARAFAC for dissolved organic matter characterization In: Wu FC, Xing B (Ed), Natural Organic Matter and Its Significance in the Environment, Science Press, Beijing, pp 90-104

Osburn CL, Morris DP, Thorn KA, Moeller RE (2001) Chemical and optical changes in freshwater dissolved organic matter exposed to solar radiation. Biogeochemistry 54:251–278

Osburn CL, O'Sullivan DW, Boyd TJ (2009) Increases in the longwave photobleaching of chromophoric dissolved organic matter in coastal waters. Limnol Oceanogr 54:145–159

Otto L (1967) Investigations on optical properties and water-masses of the Southern North Sea. Neth J Sea Res 3:532–552

Palenik B, Morel FMM (1988) Dark production of H2O2 in the Sargasso Sea. Limnol Oceanogr 33:1606–1611

Parlanti P, Worz K, Geoffroy L, Lamotte M (2000) Dissolved organic matter fluorescence spectroscopy as a tool of estimate biological activity in a coastal zone submitted to anthropogenic inputs. Org Geochem 31:1765–1781

Patel-Sorrentino N, Mounier S, Benaim JY (2002) Excitation- emission fluorescence matrix to study pH influence on organic matter fluorescence in the Amazon basin rivers. Water Res 36(10):2571–2581

Patel-Sorrentino N, Mounier S, Lucas Y, Benaim JY (2004) Effects of UV-visible irradiation on natural organic matter from the Amazon basin. Sci Total Environ 321:231–239

Peña-Méndez EM, Havel J, Patočka J (2005) Humic substances-compounds of still unknown structure: applications in agriculture, industry, environment, and biomedicine. J Appl Biomed (Rev) 3:13–24

Persson T, Wedborg M (2001) Multivariate evaluation of the fluorescence of aquatic organic matter. Anal Chim Acta 434:179–192

Petersen HT (1989) Determination of an Isochrysis galbana algal bloom by L-tryptophan fluorescence. Mar Pollut Bull 20:447–451

Peuravuori J, Pihlaja K (1999) Structural characterization of humic substances. In: Keskitalo J, Eloranta P (eds) Limnology of humic waters. Backhuy Publishers, Leiden, pp 22–39

Poiger T, Field JA, Field TM, Giger W (1996) Occurrence of fluorescent whitening agents in sewage and river water determined by solid-phase extraction and high-performance liquid chromatography. Environ Sci Technol 30:2220–2226

Poiger T, Kari FG, Giger W (1999) Fate of fluorescent whitening agents in the River Glatt. Environ Sci Technol 33:533–539

Provenzano MR, Orazio VD, Jerzykiewicz M, Senesi N (2004) Fluorescence behaviour of Zn and Ni complexes of humic acids from different sources. Chemosphere 55:885–892

Pullin MJ, Cabaniss SE (1995) Rank analysis of the pH dependent synchronous fluorescence spectra of six standard humic substances. Environ Sci Technol 29:1460–1467

Qian J, Mopper K, Kieber DJ (2001) Photochemical production of the hydroxyl radical in Antarctic waters. Deep-Sea Res I 48:741–759

Randall CE, Harvey VL, Manney GL, Orsolini Y, Codrescu M, Sioris C, Brohede S, Haley CS, Gordley LL, Zawdony JM, Russell JM (2005) Stratospheric effects of energetic particle precipitation in 2003–2004. Geophys Res Lett:LO5082. doi:101029/2004GL022003

Reche I, Pace ML, Cole JJ (1999) Relationship of trophic and chemical conditions to photobleaching of dissolved organic matter in lake ecosystems. Biogeochemistry 44:259–280

Reynolds DM (2003) Rapid and direct determination of tryptophan in water using synchronous fluorescence spectroscopy. Water Res 37:3055–3060

Richard C, Canonica S (2005) Aquatic phototransformation of organic contaminants induced by coloured dissolved natural organic matter. Hdb Env Chem 2(Part M):299–323

Richardson SD (2007) Water analysis: emerging contaminants and current issue. Anal Chem 79:4295–4324

Rochelle-Newall EJ, Fisher TR (2002) Production of chromophoric dissolved organic matter fluorescence in marine and estuarine environments: an investigation into the role of phytoplankton. Mar Chem 77:7–21

Rochelle-Newall EJ, Fisher TR, Fan C, Glibert PM (1999) Dynamics of chromophoric dissolved organic matter and dissolved organic carbon in experimental mesocosms. Int J Remote Sens 20:627–641

Rosenstock B, Simon M (2003) Consumption of dissolved amino acids and carbohydtares by limnetic bacterioplankton according to molecular weight fractions and proportions bound to humic matter. Microb Ecol 45:433–443

Rosenstock B, Zwisler W, Simon M (2005) Bacterial consumption of humic and non-humic low and high molecular weight DOM and the effect of solar irradiation on the turnover of labile DOM in the Southern Ocean. Microb Ecol 50:90–101

Ross RT, Lee C, Davis CM, Ezzeddine BM, Fayyad EA, Leurgans SE (1991) Resolution of the fluorescence spectra of plant pigment complexes using trilinear models. Biochim Biophys Acta 1056:317–320

Santín C, Yamashita Y, Otero XL, Álvarez MÁ, Jaffé R (2009) Characterizing humic substances from estuarine soils and sediments by excitation-emission matrix spectroscopy and parallel factor analysis. Biogeochemistry 96:131–147

Schwede-Thomas SB, Chin Y, Dria KJ, Hatcher P, Kaiser E, Sulzberger B (2005) Characterizing the properties of dissolved organic matter isolated by XAD and C-18 solid phase extraction and ultrafiltration. Aquat Sci 67:61–71

Seitzinger SP, Hartnett H, Lauck R, Mazurek M, Minegishi T, Spyres G, Styles R (2005) Molecular-level chemical characterization and bioavalability of dissolved organic matter in stream water using electrospray-ionization mass spectrometry. Limnol Oceanogr 50:1–12

Senesi N (1990a) Molecular and quantitative aspects of the chemistry of fulvic acid and its interactions with metal ions and organic chemicals. Part II: The fluorescence spectroscopy approach. Anal Chim Acta 232:77–106

Senesi N (1990b) Molecular and quantitative aspects of the chemistry of fulvic acid and its interactions with metal ions and organic chemicals. Part I: The electron spin resonance approach. Anal Chim Acta 232:51–75

Seredyńska-Sobecka B, Baker A, Lead JR (2007) Characterisation of colloidal and particulate organic carbon in freshwaters by thermal fluorescence quenching. Water Res 41:3069–3076

Seritti A, Russo D, Nannicini L, del Vecchio R (1998) DOC, absorption and fluorescence properties of estuarine and coastal waters of the northern Tyrrhenian Sea. Chem Speciat Bioavailab 10:95–105

Shank GC, Zepp RG, Vähätalo A, Lee R, Bartels E (2010) Photobleaching kinetics of chromophoric dissolved organic matter derived from mangrove leaf litter and floating Sargassum colonies. Mar Chem 119:162–171

Sholkovitz ER (1976) Flocculation of dissolved organic and inorganic matter during the mixing of river water and seawater. Geochim Cosmochim Acta 40:831–845

Shriner RL, Horne WH, Cox RFB (1943) p-Nitrophenyl Isocyanate. Org Synth Coll 2:453–455

Singh S, D'Sa EJ, Swenson EM (2010) Chromophoric dissolved organic matter (CDOM) variability in Barataria Basin using excitation-emission matrix (EEM) fluorescence and parallel factor analysis (PARAFAC). Sci Total Environ 408:3211–3222

Skoog A, Wedborg M, Fogelqvist E (1996) Photobleaching of fluorescence and the organic carbon concentration in a coastal environment. Mar Chem 55:333–345

Smith DS, Kramer JR (1999) Fluorescence analysis for multisite aluminium binding to natural organic matter. Environ Int 25:307–314

Southworth BA, Voelker BM (2003) Hydroxyl radical production via the photo-Fenton reaction in the presence of fulvic acid. Environ Sci Technol 37:1130–1136

Spencer RGM, Bolton L, Baker A (2007) Freeze/thaw and pH effects on freshwater dissolved organic matter fluorescence and absorbance properties from a number of UK locations. Water Res 41:2941–2950

Stabenau ER, Zika RG (2004) Correlation of the absorption coefficient with a reduction in mean mass for dissolved organic matter in southwest Florida river plumes. Mar Chem 89:55–67

Stedmon CA, Markager S (2005a) Tracing the production and degradation of autochthonous fractions of dissolved organic matter by fluorescence analysis. Limnol Oceanogr 50:1415–1426

Stedmon CA, Markager S (2005b) Resolving the variability in dissolved organic matter fluorescence in a temperate estuary and its catchment using PARAFAC analysis. Limnol Oceanogr 50:686–697

Stedmon CA, Markager S, Bro R (2003) Tracing dissolved organic matter in aquatic environments using a new approach to fluorescence spectroscopy. Mar Chem 82(3–4):239–254

Stedmon CA, Markager S, Tranvik L, Kronberg L, Slätis T, Martinsen W (2007a) Photochemical production of ammonium and transformation of dissolved organic matter in the Baltic Sea. Mar Chem 104:227–240

Stedmon CA, Thomas DN, Granskog M, Kaartokallio H, Papaditriou S, Kuosa H (2007b) Characteristics of dissolved organic matter in Baltic coastal sea ice: allochthonous or autochthonous origins? Environ Sci Technol 41:7273–7279

Steelink C (2002) Investigating humic acids in soils. Anal Chem 74:328A–333A

Stewart AJ, Wetzel RG (1980) Fluorescence: absorbance ratios-a molecular-weight tracer of dissolved organic matter. Limnol Oceanogr 25:559–564

Stoll J-MA, Giger W (1998) Mass balance for detergent-derived fluorescent whitening agents in surface waters of Switzerland. Water Res 32:2041–2050

Stoll J-MA, Ulrich MM, Giger W (1998) Dynamic behavior of fluorescent whitening agents in Greifensee: field measurements combined with mathematical modeling of sedimentation and photolysis. Environ Sci Technol 32:1875–1881

Sugiyama Y, Anegawa A, Inokuchi H, Kumagai T (2005) Distribution of dissolved organic carbon and dissolved fulvic acid in mesotrophic Lake Biwa, Japan. Limnology 6:161–168

Sun M, Lee C, Aller RC (1993) Anoxic and oxic degradation of 14C-labeled chloropigments and a 14C-labeled diatom in Long Island Sound sediments. Limnol Oceanogr 38:1438–1451

Sun Q, Lu R, Yu A (2012) Structural heterogeneity in the collision complex between organic dyes and tryptophan in aqueous solution. J Phys Chem B 116:660–666

Suzuki Y, Nagao S, Nakaguchi Y, Matsunaga T, Muraoka S, Hiraki K (1997) Spectroscopic properties of fluorescent substances in natural waters. Geochemistry 31:171–180 (in Japanese)

Świetlik J, Sikorska E (2004) Application of fluorescence spectroscopy in the studies of natural organic matter fractions reactivity with chlorine dioxide and ozone. Water Res 38:3791–3799

Tadrous PJ (2000) Methods for imaging the structure and function of living tissues and cells: 2. Fluorescence lifetime imaging. J Pathol 191:229–234

Taylor JR, Fang MM, Nie S (2000) Probing specific sequences on single DNA molecules with bioconjugated fluorescent nanoparticles. Ana Chem 72:1979–1986

Thomas DN, Lara RJ (1995) Photodegradation of algal derived dissolved organic carbon. Mar Ecol Prog Ser 116:309–310

Thurman EM (1985) Humic substances in groundwater. In: Aiken GR, McKnight DM, Wershaw RL, MacCarthy P (eds) Humic substances in soil, sediment, and water: geochemistry, isolation, and characterization, Wiley, New York, pp 87–103

Tranvik LJ (1992) Allochthonous disolved organic matter as an energy source for pelagic bacteria and the concept of the microbial loop. Hydrobiologia 229:107–114

Uyguner CS, Bekbolet M (2005) Evaluation of humic acid photocatalytic degradation by UV-vis and fluorescence spectroscopy. Catal Today 101:267–274

Vacher L (2004) Etude par fluorescence des propriétés de la matière organique dissoute dans les systèmes estuariens. Cas des estuaires de la Gironde et de la Seine, PhD thesis, Université Bordeaux 1, pp 255. http://archives.disvu.ubordeaux1.fr/proprietes.html?numero_ordre=2923

Vähätalo AV, Järvinen M (2007) Photochemically produced bioavailable nitrogen from biologically recalcitrant dissolved organic matter stimulates production of a nitrogen-limited microbial food web in the Baltic Sea. Limnol Oceanogr 52:132–143

Vähätalo AV, Wetzel RG (2004) Photochemical and microbial decomposition of chromophoric dissolved organic matter during long (months-years) exposures. Mar Chem 89:313–326

Vähätalo AV, Salkinoja-Saonen M, Taalas P, Salnen K (2000) Spectrum of the quantum yield for photochemical mineralization of dissolved organic carbon in a humic lake. Limnol Oceanogr 45:664–676

Valentine RL, Zepp RG (1993) Formation of carbon monoxide from the photodegradation of terrestrial dissolved organic carbon in natural waters. Environ Sci Technol 27:409–412

Van Heemst JDH, Megens L, Hatcher PG, de Leeuw JW (2000) Nature, origin and average age of estuarine ultrafiltered dissolved organic matter as determined by molecular and carbon isotope characterization. Org Geochem 31:847–857

Vione D, Lauri V, Minero C, Maurino V, Malandrino M, Carlotti ME, Olariu RI, Arsene C (2009) Photostability and photolability of dissolved organic matter upon irradiation of natural water samples under simulated sunlight. Aquat Sci 71:34–45

Visser SA (1984) Fluorescence phenomena of humic matter of aquatic origin and microbial cultures. In: Christman RF, Gjessing ET (eds) Aquatic and Terrestrial Humic materials. Ann Arbor Science, Ann Arbor, pp 183–202

Vodacek A, Philpot WD (1987) Environmental effects on laser induced fluorescence spectra of natural waters. Remote Sens Environ 21:83–95

Vodacek A, Hoge F, Swift RN, Yungel JK, Peltzer ET, Blough NV (1995) The use of in situ and airborne fluorescence measurements to determine UV absorption coefficients and DOC concentrations in surface waters. Limnol Oceanogr 40:411–415

Vodacek A, Blough NV, DeGrandpre D, Peltzer ET, Nelson RK (1997) Seasonal variation of CDOM and DOC in the Middle Atlantic Bight: terrestrial input and photooxidation. Limnol Oceanogr 42:674–686

Volk CJ, Volk CB, Kaplan LA (1997) Chemical composition of biodegradable dissolved organic matter in streamwater. Limnol Oceanogr 42:39–44

Waiser MJ, Robarts RD (2004) Photodegradation of DOC in a shallow prairie wetland: evidence from seasonal changes in DOC optical properties and chemical characteristics. Biogeochemistry 69:263–284

Wakeham SG, Lee C (1993) Production, transport, and alteration of particulate organic matter in the marine water column. In: Engel MH, Macko SA (eds) Org Geochem. Plenum Press, New York, pp 145–169

Wakeham SG, Lee C, Hedges JI, Hernes PJ, Peterson ML (1997) Molecular indicators of diagenetic status in marine organic matter. Geochim Cosmochim Acta 61:5363–5369

Wang GS, Liao CH, Wu FJ (2001) Photodegradation of humic acids in the presence of hydrogen peroxide. Chemosphere 42:379–387

Wang Z-G, Liu W-Q, Zhao N-J, Li H-B, Zhang Y-J, Si-Ma W-C, Liu J-G (2007) Composition analysis of colored dissolved organic matter in Taihu Lake based on three dimension excitation-emission fluorescence matrix and PARAFAC model, and the potential application in water quality monitoring. J Environ Sci 19:787–791

Wedborg M, Persson T, Larsson T (2007) On the distribution of UV-blue fluorescent organic matter in the Southern Ocean. Deep-Sea Res I 54:1957–1971

Wehry EL (1973) Effects of molecular structure and molecular environment on fluorescence. In: Guilbault GG (ed) Practical fluorescence: theory, methods, and techniques. Marcel Dekker, New York, pp 79–136

Weishaar JL, Aiken GR, Bergamaschi BA, Fram MS, Fujii R, Mopper K (2003) Evaluation of specific ultraviolet absorbance as an indicator of the chemical composition and reactivity of dissolved organic carbon. Environ Sci Technol 37:4702–4708

Weiss RF, Carmack EC, Koropalov VM (1991) Deep-water renewal and biological production in Lake Baikal. Nature 349:665–669

Westerhoff P, Chen W, Esparza M (2001) Fluorescence analysis of a standard fulvic acid and tertiary treated wastewater. J Environ Qual 30:2037–2046

Wetzel RG (1992) Gradient-dominated ecosystems: Sources and regulatory functions of dissolved organic matter in freshwater ecosystems. Hydrobiologia 229:181–198

Wetzel RG, Hatcher PG, Bianchi TS (1995) Natural photolysis by ultraviolet irradiance of recalcitrant dissolved organic matter to simple substrates for rapid bacterial metabolism. Limnol Oceanogr 40:1369–1380

White EM, Vaughan PP, Zepp RG (2003) Role of the photo-Fenton reaction in the production of hydroxyl radicals and photobleaching of colored dissolved organic matter in a coastal river of the southeastern United States. Aquat Sci 65:402–414

Willey JD (1984) The effect of seawater magnesium on natural fluorescence during estuarine mixing, and implications for tracer applications. Mar Chem 15:19–45

Willey JD, Atkinson LP (1982) Natural fluorescence as a tracer for distinguishing between Piedmont and coastal plain river waters in the nearshore waters of Georgia and North Carolina. Estuarine Coastal Shelf Sci 14:49–59

Winter AR, Fish TAE, Playle RC, Smith DS, Curtis PJ (2007) Photodegradation of natural organic matter from diverse freshwater sources. Aquat Toxicol 84:215–222

Wolfbeis OS (1985) The fluorescence of organic natural products In: Schulman SG (ed) Molecular luminescence spectroscopy, methods and applications V 77: Part I. Wiley, New York, pp 167–370

Wu FC, Tanoue E (2001a) Molecular mass distribution and fluorescence characteristics of dissolved organic ligands for copper (II) in Lake Biwa, Japan. Org Geochem 32:11–20

Wu FC, Tanoue E (2001b) Geochemical characterization of organic ligands for copper (II) in different molecular size fractions in Lake Biwa, Japan. Org Geochem 32:1311–1318

Wu FC, Midorikawa T, Tanoue E (2001) Fluorescence properties of organic ligands for copper(II) in Lake Biwa and its rivers. Geochem J 35:333–346

Wu FC, Evans RD, Dillon PJ (2002) Fractionation and characterization of fulvic acid by immobilized metal ion affinity chromatography. Anal Chim Acta 452:85–93

Wu FC, Tanoue E, Liu CQ (2003a) Fluorescence and amino acid characteristics of molecular size fractions of DOM in the waters of Lake Biwa. Biogeochem 65:245–257

Wu FC, Evans RD, Dillon PJ (2003b) High-performance liquid chromatographic fractionation and characterization of fulvic acid. Anal Chim Acta 464:47–55

Wu FC, Cai YR, Evans RD, Dillon PJ (2004a) Complexation between Hg(II) and dissolved organic matter in stream waters: an application of fluorescence spectroscopy. Biogeochemistry 71:339–351

Wu FC, Evans RD, Dillon PJ, Schiff S (2004b) Molecular size distribution characteristics of the metal-DOM complexes in stream waters by high-performance size-exclusion chromatography and high-resolution inductively coupled plasma mass spectrometry. J Anal Atomic Spectrom 19:979–983

Wu FC, Mills RB, Evans RD, Dillon PJ (2004c) Kinetics of metal-fulvic acid complexation using a stopped-flow technique and three-dimensional excitation emission fluorescence spectrometer. Anal Chem 76:110–113

Wu FC, Mills RB, Evans RD, Dillon PJ (2005) Photodegradation-induced changes in dissolved organic matter in acidic waters. Can J Fish Aqua Sci 62:1019–1027

Wu FC, Liu CQ, Wang FY, Xing BS, Fu PQ, Mostofa KMG (2007) Natural organic matter and its role in the environment water-rock interaction. In: Bullen TD, Wang Y (eds) Proceedings of the 12th international symposium on water-rock interaction WRI-12, Kunming, China, 31 July–5 August 2007, vol 1. Taylor & Francis Group, London, pp 37–40

Wu J, Zhang H, He P-J, Shao L-M (2011) Insight into the heavy metal binding potential of dissolved organic matter in MSW leachate using EEM quenching combined with PARAFAC analysis. Water Res 45:1711–1719

Xie HX, Zafiriou OC, Cai WJ, Zepp RG, Wang YC (2004) Photooxidation and its effects on the carboxyl content of dissolved organic matter in two coastal rivers in the Southeastern United States. Environ Sci Technol 38:4113–4119

Yamaji N, Hayakawa K, Takada H (2010) Role of photodegradation in the fate of fluorescent whitening agents (FWAs) in lacustrine environments. Environ Sci Technol 44:8791

Yamashita Y, Jaffé R (2008) Characterizing the interactions between trace metals and dissolved organic matter using excitation-emission matrix and parallel factor analysis. Environ Sci Technol 42:7374–7379

Yamashita Y, Tanoue E (2003a) Chemical characterization of protein-like fluorophores in DOM in relation to aromatic amino acids. Mar Chem 82:255–271

Yamashita Y, Tanoue E (2003b) Distribution and alteration of amino acids in bulk DOM along a transect from bay to oceanic waters. Mar Chem 82:145–160

Yamashita Y, Tanoue E (2004) In situ production of chromophoric dissolved organic matter in coastal environments. Geophys Res Lett 31:L14302. doi:101029/2004GL019734

Yamashita Y, Tanoue E (2008) Production of bio-refractory fluorescent dissolved organic matter in the ocean interior. Nature Geosci:579–582. doi:101038/ngeo279

Yamashita Y, Tsukasaki A, Nishida T, Tanoue E (2007) Vertical and horizontal distribution of fluorescent dissolved organic matter in the Southern Ocean. Mar Chem 106:498–509

Yamashita Y, Jaffé R, Maie N, Tanoue E (2008) Assessing the dynamics of dissolved organic matter (DOM) in coastal environments by excitation emission matrix fluorescence and parallel factor analysis (EEM-PARAFAC). Limnol Oceanogr 53:1900–1908

Yamashita Y, Cory RM, Nishioka J, Kuma K, Tanoue E, Jaffe R (2010) Fluorescence characteristics of dissolved organic matter in the deep waters of the Okhotsk Sea and the northwestern North Pacific Ocean. Deep-Sea Res II 57:1478–1485

Yamashita Y, Panton A, Mahaffey C, Jaffé R (2011) Assessing the spatial and temporal variability of dissolved organic matter in Liverpool Bay using excitation–emission matrix fluorescence and parallel factor analysis. Ocean Dyn 61:569–579. doi:10.1007/s10236-010-0365-4

Ying GG (2006) Fate, behavior and effects of surfactants and their degradation products in the environment. Environ Int 32:417–431

Yoshioka T, Mostofa KMG, Konohira E, Tanoue E, Hayakawa K, Takahashi M, Ueda S, Katsuyama M, Khodzher T, Bashenkhaeva N, Korovyakova I, Sorokovikova L, Gorbunova L (2007) Distribution and characteristics of molecular size fractions of freshwater dissolved organic matter in watershed environments: its implication to degradation. Limnology 8:29–44

Yu G-H, Luo Y-H, Wu M-J, Tang Z, Liu D-Y, Yang X-M, Shen Q-R (2010) PARAFAC modeling of fluorescence excitation–emission spectra for rapid assessment of compost maturity. Bioresour Technol 101:8244–8251

Yue L, Wu FC, Liu C, Li W, Fu P, Bai Y, Wang L, Yin Z, Lü Z (2006) Relationship between fluorescence characteristics and molecular weight distribution of natural dissolved organic matter in Lake Hongfeng and Lake Baihua, China. Chin Sci Bull 51:89–96

Zanardi-Lamardo E, Moore C, Zika RG (2004) Seasonal variation in molecular mass and optical properties of chromophoric dissolved organic material in coastal waters of southwest Florida. Mar Chem 89:37–54

Zepp RG, Scholtzhauer PF (1981) Comparisson of photochemical behavior of various humic substances in water: III Spectroscopic properties of humic substances. Chemosphere 10:479–486

Zepp RG, Braun AM, Hoigne J, Leenheer JA (1987) Photoproduction of hydrated electrons from natural organic solutes in aquatic environments. Environ Sci Technol 21:485–490

Zepp RG, Faust BC, Hoigné J (1992) Hydroxyl radical formation in aqueous reactions (pH 3–8) of iron(II) with hydrogen peroxide: the Photo-Fenton reaction. Environ Sci Technol 26:313–319

Zhang Y, Zhu L, Zeng X, Lin Y (2004) The biogeochemical cycling of phosphorus in the upper ocean of the East China Sea. Est Coast Shelf Sci 60:369–379

Zhang X, Minear RA, Barrett SE (2005) Characterization of high molecular weight disinfection byproducts from chlorination of humic substances with/without coagulation pretreatment using UF-SEC-ESI-MS/MS. Environ Sci Technol 39:963–972

Zhang Y, van Dijk MA, Liu M, Zhu G, Qin B (2009a) The contribution of phytoplankton degradation to chromophoric dissolved organic matter (CDOM) in eutrophic shallow lakes: field and experimental evidence. Water Res 43:4685–4697

Zhang Y, Liu M, Qin B, Feng S (2009b) Photochemical degradation of chromophoric-dissolved organic matter exposed to simulated UV-B and natural solar radiation. Hydrobiologia 627:159–168

Zhang DY, Pan XL, Mostofa KMG, Chen X, Mu G, Wu FC, Liu J, Song WJ, Yang JY, Liu Y, Fu QL (2010) Complexation between Hg(II) and biofilm extracellular polymeric substances: an application of fluorescence spectroscopy. J Hazard Matter 175(1–3):359–365

Zhao WH, Wang JT, Cui X, Ji NY (2006) Research on fluorescence excitation and emission matrix spectra of dissolved organic matter in phytoplankton growth process. Chin High Technol Lett 16:425–430 (in Chinese with English abstract)

Zhao W, Wang J, Chen M (2009) Three-dimensional fluorescence characteristics of dissolved organic matter produced by Prorocentrum donghaiense Lu. Chin J Oceanol Limnol 27:564–569. doi:10.1007/s00343-009-9141-z

Zimmerman JTF, Rommets JW (1974) Natural fluorescence as a tracer in the Dutch Wadden Sea and the adjacent North Sea. Neth J Sea Res 8:117–125

Zinder SH (1993) Physiological ecology of methanogens. In: Ferry JG (ed) Methanogenesis: ecology, physiology, biochemistry and genetics. Chapman and Hall, New York, pp 128–206

Zsolnay A, Baigar E, Jimenez M, Steinweg B, Saccomandi F (1999) Differentiating with fluorescence spectroscopy the sources of dissolved organic matter in soils subjected to drying. Chemosphere 38:45–50

Photosynthesis in Nature: A New Look

Khan M. G. Mostofa, Cong-qiang Liu, Xiangliang Pan, Takahito Yoshioka, Davide Vione, Daisuke Minakata, Kunshan Gao, Hiroshi Sakugawa and Gennady G. Komissarov

1 Introduction

Photosynthesis is a fundamental process on the Earth's surface that can convert the sunlight energy to chemical energy that can be used by essentially all forms all life (Komissarov 2003; Krauß 2003). The outstanding English chemist Joseph Priestley

K. M. G. Mostofa (✉) · C. Liu
State Key Laboratory of Environmental Geochemistry, Institute of Geochemistry, Chinese Academy of Sciences, Guiyang 550002, China
e-mail: mostofa@vip.gyig.ac.cn

X. L. Pan
Xinjiang Key Laboratory of Water Cycle and Utilization in Arid Zone, Xinjiang Institute of Ecology and Geography, Chinese Academy of Sciences, Urumqi 830011, People's Republic of China

T. Yoshioka
Field Science Education and Research Center, Kyoto University, Kitashirakawa Oiwake-cho, Sakyo-ku, Kyoto 606-8502, Japan

D. Vione
Dipartimento Chim Analit, University Turin, I-10125 Turin, Italy

Centro Interdipartimentale NatRisk, I-10095 Grugliasco, (TO), Italy

D. Minakata
School of Civil and Environmental Engineering, Brook Byers Institute for Sustainable Systems Georgia Institute of Technology, 828 West Peachtree Street, Suite 320, Atlanta, GA 30332, USA

K. Gao
State Key Laboratory of Marine Environmental Science, Xiamen University, Xiamen, Fujian, China

H. Sakugawa
Graduate School of Biosphere Science, Department of Environmental Dynamics and Management, Hiroshima University, 1-7-1, Kagamiyama, Higashi-Hiroshima 739-8521, Japan

G. G. Komissarov
Semenov Institute of Chemical Physics, Russian Academy of Sciences, Moscow 117977, Russia

K. M. G. Mostofa et al. (eds.), *Photobiogeochemistry of Organic Matter*, Environmental Science and Engineering, DOI: 10.1007/978-3-642-32223-5_7,

in 1771 and 1772 firstly hypothesised on photosynthesis that plants can restore to the air whatever breathing animals and burning candles remove. Jan Ingenhousz in 1779 showed that light is essential to the plant process that somehow purifies air fouled by candles or animals. Based on the experiments, he concluded that plants are dependent on light and their green parts for nutrients and energy.

The experiments conducted by J. Senebier and N. Th. de Saussure revealed that the initial substances of photosynthesis are carbon dioxide (CO_2) and water (H_2O) (de Saussure 1804; Bay 1931). It has been shown by de Saussure (1804) that H_2O is a reactant in photosynthesis. The CO_2 cleavage hypothesis readily accounted for the deceptively simple overall photosynthesis equation ($CO_2 + H_2O + h\upsilon \rightarrow CH_2O + O_2$) (de Saussure 1804). The C:2H:O proportion in the reaction made people assumed that carbon from the photodecomposition of CO_2 can recombine with the elements of water. In 1905 the British scientist F. Blackmann discovered that photosynthesis consists of a light reaction, which is rapid, and a slower dark reaction (Blackman 1905; Blackman and Matthaei 1905). In 1924, O. Warburg and T. Uyesugi explained the result of Blackman as showing that photosynthesis has two classes of reactions: light and dark reactions (Warburg and Uyesugi 1924). In 1922 the German Scientists O. Warburg and E. Negelein revealed the minimum quantum requirement (i.e., minimum number of photons) to be 3–4 per oxygen molecule evolved during the overall process of photosynthesis (Warburg and Negelein 1922). This was later shown to be in error by a factor of 2–3 (Govindjee 1999). Warburg then was awarded the 1931 Nobel Prize in Physiology and Medicine for his discoveries concerning respiration. In 1937 the British scientist R. Hill provided the biochemical proof of the existence of these light and dark phases (Hill 1937, 1939).

In 1931 the American microbiologist van Niel showed that the photosynthetic processes in various pigmented organisms can be interpreted as special cases of a general process expressed as follows (van Niel 1931):

$$2H_2A + CO_2 \xrightarrow[\text{[Radiant energy]}]{\text{Pigments and}} CH_2O + H_2O + 2A \tag{1.1}$$

where light energy is used to photodecompose a hydrogen donor (H_2A) whilst carbon dioxide is reduced anaerobically to cell substance in the dark, using enzymatic reactions (van Niel 1931). According to this generalization, H_2A is water in the case of plants, whilst H_2A is H_2S (or $Na_2S_2O_3$, Na_2SO_3, S, molecular hydrogen, organic substrates and so on) in green and purple sulfur bacteria. Therefore, O_2 is the by-product of plant photosynthesis and elemental sulfur or other compounds are the by-products of bacterial photosynthesis (van Niel 1931, 1936; Roelofsen 1935; Muller 1933).

van Niel in 1941 postulated that the photoinduced reaction in the photosynthetic processes of green bacteria, purple bacteria, and green plants represents, in all cases, a photodecomposition of water (van Niel 1941). The scientist Hill then demonstrated that isolated chloroplasts can evolve oxygen but cannot assimilate CO_2 (Hill 1939, 1951). In 1941, two Soviet and several American scientists discovered that oxygen released by higher plants and algae in photosynthesis is originated from H_2O and not from CO_2 (Ruben et al. 1941; Vinogradov and Teĭs 1941).

Calvin and his colleagues (A. Benson and J. Bassham) during the period of 1947–1956 depicted the mechanism that carbon travels through a plant during photosynthesis, starting from its absorption as atmospheric carbon dioxide to its conversion into carbohydrates and other organic compounds (Bassham et al. 1956; Calvin 1956). They showed that sunlight can act on chlorophyll in a plant to fuel the production of organic compounds, rather than on CO_2 as was previously believed. Calvin was then awarded the Nobel Prize in Chemistry in 1961 for his Calvin cycle (sometimes termed as Calvin-Benson-Bassham Cycle).

Since then, a lot of studies have been conducted on photosynthesis regarding release of electrons from chlorophylls, characterization of the primary reaction centers of photosystems (PSI and PSII), occurrence and formation of chlorophyll dimer in PSI and PSII, functions and roles of PSI and PSII, endogenous formation of hydrogen peroxide (H_2O_2) in photosynthetic cells, release of O_2, and so on (Hill 1937, 1939; van Niel 1931; Bach 1894; Arnon 1949, 1959, 1961, 1971; Mehler 1951; Asada et al. 1974; Chance et al. 1979; Halliwell 1981; Holland et al. 1982; Boekema et al. 1987, 2001 Shipman et al. 1976; Hynninen and Lötjönen 1993; Krauss et al. 1993; Krauß et al. 1996; Shubin et al. 1993; Golbeck 1994; Kruip et al. 1994; Boussaad et al. 1997; Brettel 1997; Wilhelm et al. 1996, 1997, 1999; Klukas et al. 1999; Halliwell and Gutteridge 1999; López-Huertas et al. 1999; Stewart et al. 2000; Jordan et al. 2001; Baker and Graham 2002; Ben-Shem et al. 2003; Catalan et al. 2004; Dashdorj et al. 2004; Germano et al. 2004; Diner and Rappaport 2002; del Río et al. 2006; Li et al. 2006; Krasnovsky 2007; Krieger-Liszkay et al. 2008; Rappaport and Diner 2008; Amunts et al. 2010; Müller et al. 2010; Nilsson Lill 2011; Umena et al. 2011).

Moreover, release of O_2 during photosynthesis still remains under debate because it is considered to be originated either from H_2O (Dashdorj et al. 2004; Germano et al. 2004; Rappaport and Diner 2008; Müller et al. 2010; Takahashi et al. 1987; Periasamy et al. 1978) or from H_2O_2 (Komissarov 1994, 2003; Bach 1893; Velthuys and Kok 1978; Asada and Badger 1984; Asada and Takahashi 1987; Mano et al. 1987; Renger 1987; Anan'ev and Klimov 1988; Bader and Schmid 1988, 1989; Schroeder 1989; Schröder and Åkerlund 1990; Miyake and Asada 1992; Bader 1994, 1995; Yin et al. 2006; Mostofa et al. 2009; Kuznetsov et al. 2010; Bernardini et al. 2011; Mostofa et al. 2009). The scientist Bach has been the first to show that plants actively accumulate H_2O_2 upon illumination (Bach 1894). The major generation sites of reactive oxygen species (ROS) are the PSI and PSII photosystems in chloroplast thylakoids in higher plants. Here, photoreduction of O_2 to H_2O_2 in PSI has firstly been discovered by Mehler (Mehler 1951). Subsequently, the primary reduced product has been identified as the superoxide anion ($O_2^{\bullet -}$), the disproportionation of which can produce H_2O_2 and O_2 (Asada et al. 1974). Recently, H_2O_2 instead of H_2O has been proposed to react with CO_2 in photosynthesis, whereas H_2O_2 is used as an intermediate to release O_2 (Komissarov 1994, 2003; Velthuys and Kok 1978; Mano et al. 1987; Renger 1987; Anan'ev and Klimov 1988; Bader and Schmid 1988,

1989; Schroeder 1989; Schröder and Åkcrlund 1990; Miyake and Asada 1992; Bader 1994; 1995; Yin et al. 2006; Mostofa et al. 2009; Kuznetsov et al. 2010; Bernardini et al. 2011). Komissarov (1994, 1995, 2003) has been summarizing on the new hypothesis concerning the photosynthetic reaction, according to which the interaction between CO_2 in air and H_2O_2 in aqueous media (instead of H_2O as for the earlier concept) may form carbohydrate in plants.

Photosynthesis is significantly affected by several factors such as seasonal variation in sunlight and UV radiation (Marshall and Orr 1928; Barker 1935; Jenkin 1937; Rabinowitch 1951; Nielsen 1951, 1952; Aro et al. 1993; Melis 1999; Andersson and Aro 2001; Han et al. 2001; Nishiyama et al. 2001, 2009; Sinha et al. 2001a; Adir et al. 2003; Rastogi et al. 2010; Jiang and Qiu 2011; Sobek et al. 2007; Zhang et al. 2010), occurrence of CO_2 forms (CO_2, H_2CO_3, HCO_3^-, and CO_3^-) (Wong et al. 1975; O'Leary 1981; Cooper and McRoy 1988; Farquhar et al. 1989; Raven and Farquhar 1990; Fogel et al. 1992; Rau et al. 1992; Francois et al. 1993; Raven et al. 1993, 2002; Jasper and Hayes 1994; Laws et al. 1995; Yoshioka 1997; Hu et al. 2012), variations in temperature (Sobek et al. 2007; Mortain-Bertrand et al. 1988; Davison 1991; Wilen et al. 1995; Lesser and Gorbunov 2001; Baulch et al. 2005; Doyle et al. 2005; Yoshiyama and Sharp 2006; Ogweno et al. 2008; Higuchi et al. 2009; Bouman et al. 2010), effects of water stress (drought) and precipitation/rainfall (Munns et al. 1979; Jones and Turner 1978; Matsuda and Riazi 1981; Kaiser 1987; Asada 1992; Hopkins and Hüner 1995; Aziz and Larher 1998; Li and van Staden 1998; Nam et al. 1998; Cornic 2000; Wilson et al. 2000; Lawlor 2002; Velikova and Tsonev 2003; Flexas et al. 2004; Hassan 2006; Liu et al. 2006; Ohashi et al. 2006; Fariduddin et al. 2009), effects of the contents and nature of DOM and POM (Haan 1974; de Haan 1977; Stabel et al. 1979; Jackson and Hecky 1980; Wright 1988; Lindell et al. 1995; Brussaard et al. 1996; Brussaard et al. 2005; Brussaard et al. 2007; Carpenter et al. 1998; Igarashi et al. 1998; Reche et al. 1998; Rengefors and Legrand 2001; Sukenik et al. 2002; de Lange et al. 2003; Hanson et al. 2003; Houser et al. 2003; Druon et al. 2010), variation in nutrient contents (Yoshiyama and Sharp 2006; Martinez and Cerda 1989; Vıllora et al. 2000; Parkhill et al. 2001; Smith 2003; Kaneko et al. 2004; Sterner et al. 2004; Turhan and Eris 2005; Huszar et al. 2006; Liu et al. 2007; Wang and Han 2007; Nõges et al. 2008; McCarthy et al. 2009; Mohlin and Wulff 2009; Achakzai et al. 2010; Bybordi 2010; Tunçtürk et al. 2011), variation in trace metal ions with effects on aquatic microorganisms (Zhang et al. 2010; Crist et al. 1981; Zhou and Wangersky 1985, 1989; Simkiss and Taylor 1989; Xue and Sigg 1990; Tessier and Turner 1995; Sunda and Huntsman 1998; Burda et al. 2003; Koukal et al. 2003; Mylon et al. 2003; Sigfridsson et al. 2004; Berden-Zrimec et al. 2007; Lamelas and Slaveykova 2007; Hopkinson and Barbeau 2008; Lamelas et al. 2009; Pan et al. 2009), effect of salinity or salt stress (Liu et al. 2007; Bybordi 2010; Tunçtürk et al. 2011; Satoh et al. 1983; Ahel et al. 1996; Moisander et al. 2002; Marcarelli et al. 2006; Segal et al. 2006; Demetriou et al. 2007; Allakhverdiev and Murata 2008; Melgar et al. 2008; Pandey and Yeo 2008; Pandey et al. 2009; Bybordi et al. 2010a, b, c), effects of toxic pollutants on aquatic microorganisms (Berden-Zrimec et al. 2007; Mayer et al. 1997; Halling-Sørensen et al. 2000; Katsumata et al. 2005, 2006 Kvíderová and Henley 2005; Zrimec et al. 2005; Pan et al. 2009; Yates and Rogers 2011), effect of size-fractionated phytoplankton (Malone 1980; Chisholm 1992; Li 1994; Tarran et al. 2001; Hansen and Hjorth 2002; Stibor and Sommer 2003; Tittel et al. 2003; Cermeno et al. 2005; Unrein et al. 2007; Zubkov et al. 2007; Zubkov and Tarran 2008), and effects of global warming (Mostofa et al. 2009; Baulch et al. 2005; Yates and Rogers 2011; Morris and Hargreaves 1997; Cooke et al. 2006; Huisman et al. 2006; Llewellyn 2006; Richardson 2007; Malkin et al. 2008; Prince et al. 2008; Davis et al. 2009; Castle and Rodgers 2009; Mostofa and Sakugawa 2009; Etheridge 2010; Keeling et al. 2010). These factors have been assessed in recent studies and are vital to understanding and solving the debate about the occurrence of photosynthesis in terrestrial plants and aquatic microorganisms.

This chapter will give a general overview on photosynthesis, its key biogeochemical functions, the functions of photosystems (I and II) in organisms during photosynthesis, and will describe a new hypothesis for photosynthesis that adopts H_2O_2 instead of H_2O. It will also address the debates/questions regarding O_2 release from PSI and PSII during photosynthesis. Finally, it will extensively discuss the key factors that may significantly influence the photosynthetic activities of organisms, including higher plants.

2 Photosynthesis in Natural Waters

Photosynthesis is typically defined as a combination of photoinduced and biological processes that can convert carbon dioxide (CO_2) and hydrogen peroxide (H_2O_2: photoinduced generation from dissolved oxygen in water) into organic compounds (e.g. carbohydrates) and O_2 using the sunlight energy. These processes take place in photosynthetic cells of higher plants, cyanobacteria (or algae) and bacteria. Carbohydrates are then used for metabolic activities within the cell, and the whole process is termed as the oxygenic photosynthesis. It should be noted that cyanobacteria are not bacterial but generally referred to as algae. The chloroplast pigments of all cyanobacteria and aquatic higher plants have absorption bands in the blue region of the spectrum, such as the chlorophyll Soret band, and carotenoid bands (Kirk 1976). The action spectrum of photosynthesis in green algae, brown algae, diatoms and euglenas has two broad and intense peaks in the range from 400 to 500 nm of wavelength and in the region from 670 to 700 nm, respectively (Kirk 1976; Haxo and Blinks 1950; Mann and Myers 1968; Kirk and Reade 1970; Iverson and Curl 1973; Telfer et al. 1990; Schelvis et al. 1992; Durrant et al. 1995; Renger and Marcus 2002; Zhang et al. 2009). Photons of light initiate photosynthesis through releases of electrons across a membrane. It is catalysed by a special type of membrane-bound pigment-protein complexes called photosynthetic reaction centers (RCs). They are composed of photosystem I (PSI) and photosystem II (PSII), which will be discussed in the next sections. Oxygenic photosynthesis is caused by cooperation of PSI and PSII RCs and generally occurs in higher plants, bacteria and cyanobacteria. Cyanobacteria, in contrast to higher plants, are well enriched with PSI as compared with PSII: the PSI/PSII ratio is about unity in higher plants, but it is much higher in cyanobacteria, varying between 3 and 5.5 (Rakhimberdieva et al. 2001). On the other hand, either PSI or PSII RCs are used to convert light energy in anoxygenic photosynthesis, which typically occurs in many bacteria. Anoxygenicphotosynthesis is a process where uptake of light energy occurs without the release of O_2. Anoxygenic species can utilize hydrogen sulfide (H_2S) or other species as sources of reductants, giving various forms of sulfur as by-products. It is noted that green bacteria can use H_2S, while purple sulfur bacteria (*Thiorhodaceae*) can use various reduced sulfur compounds including $Na_2S_2O_3$, Na_2SO_3, S and H_2S, molecular hydrogen (H_2) and organic substances during photosynthesis (van Niel 1931; 1936; Roelofsen 1935; Muller 1933). Anoxygenic species are mostly equipped with variety of bacteriochlorophylls.

The chlorophyll absorption bands at the red end of the spectrum are only of limited use in water ecosystems, because of the rapid attenuation of red light by water (Kirk 1976). Therefore, the ability of many cyanobacteria and aquatic higher plants to photosynthesize and grow are markedly affected by the availability of blue light, which is in turn highly dependent on the concentration of yellow substance within water (Kirk 1976). All natural waters generally contain a significant amount of yellow substances that absorb light in the blue and ultraviolet (Hutchinson 1957; Kalle 1966; Jerlov 1968; Morel et al. 2007). Yellow substances originate generally from the occurrences of both allochthonous humic substances (fulvic and humic acids) of terrestrial plant origin and autochthonous fulvic acids of algal or phytoplankton origin, which absorb light in the blue and ultraviolet range (see also chapters "Dissolved Organic Matter in Natural Waters" and "Fluorescent Dissolved Organic Matter in Natural Waters") (Mostofa et al. 2009; Mostofa et al. 2009; Zhang et al. 2009; Hutchinson 1957; Kalle 1966; Jerlov 1968; Parlanti et al. 2000).

2.1 Biogeochemical Functions of Photosynthesis

The different functions of photosynthesis can be summarized as follows: (i) Photosynthetic oxygen production by cyanobacteria can lead to oxygenation of the atmosphere and oceans, in turn allowing aerobic respiration and the evolution of large, complex and ultimately intelligent organisms (Catling et al. 2005). Oxygenic photosynthesis has evolved hundreds of millions of years before the atmosphere became permanently oxygenated. Therefore, biogenic oxygen production started very early in Earth's history, before the start of the geological record, leading to an Archaean (greater than 2.5 Ga, gigaannum: 10^9 years) atmosphere that was highly oxygenated (Ohmoto 1997; Catling and Claire 2005; Buick 2008). (ii) Photosynthesis is the only process that can balance the biosphere by converting atmospheric CO_2 into organic/biological matter, at the same time by releasing O_2 into the atmosphere. (iii) All forms of life in the biosphere are dependent on food and primarily on vegetables and terrestrial plants, the matter of which is produced through photosynthesis. (iv) Plant litter materials or biomass, developed initially through photosynthesis, represent the largest pool of terrestrial carbon. It is currently estimated at approximately 1500–2000 Pg of C that are stored in the world's soils (Schlesinger 1997; CAST 2004). Upon microbial processing, this material can produce soil organic matter or allochthonous dissolved organic matter (DOM), including humic substances (fulvic and humic acids) and inorganic components such as nutrients and various elements (see also chapter "Dissolved Organic Matter in Natural Waters") (Mostofa et al. 2009; Nakane et al. 1997; Uchida et al. 2000; Kögel-Knabner 2002; Grandy and Neff 2008; Moore et al. 2008; Braakhekke et al. 2011; Spence et al. 2011; Tu et al. 2011). These chemical components are ultimately released into the water ecosystem and undergo photoinduced and microbial degradation. Their end-products are CO_2, H_2O_2 and dissolved inorganic carbon (DIC: generally defined as dissolved CO_2, H_2CO_3, HCO_3^-, and CO_3^{2-}), which can fuel/accelerate the primary production (see also chapter "Photoinduced and Microbial Degradation of Dissolved Organic Matter in Natural Waters" and "Impacts of Global Warming on Biogeochemical Cycles in Natural Waters") (Mostofa et al. 2009; Jones 1992; Hessen and Tranvik 1998; Jansson et al. 2000; Meili et al. 2000; Grey et al. 2001; Hernes and Benner 2003; Tranvik et al. 2009; Ballaré et al. 2011; Zepp et al. 2011). (v) Photosynthesis is the key process for primary and secondary production and uses natural sunlight in aquatic ecosystems. Aquatic microorganisms that are key components of the Earth's biosphere can produce more than 50 % of the biomass of our planet through photosynthesis, using allochthonous DOM and nutrients. Therefore, aquatic environments can incorporate at least the same amount of atmospheric carbon dioxide (CO_2) as terrestrial ecosystems (de Haan 1974, 1977; Tranvik 1988; Häder et al. 2003; Zepp et al. 2007). Life is mostly composed of the elements carbon, hydrogen, nitrogen, oxygen, sulfur and phosphorus, which make up nucleic acids (e.g. DNA), proteins and lipids and can thus form the bulk of living matter (Wolfe-Simon et al. 2011). (vi) Aquatic microorganisms (e.g. algae or phytoplankton cells) can produce autochthonous DOM, including autochthonous fulvic acids, CO_2 and nutrients under both photoinduced and microbial respiration or assimilation (see also chapters "Dissolved Organic Matter in Natural Waters", "Photoinduced and Microbial Degradation of Dissolved Organic Matter in Natural Waters", "Fluorescent Dissolved Organic Matter in Natural Waters", and "Impacts of Global Warming on Biogeochemical Cycles in Natural Waters") (Mostofa et al. 2009; Mostofa et al. 2009; Zhang et al. 2009; Tranvik et al. 2009; Biddanda and Benner 1997; Gobler et al. 1997; Kritzberg et al. 2006; Mostofa et al. 2011). These compounds can be used by aquatic microorganisms for their further photosynthetic activity and can, therefore, promote the primary production (see also chapters "Dissolved Organic Matter in Natural Waters", and "Impacts of Global Warming on Biogeochemical Cycles in Natural Waters") (Hessen and Tranvik 1998; Cole et al. 1982). (vii) Photosynthesis is the dominant energy mobilization process for secondary production in natural waters, where organic carbon fixed by primary producers is consumed directly by grazing or is recycled via the microbial loop (Wetzel 2001). (viii) The primary producers in freshwater and marine ecosystems can constitute the basis of the intricate food webs, providing energy for the primary and secondary consumers. Therefore, they are important contributors for the production of the human staple diet in the form of crustaceans, fish, and mammals derived from the sea (Häder et al. 2007). (xi) Cyanobacteria (e.g., mostly *Anabaena* and *Nostoc* spp.) that grow through photosynthesis are a rich source of potentially useful marine natural products (secondary metabolites) that have specific activities such as anti-HIV, anticancer, antifungal, antimalarial, antifoulants, anti-inflammatory, antituberculosis, and antimicrobial (Moore 1996; Burja et al. 2001; Singh et al. 2002; Blunt et al. 2007). For example cyanovirin (CV-N, cyanovirin-N), a 101 amino acid protein extracted from *Nostoc ellipsosporum* has potent activity against a wide range of immunodeficiency viruses such as HIV-1, M-and T-tropic stains of HIV-1, HIV-2, SIV (simian) and FIV (feline) (Burja et al. 2001). (x) Marine microorganisms could be used as sources of natural bioactive molecules that play a photoprotective role, which could be used in commercial applications (Rastogi et al. 2010). A number of photoprotective compounds such as melanins, mycosporines, mycosporine-like amino acids (MAAs), scytonemin, parietin, usnic acid, carotenoids, phycobiliproteins, phenylpropanoids and flavonoids and several other UV-absorbing substances of unknown chemical structure are produced by different microorganisms (Rastogi et al. 2010; Blunt et al. 2007; Jeffrey et al. 1999; Gauslaa and McEvoy 2005; Sinha et al. 2007b; Coesel et al. 2008; Klisch and Häder 2008; Hylander et al. 2009; Lee and Shiu 2009; Ingalls et al. 2010).

3 New Hypothesis for Photosynthesis Using H_2O_2 Instead of H_2O

Studies demonstrate that the reaction of CO_2 and H_2O_2 (instead of H_2O) can cause photosynthesis of organisms in photosynthetic cell in new hypothesis (Komissarov 1994, 1995, 2003; Velthuys and Kok 1978; Mano et al. 1987; Renger 1987; Anan'ev and Klimov 1988; Bader and Schmid 1988, 1989; Schroeder 1989; Schröder and Åkerlund 1990; Miyake and Asada 1992; Bader 1994; Mostofa et al. 2009; Kuznetsov et al. 2010; Bernardini et al. 2011; Mostofa et al. 2009). According to these studies, the reaction of CO_2 and H_2O_2 (instead of H_2O) can cause photosynthesis of organisms by either simultaneous photoinduced formation of H_2O_2 in chlorophylls bound in photosynthetic cell or photoinduced and microbial formation of H_2O_2 and CO_2 from dissolved organic matter (DOM) and particulate organic matter (POM) in aqueous media.

The general photosynthetic reaction can be expressed as follows (Eqs. 3.1, 3.2):

$$xCO_{2(H_2O)} + yH_2O_{2(H_2O)} \xrightarrow{h\nu} C_x(H_2O)_y + O_2 + E(\pm) \tag{3.1}$$

$$2H_2O_2 + \text{ photoinduced/biological processes } \rightarrow O_2 + 2H_2O/\text{unknown oxidants} \tag{3.2}$$

Carbohydrate $C_x(H_2O)_y$ is formed (Eq. 3.1), where x and y are whole numbers that differ depending on the specific carbohydrate that is being produced. The release of O_2 in photosynthesis is the fundamental photoinduced conversion reaction, which under the present hypothesis is supposed to involve H_2O_2 either via disproportionation or upon biological processes (Eq. 3.2) (Komissarov 2003; Buick 2008; Moffett and Zafiriou 1990; Liang et al. 2006).

The release of O_2 from H_2O_2 instead of H_2O can be understood from several mechanistic approaches: (i) Mechanism for oxygen release from H_2O_2 instead of H_2O; (ii) Effective oxidation of H_2O_2 instead of H_2O in releasing photosynthetic O_2 (iii) generation of H_2O_2 from DOM and POM; (iv) photoinduced generation of H_2O_2 from ultrapure H_2O; (v) Endogenous H_2O_2 in the photosynthetic cell and effects of exogenous H_2O_2; (vi) H_2O_2 formation in water, lipid and protein environments in the presence of Chlorophyll; and (vii) Occurrence of H_2O_2 and its effects on photosynthesis.

3.1 Mechanism for Oxygen Release from H_2O_2 Instead of H_2O

Experimental studies show that various cyanobacteria may release O_2 from the decomposition of H_2O_2 during photosynthesis (Komissarov 1994, 1995, 2003; Velthuys and Kok 1978; Mano et al. 1987; Renger 1987; Anan'ev and Klimov 1988; Bader and Schmid 1988, 1989; Schroeder 1989; Schröder and Åkerlund 1990; Miyake and Asada 1992; Bader 1994). Based on the mechanism given by Komissarov (1994, 1995, 2003) and the mechanism of H_2O_2 photodecomposition by earlier studies (Christensen et al. 1982; Bielski et al. 1985), the release of O_2 from H_2O_2 can be generalized as follows (Eqs. 3.3–3.6 or 3.7–3.10):

$$H_2O_2 \rightarrow H^+ + HO_2^- \tag{3.3}$$

$$HO_2^- + h \rightarrow HO_2 \tag{3.4}$$

$$HO_2 \rightarrow H^+ + O_2^- \tag{3.5}$$

$$O_2^- + h \rightarrow O_2 \tag{3.6}$$

or

$$H_2O_2 + HO^\bullet \rightarrow HO_2^\bullet + H_2O \quad k = 3.0 \times 10^7\,M^{-1}s^{-1} \tag{3.7}$$

$$HO_2^\bullet + HO_2^\bullet \rightarrow H_2O_2 + O_2 \quad k = 8.3 \times 10^5\,M^{-1}s^{-1} \tag{3.8}$$

$$HO_2^\bullet + O_2^{\bullet -} \rightarrow HO_2^- + O_2 \quad k = 9.7 \times 10^7\,M^{-1}s^{-1} \tag{3.9}$$

$$HO_2^\bullet \leftrightarrow O_2^{\bullet -} + H^+ \quad k = 1.58 \times 10^{-5}\,M^{-1}s^{-1} \tag{3.10}$$

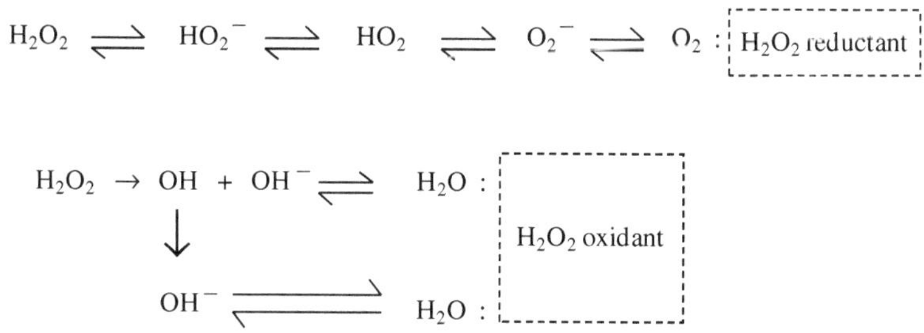

Fig. 1 Electron transfer and proton transfer reactions in the reduction of O_2 from H_2O_2 to H_2O, demonstrating the intermediates involved. *Data source* Moffett and Zafiriou (1990)

where *h* is an electron vacancy (hole), generated in the pigment under the effect of light. The generation of a single molecule of oxygen from water requires at least four light quanta, each of which generates an 'electron–hole' couple. The electron is used in the reaction ($H^+ + e \rightarrow H$) required for the subsequent fixation of CO_2.

Hylakoid particle preparations of the filamentous cyanobacterium *Oscillatoria chalybea* in laboratory experiments using labeled $^{16}O_2$ and $^{18}O_2$ show the occurrence of at least three types of oxygen uptake: one is associated with PSII and the S-state system, whereas the other two types apparently belong to the respiratory pathway. The S-state system is involved in $^{18}O_2$ production from H_2O_2 (Bader and Schmid 1988, 1989). Comparison of the relaxation kinetics of the first two flashes of a sequence with the steady-state signals as well as the detailed analysis of the mass spectrometric signals reveal that O_2 is evolved by various cyanobacteria through the decomposition of H_2O_2, which requires only two light quanta (Bader 1994).

The release of O_2 from H_2O_2 is confirmed by the redox behavior of H_2O_2 in water (Moffett and Zafiriou 1990; Rose and Waite 2003; Jeong and Yoon 2005). When H_2O_2 acts as a reductant, O from H_2O_2 is transformed into O_2 (Moffett and Zafiriou 1990). When H_2O_2 acts as an oxidant, O from H_2O_2 is converted into H_2O (Moffett and Zafiriou 1990). The chain reactions of H_2O_2 as reductant and oxidant are schematically depicted below (Fig. 1) (Moffett and Zafiriou 1990):

The detailed mechanism for the release of O_2 in the first scheme can be generalized using the reduction of Fe^{3+} (or Cu^{2+}) by H_2O_2 in the following ways (Eqs. 3.11–3.15) (Bielski et al. 1985; Hardwick 1957; Moffett and Zika 1987a, b; Marianne and Sulzberger 1999):

$$HOOH \leftrightarrow H^+ + HO_2^- \tag{3.11}$$

$$Fe^{3+} + HO_2^- \rightarrow Fe^{2+} + HO_2^{\bullet} \tag{3.12}$$

$$HO_2^{\bullet} \leftrightarrow H^+ + O_2^{\bullet -} \quad k = 1.58 \times 10^{-5}\,M^{-1}s^{-1} \tag{3.13}$$

$$Fe^{3+} + O_2^{\bullet -} \rightarrow Fe^{2+} + O_2 \quad k = 1.5 \times 10^{8}\,M^{-1}s^{-1} \tag{3.14}$$

$$H_2O_2 + Fe^{2+} \rightarrow Fe^{3+} + HO^{\bullet} + OH^- \quad k = 63\,M^{-1}s^{-1} \tag{3.15}$$

In the reactions above, release of O_2 occurs not from H_2O but from H_2O_2.

Correspondingly, photosynthetic O_2 evolution would involve different stages that carry out a gradual accumulation of oxidizing equivalents in the Mn-containing water-oxidizing complex (WOC) (Samuilov et al. 2001). The WOC can exist in different oxidation states (S_n, where high n indicates the most oxidised states), which can be probed by addition of different redox-active molecules. The interaction of H_2O_2 with the S states of the WOC is depicted in the scheme that follows (Velthuys and Kok 1978; Mano et al. 1987; Samuilov et al. 2001; Latimer 1952; Ilan et al. 1976; Samuilov 1997):

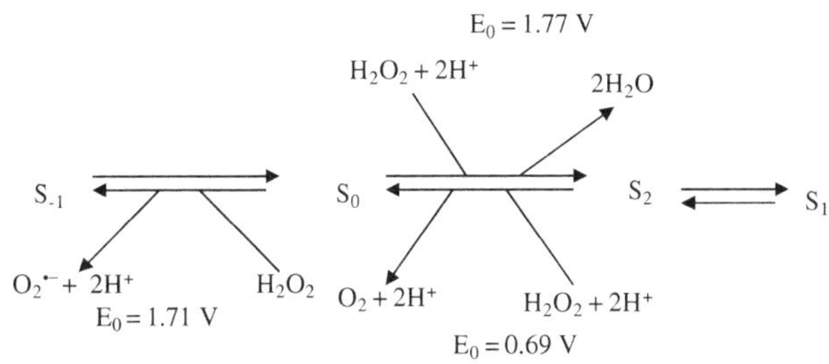

These studies suggest that H_2O_2 is an evolutionary precursor of H_2O as the electron donor for PSII in cyanobacteria (Bader 1994; Samuilov 1997; Blankenship and Hartman 1998).

The release of O_2 from H_2O_2 instead of H_2O can be justified by the rapid formation of H_2O_2 and of highly reactive chemical forms collectively denoted as 'reactive oxygen species (ROS)'. Both H_2O_2 and ROS are formed from O_2 when it is exposed to high-energy or electron-transfer chemical reactions, which can be expressed as follows (Chance et al. 1979; Koppenol 1976; Klotz 2002; Apel and Hirt 2004):

$$3O_2 + h\upsilon \rightarrow {}^1O_2 \rightarrow {}^1O_2 \tag{3.16}$$

$$3O_2 + e^- + h\upsilon \rightarrow O_2{}^{\bullet-} \xrightarrow{H+} HO_2{}^{\bullet} \tag{3.17}$$

$$O_2{}^{\bullet-} + e^- + h\upsilon \rightarrow O_2{}^{2-} \xrightarrow{2H+} H_2O_2 \tag{3.18}$$

$$O_2{}^{2-} + e^- + h\upsilon \rightarrow O_2{}^{3-} \quad \rightarrow O^- \xrightarrow{H^+} HO^{\bullet}; \quad \downarrow 2H^+ \; H_2O \tag{3.19}$$

$$O^- + e^- + h\upsilon \rightarrow O^{2-} \xrightarrow{2H+} H_2O \tag{3.20}$$

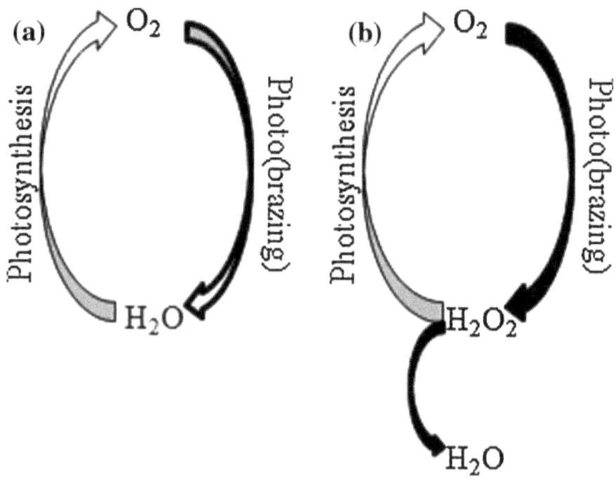

Fig. 2 The relationship between photosynthesis and (photo) bleeding within the framework of the conventional considerations regarding photosynthesis (**a**) and in accordance with the concept proposed by the author of the article (**b**). *Data source* Komissarov (2003)

Singlet oxygen ($^1O_2$) and superoxide radical ion $\left(O_2^{\bullet -}\right)$ are formed from the triplet state of O_2 $\left(^3O_2\right)$ in the presence of light (Eqs. 3.16, 3.17). The radical ion $O_2^{\bullet -}$ then reacts with an hydrogen ion (H^+) to form the perhydroxyl radical ($HO_2^{\bullet}$) (Eq. 3.17). The species $O_2^{\bullet -}$ can also accept one more electron (e^-) to form peroxide ion (O_2^{2-}), which then combines with H^+ to generate hydrogen peroxide (H_2O_2) (Eq. 3.18). Further acceptance of one e^- by O_2^{2-} can form O_2^{3-}, which can then produce H_2O and an oxene ion (O^-) in the presence of H^+ (Eq. 3.19). The ion radical O^- can produce the hydroxyl radical in the presence of H^+ (Eq. 3.19). Further acceptance of one e^- by O^- can yield the oxide ion (O^{2-}), which finally gives H_2O in the presence of H^+ (Eq. 3.20). This result shows that formation of water from O_2 is relatively more difficult than the process involving H_2O_2.

In the new hypothesis, the relationship between the fundamental biological process and breathing is complicated because the final product in breathing is water, which would not dissociate during photosynthesis (Fig. 2b) (Komissarov 2003). This is not contemplated in the conventional view of photosynthesis, which is illustrated in Fig. 2a. Breathing is followed from right to left in both equations.

However, breathing is accompanied by the formation of endogenous H_2O_2 that is not only a source of O_2, injected into the atmosphere, but also of hydrogen used in the synthetic processes of growth (Komissarov 2003).

Mass spectrometric examination of photosynthetic generation of O_2 using H_2O_2, marked with heavy isotopic oxygen $\left(H_2{}^{18}O_2\right)$, suggests that H_2O_2 is the source of the entire amount of generated O_2 (Mano et al. 1987). Experimental studies using ^{18}O-labeled H_2O_2 $\left(H_2{}^{18}O_2\right)$ and O_2 $\left(^{18}O_2\right)$ added to seawater also suggest that photoinduced oxidation can produce $^{18}O_2$and H_2O (Moffett and Zafiriou 1990), whereas label transfer is governed by the mass balance (Eq. 3.21):

$$-\Delta H_2{}^{18}O_2 = \Delta H_2{}^{18}O + \Delta {}^{18}O_2 \tag{3.21}$$

Similarly, catalytic epoxidation experiments using the ^{18}O labels in an acetone/water $\left(H_2{}^{18}O\right)$ solvent demonstrate that no ^{18}O coming from water $\left(H_2{}^{18}O\right)$

is incorporated into epoxide products, even though oxygen exchange is observed between the Mn^{IV} catalyst species and $H_2{}^{18}O$. Therefore, one can conclude that O_2 transfer does not proceed by the well-known oxygen-rebound mechanism (Yin et al. 2006). Experiments using labeled dioxygen, $^{18}O_2$, and hydrogen peroxide, $H_2{}^{18}O_2$, confirm that an oxygen atom is transferred directly from the $H_2{}^{18}O_2$ oxidant to the olefin substrate in the predominant pathway (Yin et al. 2006). Moreover, some recent experiments show that photoinduced H_2O oxidation occurs in the presence of inorganic catalysts (Kuznetsov et al. 2010; Bernardini et al. 2011). This result does not imply that H_2O is oxidized, but rather that $O_2{}^{\bullet-}$ and then H_2O_2 are produced photolytically. H_2O_2 is then photolytically decomposed into O_2 and H_2O.

Biological release of O_2 is observed using catalase for the decomposition of H_2O_2 in aqueous media, a process that can be depicted as follows (Eqs. 3.22, 3.23) (Moffett and Zafiriou 1990):

$$\text{HOOH} + \text{Catalase} \rightarrow \text{HOOH}^* + \text{Catalase}^\# \tag{3.22}$$

$$2\text{HOOH}^* + \text{Catalase}^\# \rightarrow \text{H–O–H} + O_2 + \text{Catalase} \tag{3.23}$$

In the above reactions, catalase enzymatically activates $HOOH^*$ to use them as oxidants (electron acceptors) and reductants (electron donors) (Eq. 3.22). Afterwards, disproportionation of activated $HOOH^*$ converts them into H_2O and O_2 (Eq. 3.23). Therefore, H_2O_2 can release O_2 under both photoinduced and microbial decomposition processes. The widespread occurrence of such a process justifies the hypothesis that the release of photosynthetic O_2 may occur from H_2O_2 instead of H_2O. Note that the contribution percentage decay of H_2O_2 is 65–80 % by catalase enzyme and 20–35 % by peroxidase enzyme, as estimated by isotopic measurements in seawater (Moffett and Zafiriou 1990).

Based on the current evidence, it is hypothesized that oxygenic photosynthesis has evolved by the end of the 'Great Oxidation Event' *ca.* 2.4 Ga ago. It has permanently raised atmospheric oxygen above the levels produced by photolysis of water (Buick 2008). The latter process can produce primarily H_2O_2, which might be source of oxygenic photosynthesis.

3.2 *Effective Oxidation of H_2O_2 Instead of H_2O in Releasing Photosynthetic O_2*

The oxidation of water to molecular oxygen is described by the equation (Rappaport and Diner 2008): $2H_2O \rightarrow O_2 + 4H^+ + 4e^-$, where at pH 7.0 the midpoint potential of the $O_2/2H_2O$ couple is 810 mV. Water is a very stable molecule and its oxidation requires the successive absorption of four photons and their photoinduced conversion into electrochemical energy. The energy of the quantum of a visible light is relatively small, such as 1.8 eV at the maximum absorption of chlorophyll (Komissarov 2003).

The value of standard electrode potential of the reaction of O_2 formation from H_2O_2 (Eq. 3.19) is significantly lower than for H_2O (Eqs. 3.24, 3.25) (Komissarov 2003):

$$H_2O_2 = O_2 + 2H^+ + 2e^-, \quad E^0 = -0.69\ V \tag{3.24}$$

$$H_2O = 1/2O_2 + 2H^+ + 2e^-, E^0 = -1.23\ V \tag{3.25}$$

Therefore, in vivo formation of oxygen would be preferable from hydrogen peroxide than from water.

3.3 Generation of H_2O_2 from DOM and POM

The most important source of H_2O_2 is the photoinduced generation from DOM and POM (e.g. algae) under solar illumination in natural waters. The mechanism has been discussed in earlier chapters (see "Photoinduced and Microbial Generation of Hydrogen Peroxide and Organic Peroxides in Natural Waters" and "Chlorophylls and Their Degradation in Nature"). In addition, DOM can also produce H_2O_2 under dark incubation. Algae or phytoplankton can produce H_2O_2 from superoxide radical anion $(O_2^{\bullet-})$, which can be formed either by photoinduced generation of electrons from Chlorophyll bound in microorganisms, or via autochthonous DOM. In the latter case, H_2O_2 generation can take place under photo- and microbial respiration (assimilations) of phytoplankton (see chapter "Photoinduced and Microbial Generation of Hydrogen Peroxide and Organic Peroxides in Natural Waters" and "Chlorophylls and Their Degradation in Nature"). Overall, production of H_2O_2 from various sources can be depicted as follows (Fig. 3).

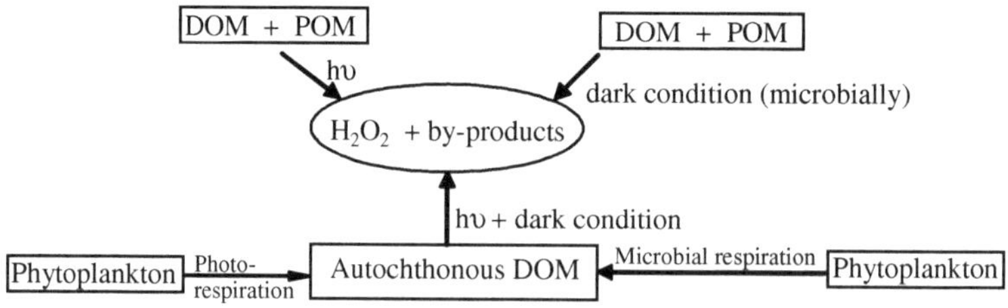

Fig. 3 Production of H_2O_2 from various sources in natural waters

3.4 Endogenous H_2O_2 in the Photosynthetic Cell and Effects of Exogenous H_2O_2

Endogenous H_2O_2 is formed in photosynthetic cells of organisms through production of superoxide radical ion $(O_2^{\bullet-})$ from whole bacteria of several species, from phagocytic cells, from spermatozoa as well as peroxisomes, mitochondria and chloroplasts (Komissarov 2003; Bach 1894; Chance et al. 1979; Halliwell 1981; Holland et al. 1982; Wilhelm et al. 1996, 1997, 1999; Halliwell and Gutteridge 1999; López-Huertas et al. 1999; Baker and Graham 2002; del Río et al. 2006; Krieger-Liszkay et al. 2008; Lyubimov and Zastrizhnaya 1992a, b; Turrens 1997; Karuppanapandian et al. 2011). H_2O_2 is also detected in the lens of the human eye and cataracts, aqueous humor and urine, in expired human breath and rat breath. Furthermore, increased H_2O_2 concentrations are also observed in patients with the adult respiratory distress syndrome, in patients with a cardiopulmonary bypass, in people exposed to ozone, in alveolar and peritoneal macrophages isolated from rats exposed to hypoxia, and in the breath of smokers (Wilhelm et al. 1996, 1997; Bhuyan and Bhuyan 1977; Spector and Garner 1981; Williams and Chance 1983; Ramachandran et al. 1991; Wilson et al. 1993; Nowak et al. 1996; Madden et al. 1997).

It has also been observed that oral bacteria may produce H_2O_2 (Thomas and Pera 1983) and that several enzymes, including glycollate and urate oxidases, can produce H_2O_2. It is calculated that 82 nM of H_2O_2 is produced per g of tissue per min in perfused livers isolated from normally fed rats (Chance et al. 1979). The H_2O_2 production rate is increased with inclusion of glycollate or urate in the perfusion medium. H_2O_2 is a precursor of $HO^\bullet$, a strong oxidizing agent, which is mostly formed either in the Fenton-type reaction in the presence of transition metals or via the Haber–Weiss reaction in the presence of superoxide and iron (Fong et al. 1976). Catalase, the enzyme that metabolizes H_2O_2 to H_2O and O_2 is detected in liver, kidney, blood, mucous membranes and other highly vascularized tissues (Sohal et al. 1994; Matutte et al. 2000). Correspondingly, detoxification of H_2O_2 by catalase has also been observed in the rabbit iris-ciliary body and in cultured lens epithelial cells (Delamere and Williams 1985; Giblin et al. 1990).

The radical $O_2^{\bullet-}$ can rapidly produce H_2O_2 and O_2 by the following reaction (Eq. 3.26) (Koppenol 1976):

$$2O_2^{\bullet-} + 2H^+ \rightarrow H_2O_2 + O_2 \quad k = 4.5 \times 10^5 M^{-1} s^{-1} \text{at pH 7 and } 22^\circ C \qquad (3.26)$$

although the reaction between $O_2^{\bullet-}$ and $HO_2^\bullet$ is much faster.
Similarly, $HO^\bullet$ can react with O_2^- to produce H_2O and O_2 (Eq. 3.27) (Koppenol 1976):

$$HO^\bullet + O_2^- + H^+ \rightarrow O_2 + H_2O \qquad (3.27)$$

Several studies have proposed that $^1O_2$ is formed in the cells or in PSII (Halliwell and Gutteridge 1999; Krieger-Liszkay et al. 2008; Kautsky et al. 1931; Durrant et al. 1990; Vass et al. 1992; Macpherson et al. 1993; Hideg et al. 1994; Keren et al. 1997; Fufezan et al. 2002; Krieger-Liszkay 2005). The chlorophyll (Chl) triplet state can produce the very reactive $^1O_2$ upon reaction with ground state $^3O_2$, if it is not efficiently quenched (Krieger-Liszkay et al. 2008). The lifetime of $^1O_2$ in a cell is estimated into approximately 3 s (Skovsen et al. 2005; Hatz et al. 2007).

The reactive transient $^1O_2$ is also formed from superoxide anion (O_2^-) in the following process (3.28) (Koppenol 1976):

$$O_2^- \rightarrow {}^1O_2 + e^- \text{ where } E_0 = +0.65 \text{ V} \tag{3.28}$$

In addition, any sensitizer (e.g. photoactive organic matter) can photolytically produce $^1O_2$ via the following processes (Eqs. 3.29, 3.30) (Braun and Oliveros 1990):

$$\text{Sens} + h\upsilon \rightarrow {}^1\text{Sens}^* \rightarrow {}^3\text{Sens}^* \tag{3.29}$$

$${}^3\text{Sens}^* + O_2 \rightarrow \text{Sens} + {}^1O_2 \tag{3.30}$$

where Sens is the sensitizer that can absorb photons and is promoted to the singlet excited state $\left({}^1\text{Sens}^*\right)$. The latter can undergo intersystem crossing (ISC) and be converted into the triplet state $\left({}^3\text{Sens}^*\right)$ (Eq. 3.29), which can react with O_2 to produce $^1O_2$ (Eq. 3.30).

On the other hand, deactivation of $^1O_2$ involves two major processes such as energy-transfer quenching and charge-ransfer quenching,through any acceptor or sensitizer (Eqs. 3.31, 3.32) (Braun and Oliveros 1990; Halliwell and Gutteridge 2007):

$${}^1O_2 + \text{Sens} \rightarrow \text{SensO}_2 \tag{3.31}$$

$$\text{SensO}_2 \rightarrow O_2 + \text{Sens} \tag{3.32}$$

The H_2O_2 concentration in plant cells is approximately 0.5–1 μmol per milligram of Chl, including Chl of photosynthetic antennae (Lyubimov and Zastrizhnaya 1992a). Therefore, the amount of H_2O_2 is much higher than the Chl content in the composition of so called oxygen-evolving complexes in chloroplasts (Lobanov et al. 2008). Experimental studies have shown that the content of H_2O_2 can increase during ontogenesis of both the whole plant and populations of protoplasts of separate leaves in the dark, and the light-dependent component of peroxide formation increases regardless of the metabolic type of the plant antennae (Lyubimov and Zastrizhnaya 1992b). It is known that each molecule of the chlorophyll absorbs light quanta ~1 time per second, even at the maximum intensity of daylight (Komissarov 2003). Synthetic Chl, metal complexes of porphyrins and phthalocyanines are photoactive and can produce H_2O_2 under irradiation in aqueous solutions saturated with O_2 (Lobanov et al. 2008; Hong et al. 1987; Bazanov et al. 1999; Premkumar and Ramaraj 1999).

Lower volatility of H_2O_2 compared to H_2O may cause the green leaves to be a unique concentrator of H_2O_2 (Komissarov 2003). It is shown that the heat of vapour formation of pure H_2O_2 is 12.3 kcal mole^{-1}, whilst that of water is 10.5 kcal mole^{-1} (Shamb et al. 1958). Transpiration (evaporation of water by plants) may evidently play the same function of H_2O_2 concentrator in addition to the protection of plants against overheating. For each kg of water, absorbed by the roots from soil, only 1 g is used by the plant for the construction of tissue. Therefore, the transpiration process may enhance the total contents of H_2O_2 in the plant cells. Terrestrial plants can receive high concentrations of rainwater H_2O_2 (0–199,000 nM: see Table 2 in chapter "Photoinduced and Microbial Generation of Hydrogen Peroxide and Organic Peroxides in Natural Waters"), which is a vital source of exogenous H_2O_2 and is susceptible to promote photosynthesis in plants and algae (Komissarov 1994, 1995, 2003; Mostofa et al. 2009). Experimental studies demonstrate that H_2O_2 concentrations (up to 10^{-5} M) in culture media can stimulate plant growth (Komissarov 1994, 1995, 2003). In addition, H_2O_2 can inhibit growth at concentrations as low as 10^{-5}–10^{-4} M under the conditions of a dialysis culture (Samuilov et al. 2001). H_2O_2 can inhibit the photosynthetic electron transport in cells of cyanobacteria (Samuilov et al. 2001, 2004) and can also destroy the function of the oxygen-evolving complex (OEC) in some chloroplasts and photosystem II preparations. In such a case it would cause the release of manganese from cyanobacterial cells, which inhibits the OEC activity.

3.5 H_2O_2 Formation in Water, Lipid and Protein Environments in the Presence of Chlorophyll

Chlorophyll can produce H_2O_2 in aqueous solution under acidic and alkaline pH conditions (pH = 3.8–12.4) under visible light irradiation (Lobanov et al. 2008). The mechanism behind the production of H_2O_2 from illuminated Chl can be illustrated as follows (3.33–3.39) (Lobanov et al. 2008; Parmon 1985; Bruskov2002):
At pH < 7

$$\mathrm{Chl} + h\upsilon \rightarrow \mathrm{Chl}^{*} \tag{3.33}$$

$$\mathrm{Chl}^{*} + O_2 \rightarrow \mathrm{Chl}^{+} + O_2^{\bullet -} \tag{3.34}$$

$$O_2^{\bullet -} + H^{+} \rightarrow HO_2^{\bullet} \quad k = 5 \times 10^{10}\,\mathrm{Ms}^{-1} \tag{3.35}$$

$$HO_2^{\bullet} + O_2^{\bullet -} + H^{+} \rightarrow H_2O_2 + O_2 \quad k = 9 \times 10^{7}\,\mathrm{Ms}^{-1} \tag{3.36}$$

$$HO_2^{\bullet} + HO_2^{\bullet} \rightarrow H_2O_2 \quad k = 2 \times 10^{6}\,\mathrm{Ms}^{-1} \tag{3.37}$$

At pH > 7

$$H_2O_2 + Fe^{2+}\ (\text{or other metals}) \rightarrow HO^{\bullet} + HO^{-} + Fe^{3+} \tag{3.38}$$

$$HO^{\bullet} + HO^{\bullet} \rightarrow H_2O_2 \quad k = 6 \times 10^{9}\,\mathrm{Ms}^{-1} \tag{3.39}$$

The electron donor for the conversion $O_2 \rightarrow O_2^{\bullet -}$ (redox potential $\phi^{\circ} = -0.12$ V) can be Chl in the singlet or triplet excited state (the S1 and T1), with $\phi^{\circ} = -1.14$ and -1.54 V, respectively (Lobanov et al. 2008). The occurrence of reaction (Eq. 3.39) is confirmed by the addition of 1 M ethanol as a scavenger of $HO^{\bullet}$ into the water suspension of silica gel with immobilized Chl inhibits the formation of H_2O_2 in the alkaline medium with pH 12.4 (Lobanov et al. 2008; Bruskov and Masalimov 2002). Formation of H_2O_2 from Chl can generally be expressed as follows (Eq. 3.40) (Lobanov et al. 2008): at pH < 7,

$$1/2O_2 + H^+ + Chl + h\upsilon \rightarrow 1/2H_2O_2 + Chl^+ \quad (3.40)$$

where redox potentials ($\Delta\varphi^\circ$) and Gibbs energy changes (ΔG^0) for the reduction of O_2 to H_2O_2 with simultaneous oxidation of Chl to the radical cation ($T = 298$ K) are −0.03 V and 5.8 kJ for H_2O_2 generation, 1.83 V and −353 kJ for the singlet excited state of Chl, as well as 1.23 V and −237 kJ for the triplet excited state of Chl, respectively.Similarly at pH > 7 (Eq. 3.41),

$$1/2O_2 + Chl + h\upsilon \rightarrow 1/2HO_2^- + Chl^+ \quad (3.41)$$

where $\Delta\varphi^\circ$ and ΔG^0 for the reduction of O_2 to HO_2^- with simultaneous oxidation of Chl to the cation radical ($T = 298$ K) are −0.80 V and 154 kJ for HO_2^- generation, 1.06 V and −204 kJ for the singlet excited state of Chl, and 0.46 V and −89 kJ for the triplet excited state of Chl, respectively (Lobanov et al. 2008).

In addition, H_2O_2 is significantly formed photolytically in aqueous mixtures of Chl and either micelles of cetyltrimethylammonium bromide (CTAB) or macromolecules of bovine serum albumin (BSA) in a noncovalent complex. Insuch a case, Chl acts as a photocatalyst (Lobanov et al. 2008). The Chl may affect the donors of electron density, polarize chemical bonds, and stabilize reaction intermediates (similar to enzyme–substrate complexes) by the occurrence of N-, O-, and S-containing functional groups bound in proteins and lipids (Lobanov et al. 2008).

Under certain physiological conditions such as exposure to high light intensity or drought, reduction of O_2 in photosynthetic organisms can produce reactive oxygen species (ROS), such as $O_2^{\bullet-}$, H_2O_2 or $^1O_2$. These species can lead to the closure of the stomata and cause low CO_2 concentrations in the chloroplasts (Krieger-Liszkay et al. 2008; Asada 1992, 2006 Halliwell and Gutteridge 1990; Hideg et al. 2001, 2002; Trebst et al. 2002). It is shown that a key ROS in UV-irradiated leaves is $O_2^{\bullet-}$, whilst $^1O_2$ is minor (Hideg et al. 2002). Therefore, H_2O_2 may be produced in the plant cells via $O_2^{\bullet-}$. Under such conditions, the plastoquinone pool can be in a very highly reduced state that would allow photoinhibition, i.e. the light induced loss of PSII activity (Adir et al. 2003). The $HO^\bullet$ produced photolytically from H_2O_2 or $^1O_2$ and ROS itself can react with proteins, pigments, nucleic acids and lipids, and could also be connected to the light-induced loss of PSII activity, to the degradation of the D1 polypeptide (PSII reaction centre polypeptide) and to pigment bleaching (Krieger-Liszkay et al. 2008; Aro et al. 1993; Nishiyama et al. 2001, 2004; Vass et al. 1992; Hideg et al. 1994; Keren et al. 1997; Halliwell and Gutteridge 1990; Sopory et al. 1990; Prasil et al. 1992; Hideg et al. 1998; Okada et al. 1996; 2006; Allakhverdiev and Murata 2004; Nixon et al. 2005; Hideg et al. 2007; Aro 2007; Tyystjärvi 2008). Such reactions are often observed in water, where photoinduced generation of $HO^\bullet$ either from H_2O_2 (both upon direct photolysis by sunlight and photo-Fenton reaction) or other sources (e.g. $NO_2^-$and NO_3^-) can decompose the DOM components

(Draper and Crosby 1981; Zepp et al. 1992; Wang et al. 2001; White et al. 2003; Nakatani et al. 2007; Vione et al. 2006, 2009a, b).

3.6 Occurence of H_2O_2 and its Effect on Photosynthesis

In support of the involvement of H_2O_2 in the photosynthetic reaction, several H_2O_2-related phenomena have been observed in natural waters, which can be classified as follows (Mostofa et al. 2009). First, the correlation between carbon production and photolytically formed H_2O_2 concentration, suggesting a link between hydrogen peroxide and organic matter photosynthesis in lake water (Anesio et al. 2005). Second, Chl *a* production in the epilimnetic layer (5–10 m) is typically observed to increase with a decrease in total CO_2 contents (Talling 2006), suggesting that photosynthesis is highest at the epilimnetic layer (5–10 m) than in the uppermost epilimnion (0–1 m). Correspondingly, the O_2 and Chl *a* contents reach a minimum when the water temperature become highest during the summer stratification period (Talling 2006), suggesting that photoinduced degradation or assimilation of Chl *a* may be responsible for the decrease in Chl *a* at the uppermost layer. Here O_2 may be involved in the production of free radicals (H_2O_2 or $HO^{\bullet}$) that could inhibit photosynthesis (Mostofa and Sakugawa 2009; Moffett and Zafiriou 1990). This result is similar to earlier studies where photosynthesis was observed to be less effective in the uppermost layer (1 m) compared to the subsequent epilimnion (3 m) (Nozaki et al. 2002). A ratio of variable to maximal fluorescence (Fv/Fm) of phytoplankton productivity showed a decrease as irradiance increased during the morning and an increase as irradiance declined in the afternoon. These results may be associated with both photoprotective strategies in the antennae of PSII and photo damage of PSII reaction centers (Zhang et al. 2008). Conversely, H_2O_2 usually increases gradually starting in the morning, reaches a maximum at noon and then gradually decreases in the afternoon (Mostofa and Sakugawa 2009). It is therefore suggested that high production of H_2O_2 and subsequent photoinduced generation of $HO^{\bullet}$ at noon is susceptible to damage the PSII reaction centers.

Third, H_2O_2 may be concentrated by particulate organic matter or small fungi through rapid transpiration (Komissarov 1994, 1995, 2003). This hypothesis can be supported by observation of relatively low production of H_2O_2 in unfiltered samples compared to filtered ones during irradiation (Moffett and Zafiriou 1990; Cooper et al. 1988; Petasne and Zika 1997). An increase in the growth rate of plants and mycelial fungi is detected when the H_2O_2 concentration increases up to an optimum level, from 1 nM to 10 M, and the growth rate decreases when H_2O_2 approaches 1 mM (Komissarov 2003; Ivanova et al. 2005). High levels of H_2O_2 may photolytically produce $HO^{\bullet}$, a strong oxidizing agent, that may cause ecophysiological disorders in plants, decrease the CO_2 assimilation rate and affect stomatal conductance, fluorescence and needle life span (Kume et al. 2000; Kobayashi et al. 2002). In natural waters, $HO^{\bullet}$ that is produced photolytically from

H_2O_2 can degrade phytoplankton cells, thereby decreasing photosynthesis. The synergistic effect of high contents of H_2O_2 combined with elevated seawater temperature (27–31 °C) can result in a 134 % increase in respiration rates of the coral *Galaxea fascicularis*, which can surpass the effect of either H_2O_2 or high seawater temperature alone (Higuchi et al. 2009). A possible explanation is that an increase in growth of plant species with increasing H_2O_2 might enhance carbohydrate production, and therefore enhance the activity throughout the food web.

4 Functions of Photosystems (I and II) in Organisms During Photosynthesis

Photosynthesis is primarily initiated by the light-induced release of electrons across a membrane, which is catalyzed by two multisubunits, special type of membrane-bound pigment-protein complexes called photosynthetic reaction centres (RCs). They are photosystem I (PSI) and photosystem II (PSII) (Krauß 2003; Golbeck 1994; Brettel 1997; Li et al. 2006; Rappaport and Diner 2008; Müller et al. 2010; Nilsson Lill 2011; Umena et al. 2011; Renger and Holzwarth 2005; Fromme 2008; Holzwarth 2008). PSI of higher plants and algae (named PSI-200) consists of the PSI core complex and the peripheral light-harvesting complex LHCI. In cyanobacteria, it only consists of the PSI core (Schlodder et al. 2011). The PSI core complexes in cyanobacteria are organized preferentially as trimers, whereas PSI in higher plants and algae is present only as a monomer (Boekema et al. 1987, 2001; Shubin et al. 1993; Kruip et al. 1994; Jordan et al. 2001; Amunts et al. 2010).

By studying the crystal structure of cyanobacterial PSI it has been shown that it is composed of 128 cofactors including approximately 96–100 Chl molecules, two phylloquinones, three [Fe_4S_4] clusters, 22 carotenoids, four lipids and a putative Ca^{2+} ion (Fig. 4) (Krauß 2003; Krauss et al. 1993; Krauß et al. 1996; Klukas et al. 1999; Jordan et al. 2001; Ben-Shem et al. 2003; Müller et al. 2010; Webber and Lubitz 2001). The PSI antenna consists of 90 Chls, of which 79 are bound to a heterodimeric core formed by subunits PsaA and PsaB, with 2 × 11 transmembrane α-helices (Krauß 2003). The cofactors in the RC of PSI form two quasi-symmetric branches (Fig. 4), diverging from a Chl *a*/Chl *a* pair (ec1A/ec1B) traditionally called P700 (Jordan et al. 2001; Müller et al. 2010). In each branch there is a pair of Chl *a* molecules (ec2A/ec3A or ec2B/ec3B) and a phylloquinone (PhQA or PhQB) and then the branches join again at the FX iron-sulfur (FeS) cluster (Müller et al. 2010). The carotenoids have a dual function in light harvesting and photoprotection. The organic cofactors of the electron transfer chain are bound to PsaA/PsaB and arranged in two branches of three Chl and one phylloquinone molecule each, related by a pseudo-C2 axis (Krauß 2003). These studies show that the PSI reaction center or primary donor P700 in PSI is composed of six chlorophyll (Chl) *a* cofactors: the P700 special pair Chls (analogous to the special pair bacteriochlorophylls in purple bacterial reaction centers), two accessory Chls

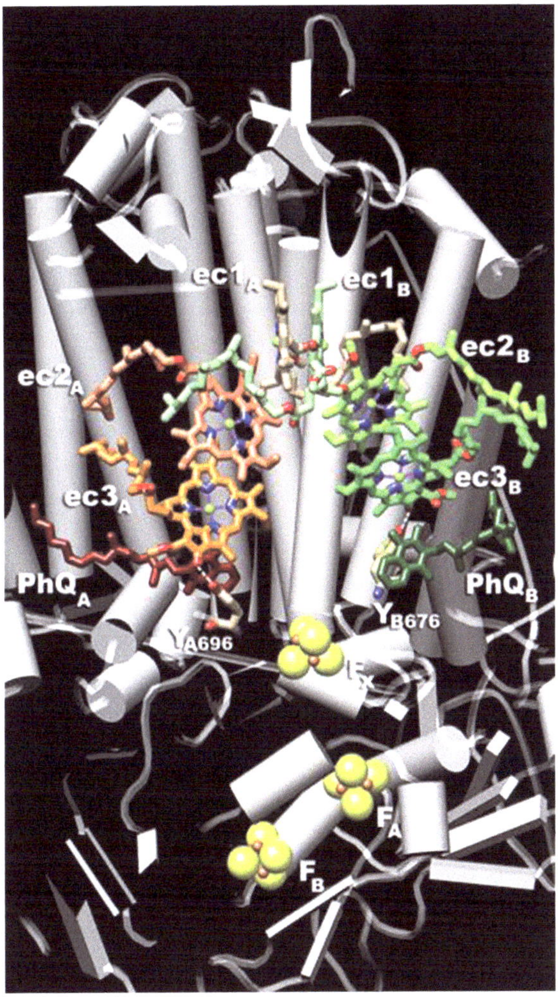

Fig. 4 Organization of the ET cofactors in the RC of PSI, based on the X-ray crystal structure of cyanobacterial PSI [1JB0] (Jordan et al. 2001), and using the nomenclature suggested by Redding and van der Est (Redding and van der Est 2006) (Figure is generated using UCSF Chimera). *Data source* Müller et al. (2010)

(analogous to the accessory bacteriochlorophylls), and two chlorophylloid. Based on the crystal structures, it is generally assumed that the PSI core complexes, particularly the cofactor arrangement in the reaction centre, are similar in all organisms and plants (Jordan et al. 2001; Ben-Shem et al. 2003).

On the other hand, crystal structure analysis of cyanobacterial photosystem II (PSII) demonstrates that PSII monomer contains 20 subunits with a total molecular mass of 350 kDa (Umena et al. 2011). It is composed of 19 protein subunits, 32–36 Chl molecules (35 Chls for *T. vulcanus*) (Umena et al. 2011) including chlorophyll *a* dimer ($P_{D1}P_{D2}$) and monomers (Chl_{D1} and Chl_{D2}), two pheophytins *a* ($Pheo_{D1}$ and $Pheo_{D2}$), 11 ß-carotenes, more than 20 lipids, two plastoquinones Q_A and Q_B, two haem irons, one non–haem iron, a tetranuclear manganese cluster forming $Mn_4CaO_5(H_2O)_4$ or $Mn_4CaO_4(OH)(H_2O)_4$, three or four calcium atoms

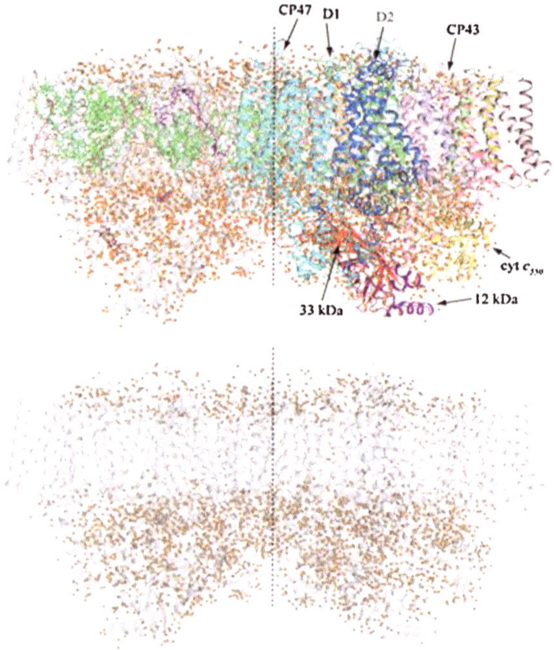

Fig. 5 Overall structure of PSII dimer from *Thermosynechococcus vulcanus* at a resolution of 1.9Å. View from the direction perpendicular to the membrane normal. **a** Overall structure. The protein subunits are coloured individually in the right hand monomer and in light *grey* in the left-hand monomer, and the cofactors are coloured in the left-hand monomer and in light *grey* in the right-hand monomer. Orange balls represent water molecules. **b** Arrangement of water molecules in the PSII dimer. The protein subunits are coloured in light *grey* and all other cofactors are omitted. The central broken lines are the noncrystallographic two-fold axes relating the two monomers. *Data source* Umena et al. (2011)

(one of which is in the Mn_4Ca cluster), three Cl^- ions (two of which are near the Mn_4CaO_5 cluster), one bicarbonate ion and more than 15 detergents (Fig. 5) (Krauß 2003; Nilsson Lill 2011; Umena et al. 2011; Zouni et al. 2001; Kamiya and Shen 2003; Ferreira et al. 2004; Loll et al. 2005; Murray et al. 2008; Kawakami et al. 2009; Guskov et al. 2009; Biesiadka et al. 2004). PSII reaction center or primary donor P680 in PSII is an approximately C_2-symmetric structure formed by polypeptides (D1 and D2) and six chlorin cofactors: four chlorophyll *a* and two pheophytin *a* ($Pheo_{D1}$ and $Pheo_{D2}$) (Fig. 5) (Nilsson Lill 2011; Umena et al. 2011). Each PSII monomer consists of more than 1,300 water molecules, yielding a total of 2,795 water molecules in the dimer (Umena et al. 2011). The water molecules are organized into two layers located on the surfaces of the stromal and lumenal sides, respectively, with the latter having more water molecules than the former (Umena et al. 2011). A few water molecules are detected within the membrane region, most of them serving as ligands to chlorophylls (Umena et al. 2011).

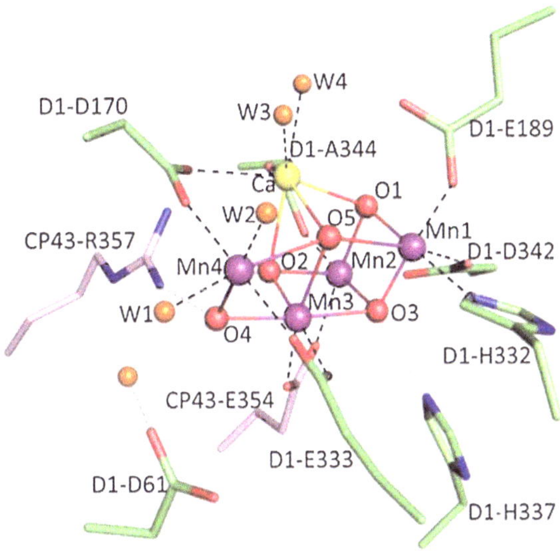

Fig. 6 Structure of the Mn_4CaO_5 cluster. Stereo view of the Mn_4CaO_5 cluster and its ligand environment. The distances shown are the average distances between the two monomers. Manganese, *purple*; calcium, *yellow*; oxygen, *red*; D1, *green*; CP43, *pink*. *Data source* Umena et al. (2011)

$Mn_4CaO_5(H_2O)_4$ or $Mn_4CaO_4(OH)(H_2O)_4$ is formed through five oxygen atoms that act as oxo bridges linking the five metal atoms, and four water molecules that are bound to the Mn_4CaO_5 cluster and can generate O_2 (Fig. 6) (Umena et al. 2011; Yamanaka et al. 2012). Among the five metal and five oxygen atoms, three Mn, one Ca and four O form a cubane-like structure in which Ca and Mn occupy four corners and the O atoms occupy the other four. The fourth manganese (Mn4) is located outside the cubane; it is linked to Mn1 and Mn3 within the cubane by O5, and to O4 by a di-μ-oxo bridge (Umena et al. 2011). In this way, every two adjacent Mn atoms are linked by di-μ-oxo bridges: Mn1 and Mn2 via O1 and O3, Mn2 and Mn3 via O_2 and O3, and Mn3 and Mn4 via O4 and O5. The calcium is linked to all four Mn by oxo bridges: to Mn1 via the di-μ-oxo bridge formed by O1 and O5, to Mn2 via O1 and O2, to Mn3 via O2 and O5, and to Mn4 via the mono-μ-oxo bridge formed by O5 (Umena et al. 2011). It is also shown that four water molecules (W1 to W4) are associated with the Mn_4CaO_5 cluster, of which W1 and W2 are coordinated to Mn4 with respective distances of 2.1 and 2.2 Å, and W3 and W4 are coordinated to Ca with a distance of 2.4 Å. This suggests that some of the four waters may serve as the substrates for water oxidation (Umena et al. 2011).

Several studies are conducted to evaluate the functions of the PSI and PSII (Jordan et al. 2001; Dashdorj et al. 2004; Germano et al. 2004; Diner and Rappaport 2002; Li et al. 2006; Rappaport and Diner 2008; Müller et al. 2010; Nilsson Lill 2011; Schlodder et al. 2007, 2011; Nanba and Satoh 1987; Dekker and van Grondelle 2000; Greenfield and Wasielewski 1996; Klug et al. 1998; Prokhorenko and Holzwarth 2000; Byrdin et al. 2002; Yoder et al. 2002; Holzwarth et al. 2006).

4.1 Debates/Questions Regarding O_2-Releases from PSI and PSII

Some key issues on the debate concerning the details of electron- and O_2-release from PSI and PSII will be discussed in the following parts.

First, an electron is released upon excitation by light, either producing the charge-separated state $P680^+H_A^-$ from Chl molecules (P680), or accompanied by no charge separation (or by considerable protein relaxation) (Dashdorj et al. 2004; Germano et al. 2004; Rappaport and Diner 2008; Müller et al. 2010; Takahashi et al. 1987; Periasamy et al. 1978). Accordingly, after release of an electron by PSI or PSII upon excitation by light, is it possible to accept the same component of PSI or PSII? From the point of view of aquatic humic substances (fulvic and humic acids) or CDOM (DOM or FDOM, fluorescent dissolved organic matter), the answer is no. The secondary component (dissolved O_2 in water) can accept the electron to produce super oxide radical anion ($O_2^{\bullet-}$) and then H_2O_2 (Eqs. 3.36–3.40). The detailed mechanism for H_2O_2 production from DOM (or FDOM or CDOM) is extensively discussed in chapter "Photoinduced and Microbial Generation of Hydrogen Peroxide and Organic Peroxides in Natural Waters".

Second, which and how many Chl molecules are taking part to the primary donor sites in PSI and PS II? (Stewart et al. 2000; Jordan et al. 2001; Diner and Rappaport 2002; Li et al. 2006; Müller et al. 2010; Durrant et al. 1995; Dekker and van Grondelle 2000; van Gorkom and Schelvis 1993) The answer is that the first electron is released from the functional or chromophoric group bound to PSI or PSII, which is the easiest way to do it upon excitation by light. Subsequent electron releases occur in succession from the functional groups (for an analogy, see CDOM and FDOM, chapter "Colored and Chromophoric Dissolved Organic Matter in Natural Waters" and "Fluorescent Dissolved Organic Matter in Natural Waters"). It has been shown by fluorescence spectroscopy that longer-wavelength excitation is usually the first to take place, followed by the others. Therefore, Chl dimers or Chl molecules (generally with emission wavelengths >675 nm) bound to PSI or PSII are primarily responsible for excitation of electrons. In contrast, proteins or aromatic amino acid residues (generally having shorter emission wavelengths: <370 nm) are not excited in presence of Chl molecules upon irradiation (see also chapter "Fluorescent Dissolved Organic Matter in Natural Waters").

Third, why are PSI and PSII formed by a number of Chl molecules in their structure? It is assumed here that Chl *a* (or dimer Chl *a*) molecules are extremely photosensitive and can be excited by a small light intensity. Continuous H_2O_2 generation in the presence of little light is enabled by the occurrence of high numbers of Chl *a* molecules in PSI and PSII, which at the same time can contribute to the continuous photosynthesis in organisms and plants under light conditions. The factors affecting the generation of H_2O_2 (e.g. high or low light intensity, pH, nutrients and so on) can affect photosynthesis and induce structural modifications in PSI and PSII. For example, under intense light conditions there is an elevated production of H_2O_2, the excess of which can be photolytically converted into $HO^\bullet$. The hydroxyl radical can then degrade the proteins or amino acid residues. Such an effect can reduce the contents of proteins or amino acid residues, which are often observed in PSI and PSII (Neufeld et al. 2004; Shutova et al. 2005). The decomposition of proteins or amino acids (e.g. tryptophan) is also generally observed in sunlit water environments because of the effects of $HO^{\bullet}$ and other ROS (Mostofa et al. 2007, 2010, 2011; Moran et al. 2000).

Fourth, are there any O_2 or H_2O_2 molecules that may remain undetected among the 1,300 water molecules found in PSII? It is consistent to detect O_2 and H_2O_2 molecules in the PSII structure, which have often been observed in earlier studies. The occurrence of a large number of H_2O molecules suggests that O_2 may remain and be dissolved in those water molecules. Furthermore, H_2O_2 may be produced photolytically from O_2 as discussed before. Two facts may be responsible for not detecting O_2 or H_2O_2: (i) O_2 and H_2O_2 may disappear during the primary processing of the photosynthetic cells before examination; and (ii) former studies did not focus on the occurrence of H_2O_2 in PSII. In a recent study, it has been assumed that H_2O_2 may be "lost" amongst 1,300 H_2O molecules (Umena et al. 2011). The most likely reasons would be the structural similarity and the fact that H_2O_2 occurrence in the PSII structure was not expected. It should be noted that H_2O_2 may be decomposed to H_2O during the processing of photosynthetic cells for the determination of PS crystal structure.

It has been shown that two H_2O molecules in four reaction-center Chls are linked through H-bonding between water ligand and Chl_{D1} (Umena et al. 2011), and it may well be H_2O_2 that can make H-bonding in the proposed structure. It is also shown that two balls labeled I and II represent a single water molecule, disordered at two different positions separated by 1.8 Å. Position-I is able to H-bond to YD (redox-active tyrosine residue located at D2-tyr 160), whereas position-II is not able to H-bond to YD (Supplementary part) (Umena et al. 2011). On this basis, it can be assumed that H_2O_2 may occur in that structure instead of H_2O. Note that the bond length of O–O in H_2O_2 is 1.49 Å, which is larger than in the ground (triplet) state of molecular oxygen ($^3O_2$, 1.21 Å) (Abrahams et al. 1951). Among the 1300 H_2O molecules in each PSII monomer, a few of them are detected as disordered (Umena et al. 2011), a case in which the probability to mistakenly detect H_2O instead of H_2O_2 is relatively high. Future studies will be important to find out any presence of H_2O_2 instead of H_2O in the crystal structure of PSII.

The first two questions will be discussed comprehensively in the next section.

4.2 Mechanism for Electron Transfer and O_2-Release in Photosystem II Reaction Centers

Upon excitation by light, the electron release takes place at the central part of the reaction center (RC), at the primary donor P700 in PSI or P680 in PSII (Figs. 4, 5) (Müller et al. 2010; Nilsson Lill 2011; Umena et al. 2011). It is suggested that the

(a)

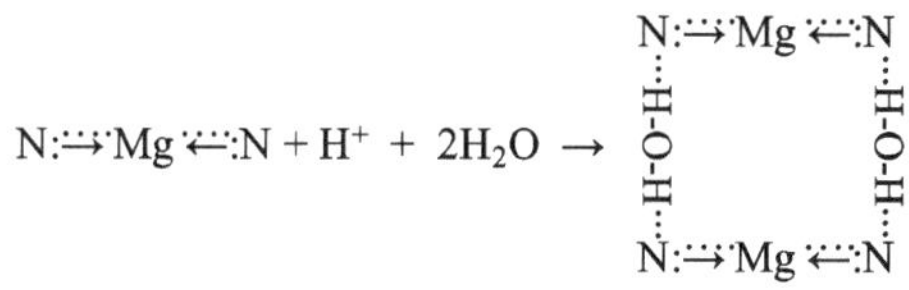

(b)

Fig. 7 The possible resonance configuration of Mg with π-electrons of two N-atoms located in the chlorophyll *a* structure (**a**) and chlorophyll *a* dimer (**b**). Only the two N-atoms in porphyrin ring with Mg are presented in the structure to simplify the resonance structure

primary electron release in PSII involves the chlorophyll *a* dimer (Boussaad et al. 1997; Nilsson Lill 2011). This can be justified by the theory of excitation of multiple functional groups bound to macromolecular organic substances (e.g. fulvic acids or humic acids). Light excitation is expected to induce first the release of the electron less strongly bound in the relevant functional groups, and then of the subsequent ones (see chapter "Colored and Chromophoric Dissolved Organic Matter in Natural Waters").

It is hypothesized that the first electron is released from the π-bonding system formed between two N-atoms in the porphyrin ring and Mg. In fact, Mg ($1s^2 2s^2 2p^6 3s^1 3p_x{}^1 3p_y{}^0 3p_z{}^0$) can form two covalent bonds with two N-atoms of the porphyrin ring using $3s^1$ and $3p_x{}^1$ orbitals, whilst other two empty $3p_y{}^0$ and $3p_z{}^0$ orbitals can accept the π-electrons from the remaining two N-atoms. The π-bonding systems among these orbitals ($3p_y$ and $3p_z$) can interchange with one another because of the similar energy levels. Therefore, one can have resonance configuration upon exchange of electrons between the orbitals and Mg (Fig. 7a). Chl *a* dimer is formed through hydrogen bonding via H_2O bridges, and H_2O is the key component in the formation of such dimers (Shipman et al. 1976; Hynninen and Lötjönen 1993; Boussaad et al. 1997; Catalan et al. 2004). It is supposed that hydrogen (H)-bonding is formed between the non-bonding π-electrons of two N-atoms in the porphyrin ring. The latter is also a resonance structure where electrons can move through the whole Chl *a* dimer (Fig. 7b).

The formation of H-bonds through H_2O bridges is suggested by earlier studies (Shipman et al. 1976), and can be justified by the shift of the π-bonding system in H–N–Mg–N–H (Fig. 7b). This system can assist the release of electrons in a much easier way than the single N–Mg–N system (Fig. 7a). Based on multimer model studies one obtains equal site energies and inhomogeneous widths for all pigments, which leads to similar distances and to nearest-neighbor dipole–dipole interactions between the central chlorin cofactors (Durrant et al. 1995; Renger and Marcus 2002; Barter et al. 2003). This may result into two wavelength positions for the electronic states in the reaction center (RC): uncoupled Chls can absorb at 670 nm, and electronically coupled chlorins (the central cofactors) or Chl dimers can absorb between 676 and 684 nm (Telfer et al. 1990; Durrant et al. 1995; Renger and Marcus 2002). Red shifts are commonly observed in in vitro Chl *a* systems, such as thin films, monolayers and colloidal dispersions, used as models for the in vivo system (Katz et al. 1991). It is known that red shifts occur when the release of electrons takes place in the functional groups that is bound to the component system (see also chapter "Colored and Chromophoric Dissolved Organic Matter in Natural Waters", "Fluorescent Dissolved Organic Matter in Natural Waters") (Mostofa et al. 2009; Senesi 1990). Note that Chl *a* has a broad absorption spectrum and can form dimers or aggregates through self assembly, which typically leads to changes in its optical properties (Shipman et al. 1976; Hynninen and Lötjönen 1993; Closs et al. 1963; Katz et al. 1963; Fong 1974; Shipman et al. 1975; Katz 1990, 1994; Frackowiak et al. 1994). Formation of the dimer often occurs through H-bonding in the N-heterocyclic base pair (Catalan et al. 2004), which can support the occurrence of H-bonding between N and H_2O (Fig. 7b).

Two possible hydrogen bonds were also discussed in earlier studies. First, formation of H-bonds might occur between central Mg and H_2O according to the Mg…OH_2 interaction (Hynninen and Lötjönen 1993). Second, the keto carbonyl group of Chl a may participate in the formation of Chl a dimers, either through coordination with Mg or through H-bonding of the H–X type, where X = O, N and S (Shipman et al. 1976; Closs et al. 1963; Katz et al. 1963; Fong 1974; Shipman et al. 1975; Katz 1990). However, these two previous assumptions are not possible electronically because the outer shells of Mg are entirely full, after bonding with two covalent bonds and two unpaired π–electron systems with four N-atoms of the Chl *a*. Therefore, Mg has less probability to accept further electrons or H-bonding with other groups. Moreover, the formation of such proposed bonding systems is not consistent with the easiest way of electron release via absorption in the longer wavelength region.

Crystal structures of the reaction center have identified two chlorophyll monomers forming a dimer with a partial structural overlap, which are thus stabilized by van der Waals interactions (Nilsson Lill 2011). The structure of the chlorophyll dimer has been optimized using dispersion-corrected density functional theory (B3LYP-DCP) and it has been found that the dimerization energy is approximately -17 kcal mol^{-1} (Nilsson Lill 2011). Electrons may be rapidly released from these resonance configurations upon irradiation of the Chl *a* dimmer, according to the proposed dimer formation (Fig. 6). This can be understood from the interaction mechanism between the functional group [$-CH_2-(NH_3^+)-CH-COO^-$] in tryptophan [$C_8H_5(NH)$-$CH_2(NH_3^+)CHCOO^-$] and metal ions, where the functional group [$-CH_2-(NH_3^+)-CH-COO^-$] can display resonance configuration that is responsible for the longer wavelength fluorescence emission spectra (see chapter "Complexation of Dissolved Organic Matter With Trace Metal Ions in Natural Waters").

PSII acts as one component and upon irradiation, the released electron may not accept the same component of PSII that can be understood from aquatic ecosystem. For example, in aqueous media fulvic acid or humic acid upon irradiation can donate the electron to O_2 and form $O_2^{\bullet-}$ and then H_2O_2, which is a well-accepted mechanism by all aquatic scientists. Therefore, it is hypothesized that the released electron in PSII may react with other components present in the cells, the most efficient of which is O_2 that can form $O_2^{\bullet-}$ and then H_2O_2. The latter species are often detected in cells as discussed in the earlier sections. It is also established that H_2O_2 formation is the primary step of many photoinduced processes in aqueous solution that finally lead to the formation of the $HO^{\bullet}$ radical (see chapter "Photoinduced and Microbial Generation of Hydrogen Peroxide and Organic Peroxides in Natural Waters").

Upon excitation, an electron is transferred from the Chls to the Pheo HA, producing the charge-separated state $P680^{+}H_A^{-}$ as assumed by earlier studies (Germano et al. 2004; Rockley et al. 1975; Thurnauer et al. 1975; Shuvalov and Klevanik 1983; Kirmaier and Holten 1987; Holzapfel et al. 1990). Similarly, in PSI a primary charge separation occurs in the P700 reaction center that can lead to the reduction of A_0 (two chlorophylloid primary electron acceptors), creating the radical ion pair $P700^{+}A_0^{-}$ (Krauß 2003; Brettel 1997; Müller et al. 2010; Webber and Lubitz 2001; Fromme et al. 2001). However, no concrete evidence has been found for the formation of these types of radicals in PSI or PSII. Rather, experimental studies support the idea that primary electron transfer reactions are accompanied by molecular readjustments or reorganizations involving pigments and proteins, or the interaction of pigment-protein complexes in the reaction center (Dashdorj et al. 2004; Kleinfeld et al. 1984; Woodbury and Parson 1984; Kirmaier et al. 1985a, b; Holten et al. 1986; Kirmaier et al. 1986; Tiede et al. 1987; Mullineaux et al. 1993; Savikhin et al. 2001; Karapetyan 2004).

It is also observed that chlorophyll-binding PsbS protein (22-kD protein of PSII), which belongs to the family of light-harvesting proteins, can contribute only to quenching but not to light harvesting (Li et al. 2000, 2002; Aspinall-O'Dea et al. 2002; Bergantino et al. 2003). Indeed, the degree of fluorescence quenching in vivo can correlate with the content of PsbS (Li et al. 2004). Dissipation of energy in PSI trimers of cyanobacteria takes place with a contribution of the long-wavelength chlorophyll, and the excited state of which is quenched by the cation radical of P700 or by P700 in its triplet state (Karapetyan 2004). The low fluorescence yield of Chls in light-harvesting antenna complexes is indicative of an additional pathway of energy dissipation in oligomers, which would protect the PSII complex of cyanobacteria against photodestruction (Karapetyan 2004).

It can thus be hypothesized that excitation followed by charge transfer could produce $P680^{+}O_2^{\bullet-}$ instead of $P680^{+}H_A^{-}$. O_2 is the primary acceptor for excited electrons in aquatic media and is involved in the production of H_2O_2 as discussed earlier. This result is supported by Laser flash photolysis studies, in which a charge-transfer excited state has not been detected from the spectra. Recovery kinetics, including observation of both triplet decay and ground-state folding reactions, show that the flash transient obtained from the pinned form consists of a triplet and of a ground state moiety in the unpinned configuration (Periasamy et al. 1978). Experimental optical data and structure-based simulations showed nanosecond absorption dynamics at ~685 nm, after excitation of PS I from *Synechocystis* sp. PCC 6803. It is suggested that the electrochromic shift of absorption bands of the Chl *a* pigments may occur around the secondary electron acceptor, through considerable protein relaxation (Dashdorj et al. 2004; Savikhin et al. 2001).

A recent study has shown that the PSII monomer consists of 1300 H_2O molecules, a few of which have been detected as disordered (Umena et al. 2011). H_2O_2 was not considered as a component of PSII structure in that study. Concurrently, four successive photoinduced turnovers provide the WOC with four oxidising equivalents and drive it through an S-state cycle, with S-states ranging from S_0 to S_4 and O_2 is being released on the S_3 to S_4 transition.

Now the questions are: how is it possible for H_2O to undergo photodissociation through four successive photoinduced turnovers, needing energy in the presence of H_2O_2 that can easily be decomposed and produce O_2? How can H_2O in a cell accept four consecutive electrons in the presence of many additional components including O_2 that can more easily accept electrons? Under these conditions, the easiest pathway would be the addition of one electron to O_2 with formation of $O_2^{\bullet-}$ and then of H_2O_2. This is a well established mechanism in water media and could take place in photosynthetic cells as well. Note that the main radiation absorbers in natural waters are chromophoric (or colored) DOM (CDOM) (10–98 %), phytoplankton or chlorophyll (32–85 %), H_2O (0.3–9 % in the red portion of the visible spectrum, depending on water being clear or turbid) and so on (see chapter "Colored and Chromophoric Dissolved Organic Matter in Natural Waters"). It is entirely impractical to consider that H_2O can accept four successive electrons under light condition in the presence of O_2 or other organic components in a photosynthetic cell and there is no evidence in that regards.

It is therefore theorized that

> if H_2O would decompose by the reaction with CO_2 in photosynthesis, then all H_2O would convert into O_2 by organisms and plants after the origin of life on earth to date and no H_2O would remain in the biosphere. Instead of H_2O, photoinduced generation of H_2O_2 from dissolved O_2 in water bound in photosynthetic cells (3.33–3.39) is reacted with CO_2 in photosynthesis that can limit the photosynthesis under light condition.
> Then further conversion of H_2O_2 to O_2 either through photosynthesis [$xCO_2(H_2O) + yCO_2(H_2O) \rightarrow C_x(H_2O)_y + O_2 + E\ (\pm)$] or both photolytically ($2H_2O_2 + h\upsilon \rightarrow O_2 +$ unknown oxidant) and biologically ($2H_2O_2 +$ catalases/peroxidases $\rightarrow O_2 + 2H_2O$) may balance the environment.

This can be supported by the observation of several phenomena:

(i) Formation and occurences of H_2O_2 in photosynthetic cells of organisms through production of $O_2^{\bullet-}$ from whole bacteria of several species, from phagocytic cells, from spermatozoa as well as peroxisoms, mitochondria and chloroplasts (Komissarov 2003; Bach 1894; Chance et al. 1979; Halliwell 1981; Holland et al. 1982; Wilhelm et al. 1996, 1997, 1999; Halliwell and Gutteridge 1999; López-Huertas et al. 1999; Baker and Graham 2002; del Río et al. 2006; Krieger-Liszkay et al. 2008; Lyubimov and Zastrizhnaya 1992a, b; Turrens 1997; Karuppanapandian et al. 2011). (ii) Releases of O_2 from H_2O_2 during photosynthesis are evidenced in earlier studies (Komissarov 1994, 2003; Velthuys and Kok 1978; Asada and Badger 1984; Asada and Takahashi 1987; Mano et al. 1987; Renger 1987; Anan'ev and Klimov 1988; Bader and Schmid 1988, 1989; Schroeder 1989; Schröder and Åkerlund 1990; Miyake and Asada 1992; Kuznetsov et al. 2010; Bernardini et al. 2011; Yin et al. 2006). (iii) The

O_2-releases [H_2O_2 + light or enzymes (catalases/peroxidases) $\rightarrow$ O_2 + $2H_2O$ or other components] and their reused in H_2O_2 generation in photosynthetic organisms (O_2 + Chl + H^+ + $h\upsilon$ $\rightarrow$ H_2O_2) can balance the O_2 level in the environments in new photosynthetic reaction. But this does not occur in old photosynthetic reaction. (iv) Conversion of H_2O_2 to O_2 occurs at a higher extent in biological systems than in photoinduced decomposition processes (Moffett and Zafiriou 1990). For instance, the Cu(II) bathocuproinedisulfonic acid complex ($Cubc_2$) can convert H_2O_2 to O_2 via the reaction (Eq. 4.1) (Moffett et al. 1985):

$$2Cu(II)bc_2 + H_2O_2 \rightarrow 2Cu(I)bc_2 + O_2 + 2H^+ \quad (4.1)$$

Unconvincing evidence has been found for S_0 to S_4 transitions, and four successive transitions are needed for H_2O decomposition (Rappaport and Diner 2008; Kok et al. 1970; Joliot and Kok 1975; Krishtalik 1986, 1990).

(v) It is hypothesized that the O–O bond formation occurs when O5 in Mn_4CaO_5 cluster provides one O atom via formation of hydroxide ion in the S_1 state (Umena et al. 2011; Saito et al. 2012). A major issuer is then if it is possible to break down O5 in the Mn_4CaO_5 cluster. The problem is that, were it possible, probably the entire PSII system would be broken down. Furthermore, each Mn atom in the Mn_4CaO_5 cluster is in octrahedral form with six ligands, and it is also paramagnetic with 5 unpaired electrons in its outer *d*-orbitals ($Mn^{2+} = 1s^22s^22p^63s^23p^64s^03d^5$). The result is that Mn could carry out strong H-bonding with other components. Such an effect enables a second coordination sphere by D1-Asp 61, D1-His 337 and CP43-Arg 357, in addition to the direct ligands. Therefore, these three residues might be responsible for maintaining the oxygen-evolving activity (Umena et al. 2011; Nixon and Diner 1994; Chu et al. 1995; Hwang et al. 2007; Service RJ, Hillier W, Debus RJ 2010). D1-Asp 61 is located at the entrance of a proposed proton exit channel involving a chloride ion (Cl^-) in Mn_4CaO_5 (Umena et al. 2011; Kawakami et al. 2009; Guskov et al. 2009; Murray and Barber 2007; Ho and Styring 2008). This residue may facilitate proton exit from the Mn cluster. Proton releases from Mn_4CaO_5 may play a key role in the formation of H_2O_2 via $O_2^{\bullet-}$ and $HO_2^{\bullet}$.

(vi) Finally, the occurrence of about 1,300 water molecules in the PSII monomer, located at the luminal and stromal sides (Umena et al. 2011), could allow the inclusion of a lot of dissolved O_2 molecules. They could add electrons after they are released from Chl molecules upon excitation by light. Crystal structures of PSI or PSII do not include any information about dissolved O_2, and issue that will need further studies to be clarified.

5 Factors Affecting the Photosynthesis of Organisms

Cyanobacteria carry out oxygenic photosynthesis using a photosynthetic system similar to that observed in chloroplasts of higher plants. Therefore, cyanobacteria can be used in model studies to understand the effects of various environmental factors

(Allakhverdiev and Murata 2008; Pfenning 1978; Öquist et al. 1995). However, the anti-oxidant systems in cyanobacteria are significantly different from those of higher plants (Asada 2006; Demmig-Adams and Adams III 1992, 2002). This can vary the effects of various environmental stresses on cyanobacteria, bacteria and higher plants.

Studies show that terrestrial plants are adapted to their annual life cycles of growth, reproduction and senescence. Compared to the annual climate cycle, phytoplankton biomass can turn over around 100 times a year as a result of fast growth and equally fast consumption by grazers (Calbet and Landry 2004; Behrenfeld et al. 2006; Winder and Cloern 2010). It has been observed that the timing of these life-history transitions can vary among species and among regions with variation in temperature and sunlight intensity (Winder and Cloern 2010; Myneni et al. 1997; Menzel and Fabian 1999; Peñuelas and Filella 2001; Jolly et al. 2005; White et al. 2009; Richardson et al. 2010). Correspondingly, annual phytoplankton cycles can differ across ecosystems, because of year to year variability and with changes in the climate system (Winder and Cloern 2010; Garcia-Soto and Pingree 2009; Thackeray et al. 2008; Paerl and Huisman 2008; McQuatters-Gollop et al. 2008; Cloern and Jassby 2008; Winder and Schindler 2004; Edwards and Richardson 2004; Scheffer 1991; Pratt 1959). These periodic cycles can be linked with annual fluctuations of mixing, temperature, light, precipitation and with other drivers of population variability, including human disturbance. There are also effects from periodic weather events and strong trophic coupling between phytoplankton and their consumers (Winder and Cloern 2010; Smetacek 1985; Sommer et al. 1986; Cloern 1996).

Cyanobacteria can control a variety of environmental stressors such as UV light, heat, cold, drought, salinity, nitrogen starvation, photo-oxidation, anaerobiosis and osmotic stress, by developing a number of defence mechanisms (Fay 1992; Tandeau de Marsac and Houmard 1993; Sinha and Häder 1996). The most important one is the production of photoprotective compounds such as mycosporine-like amino acids (MAAs) and scytonemin (Sinha et al. 1998, 1999a, b; 2001); availability of enzymes such as superoxide dismutase, catalase and peroxidase (Burton and Ingold 1984; Canini et al. 2001); repair of DNA damage (Sinha and Häder 2002) and synthesis of shock proteins (Sinha and Häder 1996; Borbely and Suranyi 1988; Bhagwat and Apte 1989).

Organisms are thus affected by several factors that could either increase or decrease their photosynthetic and respiratory activities (Doyle et al. 2005; Nozaki et al. 2002; Shimura and Ichimura 1973; Pope 1975; Pick and Lean 1987; Babin et al. 1996; Shapiro 1997; Hyenstrand et al. 1998; Elser 1999; Dokulil and Teubner 2000; MacIntyre et al. 2000; Xie et al. 2003; Qu et al. 2004; Tank et al. 2005; Wängberg et al. 2006; Sobrino et al. 2008). The key factors affecting these activities are mostly documented on the basis of the growth and development of organisms. Such factors are: (i) seasonal variation in sunlight and UV radiation, which affect photosynthesis; (ii) occurrence of CO_2 forms (dissolved CO_2, carbonic acid, bicarbonate, carbonate); (iii) variation in temperature; (iv) water stress (drought) and precipitation/rainfall; (v) contents and nature of DOM and POM; (vi) nutrient availability; (vii) variation in trace metal ions; (viii) salinity or salt stress; (ix) presence of toxic pollutants; (x) effect of size-fractionated phytoplankton; (xi) global warming.

5.1 Seasonal Variation in Sunlight and UV Radiation on Photosynthesis

Solar radiation is the key driving force for the occurrence of photosynthesis in natural waters (Sinha et al. 2001; Rastogi et al. 2010; Jiang and Qiu 2011; Sobek et al. 2007). Exposure of photosynthetic organisms to strong light (or UV light) can significantly inhibit the PSII activity, with resulting photoinhibition of or photodamage to PS II (Aro et al. 1993; Melis 1999; Andersson and Aro 2001; Han et al. 2001; Nishiyama et al. 2001, 2008; Adir et al. 2003). Photoinhibition of photosynthesis is a process by which excessive irradiance, absorbed by the leaves, can inactivate or impair the chlorophyll-containing reaction centers of chloroplasts, thus inhibiting photosynthesis (Bertamini et al. 2006). Because of the differences among the organisms, the effects of light can be classified into two sections (aquatic microorganisms and higher plants) for their better understanding.

Effects of Sunlight on Aquatic Microorganisms

Cyanobacteria or phytoplankton cells can utilize photosynthetically active radiation (PAR, 400–700 nm) to drive photosynthesis within the euphotic zone (see also global warming chapter "Impacts of Global Warming on Biogeochemical Cycles in Natural Waters") (Smith and Baker 1979; Abboudi et al. 2008; Li et al. 2011). Solar UV-A radiation (315–400 nm) acts as an additional source of energy for photosynthesis to enhance the CO_2 fixation in tropical marine phytoplankton (Li et al. 2011; Gao et al. 2007, 2007). However, UV-A does not bring any enhancement to carbon fixation in pelagic water (Li et al. 2011). The cells of aquatic microorganisms can be exposed to ultraviolet radiation (UVR, 280–400 nm), which can penetrate up to 60 m into the pelagic water column (Smith and Baker 1979). Furthermore, depletion of the stratospheric ozone layer can cause additional penetration of UV radiation in the Arctic and Antarctic regions. Such a phenomenon has detrimental effects on the processes involved in primary production (see also chapter "Impacts of Global Warming on Biogeochemical Cycles in Natural Waters") (Huisman et al. 2006; Häder et al. 2007; Zhang et al. 2007). Solar UV-B (280–315 nm), and partly UV-A (315–400 nm) can reduce growth and photosynthetic rates, increase permeability of cell membranes, damage proteins or DNA molecules, pigments, and even lead to cell death (see also chapter "Impacts of Global Warming on Biogeochemical Cycles in Natural Waters") (Jiang and Qiu 2011; Wängberg et al. 2006; Behrenfeld et al. 1993; Sass et al. 1997; Campbell et al. 1998; Rajagopal et al. 2000; Helbling et al. 2001; He and Häder 2002; Buma et al. 2003; Sobrino et al. 2004; Litchman and Neale 2005; Wu et al. 2005; Bouchard et al. 2006; Agusti and Llabrés 2007; Rath and Adhikary 2007; Gao et al. 2008; Pattanaik et al. 2008; Jiang and Qiu 2005).

It has been shown that, ranging from coastal (case 1) to pelagic (case 2) surface seawaters, UV-B can cause similar inhibition whilst the inhibition of photosynthesis by UV-A (315–400 nm) increases when passing from coastal to offshore waters (Li et al. 2011). UV-B inhibits photosynthesis up to 27 % and UV-A up to 29 %. It has also been shown that the daily integrated inhibition by UV-A can reach 4.3 % and 13.2 %, whilst that by UV-B can reach 16.5 % and 13.5 % in coastal and offshore waters, respectively (Li et al. 2011). Additionally, exclusion of UV radiation can increase photosynthesis by 10–65 % in algae from the Mediterranean, 17–46 % in intertidal algae from southern Chile, and 15–20 % in algae (*Laminaria Saccharina*) from the North Sea (Hanelt et al. 1997; Jiménez et al. 1998; Gómez et al. 2004).

UV-stimulated inorganic carbon acquisition is often observed in phytoplankton species (Beardall et al. 2009; Wu and Gao 2009). Phytoplankton cells grown in nutrient replete conditions are more resistant to solar UV radiation, and also their contents of UV-absorbing compounds increases (Marcoval et al. 2008). Microplankton (>20 m) are more plentiful in coastal waters, while picoplankton (<2 m) are more abundant in open oceans (Marañón et al. 2001; Ho et al. 2008). In terms of their responses to UV, large cells are capable of synthesizing and accumulating UV-absorbing compounds that play a protective role against UV. These screening compounds are not found in picoplankton cells (Raven 1991; Garcia-Pichel 1994) that, therefore, would be more sensitive to solar UV. This issue is partially offset by a much faster repair process of damaged DNA (Helbling et al. 2001; Callieri et al. 2001). Because taxonomic composition, accumulation of UV-absorbing compounds and nutrient availability are typically different, physiological responses of phytoplankton assemblages to solar UV can differ geographically from coastal to pelagic waters (Li et al. 2011).

Cyanobacteria are important and ubiquitous prokaryotes that populate terrestrial and aquatic habitats, and they are important contributors to global photosynthetic biomass production (Whitton and Potts 2000). Enhanced UV-B radiation can affect cyanobacterial growth, photosynthetic efficiency, pigments, morphology, as well as cell size and shape. Anyway, different responses are observed in different species exposed to different UV doses (Wu et al. 2005; Rath and Adhikary 2007; Pattanaik et al. 2008; Jiang and Qiu 2005; Harrison and Smith 2009). It has also been shown that exposure to UV radiation can reduce the activity of alkaline phosphatase, a common extracellular enzyme, by up to 57 %. Interestingly, it is more often decreased under ultraviolet A than ultraviolet B exposure (Tang et al. 2005). As already mentioned, algal nutritional status can influence the UV radiation sensitivity but, on the other hand, UV radiation can inhibit uptake and assimilation of inorganic nutrients (Harrison and Smith 2009). This is likely caused by the rapid UV radiation-induced changes of nitrate into $HO^{\bullet}$ and $^{\bullet}NO_2/NO_2^-$, which may reduce the availability of NO_3^- for primary production (see chapter "Photoinduced Generation of Hydroxyl Radical in Natural Waters").

It is estimated that, depending on location, ambient UV radiation can reduce carbon fixation rates up to 65 % in surface waters of the Antarctic region, down to undetectable levels at 36 m (Boucher and Prezelin 1996). On average, up to 42 % of primary production inhibition in the water column is carried out by UV radiation on a daily basis outside the ozone hole (Wängberg et al. 2006; Harrison and Smith 2009; Helbling et al. 1992; Smith et al. 1992; Holm-Hansen et al. 1993; Bertoni et al. 2011). In contrast, during a ozone hole depletion event, the inhibition is increased to ~50 %. This can be supported by the experimental observations that UV-B radiation can inhibit the oxygen-evolving complex of PSII in *M. aeruginosa* (Jiang and Qiu 2011). The whole electron-transport activities are significantly varied: the transfer from water to methyl viologen being inhibited by 27.9 % under UV-B, that from diphenylcarbazide to methyl viologen by 13.3 % (Jiang and Qiu 2011).

Cyanobacterial blooms in freshwater have apparently increased over the last few decades all over the world (Xu et al. 2000; Chen et al. 2003; McCarthy et al. 2007). UV-B influences the CO_2-uptake mechanism of M. *aeruginosa*, and this cyanobacterium has many adaptive strategies to cope with prolonged UV-B exposure (Jiang and Qiu 2005; Song and Qiu 2007). It has been shown that maximum quantum yield and maximum electron transport rate in seaweeds collected from the Red Sea decreased largely due to the combined effects of increased irradiance (PAR) and presence of UV radiation (Figueroa et al. 2009). A 33-kDa protein of the water-splitting complex is sensitive to UV-B. Therefore, its degradation contributes importantly to the decline of the electron transport rate (Jiang and Qiu 2011; Prabha and Kulandaivelu 2002). Short-term UV-B exposure can severely inhibit photosynthetic capability, which could be quickly restored upon exposure to PAR (Jiang and Qiu 2011). Quite surprisingly, UV-A can assist the photo repair of UV-damaged DNA and enhance carbon fixation under reduced levels of solar radiation or fast mixing conditions (Gao et al. 2007, 2007; Karentz et al. 1991; Barbieri et al. 2002; Helbling et al. 2003). Recent study reveals that the PSII of *M. aeruginosa* FACHB 854 is more sensitive to UV-B exposure than PSI, and the oxygen-evolving complex of PS II is an important target for UV-B damage (Jiang and Qiu 2011).

The mechanisms behind the photoinhibition effects of strong sunlight, UV light or high irradiance (drought/heat stress) on aquatic microorganisms are presumably involving two facts: First, there are direct effects in which a high number of electrons is released from chlorophylls (Chl) (P680) in PSII of microorganisms, upon excitation by strong light or strong UV light (Eq. 5.1). The release of many electrons can produce elevated amounts of reactive oxygen species (ROS) such as $^1O_2$, $O_2^{\bullet-}$, H_2O_2 and $HO^{\bullet}$ (Eq. 5.2). Among the ROS, H_2O_2 can be used in photosynthesis whilst the remaining ROS including H_2O_2 can react with the Chl^+ ($P680^+$) functional groups bound to PSII, killing the cells (Eq. 5.3). These reactions can be schematically depicted as follows:

$$\text{Chl (or P680)} + h\upsilon \rightarrow \text{Chl}^+ \left(\text{or P680}^+\right) + e^- \tag{5.1}$$

$$e^- + O_2 + h\upsilon \rightarrow {}^1O_2/O_2^{\bullet-}/H_2O_2 + h\upsilon \rightarrow HO^{\bullet} \tag{5.2}$$

$$HO^{\bullet}\left({}^1O_2/O_2^{\bullet-}/H_2O_2\right) + \text{Chl}^+\left(\text{or P680}^+\right) \rightarrow \text{Chl}^+\left(\text{or P680}^+\right)\text{ damage} \tag{5.3}$$

ROS production in cells of aquatic microorganisms has generally been detected in earlier studies, which are extensively discussed in earlier sections. The process is supported by the earlier observation that chlorophylls can easily undergo photooxidation, involving attack of singlet oxygen and enzymatic degradation (Brown SB and Hendry 1991; Gossauer and Engel 1996). Experimental studies show that H_2O_2 can affect the cyanobacterium at 10 times lower concentrations than green alga and diatom, and a strong light-dependent toxicity can enhance the difference (Drábková et al. 2007).

Second, indirect effects can be operational by which UV or strong light can produce a significant amount of strong oxidizing agents. For instance, $HO^{\bullet}$ can be photolytically generated in the presence of H_2O_2 (photo-Fenton raction or direct photodissociation), hydrogen peroxide being produced by DOM (of both algal and terrestrial origin). The hydroxyl radical can also be photoproduced by other chemical species such as NO_2^- and NO_3^- (see the chapters "Dissolved Organic Matter in Natural Waters", "Photoinduced and Microbial Generation of Hydrogen Peroxide and Organic Peroxides in Natural Waters", "Photoinduced Generation of Hydroxyl Radical in Natural Waters" and "Photoinduced and Microbial Degradation of Dissolved Organic Matter in Natural Waters" for a detailed description). The $HO^{\bullet}$ radical would subsequently react with the functional groups present in the cells of aquatic microorganisms. The indirect effect may significantly affect waters with high contents of DOM and POM, which are usually associated to elevated production of photo- and microbial products and, as a consequence, to high photosynthesis and high primary production. Moreover, it has been shown that the production of $HO^{\bullet}$ during an ozone hole (151 Dobson units) is enhanced by at least 20 %, mostly from nitrate photolysis and to a lesser extent from DOM photoinduced reactions, in Antarctic seawater. Similar results have been observed for Arctic water (see chapters "Photoinduced and Microbial Generation of Hydrogen Peroxide and Organic Peroxides in Natural Waters" and "Photoinduced and Microbial Degradation of Dissolved Organic Matter in Natural Waters" for detailed description) (Rex et al. 1997; Qian et al. 2001; Randall et al. 2005).

Note that cyanobacteria (or phytoplankton) can produce autochthonous DOM including autochthonous fulvic acids, which are very efficient in the production of H_2O_2 (and of $HO^{\bullet}$ as a consequence under irradiation). Regeneration of autochthonous DOM and nutrients $\left(NO_3^-, NO_2^-, PO_4^{3-} \text{and } NH_4^+\right)$ occurs during the photoinduced and microbial assimilation of cyanobacteria or phytoplankton, and simultaneously also from the photoinduced degradation of DOM in natural waters (see chapter "Dissolved Organic Matter in Natural Waters", "Photoinduced and Microbial Generation of Hydrogen Peroxide and Organic Peroxides in Natural Waters", "Photoinduced Generation of Hydroxyl Radical in Natural Waters", and "Impacts of Global Warming on Biogeochemical Cycles in Natural Waters" for detailed description). High solar irradiation generally induces the production of large amounts of H_2O_2 and $HO^{\bullet}$, from DOM or NO_2^- and NO_3^- in aqueous media (see also the chapters "Photoinduced and Microbial Generation of Hydrogen Peroxide and Organic Peroxides in Natural Waters" and "Photoinduced Generation of Hydroxyl Radical in Natural Waters") (Mostofa and Sakugawa 2009; Takeda et al. 2004). Moreover, light plays a significant role in the cycling of terrestrially-derived DOM and (to a certain extent) of autochthonous DOM. It can potentially increase metabolism of both terrestrially and microbially derived DOM in natural waters (Hiriart-Baer et al. 2008). Low light levels, due to increased CDOM, do not have significant effects on the benthic microfloral community at mid-shelf locations (Darrow et al. 2003).

Enhanced solar UV-A (315–400 nm) and/or UV-B radiation (280–315 nm) can reduce growth and photosynthetic rates, inhibit pigment production, increase permeability of cell membranes, damage proteins or DNA molecules, and even lead to cell death (see chapter "Impacts of Global Warming on Biogeochemical Cycles in Natural Waters" for more references) (Jiang and Qiu 2011). At normal ozone concentrations (i.e. 344 Dobson Units), UV radiation can reduce primary productivity in surface waters by as much as 50 % (see chapter "Impacts of Global Warming on Biogeochemical Cycles in Natural Waters" for more references) (Cullen and Neale 1994). A normal level of UV radiation also reduces phytoplankton production by 57 % at a depth of 1 m, while such inhibition decreases to <5 % at 30 m, at 50ºS in mid December (Arrigo 1994). Such effects on aquatic organisms might be caused directly by UV radiation and indirectly through high production of $HO^{\bullet}$ in epilimnetic (upper layer) waters. Both effects are able to alter the structural configuration of organisms with release of many organic substances in epilimnetic (surface layer) waters (see chapter "Impacts of Global Warming on Biogeochemical Cycles in Natural Waters" for more references) (Mostofa et al. 2009; Mostofa et al. 2009; Rastogi et al. 2010; Ingalls et al. 2010). Some studies also hypothesize that the primary target of photodamage to PSII by strong light is the PSII reaction center. A primary event in photoinhibition could be the damage to the D1 protein, which activates its rapid degradation by several proteases (Aro et al. 1993; Andersson and Aro 2001; Nishiyama et al. 2008; Kanervo et al. 1993; Tyystjärvi et al. 2001). Studies show that hydroperoxides (H_2O_2 and organic peroxides, ROOH) are often considered as indicators of membrane damage (see also chapter "Photoinduced and Microbial Generation of Hydrogen Peroxide and Organic Peroxides in Natural Waters") (Hagege et al. 1990a, b).

Effects of Sunlight on Higher Plants

High irradiance can affect the PSII activity, with negative effect on the PSII-mediated electron transport rate, disarrangement of PSII system, degradation of the D1 protein and/or its metabolism in a circadian-dependent manner (the same issue could also involve various polypeptides) (Aro et al. 1993; Pandey and Yeo 2008; Maslenkova et al. 1995; Rintamäki et al. 1995; Baena–González and Aro 2002; Booij-James et al. 2002; Hofman et al. 2002; Henmi et al. 2003, 2004; Nováková et al. 2004; Porta et al. 2004; Suzuki et al. 2004; Szilárd et al. 2007). The final result is a decrease of the photosynthetic capacity of plants. The decrease in photosynthetic efficiency is mostly associated with three facts: First, the decline in the enzymatic reactions of the Calvin–Benson cycle (Friedrich and Huffaker 1980); second, the decrease in the light reactions, i.e. the photoinduced reactions of PSI and PSII (Grover and Mohanty 1992; Wingler et al. 2004); and third, the changes in the structure of chloroplasts (Tang et al. 2005). UV-B sensitivity depends on the oxidation state of the water-splitting complex of PS II in higher plant such as spinach (Szilárd et al. 2007). It has been shown that ROS produced endogeneously under high-irradiance conditions can cause more deleterious effect in the decrease of PSII-mediated electron transfer rate, compared with exogenously applied H_2O_2 and $^{\bullet}OH$ stresses (Pandey and Yeo 2008). Strong illumination of thylakoid membranes in the absence of an acceptor can results in oxygen accepting electrons and subsequently producing reactive oxygen species, ROS (Pandey and Yeo 2008).

The photoproduction rate of ROS is largely enhanced under conditions where photon intensity is in excess of that required for the CO_2 assimilation (Asada 2006). It has been shown that the quantum yield of PSII is increased more rapidly than CO_2 assimilation in 20 % O_2, which can result from the electron flux through the water–water cycle (Makino et al. 2002). This flux can reach a maximum just after illumination, and can rapidly produce non-photoinduced quenching. With increasing CO_2 assimilation, the electron flux of water–water cycle and the non-photoinduced quenching is decreased (Makino et al. 2002). The cyclic electron flow around PSI can produce non-photoinduced quenching, which remains at elevated levels upon switching to low oxygen (2 % O_2) (Makino et al. 2002). The water–water cycle is thus believed to dissipate the energy of excess photons (Asada 1999, 2000, 2006; Foyer and Noctor 2000; Osmond 1997; Osmond and Grace 1995;). Such a cycle is defined as the process of the electron flow from water in PSII to water in PSI (Asada 1999). In addition, H_2O_2 and ROS can directly be produced by excited PSII under photoinhibitory conditions that trigger the turnover of the D1 protein (see also earlier sections) (Aro et al. 1993; Prasil et al. 1992; Bradley et al. 1991). ROS can influence the outcome of photodamage, primarily via inhibition of translation of the *psbA* gene, which encodes the precursor of the D1 protein (Nishiyama et al. 2001). The rate of photo-damage is proportional to irradiance (Pandey and Yeo 2008).

The mechanism behind the high irradiance (or heat stress or high temperature or drought) effect on higher plant is the similar to that explained before for cyanobacteria or phytoplankton in aqueous media. However, in higher plants the reaction centers of PSI and PSII in chloroplast thylakoids are the major ROS generation site. Photoreduction of O_2 to H_2O_2 occurs in PSI (Mehler 1951): the primary reduced species is the superoxide radical anion $\left(O_2{}^{\bullet-}\right)$, and its disproportionation produces H_2O_2 and O_2 (Asada et al. 1974). Correspondingly, ground (triplet) state oxygen $\left({}^3O_2\right)$ in PSII is excited to singlet state $\left({}^1O_2\right)$ by the triplet state of chlorophyll (Hideg et al. 1998; Telfer et al. 1994). The mechanism behind the photoreduction of O_2 in PSI of higher plants according to Asada (Asada 2006) and other studies (Lobanov et al. 2008; Parmon 1985; Bruskov and Masalimov 2002) can be expressed as follows (Eqs. 5.4–5.11):

$$\text{P680 or P700} + h\upsilon \rightarrow e^- + \text{P680}^+ \text{or P700}^+ \tag{5.4}$$

$$O_{2(aq)} + e^- + h\upsilon \rightarrow O_2{}^{\bullet-}\ (\text{PSI}) \tag{5.5}$$

$$2O_2{}^{\bullet-} + 2H^+ \rightarrow H_2O_2 + O_2 \tag{5.6}$$

$$H_2O_2 + 2\text{AsA} \rightarrow 2H_2O + 2\text{MDA} \tag{5.7}$$

In this modified mechanism, the electron is originated mostly from photoinduced excitation of both P680 and P700 (Eq. 5.4). Dissolved O_2 in water is thus reduced photolytically, differently from the results of earlier studies. The disproportionation of O_2^- to H_2O_2 and O_2 is catalyzed by superoxide dismutase (Eq. 5.6). H_2O_2 is then reduced to H_2O by ascorbate (AsA), a process that is catalyzed by ascorbate peroxidase (APX). AsA is oxidized to monodehydroascorbate radical, MDA (Eq. 5.7).

Additional electron pathways in chloroplasts that protect the photosynthetic apparatus from photo-oxidative stress are the Mehler reaction, xanthophyll cycle–dependent energy, the cyclic electron flow around PSI, the cyclic electron flow within PSII, and antioxidant metabolism (Mehler 1951; Heber et al. 1978; Verhoeven et al. 1997; Miyake and Yokota 2001; Miyake et al. 2002; Hirotsu et al. 2004). Nitrate assimilation is referred to as alternative electron flow (Makino et al. 2002). The Mehler reaction implies that the photoreduction of O_2 at PSI can produce superoxide radical $\left(O_2^{\bullet-}\right)$, which disproportionates to H_2O_2 (Mehler 1951; Asada 2006). It is estimated that the maximum rate of O_2 photoreduction is approximately 7.5 mmol $O_2^{\bullet-}$ (mol Chl)$^{-1}$ s^{-1} (30 mol (mg Chl)$^{-1}$ h^{-1}) in washed thylakoids, which corresponds to 5–10 % of the rate of total electron transport (Asada et al. 1974). It has also been observed that the $O_2^{\bullet-}$ reduction rate can reach a maximum around 2.0 kPa O_2 (Heber and French 1968; Takahashi and Asada 1982).

5.2 CO_2 Forms Used in Phytoplankton Photosynthesis

CO_2 and DIC (CO_2, H_2CO_3, HCO_3^-, and CO_3^{2-}) can be produced either photolytically or microbially from both DOM and POM (e.g. alage or phytoplankton) in natural waters (see also chapter "Photoinduced and Microbial Degradation of Dissolved Organic Matter in Natural Waters" and "Impacts of Global Warming on Biogeochemical Cycles in Natural Waters") (Jones 1992; Jansson et al. 2000; Meili et al. 2000; Grey et al. 2001; Hernes and Benner 2003; Tranvik et al. 2009; Ballaré et al. 2011; Zepp et al. 2011; Miller and Zepp 1995; Graneli et al. 1996; Granéli et al. 1998; Bertilsson and Tranvik 2000; Ma and Green 2004; Xie et al. 2004; Fu et al. 2007). This production varies seasonally and spatially depending on several factors such as contents of DOM and POM, solar intensity, water temperature and other geological and environmental conditions (White et al. 2010).

Gaseous CO_2 is rapidly dissolved in waters (Liu et al. 2010):

$$CO_2 + H_2O \leftrightarrow H_2CO_3 \leftrightarrow H^+ + HCO_3^- \leftrightarrow 2H^+ + CO_3^{2-} \tag{5.8}$$

where the reaction (Eq. 5.8) is an equilibrium mixture of dissolved carbon dioxide ($[CO_2]_{aq}$), carbonic acid (H_2CO_3), bicarbonate $\left(HCO_3^-\right)$ and carbonate $\left(CO_3^{2-}\right)$ ions with the pKa of 6.3 and 10.3 for $H_2CO_3 \leftrightarrow H^+ + HCO_3^-$ and $HCO_3^- \leftrightarrow H^+ + CO_3^{2-}$, respectively (Liu et al. 2010; Appelo and Postma

2005). The proportion of each species depends on pH: at high pH the reaction shifts to the right hand side of (Eq. 5.8) and HCO_3^- dominates at pH between 7 and 9, approximately 95 % of the carbon in water. At pH > 10.5, CO_3^{2-} predominates (Dreybrodt 1988). The equilibrium constants for this system are altered by the salinity of the medium: the values for seawater are an order of magnitude higher than those of freshwater toward the right-hand-side of the reaction (Raven et al. 2002; Millero and Roy 1997).

It is well known that the stable carbon isotope composition ($\delta^{13}C$ value) of organic matter, produced either by phytoplankton or terrestrial plants during photosynthesis, is significantly varied depending on the taxon-specific photosynthetic pathways (such as C_3, C_4, and crassulacean acid metabolism, CAM). It also varies depending on: variety of phytoplankton; diffusion of CO_2; incorporation of CO_2 by phosphoenolpyruvate carboxylase or Ribulose Bisphosphate Carboxylase-Oxygenase (Rubisco), and respiration; sources and interconversion of CO_2 and HCO_3^- (depending on a variety of environmental conditions including light intensity, temperature, DOM and POM contents, water depth, atmospheric CO_2 concentration and so on) (O'Leary 1981; Cooper and McRoy 1988; Farquhar et al. 1989; Raven and Farquhar 1990; Yoshioka 1997; Raven et al. 2002; Hu et al. 2012). Note that the $\delta^{13}C$ values of $[CO_2]_{aq}$ and DIC are −16.5 to −14.5 ‰ and −7.4 to −4.5 ‰, respectively (Yoshioka 1997). The values of $\delta^{13}C$ of organic matter in marine macroalgae and seagrass collected from the natural environment can vary from –2.7 ‰ to –35.3 ‰ (Raven et al. 2002; Hu et al. 2012; Beardall 2003; Hemminga and Mateo 1996; Raven 1997; Dunton 2001). Plants with C_4 characteristics show $\delta^{13}C$ values of −6 to −19 ‰ whilst plants with C_3 characteristics exhibit $\delta^{13}C$ values of −24 to −34 ‰ (Smith and Epstein 1971).

Such variation in the $\delta^{13}C$ value can be caused by (Farquhar et al. 1989; Raven and Farquhar 1990): (i) the isotope fractionation factor (α), which is the ratio of the reaction rates of $^{12}CO_2$ and $^{13}CO_2$ with Rubisco ($\alpha = 1.029$ for gaseous CO_2 and $\alpha = 1.030$ for dissolved CO_2); (ii) the relative contribution of phosphoenolpyruvate carboxylase (PEPC) activity to the photosynthetic carbon assimilation; and (iii) the supply of CO_2 to Rubisco is restricted by the boundary layer, stomata, and intercellular gas spaces that can differ for CO_2 diffusion in the gas phase ($\alpha = 1.0044$), and in the aqueous phase ($\alpha = 1.0007$).

The $\delta^{13}C$ values of POM are varied spatially and seasonally. They increase with increasing pH of lake water, which may reflect a shift by phytoplankton from using CO_2 to using HCO_3^- for photosynthesis (Zohary et al. 1994; Doi et al. 2006). The pH is increased with increasing water temperature during the time span of the summer stratification period, which may be connected with photoinduced degradation of DOM and POM (see also chapter "Photoinduced and Microbial Degradation of Dissolved Organic Matter in Natural Waters") (Kopáček et al. 2003; Mostofa et al. 2005). Photoinduced generation of H_2O_2 $\left(2O_2^{\bullet-} + 2H^+ \rightarrow H_2O_2 + O_2\right)$ (Mostofa and Sakugawa 2009; Fujiwara et al. 1993) might be one of the key factors for enhancing alkalinity or pH in waters. Therefore, uptake of HCO_3^- for phytoplankton photosynthesis at high pH might be the effect of its dominant presence in waters. A significant increase in the

$\delta^{13}C$ value in the phytoplankton bloom season suggests that phytoplankton photosynthesis may be limited by CO_2 depletion (Takahashi et al. 1990). It has been observed that aqueous CO_2, $[CO_2]_{aq}$, determined in freshwater and marine waters is relatively low (0.13–35 M) in freshwater and relatively higher (5–120 M) in seawater (Fogel et al. 1992; Francois et al. 1993; Yoshioka 1997; Takahashi et al. 1990; Herczeg and Fairbanks 1987). All aquatic phototrophs are depleted in $\delta^{13}C$ relative to dissolved inorganic carbon (DIC), because Rubisco discriminates against ^{13}C (Hu et al. 2012).

The spatial and temporal variability of $\delta^{13}C$ values in aquatic organisms depends on several factors such as isotopic shifts in available inorganic carbon, resulting from light-induced HCO_3^- utilization, variation in solar intensity, differences in water temperature, internal recycling of respiratory CO_2, photoinduced generation of DIC from DOM and POM, and dissolution of sedimentary carbonate (Yoshioka 1997; Raven et al. 2002; Jones 1992; Ma and Green 2004; Xie et al. 2004; White et al. 2010; Liu et al. 2010; Dreybrodt 1988; Hemminga and Mateo 1996; Campbell and Fourqurean 2009). It is shown that $[CO_2]_{aq}$ concentration is inversely correlated with the $\delta^{13}C$ of organic matter produced by phytoplankton (Rau et al. 1992; Freeman and Hayes 1992). The carbon isotope fractionation of phytoplankton could be a useful indicator for the assessment of its growth rate and of CO_2 availability (Fogel et al. 1992; Takahashi et al. 1991). Phytoplankton can actively transport CO_2 by a carbon-concentrating mechanism (CCM) that can affect its $\delta^{13}C$ value (Yoshioka 1997; Sharkey and Berry 1985; Bums and Beardall 1987; Thielmann et al. 1990). Correspondingly, ß-carboxylation catalysed by phosphoenolpyruvate carboxylase and phosphoenolpyruvate carboxykinase can affect the $\delta^{13}C$ of phytoplankton (Descolas-Gros and Fontugne 1985; Falkowski 1991).

To understand the mechanism behind the uptake of CO_2 or HCO_3^-, a fractionation equation was developed for plant photosynthesis (O'Leary 1981; Farquhar et al. 1989; Raven et al. 1993) and phytoplankton photosynthesis (Fogel et al. 1992; Rau et al. 1992; Francois et al. 1993; Jasper and Hayes 1994; Laws et al. 1995; Yoshioka 1997; Berry 1988).

5.2.1 Basic Equation for Expressing Photosynthetic Carbon Isotope Fractionation

The photosynthetic carbon isotope fractionation is initially derived based on the land C_3 plants (O'Leary 1981; Farquhar et al. 1989; Yoshioka 1997). The photosynthetic process for uptake of carbon can be depicted as follows (Yoshioka 1997)::

$$[CO_2]_{out} \underset{k_3}{\overset{k_1}{\longleftrightarrow}} [CO_2]_{in} \longrightarrow k_2 \text{ organic carbon} \tag{5.9}$$

where k_i is the rate constant for process i. Processes 1 and 3 are the diffusive influx and efflux of CO_2, respectively, whilst process 2 is the carboxylation step by

Rubisco. At steady state, or $d[CO_2]_{in}/dt = 0$, the overall fractionation factor (α) can be written as

$$\alpha = 1 + \Delta k_1 + (\Delta k_2 - \Delta k_1)\frac{Ci}{Ce} \tag{5.10}$$

where Ce and Ci are the CO_2 concentrations in air and at the carboxylation site, respectively, and $\Delta k_i = \alpha_i - 1$. In the equation (O'Leary 1981), subscripts for efflux and carboxylation steps are 2 and 3, respectively, and $E_i = 1 + \Delta k_i$:

$$\alpha = E_1(E_3/E_2 + k_3/k_2)(1 + \frac{k_3}{k_2}) \tag{5.11}$$

When $a = \Delta k_1$, $b = \Delta k_2$ and CO_2 concentrations in air and intercellular leaf spaces are denoted in partial pressure p_a and p_i, respectively, then (Eq. 5.10) can be modified into Farquhar's equation:

$$\Delta = \alpha - 1 = a + (b - a)\frac{p_i}{p_a} \tag{5.12}$$

On the other hand, the fractionation equation for passive diffusion-phytoplankton photosynthesis is substantially similar to that of land C_3 plants (Eq. 5.10). The CO_2 diffusion must be considered in the aqueous phase and Ce denotes the CO_2 concentration in bulk solution or $[CO_2]_{aq}$. The term 'CO_2 demand' = 'Ce – Ci' has been introduced into the new model (Rau et al. 1992). The relationship between the $\delta^{13}C$ value of POM and $[CO_2]_{aq}$ can be determined using the fractionation equation that includes the (Ce–Ci) term:

$$\varepsilon_p = \varepsilon_1 + \left(1 - \frac{Ce - Ci}{Ce}\right)(\varepsilon_2 - \varepsilon_1) \tag{5.13}$$

where (Ce – Ci) = 7–8 μM in southwestern Indian Ocean. When (Ce – Ci) is constant, the (Eq. 5.10) at infinite Ce can be expressed as:

$$\alpha = 1 + \Delta k_2 \tag{5.14}$$

This implies that the overall fractionation can reach a maximum value, which corresponds to that of Rubisco (α – 1.027–1.029, or Δk_2 = 0.027 – 0.029) at high Ce (Roeske and O'Leary 1984; Farquhar and Richards 1984). Furthermore, (Ce – Ci) may increase with increasing Ce as found in a culture study of *Skeletonema costatum* and *Emiliania huxley*, which introduces the possibility of β-carboxylation at high Ce (Hinga et al. 1994). Interestingly, the activity of the PEPCKase of *S.costatum* can increase to >50 % of Rubisco activity at the end of growth (Descolas-Gros and Fontugne 1985, 1990).

The low fractionation observed at high Ce is possibly due to β-carboxylation (Goericke and Fry 1994), particularly in the case of PEPCKase-mediating β-carboxylation. The latter shows similar discrimination against $^{13}CO_2$ as that of Rubisco (Arnelle and O'Leary 1992). Active transport by CCM may contribute to a fractionation at high Ce, which is lower than that given by the fractionation equation (Yoshioka 1997). It is shown that passive CO_2 diffusion is efficient to sustain maximum growth of *Phaeodactylum tricornutum*, which does not require active transport of inorganic carbon at $[CO_2]_{aq} = 10$ M (Laws et al. 1995). This study also shows that maximum growth rate is expected when the CO_2 influx is equal to the growth rate (Laws et al. 1995). In that case, Ci = 0 and also the growth rate (photosynthetic activity) is zero or even negative, because of the oxygenase activity of Rubisco (Yoshioka 1997). The contradiction may occur because the growth rate is not independent of Ce and Ci. Therefore, diffusive transport of CO_2 can operate together with active transport (Yoshioka 1997), and CCM possibly requires an energy expenditure (Berry 1988). However, it is difficult to identify the relative contribution of active transport to the total CO_2 influx from the ealier fractionation equations. In the derivation of (Eq. 5.10), it is assumed that the resistance to CO_2 diffusion is similar in either direction across the cell membrane, or $k_1 = k_3$ (Francois et al. 1993). This assumption originally came from the expectation that resistance to CO_2 diffusion through the stoma of a plant leaf would be the same in both directions (O'Leary 1981). Aquatic phytoplankton may have a CCM with different values for this resistance ($k_1 \neq k_3$), probably ($k_1 > k_3$), and thus the fractionation equation can be rewritten as:

$$\alpha = 1 + \Delta k_1 + (\Delta k_2 - \Delta k_1)\frac{k_3 \mathrm{Ci}}{k_1 \mathrm{Ce}} \tag{5.15}$$

which may provide some measure of the contribution of active transport. It is generally assumed that the resistances to CO_2 diffusion in both directions across the cell membrane are the same (symmetric permeability). A fractionation equation is required to express the decrease in fractionation with increasing contribution of active transport (*f*), as some function *f* (Yoshioka 1997). Basically, *f* and $k_1 \neq k_3$ may have the same importance for CO_2 acquisition by phytoplankton. Therefore, active transport of inorganic carbon by CCM may be linked (as a homologue) to the asymmetric permeability of the cell membrane for CO_2.

Deviation of Fractionation Equations Involving Active Transport (*Yoshioka 1997*)

Various phytoplankton species can actively transport CO_2 and HCO_3^- in aqueous media (Bums and Beardall 1987). However, they depend on two phenomena: (i) the occurrence of internal and external carbonic anhydrase (CA), which can catalyse the equilibrium between CO_2 and HCO_3^- and can affect the determination of the inorganic carbon species crossing the cell membrane; (ii) the difference in inorganic carbon species can substantially vary the fractionation factor of the substrate for photosynthesis. It is shown that fractionation between $[CO_2]_{aq}$ and HCO_3^- can differ by at most 10 ‰ in both equilibrium- and CA-catalyzed reactions (Deines et al. 1974; Paneth and O'Leary 1985). Considering these phenomena, it is important to develop the fractionation equations for two cases in which transported carbon has the $\delta^{13}C$ value of either bulk $[CO_2]_{aq}$ or HCO_3^-.

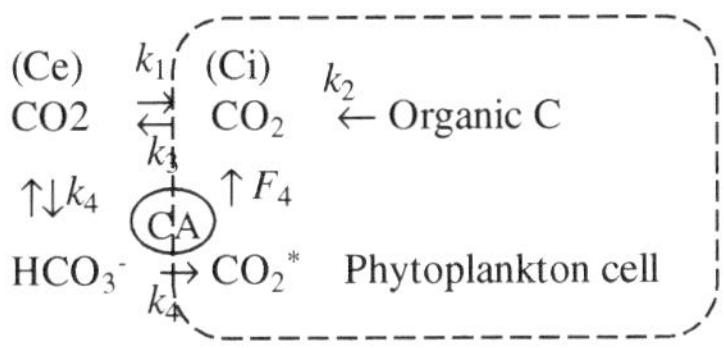

Fig. 8 Schematic presentation of the active transport of CO_2. The $\delta^{13}C$ of the actively transported carbon (CO_2^*) is assumed to be the same as that of the CO_2 in the medium (Ce). *Data source* Yoshioka (1997)

(1) ***Active transport of CO_2***. The $\delta^{13}C$ value of actively transported inorganic carbon is assumed to be the same as that of Ce (Fig. 8). Extracellular CA may contribute to the conversion of HCO_3^- to CO_2 at the cell surface.

At steady state:

$$\frac{dCi}{dt} = k_1Ce + F_4 - (k_2 - k_3)Ci = 0 \tag{5.16}$$

where F_4 is the is the flux of actively transported CO_2.
The relative contribution of active transport (f) can be defined by:

$$f = \frac{F_4}{k_1Ce + F_4} \tag{5.17}$$

If $0 \leq f < 1$, (Eq. 5.17) can be rewritten as:

$$\frac{dCi}{dt} = \frac{1}{1-f}k_1Ce - (k_2 + k_3)Ci = 0 \tag{5.18}$$

Overall, fractionation becomes:

$$\alpha = 1 + \Delta k_1 + (\Delta k_2 - \Delta k_1)(1 - f)\frac{Ci}{Ce} \tag{5.19}$$

By assuming the same f value for $^{12}CO_2$ and $^{13}CO_2$, and $\Delta k_1 = \Delta k_3$, (Eq. 5.19) becomes the same as (Eq. 5.15) when k_1/k_3 is substituted for $(1 - f)$. This supports the expectation that active transport might be linked with the asymmetric permeability of the cell membrane for CO_2. Leakiness, X (the ratio of efflux to influx of DIC) (Berry 1988), can be expressed as follows:

$$X - 1 + \frac{k_3Ci}{k_1Ce\,F_4}(1 - f)\frac{Ci}{Ce} \tag{5.20}$$

When all carbon is transported by active transport ($f = 1$), k_1Ce would be zero. In that case, one cannot substitute $f = 1$ in (Eq. 5.19), because the denominator in (Eq. 5.18) becomes zero. Then, α becomes:

X is not zero, but

$$\begin{aligned}\alpha &= 1 + \frac{\Delta k_2 - \Delta k_1}{\Delta k_1 + 1}\frac{k_3Ci}{F_4} \\ &= 1 + (\Delta k_2 - \Delta k_1)\frac{k_3Ci}{F_4}\end{aligned} \tag{5.21}$$

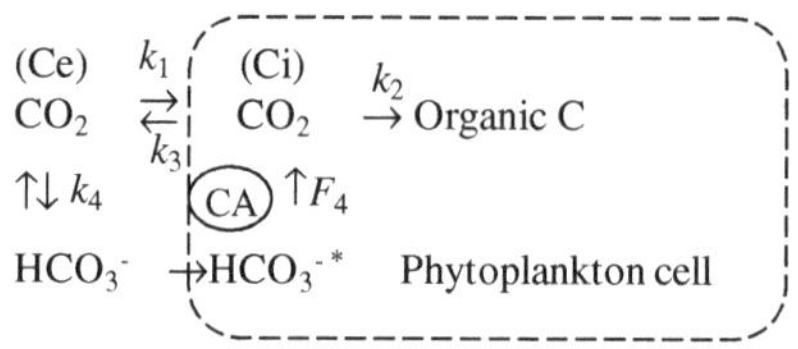

Fig. 9 Schematic presentation of the active transport of HCO_3^-. The $\delta^{13}C$ of the actively transported carbon (HCO_3^{-*}) is assumed to be the same as that of HCO_3^- in the medium. *Data source* Yoshioka (1997)

$$X = \frac{k_3 \mathrm{Ci}}{F_4} \tag{5.25}$$

(2) ***Active transport of*** $\boldsymbol{HCO_3^-}$: The transported carbon has the same $\delta^{13}C$ value as HCO_3^-, as depicted in the scheme shown in (Fig. 9). The overall fractionation equation is substantially different from Eq. (5.19), although the steady-state for Ci is denoted by a similar term as Eq. (5.16), which can be written as:

$$\alpha = \frac{(\Delta k_1 + 1)(\Delta k_3 + 1)(1 - X) + (\Delta k_1 + 1)(\Delta k_2 + 1)X}{(\Delta k_3 + 1)(1 - f) + (\Delta k_1 + 1)(\Delta k_3 + 1)(\Delta k_4 + 1)f} \tag{5.23}$$

where Δk_4 denotes the fractionation in the CO_2—HCO_3^- dissociation process. Note that f and X are the same as those in the active transport of CO_2.

Considering that the second- and third-order terms of Δk_i are negligible, and $\Delta k_1 = \Delta k_3$, then α can be approximated as follows:

$$\alpha = 1 + \Delta k_1(1 - f) + (\Delta k_2 - \Delta k_1)(1 - f)\frac{\mathrm{Ci}}{\mathrm{Ce}} - \Delta k_{4f} \tag{5.24}$$

When $f = 1$, α becomes:

$$\alpha = 1 + (\Delta k_2 - \Delta k_1)\frac{k_3 \mathrm{Ci}}{F_4} - \Delta k_4 \tag{5.25}$$

which implies that the overall fractionation decreases by ($\Delta k_1 + \Delta k_4$) when all carbon derives from the active transport of HCO_3^- ($f = 1$), compared to the passive diffusion model (Eq. 5.15). It can be deduced from (Eq. 5.24) that all fractionation steps, including overall fractionation would be affected by *f*. The difference between (Eqs. 5.19 and 5.24) or ($\Delta k_1 + \Delta k_4$) corresponds to the difference in $\delta^{13}C$ values between CO_2 and HCO_3^-. These equations indicate that the overall fractionation from $[CO_2]_{aq}$ to organic carbon may be less than unity under some conditions (Yoshioka 1997).

From a reanalysis of Hinga's data (Hinga et al. 1994) one gets that the active transport of CO_2 for *S. costatum* can contribute ~ 30–40 % of the total carbon influx. The relative contribution of active transport can reach 25–35 %, without any change in CO_2 demand for an uptake of 10 % of the total carbon mediated by β-carboxylation

(Yoshioka 1997). Finally, carbon assimilation by various kinds of phytoplankton, such as *S. costatum*, *Microcystis* spp. and others (Fogel et al. 1992; Francois et al. 1993; Yoshioka 1997; Takahashi et al. 1990; Herczeg and Fairbanks 1987; Hinga et al. 1994) may operate under almost constant CO_2 demand, amounting on average to 4.4 μM in seawater and 0.29 μM in freshwater (Yoshioka 1997). Phytoplankton photosynthesis is largely dependent on habitats (either seawater or freshwater), and on phytoplankton species that have variable efficiency for CCM. The process involves either active transport of HCO_3^-, or coupled dehydration of HCO_3^- by a cell-surface carbonic anhydrase and CO_2 transport (MacIntyre et al. 2000; Badger and Price 1992; Tortell et al. 1997; Berman-Frank et al. 1998; Nimer et al. 1999).

5.3 Variation in Temperature

Temperature, driven by solar radiation, is one of the key factors for variating the primary production by photosynthesis in natural waters (Sobek et al. 2007; Mortain-Bertrand et al. 1988; Davison 1991; Wilen et al. 1995; Lesser and Gorbunov 2001; Baulch et al. 2005; Doyle et al. 2005; Yoshiyama and Sharp 2006; Ogweno et al. 2008; Bouman et al. 2010; Fu et al. 2007). This effect can be discussed, based on aquatic microorganisms and higher plants.

Temperature Effects on Aquatic Microorganisms

Cyanobacteria, the most ancient life forms on earth, are unusual prokaryotic microorganisms that are able to perform oxygenic photosynthesis. Optimum growth, with respect to optimal temperatures, is in this case influenced by their ability to tolerate temperature stress, such extreme cold in Antarctica (where temperatures never exceed −20 °C) and in water pockets of Antarctic lake ice, where temperatures are always below 0 °C. At the opposite end of the variation scale there are extremely high temperatures such as 55–60 °C and even the case of hot springs, where temperatures reach 70 °C (Schopf et al. 1965; Meeks and Castenholz 1971; Margulis 1975; Priscu et al. 1998; Psenner and Sattler 1998; Ward et al. 1998).

At ambient water temperature (WT), the primary excitation requires 2–3 ps, and the subsequent electron transfer to the primary quinone QA exhibits multiphasic kinetics (80–300 ps) (Dashdorj et al. 2004). It is commonly considered that that the primary excitation occurs within 1–3 ps after the creation of the electronically excited special pair P700* (Brettel 1997; Dashdorj et al. 2004). The state of thylakoid membranes in cyanobacteria plays a prominent role in the tolerance of the photosynthetic machinery to environmental stresses, such as cold (chilling) (Wada et al. 1990; Murata et al. 1992).

At low temperatures, ultrafast time-resolved spectroscopy suggests multiexponential evolution of the excited state and of photoproduct populations, even when excitation takes place in the red edge of the absorption spectrum (Germano et al. 2004). The different time components observed at low temperatures are generally recognized to produce charge separation. The latter can either take place through direct excitation of the primary donor by 1–5 ps (Prokhorenko and Holzwarth 2000; Tang et al. 1990; Germano M et al. 1995; Groot et al. 1997; Konermann et al. 1997; Greenfield et al. 1999), or be slowed down by energy transfer to the primary donor in tens or hundreds of picoseconds (Groot et al. 1997; Greenfield et al. 1999). However, calculations based on structural information, from both the crystallographic structure and a model, predict subpicosecond excitation energy equilibration among the six central cofactors (Durrant et al. 1995; Renger and Marcus 2002; Zouni et al. 2001; Kamiya and Shen 2003; Svensson et al. 1996; Leegwater et al. 1997). Electron transfer thus occurs from other Chls, and the slower components observed in the tens of picoseconds timescale at low temperatures are due to secondary electron transfer (Prokhorenko and Holzwarth 2000). A model study has shown that the ~67 % variability of observed primary production indicates that estuarine production is mainly controlled by light availability and temperature (Yoshiyama and Sharp 2006). Bacterial abundance (12×10^6 mL^{-1}) and production (1.7 g C L^{-1} h^{-1}) depend on temperature. During late spring and summer, at constantly higher temperatures, bacterial production can correlate positively with readily utilisable substrates and humic compounds (Freese et al. 2007).

High surface temperatures and heavy precipitation in late spring and summer can give rise to a highly-stratified water column that can stimulate a series of phytoplankton blooms. During winter in Tokyo bay, a weakly-stratified and deeply-mixed water column can lead to a rapid decline in phytoplankton biomass under light-limited growth conditions (Bouman et al. 2010). The effect of high WT can be a decrease in PSII efficiency, which can ultimately cause cell stress (Lesser and Gorbunov 2001).

At highly elevated WT, several effects on phytoplankton can take place such as disorganization of thylakoid membranes, disrupted electron flow to the dark reactions of photosystem II, elevated concentrations of damaging oxygen and hydroxyl radicals, and the loss of the D1 repair protein (Goulet et al. 2005). The mechanism behind the changes in photosynthetic efficiency caused by WT, driven by natural solar intensity, mostly follows a similar mechanism as sunlight effects (see the earlier section). However, WT can cause photosynthetic efficiency to be either enhanced or decreased, an issue that involves three facts: First, at low WT (lower than 12 °C, including chilling stress that generally refers to nonfreezing temperatures at 0–12 °C) the key reactants such as CO_2, H_2O_2 and DIC (generated both photolytically and microbially from DOM and POM) are quite low at low sunshine in natural surface waters. Low availability of these reactants can decrease the photosynthetic efficiency of aquatic microorganisms in natural waters.

Second, at moderate WT (approximately 12–25 °C) and with an increase in WT, the key reactants are significantly increased, usually also because of enhanced sunlight intensity. This effect may greatly enhance photosynthesis at optimum WT and, as a consequence, primary production in waters. It has been shown that the Chl *a* concentrations at the epilimnion are well correlated with WT in lakes, but those correlations are not observed in the deeper layers (Fu et al. 2010; Mostofa KMG et al., unpublished data). This suggests that an optimum water temperature, driven by solar intensity, may play a significant role in the origin of Chl *a* or in enhancing phytoplankton biomass in natural waters.

At highly elevated WT (approximately >25–50 °C), photoinduced and microbial degradation of DOM and POM is extremely enhanced, with extremely high generation of H_2O_2, CO_2 and DIC. It has been shown that $[CO_2]_{aq}$ is significantly higher (~10–120 M) at 25 °C than at 15 °C (~5–110 M) or at 9 °C (~5–50 M) in marine waters (Hinga et al. 1994). This effect can cause extremely high photosynthesis and high primary production. This can be supported by the synergistic effect of high H_2O_2, combined with high seawater temperature, which can cause a 134 % increase in respiration rates. Such an increase surpassed the effect of either H_2O_2 or high seawater temperature alone (Higuchi et al. 2009). High temperature, driven by strong solar intensity, is responsible for high production of H_2O_2 (see also chapter "Photoinduced and Microbial Generation of Hydrogen Peroxide and Organic Peroxides in Natural Waters") (Mostofa and Sakugawa 2009), which is directly linked with photosynthesis. Simultaneously, this process can also generate a high amount of ROS such as $O_2^{\bullet-}$, $^1O_2$, H_2O_2, and $HO^{\bullet}$. The latter is a strong oxidizing agent, produced either from H_2O_2 (via direct photo-dissociation by sunlight or photo-Fenton reaction) or other sources, such as the direct photolysis of NO_2^- and NO_3^- (see the chapters "Photoinduced and Microbial Generation of Hydrogen Peroxide and Organic Peroxides in Natural Waters", "Photoinduced Generation of Hydroxyl Radical in Natural Waters" and "Photoinduced and Microbial Degradation of Dissolved Organic Matter in Natural Waters"). This effect can significantly degrade algal or phytoplankton cells, thereby decreasing the photosynthetic efficiency. All these processes should be able to significantly promote the photosynthetic efficiency in waters with high contents of DOM and POM.

Temperature Effects on Higher Plants

Plants need an optimum temperature for photosynthesis. The stress represented by extremely high- or low-temperature has a significantly negative effect on the growth and productivity of plants (Allen and Ort 2001; Adams et al. 2002; Adams Iii et al. 2004; Öquist and Huner 2003; Yang et al. 2009). It has been shown that suboptimal and above-optimal temperatures can promote photoinhibition, caused by an over-excitation of photosystems (Powles 1984; Öquist et al. 1993; Huner et al. 1998). Effects of temperature on the photosynthesis of plants have been discussed as follows: First, low temperature stress or chilling stress (generally at 0–12 °C) can highly inhibit growth and development of most plants, and in particular of those coming from tropical and subtropical regions (Allen and Ort 2001; Yang et al. 2009; D'Ambrosio et al. 2006).

The chilling stress or lower temperatures can affect several physiological functions and induce water deficiency. Commonly observed effects are decrease of leaf water potential, of electron transport rate, of total Chl contents, of CO_2 uptake and of the carotenoid content; stomatal closure; inhibition of thylakoid electron transport and photophosphorylation; Rubisco inactivation; inhibition of carbohydrate metabolism; and finally, a significant decrease of the maximum quantum efficiency of PSI and PSII primary photochemistry (Allen and Ort 2001; Yang et al. 2009; D'Ambrosio et al. 2006; Berry and Bjorkman 1980; Eamus 1986; Sage and

Sharkey 1987; Huner et al. 1993; Ebrahim et al. 1998; Sundar and Ramachandra Reddy 2001; Caramori et al. 2002; Kudoh and Sonoike 2002; Yu et al. 2002; Huang and Guo 2005). The latter effect can limit the photosynthetic rates or processes of chilling-sensitive plants.

It is also shown that low temperatures can inhibit the enzymes of carbon assimilation, such as fructose-1,6-bisphosphatase and sedoheptulose-1,7-bisphosphatase (D'Ambrosio et al. 2006; Sassenrath et al. 1990; Sassenrath and Ort 1990). It has also been shown that the O_2-induced inhibition of photosynthesis can increase with temperature, from 12.2 % at 5 °C to 33.5 % at 35 °C (D'Ambrosio et al. 2006). Plants of *B. vulgaris* exposed to low temperatures (5–15 °C) also show a significant stimulation of CO_2 assimilation at 2 % O_2 concentration (D'Ambrosio et al. 2006). The inhibition of photosynthesis (photorespiration) at high temperatures is generally caused by the increase of the ratio oxygenase/carboxylase activity of Rubisco (Sage and Sharkey 1987).

It has been observed that low night temperature under chilling conditions (mostly affected at 5 °C) can increase photoinhibition of photosynthesis with a marked loss of D1 and 33 kDa proteins in various plants (Yang et al. 2009; Sundar and Ramachandra Reddy 2001; Lidon et al. 2001; Bertamini et al. 2006). This can be due to accumulation of soluble sugars and reduced orthophosphate cycling from the cytosol back to the chloroplast. Therefore, it limits the ATP synthesis needed for Rubisco regeneration (Ebrahim et al. 1998; Hurry et al. 1998). Inhibition of photosynthetic electron transport is susceptible to lessen net photosynthesis in some chilling-sensitive plant species, despite relatively minimal reductions in the ratio of variable to maximum chlorophyll (Chl) fluorescence (F_v/F_m). Such an effect is due to the net photoinactivation of PSI rather than PSII (Bertamini et al. 2006; Tjus et al. 1998; Sonoike 1999). A significant decrease of electron transport rate under chilling conditions might cause a low temperature-induced limitation of carbon metabolism. Furthermore, sinks of electrons can result in alternative processes to CO_2 fixation (D'Ambrosio et al. 2006; Huner et al. 1993; Osmond 1981; Hendrickson et al. 2003, 2004). The decrease of electron transport in PSII (D'Ambrosio et al. 2006) is susceptible to decrease in the photoinduced generation of $O_2^{\bullet -}$ and then H_2O_2, which is directly liked to the occurrences of photosynthesis. The decrease in the contents of H_2O_2 production at chilling conditions can decrease the photosynthesis that subsequently decreases the growth and development of plants. This effect is mostly responsible for other physiological changes in plants at chilling stress.

It has also been observed that a significant increase of the proportion of electron flow in chilling conditions can occur in non-assimilative processes in some plants, such as maize and grapevine leaves (Fryer et al. 1998; Flexas et al. 1999). These studies suggest that a higher electron flow could reach O_2, by the Mehler reaction, as an alternative acceptor to CO_2 at low temperatures. This effect can enhance the production of ROS such as $O_2^{\bullet -}$ and H_2O_2, which may not be used in photosynthesis because of CO_2 shortage and other still unknown reasons. In contrast, H_2O_2 and photogenerated $HO^{\bullet}$ can damage the cells. Coherently, damage of chlorophyll-protein complexes and pigments in has been observed in plant cells under chilling condition (Powles 1984; Kudoh and Sonoike 2002; Bongi and Long 1987; Garstka et al. 2007). The decrease of the carotenoid content at lower temperatures in *B. vulgaris* can enhance damage by ROS, because of the important photoprotective function of carotenoids in scavenging highly destructive singlet oxygen. Furthermore, they can prevent $^1O_2$ formation by reacting with the chlorophyll triplet state (Havaux et al. 1998). Low temperature stress can also enhance photodamage to PS II under strong light (Wada et al. 1990; Murata et al. 1992; Öquist et al. 1993; Öquist and Huner 1991), and repair of PS II under low-temperature stress conditions is inhibited both in *Synechocystis* and plants (Gombos et al. 1994; Wada et al. 1994; Moon et al. 1995; Alia et al. 1998).

At higher temperature (>25 °C) caused by heat stress or drought stress, photosynthetic efficiency is significantly altered and can lead to decreased growth and development of plants (D'Ambrosio et al. 2006; Pastenes and Horton 1996; Pastenes and Horton 1996; Salvucci and Crafts-Brandner 2004; Sharkey 2005). The effect of high temperature on organisms is expected to become more and more significant. The global mean temperature has increased by 0.6 °C from 1990 to 2000 and is projected to increase by another 1.4 to over 5 °C by 2100 (see chapter "Impacts of Global Warming on Biogeochemical Cycles in Natural Waters" for detailed description). Heat stress can induce several processes such as: saturation of electron transport rate and disruption of its activity; decrease of stomatal conductance; increase in increase in O_2-consuming photorespiration and non-photoinduced quenching; decreased affinity of the enzyme for CO_2; decrease in CO_2 fixation; inactivation of the oxygen-evolving enzymes of PSII; increase in the activity of antioxidant enzymes such as superoxide dismutase, ascorbate peroxidase, guaiacol peroxidase, and catalase; decrease in PSII activity, and finally of photosynthetic capacity (Ogweno et al. 2008; D'Ambrosio et al. 2006; Pastenes and Horton 1996; Pastenes and Horton 1996; Salvucci and Crafts-Brandner 2004; Sharkey 2005; Schuster and Monson 1990; Heckathorn et al. 2002; Mazorra et al. 2002; Barua et al. 2003; Núñez et al. 2003; El-Shintinawy et al. 2004; Rivero et al. 2004; Cao et al. 2005).

Moderate heat stress can cause increased thylakoid proton conductance and increased cyclic electron flow around PSI (Pastenes and Horton 1996; Bukhov et al. 1999, 2000; Bukhov and Carpentier 2000; Egorova and Bukhov 2002). PSI-mediated cyclic electron flow can occur via either of two routes: the first is antimycin A-sensitive and involves ferredoxin plastoquinone reductase; the second one involves the NAD(P)H dehydrogenase complex (Bukhov et al. 2000; Thomas et al. 1986; Boucher et al. 1990; Joët et al. 2001).

It has also been shown that high temperatures stress (often above 45 °C) can damage PSII (Terzaghi et al. 1989; Thompson et al. 1989; Gombos et al. 1994; Çjánek et al. 1998; Yamane et al. 1998). Furthermore, photorespiration increases with increasing temperature, faster than photosynthesis (Schuster and Monson 1990). High leaf temperatures can reduce plant growth, and it is estimated that up to 17 % decrease in crop yield can occur for each degree Celsius increase of average temperature during the growing season (Lobell and Asner 2003). Additionally, leaves with low transpiration rates (e.g. oak leaves) can suffer frequent high-temperature episodes when leaf temperature can exceed the air temperature by as much as 15 °C (Singsaas and Sharkey 1998; Hanson et al. 1999; Singsaas et al. 1999). Rubisco can produce hydrogen peroxide as a result of oxygenase side reactions, which can increase substantially with temperature (Sharkey 2005).

Moreover, an increase in temperature can induce sinks of electron transport different from CO_2 assimilation, and photorespiration is increased at 30–35 °C (D'Ambrosio et al. 2006). The O_2-independent electron transport can account for up to 20 % of the total PSII electron transport in wild watermelon leaves (Miyake and Yokota 2000, 2001). The electron flux in PSII that exceeds the flux required for the cycles of photosynthetic carbon reduction and photorespiratory carbon oxidation, can induce photoreduction of O_2 in the water–water cycle (Miyake and Yokota 2000, 2001). It has been shown that the greater partitioning of reductive power to non-assimilative processes consuming O_2 (photorespiration, Mehler reaction and chlororespiration) with respect to CO_2 assimilation allows keeping the PSII efficiency factor unmodified at temperatures as high as 35 °C (D'Ambrosio et al. 2006).

The unsaturation of fatty acids can protect PSII from the inhibition of the activity that is caused by strong light at low temperatures (Wada et al. 1990; Murata et al. 1992), and can accelerate the repair of photodamaged PSII (Gombos et al. 1994; Wada et al. 1994; Moon et al. 1995). After photodamage to PSII in *Synechocystis* at low temperatures (0–10 °C), activity recovery can reach up to 50 % of the original level in the darkness at moderate temperatures, without the de novo synthesis of D1 protein (Nishiyama et al. 2008).

High-temperature stress can disrupt the cellular metabolic homeostasis and promote the production of reactive oxygen species (H_2O_2, $^1O_2$, $O_2^{\bullet-}$, and $HO^{\bullet}$) (Mittler 2002). Oxidative stress occurs in any plant cell when there is an imbalance between production of ROS and antioxidant defense (Apel and Hirt 2004; Mittler 2002; Scandalios 2002). The consequence is a decrease of the net photosynthetic efficiency that affects various plant activities (Ogweno et al. 2008; Apel and Hirt 2004; García-Ferris and Moreno 1994; Alscher et al. 1997; Anderson 2002; Irihimovitch and Shapira 2000; Pfannschmidt 2003). Calvin-cycle enzymes within chloroplasts are particularly sensitive to high levels of H_2O_2, which decreases CO_2 fixation and foliar biomass (Willekens et al. 1997; Zhou et al. 2004, 2006). The mechanism behind the decline of plant photosynthesis by high-temperature stress, driven by high irradiance or drought or heat stress, is similar to that of high irradiance as mentioned earlier.

5.4 Effects of Water Stress (Drought) and of Precipitation/Rainfall

Water stress or drought stress can significantly affect plant photosynthesis and decrease their growth, development and productivity (Li and van Staden 1998; Hassan 2006; Liu et al. 2006; Ohashi et al. 2006; Fariduddin et al. 2009). Water or drought stress can stimulate changes in water balance, leaf area expansion, absorption of photosynthetically active radiation, stomatal closure that reduces the internal CO_2 concentration, integrity of membranes and proteins, metabolic dysfunction, damage at the cellular and subcellular membrane levels via lipid peroxidation, loss of activity of membrane-based enzymes, chloroplast capacity, and PSII activities (Jones and Turner 1978; Matsuda and Riazi 1981; Kaiser 1987; Asada 1992; Hopkins and Hüner 1995; Aziz and Larher 1998; Nam et al. 1998; Cornic 2000; Wilson et al. 2000; Lawlor 2002; Velikova and Tsonev 2003; Flexas et al. 2004; Hassan 2006; Fariduddin et al. 2009; Munns et al. 1979). The final result is a decline in net photosynthesis. The drought stress can reduce stomatal conductance and lead to decreased carbon assimilation, with consequently low biomass production (Fariduddin et al. 2009; Medrano et al. 2002). Decrease in photosynthetic efficiency is generally attributed to reduced CO_2 supply resulting from stomatal closure (Hsiao 1973). A decrease in nitrate reductase activity can inhibit protein synthesis, inactivate enzymes, and reduce the flux of nitrate to the leaf (Fariduddin et al. 2009; Morilla et al. 1973; Shaner and Boyer 1976). The rapid loss of nitrate reductase activity could be part of a biochemical adaptation to water deficit, shutting off the nitrate assimilation pathway and preventing accumulation of nitrite and ammonium (Huffaker et al. 1970).

Cell membranes, which are structurally composed of large amounts of polyunsaturated fatty acid, are highly susceptible to react photolytically with possible changes in membrane fluidity, permeability, and cellular metabolic functions (Bandyopadhyay et al. 1999). The elevation in the antioxidant system defences can detoxify the reactive oxygen species generated by drought stress and can thereby recover the altered physiological performance of stressed plants (Fariduddin et al. 2009).

Water (drought) stress and high temperature together can cause a marked decrease of PSII activity that, together with other functions, can lead to a significant decrease in the net photosynthetic rate of plants (Hassan 2006; Flagella et al. 1998; Hassan et al. 1998; Yordanov et al. 1997, 1999, 2000). It has been shown that this effect may be caused by stomatal and non-stomatal limitations. Stomatal closure usually occurs before inhibition of photosynthesis and restricts CO_2 availability at the assimilation sites in chloroplast. In contrast, non-stomatal limitation of photosynthesis has been attributed to reduced carboxylation efficiency, reduced ribulose-1,5-bisphosphate (RuBP) regeneration, or inhibited chloroplast activity (Wise et al. 1992; Lawlor 1995; Shangguan et al. 1999). Conversely, water stress mostly causes a progressive suppression of photosynthetic carbon assimilation in desiccation-tolerant and intolerant wheat plants (Deltoro et al. 1998).

The mechanism behind the water (drought) stress effect of decreasing photosynthesis is similar to that of high-irradiance/high temperature stress. It occurs particularly in tropical and subtropical regions as mentioned before. Moreover, water stress or drought in low temperature regions can decrease the water content of plant cells that contain dissolved O_2. Shortage of dissolved O_2 in response to water stress can decrease the photoinduced generation of H_2O_2, which is directly linked to photosynthesis. This effect can decrease photosynthesis and cause decline in growth or death of organisms.

The water stress can shift the temperature threshold towards higher values and cause an increase of the heat resistance (Yordanov et al. 1997, 2000; Havaux

1992). Some desiccation-tolerant cells accumulate large amounts of the disaccharides trehalose and/or sucrose. Of these, mostly trehalose can prevent damage from dehydration, not only by inhibiting fusion between adjacent membrane vesicles during drying, but also by maintaining membrane lipids in the fluid phase in the absence of water (Singh et al. 2002; Crowe et al. 1987, 1992; Leslie et al. 1994). Trehalose can depress the phase transition temperature of the dry lipids after desiccation and maintain them in the liquid crystal state (Crowe et al. 1992; Leslie et al. 1994). The increasing activities of catalase, peroxidase and superoxide dismutase and the level of proline can constitute a natural endogenous defense system that increases the response to water stress (Fariduddin et al. 2009; Schützendübel and Polle 2002).

Rainfall can greatly increase photosynthesis, particularly by increasing various physiological phenomena such as leaf water potential, net photosynthetic rate, stomatal conductance, and transpiration (Souch and Stephens 1998; Smit and Rethman 2000; Morris et al. 2004; Li et al. 2007). Moreover, it can significantly enhance the sudden growth of plants all over the world at the beginning of summer season and at the end of winter season. The consequence is an increase of annual biomass production and a simultaneous increase of the production of various food and crops. Maximum photoinduced efficiency of PSII is significantly increased with an increase in rainfall (Li et al. 2007). Among other issues, this might also be caused by the occurrence of H_2O_2 and nutrients in rain water. The supply of exogenous H_2O_2 from rainfall (up to 200 μM, see chapter "Photoinduced and Microbial Generation of Hydrogen Peroxide and Organic Peroxides in Natural Waters") could enhance photosynthesis and make PSII reach its maximum photoinduced efficiency. On the other hand, leaf wetness causes not only instantaneous suppression of photosynthesis but also chronic damage to the photosynthetic apparatus (Ishibashi and Terashima 1995). Interestingly, a direct link has been observed between rainwater H_2O_2 content and the rate of photosynthesis (Komissarov 1995, 2003; Mostofa et al. 2009). However, high concentrations of H_2O_2 (50–100 M) in the presence of iron (Fe) and oxalate can generate $HO^{\bullet}$ that would decrease plant productivity and growth (Kobayashi et al. 2002).

In the case of aquatic microorganisms, drought stress or absence of rainfall for a longer period can significantly affect photosynthesis. In this case, similar mechanisms are followed as for high light irradiance as explained in the earlier section.

5.5 *Effects of the Contents and Nature of DOM and POM*

Organic matter (OM) consisting of DOM and POM is one of the key factors that can produce nutrients (NO_2^-, NO_3^- and PO_4^{3-}) and various photo- and microbial products, such as H_2O_2, CO_2, DIC, LMW DOM, and so on (see also chapters "Dissolved Organic Matter in Natural Waters", "Photoinduced and Microbial Degradation of Dissolved Organic Matter in Natural Waters", "Chlorophylls and Their Degradation in Nature", and "Impacts of Global Warming on

Biogeochemical Cycles in Natural Waters") (Mostofa et al. 2009; Mostofa and Sakugawa 2009; Zhang et al. 2009; Tranvik et al. 2009; Zepp et al. 1987, 2011; Mostofa et al. 2011; Graneli et al. 1996; Granéli et al. 1998; Ma and Green 2004; White et al. 2010; Liu et al. 2010; Fu et al. 2010; Palenik et al. 1987; Cooper and Lean 1992; Bushaw et al. 1996; Molot et al. 2005; Kim et al. 2006; Johannessen et al. 2007; Borges et al. 2008; Li et al. 2008; Kujawinski et al. 2009; Lohrenz et al. 2010; Omar et al. 2010; Cai 2011). Such processes can influence photosynthesis directly or indirectly in water.

Photoinduced degradation of DOM and POM (POM includes e.g. phytoplankton) can be summarized as follows (Eq. 5.29):

$$\begin{aligned}&\text{DOM} + \text{POM} + h\upsilon \rightarrow \text{H}_2\text{O}_2 + \text{CO}_2 + \text{DIC} + \text{LOW DOM} + \text{NO}_3^- + \text{NO}_2^- \\ &+\text{PO}_4^{3-} + \text{autochthonous DOM} + \text{other species}\end{aligned} \tag{5.29}$$

where DIC is usually defined as the sum of an equilibrium mixture of dissolved

$$CO_2, H_2CO_3,\ HCO_3^-,\ \text{and } CO_3^{2-}.$$

Microbial degradation of DOM and POM could be indicated as (5.30):

$$\begin{aligned}&\text{DOM} + \text{POM} + \text{microbes} \rightarrow \text{H}_2\text{O}_2 + \text{CO}_2 + \text{DIC} + \text{LOW DOM} \\ &+ \text{NO}_3^- + \text{PO}_4^{3-} + \text{autochthonous DOM} + \text{other species}\end{aligned} \tag{5.30}$$

Products of these reactions are extensively discussed in chapter "Dissolved Organic Matter in Natural Waters", "Photoinduced and Microbial Generation of Hydrogen Peroxide and Organic Peroxides in Natural Waters" and "Photoinduced and Microbial Degradation of Dissolved Organic Matter in Natural Waters", "Colored and Chromophoric Dissolved Organic Matter in Natural Waters", "Fluorescent Dissolved Organic Matter in Natural Waters". The compounds H_2O_2, CO_2, DIC, and nutrients (NO_3^- and PO_4^{3-}) are primarily responsible for an increase in photosynthetic efficiency in water (Eq. 3.1). Therefore photosynthesis depends on the contents and chemical nature of allochthonous OM (of terrestrial vascular plant origin) and autochthonous OM (of algal or phytoplankton origin). They are characterized by a large variation in different water environments (see chapter "Dissolved Organic Matter in Natural Waters"). Dependence of photosynthesis on OM (DOM and POM) is supported by several observation reported below.

First, DOM contents can affect photosynthesis in the water column. DOM can limit productivity and affect epilimnetic and hypolimnetic respiration (Jackson and Hecky 1980; Carpenter et al. 1998; Hanson et al. 2003; Houser et al. 2003; Druon et al. 2010). Both DOM and POM can limit light penetration in deeper water, thus shoaling the euphotic zone (Bertilsson and Tranvik 2000; Laurion et al. 2000; Hayakawa and Sugiyama 2008; Effler et al. 2010). The vertical attenuation coefficient for downward irradiance of PAR (K_{PAR}) is strongly dependent on water color (Eloranta 1978; Jones and Arvola 1984), which subsequently depends on DOC concentration (Jones and Arvola 1984). Elevated DOM may decrease the efficiency of photosynthesis and growth in deeper waters and produce surface Chl *a* maxima in the upper epilimnion (0–8 m). Such an effect has been observed in the lakes Hongfeng, Baihua and Kinneret, and is quite different from Lake Biwa (0–20 m) and Lake Baikal where DOM contents are relatively low (see also chapters "Dissolved Organic Matter in Natural Waters" and "Chlorophylls and Their Degradation in Nature") (Fu et al. 2010; Mostofa KMG et al., unpublished data; Hayakawa 2004; Yacobi 2006). Waters with high contents of DOM and POM are responsible for the occurrence of toxic algal blooms through high photosynthesis. The latter would be linked to elevated amounts of photo- and microbial products, provided that algal growth is limited by nutrient availability and not by light, and would also be affected by global warming (see later).

The second issue is the dependence of photosynthesis on allochthonous DOM. It has been shown that photosynthetic primary production is significantly dependent on allochthonous humic substances (fulvic and humic acids) in natural waters. It has been observed an increase of bacterial biomass with high humic contents (Jones 1992; Tranvik 1988; 1989; Hessen 1985; Tranvik and Höfle 1987). Typhoon-enhanced terrestrial discharges can elevate Chl *a* concentrations by four times and shift phytoplankton composition (spectral class-based), from an initial dominance of diatoms and green microalgae to the dominance of blue green microalgae (cyanobacteria are increased by more than 200 %) and cryptophytes (Blanco et al. 2008). This enhancement is likely caused either by favorable nutrient availability (Blanco et al. 2008) or by high input of allochthonous DOM including humic substances. A higher ratio of bacterial production to primary production has been observed in a humic lake compared with a clear-water lake, suggesting that the bacterioplankton of the humic lake utilized allochthonous substrates in addition to substrates originating from autochthonous primary production (Tranvik 1989). Moreover, a isolated (ca. *Pseudomonas sp.*) bacterial cell does not utilize fulvic acid, but in the presence of added lactate fulvic acid is partially degraded and causes an increase in the cell yield because of co-metabolism (Stabel et al. 1979; Wright 1988; de Haan 1974). Bacteria (ca. *Arthrobacter* sp.) can utilize fulvic acid, but this is only partially degraded and produces a small cell yield compared to e.g. benzoate. However, in media containing benzoate and fulvic acid, bacteria have higher growth rate and cell yield compared to media with only benzoate or fulvic acid (de Haan 1977). The fluctuations in the content of fulvic acids and the amount of benzoate-oxidizing bacteria suggest that the priming effect might be more important than co-metabolism during the decomposition of fulvic acids in lake water (de Haan 1977). The mechanism behind this phenomenon is, presumably, the acceleration of the photoinduced degradation of fulvic acid in the presence of benzoate. It may cause enhanced production of biologically labile substrates that subsequently increase bacterial production. Benzoate (C_6H_5-COONa) may photolytically release electrons (e_{aq}^-) in aqueous solutions of fulvic acid (Fujiwara et al. 1993; Zepp et al. 1987; Assel et al. 1998; Richard and Canonica 2005), an effect that might lead to the production of hydrogen peroxide in natural waters (Mostofa and Sakugawa 2009; Fujiwara et al. 1993).

The generation of hydrogen peroxide (H_2O_2) upon irradiation of ultra-filtered river DOM is substantially increased, from 15 to 368 nM h^{-1}, with increasing salinity at circumneutral pH values (Osburn et al. 2009). Production of $HO^\bullet$ from H_2O_2

either by direct photoinduced reaction ($H_2O_2 + h\upsilon \rightarrow 2HO^\bullet$) or by photo-Fenton processes is susceptible to decompose DOM in aqueous solution (Zepp et al. 1992; Zellner et al. 1990; Goldstein and Rabani 2008). These photoinduced effects are associated with two impacts on growth of primary production: (i) photoinduced generation of $HO^\bullet$ has direct negative effects on bacterial growth and/or indirect effects, because of the loss of bioavailable DOM associated to ROS mineralization (Scully et al. 2003a). Correspondingly, extracellular enzymes (e.g., phosphatase and glucosidase) can be inactivated in natural waters by secondary photoinduced processes that can lead to a reduction of the substrate uptake by bacteria (Scully et al. 2003b; Ortega-Retuerta et al. 2007). (ii) Studies of abundance and growth in the presence of humic substances indicate that bacteria are the most significant utilizers of allochthonous DOM. This issue is apparently made easier by DOM photolysis under natural sunlight, with production of lower molecular weight, and biologically labile organic products (Miller and Zepp 1995; Strome and Miller 1978; Amador et al. 1989; Kieber et al. 1989; Moran and Zepp 1997). This photoinduced effect can be supported by the observation that DOM photobleaching is accompanied by bacterial growth in humic lakes with significant amounts of chromophoric DOM (Lindell et al. 1995; Reche et al. 1998; de Lange et al. 2003). Thus, humic substances in lakes may serve as a substrate for bacterioplankton and lead to enhanced microbial production. Such stimulation of bacterioplankton productivity could influence food chains in two ways (Jones 1992): firstly, by providing an alternative base (in addition to autotrophic primary production) for the energetic and nutritional support of consumer organisms, of course if bacterial production can be effectively grazed; secondly, by increasing bacterial demand for limiting nutrients at the expense of phytoplankton, thereby depressing autotrophic primary production (Jones 1992).

A further issue is the dependence of photosynthesis on autochthonous DOM. Autochthonous DOM or unknown compounds produced by the cyanobacterium *Trichormus doliolum* or filtrates of dinoflagellate *Peridinium aciculiferum* or *Prorocentrum lima* can inhibit the PSII in other cyanobacteria, decreasing the photosynthetic efficiency (Igarashi et al. 1998; Rengefors and Legrand 2001; Sukenik et al. 2002; Windust et al. 1996; von Elert and Juttner 1997; Sugg and VanDolah 1999). Compounds produced by the cyanobacterium *Microcystis* sp. can inhibit carbonic anhydrase activity of the dinoflagellate *P. gatunense*, leading to CO_2 limitation and inhibition of photosynthesis (Sukenik et al. 2002). When tested as a pure compound, okadaic acid produced by the dinoflagellate *Prorocentrum lima* could inhibit the growth of three microalgal species (Windust et al. 1996), possibly because okadaic acid is a potent phosphatase inhibitor (Bialojan and Takai 1988). Also microcystins produced by the cyanobacterium *Microcystis aeruginosa* can inhibit phosphatase (Dawson 1998). Microalgal compounds have been shown to damage red blood cell membranes, which suggest that competing phytoplankton could be similarly affected (Igarashi et al. 1998). On the other hand, autochthonous DOM released by phytoplankton can be utilized with high efficiency by heterotrophic bacteria and can thus stimulate heterotrophic growth and nutrient cycling (Brussaard et al. 1996, 2005, 2007; Gobler et al. 1997; Fuhrman 1992; Bratbak et al. 1998; Middelboe 2003).

Interestingly, the viral lysis of an *Aureococcus anophagefferens* bloom can release approximately 500 g C L^{-1} that can support bacterial demands for both carbon and nutrients (Gobler et al. 1997). It has been shown that >62 % of a bacterial lysate is metabolized by other bacteria following viral lysis within a few days, with a correspondent bacterial growth efficiency of 45 % (Middelboe 2003). Fatty acids potentially produced by microalgae have been shown to increase permeability of the plasma membranes of chlorophytes and cyanobacteria, which might be connected with an increase of photosynthesis (Wu et al. 2006). Photoproduction of biologically labile substrates from CDOM could potentially stimulate the growth of biomass in Hudson Bay coastal waters (Granskog et al. 2007).

Bacterial biomass exhibits high values during the summer season and lower ones during winter in lakes with different water color (Wright 1984; Arvola and Kankaala 1989; Jones 1990). In winter, the bacterioplankton in humic lakes may primarily consist of a dormant, substrate-limited community that may sustain only a small number of microzooplankton grazers (Jones 1992; Wright 1984). During the spring and summer season fresh inputs of labile allochthonous DOM and autochthonous DOM, possibly with enhanced photoinduced activity, stimulate an increase in bacterial production (Jones 1992). In turn, a rapid development is promoted of grazing flagellates until a quasi steady-state is reached, resulting into an active, grazer-controlled bacterioplankton (Wright 1984).

Currently, model results reveal that the progressive release of dissolved organic nitrogen (DON) in the ocean's upper layer during the summer season increases the regenerated primary production by 30–300 % (Druon et al. 2010). This in turn enhances the dissolved organic carbon (DOC) production, mainly deriving from phytoplankton exudation in the upper layer, and the solubilization of particulate organic matter (POM) deeper in the water column (Druon et al. 2010). A microcosm experimental study on summer carbon metabolism in a humic lake has shown that DOC is 80–85 % of total carbon, while 75 % of POC is detritus. Bacterial biomass and production can exceed those of phytoplankton (Hessen et al. 1990). It has been shown that most of the zooplankton body carbon (46–82 %) is apparently derived from direct ingestion of the large detrital carbon pool. The loop of ingestion and defecation is important, giving a detritus particle turnover rate of 0.39 d^{-1}, and suggests that carbon cycling in humic lakes is essentially different from that in clear-water lakes (Hessen et al. 1990).

Finally, both autochthonous and allochthonous DOM contribute to the production of photo- and microbial products (CO_2, DIC, H_2O_2 and so on) and to the photoinduced generation of the reactive oxygen species (ROS) such as $O_2^{\bullet -}$, H_2O_2 and $HO^{\bullet}$ in photosynthesis. Negative effects of photoproducts on bacterial growth are linked with phototransformation of algal-derived autochthonous DOM (Ortega-Retuerta et al. 2007; Tranvik and Bertilsson 2001). This can be supported by the highly photosensitive and photodegradable nature of autochthonous DOM of algal/phytoplankton origin compared to allochthonous DOM (Mostofa et al. 2009; Johannessen et al. 2007). Penetration of sunlight to deep water significantly depends on the DOM contents, and high-DOM lakes are characterized by shoaling of the euphotic zone (Laurion et al. 2000).

Photosynthetic efficiency of phytoplankton decreases as irradiance increases during the morning, and increases as irradiance declines in the afternoon. These trends are associated with photoprotective strategies in the antennae of PSII and photodamage of PSII reaction centers (Zhang et al. 2008). Conversely, H_2O_2 usually shows strong diurnal variation and its concentrations increases gradually from the morning, reaches a maximum at noon and then gradually decreases in the afternoon (Mostofa and Sakugawa 2009). Therefore, high production of H_2O_2 and the subsequent $HO^{\bullet}$ photogeneration (either direct of photo-Fenton mediated) at noon is susceptible to damages PSII reaction centers.

In addition, autochthonous DOM can produce relatively high amounts of ROS that can inhibit primary production. The daily estimated net CO_2 fluxes (due to all processes) are much smaller than daylight photosynthetic rates (^{14}C uptake) and sometimes go in the opposite direction (Kelly et al. 2001). This indicates that CO_2 fixation measured by ^{14}C uptake is largely offset, and sometimes exceeded, by CO_2 production. Allochthonous DOC degradation could account for only a part of this CO_2 production and the remainder presumably comes from the respiration of photosynthetically fixed carbon (Kelly et al. 2001). The average rates of net epilimnetic CO_2 fixation, or net epilimnetic production (NEP) range from 20 to 60 % of ^{14}C uptake (Kelly et al. 2001). This is similar to previous estimates of the relationship between net 24 h and daylight photosynthetic fixation (Berman and Pollingher 1974). Note that NEP is a *community* parameter, including the respiration of grazers, sediment bacteria and so on. Therefore, is not the same as the term "net photosynthesis" that refers only to the photosynthesis and respiration of algae (Kelly et al. 2001).

5.6 Variation in Nutrient Contents

Photosynthesis of organisms is dependent on the contents of nutrients that can either enhance or decrease its efficiency (Parkhill et al. 2001; Liu et al. 2007; Bybordi 2010). The effects of nutrients on photosynthesis can be classified in two ways depending on the types of organisms.

Nutrients Effects on Aquatic Microorganisms

The effect of nutrients on photosynthesis in water may be a stimulation of primary production (Chl *a*), or not (Yoshiyama and Sharp 2006; Parkhill et al. 2001; Smith 2003; Kaneko et al. 2004; Sterner et al. 2004; Huszar et al. 2006; Nõges et al. 2008; McCarthy et al. 2009; Mohlin and Wulff 2009; Canfield 1983; Auclair et al. 1985; Ferris and Tyler 1985; Steinberg and Muenster 1985; Francko 1986; Jones et al. 1988; Lewis 1990; Salas and Martino 1991; Cullen et al. 1992; Sarnelle et al. 1998; Brown et al. 2000; Guildford and Hecky 2000; Jones 2000). Observing the uptake of nutrients during primary production or algal productivity is a complex issue, because of the many factors involved for the demand and supply of N and P in water. Such factors can be classified as follows: (i) Nutrients (NO_3^-, NO_2^-, NH_4^+ and PO_4^{3-}) are mostly released during photoinduced and microbial assimilation or respiration of algal/phytoplankton biomass (Mostofa et al. 2011; Kopáček et al. 2003; Li et al. 2008; Mallet et al. 1998; Carrillo et al. 2002; Lehmann et al. 2004; Fu et al. 2005). (ii) Formation of N-containing (NH_4^+ or NO_2^-) and P-containing inorganic compounds (PO_4^{3-}) typically occurs upon degradation of dissolved organic nitrogen (DON) and dissolved organic phosphorus (DOP) in natural waters (Mostofa et al. 2011; Kim et al. 2006; Li et al. 2008; Bronk 2002; Zhang et al. 2004; Vähätalo and Järvinen 2007; Haaber and Middelboe 2009). The degradation of *Phaeocystis pouchetii* lysates is associated with significant regeneration of inorganic N and P and produces 148 g N L^{-1} and 7 g P L^{-1}, which corresponds to 78 % and 26 % of lysate N and P being mineralized to NH_4^+ and PO_4^{3-}, respectively (Haaber and Middelboe 2009). Contribution of nutrients through viral lysis might be an important mechanism that promotes heterotrophic nutrient cycling and stimulates primary production (Haaber and Middelboe 2009, 2008; Brussaard et al. 2008). (iii) NO_3^- and NO_2^- can be regenerated by oxidation of ammonia in nitrification ($NH_4^+ + 2O_2 \rightarrow NO_3^- + 2H^+ + H_2O$) and of dissolved organic nitrogen (DON) in lake waters (Lehmann et al. 2004; Mack and Bolton 1999; Kopáček et al. 2004; Minero et al. 2007). (iv) NO_2^- and NO_3^- are preferentially detected in epilimnetic water rather than the hypolimnion (Mostofa KMG et al., unpublished data; Kim et al. 2006; Li et al. 2008; Lehmann et al. 2004; Kopáček et al. 2004; Minero et al. 2007), and they are also involved in photoinduced generation of $HO^•$ that is able to degrade DOM in the epilimnion (see also chapters "Photoinduced Generation of Hydroxyl Radical in Natural Waters" and "Photoinduced and Microbial Degradation of Dissolved Organic Matter in Natural Waters") (Mostofa et al. 2009; Nakatani et al. 2007; Takeda et al. 2004; Zellner et al. 1990; Mopper and Zhou 1990). Furthermore, the NO_2^- ion is generally observed at low concentration during the summer season (Mostofa KMG et al., unpublished data; Kim et al. 2006; Li et al. 2008), and possibly it is photolytically more active in production of $HO^•$ than in NO_3^- (see also chapters "Photoinduced Generation of Hydroxyl Radical in Natural Waters" and "Photoinduced and Microbial Degradation of Dissolved Organic Matter in Natural Waters"). It is also a rather photolabile compound in surface waters, undergoing faster direct photolysis in lake than in ultrapure water. This effect is linked to the scavenging of photogenerated transients by DOM, which finally prevents the recombination of photogenerated, transient nitrogen species back into nitrite (Vione et al. 2009a).

Primary production or Chl *a* often increases with increasing total phosphorus (TP) and nutrients, suggesting that uptake of P and nutrients takes place during primary production (Doyle et al. 2005; Huszar et al. 2006; Nõges et al. 2008; McCarthy et al. 2009; Mohlin and Wulff 2009; Guildford and Hecky 2000; Lehmann et al. 2004; Schindler 1974, 2006; Havens et al. 1995; Smith et al. 1995). Chl *a* is significantly correlated with total P in marine environments, but total P concentration in marine sites is relatively higher compared to freshwater (Guildford and Hecky 2000). Uptake of phosphorus during phytoplankton growth is greatly stimulated in presence of humic substances, but the phosphate uptake is inhibited by toxic compounds (Kaneko et al. 2004; Auclair et al. 1985; Steinberg and Muenster 1985; Francko 1986; Jones et al. 1988). Humic substances can easily undergo photoinduced decomposition into several photoproducts such as CO_2, H_2O_2, DIC, LMW DOM, and so on. These compounds are directly and indirectly linked with photosynthesis and can stimulate primary production.

Primary producers or phytoplankton (or Chl *a*) depend on the total nitrogen (TN) as they can uptake both inorganic and organic N forms such as urea, NH_4^+, and NO_3^- (McCarthy et al. 2007, 2009; Walsh and Dugdale 1971; Kappers 1980; Syrett 1981; Dugdale et al. 1990; Probyn 1992; Blomqvist et al. 1994; Berg et al. 2003; Giani et al. 2005; Rolland et al. 2005; Heil et al. 2007). No evidence has been found for a control of Chl *a* by TN in lake and marine environments (Guildford and Hecky 2000). However, TN (mostly NO_3^- and NH_4^+) can limit primary production in most cases where nutrients are limiting (Huszar et al. 2006; Sarnelle et al. 1998; Barica et al. 1980; Smith 1982; Elser et al. 1990; Aldridge et al. 1995; Levine et al. 1997; Philips et al. 1997; Lewis 1996, 2002).

The nutrients-ratio theory predicts that cyanobacteria will dominate in lakes with low TN:TP ratios, due to their superior ability to compete for dissolved N and, in some cases, to fix atmospheric N (Smith 1983). Recent studies show that primary production or cyanobacteria do not follow this predicted theory in a variety of waters, with either high or low TN:TP ratio (Nõges et al. 2008; McCarthy et al. 2009; Xie et al. 2003; Smith et al. 1995; Smith 1983; Smith and Bennett 1999; Downing et al. 2001). The TN:TP ratio theory can not consistently predict cyanobacterial dominance in a variey of waters. Indeed, recent studies show that nutrients such as PO_4^{3-} and NO_3^- are significantly produced from either POM (e.g. phytoplankton) or allochthonous and autochthonous DOM (see chapters "Dissolved Organic Matter in Natural Waters", "Photoinduced and Microbial Degradation of Dissolved Organic Matter in Natural Waters" "Impacts of Global Warming on Biogeochemical Cycles in Natural Waters"). Correspondingly, waters with extreme eutrification are composed of excess PO_4^{3-} that does not follow this theory at all. This can be justified by the observation that primary production is probably not limited by nutrient availability, because of the high nutrient loadings in natural water (McCarthy et al. 2007, 2009; Heath 1992). Primary productivity within a plume appears to rely upon recycled nutrients, with organic fractions representing the majority of the nutrient pool (Davies 2004). Furthermore, remineralized nutrients from the declining chlorophyll bloom in surface waters are taken up by heterotrophic bacteria in the water-column and by benthic microalgae in sediments (Darrow et al. 2003). Variations in DOM and POM contents can greatly modify the contents of nutrients, and additional factors would be involved into the variations of primary production.

Based on these studies, photosynthesis dependence on nutrients is quite complex in natural waters. First, photosynthesis does not depend on nutrients in waters with high contents of DOM and POM, particularly in lakes, estuarine and coastal waters. High content of DOM and POM can often supply the nutrients (NO_3^- and PO_4^{3-}) under both photoinduced and microbial assimilation or degradation, thus the nutrients in excess have no effects on primary production. Second, photosynthesis may depend on nutrients in waters with low contents of DOM and POM.

This effect is the opposite as the previous one, and is most likely accounted for by the low production of nutrients from low contents of DOM and POM.

On the other hand, a decrease in PSII efficiency with changes in cellular physiology of microalgae can result into nutrient (and mostly nitrogen) stress, ultimately followed by a cell stress (Parkhill et al. 2001; Babin et al. 1996; Cullen et al. 1992; Geider et al. 1993; Graziano et al. 1996). These studies thus show that a decrease of photosynthetic efficiency is caused by nutrient stress. Nitrogen stress is found to reduce the maximum quantum yield of carbon fixation (Babin et al. 1996). The mechanism behind the N-containing (NO_2^- and NO_3^-) nutrient stress is presumably that the strong oxidizing agent $HO^\bullet$, photogenerated from both NO_2^- and NO_3^-, could react with the functional groups bound to PSII and can damage the cells. The result is a decline of the overall photosynthetic efficiency that suppresses the primary production. In addition, the synergic effect of UV radiation due to depletion of the stratospheric ozone layer in combination with N-containing nutrient stress can generate extremely high contents of $HO^\bullet$, which can kill aquatic microorganisms. Note that in Antarctic seawater during an ozone hole event, the production of $HO^\bullet$ is enhanced by at least 20 %. Such enhancement would mostly come from nitrate photolysis and to a minor extent from DOM photoinduced reactions (see also chapter "Photoinduced Generation of Hydroxyl Radical in Natural Waters") (Qian et al. 2001).

Nutrients Effects on Higher Plants

Plant growth is enhanced at 200 mg l^{-1} N (as NH_4NO_3) in cvs. (cultivars) 'Licord' and 'Okapi', but it is reduced when the N concentration increases up to 300 mg l^{-1} (Bybordi 2010). Nitrate reductase (NR), a substrate inducible enzyme, is slightly inhibited by salinity in tomato roots, while leaf NR is decreased sharply (Cramer and Lips 1995). In the leaves of tomatoes and cucumbers, NR activity can increase with exogenous nitrate concentration (Martinez and Cerda 1989). NR is decreased in leaves under salinization, which can subsequently decrease NO_3^- uptake by plants under salt stress (Bybordi 2010; Cramer and Lips 1995; Lacuesta et al. 1990; Abd-El Baki et al. 2000). The decreased of nitrate is accompanied by a high Cl^- uptake and low rate of xylem exudation in high osmotic conditions, by either NaCl or other nutrients (Parida et al. 2004; Tabatabaie et al. 2004). Reduced nitrate uptake or translocation can lead to lower NO_3^- concentration in leaves, which can consequently decrease the NR activity under saline conditions.

Several factors (e.g. salinity) can modify the uptake of some nutrients (e.g. Fe, Mn, Cu, Zn, K, etc.), and either increase or decrease their contents in various parts of most plants (Vıllora et al. 2000; Turhan and Eris 2005; Wang and Han 2007; Achakzai et al. 2010; Tunçtürk et al. 2011).

5.7 Effects of Trace Elements on Aquatic Microorganisms

Aquatic organisms that carry out photosynthesis are largely affected by trace elements, and PSII is thought to be the primary and most sensitive site of inhibition

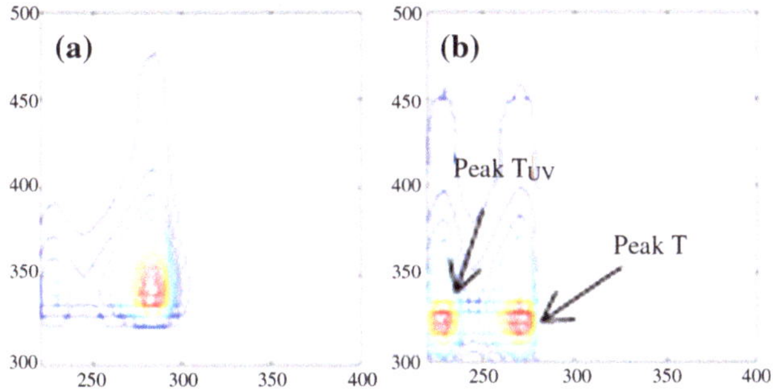

Fig. 10 Two fluorescent components (**a**, **b**) of lake green algae isolated and resuspensions in aqueous media (Milli-Q waters) identified using PARAFAC modeling on the respective EEM data. *Data source* Mostofa KMG et al. (unpublished data)

(Zhang et al. 2010; Crist et al. 1981; Zhou and Wangersky 1985, 1989; Simkiss and Taylor 1989; Xue and Sigg 1990; Tessier and Turner 1995; Sunda and Huntsman 1998; Burda et al. 2003; Koukal et al. 2003; Mylon et al. 2003; Sigfridsson et al. 2004; Berden-Zrimec et al. 2007; Lamelas and Slaveykova* 2007; Hopkinson and Barbeau 2008; Lamelas et al. 2009; Pan et al. 2009). Various trace elements detected in phytoplankton are N, P, S, K, Mg, Ca, Sr, Fe, Mn, Zn, Cu, Co, Cd, Ni, and Mo (Quigg et al. 2003, 2011; Finkel et al. 2006). Study shows that many elements (Fe, Mn, Zn, Cu, Co, and Mo) are enriched relative to P by about two to three orders of magnitude under irradiances that are limiting for growth, and net steady-state uptake of element: P is often elevated under lower irradiances (Finkel et al. 2006). Cyanobacteria or phytoplankton cells can form complexes with or uptake trace metals, either directly or in the presence of humic acids (Zhou and Wangersky 1985, 1989; Xue and Sigg 1990; Koukal et al. 2003; Mylon et al. 2003; Lamelas and Slaveykova* 2007; Lamelas et al. 2009). The latter can substantially enhance the metal ion uptake. Bacteria, algae (or phytoplankton cells) and their exudates are composed of a mosaic of functional groups (e.g. amino, phosphoryl, sulfhydryl, and carboxylic), and the net charge on the cell wall is dependent on the pH of the medium (see also chapter "Complexation of Dissolved Organic Matter With Trace Metal Ions in Natural Waters" for detailed description) (Mostofa et al. 2009; Zhang et al. 2009, 2010; Mostofa et al. 2011; Filella 2008). Cyanobacteria or phytoplankton are composed of two fluorescent components that can be identified using parallel factor (PARAFAC) analysis on the excitation-emission maxima (EEM) spectra of their resuspensions in pure water (Fig. 10) (Mostofa KMG et al. unpublished data). The EEM spectra of these two fluorescent components identify functional groups bound to tryptophan or protein-like components (Fig. 10; see the chapter "Fluorescent Dissolved Organic Matter in Natural Waters" for detailed description) (Mostofa KMG et al. unpublished data). Furthermore, some trace metal ions (e.g., Th^{4+} and U) form complexes at the surface of particulate matter with an organic ligand that might be a nonmetal-specific chelator originating from the cell surface of microorganisms

(Hirose 2004). In addition, autochthonous DOM originating from phytoplankton or algal biomass may contain amino and sulfidic functional groups in its molecular structure, which may form complexes with trace metals in water (Xue and Sigg 1993; Xue et al. 1995).

Fe uptake by phytoplankton is significantly enhanced in the presence of humic substances (Provasoli 1963; Prakash et al. 1973), which is presumably caused by improved metal chelation in aqueous solution (Anderson and Morel 1982). Under low-Fe conditions, Fe allocation in the diatoms *Thalassiosira weissflogii* and *Thalassiosira oceanica* is localized in photosynthetic light–harvesting and electron-transport proteins (Strzepek and Harrison 2004). Increased iron quotas and lowered iron-use efficiencies are often observed in phytoplankton, in response to decreased light levels (Hopkinson and Barbeau 2008; Strzepek and Harrison 2004; Sunda and Huntsman 1997). Iron requirements by phytoplankton increase as available light for photosynthesis decreases, which can lead to the hypothesis that phytoplankton may be colimited by iron and light in low-light environments (Sunda and Huntsman 1997). In an iron–light colimited state growth and photosynthesis are ultimately limited by light processing, whilst production of photosynthetic proteins able to harvest and process light is constrained by iron availability (Hopkinson and Barbeau 2008). Iron– light colimitation may occur in low-iron regions with deep mixed layers, such as the Southern Ocean, or even in macronutrient-limited and stratified waters, near the base of the euphotic zone (Sunda and Huntsman 1997). An iron–light colimitation is observed during winter in the subarctic North Pacific. Here a deep mixed layer (80 m), low incident irradiance, and lack of available iron are all combined to limit photosynthesis, which maintains low phytoplankton biomass (Maldonado et al. 1999). Iron can limit growth in an area with a relatively shallow (40 m) mixed layer in the Subantarctic Front. However light, in conjunction with iron, can control growth in an area with deeper (90 m) mixed layers (Boyd et al. 2001). Iron–light colimitation should also be a factor influencing phytoplankton growth during the North Atlantic spring bloom (Moore et al. 2006).

Availability of iron alone has also been implicated as an important factor in the bloom of some harmful algal species (Bruland et al. 2001; Maldonado et al. 2002), whilst an increase in the toxicity of *Microcystis aeruginosa* has been observed when iron is limited (Lukač and Aegerter 1993). Iron deficiency can affect the electron transfer rate in *Pisum sativum* chloroplasts (Muthuchelian et al. 2001), and stable organic Fe(III) complexes (FeL) photolytically produce dissolved inorganic iron at a higher extent than thermal decomposition and cell-surface reduction of FeL. Such a process can facilitate phytoplankton uptake of iron in the ocean (Fan 2008). On the other hand, during nighttime the reactive oxygen species (H_2O_2 and $O_2^{\bullet -}$) produced by reductases on cell surfaces react with FeL, producing Fe(II). Such a process slows down the oxidation of Fe(II) and the subsequent formation of FeL, thereby maintaining significant levels of bio-available Fe (Fan 2008).

A significant effect of toxic metals on photosynthesis is observed, and the relevant photosynthetic efficiency can be either enhanced or suppressed in natural waters (Zhang et al. 2010; Burda et al. 2003; Koukal et al. 2003; Sigfridsson et al. 2004; Berden-Zrimec et al. 2007; Pan et al. 2009; Mayer et al. 1997; Horton and

Bowyer 1990; Prasad et al. 1991; Barraza and Carballeira 1999; Susplugas et al. 2000; Appenroth et al. 2001; Franklin et al. 2001; Drinovec et al. 2004; Miller-Morey and van Dolah 2004; Shanker et al. 2005; Alam et al. 2007; Hayat et al. 2007; Perales-Vela et al. 2007; Ali et al. 2008; Hasan et al. 2008; Vernay et al. 2008). The esterase activity in several species of marine and freshwater cyanobacteria can be either enhanced or suppressed by copper (Franklin et al. 2001), and antimony (Sb) exposure at concentrations ranging from 1.0 to 10.0 mg L^{-1} inhibits O_2 evolution (Zhang et al. 2010). A decrease in photosynthetic efficiency is caused by the reduction of phytoplankton enzyme activity, which may be a general indicator of cell stress. The stimulating action of Cu for a definite concentration level (e.g. 0.02 mg Cu L^{-1}) on PSII system is often observed in natural waters (Franklin et al. 2001; Burda et al. 2002; Schaffer and Sebetich 2004).

Toxicity of Cd and Zn to the green alga *Pseudokirchneriella subcapitatais* can be significantly ($p < 0.05$) reduced in the presence of humic acids (soil and peat), but not in the presence of Suwannee River fulvic acid (SRFA) (Koukal et al. 2003). It is postulated that humic acid can reduce Cd and Zn toxicity in two different ways (Koukal et al. 2003): (i) Humic acid is capable of decreasing the amount of free metal ions through complex formation with the metal. Humic acid has high molecular weight and is relatively stable with regard to metal-exchange reactions, which can make the metals less bioavailable. (ii) Humic acid can be adsorbed onto algal surfaces, shielded the cells from free Cd and Zn ions. On the other hand, several hypotheses have been advanced to explain why SRFA is unable to reduce metal toxicity (Koukal et al. 2003): (i) Cd- and Zn-SRFA complexes are thought to be labile (i.e. to undergo rapid dissociation); (ii) SRFA can coagulate, presumably during equilibration, which can alter their metal complexing behavior; and (iii) SRFA has a low ability to adsorb on cell membranes at pH > 7.

For better understanding the mechanism of metal toxicity to organisms, it is interesting to have a look at the outer-shell electronic configurations of toxic metals:

As^{33}: $1s^2 2s^2 2p^6 3s^2 3p^6 3d^{10} 4s^2 4p^3$ and As^{3+}: $1s^2 2s^2 2p^6 3s^2 3p^6 3d^{10} 4s^2 4p^0$;

Sb^{51}: $1s^2 2s^2 2p^6 3s^2 3p^6 3d^{10} 4s^2 4p^6 4d^{10} 5s^2 5p^3$ and Sb^{3+}: $1s^2 2s^2 2p^6 3s^2 3p^6 3d^{10} 4s^2 4p^6 4d^{10} 5s^2 5p^0$;

Zn^{30}: $1s^2 2s^2 2p^6 3s^2 3p^6 3d^{10} 4s^2$ and Zn^{2+}: $1s^2 2s^2 2p^6 3s^2 3p^6 3d^{10} 4s^0$;

Cd^{48}: $1s^2 2s^2 2p^6 3s^2 3p^6 3d^{10} 4s^2 4p^6 4d^{10} 5s^2$ and
Cd^{2+}: $1s^2 2s^2 2p^6 3s^2 3p^6 3d^{10} 4s^2 4p^6 4d^{10} 5s^0$;

Cr^{24}: $1s^2 2s^2 2p^6 3s^2 3p^6 4s^1 3d^5$ and Cr^{4+}: $1s^2 2s^2 2p^6 3s^2 3p^6 4s^1 3d^1$

Cu^{29}: $1s^2 2s^2 2p^6 3s^2 3p^6 4s^1 3d^{10}$ and Cu^{2+}: $1s^2 2s^2 2p^6 3s^2 3p^6 4s^1 3d^8$

These metal ions have empty *s*-, *p*- or *d*-orbitals in the outer shell, which allows them to be involved in a strong π–electron bonding system through donation of electrons from the functional groups of PSII (e.g. N- and S-containing carboxylic, amino, thio and hydroxyl groups) (see chapter "Complexation of Dissolved Organic Matter With Trace Metal Ions in Natural Waters" for detailed discussion)

(Mostofa et al. 2009, 2011). After formation of the strong π–electron bonding system in the metal-protein (or organism) complex, the remaining outer-shell electrons (e.g. $4s^2$ for As) are loosely bound and can easily move (see chapter "Complexation of Dissolved Organic Matter With Trace Metal Ions in Natural Waters" for detailed explanation). After complex formation between metals and proteins (or amino acids) in PSII, the normal cells metabolism can be disrupted by electrons in the outer shell of the metal ion itself, or via $HO^{\bullet}$ formation in Fenton or Fenton-like or other unknown processes, finally leading to cell death.

Interestingly, As-protein complexes may be accumulated in the human skin and, when the skin is exposed to natural sunlight (mostly UV-light), irradiation induces the formation of $HO^{\bullet}$ or other reactive oxygen species (e.g. $O_2^{\bullet -}$ and H_2O_2). These species can cause damage to DNA and finally induce cancer in the human body. Coherently, it has been suggested that DNA damage induced by methylated trivalent arsenicals is mediated by reactive oxygen species (Nesnow et al. 2002). Furthermore, arsenite can play a role in the enhancement of UV-induced skin cancers (Rossman et al. 2004). The carcinogenic effects may be connected with accumulation of As^{3+} or Sb^{3+} and other toxic metals. As^{3+} or Sb^{3+} have two electrons in the outer shell, while their inner shells are entirely filled with electrons. This situation makes the outer-shell electrons of metal-protein complexes highly mobile.

As and cigarette smoke are synergistic, producing an elevated risk of bladder and lung cancer in smokers (Hopenhayn-Rich et al. 1998; Steinmaus et al. 2003; Chen et al. 2004). Smoking could help promoting the excitation of electrons from the outer shells of As in As-protein complexes, which can finally induce formation of $HO^{\bullet}$ that damages DNA. While eating soil is quite unusual for humans, in some rural Bangladesh villages it is customary for pregnant women to eat *shikal* (it sounds like "chain" in English), which consists of small (2 in. × 1 in. × 1/2 in.) pellets made out of flooded soil (information source is personal experience of one of the authors). While the origin of this tradition is quite uncertain, it is noteworthy that it was observed in one of the world's regions where human exposure to As is the highest. Interestingly, intake of black soil (dark brown soil) with high contents of humic acid could reduce As levels in the human body, because complexes between As and humic acids are much stronger than As-protein or As-fulvic acids ones (see chapter "Complexation of Dissolved Organic Matter With Trace Metal Ions in Natural Waters" for detailed discussion).

Coming back to photosynthetic microorganisms, the decrease in photosynthetic efficiency can be cause by complex formation between metals and the functional groups of PSII. Two possible mechanisms can be operational. First, the formed metal-cells or metal-proteins complexes in the PSII of aquatic microorganisms can produce electrons (e^-) photolytically upon exposure to sunlight, which can finally lead to H_2O_2 generation. (see chapter "Photoinduced and Microbial Generation of Hydrogen Peroxide and Organic Peroxides in Natural Waters" for detailed mechanism) (Komissarov 2003; Mostofa and Sakugawa 2009; Mostofa et al. 2011; Fujiwara et al. 1993). Coherently, it has been observed that methyl viologen acts as an inducer of photo-oxidative stress (Donahue et al. 1997; Mano et al. 2001) and can induce the photoreduction of dioxygen (O_2) by accepting electrons from the iron-sulfur cluster Fe-S_A/Fe-S_B of PS1. This process could induce the production of superoxide radical ($O_2^{\bullet -}$) and H_2O_2 (Fujii et al. 1990). When catalytic free metals [e.g., Fe(II), Cu(I) and Mn(II)] are present near the production site of $O_2^{\bullet -}$ and H_2O_2, the strong oxidizing agent $OH^{\bullet}$ can be produced via a Fenton reaction or by direct photodissociation (Zepp et al. 1992; Nakatani et al. 2007; Zellner et al. 1990; Goldstein and Rabani 2008; Halliwell and Gutteridge 1984; Stadtman 1993). The photo-Fenton reaction could proceed as follows (Eqs. 5.31, 5.32):

$$Fe^{3+} + h\upsilon/H_2O^{\bullet}/O_2^{\bullet -} \rightarrow Fe^{2+} + O_2 + H^+ \tag{5.31}$$

$$Fe^{2+} + H_2O_2 + h\upsilon \rightarrow Fe^{3+} + HO^{\bullet} + OH^- \tag{5.32}$$

The $HO^{\bullet}$ radical thus generated kills the cells of microorganisms in natural waters. Fe concentration and pH can significantly affect both the growth and the reactive oxygen species (ROS) production in *Chattonella marina*, a harmful algal bloom species (Liu et al. 2007). The rapid photoinduced release of electrons from the outer shell of metal ions in PSII metal-protein complexes can be involved in chronic toxicity. It has recently been shown that exposure of PSII to Sb^{3+} and Cr^{4+} in *Synechocystis* sp. can increase the dissipated energy flux and decrease the performance index and the maximum quantum yield for primary photochemistry (φP_O) (Zhang et al. 2010; Pan et al. 2009). It can also cause damage to cellular components and to the overall photosynthetic driving force. The fluorescence yield at 684 nm, which is connected to the maximum quantum yield for primary photochemistry (Zhang et al. 2010) is affected by the metal-functional groups of PSII. It can be the easiest way by which electrons are released, which can subsequently result into high production of $HO^{\bullet}$ via $O_2^{\bullet -}$ or H_2O_2. Such a process can decressee photosynthetic efficiency or damage the PSII or DNA. Along with this effect, complexation between metal ions and the functional groups of PSII can inhibit electron transport and cause the overall photosynthetic efficiency to decline. Severe damage in the water-splitting site of PSII can produce an increased ratio of F_O / F_V (fluorescence level before and after flash) (Pan et al. 2009), which may occur because of $HO^{\bullet}$ photoproduction. Note that Sb^{3+} compounds are approximately ten times more toxic than Sb^{5+} ones, possibly because of the two unpaired electrons in Sb^{3+} species. In contrast, Sb^{5+} species have no outer shell electrons.

Another possible effect linked with complex formation between metals and the functional groups bound to PSII is the blockage of the normal function of electron release by PSII-bound functional groups. This can also significantly decrease the photosynthetic efficiency of aquatic organisms.

The stimulating effect of Cu^{2+} ($1s^2\ 2s^2 2p^6\ 3s^2 3p^6 4s^1 3d^8$) on PSII is presumably caused by the partially and completely filled outer shell of its *s*- and *d*-orbitals. Therefore, the complexes formed between Cu^{2+} and the PSII functional groups are not as strong as those involving other metal ions with empty orbitals. This effect can roughly recover the normal function of PSII in organisms. In addition, the electrons released from the metal-protein complexes of PSII can induce the generation of relatively low amounts of H_2O_2, which might enhance photosynthesis in aquatic organisms. Further studies will be needed to evaluate the exact mechanism behind this phenomenon. The formation of complexes between metals and the functional groups of either PSII or PSI is a relatively new hypothesis, which could greatly help improving the present understanding of the effects of metal ions on the photosynthetic efficiency of aquatic organisms.

5.8 Effect of Salinity or Salt Stress

Salinity is an important controlling factor for photosynthesis, its effect depending on the organisms such as higher plants or aquatic microorganisms. Therefore, the effect of salinity can be divided into two classes: (i) effect on aquatic microorganisms; (ii) effect on higher plants.

Effects of Salinity on Aquatic Microorganisms

Water salinity or salt stress has a significant impact on the photosynthetic capability of organisms, but the actual effect is highly dependent on the different kinds of microorganisms (Liu et al. 2007; Satoh et al. 1983; Ahel et al. 1996; Moisander et al. 2002; Marcarelli et al. 2006; Segal et al. 2006; Demetriou et al. 2007; Allakhverdiev and Murata 2008). It has been shown that salinity in marine waters can alter the PSI and PSII of organisms, which is connected to salt stress. However, some organisms such as cyanobacteria or phytoplankton can overcome the salt stress and are capable of growing under salinity conditions which would be harmful to most other organisms. The basic physiological responses of cyanobacterial cells to salt stress occur in three phases (Hagemann and Erdmann 1997). First, within seconds an increase of the ambient concentration of NaCl can cause an influx of Na^+ and Cl^- ions into the cytoplasm. Second, within an hour it starts the replacement of Na^+ by K^+ ions, leading to a decline in the toxic effects of high Na^+ concentrations. Third, within several hours the cells become acclimatized to the elevated ion concentrations. During this phase, the synthesis or uptake of compatible solutes/components mitigates the toxic effects of salts and preserves the structures of complex proteins and cell membranes (Bhagwat and Apte 1989; Reed et al. 1985; Hagemann et al. 1990, 1991; Hayashi and Murata 1998; Chen and Murata 2002). The cyclic electron transport via PSI is also activated (Joset et al. 1996; Hibino et al. 1996). However, over a longer period of time, such as within several days, salt stress can inhibit cell division (Ferjani et al. 2003).

The increase in the intracellular concentrations of Na^+ and Cl^- ions can cause irreversible inactivation of the oxygen-evolving machinery and of the electron-transport activity of PSI (Kuwabara and Murata 1983; Miyao and Murata 1983; Murata and Miyao 1985; Allakhverdiev et al. 2000a, b). For instance, incubation of *Synechococcus* cells with 0.5 M NaCl can suppress the reduction of $P700^+$ (Allakhverdiev et al. 2000a). Because $P700^+$ is reduced by plastocyanin, it is suggested that the association of this compound with the PSI complex is disturbed by the presence of NaCl (Allakhverdiev et al. 2000a, b).

In cyanobacteria, the oxygen-evolving machinery of PSII located on the luminal side of thylakoid membranes is stabilized by three extrinsic proteins. They are PsbO (33-kD protein), PsbV (cytochrome c_{550}), and PsbU (Allakhverdiev and Murata 2008; Shen et al. 1998; Nishiyama et al. 1999). Cyt c_{550} and PsbU are loosely bound to the donor side of the core complex of PSII (Nishiyama et al. 1997, 1999). These proteins could be easily dissociated from the cyanobacterial PSII complex in the presence of elevated concentrations of NaCl (Shen et al. 1998, 1992). Moreover, pulse-chase experiments revealed that salt stress can inhibit the de novo synthesis of D1 in *Synechococcus* (Ohnishi and Murata 2006).

Light is an important factor in restoring the activity of PSII and PSI during dark incubation of cyanobacterial cells under salt stress (Allakhverdiev et al. 2005). When light is applied to *Synechococcus* cells, protein synthesis occurs for the recovery of the photosystems from salt stress (Allakhverdiev and Murata 2008; Hagemann et al. 1991; Allakhverdiev et al. 1999, 2005). Weak light at 70 mE m^{-2} s^{-1} is sufficient to generate ATP, which seems to support recovery (Allakhverdiev and Murata 2008). Such conditions are sufficient to induce the necessary excitation, because of the formation of complexes between cations (e.g. Na^+ and other cations from salts) and the functional groups bound to PSII and PSI. Recent studies of PSII photoinhibition in cyanobacteria suggest that oxidative stress due to reactive oxygen species (ROS) can inhibit protein synthesis and the repair of PSII. However, it does not stimulate photodamage to PSII (Nishiyama et al. 2005, 2006; Takahashi and Murata 2008; Murata et al. 2007). Note that salinity in marine waters is accounted for various salts including NaCl (86 %), but comparison of river and sea water shows that Na^+, Ca^{2+}, Mg^{2+}, K^+, HCO_3^-, Cl^- and SO_4^{2-} in the sea are typically 1,670, 27, 330, 170, 2.4, 2,400 and 245 times, respectively, higher than in rivers (see chapter "Complexation of Dissolved Organic Matter With Trace Metal Ions in Natural Waters" for more discussion). Also note that the occurrence of these salts can cause changes in the absorption properties of chromophoric dissolved organic matter (CDOM), and in the fluorescence properties of fluorescent dissolved organic matter (FDOM). A change in the optical properties (generally shifting from shorter towards longer wavelengths) and in the complexation behavior of both CDOM and FDOM can be linked to an enhanced photoinduced generation of H_2O_2 (see chapters "Photoinduced and Microbial Generation of Hydrogen Peroxide and Organic Peroxides in Natural Waters" "Colored and Chromophoric Dissolved Organic Matter in Natural Waters", "Fluorescent Dissolved Organic Matter in Natural Waters", and "Complexation of Dissolved Organic Matter With Trace Metal Ions in Natural Waters", respectively for their detailed discussion).

A proposed mechanism for the decline of photosynthesis of microorganisms is that cations (e.g. Na^+, Ca^{2+}, Mg^{2+}, Sr^{2+}) of various salts occurring in marine waters can form complexes with functional groups bound to microorganisms (or with their PSII). This complexation can decrease photosynthesis, either by inducing high photoinduced formation of $HO^•$ from H_2O_2, which would damage PSII, or by blocking the normal function of electron release by the PSII functional groups. Either effect could alter the normal function of PSII, as extensively discussed in an earlier section (see the "effect of trace metal ions" section). In addition, complexes of trace metal ions in marine waters with autochthonous DOM of algal/phytoplankton origin and with terrestrial DOM of vascular plant origin can induce rapid photoinduced excitation of electrons (e^-). Such a process can produce $O_2^{•-}$, H_2O_2 and $HO^•$ that can subsequently either decompose the proteins and the functional groups bound to microorganisms, decreasing their photosynthetic efficiency, or transform the DOM components with production of a number of photoproducts such as CO_2, DIC, H_2O_2 and LMW DOM. These products are directly linked with an enhancement of photosynthesis and might account for algal blooms, particularly in coastal marine environments.

This mechanism is supported by earlier studies, showing that the inhibition of electron flow on the oxidizing (water) side of PSII causes photoinhibition. Moreover, photoactivation and dark-inactivation of electron flow on the reducing side of PSI is completely inhibited at high salinity (Satoh et al. 1983). It is known that photobleaching of carotenoids and Chl can take place when the oxidizing side of PSII is inhibited (Yamashita et al. 1969; Katoh 1972). It has also been observed that inhibition on the reducing side of PSI can give rise to strong reductants, which can also destroy the reaction centers of both PSI and PSII (Satoh and Fork 1982a, b). Photobleaching of carotenoids and Chl might be caused by $HO^•$ or other oxidants, generated photolytically by the above mechanism, in analogy with the well-known phenomena concerning DOM photobleaching in natural waters (see chapters "Photoinduced Generation of Hydroxyl Radical in Natural Waters", "Photoinduced and Microbial Degradation of Dissolved Organic Matter in Natural Waters", "Colored and Chromophoric Dissolved Organic Matter in Natural Waters", and "Fluorescent Dissolved Organic Matter in Natural Waters" for detailed discussion).

Effects of Salinity on Higher Plants

Salinity of soil or water is one of the key environmental factors that limit plant growth and productivity, particularly in arid, semi-arid and freshwater land near coastal regions. Salinity can have a two-fold effect on plants: (i) osmotic stress, by which salt in the soil can reduce the availability of water to the roots, and (ii) ionic stress due to the salt taken up by the plant that can be accumulated to toxic levels in certain tissues (Munns et al. 1995). Reduction of photosynthesis caused by salt stress has an impact on several physiological responses, such as inhibition of growth and development, modification of ion balance, mineral nutrition, water status, stomatal behavior, decrease in photosynthetic efficiency and in chlorophyll content (which leads to a corresponding reduction of light absorption by leaves), decrease of carotenoids, carbon allocation and utilization, net carbon dioxide exchange, respiration and protein synthesis, and finally, induction of cell expansion in both roots and leaves in salt-sensitive plants (Bybordi 2010; Tunçtürk et al. 2011; Melgar et al. 2008; Pandey and Yeo 2008; Pandey et al. 2009; Bybordi et al. 2010a, b; Flowers et al. 1977; Munns and Termaat 1986; Zidan et al. 1990; Ashraf and Wu 1994; Neumann et al. 1994; Evans 1996; Jungklang et al. 2003; Meloni et al. 2003; Qiu and Lu 2003; Lee et al. 2004; Pal et al. 2004; Suwa et al. 2006; Ali et al. 2007; Desingh and Kanagaraj 2007; Šiler et al. 2007; Ahmed et al. 2008). It has been shown that some physiological responses (e.g. chlorophyll and carotenoids) are initially increased at moderate NaCl levels, but they are generally decreased by increasing salinity. It has also been observed that cations or metal ions in all plant parts are typically increased with an increase in salt stress.

The effects of salinity are mostly linked to a decrease in stomatal conductance and/or to the non-stomatal limitation related to carbon fixation (Bongi and Loreto 1989; Brugnoli and Björkman 1992; Delfine et al. 1998, 1999; Centritto et al. 2003). It is suggested that stomatal limitation prevails at intermediate salinity levels, while the non-stomatal limitations predominate under severe salt stress conditions (Bongi and Loreto 1989). The photosynthetic rate, PSII efficiency, root and shoot growth of *Centaurium erythraea* is increased or remains the same at moderate salt levels (50–200 mM NaCl), but it is decreased significantly at high salt concentration (400 mM NaCl). Root growth is more adversely affected by increasing NaCl concentration than shoot growth (Šiler et al. 2007). Chlorophyll contents are decreased under elevated salinity conditions for some salt-sensitive plant species, but they are not modified at moderate salt levels (Jungklang et al. 2003; Lee et al. 2004; Šiler et al. 2007; Delfine et al. 1998, 1999; Ashraf et al. 2002). This suggests that the decline of chlorophyll content depends on the salinity level, on the time of exposure to salts and on the plant species. Salinity can rapidly inhibit root growth and subsequently decrease the uptake of water and essential mineral nutrients from soil (Neumann 1997). An increase of NaCl concentration in solution can reduce N and NO_3 concentrations in leaves, when plants are treated with NaCl and NH_4NO_3 (Bybordi 2010). An apparent increase in salt tolerance is observed when N levels, supplied under saline conditions, exceed the optimum ones observed under non-saline conditions (Bybordi et al. 2010a; Papadopoulos and Rendig 1983). This indicates that increased fertilization, especially by N, may improve the deleterious effect of salinity (Ravikovitch and Porath 1967).

A contribution to salt stress in salt-sensitive plants may derive from the fact that an increase of salinity can enhance the metal ion contents in plant cells, because metal ions can form complexes with PSII functional groups. As already mentioned, such a complexation may cause either a high production of photoinduced electrons (e^-) and of superoxide anion ($O_2^{\bullet -}$), H_2O_2 and $HO^\bullet$, which can damage PSII, or block further photoinduced generation of electrons from PSII itself.

Conversely, the plant growth at moderate levels of NaCl might also be favored by photoinduced generation of H_2O_2 from PSII-metal complexes. If moderate, such H_2O_2 levels could be favourable to photosynthesis as discussed before (Eq. 3.1). The balance is delicate, however, because excessive salt can cause high production of H_2O_2 and $HO^\bullet$ that can damage the PSII. These proposed mechanisms can be justified by the observation of several physiological functions caused by salt stress, such as: (i) salinity stress in plants can produce reactive oxygen species (ROS) such as $O_2^{\bullet-}$, H_2O_2 and $HO^{\bullet}$, particularly in chloroplasts and mitochondria (Mittler 2002; Masood et al. 2006). Plants possess a number of antioxidant enzymes such as superoxide dismutase (SOD), ascorbate peroxidase (APX) and glutathione reductase (GR) for protection against the damaging effects of ROS (Asada 1992; Prochazkova and Wilhelmova 2007), but ROS-linked salinity stress can cause membrane disorganization, metabolic toxicity and attenuated nutrients (Frommer et al. 1999; Zhu 2000; Costa et al. 2005) These initial effects can then induce more catastrophic events in plants. Excessive salt stress can eventually cause photoinhibition and photodamage of PSII (Krause and Weis 1991; Belkhodja et al. 1994). (ii) Strong salt stress in salt-sensitive species can severely reduce the potential of electron transport in PSII (Jungklang et al. 2003). (iii) Salinity can increase or decrease uptake of some nutrients (e.g. Fe, Mn, Cu, Zn, K, etc.) depending on the plant species (Vıllora et al. 2000; Turhan and Eris 2005; Wang and Han 2007; Achakzai et al. 2010; Tunçtürk et al. 2011; Greenway and Munns 1980; Martinez et al. 1987; Cornillon and Palloix 1997; Alpaslan et al. 1998). The increase in these metals can enhance complexation with the PSI and PSII functional groups, leading to ROS production. High Na^+ content is generally responsible for alteration of the nutrient balance, which can cause specific ion toxicity in addition to disturbing the osmotic regulation (Greenway and Munns 1980). (iv) Due to the complex formation between metals and PSII functional groups, electron excitation at low irradiance can induce effective generation of H_2O_2 and ROS. This can be justified by the in vivo observation of ROS generation inside PSII membranes. Salt stress may thus damage the photosynthetic activity of PSII even at low irradiance (Pandey et al. 2009). (v) Complexation of trace metal ions with functional groups bound to PSII under salinity conditions can enhance electron release and, as a consequence, ROS production (see chapter "Complexation of Dissolved Organic Matter With Trace Metal Ions in Natural Waters"). Such effects are able to photodamage PSII in *Chlamydomonas reinhardtii*, barley leaves (*Hordeum vulgare*), sorghum (*Sorghum bicolor*), rye (*Secale cereal*), and *Spirulina platensis* (Neale and Melis 1989; Sharma and Hall 1991; Hertwig et al. 1992; Lu and Zhang 1999).

Chl content in salt-tolerant plants would either remain the same or be significantly enhanced with increasing salinity (Qiu and Lu 2003; Brugnoli and Björkman 1992), and accumulation of compatible solutes (e.g. proline, betaine, polyols, sugar alcohols, and soluble sugars) in many plants can increase the tolerance of PSI and PSII to salt stress (Chen and Murata 2002; Fulda et al. 1999; Zhu 2002; Reed and Stewart 1988). The increase of Na^+ and Cl^- ions in both leaves and roots is accompanied with an increase in proline and soluble sugars which could play a role in salt tolerance (Melgar et al. 2008; Ahmed et al. 2008).

While functioning in an otherwise similar way as non-tolerant plants, salt-tolerant plant species may supply relatively low amounts of salt ions to leaves through roots. The consequence may be the occurrence of relatively low contents of H_2O_2. If the latter be present in moderate amount, it would mostly be used in photosynthesis and would not produce dangerous levels of $HO^{\bullet}$. Therefore, the plant may maintain normal photosynthesis in the presence of high salt levels. Salt tolerance in canola is associated with the ability to reduce uptake and/or transport of saline ions (Bybordi 2010).

In addition, resistance to salinity may occur when a plant is capable of producing large amounts of antioxidant enzymes such as superoxide dismutase (SOD), ascorbate peroxidase (APX) and glutathione reductase (GR) (Asada 1992; Prochazkova and Wilhelmova 2007; Mittova et al. 2002; Gossett et al. 1994; Pastori and Trippi 1993). These enzymes can significantly scavenge free radicals under stress conditions.

Elevated levels of GR are capable of increasing the amount of $NADP^+$, which accepts electrons from the photosynthetic electron transport chain (Peltzer et al. 2002; Reddy et al. 2004). The activity of antioxidant enzymes under saline conditions are typically increased in the case of salt-tolerant cotton varieties, shoot cultures of rice, cucumber, wheat shoot and pea (Bybordi et al. 2010b, c; Meloni et al. 2003; Desingh and Kanagaraj 2007; Fadzilla et al. 1997; Lechno et al. 1997; Hernandez et al. 1999; Meneguzzo et al. 1999). Due to salinity stress, plants can accumulate osmolytes such as proline and glycine betaine, which are known to protect macromolecules by stabilizing protein structure during dehydration and/ or by scavenging ROS produced under stress conditions (Desingh and Kanagaraj 2007; McNeil et al. 2001; Zhu 2001; Matysik et al. 2002; Rontein et al. 2002). Tolerance of photosystems to salt stress can be enhanced by genetically engineered increase in the unsaturation of fatty acids in membrane lipids, and by intracellular synthesis of compatible solutes (e.g. glucosylglycerol and glycinebetaine) (Allakhverdiev and Murata 2008). When photosynthetic organisms are exposed to salt stress, fatty acids of membrane lipids are desaturated (Huflejt et al. 1990). Therefore, unsaturation of fatty acid in membrane lipids can enhance tolerance of PSI and PSII to salt stress (Allakhverdiev and Murata 2008).

Enhanced tolerance of PSII to salt stress upon unsaturation of membrane lipids is probably accounted for by the fact that unsaturated fatty acids are generally capable of surrounding the cations [e.g. Na^+/H^+ antiporter(s) and/or H^+-ATPase(s)] with their electron-rich double bonds. An increase in the levels of the antiport system components can decrease the concentration of Na^+ ions in the cytosol, which may protect PSII and PSI against NaCl-induced inactivation (Allakhverdiev and Murata 2008).

5.9 Effects of Toxic Pollutants on Aquatic Microorganisms

Environmentally-occurring toxic organic pollutants can decrease the efficiency of photosynthesis, most presumably by adversely affecting the PSII (Berden-Zrimec et al. 2007; Mayer et al. 1997; Halling-Sørensen et al. 2000; Katsumata et al. 2005, 2006; Kvíderová and Henley 2005; Zrimec et al. 2005; Pan et al. 2009; Yates and Rogers 2011). Some antibiotics (e.g. ampicillin, streptomycin, levofloxacin hydrochloride, mecillinam, trimethoprim, ciprofloxacin), phenols (e.g. 3,5-dichlorophenol), pesticides and herbicides (e.g. DCMU or diuron, simazine, atrazine) are highly toxic to microorganisms such as cyanobacteria or phytoplankton cells (Berden-Zrimec et al. 2007; Halling-Sørensen et al. 2000; Katsumata et al. 2005, 2006 Kvíderová and Henley 2005; Zrimec et al. 2005; Pan et al. 2009; Yates and Rogers 2011; DeLorenzo et al. 2001). The toxic organic compounds are effective in inhibiting O_2 evolution and PSII activity of microorganisms. The inhibitory effect on PSII is often increased with increasing concentration of toxic compounds. Levofloxacin hydrochloride, one of the most commonly used fluoroquinolone antibiotics, can decrease the density of the active photosynthetic reaction centers of *Synechocystis* sp., inhibit electron transport, and increase the dissipated energy flux per reaction center. All these effects together are able to decrease the photosynthetic efficiency (Pan et al. 2009).

The adverse effect on photosynthesis is thought to be caused by two facts. First, the molecular structures of organic contaminants are mostly composed of N-, S-, O-, and/or P-containing functional groups, which are susceptible to form H-bonding with the functional groups of PSII. This effect can decrease the efficiency of electron release from PSII. It has in fact been demonstrated that the herbicide DCMU can directly block the electron transport in PSII (Berden-Zrimec et al. 2007; Tissut et al. 1987; Behrenfeld et al. 1998). The second issue is that N-, S-, O-, or P-containing functional groups can release electrons upon excitation by light, which can produce ROS such as $O_2^{\bullet-}$, H_2O_2 and $HO^{\bullet}$. These oxidizing species can damage the PSII system, thereby reducing the photosynthetic efficiency as a whole.

KCN (an inhibitor of mitochondrial respiration) and 3-(3,4-dichlorophenyl)-1,1-dimethylurea (an inhibitor of photosynthesis) had no significant effects on ROS production. In contrast, vitamin K3 (a plasma membrane electron shuttle) can enhance ROS production and its antagonist, dicumarol, can decrease it (Liu et al. 2007). Photosynthetic organisms can produce ROS by activating various oxidases and peroxidases, in response to environmental stresses such as pathogens, drought, light intensity, an increase in temperature from 7 °C to 30 °C, and contaminants such as paraquat (Peng and Kuc 1992; Moran et al. 1994; Karpinski et al. 1997; Iturbe-Ormaetxe et al. 1998; Twiner and Trick 2000).

5.10 *Effect of Size-Fractionated Phytoplankton*

Planktonic algae of <5 μm in size are major fixers of inorganic carbon in the ocean and dominate phytoplankton biomass in post-bloom, stratified oceanic temperate waters (Li 1994; Tarran et al. 2001). Large and small algae are viewed as having a critical growth dependence on inorganic nutrients. The latter can be assimilated at lower ambient concentrations due to the higher surface-area-to-volume ratios of small vs. larger organisms (Malone 1980; Chisholm 1992; Zubkov and Tarran 2008). Experimental studies that adopted phosphate tracer suggest that small algae can uptake inorganic phosphate indirectly, possibly through feeding on bacterioplankton (Hansen and Hjorth 2002; Stibor and Sommer 2003; Tittel et al. 2003; Unrein et al. 2007; Zubkov et al. 2007; Jones 2000; Bird and Kalff 1986; Arenovski et al. 1995; Rothhaupt 1996; Thingstad et al. 1996; Caron 2000). Inorganic phosphate and other nutrients (e.g. NO_3^-) can originate mostly from two processes: (i) photoinduced and microbial assimilations of algae (or cyanobacteria), and (ii) photoinduced and microbial degradation of allochthonous DOM of plant origin and autochthonous DOM of algal/phytoplankton origin (see also chapters "Dissolved Organic Matter in Natural Waters", "Photoinduced and Microbial Degradation of Dissolved Organic Matter in Natural Waters", and "Impacts of Global Warming on Biogeochemical Cycles in Natural Waters"). Small algae can carry out 40–95 % of the bacterivory activity in the euphotic layer of the temperate North Atlantic Ocean in summer, and 37–70 % in the surface waters of the tropical North-East Atlantic Ocean (Zubkov and Tarran 2008). This reveals that the smallest algae have less dependence on dissolved inorganic nutrients (Zubkov and Tarran 2008).

The volume of planktonic bacteria increases as water temperature decreases (Albright and McCrae 1987; Chrzanowski et al. 1988; Bjørnsen et al. 1989), and thus the occurrence of larger cells in the hypolimnion is linked to its low temperature (Wiebe et al. 1992; Callieri et al. 2009; Bertoni et al. 2010). Bacterial cells are often observed to be approximately 30 % larger in the Arctic Ocean and Antarctic coastal waters than in temperate regimes (Straza et al. 2009).

The mechanism behind this size shift is presumably that low temperature in hypolimnion and Arctic or Antactic regions can protect against cell degradation, whereas microbial assimilations of planktonic bacteria cells can release both nutrients and autochthonous DOM. Correspondingly, high temperatures along with solar radiation and surface waters mixing by strong waves are effective in photolytically and microbially releasing nutrients and autochthonous DOM. These two effects could be responsible for the occurrence of large cells in low temperature regions including the hypolimnion. The photoinduced and microbial releases of nutrients, photo-/microbial products, and autochthonous DOM from algae/phytoplankton (Mostofa et al. 2009; Zhang et al. 2009; Tranvik et al. 2009; Zepp et al. 2011; Ma and Green 2004; White et al. 2010; Liu et al. 2010; Mostofa et al. 2005; Bushaw et al. 1996; Molot et al. 2005; Johannessen et al. 2007; Borges et al. 2008; Kujawinski et al. 2009; Lohrenz et al. 2010; Omar et al. 2010; Cai 2011) are responsible for low photosynthesis in most upper surface layers.

It has also been observed that lower photosynthesis in the shallower epilimnion (1 m) than in the deeper epilimnion (3 m) (Nozaki et al. 2002) might be the effect of higher photoinduced degradation of algae near the water surface. This effect, coupled with strong wind mixing and turbulence can decrease the size structure of phytoplankton or algae in the upper epilimnion, thereby decreasing the photosynthetic efficiency (Cermeno et al. 2005; Nozaki et al. 2002). Note that physical mixing in the surface mixing zone is an important factor for promoting the photoinduced degradation of DOM (see chapter "Complexation of Dissolved Organic Matter With Trace Metal Ions in Natural Waters").

5.11 *Effects of Global Warming*

Global warming causes an increase in water temperature, lengthens the summer season, extends the surface water mixing zone and increases the stability of the vertical stratification in large parts of lakes and oceans. An increase in photoinduced and microbial degradation rates of OM (DOM and POM) by global warming may affect water transparency and generation of photo- and microbial products (H_2O_2, CO_2, DIC, and so on), may modify seasonal patterns in chlorophyll or primary production, contents of nutrients (NO_2^-, NO_3^-, and PO_4^{3-}), carbon cycling, pH values, microbial food web stimulation that varies seasonally on a variety of time scales, and the depth of the mixing layer (see also chapters "Dissolved Organic Matter in Natural Waters", "Chlorophylls and Their Degradation in Nature", and "Impacts of Global Warming on Biogeochemical Cycles in Natural Waters") (Mostofa et al. 2009; Baulch et al. 2005; Morris and Hargreaves 1997; Cooke et al. 2006; Huisman et al. 2006; Malkin et al. 2008; Davis et al. 2009; Castle and Rodgers 2009; Mostofa and Sakugawa 2009; Keeling et al. 2010; Zepp et al. 2011; Granéli et al. 1998). Two phenomena can result from this. First, in water with high contents of OM, photoinduced and microbial processes that correspond to high photosynthesis may be prolonged, thereby causing the prolongation of the primary productivity (Malkin et al. 2008). This may eventually result into toxic or harmful algal bloom in natural waters. Second, low photosynthesis could take place in waters with low contents of OM, causing low production of photo- and microbial products. This can subsequently reduce the vertical mixing and suppress the upward flux of nutrients, leading to a decrease in primary production in oceans (Huisman et al. 2006).

An increase in water temperature by global warming can also decrease the concentration of dissolved oxygen (O_2) in natural waters (Keeling et al. 2010; Epstein et al. 1993; Garcia et al. 1998; Sarmiento et al. 1998; Plattner et al. 2001; Bopp et al. 2002; Keeling and Garcia 2002; Matear and Hirst 2003). This could enhance the growth of cyanobacteria and other algae (Epstein et al. 1993) and/or decrease the growth of other organisms (Keeling et al. 2010). The decrease of dissolved O_2 in the upper surface layer would occur because of decreased O_2 solubility in warmer water and due to photoinduced generation of superoxide radical ion ($O_2^{\bullet-}$) and H_2O_2 (see chapter "Photoinduced and Microbial Generation of Hydrogen Peroxide and Organic Peroxides in Natural Waters" for detailed mechanism) by the effect of global warming. An increase in O_2 can enhance the production of H_2O_2 (Moffett and Zafiriou 1990) and different algae can show unalike responses to O_2 concentration (Pope 1975). This may for instance be linked to eutrophication from excess algal growth. The most prominent symptoms of eutrophication are oxygen depletion in bottom waters and harmful algal blooms (Richardson and Jorgensen 1996).

The decrease of dissolved O_2 in deeper waters would be caused by the decrease in vertical mixing of the water column due to the longer stratification period as a consequence of global warming. This effect can reduce the primary production as well as to survival of organisms in deeper water layers, particularly in lakes and oceans. Earlier studies did not provide any clear mechanisms about the decrease of dissolved O_2, which includes changes in ocean circulation rates (Bindoff and McDougall 2000; Shaffer et al. 2000; Emerson et al. 2001; Keller et al. 2001), in preformed values (Garcia et al. 1998), in Redfield ratios (Pahlow and Riebesell 2000), and in biological production (Emerson et al. 2001). The abundance and geographic distribution of toxin-producing algae is significantly increasing with respect to global warming and because of increased anthropogenic input of nutrients to aquatic environments (Shumway 1990; Harvell et al. 1999; Haines et al. 2000; vanDolah 2000; Shumway et al. 2003; Phlips et al. 2004; Yan and Zhou 2004; Luckas et al. 2005). The effects of global warming on waters are extensively discussed in the global warming chapter, which can help understanding the overall effects on photosynthesis and other key biogeochemical issues (see chapter "Impacts of Global Warming on Biogeochemical Cycles in Natural Waters").

5.11.1 Harmful Algal Blooms

The harmful algal blooms are presumably an effect of global warming on waters with high content of DOM and POM, as extensively discussed in earlier sections. Harmful algal blooms can cause loss of phytoplankton competitor motility, inhibition of photosynthesis, inhibition of enzymes, membrane damage, large fish kills, shellfish poisoning, deaths of livestock and wildlife, death of coral reefs and, finally, illness or even death in humans (Yates and Rogers 2011; Llewellyn 2006; Richardson 2007; Prince et al. 2008; Etheridge 2010; Harvell et al. 1999; Hallock and Schlager 1986; Hallegraeff 1993; Negri et al. 1995; Braun and Pfeiffer 2002; Landsberg 2002; Legrand et al. 2003). Autochthonous organic compounds (e.g. autochthonous fulvic acids) and nutrients are generally produced by algae or phytoplankton, either biologically (also termed allelopathy: a biological phenomenon by which an organism can produce various types of biochemicals, which can influence growth, survival, death, and reproduction of other organisms) or photolytically (see also chapters "Dissolved Organic Matter in Natural Waters", "Photoinduced and Microbial Degradation of Dissolved Organic Matter in Natural Waters", "Colored and Chromophoric Dissolved Organic Matter in Natural Waters" and "Fluorescent Dissolved Organic Matter in Natural Waters" for detailed description) (Mostofa et al. 2009, 2011; Prince et al. 2008; Zhang et al. 2009). In addition, various types of algae such as diatoms, dinoflagellates and cyanobacteria can produce toxins that can cause death of higher organisms (Castle and Rodgers 2009; Falconer 1993). Blooms of a red-tide dinoflagellate such as *Karenia brevis*, occurring in the coastal seawaters, and *Prymnesium parvum*, also known as golden algae, can produce neurotoxic compounds (brevetoxins) that can kill fish and accumulate in shellfish (Yates and Rogers 2011; Landsberg 2002; Southard et al. 2010; Tester et al. 1991). Moreover, autochthonous compounds and toxins produced during toxic algal blooms are susceptible to decrease the photosynthetic efficiency of natural waters.

6 Scope of the Future Researches

The mechanisms of the photosynthetic reaction and the changes of the photosynthetic efficiency of organisms are affected by the different factors discussed in this study. A number of issues may greatly assist to improve the present understanding of photosynthesis. For example, formation of complexes between metal ions and the functional groups of PSII or PSI is a new suggestion about the processes that might occur in aquatic environments. Earlier studies did not place much attention on the complexation theory, which may greatly assist a better understanding of similar researches. The effects of organic matter (DOM and POM) and of other factors on photosynthesis are important for understanding the mechanisms of the release of electrons and O_2, as well as other vital phenomena. The complexation theory may provide a better understanding of the molecular basis for the mechanisms of photosystem tolerance to salt or metal ions stress. If proven, such a theory may greatly help the introduction, by plant breeding and genetic engineering, of salt tolerance in crop plants.

The need for additional studies in photosynthesis can be summarized as follows: (i) Observations are required of the effects of diverse toxic and non-toxic organic substances and metals on efficiency of the photosynthesis of various microorganisms; (ii) The mechanism of relase of O_2 from H_2O_2 during photosynthesis should be substantiated; (iii) Special attention should be paid to the photosystems crystal structure, to find out any presence of H_2O_2 (or O_2) instead of H_2O. Concurrently, further studies are needed that take special attention during sample preparation, to avoid the possible decomposition of H_2O_2. Such studies would help solving the debate concerning the process of oxygen release; (iv) A better understanding is required of the effect on photosynthesis of exogenous H_2O_2, produced from allochthonous DOM (humic substances including fulvic and humic acids); (v) The effect of autochthonous DOM (autochthonous fulvic acids of algal/phytoplankton origin) on photosynthesis also requires better understanding; (vi) Crystal structures of PSI or PSII do not include any information about dissolved O_2, but they are composed of about 1,300 water molecules (Umena et al. 2011) and issue that will need further studies to be clarified. Finally, it is important to remember during the sample processing of any photosynthesis experiments that H_2O_2 is rapidly decomposed microbilly and it is also rapidly produced under light condition.

7 Nomenclature

CCM	Carbon-concentrating mechanism
Chl	Chlorophyll
CO_2	Carbon dioxide
DIC	Dissolved inorganic carbon = dissolved CO_2, H_2CO_3, HCO_3^-, and CO_3^{2-}
DOM	Dissolved organic matter

Ga	Gigaannum: 10^9 years
H_2O	Water
H_2O_2	Hydrogen peroxide
MAAs	Mycosporine-like amino acids
$^1O_2$	Singlet state of oxygen
$^3O_2$	Triplet state of oxygen
$O_2^{\bullet -}$	Superoxide radical anion
OEC	Oxygen-evolving complex
POM	Particulate organic matter
PSI	Photosystem I
PSII	Photosystem II
ROS	Reactive oxygen species
Rubisco	Ribulose Bisphosphate Carboxylase-Oxygenase
UV	Ultraviolet
WOC	Water-oxidizing complex

Problems

1. Define oxygenic photosynthesis and how does it differ from anoxygenic photosynthesis. Explain three key functions of photosynthesis in aquatic environments.
2. Define the key photosynthetic reactions under the hypotheses of H_2O and H_2O_2 involvement.
3. Which and how many Chl molecules can participate in the primary donor sites in PSI and PS II under illumination?
4. Explain the mechanism of H_2O_2 formation from chlorophyll bound in photosynthetic cells and ultrapure water under illumination.
5. Why are PSI and PSII composed of a number of chlorophyll molecules in their structures?
6. Explain the mechanism of electron transfer and O_2 release from PSII during photosynthesis.
7. Distinguish the various factors that influence photosynthesis. How do seasonal light cycle and temperature affect photosynthesis?
8. Why do precipitation/rainfall substantially enhance plant photosynthesis?
9. Explain the mechanism for the occurrence of algal (cyanobacterial) bloom in waters with high contents of DOM and POM. How does global warming accelerate the algal bloom in natural waters?
10. How do trace metal ions become toxic during phytoplankton photosynthesis? Explain the mechanism.
11. Explain how does salinity affect both plant and phytoplankton photosynthesis.
12. How can UV radiation affect phytoplankton photosynthesis?
13. How does metal toxicity impact organisms or induce cancer in humans?

Acknowledgments We thank Dr. Li Xiao-Dong of Institute of Geochemistry, Chinese Academy of Sciences, China for his assistance during the manuscript preparation. This work was financially supported by the Institute of Geochemistry, the Chinese Academy of Sciences, China. This work was partly supported by the Xinjiang Institute of Ecology and Geography, Chinese Academy of Sciences, China; Kyoto University, Japan; University of Turin, Italy; Brook Byers Institute for Sustainable Systems, Georgia Institute of Technology, United States and National Science Foundation of the United States award 0854416; Xiamen University, China; Hiroshima University, Japan; Semenov Institute of Chemical Physics, Russian Academy of Sciences, Russia. We also acknowledge the Copyright (1990) by the Association for the Sciences of Limnol Oceanogr, Inc.; PNAS Copyrighted (2010) by the National Academy of Sciences of USA; reprinted by permission from Macmillan Publishers Ltd: [Nature] (Jordan P, Fromme P, Witt HT, Klukas O, Saenger W, Krauß N, Three dimensional structure of cyanobacterial photosystem I at 25 Å – resolution, 411, 909–917), copyright (2001); reprinted by permission from Macmillan Publishers Ltd: [Nature] (Umena Y, Kawakami K, Shen J-R, Kamiya N, Crystal structure of oxygen-evolving photosystem II at a resolution of 1.9Å, 473, 55–61, Copyright (2011); Copyright (2008) by Pleiades Publishing, Ltd with Original Russian Text (2008) by AV Lobanov, NA Rubtsova, YuA Vedeneeva, GG Komissarov published in Doklady Akademii Nauk; and Yoshioka T, Phytoplanktonic carbon isotope fractionation: equations accounting for CO_2-concentrating mechanisms, J Plankton Res, 1997, 19 (10), 1455–1476, by permission of Oxford University Press.

References

Abboudi M, Surget S, Rontani J-F, Sempéré R, Joux F (2008) Physiological alteration of the marine bacterium *Vibrio angustum* s14 exposed to simulated sunlight during growth. Curr Microbiol 57:412–417

Abd-El Baki GK, Siefritz F, Man HM, Weiner H, Kaldenhoff R, Kaiser W (2000) Nitrate reductase in Zea mays L. under salinity. Plant, Cell Environ 23:515–521

Abrahams S, Collin R, Lipscomb W (1951) The crystal structure of hydrogen peroxide. Acta Cryst 4:15–20

Achakzai AKK, Kayani S, Hanif A (2010) Effect of salinity on uptake of micronutrients in sunflower at early vegetative stage. Pak J Bot 42:129–139

Adams Iii WW, Zarter CR, Ebbert V, Demmig-Adams B (2004) Photoprotective strategies of overwintering evergreens. Bioscience 54:41–49

Adams W III, Demmig-Adams B, Rosenstiel T, Brightwell A, Ebbert V (2002) Photosynthesis and photoprotection in overwintering plants. Plant Biol 4:545–557

Adir N, Zer H, Shochat S, Ohad I (2003) Photoinhibition–a historical perspective. Photosynth Res 76:343–370

Agusti S, Llabrés M (2007) Solar radiation-induced Mortality of marine pico-phytoplankton in the oligotrophic ocean. Photochem Photobiol 83:793–801

Ahel M, Barlow R, Mantoura R (1996) Effect of salinity gradients on the distribution of phytoplankton pigments in a stratified estuary. Mar Ecol Prog Ser 143:289–295

Ahmed CB, Rouina BB, Boukhris M (2008) Changes in water relations, photosynthetic activity and proline accumulation in one-year-old olive trees (*Olea europaea* L. cv. Chemlali) in response to NaCl salinity. Acta Physiol Plant 30:553–560

Alam MM, Hayat S, Ali B, Ahmad A (2007) Effect of 28-homobrassinolide treatment on nickel toxicity in Brassica juncea. Photosynthetica 45:139–142

Albright L, McCrae S (1987) Annual cycle of bacterial specific biovolumes in Howe sound, a Canadian west coast fjord sound. Appl Environ Microbiol 53:2739–2744

Aldridge FJ, Phlips EJ, Schelske CL (1995) The use of nutrient enrichment bioassays to test for spatial and temporal distribution of limiting factors affecting phytoplankton dynamics in Lake Okeechobee, Florida. Ergebn Limnol 45:177–190

Ali B, Hayat S, Ahmad A (2007) 28-Homobrassinolide ameliorates the saline stress in chickpea (*Cicer arietinum* L.). Environ Exp Bot 59:217–223

Ali B, Hasan S, Hayat S, Hayat Q, Yadav S, Fariduddin Q, Ahmad A (2008) A role for brassinosteroids in the amelioration of aluminium stress through antioxidant system in mung bean (*Vigna radiata* L. Wilczek). Environ Exp Bot 62:153–159

Alia Hayashi H, Chen T, Murata N (1998) Transformation with a gene for choline oxidase enhances the cold tolerance of Arabidopsis during germination and early growth. Plant, Cell Environ 21:232–239

Allakhverdiev SI, Murata N (2004) Environmental stress inhibits the synthesis de novo of proteins involved in the photodamage–repair cycle of photosystem II in *Synechocystis* sp. PCC 6803. Biochim Biophys Acta 1657:23–32

Allakhverdiev SI, Murata N (2008) Salt stress inhibits photosystems II and I in cyanobacteria. Photosynth Res 98:529–539

Allakhverdiev SI, Nishiyama Y, Suzuki I, Tasaka Y, Murata N (1999) Genetic engineering of the unsaturation of fatty acids in membrane lipids alters the tolerance of *Synechocystis* to salt stress. PNAS 96:5862

Allakhverdiev SI, Sakamoto A, Nishiyama Y, Inaba M, Murata N (2000a) Ionic and osmotic effects of NaCl-induced inactivation of photosystems I and II in Synechococcus sp. Plant Physiol 123:1047–1056

Allakhverdiev SI, Sakamoto A, Nishiyama Y, Murata N (2000b) Inactivation of photosystems I and II in response to osmotic stress in Synechococcus. Contribution of water channels. Plant Physiol 122:1201–1208

Allakhverdiev SI, Klimov VV, Hagemann M (2005) Cellular energization protects the photosynthetic machinery against salt-induced inactivation in *Synechococcus*. Biochim Biophys Acta 1708:201–208

Allen DJ, Ort DR (2001) Impacts of chilling temperatures on photosynthesis in warm-climate plants. Trends Plant Sci 6:36–42

Alpaslan M, Guenes A, Taban S, Erdal I, Tarakcioglu C (1998) Variations in calcium, phosphorus, iron, copper, zinc and manganese contents of wheat and rice varieties under salt stress. Turkish Journal of Agriculture and Forestry 22:227–234

Alscher RG, Donahue JL, Cramer CL (1997) Reactive oxygen species and antioxidants: relationships in green cells. Physiol Plant 100:224–233

Amador JA, Alexander M, Zika RG (1989) Sequential photochemical and microbial degradation of organic molecules bound to humic acid. Appl Environ Microbiol 55:2843–2849

Amunts A, Toporik H, Borovikova A, Nelson N (2010) Structure determination and improved model of plant photosystem I. J Biol Chem 285:3478

Anan'ev G, Klimov VV (1988) Doklady Akademii Nauk SSSR 298:1007–1011

Anderson JA (2002) Catalase activity, hydrogen peroxide content and thermotolerance of pepper leaves. Sci Horticult 95:277–284

Anderson MA, Morel FMM (1982) The influence of aqueous iron chemistry on the uptake of iron by the coastal diatom Thalassiosira weissflogii. Limnol Oceanogr 27:789–813

Andersson B, Aro EM (2001) Photodamage and D1 protein turnover in photosystem I photodamage and D1 protein turnover in photosystem II In: AroE-Mand Andersson B(Eds), Regulation of photosynthesis, Kluwer Academic Publishers, Dordrecht, pp 377–393

Anesio AM, Granéli W, Aiken GR, Kieber DJ, Mopper K (2005) Effect of humic substance photodegradation on bacterial growth and respiration in lake water. Appl Environ Microbiol 71:6267–6275

Apel K, Hirt H (2004) Reactive oxygen species: metabolism, oxidative stress, and signal transduction. Annu Rev Plant Biol 55:373–399

Appelo CAJ, Postma D (2005) Geochemistry, ground water, and pollution. 2nd ed. A.A. Balkema, Leiden, the Netherlands

Appenroth KJ, Stöckel J, Srivastava A, Strasser R (2001) Multiple effects of chromate on the photosynthetic apparatus of Spirodela polyrhiza as probed by OJIP chlorophyll a fluorescence measurements. Environ Pollut 115:49–64

Arenovski AL, Lim EL, Caron DA (1995) Mixotrophic nanoplankton in oligotrophic surface waters of the Sargasso Sea may employ phagotrophy to obtain major nutrients. J Plankton Res 17:801–820
Arnelle DR, O'Leary MH (1992) Binding of carbon dioxide to phosphoenolpyruvate carboxykinase deduced from carbon kinetic isotope effects. Biochemistry 31:4363–4368
Arnon DI (1949) Copper enzymes in isolated chloroplasts polyphenoloxidase in Beta vulgaris. Plant Physiol 24:1
Arnon DI (1959) Conversion of light into chemical energy in photosynthesis. Nature 184:10–21
Arnon DI (1961) Cell-free photosynthesis and the energy conversion process. In: McElroy WD, Glass B (eds) Light and life. The Johns Hopkins Press, Baltimore Maryland, pp 489–566
Arnon DI (1971) The light reactions of photosynthesis. Nat Acad Sci USA 68:2883–2892
Vass I, Aro, E.M. (2007) Photoinhibition of photosynthetic electron transport. In: Renger G (Ed), Primary processes in photosynthesis, comprehensive series in photochemical and photobiological sciences RSC Publishing, The Royal Society of Chemistry, Cambridge, UK, pp 393–425 part 1
Aro EM, Virgin I, Andersson B (1993) Photoinhibition of photosystem II. Inactivation, protein damage and turnover. Biochim Biophys Acta 1143:113–134
Arrigo KR (1994) Impact of ozone depletion on phytoplankton growth in the Southern Ocean: large-scale spatial and temporal variability. Marine Ecology-Progress Series 114:1–1
Arvola L, Kankaala P (1989) Winter and spring variability in phyto- and bacterioplankton in lakes with different water colour. Aqua Fenn 19:29–39
Asada K (1992) Ascorbate peroxidase–a hydrogen peroxide-scavenging enzyme in plants. Physiol Plant 85:235–241
Asada K (1999) The water–water cycle in chloroplasts: scavenging of active oxygens and dissipation of excess photons. Annu Rev Plant Biol 50:601–639
Asada K (2000) The water–water cycle as alternative photon and electron sinks. Phil Trans R Soc Lond B 355:1419–1431
Asada K (2006) Production and scavenging of reactive oxygen species in chloroplasts and their functions. Plant Physiol 141:391–396
Asada K, Badger MR (1984) Photoreduction of $^{18}O_2$ and $H_2{}^{18}O_2$ with concomitant evolution of 16O2 in intact spinach chloroplasts: evidence for scavenging of hydrogen peroxide by peroxidase. Plant Cell Physiol 25:1169–1179
Asada K, Takahashi M (1987) Production and scavenging of active oxygen in photosynthesis. In: Kyle DJ, Osmond CB, Arntzen CJ (eds) Photoinhibition. Elsevier, Amsterdam, pp 227–287
Asada K, Kiso K, Yoshikawa K (1974) Univalent reduction of molecular oxygen by spinach chloroplasts on illumination. J Biol Chem 249:2175–2181
Ashraf M, Wu DL (1994) Breeding for salinity tolerance in plants. Crit Rev Plant Sci 13:17–42
Ashraf M, Karim F, Rasul E (2002) Interactive effects of gibberellic acid (GA 3) and salt stress on growth, ion accumulation and photosynthetic capacity of two spring wheat (Triticum aestivum L.) cultivars differing in salt tolerance. Plant Growth Regul 36:49–59
Aspinall-O'Dea M, Wentworth M, Pascal A, Robert B, Ruban A, Horton P (2002) In vitro reconstitution of the activated zeaxanthin state associated with energy dissipation in plants. PNAS 99:16331
Assel M, Laenen R, Laubereau A (1998) Ultrafast electron trapping in an aqueous NaCl-solution. Chem Phys Lett 289:267–274
Auclair J, Brassard P, Couture P (1985) Total dissolved phosphorus: effects of two molecular weight fractions on phosphorus cycling in natural phytoplankton communities. Water Res 19:1447–1453
Aziz A, Larher F (1998) Osmotic stress induced changes in lipid composition and peroxidation in leaf discs of *Brassica napus* L. J Plant Physiol 153:754–762
Babin M, Morel A, Claustre H, Bricaud A, Kolber Z, Falkowski PG (1996) Nitrogen-and irradiance-dependent variations of the maximum quantum yield of carbon fixation in eutrophic, mesotrophic and oligotrophic marine systems. Deep Sea Res Part I 43:1241–1272
Bach A (1893) Mon Sci 7:669
Bach A (1894) Mon Sci 8:241

Bader KP (1994) Physiological and evolutionary aspects of the O_2/H_2O_2-cycle in cyanobacteria. Biochim Biophys Acta 1188:213–219
Bader K, Schmid G (1988) Mass spectrometric analysis of a photosystem-II-mediated oxygen uptake phenomenon in the filamentous cyanobacterium, Oscillatoria chalybea. Biochim Biophys Acta 936:179–186
Bader K, Schmid G (1989) Photosynthetic and respiratory oxygen gas exchange measured by mass spectrometry in the filamentous cyanobacterium Oscillatoria chalybea in dependence on the nitrogen source in the growth medium. Biochim Biophys Acta 974:303–310
Badger MR, Price GD (1992) The CO_2 concentrating mechanism in cyanobactiria and microalgae. Physiol Plant 84:606–615
Baena-González E, Aro EM (2002) Biogenesis, assembly and turnover of photosystem II units. Phil Trans R Soc Lond B 357:1451-1460
Baker A, Graham IA (2002) Plant peroxisomes: biochemistry, cell biology, and biotechnological applications. Kluwer Academic Publishers, Dordrecht
Ballaré CL, Caldwell MM, Flint SD, Robinson S, Bornman JF (2011) Effects of solar ultraviolet radiation on terrestrial ecosystems. Patterns, mechanisms, and interactions with climate change. Photochem Photobiol Sci 10:226–241
Bandyopadhyay U, Das D, Banerjee RK (1999) Reactive oxygen species: oxidative damage and pathogenesis. Curr Sci 77:658–666
Barbieri ES, Villafane VE, Helbling EW (2002) Experimental assessment of UV effects on temperate marine phytoplankton when exposed to variable radiation regimes. Limnol Oceanogr 47:1648–1655
Barica J, Kling H, Gibson J (1980) Experimental manipulation of algal bloom composition by nitrogen addition. Can J Fish Aquat Sci 37:1175–1183
Barker HA (1935) Photosynthesis in diatoms. Arch Microbiol 6:141–156
Barraza J, Carballeira A (1999) Chlorophyll fluorescence analysis and cadmium-copper bioaccumulation in Ulva rigida (C. Agardh). Boletín del Instituto Español de Oceanografía 15:395–399
Barter LMC, Durrant JR, Klug DR (2003) A quantitative structure-function relationship for the photosystem II reaction center: supermolecular behavior in natural photosynthesis. PNAS 100:946–951
Barua D, Downs CA, Heckathorn SA (2003) Variation in chloroplast small heat-shock protein function is a major determinant of variation in thermotolerance of photosynthetic electron transport among ecotypes of Chenopodium album. Funct Plant Biol 30:1071–1079
Bassham JA, Barker SA, Calvin M, Quarck UC (1956) Intermediates in the photosynthetic cycle. Biochim Biophys Acta 21:376–377
Baulch H, Schindler D, Turner M, Findlay D, Paterson M, Vinebrooke R (2005) Effects of warming on benthic communities in a boreal lake: implications of climate change. Limnol Oceanogr 50:1377–1392
Bay JC (1931) Jean Senebier. Plant Physiol 6:188
Bazanov MI, Berezin BD, Berezin DB et al (1999) Uspekhi khimii porfirinov (Progress in the Chemistry of Porphyrins). NII khimii SPbGU, St Petersburg
Raven J, Beardall, J (2003) Carbohydrate metabolism and respiration in algae. In: Larkum AWD, Douglas SE, Raven JA (Eds), Photosynthesis in Algae Advances in Photosynthesis and Respiration, Springer, Dordrecht, 14:205-224
Beardall J, Sobrino C, Stojkovic S (2009) Interactions between the impacts of ultraviolet radiation, elevated CO2, and nutrient limitation on marine primary producers. Photochem Photobiol Sci 8:1257–1265
Behrenfeld M, Hardy J, Gucinski H, Hanneman A, Lee H, Wones A (1993) Effects of ultraviolet-B radiation on primary production along latitudinal transects in the South Pacific Ocean. Mar Environ Res 35:349–363
Behrenfeld MJ, Prasil O, Kolber ZS, Babin M, Falkowski PG (1998) Compensatory changes in photosystem II electron turnover rates protect photosynthesis from photoinhibition. Photosynth Res 58:259–268

Behrenfeld MJ, O'Malley RT, Siegel DA, McClain CR, Sarmiento JL, Feldman GC, Milligan AJ, Falkowski PG, Letelier RM, Boss ES (2006) Climate-driven trends in contemporary ocean productivity. Nature 444:752–755

Belkhodja R, Morales F, Abadia A, Gomez-Aparisi J, Abadia J (1994) Chlorophyll fluorescence as a possible tool for salinity tolerance screening in barley (*Hordeum vulgare* L.). Plant Physiol 104:667–673

Ben-Shem A, Frolow F, Nelson N (2003) Crystal structure of plant photosystem I. Nature 426:630–635

Berden-Zrimec M, Drinovec L, Zrimec A, Tišler T (2007) Delayed fluorescence in algal growth inhibition tests. Central Eur J Biol 2:169–181

Berg G, Balode M, Purina I, Bekere S, Béchemin C, Maestrini S (2003) Plankton community composition in relation to availability and uptake of oxidized and reduced nitrogen. Aquatic Microb Ecol 30:263–274

Bergantino E, Segalla A, Brunetta A, Teardo E, Rigoni F, Giacometti GM, Szabò I (2003) Light-and pH-dependent structural changes in the PsbS subunit of photosystem II. PNAS 100:15265

Berman T, Pollingher U (1974) Annual and seasonal variations of phytoplankton, chlorophyll, and photosynthesis in Lake Kinneret. Limnol Oceanogr:31-54

Berman-Frank I, Erez J, Kaplan A (1998) Changes in inorganic carbon uptake during the progression of a dinoflagellate bloom in a lake ecosystem. Can J Bot 76:1043–1051

Bernardini G, Zhao C, Wedd AG, Bond AM (2011) Ionic liquid-enhanced photooxidation of water using the polyoxometalate Anion [P2W18O62] 6–as the sensitizer. Inorg Chem 50:5899–5909

Berry JA (1988) Studies of mechanisms affecting the fractionation of carbon isotopes in photosynthesis In: Rundel PW, Ehleringer JR, Nagy KA (Eds), Stable Isotopes in Ecological Research Ecological Studie, VoL 68 Springer-Verlag, New York, pp 82-94.

Berry J, Bjorkman O (1980) Photosynthetic response and adaptation to temperature in higher plants. Annu Rev Plant Physiol 31:491–543

Bertamini M, Muthuchelian K, Rubinigg M, Zorer R, Velasco R, Nedunchezhian N (2006) Low-night temperature increased the photoinhibition of photosynthesis in grapevine (*Vitis vinifera* L. cv. Riesling) leaves. Environ Exp Bot 57:25–31

Bertilsson S, Tranvik LJ (2000) Photochemical transformation of dissolved organic matter in lakes. Limnol Oceanogr 45:753–762

Bertoni R, Callieri C, Corno G, Rasconi S, Caravati E, Contesini M (2010) Long-term trends of epilimnetic and hypolimnetic bacteria and organic carbon in a deep holo-oligomictic lake. Hydrobiologia 644:279–287

Bertoni R, Jeffrey WH, Pujo-Pay M, Oriol L, Conan P, Joux F (2011) Influence of water mixing on the inhibitory effect of UV radiation on primary and bacterial production in Mediterranean coastal water. Aquat Sci 73:377–387

Bhagwat A, Apte SK (1989) Comparative analysis of proteins induced by heat shock, salinity, and osmotic stress in the nitrogen-fixing cyanobacterium Anabaena sp. strain L-31. J Bacteriol 171:5187–5189

Bhuyan KC, Bhuyan DK (1977) Regulation of hydrogen peroxide in eye humors effect of 3-amino-1H-1, 2, 4-triazole on catalase and glutathione peroxidase of rabbit eye. Biochim Biophys Acta 497:641–651

Bialojan C, Takai A (1988) Inhibitory effect of a marine-sponge toxin, okadaic acid, on protein phosphatases. Specificity and kinetics. Biochem J 256:283

Biddanda B, Benner R (1997) Carbon, nitrogen, and carbohydrate fluxes during the production of particulate and dissolved organic matter by marine phytoplankton. Limnol Oceanogr 42:506–518

Bielski BHJ, Cabelli DE, Arudi RL, Ross AB (1985) Reactivity of HO/O radicals in aqueous solution. J Phys Chem Ref Data 14:1041–1100

Biesiadka J, Loll B, Kern J, Irrgang K-D, Zouni A (2004) Crystal structure of cyanobacterial photosystem II at 3.2 Å resolution: a closer look at the Mn-cluster. Phys Chem Chem Phys 6(20):4733–4736

Bindoff NL, McDougall TJ (2000) Decadal changes along an Indian Ocean section at 32 S and their interpretation. J Phys Oceanogr 30:1207–1222

Bird DF, Kalff J (1986) Bacterial grazing by planktonic lake algae. Science 231:493–495

Bjørnsen PK, Riemann B, Pock-Steen J, Nielsen TG, Horsted SJ (1989) Regulation of bacterioplankton production and cell volume in a eutrophic estuary. Appl Environ Microbiol 55:1512–1518

Blackman F (1905) Optima and limiting factors. Ann Bot 19:281–295

Blackman FF, Matthaei GLC (1905) Experimental researches in vegetable assimilation and respiration. IV—a quantitative study of carbon-dioxide assimilation and leaf-temperature in natural illumination. Proc R Soc London 76:402–460

Blanco AC, Nadaoka K, Yamamoto T (2008) Planktonic and benthic microalgal community composition as indicators of terrestrial influence on a fringing reef in Ishigaki Island, Southwest Japan. Mar Environ Res 66:520–535

Blankenship RE, Hartman H (1998) The origin and evolution of oxygenic photosynthesis. Trends Biochem Sci 23:94–97

Blomqvist P, Petterson A, Hyenstrand P (1994) Ammonium-nitrogen: a key regulatory factor causing dominance of non-nitrogen-fixing cyanobacteria in aquatic systems. Arch Hydrobiol 132:141–164

Blunt JW, Copp BR, Hu WP, Munro MHG, Northcote PT, Prinsep MR (2007) Marine natural products. Nat Prod Rep 24:31–86

Boekema E, Dekker J, van Heel M, Rögner M, Saenger W, Witt I, Witt H (1987) Evidence for a trimeric organization of the photosystem I complex from the thermophilic cyanobacterium *Synechococcus* sp. FEBS Lett 217:283–286

Boekema EJ, Jensen PE, Schlodder E, van Breemen JFL, van Roon H, Scheller HV, Jan P (2001) Green plant photosystem I binds light-harvesting complex I on one side of the complex. Biochemistry 40:1029–1036

Bongi G, Long S (1987) Light-dependent damage to photosynthesis in olive leaves during chilling and high temperature stress. Plant, Cell Environ 10:241–249

Bongi G, Loreto F (1989) Gas-exchange properties of salt-stressed olive (*Olea europea* L.) leaves. Plant Physiol 90:1408–1416

Booij-James IS, Swegle WM, Edelman M, Mattoo AK (2002) Phosphorylation of the D1 photosystem II reaction center protein is controlled by an endogenous circadian rhythm. Plant Physiol 130:2069–2075

Bopp L, Le Quéré C, Heimann M, Manning AC, Monfray P (2002) Climate-induced oceanic oxygen fluxes: Implications for the contemporary carbon budget. Glob Biogeochem Cy 16:1022

Borbely G, Suranyi G (1988) Cyanobacterial heat-shock proteins and stress responses. Methods Enzymol 167:622–629

Borges A, Ruddick K, Schiettecatte LS, Delille B (2008) Net ecosystem production and carbon dioxide fluxes in the Scheldt estuarine plume. BMC Ecol 8, doi:101186/1472-6785-8-15

Bouchard JN, Roy S, Campbell DA (2006) UVB Effects on the photosystem II-D1 protein of phytoplankton and natural phytoplankton communities. Photochem Photobiol 82:936–951

Boucher NP, Prezelin BB (1996) Spectral modeling of UV Inhibition of In Situ Antarctic primary production using a field-derived biological weighting function. Photochem Photobiol 64:407–418

Boucher N, Harnois J, Carpentier R (1990) Heat-stress stimulation of electron flow in a photosystem I submembrane fraction. Biochem Cell Biol 68:999–1004

Bouman HA, Nakane T, Oka K, Nakata K, Kurita K, Sathyendranath S, Platt T (2010) Environmental controls on phytoplankton production in coastal ecosystems: a case study from Tokyo Bay. Estuar Coast Shelf Sci 87:63–72

Boussaad S, Tazi A, Leblanc R (1997) Chlorophyll a dimer: a possible primary electron donor for the photosystem II. PNAS 94:3504

Boyd P, Crossley A, DiTullio G, Griffiths F, Hutchins D, Queguiner B, Sedwick P, Trull T (2001) Control of phytoplankton growth by iron supply and irradiance in the subantarctic Southern Ocean: experimental results from the SAZ project. J Geophys Res 106:573–531

Braakhekke MC, Beer C, Hoosbeek MR, Reichstein M, Kruijt B, Schrumpf M, Kabat P (2011) SOMPROF: a vertically explicit soil organic matter model. Ecol Model 222:1712–1730
Bradley RL, Long KM, Frasch WD (1991) The involvement of photosystem II-generated H_2O_2 in photoinhibition. FEBS Lett 286:209–213
Bratbak G, Jacobsen A, Heldal M (1998) Viral lysis of phaeocystis pouchetii and bacterial secondary production. Aquatic Microb Ecol 16:11–16
Braun AM, Oliveros E (1990) Applications of singlet oxygen reactions: mechanistic and kinetic investigations. Pure Appl Chem 62:1467–1476
Braun A, Pfeiffer T (2002) Cyanobacterial blooms as the cause of a Pleistocene large mammal assemblage. Paleobiology 28:139–154
Brettel K (1997) Electron transfer and arrangement of the redox cofactors in photosystem I. Biochim Biophys Acta 1318:322–373
Bronk DA (2002) Dynamics of DON. In: Carlson CA (ed) Hansell DA. Academic Press, Biogeochemistry of Marine Dissolved Organic Matter, pp 153–249
Brown SB HJ, Hendry GAF (1991) Chlorophyll breakdown. In: Scheer H (Ed), Chlorophylls, pp 465–489
Brown CD, Hoyer MV, Bachmann RW, Canfield DE Jr (2000) Nutrient-chlorophyll relationships: an evaluation of empirical nutrient-chlorophyll models using Florida and north-temperate lake data. Can J Fish Aquat Sci 57:1574–1583
Brugnoli E, Björkman O (1992) Growth of cotton under continuous salinity stress: influence on allocation pattern, stomatal and non-stomatal components of photosynthesis and dissipation of excess light energy. Planta 187:335–347
Bruland KW, Rue EL, Smith GJ (2001) Iron and macronutrients in California coastal upwelling regimes: implications for diatom blooms. Limnol Oceanogr 46:1661–1674
Brussaard C, Gast G, van Duyl F, Riegman R (1996) Impact of phytoplankton bloom magnitude on a pelagic microbial food web. Mar Ecol Prog Ser 14:211–221
Brussaard C, Mari X, Bleijswijk JDLV, Veldhuis M (2005) A mesocosm study of *Phaeocystis globosa* (*Prymnesiophyceae*) population dynamics: II. Significance for the microbial community. Harmful Algae 4:875–893
Brussaard CPD, Bratbak G, Baudoux AC, Ruardij P (2007) Phaeocystis and its interaction with viruses. Biogeochemistry 83:201–215
Brussaard CPD, Wilhelm SW, Thingstad F, Weinbauer MG, Bratbak G, Heldal M, Kimmance SA, Middelboe M, Nagasaki K, Paul JH (2008) Global-scale processes with a nanoscale drive: the role of marine viruses. ISME J 2:575–578
Buick R (2008) When did oxygenic photosynthesis evolve? Phil Trans R Soc B 363:2731–2743
Bukhov NG, Carpentier R (2000) Heterogeneity of photosystem II reaction centers as influenced by heat treatment of barley leaves. Physiol Plant 110:279–285
Bukhov NG, Wiese C, Neimanis S, Heber U (1999) Heat sensitivity of chloroplasts and leaves: leakage of protons from thylakoids and reversible activation of cyclic electron transport. Photosynth Res 59:81–93
Bukhov NG, Samson G, Carpentier R (2000) Nonphotosynthetic reduction of the intersystem electron transport chain of chloroplasts following heat stress. Steady-state Rate. Photochem Photobiol 72:351–357
Buma AGJ, Boelen P, Jeffrey WH (2003) UVR-induced DNA damage in aquatic organisms. In: Helbling EW, Zagarese HE (eds) UV Effects in aquatic organisms and ecosystems. The Royal Society of Chemistry, Cambridge, pp 291–327
Bums B, Beardall J (1987) Utilization of inorganic carbon by marine micToalgae. Exp Mar Biol Ecol 107:75–86
Burda K, Kruk J, Strzalka K, Schmid G (2002) Stimulation of oxygen evolution in photosystem II by copper (II) ions. Z Naturforsch C 57:853–857
Burda K, Kruk J, Schmid GH, Strzalka K (2003) Inhibition of oxygen evolution in photosystem II by Cu (II) ions is associated with oxidation of cytochrome b559. Biochem J 371:597
Burja AM, Banaigs B, Abou-Mansour E, Grant Burgess J, Wright PC (2001) Marine cyanobacteria–a prolific source of natural products. Tetrahedron 57:9347–9377

Burton GW, Ingold K (1984) Beta-carotene: an unusual type of lipid antioxidant. Science 224:569–573

Bushaw KL, Zepp RG, Tarr MA, Schulz-Jander D, Bourbonniere RA, Hodson RE, Miller WL, Bronk DA, Moran MA (1996) Photochemical release of biologically available nitrogen from aquatic dissolved organic matter. Nature 381:404–407

Bybordi A (2010) Effects of salinity and n on the growth, photosynthesis and N status of Canola (Brassica napus L.). Not Sci Biol 2:92–97

Bybordi A, Jalal Tabatabaei S, Ahmedov A (2010a) Effects of salinity on fatty acid composition of canola (*Brassica napus* L.). Int J Food Agric Environ 8:113–115

Bybordi A, Jalal Tabatabaei S, Ahmedov A (2010b) Effect of salinity on the growth and peroxidase and IAA oxidase activities in canola. Int J Food Agric Environ 8:109–112

Bybordi A, Jalal Tabatabaei S, AHMEDOV A (2010c) The influence of salinity stress on antioxidant activity in canola cultivars (Brassica napus L.). Int J Food Agric Environ 8:122–127

Byrdin M, Jordan P, Krauss N, Fromme P, Stehlik D, Schlodder E (2002) Light harvesting in photosystem I: modeling based on the 2.5 Å structure of photosystem I from Synechococcus elongatus. Biophys J 83:433–457

Cai WJ (2011) Coastal ocean carbon paradox: CO_2 sinks or sites of terrestrial carbon incineration. Annu Rev Mar Sci 3:123–145

Calbet A, Landry MR (2004) Phytoplankton growth, microzooplankton grazing, and carbon cycling in marine systems. Limnol Oceanogr 49:51–57

Callieri C, Morabito G, Huot Y, Neale PJ, Litchman E (2001) Photosynthetic response of pico-and nanoplanktonic algae to UVB, UVA and PAR in a high mountain lake. Aquat Sci 63:286–293

Callieri C, Corno G, Caravati E, Rasconi S, Contesini M, Bertoni R (2009) Bacteria, Archaea, and Crenarchaeota in the epilimnion and hypolimnion of a deep holo-oligomictic lake. Appl Environ Microbiol 75:7298–7300

Calvin M (1956) The photosynthetic cycle. Bull Soc Chim Biol 38:1233–1244

Campbell JE, Fourqurean JW (2009) Interspecific variation in the elemental and stable isotope content of seagrasses in South Florida. Mar Ecol Prog Ser 387:109–123

Campbell D, Eriksson MJ, Öquist G, Gustafsson P, Clarke AK (1998) The cyanobacterium synechococcus resists UV-B by exchanging photosystem II reaction-center D1 proteins. PNAS 95:364

Canfield DE Jr (1983) Prediction of chlorophyll-a concentrations in Florida lakes: the importance of phosphorus and nitrogen. Water Res Bull 19:255–262

Canini A, Leonardi D, Caiola MG (2001) Superoxide dismutase activity in the cyanobacterium microcystis aeruginosa after surface bloom formation. New Phytol 152:107–116

Cao S, Xu Q, Cao Y, Qian K, An K, Zhu Y, Binzeng H, Zhao H, Kuai B (2005) Loss-of-function mutations in DET2 gene lead to an enhanced resistance to oxidative stress in arabidopsis. Physiol Plant 123:57–66

Caramori LPC, Caramori PH, Manetti Filho J (2002) Effect of leaf water potential on cold tolerance of coffea arabica L. Braz Arch Biol Technol 45:439–443

Caron D (2000) Symbiosis and mixotrophy among pelagic microorganisms. In: Kirchman DL (ed) Microbial ecology of the oceans. Wiley-Liss, New York, pp 495–523

Carpenter SR, Cole JJ, Kitchell JF, Pace ML (1998) Impact of dissolved organic carbon, phosphorus, and grazing on phytoplankton biomass and production in experimental lakes. Limnol Oceanogr 43:73–80

Carrillo P, Medina-Sánchez JM, Villar-Argaiz M (2002) The interaction of phytoplankton and bacteria in a high mountain lake: importance of the spectral composition of solar radiation. Limnol Oceanogr 47:1294–1306

CAST-2004 (2004) Emissions and mitigation of agricultural greenhouse gases. Climate change and greenhouse gas mitigation: challenges and opportunities for agriculture. Council for Agricultural Science and Technology (CAST), Ames

Castle JW, Rodgers JH Jr (2009) Hypothesis for the role of toxin-producing algae in phanerozoic mass extinctions based on evidence from the geologic record and modern environments. Environ Geosci 16:1–23

Catalan J, Perez P, del Valle J, de Paz J, Kasha M (2004) H-bonded N-heterocyclic base-pair phototautomerizational potential barrier and mechanism: the 7-azaindole dimer. PNAS 101:419–422

Catling DC, Claire MW (2005) How Earth's atmosphere evolved to an oxic state: a status report. Earth Planet Sci Lett 237:1–20

Catling DC, Glein CR, Zahnle KJ, McKay CP (2005) Why O2 Is required by complex life on habitable planets and the concept of planetary" Oxygenation Time". Astrobiology 5:415–438

Centritto M, Loreto F, Chartzoulakis K (2003) The use of low [CO_2] to estimate diffusional and non-diffusional limitations of photosynthetic capacity of salt-stressed olive saplings. Plant, Cell Environ 26:585–594

Cermeno P, Estévez-Blanco P, Maranón E, Fernández E (2005) Maximum photosynthetic efficiency of size-fractionated phytoplankton assessed by 14 C uptake and fast repetition rate fluorometry. Limnol Oceanogr 50:1438–1446

Chance B, Sies H, Boveris A (1979) Hydroperoxide metabolism in mammalian organs. Physiol Rev 59:527–605

Chen THH, Murata N (2002) Enhancement of tolerance of abiotic stress by metabolic engineering of betaines and other compatible solutes. Curr Opin Plant Biol 5:250–257

Chen Y, Qin B, Teubner K, Dokulil MT (2003) Long-term dynamics of phytoplankton assemblages: microcystis-domination in Lake Taihu, a large shallow lake in China. J Plankton Res 25:445–453

Chen CL, Hsu LI, Chiou HY, Hsueh YM, Chen SY, Wu MM, Chen CJ (2004) Ingested arsenic, cigarette smoking, and lung cancer risk. JAMA 292:2984–2990

Chisholm SW (1992) Phytoplankton size. In: Falkowski PG, Woodhead AD (eds) Primary productivity and biogeochemical cycles in the sea. Plenum Press, New York, pp 213–237

Christensen H, Sehested K, Corfitzen H (1982) Reactions of hydroxyl radicals with hydrogen peroxide at ambient and elevated temperatures. J Phys Chem 86:1588–1590

Chrzanowski TH, Crotty R, Hubbard G (1988) Seasonal variation in cell volume of epilimnetic bacteria. Microb Ecol 16:155–163

Chu HA, Nguyen AP, Debus RJ (1995) Amino acid residues that influence the binding of manganese or calcium to photosystem II. 1. The lumenal interhelical domains of the D1 polypeptide. Biochemistry 34:5839–5858

Çjánek M, Stroch M, Lachetová I, Kalina J, Spunda V (1998) Characterization of the photosystem II inactivation of heat-stressed barley leaves as monitored by the various parameters of chlorophyll a fluorescence and delayed fluorescence. J Photochem Photobiol, B 47:39–45

Cloern JE (1996) Phytoplankton bloom dynamics in coastal ecosystems: a review with some general lessons from sustained investigation of San Francisco Bay, California. Rev Geophys 34:127–168

Cloern JE, Jassby AD (2008) Complex seasonal patterns of primary producers at the land–sea interface. Ecol Lett 11:1294–1303

Closs G, Katz J, Pennington F, Thomas M, Strain H (1963) Nuclear magnetic resonance spectra and molecular association of chlorophylls a and b, methyl chlorophyllides, pheophytins, and methyl pheophorbides. JACS 85:3809–3821

Coesel S, Oborník M, Varela J, Falciatore A, Bowler C (2008) Evolutionary origins and functions of the carotenoid biosynthetic pathway in marine diatoms. PLoS ONE 3:e2896

Cole JJ, Likens GE, Strayer DL (1982) Photosynthetically produced dissolved organic carbon: an important carbon source for planktonic bacteria. Limnol Oceanogr 27:1080–1090

Cooke SL, Williamson CE, Hargreaves BR, Morris DP (2006) Beneficial and detrimental interactive effects of dissolved organic matter and ultraviolet radiation on zooplankton in a transparent lake. Hydrobiologia 568:15–28

Cooper W, Lean D (1992) Hydrogen peroxide dynamics in marine and fresh water systems. Encyclop Earth System Sci, Academic Press Inc 2:527–535

Cooper L, McRoy C (1988) Stable carbon isotope ratio variations in marine macrophytes along intertidal gradients. Oecologia 77:238–241

Cooper WJ, Zika RG, Petasne RG, Plane JMC (1988) Photochemical formation of hydrogen peroxide in natural waters exposed to sunlight. Environ Sci Technol 22:1156–1160

Cornic G (2000) Drought stress inhibits photosynthesis by decreasing stomatal aperture-not by affecting ATP synthesis. Trends Plant Sci 5:187–188
Cornillon P, Palloix A (1997) Influence of sodium chloride on the growth and mineral nutrition of pepper cultivars. J Plant Nutrition 20:1085–1094
Costa PHA, Neto ADA, Bezerra MA, Prisco JT, Gomes-Filho E (2005) Antioxidant-enzymatic system of two sorghum genotypes differing in salt tolerance. Brazilian J Plant Physiol 17:353–361
Cramer M, Lips S (1995) Enriched rhizosphere CO_2 concentrations can ameliorate the influence of salinity on hydroponically grown tomato plants. Physiol Plant 94:425–432
Crist RH, Oberholser K, Shank N, Nguyen M (1981) Nature of bonding between metallic ions and algal cell walls. Environ Sci Technol 15:1212–1217
Crowe JH, Spargo BJ, Crowe LM (1987) Preservation of dry liposomes does not require retention of residual water. PNAS 84:1537–1540
Crowe JH, Hoekstra FA, Crowe LM (1992) Anhydrobiosis. Annu Rev Physiol 54:579–599
Cullen JJ, Neale PJ (1994) Ultraviolet radiation, ozone depletion, and marine photosynthesis. Photosynth Res 39:303–320
Cullen JJ, Yang X, MacIntyre HL (1992) Nutrient limitation of marine photosynthesis. In: Woodhead AD (ed) Falkowski PG. Primary Productivity and Biogeochemical Cycles in the Sea Plenum Press, New York, pp 69–88
D'Ambrosio N, Arena C, de Santo AV (2006) Temperature response of photosynthesis, excitation energy dissipation and alternative electron sinks to carbon assimilation in *Beta vulgaris* L. Environ Exp Bot 55:248–257
Darrow BP, Walsh JJ, Vargo GA, Masserini RT, Fanning KA, Zhang JZ (2003) A simulation study of the growth of benthic microalgae following the decline of a surface phytoplankton bloom. Continent Shelf Res 23:1265–1283
Dashdorj N, Xu W, Martinsson P, Chitnis PR, Savikhin S (2004) Electrochromic shift of chlorophyll absorption in photosystem I from Synechocystis sp. PCC 6803: a probe of optical and dielectric properties around the secondary electron acceptor. Biophys J 86:3121–3130
Davies P (2004) Nutrient processes and chlorophyll in the estuaries and plume of the Gulf of Papua. Continent Shelf Res 24:2317–2341
Davis TW, Berry DL, Boyer GL, Gobler CJ (2009) The effects of temperature and nutrients on the growth and dynamics of toxic and non-toxic strains of microcystis during cyanobacteria blooms. Harmful Algae 8:715–725
Davison IR (1991) Environmental effects on algal photosynthesis: temperature. J Phycol 27:2–8
Dawson R (1998) The toxicology of microcystins. Toxicon 36:953–962
de Haan H (1974) Effect of a fulvic acid fraction on the growth of a Pseudomonas from Tjeukemeer (The Netherlands). Freshwater Biol 4:301–310
de Haan H (1977) Effect of benzoate on microbial decomposition of fulvic acids in Tjeukemeer (The Netherlands). Limnol Oceanogr 22:38–44
de Lange HJ, Morris DP, Williamson CE (2003) Solar ultraviolet photodegradation of DOC may stimulate freshwater food webs. J Plankton Res 25:111–117
de Saussure N-T (1804) Recherches chimiques sur la vegetation. Chez la Ve Nyon, Paris
Deines P, Langmuir D, Harmon RS (1974) Stable carbon isotope ratios and the existence of a gas phase in the evolution of carbonate ground waters. Geochim Cosmochim Acta 38:1147–1164
Dekker JP, van Grondelle R (2000) Primary charge separation in photosystem II. Photosynth Res 63:195–208
del Río L, Sandalio LM, Corpas FJ, Palma JM, Barroso JB (2006) Reactive oxygen species and reactive nitrogen species in peroxisomes. Production, scavenging, and role in cell signaling. Plant Physiol 141:330–335
Delamere NA, Williams RN (1985) Detoxification of hydrogen peroxide by the rabbit iris-ciliary body. Exp Eye Res 40:805–811
Delfine S, Alvino A, Zacchini M, Loreto F (1998) Consequences of salt stress on conductance to CO_2 diffusion, Rubisco characteristics and anatomy of spinach leaves. Funct Plant Biol 25:395–402

Delfine S, Alvino A, Villani MC, Loreto F (1999) Restrictions to carbon dioxide conductance and photosynthesis in spinach leaves recovering from salt stress. Plant Physiol 119:1101–1106
DeLorenzo ME, Scott GI, Ross PE (2001) Toxicity of pesticides to aquatic microorganisms: a review. Environ Toxicol Chem 20:84–98
Deltoro VI, Calatayud A, Gimeno C, Abadía A, Barreno E (1998) Changes in chlorophyll a fluorescence, photosynthetic CO_2 assimilation and xanthophyll cycle interconversions during dehydration in desiccation-tolerant and intolerant liverworts. Planta 207:224–228
Demetriou G, Neonaki C, Navakoudis E, Kotzabasis K (2007) Salt stress impact on the molecular structure and function of the photosynthetic apparatus—the protective role of polyamines. Biochim Biophys Acta 1767:272–280
Demmig-Adams B, Adams WW III (2002) Antioxidants in photosynthesis and human nutrition. Science 298:2149–2153
Demmig-Adams B, Adams Iii W (1992) Photoprotection and other responses of plants to high light stress. Annu Rev Plant Biol 43:599–626
Descolas-Gros C, Fontugne M (1985) Carbon fixation in marine phytoplankton: carboxylase activities and stable carbon-isotope ratios; physiological and paleoclimatological aspects. Mar Biol 87:1–6
Descolas-Gros C, Fontungne M (1990) Stable carbon isotope fractionation by marine phytoplankton during photosynthesis. Plant, Cell Environ 13:207–218
Desingh R, Kanagaraj G (2007) Influence of salinity stress on photosynthesis and antioxidative systems in two cotton varieties. Gen Appl Plant Physiol 33:221–234
Diner BA, Rappaport F (2002) Structure, dynamics, and energetics of the primary photochemistry of photosystem II of oxygenic photosynthesis. Annu Rev Plant Biol 53:551–580
Doi H, Zuykova EI, Kikuchi E, Shikano S, Kanou K, Yurlova N, Yadrenkina E (2006) Spatial changes in carbon and nitrogen stable isotopes of the plankton food web in a saline lake ecosystem. Hydrobiologia 571:395-400
Dokulil MT, Teubner K (2000) Cyanobacterial dominance in lakes. Hydrobiologia 438:1–12
Donahue JL, Okpodu CM, Cramer CL, Grabau EA, Alscher RG (1997) Responses of antioxidants to paraquat in pea leaves (relationships to resistance). Plant Physiol 113:249–257
Downing JA, Watson SB, McCauley E (2001) Predicting cyanobacteria dominance in lakes. Can J Fish Aquat Sci 58:1905–1908
Doyle SA, Saros JE, Williamson CE (2005) Interactive effects of temperature and nutrient limitation on the response of alpine phytoplankton growth to ultraviolet radiation. Limnol Oceanogr 50:1362–1367
Drábková M, Admiraal W, Maršálek B (2007) Combined exposure to hydrogen peroxide and light selective effects on cyanobacteria, Green Algae, and diatoms. Environ Sci Technol 41:309–314
Draper WM, Crosby DG (1981) Hydrogen peroxide and hydroxyl radical: intermediates in indirect photolysis reactions in water. J Agric Food Chem 29:699–702
Dreybrodt W (1988) Processes in karst systems: physics, chemistry, and geology. Springer, Berlin
Drinovec L, Drobne D, Jerman I, Zrimec A (2004) Delayed fluorescence of Lemna minor: a biomarker of the effects of copper, cadmium, and zinc. Bull Environ Contam Toxicol 72:896–902
Druon J, Mannino A, Signorini S, McClain C, Friedrichs M, Wilkin J, Fennel K (2010) Modeling the dynamics and export of dissolved organic matter in the Northeastern US continental shelf. Estuar Coast Shelf Sci 88:488–507
Dugdale R, Wilkerson F, Morel A (1990) Realization of new production in coastal upwelling areas: a means to compare relative performance. Limnol Oceanogr 35:822–829
Dunton KH (2001) δ15 N and δ13C measurements of Antarctic Peninsula fauna: trophic relationships and assimilation of benthic seaweeds. Am Zoolog 41:99–112
Durrant J, Giorgi L, Barber J, Klug D, Porter G (1990) Characterisation of triplet states in isolated Photosystem II reaction centres: oxygen quenching as a mechanism for photodamage. Biochim Biophys Acta 1017:167–175

Durrant JR, Klug DR, Kwa S, van Grondelle R, Porter G, Dekker JP (1995) A multimer model for P680, the primary electron donor of photosystem II. PNAS 92:4798
Eamus D (1986) The responses of leaf water potential and leaf diffusive resistance to abscisic acid, water stress and low temperature in Hibiscus esculentus: the effect of water stress and ABA pre-treatments. J Exp Bot 37:1854–1862
Ebrahim MKH, Vogg G, Osman MNEH, Komor E (1998) Photosynthetic performance and adaptation of sugarcane at suboptimal temperatures. J Plant Physiol 153:587–592
Edwards M, Richardson AJ (2004) Impact of climate change on marine pelagic phenology and trophic mismatch. Nature 430:881–884
Effler SW, Perkins MG, Peng F, Strait C, Weidemann AD, Auer MT (2010) Light-absorbing components in Lake Superior. J Great Lakes Res 36:656–665
Egorova E, Bukhov N (2002) Effect of elevated temperatures on the activity of alternative pathways of photosynthetic electron transport in intact barley and maize leaves. Russian J Plant Physiol 49:575–584
Eloranta P (1978) Light penetration in different types of lakes in Central Finland. Ecography 1:362–366
Elser JJ (1999) The pathway to noxious cyanobacteria blooms in lakes: the food web as the final turn. Freshwater Biol 42:537–543
Elser JJ, Marzolf ER, Goldman CR (1990) Phosphorus and nitrogen limitation of phytoplankton growth in the freshwaters of North America: a review and critique of experimental enrichments. Can J Fish Aquat Sci 47:1468–1477
El-Shintinawy F, Ebrahim M, Sewelam N, El-Shourbagy M (2004) Activity of photosystem 2, lipid peroxidation, and the enzymatic antioxidant protective system in heat shocked barley seedlings. Photosynthetica 42:15–21
Emerson S, Mecking S, Abell J (2001) The biological pump in the subtropical North Pacific Ocean: nutrient sources, redfield ratios, and recent changes. Glob Biogeochem Cy 15:535–554
Epstein PR, Ford TE, Colwell RR (1993) Marine ecosystems: emerging diseases as indicators of change. Lancet 342:1216–1219
Etheridge SM (2010) Paralytic shellfish poisoning: seafood safety and human health perspectives. Toxicon 56:108–122
Evans JR (1996) Developmental constrains on photosynthesis: effects of light and nutrition In: Baker NR (Ed), Photosynthesis and Environment, Kluwer Academic Publishers, Dordrecht, The Netherlands, pp 281-304
Fadzilla NM, Finch RP, Burdon RH (1997) Salinity, oxidative stress and antioxidant responses in shoot cultures of rice. J Exp Bot 48:325–331
Falconer IR (1993) Algal toxins in seafood and drinking water. Academic Press
Falkowski PG (1991) Species variability in the fractionation of 13C and 12C by marine phytoplankton. J Plankton Res 13:21–28
Fan SM (2008) Photochemical and biochemical controls on reactive oxygen and iron speciation in the pelagic surface ocean. Mar Chem 109:152–164
Fariduddin Q, Khanam S, Hasan S, Ali B, Hayat S, Ahmad A (2009) Effect of 28-homobrassinolide on the drought stress-induced changes in photosynthesis and antioxidant system of Brassica juncea L. Acta Physiol Plant 31:889–897
Farquhar G, Richards R (1984) Isotopic composition of plant carbon correlates with water-use efficiency of wheat genotypes. Funct Plant Biol 11:539–552
Farquhar GD, Ehleringer JR, Hubick KT (1989) Carbon isotope discrimination and photosynthesis. Annu Rev Plant Biol 40:503–537
Fay P (1992) Oxygen relations of nitrogen fixation in cyanobacteria. Microbiol Rev 56:340
Ferjani A, Mustardy L, Sulpice R, Marin K, Suzuki I, Hagemann M, Murata N (2003) Glucosylglycerol, a compatible solute, sustains cell division under salt stress. Plant Physiol 131:1628–1637
Ferreira KN, Iverson TM, Maghlaoui K, Barber J, Iwata S (2004) Architecture of the photosynthetic oxygen-evolving center. Science 303:1831–1838

Ferris J, Tyler P (1985) Chlorophyll-total phosphorus relationships in Lake Burragorang, New South Wales, and some other Southern Hemisphere lakes. Mar Freshwater Res 36:157–168

Figueroa FL, Martínez B, Israel A, Neori A, Malta E, Ang P Jr, Inken S, Marquardt R, Rachamim T, Arazi U (2009) Acclimation of red sea macroalgae to solar radiation: photosynthesis and thallus absorptance. Aquat Biol 7:159–172

Filella M (2008) NOM site binding heterogeneity in natural waters: discrete approaches. J Mol Liquids 143:42–51

Finkel Z, Quigg A, Raven J, Reinfelder J, Schofield O, Falkowski P (2006) Irradiance and the elemental stoichiometry of marine phytoplankton. Limnol Oceanogr 51:2690–2701

Flagella Z, Campanile R, Stoppelli M, de Caro A, Di Fonzo N (1998) Drought tolerance of photosynthetic electron transport under CO2-enriched and normal air in cereal species. Physiol Plant 104:753–759

Flexas J, Badger M, Chow WS, Medrano H, Osmond CB (1999) Analysis of the relative increase in photosynthetic O_2 uptake when photosynthesis in grapevine leaves is inhibited following low night temperatures and/or water stress. Plant Physiol 121:675–684

Flexas J, Bota J, Loreto F, Cornic G, Sharkey T (2004) Diffusive and metabolic limitations to photosynthesis under drought and salinity in C3 plants. Plant Biol 6:269–279

Flowers T, Troke P, Yeo A (1977) The mechanism of salt tolerance in halophytes. Annu Rev Plant Physiol 28:89–121

Fogel M, Cifuentes L, Velinsky D, Sharp J (1992) Relationship of carbon availability in estuarine phytoplankton to isotopic composition. Mar Ecol Prog Ser 82:291–300

Fong FK (1974) Molecular basis for the photosynthetic primary process. PNAS 71:3692

Fong KL, McCay PB, Poyer JL (1976) Evidence for superoxide-dependent reduction of Fe3 + and its role in enzyme-generated hydroxyl radical formation. Chem Biolog Interact 15:77

Foyer CH, Noctor G (2000) Tansley Review No. 112. New Phytol 146:359–388

Frackowiak D, Zelent B, Malak H, Planner A, Cegielski R, Leblanc R (1994) Fluorescence of aggregated forms of Chl *a* in various media. J Photochem Photobiol, A 78:49–55

Francko DA (1986) Epilimnetic phosphorus cycling: Influence of humic materials and iron on coexisting major mechanisms. Can J Fish Aquat Sci 43:302–310

Francois R, Altabet M, Goericke R, McCorkleJ C, Brunet C, Poisson A (1993) Changes in the 8I3C of surface water participate organic matter across the subtropical convergence in the SW Indian Ocean. Glob Biogeochem Cy 7:627–644

Franklin NM, Stauber JL, Lim RP (2001) Development of flow cytometry-based algal bioassays for assessing toxicity of copper in natural waters. Environ Toxicol Chem 20:160–170

Freeman KH, Hayes J (1992) Fractionation of carbon isotopes by phytoplankton and estimates of ancient CO2 levels. Glob Biogeochem Cy 6:185–198

Freese HM, Görs S, Karsten U, Schumann R (2007) Dissolved inorganic nutrients and organic substrates in the River Warnow (North-Eastern Germany)-Utilisation by bacterioplankton. Limnologica 37:264–277

Friedrich JW, Huffaker RC (1980) Photosynthesis, leaf resistances, and ribulose-1, 5-bisphosphate carboxylase degradation in senescing barley leaves. Plant Physiol 65:1103

Fromme P (2008) Photosynthetic Protein Complexes: A Structural Approach, Wiley-VCH, Weinheim, 1, pp XXVI–360

Fromme P, Jordan P, Krauß N (2001) Structure of photosystem I. Biochim Biophys Acta 1507:5–31

Frommer WB, Ludewig U, Rentsch D (1999) Taking transgenic plants with a pinch of salt. Science 285:1222–1223

Fryer MJ, Andrews JR, Oxborough K, Blowers DA, Baker NR (1998) Relationship between CO_2 assimilation, photosynthetic electron transport, and active O_2 metabolism in leaves of maize in the field during periods of low temperature. Plant Physiol 116:571–580

Fu FX, Zhang Y, Leblanc K, Sanudo-Wilhelmy SA, Hutchins DA (2005) The biological and biogeochemical consequences of phosphate scavenging onto phytoplankton cell surfaces. Limnol Oceanogr 50:1459–1472

Fu FX, Warner ME, Zhang Y, Feng Y, Hutchins DA (2007) Effects of increased temperature and CO2 on photosynthesis, growth, and elemental ratios in marine synechococcus and prochlorococcus (cyanobacteria). J Phycol 43:485–496

Fu P, Mostofa KMG, Wu F, Liu CQ, Li W, Liao H, Wang L, Wang J, Mei Y (2010) Excitation-emission matrix characterization of dissolved organic matter sources in two eutrophic lakes (Southwestern China Plateau). Geochem J 44:99–112

Fufezan C, Rutherford AW, Krieger-Liszkay A (2002) Singlet oxygen production in herbicide-treated photosystem II. FEBS Lett 532:407–410

Fuhrman J (1992) Bacterioplankton roles in cycling of organic matter: the microbial food web. In: Falkowski PG, Woodhead AD (eds) Primary productivity and biogeochemical cycles in the sea. Plenum Press, New York, pp 361–382

Fujii T, Yokoyama E, Inoue K, Sakurai H (1990) The sites of electron donation of photosystem I to methyl viologen. Biochim Biophys Acta 1015:41–48

Fujiwara K, Ushiroda T, Takeda K, Kumamoto YI, Tsubota H (1993) Diurnal and seasonal distribution of hydrogen peroxide in seawater of the Seto Inland Sea. Geochem J 27:103–115

Fulda S, Huckauf J, Schoor A, Hagemann M (1999) Analysis of stress responses in the cyanobacterial strains *Synechococcus* sp. PCC 7942, *Synechocystis* sp. PCC 6803, and *Synechococcus* sp. PCC 7418: osmolyte accumulation and stress protein synthesis. J Plant Physiol 154:240–249

Gao K, Li G, Helbling EW, Villafane VE (2007a) Variability of UVR effects on photosynthesis of summer phytoplankton assemblages from a Tropical Coastal Area of the South China Sea†. Photochem Photobiol 83:802–809

Gao K, Wu Y, Li G, Wu H, Villafane VE, Helbling EW (2007b) Solar UV radiation drives CO2 fixation in marine phytoplankton: a double-edged sword. Plant Physiol 144:54–59

Gao K, Li P, Watanabe T, Walter Helbling E (2008) Combined effects of ultraviolet radiation and temperature on morphology, photosynthesis, and DNA of Arthrospira (spirulina) platensis (Cyanophyta). J Phycol 44:777–786

Garcia H, Cruzado A, Gordon L, Escanez J (1998) Decadal-scale chemical variability in the subtropical North Atlantic deduced from nutrient and oxygen data. J Geophys Res 103:2817–2830

García-Ferris C, Moreno J (1994) Oxidative modification and breakdown of ribulose-1, 5-bisphosphate carboxylase/oxygenase induced in Euglena gracitis by nitrogen starvation. Planta 193:208–215

Garcia-Pichel F (1994) A model for internal self-shading in planktonic organisms and its implications for the usefulness of ultraviolet sunscreens. Limnol Oceanogr 39:1704–1717

Garcia-Soto C, Pingree RD (2009) Spring and summer blooms of phytoplankton (SeaWiFS/MODIS) along a ferry line in the Bay of Biscay and western English Channel. Continent Shelf Res 29:1111–1122

Garstka M, Venema JH, Rumak I, Gieczewska K, Rosiak M, Koziol-Lipinska J, Kierdaszuk B, Vredenberg WJ, Mostowska A (2007) Contrasting effect of dark-chilling on chloroplast structure and arrangement of chlorophyll protein complexes in pea and tomato: plants with a different susceptibility to non-freezing temperature. Planta 226:1165–1181

Gauslaa Y, McEvoy M (2005) Seasonal changes in solar radiation drive acclimation of the sun-screening compound parietin in the lichen Xanthoria parietina. Basic Appl Ecol 6:75–82

Geider RJ, Greene RM, Kolber Z, MacIntyre HL, Falkowski PG (1993) Fluorescence assessment of the maximum quantum efficiency of photosynthesis in the western North Atlantic. Deep Sea Res Part I 40:1205–1224

Germano M VM, Martin JL, Aartsma TJ, van Gorkom HJ (1995) Femtosecond absorbance changes in PS II reaction centers. In: Mathis P (Ed), Photosynthesis: From Light to Biosphere, Kluwer Academic Publishers, Dordrecht, The Netherlands, 1:503–506

Germano M, Gradinaru C, Shkuropatov AY, van Stokkum I, Shuvalov V, Dekker J, van Grondelle R, van Gorkom H (2004) Energy and electron transfer in photosystem II reaction centers with modified pheophytin composition. Biophys J 86:1664–1672

Giani A, Bird DF, Prairie YT, Lawrence JF (2005) Empirical study of cyanobacterial toxicity along a trophic gradient of lakes. Can J Fish Aquat Sci 62:2100–2109

Giblin FJ, Reddan JR, Schrimscher L, Dziedzic DC, Reddy VN (1990) The relative roles of the glutathione redox cycle and catalase in the detoxification of H_2O_2 by cultured rabbit lens epithelial cells. Exp Eye Res 50:795–804

Gobler CJ, Hutchins DA, Fisher NS, Cosper EM, Sanudo-Wilhelmy SA (1997) Release and bioavailability of C, N, P, Se, and Fe following viral lysis of a marine chrysophyte. Limnol Oceanogr 42:1492–1504

Goericke R, Fry B (1994) Variations of marine plankton ^{13}C with latitude, temperature, and dissolved CO ~ 2 in the world ocean. Glob Biogeochem Cy 8:85–90

Golbeck JH (1994) Photosystem I in cyanobacteria. In: Bryant DA (ed) The molecular biology of cyanobacteria. Kluwer Academic Publishers, Dordrecht, pp 319–360

Goldstein S, Rabani J (2008) Polychromatic UV photon irradiance measurements using chemical actinometers based on NO3 – and H2O2 excitation: applications for industrial photoreactors. Environ Sci Technol 42:3248–3253

Gombos Z, Wada H, Murata N (1994a) The recovery of photosynthesis from low-temperature photoinhibition is accelerated by the unsaturation of membrane lipids: a mechanism of chilling tolerance. PNAS 91:8787–8791

Gombos Z, Wada H, Hideg E, Murata N (1994b) The unsaturation of membrane lipids stabilizes photosynthesis against heat stress. Plant Physiol 104:563–567

Gómez I, López-Figueroa F, Ulloa N, Morales V, Lovengreen C, Huovinen P, Hess S (2004) Patterns of photosynthesis in 18 species of intertidal macroalgae from southern Chile. Mar Ecol Prog Ser 270:103–116

Gossauer A, Engel N (1996) Chlorophyll catabolism–structures, mechanisms, conversions. J Photochem Photobiol, B 32:141–151

Gossett DR, Millhollon EP, Lucas M (1994) Antioxidant response to NaCl stress in salt-tolerant and salt-sensitive cultivars of cotton. Crop Sci 34:706–714

Goulet TL, Cook CB, Goulet D (2005) Effect of short-term exposure to elevated temperatures and light levels on photosynthesis of different host-symbiont combinations in the aiptasia pallidal symbiodinium symbiosis. Limnol Oceanogr 50:1490–1498

Govindjee (1999) On the requirement of minimum number of four versus eight quanta of light for the evolution of one molecule of oxygen in photosynthesis: a historical note. Photosynth Res 59:249–254

Grandy AS, Neff JC (2008) Molecular C dynamics downstream: the biochemical decomposition sequence and its impact on soil organic matter structure and function. Sci Total Environ 404:297–307

Graneli W, Lindell M, Tranvik L (1996) Photo-oxidative production of dissolved inorganic carbon in lakes of different humic content. Limnol Oceanogr 41:698–706

Granéli W, Lindell M, de Faria BM, de Assis Esteves F (1998) Photoproduction of dissolved inorganic carbon in temperate and tropical lakes–dependence on wavelength band and dissolved organic carbon concentration. Biogeochemistry 43:175–195

Granskog MA, Macdonald RW, Mundy CJ, Barber DG (2007) Distribution, characteristics and potential impacts of chromophoric dissolved organic matter (CDOM) in Hudson Strait and Hudson Bay, Canada. Continent Shelf Res 27:2032–2050

Graziano L, Geider R, Li W, Olaizola M (1996) Nitrogen limitation of North Atlantic phytoplankton: Analysis of physiological condition in nutrient enrichment experiments. Aquatic Microb Ecol 11:53–64

Greenfield SR, Wasielewski MR (1996) Excitation energy transfer and charge separation in the isolated photosystem II reaction center. Photosynth Res 48:83–97

Greenfield S, Seibert M, Wasielewski MR (1999) Time-resolved absorption changes of the pheophytin Q x band in isolated photosystem II reaction centers at 7 K: energy transfer and charge separation. J Phys Chem B 103:8364–8374

Greenway H, Munns R (1980) Mechanisms of salt tolerance in nonhalophytes. Annu Rev Plant Physiol 31:149–190

Grey J, Jones RI, Sleep D (2001) Seasonal changes in the importance of the source of organic matter to the diet of zooplankton in Loch Ness, as indicated by stable isotope analysis. Limnol Oceanogr 46:505–513
Groot ML, van Mourik F, Eijckelhoff C, van Stokkum IHM, Dekker JP, van Grondelle R (1997) Charge separation in the reaction center of photosystem II studied as a function of temperature. PNAS 94:4389
Grover A, Mohanty P (1992) Leaf senescence-induced alterations in structure and function of higher plant chloroplasts. In: Abrol YP, Mohanty P, Govindjee (Eds), Photosynthesis: Photoreactions to Plant Productivity, Kluwer Academic Publishers, Dordrecht, pp 225–255
Guildford SJ, Hecky RE (2000) Total nitrogen, total phosphorus, and nutrient limitation in lakes and oceans: is there a common relationship? Limnol Oceanogr 45:1213–1223
Guskov A, Kern J, Gabdulkhakov A, Broser M, Zouni A, Saenger W (2009) Cyanobacterial photosystem II at 2.9-Å resolution and the role of quinones, lipids, channels and chloride. Nature Struct Mol Biol 16:334–342
Haaber J, Middelboe M (2009) Viral lysis of Phaeocystis pouchetii: implications for algal population dynamics and heterotrophic C, N and P cycling. ISME J 3:430–441
Häder DP, Kumar H, Smith RC, Worrest RC (2003) Aquatic ecosystems: effects of solar ultraviolet radiation and interactions with other climatic change factors. Photochem Photobiol Sci 2:39–50
Häder DP, Kumar H, Smith R, Worrest R (2007) Effects of solar UV radiation on aquatic ecosystems and interactions with climate change. Photochem Photobiol Sci 6:267–285
Hagege D, Kevers C, Boucaud J, Duyme M, Gaspar T (1990a) Polyamines, phospholipids, and peroxides in normal and habituated sugar beet calli. J Plant Physiol 136:641–645
Hagege D, Nouvelot A, Boucaud J, Gaspar T (1990b) Malondialdehyde titration with thiobarbiturate in plant extracts: avoidance of pigment interference. Phytochem Anal 1:86–89
Hagemann M, Erdmann N (1997) Environmental stresses. In: Rai AK (ed) Cyanobacterial nitrogen metabolism and environmental biotechnology. Springer, Heidelberg, pp 156–221
Hagemann M, Wölfel L, Krüger B (1990) Alterations of protein synthesis in the cyanobacterium Synechocystis sp. PCC 6803 after a salt shock. J Gen Microbiol 136:1393–1399
Hagemann M, Techel D, Rensing L (1991) Comparison of salt-and heat-induced alterations of protein synthesis in the cyanobacterium *Synechocystis* sp. PCC 6803. Arch Microbiol 155:587–592
Haines A, McMichael AJ, Epstein PR (2000) Environment and health: 2. Global climate change and health. Can Med Assoc J 163:729–734
Hallegraeff GM (1993) A review of harmful algal blooms and their apparent global increase*. Phycologia 32:79–99
Halling-Sørensen B, Lützhøft HCH, Andersen H, Ingerslev F (2000) Environmental risk assessment of antibiotics: comparison of mecillinam, trimethoprim and ciprofloxacin. J Antimicrob Chemother 46:53–58
Halliwell B (1981) Free radicals, oxygen toxicity and aging. Age pigments. Elsevier, Amsterdam pp. 1–62
Halliwell B, Gutteridge J (1984) Oxygen toxicity, oxygen radicals, transition metals and disease. Biochem J 219:1–14
Halliwell B, Gutteridge J (1990) Role of free radicals and catalytic metal ions in human disease: an overview. Methods Enzymol 186:1
Halliwell B, Gutteridge JMC (1999) Free radicals in biology and medicine, 3rd edn. Oxford University Press, Oxford
Halliwell B, Gutteridge JMC (2007) Free radicals in biology and medicine. Free Rad Biol Med 10:449–450
Hallock P, Schlager W (1986) Nutrient excess and the demise of coral reefs and carbonate platforms. Palaios 1:389–398
Han T, Sinha RP, Häder DP (2001) UV-A/blue light-induced reactivation of photosynthesis in UV-B irradiated cyanobacterium, *Anabaena* sp. J Plant Physiol 158:1403–1413

Hanelt D, Wiencke C, Nultsch W (1997) Influence of UV radiation on the photosynthesis of arctic macroalgae in the field. J Photochem Photobiol, B 38:40–47
Hansen PJ, Hjorth M (2002) Growth and grazing responses of Chrysochromulina ericina (Prymnesiophyceae): The role of irradiance, prey concentration and pH. Mar Biol 141:975–983
Hanson DT, Swanson S, Graham LE, Sharkey TD (1999) Evolutionary significance of isopreneemission from mosses. Am J Bot 86:634–639
Hanson PC, Bade DL, Carpenter SR, Kratz TK (2003) Lake metabolism: relationships with dissolved organic carbon and phosphorus. Limnol Oceanogr 48:1112–1119
Hardwick T (1957) The rate constant of the reaction between ferrous ions and hydrogen peroxide in acid solution. Can J Chem 35:428–436
Harrison JW, Smith REH (2009) Effects of ultraviolet radiation on the productivity and composition of freshwater phytoplankton communities. Photochem Photobiol Sci 8:1218–1232
Harvell C, Kim K, Burkholder J, Colwell R, Epstein PR, Grimes D, Hofmann E, Lipp E, Osterhaus A, Overstreet RM (1999) Emerging marine diseases–climate links and anthropogenic factors. Science 285:1505–1510
Hasan S, Hayat S, Ali B, Ahmad A (2008) 28-Homobrassinolide protects chickpea (Cicer arietinum) from cadmium toxicity by stimulating antioxidants. Environ Pollut 151:60–66
Hassan I (2006) Effects of water stress and high temperature on gas exchange and chlorophyll fluorescence in Triticum aestivum L. Photosynthetica 44:312–315
Hassan I, Bendert J, Weigel H-J (1998) Effects of O_3 and water stress on growth and physiology of Egyptian cultivars of tomatoes. Gartenbauwissenschaft 76:122–135
Hatz S, Lambert JDC, Ogilby PR (2007) Measuring the lifetime of singlet oxygen in a single cell: addressing the issue of cell viability. Photochem Photobiol Sci 6:1106–1116
Havaux M (1992) Stress tolerance of photosystem II in vivo: antagonistic effects of water, heat, and photoinhibition stresses. Plant Physiol 100:424–432
Havaux M, Tardy F, Lemoine Y (1998) Photosynthetic light-harvesting function of carotenoids in higher-plant leaves exposed to high light irradiances. Planta 205:242–250
Havens KE, Bierman VJ Jr, Flaig EG, Hanlon C, James RT, Jones BL, Smith VH (1995) Historical trends in the Lake Okeechobee ecosystem. VI Synthesis Arch Hydrobiol 107(Suppl):101–111
Haxo F, Blinks L (1950) Photosynthetic action spectra of marine algae. J Gen Physiol 33:389–422
Hayakawa K (2004) Seasonal variations and dynamics of dissolved carbohydrates in Lake Biwa. Org Geochem 35:169–179
Hayakawa K, Sugiyama Y (2008) Spatial and seasonal variations in attenuation of solar ultraviolet radiation in Lake Biwa, Japan. J Photochem Photobiol, B 90:121–133
Hayashi H, Murata N (1998) Genetically engineered enhancement of salt tolerance in higher plants. In: Sato K, Murata N (eds) Stress responses of photosynthetic organisms Molecular mechanisms and molecular regulation. Elsevier, Amsterdam, pp 133–148
Hayat S, Ali B, Aiman Hasan S, Ahmad A (2007) Brassinosteroid enhanced the level of antioxidants under cadmium stress in *Brassica juncea*. Environ Exp Bot 60:33–41
He YY, Häder DP (2002) UV-B-induced formation of reactive oxygen species and oxidative damage of the cyanobacterium *Anabaena* sp.: protective effects of ascorbic acid and N-acetyl-L-cysteine. J Photochem Photobiol, B 66:115–124
Heath RT (1992) Nutrient dynamics in Great Lakes coastal wetlands: future directions. J Great Lakes Res 18:590–602
Heber U, French C (1968) Effects of oxygen on the electron transport chain of photosynthesis. Planta 79:99–112
Heber U, Egneus H, Hanck U, Jensen M, Köster S (1978) Regulation of photosynthetic electron transport and photophosphorylation in intact chloroplasts and leaves of Spinacia oleracea L. Planta 143:41–49
Heckathorn SA, Ryan SL, Baylis JA, Wang D, Hamilton EW III, Cundiff L, Luthe DS (2002) In vivo evidence from an agrostis stolonifera selection genotype that chloroplast small

heat-shock proteins can protect photosystem II during heat stress. Funct Plant Biol 29:935–946
Heil CA, Revilla M, Glibert PM, Murasko S (2007) Nutrient quality drives differential phytoplankton community composition on the southwest Florida shelf. Limnol Oceanogr 52:1067–1078
Helbling EW, Villafane V, Ferrario M, Holm-Hansen O (1992) Impact of natural ultraviolet radiation on rates of photosynthesis and on specific marine phytoplankton species. Mar Ecol Prog Ser 80:89–100
Helbling EW, Buma AGJ, de Boer MK, Villafañe VE (2001) In situ impact of solar ultraviolet radiation on photosynthesis and DNA in temperate marine phytoplankton. Mar Ecol Prog Ser 211:43–49
Helbling EW, Gao K, Gonçalves RJ, Wu H, Villafañe VE (2003) Utilization of solar UV radiation by coastal phytoplankton assemblages off SE China when exposed to fast mixing. Mar Ecol Prog Ser 259:59–66
Hemminga M, Mateo M (1996) Stable carbon isotopes in seagrasses: variability in ratios and use in ecological studies. Mar Ecol Prog Ser 140:285–298
Hendrickson L, Ball MC, Osmond CB, Furbank RT, Chow WS (2003) Assessment of photoprotection mechanisms of grapevines at low temperature. Funct Plant Biol 30:631–642
Hendrickson L, Förster B, Furbank RT, Chow WS (2004) Processes contributing to photoprotection of grapevine leaves illuminated at low temperature. Physiol Plant 121:272–281
Henmi T, Yamasaki H, Sakuma S, Tomokawa Y, Tamura N, Shen JR, Yamamoto Y (2003) Dynamic interaction between the D1 protein, CP43 and OEC33 at the lumenal side of photosystem II in spinach chloroplasts: evidence from light-induced cross-linking of the proteins in the donor-side photoinhibition. Plant Cell Physiol 44:451–456
Henmi T, Miyao M, Yamamoto Y (2004) Release and reactive-oxygen-mediated damage of the oxygen-evolving complex subunits of PSII during photoinhibition. Plant Cell Physiol 45:243–250
Herczeg AL, Fairbanks RG (1987) Anomalous carbon isotope fractionation between atmospheric CO2 and dissolved inorganic carbon induced by intense photosynthesis. Geochim Cosmochim Acta 51:895–899
Hernandez J, Campillo A, Jimenez A, Alarcon J, Sevilla F (1999) Response of antioxidant systems and leaf water relations to NaCl stress in pea plants. New Phytol 141:241–251
Hernes PJ, Benner R (2003) Photochemical and microbial degradation of dissolved lignin phenols: Implications for the fate of terrigenous dissolved organic matter in marine environments. J Geophys Res 108:3291. doi:101029/2002JC001421
Hertwig B, Streb P, Feierabend J (1992) Light dependence of catalase synthesis and degradation in leaves and the influence of interfering stress conditions. Plant Physiol 100:1547–1553
Hessen DO (1985) The relation between bacterial carbon and dissolved humic compounds in oligotrophic lakes. FEMS Microbiol Lett 31:215–223
Hessen DO, Tranvik LJ (1998) Aquatic humic matter: From molecular structure to ecosystem stability. In: Hessen DO, Tranvik LJ (Eds), Squatic Humic Substances: Ecology and Biogeochemistry, Ecological Studies 133, Springer, pp 333-342
Hessen DO, Andersen T, Lyche A (1990) Carbon metabolism in a humic lake: pool sizes and cycling through zooplankton Limnol Oceanogr 35:84–99
Hibino T, Lee BH, Rai A, Ishikawa H, Kojima H, Tawada M, Shimoyama H, Takabe T (1996) Salt enhances photosystem I content and cyclic electron flow via NAD (P) H dehydrogenase in the halotolerant cyanobacterium Aphanothece halophytica. Funct Plant Biol 23:321–330
Hideg É, Spetea C, Vass I (1994) Singlet oxygen production in thylakoid membranes during photoinhibition as detected by EPR spectroscopy. Photosynth Res 39:191–199
Hideg É, Kálai T, Hideg K, Vass I (1998) Photoinhibition of photosynthesis in vivo results in singlet oxygen production detection via nitroxide-induced fluorescence quenching in broad bean leaves. Biochemistry 37:11405–11411
Hideg É, Ogawa K, Kálai T, Hideg K (2001) Singlet oxygen imaging in Arabidopsis thaliana leaves under photoinhibition by excess photosynthetically active radiation. Physiol Plant 112:10–14

Hideg É, Barta C, Kálai T, Vass I, Hideg K, Asada K (2002) Detection of singlet oxygen and superoxide with fluorescent sensors in leaves under stress by photoinhibition or UV radiation. Plant Cell Physiol 43:1154–1164
Hideg E, Kós PB, Vass I (2007) Photosystem II damage induced by chemically generated singlet oxygen in tobacco leaves. Physiol Plant 131:33–40
Higuchi T, Fujimura H, Arakaki T, Oomori T (2009) The synergistic effects of hydrogen peroxide and elevated seawater temperature on the metabolic activity of the coral Galaxea fascicularis. Mar Biol 156:589–596
Hill R (1937) Oxygen evolved by isolated chloroplasts. Nature 139:i939
Hill R (1939) Oxygen produced by isolated chloroplasts. Proc R Soc London 127:192–210
Hill R (1951) Reduction by cliloroplasts. Symp Soc Exp Biol 5:223–231
Hinga KR, Arthur MA, Pilson MEQ, Whitaker D (1994) Carbon isotope fractionation by marine phytoplankton in culture: the effects of CO_2 concentration, pH, temperature, and species. Glob Biogeochem Cy 8:91–102
Hiriart-Baer VP, Diep N, Smith REH (2008) Dissolved organic matter in the Great Lakes: role and nature of allochthonous material. J Great Lakes Res 34:383–394
Hirose K (2004) Chemical speciation of thorium in marine biogenic particulate matter. TSWJ 4:67–76
Hirotsu N, Makino A, Ushio A, Mae T (2004) Changes in the thermal dissipation and the electron flow in the water–water cycle in rice grown under conditions of physiologically low temperature. Plant Cell Physiol 45:635–644
Ho FM, Styring S (2008) Access channels and methanol binding site to the CaMn4 cluster in photosystem II based on solvent accessibility simulations, with implications for substrate water access. Biochim Biophys Acta 1777:140–153
Ho AYT, Xu J, Yin K, Yuan X, He L, Jiang Y, Lee JHW, Anderson DM, Harrison PJ (2008) Seasonal and spatial dynamics of nutrients and phytoplankton biomass in Victoria Harbour and its vicinity before and after sewage abatement. Mar Pollut Bull 57:313–324
Hofman P, Haisel D, Komenda J, Vágner M, Tichá I, Schäfer C, Čapková V (2002) Impact of in vitro cultivation conditions on stress responses and on changes in thylakoid membrane proteins and pigments of tobacco during ex vitro acclimation. Biol Plant 45:189–195
Holland MK, Alvarez JG, Storey BT (1982) Production of superoxide and activity of superoxide dismutase in rabbit epididymal spermatozoa. Biol Reprod 27:1109–1118
Holm-Hansen O, Helbling EW, Lubin D (1993) Ultraviolet radiation in Antarctica: inhibition of primary production. Photochem Photobiol 58:567–570
Holten D, Kirmaier C, Parson WW (1986) Picosecond measurements of electron transfer in bacterial photosynthetic reaction centers. Am Chem Soc 321:205–218
Holzapfel W, Finkele U, Kaiser W, Oesterhelt D, Scheer H, Stilz HU, Zinth W (1990) Initial electron-transfer in the reaction center from rhodobacter sphaeroides. PNAS 87:5168
Holzwarth AR (2008) Ultrashort Laser Pulses in Biology and Medicine. In: Gilch P, Zinth W (eds) Braun M. Springer, Dordrecht, pp 141–164
Holzwarth AR, Müller MG, Niklas J, Lubitz W (2006) Ultrafast transient absorption studies on photosystem I reaction centers from Chlamydomonas reinhardtii. 2: mutations near the P700 reaction center chlorophylls provide new insight into the nature of the primary electron donor. Biophys J 90:552–565
Hong AP, Bahnemann DW, Hoffmann MR (1987) Cobalt (II) tetrasulfophthalocyanine on titanium dioxide: a new efficient electron relay for the photocatalytic formation and depletion of hydrogen peroxide in aqueous suspensions. J Phys Chem 91:2109–2117
Hopenhayn-Rich C, Biggs ML, Smith AH (1998) Lung and kidney cancer mortality associated with arsenic in drinking water in Cordoba, Argentina. Int J Epidemiol 27:561–569
Hopkins WG, Hüner NPA (1995) Introduction to Plant Physiol, vol 355. Wiley, New York
Hopkinson BM, Barbeau KA (2008) Interactive influences of iron and light limitation on phytoplankton at subsurface chlorophyll maxima in the eastern North Pacific. Limnol Oceanogr 53:1303–1318
Horton P, Bowyer TR (1990) Chlorophyll fluorescence transients. In: Harwood JL, Bowyer TR (eds) Methods in plant biochemistry Lipids. Membranes and Aspects of Photobiology, Academic, San Diego, pp 259–295

Houser JN, Bade DL, Cole JJ, Pace ML (2003) The dual influences of dissolved organic carbon on hypolimnetic metabolism: organic substrate and photosynthetic reduction. Biogeochemistry 64:247–269

Hsiao TC (1973) Plant responses to water stress. Annu Rev Plant Physiol 24:519–570

Hu XP, Burdige DJ, Zimmerman RC (2012) delta C-13 is a signature of light availability and photosynthesis in seagrass. Limnol Oceanogr 57:441–448

Huang M, Guo Z (2005) Responses of antioxidative system to chilling stress in two rice cultivars differing in sensitivity. Biol Plant 49:81–84

Huffaker RCRT, Kleinkopf GE, Cox EL (1970) Effects of mild water stress on enzymes of nitrate assimilation and of the carboxylative phase of photosynthesis in barley. Crop Sci 10:471–474

Huflejt ME, Tremolieres A, Pineau B, Lang JK, Hatheway J, Packer L (1990) Changes in membrane lipid composition during saline growth of the fresh water cyanobacterium Synechococcus 6311. Plant Physiol 94:1512

Huisman J, Thi NNP, Karl DM, Sommeijer B (2006) Reduced mixing generates oscillations and chaos in the oceanic deep chlorophyll maximum. Nature 439:322–325

Huner NPA, Öquist G, Hurry VM, Krol M, Falk S, Griffith M (1993) Photosynthesis, photoinhibition and low temperature acclimation in cold tolerant plants. Photosynth Res 37:19–39

Huner N, Öquist G, Sarhan F (1998) Energy balance and acclimation to light and cold. Trends Plant Sci 3:224–230

Hurry V, Huner N, Selstam E, Gardeström P, Öquist G (1998) Photosynthesis at low growth temperatures. In: Raghavendra AS (ed) Photosynthesis a comprehensive treatise. Cambridge University Press, Cambridge, pp 238–249

Huszar VLM, Caraco NF, Roland F, Cole J (2006) Nutrient-chlorophyll relationships in tropical-subtropical lakes: do temperate models fit? Biogeochemistry 79:239–250

Hutchinson G (1957) A treatise on limnology, v. 1. Wiley eutrophication of the St Lawrence Great Lakes. J Fish Res Bd Can 29:1451–1462

Hwang HJ, Dilbeck P, Debus RJ, Burnap RL (2007) Mutation of arginine 357 of the CP43 protein of photosystem II severely impairs the catalytic S-state cycle of the H2O oxidation complex. Biochemistry 46:11987–11997

Hyenstrand P, Blomqvist P, Pettersson A (1998) Factors determining cyanobacterial success in aquatic systems-a literature review. Ergebn Limnol 51:41–62

Hylander S, Boeing WJ, Granéli W, Karlsson J, Von Einem J, Gutseit K, Hansson LA (2009) Complementary UV protective compounds in zooplankton. Limnol Oceanogr 54:1883–1893

Hynninen PH, Lötjönen S (1993) Hydrogen bonding of water to chlorophyll *a* and its derivatives as detected by ^{1}H-NMR spectroscopy. Biochim Biophys Acta 1183:381–387

Igarashi T, Aritake S, Yasumoto T (1998) Biological activities of prymnesin-2 isolated from a red tide alga prymnesium parvum. Nat Toxins 6:35–41

Ilan Y, Czapski G, Meisel D (1976) The one-electron transfer redox potentials of free-radicals. 1. Oxygen/superoxide system. Biochim Biophys Acta 430:209–224

Ingalls AE, Whitehead K, Bridoux MC (2010) Tinted windows: the presence of the UV absorbing compounds called mycosporine-like amino acids embedded in the frustules of marine diatoms. Geochim Cosmochim Acta 74:104–115

Irihimovitch V, Shapira M (2000) Glutathione redox potential modulated by reactive oxygen species regulates translation of Rubisco large subunit in the chloroplast. J Biol Chem 275:16289–16295

Ishibashi M, Terashima I (1995) Effects of continuous leaf wetness on photosynthesis: adverse aspects of rainfall. Plant, Cell Environ 18:431–438

Iturbe-Ormaetxe I, Escuredo PR, Arrese-Igor C, Becana M (1998) Oxidative damage in pea plants exposed to water deficit or paraquat. Plant Physiol 116:173–181

Ivanova A, Aslanidi K, Karpenko YV, Belozerskaya T (2005) The effect of hydrogen peroxide on the growth of microscopic mycelial fungi isolated from habitats with different levels of radioactive contamination. Microbiology 74:655–663

Iverson RL, Curl H Jr (1973) Action spectrum of photosynthesis for skeletonema costatum obtained with carbon-14. Physiol Plant 28:498–502

Jackson TA, Hecky RE (1980) Depression of primary productivity by humic matter in lake and reservoir waters of the boreal forest zone. Can J Fish Aquat Sci 37:2300–2317
Jansson M, Bergström AK, Blomqvist P, Drakare S (2000) Allochthonous organic carbon and phytoplankton/bacterioplankton production relationships in lakes. Ecology 81:3250–3255
Jasper JP, Hayes J (1994) Photosynthetic fractionation of and concentrations of dissolved COz in the central equatorial Pacific. Paleoceanography 9:781–798
Jeffrey S, MacTavish H, Dunlap W, Vesk M, Groenewoud K (1999) Occurrence of UVA-and UVB-absorbing compounds in 152 species (206 strains) of marine microalgae. Mar Ecol Prog Ser 189:35–51
Jenkin PM (1937) Oxygen production by the diatom coscinodiscus excentricus Ehr. In relation to submarine illumination in the english channel. J Mar Biol Assoc UK 22:301–343
Jeong J, Yoon J (2005) pH effect on OH radical production in photo/ferrioxalate system. Water Res 39:2893–2900
Jerlov NG (1968) Optical oceanography, vol 5. Elsevier, Amsterdam
Jiang H, Qiu B (2005) Photosynthetic adaptation of a bloomforming cyanobacterium microcystis aeruginosa (Cyanophyceae) to prolonged UV-B exposure. J Phycol 41:983–992
Jiang H, Qiu B (2011) Inhibition of photosynthesis by UV-B exposure and its repair in the bloom-forming cyanobacterium microcystis aeruginosa. J Appl Phycol 23:691–696
Jiménez C, Figueroa F, Salles S, Aguilera J, Mercado J, Viñegla B, Flores-Moya A, Lebert M, Häder DP (1998) Effects of solar radiation on photosynthesis and photoinhibition in red macrophytes from an intertidal system of southern Spain. Bot Mar 41:329-338
Joët T, Cournac L, Horvath EM, Medgyesy P, Peltier G (2001) Increased sensitivity of photosynthesis to antimycin A induced by inactivation of the chloroplast ndhB gene. Evidence for a participation of the NADH-dehydrogenase complex to cyclic electron flow around photosystem I. Plant Physiol 125:1919–1929
Johannessen SC, Peña MA, Quenneville ML (2007) Photochemical production of carbon dioxide during a coastal phytoplankton bloom. Estuar Coast Shelf Sci 73:236–242
Joliot P, Kok B (1975) Oxygen evolution in photosynthesis. In: Govindjee (ed) Bioenergetics of photosynthesis. Academic Press, New York, pp 387–412
Jolly WM, Nemani R, Running SW (2005) A generalized, bioclimatic index to predict foliar phenology in response to climate. Glob Change Biol 11:619–632
Jones R (1990) Phosphorus transformations in the epilimnion of humic lakes: biological uptake of phosphate. Freshwater Biol 23:323–337
Jones RI (1992) The influence of humic substances on lacustrine planktonic food chains. Hydrobiologia 229:73–91
Jones RI (2000) Mixotrophy in planktonic protists: an overview. Freshwater Biol 45:219–226
Jones RI, Arvola L (1984) Light penetration and some related characteristics in small forest lakes in Southern Finland. Verh int Ver Limnol 22:811–816
Jones MM, Turner NC (1978) Osmotic adjustment in leaves of sorghum in response to water deficits. Plant Physiol 61:122–126
Jones R, Salonen K, Haan H (1988) Phosphorus transformations in the epilimnion of humic lakes: abiotic interactions between dissolved humic materials and phosphate. Freshwater Biol 19:357–369
Jordan P, Fromme P, Witt HT, Klukas O, Saenger W, Krauß N (2001) Three-dimensional structure of cyanobacterial photosystem I at 2.5 Å resolution. Nature 411:909–917
Joset F, Jeanjean R, Hagemann M (1996) Dynamics of the response of cyanobacteria to salt stress: deciphering the molecular events. Physiol Plant 96:738–744
Jungklang J, Usui K, Matsumoto H (2003) Differences in physiological responses to NaCl between salt-tolerant Sesbania rostrata Brem. & Oberm. and non-tolerant Phaseolus vulgaris L. Weed Biol Manag 3:21–27
Kaiser WM (1987) Effects of water deficit on photosynthetic capacity. Physiol Plant 71:142–149
Kalle K (1966) The problem of the gelbstoff in the sea. Oceanogr Mar Biol 4:91–104
Kamiya N, Shen JR (2003) Crystal structure of oxygen-evolving photosystem II from thermosynechococcus vulcanus at 3.7-Å resolution. PNAS 100:98–103

Kaneko H, Shimada A, Hirayama K (2004) Short-term algal toxicity test based on phosphate uptake. Water Res 38:2173–2177

Kanervo E, Mäenpää P, Aro EM (1993) D1 Protein degradation and *psbA* transcript levels in *Synechocystis* PCC 6803 during photoinhibition in vivo. J Plant Physiol 142:669–675

Kappers FI (1980) The cyanobacterium Microcystis aeruginosa Kg. and the nitrogen cycle of the hypertrophic Lake Brielle (The Netherlands). In: Barica J, Mur L (eds) Hypertrophic ecosystems. Dr W Junk, The Hague, pp 37–43

Karapetyan N (2004) Interaction of pigment—protein complexes within aggregates stimulates dissipation of excess energy. Biochemistry (Moscow) 69:1299–1304

Karentz D, Cleaver JE, Mitchell DL (1991) Cell survival characteristics and molecular responses of antarctic phytoplankton to ultraviolet-B radiation. J Phycol 27:326–341

Karpinski S, Escobar C, Karpinska B, Creissen G, Mullineaux PM (1997) Photosynthetic electron transport regulates the expression of cytosolic ascorbate peroxidase genes in Arabidopsis during excess light stress. Plant Cell 9:627–640

Karuppanapandian T, Moon JC, Kim C, Manoharan K, Kim W (2011) Reactive oxygen species in plants: their generation, signal transduction, and scavenging mechanisms. Aust J Crop Sci 5:709

Katoh S (1972) Inhibitors of electron transport associated with photosystem II in chloroplasts. Plant Cell Physiol 13:273–286

Katsumata M, Koike T, Nishikawa M, Tsuchiya H (2005) Influences of herbicides and mercury on Blue-Green Alga spirulina platensis-analysis of long-term behavior of S. platensis delayed fluorescence. J Jpn Soc Water Environ 28:23–28

Katsumata M, Koike T, Nishikawa M, Kazumura K, Tsuchiya H (2006) Rapid ecotoxicological bioassay using delayed fluorescence in the green alga Pseudokirchneriella subcapitata. Water Res 40:3393–3400

Katz JJ (1990) Green thoughts in a green shade. Photosynth Res 26:143–160

Katz J (1994) Long wavelength chlorophyll. Spectrum 7:1–9

Katz J, Closs G, Pennington F, Thomas M, Strain H (1963) Infrared spectra, molecular weights, and molecular association of chlorophylls a and b, methyl chlorophyllides, and pheophytins in various solvents. JACS 85:3801–3809

Katz JJ, Bowman MK, Michalski TJ, Worcester DL (1991) Chlorophyll aggregation: chlorophyll/water micelles as models for in vivo long-wavelength chlorophyll. In: Scheer H (ed) Chlorophylls. CRC Press, Boca Raton, pp 211–235

Kautsky H, de, Bruijn (1931) Die Aufklärung der Photoluminescenztilgung fluorescierender Systeme durch Sauerstoff: Die Bildung aktiver, diffusionsfa¨higer Sauerstoffmoleküle durch Sensibilisierung. Naturwissenschaften 19:1043 doi:101007/BF01516190

Kawakami K, Umena Y, Kamiya N, Shen JR (2009) Location of chloride and its possible functions in oxygen-evolving photosystem II revealed by X-ray crystallography. PNAS 106:8567

Keeling RF, Garcia HE (2002) The change in oceanic O_2 inventory associated with recent global warming. PNAS 99:7848–7853

Keeling RF, Körtzinger A, Gruber N (2010) Ocean deoxygenation in a warming world. Annu Rev Mar Sci 2:199–229

Keller K, Slater RD, Bender M, Key RM (2001) Possible biological or physical explanations for decadal scale trends in North Pacific nutrient concentrations and oxygen utilization. Deep Sea Res Part II 49:345–362

Kelly CA, Fee E, Ramlal PS, Rudd JWM, Hesslein RH, Anema C, Schindler EU (2001) Natural variability of carbon dioxide and net epilimnetic production in the surface waters of boreal lakes of different sizes. Limnol Oceanogr 46:1054–1064

Keren N, Berg A, van Kan PJM, Levanon H, Ohad I (1997) Mechanism of photosystem II photoinactivation and D1 protein degradation at low light: the role of back electron flow. PNAS 94:1579–1584

Kieber DJ, McDaniel J, Mopper K (1989) Photochemical source of biological substrates in sea water: implications for carbon cycling. Nature 341:637–639

Kim C, Nishimura Y, Nagata T (2006) Role of dissolved organic matter in hypolimnetic mineralization of carbon and nitrogen in a large, monomictic lake. Limnol Oceanogr 51:70–78
Kirk J (1976) Yellow substance (Gelbstoff) and its contribution to the attenuation of photosynthetically active radiation in some inland and coastal south-eastern Australian waters. Mar Freshwater Res 27:61–71
Kirk JTO, Reade J (1970) The action spectrum of photosynthesis in Euglena Gracilis at different stages of chloroplast development. Aust J Biol Sci 23:33–42
Kirmaier C, Holten D (1987) Primary photochemistry of reaction centers from the photosynthetic purple bacteria. Photosynth Res 13:225–260
Kirmaier C, Holten D, Parson WW (1985a) Temperature and detection-wavelength dependence of the picosecond electron-transfer kinetics measured in Rhodopseudomonas sphaeroides reaction centers. Resolution of new spectral and kinetic components in the primary charge-separation process. Biochim Biophys Acta 810:33–48
Kirmaier C, Holten D, Parson WW (1985b) Picosecond-photodichroism studies of the transient states in Rhodopseudomonas sphaeroides reaction centers at 5 K. Effects of electron transfer on the six bacteriochlorin pigments. Biochim Biophys Acta 810:49–61
Kirmaier C, Blankenship RE, Holten D (1986) Formation and decay of radical-pair state P + I-in chloroflexus aurantiacus reaction centers. Biochim Biophys Acta 850:275–285
Kleinfeld D, Okamura M, Feher G (1984) Electron-transfer kinetics in photosynthetic reaction centers cooled to cryogenic temperatures in the charge-separated state: evidence for light-induced structural changes. Biochemistry 23:5780–5786
Klisch M, Häder DP (2008) Mycosporine-like amino acids and marine toxins: the common and the different. Mar Drugs 6:147–163
Klotz LO (2002) Oxidant-induced signaling: effects of peroxynitrite and singlet oxygen. Biol Chem 383:443–456
Klug DR, Durrant JR, Barber J (1998) The entanglement of excitation energy transfer and electron transfer in the reaction centre of photosystem II. Philos T Roy Soc A 356:449–464
Klukas O, Schubert WD, Jordan P, Krauß N, Fromme P, Witt HT, Saenger W (1999) Photosystem I, an improved model of the stromal subunits PsaC, PsaD, and PsaE. J Biol Chem 274:7351–7360
Kobayashi T, Natanani N, Hirakawa T, Suzuki M, Miyake T, Chiwa M, Yuhara T, Hashimoto N, Inoue K, Yamamura K (2002) Variation in CO2 assimilation rate induced by simulated dew waters with different sources of hydroxyl radical (· OH) on the needle surfaces of Japanese red pine (Pinus densiflora Sieb. et Zucc.). Environ Pollut 118:383–391
Kögel-Knabner I (2002) The macromolecular organic composition of plant and microbial residues as inputs to soil organic matter. Soil Biol Biochem 34:139–162
Kok B, Forbush B, McGloin M (1970) Cooperation of charges in photosynthetic O2 evolution–I. A linear four step mechanism. Photochem Photobiol 11:457–475
Komissarov G (1994) Photosynthesis: a new look. Science in Russia 52–55
Komissarov G (1995) Photosynthesis as a physicochemical process. Chem Phys Rep 14:1723–1732
Komissarov G (2003) Photosynthesis: the physical-chemical approach. J Adv Chem Phys 2:28–61
Konermann L, Gatzen G, Holzwarth AR (1997) Primary processes and structure of the photosystem ii reaction center. 5. Modeling of the fluorescence kinetics of the D1–D2-cyt-b 559 complex at 77 K. J Phys Chem B 101:2933–2944
Kopáček J, Hejzlar J, Kaňa J, Porcal P, Klementová Š (2003) Photochemical, chemical, and biological transformations of dissolved organic carbon and its effect on alkalinity production in acidified lakes. Limnol Oceanogr 48:106–117
Kopáček J, Brzáková M, Hejzlar J, Nedoma J, Porcal P, Vrba J (2004) Nutrient cycling in a strongly acidified mesotrophic lake. Limnol Oceanogr 49:1202–1213
Koppenol W (1976) Reactions involving singlet oxygen and the superoxide anion
Koukal B, Guéguen C, Pardos M, Dominik J (2003) Influence of humic substances on the toxic effects of cadmium and zinc to the green alga Pseudokirchneriella subcapitata. Chemosphere 53:953–961

Krasnovsky A (2007) Primary mechanisms of photoactivation of molecular oxygen. History of development and the modern status of research. Biochemistry (Moscow) 72:1065–1080

Krause G, Weis E (1991) Chlorophyll fluorescence and photosynthesis: the basics. Annu Rev Plant Biol 42:313–349

Krauß N (2003) Mechanisms for photosystems I and II. Curr Opin Chem Biol 7:540–550

Krauss N, Hinrichs W, Witt I, Fromme P, Pritzkow W, Dauter Z, Betzel C, Wilson KS, Witt HT, Saenger W (1993) Three-dimensional structure of system I of photosynthesis at 6Å resolution. Nature 361:326–331

Krauß N, Schubert WD, Klukas O, Fromme P, Witt HT, Saenger W (1996) Photosystem I at 4 Å resolution represents the first structural model of a joint photosynthetic reaction centre and core antenna system. Nature struct Mol Biol 3:965–973

Krieger-Liszkay A (2005) Singlet oxygen production in photosynthesis. J Exp Bot 56:337–346

Krieger-Liszkay A, Fufezan C, Trebst A (2008) Singlet oxygen production in photosystem II and related protection mechanism. Photosynth Res 98:551–564

Krishtalik LI (1986) Energetics of multielectron reactions. Photosynthetic oxygen evolution. Biochim Biophys Acta 849:162–171

Krishtalik L (1990) Activation energy of photosynthetic oxygen evolution: an attempt at theoretical analysis. Bioelectrochem Bioenerg 23:249–263

Kritzberg ES, Cole JJ, Pace MM, Granéli W (2006) Bacterial growth on allochthonous carbon in humic and nutrient-enriched lakes: Results from whole-lake 13 C addition experiments. Ecosystems 9:489–499

Kruip J, Bald D, Boekema E, Rögner M (1994) Evidence for the existence of trimeric and monomeric Photosystem I complexes in thylakoid membranes from cyanobacteria. Photosynth Res 40:279–286

Kudoh H, Sonoike K (2002) Irreversible damage to photosystem I by chilling in the light: cause of the degradation of chlorophyll after returning to normal growth temperature. Planta 215:541–548

Kujawinski EB, Longnecker K, Blough NV, Vecchio RD, Finlay L, Kitner JB, Giovannoni SJ (2009) Identification of possible source markers in marine dissolved organic matter using ultrahigh resolution mass spectrometry. Geochim Cosmochim Acta 73:4384–4399

Kume A, Tsuboi N, Satomura T, Suzuki M, Chiwa M, Nakane K, Sakurai N, Horikoshi T, Sakugawa H (2000) Physiological characteristics of Japanese red pine, Pinus densiflora Sieb. et Zucc., in declined forests at Mt. Gokurakuji in Hiroshima prefecture. Japan. Trees 14:305–311

Kuwabara T, Murata N (1983) Quantitative analysis of the inactivation of photosynthetic oxygen evolution and the release of polypeptides and manganese in the photosystem II particles of spinach chloroplasts. Plant Cell Physiol 24:741–747

Kuznetsov AE, Geletii YV, Hill CL, Musaev DG (2010) Insights into the mechanism of O_2 formation and release from the Mn4O4L6 "Cubane" cluster. J Phys Chem A 114:11417–11424

Kvíderová J, Henley WJ (2005) The effect of ampicillin plus streptomycin on growth and photosynthesis of two halotolerant chlorophyte algae. J Appl Phycol 17:301–307

Lacuesta M, Gonzalez-Moro B, González-Murua C, Muñoz-Rueda A (1990) Temporal study of the effect of phosphinothricin on the activity of glutamine synthetase, glutamate dehydrogenase and nitrate reductase in medicago sativa L. J Plant Physiol 136:410–414

Lamelas C, Slaveykova* VI (2007) Comparison of Cd (II), Cu (II), and Pb(II) biouptake by green algae in the presence of humic acid. Environ Sci Technol 41:4172-4178

Lamelas C, Pinheiro JP, Slaveykova VI (2009) Effect of humic acid on Cd (II), Cu (II), and Pb(II) uptake by freshwater algae: kinetic and cell wall speciation considerations. Environ Sci Technol 43:730–735

Landsberg JH (2002) The effects of harmful algal blooms on aquatic organisms. Rev Fish Sci 10:113–390

Latimer WM (1952) The oxidation states of the elements and their potentials in aqueous solution. Prentice-Hall, London

Laurion I, Ventura M, Catalan J, Psenner R, Sommaruga R (2000) Attenuation of ultraviolet radiation in mountain lakes: factors controlling the among-and within-lake variability. Limnol Oceanogr 45:1274–1288

Tezara W, Lawlor, DW (1995) Effects of heat stress on the biochemistry and physiology of photosynthesis in sunflower. In: Mathis P (Ed), Photosynthesis: from Light to Biosphere, Kluwer Academic Publ, Dordrecht – Boston – London, vol IV, pp 625-628

Lawlor DW (2002) Limitation to photosynthesis in water-stressed leaves: stomata vs. metabolism and the role of ATP. Ann Bot 89:871–885

Laws EA, Popp BN, Bidigare RR, Kennicutt MC, Macko SA (1995) Dependence of phytoplankton carbon isotopic composition on growth rate and $[CO_2)]_{aq}$: Theoretical considerations and experimental results. Geochim Cosmochim Acta 59:1131–1138

Lechno S, Zamski E, Tel-Or E (1997) Salt stress-induced responses in cucumber plants. J Plant Physiol 150:206–211

Lee TM, Shiu CT (2009) Implications of mycosporine-like amino acid and antioxidant defenses in UV-B radiation tolerance for the algae species Ptercladiella capillacea and Gelidium amansii. Mar Environ Res 67:8–16

Lee G, Carrow RN, Duncan RR (2004) Photosynthetic responses to salinity stress of halophytic seashore paspalum ecotypes. Plant Sci 166:1417–1425

Leegwater JA, Durrant JR, Klug DR (1997) Exciton equilibration induced by phonons: theory and application to PS II reaction centers. J Phys Chem B 101:7205-7210

Legrand C, Rengefors K, Fistarol GO, Graneli E (2003) Allelopathy in phytoplankton-biochemical, ecological and evolutionary aspects. Phycologia 42:406–419

Lehmann MF, Bernasconi SM, McKenzie JA, Barbieri A, Simona M, Veronesi M (2004) Seasonal variation of the $\delta^{13}C$ and $\delta^{15}N$ of particulate and dissolved carbon and nitrogen in Lake Lugano: constraints on biogeochemical cycling in a eutrophic lake. Limnol Oceanogr 49:415–429

Leslie SB, Teter SA, Crowe LM, Crowe JH (1994) Trehalose lowers membrane phase transitions in dry yeast cells. Biochim Biophys Acta 1192:7–13

Lesser MP, Gorbunov MY (2001) Diurnal and bathymetric changes in chlorophyll fluorescence yields of reef corals measured in situ with a fast repetition rate fluorometer. Mar Ecol Prog Ser 212:69–77

Levine SN, Shambaugh A, Pomeroy SE, Braner M (1997) Phosphorus, nitrogen, and silica as controls on phytoplankton biomass and species composition in Lake Champlain (USA-Canada). J Great Lakes Res 23:131–148

Lewis WM Jr (1990) Comparisons of phytoplankton biomass in temperate and tropical lakes. Limnol Oceanogr 35:1838–1845

Lewis WM Jr (1996) Tropical lakes: how latitude makes a difference. In: Schiemer F, Boland KT (eds) Perspectives in tropical limnology. SPB Academic Publ, Amsterdam, pp 43–64

Lewis WM Jr (2002) Causes for the high frequency of nitrogen limitation in tropical lakes. Verh Int Verein Theor Angew Limnol 28:210–213

Li WKW (1994) Primary production of prochlorophytes, cyanobacteria, and eucaryotic ultraphytoplankton: measurements from flow cytometric sorting. Limnol Oceanogr 39:169–175

Li L, van Staden J (1998) Effects of plant growth regulators on the antioxidant system in callus of two maize cultivars subjected to water stress. Plant Growth Regul 24:55–66

Li XP, Bjorkman O, Shih C, Grossman AR, Rosenquist M, Jansson S, Niyogi KK (2000) A pigment-binding protein essential for regulation of photosynthetic light harvesting. Nature 403:391–395

Li XP, Gilmore AM, Niyogi KK (2002) Molecular and global time-resolved analysis of a psbsgene dosage effect on pH-and xanthophyll cycle-dependent nonphotochemical quenching in photosystem II. J Biol Chem 277:33590–33597

Li XP, Gilmore AM, Caffarri S, Bassi R, Golan T, Kramer D, Niyogi KK (2004) Regulation of photosynthetic light harvesting involves intrathylakoid lumen pH sensing by the PsbS protein. J Biol Chem 279:22866–22874

Li Y, van Der Est A, Lucas MG, Ramesh V, Gu F, Petrenko A, Lin S, Webber AN, Rappaport F, Redding K (2006) Directing electron transfer within photosystem I by breaking H-bonds in the cofactor branches. PNAS 103:2144–2149

Li YG, Jiang GM, Liu MZ, Niu SL, Gao LM, Cao XC (2007) Photosynthetic response to precipitation/rainfall in predominant tree (Ulmus pumila) seedlings in Hunshandak Sandland, China. Photosynthetica 45:133–138

Li W, Wu F, Liu C, Fu P, Wang J, Mei Y, Wang L, Guo J (2008) Temporal and spatial distributions of dissolved organic carbon and nitrogen in two small lakes on the Southwestern China Plateau. Limnology 9:163–171

Li G, Gao K, Gao G (2011) Differential impacts of solar UV radiation on photosynthetic carbon fixation from the coastal to offshore surface waters in the South China Sea. Photochem Photobiol 87:329–334

Liang MC, Hartman H, Kopp RE, Kirschvink JL, Yung YL (2006) Production of hydrogen peroxide in the atmosphere of a Snowball earth and the origin of oxygenic photosynthesis. PNAS 103:18896–18899

Lidon F, Loureiro A, Vieira D, Bilho E, Nobre P, Costa R (2001) Photoinhibition in chilling stressed wheat and maize. Photosynthetica 39:161–166

Lindell MJ, Graneli W, Tranvik LJ (1995) Enhanced bacterial growth in response to photochemical transformation of dissolved organic matter. Limnol Oceanogr 40:195–199

Litchman E, Neale PJ (2005) UV effects on photosynthesis, growth andacclimation of an estuarine diatom and cryptomonad. MEPS 300:53–62

Liu WJ, Yuan S, Zhang NH, Lei T, Duan HG, Liang HG, Lin HH (2006) Effect of water stress on photosystem 2 in two wheat cultivars. Biol Plant 50:597–602

Liu W, Au DWT, Anderson DM, Lam PKS, Wu RSS (2007) Effects of nutrients, salinity, pH and light: dark cycle on the production of reactive oxygen species in the alga Chattonella marina. J Exp Mar Biol Ecol 346:76–86

Liu Z, Dreybrodt W, Wang H (2010) A new direction in effective accounting for the atmospheric CO_2 budget: considering the combined action of carbonate dissolution, the global water cycle and photosynthetic uptake of DIC by aquatic organisms. Earth Sci Rev 99:162–172

Llewellyn LE (2006) Saxitoxin, a toxic marine natural product that targets a multitude of receptors. Nat Prod Rep 23:200–222

Lobanov A, Rubtsova N, Vedeneeva YA, Komissarov G (2008) Photocatalytic activity of chlorophyll in hydrogen peroxide generation in water. Doklady Chem 421:190–193

Lobell DB, Asner GP (2003) Climate and management contributions to recent trends in US agricultural yields. Science 299:1032–1032

Lohrenz SE, Cai WJ, Chen F, Chen X, Tuel M (2010) Seasonal variability in air-sea fluxes of CO_2 in a river-influenced coastal margin. J Geophys Res 115:C10034

Loll B, Kern J, Saenger W, Zouni A, Biesiadka J (2005) Towards complete cofactor arrangement in the 3.0 Å resolution structure of photosystem II. Nature 438:1040–1044

López-Huertas E, Corpas F, Sandalio L, del Rio L (1999) Characterization of membrane polypeptides from pea leaf peroxisomes involved in superoxide radical generation. Biochem J 337:531

Lu CM, Zhang JH (1999) Effects of salt stress on PSII function and photoinhibition in the cyanobacterium Spirulina platensis. J Plant Physiol 155:740–745

Luckas B, Dahlmann J, Erler K, Gerdts G, Wasmund N, Hummert C, Hansen P (2005) Overview of key phytoplankton toxins and their recent occurrence in the North and Baltic Seas. Environ Toxicol 20:1–17

Lukač M, Aegerter R (1993) Influence of trace metals on growth and toxin production of *Microcystis aeruginosa*. Toxicon 31:293–305

Lyubimov V, Zastrizhnaya OM (1992a) Fiziol Rast 39:701–706

Lyubimov V, Zastrizhnaya O (1992b) Role of hydrogen peroxide in photorespiration of C-4 plants. Soviet Plant Physiol 39:454–460

Ma X, Green SA (2004) Photochemical transformation of dissolved organic carbon in Lake Superior–an In situ experiment. J Great Lakes Res 30:97–112

MacIntyre HL, Kana TM, Geider RJ (2000) The effect of water motion on short-term rates of photosynthesis by marine phytoplankton. Trends Plant Sci 5:12–17

Mack J, Bolton JR (1999) Photochemistry of nitrite and nitrate in aqueous solution: a review. J Photochem Photobiol, A 128:1–13

Macpherson AN, Telfer A, Barber J, Truscott TG (1993) Direct detection of singlet oxygen from isolated photosystem II reaction centres. Biochim Biophys Acta 1143:301–309

Madden MC, Hanley N, Harder S, Velez G, Raymer JH (1997) Increased amounts of hydrogen peroxide in the exhaled breath of ozone-exposed human subjects. Inhal Toxicol 9:317–330

Makino A, Miyake C, Yokota A (2002) Physiological functions of the water–water cycle (Mehler reaction) and the cyclic electron flow around PSI in rice leaves. Plant Cell Physiol 43:1017–1026

Maldonado MT, Boyd PW, Harrison PJ, Price NM (1999) Co-limitation of phytoplankton growth by light and Fe during winter in the NE subarctic Pacific Ocean. Deep Sea Res Part II 46:2475–2485

Maldonado MT, Hughes MP, Rue EL, Wells ML (2002) The effect of Fe and Cu on growth and domoic acid production by Pseudo-nitzschia multiseries and Pseudo-nitzschia australis. Limnol Oceanogr 47:515–526

Malkin SY, Guildford SJ, Hecky RE (2008) Modeling the growth response of Cladophora in a Laurentian great lake to the exotic invader dreissena and to lake warming. Limnol Oceanogr 53:1111–1124

Mallet C, Charpin M, Devaux J (1998) Nitrate reductase activity of phytoplankton populations in eutrophic Lake Aydat and meso-oligotrophic Lake Pavin: a comparison. Hydrobiologia 373:135–148

Malone T (1980) Algal size. In: Morris I (ed) The physiological ecology of phytoplankton. Blackwell Scientific Publications, Oxford, pp 433–463

Mann JE, Myers J (1968) Photosynthetic enhancement in the diatom phaeodactylum tricornutum. Plant Physiol 43:1991

Mano J, Takahashi M, Asada K (1987) Oxygen evolution from hydrogen peroxide in photosystem II: flash-induced catalytic activity of water-oxidizing photosystem II membranes. Biochemistry 26:2495–2501

Mano J, Ohno C, Domae Y, Asada K (2001) Chloroplastic ascorbate peroxidase is the primary target of methylviologen-induced photooxidative stress in spinach leaves: its relevance to monodehydroascorbate radical detected with in vivo ESR. Biochim Biophys Acta 1504:275–287

Marañón E, Holligan PM, Barciela R, González N, Mouriño B, Pazó MJ, Varela M (2001) Patterns of phytoplankton size structure and productivity in contrasting open-ocean environments. Mar Ecol Prog Ser 216:43–56

Marcarelli AM, Wurtsbaugh WA, Griset O (2006) Salinity controls phytoplankton response to nutrient enrichment in the Great Salt Lake, Utah, USA. Can J Fish Aquat Sci 63:2236–2248

Marcoval MA, Villafaņe VE, Helbling EW (2008) Combined effects of solar ultraviolet radiation and nutrients addition on growth, biomass and taxonomic composition of coastal marine phytoplankton communities of Patagonia. J Photochem Photobiol, B 91:157–166

Margulis L (1975) Symbiotic theory of the origin of eukaryotic organelles criteria for proof. Symp Soc Exp Biol 29:21–38

Marianne E, Sulzberger B (1999) Atrazine degradation in irradiated iron/oxalate systems: effects of pH and oxalate. Environ Sci Technol 33:2418–2424

Marshall SM, Orr A (1928) The photosynthesis of diatom cultures in the sea. J Mar Biol Assoc UK 15:321–360

Martinez V, Cerda A (1989) Nitrate reductase activity in tomato and cucumber leaves as influenced by NaCl and N source. J Plant Nutrition 12:1335–1350

Martinez V, Cerda A, Fernandez F (1987) Salt tolerance of four tomato hybrids. Plant Soil 97:233–241

Maslenkova L, Gambarova N, Zeinalov Y (1995) NaCl-induced changes in oxygen evolving activity and thylakoid membrane patterns of barley plants. Adaptation to salinity. Bulg J Plant Physiol 21:29–35

Masood A, Shah NA, Zeeshan M, Abraham G (2006) Differential response of antioxidant enzymes to salinity stress in two varieties of *Azolla* (*Azolla pinnata* and *Azolla filiculoides*). Environ Exp Bot 58:216–222

Matear R, Hirst A (2003) Long-term changes in dissolved oxygen concentrations in the ocean caused by protracted global warming. Glob Biogeochem Cy 17:1125. doi:101029/2002GB001997

Matsuda K, Riazi A (1981) Stress-induced osmotic adjustment in growing regions of barley leaves. Plant Physiol 68:571

Matutte B, Awe SO, Ameh FA, Leday AM, Rice JC, Opere CA, Ohia SE (2000) Role of catalase in pre-and postjunctional responses of mammalian irides to hydrogen peroxide. J Ocul Pharmacol Ther 16:429–438

Matysik J, Bhalu B, Mohanty P (2002) Molecular mechanisms of quenching of reactive oxygen species by proline under stress in plants. Curr Sci 82:525–532

Mayer P, Cuhel R, Nyholm N (1997) A simple in vitro fluorescence method for biomass measurements in algal growth inhibition tests. Water Res 31:2525–2531

Mazorra L, Nunez M, Hechavarria M, Coll F, Sánchez-Blanco MJ (2002) Influence of brassinosteroids on antioxidant enzymes activity in tomato under different temperatures. Biol Plant 45:593–596

McCarthy MJ, Lavrentyev PJ, Yang L, Zhang L, Chen Y, Qin B, Gardner WS (2007) Nitrogen dynamics and microbial food web structure during a summer cyanobacterial bloom in a subtropical, shallow, well-mixed, eutrophic lake (Lake Taihu, China). Hydrobiologia 581:195–207

McCarthy MJ, James RT, Chen Y, East TL, Gardner WS (2009) Nutrient ratios and phytoplankton community structure in the large, shallow, eutrophic, subtropical Lakes Okeechobee (Florida, USA) and Taihu (China). Limnology 10:215–227

McNeil SD, Nuccio ML, Ziemak MJ, Hanson AD (2001) Enhanced synthesis of choline and glycine betaine in transgenic tobacco plants that overexpress phosphoethanolamine N-methyltransferase. PNAS 98:10001–10005

McQuatters-Gollop A, Mee LD, Raitsos DE, Shapiro GI (2008) Non-linearities, regime shifts and recovery: the recent influence of climate on Black Sea chlorophyll. J Mar Systems 74:649–658

Medrano H, Escalona J, Bota J, Gulias J, Flexas J (2002) Regulation of photosynthesis of C3 plants in response to progressive drought: stomatal conductance as a reference parameter. Ann Bot 89:895–905

Meeks JC, Castenholz RW (1971) Growth and photosynthesis in an extreme thermophile, Synechococcus lividus (Cyanophyta). Arch Microbiol 78:25–41

Mehler AH (1951) Studies on reactions of illuminated chloroplasts: I. Mechanism of the reduction of oxygen and other hill reagents. Arch Biochem Biophys 33:65–77

Meili M, Jonsson A, Jansson M (2000) Seasonal dynamics of plankton and carbon stable isotopes (δ13C) in a large humic lake (Örträsket, N Sweden). Verh Int Ver Limnol 27:1940–1942

Melgar J, Syvertsen J, Martínez V, García-Sánchez F (2008) Leaf gas exchange, water relations, nutrient content and growth in citrus and olive seedlings under salinity. Biol Plant 52:385–390

Melis A (1999) Photosystem-II damage and repair cycle in chloroplasts: what modulates the rate of photodamage in vivo? Trends Plant Sci 4:130–135

Meloni DA, Oliva MA, Martinez CA, Cambraia J (2003) Photosynthesis and activity of superoxide dismutase, peroxidase and glutathione reductase in cotton under salt stress. Environ Exp Bot 49:69–76

Meneguzzo S, Navam-Izzo F, Izzo R (1999) Antioxidative responses of shoots and roots of wheat to increasing NaCI concentrations. J Plant Physiol 155:274–280

Menzel A, Fabian P (1999) Growing season extended in Europe. Nature 397:659–659

Middelboe M (2003) Microbial disease in the sea: Effects of viruses on marine carbon and nutrient cycling. In: Ostfeld RS, Keesing F, Eviner VT (Eds), Infectious disease ecology: effects of ecosystems on disease and of disease on ecosystems, Princeton University Press: Princeton, NJ, pp 242–259

Middelboe M (2008) Microbial disease in the sea: Effects of viruses on marine carbon and nutrient cycling. In: Ostfeld RS, Keesing F, Eviner VT (Eds), Infectious disease ecology:

effects of ecosystems on disease and of disease on *ecosystems*, Princeton University Press: Princeton, NJ, pp 242–259
Miller WL, Zepp RG (1995) Photochemical production of dissolved inorganic carbon from terrestrial organic matter: Significance to the oceanic organic. Geophys Res Lett 22:417–420
Miller-Morey JS, van Dolah FM (2004) Differential responses of stress proteins, antioxidant enzymes, and photosynthetic efficiency to physiological stresses in the Florida red tide dinoflagellate, *Karenia brevis*. Comp Biochem Physiol Part C 138:493–505
Millero FJ, Roy RN (1997) A chemical equilibrium model for the carbonate system in natural waters. Croat Chem Acta 70:1–38
Minero C, Chiron S, Falletti G, Maurino V, Pelizzetti E, Ajassa R, Carlotti ME, Vione D (2007) Photochemincal processes involving nitrite in surface water samples. Aquat Sci 69:71–85
Mittler R (2002) Oxidative stress, antioxidants and stress tolerance. Trends Plant Sci 7:405–410
Mittova V, Tal M, Volokita M, Guy M (2002) Salt stress induces up-regulation of an efficient chloroplast antioxidant system in the salt-tolerant wild tomato species Lycopersicon pennellii but not in the cultivated species. Physiol Plant 115:393–400
Miyake C, Asada K (1992) Thylakoid-bound ascorbate peroxidase in spinach chloroplasts and photoreduction of its primary oxidation product monodehydroascorbate radicals in thylakoids. Plant Cell Physiol 33:541–553
Miyake C, Yokota A (2000) Determination of the rate of photoreduction of O2 in the water–water cycle in watermelon leaves and enhancement of the rate by limitation of photosynthesis. Plant Cell Physiol 41:335–343
Miyake C, Yokota A (2001) Cyclic flow of electrons within PSII in thylakoid membranes. Plant Cell Physiol 42:508–515
Miyake C, Yonekura K, Kobayashi Y, Yokota A (2002) Cyclic electron flow within PSII functions in intact chloroplasts from spinach leaves. Plant Cell Physiol 43:951–957
Miyao M, Murata N (1983) Partial disintegration and reconstitution of the photosynthetic oxygen evolution system. Binding of 24 kilodalton and 18 kilodalton polypeptides. Biochim Biophys Acta 725:87–93
Moffett JW, Zafiriou OC (1990) An Investigation of hydrogen peroxide chemistry in surface waters of Vineyard Sound with $H_2{}^{18}O_2$ and $^{18}O_2$. Limnol Oceanogr 35:1221–1229
Moffett JW, Zika RG (1987a) Photochemistry of a copper complexes in sea water. In: Zika RG, Cooper WJ (eds) Photochemistry of environmental aquatic systems, ACS Sym Ser 327. Am Chem Soc, Washington DC, pp 116–130
Moffett JW, Zika RG (1987b) Reaction kinetics of hydrogen peroxide with copper and iron in seawater. Environ Sci Technol 21:804–810
Moffett JW, Zika RG, Petasne RG (1985) Evaluation of bathocuproine for the spectro-photometric determination of copper (I) in copper redox studies with applications in studies of natural waters. Anal Chim Acta 175:171–179
Mohlin M, Wulff A (2009) Interaction effects of ambient UV radiation and nutrient limitation on the toxic cyanobacterium Nodularia spumigena. Microb Ecol 57:675–686
Moisander P, McClinton E, Paerl H (2002) Salinity effects on growth, photosynthetic parameters, and nitrogenase activity in estuarine planktonic cyanobacteria. Microb Ecol 43:432–442
Molot LA, Hudson JJ, Dillon PJ, Miller SA (2005) Effect of pH on photo-oxidation of dissolved organic carbon by hydroxyl radicals in a coloured, softwater stream. Aquat Sci 67:189–195
Moon BY, Higashi S, Gombos Z, Murata N (1995) Unsaturation of the membrane lipids of chloroplasts stabilizes the photosynthetic machinery against low-temperature photoinhibition in transgenic tobacco plants. PNAS 92:6219–6223
Moore R (1996) Cyclic peptides and depsipeptides from cyanobacteria: a review. J Ind Microbiol Biotechnol 16:134–143
Moore CM, Mills MM, Milne A, Langlois R, Achterberg EP, Lochte K, Geider RJ, La Roche J (2006) Iron limits primary productivity during spring bloom development in the central North Atlantic. Glob Change Biol 12:626–634
Moore TR, Paré D, Boutin R (2008) Production of dissolved organic carbon in Canadian forest soils. Ecosystems 11:740–751

Mopper K, Zhou X (1990) Hydroxyl radical photoproduction in the sea and its potential impact on marine processes. Science 250:661–664
Moran MA, Zepp RG (1997) Role of photoreactions in the formation of biologically labile compounds from dissolved organic matter. Limnol Oceanogr 42:1307–1316
Moran JF, Becana M, Iturbe-Ormaetxe I, Frechilla S, Klucas RV, Aparicio-Tejo P (1994) Drought induces oxidative stress in pea plants. Planta 194:346–352
Moran MA, Sheldon WM Jr, Zepp RG (2000) Carbon loss and optical property changes during long-term photochemical and biological degradation of estuarine dissolved organic matter. Limnol Oceanogr 45:1254–1264
Morel A, Gentili B, Claustre H, Babin M, Bricaud A, Ras J, Tieche F (2007) Optical properties of the" clearest" natural waters. Limnol Oceanogr 52(1):217–229
Morilla CA, Boyer J, Hageman R (1973) Nitrate reductase activity and polyribosomal content of corn (Zea mays L.) having low leaf water potentials. Plant Physiol 51:817–824
Morris DP, Hargreaves BR (1997) The role of photochemical degradation of dissolved organic carbon in regulating the UV transparency of three lakes on the Pocono Plateau. Limnol Oceanogr 42:239–249
Morris J, Ningnan Z, Zengjiang Y, Collopy J, Daping X (2004) Water use by fast-growing Eucalyptus urophylla plantations in southern China. Tree Physiol 24:1035–1044
Mortain-Bertrand A, Descolas-Gros C, Jupin H (1988) Growth, photosynthesis and carbon metabolism in the temperate marine diatom Skeletonema costatum adapted to low temperature and low photon-flux density. Mar Biol 100:135–141
Mostofa KMG LC, Wu FC, Fu PQ, Ying WL, Yuan J (2009) Overview of key biogeochemical functions in lake ecosystem: Impacts of organic matter pollution and global warming. Proceedings of the 13 th World Lake Conf Wuhan, China, 1-5 Nov 2009, Keynote Speech, pp 59-60
Mostofa KMG, Sakugawa H (2009) Spatial and temporal variations and factors controlling the concentrations of hydrogen peroxide and organic peroxides in rivers. Environ Chem 6:524–534
Mostofa KMG, Yoshioka T, Konohira E, Tanoue E, Hayakawa K, Takahashi M (2005) Three-dimensional fluorescence as a tool for investigating the dynamics of dissolved organic matter in the Lake Biwa watershed. Limnology 6:101–115
Mostofa KMG, Yoshioka T, Konohira E, Tanoue E (2007) Photodegradation of fluorescent dissolved organic matter in river waters. Geochem J 41:323–331
Mostofa K, Wu FC, Yoshioka T, Sakugawa H, Tanoue E (2009) Dissolved organic matter in the aquatic environments. In: Wu FC, Xing B (eds) Natural organic matter and its significance in the environment. Science Press, Beijing, pp 3–66
Mostofa KMG, Wu F, Liu CQ, Fang WL, Yuan J, Ying WL, Wen L, Yi M (2010) Characterization of Nanming river (Southwestern China) sewerage-impacted pollution using an excitation-emission matrix and PARAFAC. Limnology 11:217–231
Mostofa KMG, Wu F, Liu CQ, Vione D, Yoshioka T, Sakugawa H, Tanoue E (2011) Photochemical, microbial and metal complexation behavior of fluorescent dissolved organic matter in the aquatic environments. Geochem J 45:235–254
Muller F (1933) On the metabolism of the purple sulfur bacteria in organic media. Arch Mikrobiol 4:131–166
Müller MG, Slavov C, Luthra R, Redding KE, Holzwarth AR (2010) Independent initiation of primary electron transfer in the two branches of the photosystem I reaction center. PNAS 107:4123–4128
Mullineaux CW, Pascal AA, Horton P, Holzwarth AR (1993) Excitation-energy quenching in aggregates of the LHC II chlorophyll-protein complex: a time-resolved fluorescence study. Biochim Biophys Acta 1141:23–28
Munns R, Termaat A (1986) Whole-plant responses to salinity. Funct Plant Biol 13:143–160
Munns R, Brady C, Barlow E (1979a) Solute accumulation in the apex and leaves of wheat during water stress. Funct Plant Biol 6:379–389
Munns R, Brady C, Barlow E (1979b) Solute accumulation in the apex and leaves of wheat during water stress. Aust J Plant Physiol 6:379–389

Munns R, Schachtman D, Condon A (1995) The significance of a two-phase growth response to salinity in wheat and barley. Funct Plant Biol 22:561–569
Murata N, Miyao M (1985) Extrinsic membrane proteins in the photosynthetic oxygen-evolving complex. Trends Biochem Sci 10:122–124
Murata N, Ishizaki-Nishizawa O, Higashi S, Hayashi H, Tasaka Y, Nishida I (1992) Genetically engineered alteration in the chilling sensitivity of plants. Nature 356:710–713
Murata N, Takahashi S, Nishiyama Y, Allakhverdiev SI (2007) Photoinhibition of photosystem II under environmental stress. Biochim Biophys Acta 1767:414–421
Murray JW, Barber J (2007) Structural characteristics of channels and pathways in photosystem II including the identification of an oxygen channel. J Struct Biol 159:228–237
Murray JW, Maghlaoui K, Kargul J, Ishida N, Lai TL, Rutherford AW, Sugiura M, Boussac A, Barber J (2008) X-ray crystallography identifies two chloride binding sites in the oxygen evolving centre of Photosystem II. Energy Environ Sci 1:161–166
Muthuchelian K, Bertamini M, Nedunchezhian N (2001) Iron deficiency induced changes on electron transport rate in Pisum sativum chloroplasts. Biol Plant 44:313–316
Mylon SE, Twining BS, Fisher NS, Benoit G (2003) Relating the speciation of Cd, Cu, and Pb in two connecticut rivers with their uptake in algae. Environ Sci Technol 37:1261–1267
Myneni RB, Keeling C, Tucker C, Asrar G, Nemani R (1997) Increased plant growth in the northern high latitudes from 1981 to 1991. Nature 386:698–702
Nakane K, Kohno T, Horikoshi T, Nakatsubo T (1997) Soil carbon cycling at a black spruce (Picea mariana) forest stand in Saskatchewan, Canada. J Geophys Res 103:28–28
Nakatani N, Ueda M, Shindo H, Takeda K, Sakugawa H (2007) Contribution of the photo-Fenton reaction to hydroxyl radical formation rates in river and rain water samples. Anal Sci 23:1137–1142
Nam N, Subbarao G, Johansen C, Chauhan Y (1998) Importance of canopy attributes in determining dry matter accumulation of pigeonpea under contrasting moisture regimes. Crop Sci 38:955–961
Nanba O, Satoh K (1987) Isolation of a photosystem II reaction center consisting of D1 and D2 polypeptides and cytochrome b-559. PNAS 84:109–112
Neale PJ, Melis A (1989) Salinity-stress enhances photoinhibition of photosynthesis in *chlamydomonas reinhardtii*. J Plant Physiol 134:619–622
Negri AP, Jones GJ, Hindmarsh M (1995) Sheep mortality associated with paralytic shellfish poisons from the cyanobacterium *Anabaena circinalis*. Toxicon 33:1321–1329
Nesnow S, Roop BC, Lambert G, Kadiiska M, Mason RP, Cullen WR, Marc J (2002) DNA damage induced by methylated trivalent arsenicals is mediated by reactive oxygen species. Chem Res Toxicol 15:1627–1634
Neufeld S, Zinchenko V, Stephan D, Bader K, Pistorius E (2004) On the functional significance of the polypeptide PsbY for photosynthetic water oxidation in the cyanobacterium Synechocystis sp. strain PCC 6803. Mol Gen Genomics 271:458–467
Neumann P (1997) Salinity resistance and plant growth revisited. Plant, Cell Environ 20:1193–1198
Neumann P, Azaizeh H, Leon D (1994) Hardening of root cell walls: a growth inhibitory response to salinity stress. Plant, Cell Environ 17:303–309
Nielsen ES (1951) The marine vegetation of the Isefjord. A study on ecology and production. Ser Plankton 5:1–114
Nielsen E (1952a) The use of radioactive carbon (C14) for measuring organic production in the sea. J Cons Int Explor Mer 18:117–140
Nielsen ES (1952b) On detrimental effects of high light intensities on the photosynthetic mechanism. Physiol Plant 5:334–344
Nilsson Lill S (2011) On the dimerization of chlorophyll in photosystem II. Phys Chem Chem Phys 13:16022–16027
Nimer NA, Brownlee C, Merrett MJ (1999) Extracellular carbonic anhydrase facilitates carbon dioxide availability for photosynthesis in the marine dinoflagellateprorocentrum micans. Plant Physiol 120:105–112

Nishiyama Y, Los DA, Hayashi H, Murata N (1997) Thermal protection of the oxygen-evolving machinery by PsbU, an extrinsic protein of photosystem II, in *Synechococcus* species PCC 7002. Plant Physiol 115:1473–1480
Nishiyama Y, Los DA, Murata N (1999) PsbU, a protein associated with photosystem II, is required for the acquisition of cellular thermotolerance in *Synechococcus* species PCC 7002. Plant Physiol 120:301–308
Nishiyama Y, Yamamoto H, Allakhverdiev SI, Inaba M, Yokota A, Murata N (2001) Oxidative stress inhibits the repair of photodamage to the photosynthetic machinery. EMBO J 20:5587–5594
Nishiyama Y, Allakhverdiev SI, Yamamoto H, Hayashi H, Murata N (2004) Singlet oxygen inhibits the repair of photosystem II by suppressing thc translation elongation of the D1 protein in Synechocystis sp. PCC 6803. Biochemistry 43:11321–11330
Nishiyama Y, Allakhverdiev SI, Murata N (2005) Inhibition of the repair of photosystem II by oxidative stress in cyanobacteria. Photosynth Res 84:1–7
Nishiyama Y, Allakhverdiev SI, Murata N (2006) A new paradigm for the action of reactive oxygen species in the photoinhibition of photosystem II. Biochim Biophys Acta 1757:742–749
Nishiyama Y, Allakhverdiev SI, Murata N (2008) Regulation by environmental conditions of the repair of photosystem II in cyanobacteria. In: Demmig-Adams B, Adams III WW, Mattoo AK (Eds), Photoprotection, Photoinhibition, Gene Regulation, and Environment, Springer, pp 193-203
Nishiyama Y, Allakhverdiev SI, Murata N (2009) Regulation by environmental conditions of the repair of photosystem II in cyanobacteria. In: Demmig-Adams B, Adams III WW, Mattoo AK (Eds), Photoprotection, photoinhibition, gene regulation, and environment, Springer, pp 193–203
Nixon P, Diner B (1994) Analysis of water-oxidation mutants constructed in the cyanobacterium Synechocystis sp. PCC 6803. Biochem Soc Trans 22:338–343
Nixon PJ, Barker M, Boehm M, de Vries R, Komenda J (2005) FtsH-mediated repair of the photosystem II complex in response to light stress. J Exp Bot 56:357–363
Nõges T, Laugaste R, Noges P, Tonno I (2008) Critical N: P ratio for cyanobacteria and N 2-fixing species in the large shallow temperate lakes Peipsi and Võrtsjärv, North-East Europe. Hydrobiologia 599:77–86
Nováková M, Matějová E, Sofrová D (2004) Cd 2 + effect on photosynthetic apparatus in synechococcus elongatus and spinach (Spinacia oleracea L.). Photosynthetica 42:425–430
Nowak D, Antczak A, Krol M, Pietras T, Shariati B, Bialasiewicz P, Jeczkowski K, Kula P (1996) Increased content of hydrogen peroxide in the expired breath of cigarette smokers. Eur Respir J 9:652–657
Nozaki K, Morino H, Munehara H, Sideleva VG, Nakai K, Yamauchi M, Kozhova OM, Nakanishi M (2002) Composition, biomass, and photosynthetic activity of the benthic algal communities in a littoral zone of Lake Baikal in summer. Limnology 3:175–180
Núñez M, Mazzafera P, Mazorra L, Siqueira W, Zullo M (2003) Influence of a brassinosteroid analogue on antioxidant enzymes in rice grown in culture medium with NaCl. Biol Plant 47:67–70
Ogweno JO, Song XS, Shi K, Hu WH, Mao WH, Zhou YH, Yu JQ, Nogués S (2008) Brassinosteroids alleviate heat-induced inhibition of photosynthesis by increasing carboxylation efficiency and enhancing antioxidant systems in Lycopersicon esculentum. J Plant Growth Regul 27:49–57
Ohashi Y, Nakayama N, Saneoka H, Fujita K (2006) Effects of drought stress on photosynthetic gas exchange, chlorophyll fluorescence and stem diameter of soybean plants. Biol Plant 50:138–141
Ohmoto H (1997) When did the Earth's atmosphere become oxic. Geochem News 93:26
Ohnishi N, Murata N (2006) Glycinebetaine counteracts the inhibitory effects of salt stress on the degradation and synthesis of D1 protein during photoinhibition in *Synechococcus* sp. PCC 7942. Plant Physiol 141:758–765
Okada K, Ikeuchi M, Yamamoto N, Ono T, Miyao M (1996) Selective and specific cleavage of the D1 and D2 proteins of photosystem II by exposure to singlet oxygen: factors responsible for the susceptibility to cleavage of the proteins. Biochim Biophys Acta 1274:73–79

O'Leary MH (1981) Carbon isotope fractionation in plants. Phytochemistry 20:553–567
Omar A, Olsen A, Johannessen T, Hoppema M, Thomas H, Borges A (2010) Spatiotemporal variations of fCO_2 in the North Sea. Ocean Sci 6:77–89
Öquist G, Huner N (1991) Effects of cold acclimation on the susceptibility of photosynthesis to photoinhibition in Scots pine and in winter and spring cereals: a fluorescence analysis. Func Ecol 5:91–100
Öquist G, Huner NPA (2003) Photosynthesis of overwintering evergreen plants. Annu Rev Plant Biol 54:329–355
Öquist G, Hurry V, Huner N (1993) The temperature dependence of the redox state of QA and susceptibility of photosynthesis to photoinhibition. Plant Physiol Biochem 31:683–691
Öquist G, Campbell D, Clarke AK, Gustafsson P (1995) The cyanobacterium Synechococcus modulates photosystem II function in response to excitation stress through D1 exchange. Photosynth Res 46:151–158
Ortega-Retuerta E, Pulido-Villena E, Reche I (2007) Effects of dissolved organic matter photoproducts and mineral nutrient supply on bacterial growth in Mediterranean inland waters. Microb Ecol 54:161–169
Osburn CL, Retamal L, Vincent WF (2009) Photoreactivity of chromophoric dissolved organic matter transported by the Mackenzie river to the Beaufort Sea. Mar Chem 115:10–20
Osmond C (1981) Photorespiration and photoinhibition: some implications for the energetics of photosynthesis. Biochim Biophys Acta 639:77–98
Osmond B (1997) Too many photons: photorespiration, photoinhibition and photooxidation. Trends Plant Sci 2:119–121
Osmond C, Grace S (1995) Perspectives on photoinhibition and photorespiration in the field: quintessential inefficiencies of the light and dark reactions of photosynthesis? J Exp Bot 46:1351
Paerl HW, Huisman J (2008) Blooms like it hot. Science 320:57–58
Pahlow M, Riebesell U (2000) Temporal trends in deep ocean redfield ratios. Science 287:831–833
Pal M, Singh D, Rao L, Singh K (2004) Photosynthetic characteristics and activity of antioxidant enzymes in salinity tolerant and sensitive rice cultivars. Ind J Plant Physiol 9:407–412
Palenik B, Zafiriou O, Morel F (1987) Hydrogen peroxide production by a marine phytoplankter. Limnol Oceanogr 32:1365–1369
Pan X, Chen X, Zhang D, Wang J, Deng C, Mu G, Zhu H (2009a) Effect of chromium(VI) on photosystem II activity and heterogeneity of synechocystis sp. (Cyanophyta): studied with in vivo chlorophyll fluorescence tests 1. J Phycol 45:386–394
Pan X, Zhang D, Chen X, Mu G, Li L, Bao A (2009b) Effects of levofloxacin hydrochlordie on photosystem II activity and heterogeneity of *Synechocystis* sp. Chemosphere 77:413–418
Pandey D, Yeo UD (2008) Stress-induced degradation of D1 protein and its photoprotection by DCPIP in isolated thylakoid membranes of barley leaf. Biol Plant 52:291–298
Pandey D, Choi I, Yeo UD (2009) Photosystem 2-activity and thylakoid membrane polypeptides of in vitro cultured chrysanthemum as affected by NaCl. Biol Plant 53:329–333
Paneth P, O'Leary MH (1985) Carbon isotope effect on dehydration of bicarbonate ion catalyzed by carbonic anhydrase. Biochemistry 24:5143–5147
Papadopoulos I, Rendig V (1983) Interactive effects of salinity and nitrogen on growth and yield of tomato plants. Plant Soil 73:47–57
Parida AK, Das A, Mittra B (2004) Effects of salt on growth, ion accumulation, photosynthesis and leaf anatomy of the mangrove, Bruguiera parviflora. Trees 18:167–174
Parkhill JP, Maillet G, Cullen JJ (2001) Fluorescence-based maximal quantum yield for PSII as a diagnostic of nutrient stress. J Phycol 37:517–529
Parlanti E, Wörz K, Geoffroy L, Lamotte M (2000) Dissolved organic matter fluorescence spectroscopy as a tool to estimate biological activity in a coastal zone submitted to anthropogenic inputs. Org Geochem 31:1765–1781
Parmon V (1985) In: Fotokataliticheskoe preobrazovanie solnechnoi energii, Ch 2 Molekulyarnye sistemy dlya razlozheniya vody (Photocatalytic Sunlight Conversion, part 2: Molecular Systems for Water Decomposition). Nauka, Novosibirsk

Pastenes C, Horton P (1996a) Effect of high temperature on photosynthesis in beans (I. Oxygen evolution and chlorophyll fluorescence). Plant Physiol 112:1245–1251
Pastenes C, Horton P (1996b) Effect of high temperature on photosynthesis in beans (II. CO2 assimilation and metabolite contents). Plant Physiol 112:1253–1260
Pastori G, Trippi V (1993) Antioxidative protection in a drought-resistant maize strain during leaf senescence. Physiol Plant 87:227–231
Pattanaik B, Roleda MY, Schumann R, Karsten U (2008) Isolate-specific effects of ultraviolet radiation on photosynthesis, growth and mycosporine-like amino acids in the microbial mat-forming cyanobacterium microcoleus chthonoplastes. Planta 227:907–916
Peltzer D, Dreyer E, Polle A (2002) Differential temperature dependencies of antioxidative enzymes in two contrasting species: Fagus sylvatica and Coleus blumei. Plant Physiol Biochem 40:141–150
Peng M, Kuc J (1992) Peroxidase-generated hydrogen peroxide as a source of antifungal activity in vitro and on tobacco leaf disks. Phytopathology 82:696–699
Peñuelas J, Filella I (2001) Responses to a warming world. Science 294:793–795
Perales-Vela HV, González-Moreno S, Montes-Horcasitas C, Cañizares-Villanueva RO (2007) Growth, photosynthetic and respiratory responses to sub-lethal copper concentrations in *Scenedesmus incrassatulus* (Chlorophyceae). Chemosphere 67:2274–2281
Periasamy N, Linschitz H, Closs G, Boxer S (1978) Photoprocesses in covalently linked pyrochlorophyllide dimer: triplet state formation and opening and closing of hydroxylic linkages. PNAS 75:2563
Petasne RG, Zika RG (1997) Hydrogen peroxide lifetimes in south Florida coastal and offshore waters. Mar Chem 56:215–225
Pfannschmidt T (2003) Chloroplast redox signals: how photosynthesis controls its own genes. Trends Plant Sci 8:33–41
Pfenning N (1978) General physiology and ecology of photosynthetic bacteria. In: Sistrom WR (ed) Clayton RK. The photosynthetic bacteria Plenum Press, New York, pp 3–18
Philips EJ, Cichra M, Havens K, Hanton C, Badylak S, Rueter B, Randall M, Hansen P (1997) Relationships between phytoplankton dynamics and the availability of light and nutrients in a shallow sub-tropical lake. J Plankton Res 19:319–342
Phlips E, Badylak S, Youn S, Kelley K (2004) The occurrence of potentially toxic dinoflagellates and diatoms in a subtropical lagoon, the Indian River Lagoon, Florida, USA. Harmful Algae 3:39–49
Pick FR, Lean DRS (1987) The role of macronutrients (C, N, P) in controlling cyanobacterial dominance in temperate lakes. New Zeal J Mar Freshwater Res 21:425–434
Plattner GK, Joos F, Stocker T, Marchal O (2001) Feedback mechanisms and sensitivities of ocean carbon uptake under global warming. Tellus B 53:564–592
Pope DH (1975) Effects of light intensity, oxygen concentration, and carbon dioxide concentration on photosynthesis in algae. Microb Ecol 2:1–16
Porta NL, Bertamini M, Nedunchezhian N, Muthuchelian K (2004) High irradiance induced changes of photosystem 2 in young and mature needles of cypress (Cupressus sempervirens L.). Photosynthetica 42:263–271
Powles SB (1984) Photoinhibition of photosynthesis induced by visible light. Annu Rev Plant Physiol 35:15–44
Prabha GL, Kulandaivelu G (2002) Induced UV-B resistance against photosynthesis damage by adaptive mutagenesis in Synechococcus PCC 7942. Plant Sci 162:663–669
Prakash A, Rashid M, Jensen A, Rao DVS (1973) Influence of humic substances on the growth of marine phytoplankton: diatoms. Limnol Oceanogr 18:516–524
Prasad S, Singh J, Rai L, Kumar H (1991) Metal-induced inhibition of photosynthetic electron transport chain of the cyanobacterium *Nostoc muscorum*. FEMS Microbiol Lett 82:95–100
Prasil O, Adir N, Ohad I (1992) Dynamics of photosystem II: mechanism of photoinhibition and recovery process. In: Barber J (ed) Topics in photosynthesis, the photosystems: structure, function and molecular biology. Elsevier, Amsterdam, pp 220–250

Pratt DM (1959) The phytoplankton of Narragansett Bay. Limnol Oceanogr 4:425–440
Premkumar J, Ramaraj R (1999) Photoreduction of dioxygen to hydrogen peroxide at porphyrins and phthalocyanines adsorbed Nafion membrane. J Mol Catal 142:153–162
Prince EK, Myers TL, Kubanek J (2008) Effects of harmful algal blooms on competitors: allelopathic mechanisms of the red tide dinoflagellate" Karenia Brevis". Limnol Oceanogr 53:531–541
Priscu JC, Fritsen CH, Adams EE, Giovannoni SJ, Paerl HW, McKay CP, Doran PT, Gordon DA, Lanoil BD, Pinckney JL (1998) Perennial Antarctic lake ice: an oasis for life in a polar desert. Science 280:2095–2098
Probyn T (1992) The inorganic nitrogen nutrition of phytoplankton in the southern Benguela: new production, phytoplankton size and implications for pelagic foodwebs. S Afr J Mar Sci 12:411–420
Prochazkova D, Wilhelmova N (2007) Leaf senescence and activities of the antioxidant enzymes. Biol Plant 51:401–406
Prokhorenko VI, Holzwarth AR (2000) Primary processes and structure of the photosystem II reaction center: a photon echo study. J Phys Chem B 104:11563–11578
Provasoli L (1963) Organic regulation of phytoplankton fertility. In: Hill MN (Ed), The composition of seawater. Comparative and descriptive oceanography. The sea: ideas and observations on progress in the study of the seas, 2, New York, pp 165-219
Psenner R, Sattler B (1998) Life at the freezing point. Science 280:2073–2074
Qian J, Mopper K, Kieber DJ (2001) Photochemical production of the hydroxyl radical in Antarctic waters. Deep Sea Res Part I 48:741–759
Qiu N, Lu C (2003) Enhanced tolerance of photosynthesis against high temperature damage in salt-adapted halophyte atriplex centralasiatica plants. Plant, Cell Environ 26:1137–1145
Qu W, Su C, West R, Morrison R (2004) Photosynthetic characteristics of benthic microalgae and seagrass in Lake Illawarra, Australia. Hydrobiologia 515:147–159
Quigg A, Finkel ZV, Irwin AJ, Rosenthal Y, Ho TY, Reinfelder JR, Schofield O, Morel FMM, Falkowski PG (2003) The evolutionary inheritance of elemental stoichiometry in marine phytoplankton. Nature 425:291–294
Quigg A, Irwin AJ, Finkel ZV (2011) Evolutionary inheritance of elemental stoichiometry in phytoplankton. Proc R Soc Lond B 278:526–534
Rabinowitch EI (1951) Photosynthesis and related processes, vol II, Part 1, Interscience Publ Inc, New York
Rajagopal S, Murthy S, Mohanty P (2000) Effect of ultraviolet-B radiation on intact cells of the cyanobacterium Spirulina platensis: characterization of the alterations in the thylakoid membranes. J Photochem Photobiol, B 54:61–66
Rakhimberdieva MG, Boichenko VA, Karapetyan NV, Stadnichuk IN (2001) Interaction of phycobilisomes with photosystem II dimers and photosystem I monomers and trimers in the cyanobacterium Spirulina platensis. Biochemistry 40:15780–15788
Ramachandran S, Morris SM, Devamanoharan P, Henein M, Varma SD (1991) Radioisotopic determination of hydrogen peroxide in aqueous humor and urine. Exp Eye Res 53:503–506
Randall C, Harvey V, Manney G, Orsolini Y, Codrescu M, Sioris C, Brohede S, Haley C, Gordley L, Zawodny J (2005) Stratospheric effects of energetic particle precipitation in 2003–2004. Geophys Res Lett 32:L05802
Rappaport F, Diner BA (2008) Primary photochemistry and energetics leading to the oxidation of the (Mn) 4Ca cluster and to the evolution of molecular oxygen in Photosystem II. Coord Chem Rev 252:259–272
Rastogi RP, Sinha RP, Singh SP, Häder DP (2010) Photoprotective compounds from marine organisms. J Ind Microbiol Biotechnol 37:537–558
Rath J, Adhikary SP (2007) Response of the estuarine cyanobacterium Lyngbya aestuarii to UV-B radiation. J Appl Phycol 19:529–536
Rau G, Takahashi T, Des Marais D, Repeta D, Martin J (1992) The relationship between δI3C of organic matter and $[CO]_{(aq)}]$ in ocean surface water data from a JGOFS site in the northeast Atlantic Ocean and a model. Geochim Cosmochim Acta 56:1413–1419

Raven J (1991) Responses of aquatic photosynthetic organisms to increased solar UVB. J Photochem Photobiol B: Biol 9:239–244
Raven J (1997) Inorganic carbon acquisition by marine autotrophs. Adv Bot Res 27:85–209
Raven JA, Farquhar GD (1990) The influence of N metabolism and organic acid synthesis on the natural abundance of isotopes of carbon in plants. New Phytol 116:505–529
Raven J, Johnston A, Turpin D (1993) Influence of changes in CO_2 concentration and temperature on marine phytoplankton $^{13}C/^{12}C$ ratios: an analysis of possible mechanisms. Glob Planet Change 8:1–12
Raven JA, Johnston AM, Kübler JE, Korb R, McInroy SG, Handley LL, Scrimgeour CM, Walker DI, Beardall J, Vanderklift M (2002) Mechanistic interpretation of carbon isotope discrimination by marine macroalgae and seagrasses. Funct Plant Biol 29:355–378
Ravikovitch S, Porath A (1967) The effect of nutrients on the salt tolerance of crops. Plant Soil 26:49–71
Reche I, Pace M, Cole J (1998) Interactions of photobleaching and inorganic nutrients in determining bacterial growth on colored dissolved organic carbon. Microb Ecol 36:270–280
Redding K, van der Est A (2006) The directionality of electron transfer in photosystem I. In: Golbeck JH (Ed), Photosystem I: The Light-Driven Plastocyanin:Ferredoxin Oxidoreductase, Springer, The Netherlands, Vol 24, pp 413–437
Reddy AR, Chaitanya KV, Vivekanandan M (2004) Drought-induced responses of photosynthesis and antioxidant metabolism in higher plants. J Plant Physiol 161:1189–1202
Reed R, Stewart W (1988) The responses of cyanobacteria to salt stress. In: Rogers LJ, Gallan LJ (Eds) Biochemistry of the algae and cyanobacteria, Clarendon Press, Oxford, 12:217–231
Reed RH, Warr SRC, Richardson DL, Moore DJ, Stewart WDP (1985) Multiphasic osmotic adjustment in a euryhaline cyanobacterium. FEMS Microbiol Lett 28:225–229
Rengefors K, Legrand C (2001) Toxicity in peridinium aciculiferum-an adaptive strategy to outcompete other winter phytoplankton? Limnol Oceanogr 46:1990–1997
Renger G (1987) Mechanistic aspects of photosynthetic water cleavage. Photosynthetica 21:203–224
Renger G, Holzwarth A (2005) Primary electron transfer. In: Wydrzynski TJ, Satoh K (eds) Photosystem II: the light-driven water: plastoquinone oxidoreductase. Springer Netherland, Berlin, pp 139–175
Renger T, Marcus R (2002) Photophysical properties of PS-2 reaction centers and a discrepancy in exciton relaxation times. J Phys Chem B 106:1809–1819
Rex M, Harris NRP, von der Gathen P, Lehmann R, Braathen GO, Reimer E, Beck A, Chipperfield MP, Alfier R, Allaart M (1997) Prolonged stratospheric ozone loss in the 1995–96 arctic winter. Nature 389:835–838
Richard C, Canonica S (2005) Aquatic phototransformation of organic contaminants induced by coloured dissolved natural organic matter. Environ Photochem Part II 2M:299–323
Richardson SD (2007) Water analysis: emerging contaminants and current issues. Anal Chem 79:4295–4324
Richardson K, Jorgensen BB (1996) Eutrophication: definition, history, and effects. In: Richardson K (ed) Jorgensen BB. American Geophysical Union, Eutrophication in coastal marine ecosystems, pp 1–19
Richardson AD, Andy Black T, Ciais P, Delbart N, Friedl MA, Gobron N, Hollinger DY, Kutsch WL, Longdoz B, Luyssaert S (2010) Influence of spring and autumn phenological transitions on forest ecosystem productivity. Phil Trans R Soc B 365:3227–3246
Rintamäki E, Kettunen R, Tyystjärvi E, Aro EM (1995) Light-dependent phosphorylation of D1 reaction centre protein of photosystem II: hypothesis for the functional role in vivo. Physiol Plant 93:191–195
Rivero R, Ruiz J, Romero L (2004) Oxidative metabolism in tomato plants subjected to heat stress. J Hort Sci Biotechnol 79:560–564
Rockley MG, Windsor MW, Cogdell RJ, Parson WW (1975) Picosecond detection of an intermediate in the photochemical reaction of bacterial photosynthesis. PNAS 72:2251
Roelofsen PA (1935) On photosynthesis of the Thiorhodaceae. vol 3, de Voorpost

Roeske C, O'Leary MH (1984) Carbon isotope effects on enzyme-catalyzed carboxylation of ribulose bisphosphate. Biochemistry 23:6275–6284

Rolland A, Bird DF, Giani A (2005) Seasonal changes in composition of the cyanobacterial community and the occurrence of hepatotoxic blooms in the eastern townships, Quebec, Canada. J Plankton Res 27:683–694

Rontein D, Basset G, Hanson AD (2002) Metabolic engineering of osmoprotectant accumulation in plants. Metabol Eng 4:49–56

Rose AL, Waite TD (2003) Predicting iron speciation in coastal waters from the kinetics of sunlight-mediated iron redox cycling. Aquat Sci 65:375–383

Rossman TG, Uddin AN, Burns FJ (2004) Evidence that arsenite acts as a cocarcinogen in skin cancer. Toxicol Appl Pharmacol 198:394–404

Rothhaupt KO (1996) Laboratory experiments with a mixotrophic chrysophyte and obligately phagotrophic and phototrophic competitors. Ecology 77:716–724

Ruben S, Randall M, Kamen M, Hyde JL (1941) Heavy oxygen (O18) as a tracer in the study of photosynthesis. JACS 63:877–879

Sage RF, Sharkey TD (1987) The effect of temperature on the occurrence of O_2 and CO_2 insensitive photosynthesis in field grown plants. Plant Physiol 84:658–664

Saito T, Yamanaka S, Kanda K, Isobe H, Takano Y, Shigeta Y, Umena Y, Kawakami K, Shen JR, Kamiya N (2012) Possible mechanisms of water splitting reaction based on proton and electron release pathways revealed for CaMn4O5 cluster of PSII refined to 1.9 Å X-ray resolution. Int J Quant Chem 253:253–2766

Salas HJ, Martino P (1991) A simplified phosphorus trophic state model for warm-water tropical lakes. Water Res 25:341–350

Salvucci ME, Crafts-Brandner SJ (2004) Inhibition of photosynthesis by heat stress: the activation state of Rubisco as a limiting factor in photosynthesis. Physiol Plant 120:179–186

Samuilov V (1997) Photosynthetic oxygen: the role of H_2O_2. A review. Biochemistry 62:451–454

Samuilov V, Bezryadnov D, Gusev M, Kitashov A, Fedorenko T (2001) Hydrogen peroxide inhibits photosynthetic electron transport in cells of cyanobacteria. Biochemistry (Moscow) 66:640–645

Samuilov V, Timofeev K, Sinitsyn S, Bezryadnov D (2004) H_2O_2-induced inhibition of photosynthetic O_2 evolution by Anabaena variabilis cells. Biochemistry (Moscow) 69:926–933

Sarmiento JL, Hughes TMC, Stouffer RJ, Manabe S (1998) Simulated response of the ocean carbon cycle to anthropogenic climate warming. Nature 393:245–249

Sarnelle O, Cooper SD, Wiseman S, Mavuti KM (1998) The relationship between nutrients and trophic-level biomass in turbid tropical ponds. Freshwater Biol 40:65–75

Sass L, Spetea C, Máté Z, Nagy F, Vass I (1997) Repair of UV-B induced damage of photosystem II via de novo synthesis of the D1 and D2 reaction centre subunits in Synechocystis sp. PCC 6803. Photosynth Res 54:55–62

Sassenrath G, Ort D (1990) The relationship between inhibition of photosynthesis at low temperature and the inhibition of photosynthesis after rewarming in chill-sensitive tomato. Plant Physiol Biochem (Paris) 28:457–465

Sassenrath GF, Ort DR, Portis AR (1990) Impaired reductive activation of stromal bisphosphatases in tomato leaves following low-temperature exposure at high light. Arch Biochem Biophys 282:302–308

Satoh K, Fork D (1982a) The light-induced decline in chlorophyll fluorescence as an indication of photoinhibition in intact Bryopsis chloroplasts illuminated under anaerobic conditions. Photobiochem Photobiophys 4:153–162

Satoh K, Fork DC (1982b) Photoinhibition of reaction centers of photosystems I and II in intact Bryopsis chloroplasts under anaerobic conditions. Plant Physiol 70:1004–1008

Satoh K, Smith CM, Fork DC (1983) Effects of salinity on primary processes of photosynthesis in the red alga Porphyra perforata. Plant Physiol 73:643

Savikhin S, Xu W, Martinsson P, Chitnis PR, Struve WS (2001) Kinetics of charge separation and $A_0^- \rightarrow A_1$ electron transfer in photosystem I reaction centers. Biochemistry 40:9282–9290

Scandalios JG (2002) The rise of ROS. Trends Biochem Sci 27:483–486

Schaffer J, Sebetich M (2004) Effects of aquatic herbicides on primary productivity of phytoplankton in the laboratory. Bull Environ Contam Toxicol 72:1032–1037
Scheffer M (1991) Should we expect strange attractors behind plankton dynamics–and if so, should we bother? J Plankton Res 13:1291–1305
Schelvis JPM, Vannoort PI, Aartsma TJ, Vangorkom HJ (1992) Energy and electron-transfer in D1–D2-CYTB559-complexes studied with picosecond transient absorption-spectroscopy. Photosynth Res 34:137–137
Schindler DW (1974) Eutrophication and recovery in experimental lakes: implications for lake management. Science 184:897
Schindler DW (2006) Recent advances in the understanding and management of eutrophication. Limnol Oceanogr 51:356–363
Schlesinger WH (1997) Biogeochemistry: an analysis of global change, vol Ed. 2. Academic press, New York
Schlodder E, Shubin VV, El-Mohsnawy E, Roegner M, Karapetyan NV (2007) Steady-state and transient polarized absorption spectroscopy of photosytem I complexes from the cyanobacteria arthrospira platensis and thermosynechococcus elongatus. Biochim Biophys Acta 1767:732–741
Schlodder E, Martin Hussels M, Çetin M, Karapetyan NV, Brecht M (2011) Fluorescence of the various red antenna states in photosystem I complexes from cyanobacteria is affected differently by the redox state of P700. Biochim Biophys Acta 1807:1423–1431
Schopf JW, Barghoorn ES, Maser MD, Gordon RO (1965) Electron microscopy of fossil bacteria two billion years old. Science 149:1365–1367
Schröder W, Åkerlund H (1990) Hydrogen peroxide production in photosystem II preparations. In: Baltscheffsky C (Ed), Current research in photosynthesis, vol 1 Kluwer, Dordrecht pp 901–904
Schroeder W (1989) The photosynthetic oxygen evolving system, hydrogen peroxide interaction and protein components. University of Lund
Schuster W, Monson R (1990) An examination of the advantages of C3–C4 intermediate photosynthesis in warm environments. Plant, Cell Environ 13:903–912
Schützendübel A, Polle A (2002) Plant responses to abiotic stresses: heavy metal-induced oxidative stress and protection by mycorrhization. J Exp Bot 53:1351–1365
Scully NM, Cooper WJ, Tranvik LJ (2003a) Photochemical effects on microbial activity in natural waters: the interaction of reactive oxygen species and dissolved organic matter. FEMS Microbiol Ecol 46:353–357
Scully NM, Tranvik LJ, Cooper WJ (2003b) Photochemical effects on the interaction of enzymes and dissolved organic matter in natural waters. Limnol Oceanogr 48:1818–1824
Segal RD, Waite AM, Hamilton DP (2006) Transition from planktonic to benthic algal dominance along a salinity gradient. Hydrobiologia 556:119–135
Senesi N (1990) Molecular and quantitative aspects of the chemistry of fulvic acid and its interactions with metal ions and organic chemicals: part II. The fluorescence spectroscopy approach. Anal Chim Acta 232:77–106
Service RJ, Hillier W, Debus RJ (2010) Evidence from FTIR Difference Spectroscopy of an Extensive Network of Hydrogen Bonds near the Oxygen-Evolving Mn4Ca Cluster of Photosystem II Involving D1-Glu65, D2-Glu312, and D1-Glu329. Biochemistry 49:6655-6669
Shaffer G, Leth O, Ulloa O, Bendtsen J, Daneri G, Dellarossa V, Hormazabal S, Sehlstedt PI (2000) Warming and circulation change in the eastern South Pacific Ocean. Geophys Res Lett 27:1247–1250
Shamb U, Setterfield C, Wentworth R (1958) Hydrogen peroxide [Russian translation]. Izd-vo Inostr, Lit, Moscow
Shaner DL, Boyer JS (1976) Nitrate reductase activity in maize (Zea mays L.) Leaves: II. regulation by nitrate Flux at low leaf water potential 1. Plant Physiol 58:505–509
Shangguan Z, Shao M, Dyckmans J (1999) Interaction of osmotic adjustment and photosynthesis in winter wheat under soil drought. J Plant Physiol 154:753–758
Shanker AK, Cervantes C, Loza-Tavera H, Avudainayagam S (2005) Chromium toxicity in plants. Environ Int 31:739–753

Shapiro J (1997) The role of carbon dioxide in the initiation and maintenance of blue-green dominance in lakes. Freshwater Biol 37:307–323

Sharkey TD (2005) Effects of moderate heat stress on photosynthesis: importance of thylakoid reactions, rubisco deactivation, reactive oxygen species, and thermotolerance provided by isoprene. Plant, Cell Environ 28:269–277

Sharkey TD, Berry JA (1985) Carbon isotope fractionation of algae as influenced by inducible CO_2 concentrating mechanism. In: Berry JA (ed) Lucas WJ. Inorganic Carbon Uptake by Aquatic Photosynthetic Organisms American Society of Plant Physiologists, Rockville, pp 389–401

Sharma PK, Hall DO (1991) Interaction of salt stress and photoinhibition on photosynthesis in barley and sorghum. J Plant Physiol 138:614–619

Shen JR, Ikeuchi M, Inoue Y (1992) Stoichiometric association of extrinsic cytochrome c550 and 12 kDa protein with a highly purified oxygen-evolving photosystem II core complex from *Synechococcus vulcanus*. FEBS Lett 301:145–149

Shen JR, Qian M, Inoue Y, Burnap RL (1998) Functional characterization of Synechocystis sp. PCC 6803 Δ psb U and Δ psb V mutants reveals important roles of cytochrome c-550 in cyanobacterial oxygen evolution. Biochemistry 37:1551–1558

Shimura S, Ichimura S (1973) Selective transmission of light in the ocean waters and its relation to phytoplankton photosynthesis. J Oceanogr 29:257–266

Shipman LL, Janson TR, Ray GJ, Katz JJ (1975) Donor properties of the three carbonyl groups of chlorophyll *a*: ab initio calculations and ^{13}C magnetic resonance studies. PNAS 72:2873

Shipman LL, Cotton TM, Norris JR, Katz JJ (1976) New proposal for structure of special-pair chlorophyll. PNAS 73:1791

Shubin V, Tsuprun V, Bezsmertnaya I, Karapetyan N (1993) Trimeric forms of the photosystem I reaction center complex pre-exist in the membranes of the cyanobacterium Spirulina platensis. FEBS Lett 334:79–82

Shumway SE (1990) A review of the effects of algal blooms on shellfish and aquaculture. J World Aquaculture Soc 21:65–104

Shumway SE, Allen SM, Dee Boersma P (2003) Marine birds and harmful algal blooms: sporadic victims or under-reported events? Harmful Algae 2:1–17

Shutova T, Nikitina J, Deikus G, Andersson B, Klimov V, Samuelsson G (2005) Structural dynamics of the manganese-stabilizing protein effect of pH, calcium, and manganese. Biochemistry 44:15182–15192

Shuvalov V, Klevanik A (1983) The study of the state [P870 + B800-] in bacterial reaction centers by selective picosecond and low-temperature spectroscopies. FEBS Lett 160:51–55

Sigfridsson KGV, Bernat G, Mamedov F, Styring S (2004) Molecular interference of Cd^{2+} with photosystem II. Biochim Biophys Acta 1659:19–31

Šiler B, Mišic D, Filipovic B, Popovic Z, Cvetic T, Mijovic A (2007) Effects of salinity on in vitro growth and photosynthesis of common centaury (Centaurium erythraea Rafn.). Arch Biol Sci 59:129–134

Simkiss K, Taylor MG (1989) Metal fluxes across the membranes of aquatic organisms. Rev Aquat Sci 1:173–188

Singh SC, Sinha RP, Hader D (2002) Role of lipids and fatty acids in stress tolerance in cyanobacteria. Acta protozool 41:297–308

Singsaas E, Sharkey T (1998) The regulation of isoprene emission responses to rapid leaf temperature fluctuations. Plant, Cell Environ 21:1181–1188

Singsaas EL, Laporte MM, Shi JZ, Monson RK, Bowling DR, Johnson K, Lerdau M, Jasentuliytana A, Sharkey TD (1999) Kinetics of leaf temperature fluctuation affect isoprene emission from red oak (Quercus rubra) leaves. Tree Physiol 19:917–924

Sinha RP, Häder DP (1996) Response of a rice field cyanobacterium *Anabaena* sp. to physiological stressors. Environ Exp Bot 36:147–155

Sinha RP, Häder DP (2002) UV-induced DNA damage and repair: a review. Photochem Photobiol Sci 1:225–236

Sinha R, Klisch M, Gröniger A, Häder DP (1998) Ultraviolet-absorbing/screening substances in cyanobacteria, phytoplankton and macroalgae. J Photochem Photobiol, B 47:83–94

Sinha RP, Klisch M, Häder DP (1999a) Induction of a mycosporine-like amino acid (MAA) in the rice-field cyanobacterium Anabaena sp. by UV irradiation. J Photochem Photobiol, B 52:59–64
Sinha R, Klisch M, Vaishampayan A, Hader D (1999b) Biochemical and spectroscopic characterization of the cyanobacterium Lyngbya sp. inhabiting Mango (Mangifera indica) trees: presence of an ultraviolet-absorbing pigment, scytonemin. Acta Protozool 38:291–298
Sinha RP, Klisch M, Gröniger A, Häder DP (2001a) Responses of aquatic algae and cyanobacteria to solar UV-B. Plant Ecol 154:219–236
Sinha RP, Klisch M, Walter Helbling E, Häder DP (2001b) Induction of mycosporine-like amino acids (MAAs) in cyanobacteria by solar ultraviolet-B radiation. J Photochem Photobiol, B 60:129–135
Sinha RP, Singh SP, Häder DP (2007) Database on mycosporines and mycosporine-like amino acids (MAAs) in fungi, cyanobacteria, macroalgae, phytoplankton and animals. J Photochem Photobiol, B 89:29–35
Skovsen E, Snyder JW, Lambert JDC, Ogilby PR (2005) Lifetime and diffusion of singlet oxygen in a cell. J Phys Chem B 109:8570–8573
Smetacek V (1985) The annual cycle of Kiel Bight plankton: a long-term analysis. Estuaries Coasts 8:145–157
Smit G, Rethman N (2000) The influence of tree thinning on the soil water in a semi-arid savanna of southern Africa. J Arid Environ 44:41–59
Smith VH (1982) The nitrogen and phosphorus dependence of algal biomass in lakes: an empirical and theoretical analysis. Limnol Oceanogr 27:1101–1112
Smith VH (1983) Low nitrogen to phosphorus ratios favor dominance by blue-green algae in lake phytoplankton. Science 221:669–671
Smith VH (2003) Eutrophication of freshwater and coastal marine ecosystems a global problem. Environ Sci Pollut Res 10:126–139
Smith RC, Baker KS (1979) Penetration of UV-B and biologically effective dose-rates in natural waters. Photochem Photobiol 29:311–323
Smith VH, Bennett S (1999) Nitrogen: phosphorus supply ratios and phytoplankton community structure in lakes: Nutrient ratios. Arch Hydrobiol 146:37–53
Smith BN, Epstein S (1971) Two categories of $^{13}C/^{12}C$ ratios for higher plants. Plant Physiol 47:380
Smith RC, Prezelin B, Baker K, Bidigare R, Boucher N, Coley T, Karentz D, MacIntyre S, Matlick H, Menzies D (1992) Ozone depletion: ultraviolet radiation and phytoplankton biology in Antarctic waters. Science 255:952–959
Smith VH, Bierman V Jr, Jones B, Havens K (1995) Historical trends in the Lake Okeechobee ecosystem IV nitrogen: phosphorus ratios, cyanobacterial dominance, and nitrogen fixation potential. Arch Hydrobiol 107(Suppl):71–88
Sobek S, Tranvik LJ, Prairie YT, Kortelainen P, Cole JJ (2007) Patterns and regulation of dissolved organic carbon: An analysis of 7,500 widely distributed lakes. Limnol Oceanogr 52:1208–1219
Sobrino C, Montero O, Lubián LM (2004) UV-B radiation increases cell permeability and damages nitrogen incorporation mechanisms inNannochloropsis gaditana. Aquat Sci 66:421–429
Sobrino C, Ward ML, Neale PJ (2008) Acclimation to elevated carbon dioxide and ultraviolet radiation in the diatom" Thalassiosira pseudonana": Effects on growth, photosynthesis, and spectral sensitivity of photoinhibition. Limnol Oceanogr 53:494–505
Sohal R, Agarwal S, Candas M, Forster MJ, Lal H (1994) Effect of age and caloric restriction on DNA oxidative damage in different tissues of C57BL/6 mice. Mech Ageing and Dev 76:215–224
Sommer U, Gliwicz ZM, Lampert W, Duncan A (1986) The PEG-model of seasonal succession of planktonic events in fresh waters. Arch Hydrobiol 106:433–471
Song Y, Qiu B (2007) The CO2-concentrating mechanism in the bloom-forming cyanobacterium Microcystis aeruginosa (Cyanophyceae) and effects of UVB radiation on its operation1. J Phycol 43:957–964
Sonoike K (1999) The different roles of chilling temperatures in the photoinhibition of photosystem I and photosystem II. J Photochem Photobiol, B 48:136–141

Sopory S, Greenberg B, Mehta R, Edelman M, Mattoo A (1990) Free radical scavengers inhibit light-dependent degradation of the 32 kDa photosystem II reaction center protein. Z Naturforsch 45:412–417
Souch C, Stephens W (1998) Growth, productivity and water use in three hybrid poplar clones. Tree Physiol 18:829–835
Southard GM, Fries LT, Barkoh A (2010) Prymnesium parvum: the texas experience. J Am Water Resour Assoc 46:14–23
Spector A, Garner WH (1981) Hydrogen peroxide and human cataract. Exp Eye Res 33:673–681
Spence A, Simpson AJ, Mcnally DJ, Moran BW, McCaul MV, Hart K, Paull B, Kelleher BP (2011) The degradation characteristics of microbial biomass in soil. Geochim Cosmochim Acta 75:2571–2581
Stabel H, Moaledj K, Overbeck J (1979) On the degradation of dissolved organic molecules from plussee by oligocarbophilic bacteria. Arch Hydrobiol Ergebn Limnol 12:95–104
Stadtman E (1993) Oxidation of free amino acids and amino acid residues in proteins by radiolysis and by metal-catalyzed reactions. Annu Rev Biochem 62:797–821
Steinberg C, Muenster U (1985) In: Aiken GR, McKnight DM, Wershaw RL, MacCarthy P (eds) Humic Substances in Soil, Sediment, and Water: Geochemistry. Isolation and Characterization, Wiley, NY, pp 104–145
Steinmaus C, Yuan Y, Bates MN, Smith AH (2003) Case-control study of bladder cancer and drinking water arsenic in the western United States. Am J Epidemiol 158:1193–1201
Sterner RW, Smutka TM, McKay RML, Xiaoming Q, Brown ET, Sherrell RM (2004) Phosphorus and trace metal limitation of algae and bacteria in Lake superior. Limnol Oceanogr 49:495–507
Stewart DH, Nixon PJ, Diner BA, Brudvig GW (2000) Assignment of the Q y absorbance bands of Photosystem II chromophores by low-temperature optical spectroscopy of wild-type and mutant reaction centers. Biochemistry 39:14583–14594
Stibor H, Sommer U (2003) Mixotrophy of a photosynthetic flagellate viewed from an optimal foraging perspective. Protist 154:91–98
Straza TRA, Cottrell MT, Ducklow HW, Kirchman DL (2009) Geographic and phylogenetic variation in bacterial biovolume as revealed by protein and nucleic acid staining. Appl Environ Microbiol 75:4028–4034
Strome D, Miller M (1978) Photolytic changes in dissolved humic substances. Int Ver Theor Angew Limnol 20:1248–1254
Strzepek RF, Harrison PJ (2004) Photosynthetic architecture differs in coastal and oceanic diatoms. Nature 431:689–692
Sugg LM, VanDolah FM (1999) No evidence for an allelopathic role of okadaic acid among ciguatera-associated dinoflagellates J Phycol 35:93–103
Sukenik A, Eshkol R, Livne A, Hadas O, Rom M, Tchernov D, Vardi A, Kaplan A (2002) Inhibition of growth and photosynthesis of the dinoflagellate peridinium gatunense by Microcystis sp. (cyanobacteria): a novel allelopathic mechanism. Limnol Oceanogr 47:1656–1663
Sunda WG, Huntsman SA (1997) Interrelated influence of iron, light and cell size on marine phytoplankton growth. Nature 390:389–392
Sunda WG, Huntsman SA (1998) Processes regulating cellular metal accumulation and physiological effects: phytoplankton as model systems. Sci Total Environ 219:165–181
Sundar D, Ramachandra Reddy A (2001) Low night temperature-induced changes in photosynthesis and rubber accumulation in guayule (Parthenium argentatum Gray). Photosynthetica 38:421–427
Susplugas S, Srivastava A, Strasser RJ (2000) Changes in the photosynthetic activities during several stages of vegetative growth of *Spirodela polyrhiza*: effect of chromate. J Plant Physiol 157:503–512
Suwa R, Nguyen NT, Saneoka H, Moghaieb R, Fujita K (2006) Effect of salinity stress on photosynthesis and vegetative sink in tobacco plants. Soil Sci Plant Nutr 52:243–250
Suzuki T, Tada O, Makimura M, Tohri A, Ohta H, Yamamoto Y, Enami I (2004) Isolation and characterization of oxygen-evolving photosystem II complexes retaining the PsbO, P and Q proteins from Euglena gracilis. Plant Cell Physiol 45:1168–1175

Svensson B, Etchebest C, Tuffery P, van Kan P, Smith J, Styring S (1996) A model for the photosystem II reaction center core including the structure of the primary donor P680. Biochemistry 35:14486–14502

Syrett P (1981) Nitrogen metabolism of microalgae. Can Bull Fish Aquat Sci 210:182–210

Szilárd A, Sass L, Deák Z, Vass I (2007) The sensitivity of photosystem II to damage by UV-B radiation depends on the oxidation state of the water-splitting complex. Biochim Biophys Acta 1767:876–882

Tabatabaie S, Gregory P, Hadley P (2004) Uneven distribution of nutrients in the root zone affects the incidence of blossom end rot and concentration of calcium and potassium in fruits of tomato. Plant Soil 258:169–178

Takahashi M, Asada K (1982) Dependence of oxygen affinity for Mehler reaction on photochemical activity of chloroplast thylakoids. Plant Cell Physiol 23:1457–1461

Takahashi S, Murata N (2008) How do environmental stresses accelerate photoinhibition? Trends Plant Sci 13:178–182

Takahashi Y, Hansson O, Mathis P, Satoh K (1987) Primary radical pair in the photosystem II reaction centre. Biochim Biophys Acta 893:49–59

Takahashi K, Yoshioka T, Wada E, Sakamoto M (1990) Temporal variations in carbon isotope ratio of phytoplankton in a eutrophic lake. J Plankton Res 12:799–808

Takahashi K, Wada E, Sakamoto M (1991) Relationship between carbon isotope discrimination and the specific growth rate of green alga chlamydomonas reinhardtii. Jpn J Limnol 52:105–112

Takeda K, Takedoi H, Yamaji S, Ohta K, Sakugawa H (2004) Determination of hydroxyl radical photoproduction rates in natural waters. Anal Sci 20:153–158

Talling J (2006) Interrelated seasonal shifts in acid–base and oxidation–reduction systems that determine chemical stratification in three dissimilar English Lake Basins. Hydrobiologia 568:275–286

Tandeau de Marsac N, Houmard J (1993) Adaptation of cyanobacteria to environmental stimuli: new steps towards molecular mechanisms. FEMS Microbiol Lett 104:119–189

Tang D, Jankowiak R, Seibert M, Yocum C, Small G (1990) Excited-state structure and energy-transfer dynamics of two different preparations of the reaction center of photosystem II: a hole-burning study. J Phys Chem 94:6519–6522

Tang Y, Wen X, Lu C (2005) Differential changes in degradation of chlorophyll-protein complexes of photosystem I and photosystem II during flag leaf senescence of rice. Plant Physiol Biochem 43:193–201

Tank SE, Xenopoulos MA, Hendzel LL (2005) Effect of ultraviolet radiation on alkaline phosphatase activity and planktonic phosphorus acquisition in Canadian boreal shield lakes. Limnol Oceanogr 50:1345–1351

Tarran G, Zubkov M, Sleigh M, Burkill P, Yallop M (2001) Microbial community structure and standing stocks in the NE Atlantic in June and July of 1996. Deep Sea Res Part II 48:963–985

Telfer A, He WZ, Barber J (1990) Spectral resolution of more than one chlorophyll electron donor in the isolated Photosystem II reaction centre complex. Biochim Biophys Acta 1017:143–151

Telfer A, Bishop SM, Phillips D, Barber J (1994) Isolated photosynthetic reaction center of photosystem II as a sensitizer for the formation of singlet oxygen. Detection and quantum yield determination using a chemical trapping technique. J Biol Chem 269:13244–13253

Terzaghi WB, Fork DC, Berry JA, Field CB (1989) Low and High Temperature Limits to PSII: a survey using trans-parinaric acid, Delayed light emission, and Fo chlorophyll fluorescence. Plant Physiol 91:1494–1500

Tessier A, Turner DR (1995) Metal laspeciation and bioavaibility in aquatic systems. IUPAC series on analytical and physical chemistry of environmental systems, Wiley and Sons Ltd, Chichester, vol 3, pp 696

Tester PA, Stumpf RP, Vukovich FM, Fowler PK, Turner JT (1991) An expatriate red tide bloom: transport, distribution, and persistence. Limnol Oceanogr 36:1053–1061

Thackeray S, Jones I, Maberly S (2008) Long-term change in the phenology of spring phytoplankton: species-specific responses to nutrient enrichment and climatic change. J Ecol 96:523–535

Thielmann J, Tolbert NE, Goyal A, Senger H (1990) Two systems for concentrating CO_2 and bicarbonate during photosynthesis by *Scenedesmus*. Plant Physiol 92:622–629

Thingstad TF, Havskum H, Garde K, Riemann B (1996) On the strategy of" eating your competitor": a mathematical analysis of algal mixotrophy. Ecology 77:2108–2118

Thomas EL, Pera KA (1983) Oxygen metabolism of Streptococcus mutans: uptake of oxygen and release of superoxide and hydrogen peroxide. J Bacteriol 154:1236–1244

Thomas PG, Dominy PJ, Vigh L, Mansourian AR, Quinn PJ, Williams WP (1986) Increased thermal stability of pigment-protein complexes of pea thylakoids following catalytic hydrogenation of membrane lipids. Biochim Biophys Acta 849:131–140

Thompson LK, Blaylock R, Sturtevant JM, Brudvig GW (1989) Molecular basis of the heat denaturation of photosystem II. Biochemistry 28:6686–6695

Thurnauer MC, Katz JJ, Norris JR (1975) The triplet state in bacterial photosynthesis: Possible mechanisms of the primary photo-act. PNAS 72:3270

Tiede DM, Kellogg E, Breton J (1987) Conformational changes following reduction of the bacteriopheophytin electron acceptor in reaction centers of *Rhodopseudomonas viridis*. Biochim Biophys Acta 892:294–302

Tissut M, Taillandier G, Ravanel P, Benoit-Guyod JL (1987) Effects of chlorophenols on isolated class A chloroplasts and thylakoids: a QSAR study. Ecotoxicol Environ Saf 13:32–42

Tittel J, Bissinger V, Zippel B, Gaedke U, Bell E, Lorke A, Kamjunke N (2003) Mixotrophs combine resource use to outcompete specialists: implications for aquatic food webs. PNAS 100:12776

Tjus SE, Møller BL, Scheller HV (1998) Photosystem I is an early target of photoinhibition in barley illuminated at chilling temperatures. Plant Physiol 116:755–764

Tortell PD, Reinfelder JR, Morel FMM (1997) Active uptake of bicarbonate by diatoms. Nature 390:243–244

Tranvik LJ (1988) Availability of dissolved organic carbon for planktonic bacteria in oligotrophic lakes of differing humic content. Microb Ecol 16:311–322

Tranvik LJ (1989) Bacterioplankton growth, grazing mortality and quantitative relationship to primary production in a humic and a clear water lake. J Plankton Res 11:985–1000

Tranvik LJ, Bertilsson S (2001) Contrasting effects of solar UV radiation on dissolved organic sources for bacterial growth. Ecol Lett 4:458–463

Tranvik LJ, Höfle MG (1987) Bacterial growth in mixed cultures on dissolved organic carbon from humic and clear waters. Appl Environ Microbiol 53:482–488

Tranvik LJ, Downing JA, Cotner JB, Loiselle SA, Striegl RG, Ballatore TJ, Dillon P, Knoll L, Kutser T, Larsen S (2009) Lakes and reservoirs as regulators of carbon cycling and climate. Limnol Oceanogr 54:2298–2314

Trebst A, Depka B, Holländer-Czytko H (2002) A specific role for tocopherol and of chemical singlet oxygen quenchers in the maintenance of photosystem II structure and function in Chlamydomonas reinhardtii. FEBS Lett 516:156–160

Tu CL, Liu CQ, Lu XH, Yuan J, Lang YC (2011) Sources of dissolved organic carbon in forest soils: evidences from the differences of organic carbon concentration and isotope composition studies. Environ Earth Sci 63:723–730

Tunçtürk M, Tunçtürk R, Yildirim B, Çiftçi V (2011) Changes of micronutrients, dry weight and plant development in canola (Brassica napus L.) cultivars under salt stress. African J Biotechnol 10:3726–3730

Turhan E, Eris A (2005) Changes of micronutrients, dry weight, and chlorophyll contents in strawberry plants under salt stress conditions. Commun Soil Sci Plant Anal 36:1021–1028

Turrens JF (1997) Superoxide production by the mitochondrial respiratory chain. Biosci Rep 17:3–8

Twiner MJ, Trick CG (2000) Possible physiological mechanisms for production of hydrogen peroxide by the ichthyotoxic flagellate Heterosigma akashiwo. J Plankton Res 22:1961–1975

Tyystjärvi E (2008) Photoinhibition of photosystem II and photodamage of the oxygen evolving manganese cluster. Coord Chem Rev 252:361–376
Tyystjärvi T, Herranen M, Aro EM (2001) Regulation of translation elongation in cyanobacteria: membrane targeting of the ribosome nascent-chain complexes controls the synthesis of D1 protein. Mol Microbiol 40:476–484
Uchida M, Nakatsubo T, Kasai Y, Nakane K, Horikoshi T (2000) Altitudinal differences in organic matter mass loss and fungal biomass in a subalpine coniferous forest, Mt. Fuji Japan. Arct Antar Alp Res 32:262–269
Umena Y, Kawakami K, Shen JR, Kamiya N (2011) Crystal structure of oxygen-evolving photosystem II at a resolution of 1.9 Å. Nature 473:55–60
Unrein F, Massana R, Alonso-Sáez L, Gasol JM (2007) Significant year-round effect of small mixotrophic flagellates on bacterioplankton in an oligotrophic coastal system. Limnol Oceanogr 52:456–469
Vähätalo AV, Järvinen M (2007) Photochemically produced bioavailable nitrogen from biologically recalcitrant dissolved organic matter stimulates production of a nitrogen-limited microbial food web in the Baltic Sea. Limnol Oceanogr 52:132–143
van Dolah FM (2000) Marine algal toxins: origins, health effects, and their increased occurrence. Environ Health Perspect 108:133
van Gorkom HJ, Schelvis JPM (1993) Kok's oxygen clock: what makes it tick? The structure of P680 and consequences of its oxidizing power. Photosynth Res 38:297–301
van Niel C (1931) On the morphology and physiology of the purple and green sulfur bacteria. Arch Mikrobiol 3:1–112
van Niel CB (1936) On the metabolism of the thiorhodaceae. Arch Microbiol 7:323–358
van Niel C (1941) The bacterial photosyntheses and their importance for the general problem of photosynthesis. Advan Enzymol 1:263–328
Vass I, Styring S, Hundal T, Koivuniemi A, Aro E, Andersson B (1992) Reversible and irreversible intermediates during photoinhibition of photosystem II: stable reduced QA species promote chlorophyll triplet formation. PNAS 89:1408–1412
Yordanov I, Velikova, V, Tsonev, T (2003) Plant responses to drought and stress tolerance. Bulg J Plant Physiol (special issue):187–286
Velthuys B, Kok B (1978) Photosynthetic oxygen evolution from hydrogen peroxide. Biochim Biophys Acta 502:211–221
Verhoeven AS, Demmig-Adams B, Adams WW III (1997) Enhanced employment of the xanthophyll cycle and thermal energy dissipation in spinach exposed to high light and N stress. Plant Physiol 113:817–824
Vernay P, Gauthier-Moussard C, Jean L, Bordas F, Faure O, Ledoigt G, Hitmi A (2008) Effect of chromium species on phytochemical and physiological parameters in *Datura innoxia*. Chemosphere 72:763–771
Vıllora G, Moreno DA, Pulgar G, Romero L (2000) Yield improvement in zucchini under salt stress: determining micronutrient balance. Sci Horticult 86:175–183
Vinogradov A, Teis RV (1941) Isotopic composition of oxygen of different origin. CR Acad Sci, URSS 33:490–493
Vione D, Falletti G, Maurino V, Minero C, Pelizzetti E, Malandrino M, Ajassa R, Olariu RI, Arsene C (2006) Sources and sinks of hydroxyl radicals upon irradiation of natural water samples. Environ Sci Technol 40:3775–3781
Vione D, Minella M, Minero C, Maurino V, Picco P, Marchetto A, Tartari G (2009a) Photodegradation of nitrite in lake waters: role of dissolved organic matter. Environ Chem 6:407–415
Vione D, Khanra S, Man SC, Maddigapu PR, Das R, Arsene C, Olariu RI, Maurino V, Minero C (2009b) Inhibition vs. enhancement of the nitrate-induced phototransformation of organic substrates by the •OH scavengers bicarbonate and carbonate. Water Res 43:4718–4728
von Elert E, Juttner F (1997) Phosphorus limitation and not light controls the extracellular release of allelopathic compounds by Trichormus doliolum (Cyanobacteria). Limnol Oceanogr 42:1796–1802

Wada H, Combos Z, Murata N (1990) Enhancement of chilling tolerance of a cyanobacterium by genetic manipulation of fatty acid desaturation. Nature 347:200–203
Wada H, Gombos Z, Murata N (1994) Contribution of membrane lipids to the ability of the photosynthetic machinery to tolerate temperature stress. PNAS 91:4273–4277
Walsh JJ, Dugdale RC (1971) A simulation model of the nitrogen flow in the Peruvian upwelling system. Invest Pesq 35:309–330
Wang XS, Han JG (2007) Effects of NaCl and silicon on ion distribution in the roots, shoots and leaves of two alfalfa cultivars with different salt tolerance. Soil Sci Plant Nutr 53:278–285
Wang GS, Liao CH, Wu FJ (2001) Photodegradation of humic acids in the presence of hydrogen peroxide. Chemosphere 42:379–387
Wängberg SÅ, Andreasson KIM, Garde K, Gustavson K, Henriksen P, Reinthaler T (2006) Inhibition of primary production by UV-B radiation in an arctic bay–model calculations. Aquat Sci 68:117–128
Warburg O, Negelein E (1922) Über den Energieumsatz bei der Kohlensäureassimilation. Naturwissenschaften 10:647–653
Warburg O, Uyesugi T (1924) Über die blackmansche reaktion. Biochem Z 146:486–492
Ward DM, Ferris MJ, Nold SC, Bateson MM (1998) A natural view of microbial biodiversity within hot spring cyanobacterial mat communities. Microbiol Mol Biol Rev 62:1353–1370
Webber AN, Lubitz W (2001) P700: the primary electron donor of photosystem I. Biochim Biophys Acta 1507:61–79
Wetzel R (2001) Limnology. 3rd ed Academic Press
White EM, Vaughan PP, Zepp RG (2003) Role of the photo-Fenton reaction in the production of hydroxyl radicals and photobleaching of colored dissolved organic matter in a coastal river of the southeastern United States. Aquat Sci 65:402–414
White MA, de Beurs KM, Didan K, Inouye DW, Richardson AD, Jensen OP, O'KEEFE J, Zhang G, Nemani RR, van Leeuwen WJD (2009) Intercomparison, interpretation, and assessment of spring phenology in North America estimated from remote sensing for 1982–2006. Glob Change Biol 15:2335–2359
White EM, Kieber DJ, Sherrard J, Miller WL, Mopper K (2010) Carbon dioxide and carbon monoxide photoproduction quantum yields in the Delaware Estuary. Mar Chem 118:11–21
Whitton BA, Potts M (2000) Introduction to cuanobacteria. In: Whitton BA, Potts M (eds) The ecology of cyanobacteria. Kluwer Academic Publishers, Netherlands, pp 1–11
Wiebe W, Sheldon W Jr, Pomeroy L (1992) Bacterial growth in the cold: evidence for an enhanced substrate requirement. Appl Environ Microbiol 58:359–364
Wilen RW, Sacco M, Gusta LV, Krishna P (1995) Effects of 24-epibrassinolide on freezing and thermotolerance of bromegrass (Bromus inermis) cell cultures. Physiol Plant 95:195–202
Wilhelm J, Sojkova J, Herget J (1996) Production of hydrogen peroxide by alveolar macrophages from rats exposed to subacute and chronic hypoxia. Physiol Res 45:185
Wilhelm J, Frydrychová M, Hezinová A, Vizek M (1997) Production of hydrogen peroxide by peritoneal macrophages from rats exposed to subacute and chronic hypoxia. Physiol Res 46:35
Wilhelm J, Frydrychova M, Vizek M (1999) Hydrogen peroxide in the breath of rats: the effects of hypoxia and paraquat. Physiol Res 48:445–450
Willekens H, Chamnongpol S, Davey M, Schraudner M, Langebartels C, van Montagu M, Inzé D, van Camp W (1997) Catalase is a sink for H_2O_2 and is indispensable for stress defence in C3 plants. The EMBO journal 16:4806–4816
Williams MD, Chance B (1983) Spontaneous chemiluminescence of human breath. Spectrum, lifetime, temporal distribution, and correlation with peroxide. J Biol Chem 258:3628–3631
Wilson WC, Laborde PR, Benumof JL, Taylor R, Swetland JF (1993) Reperfusion injury and exhaled hydrogen peroxide. Anesth Analg 77:963–970
Wilson KB, Baldocchi DD, Hanson PJ (2000) Quantifying stomatal and non-stomatal limitations to carbon assimilation resulting from leaf aging and drought in mature deciduous tree species. Tree Physiol 20:787–797
Winder M, Cloern JE (2010) The annual cycles of phytoplankton biomass. Phil Trans R Soc B 365:3215–3226

Winder M, Schindler DE (2004) Climatic effects on the phenology of lake processes. Glob Change Biol 10:1844–1856

Windust A, Wright J, McLachlan J (1996) The effects of the diarrhetic shellfish poisoning toxins, okadaic acid and dinophysistoxin-1, on the growth of microalgae. Mar Biol 126:19–25

Wingler A, Marès M, Pourtau N (2004) Spatial patterns and metabolic regulation of photosynthetic parameters during leaf senescence. New Phytol 161:781–789

Wise RR, Ortiz-Lopez A, Ort DR (1992) Spatial distribution of photosynthesis during drought in field-grown and acclimated and nonacclimated growth chamber-grown cotton. Plant Physiol 100:26–32

Wolfe-Simon F, Blum JS, Kulp TR, Gordon GW, Hoeft SE, Pett-Ridge J, Stolz JF, Webb SM, Weber PK, Davies PCW (2011) A bacterium that can grow by using arsenic instead of phosphorus. Science 332:1163

Wong W, Sackett WM, Benedict CR (1975) Isotope fractionation in photosynthetic bacteria during carbon dioxide assimilation. Plant Physiol 55:475

Woodbury NWT, Parson WW (1984) Nanosecond fluorescence from isolated photosynthetic reaction centers of *Rhodopseudomonas sphaeroides*. Biochim Biophys Acta 767:345–361

Wright RT (1984) Dynamics of pools of dissolved organic carbon. In: Hobbie JE, Williams PJleB (Eds), Heterotrophic Activity in the Sea, Proc NATO ARI, Cascais, Portugal, 1981 Plenum, NY, pp 121-154

Wright RT (1988) A model for short-term control of the bacterioplankton by substrate and grazing. Hydrobiologia 159:111–117

Wu H, Gao K (2009) Ultraviolet radiation stimulated activity of extracellular carbonic anhydrase in the marine diatom Skeletonema costatum. Funct Plant Biol 36:137–143

Wu H, Gao K, Villafañe VE, Watanabe T, Helbling EW (2005) Effects of solar UV radiation on morphology and photosynthesis of filamentous cyanobacterium Arthrospira platensis. Appl Environ Microbiol 71:5004–5013

Wu JT, Chiang YR, Huang WY, Jane WN (2006) Cytotoxic effects of free fatty acids on phytoplankton algae and cyanobacteria. Aquat Toxicol 80:338–345

Xie L, Xie P, Li S, Tang H, Liu H (2003) The low TN: TP ratio, a cause or a result of *Microcystis* blooms? Water Res 37:2073–2080

Xie H, Zafiriou OC, Cai WJ, Zepp RG, Wang Y (2004) Photooxidation and its effects on the carboxyl content of dissolved organic matter in two coastal rivers in the southeastern United States. Environ Sci Technol 38:4113–4119

Xu L, Lam P, Chen J, Xu J, Wong B, Zhang Y, Wu R, Harada K (2000) Use of protein phosphatase inhibition assay to detect microcystins in Donghu Lake and a fish pond in China. Chemosphere 41:53–58

Xue H, Sigg L (1990) Binding of Cu (II) to algae in a metal buffer. Water Res 24:1129–1136

Xue HB, Sigg L (1993) Free cupric ion concentration and Cu (II) speciation in a eutrophic lake. Limnol Oceanogr 38:1200–1213

Xue HB, Kistler D, Sigg L (1995) Competition of copper and zinc for strong ligands in a eutrophic lake. Limnol Oceanogr 40:1142–1152

Bruskov V, Masalimov ZK (2002) Chernikov A Heat-induced generation of reactive oxygen in water. Dokl Biochem Biophys 384:181–184

Yacobi YZ (2006) Temporal and vertical variation of chlorophyll a concentration, phytoplankton photosynthetic activity and light attenuation in Lake Kinneret: possibilities and limitations for simulation by remote sensing. J Plankton Res 28:725–736

Yamanaka S, Saito T, Kanda K, Isobe H, Umena Y, Kawakami K, Shen JR, Kamiya N, Okumura M, Nakamura H (2012) Structure and reactivity of the mixed-valence $CaMn_4O_5(H_2O)_4$ and $CaMn_4O_4(OH)(H_2O)_4$ clusters at oxygen evolution complex of photosystem II Hybrid DFT (UB3LYP and UBHandHLYP) calculations. Int J Quantum Chem 112:321–343

Yamane Y, Kashino Y, Koike H, Satoh K (1998) Effects of high temperatures on the photosynthetic systems in spinach: oxygen-evolving activities, fluorescence characteristics and the denaturation process. Photosynth Res 57:51–59

Yamashita K, Konishi K, Itoh M, Shibata K (1969) Photo-bleaching of carotenoids related to the electron transport in chloroplasts. Biochim Biophys Acta 172:511–524
Yan T, Zhou MJ (2004) Environmental and health effects associated with harmful algal bloom and marine algal toxins in China. Biomed Environ Sci 17:165–176
Yang J, Kong Q, Xiang C (2009) Effects of low night temperature on pigments, chl a fluorescence and energy allocation in two bitter gourd (Momordica charantia L.) genotypes. Acta Physiol Plant 31:285–293
Yates BS, Rogers WJ (2011) Atrazine selects for ichthyotoxic prymnesium parvum, a possible explanation for golden algae blooms in lakes of Texas, USA. Ecotoxicology 20:2003–2010
Yin C, Berninger F, Li C (2006a) Photosynthetic responses of Populus przewalski subjected to drought stress. Photosynthetica 44:62–68
Yin G, Buchalova M, Danby AM, Perkins CM, Kitko D, Carter JD, Scheper WM, Busch DH (2006b) Olefin epoxidation by the hydrogen peroxide adduct of a novel non-heme manganese (IV) complex: demonstration of oxygen transfer by multiple mechanisms. Inorg Chem 45:3467–3474
Yoder LM, Cole AG, Sension RJ (2002) Structure and function in the isolated reaction center complex of photosystem II: energy and charge transfer dynamics and mechanism. Photosynth Res 72:147–158
Yordanov I, Tsonev T, Goltsev V, Kruleva L, Velikova V (1997) Interactive effect of water deficit and high temperature on photosynthesis in sunflower and maize plants.1. Changes in parameters of chlorophyll fluorescence induction kinetics and fluorescence quenching. Photosynth Res 33:391–402
Yordanov I, Velikova V, Tsonev T (1999) Influence of drought, high temperature, and carbamide cytokinin 4-PU-30 on photosynthetic activity of bean plants. 1. Changes in chlorophyll fluorescence quenching. Photosynthetica 37:447–457
Yordanov I, Velikova V, Tsonev T (2000) Plant responses to drought, acclimation, and stress tolerance. Photosynthetica 38:171–186
Yoshioka T (1997) Phytoplanktonic carbon isotope fractionation: equations accounting for CO2-concentrating mechanisms. J Plankton Res 19:1455–1476
Yoshiyama K, Sharp JH (2006) Phytoplankton response to nutrient enrichment in an urbanized estuary: apparent inhibition of primary production by overeutrophication. Limnol Oceanogr 51:424–434
Yu CW, Murphy TM, Sung WW, Lin CH (2002) H2O2 treatment induces glutathione accumulation and chilling tolerance in mung bean. Funct Plant Biol 29:1081–1087
Zellner R, Exner M, Herrmann H (1990) Absolute OH quantum yields in the laser photolysis of nitrate, nitrite and dissolved H2O2 at 308 and 351 nm in the temperature range 278–353 K. J Atmos Chem 10:411–425
Zepp RG, Skurlatov Y, Pierce J (1987a) Algal-induced decay and formation of hydrogen peroxide in water: its possible role in oxidation of anilines by algae. In: Zika RG, Cooper WJ (eds) Photochemistry of environmental aquatic systems, ACS Symp Ser 327. Am Chem Soc, Washington DC, pp 213–224
Zepp RG, Braun AM, Hoigne J, Leenheer JA (1987b) Photoproduction of hydrated electrons from natural organic solutes in aquatic environments. Environ Sci Technol 21:485–490
Zepp RG, Faust BC, Hoigne J (1992) Hydroxyl radical formation in aqueous reactions (pH 3–8) of iron (II) with hydrogen peroxide: the photo-Fenton reaction. Environ Sci Technol 26:313–319
Zepp R, Erickson Iii D, Paul N, Sulzberger B (2007) Interactive effects of solar UV radiation and climate change on biogeochemical cycling. Photochem Photobiol Sci 6:286–300
Zepp R, Erickson D III, Paul N, Sulzberger B (2011) Effects of solar UV radiation and climate change on biogeochemical cycling: interactions and feedbacks. Photochem Photobiol Sci 10:261–279
Zhang Y, Zhu L, Zeng X, Lin Y (2004) The biogeochemical cycling of phosphorus in the upper ocean of the East China Sea. Estuar Coast Shelf Sci 60:369–379

Zhang YL, Lou Zhang E, Liang Liu M, Wang X, Qiang Qin B (2007) Variation of chromophoric dissolved organic matter and possible attenuation depth of ultraviolet radiation in Yunnan Plateau lakes. Limnology 8:311–319

Zhang M, Kong F, Wu X, Xing P (2008) Different photochemical responses of phytoplankters from the large shallow Taihu Lake of subtropical China in relation to light and mixing. Hydrobiologia 603:267–278

Zhang Y, van Dijk MA, Liu M, Zhu G, Qin B (2009) The contribution of phytoplankton degradation to chromophoric dissolved organic matter (CDOM) in eutrophic shallow lakes: field and experimental evidence. Water Res 43:4685–4697

Zhang D, Pan X, Mu G, Wang J (2010a) Toxic effects of antimony on photosystem II of *Synechocystis* sp. as probed by in vivo chlorophyll fluorescence. J Appl Phycol 22:479–488

Zhang J, Hudson J, Neal R, Sereda J, Clair T, Turner M, Jeffries D, Dillon P, Molot L, Somers K (2010b) Long-term patterns of dissolved organic carbon in lakes across eastern Canada: evidence of a pronounced climate effect. Limnol Oceanogr 55:30–42

Zhang D, Pan X, Mostofa KMG, Chen X, Mu G, Wu F, Liu J, Song W, Yang J, Liu Y (2010c) Complexation between Hg(II) and biofilm extracellular polymeric substances: an application of fluorescence spectroscopy. J Hazard Mater 175:359–365

Zhou X, Wangersky PJ (1985) Copper complexing capacity in cultures of Phaeodactylum tricornutum: diurnal changes. Mar Chem 17:301–312

Zhou X, Wangersky PJ (1989) Production of copper-complexing organic ligands by the marine diatom Phaeodactylum tricornutum in a cage culture turbidostat. Mar Chem 26:239–259

Zhou Y, Yu J, Huang L, Nogués S (2004) The relationship between CO_2 assimilation, photosynthetic electron transport and water–water cycle in chill-exposed cucumber leaves under low light and subsequent recovery. Plant, Cell Environ 27:1503–1514

Zhou YH, Yu JQ, Mao WH, Huang LF, Song XS, Nogués S (2006) Genotypic variation of rubisco expression, photosynthetic electron flow and antioxidant metabolism in the chloroplasts of chill-exposed cucumber plants. Plant Cell Physiol 47:192–199

Zhu JK (2000) Genetic analysis of plant salt tolerance using Arabidopsis. Plant Physiol 124:941–948

Zhu JK (2001) Plant salt tolerance. Trends Plant Sci 6:66–71

Zhu JK (2002) Salt and drought stress signal transduction in plants. Annu Rev Plant Biol 53:247

Zidan I, Azaizeh H, Neumann PM (1990) Does salinity reduce growth in maize root epidermal cells by inhibiting their capacity for cell wall acidification? Plant Physiol 93:7

Zohary T, Erez J, Gophen M, Berman-Frank I, Stiller M (1994) Seasonality of stable carbon isotopes within the pelagic food web of Lake Kinneret. Limnol Oceanogr 39:1030–1043

Zouni A, Witt HT, Kern J, Fromme P, Krauû N, Saenger W, Orth P (2001) Crystal structure of photosystem II from Synechococcus elongatusat 3.8 Å resolution. Nature 409:739–743

Zrimec A, Drinovec L, Berden-Zrimec M (2005) Influence of chemical and physical factors on long-term delayed fluorescence in Dunaliella tertiolecta. Electromagn Biol Med 24:309–318

Zubkov MV, Tarran GA (2008) High bacterivory by the smallest phytoplankton in the North Atlantic Ocean. Nature 455:224–226

Zubkov MV, Mary I, Woodward EMS, Warwick PE, Fuchs BM, Scanlan DJ, Burkill PH (2007) Microbial control of phosphate in the nutrient-depleted North Atlantic subtropical gyre. Environ Microbiol 9:2079–2089

"*If H_2O would decompose by the reaction with CO_2* in *photosynthesis, then all H_2O would convert into O_2 by organisms and plants after the origin of life on earth to date and no H_2O would remain in the biosphere.*

Instead of H_2O, *photoinduced generation of H_2O_2 from dissolved O_2 in water bound in photosynthetic cells is reacted with CO_2 in photosynthesis that can limit the photosynthesis under light condition.*

Then further conversion of H_2O_2 *to O_2 either through photosynthesis [$xCO_{2(H2O)} + yH_2O_{2(H2O)} \rightarrow C_x(H_2O)_y + O_2 + E\ (\pm)$] or both photolytically ($2H_2O_2 + h\upsilon \rightarrow O_2 +$ unknown oxidant) and biologically ($2H_2O_2 +$ catalases/ peroxidases $\rightarrow O_2 + 2H_2O$) may balance the environment.*"

Chlorophylls and their Degradation in Nature

Khan M. G. Mostofa, Cong-qiang Liu, Xiangliang Pan, Davide Vione, Kazuhide Hayakawa, Takahito Yoshioka and Gennady G. Komissarov

1 Introduction

Phytoplankton are responsible for approximately 40–50 % of the total primary production on Earth. They contribute to controlling the total CO_2 concentration and pH of the ocean, which together with physical processes (e.g. solar energy input, sea–air heat exchanges, upwelling of subsurface waters and mixed layer thickness) dictates the air-to-sea CO_2 gas exchanges (Longhurst et al. 1995; Field et al. 1998; Takahashi et al. 2002; Falkowski et al. 2004). The global net primary production from phytoplankton is 45–50 Gt C $year^{-1}$, whilst from land plants

K. M. G. Mostofa (✉) · C. Q. Liu
State Key Laboratory of Environmental Geochemistry, Institute of Geochemistry, Chinese Academy of Sciences, Guiyang 550002, China
e-mail: mostofa@vip.gyig.ac.cn

X. L Pan
Xinjiang Key Laboratory of Water Cycle and Utilization in Arid Zone, Xinjiang Institute of Ecology and Geography, Chinese Academy of Sciences, Urumqi 830011, People Republic of China

D. Vione
Dipartimento di Chimica, University of Turin,-I 10125 Turin, Italy

Centro Interdipartimentale NatRisk, I-10095 Grugliasco, (TO), Italy

K. Hayakawa
Lake Biwa Environmental Research Institute, Shiga Prefecture, Ohtsu 520-0806, Japan

T. Yoshioka
Field Science Education and Research Center, Kyoto University, Kitashirakawa Oiwake-cho, Sakyo-ku, Kyoto 606-8502, Japan

G. G. Komissarov
Semenov Institute of Chemical Physics, Russian Academy of Sciences, Moscow 117977, Russia

K. M. G. Mostofa et al. (eds.), *Photobiogeochemistry of Organic Matter*, Environmental Science and Engineering, DOI: 10.1007/978-3-642-32223-5_8,

it is of 45–68 Gt C year^{-1} and from coastal vegetation it is of 1.9 Gt C year^{-1} (Longhurst et al. 1995; Box 2004; Haberl et al. 2007).

Since the development of techniques for Chl *a* detection in water in the decade of 1930 and 1940s (Harvey 1934, 1939), a number of research works has been published to develop analytical methodologies (Richards and Thompson 1952; Parsons and Strickland 1963; Jeffrey and Humphrey 1975), to elucidate Chl *a* origin (Fennel and Boss 2003; Letelier et al. 2004; Huisman et al. 2006) and to understand its photoinduced degradation into various pheopigments (Welschmeyer and Lorenzen 1985; Barlow et al. 1993; Stephens et al. 1997). An additional issue is the production of autochthonous DOM by photoinduced degradation of Chl *a* or phytoplankton biomass, under both photoinduced and microbial (bacterial) metabolism/assimilation/respiration (Kirchman et al. 1991, 1995; Tranvik 1993; Nelson et al. 1998, 2004; Hart et al. 2000; Parlanti et al. 2000; Carrillo et al. 2002; Rochelle-Newall and Fisher 2002; Nieto-Cid et al. 2006; Mostofa et al. 2009; Zhang et al. 2009).

The spatial variability of the net primary productivity over the globe is substantially high, varying from about 1,000 g C m^{-2} for evergreen tropical rain forests to less than 30 g C m^{-2} for deserts (Scurlock et al. 1999). On the other hand, chlorophyll *a* (Chl *a*) concentrations vary from 0.0 to 2,080 μg L^{-1} in a variety of natural waters. Such a variability in Chl *a* concentration can produce either a surface/subsurface Chl *a* maximum (SCM) or a deep Chl *a* maximum (DCM) in natural waters (Huisman et al. 1999, 2006; Riley et al. 1949; Bainbridge 1957; Steele and Yentsch 1960; Anderson 1969; Derenbach et al. 1979; Dortch 1987; Viličić et al. 1989; Bjørnsen and Nielsen 1991; Donaghay et al. 1992; Huisman and Weissing 1995; Djurfeldt 1994; Gentien et al. 1995; Odate and Furuya 1998; Dekshenieks et al. 2001; Franks and Jaffe 2001; Klausmeier and Litchman 2001; Diehl 2002; Rines et al. 2002; Yoshiyama and Nakajima 2002; Arístegui Ruiz et al. 2003; Hodges and Rudnick 2004; Matondkar et al. 2005; Weston et al. 2005; Lund-Hansen et al. 2006; Beckmann and Hense 2007; Hense and Beckmann 2008; Hopkinson and Barbeau 2008; Whitehouse et al. 2008; Yoshiyama et al. 2009; Lu et al. 2010; Martin et al. 2010; Ryabov et al. 2010; Velo-Suárez et al. 2010).

The high variation in Chl *a* content is generally used as a universal signature of cyanobacteria (algae), or of phytoplankton bloom or eutrophication in a variety of waters (Fielding and Seiderer 1991; Ondrusek et al. 1991; Williams and Claustre 1991; Millie et al. 1993; Jeffrey et al. 1999; Bianchi et al. 1993, 2002, Blanco et al. 2008; Kasprzak et al. 2008). Variations in Chl *a* concentrations or primary production is entirely dependent on various environmental factors in natural waters, which have been extensively discussed before (see also chapter "Photosynthesis in Nature: A New Look").

It has been found that Chl *a* bound to phytoplankton can be degraded by photoinduced and microbial processes, thereby producing a number of pigments and colourless organic compounds in natural waters (Welschmeyer and Lorenzen 1985; Barlow et al. 1993; Stephens et al. 1997; Zhang et al. 2009; Bianchi et al. 2002; Schulte-Elte et al. 1979; Falkowski and Sucher 1981; Pietta et al. 1981; Mantoura and Llewellyn 1983; Keely and Maxwell 1991; Nelson 1993; Sun et al. 1993; Rontani et al. 1995, 1998, 2003, 2011; Rontani and Marchand 2000; Yacobi et al. 1996; Cuny et al.

1999; Marchand and Rontani 2001; Rontani 2001; Lemaire et al. 2002; Rontani and Volkman 2003; Marchand et al. 2005; Christodoulou et al. 2009; Christodoulou et al. 2010). Chl can also be degraded in higher plants, which for instance causes the colour change in leaves from green to yellow or red that is naturally observed in autumn. However, degradation can also occur as a consequence of cell death caused by external factors, such as injuries due to low or high temperature, pathogen attack, as well as phenomena taking place during various phases of the life cycle of plants (Hendry et al. 1987; Takamiya et al. 2000). Conversion of Chls to pheophytins can take place during discolouration of green vegetable upon processing by several chemicals, photoinduced or enzymatic reactions including simultaneous actions of enzymes, weak acids or changes in pH, oxygen, light and heat (Blair and Ayres 1943; Gupte et al. 1964; Hayakawa and Timbers 1977; Minguez-Mosquera et al. 1989; Mangos and Berger 1997; Koca et al. 2007). Moreover, the key PSII degradation reactions of Chls are photooxidation, involving attack of singlet oxygen or $HO^{\bullet}$ via H_2O_2, and enzymatic degradation (Takamiya et al. 2000; Brown et al. 1991; Hörtensteiner 2006; Kräutler and Hörtensteiner 2006; Moser et al. 2009; Hörtensteiner and Kräutler 2011; Gálvez et al. 1988).

This chapter will give an overview of the various kinds of Chl, their properties, functions, and techniques for their precise determination. It extensively discusses the distribution of Chl *a* providing information about SCM and DCM depths, the formation mechanisms of such maxima as well as the changes of Chl *a* concentrations in a variety of natural waters, under both field and experimental conditions. It also discusses the degradation and degradation mechanisms of Chl *a* bound to aquatic microorganisms and higher plants, as well as the modifications taking place during food processing. Finally, an explanation will be provided of how Chl *a* acts as a universal signature of phytoplankton bloom, and of the possible actions to be adopted for the management of eutrophication by controlling primary production or Chl *a*.

2 Chlorophylls (Pigments) in Phytoplankton

Photosynthetic organisms can collect light energy with their light-harvesting systems that are composed of core and peripheral antenna complexes (Green and Durnford 1996). Core antenna complexes of oxygen-evolving photosynthetic organisms have Chl *a* as pigment. In contrast peripheral antenna complexes, particularly for photosystem II (PSII), have various pigments depending on the group of photosynthetic organisms. They are Chl *b*, Chl *c* (made up of c_1, c_2 and c_3), Chl *d*, phycobilins, fucoxanthin, zeaxanthin (carotenoids), echinenone, peridinin, and so on (Bianchi et al. 2002; Woodward et al. 1960, 1990; Dougherty et al. 1966; Fleming 1967; Wu and Rebeiz 1985; Jeffrey and Wright 1987; Verne-Mismer et al. 1988, 1990; Fookes and Jeffrey 1989; Rowan 1989; Grossman et al. 1995; Miyashita et al. 1996, 1997; Motilva 2008).

Chl *b* is detected in various forms such as: divinyl Chl *b*, with two vinyl groups at R_1 and R_2 positions; monovinyl Chl *b*, with vinyl at R_1 and ethyl at

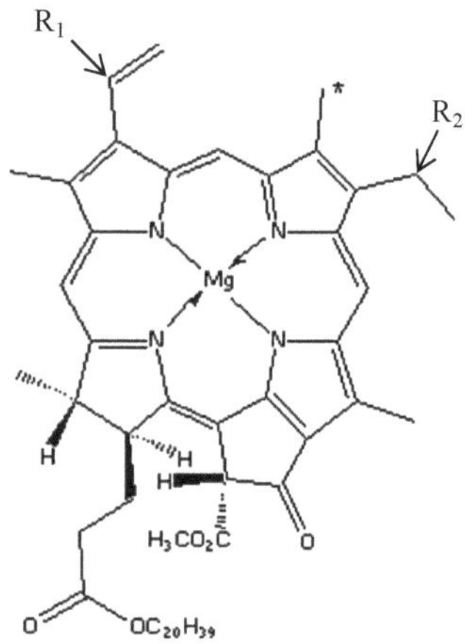

Fig. 1 Molecular structure of chlorophyll *a* and chlorophyll *b* with some medication. *Data source* Clarke et al. (1976)

R_2, and meso-Chl *b* with two ethyl groups at R_1 and R_2 (Fig. 1) (Clarke et al. 1976). Moreover, Chl *c* composed of a 17-acrylic side-chain has been isolated from two Moroccan oil shales of Cretaceous age (Verne-Mismer et al. 1988). Petroporphyrins, showing both a free C-7 position and a C-13 to C-17 substitution pattern typical of Chl *c* fossils, have been isolated from two Cretaceous Moroccan oil shales and may arise from a still unknown Chl, 7-formylchlorophyll *c* or Chl c_3 (Verne-Mismer et al. 1990). Chl c_3, isolated from a microalga, has tentatively been identified as 7-demethyl-7-methoxycarbonyl chlorophyll (Fookes and Jeffrey 1989). Chl c_2, isolated from the coccolithophorid *Emiliania huxleyi* (Prymnesiophyceae), is present in approximately equal proportions as Chl c_3. It has absorption maxima at 447, 579 and 628 nm, whilst no Chl c_1 has been detected (Jeffrey and Wright 1987). Cells of the new prokaryote have an absorption maximum in the red region (714–718 nm) due to Chl *d* absorption, but they do not show any characteristic absorption peak of Chl *a* at around 680 nm (Miyashita et al. 1997). Chl *f* is [2-formyl]-chlorophyll *a* ($C_{55}H_{70}O_6N_4Mg$), and its in vitro absorption (706 nm) and fluorescence (722 nm) maxima are red-shifted compared to all other Chls from oxygenic phototrophs (Chen et al. 2010).

Uncoupled Chls in PSII of cyanobacteria or phytoplankton and red algae can absorb at shorter wavelength, e.g. 670 nm (close to their site energy), whilst electronically coupled chlorins (the central cofactors) or Chl dimers can absorb between 676 and 684 nm (Zhang et al. 2009; Telfer et al. 1990; Durrant et al. 1995; Renger and Marcus 2002). Formation of Chl dimers and their light-induced excitations are extensively discussed in the photosynthesis chapter “Photosynthesis in Nature: A New Look”. Upon 440-nm excitation of Chl *a* bound to *Synechocystis* cells, a typical PSII fluorescence at 685 nm has been observed both at room temperature and at 77 K (Satoh et al. 2001). Two additional fluorescence components have been detected at 695 and 722 nm, whilst no fluorescence peak for Chl *b* has been observed in these cyanobacteria (Satoh et al. 2001). Red shifts are commonly observed in in vitro Chl *a* systems, including thin films, monolayers and colloidal dispersions used as models for the in vivo systems (Katz et al. 1991). Red shifts generally occur when electron releases follow the easiest way in the functional groups bound to the component system (see also chapters "Colored and Chromophoric Dissolved Organic Matter (CDOM) in Natural Waters" and "Fluorescent Dissolved Organic Matter in Natural Waters") (Mostofa et al. 2009; Senesi 1990). Chl *b* found in chlorophytes and prochlorophytes can absorb sunlight at around 470 nm (highest peak) and 650 nm (small peak) (Satoh et al. 2001). The Chl *c* isolated from *Peridinium gatunense* showed two peaks at 448–449 and 634–635 nm (Yacobi et al. 1996).

All pigments can bind to their specific proteins to form pigment-protein complexes (Cogdell et al. 1996; Pearlstein 1996). Complexation can provide the easiest way of electron release, as depicted in other chapters (see "Photosynthesis in Nature: A New Look" and "Complexation of Dissolved Organic Matter with Trace Metal ions in Natural Waters"). The Chl *b* content of the light-harvesting complex (LHC) of PSII in higher plants is highly preserved, approximately between 45 and 50 % or in the approximate ratio of 3:1 of Chl *a* to Chl *b* (Anderson 1986; von Elbe and Schwartz 1996). Conversely, the contents of Chl *b* in cyanobacteria are variable and relatively low (1.4–10.6 % or more) (Bianchi et al. 2002; Satoh et al. 2001). Experimental and other observation have shown that Chl *a* molecules can bind to LHC of PSII at Chl *b* binding sites (Thornber and Highkin 1974; Terao and Katoh 1989; Murray and Kohorn 1991; Paulsen et al. 1993; Polle et al. 2000). Correspondingly, Chl *b* is vital for the stability of LHC of PSII in the thylakoid membrane (Murray and Kohorn 1991; Bellemare et al. 1982). The core antenna complexes of chlorophytes have Chl *a* and do not bind Chl *b*, despite its presence (Satoh et al. 2001; Anderson et al. 1978).

2.1 Properties and Functions of Chlorophyll

Chlorophyll (Chl) *a* has a methyl group at the C-3 carbon (molecular formula $C_{55}H_{72}MgN_4O_5$), while Chl *b* has the same chemical structure as Chl *a* but with a $-CH_3$ group replaced by a –CHO one, providing the molecular formula $C_{55}H_{70}MgN_4O_6$ (Fig. 1) (Clarke et al. 1976). The correct gross structure of Chl has been suggested at first by Fischer (Fischer and Wenderoth 1940) and verified in a synthesis by Woodward (Woodward et al. 1960; Woodward 1961). The relative configuration of the methyl and propionic ester groups on the D ring in the structure was shown to be *trans* by Ficken and his colleagues (Ficken et al. 1956). The stereochemistry and absolute configuration of the phytyl group is 2′-*trans*-7′*R*,11′*R*, as discovered in 1959 (Burrell et al. 1959; Crabbe et al. 1959). The relative configuration at C_{10} is such that the methoxycarbonyl group is *trans* to the propionic ester side chain on C_7 (Closs et al. 1963; Wolf et al. 1967). In addition to their structural differences, Chl *a* is observed to be thermally less stable than Chl *b* (Buckle and Edwards 1970; Lajollo et al. 1971; Schwartz and von Elbe 1983; Canjura et al. 1991; Schwartz and Lorenzo 1991).

Functions of Chls and their degradation products can be discriminated as follows: (i) Chl is an efficient visible-light photosensitizer and a key component required for the absorption of sunlight. It is essential for the occurrence of photosynthesis as it is involved into the initiation of electron release in aquatic microorganisms and higher plants (see also chapter "Photosynthesis in Nature: A New Look") (Hörtensteiner and Kräutler 2011; Foote 1976; Kimball 1979; Knox and Dodge 1985). Under specific conditions (e.g. high light conditions, high temperature, drought and so on) Chl can significantly produce reactive oxygen species (ROS) such as singlet oxygen ($^1O_2$), superoxide radical anion ($O_2^{\bullet -}$), hydrogen peroxide (H_2O_2) and $HO^\bullet$ (see chapter "Photosynthesis in Nature: A New Look"). In turn, these species can degrade Chl and cause cell death (Rontani 2001; Hörtensteiner and Kräutler 2011; Marshall et al. 2002; Oda et al. 1998). ROS also play a role during senescence of photosynthetic cells or fruit ripening. Strong light gradients cause unbalanced excitation of the two photosystems and reduce photosynthetic efficiency (Dietzel et al. 2011). (ii) According to the Treibs hypothesis, petroporphyrins can originate from Chl (Treibs 1936; Liang et al. 1993). (iii) Chl, some of its synthetic analogues, metal complexes of porphyrins and phthalocyanines are all photoactive. As key components bound to organisms they can cause production of H_2O_2 in vivo under light, in aqueous solutions saturated with dioxygen (Hong et al. 1987; Bazanov et al. 1999; Premkumar and Ramaraj 1999; Lobanov et al. 2008). (iv) Chls, the pigments responsible for green color in fruits and vegetables, are highly susceptible to degradation during processing. This can result into changes from bright green to olive brown or other colors, during storage and processing in the agriculture and food industry (Schwartz and von Elbe 1983; Sweeney and Martin 1961). Color, the major quality attribute of vegetables and fruits, is a key factor in the commercial value of food to the consumer and can be highly affected by Chl breakdown as an important catabolic process of leaf senescence and fruit ripening (Takamiya et al. 2000; Hörtensteiner and Kräutler 2011; Schwartz and von Elbe 1983; Steet and Tong 1996). (v) The colorless "nonfluorescent Chl catabolites (NCC)" found in ripening fruits (e.g. apples and pears) can act as antioxidants, in a similar way as bilirubin (Moser et al. 2009; Stocker et al. 1987; Barañano et al. 2002). It has been shown that the rates of formation of hydroperoxides of linoleic acid in the presence of NCC is significantly reduced. The observed effect is a function of time and of the concentration of the added antioxidants. Moreover, the (concentration-dependent) peroxy radical scavenging effect of NCC is only slightly inferior to that of bilirubin (Moser et al. 2009; Stocker et al. 1987; Müller et al. 2007). (vi) Chl *a* is generally used to estimate the primary biomass production or the phytoplankton/cyanobacterial biomass or bloom in natural waters. In contrast, carotenoids and the degradation intermediates xanthophylls could be effective biomarkers of different classes of phytoplankton (Fielding and Seiderer 1991; Ondrusek et al. 1991; Williams and Claustre 1991; Millie et al. 1993; Jeffrey et al. 1999; Bianchi et al. 1993, 2002; Kasprzak et al. 2008). Therefore, Chl *a* and its degradation products could be useful indicators of the fate and composition of phytoplankton species and of transformation and degradation of phytoplanktonic carbon. As a key characteristic of phototrophic organisms, they can be used as a criterion in the classification of autotrophic bacteria and cyanobacteria or algae (Williams and Claustre 1991; Marchand et al. 2005; Rowan 1989; Liang et al. 1993; Downs and Lorenzen 1985; Trüper 1987; Volkman et al. 1988; Vaulot et al. 1990; Veldhuis and Kraay 1990; Wilhelm et al. 1991; Brunet et al. 1992; Head and Horne 1993; Soma et al. 1993). Similarly, phaeopigments (Chl degradation products) represent the dominant form of plant pigments in marine sediments (Brown et al. 1991; Baker and Louda 1983; Furlong and Carpenter 1988; Leavitt and Carpenter 1990; Bianchi and Findlay 1991; Bianchi et al. 1993; Jeffrey et al. 1997). Chl *b* is used as a biomarker for chlorophytes (Bianchi et al. 2002). (vii) Primary production (e.g. algae) is substantially high in ice bed (0.1–1,000 $\mu g\ L^{-1}$) and can provide food resources for organisms in higher trophic levels, in seasons and regions where the water-column biological production is low or negligible (Palmisano et al. 1985; Garrison et al. 1986; Wheeler et al. 1996; Mock and Gradinger 1999; Lizotte 2001). (viii) The specific Chl *a* content per unit of phytoplankton biomass is typically decreased with increasing phytoplankton standing stocks, and with Chl *a* concentration in natural waters and also in laboratory cultures of certain species (Kasprzak et al. 2008; Desortová 1981; Shlgren 1983; Wojciechowska 1989; Watson et al. 1992; Talling 1993; Chow-Fraser et al. 1994; Schmid et al. 1998; Felip and Catalan 2000; Sandu et al. 2003; Kiss et al. 2006). Such a trend might reflect several phenomena such as: degradation of Chl *a* bound in phytoplankton; lake trophic status; phytoplankton community structure; size frequency distribution of algal cells; and seasonal shifts within the plankton community (Bianchi et al. 2002; Bursche 1961; Nusch and Palme 1975; Harris 1986; Watson and McCauley 1988; Arnott and Vanni 1993; Fu et al. 2010; Mostofa KMG et al. unpublished data). (ix) Chloropigments (Chl *a* and carotenoids) and their degradation products could be important determinants of UV and PAR attenuation in natural waters, due to their efficient radiation absorption (see also chapter "Colored and Chromophoric Dissolved Organic Matter (CDOM) in Natural Waters") (Zhang et al. 2009; Devlin et al. 2009; Zhang and Qin 2007; Dupouy et al. 2010; Zhang et al. 2007). (x) The ultimate degradation products of Chls and pigments are colorless (Zhang et al. 2009; Marchand et al. 2005; Mostofa K et al. unpublished data; Wakeham and Lee 1993; Mostofa K et al. unpublished data; Meyers 1997). They may contribute to autochthonous DOM and, therefore, to DOM dynamics in natural waters. Lipids, one of the three major classes of organic matter in algal material, are often used as biomarkers because of their lower labilility compared to proteins and carbohydrates (Mostofa et al. 2009; Sun et al. 2002; Wakeham 1995; Volkman 1986).

2.2 Determination of Chls and Other Pigments

For the measurement of Chls, and particularly of Chl *a*, Chl *b* and Chl *c*, various absorption peaks have been used. Absorption peaks have small variations depending on the phytoplankton species (Goedheer 1970; Prezelin 1981; Aguirre-Gomez

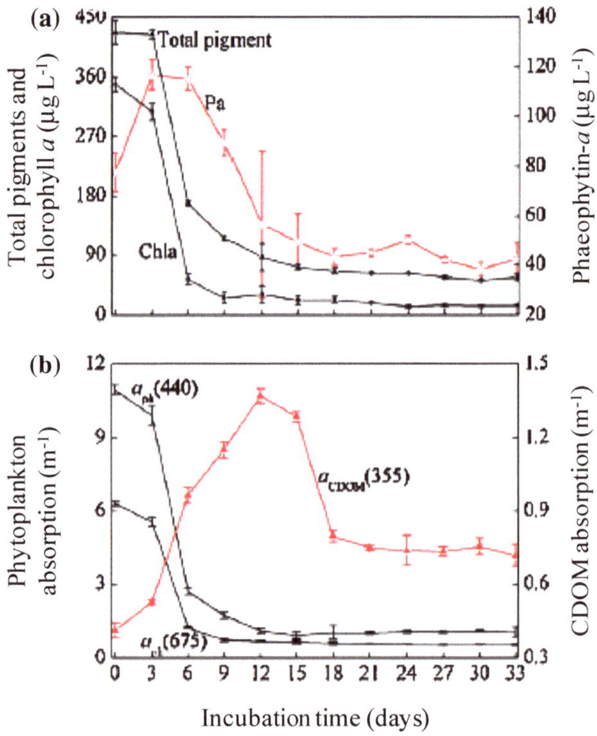

Fig. 2 Changes in the (**a**) mean concentrations of total pigment, chlorophyll *a* and Phaeophytin-*a* (*P*a); and (**b**) phytoplankton pigment absorption at the Chl *a* absorption maxima at 440 and 675 nm and CDOM absorption at 355 nm during the degradation experiment period (0–33 days). *Error bar* indicates the means and standard deviations (n = 3). *Data source* Zhang et al. (2009)

et al. 2001; Pérez et al. 2007). For Chl *a*, peaks that are often used are those at 412–425, 435–455, 618–623 and 662–675 nm, respectively (Zhang et al. 2009; Goedheer 1970; Prezelin 1981; Aguirre-Gomez et al. 2001; Pérez et al. 2007). The in vivo absorption spectra of the brown alga *Laminaria digitata* have Chl *a* peaks at 418, 437, 618 and 673 nm. Moreover, absorption peaks of *Glenodinium* sp. occur at 419, 437, 618 and 675 nm, and absorption peaks (average) of three different groups of algae are located at 412, 435, 623 and 675 nm (Goedheer 1970; Prezelin 1981; Hoepffner and Sathyendranath 1991). The structural configuration of PSI and PSII in the reaction center shows that they may have two wavelength positions: uncoupled Chls can absorb at 670 nm (close to their site energy), and electronically coupled chlorins (the central cofactors) or Chl dimers can absorb between 676 and 684 nm (see also chapter "Photosynthesis in Nature: A New Look") (Telfer et al. 1990; Durrant et al. 1995; Renger and Marcus 2002).

Microbial degradation experiments show that absorbance of Chl *a* in the shorter wavelength region (~440 nm) disappears relatively faster compared to the longer wavelength region (~675 nm). Therefore, only the 675 nm absorption peak remains visible in the suspension if degradation time is long enough (Fig. 2; see also chapter "Colored and Chromophoric Dissolved Organic Matter (CDOM) in Natural Waters") (Zhang et al. 2009). Absorption peaks in the shorter wavelength region are generally accounted for by various substances such as proteins, amino acids and other organic components bound to PSI and PSII. These compounds are all susceptible to undergo microbial decompositon (see chapters "Photoinduced and Microbial Degradation of Dissolved Organic Matter in Natural Waters, "Colored and Chromophoric Dissolved Organic Matter (CDOM) in Natural Waters", "Fluorescent Dissolved Organic Matter in Natural Waters"). On the other hand, Chls that absorb radiation in the longer wavelength region are susceptible to undergo photochemical decomposition (see chapters "Photoinduced and Microbial Degradation of Dissolved Organic Matter in Natural Waters", "Colored and Chromophoric Dissolved Organic Matter (CDOM) in Natural Waters", "Fluorescent Dissolved Organic Matter in Natural Waters"). Absorbance in the longer wavelength regions (>600 nm) is generally linked to the easiest way of electron release from the functional groups bound to the parent molecule. Chl molecules are thus responsible for the absorption peaks located at $\lambda > 600$ nm. Interestingly, longer wavelength absorption peaks (>600 nm) are often observed for some functional groups that are present in terrestrial humic substances (fulvic and humic acids) in riverine ecosystems (see chapter "Colored and Chromophoric Dissolved Organic Matter (CDOM) in Natural Waters"). Therefore, changes in functional groups or molecules bound to PSI and PSII, which take place through either photoinduced or microbial processes, may affect the absorption peaks. Note that peaks appearing in the green region (500–600 nm) are small compared to those located in the blue (<500 nm) and red (>600 nm) regions (Aguirre-Gomez et al. 2001).

Considering the previously reported findings, the following suggestions can be followed for Chl determination: First, measurement of Chl *a* should be conducted only at a single wavelength, not at several ones. The most suitable is at around 665–675 nm, and absence of light should be ensured during sample processing and measuring. Second, Chl *b* should be measured only at around 643–650 nm. In earlier studies, the measurement of Chl *b* has been carried out using its absorption peaks at 465–470 or 483, 585–595 and 643–650 nm, but only the latter provides sufficiently accurate results (Satoh et al. 2001; Aguirre-Gomez et al. 2001; Bidigare et al. 1989; Millie et al. 1997). Some differences in absorption wavelengths in Chl *b* can be caused by the occurrence of various forms of this Chl, as mentioned before. The third issue is that Chl *c* should be detected at 630–639 nm, although earlier studies have adopted absorption peaks at 465–470, 589, and 630–639 nm (Bidigare et al. 1989; Millie et al. 1995, 1997). The many absorption peaks used in earlier studies, in particular at short wavelengths, should not be adopted for the measurement of any Chl molecule. The reason is that absorbance at shorter wavelengths has been observed for other pigments that could interfere with Chl determination, such as hycoerythrin (detected at 543–550 and 566–568 nm) (Payri et al. 2001; Smith and Alberte 1994); phycoerythrocyanin (~550 nm and ~575 nm) (Millie et al. 2002); phycocyanin (625–630 nm) (Payri et al. 2001; Millie et al. 2002); fucoxanthin (521–531 nm) (Bidigare et al. 1989); and different carotenoids (490–495 nm) (Millie et al. 1997; Owens et al. 1987). Furthermore, CDOM absorbs radiation in lower wavelength regions (250–500 nm) because of the functional groups present in allochthonous and autochthonous organic substances (see chapter "Colored and Chromophoric Dissolved Organic Matter (CDOM) in Natural Waters").

3 Distribution of Chlorophyll (Chl *a*)

Chl *a* concentrations are significantly varied in the water column, where a given set of parameters may lead to either a surface or a subsurface Chl *a* maximum (SCM), or to a deep Chl *a* maximum (DCM) (Huisman et al. 2006; Riley et al. 1949; Bainbridge 1957; Steele and Yentsch 1960; Anderson 1969; Derenbach et al. 1979; Dortch 1987; Viličić et al. 1989; Bjørnsen and Nielsen 1991; Donaghay et al. 1992; Huisman and Weissing 1995; Djurfeldt 1994; Gentien et al. 1995; Odate and Furuya 1998; Huisman et al. 1999; Dekshenieks et al. 2001; Franks and Jaffe 2001; Klausmeier and Litchman 2001; Diehl 2002; Rines et al. 2002; Yoshiyama and Nakajima 2002; Arístegui Ruiz et al. 2003; Hodges and Rudnick 2004; Matondkar et al. 2005; Weston et al. 2005; Lund-Hansen et al. 2006; Beckmann and Hense 2007; Hense and Beckmann 2008; Hopkinson and Barbeau 2008; Whitehouse et al. 2008; Yoshiyama et al. 2009; Lu et al. 2010; Martin et al. 2010; Ryabov et al. 2010; Velo-Suárez et al. 2010). The location of the maximum is entirely determined by the environmental conditions. The cited studies have shown that SCM and DCM of phytoplankton can occur in a variety of conditions in lake and marine waters. They can range in the vertical dimension from centimeters to a few meters, and have been observed to extend horizontally for kilometers.

3.1 Surface or Subsurface Chl a *Maximum*

The surface or subsurface Chl *a* maximum (SCM) is detected in the surface layer, which varies in different waters and may range between 0–25 m in lakes and 0–30 m or more in seawater (Fig. 3a; Table 1) (Fu et al. 2010; Mostofa K et al. unpublished data; Apollonio 1980; Vicente and Miracle 1984; Kimor et al. 1987; Pedros-Alio et al. 1987; Millán-Núñez et al. 1996; Gomes et al. 2000; Guildford and Hecky 2000; Li and Harrison 2001; Echevin et al. 2004; Koné et al. 2005; Camacho 2006; Ediger et al. 2006; Parab et al. 2006; Roy et al. 2006; Satoh et al. 2006; Sawatzky et al. 2006; Yacobi 2006; Norrbin et al. 2009; Xiu et al. 2009; Zhu et al. 2009; Hamilton et al. 2010). According to these studies, SCM can be defined as a zone of maximum photosynthetic activity that shows the highest Chl *a* contents. It occurs in the upper surface layer of the euphotic zone in the presence of strong light, high DOM contents and nutrients, and under high temperature as well as low or high turbulence. It is a remarkable feature of highly turbid water in the surface layer of stagnant natural waters, particularly in lakes and oceans. High variation with depth of SCM in seawater is presumably caused by an increase of the surface-water mixing zone, due to strong wind and wave compared

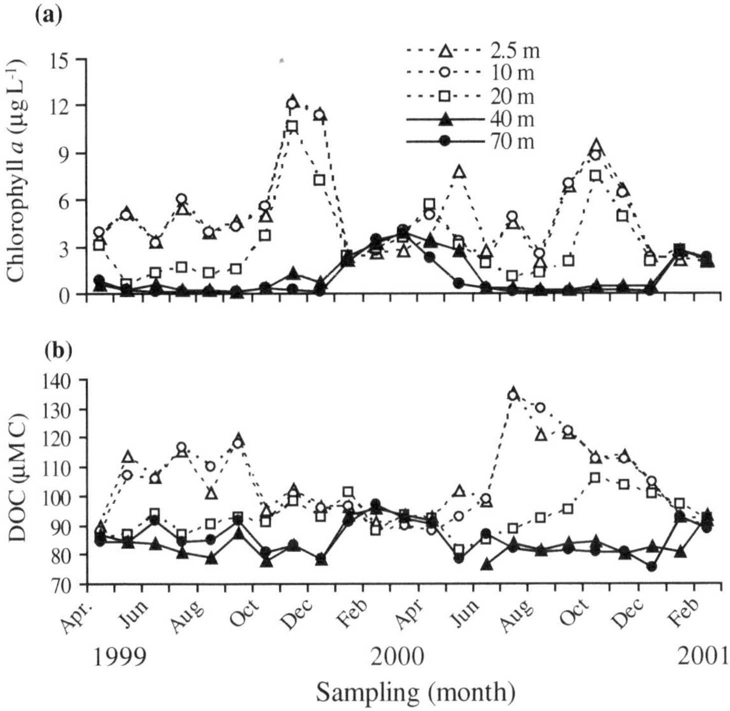

Fig. 3 Vertical changes in the chlorophyll *a* (**a**) and dissolved organic carbon (DOC) concentrations (**b**) in monthly collected samples from Lake Biwa and *Error bars* indicate the standard deviation *Data source* Mostofa et al. (2005), Mostofa KMG et al. (unpublished data)

to conditions in lakes. An upper-surface mixed layer commonly occurs in lakes and oceans, due to mechanical perturbation of surface waters (e.g. by wind, waves and storms) (Deuser 1987; Venrick 1993; Law et al. 2003; Moum et al. 1989; Brainerd and Gregg 1995). It is characterized by strong turbulent mixing, up to a depth of approximately 30–200 m or more. Note that few studies have reported the occurrence of DCM (or subsurface Chl a maximum) at a depth of 5–25 m or more (Table 1) (Parab et al. 2006; Sawatzky et al. 2006; Xiu et al. 2009; Hamilton et al. 2010; Fee 1976; Sommaruga and Augustin 2006). Considering the surface mixing zone of the water column, it might be supposed to have a similar meaning as the surface Chl *a* maximum (SCM). A high content of Chl *a* at a depth of 5–15 m may be due to the occurrence of strong photoinduced degradation of Chl *a* in the upper surface layer, e.g. at 0–4 m depth. Note that subsurface Chl *a* maxima have been considered as DCM in several earlier studies, while in this chapter a similar meaning (SCM) is adopted for the subsurface Chl *a* maximum (at e.g. 5–15 m depth) and for the surface Chl *a* maximum (0–30 m depth). SCM should thus be well differentiated from DCM to avoid any confusion. Such a rationalization could be useful to avoid confusion between SCM and DCM in future studies.

Table 1 Variation in chlorophyll *a* concentrations along with pH, water temperature and dissolved organic carbon (DOC) concentrations in a variety of natural waters

Sampling	pH	Water temperature (°C)	Chl *a* Epilimnion ($\mu g\ L^{-1}$)	Hypolimnion	Chl *a* and depth in SCM and DCM ($\mu g\ L^{-1}$; m)	DOC Epilimnion (μM C)	Hypolimnion	Reference
Streams and rivers								
Streams (n = 9)	–	–	0.0–12.7	–	–	–	–	Gao et al. (2004)
Temperate streams (USA)	–	–	0.4–170	–	–	–	–	van Nieuwenhuyse and Jones (1996)
Ozark Streams (USA)	–	–	0.5–44.6	–	–	–	–	Lohman and Jones (1999)
Chalk stream (UK)	–	–	0.5–17.0	–	–	–	–	Palmer-Felgate et al. (2008)
La Trobe River Streams (Victoria, Australia)	–	–	~<65	–	–	–	–	Chessman (1985)
Streams and Rivers (USA)	–	–	1.0–97.0	–	–	–	–	Royer et al. (2008)
Streams and Rivers (Illinois, USA)	–	6.0–27.0	~00.0–18.0	–	–	–	–	Morgan et al. (2006)
Streams and Rivers (Paraná River basin: Argentina, Brazil and Paraguay)	6.74–7.78	11.7–21.5	0.14–216	–	–	–	–	Devercelli and Peruchet (2008)
Red River and its basin (USA)	–	–	0.1–263	–	–	–	–	Longing and Haggard (2010)
Rideau River (Ontario, Canada)	–	–	<27	–	–	–	–	Basu and Pick (1997)
Yukon River (Canada)	7.4–7.9	–	0.20–5.07	–	–	508–2835	–	Guéguen et al. (2006)

(continued)

Table 1 (continued)

Sampling	pH	Water temperature (°C)	Chl *a* Epilimnion ($\mu g\ L^{-1}$)	Hypolimnion	Chl *a* and depth in SCM and DCM ($\mu g\ L^{-1}$; m)	DOC Epilimnion (μM C)	Hypolimnion	Reference
River Avon (Warwickshire, UK)	–	–	~<100	–	–	–	–	Foster et al. (1997)
River Alne (Warwickshire, UK)	–	–	~<280	–	–	–	–	Foster et al. (1997)
River Arrow (Warwickshire, UK)	–	–	~<240	–	–	–	–	Foster et al. (1997)[a]
Jacupiranguinha and Pariquera-Açu Rivers	–	–	1.4–12.0	–	–	–	–	Calijuri et al. (2008)
Lakes and reservoirs								
Lake Hongfeng (Southwest China)	–	7.0–31.0	2.7–47.8 (0–8 m)	1.4–32.7 (10–25 m)	43.6–47.8; SCM (4–5 m)	170–250	134–237	Fu et al. (2010)
Lake Baihua (Southwest China)	–	6.5–25	4.1–65.5 (0–8 m)	1.4–41.0 (10–25 m)	58.7–65.5; SCM (0–2 m)	169–330	157–303	Fu et al. (2010)
Subtropical and urban shallow Lakes (Wuhan, China)	6.8–9.1	3.7–31.6	2.1–189.8	–	–	–	–	Lv et al. (2011)
Lakes (38 Chinese lakes)	7.31–9.73 (8.67 ± 0.43)	13.3–28.3 (23.2 ± 2.69	0.01–133.22	–	–	–	–	Zhang et al. (2007)
Lakes (3 Chinese lakes)	–	16.5–16.7 (mean)	0.85 ± 0.17–9.67 ± 2.25	–	–	–	–	Pan et al. (2009)
Lake Taihu (China)	–	5.0–30.0	~5.0–30.0	–	–	–	–	James et al. (2009)[a], Liu et al. (2011)

(continued)

Table 1 (continued)

Sampling	pH	Water temperature (°C)	Chl *a* Epilimnion (μg L^{-1})	Hypolimnion	Chl *a* and depth in SCM and DCM (μg L^{-1}; m)	DOC Epilimnion (μM C)	Hypolimnion	Reference
Lake Baiyangdian (China)	–	–	0.0–66.93	–	–	–	–	Wang et al. (2012)
Lake Biwa: (Japan)	–	–	0.5–12.3 (0–20 m)	0.1–3.9 (40–80 m)	9.4–12.3; SCM (2.5–10 m)	81.5–135.1	75.6–96.9	Mostofa et al. (unpublished data)
Lake Kizaki (Japan)	7.68–9.55	8.40–21.49	15.3–82.1 (2 m)	–	–	–	–	Yoshioka (1997)
Lakes (Japan)	–	–	0.2–189	–	–	–	–	Aizaki et al. (1981)
Lake Baikal (Russia)	–	0-17.9 (s)	0.5–5.8	–	<5.8; SCM (~10 m)	–	–	Satoh et al. (2006)[a], Yuma et al. (2006)
Lake Baikal (Southern Basin) during late winter	–	0.40–0.84 (average)	0.67–2.0 (0–25 m)	–	–	–	–	Straškrábová et al. (2005)
Lake Victoria (Africa)	–	–	~15–110	–	–	–	–	Silsbe et al. (2006)[a]
Lake Victoria (Africa)	–	–	4.7–78.5	–	–	–	–	Guildford and Hecky (2000)
Lake Tanganyika (Africa)	–	23–37 (surface layer)	0.1–4.5	–	–	–	–	Yuma et al. (2006)
Lake Malawi (Africa)	–	–	0.03–18.7	–	–	–	–	Guildford and Hecky (2000)
Lakes (16 shallow Danish lakes)	–	–	33–276	–	–	–	–	Windolf et al. (1996)
Lakes (Lake Ånnsjön, Erken and Balaton, Sweden)	7.3–8.4	6–11.3	1.2–10 (1 m)	–	–	–	–	Kahlert (2002)
Lakes (Stechlin, Kleiner Väter, Großer Väter, Tiefwaren & Feldberger Haus)	–	–	0.7–175.9	–	–	–	–	Kasprzak et al. (2008)

(continued)

Table 1 (continued)

Sampling	pH	Water temperature (°C)	Chl *a* Epilimnion (μg L^{-1})	Hypolimnion	Chl *a* and depth in SCM and DCM (μg L^{-1}; m)	DOC Epilimnion (μM C)	Hypolimnion	Reference
Lakes (Förchensee, Brunnensee, Klostersee, Langbürgenersee, Thalersee, Bansee)	–	–	0.2–53.4	–	–	–	–	Striebel et al. (2008)
Lake Krankesjön (Sweden)	–	–	10–60 (0–2 m)	–	–	–	–	Blindow et al. (2006)
Lake Börringesjön (Sweden)	–	–	60–145 (0–2 m)	–	–	–	–	Blindow et al. (2006)
Lake La Caldera (Southern Spain)	–	1.2–12.4	0.14–2.85	–	–	42	–	Carrillo et al. (2002)
Lake Cisó (Spain)	–	–	<850	–	850; SCM or DCM? (1.0–1.5 m)	–	–	Pedros-Alio et al. (1987)
Lake Arcas (Spain)	–	–	<298	–	298; SCM or DCM? (8-9 m)	–	–	Camacho (1997)
Lake El Tobar (Spain)	–	–	< 90	–	90; SCM or DCM? (11–12 m)	–	–	Miracle et al. (1993); Camacho (2006)
Lake La Cruz (Spain)	–	–	<25	–	25; SCM or DCM? (10–12 m)	–	–	Rojo and Miracle (1987), Dasí and Miracle (1991)
Lake La Parra (Spain)	–	–	<15	–	15; SCM or DCM? (9 m)	–	–	Camacho et al. (2003)
Lake Lagunillo del Tejo (Spain)	–	–	<15	–	15; SCM or DCM? (4–11 m)	–	–	Vicente and Miracle (1984)
Lakes (26 lakes in Austria, Italy and Spain)	–	–	0.3–7.97	–	–	–	–	Laurion et al. (2002)
Amazon flood plain lakes (shallow lakes, 1–3 m)	–	–	8.2–89.2	–	–	–	–	de Moraes Novo et al. (2006)

(continued)

Table 1 (continued)

Sampling	pH	Water temperature (°C)	Chl *a* Epilimnion (μg L^{-1})	Hypolimnion	Chl *a* and depth in SCM and DCM (μg L^{-1}; m)	DOC Epilimnion (μM C)	Hypolimnion	Reference
Lakes (New Zealand, n = 11)	–	–	0.39–4.38	–	–	25-833	–	Rae et al. (2001)
Lake Taupo (New Zealand)	–	11.1–20.6 (s); 10.6–11.1 (b)	0.75 (annual mean)	–	~3–6; SCM/DCM? (~10, 40–85 m)	–	–	Hamilton et al. (2010)[a]
Lake Rotoma (New Zealand)	–	11.2–22.1 (s); 10.7-12.6 (b)	0.93 (annual mean)	–	~2–8; SCM/DCM? (~ 10-20, 75 m)	–	–	Hamilton et al. (2010)[a]
Lake Tarawera (New Zealand)	–	11.0–21.7 (s); 10.9–11.7 (b)	1.26 (annual mean)	–	~1–5.8; SCM/DCM? (~10–25; 25–30 m)	–	–	Hamilton et al. (2010)[a]
Lake Rotoiti (New Zealand)	–	11.1–22.0 (s); 10.7–13.5 (b)	8.7 (annual mean)	–	~10-30; SCM/DCM? (~0–10, 50–90 m)	–	–	Hamilton et al. (2010)[a]
Lake Superior (USA)	–	–	0.57–1.3	–	–	110-119	–	Biddanda et al. (2001), Guildford and Hecky (2000)
Lake Superior (USA)	–	6.0-18.0	0.10–1.82 (0–80 m)	–	<1.82; DCM (23–35 m)	–	–	Barbiero and Tuchman (2004)
Lake Michigan (USA)	–	~3.0–24.0	~0.5–8.0 (0–70 m)	–	<~4: SCM? (15 m) or <~8.0: DCM (25–70 m)	–	–	Fahnenstiel and Scavia (1987)[a]
Lake Josephine (USA)	–	–	11.57	–	–	545	–	Biddanda et al. (2001)
Lake Johanna (USA)	–	–	20.2	–	–	484	–	Biddanda et al. (2001)
Lake Eagle (USA)	–	–	25.16	–	–	643	–	Biddanda et al. (2001)
Lake Medicine (USA)	–	–	40.49	–	–	615	–	Biddanda et al. (2001)
Lake Christmas (USA)	–	–	1.37	–	–	551	–	Biddanda et al. (2001)

(continued)

Table 1 (continued)

Sampling	pH	Water temperature (°C)	Chl *a* Epilimnion (μg L^{-1})	Hypolimnion	Chl *a* and depth in SCM and DCM (μg L^{-1}; m)	DOC Epilimnion (μM C)	Hypolimnion	Reference
Lake Turtle (USA)	–	–	2.47	–	–	600	–	Biddanda et al. (2001)
Lake Minnetonka (USA)	–	–	3.35	–	–	712	–	Biddanda et al. (2001)
Lake Round (USA)	–	–	8.74	–	–	537	–	Biddanda et al. (2001)
Lake Owasso (USA)	–	–	4.25	–	–	695	–	Biddanda et al. (2001)
Lake Mitchell (USA)	–	–	52.7	–	–	784	–	Biddanda et al. (2001)
Lake Okeechobee (USA)	–	14.8–30.2 (monthly mean)	10.0–60.0	–	–	–	–	James et al. (2009)[a]
Lake Tahoe (USA)	–	~5.0–16.0	0.2–0.9 (0–100 m)	–	~0.9; DCM (40–60 m)	–	–	Winder et al. (2009)[a]
Lake Tahoe (USA)	–	~4–14	~0.1–0.7 (0–500 m)	–	~0.3–0.7; DCM (100, 320, 350 m)	–	–	Kiefer et al. (1972)[a]
Florida Lakes (n = 438)	–	–	2–265	–	–	–	–	Bachmann et al. (2003)
Yellow Belly Lake (USA)	–	~3–17 (4–9 at DCM)	0.7–3 (0–20 m)	–	~2–3; SCM? (~8–15 m)	–	–	Sawatzky et al. (2006)
Redfish Lake (USA)	–	–	<4.8	–	2.2–4.8; DCM (~18–35 m)	–	–	Gross et al. (1997)[a]
Lakes of the Experimental Lakes Area, northwestern Ontario (Canada)	–	4.0–20.0	~1.0–327 (0–15 m)	–	311; SCM or DCM ? (5–7 m)	–	–	Fee (1976)
Quebec lakes (n = 8), Canada	–	–	1.5–6.9 (0.5–1.0 m)	–	–	233-625	–	McCallister and Del Giorgio (2008)

(continued)

Table 1 (continued)

Sampling	pH	Water temperature (°C)	Chl *a* Epilimnion (μg L^{-1})	Hypolimnion	Chl *a* and depth in SCM and DCM (μg L^{-1}; m)	DOC Epilimnion (μM C)	Hypolimnion	Reference
Lakes (small sizes in Northwestern Ontario)	–	–	0.3–7.9	–	–	–	–	Guildford and Hecky (2000)
Lakes (large sizes in Northwestern Ontario)	–	–	0.7–18.1	–	–	–	–	Guildford and Hecky (2000)
Alpine lake (Gossenköllesee)	–	–	0.3–8.5 (0–9 m)	–	8.5; DCM? (9 m)	10-54.	–	Sommaruga and Augustin (2006)
High Arctic Lake (Canada)	~7.0–8.2	~1.0–8.45	0.03–0.23	–	–	–	–	Antoniades et al. (2009)
Atazar Reservoir (Spain)	–	–	<22	–	< 22; SCM Or DCM (0–10 m)	–	–	Almodovar et al. (2004)
La Concepción Reservoir (Spain)			<58		<58; SCM Or DCM? (8–10 m)			Gálvez et al. (1988)
Forata Reservoir (Spain)			<15		15; SCM or DCM? (8 m)			Dasí et al. (1998)
Taechung Reservoir (South Korea)	–	–	2–173	–	–	–	–	An and Park (2002)
Stanford Reservoir (UK)	–	–	~0.0–919	–	–	–	–	Foster et al. (1997)[a]
Lower Bittel Reservoir (UK)	–	–	~<120	–	–	–	–	Foster et al. (1997)[a]
Draycote Reservoir (UK)	–	–	<200	–	–	–	–	Foster et al. (1997)[a]
Gorky Reservoir (Russia)	7.8–8.2	19.7–21.9	5.9–28.0	–	–	–	–	Mineeva et al. (2008)
Cheboksary Reservoir (Russia)	7.6–8.4	19.9–24.0	4.2–72.4	–	–	–	–	Mineeva et al. (2008)
Cheboksary Reservoir (Russia)	–	11.0–24.0	6.6 ± 0.7–239.8 ± 68.2	–	–	–	–	Mineeva and Abramova (2009)

(continued)

Table 1 (continued)

Sampling	pH	Water temperature (°C)	Chl *a* Epilimnion ($\mu g\ L^{-1}$)	Hypolimnion	Chl *a* and depth in SCM and DCM ($\mu g\ L^{-1}$; m)	DOC Epilimnion (μM C)	Hypolimnion	Reference
Upper Volga Chain of Reservoirs (Russia)	–	–	0.4–54.5	–	–	–	–	Sigareva and Pyrina (2006)
Estuaries								
Upper reach, Pearl River Estuary (China)	–	–	15.5–40.0	–	–	199–473	–	He et al. (2010)
Lower reach, Pearl River Estuary (China)	–	–	1.5–9.3	–	–	84–161	–	He et al. (2010)
Mixing zone, Pearl River Estuary (China)	–	–	1.2–14.6	–	–	165–278	–	He et al. (2010)
Changjiang (Yangtze River) Estuary (China)	–	18.0 (mean)	0.4–11.0	–	11.0; SCM (surface)	–	–	Zhu et al. (2009)
Patuxent River Estuary, USA	7.6–8.2	0.0–28.6	3.7–33.3	–	–	–	–	Stross and Stottlemyer (1965)
Rhine Estuary (Germany, Italy, Austria, Switzerland, France, Netherlands)	–	–	0.5–5.1	–	–	142–258	–	Abril et al. (2002)
Gironde Estuary	–	–	0.3–3.5	–	–	92–208	–	Abril et al. (2002)
Thames Estuary	–	–	1.5–5.1	–	–	217–417	–	Abril et al. (2002)
Elbe Estuary	–	–	3.2–8.9	–	–	258–367	–	Abril et al. (2002)
Ems Estuary	–	–	4.5–5.4	–	–	425–592	–	Abril et al. (2002)
Sado Estuary	–	–	1.6–13.8	–	–	300–525	–	Abril et al. (2002)
Douro Estuary	–	–	2.0–5.0	–	–	158–208	–	Abril et al. (2002)
Loire Estuary	–	–	5.1–60	–	–	200–292	–	Abril et al. (2002)

(continued)

Table 1 (continued)

Sampling	pH	Water temperature (°C)	Chl *a* Epilimnion (μg L^{-1})	Hypolimnion	Chl *a* and depth in SCM and DCM (μg L^{-1}; m)	DOC Epilimnion (μM C)	Hypolimnion	Reference
Scheldt Estuary	–	–	1.2–220	–	–	183–517	–	Abril et al. (2002)
European Estuaries (n = 9)	–	–	0.2–220	–	–	–	–	Lemaire et al. (2002)
Estuaries: Chesapeake Bay	–	–	9–77.4 (surface)	–	–	–	–	Gitelson et al. (2007)
Delaware Estuary	–	–	15–60	–	–	–	–	Pennock (1985)
The Exe Estuary SPA	–	–	>101	–	–	–	–	Langston et al. (2003)
North Carolina estuaries (n = 6)	–	–	0–184	–	–	–	–	Mallin (1994)
Neuse River Estuary	–	~10.0–30.0	~0.0–80	–	–	–	–	Gaulke et al. (2010)
Estuary, Bedford Basin (Canada)	–	–	0.58–18.02	–	–	–	–	Craig et al. (2012)
Temperate Estuaries (n = 7)	–	–	4.0–23.0	–	–	–	–	Hauxwell et al. (2003)
Coastal and open oceans								
Yellow Sea (Southern region)	–	~9.0–20.0	0.06–152	–	–	–	–	Li et al. (2007)
East China Sea	–	15–24.0	~0.06–3.2 (3–70 m)	–	–	–	–	Hung et al. (2000), Gong et al. (2000)
East China Sea surrounding Cheju Island	–	12.0–28	0.08–4.14	–	–	–	–	Kim et al. (2009)
Subtropical coastal waters (Hong Kong)	–	15.0–30.6	0.6–14.5 (0.0–10 m)	–	–	–	–	Chen et al. (2011)

(continued)

Table 1 (continued)

Sampling	pH	Water temperature (°C)	Chl *a* Epilimnion (μg L^{-1})	Hypolimnion	Chl *a* and depth in SCM and DCM (μg L^{-1}; m)	DOC Epilimnion (μM C)	Hypolimnion	Reference
Bay of Bengal, Inshore to offshore waters (Bangladesh-India)	–	–	0.1–2.5 (0–250 m)	–	2.5: SCM (0-10 m); 0.5-0.8: DCM (60-80 m)	–	–	Gomes et al. (2000)[a]
Southwest coastal waters (India)	–	23.8–27.8	0.04–8.3 (0–45 m)	–	8.3; SCM (5 m)	–	–	Roy et al. (2006)
Concepción Bay (Chile)	–	–	1.0–25	–	–	–	–	Gonzalez et al. (1989), Ahumada et al. (1991)
Mejillones Bay (Chile)	–	–	1.0–35	–	–	–	–	Iriarte and González (2004)
Inner Sea of Chiloé and austral fjord (Chile)	–	–	0.1–40	–	–	–	–	Dellarossa (1998), Pizarro et al. (2000)
Coastal waters, Inner Sea of Chiloé (Chile)	–	~10.0–17.0	0.0–32	–	–	–	–	Iriarte et al. (2007)
Upwelling seawaters of northern Chile	–	14.4–22.1	0.2–16.8	–	–	–	–	Morales et al. (1996)
The Kattegat, shallow marginal sea	–	~14.7	1.27–2.98	–	–	–	–	Carstensen et al. (2004)
Chesapeake Bay (USA)	–	–	0.2–23.3	–	–	100–341	–	Rochelle-Newall and Fisher (2002)
Scotian Shelf	–	–	0.5–5.6	–	–	–	–	Guildford and Hecky (2000)
Continental Slope	–	–	00.01–1.1	–	–	–	–	Guildford and Hecky (2000)
Southwest Florida Shelf (Shark River: Sts 26-30)	–	–	3.66–4.57	–	–	391–671	–	Clark et al. (2004)

(continued)

Table 1 (continued)

Sampling	pH	Water temperature (°C)	Chl *a* Epilimnion (μg L^{-1})	Hypolimnion	Chl *a* and depth in SCM and DCM (μg L^{-1}; m)	DOC Epilimnion (μM C)	Hypolimnion	Reference
Southwest Florida Shelf (Caloosahatchee River: Sts 51-55)	–	–	4.41–11.60	–	–	419–1425	–	Clark et al. (2004)
Southwest Florida Shelf (Charlotte Harbor/ Peace River: Sts 62-68)	–	–	0.97–15.54	–	–	165–966	–	Clark et al. (2004)
Southwest Florida Shelf (Gulf of Mexico: St 73)	–	–	0.38	–	–	146	–	Clark et al. (2004)
California Current System, seawater	–	~12–22	0.06–15.23 (mean)	–	<7.33; SCM (0–11 m)	–	–	Millán-Núñez et al. (1996)[a]
California Current System, seawater	–	~12–22	0.06–15.23 (mean)	–	~1.17–6.45; DCM (31–78 m)	–	–	Millán-Núñez et al. (1996)[*]
Scotia Sea (near South Georgia)	–	–	0.06–14.6	–	–	–	–	Holm-Hansen et al. (2004)
Gulf of St. Lawrence (Canada)	–	–	0.13–10.88	–	–	–	–	Doyon et al. (2000)
Gulf of Mexico (North-central location)	–	21.1–32.0	0.7–5.9 ± 1.5 (~<20 m)	–	–	–	–	Grippo et al. (2010)
Bohai Sea at station (38.1 N, 119.5E)	–	~24.7–25.6	~1–2 (~0–20 m)	–	~1.9; SCM or DCM? (5–6 m)	–	–	Xiu et al. (2009)
Sargasso Sea	–	–	0.03–0.5	–	–	–	–	Guildford and Hecky (2000)
Baltic Sea	–	12.0–24.0	1.0–12.5	–	–	–	–	Seppälä et al. (2007)
Bering Sea (Southeast region)	–	1.5–10.5	0.40–4.45 (4–30 m)	–	–	–	–	Olson and Strom (2002)

(continued)

Table 1 (continued)

Sampling	pH	Water temperature (°C)	Chl *a* Epilimnion ($\mu g\ L^{-1}$)	Hypolimnion	Chl *a* and depth in SCM and DCM ($\mu g\ L^{-1}$; m)	DOC Epilimnion (μM C)	Hypolimnion	Reference
Alboran Sea (North-western region) or W-Mediterranean)	–	14.0–18.0	0.4 ± 0.3–148 ± 63.1 (0–200 m)	–	–	–	–	Reul et al. (2005)
Arabian Sea	–	~ 20–30	~ 0.0–2.1	–	–	–	–	Kinkade et al. (2001)
Arabian Sea (Eastern region)	–	20.6–29.4	0.17–2,080	–	7.52; SCM or DCM? (~10–15 m)	–	–	Parab et al. (2006)
Black Sea (Southwestern region)	–	~8.0–18.05	0.0–1.7	–	~1.7; SCM (0–15 m)	–	–	Ediger et al. (2006)[a]
Eastern North Pacific marine water	–	~11–28	~0.1–0.31	–	~ < 0.33–1.0; DCM (~60–75 m)	–	–	Hopkinson and Barbeau (2008)[a]
Central Atlantic Ocean	–	–	0.02–0.88	–	0.23–0.88; DCM (50–139 m)	–	–	Planas et al. (1999)
Atlantic subtropical gyres	–	–	0.06 ± 0.01–0.09 ± 0.01 (surface)	–	0.29 ± 0.01–0.34 ± 0.02; DCM (93 ± 3–119 ± 4 m)	–	–	Pérez et al. (2006)
Atlantic Ocean	–	–	0.1–4.0	–	–	–	–	Gibb et al. (2000)
Atlantic Ocean	–	13.8–27.3 (mean)	0.13–0.96 (mean)	–	–	–	–	Calbet et al. (2009)
North Atlantic Ocean	–	–	0.0–17.0	–	<17; SCM (0.0-30 m); DCM (40-100 m)	–	–	Li and Harrison (2001)[a]
North Atlantic Sub-tropical Gyral East province	–	–	~ 0.0–0.6	–	<0.6; DCM (80-110 m)	–	–	Li and Harrison (2001)[a]
North Pacific Subtropical Gyre	–	–	<0.12–1.08	–	0.12–1.08; DCM (100-136 m)	–	–	Letelier et al. (2004)
Western equatorial Pacific Ocean	–	–	0.1–0.4	–	0.4; DCM (74–96 m)	–	–	Mackey et al. (1995)

(continued)

Table 1 (continued)

Sampling	pH	Water temperature (°C)	Chl *a* Epilimnion (μg L⁻¹)	Hypolimnion	Chl *a* and depth in SCM and DCM (μg L⁻¹; m)	DOC Epilimnion (μM C)	Hypolimnion	Reference
Pacific Ocean (North-west region)	–	2.4-14.8	0.37–17.0	–	–	–	–	Isada et al. (2010)
Pacific Ocean (western subarctic region)	–	~3.0–8.0	~ 0.1–30 (0–200 m)	–	–	–	–	Sasaoka et al. (2002)[a]
Arctic and Antarctic ice waters								
Barents and Greenland Sea ice (Arctic Ocean)	–	–	<86 (ice)	–	–	–	–	Mock and Gradinger (1999)
Dumbell Bay, Arctic Ocean	–	(–)1.7 to (+)1.9	0.5–8.2 (1–25 m)	–	8.2; SCM (5 m)	–	–	Apollonio (1980)
Arctic coastal waters (Summer and Winter)	–	(–)1.8 ± 0 to (+)4.1 ± 1.3	0.06 ± 0.01–0.48 ± 0.11 (2 m)	–	–	–	–	Cottrell and Kirchman (2009)
Barents Sea (Arctic Ocean)	–	–	<0.11–2.27	–	<2.27; SCM? (2.5–32 m)	–	–	Norrbin et al. (2009)
Central Arctic Ocean	–	–	0.1–297 (ice)	0.1–5.2 (water)	–	70-208	–	Wheeler et al. (1996)
Arctic Ocean	–	–	0.2–7.8	–	–	–	–	Guildford and Hecky (2000)
South Shetland Islands (Antarctica)	–	0.0–2.4	<0.6–4.0 (<70 m)	–	–	–	–	Hewes et al. (2009)
Gerlache and south Bransfield Straits (Antarctic Peninsula)	–	–	~<25 (0–70 m)	–	–	–	–	Varela et al. (2002)
Coastal seawater ice, Antarctic Ocean	–	(–)2 to (+)1.0	0.45–4.03 (surface)	–	–	–	–	Verlencar et al. (1990)

(continued)

Table 1 (continued)

Sampling	pH	Water temperature (°C)	Chl *a* Epilimnion ($\mu g\ L^{-1}$)	Hypolimnion	Chl *a* and depth in SCM and DCM ($\mu g\ L^{-1}$; m)	DOC Epilimnion (µM C)	Hypolimnion	Reference
Oceanic seawater, Antarctic Ocean	–	(–)2 to (+)0.5	0.19–0.43 (surface)	–	–	–	–	Verlencar et al. (1990)
Antarctica ice seawater (control and light perturbation experiment)	–	–	3.54 ± 1.00–111 ± 30	–	–	–	–	Palmisano et al. (1985)
Southern Ocean (Antarctic marine waters)	–	(–) (0.3 to 1.4)	~0.87–30.0 (experimental growth)	–	–	–	–	Spies (1987)
Southern Ocean (Antarctic marine waters)	–	–	10.0–50.0 (experimental growth)	–	–	–	–	Sakshaug and Holm-Hansen (1986)

SCM surface Chlorophyll *a* maximum; *DCM* deep chlorophyll *a* maximum; *S* surface tempertaure; *b* bottom temperature; *?* remark either SCM or DCM which reasonably formed in water column

[a]Some data used approximately which are taken from the graphs of the related paper

SCM is often observed in coastal seawater and subsequently decreases in the offshore direction, whereas DCM is increased along the offshore direction with its enhanced depth (Millán-Núñez et al. 1996; Hayward et al. 1995; Maranón et al. 2004). Chl *a* values are quite high in SCM in coastal seawater, and decrease much more rapidly in the offshore direction (Millán-Núñez et al. 1996; Echevin et al. 2004). The occurrence of SCM in coastal seawater is possibly responsible for the high contents of DOM and nutrients, which is a general phenomenon in coastal environments. High contents of DOM can protect surface waters against sunlight exposure (Laurion et al. 2000; Hayakawa and Sugiyama 2008), and the photoproducts generated from photoinduced degradation of DOM and POM can enhance photosynthesis. The consequence is SCM formation in surface waters. For similar reasons, SCM is limited at the epilimnion (~0–5 m) in Lake Kinneret, Lake Hongfeng (4–5 m), Lake Baihua (0–2 m); Lake Biwa (2.5–10 m) and Lake Baikal (~0–30 m, with a peak at 10 m). DOC concentrations are quite higher (258–485 μM C) in Lake Kinneret compared to Lake Hongfeng (134–250 μM C at 0–25 m), Lake Baihua (157–330 μM C at 0–25 m), Lake Biwa (76–135 μM C at 0–80 m), and Lake Baikal (88–142 at 0–1,620 m) (Table 1) (Fu et al. 2010; Mostofa KMG et al. unpublished data; Satoh et al. 2006; Yacobi 2006; Berman et al. 1995; Yoshioka et al. 2002, 2007; Annual Report 2004; Sugiyama et al. 2004; Yuma et al. 2006; Mostofa et al. 2005). The overall penetration depth in Lake Kinneret was on average 1.77 m, and the uppermost layer is supposed to be representative of the entire euphotic zone (Yacobi 2006). Therefore, it is suggested that DOM and mechanical perturbation of surface waters (e.g., by wind, waves and storms), which also depends on the depth and size of the water ecosystem (particularly for lakes), are the two key factors for SCM formation.

Lakes having high water temperature (WT) often exhibit the SCM in the epilimnion, such as Lake Hongfeng (10.9–47.8 μg L^{-1} at 4–5 m and 15.3–31.0 °C), Lake Baihua (15.0–65.5 μg L^{-1} at 0–2 m and 15.3–31.0 °C), Lake Biwa (2.0–12.3 μg L^{-1} at 2.5–10 m and 11.6–28.7 °C), Lake Kinneret (95 % of Chl *a* at 0–5 m and 15–30 °C), Lake Baikal (0.7–5.8 μg L^{-1} at 0–30 m and 16.5–17.9 °C); Lake Malawi (0.03–18.7 μg L^{-1} at upper mixing layer and ~40 °C); Lake Victoria (4.7–78.5 μg L^{-1} at upper mixing layer and 25–29 °C) and Lakes of Experimental Lakes Area (<311–327 μg L^{-1} at 5–7 m and 4–20 °C) (Table 1) (Fu et al. 2010; Mostofa K et al. unpublished data; Guildford and Hecky 2000; Satoh et al. 2006; Yacobi 2006; Fee 1976; Berman et al. 1995; Mostofa et al. 2005). Therefore, high contents of DOM in surface water under high WT, driven by strong sunlight, can photochemically decompose DOM and POM to produce high amounts of DIC, CO_2, and H_2O_2. These species are directly linked with occurrence of high photosynthesis and high primary production. Moreover, in mesocosm experiments it has been observed that increasing DOM concentrations from ~10 mg C. L^{-1} to ~20 mg C. L^{-1} had a negative effect on total phytoplankton growth. The most likely explanation is the reduction of irradiance because of radiation absorption by DOM (Klug 2002).

DOC concentrations are relatively low in the offshore direction, which may afford easy penetration of sunlight (UVR) that can reach the deeper layers. This issue may increase the photosynthetic layer depth in the water column, i.e. increase the DCM depth. A significant contribution to Chl *a* may come from phytoplankton in deeper layers, in the case of a low-DOC lake water when UV attenuation increases with Chl *a* concentrations (Laurion et al. 2000). Moreover, the mixing depth can play an important role in SCM or DCM formation in lakes or oceans. A low mixing depth can often induce SCM formation, whilst high mixing depth can cause DCM formation. For example, SCM formation (~0–15 m) occurs when mixed-layer depth is low (3–15 m), whilst DCM formation (~65 m) takes place when the mixed-layer depth is high (e.g. 33 m in East China Sea) (Hung et al. 2000).

The Chl *a* concentrations in Lake Biwa are several times (~15–24) higher at the epilimnion (0–10 m), compared to those of deeper layers (40 and 70 m) during the summer stratification period (Mostofa KMG et al. unpublished data). SCM is often observed during autumn, e.g. November in 1999 and October in 2000 at the epilimnion. Chl concentration largely fluctuates and it is lower during the summer stratification period compared to early spring and autumn seasons (Fig. 3a) (Mostofa KMG et al. unpublished data). The low Chl *a* contents in SCM and its fluctuation during the summer stratification period is presumably caused by photodegradation induced by strong sunlight, coupled with high WT (maximum 28.7 °C). However, an early bloom in 2000 compared to 1999 was probably caused by a longer summer period. WT was 26.8 to 21.9 °C during September–October in 1999, which is lower compared to 2000 (WT: 23.6 to 19.5 °C at the same time).

Moreover, reduction of water clarity through eutrophication can cause a shift in phytoplankton distributions, from a DCM in spring or summer to a SCM within the surface mixed layer. This may happen when the depth of the euphotic zone is consistently shallower than the depth of the surface mixed layer (Hamilton et al. 2010). Such a SCM, which is susceptible to occur because of high primary production in spring or summer, is initially caused by the photoinduced generation of photoproducts in waters. Simultaneously, the decrease of primary production because of photoinduced degradation does not predominantly occur during that period. Therefore, the new primary production may prevail over photoinduced degradation processes. The DOM that is generated as a consequence of the high primary production can substantially absorb sunlight and cause the depth of the euphotic zone to be shallow.

These results may suggest that two important phenomena account for the occurrence of SCM and DCM in natural waters: First, waters with high contents of DOM and POM can have intense solar radiation in the surface layer. In contrast, photoproducts (DIC, CO_2, H_2O_2, and so on) are generated photochemically under high WT (caused by strong sunlight) from DOM and POM. They are responsible for the occurrence of high photosynthesis, with the consequence that high primary production can form SCM in surface waters. The second phenomenon is that water with low contents of DOM and POM lets sunlight to penetrate in the deeper water layer. Photoinduced or microbial products (DIC, CO_2, H_2O_2, nutrients, and so on) are generated from DOM and POM and are responsible for occurrence of photosynthesis. As a consequence, enhanced primary production at depth can produce DCM in deep water. The two described phenomena are extensively discussed in the next sections.

3.1.1 Mechanism of SCM Formation

SCM is driven by sunlight and it is formed during the summer stratification period in waters with high contents of DOM and POM and high temperature. High contents of DOM (of both allochthonous and autochthonous origin) and POM (e.g. algae) along with Chl *a* or phytoplankton, together with incident light wavelengths or solar zenith angle are the main limiting factors for sunlight in the surface layer (see also chapter "Colored and Chromophoric Dissolved Organic Matter (CDOM) in Natural Waters" for detailed description) (Laurion et al. 2000; Hayakawa and Sugiyama 2008; Markager and Vincent 2000; Belzile et al. 2002; Shank et al. 2005; Zhao et al. 2009). High contents of DOM and POM are thus responsible for having most of the sunlight intensity in the upper surface layer. Therefore, most of the photoinduced degradation processes would occur in the surface layers or in epilimnion. OM including DOM and POM is one of the key factors that can produce nutrients (NO_3^-, NH_4^+ and PO_4^{3-}) and various photo- and microbial products (H_2O_2, CO_2, DIC, LMW DOM, and so on) (see also chapters "Dissolved Organic Matter in Natural Waters , Photoinduced and Microbial Degradation of Dissolved Organic Matter in Natural Waters ", "Photosynthesis in Nature: A New Look" and "Impacts of Global Warming on Biogeochemical Cycles in Natural Waters") (Zepp et al. 1987; Palenik et al. 1987; Palenik and Morel 1988; Cooper and Lean 1992; Miller and Zepp 1995; Bushaw et al. 1996; Graneli et al. 1996, 1998; Miller and Moran 1997; Sarthou et al. 1997; Gao and Zepp 1998; Jørgensen et al. 1998; Bertilsson et al. 1999; Bertilsson and Tranvik 2000; Anesio and Granéli 2003; Scully et al. 2003; Obernosterer and Benner 2004; Ma and Green 2004; Croot et al. 2005; Molot et al. 2005; Johannessen et al. 2007; Kujawinski et al. 2009; Mostofa and Sakugawa 2009; Finlay et al. 2009; Stets et al. 2009; Jiao et al. 2010; Liu et al. 2010; Lohrenz et al. 2010; Omar et al. 2010; White et al. 2010; Zepp et al. 2011; Borges et al. 2008). All these species can influence photosynthesis directly and indirectly in waters.

Photoinduced degradation of DOM and POM (e.g. degradation of phytoplankton) can be described as follows:

$$\begin{aligned} \text{DOM} + \text{POM} + h\upsilon \rightarrow\ & H_2O_2 + CO_2 + \text{DIC} + \text{LOW DOM} + NO_3^- \\ & + NO_2^- + PO_4^{3-} + \text{autochthonous DOM} + \text{others} \end{aligned} \tag{3.1}$$

where DIC is usually defined as the sum of an equilibrium mixture of dissolved CO_2, H_2CO_3, HCO_3^-, and CO_3^{2-}.

Microbial degradation of DOM and POM is as follows:

$$\begin{aligned} \text{DOM} + \text{POM} + \text{microbes} \rightarrow\ & H_2O_2 + CO_2 + \text{DIC} + \text{LOW DOM} + NO_3^- \\ & + PO_4^{3-} + \text{autochthonous DOM} + \text{others} \end{aligned} \tag{3.2}$$

The mechanism behind the formation of SCM might be that H_2O_2, photogenerated intracellularly in a photosynthetic cell or extracellularly from DOM and POM in surface waters can induce photosynthesis in the presence of CO_2 or DIC (dissolved CO_2, H_2CO_3, HCO_3^-, CO_3^{2-}). Dependence of photosynthesis by aquatic microorganisms on OM (DOM and POM) is extensively documented in the literature (see

Sect. 5.5, chapter "Photosynthesis in Nature: A New Look"). It has recently been shown that dissolved O_2 is significantly related to benthic or sestonic Chl concentration (Heiskary and Markus 2003; Miltner 2010). Moreover, a 10 mg L^{-1} difference between daytime and nighttime dissolved O_2 concentrations was observed at an enriched site, where benthic Chl *a* levels exceeded 500 mg m^{-2} (Sabater et al. 2000). Variation in dissolved O_2 concentration forced by algal respiration is an important link between increasing nutrients and decreasing biological quality, as shown in a study of medium to large rivers (Heiskary and Markus 2003). These findings are consistent with the hypothesis that photoinduced formation of H_2O_2 from dissolved O_2 may be involved in SCM formation or primary production. Correspondingly, when cyanobacterial blooms are accumulated as scums in surface waters, prolonged exposure to UV radiation can cause enhanced carotenoid production, which can subsequently increase Chl *a*-specific photosynthetic production of O_2 (Jeffrey et al. 1997).

3.2 Deep Chl *a* Maximum

The Deep Chl *a* maximum (DCM) is a well-known phenomenon occurring in the presence of maximal Chl *a* contents in the deeper layer of the euphotic zone of the water column (Table 1) (Fennel and Boss 2003; Letelier et al. 2004; Huisman et al. 2006; Steele and Yentsch 1960; Anderson 1969; Klausmeier and Litchman 2001; Hodges and Rudnick 2004; Beckmann and Hense 2007; Hense and Beckmann 2008; Ryabov et al. 2010; Pérez et al. 2007; Gomes et al. 2000; Camacho 2006; Sawatzky et al. 2006; Fee 1976; Kiefer et al. 1972; Cullen 1982; Moll and Stoermer 1982; Abbott et al. 1984; Pick et al. 1984; Steinhart et al. 1994; Varela et al. 1994; Budy et al. 1995; Ediger and Yilmaz 1996; Gross et al. 1997; Goericke and Welschmeyer 1998; Marañón et al. 2000; Wurtsbaugh et al. 2001; Cuny et al. 2002; Pérez et al. 2002; Tittel et al. 2003; Barbiero and Tuchman 2004; Chapin et al. 2004; Holm-Hansen and Hewes 2004; Park et al. 2004; Ghai et al. 2010; Johnson et al. 2010; Harrison and Smith 2011; Mellard et al. 2011). According to these studies, DCM can be defined as a zone of maximum photosynthetic activity with highest Chl *a* contents. It is usually a region lacking a pronounced density gradient, generally occurring in or below the thermocline (the metalimnion). It is a stable and common feature occurring in the presence of sufficient light and nutrients under low temperature and low turbulence, and it is a remarkable characteristic of clear water with low nutrients in the deep layer, particularly in lakes and oceans. The DCM is a stable feature in tropical waters whilst it is a seasonal phenomenon in the Mediterranean and other temperate waters, following seasonal changes in incident light intensity and nutrient conditions (Letelier et al. 2004; Huisman et al. 2006; Ghai et al. 2010). The DCM is found to vary from 20 to 350 m in lakes and from 30 to 139 m in oceans (Table 1).

DCM is entirely different in Lake Superior and Lake Michigan. It is observed in the upper hypolimnion at a depth of 23–35 m in Lake Superior, whilst its depth in Lake Michigan changes seasonally. Depth varies from 15 to 30 m during early thermal stratification primarily in June, to 25–50 m by mid-stratification in July, and finally reaches 40–70 m in August (Barbiero and Tuchman 2004; Moll et al. 1984; Fahnenstiel and Scavia 1987). It has been shown that WT is relatively higher (3–24 °C) in Lake Michigan than in Lake Superior (6–18 °C), and high WT along with DOM contents may affect the DCM depth variation in Lake Michigan. Redfish Lake and other Sawtooth Valley (Idaho) lakes had DCM with mean Chl *a* peaks reaching 240–1,000 % of the mean epilimnetic Chl *a* concentrations. The DCM can be present at low light levels and account for 36.72 % of the lake primary production (Gross et al. 1997). The Sawtooth Valley lakes have DCM Chl values that can be up to 10 times higher in the metalimnia and hypolimnia than in the epilimnia (Steinhart et al. 1994; Budy et al. 1995). The DCM in the Sawtooth Valley lakes are located at depths where the light levels are near or below 1 % of surface light (Gross et al. 1997).

Seasonal changes in mixing and light intensity can produce a seasonal reset of Chl distributions, which can alter the DCM or SCM formation and ablation as a regime shift (Hense and Beckmann 2008; Hamilton et al. 2010; Abbott et al. 1984; Vincent 1983; Vincent et al. 1984; Marshall and Peters 1989; Bayley et al. 2007; Carpenter et al. 2003). Three different 'regimes' can occur during the seasonal occurrence of a DCM in Lake Tahoe, with transitions alternatively controlled by diffusion, nutrient supply and light (Abbott et al. 1984). Seasonal changes of DCM in the water column can depend on the depth of light penetration, which can largely affect DCM depth during the summer stratification period (Hamilton et al. 2010). Seasonal variations in the water-column attenuation coefficient of the photosynthetically available radiation (PAR) can shift the 1 % sea-surface PAR depth from approximately 105 m in winter to 121 m in summer, in the North Pacific Subtropical Gyre (Letelier et al. 2004). Such a seasonal depth shift of isolumes (constant daily integrated photon flux strata) can also be increased to 31 m due to the added effect of changes in sea-surface PAR (Letelier et al. 2004). Such a discrepancy can induce a significant deepening of the DCM during the summer period, with a concomitant increase in Chl *a* (Letelier et al. 2004).

The DCM phytoplankton contains higher amounts of phosphorus than for the epilimnion, which is likely caused by the rapid photochemical degradation of SCM phytoplankton in epilimnion. Nutrient-rich DCM might be useful as a food source for grazers, including deep-living calanoid copepods that may have a substantial impact on total lake phytoplankton productivity (Barbiero and Tuchman 2004; Moll et al. 1984). The DCM also releases the new DOM and nutrients in the hypolimnion under microbial assimilation (Rochelle-Newall and Fisher 2002; Maurin et al. 1997; Yamashita and Tanoue 2008). Phytoplankton from DCM do not show marked differences from epilimnetic communities in taxonomy or nutrient status, but can exhibit substantially higher photosynthetic impairment under UVR exposure (Harrison and Smith 2011). This suggests that epilimnetic phytoplankton can be acclimated to in situ light conditions in a spectrally-specific manner, and that ultraviolet-A radiation may be a stronger stressor than ultraviolet-B or photosynthetically active radiation in the mixed layers of lakes (Harrison and Smith 2011). DCM has varying characteristics that suggest multiple processes contributing to its formation and maintenance in lakes and oceans (Anderson 1969; Steele 1964; Hobson and Lorenzen 1972).

3.2.1 Mechanism of DCM Formation

It has been shown that DCM is generally developed in clear water at low temperature. The main effect of these conditions result is the penetraton of radiation into deep water, in which case photosynthesis can enhance the primary production and produce the DCM in deeper water. The mechanism behind DCM formation is presumably that H_2O_2 and HCO_3^- produced in the DCM water layer are susceptible to take part in phytoplankton photosynthesis. It has been shown that DIC (dissolved CO_2, H_2CO_3, HCO_3^-, CO_3^{2-}) is mostly produced from particulate organic matter (POM: e.g. algae or cyanobacteria) and DOM microbiologically in natural waters as well as under in situ experiments (Ma and Green 2004; Finlay et al. 2009; Stets et al. 2009; Jiao et al. 2010). Correspondingly, most H_2O_2 can be produced either from algae (cyanobacteria or phytoplankton or biota) or from DOM, by several biological or photochemical processes (see also chapter "Photoinduced and Microbial Generation of Hydrogen Peroxide and Organic Peroxides in Natural Waters" for more references and description) (Palenik et al. 1987; Palenik and Morel 1988; Cooper and Lean 1992; Sarthou et al. 1997; Croot et al. 2005; Mostofa and Sakugawa 2009; Zepp et al. 1987; Angel et al. 1999; Wentworth et al. 2000; Wentworth et al. 2001; Moreno 2012; Moffett and Zafiriou 1990). Such processes are: (i) extracellular phenomena, (ii) biological processes such as glycolate oxidation during photorespiration, (iii) enzymatic reduction of oxygen at the cell surface, and (iv) microbial degradation of DOM under dark incubation. Most phytoplankton cells have the enzyme superoxide dismutase (SOD), which can catalyse the conversion of superoxide to H_2O_2. This is one of the many biological reactions that produce H_2O_2 in seawater (Croot et al. 2005).

In a field study, dark production of H_2O_2 was highest at 40–60 m depth and the corresponding DCM was detected at 90 m. The finding suggests that photosynthesis, which causes the DCM may reduce the dark production of H_2O_2 at 90 m depth (Palenik and Morel 1988). Simultaneously, the increase in pigment production caused by phytoplankton under the low-light conditions of the DCM layer (Steele 1964; Hobson and Lorenzen 1972; Kiefer et al. 1976) may lead to high contents of H_2O_2 and contribute to DCM formation. Note that pigments made up of Chls can rapidly absorb light energy upon irradiation. Radiation absorption can excite an electron to form the superoxide radical anion ($O_2^{\bullet-}$) and then H_2O_2 (see chapters "Photosynthesis in Nature: A New Look" and "Photoinduced and Microbial Generation of Hydrogen Peroxide and Organic Peroxides in Natural Waters"). The H_2O_2 concentration increase at the depth of the Chl maximum is possibly due to biological production (Croot et al. 2005). The formation of H_2O_2 by phytoplankton in the DCM layer can be supported by the observation that *Chattonella marina*, a harmful algal bloom species, is capable of producing reactive oxygen species (ROS) including $O_2^{\bullet-}$, H_2O_2, and $HO^{\bullet}$ at levels 100 times higher than those produced by most algae (Marshall et al. 2002; Oda et al. 1998). ROS are often produced as by-products of various metabolic pathways localized in mitochondria, chloroplasts, and peroxisomes (see also chapter "Photosynthesis in Nature: A New Look") (Apel and Hirt 2004). The presence of the cyanobacterium *Microcystis* sp. can produce a buildup of apoptosis-inducing ROS in the competing dinoflagellate *Peridinium gatunense* (Vardi et al. 2002). A distinct H_2O_2 maximum at depth in the Southern Ocean can correspond to a DCM, which also suggest a significant biological source of H_2O_2 (Sarthou et al. 1997; Croot et al. 2005). The decay of H_2O_2 apparently follows first-order kinetics (Petasne and Zika 1997; Yuan and Shiller 2001) and is biologically mediated by small microorganisms (Petasne and Zika 1997).

Filtration of seawater to remove the biota typically produces a dramatic reduction in the decay rate of H_2O_2 (Moffett and Zafiriou 1990; Petasne and Zika 1997; Fujiwara et al. 1993), whilst the amount of colloidal material influences the decay rate (Yuan and Shiller 2001). H_2O_2 may be concentrated by particulate organic matter or small fungi through rapid transpiration (Komissarov 1994, 1995, 2003). The decay process of H_2O_2 can be explained in two ways: one is the uptake possible of H_2O_2 by microorganisms during photosynthesis, the other is the decomposition of H_2O_2 by catalases and peroxidases bound to microorganisms. Catalases and peroxidase can enzymatically activate H_2O_2 to detoxify it to H_2O (see also chapter "Photoinduced and Microbial Generation of Hydrogen Peroxide and Organic Peroxides in Natural Waters") (Moffett and Zafiriou 1990). Moreover, conversion of H_2O_2 to H_2O by catalases and peroxidases could play a key role in photosynthesis and needs further study to clarify the possible links. Note that dark reduction of CO_2 may take place because of the electrons that are released by organic molecules and sulfide (Jagannathan and Golbeck 2009). Some important phenomena relevant to this context are extensively discussed in the photosynthesis chapter ("Photosynthesis in Nature: A New Look").

3.3 Changes in the Chl *a* Concentrations in Natural Waters

Chl *a* concentrations undergo significant variations in the water column, which can be seasonal, spatial and temporal depending on various factors that characterize water (Bianchi et al. 2002; Sommaruga and Augustin 2006; Biggs 2000; de Moraes Novo et al. 2006; Duan and Bianchi 2006; Lewis et al. 2010).

Streams and Rivers

Chl *a* concentrations range from 0.0 to 280 $\mu g\ L^{-1}$ in streams and rivers (Table 1) (Miltner 2010; Chessman 1985; Lohman and Jones 1999; van Nieuwenhuyse and Jones 1996; Basu and Pick 1997; Gao et al. 2004; Guéguen et al. 2006; Morgan et al. 2006; Devercelli and Peruchet 2008; Palmer-Felgate et al. 2008; Royer et al. 2008; Longing and Haggard 2010; Calijuri et al. 2008). The highest Chl *a* concentrations in freshwater riverine ecosystems are in the order of <280 $\mu g\ L^{-1}$ in River Alne (Warwickshire, UK); <263 $\mu g\ L^{-1}$ in Red River and its basin (USA); <240 $\mu g\ L^{-1}$ in River Arrow (Warwickshire, UK); <216 $\mu g\ L^{-1}$ in Paraná River basin (South America); <170 $\mu g\ L^{-1}$ in temperate streams (USA); <100 $\mu g\ L^{-1}$

in River Avon (Warwickshire, UK); <97 μg L^{-1} in streams and rivers (USA); <65 μg L^{-1} in La Trobe River Streams (Victoria, Australia); <44.6 μg L^{-1} in Ozark Streams (Missouri, USA); <27 μg L^{-1} in Rideau River (Ontario, Canada); <18.0 μg L^{-1} in sStreams and rivers (Illinois, USA); <17 μg L^{-1} in Chalk stream (UK); and 0.0–12.7 μg L^{-1} in other studied systems (Table 1).

Chl *a* mostly results from in-channel production rather than from tributary or outside inputs. Chl *a* concentrations in the Pearl River are high only during summer low-flow periods and are often controlled by temperature and by CDOM concentration (Duan and Bianchi 2006). Lower phytoplankton biomass (dominated by chlorophytes) in the Pearl River is likely linked with intense shading by CDOM and lower availability of nutrient inputs (Duan and Bianchi 2006). High concentrations of Chl *a* (0.4–170 μg L^{-1}) are strongly correlated with high contents of phosphorus (5–1,030 μg L^{-1}) in temperate streams (van Nieuwenhuyse and Jones 1996). Chl *a* concentrations in the lower Mississippi River are high in summer low-flow periods and also during interims of winter and spring. They are not coupled with physical variables or nutrients, likely due to a combination of in situ production and inputs from reservoirs, navigation locks and oxbow lakes in the upper Mississippi River and Missouri River (Duan and Bianchi 2006). The high, diatom-dominated phytoplankton biomass in the lower Mississippi River is likely the result of decreasing total suspended solids (because of increased damming in the watershed) and increasing nutrients (due to enhanced agricultural runoff) over the past few decades (Duan and Bianchi 2006).

Lakes and Reservoirs

Chl *a* concentrations are significantly variable, from 0.01 to 850 μg L^{-1} in a variety of lakes (Table 1) (Carrillo et al. 2002; Kasprzak et al. 2008; Fu et al. 2010; Mostofa KMG et al. unpublished data; Vicente and Miracle 1984; Pedros-Alio et al. 1987; Guildford and Hecky 2000; Camacho 2006; Satoh et al. 2006; Sawatzky et al. 2006; Hamilton et al. 2010; Fee 1976; Sommaruga and Augustin 2006; Yuma et al. 2006; Kiefer et al. 1972; Gross et al. 1997; Barbiero and Tuchman 2004; Fahnenstiel and Scavia 1987; de Moraes Novo et al. 2006; Aizaki et al. 1981; Rojo and Miracle 1987; Dasí and Miracle 1991; Miracle et al. 1993; Windolf et al. 1996; Camacho 1997; Yoshioka 1997, Biddanda et al. 2001; Kahlert 2002; Laurion et al. 2002; Bachmann et al. 2003; Camacho et al. 2003; Straškrábová et al. 2005; Blindow et al. 2006; Silsbe et al. 2006; McCallister and del Giorgio 2008; Striebel et al. 2008; Antoniades et al. 2009; James et al. 2009; Pan et al. 2009; Winder et al. 2009; Lv et al. 2011; Wang et al. 2012; Liu et al. 2011; Zhang et al. 2007; Rae et al. 2001). These studies demonstrate that the highest detected Chl *a* concentrations can be ordered as follows: <850 μg L^{-1} in Lake Ciśó (Spain); <327 μg L^{-1} in lakes of the Experimental Lakes Area (northwestern Ontario, Canada); <298 μg L^{-1} in Lake Arcas (Spain); <276 μg L^{-1} in several shallow Danish lakes; <265 μg L^{-1} in numerous Florida Lakes; <189.8 μg L^{-1} in Subtropical and urban shallow Lakes (Wuhan, China); <189 μg L^{-1} in several lakes in Japan; <175.9 μg

L^{-1} in several lakes in Germany; <145 μg L^{-1} in Lake Börringesjön (Sweden); <133.22 μg L^{-1} in several Chinese lakes; <~110 μg L^{-1} in Lake Victoria; <90 μg L^{-1} in Lake El Tobar (Spain); <89.2 μg L^{-1} in Amazon flood plain lakes (shallow lakes: 1–3 m depth); <82.1 μg L^{-1} in Lake Kizaki (Japan); <66.93 μg L^{-1} in Lake Baiyangdian (China); <65.5 μg L^{-1} in Lake Hongfeng and Lake Baihua (China); <60 μg L^{-1} in Lake Okeechobee (USA) and Lake Krankesjön (Sweden); 53.4 μg L^{-1} in Lake Bansee (Germany); 52.7 μg L^{-1} in Lake Mitchell (USA); 40.49 μg L^{-1} in Lake Medicine (Canada); <30 μg L^{-1} in Lake Taihu (China); 26.9 μg L^{-1} in Lake Thalersee (Germany); 25.16 μg L^{-1} in Lake Eagle (Canada); <25 μg L^{-1} in Lake La Cruz (Spain); 20.2 μg L^{-1} in Lake Johanna; <18.7 μg L^{-1} in Lake Malawi (Africa); <18.1 μg L^{-1} in large Northwestern Ontario lakes; <15 μg L^{-1} in Lake La Parra and Lake Lagunillo del Tejo (Spain); <12.3 μg L^{-1} in Lake Biwa (Japan); 11.57 μg L^{-1} in Lake Josephine (USA); 0.03–10.0 μg L^{-1} in other lakes studied including relatively low Chl *a* concentrations in some famous lakes such as in Lake Superior (<0.73 μg L^{-1}); Lake Michigan (< ~ 8.0 μg L^{-1}), Lake Baikal (<5.8 μg L^{-1}) and Lake Tanganyika (<4.5 μg L^{-1}) (Table 1).

In Lake Biwa, Chl *a* concentration ranged from 2.1 to 12.3 μg L^{-1} in the upper epilimnion (2.5 and 10 m), from 0.5 to 10.7 μg L^{-1} in the deeper epilimnion (20 m), and from 0.1 to 3.3 μg L^{-1} in the hypolimnion (40 and 70 m) during the summer stratification period (Fig. 3a) (Mostofa KMG et al. unpublished data). From January to March, Chl *a* (2.0–4.0 μg L^{-1}) was almost uniformly distributed throughout the entire water column, due to vertical mixing during the overturn period (Fig. 3a) (Mostofa et al. 2005).

The summer maximum of *Microcystis* biomass in Lake Taihu peaked at 112.0 mg L^{-1} in August 1998, which accounted for 94.5 % of the total phytoplankton biomass. In contrast, Chl *a* concentrations varied from approximately 5–30 μg L^{-1} (Table 1) (James et al. 2009; Liu et al. 2011). It has also been shown that the annual cycles of WT showed a regular summer peak each year in lake Taihu, in accordance with fluctuations in *Microcystis* biomass. WT reached almost up to 30 °C during summer and declined to 5 °C by January (Liu et al. 2011). However, WT is relatively high (14.5–30.2 °C) in Lake Okeechobee (USA) that showed substantially high contents of Chl *a* (10–60 μg L^{-1}, Table 1) (James et al. 2009).

In two eutrophic lakes, e.g. Lake Hongfeng and Baihua (Southwestern China), Chl *a* concentration showed the highest level (44–66 μg L^{-1}) in the epilimnion (0–6 m) in July, during the summer stratification period (Fu et al. 2010). WT and DOC concentrations for these two lake waters were 25–31 °C and 134–330 μM C, respectively. Similarly, in the warm monomictic Lake Kinneret (Israel) Chl *a* concentrations exhibited a maximum at the epilimnion (0–5 m) during the spring season (April–May) (Yacobi 2006; Berman et al. 1995). WT and lake DOC concentrations were 15–30 °C and 258–485 μM C, respectively (Yacobi 2006; Annual Report 2004). On the other hand, in water of monomictic Lake Biwa Chl *a* maximum was observed in the epilimnion during the autumn season: November 1999 (12.3 μg L^{-1}) and October 2000 (9.4 μg L^{-1}) (Mostofa KMG et al. unpublished data). Moreover, WT and DOC concentrations were 17.0–19.5 °C and 76–135 μM C, respectively (Mostofa et al. 2005; Mostofa KMG et al. unpublished data).

Chl *a* concentrations are greatly variable, ranging from 0.01 to 133.22 μg L^{-1} in 38 Chinese lakes. Most of them are mesotrophic (TN = 0.31–2.30 mg L^{-1}; TP = 0.01–0.11 mg L^{-1}), five lakes are oligotrophic (TN < 0.31 mg L^{-1}; TP < 0.01 mg L^{-1}), and another four lakes are eutrophic (TN > 2.30 mg L^{-1}; TP > 0.11 mg L^{-1}) with algal blooms during the summer period (Zhang et al. 2007). The TN:TP ratio ranged from 2:1 to 253:1 for all 38 lakes (Zhang et al. 2007). Chl *a* concentrations significantly varied (10–145 μg L^{-1}) in two Swedish lakes. In Lake Börringesjön the highest concentration (145 μg L^{-1}) has been found in September, when light attenuation ranged from 4.61 to 7.81 m^{-1} (Blindow et al. 2006). Chl *a* concentrations were low (0.3–1.2 μg L^{-1}) in an alpine lake during the ice-cover period, but after ice-break the values increased particularly in the deep layers. The maximum was observed at 9 m depth (8.5 μg L^{-1}), whilst DOC concentrations in the water column ranged from 10 to 54 μM C (Sommaruga and Augustin 2006). Chl *a* concentrations were also very low (0.14–2.85 μg L^{-1}) in lake water with low WT (1.2–12.4 °C) and low DOC concentrations (such as ~42 μM C) (Carrillo et al. 2002). In Bohai Sea the vertical distribution of Chl *a* and water temperature at depth 0–20 m was approximately 1–2 μg L^{-1} and 24.7–25.6 °C. The diffuse attenuation coefficient increased with depth, producing a DCM at around 5–6 m depth (Xiu et al. 2009). High temperature and other factors suggest that this low variation of Chl *a* (1–2 μg L^{-1}) might be caused by high photoinduced decomposition of Chl *a* in the surface layer (0–5 m). This result is not accounted for by DCM, rather it can be considered as SCM or mixed layer depth.

On the other hand, Chl *a* concentrations in reservoirs are substantially high, ranging from approximately 0.0–919 μg L^{-1} (Gálvez et al. 1988; Foster et al. 1997; Dasí et al. 1998; An and Park 2002; Almodovar et al. 2004; Sigareva and Pyrina 2006; Mineeva et al. 2008; Mineeva and Abramova 2009). The highest Chl *a* concentrations were detected in several UK reservoirs, such as <120–919 μg L^{-1}; Chl *a* was then found at <54.5–239.8 ± 68.2 μg L^{-1} in several Russia's reservoirs and <173 μg L^{-1} in Taechung Reservoir (South Korea) (Table 1). The Chl *a* concentrations in Gorky Reservoir varied from 6.3 to 28.0 μg L^{-1} in both right and left banks, and from 5.9 to 20.6 μg L^{-1} in riverbed with variation of water temperature (WT) from 19.7 to 21.9 °C. In Cheboksary reservoir, Chl *a* concentrations were 4.2–72.4 and 6.6 ± 0.7 - 239.8 ± 68.2 μg L^{-1}, respectively, with variation of WT from 11.0 to 24.0 °C (Table 1) (Mineeva et al. 2008; Mineeva and Abramova 2009). The peak Chl *a* levels in Stanford reservoir exceeded 916 μg L^{-1} in June and July, but they remained below 25 μg L^{-1} for the remainder of the sampling period (Foster et al. 1997).

Estuaries

The Chl *a* concentrations are quite high (0.0–220 μg L^{-1}) in estuaries (Table 1) (Lemaire et al. 2002; Zhu et al. 2009; Stross and Stottlemyer 1965; Pennock 1985; Abril et al. 2002; Hauxwell et al. 2003; Langston et al. 2003; Gitelson et al. 2007; He et al. 2010; Craig et al. 2012; Mallin 1994; Gaulke et al. 2010). The highest Chl *a* concentrations are <220 μg L^{-1} in European estuaries; <184 μg L^{-1} in North Carolina estuaries; >101 μg L^{-1} in the Exe Estuary SPA; <80 μg L^{-1} in Neuse River Estuary; <77.4 μg L^{-1} in estuaries of Chesapeake Bay; <60 μg

L^{-1} in Delaware and Loire Estuaries; <40.0 μg L^{-1} in Pearl River Estuary; <33.3 μg L^{-1} in Patuxent River Estuary; <23.0 μg L^{-1} in several temperate estuaries; <18.02 μg L^{-1} in estuary of Bedford Basin, Canada; <13.8 μg L^{-1} in Sado Estuary; <11.0 μg L^{-1} in Changjiang (Yangtze River) Estuary; <8.9 μg L^{-1} in Elbe Estuary; and 0.3–5.4 μg L^{-1} in all other estuaries studied (Table 1). Such high contents of Chl *a* in estuaries are indicative of highly productive waters, which might be cause by several factors: (i) Estuarine waters contain high contents of DOM, such as 84–525 μM C, which are mostly originated from terrestrial DOM along with the autochthonous DOM and land-derived nutrients (Table 1; see also in chapter "Dissolved Organic Matter in Natural Waters") (Hauxwell et al. 2003; Monbet 1992). Water with high contents of DOM can significantly enhance primary production in estuaries, along with factors that have been discussed previously (see also chapter "Photosynthesis in Nature: A New Look"). (ii) Tidally-driven resuspension along with other associated processes (e.g. tidal mixing, current velocity, light penetration, and sediment resuspension) can influence the variability of suspended particulate matter in estuaries (Monbet 1992; Nichols and Biggs 1985; Allen et al. 1980; Schubel 1971). Estuaries with a low tidal range have maximum suspended sediment load, on the order of 100–200 mg L^{-1}. In contrast, systems with high tidal ranges have sediment concentrations of about 1,000–10,000 mg L^{-1} (Nichols and Biggs 1985). Comparative data analysis from 40 microtidal and macrotidal estuaries shows that mean annual Chl *a* levels are significantly lower in systems with high tidal energy (Monbet 1992). In contrast, nitrogen concentrations are equal to nitrogen levels in the microtidal systems (Monbet 1992). The mechanism behind these phenomena is presumably that strong tidal wave along with strong wind mixing can produce high concentrations of H_2O_2, DIC, nutrients, and so on. These species can be produced either photochemically or microbially from DOM and POM, and can strongly influence photosynthesis and primary production as discussed in an earlier chapter (see "Photosynthesis in Nature: A New Look").

Coastal and Open Oceanic Environments

The Chl *a* concentrations undergo higher variations, from 0.02 to 2080 μg L^{-1} in the waters of coastal and open oceans compared to those of lakes and estuaries (Table 1) (Letelier et al. 2004; Rochelle-Newall and Fisher 2002; Hopkinson and Barbeau 2008; Wheeler et al. 1996; Millán-Núñez et al. 1996; Gomes et al. 2000; Guildford and Hecky 2000; Li and Harrison 2001; Ediger et al. 2006; Parab et al. 2006; Roy et al. 2006; Norrbin et al. 2009; Xiu et al. 2009; Hung et al. 2000; Ahumada et al. 1991; Morales et al. 1996; Dellarossa 1998; Planas et al. 1999; Doyon et al. 2000; Gibb et al. 2000; Gong et al. 2000; Pizarro et al. 2000; Kinkade et al. 2001; Olson and Strom 2002; Sasaoka et al. 2002; Carstensen et al. 2004; Clark et al. 2004; Reul et al. 2005; Holm-Hansen et al. 2004; Pérez et al. 2006; Iriarte et al. 2007; Li et al. 2007; Seppälä et al. 2007; Calbet et al. 2009; Kim et al.

2009; Grippo et al. 2010; Isada et al. 2010; Chen et al. 2011; Iriarte and González 2004; Gonzalez et al. 1989; Mackey et al. 1995). Detected Chl *a* concentrations are as high as 2080 μg L^{-1} in Arabian Sea, <152 μg L^{-1} in Yellow Sea, <148 ± 63.1 μg L^{-1} in north-western Alboran Sea or W-Mediterranean (0–200 m depth), <40 μg L^{-1} in Chiloé and austral fjord (Chile), <35 μg L^{-1} in Concepción and Mejillones Bay (Chile), <30 μg L^{-1} in western subarctic waters of the Pacific Ocean (0–200 m depth), <23.3 μg L^{-1} in Chesapeake Bay (USA), <17 μg L^{-1} in Northwest Pacific Ocean, <17 μg L^{-1} in North Atlantic Ocean, <16.8 μg L^{-1} in upwelling seawater of northern Chile, <15.54 μg L^{-1} in Southwest Florida Shelf, <15.23 μg L^{-1} in California Current System, <14.6 μg L^{-1} in Scotia Sea (near South Georgia), <14.5 μg L^{-1} in Subtropical coastal waters (Hong Kong: 0–10 m depth), <12.5 μg L^{-1} in Baltic Sea, <11.6 μg L^{-1} in Southwest Florida Shelf (Caloosahatchee River: Sts 51–55), <10.88 μg L^{-1} in Gulf of St. Lawrence (Canada), <8.3 μg L^{-1} in Southwest coastal waters (India), and 0.0–4.45 μg L^{-1} in rest of the coastal and other oceans (Table 1). Very low values have been found in Southeast Bering Sea (4.45 μg L^{-1}), Atlantic Ocean (<4.0 μg L^{-1}) and East China Sea (<4.14 μg L^{-1}) (Table 1).

Extremely high Chl *a* concentrations at the surface of eastern Arabian Sea (the highest ever observed in natural water) are responsible for the surface growth of *Trichodesmium* spp. (Parab et al. 2006). This effect is probably linked to high water temperature (20.6–29.4 °C) (Parab et al. 2006) and relatively high DOC contents, varying from 80 to 300 μM C (Menzel 1964; Dileep Kumar et al. 1990; Breves et al. 2003). High contents of Chl *a* in Yellow seawater are also presumably caused by the occurrence of high contents of DOM (129–268 μM C) (Xia et al. 2010) and relatively high water temperature (9–20 °C) (Li et al. 2007) driven by solar irradiance. High contents of Chl *a* are generally detected in coastal seawaters, probably due to high terrestrial input of DOM and POM. Both DOM and POM can produce DIC, CO_2 and H_2O_2 upon photoinduced or microbial respiration/degradation, which are responsible for high photosynthesis and high primary production (see chapter "Photosynthesis in Nature: A New Look" for detailed mechanisms).

In the Baltic Sea, the Chl *a* concentrations are highest in the water column during the spring bloom in late April and during the cyanobacterial bloom in August, which are the two major bloom events (Bianchi et al. 2002). In contrast Chl *a* con centration is low during the summer period, despite the extensive development of cyanobacterial surface blooms (Bianchi et al. 2002). The contents of Chl *a* vary from 0.3 to 13.5 nmol L^{-1}, whilst those of Chl *b* vary from 0.05 to 0.92 nmol L^{-1} (Bianchi et al. 2002). Chl *a* is approximately 15 times higher than Chl *b* in the Baltic Sea.

The observed, relatively low concentrations of Chl *a* in oceanic environments are presumably due to several facts: (i) Low contents of DOM and POM, particularly in open Oceanic environments, may cause the occurrence of low contents of CO_2, DIC, H_2O_2, nutrients, and so on. They are responsible for low photosynthesis and low primary production, as extensively discussed in the photosynthesis chapter (see chapter "Photosynthesis in Nature: A New Look"). In contrast, high contents of organic matter (DOM and POM) in coastal waters are responsible for the higher observed contents of Chl *a* compared to the open ocean (Clark et al. 2004). It is generally known that DOM and POM (e.g. phytoplankton) can release NO_3^- and PO_4^{3-}, by either photoinduced or microbial assimilation/respiration in waters (see chapters “Dissolved Organic Matter in Natural Waters, Photoinduced and Microbial Degradation of Dissolved Organic Matter in Natural Waters” and “Photosynthesis in Nature: A New Look”).

(ii) Strong wind and wave mixing along with the solar (UV and PAR) radiation may degrade Chl *a*, DNA or biomolecules bound to PSI and PSII of microorganisms. The effect would be more marked in the open ocean compared to coastal waters. This issue would be supported by the observation that UV-B radiation (280–315 nm) can inhibit photosynthetic carbon fixation by tropical phytoplankton assemblages in coastal to pelagic surface seawaters (Li et al. 2011). The inhibition of photosynthesis by UV-A (315–400 nm) increases from coastal to offshore waters (Li et al. 2011). It has also been shown that UV-B inhibits photosynthesis by up to 27 % and UV-A by up to 29 % (Li et al. 2011). In East China Sea, lower concentration of Chl *a* (0.06–0.07 $\mu g\ L^{-1}$: Kuroshio sites) has been detected in the open ocean, with high water temperature (23.9–24.0 °C) and low NO_3^- (<0.1 μM), than in coastal seawater (0.43–2.44 $\mu g\ L^{-1}$) (Hung et al. 2000). The latter had low temperature (16.3–18.9 °C) and high NO_3^- (<0.4–6.0 μM) (Hung et al. 2000).

Similarly and interestingly, Chl *a* concentrations are largely variable (0.06–1,000 $\mu g\ L^{-1}$) and substantially high (occasionally >1,000 $\mu g\ L^{-1}$) in ice-covered Antarctic and Arctic Oceans (Table 1) (Palmisano et al. 1985; Garrison et al. 1986; Wheeler et al. 1996; Mock and Gradinger 1999; Apollonio 1980; Guildford and Hecky 2000; Norrbin et al. 2009; Sakshaug and Holm-Hansen 1986; Spies 1987; Verlencar et al. 1990; Varela et al. 2002; Cottrell and Kirchman 2009; Hewes et al. 2009). The highest Chl *a* concentrations, reaching values higher than 1,000 $\mu g\ L^{-1}$, have been detected in bottom-ice communities of Antarctica Ocean. Otherwise, Chl *a* is largely variable: it reached <297 $\mu g\ L^{-1}$ in the ice undersurface; <5.2 in the water column of central Arctic Ocean; <86 $\mu g\ L^{-1}$ in Barents and Greenland Sea ice (Arctic Ocean); <25 $\mu g\ L^{-1}$ in Gerlache and south Bransfield Straits (Antarctic Peninsula); <8.2 $\mu g\ L^{-1}$ in Dumbell Bay (Arctic Ocean); <4.03 $\mu g\ L^{-1}$ in ocean seawater (Antarctic Ocean); <4.0 $\mu g\ L^{-1}$ in South Shetland Islands (Antarctica), 0.10–2.27 in other ice seawater; and finally 111 ± 30 $\mu g\ L^{-1}$ in incubation experimental studies using Antarctic ice seawater (Table 1).

It has been shown that Chl *a* varies significantly, from 0.1 to 297 $\mu g\ L^{-1}$ in ice undersurface and from 0.1 to 5.2 $\mu g\ L^{-1}$ in the water column (Wheeler et al. 1996). The values of Chl *a* can increase in the range of the potential phytoplankton standing stock (25–50 $\mu g\ L^{-1}$) in Antarctic marine waters, Southern Ocean (Sakshaug and Holm-Hansen 1986; Spies 1987). Similarly, Chl *a* contents in bottom ice communities reach 300–400 mg m^{-2} (Steemann-Nielsen 1962; Palmisano and Sullivan 1983). Such numerous algal communities are presumably the consequence of several phenomena: (i) Algal growth may be prolonged due to low temperature and low solar irradiance, which are unable to form $O_2^{\bullet -}$ and subsequently H_2O_2 or $HO^{\bullet}$. This phenomenon can protect algal cells from death, allowing high primary production caused by accumulation of algal species in the ice bed. Interestingly, observation of a series of ice age classes indicated that older ice has higher concentrations of particulate organic carbon, Chl and algal cells (Gleitz and Thomas 1993). Substantial increases have been observed for the abundance of *Chaetoceros neogracile, F. cylindrus,* and *Nitzschia lecointei,* implying growth of these algae (Gleitz and Thomas 1993). The abundance of other species (*F. kerguelensis, Dactyliosolen*) decreased with the age of the sea ice, implying that they can possibly accumulate in ice but are selected against over time (Gleitz and Thomas 1993). Correspondingly, algal pigment signatures in sea ice also suggest that older ice is more diatom-dominated (Lizotte et al. 1998). Lower concentrations of Chl *a*, which have been observed in a light perturbation experiment (3.54 ± 1.00 to 14.2 ± 12.4 μg L^{-1}) compared to the control experiment (5.21 ± 2.33 to 111 ± 30 μg L^{-1}) in Antarctica ice seawater (Palmisano et al. 1985) can also support the above phenomena. The occurrence of more elevated concentrations of dissolved O_2 in Arctic and Antarctic Oceans compared to tropical and subtropical waters (Codispoti and Christensen 1985; Falkner et al. 2005; Garcia et al. 2005; Schmittner et al. 2007; Araoye 2009; Abowei 2010; Keeling et al. 2010) are also responsible for the rapid formation of H_2O_2 under low irradiance. These phenomena can support high photosynthesis in seawater ice. (ii) The occurrence of the ozone hole, and a corresponding increase in UV-B exposure, can cause unequivocal increase of direct or indirect oxidative damage, either directly or indirectly through formation of ROS. It has been shown that the latter can alter biomolecules (lipids, DNA, amino acids, proteins, Chls) and can affect photosynthetic efficiency, reproduction and development in Antarctic marine organisms (see also chapter "Photosynthesis in Nature: A New Look") (Bidigare 1989; Smith et al. 1992; Arrigo 1994; Lesser et al. 2001, 2004; Lesser and Barry 2003; Karentz et al. 2004; Leu et al. 2007; Lister et al. 2010; Cullen and Neale 1997). The effects of the ozone hole and of the corresponding UV-B exposure is largely mitigated by sea ice coverage, in the case of aquatic organisms that live beneath the ice cover (Moreno 2012; Karentz et al. 2004; Lister et al. 2010; Tremblay et al. 2006; Perovich 1993; Trodahl and Buckley 1989). (iii) Intracellular and extracellular production of H_2O_2 from algae (or phytoplankton species) can take place under light conditions in the ice layer (see also chapter "Photosynthesis in Nature: A New Look") (Hong et al. 1987; Bazanov et al. 1999; Premkumar and Ramaraj 1999; Lobanov et al. 2008; Palenik et al. 1987; Palenik and Morel 1988; Komissarov 2003), and could enhance photosynthesis. A further enhancement effect could be caused by relatively high amounts of DIC, H_2O_2 and nutrients produced from DOM and POM, either by microbial or photoinduced processes in Arctic and Antarctic Oceans.

Photosynthesis could rapidly occur under low irradiance conditions in the presence of large amounts of algae (or phytoplankton), and if H_2O_2, DIC and nutrients are available. It has been shown that nutrient concentrations (e.g. nitrate) are considerably high (2–12 μM) in the Arctic Ocean (Tremblay et al. 2006). In the Antarctic Ocean, Chl *a* concentrations in coastal surface seawater ice are high (0.45–4.03 μg L^{-1}), and at the same time there are low contents of NH_4^+ (0.05–2.21 μM), NO_3^- (7.82–23.1 μM), and PO_4^{3-} (0.60–3.0 μM) compared to those of oceanic offshore waters (Table 1) (Verlencar et al. 1990). In contrast, Chl *a* concentrations are relatively low (0.19–0.43 μg L^{-1}) in the presence of rather elevated amounts of NH_4^+ (0.14–1.36 μM), NO_3^- (22.55–29.50 μM), and PO_4^{3-} (1.71–2.35 μM), even in the presence of similar water temperatures (Table 1) (Verlencar et al. 1990). This result can imply that nutrients have limited influence on photosynthesis in offshore seawater. A more important effect could be that high contents of algae (or phytoplankton species) in coastal Antarctic seawater ice can absorb irradiance by Chl *a* bound to PSI and PSII. A possible consequence would be intracellular or extracellular H_2O_2 formation, which could directly affect photosynthesis. This effect could be more important in coastal seawater ice than in offshore oceanic seawater ice. The covariation of dissolved nitrate and phosphate maintained by ocean circulation (Weber and Deutsch 2010) might be a factor that affects photosynthesis in offshore regions. However, future studies will be required to provide evidence for this mechanism.

4 Factors Controlling Chl *a* in Natural Waters

There are a numbers of environmental factors that substantially influence Chl *a* concentrations or primary production in natural waters. The key factors affecting photosynthetic and respiratory activities can be detected based on the growth and development of organisms. They are: (i) seasonal variation in sunlight and UV radiation, which affect photosynthesis; (ii) occurrence of CO_2 forms; (iii) variation in temperature; (iv) effects of water stress (drought) and precipitation/rainfall; (v) effects of the amount and nature of DOM and POM; (vi) variation in nutrient contents; (vii) variation in trace metal ions; (viii) effect of salinity or salt stress; (ix) effects of toxic pollutants on aquatic microorganisms; (x) effect of size-fractionated phytoplankton; (xi) effects of global warming. These factors are similar to those affecting primary production or cyanobacterial bloom, which the exception of the effect of global warming (see chapter "Photosynthesis in Nature: A New Look" and "Impacts of Global Warming on Biogeochemical Cycles in Natural Waters").

4.1 Effects of Global Warming

Global warming can affect the heat budget and other physical processes of a water body, and can subsequently alter the stratification and mixed layer depths (Huisman et al. 2006; Schindler 1997; Magnuson et al. 1997). Such changes, along with global warming-induced changes in the seasonal light cycle, can alter the seasonal patterns of Chl contents (or primary production), phytoplankton composition and nutrient concentrations in SCM and DCM (Huisman et al. 2006; Walsby et al. 1997; O'Reilly et al. 2003; Verburg et al. 2003; Baulch et al. 2005; Fu et al. 2007; Jöhnk et al. 2008; Castle and Rodgers 2009; Davis et al. 2009; Paerl and Huisman 2009). Correspondingly, an extension of the summer season due to global warming may prolong the photochemical processes, with high production of photoproducts, pH alteration, and microbial food web stimulation (Baulch et al. 2005; Morris and Hargreaves 1997; Cooke et al. 2006; Malkin et al. 2008). These issues can result into high photosynthesis, thereby enhancing phytoplankton productivity in lakes and oceans. These phenomena will particularly affect the Arctic and Antarctic regions.

Climate models predict that global warming will increase the stability of the vertical stratification in large parts of lakes and oceans (Huisman et al. 2006; Sarmiento et al. 1998, 2004; Bopp et al. 2001, 2005; Schmittner 2005). This will subsequently reduce vertical mixing and suppress the upward flux of nutrients, leading to a decrease in primary production. However, increased stability of the water column might also increase the photochemical degradation of DOM, and cause high photosynthesis via high temperature and longer summer season. Reduced vertical mixing can generate oscillations and chaos in phytoplankton biomass, size and species composition of DCM (Huisman et al. 2006; Barbiero and Tuchman 2004; Winder et al. 2009). These perturbations are generated by the difference in timescale between the sinking flux of phytoplankton and the upward flux of nutrients. Increasing background light attenuation can increase light limitation, shifting phytoplankton towards the surface and generally decreasing DCM depth and total biomass, particularly in the mixed layer (Mellard et al. 2011). Climate warming may promote the growth of toxic, rather than non-toxic, phytoplankton populations (Davis et al. 2009). Therefore, changes induced by global warming can significantly impact the SCM, DCM, species composition, nutrients dynamics, and carbon cycle. This issue is also extensively discussed in other chapters (see chapters "Photosynthesis in Nature: A New Look" and "Impacts of Global Warming on Biogeochemical Cycles in Natural Waters").

5 Degradation of Chl

It has been shown that terrestrial plants adapt their annual life cycles of growth, reproduction and senescence to the annual climate cycle with period of one year. In contrast, phytoplankton biomass can turn over around 100 times each year as a result of fast growth and equally fast consumption by grazers (Calbet and Landry 2004; Behrenfeld et al. 2006, Winder and Cloern 2010). Therefore, the significance of the degradation of Chl *a* bound to higher plants and aquatic microorganisms shows characteristic differences.

5.1 Degradation of Chl a in Aquatic Microorganisms

Chl *a* bound to phytoplankton or cyanobacteria can be degraded by both photoinduced and microbial degradation processes and can produce chlorophyllide *a*, pheophorbide *a*, pheophytin *a*, and pyropheophytin *a* in aqueous media

(Welschmeyer and Lorenzen 1985; Stephens et al. 1997; Zhang et al. 2009; Bianchi et al. 2002; Schulte-Elte et al. 1979; Falkowski and Sucher 1981; Pietta et al. 1981; Mantoura and Llewellyn 1983; Keely and Maxwell 1991; Nelson 1993; Sun et al. 1993; Rontani et al. 1995; Rontani et al. 1998, 2003, 2011; Rontani and Marchand 2000; Yacobi et al. 1996; Cuny et al. 1999; Marchand and Rontani 2001; Rontani 2001; Lemaire et al. 2002; Rontani and Volkman 2003; Marchand et al. 2005; Christodoulou et al. 2009; Christodoulou et al. 2010; Rontani et al. 2000). Photosynthetically active radiation (PAR, 400–700 nm) and UV radiation (UV-B: 280–315 nm and UV-A: 315–400 nm) are responsible for the degradation of Chls, of PSI, and of PSII bound to phytoplankton species, either directly or through photoinduced generation of ROS in the natural environment (see also chapter "Photosynthesis in Nature: A New Look") (Schulte-Elte et al. 1979; Nelson 1993; Rontani et al. 1995; Nelson and Wakeham 1989; Rontani et al. 1994; Sinha and Häder 2002; Häder and Sinha 2005; Rath and Adhikary 2007; Gao et al. 2008; Pattanaik et al. 2008; Jiang and Qiu 2011). It has also been shown that the degradation rates of Chl *a* bound to algae are several times higher than those of sediment TOC or of algae themselves (Leavitt and Carpenter 1990; Westrich and Berner 1984; Garber 1984; Henrichs and Doyle 1986). The photodegradation of different lipid compounds in killed cells of *Phaeodactylum tricornutum* and *Dunaliella* sp. shows that Chl phytyl chain is degraded to 6,10,14-trimethylpentadecan-2-one and 3-methylidene-7,11,15-trimethylhexadecan-1,2-diol, sterols to 5α- and 6α/6β-hydroxysterols, carotenoids to loliolide and *iso*-loliolide, and unsaturated fatty acids to C_7–$C_{11}\omega$-oxocarboxylic and α,ω-dicarboxylic acids (Rontani et al. 1998). After elimination of insufficiently specific photoproducts, the compounds 3-methylidene-7,11,15-trimethylhexadecan-1,2-diol, 5α- and 6α/6β-hydroxysterols, C_7–$C_{11}\omega$-oxocarboxylic and α, ω-dicarboxylic acids (with C_9 as the most abundant species) have been selected to constitute a "pool" of useful indicators of photooxidative alteration of phytoplankton (Rontani et al. 1998).

Irradiation of killed non-axenic cells of *Emiliania huxleyi* (Prymnesiophyceae) under PAR and UV radiation can degrade most of the unsaturated lipid components, such as Chls, unsaturated fatty acids and brassicasterol (Christodoulou et al. 2010). Exposure to UV radiation can also induce photosensitized stereomutation (cis–trans isomerization) of the double bonds of some lipids (e.g. monounsaturated fatty acids and Chl phytyl side-chain) and of some of their oxidation products. These processes yield (after reduction) some compounds (e.g. 9-hydroxyoctadec-cis-10-enoic and 10-hydroxyoctadeccis-8-enoic acids arising from oleic acid oxidation and 11-hydroxyoctadec-cis-12-enoic and 12-hydroxyoctadec-cis-10-enoic acids arising from cis-vaccenic acid oxidation), which are sufficiently specific to act as tracers of UV-induced in situ photodegradation (Christodoulou et al. 2010). The abiotic degradation processes can act on most of the unsaturated lipid components of senescent phytoplankton, such as sterols, unsaturated fatty acids, Chl phytyl side-chain, carotenoids, alkenones and alkenes (Rontani et al. 1998; Rontani 2001, 2008; Christodoulou et al. 2010). In phytodetritus, the visible light-dependent degradation rates are 3–4 times higher for the Chl tetrapyrrolic structure than for the phytyl side-chain (Cuny et al. 1999; Cuny and Rontani 1999).

Planktonic lipids are more susceptible to biodegradation than terrestrial lipids. Moreover, biodegradation is more intense in sinking particulate organic matter (POM) than in suspended POM (Rontani et al. 2011). Simultaneously, there would be efficient transfer of singlet oxygen from suspended and senescent phytoplankton cells to associated bacteria, with subsequent inhibition of heterotrophic degradation (Rontani et al. 2011). The in vitro enzymatic degradation of Chl *a* in several species of marine phytoplankton can produce chlorophyllide *a*, pheophorbide *a*, pheophytin *a*, and pyropheophytin *a* (Owens and Falkowskit 1982). In some species, Chl *a* can be degraded to products that do not absorb visible light. It has also been observed that losses of phytol and Mg^{2+} are catalysed by chlorophyllase and by a magnesium-releasing enzyme, respectively. Both enzymes are activated by cell disintegration (Owens and Falkowskit 1982). Phaeophytin *a*, pyrophaeophytin *a*, phaeophorbide *a*, and pyrophaeophorbide *a* are the phaeopigments found in largest amount in both sediments and water column (Furlong and Carpenter 1988). Tetrapyrrole derivatives of chloropigments (phaeopigments) are formed as a result of bacterial or autolytic cell lysis, and of metazoan grazing activities (Welschmeyer and Lorenzen 1985; Sanger and Gorham 1970; Shuman and Lorenzen 1975; Bianchi et al. 1988, 1991). Further degradation may produce several colorless organic substances (Brown et al. 1991; Westrich and Berner 1984; Henrichs and Doyle 1986).

From the differences between anoxic and oxic decomposition in incubation experiments, together with naturally observed concentration profiles, it can be inferred that Chl *a* in natural sediments can be degraded during the oscillation between oxic and anoxic conditions caused by physical and biological mixing processes (Ming-Yi et al. 1993). Oscillation experiments (oxic vs. anoxic and anoxic vs. oxic) also suggest that the activity of aerobic organisms may be an important factor that affects Chl *a* degradation (Ming-Yi et al. 1993). Examination of the effects of meiofauna on Chl *a* degradation under oxic conditions, implies that microorganisms may play a stronger role in Chl *a* degradation than meiofauna (Ming-Yi et al. 1993). The relative temperature independence of anoxic degradation and temperature dependence of oxic degradation suggest that anoxic degradation may be largely controlled by chemical factors, while oxic degradation may be more strongly controlled by biophysical and biochemical processes (Ming-Yi et al. 1993).

It is shown that the maximum DOM production lags in time relative to Chl *a* concentration in surface waters, whilst Chl *a* concentrations were relatively low and fluctuated during the summer stratification period in Lake Biwa (Fig. 3a and b) (Zhang et al. 2009; Mostofa KMG et al. unpublished data; Mostofa et al. 2005; Sasaki et al. 2005; Hanamachi et al. 2008). The summertime fluctuation of Chl *a* is possibly linked to its photoinduced degradation, which can contribute to the DOC increase in the surface water of Lake Biwa (Fig. 3a and b) (Mostofa KMG et al. unpublished data; Mostofa et al. 2005). The release of DOM from algae or phytoplankton might be one of the key causes for the decrease of Chl *a* or of the primary production in the surface layer, during the summer season. It is shown that both 'labile' and 'refractory' fractions of DOM are produced during phytoplankton or algal biomass degradation. However, the 'labile' fraction of organic matter, such as glucose, is rapidly decomposed within a few days and the 'refractory' fraction is decomposed more slowly (Mostofa et al. 2009; Zhang et al. 2009; Mostofa KMG et al. 2008; Ogawa et al. 2001).

Low concentrations of Chl *a* during the summer stratification period in upper surface waters might be the effect of photoinduced degradation of Chl *a* by sunlight. Degradation of Chl *a* presumably involves two facts. First of all, cyanobacteria can generate internally reactive oxygen species (ROS) such as superoxide radical anion (O_2^-), hydrogen peroxide (H_2O_2) and hydroxyl radical ($HO^\bullet$) in PSII, which can all be involved into cells decomposition (see chapter "Photosynthesis in Nature: A New Look" for a detailed description). The second fact is the photoinduced generation of ROS from DOM (of both allochthonous and autochthonous origin), NO_2^- and NO_3^- (see also chapters "Photoinduced and Microbial Generation of Hydrogen Peroxide and Organic Peroxides in Natural Waters" and "Photoinduced Generation of Hydroxyl Radical in Natural Waters"). These ROS can decompose Chl *a* that is found outside the cells (see chapter "Photosynthesis in Nature: A New Look"). H_2O_2 involvement can be justified by the observation that autoxidation is substantially enhanced in the presence of a peroxide or hydroperoxide initiator (Fossey et al. 1996; Wilson et al. 2000; Kwan and Voelker 2003). Dissolved O_2 is substantially varied (from 6.0 to 12.0 mg L^{-1}) in a variety of surface waters, whereas the saturated dissolved O_2 concentration in pure water is 7.5 mg L^{-1} at 30 °C (Falkner et al. 2005; Garcia et al. 2005; Schmittner et al. 2007; Araoye 2009; Abowei 2010; Keeling et al. 2010; Hatcher 1987). High contents are generally found at low temperature, particularly in the Arctic and Antarctic Oceans. Such high contents of dissolved O_2 prompt the rapid absorption of electrons released from either chromophoric DOM (CDOM) or POM (e.g. phytoplankton or algae) upon light illumination, which enhances production of $O_2^{\bullet-}$ and H_2O_2. Dissolved O_2 in water is the ultimate electron acceptor upon illumination by light, forming $O_2^{\bullet-}$ that is a long-suspected first intermediate in photoinduced reactions that take place in natural surface waters (Baxter and Carey 1983; Bielski et al. 1985; Petasne and Zika 1987; Micinski et al. 1993). The involvement of dissolved O_2 in H_2O_2 production can be justified by the experimental observation that 5–40 % of the oxygen produced by photosynthetically active organisms can be fixed through photochemical reactions in natural waters (Laane et al. 1985).

Experimental studies show that H_2O_2 can affect cyanobacteria at concentration values that are 10 times lower than for green algae and diatoms. Strong light-dependent toxicity can enhance the difference, for which reason H_2O_2 can act as a limiting factor for cyanobacterial growth (Drábková et al. 2007). H_2O_2 concentrations of approximately 2–8 μM, which are produced during light exposure of aquatic macrophyte leachates or DOM, can inhibit microbial growth or bacterial carbon production (Farjalla et al. 2001; Anesio et al. 2005). The addition of 0.1 μM H_2O_2 to humic lake water can inhibit BCP by as much as 40 % (Xenopoulos and Bird 1997). Photobleaching and CO_2 production in irradiated waters can be significantly decreased upon addition of ROS scavengers, whilst post-irradiation bacterial growth in samples containing a ROS scavenger can be significantly increased Scully et al. (2003). The decrease of ROS activity (CO_2 production) can likely cause an accumulation of bioavailable DOM and enhance microbial processes (Scully et al. 2003). Chl *a* is more susceptible to photochemical decomposition than zeaxanthin in the epilimnion, because zeaxanthin is generally a more stable compound. It is photo-resistant and is found in higher contents than Chl *a* during the summer period (Bianchi et al. 2002; Rowan 1989, 2000). Photoresistance of carotenoids such as zeaxanthin and *ß*, *ß*-carotene involves quenching of singlet oxygen, which prevents photooxidation reactions (Rowan 1989; Jeffrey et al. 1997).

5.2 Degradation of Chl *a* in Higher Plants

Degradation of Chl can have two visible effects on plant leaves (Hendry et al. 1987; Takamiya et al. 2000; Matile et al. 1996; Amir-Shapira et al. 1987; Merzlyak et al. 1999; Park et al. 2007; Pruzinská et al. 2005; Zimmermann and Zentgraf 2005; Kratsch and Wise 2000; Karuppanapandian et al. 2011; Hillman et al. 1994). The first is the colour change from green to yellow or red, which naturally occurs during the season change in autumn and is the most conspicuous and rapid event. The second is cell death caused by external factors, such as injuries sustained by low or high temperature, pathogen attack during various phases of the life cycle of plants, and so on. It has been estimated that approximately 1.2 billion tons of Chl is degraded globally each year (Hendry et al. 1987). The conversions of Chl to chlorophyllide and of pheophytin to pheophorbide in coleslaw, cucumbers and brined olives are the result of chlorophyllase activity (Heaton et al. 1996). Chl *a* in crude extracts of *Chenopodium album* (white goose foot) in the dark can produce chlorophyllide *a*, pheophorbide *a*, 13^2-hydroxychlorophyllide *a* and pyropheophorbide *a*, the increase of which is accompanied by a concomitant decrease in levels of Chl *a* (Shioi et al. 1991). Chl *a* is degraded in a crude extract of *C. album* via enzymatically catalyzed reactions (Shioi et al. 1991).

Chl of detached rice leaves undergoes an initial long lag that lasts for one whole day, after which it is rapidly degraded in the second and third days during experiments conducted under total darkness at 30 °C (Okada et al. 1992). Light only has a weak protecting effect on soluble proteins, and ribulose-1,5-bisphosphate carboxylase/oxygenase rapidly disappeared under illumination with weak white light (Okada et al. 1992). In an in vitro system of extracted broccoli florets, Chl *a* is degraded initially to chlorophyllide *a* or 13^2-hydroxychlorophyll *a*. Subsequently, chlorophyllide *a* is degraded to pyrophaeophorbide *a* through 13^2-hydroxychlorophyll *a* (Yamauchi et al. 1997). Finally, 13^2-hydroxychlorophyll *a* and pyrophaeophorbide *a* can be degraded to colourless, low molecular weight compounds.

5.3 Degradation of Chl During Food Processing

It is well-known that blanching can inactivate chlorophyllase and enzymes, producing a subsequent decrease in the photosynthetic capacity that is responsible for senescence and rapid loss of green colour. The discolouration of green vegetable during processing is caused by conversion of Chls to pheophytins, which is also influenced by pH (Blair and Ayres 1943; Gupte et al. 1964; Minguez-Mosquera et al. 1989; Koca et al. 2007). Chl degradation reactions can be caused by several chemical, photoinduced or enzymatic processes, including simultaneous actions of enzymes, weak acids or pH changes, oxygen, light and heat. Such processes can lead to the formation of a large number of degradation products (Hayakawa and Timbers 1977; Koca et al. 2007). Major chemical degradation processes are pheophytinization, epimerization, pyrolysis, as well as hydroxylation, oxidation or photoinduced oxidation (Mangos and Berger 1997).

The green colour of vegetables can be altered to an olive green under mild acidic conditions, whereas hydrogen ions can transform Chls to their corresponding pheophytins by substitution of the magnesium ion in the porphyrin ring (Minguez-Mosquera et al. 1989; Gold and Weckel 1958; Gunawan and Barringer 2000). Preferential degradation of Chl *b* in the degreening of 'Satsuma' mandarin (*Citrus unshiu* Marc.) is found in ethylene-treated fruits and in fruits ripening on the tree. In contrast, Chl *a* is predominantly degraded in non-treated fruits (Keishi 1979). Methyl jasmonate and ethylene can markedly enhance the mRNA levels and chlorophyllase activity, which presumably accelerates leaf senescence and fruit ripening (Drazkiewicz 1994; Smart 1994; Creelman and Mullet 1997; Jacob-Wilk et al. 1999; Tsuchiya et al. 1999). Stimulatory effects by methyl jasmonate and ethylene also indicate that chlorophyllases are key enzymes for senescence or ripening.

5.4 Mechanism for Degradation of Chl

The key PSII reactions of Chls are photooxidation, involving attack of $^1O_2$, $HO^\bullet$ or H_2O_2, and enzymatic degradation (see also chapter "Photosynthesis in Nature: A New Look") (Takamiya et al. 2000; Brown et al. 1991; Gossauer and Engel 1996; Hörtensteiner 2006; Kräutler and Hörtensteiner 2006; Moser et al. 2009; Hörtensteiner and Kräutler 2011). The processes occurring under high irradiance or UV light and high temperature have been documented in the photosynthesis chapter (see chapter "Photosynthesis in Nature: A New Look"). Three Chl catabolic enzymes, such as chlorophyllase, pheophorbide *a* oxygenase, and red Chl catabolite reductase (RCCR) are susceptible to play key roles into Chl degradation, either during leaf senescence and fruit ripening or in response to pathogens and wounding (Hörtensteiner 2006; Hörtensteiner and Kräutler 2011; Kariola et al. 2005; Azoulay Shemer et al. 2008).

The mechanism responsible for the degreening of plants and the degradation of Chl involves enzymatic reactions in two phases, through several chain reactions (Fig. 4) (Takamiya et al. 2000; Hörtensteiner 2006; Kräutler and Hörtensteiner 2006; Moser et al. 2009; Hörtensteiner and Kräutler 2011). In the first phase, Chl degradation is caused by the removal of the phytol tail (dephytylation) and of the central Mg atom (magnesium dechelatase). Dephytylation occurs first by

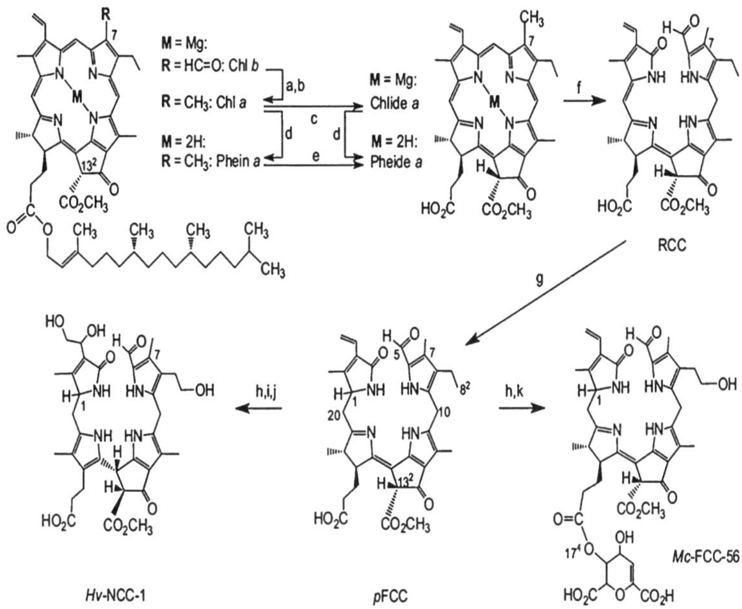

Fig. 4 Representative structural outline of major catabolites delineating the main paths of chlorophyll breakdown in higher plants (Kräutler and Hörtensteiner 2006; Moser et al. 2009): Chls are degraded in the chloroplast by enzyme-catalyzed processes via pheophorbide (Pheide) *a* and the red chlorophyll catabolite (RCC) to give primary fluorescent chlorophyll catabolites (*p*FCC, or its C1-epimer, epipFCC). The relevant enzymes involved in this part are: (**a**) Chl *b* reductase; (**b**) 7-hydroxymethyl Chl reductase; (**c**) chlorophyllase (CLH);)**d**) 85magnesium dechelatase; (**e**) pheophytinase (PPH); (**f**) Pheide *a* oxygenase (PAO); (**g**) RCC reductase (RCCR). *p*FCCs are modified further by unidentified hydroxylating enzymes (**h**, **i**). When carrying a free propionic acid group, FCCs are transported into the vacuole, where they are suggested to isomerize by a spontaneous, acid catalyzed reaction (**j**) to the corresponding nonfluorescent chlorophyll catabolites (NCCs), such as Hv-NCC-1 (the main tetrapyrrolic catabolite found in senescent leaves of barley, Hordeum vulgare). Else, they are esterified by unknown enzymes at the propionic acid group (**k**) to give 'persistent' hypermodified FCCs, such as Mc-FCC-56 (the main FCC in peels of ripe bananas, Musa acuminata, cavendish cultivar). Relevant atom numbering is specified *Data source* Hörtensteiner and Kräutler (2011)

hydrolysis of a phytol residue in ring IV, catalyzed by the enzyme chlorophyllase that converts Chl to phytol and chlorophyllide (Drazkiewicz 1994; Jacob-Wilk et al. 1999; Tsuchiya et al. 1999; Willstätter and Stoll 1913; Trebitsh et al. 1993; Schelbert et al. 2009). The 'Mg dechelatase' subsequently occurs in chlorophyllide, by displacement with 2 H^+ (dechelation) that produces pheophorbide. The latter is subsequently cleaved by an oxygenase enzyme and converted into red Chl catabolite (RCC). Subsequent reduction can produce colorless primary fluorescent Chl catabolite (*p*FCC). The in vivo and in vitro accumulation of pheopigments during Chl degradation in algae and higher plants suggests the presence of a 'magnesium dechelatase' enzyme (Owens and Falkowskit 1982; Shioi et al.

1991; Janave 1997; Maeda et al. 1998; Ziegler et al. 1988). Pheophorbide *a* oxygenase is thought to catalyze the reaction that produces RCC in various leaves and fruits (Fig. 4) (Hörtensteiner 2006; Kräutler et al. 1997; Mühlecker et al. 1997; Hörtensteiner et al. 1998). Pheophytinase, a chloroplast-located and senescence-induced hydrolase that is widely distributed in algae and land plants can also specifically dephytylate the Mg-free Chl pigment, pheophytin (phein), yielding pheophorbide (Schelbert et al. 2009).

In the second phase, *p*FCC-modifying reactions produce FCCs that are imported into the vacuole by a primary active transport process. FCCs are further converted to nonfluorescent Chl catabolites (NCCs) by an acid-catalyzed isomerization, taking place inside the vacuole (Fig. 4) (Hörtensteiner 2006; Moser et al. 2009; Hörtensteiner and Kräutler 2011; Hinder et al. 1996; Kräutler 2003; Christ et al. 2012). Transfer of catabolites from senescent chloroplasts to the vacuole is mediated by primary activated transport processes (Hörtensteiner and Kräutler 2011). Note that the vacuole is a membrane-bound organelle within the cell cytoplasm. It occurs in plant cells and other microorganisms and can store water, salts, minerals, nutrients, proteins, pigments and enzymes. It is involved in growth, protection, waste disposal and structural support and tends to be very large in mature plant cells. Degradation products and enzymes involved in the described reactions have been identified in leaves and fruits (Hörtensteiner and Kräutler 2011; Hörtensteiner et al. 1995, 1998; Hinder et al. 1996; Christ et al. 2012; Kräutler et al. 1991; Matile et al. 1992; Ginsburg and Matile 1993; Mühlecker and Kräutler 1996; Matile et al. 1999).

A process that is closely coupled with the oxygenase reaction is a reduction of the δ-methine bridge of the RCC by a stromal enzyme, termed RCC reductase (RCCR). The reaction yields colorless fluorescent products (Fig. 4) (Hörtensteiner 2006; Rodoni et al. 1997; Wüthrich et al. 2000; Oberhuber and Kräutler 2002; Oberhuber et al. 2008). RCCR has been purified and cloned recently in barley and *Arabidopsis* (Wüthrich et al. 2000).

Spectroscopic analysis shows that *p*FCC has been identified from senescent leaves of various plants (Matile et al. 1996; Mühlecker et al. 1997, 2000; Kräutler and Matile 1999). The *p*FCC is converted to FCCs by several modifications depending on the plants, such as demethylation and hydroxylation (Hörtensteiner 2006; Hörtensteiner and Kräutler 2011; Matile et al. 1992). Modified FCCs are transported to the central vacuole by ATP-dependent translocator(s) in the tonoplast. They are non-enzymatically converted to NCCs by rearrangement of double bonds, in the pyrrole IV ring and adjacent g-methine bridge (Fig. 4) (Hörtensteiner 2006; Moser et al. 2009; Hörtensteiner and Kräutler 2011; Hinder et al. 1996; Kräutler 2003; Christ et al. 2012; Matile et al. 1999). The *p*FCC and all fluorescent Chl catabolites have the same absorption spectrum, with a major peak at around 320 nm and a shoulder at around 360 nm (Takamiya et al. 2000). In contrast, NCCs have an absorption maximum at 316 nm with no shoulder (Takamiya et al. 2000). Finally, three degradation products of monopyrrole derivatives such as hematinic acid, methyl ethyl maleimide and methyl vinyl maleimide aldehyde have been detected in senescent leaves and cotyledons of barley, spinach, pea and cucumber (Suzuki and Shioi 1999).

Senescent mes16 mutants exhibit a strong UV-excitable fluorescence, which is due to accumulation of FCCs. This derives, at least in part, from the fact that FCC isomerization to the respective NCC in the presence of an intact C132-carboxymethylester is slower than with a free carboxylic acid group (Christ et al. 2012; Oberhuber et al. 2008). The most likely reason is differences in the vacuolar pH, which determine the rate of FCC-to-NCC isomerization. Therefore, whether a plant can accumulate FCCs or NCCs might depend on the presence/absence of $O13^4$-demethylation and/or on the vacuolar pH (Christ et al. 2012). Accumulation of 'hypermodified' FCCs (*h*FCCs) in ripening bananas (*Musa acuminata*, Cavendish cultivar) can indicate a new role of Chl catabolites. Moreover, *h*FCCs are a group of unprecedented FCC-esters, and their accumulation in the peels of ripening bananas is rationalized by the corresponding deactivation of the natural, acid-induced (FCC-to-NCC) isomerization (Moser et al. 2008). Such isomerization occurs rapidly in weakly acidic solution (at pH 4.9) and at ambient temperature in aqueous solution. It also occurs in the vacuoles of senescent leaves, in senescent leaves of banana plants and of the peace lily (*Spathiphyllum wallisii*) (Matile et al. 1988; Matile 1997; Oberhuber et al. 2003; Moser et al. 2009; Banala et al. 2010; Kräutler et al. 2010). The *h*FCCs are esterified at the C17-propionic acid side chain, but they are not isomerized to NCCs in some senescing leaves and in ripening banana fruits (Moser et al. 2009; Banala et al. 2010; Kräutler et al. 2010).

The conversion of FCCs to NCCs in vacuole is partly due to either Fenton-type or photo-Fenton type reactions that can generate the $HO^\bullet$, a strong oxidizing agent. This issue is supported by the observation of hydroxylated NCC products or of products with OH-containing other functional groups in place of CH_3 (R_1 or R_3 positions) (Moser et al. 2009; Hörtensteiner and Kräutler 2011; Müller et al. 2007; Pruzinská et al. 2005; Christ et al. 2012; Kräutler et al. 1991; Mühlecker and Kräutler 1996; Oberhuber et al. 2003; Kräutler et al. 1992; Curty and Engel 1996; Berghold et al. 2004; Berghold et al. 2006). Further evidence is the occurrence of the reactions under acidic conditions (pH 4.9), which is vital for obtaining sufficiently high efficiency of Fenton or photo-Fenton reactions. Note that Fenton reaction occurs in an aqueous solution of H_2O_2 and ferrous or ferric salts, which can produce $HO^\bullet$ (see also 'Photoinduced Generation of Hydroxyl Radical in Natural Waters") (Fenton 1894; Barb et al. 1951; Zepp et al. 1992; Kwan and Voelker 2002). The efficiency of the Fenton reaction is highest at pH 3, whilst the photo-Fenton process takes place in the presence of light. The occurrence of various salts, minerals, proteins, FCCs, water and so on in vacuole may favor such type of reactions. The reduction of the rate of formation of hydroperoxides of linoleic acid (induced by H_2O_2) in the presence of NCC may also support the occurrence of such reactions in vacuole (Moser et al. 2009; Müller et al. 2007). High production rates of H_2O_2 in vacuole can be due either to light-sensitive FCCs or from the complexes of FCCs with metal ions present in vacuole. Upon irradiation, such compounds yield electrons (e^-) that can subsequently produce superoxide radical anions ($O_2^{\bullet-}$), H_2O_2, and finally $HO^\bullet$ from H_2O_2. The latter process can take place by either direct photodissociation ($H_2O_2 + h\upsilon \rightarrow HO^\bullet$) or upon Fenton and photo-Fenton reactions. Such processes are discussed in detail in other chapters (see chapters "Photoinduced and Microbial Generation of Hydrogen Peroxide and Organic Peroxides in Natural Waters",

Photoinduced Generation of Hydroxyl Radical in Natural Waters", and "Complexation of Dissolved Organic Matter with Trace Metal ions in Natural Waters"). The transformation of FCCs to NCCs can be depicted shortly as below (Eqs. 5.1–5.4):

$$\mathrm{FCCs} + h\upsilon \rightarrow \mathrm{FCCs}^{+} + e^{-} \tag{5.1}$$

$$e^{-} + O_2 \rightarrow O_2^{\bullet -} \tag{5.2}$$

$$O_2^{\bullet -} + 2H^{+} \rightarrow H_2O_2 \rightarrow HO^{\bullet} \tag{5.3}$$

$$\mathrm{FCC}^{+} + HO^{\bullet} \rightarrow \mathrm{FCC-OH} \rightarrow \mathrm{NCC-OH} \tag{5.4}$$

where FCCs upon illumination by light are excited and produce e^{-}, (Eq. 5.1) which then reacts with aqueous dissolved oxygen to generate $O_2^{\bullet -}$ (Eq. 5.2). $O_2^{\bullet -}$ then produces H_2O_2 and subsequently $HO^{\bullet}$ upon several pathways as mentioned earlier (Eq. 5.3). The $HO^{\bullet}$ radical can then react with FCC^{+} to convert it into FCCs-OH and then into NCCs-OH (Eq. 5.4). Organic peroxides (ROOH) are produced either by similar processes or by breakdown of other organic components. They can generate the organic peroxide radical ($RO^{\bullet}$) and give NCC-OR. These reactions are extensively discussed in earlier chapters (see chapters "Photoinduced and Microbial Generation of Hydrogen Peroxide and Organic Peroxides in Natural Waters and Photoinduced and Microbial Degradation of Dissolved Organic Matter in Natural Waters"). Such a mechanism can also be supported by the observation that Chl degradation of chloroplast lysate or leaf extracts can be induced by intrinsic (per)oxidation with phenolic compounds and H_2O_2, and by lipoxygenation with linolenic acid ('oxidative Chl bleaching') (Janave 1997; Johnson-Flanagan and Spencer 1996; Adachi et al. 1999). Similarly, peroxidase or oxidase activity rise in parallel to the degreening of seeds or cotyledons in some plants (Johnson-Flanagan and Spencer 1996; Adachi et al. 1999). Therefore, $HO^{\bullet}$ or $RO^{\bullet}$ may play a significant role in the transformation of FCCs to NCCs in vacuole. Chl breakdown is a prerequisite to detoxify potentially phototoxic pigments within the vacuoles, to allow the remobilization of nitrogen from Chl-binding proteins that takes place during senescence (Hörtensteiner 2006).

On othe other hand, Chl *b* is degraded to chlorophyllide *b* by chlorophyllase, then chlorophyllide *b* is converted to chlorophyllide *a* by 'Chl *b* reductase' (Schelbert et al. 2009; Ito et al. 1996; Folly and Engel 1999; Scheumann et al. 1999; Tanaka and Tanaka 2006; Rüdiger 2003). The further degradation of chlorophyllide *a* proceeds in similar ways as mentioned before.

6 Chl Acting as Universal Signature of Cyanobacteria (Algae) or Phytoplankton Dynamics

Chl *a* concentrations are very variable in waters, ranging from 0.0 to 280 μg L^{-1} in streams and rivers, 0.01–850 μg L^{-1} in lakes, 0.0–919 μg L^{-1} in reservoirs, 0.0–220 μg L^{-1} in estuaries, 0.0–2080 μg L^{-1} in coastal and marine waters, and 0.06–1,000 μg L^{-1} in ice-covered Arctic and Antarctic Oceans (Table 1). Changes in Chl *a* concentrations reflect the occurrence and features of microorganisms present in natural waters. Therefore, Chl *a* can be used to estimate the primary production or the cyanobacterial (algal) bloom in a variety of waters (Fielding and Seiderer 1991; Ondrusek et al. 1991; Williams and Claustre 1991; Millie et al. 1993; Jeffrey et al. 1999; Bianchi et al. 1993, 2002; Kasprzak et al. 2008). Chl *a* concentration is a predictor of phytoplankton biomass across a broad trophic gradient of lakes, ranging from oligotrophic to highly eutrophic. It is also the most generally used indicator of eutrophication (Blanco et al. 2008; Kasprzak et al. 2008). Concentrations of Chl *a* depend on the fractional contributions of three phytoplankton size classes (micro-, nano- and picoplankton), whereas small cells dominate at low Chl *a* concentrations and large cells at high Chl *a* concentrations (Sathyendranath et al. 2001; Brewin et al. 2010).

The specific Chl *a* content per unit of phytoplankton biomass typically decreases with an increase of phytoplankton standing stocks in filed and experimental observations (Zhang et al. 2009; Kasprzak et al. 2008; Desortová 1981; Shlgren 1983; Wojciechowska 1989; Watson et al. 1992; Talling 1993; Chow-Fraser et al. 1994; Schmid et al. 1998; Felip and Catalan 2000; Sandu et al. 2003; Kiss et al. 2006). The decreases in Chl *a* content per unit of phytoplankton biomass presumably involves two facts: First, Chl *a* bound to microorganisms is the individual component that can be rapidly degraded by either photoinduced or microbial processes (Zhang et al. 2009; Takamiya et al. 2000; Hörtensteiner 2006; Kräutler and Hörtensteiner 2006; Moser et al. 2009; Hörtensteiner and Kräutler 2011). Second, the release of autochthonous DOM from phytoplankton biomass, by either photoinduced or microbial assimilation/respiration (see also chapter "Dissolved Organic Matter in Natural Waters") (Parlanti et al. 2000; Mostofa et al. 2009; Mostofa et al. 2009; Zhang et al. 2009) may affect the decrease in the total content of Chl *a* in phytoplankton standing stocks. In addition, Chl *a* concentrations are substantially affected by the occurrence of phytoplankton species or of size-fractionated phytoplankton, which undergoes seasonal variations in different waters (Bianchi et al. 2002; Satoh et al. 2001; Goedheer 1970; Prezelin 1981; Aguirre-Gomez et al. 2001; Pérez et al. 2007; Hoepffner and Sathyendranath 1991; Parab et al. 2006; Huang et al. 2004, 2005; Buchanan et al. 2005; Qiu et al. 2010). Micro- and nano-Chl *a* are both higher than pico-Chl *a*, but pico-Chl *a* can reach 40 % of total Chl *a* in Wanshan islands in summer (Huang et al. 2005). Micro- and nano-Chl *a* in Pearl River Estuary (South China Sea) generally account for 60 % of total Chl *a*, and pico-Chl *a* account for 20 % of total Chl *a* in most samples (Qiu et al. 2010). In September, picophytoplankton is dominant except for the estuary head, where nano-phytoplankton is predominant. Pico-Chl *a* in far offshore samples accounts for 69 and 75 % of total Chl *a* (Qiu et al. 2010). Picophytoplankton typically accounts for less than 10 % of the total phytoplankton biomass during winter and early spring in Chesapeake Bay. However, it can often contribute to more than 50 % of total phytoplankton biomass in summer and early autumn, particularly in mesohaline and polyhaline waters (Buchanan et al. 2005). Variations in Chl *a* concentrations among phytoplankton species and changes in Chl *a* concentrations per unit of phytoplankton biomass are caused by environmental factors, but Chl *a* is the only parameter that allows precise and rapid determination of phytoplankton biomass or primary production in natural waters.

6.1 Possible Mechanisms For the Management of Eutrophication by Control of Primary Production

Most present studies try to correlate Chl *a* with nutrients, in order to regulate Chl *a*, primary production or photosynthesis by controlling nutrients or by other measures of flood disturbance frequency or of days available for accrual (Biggs 1985, 1995, 2000; Biggs et al. 1998, 1999; Lohman et al. 1992; Welch and Lindell 1992; Chapra 1997; Dodds et al. 1998; Chetelat et al. 1999; Huszar et al. 2006). Nutrients, particularly NO_3^- and PO_4^{3-} are produced mostly by DOM and POM (e.g. phytoplankton species or algae or cyanobacteria), via photoinduced or microbial respiration and degradation. This issue strongly suggests that regulating Chl *a* is vital for the control of DOM and POM in aquatic environments. DOM and POM are in fact the sources of all reactants such as CO_2, DIC, H_2O_2, nutrients and autochthonous DOM, which are responsible for photosynthesis and, therefore, for the primary production of Chl *a* (see chapters "Photosynthesis in Nature: A New Look" and "Impacts of Global Warming on Biogeochemical Cycles in Natural Waters"). DOM and POM along with global warming can lead to excess primary production and to photosynthesis, as shown in chapter "Impacts of Global Warming on Biogeochemical Cycles in Natural Waters". A conceptual model of primary production enhancement and three important steps for remediation, to control algal blooms are extensively discussed in Sects. 5 and 5.1 of chapter "Impacts of Global Warming on Biogeochemical Cycles in Natural Waters". The same measures can be adopted to control photosynthesis and, therefore, to limit primary production or Chl *a* concentration. This activity can reduce eutrophication in natural waters.

7 Scope of the Future Research

DOM along with POM (e.g., algae or phytoplankton) can play an important role in the formation of SCM and DCM. The mechanism behind SCM and DCM formation may pave the way for future research. Formation of H_2O_2 in DCM layer by phytoplankton might be important, and distribution of H_2O_2 as well as its formation from the existing phytoplankton in DCM could be interesting to understand the mechanism of DCM formation. Almost all of previous studies dealt with nutrients (total nitrogen, NO_3^- or NH_4^+, and total phosphate or PO_4^{3-}), but they have some problems. First, DOM and POM can release nutrients in natural waters by photoinduced or microbial respiration or degradation. Therefore, release and uptake of nutrients during photosynthesis has limited importance in waters with high contents of DOM and POM or high contents of nutrients. Second, waters with high contents of DOM and POM can produce DIC, CO_2, H_2O_2 and so on, which are directly linked to photosynthesis and, therefore, to primary production (see chapter "Photosynthesis in Nature: A New Look"). Therefore, DOM and POM should be more directly linked to Chl *a* than nutrients are. Important research needs can thus be listed as follows: (i) Investigation on the relationship between Chl *a* and DOM and POM in a variety of waters, with high and low DOM contents; (ii) Investigation on phytoplankton photosynthesis along with measurement of Chl *a*, DIC, CO_2, H_2O_2 and dissolved O_2. Note that H_2O_2 is photochemically produced from dissolved O_2 (see chapter "Photosynthesis in Nature: A New Look"). Recently, significant correlation has been found between dissolved O_2 and benthic or sestonic Chl concentration (Heiskary and Markus 2003; Miltner 2010), possibly due to H_2O_2 generation from dissolved O_2 followed by in algal production. (iii) New model studies are required, dealing with the mechanism of SCM and DCM formation and elucidating the role of organic matter (DOM and POM), solar radiation, photoinduced formation of DIC, CO_2 and H_2O_2, and water temperature. Note that solar radiation and water temperature are vital for the photoinduced generation of H_2O_2, DIC, CO_2 and nutrients from DOM and POM (see chapter "Photosynthesis in Nature: A New Look"). (iv) Investigation on photosynthesis in natural waters, with and without addition of phytoplankton/algae/cyanobacteria (POM) Such a study could elucidate the effect and role of POM on the photosynthesis, allowing a distinction between photosynthetic processes conducted by DOM and POM. (v) Study of changes of dissolved O_2 concentration along with those of H_2O_2 and Chl *a*. (vi) Elucidation of the role and contribution of H_2O_2, produced either intramolecularly or extracellularly on photosynthesis, in aquatic phytoplankton and higher plants.

8 Nomenclature

CDOM	Chromophoric dissolved organic matter
Chl	Chlorophyll
Chls	Chlorophylls
DCM	Deep chlorophyll *a* maximum
DIC	Dissolved inorganic carbon (dissolved CO_2, H_2CO_3, HCO_3^-, and CO_3^{2-})
DOM	Dissolved organic matter
FDOM	Fluorescent dissolved organic matter
H_2O_2	Hydrogen peroxide
NCC	Nonfluorescent chlorophyll catabolites
$^1O_2$	Singlet oxygen
O_2^-	Super oxide anion radical
$HO^\bullet$	Hydroxyl radical
OM	Organic matter
*P*fcc	Primary fluorescent chlorophyll catabolite
POM	Particulate organic matter
RCC	Red chlorophyll catabolite
RCCR	Red Chl catabolite reductase
ROS	Reactive oxygen species
SCM	Surface or subsurface chlorophyll *a* maximum
WT	Water temperature

Problems

(1) List the various kinds of Chl found in organisms
(2) Explain shortly the Chl *a* functions.
(3) How does the surface or subsurface Chl *a* maximum (SCM) differ from the deep Chl *a* maximum (DCM)?
(4) Explain the mechanisms of SCM and DCM formation in the water column.
(5) How does global warming affect SCM or DCM?
(6) Explain the mechanism of Chl *a* degradation
(7) How does Chl act as universal signature of cyanobacteria (algae) or phytoplankton biomass?
(8) Explain possible actions for the management of eutrophication by controlling primary production (Chl *a*).

Acknowledgments This work was financially supported by the Institute of Geochemistry and Xinjiang Institute of Ecology and Geography, the Chinese Academy of Sciences, China. This work was also partly supported by the University of Turin and Centro Interdipartimentale NatRisk, I-10095 Grugliasco (TO), Italy; Lake Biwa Environmental Research Institute and Kyoto University, Japan; and Semenov Institute of Chemical Physics, Russian Academy of Sciences, Russia. This chapter acknowledges the Copyright (2005) by the Japanese Society of Limnology; reprinted (adapted) with permission from Clarke RH, Connors RE, Schaafsma TJ, Kleibeuker JF, Platenkamp RJ, The triplet state of chlorophylls. J Am Chem Soc 98(12):3674–3677. Copyright (1976) American Chemical Society; reprinted from Water Res, 43(18), Zhang Y, van Dijk MA, Liu M, Zhu G, Qin B, The contribution of phytoplankton degradation to chromophoric dissolved organic matter (CDOM) in eutrophic shallow lakes: Field and experimental evidence, 4685–4697, Copyright (2009), with permission from Elsevier; reprinted from Biochimica et Biophysica Acta (BBA) Bioenergetics, 1807(8), Hörtensteiner S, Kräutler B, Chlorophyll breakdown in higher plants, 977–988, Copyright (2011), with permission from Elsevier; reprinted from Mar Chem, 119(1–4), Christodoulou S, Joux F, Marty J-C, Sempéré R, Rontani J-F, Comparative study of UV and visible light induced degradation of lipids in non-axenic senescent cells of Emiliania huxleyi, 139–152, Copyright (2010), with permission from Elsevier; and reprinted from Geochim Cosmochim Acta, 57(1), Ming-Yi S, Lee C, Aller RC, Laboratory studies of oxic and anoxic degradation of chlorophyll-a in Long Island Sound sediments, 147–157, Copyright (1993), with permission from Elsevier.

References

Abbott MR, Denman KL, Powell TM, Richerson PJ, Richards RC, Goldman CR (1984) Mixing and the dynamics of the deep chlorophyll maximum in Lake Tahoe. Limnol Oceanogr 29:862–878
Abowei JFN (2010) Salinity, dissolved oxygen, pH and surface water temperature conditions in Nkoro River, Niger Delta, Nigeria. Adv J Food Sci Technol 2:36–40
Abril G, Nogueira M, Etcheber H, Cabeçadas G, Lemaire E, Brogueira M (2002) Behaviour of organic carbon in nine contrasting European estuaries. Estuar Coast Shelf Sci 54:241–262
Adachi M, Nakabayashi K, Azuma R, Kurata H, Takahashi Y, Shimokawa K (1999) The ethylene-induced chlorophyll catabolism of radish (*Raphanus sativus* L.) cotyledons: production of colorless fluorescent chlorophyll catabolite (FCC) in vitro. J Jpn Soc Hort Sci 68:1139–1145
Aguirre-Gomez R, Weeks A, Boxall S (2001) The identification of phytoplankton pigments from absorption spectra. Int J Remote Sens 22:315–338

Ahumada R, Matrai P, Silva N (1991) Phytoplankton biomass distribution and relationship to nutrient enrichment during an upwelling event off Concepcion bay, Chile. Boletín Sociedad Biol Concepción 62:7–19

Aizaki M, Otsuki A, Fukushima T, Hosomi M, Muraoka K (1981) Application of Carlson's trophic state index to Japanese lakes and relationships between the index and other parameters. Verh Internat Verein Limnol 21:675–681

Allen GP, Salomon J, Bassoullet P, Du Penhoat Y, de Grandpre C (1980) Effects of tides on mixing and suspended sediment transport in macrotidal estuaries. Sediment Geol 26:69–90

Almodovar A, Nicola GG, Nuevo M (2004) Effects of a bloom of Planktothirix rubescens on the fish community of a Spanish reservoir. Limnetica 23:167–178

Amir-Shapira D, Goldschmidt EE, Altman A (1987) Chlorophyll catabolism in senescing plant tissues: in vivo breakdown intermediates suggest different degradative pathways for citrus fruit and parsley leaves. PNAS 84:1901–1905

An KG, Park SS (2002) Indirect influence of the summer monsoon on chlorophyll–total phosphorus models in reservoirs: a case study. Ecoll Model 152:191–203

Anderson G (1969) Subsurface chlorophyll maximum in the northeast Pacific Ocean. Limnol Oceanogr 14:386–391

Anderson JM (1986) Photoregulation of the composition, function, and structure of thylakoid membranes. Annu Rev Plant Physiol 37:93–136

Anderson JM, Waldron J, Thorne S (1978) Chlorophyll–protein complexes of spinach and barley thylakoids: Spectral characterization of six complexes resolved by an improved electrophoretic procedure. FEBS Lett 92:227–233

Anesio AM, Granéli W (2003) Increased photoreactivity of DOC by acidification: Implications for the carbon cycle in humic lakes. Limnol Oceanogr 48:735–744

Anesio AM, Granéli W, Aiken GR, Kieber DJ, Mopper K (2005) Effect of humic substance photodegradation on bacterial growth and respiration in lake water. Appl Environ Microbiol 71:6267–6275

Angel DL, Fiedler U, Eden N, Kress N, Adelung D, Herut B (1999) Catalase activity in macro- and microorganisms as an indicator of biotic stress in coastal water of the eastern Mediterranean Sea. Helgol Mar Res 53:209–218

Annual Report 2004 (2005) Monitoring and research in Lake Kinneret. Yigal Allon Kinneret limnological laboratory IOLR report T7/2005, pp 75–76

Antoniades D, Veillette J, Martineau MJ, Belzile C, Tomkins J, Pienitz R, Lamoureux S, Vincent WF (2009) Bacterial dominance of phototrophic communities in a High Arctic lake and its implications for paleoclimate analysis. Polar Sci 3:147–161

Apel K, Hirt H (2004) Reactive oxygen species: metabolism, oxidative stress, and signal transduction. Annu Rev Plant Biol 55:373–399

Apollonio S (1980) Primary production in Dumbell Bay in the Arctic Ocean. Mar Biol 61:41–51

Araoye PA (2009) The seasonal variation of pH and dissolved oxygen (DO_2) concentration in Asa lake Ilorin, Nigeria. Int J Phys Sci 4:271–274

Arístegui Ruiz J, Barton ED, Montero del Pino MF, García Muñoz M, Escánez J (2003) Organic carbon distribution and water column respiration in the NW Africa-Canaries Coastal Transition Zone. Aquat Microb Ecol 33:289–301

Arnott SE, Vanni MJ (1993) Zooplankton assemblages in fishless bog lakes: influence of biotic and abiotic factors. Ecology 74:2361–2380

Arrigo KR (1994) Impact of ozone depletion on phytoplankton growth in the Southern Ocean: large-scale spatial and temporal variability. Mar Ecol Prog Ser 114:1–12

Azoulay Shemer T, Harpaz-Saad S, Belausov E, Lovat N, Krokhin O, Spicer V, Standing KG, Goldschmidt EE, Eyal Y (2008) Citrus chlorophyllase dynamics at ethylene-induced fruit color-break: a study of chlorophyllase expression, post-translational processing kinetics and in situ intracellular localization. Plant Physiol 148:108–118

Bachmann RW, Hoyer MV, Canfield DE (2003) Predicting the frequencies of high chlorophyll levels in Florida Lakes from average chlorophyll or nutrient data. Lake Res Manage 19:229–241

Bainbridge R (1957) The size, shape and density of marine phytoplankton concentrations. Biol Rev 32:91–115
Baker E, Louda J (1983) Thermal aspects in chlorophyll geochemistry. Adv in Org Geochem 10:401–421
Banala S, Moser S, Müller T, Kreutz C, Holzinger A, Lütz C, Kräutler B (2010) Hypermodified fluorescent chlorophyll catabolites: source of blue luminescence in senescent leaves. Angew Chem Int Ed 49:5174–5177
Barañano DE, Rao M, Ferris CD, Snyder SH (2002) Biliverdin reductase: a major physiologic cytoprotectant. PNAS 99:16093–16098
Barb W, Baxendale J, George P, Hargrave K (1951) Reactions of ferrous and ferric ions with hydrogen peroxide. Part II—the ferric ion reaction. Trans Faraday Soc 47:591–616
Barbiero RP, Tuchman ML (2004) The deep chlorophyll maximum in Lake Superior. J Great Lakes Res 30:256–268
Barlow R, Mantoura R, Gough M, Fileman T (1993) Pigment signatures of the phytoplankton composition in the northeastern Atlantic during the 1990 spring bloom. Deep Sea Res Part II 40:459–477
Basu B, Pick F (1997) Phytoplankton and zooplankton development in a lowland, temperate river. J Plankton Res 19:237–253
Baulch H, Schindler D, Turner M, Findlay D, Paterson M, Vinebrooke R (2005) Effects of warming on benthic communities in a boreal lake: implications of climate change. Limnol Oceanogr 50:1377–1392
Baxter RM, Carey JH (1983) Evidence for photochemical generation of superoxide ion in humic waters. Nature 306:575–576
Bayley S, Creed I, Sass G, Wong A (2007) Frequent regime shifts in trophic states in shallow lakes on the Boreal Plain: alternative" unstable" states? Limnol Oceanogr 52:2002–2012
Bazanov MI, Berezin BD, Berezin DB et al (1999) Uspekhi khimii porfirinov (Progress in the Chemistry of Porphyrins). NII khimii SPbGU, St Petersburg
Beckmann A, Hense I (2007) Beneath the surface: characteristics of oceanic ecosystems under weak mixing conditions-a theoretical investigation. Prog Oceanogr 75:771–796
Behrenfeld MJ, O'Malley RT, Siegel DA, McClain CR, Sarmiento JL, Feldman GC, Milligan AJ, Falkowski PG, Letelier RM, Boss ES (2006) Climate-driven trends in contemporary ocean productivity. Nature 444:752–755
Bellemare G, Bartlett S, Chua N (1982) Biosynthesis of chlorophyll a/b-binding polypeptides in wild type and the chlorina f2 mutant of barley. J Biol Chem 257:7762–7767
Belzile C, Vincent WF, Kumagai M (2002) Contribution of absorption and scattering to the attenuation of UV and photosynthetically available radiation in Lake Biwa. Limnol Oceanogr 47:95–107
Berghold J, Eichmüller C, Hörtensteiner S, Kräutler B (2004) Chlorophyll breakdown in tobacco: on the structure of two nonfluorescent chlorophyll catabolites. Chem Biodivers 1:657–668
Berghold J, Müller T, Ulrich M, Hörtensteiner S, Kräutler B (2006) Chlorophyll breakdown in maize: on the structure of two nonfluorescent chlorophyll catabolites. Monatsh Chem 137:751–763
Berman T, Stone L, Yacobi YZ, Kaplan B, Schlichter M, Nishri A, Pollingher U (1995) Primary production and phytoplankton in Lake Kinneret: a long-term record (1972–1993). Limnol Oceanogr 40:1064–1076
Bertilsson S, Tranvik LJ (2000) Photochemical transformation of dissolved organic matter in lakes. Limnol Oceanogr 45:753–762
Bertilsson S, Stepanauskas R, Cuadros-Hansson R, Granéli W, Wikner J, Tranvik L (1999) Photochemically induced changes in bioavailable carbon and nitrogen pools in a boreal watershed. Aquat Microb Ecol 19:47–56
Bianchi TS, Findlay S (1991) Decomposition of Hudson estuary macrophytes: Photosynthetic pigment transformations and decay constants. Estuaries Coasts 14:65–73
Bianchi TS, Dawson R, Sawangwong P (1988) The effects of macrobenthic deposit-feeding on the degradation of chloropigments in sandy sediments. J Exp Mar Biol Ecol 122:243–255

Bianchi TS, Findlay S, Fontvieille D (1991) Experimental degradation of plant materials in Hudson river sediments. Biogeochemistry 12:171–187
Bianchi TS, Findlay S, Dawson R (1993a) Organic matter sources in the water column and sediments of the Hudson River Estuary: the use of plant pigments as tracers. Estuar Coast Shelf Sci 36:359–376
Bianchi TS, Dibb JE, Findlay S (1993b) Early diagenesis of plant pigments in Hudson River sediments. Estuar Coast Shelf Sci 36:517–527
Bianchi T, Rolff C, Widbom B, Elmgren R (2002) Phytoplankton pigments in Baltic Sea seston and sediments: seasonal variability, fluxes, and transformations. Estuar Coast Shelf Sci 55:369–383
Biddanda B, Ogdahl M, Cotner J (2001) Dominance of bacterial metabolism in oligotrophic relative to eutrophic waters. Limnol Oceanogr 46:730–739
Bidigare R (1989) Potential effects of UV-B radiation on marine organisms of the Southern Ocean: distributions of phytoplankton and krill during austral spring. Photochem Photobiol 50:469–477
Bidigare R, Morrow J, Kiefer D (1989) Derivative analysis of spectral absorption by photosynthetic pigments in the western Sargasso Sea. J Mar Res 47:323–341
Bielski BHJ, Cabelli DE, Arudi RL, Ross AB (1985) Reactivity of $HO_2^{\bullet}/O_2^{-}$ radicals in aqueous solution. J Phys Chem Ref Data 14:1041–1100
Biggs B (1985) Algae: a blooming nuisance in rivers. Soil Water 21:27–31
Biggs BJF (1995) The contribution of flood disturbance, catchment geology and land use to the habitat template of periphyton in stream ecosystems. Freshw Biol 33:419–438
Biggs BJF (2000) Eutrophication of streams and rivers: dissolved nutrient-chlorophyll relationships for benthic algae. J N Am Benthol Soc 19:17–31
Biggs B, Kilroy C, Lowe R (1998a) Periphyton development in three valley segments of a New Zealand grassland river: test of a habitat matrix conceptual model within a catchment. Arch Hydrobiol 143:147–177
Biggs B, Stevenson R, Lowe R (1998b) A habitat matrix conceptual model for stream periphyton. Arch Hydrobiol 143:21–56
Biggs BJF, Smith RA, Duncan MJ (1999) Velocity and sediment disturbance of periphyton in headwater streams: biomass and metabolism. J N Am Benthol Soc 18:222–241
Bjørnsen P, Nielsen T (1991) Decimeter scale heterogeneity in the plankton during a pycnocline bloom of Gyrodinium aureolum. Mar Ecol Prog Ser 73:263–267
Blair J, Ayres T (1943) Protection of natural green pigment in canning of peas. Ind Eng Chem 35:85–95
Blanco AC, Nadaoka K, Yamamoto T (2008) Planktonic and benthic microalgal community composition as indicators of terrestrial influence on a fringing reef in Ishigaki Island, Southwest Japan. Mar Environ Res 66:520–535
Blindow I, Hargeby A, Meyercordt J, Schubert H (2006) Primary production in two shallow lakes with contrasting plant form dominance: A paradox of enrichment? Limnol Oceanogr 51:2711–2721
Bopp L, Monfray P, Aumont O, Dufresne JL, Le Treut H, Madec G, Terray L, Orr JC (2001) Potential impact of climate change on marine export production. Glob Biogeochem Cy 15:81–100
Bopp L, Aumont O, Cadule P, Alvain S, Gehlen M (2005) Response of diatoms distribution to global warming and potential implications: a global model study. Geophys Res Lett 32:L19606. doi:101029/2005GL023653
Borges A, Ruddick K, Schiettecatte LS, Delille B (2008) Net ecosystem production and carbon dioxide fluxes in the Scheldt estuarine plume. BMC Ecol 8: 101186/1472-6785-8-15
Box EO (2004) Gross production, respiration and biosphere CO_2 fluxes under global warming. Trop Ecol 45:13–30
Brainerd KE, Gregg MC (1995) Surface mixed and mixing layer depths. Deep Sea Res Part I 42:1521–1543
Breves W, Reuter R, Delling N, Michaelis W (2003) Fluorophores in the Arabian Sea and their relationship to hydrographic conditions. Ocean Dyn 53:73–85

Brewin RJW, Sathyendranath S, Hirata T, Lavender SJ, Barciela RM, Hardman-Mountford NJ (2010) A three-component model of phytoplankton size class for the Atlantic Ocean. Ecoll Model 221:1472–1483
Brown SB, Houghton JD, Hendry GAF (1991) Chlorophyll breakdown. In: Scheer H (ed) Chlorophylls. CRC press, Boca Raton, pp 465–489
Brunet C, Brylinski J, Frontier S (1992) Productivity, photosynthetic pigments and hydrology in the coastal front of the Eastern English Channel. J Plankton Res 14:1541–1552
Buchanan C, Lacouture RV, Marshall HG, Olson M, Johnson JM (2005) Phytoplankton reference communities for Chesapeake Bay and its tidal tributaries. Estuar Coasts 28:138–159
Buckle K, Edwards R (1970) Chlorophyll, colour and pH changes in HTST processed green pea puree. Int J Food Sci Technol 5:173–186
Budy P, Luecke C, Wurtsbaugh WA, Gross H, Gubala C (1995) Limnology of the Sawtooth Valley lakes with respect to potential growth of juvenile Snake River sockeye salmon. Northwest Sci 69:133–150
Burrell JWK, Jackman LM, Weedon BCL (1959) Stereo-chemistry and synthesis of phytol, geraniol and nerol. Proc Chem Soc 1959:263–265
Bursche EM (1961) Änderungen im Chlorophyllgehalt und im Zellvolumen bei Planktonalgen, hervorgerufen durch unterschiedliche Lebensbedingungen. Int Rev Ges Hydrobiol 46:610–652
Bushaw KL, Zepp RG, Tarr MA, Schulz-Jander D, Bourbonniere RA, Hodson RE, Miller WL, Bronk DA, Moran MA (1996) Photochemical release of biologically available nitrogen from aquatic dissolved organic matter. Nature 381:404–407
Calbet A, Landry MR (2004) Phytoplankton growth, microzooplankton grazing, and carbon cycling in marine systems. Limnol Oceanogr 49:51–57
Calbet A, Atienza D, Henriksen CI, Saiz E, Adey TR (2009) Zooplankton grazing in the Atlantic Ocean: a latitudinal study. Deep Sea Res Part II 56:954–963
Calijuri M, Cunha D, Queiroz L, Moccellin J, Miwa A (2008) Nutrients and chlorophyll-a concentrations in tropical rivers of Ribeira do Iguape Basin, SP, Brazil. Acta Limnol Bras 20:131–138
Camacho A (1997) Ecología de los microorganismos fotosintéticos en las aguas microaerobias y anóxicas de la Laguna de Arcas. Ph D Thesis, University of Valencia, Valencia, Spain, p 360
Camacho A (2006) On the occurrence and ecological features of deep chlorophyll maxima (DCM) in Spanish stratified lakes. Limnetica 25:453–478
Camacho A, Miracle MR, Vicente E (2003) Which factors determine the abundance and distribution of picocyanobacteria in inland waters? A comparison among different types of lakes and ponds. Arch Hydrobiol 157:321–338
Canjura FL, Schwartz SJ, Nunes RV (1991) Degradation kinetics of chlorophylls and chlorophyllides. J Food Sci 56:1639–1643
Carpenter SR, Kinne O, Wieser W (2003) Regime shifts in lake ecosystems: pattern and variation. Excellence in ecology series, vol 15. Ecology Institute, Luhe, Germany
Carrillo P, Medina-Sánchez JM, Villar-Argaiz M (2002) The interaction of phytoplankton and bacteria in a high mountain lake: importance of the spectral composition of solar radiation. Limnol Oceanogr 47:1294–1306
Carstensen J, Conley DJ, Henriksen P (2004) Frequency, composition, and causes of summer phytoplankton blooms in a shallow coastal ecosystem, the Kattegat. Limnol Oceanogr 49:191–201
Castle JW, Rodgers JH Jr (2009) Hypothesis for the role of toxin-producing algae in Phanerozoic mass extinctions based on evidence from the geologic record and modern environments. Environ Geosci 16:1–23
Chapin BRK, DeNoyelles F Jr, Graham DW, Smith VH (2004) A deep maximum of green sulphur bacteria ('Chlorochromatium aggregatum') in a strongly stratified reservoir. Freshw Biol 49:1337–1354
Chapra SC (1997) Surface water-quality modeling. McGraw-Hill, New York
Chen M, Schliep M, Willows RD, Cai ZL, Neilan BA, Scheer H (2010) A red-shifted chlorophyll. Science 329:1318–1319

Chen M, Chen B, Harrison P, Liu H (2011) Dynamics of mesozooplankton assemblages in subtropical coastal waters of Hong Kong: a comparative study between a eutrophic estuarine and a mesotrophic coastal site. Cont Shelf Res 31:1075–1086

Chessman B (1985) Phytoplankton of the La Trobe River, Victoria. Mar Freshw Res 36:115–122

Chetelat J, Pick F, Morin A, Hamilton P (1999) Periphyton biomass and community composition in rivers of different nutrient status. Can J Fish Aquat Sci 56:560–569

Chow-Fraser P, Trew D, Findlay D, Stainton M (1994) A test of hypotheses to explain the sigmoidal relationship between total phosphorus and chlorophyll a concentrations in Canadian lakes. Can J Fish Aquat Sci 51:2052–2065

Christ B, Schelbert S, Aubry S, Süssenbacher I, Müller T, Kräutler B, Hörtensteiner S (2012) MES16, a member of the methylesterase protein family, specifically demethylates Fluorescent chlorophyll catabolites during chlorophyll breakdown in arabidopsis. Plant Physiol 158:628–641

Christodoulou S, Marty JC, Miquel JC, Volkman JK, Rontani JF (2009) Use of lipids and their degradation products as biomarkers for carbon cycling in the northwestern Mediterranean Sea. Mar Chem 113:25–40

Christodoulou S, Joux F, Marty JC, Sempéré R, Rontani JF (2010) Comparative study of UV and visible light induced degradation of lipids in non-axenic senescent cells of Emiliania huxleyi. Mar Chem 119:139–152

Clark CD, Hiscock WT, Millero FJ, Hitchcock G, Brand L, Miller WL, Ziolkowski L, Chen RF, Zika RG (2004) CDOM distribution and CO_2 production on the Southwest Florida Shelf. Mar Chem 89:145–167

Clarke RH, Connors RE, Schaafsma TJ, Kleibeuker JF, Platenkamp RJ (1976) The triplet state of chlorophylls. J Am Chem Soc 98:3674–3677

Closs G, Katz J, Pennington F, Thomas M, Strain H (1963) Nuclear magnetic resonance spectra and molecular association of chlorophylls a and b, methyl chlorophyllides, pheophytins, and methyl pheophorbides. J Am Chem Soc 85:3809–3821

Codispoti L, Christensen J (1985) Nitrification, denitrification and nitrous oxide cycling in the eastern tropical South Pacific Ocean. Mar Chem 16:277–300

Cogdell RJ, Fyfe PK, Barrett SJ, Prince SM, Freer AA, Isaacs NW, McGlynn P, Hunter CN (1996) The purple bacterial photosynthetic unit. Photosynth Res 48:55–63

Cooke SL, Williamson CE, Hargreaves BR, Morris DP (2006) Beneficial and detrimental interactive effects of dissolved organic matter and ultraviolet radiation on zooplankton in a transparent lake. Hydrobiologia 568:15–28

Cooper W, Lean D (1992) Hydrogen peroxide dynamics in marine and fresh water systems. Encyclop Earth Sys Sci 2:527–535

Cottrell MT, Kirchman DL (2009) Photoheterotrophic microbes in the Arctic Ocean in summer and winter. Appl Environ Microbiol 75:4958–4966

Crabbe P, Djerassi C, Eisenbraun E, Liu S (1959) Optical rotatory dispersion studies XXIX absolute configuration of phytol. Proc Chem Soc 1959:264–265

Craig SE, Jones CT, Li WKW, Lazin G, Horne E, Caverhill C, Cullen JJ (2012) Deriving optical metrics of coastal phytoplankton biomass from ocean colour. Remote Sens Environ 119:72–83

Creelman RA, Mullet JE (1997) Biosynthesis and action of jasmonates in plants. Annu Rev Plant Biol 48:355–381

Croot PL, Laan P, Nishioka J, Strass V, Cisewski B, Boye M, Timmermans KR, Bellerby RG, Goldson L, Nightingale P (2005) Spatial and temporal distribution of Fe(II) and H_2O_2 during EisenEx, an open ocean mescoscale iron enrichment. Mar Chem 95:65–88

Cullen JJ (1982) The deep chlorophyll maximum: comparing vertical profiles of chlorophyll a. Can J Fish Aquat Sci 39:791–803

Cullen JJ, Neale PJ (1997) Effect of UV on short-term photosynthesis of natural phytoplankton. Photochem Photobiol 65:264–266

Cuny P, Rontani JF (1999) On the widespread occurrence of 3-methylidene-7, 11, 15-trimethylhexadecan-1, 2-diol in the marine environment: a specific isoprenoid marker of chlorophyll photodegradation. Mar Chem 65:155–165

Cuny P, Romano JC, Beker B, Rontani JF (1999) Comparison of the photodegradation rates of chlorophyll chlorin ring and phytol side chain in phytodetritus: is the phytyldiol versus phytol ratio (CPPI) a new biogeochemical index? J Exp Mar Biol Ecol 237:271–290

Cuny P, Marty JC, Chiavérini J, Vescovali I, Raphel D, Rontani JF (2002) One-year seasonal survey of the chlorophyll photodegradation process in the northwestern Mediterranean Sea. Deep Sea Res Part II 49:1987–2005

Curty C, Engel N (1996) Detection, isolation and structure elucidation of a chlorophyll *a* catabolite from autumnal senescent leaves of *Cercidiphyllum japonicum*. Phytochemistry 42:1531–1536

Dasí M, Miracle M (1991) LDistribución vertical y variación estacional del fitoplancton de una laguna carstica meromíctica, la Laguna de la Cruz, (Cuenca, España). Limnetica 7:37–59

Dasí M, Miracle M, Camacho A, Soria J, Vicente E (1998) Summer phytoplankton assemblages across trophic gradients in hard-water reservoirs. Hydrobiologia 369:27–43

Davis TW, Berry DL, Boyer GL, Gobler CJ (2009) The effects of temperature and nutrients on the growth and dynamics of toxic and non-toxic strains of Microcystis during cyanobacteria blooms. Harmful Algae 8:715–725

de Moraes Novo EML, de Farias Barbosa CC, de Freitas RM, Shimabukuro YE, Melack JM, Filho WP (2006) Seasonal changes in chlorophyll distributions in Amazon floodplain lakes derived from MODIS images. Limnology 7:153–161

Dekshenieks MM, Donaghay PL, Sullivan JM, Rines JEB, Osborn TR, Twardowski MS (2001) Temporal and spatial occurrence of thin phytoplankton layers in relation to physical processes. Mar Ecol Prog Ser 223:61–71

Dellarossa V (1998) Producción primaria anual en sistemas de alta producción biológica. Tesis Escuela de Graduados, Universidad de Concepción, p 149

Derenbach J, Astheimer H, Hansen H, Leach H (1979) Vertical microscale distribution of phytoplankton in relation to the thermocline. Mar Ecol Prog Ser 1:187–193

Desortová B (1981) Relationship between chlorophyll-α concentration and phytoplankton biomass in several reservoirs in Czechoslovakia. Int Rev Ges Hydrobiol 66:153–169

Deuser WG (1987) Variability of hydrography and particle flux: Transient and long-term relationships. Mitt Geol-Palaeont Inst Univ Hamburg 62:179–193

Devercelli M, Peruchet E (2008) Trends in chlorophyll-a concentration in urban water bodies within different man-used basins. Ann Limnol Int J Lim 44:75–84

Devlin M, Barry J, Mills D, Gowen R, Foden J, Sivyer D, Greenwood N, Pearce D, Tett P (2009) Estimating the diffuse attenuation coefficient from optically active constituents in UK marine waters. Estuar Coast Shelf Sci 82:73–83

Diehl S (2002) Phytoplankton, light, and nutrients in a gradient of mixing depths: theory. Ecology 83:386–398

Dietzel L, Bräutigam K, Steiner S, Schüffler K, Lepetit B, Grimm B, Schöttler MA, Pfannschmidt T (2011) Photosystem II Supercomplex Remodeling Serves as an Entry Mechanism for State Transitions in Arabidopsis. Plant Cell 23:2964–2977

Dileep Kumar M, Rajendran A, Somasundar K, Haake B, Jenisch A, Shuo Z, Ittekkot V, Desai B (1990) Dynamics of dissolved organic carbon in the northwestern Indian Ocean. Mar Chem 31:299–316

Djurfeldt L (1994) The influence of physical factors on a subsurface chlorophyll maximum in an upwelling area. Estuar Coast Shelf Sci 39:389–400

Dodds WK, Jones JR, Welch EB (1998) Suggested classification of stream trophic state: distributions of temperate stream types by chlorophyll, total nitrogen, and phosphorus. Water Res 32:1455–1462

Donaghay P, Rines H, Sieburth J (1992) Simultaneous sampling of fine scale biological, chemical, and physical structure in stratified waters. Ergebn Limnol ERLIA 6(36):97–108

Dortch Q (1987) The biochemical composition of plankton in a subsurface chlorophyll maximum. Deep Sea Res Part I 34:705–712

Dougherty R, Strain H, Svec WA, Uphaus R, Katz J (1966) Structure of chlorophyll c_1. J Am Chem Soc 88:5037–5038

Downs JN, Lorenzen CJ (1985) Carbon: pheopigment ratios of zooplankton fecal pellets as an index of herbivorous feeding. Limnol Oceanogr 30:1024–1036

Doyon P, Klein B, Ingram R, Legendre L, Tremblay JE, Therriault JC (2000) Influence of wind mixing and upper-layer stratification on phytoplankton biomass in the Gulf of St. Lawrence. Deep Sea Res Part II 47:415–433

Drábková M, Admiraal W, Maršálek B (2007) Combined exposure to hydrogen peroxide and light selective effects on cyanobacteria, green algae, and Diatoms. Environ Sci Technol 41:309–314

Drazkiewicz M (1994) Chlorophyllase: occurrence, functions, mechanism of action, effects of external and internal factors (review). Photosynthetica 30:321–331

Duan S, Bianchi TS (2006) Seasonal changes in the abundance and composition of plant pigments in particulate organic carbon in the lower Mississippi and Pearl Rivers. Estuar Coasts 29:427–442

Dupouy C, Neveux J, Ouillon S, Frouin R, Murakami H, Hochard S, Dirberg G (2010) Inherent optical properties and satellite retrieval of chlorophyll concentration in the lagoon and open ocean waters of New Caledonia. Mar Pollut Bull 61:503–518

Durrant JR, Klug DR, Kwa S, van Grondelle R, Porter G, Dekker JP (1995) A multimer model for P680, the primary electron donor of photosystem II. PNAS 92:4798–4802

Echevin V, Aumont O, Tam J, Pasapera J (2004) The seasonal cycle of surface chlorophyll along the Peruvian coast: comparison between SeaWifs satellite observations and dynamical/biogeochemical coupled model simulations. Gayana (Concepción) 68:325–326

Ediger D, Yilmaz A (1996) Characteristics of deep chlorphyll maximum in the Northeastern Mediterranean with respect to environmental conditions. J Mar Sys 9:291–303

Ediger D, Soydemir N, Kideys A (2006) Estimation of phytoplankton biomass using HPLC pigment analysis in the southwestern Black Sea. Deep Sea Res Part II 53:1911–1922

Fahnenstiel GL, Scavia D (1987) Dynamics of Lake Michigan phytoplankton: the deep chlorophyll layer. J Great Lakes Res 13:285–295

Falkner KK, Steele M, Woodgate RA, Swift JH, Aagaard K, Morison J (2005) Dissolved oxygen extrema in the Arctic Ocean halocline from the North Pole to the Lincoln Sea. Deep Sea Res Part I 52:1138–1154

Falkowski PG, Sucher J (1981) Rapid, quantitative separation of chlorophylls and their degradation products by high-performance liquid chromatography. J Chromatogr 213:349–351

Falkowski PG, Katz ME, Knoll AH, Quigg A, Raven JA, Schofield O, Taylor F (2004) The evolution of modern eukaryotic phytoplankton. Science 305:354–360

Farjalla VF, Anesio AM, Bertilsson S, Granéli W (2001) Photochemical reactivity of aquatic macrophyte leachates: abiotic transformations and bacterial. Aquat Microb Ecol 24:187–195

Fee EJ (1976) The vertical and seasonal distribution of chlorophyll in lakes of the Experimental Lakes Area, northwestern Ontario: Implications for primary production estimates. Limnol Oceanogr 21:767–783

Felip M, Catalan J (2000) The relationship between phytoplankton biovolume and chlorophyll in a deep oligotrophic lake: decoupling in their spatial and temporal maxima. J Plankton Res 22:91–106

Fennel K, Boss E (2003) Subsurface maxima of phytoplankton and chlorophyll: Steady-state solutions from a simple model. Limnol Oceanogr 48:1521–1534

Fenton H (1894) Oxidation of tartaric acid in presence of iron. J Chem Soc Trans 65:899–910

Field CB, Behrenfeld MJ, Randerson JT, Falkowski P (1998) Primary production of the biosphere: integrating terrestrial and oceanic components. Science 281:237–240

Ficken GE, Johns RB, Linstead RP (1956) Chlorophyll and related compounds. Part IV. The position of the extra hydrogens in chlorophyll. The oxidation of pyrophœophorbide-*a*. J Chem Soc 2272–2280. doi:10.1039/JR9560002272

Fielding P, Seiderer L (1991) A fresh look at kelp bed phytoplankton populations in an upwelling area. Mar Ecol Prog Ser 72:167–177

Finlay K, Leavitt P, Wissel B, Prairie Y (2009) Regulation of spatial and temporal variability of carbon flux in six hard-water lakes of the northern Great Plains. Limnol Oceanogr 54:2553–2564

Fischer H, Wenderoth H (1940) Chlorophyll XCIX Optically active hemotricarboxylic imides from chlorophyll. Annalen 545:140–147

Fleming I (1967) Absolute configuration and the structure of chlorophyll. Nature 216:151–152. doi:101038/216151a101030

Folly P, Engel N (1999) Chlorophyll *b* to chlorophyll *a* conversion precedes chlorophyll degradation in *Hordeum vulgare* L. J Biol Chem 274:21811–21816

Fookes CJR, Jeffrey S (1989) The structure of chlorophyll c3, a novel marine photosynthetic pigment. J Chem Soc Chem Commun 23:1827–1828

Foote CS (1976) Photosensitized oxidation and singlet oxygen: consequences in biological systems. In: Pryor WA (ed) Free Radicals in Biology. Academic Press, New York, pp 85–133

Fossey J, Lefort D, Sorba J (1995) Free radicals in organic chemistry, vol 109. Masson, Paris, pp 1–307

Foster I, Baban S, Charlesworth S, Jackson R, Wade S, Buckland P, Wagstaff K, Harrison S (1997) Nutrient concentrations and planktonic biomass (chlorophyll *a*) behaviour in the basin of the River Avon, Warwickshire, UK Freshwater Contamination (Proceedings of Rabat Symposium S4, April–May 1997), IAHS Publ no 243

Franks PJS, Jaffe JS (2001) Microscale distributions of phytoplankton: initial results from a two-dimensional imaging fluorometer, OSST. Mar Ecol Prog Ser 220:59–72

Fu FX, Warner ME, Zhang Y, Feng Y, Hutchins DA (2007) Effects of increased temperature and CO_2 on photosynthesis, growth, and elemental ratios in marine Synechococcus and Prochlorococcus (cyanobacteria). J Phycol 43:485–496

Fu P, Mostofa KMG, Wu F, Liu CQ, Li W, Liao H, Wang L, Wang J, Mei Y (2010) Excitation-emission matrix characterization of dissolved organic matter sources in two eutrophic lakes (Southwestern China Plateau). Geochem J 44:99–112

Fujiwara K, Ushiroda T, Takeda K, Kumamoto YI, Tsubota H (1993) Diurnal and seasonal distribution of hydrogen peroxide in seawater of the Seto Inland Sea. Geochem J 27:103–115

Furlong ET, Carpenter R (1988) Pigment preservation and remineralization in oxic coastal marine sediments. Geochim Cosmochim Acta 52:87–99

Gálvez J, Niell F, Lucena J (1988) Description and mechanism of formation of a deep chlorphyll maximum due to Ceratium hirundinella(O. F. Mueller) Bergh. Arch Hydrobiol 112:143–155

Gao H, Zepp RG (1998) Factors influencing photoreactions of dissolved organic matter in a coastal river of the southeastern United States. Environ Sci Technol 32:2940–2946

Gao X, Olapade OA, Kershner MW, Leff LG (2004) Algal-bacterial co-variation in streams: a cross-stream comparison. Arch Hydrobiol 159:253–261

Gao K, Li P, Watanabe T, Walter Helbling E (2008) Combined effects of ultraviolet radiation and temperature on morphology, photosynthesis, and DNA of Arthrospira (spirulina) platensis (Cyanophyta). J Phycol 44:777–786

Garber JH (1984) Laboratory study of nitrogen and phosphorus remineralization during the decomposition of coastal plankton and seston. Estuar Coast Shelf Sci 18:685–702

Garcia HE, Boyer TP, Levitus S, Locarnini RA, Antonov J (2005) On the variability of dissolved oxygen and apparent oxygen utilization content for the upper world ocean: 1955 to 1998. Geophys Res Lett 32:L09604. doi:101029/102004GL022286

Garrison DL, Sullivan CW, Ackley SF (1986) Sea ice microbial communities in Antarctica. Bioscience 36:243–250

Gaulke AK, Wetz MS, Paerl HW (2010) Picophytoplankton: A major contributor to planktonic biomass and primary production in a eutrophic, river-dominated estuary. Estuar Coast Shelf Sci 90:45–54

Gentien P, Lunven M, Lehaître M, Duvent J (1995) In situ depth profiling of particle sizes. Deep Sea Res Part I 42:1297–1312

Ghai R, Martin-Cuadrado AB, Molto AG, Heredia IG, Cabrera R, Martin J, Verdú M, Deschamps P, Moreira D, López-García P (2010) Metagenome of the Mediterranean deep chlorophyll maximum studied by direct and fosmid library 454 pyrosequencing. ISME J 4:1154–1166

Gibb S, Barlow R, Cummings D, Rees N, Trees C, Holligan P, Suggett D (2000) Surface phytoplankton pigment distributions in the Atlantic Ocean: an assessment of basin scale variability between 50ºN and 50ºS. Prog Oceanogr 45:339–368

Ginsburg S, Matile P (1993) Identification of catabolites of chlorophyll-porphyrin in senescent rape cotyledons. Plant Physiol 102:521–527

Gitelson AA, Schalles JF, Hladik CM (2007) Remote chlorophyll-*a* retrieval in turbid, productive estuaries: Chesapeake Bay case study. Remote Sens Environ 109:464–472

Gleitz M, Thomas DN (1993) Variation in phytoplankton standing stock, chemical composition and physiology during sea-ice formation in the southeastern Weddell Sea, Antarctica. J Exp Mar Biol Ecol 173:211–230
Goedheer J (1970) On the pigment system of brown algae. Photosynthetica 4:97–106
Goericke R, Welschmeyer NA (1998) Response of Sargasso Sea phytoplankton biomass, growth rates and primary production to seasonally varying physical forcing. J Plankton Res 20:2223–2249
Gold HJ, Weckel K (1958) Degradation of chlorophyll to pheophytin during sterilization of canned green peas by heat. Food Technol 13:281–286
Gomes HR, Goes JI, Saino T (2000) Influence of physical processes and freshwater discharge on the seasonality of phytoplankton regime in the Bay of Bengal. Continent Shelf Res 20:313–330
Gong GC, Shiah FK, Liu KK, Wen YH, Liang MH (2000) Spatial and temporal variation of chlorophyll a, primary productivity and chemical hydrography in the southern East China Sea. Cont Shelf Res 20:411–436
Gonzalez H, Pantoja S, Iriarte J, Bernal P (1989) Winter-spring variability of size-fractionated autotrophic biomass in Concepcion Bay, Chile. J Plankton Res 11:1157–1167
Gossauer A, Engel N (1996) Chlorophyll catabolism–structures, mechanisms, conversions. J Photochem Photobiol 32:141–151
Graneli W, Lindell M, Tranvik L (1996) Photo-oxidative production of dissolved inorganic carbon in lakes of different humic content. Limnol Oceanogr 41:698–706
Granéli W, Lindell M, de Faria BM, de Assis Esteves F (1998) Photoproduction of dissolved inorganic carbon in temperate and tropical lakes–dependence on wavelength band and dissolved organic carbon concentration. Biogeochemistry 43:175–195
Green B, Durnford D (1996) The chlorophyll-carotenoid proteins of oxygenic photosynthesis. Annu Rev Plant Biol 47:685–714
Grippo M, Fleeger J, Rabalais N, Condrey R, Carman K (2010) Contribution of phytoplankton and benthic microalgae to inner shelf sediments of the north-central Gulf of Mexico. Cont Shelf Res 30:456–466
Gross HP, Wurtsbaugh WA, Budy P, Luecke C (1997) Fertilization of an oligotrophic lake with a deep chlorophyll maximum: predicting the effect on primary productivity. Can J Fish Aquat Sci 54:1177–1189
Grossman AR, Bhaya D, Apt KE, Kehoe DM (1995) Light-harvesting complexes in oxygenic photosynthesis: diversity, control, and evolution. Annu Rev Genet 29:231–288
Guéguen C, Guo L, Wang D, Tanaka N, Hung C–C (2006) Chemical characteristics and origin of dissolved organic matter in the Yukon River. Biogeochemistry 77:139–155
Guildford SJ, Hecky RE (2000) Total nitrogen, total phosphorus, and nutrient limitation in lakes and oceans: Is there a common relationship? Limnol Oceanogr 45:1213–1223
Gunawan MI, Barringer SA (2000) Green color degradation of blanched broccoli (Brassica oleracea) due to acid and microbial growth. J Food Process Preserv 24:253–263
Gupte S, El-Bisi H, Francis F (1964) Kinetics of thermal degradation of chlorophyll in spinach pureea. J Food Sci 29:379–382
Haberl H, Erb KH, Krausmann F, Gaube V, Bondeau A, Plutzar C, Gingrich S, Lucht W, Fischer-Kowalski M (2007) Quantifying and mapping the human appropriation of net primary production in earth's terrestrial ecosystems. PNAS 104:12942–12947
Häder DP, Sinha RP (2005) Solar ultraviolet radiation-induced DNA damage in aquatic organisms: potential environmental impact. Mutation Res 571:221–233
Hamilton DP, O'Brien KR, Burford MA, Brookes JD, McBride CG (2010) Vertical distributions of chlorophyll in deep, warm monomictic lakes. Aquat Sci 72:295–307
Hanamachi Y, Hama T, Yanai T (2008) Decomposition process of organic matter derived from freshwater phytoplankton. Limnology 9:57–69
Harris G (1986) Phytoplankton ecology: structure, function and fluctuation. The Concept of Limiting Nutrients Capman and Hall, London, pp 137–165
Harrison JW, Smith REH (2011) Deep chlorophyll maxima and UVR acclimation by epilimnetic phytoplankton. Freshwater Biol 56:980–992

Hart DR, Stone L, Berman T (2000) Seasonal dynamics of the Lake Kinneret food web: the importance of the microbial loop. Limnol Oceanogr 45:350–361
Harvey HW (1934) Amount of phytoplankton population. J Mar Biol Assoc UK 19:761–773
Harvey H (1939) Substances controlling the growth of a diatom. J mar biol Ass UK 23:499–520
Hatcher KJ (1987) Selecting an appropriate method for estimating the sediment oxygen demand rate. In: Lichtenberg JJ, Winter JA, Weber CI, Fradkin L (eds) Chemical and Biological Characterization of Sludges, Sediments, Dredge Spoils, and Drilling Muds, ASTM STP 976. American Society for Testing and Materials, Philadelphia, pp 438–449
Hauxwell J, Cebrián J, Valiela I (2003) Eelgrass Zostera marina loss in temperate estuaries: relationship to land-derived nitrogen loads and effect of light limitation imposed by algae. Mar Ecol Prog Ser 247:59–73
Hayakawa K, Sugiyama Y (2008) Spatial and seasonal variations in attenuation of solar ultraviolet radiation in Lake Biwa, Japan. J Photochem Photobiol 90:121–133
Hayakawa K, Timbers GE (1977) Influence of heat treatment on the quality of vegetables: changes in visual green color. J Food Sci 42:778–781
Hayward T, Cayan D, Franks P, Lynn R, Mantyla A, McGowan J, Smith P, Schwing F, Venrick E (1995) The state of the California Current in 1994–1995: a period of transition. Calif Coop Oceanic Fish Invest Rep 35:19–40
He B, Dai M, Zhai W, Wang L, Wang K, Chen J, Lin J, Han A, Xu Y (2010) Distribution, degradation and dynamics of dissolved organic carbon and its major compound classes in the Pearl River estuary, China. Mar Chem 119:52–64
Head E, Horne E (1993) Pigment transformation and vertical flux in an area of convergence in the North Atlantic. Deep Sea Res Part II 40:329–346
Heaton JW, Lencki RW, Alejandro G (1996) Kinetic model for chlorophyll degradation in green tissue. J Agric Food Chem 44:399–402
Heiskary S, Markus H (2003) Establishing relationships among instream nutrient concentrations, phytoplankton and periphyton abundance and composition, fish and macroinvertebrate indices, and biochemical oxygen demand in Minnesota USA rivers. Minnesota Pollution Control Agency, Environmental Outcomes Division, St Paul
Hendry GAF, Houghton JD, Brown SB (1987) Tansley review No. 11. The degradation of chlorophyll-a biological enigma. New Phytol 107:255–302
Henrichs SM, Doyle AP (1986) Decomposition of ^{14}C-labeled organic substances in marine sediments. Limnol Oceanogr 31:765–778
Hense I, Beckmann A (2008) Revisiting subsurface chlorophyll and phytoplankton distributions. Deep Sea Res Part I 55:1193–1199
Hewes C, Reiss C, Holm-Hansen O (2009) A quantitative analysis of sources for summertime phytoplankton variability over 18 years in the South Shetland Islands (Antarctica) region. Deep Sea Res Part I 56:1230–1241
Hillman J, Glidewell S, Deighton N (1994) The senescence syndrome in plants: an overview of phytogerontology. Proc R Soc Edinb B 102:447–458
Hinder B, Schellenberg M, Rodoni S, Ginsburg S, Vogt E, Martinoia E, Matile P, Hörtensteiner S (1996) How plants dispose of chlorophyll catabolites. J Biol Chem 271:27233–27236
Hobson LA, Lorenzen CJ (1972) Relationship of chlorophyll maxima to density structure in the Atlantic Ocean and Gulf of Mexico. Deep-Sea Res 19:297–306
Hodges BA, Rudnick DL (2004) Simple models of steady deep maxima in chlorophyll and biomass. Deep Sea Res Part I 51:999–1015
Hoepffner N, Sathyendranath S (1991) Effect of pigment composition on absorption properties of phytoplankton. Mar Ecol Prog Ser 73:11–23
Holm-Hansen O, Hewes CD (2004) Deep chlorophyll-a maxima (DCMs) in Antarctic waters. Polar Biol 27:699–710
Holm-Hansen O, Kahru M, Hewes C, Kawaguchi S, Kameda T, Sushin V, Krasovski I, Priddle J, Korb R, Hewitt R (2004) Temporal and spatial distribution of chlorophyll-a in surface waters of the Scotia Sea as determined by both shipboard measurements and satellite data. Deep Sea Res Part II 51:1323–1331

Hong AP, Bahnemann DW, Hoffmann MR (1987) Cobalt (II) tetrasulfophthalocyanine on titanium dioxide: a new efficient electron relay for the photocatalytic formation and depletion of hydrogen peroxide in aqueous suspensions. J Phys Chem 91:2109–2117

Hopkinson BM, Barbeau KA (2008) Interactive influences of iron and light limitation on phytoplankton at subsurface chlorophyll maxima in the eastern North Pacific. Limnol Oceanogr 53:1303–1318

Hörtensteiner S (2006) Chlorophyll degradation during senescence. Annu Rev Plant Biol 57:55–77

Hörtensteiner S, Kräutler B (2011) Chlorophyll breakdown in higher plants. Biochim Biophys Acta 1807:977–988

Hörtensteiner S, Vicentini F, Matile P (1995) Chlorophyll breakdown in senescent cotyledons of rape, *Brassica napus* L.: enzymatic cleavage of phaeophorbide *a* in vitro. New Phytol 129:237–246

Hörtensteiner S, Wüthrich KL, Matile P, Ongania KH, Kräutler B (1998) The key step in chlorophyll breakdown in higher plants. J Biol Chem 273:15335–15339

Huang L, Jian W, Song X, Huang X, Liu S, Qian P, Yin K, Wu M (2004) Species diversity and distribution for phytoplankton of the Pearl River estuary during rainy and dry seasons. Mar Pollut Bull 49:588–596

Huang B, Hong H, Ke L, Cao Z (2005) Size-fractionated phytoplankton biomass and productivity in the Zhujiang River Estuary in China. Acta Oceanol Sin 27:180–186

Huisman J, Weissing FJ (1995) Competition for nutrients and light in a mixed water column: a theoretical analysis. Am Nat 146:536–564

Huisman J, van Oostveen P, Weissing FJ (1999) Species dynamics in phytoplankton blooms: incomplete mixing and competition for light. Am Nat 154:46–68

Huisman J, Thi NNP, Karl DM, Sommeijer B (2006) Reduced mixing generates oscillations and chaos in the oceanic deep chlorophyll maximum. Nature 439:322–325

Hung CC, Wong GTF, Liu KK, Shiah FK, Gong GC (2000) The effects of light and nitrate levels on the relationship between nitrate reductase activity and $15NO_3$-uptake: Field observations in the East China Sea. Limnol Oceanogr 45:836–848

Huszar VLM, Caraco NF, Roland F, Cole J (2006) Nutrient chlorophyll relationships in tropical-subtropical lakes: do temperate models fit? Biogeochemistry 79:239–250

Iriarte JL, González HE (2004) Phytoplankton size structure during and after the 1997/98 El Nino in a coastal upwelling area of the northern Humboldt current system. Mar Ecol Prog Ser 269:83–90

Iriarte J, González H, Liu K, Rivas C, Valenzuela C (2007) Spatial and temporal variability of chlorophyll and primary productivity in surface waters of southern Chile (415–43°S). Estuar Coast Shelf Sci 74:471–480

Isada T, Hattori-Saito A, Saito H, Ikeda T, Suzuki K (2010) Primary productivity and its bio-optical modeling in the Oyashio region, NW Pacific during the spring bloom 2007. Deep Sea Res Part II 57:1653–1664

Ito H, Ohtsuka T, Tanaka A (1996) Conversion of chlorophyll *b* to chlorophyll *a* via 7-hydroxymethyl chlorophyll. J Biol Chem 271.1475–1479

Jacob-Wilk D, Holland D, Goldschmidt EE, Riov J, Eyal Y (1999) Chlorophyll breakdown by chlorophyllase: isolation and functional expression of the Chlase1 gene from ethylene-treated Citrus fruit and its regulation during development. Plant J 20:653–661

Jagannathan B, Golbeck JH (2009) Photosynthesis: microbial. In: Schaechter M (ed) Encyclopedia of microbiology, 3rd edn. Elsevier, London, pp 325–341

James RT, Havens K, Zhu G, Qin B (2009) Comparative analysis of nutrients, chlorophyll and transparency in two large shallow lakes (Lake Taihu, PR China and Lake Okeechobee, USA). Hydrobiologia 627:211–231

Janave MT (1997) Enzymic degradation of chlorophyll in Cavendish bananas: in vitro evidence for two independent degradative pathways. Plant Physiol Biochem 35:837–846

Jeffrey S, Humphrey GF (1975) New spectrophotometric equations for determining chlorophylls *a*, *b*, c_1 and c_2 in higher plants, algae and natural phytoplankton. Biochem Physiol Pflanz 167:1–194

Jeffrey S, Wright SW (1987) A new spectrally distinct component in preparations of chlorophyll c from the micro-alga Emiliania huxleyi (Prymnesiophycease). Biochim Biophys Acta 894:180–188

Jeffrey SW, Mantoura RFC, Wright SW (Eds) (1997) Phytoplankton Pigments in Oceanography: guidelines to modern methods, UNESCO Publishing

Jeffrey S, Wright S, Zapata M (1999) Recent advances in HPLC pigment analysis of phytoplankton. Mar Freshw Res 50:879–896

Jiang H, Qiu B (2011) Inhibition of photosynthesis by UV-B exposure and its repair in the bloom-forming cyanobacterium Microcystis aeruginosa. J Appl Phycol 23:691–696

Jiao N, Herndl GJ, Hansell DA, Benner R, Kattner G, Wilhelm SW, Kirchman DL, Weinbauer MG, Luo T, Chen F (2010) Microbial production of recalcitrant dissolved organic matter: long-term carbon storage in the global ocean. Nature Rev Microbiol 8:593–599

Johannessen SC, Peña MA, Quenneville ML (2007) Photochemical production of carbon dioxide during a coastal phytoplankton bloom. Estuar Coast Shelf Sci 73:236–242

Jöhnk KD, Huisman J, Sharples J, Sommeijer B, Visser PM, Stroom JM (2008) Summer heatwaves promote blooms of harmful cyanobacteria. Glob Change Biol 14:495–512

Johnson ZI, Shyam R, Ritchie AE, Mioni C, Lance VP, Murray JW, Zinser ER (2010) The effect of iron-and light-limitation on phytoplankton communities of deep chlorophyll maxima of the western Pacific Ocean. J Mar Res 68:283–308

Johnson-Flanagan AM, Spencer MS (1996) Chlorophyllase and peroxidase activity during degreening of maturing canola (*Brassica napus*) and mustard (*Brassica juncea*) seed. Physiol Plant 97:353–359

Jørgensen NOG, Tranvik L, Edling H, Granéli W, Lindell M (1998) Effects of sunlight on occurrence and bacterial turnover of specific carbon and nitrogen compounds in lake water. FEMS Microbiol Ecol 25:217–227

Kahlert M (2002) Horizontal variation of biomass and C: N: P ratios of benthic algae in lakes. Hydrobiologia 489:171–177

Karentz D, Bosch I, Mitchell D (2004) Limited effects of Antarctic ozone depletion on sea urchin development. Mar Biol 145:277–292

Kariola T, Brader G, Li J, Palva ET (2005) Chlorophyllase 1, a damage control enzyme, affects the balance between defense pathways in plants. Plant Cell 17:282–294

Karuppanapandian T, Moon JC, Kim C, Manoharan K, Kim W (2011) Reactive oxygen species in plants: their generation, signal transduction, and scavenging mechanisms. Aust J Crop Sci 5:709–725

Kasprzak P, Padisák J, Koschel R, Krienitz L, Gervais F (2008) Chlorophyll a concentration across a trophic gradient of lakes: an estimator of phytoplankton biomass? Limnologica 38:327–338

Katz JJ, Bowman MK, Michalski TJ, Worcester DL (1991) Chlorophyll aggregation: chlorophylly water micelles as models for in vivo long-wavelength chlorophyll. In: Scheer H (ed) Chlorophylls. CRC Press, Boca Raton, pp 211–235

Keeling RF, Körtzinger A, Gruber N (2010) Ocean deoxygenation in a warming world. Annu Rev Mar Sci 2:199–229

Keely BJ, Maxwell JR (1991) Structural characterization of the major chlorins in a recent sediment. Org Geochem 17:663–669

Keishi S (1979) Preferential degradation of chlorophyll b in ethylene-treated fruits of 'Satsuma'mandarin. Sci Hortic 11:253–256

Kiefer DA, Holm-Hansen O, Goldman CR, Richards R, Berman T (1972) Phytoplankton in Lake Tahoe: deep-living populations. Limnol Oceanogr 17:418–422

Kiefer D, Olson R, Holm-Hansen O (1976) Another look at the nitrite and chlorophyll maxima in the central North Pacific. Deep-Sea Res 23:1199–1208

Kim D, Choi SH, Kim KH, Shim JH, Yoo S, Kim CH (2009) Spatial and temporal variations in nutrient and chlorophyll-a concentrations in the northern East China Sea surrounding Cheju Island. Cont Shelf Res 29:1426–1436

Kimball JW (1979) Biology, 4th edn. Addison-Wesley, Reading
Kimor B, Berman T, Schneller A (1987) Phytoplankton assemblages in the deep chlorophyll maximum layers off the Mediterranean coast of Israel. J Plankton Res 9:433–443
Kinkade C, Marra J, Dickey T, Weller R (2001) An annual cycle of phytoplankton biomass in the Arabian Sea, 1994–1995, as determined by moored optical sensors. Deep Sea Res Part II 48:1285–1301
Kirchman DL, Suzuki Y, Garside C, Ducklow HW (1991) High turnover rates of dissolved organic carbon during a spring phytoplankton bloom. Nature 352:612–614
Kirchman DL, Rich JH, Barber RT (1995) Biomass and biomass production of heterotrophic bacteria along 140°W in the equatorial Pacific: effect of temperature on the microbial loop. Deep Sea Res Part II 42:603–619
Kiss G, Dévai G, Tóthmérész B, Szabó A (2006) Multivariate analysis of long-term water quality changes of shallow Lake Balaton. Verh Int Verein Theoret Angew Limnol 29:2051–2055
Klausmeier CA, Litchman E (2001) Algal games: the vertical distribution of phytoplankton in poorly mixed water columns. Limnol Oceanogr 46:1998–2007
Klug JL (2002) Positive and negative effects of allochthonous dissolved organic matter and inorganic nutrients on phytoplankton growth. Can J Fish Aquat Sci 59:85–95
Knox JP, Dodge AD (1985) Singlet oxygen and plants. Phytochemistry 24:889–896
Koca N, Karadeniz F, Burdurlu HS (2007) Effect of pH on chlorophyll degradation and colour loss in blanched green peas. Food Chem 100:609–615
Komissarov G (1994) Photosynthesis: a new look. Sci Russia 5:52–55
Komissarov G (1995) Photosynthesis as a physicochemical process. Chem Phys Rep 14:1723–1732
Komissarov G (2003) Photosynthesis: the physical-chemical approach. J Adv Chem Phys 2:28–61
Koné V, Machu E, Penven P, Andersen V, Garcon V, Fréon P, Demarcq H (2005) Modeling the primary and secondary productions of the southern Benguela upwelling system: A comparative study through two biogeochemical models. Glob Biogeochem Cy 19:GB4021, doi:101029/102004GB002427
Kratsch H, Wise R (2000) The ultrastructure of chilling stress. Plant Cell Environ 23:337–350
Kräutler B (2003) Chlorophyll breakdown and chlorophyll catabolites. In: Kadish KM, Smith KM, Guilard R (eds) The porphyrin handbook, vol 13. Elsevier Science, Oxford, pp 183–209
Kräutler B, Hörtensteiner S (2006) Chlorophyll catabolites and the biochemistry of chlorophyll breakdown. In: Grimm B, Porra R, Rüdiger W, Scheer H (eds) Chlorophylls and bacteriochlorophylls: biochemistry, biophysics, functions and applications. Springer, Dordrecht, pp 237–260
Kräutler B, Matile P (1999) Solving the riddle of chlorophyll breakdown. Acc Chem Res 32:35–43
Kräutler B, Jaun B, Matile P, Bortlik K, Schellenberg M (1991) On the enigma of chlorophyll degradation: the constitution of a secoporphinoid catabolite. Angew Chem Int Ed Engl 30:1315–1318
Kräutler B, Jaun B, Amrein W, Bortlik K, Schellenberg M, Matile P (1992) Breakdown of chlorophyll: constitution of a secoporphinoid chlorophyll catabolite isolated from senescent barley leaves. Plant Physiol Biochem 30:333–346
Kräutler B, Mühlecker W, Anderl M, Gerlach B (1997) Breakdown of Chlorophyll: Partial synthesis of a putative intermediary catabolite. Preliminary communication. Helvet Chim Acta 80:1355–1362
Kräutler B, Banala S, Moser S, Vergeiner C, Müller T, Lütz C, Holzinger A (2010) A novel blue fluorescent chlorophyll catabolite accumulates in senescent leaves of the peace lily and indicates a split path of chlorophyll breakdown. FEBS Lett 584:4215–4221
Kujawinski EB, Longnecker K, Blough NV, Vecchio RD, Finlay L, Kitner JB, Giovannoni SJ (2009) Identification of possible source markers in marine dissolved organic matter using ultrahigh resolution mass spectrometry. Geochim Cosmochim Acta 73:4384–4399
Kwan WP, Voelker BM (2002) Decomposition of hydrogen peroxide and organic compounds in the presence of dissolved iron and ferrihydrite. Environ Sci Technol 36:1467–1476

Kwan WP, Voelker BM (2003) Rates of hydroxyl radical generation and organic compound oxidation in mineral-catalyzed Fenton-like systems. Environ Sci Technol 37:1150–1158

Laane R, Gieskes W, Kraay G, Eversdijk A (1985) Oxygen consumption from natural waters by photo-oxidizing processes. Neth J Sea Res 19:125–128

Lajollo F, Tannenbaum S, Labuza T (1971) Reaction at limited water concentration. 2. Chlorophyll degradation. J Food Sci 36:850–853

Langston WJ, Chesman B, Burt G, Hawkins S, Readman J, Worsfold P (2003) characterisation of the South West European Marine sites. Summary report. Occas Publ Mar Biol Assoc UK, 111

Laurion I, Ventura M, Catalan J, Psenner R, Sommaruga R (2000) Attenuation of ultraviolet radiation in mountain lakes: Factors controlling the among-and within-lake variability. Limnol Oceanogr 45:1274–1288

Laurion I, Lami A, Sommaruga R (2002) Distribution of mycosporine-like amino acids and photoprotective carotenoids among freshwater phytoplankton assemblages. Aquat Microb Ecol 26:283–294

Law CS, Abraham ER, Watson AJ, Liddicoat MI (2003) Vertical eddy diffusion and nutrient supply to the surface mixed layer of the Antarctic Circumpolar Current. J Geophys Res108, 3272, 14, doi:10.1029/2002JC001604

Leavitt P, Carpenter S (1990) Regulation of pigment sedimentation by photo-oxidation and herbivore grazing. Can J Fish Aquat Sci 47:1166–1176

Lemaire E, Abril G, de Wit R, Etcheber H (2002) Distribution of phytoplankton pigments in nine European estuaries and implications for an estuarine typology. Biogeochemistry 59:5–23

Lesser MP, Barry TM (2003) Survivorship, development, and DNA damage in echinoderm embryos and larvae exposed to ultraviolet radiation (290–400 nm). J Exp Mar Biol Ecol 292:75–91

Lesser MP, Farrell JH, Walker CW (2001) Oxidative stress, DNA damage and p53 expression in the larvae of Atlantic cod (Gadus morhua) exposed to ultraviolet (290–400 nm) radiation. J Exp Biol 204:157–164

Lesser MP, Lamare MD, Barker MF (2004) Transmission of ultraviolet radiation through the Antarctic annual sea ice and its biological effects on sea urchin embryos. Limnol Oceanogr 49:1957–1963

Letelier RM, Karl DM, Abbott MR, Bidigare RR (2004) Light driven seasonal patterns of chlorophyll and nitrate in the lower euphotic zone of the North Pacific subtropical Gyre. Limnol Oceanogr 49:508–519

Leu E, Falk-Petersen S, Hessen DO (2007) Ultraviolet radiation negatively affects growth but not food quality of arctic diatoms. Limnol Oceanogr 52:787–797

Lewis J, William M, Mccutchan J, James H (2010) Ecological responses to nutrients in streams and rivers of the Colorado mountains and foothills. Freshw Biol 55:1973–1983

Li W, Harrison W (2001) Chlorophyll, bacteria and picophytoplankton in ecological provinces of the North Atlantic. Deep Sea Res Part II 48:2271–2293

Li HB, Lv RH, Ding T, Lin Y (2007) Impact of tidal front on the distribution of bacterioplankton in the southern Yellow Sea, China. J Mar Sys 67:263–271

Li G, Gao K, Gao G (2011) Differential impacts of solar UV radiation on photosynthetic carbon fixation from the coastal to offshore surface waters in the South China Sea. Photochem Photobiol 87:329–334

Liang YZ, Brereton RG, Kvalheim OM, Rahmani A (1993) Use of chemometric factor analysis for chromatographic integration: application to diode-array high-performance liquid chromatography of mixtures of chlorophyll a degradation products. Analyst 118:779–790

Lister KN, Lamare MD, Burritt DJ (2010) Sea ice protects the embryos of the Antarctic sea urchin Sterechinus neumayeri from oxidative damage due to naturally enhanced levels of UV-B radiation. J Exp Biol 213:1967–1975

Liu Z, Dreybrodt W, Wang H (2010) A new direction in effective accounting for the atmospheric CO_2 budget: Considering the combined action of carbonate dissolution, the global water cycle and photosynthetic uptake of DIC by aquatic organisms. Earth Sci Rev 99:162–172

Liu X, Lu X, Chen Y (2011) The effects of temperature and nutrient ratios on microcystis blooms in Lake Taihu, China: an 11-year investigation. Harmful Algae 10:337–343

Lizotte MP (2001) The contributions of sea ice algae to Antarctic marine primary production. Am Zool 41:57–73

Lizotte MP, Robinson DH, Sullivan CW (1998) Algal pigment signatures in Antarctic sea ice. In: Lizotte MP, Arrigo KR (eds) Antarctic sea ice: biological processes, interactions and variability. Antarctic Res Ser 73:93–106

Lobanov AV, Rubtsova NA, Vedeneeva YuA, Komissarov GG (2008) Photocatalytic activity of chlorophyll in hydrogen peroxide generation in water. Doklady Chem 421:190–193

Lohman K, Jones JR (1999) Nutrient-sestonic chlorophyll relationships in northern Ozark streams. Can J Fish Aquat Sci 56:124–130

Lohman K, Jones JR, Perkins BD (1992) Effects of nutrient enrichment and flood frequency on periphyton biomass in northern Ozark streams. Can J Fish Aquat Sci 49:1198–1205

Lohrenz SE, Cai WJ, Chen F, Chen X, Tuel M (2010) Seasonal variability in air-sea fluxes of CO_2 in a river-influenced coastal margin. J Geophys Res 115:C10034

Longhurst A, Sathyendranath S, Platt T, Caverhill C (1995) An estimate of global primary production in the ocean from satellite radiometer data. J Plankton Res 17:1245–1271

Longing SD, Haggard BE (2010) Distributions of median nutrient and chlorophyll concentrations across the Red River basin, USA. J Environ Qual 39:1966–1974

Lu Z, Gan J, Dai M, Cheung AYY (2010) The influence of coastal upwelling and a river plume on the subsurface chlorophyll maximum over the shelf of the northeastern South China Sea. J Mar Sys 82:35–46

Lund-Hansen LC, de Amezua Ayala PC, Reglero AF (2006) Bio-optical properties and development of a sub-surface chlorophyll maxima (SCM) in southwest Kattegat, Baltic Sea. Estuar Coast Shelf Sci 68:372–378

Lv J, Wu H, Chen M (2011) Effects of nitrogen and phosphorus on phytoplankton composition and biomass in 15 subtropical, urban shallow lakes in Wuhan, China. Limnologica 41:48–56

Ma X, Green SA (2004) Photochemical transformation of dissolved organic carbon in Lake Superior–an in situ experiment. J Great Lakes Res 30:97–112

Mackey D, Parslow J, Higgins H, Griffiths F, O'Sullivan J (1995) Plankton productivity and biomass in the western equatorial Pacific: biological and physical controls. Deep Sea Res Part II 42:499–533

Maeda Y, Kurata H, Adachi M, Shimokawa K (1998) Chlorophyll catabolism in ethylene-treated Citrus unshiu fruits. J Jpn Soc Hort Sci 67:497–502

Magnuson J, Webster K, Assel R, Bowser C, Dillon P, Eaton J, Evans H, Fee E, Hall R, Mortsch L (1997) Potential effects of climate changes on aquatic systems: Laurentian Great Lakes and Precambrian Shield Region. Hydrol Process 11:825–871

Malkin SY, Guildford SJ, Hecky RE (2008) Modeling the growth response of Cladophora in a Laurentian Great Lake to the exotic invader Dreissena and to lake warming. Limnol Oceanogr 53:1111–1124

Mallin MA (1994) Phytoplankton ecology of North Carolina estuaries. Estuar Coasts 17:561–574

Mangos TJ, Berger RG (1997) Determination of major chlorophyll degradation products. Zeitschrift für Lebensmitteluntersuchung und-Forschung A 204:345–350

Mantoura R, Llewellyn C (1983) The rapid determination of algal chlorophyll and carotenoid pigments and their breakdown products in natural waters by reverse-phase high-performance liquid chromatography. Anal Chim Acta 151:297–314

Maranón E, Cermeno P, Fernández E, Rodríguez J, Zabala L (2004) Significance and mechanisms of photosynthetic production of dissolved organic carbon in a coastal eutrophic ecosystem. Limnol Oceanogr 49:1652–1666

Marañón E, Holligan PM, Varela M, Mouriño B, Bale AJ (2000) Basin-scale variability of phytoplankton biomass, production and growth in the Atlantic Ocean. Deep Sea Res Part I 47:825–857

Marchand D, Rontani JF (2001) Characterisation of photo-oxidation and autoxidation products of phytoplanktonic monounsaturated fatty acids in marine particulate matter and recent sediments. Org Geochem 32:287–304

Marchand D, Marty JC, Miquel JC, Rontani JF (2005) Lipids and their oxidation products as biomarkers for carbon cycling in the northwestern Mediterranean Sea: results from a sediment trap study. Mar Chem 95:129–147
Markager S, Vincent WF (2000) Spectral light attenuation and the absorption of UV and blue light in natural waters. Limnol Oceanogr 45:642–650
Marshall CT, Peters RH (1989) General patterns in the seasonal development of chlorophyll a for temperate lakes. Limnol Oceanogr 34:856–867
Marshall JA, Hovenden M, Oda T, Hallegraeff GM (2002) Photosynthesis does influence superoxide production in the ichthyotoxic alga Chattonella marina (Raphidophyceae). J Plankton Res 24:1231–1236
Martin J, Tremblay JÉ, Gagnon J, Tremblay G, Lapoussière A, Jose C, Poulin M, Gosselin M, Gratton Y, Michel C (2010) Prevalence, structure and properties of subsurface chlorophyll maxima in Canadian Arctic waters. Mar Ecol Prog Ser 412:69–84
Matile P (1997) The vacuole and cell senescence. Adv Bot Res 25:87–112
Matile P, Ginsburg S, Schellenberg M, Thomas H (1988) Catabolites of chlorophyll in senescing barley leaves are localized in the vacuoles of mesophyll cells. PNAS 85:9529–9532
Matile P, Schellenberg M, Peisker C (1992) Production and release of a chlorophyll catabolite in isolated senescent chloroplasts. Planta 187:230–235
Matile P, Hortensteiner S, Thomas H, Krautler B (1996) Chlorophyll breakdown in senescent leaves. Plant Physiol 112:1403–1409
Matile P, Hörtensteiner S, Thomas H (1999) Chlorophyll degradation. Annu Rev Plant Biol 50:67–95
Matondkar P, Nair K, Ansari Z (2005) Biological characteristics of Central Indian Basin waters during the southern summer. Mar Geores Geotechnol 23:299–314
Maurin N, Amblard C, Bourdier G (1997) Phytoplanktonic excretion and bacterial reassimilation in an oligomesotrophic lake: molecular weight fractionation. J Plankton Res 19:1045–1068
McCallister SL, Del Giorgio PA (2008) Direct measurement of the $\delta^{13}C$ signature of carbon respired by bacteria in lakes: Linkages to potential carbon sources, ecosystem baseline metabolism, and CO_2 fluxes. Limnol Oceanogr 53:1204–1216
Mellard JP, Yoshiyama K, Litchman E, Klausmeier CA (2011) The vertical distribution of phytoplankton in stratified water columns. J Theor Biol 269:16–30
Menzel DW (1964) The distribution of dissolved organic carbon in the western Indian ocean. Deep-Sea Res 11:757–765
Merzlyak MN, Gitelson AA, Chivkunova OB, Rakitin VYU (1999) Non-destructive optical detection of pigment changes during leaf senescence and fruit ripening. Physiol Plant 106:135–141
Meyers PA (1997) Organic geochemical proxies of paleoceanographic, paleolimnologic, and paleoclimatic processes. Org Geochem 27:213–250
Micinski E, Ball LA, Zafiriou OC (1993) Photochemical oxygen activation: Superoxide radical detection and production rates in the eastern Caribbean. J Geophys Res 98:2299–2306
Millán-Núñez M, Alvarez-Borrego S, Trees C (1996) Relationship between deep chalrophyll maximum and surface chlorophyll concentration in the California current system. CalCOFI Rep 37:241–250
Miller WL, Moran MA (1997) Interaction of photochemical and microbial processes in the degradation of refractory dissolved organic matter from a coastal marine environment. Limnol Oceanogr 42:1317–1324
Miller WL, Zepp RG (1995) Photochemical production of dissolved inorganic carbon from terrestrial organic matter: Significance to the oceanic organic. Geophys Res Lett 22:417–420
Millie DF, Paerl HW, Hurley JP (1993) Microalgal pigment assessments using high-performance liquid chromatography: a synopsis of organismal and ecological applications. Can J Fish Aquat Sci 50:2513–2527
Millie D, Kirkpatrick G, Vinyard B (1995) Relating photosynthetic pigments and in vivo optical density spectra to irradiance for the Florida red-tide dinoflagellate Gymnodinium breve. Mar Ecol Prog Ser 120:65–75

Millie DF, Schofield OM, Kirkpatrick GJ, Johnsen G, Tester PA, Vinyard BT (1997) Detection of harmful algal blooms using photopigments and absorption signatures: A case study of the Florida red tide dinoflagellate, Gymnodinium breve. Limnol Oceanogr 42:1240–1251
Millie DF, Schofield OME, Kirkpatrick GJ, Johnsen G, Evens TJ (2002) Using absorbance and fluorescence spectra to discriminate microalgae. Eur J Phycol 37:313–322
Miltner RJ (2010) A method and rationale for deriving nutrient criteria for small rivers and streams in Ohio. Environ Manage 45:842–855
Mineeva N, Abramova N (2009) Phytoplankton pigments as ecological state indices of the Cheboksary Reservoir. Water Resour 36:560–567
Mineeva N, Litvinov A, Stepanova I, Kochetkova MY (2008) Chlorophyll content and factors affecting its spatial distribution in the Middle Volga reservoirs. Inland Water Biol 1:64–72
Minguez-Mosquera MI, Garrido-Fernandez J, Gandul-Rojas B (1989) Pigment changes in olives during fermentation and brine storage. J Agric Food Chem 37:8–11
Ming-Yi S, Lee C, Aller RC (1993) Laboratory studies of oxic and anoxic degradation of chlorophyll-a in Long Island Sound sediments. Geochim Cosmochim Acta 57:147–157
Miracle MR, Armengol-Diaz J, Dasi MJ (1993) Extreme meromixis determines strong differential planktonic vertical distributions. Internationale Verh Internat Verein Limnol 25:705–710
Miyashita H, Ikemoto H, Kurano N (1996) Chlorophyll d as a major pigment. Nature 383:402
Miyashita H, Adachi K, Kurano N, Ikemot H, Chihara M, Miyach S (1997) Pigment composition of a novel oxygenic photosynthetic prokaryote containing chlorophyll d as the major chlorophyll. Plant Cell Physiol 38:274–281
Mock T, Gradinger R (1999) Determination of Arctic ice algal production with a new in situ incubation technique. Mar Ecol Prog Ser 177:15–26
Moffett JW, Zafiriou OC (1990) An Investigation of hydrogen peroxide chemistry in surface waters of Vineyard Sound with $H_2{}^{18}O_2$ and $^{18}O_2$. Limnol Oceanogr 35:1221–1229
Moll R, Stoermer E (1982) Hypothesis relating trophic status and subsurface chlorophyll maxima of lakes. Arch Hydrobiol 94:425–440
Moll RA, Brache MZ, Peterson TP (1984) Phytoplankton dynamics within the subsurface chlorophyll maximum of Lake Michigan. J Plankton Res 6:751–766
Molot LA, Hudson JJ, Dillon PJ, Miller SA (2005) Effect of pH on photo-oxidation of dissolved organic carbon by hydroxyl radicals in a coloured, softwater stream. Aquat Sci 67:189–195
Monbet Y (1992) Control of phytoplankton biomass in estuaries: a comparative analysis of microtidal and macrotidal estuaries. Estuar Coasts 15:563–571
Morales CE, Blanco JL, Braun M, Reyes H, Silva N (1996) Chlorophyll-a distribution and associated oceanographic conditions in the upwelling region off northern Chile during the winter and spring 1993. Deep Sea Res Part I 43:267–289
Moreno CM (2012) Hydrogen peroxide production driven by UV-B in planktonic microorganisms: a photocatalytic factor in sea warming and ice melting in regions with ozone depletion? Biogeochemistry 107:1–8
Morgan AM, Royer TV, David MB, Gentry LE (2006) Relationships among nutrients, chlorophyll-, and dissolved oxygen in agricultural streams in Illinois. J Environ Qual 35:1110–1117
Morris DP, Hargreaves BR (1997) The role of photochemical degradation of dissolved organic carbon in regulating the UV transparency of three lakes on the Pocono Plateau. Limnol Oceanogr 42:239–249
Moser S, Müller T, Ebert MO, Jockusch S, Turro NJ, Kräutler B (2008) Blue luminescence of ripening bananas. Angew Chem Int Ed 47:8954–8957
Moser S, Müller T, Oberhuber M, Kräutler B (2009a) Chlorophyll catabolites–chemical and structural footprints of a fascinating biological phenomenon. Eur J Org Chem 2009:21–31
Moser S, Müller T, Holzinger A, Lütz C, Jockusch S, Turro NJ, Kräutler B (2009b) Fluorescent chlorophyll catabolites in bananas light up blue halos of cell death. PNAS 106:15538–15543
Mostofa KMGLC, Wu FC, Fu PQ, Ying WL, Yuan J (2009) Overview of key biogeochemical functions in lake ecosystem: Impacts of organic matter pollution and global warming. In: Proceedings of the 13th world lake conference Wuhan, China, 1–5 Nov 2009, Keynote speech, pp 59–60

Mostofa KMG, Sakugawa H (2009) Spatial and temporal variations and factors controlling the concentrations of hydrogen peroxide and organic peroxides in rivers. Environ Chem 6:524–534

Mostofa KMG, Yoshioka T, Konohira E, Tanoue E, Hayakawa K, Takahashi M (2005) Three-dimensional fluorescence as a tool for investigating the dynamics of dissolved organic matter in the Lake Biwa watershed. Limnology 6:101–115

Mostofa K, Wu FC, Yoshioka T, Sakugawa H, Tanoue E (2009) Dissolved organic matter in the aquatic environments. In: Wu FC, Xing B (eds) Natural organic matter and its significance in the environment. Science Press, Beijing, pp 3–66

Mostofa K, Yoshioka, T, Hayakawa, K, Tanoue, E, Konohira, E, Takahashi, M Distribution and dynamics of chlorophyll a and pheopigments in Lake Biwa: Implications to production of dissolved organic matter (unpublished data)

Mostofa KMG, Wu FC, Liu CQ, Ying WL Characterization of fluorescent dissolved organic matter originated under photoinduced and microbial assimilations of lake algae using EEM-PARAFAC (unpublished data)

Motilva M-J (2008) Chlorophylls—from functionality in food to health relevance, 5th Pigments in Food congress- for quality and health, University of Helsinki

Moum JN, Caldwell DR, Paulson CA (1989) Mixing in the equatorial surface layer and thermocline. J Geophys Res 94(C2):2005–2021

Mühlecker W, Kräutler B (1996) Breakdown of chlorophyll: constituent of nonfluorescing chlorophyll-catabolites from senescent cotyledons of the dicot rape. Plant Physiol Biochem 34:61–75

Mühlecker W, Ongania KH, Kräutler B, Matile P, Hörtensteiner S (1997) Tracking down chlorophyll breakdown in plants: elucidation of the constitution of a "fluorescent" chlorophyll catabolite. Angew Chem Int Ed Engl 36:401–404

Mühlecker W, Kräutler B, Moser D, Matile P, Hörtensteiner S (2000) Breakdown of chlorophyll: a fluorescent chlorophyll catabolite from sweet pepper (*Capsicum annuum*). Helvet Chim Acta 83:278–286

Müller T, Ulrich M, Ongania KH, Kräutler B (2007) Colorless tetrapyrrolic chlorophyll catabolites found in ripening fruit are effective antioxidants. Angew Chem Int Ed 46:8699–8702

Murray DL, Kohorn BD (1991) Chloroplasts of Arabidopsis thaliana homozygous for the ch-1 locus lack chlorophyll b, lack stable LHCPII and have stacked thylakoids. Plant Mol Biol 16:71–79

Nelson JR (1993) Rates and possible mechanism of light-dependent degradation of pigments in detritus derived from phytoplankton. J Mar Res 51:155–179

Nelson JR, Wakeham SG (1989) A phytol-substituted chlorophyll-c from Emiliania huxleyi (prymnesiophyceae). J Phycol 25:761–766

Nelson N, Siegel D, Michaels A (1998) Seasonal dynamics of colored dissolved material in the Sargasso Sea. Deep Sea Res Part I 45:931–957

Nelson NB, Carlson CA, Steinberg DK (2004) Production of chromophoric dissolved organic matter by Sargasso Sea microbes. Mar Chem 89:273–287

Nichols MM, Biggs RB (1985) Estuaries. In: Davis RA Jr (ed) Coastal sedimentary environments, 2nd edn. Springer, New York, pp 77–186

Nieto-Cid M, Alvarez-Salgado X, Pérez F (2006) Microbial and photochemical reactivity of fluorescent dissolved organic matter in a coastal upwelling system. Limnol Oceanogr 51:1391–1400

Norrbin F, Eilertsen HC, Degerlund M (2009) Vertical distribution of primary producers and zooplankton grazers during different phases of the Arctic spring bloom. Deep Sea Res Part II 56:1945–1958

Nusch E, Palme G (1975) Biologische methoden für die praxis der gewässeruntersuchung. GWF-Wasser/Abwasser 116:562–565

Oberhuber M, Kräutler B (2002) Breakdown of chlorophyll: electrochemical bilin reduction provides synthetic access to fluorescent chlorophyll catabolites. Chem Biochem 3:104–107

Oberhuber M, Berghold J, Breuker K, Hörtensteiner S, Kraütler B (2003) Breakdown of chlorophyll: a nonenzymatic reaction accounts for the formation of the colorless "nonfluorescent" chlorophyll catabolites. PNAS 100:6910–6915

Oberhuber M, Berghold J, Kräutler B (2008) Chlorophyll breakdown by a biomimetic route. Angew Chem Int Ed 47:3057–3061

Obernosterer I, Benner R (2004) Competition between biological and photochemical processes in the mineralization of dissolved organic carbon. Limnol Oceanogr 49:117–124

Oda T, Nakamura A, Okamoto T, Ishimatsu A, Muramatsu T (1998) Lectin-induced enhancement of superoxide anion production by red tide phytoplankton. Mar Biol 131:383–390

Odate T, Furuya K (1998) Well-developed subsurface chlorophyll maximum near Komahashi No. 2 Seamount in the summer of 1991. Deep Sea Res Part I 45:1595–1607

Ogawa H, Amagai Y, Koike I, Kaiser K, Benner R (2001) Production of refractory dissolved organic matter by bacteria. Science 292:917–920

Okada K, Inoue Y, Satoh K, Katoh S (1992) Effects of light on degradation of chlorophyll and proteins during senescence of detached rice leaves. Plant Cell Physiol 33:1183–1191

Olson MB, Strom SL (2002) Phytoplankton growth, microzooplankton herbivory and community structure in the southeast Bering Sea: insight into the formation and temporal persistence of an Emiliania huxleyi bloom. Deep Sea Res Part II 49:5969–5990

Omar A, Olsen A, Johannessen T, Hoppema M, Thomas H, Borges A (2010) Spatiotemporal variations of fCO_2 in the North Sea. Ocean Sci 6:77–89

Ondrusek ME, Bidigare RR, Sweet ST, Defreitas DA, Brooks JM (1991) Distribution of phytoplankton pigments in the North Pacific Ocean in relation to physical and optical variability. Deep Sea Res Part I 38:243–266

O'Reilly CM, Alin SR, Plisnier PD, Cohen AS, McKee BA (2003) Climate change decreases aquatic ecosystem productivity of Lake Tanganyika, Africa. Nature 424:766–768

Owens TG, Falkowskit PG (1982) Enzymatic degradation of chlorophyll *a* by marine phytoplankton in vitro. Phytochemistry 21:979–984

Owens TG, Gallagher JC, Alberte RS (1987) Photosynthetic light-harvesting function of violoxanthin in Nannochloropsis spp (Eustigmagtophyceae). J Phycol 23:79–85

Paerl HW, Huisman J (2009) Climate change: a catalyst for global expansion of harmful cyanobacterial blooms. Environ Microbiol Rep 1:27–37

Palenik B, Morel F (1988) Dark production of H_2O_2 in the Sargasso Sea. Limnol Oceanogr 33:1606–1611

Palenik B, Zafiriou O, Morel F (1987) Hydrogen peroxide production by a marine phytoplankter. Limnol Oceanogr 32:1365–1369

Palmer-Felgate EJ, Jarvie HP, Williams RJ, Mortimer RJG, Loewenthal M, Neal C (2008) Phosphorus dynamics and productivity in a sewage-impacted lowland chalk stream. J Hydrol 351:87–97

Palmisano A, Sullivan C (1983) Sea ice microbial communities (SIMCO). Polar Biol 2:171–177

Palmisano A, Kottmeier S, Moe RL, Sullivan C (1985) Sea ice microbial communities. IV. The effect of light perturbation on microalgae at the ice-seawater interface in McMurdo Sound, Antarctica. Mar Ecol Prog Ser 21:37–45

Pan BZ, Wang HJ, Liang XM, Wang HZ (2009) Factors influencing chlorophyll a concentration in the Yangtze-connected lakes. Fres Environ Bull 18:1894–1900

Parab SG, Prabhu Matondkar S, Gomes HR, Goes J (2006) Monsoon driven changes in phytoplankton populations in the eastern Arabian Sea as revealed by microscopy and HPLC pigment analysis. Continent Shelf Res 26:2538–2558

Park S, Chandra S, Müller-Navarra DC, Goldman CR (2004) Diel and vertical variability of seston food quality and quantity in a small subalpine oligomesotrophic lake. J Plankton Res 26:1489–1498

Park SY, Yu JW, Park JS, Li J, Yoo SC, Lee NY, Lee SK, Jeong SW, Seo HS, Koh HJ (2007) The senescence-induced staygreen protein regulates chlorophyll degradation. Plant Cell 19:1649–1664

Parlanti E, Wörz K, Geoffroy L, Lamotte M (2000) Dissolved organic matter fluorescence spectroscopy as a tool to estimate biological activity in a coastal zone submitted to anthropogenic inputs. Org Geochem 31:1765–1781

Parsons TR, Strickland JDH (1963) Discussion of spectrophotometric determination of marine-plant pigments, with revised equations for ascertaining chlorophylls and carotenoids. J Mar Res 21:155–163

Pattanaik B, Roleda MY, Schumann R, Karsten U (2008) Isolate-specific effects of ultraviolet radiation on photosynthesis, growth and mycosporine-like amino acids in the microbial mat-forming cyanobacterium Microcoleus chthonoplastes. Planta 227:907–916

Paulsen H, Finkenzeller B, Kühlein N (1993) Pigments induce folding of light-harvesting chlorophyll a/b-binding protein. Eur J Biochem 215:809–816

Payri CE, Maritorena S, Bizeau C, Rodière M (2001) Photoacclimation in the tropical coralline alga Hydrolithon onkodes (Rhodophyta, Corallinaceae) from a French Polynesian reef. J Phycol 37:223–234

Pearlstein RM (1996) Coupling of exciton motion in the core antenna and primary charge separation in the reaction center. Photosynth Res 48:75–82

Pedros-Alio C, Gasol JM, Guerrero R (1987) On the ecology of a Cryptomonas phaseolus population forming a metalimnetic bloom in Lake Cisó, Spain: Annual distribution and loss factors. Limnol Oceanogr 32:285–298

Pennock JR (1985) Chlorophyll distributions in the Delaware estuary: regulation by light-limitation. Estuar Coast Shelf Sci 21:711–725

Pérez GL, Queimaliños CP, Modenutti BE (2002) Light climate and plankton in the deep chlorophyll maxima in North Patagonian Andean lakes. J Plankton Res 24:591–599

Pérez V, Fernández E, Marañón E, Morán XAG, Zubkov MV (2006) Vertical distribution of phytoplankton biomass, production and growth in the Atlantic subtropical gyres. Deep Sea Res Part I 53:1616–1634

Pérez G, Queimalinos C, Balseiro E, Modenutti B (2007) Phytoplankton absorption spectra along the water column in deep North Patagonian Andean lakes (Argentina). Limnologica 37:3–16

Perovich D (1993) A theoretical model of ultraviolet light transmission through Antarctic sea ice. J Geophys Res 98(22579–22522):22587

Petasne RG, Zika RG (1987) Fate of superoxide in coastal sea water. Nature 325:516–518

Petasne RG, Zika RG (1997) Hydrogen peroxide lifetimes in south Florida coastal and offshore waters. Mar Chem 56:215–225

Pick F, Lean D, Nalewajko C (1984) Nutrient status of metalimnetic phytoplankton peaks. Limnol Oceanogr 29:960–971

Pietta P, Rava A, Biondi P (1981) High-performance liquid chromatography of nifedipine, its metabolites and photochemical degradation products. J Chromatogr 210:516–521

Pizarro G, Iriarte JL, Montecino V, Blanco JL, Guzman L (2000) Distribuci6n de la biomasa fitoplancto′nica y productividad primaria máxima de fiordos y canales australes (47–50º S) en octubre 1996. Cienc Tecnol Mar 23:25–48

Planas D, Agusti S, Duarte CM, Granata TC, Merino M (1999) Nitrate uptake and diffusive nitrate supply in the Central Atlantic. Limnol Oceanogr 49:116–126

Polle JEW, Benemann JR, Tanaka A, Melis A (2000) Photosynthetic apparatus organization and function in the wild type and a chlorophyll b-less mutant of Chlamydomonas reinhardtii. Dependence on carbon source. Planta 211:335–344

Premkumar J, Ramaraj R (1999) Photoreduction of dioxygen to hydrogen peroxide at porphyrins and phthalocyanines adsorbed Nafion membrane. J Mol Catal A 142:153–162

Prezelin BB (1981) Light reactions in photosynthesis. Can Bull Fish Aquat Sci 210:1–43

Pruzinská A, Tanner G, Aubry S, Anders I, Moser S, Müller T, Ongania KH, Kräutler B, Youn JY, Liljegren SJ (2005) Chlorophyll breakdown in senescent Arabidopsis leaves. Characterization of chlorophyll catabolites and of chlorophyll catabolic enzymes involved in the degreening reaction. Plant Physiol 139:52–63

Qiu D, Huang L, Zhang J, Lin S (2010) Phytoplankton dynamics in and near the highly eutrophic Pearl River Estuary, South China Sea. Cont Shelf Res 30:177–186

Rae R, Howard-Williams C, Hawes I, Schwarz AM, Vincent WF (2001) Penetration of solar ultraviolet radiation into New Zealand lakes: influence of dissolved organic carbon and catchment vegetation. Limnology 2:79–89
Rath J, Adhikary SP (2007) Response of the estuarine cyanobacterium Lyngbya aestuarii to UV-B radiation. J Appl Phycol 19:529–536
Renger T, Marcus R (2002) Photophysical properties of PS-2 reaction centers and a discrepancy in exciton relaxation times. J Phys Chem B 106:1809–1819
Reul A, Rodríguez V, Jiménez-Gómez F, Blanco J, Bautista B, Sarhan T, Guerrero F, Ruíz J, Garcia-Lafuente J (2005) Variability in the spatio-temporal distribution and size-structure of phytoplankton across an upwelling area in the NW-Alboran Sea (W-Mediterranean). Cont Shelf Res 25:589–608
Richards FA, Thompson TG (1952) The estimation and characterization of plankton populations by pigment analysis. II. A spectrophotometric method for the estimation of plankton pigments. J Mar Res 11:156–172
Riley GA, Stommel HM, Bumpus DF (1949) Quantitative ecology of the plankton of the western North Atlantic. Bull Bingham Oceanogr Coll 12:1–169
Rines J, Donaghay P, Dekshenieks M, Sullivan J, Twardowski M (2002) Thin layers and camouflage: hidden Pseudo-nitzschia spp. (Bacillariophyceae) populations in a fjord in the San Juan Islands, Washington, USA. Mar Ecol Prog Ser 225:123–137
Rochelle-Newall E, Fisher T (2002) Production of chromophoric dissolved organic matter fluorescence in marine and estuarine environments: an investigation into the role of phytoplankton. Mar Chem 77:7–21
Rodoni S, Muhlecker W, Anderl M, Krautler B, Moser D, Thomas H, Matile P, Hortensteiner S (1997) Chlorophyll breakdown in senescent chloroplasts (cleavage of pheophorbide a in two enzymic steps). Plant Physiol 115:669–676
Rojo C, Miracle MR (1987) Poblaciones fitoplanctónicas de la Laguna de la Cruz (Cuenca), una laguna cárstica meromíctica. Act VI Simp Nac Bot Crip 119–135
Rolff C (2000) Seasonal variation in delta $\delta^{15}C$ and $\delta^{15}N$ of size-fractionated plankton at a coastal station in the northern Baltic proper. Mar Ecol Prog Ser 203:47–65
Rontani JF (2001) Visible light-dependent degradation of lipidic phytoplanktonic components during senescence: a review. Phytochemistry 58:187–202
Rontani JF (2008) Photooxidative and autoxidative degradation of lipid components during the senescence of phototrophic organisms. In: Matsumoto T (ed) Phytochemistry research progress. Nova Science Publishers, pp 115–144
Rontani JF, Marchand D (2000) Photoproducts of phytoplanktonic sterols: a potential source of hydroperoxides in marine sediments? Org Geochem 31:169–180
Rontani JF, Volkman JK (2003) Phytol degradation products as biogeochemical tracers in aquatic environments. Org Geochem 34:1–35
Rontani J, Grossi V, Faure R, Aubert C (1994) "Bound" 3-methylidene-7, 11, 15-trimethylhexadecan-1, 2-diol: a new isoprenoid marker for the photodegradation of chlorophyll-*a* in seawater. Org Geochem 21:135–142
Rontani JF, Beker B, Raphel D, Baillet G (1995) Photodegradation of chlorophyll phytyl chain in dead phytoplanktonic cells. J Photochem Photobiol 85:137–142
Rontani JF, Cuny P, Grossi V (1998) Identification of a pool of lipid photoproducts in senescent phytoplanktonic cells. Org Geochem 29:1215–1225
Rontani JF, Perrote S, Cuny P (2000) Can a high chlorophyllase activity bias the use of the phytyldiol versus phytol ratio (CPPI) for the monitoring of chlorophyll photooxidation in seawater? Org Geochem 31:91–99
Rontani JF, Rabourdin A, Marchand D, Aubert C (2003) Photochemical oxidation and autoxidation of chlorophyll phytyl side chain in senescent phytoplanktonic cells: potential sources of several acyclic isoprenoid compounds in the marine environment. Lipids 38:241–254
Rontani JF, Zabeti N, Wakeham S (2011) Degradation of particulate organic matter in the equatorial Pacific Ocean: biotic or abiotic? Limnol Oceanogr 56:333–349
Rowan KS (1989) Photosynthetic pigments of algae. Cambridge University Press, New York

Roy R, Pratihary A, Mangesh G, Naqvi S (2006) Spatial variation of phytoplankton pigments along the southwest coast of India. Estuar Coast Shelf Sci 69:189–195

Royer TV, David MB, Gentry LE, Mitchell CA, Starks KM, Heatherly T, Whiles MR (2008) Assessment of chlorophyll-as a criterion for establishing nutrient standards in the streams and rivers of Illinois. J Environ Qual 37:437–447

Rüdiger W (2003) The last step of chlorophyll synthesis. The last step of chlorophyll synthesis In: Kadish KM, Smith KM, Guilard R (eds) The porphyrin handbook, Elsevier Science, Amsterdam, pp 71–108

Ryabov AB, Rudolf L, Blasius B (2010) Vertical distribution and composition of phytoplankton under the influence of an upper mixed layer. J Theor Biol 263:120–133

Sabater S, Armengol J, Comas E, Sabater F, Urrizalqui I, Urrutia I (2000) Algal biomass in a disturbed Atlantic river: water quality relationships and environmental implications. Sci Total Environ 263:185–195

Sakshaug E, Holm-Hansen O (1986) Photoadaptation in Antarctic phytopfankton: variations in growth rate, chemical composition and P versus I curves. J Plankton Res 8:459–473

Sandu C, Iacob R, Nicolescu N (2003) Chlorophyll-a determination-a reliable method for phytoplankton biomass assessment. Acta Botanica Hungarica 45:389–397

Sanger JE, Gorham E (1970) The diversity of pigments in lake sediments and its ecological significance. Limnol Oceanogr 15:59–69

Sarmiento JL, Hughes TMC, Stouffer RJ, Manabe S (1998) Simulated response of the ocean carbon cycle to anthropogenic climate warming. Nature 393:245–249

Sarmiento JL, Slater R, Barber R, Bopp L, Doney SC, Hirst A, Kleypas J, Matear R, Mikolajewicz U, Monfray P (2004) Response of ocean ecosystems to climate warming. Glob Biogeochem Cy 18:3001–3023

Sarthou G, Jeandel C, Brisset L, Amouroux D, Besson T, Donard OFX (1997) Fe and H2O2 distributions in the upper water column in the Indian sector of the Southern Ocean. Earth Planet Sci Lett 147:83–92

Sasaki H, Miyamura T, Saitoh S, Ishizaka J (2005) Seasonal variation of absorption by particles and colored dissolved organic matter (CDOM) in Funka Bay, southwestern Hokkaido, Japan. Estuar Coast Shelf Sci 64:447–458

Sasaoka K, Saitoh S, Asanuma I, Imai K, Honda M, Nojiri Y, Saino T (2002) Temporal and spatial variability of chlorophyll-a in the western subarctic Pacific determined from satellite and ship observations from 1997 to 1999. Deep Sea Res Part II 49:5557–5576

Sathyendranath S, Cota G, Stuart V, Maass H, Platt T (2001) Remote sensing of phytoplankton pigments: a comparison of empirical and theoretical approaches. Int J Remote Sens 22:249–273

Satoh S, Ikeuchi M, Mimuro M, Tanaka A (2001) Chlorophyll b expressed in cyanobacteria functions as a light-harvesting antenna in photosystem I through flexibility of the proteins. J Biol Chem 276:4293–4297

Satoh Y, Katano T, Satoh T, Mitamura O, Anbutsu K, Nakano S, Ueno H, Kihira M, Drucker V, Tanaka Y (2006) Nutrient limitation of the primary production of phytoplankton in Lake Baikal. Limnology 7:225–229

Sawatzky CL, Wurtsbaugh WA, Luecke C (2006) The spatial and temporal dynamics of deep chlorophyll layers in high-mountain lakes: effects of nutrients, grazing and herbivore nutrient recycling as growth determinants. J Plankton Res 28:65–86

Schelbert S, Aubry S, Burla B, Agne B, Kessler F, Krupinska K, Hörtensteiner S (2009) Pheophytin pheophorbide hydrolase (pheophytinase) is involved in chlorophyll breakdown during leaf senescence in Arabidopsis. Plant Cell 21:767–785

Scheumann V, Schoch S, Rüdiger W (1999) Chlorophyll *b* reduction during senescence of barley seedlings. Planta 209:364–370

Schindler DW (1997) Widespread effects of climatic warming on freshwater ecosystems in North America. Hydrol Process 11:1043–1067

Schmid H, Bauer F, Stich HB (1998) Determination of algal biomass with HPLC pigment analysis from lakes of different trophic state in comparison to microscopically measured biomass. J Plankton Res 20:1651–1661

Schmittner A (2005) Decline of the marine ecosystem caused by a reduction in the Atlantic overturning circulation. Nature 434:628–633
Schmittner A, Galbraith ED, Hostetler SW, Pedersen TF, Zhang R (2007) Large fluctuations of dissolved oxygen in the Indian and Pacific oceans during Dansgaard-Oeschger oscillations caused by variations of North Atlantic Deep Water subduction. Paleoceanography 22:PA3207
Schubel J (1971) Tidal variation of the size distribution of suspended sediment at a station in the Chesapeake Bay turbidity maximum. Neth J Sea Res 5:252–266
Schulte-Elte KH, Muller BL, Pamingle H (1979) Photooxygenation of 3, 3-dialkylsubstituted allyl alcohols. Occurrence of syn preference in the ene addition of 1O2 at E/Z-isomeric allyl alcohols. Helvet Chim Acta 62:816–829
Schwartz S, Lorenzo T (1991) Chlorophyll stability during continuous aseptic processing and storage. J Food Sci 56:1059–1062
Schwartz S, von Elbe J (1983) Kinetics of chlorophyll degradation to pyropheophytin in vegetables. J Food Sci 48:1303–1306
Scully NM, Cooper WJ, Tranvik LJ (2003a) Photochemical effects on microbial activity in natural waters: the interaction of reactive oxygen species and dissolved organic matter. FEMS Microbiol Ecol 46:353–357
Scully NM, Tranvik LJ, Cooper WJ (2003b) Photochemical effects on the interaction of enzymes and dissolved organic matter in natural waters. Limnol Oceanogr 48:1818–1824
Scurlock J, Cramer W, Olson R, Parton W, Prince S (1999) Terrestrial NPP: towards a consistent data set for global model evaluation. Ecol Appl 9:913–919
Senesi N (1990) Molecular and quantitative aspects of the chemistry of fulvic acid and its interactions with metal ions and organic chemicals: part II. The fluorescence spectroscopy approach. Anal Chim Acta 232:77–106
Seppälä J, Ylöstalo P, Kaitala S, Hällfors S, Raateoja M, Maunula P (2007) Ship-of-opportunity based phycocyanin fluorescence monitoring of the filamentous cyanobacteria bloom dynamics in the Baltic Sea. Estuar Coast Shelf Sci 73:489–500
Shank GC, Zepp RG, Whitehead RF, Moran MA (2005) Variations in the spectral properties of freshwater and estuarine CDOM caused by partitioning onto river and estuarine sediments. Estuar Coast Shelf Sci 65:289–301
Shioi Y, Tatsumi Y, Shimokawa K (1991) Enzymatic degradation of chlorophyll in Chenopodium album. Plant Cell Physiol 32:87–93
Shlgren G (1983) Comparison of methods for estimation of phytoplankton carbon. Arch Hydrobiol 98:489–508
Shuman FR, Lorenzen CJ (1975) Quantitative degradation of chlorophyll by a marine herbivore. Limnol Oceanogr 20:580–586
Sigareva L, Pyrina I (2006) Plant pigments as indicators of water transformation in the upper Volga Chain of reservoirs. Water Resour 33:436–445
Silsbe G, Hecky R, Guildford S, Mugidde R (2006) Variability of chlorophyll a and photosynthetic parameters in a nutrient-saturated tropical great lake. Limnol Oceanogr 51:2052–2063
Sinha RP, Häder DP (2002) UV-induced DNA damage and repair: a review. Photochem Photobiol Sci 1:225–236
Smart CM (1994) Gene expression during leaf senescence. New Phytol 126:419–448
Smith C, Alberte R (1994) Characterization of in vivo absorption features of chlorophyte, phaeophyte and rhodophyte algal species. Mar Biol 118:511–521
Smith RC, Prezelin B, Baker K, Bidigare R, Boucher N, Coley T, Karentz D, MacIntyre S, Matlick H, Menzies D (1992) Ozone depletion: ultraviolet radiation and phytoplankton biology in Antarctic waters. Science 255:952–959
Soma Y, Imaizumt T, Yagi K, Kasuga S (1993) Estimation of algal succession in lake water using HPLC analysis of pigments. Can J Fish Aqual Sci 50:1142–1146
Sommaruga R, Augustin G (2006) Seasonality in UV transparency of an alpine lake is associated to changes in phytoplankton biomass. Aquat Sci 68:129–141
Spies A (1987) Growth rates of Antarctic marine phytoplankton in the Weddell Sea. Mar Ecol Prog Ser 41:267–274
Steele JH (1964) A study of production in the Gulf of Mexico. J Mar Res 22:211222

Steele J, Yentsch C (1960) The vertical distribution of chlorophyll. J Mar Biol Assoc UK 39:217–226
Steemann-Nielsen E (1962) On the maximum quantity of plankton chlorophyll per surface unit of a lake or the sea. Int Rev Ges Hydrobiol 47:333–338
Steet JA, Tong CH (1996) Quantification of color change resulting from pheophytinization and nonenzymatic browning reactions in thermally processed green peas. J Agric Food Chem 44:1531–1537
Steinhart G, Gross H, Budy P, Luecke C, Wurtsbaugh W (1994) Limnological investigations and hydroacoustic surveys of Sawtooth Valley Lakes. In: Tuescher D, Taki D, Wurtsbaugh WA, Luecke C, Budy P, Gross HP, Steinhart G (eds) Snake River sockeye salmon habitat and limnological researchannual report 1993, Publ No DOE/BP-22548-2 US Department of Energy. Bonneville Power Administration, Portland, Oreg, pp 30–61
Stephens MP, Kadko DC, Smith CR, Latasa M (1997) Chlorophyll-a and pheopigments as tracers of labile organic carbon at the central equatorial Pacific seafloor. Geochim Cosmochim Acta 61:4605–4619
Stets EG, Striegl RG, Aiken GR, Rosenberry DO, Winter TC (2009) Hydrologic support of carbon dioxide flux revealed by whole-lake carbon budgets. J Geophys Res 114:G01008. doi:10 1029/102008JG000783
Stocker R, Yamamoto Y, McDonagh AF, Glazer AN, Ames BN (1987) Bilirubin is an antioxidant of possible physiological importance. Science 235:1043–1046
Straškrábová V, Izmest'yeva L, Maksimova E, Fietz S, Nedoma J, Borovec J, Kobanova G, Shchetinina E, Pislegina E (2005) Primary production and microbial activity in the euphotic zone of Lake Baikal (Southern Basin) during late winter. Glob Planet Change 46:57–73
Striebel M, Spörl G, Stibor H (2008) Light-induced changes of plankton growth and stoichiometry: experiments with natural phytoplankton communities. Limnol Oceanogr 53:513–522
Stross RG, Stottlemyer JR (1965) Primary production in the Patuxent River. Chesapeake Sci 6:125–140
Sugiyama Y, Anegawa A, Kumagai T, Harita Y, Hori T, Sugiyama M (2004) Distribution of dissolved organic carbon in lakes of different trophic types. Limnology 5:165–176
Sun MY, Lee C, Aller RC (1993) Anoxic and oxic degradation of14C-labeled chloropigments and a 14C-labeled diatom in long island sound sediments. Limnol Oceanogr 38:1438–1451
Sun MY, Cai WJ, Joye SB, Ding H, Dai J, Hollibaugh JT (2002) Degradation of algal lipids in microcosm sediments with different mixing regimes. Org Geochem 33:445–459
Suzuki Y, Shioi Y (1999) Detection of chlorophyll breakdown products in the senescent leaves of higher plants. Plant Cell Physiol 40:909–915
Sweeney J, Martin M (1961) Stability of chlorophyll in vegetables as affected by pH. Food Technol 15:263–266
Takahashi T, Sutherland SC, Sweeney C, Poisson A, Metzl N, Tilbrook B, Bates N, Wanninkhof R, Feely RA, Sabine C (2002) Global sea-air CO_2 flux based on climatological surface ocean pCO_2, and seasonal biological and temperature effects. Deep Sea Res Part II 49:1601–1622
Takamiya K, Tsuchiya T, Ohta H (2000) Degradation pathway (s) of chlorophyll: what has gene cloning revealed? Trends Plant Sci 5:426–431
Talling J (1993) Comparative seasonal changes, and inter-annual variability and stability, in a 26-year record of total phytoplankton biomass in four English lake basins. Hydrobiologia 268:65–98
Tanaka A, Tanaka R (2006) Chlorophyll metabolism. Curr Opin Plant Biol 9:248–255
Telfer A, He WZ, Barber J (1990) Spectral resolution of more than one chlorophyll electron donor in the isolated Photosystem II reaction centre complex. Biochim Biophys Acta 1017:143–151
Terao T, Katoh S (1989) Synthesis and breakdown of the apoproteins of light-harvesting chlorophyll a/b proteins in chlorophyll b-deficient mutants of rice. Plant Cell Physiol 30:571–580
Thornber JP, Highkin HR (1974) Composition of the photosynthetic apparatus of normal barley leaves and a mutant lacking chlorophyll *b*. Eur J Biochem 41:109–116

Tittel J, Bissinger V, Zippel B, Gaedke U, Bell E, Lorke A, Kamjunke N (2003) Mixotrophs combine resource use to outcompete specialists: implications for aquatic food webs. PNAS 100:12776–12781
Tranvik LJ (1993) Microbial transformation of labile dissolved organic matter into humic-like matter in seawater. FEMS Microbiol Ecol 12:177–183
Trebitsh T, Goldschmidt EE, Riov J (1993) Ethylene induces de novo synthesis of chlorophyllase, a chlorophyll degrading enzyme, in citrus fruit peel. PNAS 90:9441–9445
Treibs A (1936) Chlorophyll and hemin derivatives in organic materials. Angew Chem 49:682–686
Tremblay JE, Michel C, Hobson KA, Gosselin M, Price NM (2006) Bloom dynamics in early opening waters of the Arctic Ocean. Limnol Oceanogr 51:900–912
Trodahl H, Buckley R (1989) Ultraviolet levels under sea ice during the Antarctic spring. Science 245:194–195
Trüper HG (1987) Phototrophic bacteria (an incoherent group of prokaryotes). A taxonomic versus phylogenetic survey. Microbiologia 3:71–89
Tsuchiya T, Ohta H, Okawa K, Iwamatsu A, Shimada H, Masuda T, Takamiya K (1999) Cloning of chlorophyllase, the key enzyme in chlorophyll degradation: finding of a lipase motif and the induction by methyl jasmonate. PNAS 96:15362–15367
van Nieuwenhuyse EE, Jones JR (1996) Phosphorus-chlorophyll relationship in temperate streams and its variation with stream catchment area. Can J Fish Aquat Sci 53:99–105
Vardi A, Schatz D, Beeri K, Motro U, Sukenik A, Levine A, Kaplan A (2002) Dinoflagellate-cyanobacterium communication may determine the composition of phytoplankton assemblage in a mesotrophic lake. Curr Biol 12:1767–1772
Varela R, Cruzado A, Tintoré J (1994) A simulation analysis of various biological and physical factors influencing the deep-chlorophyll maximum structure in oligotrophic areas. J Mar Sys 5:143–157
Varela M, Fernandez E, Serret P (2002) Size-fractionated phytoplankton biomass and primary production in the Gerlache and south Bransfield Straits (Antarctic Peninsula) in Austral summer 1995–1996. Deep Sea Res Part II 49:749–768
Vaulot D, Partensky F, Neveux J, Mantoura RFC, Llewellyn CA (1990) Winter presence of prochlorophytes in surface waters of the northwestern Mediterranean Sea. Limnol Oceanogr 35:1156–1164
Veldhuis MJW, Kraay GW (1990) Vertical distribution and pigment composition of a picoplanktonic prochlorophyte in the subtropical North Atlantic: a combined study of HPLC-analysis of pigments and flow cytometry. Mar Ecol Prog Ser 68:121–127
Velo-Suárez L, Fernand L, Gentien P, Reguera B (2010) Hydrodynamic conditions associated with the formation, maintenance and dissipation of a phytoplankton thin layer in a coastal upwelling system. Cont Shelf Res 30:193–202
Venrick E (1993) Phytoplankton seasonality in the central North Pacific: the endless summer reconsidered. Limnol Oceanogr 38:1135–1149
Verburg P, Hecky RE, Kling H (2003) Ecological consequences of a century of warming in Lake Tanganyika. Science 301:505–507
Verlencar X, Somasunder K, Qasim S (1990) Regeneration of nutrients and biological productivity in Antarctic waters. Mar Ecol Prog Ser 61:41–59
Verne-Mismer J, Ocampo R, Callot H, Albrecht P (1988) Molecular fossils of chlorophyll c of the 17-nor-DPEP Series. Structure determination, synthesis, geochemical significance. Tetrah Lett 29:371–374
Verne-Mismer J, Ocampo R, Callot H, Albrecht P (1990) New chlorophyll fossils from moroccan oil shales. Porphyrins derived from chlorophyll C3 or a related pigment? Tetrah Lett 31:1751–1754
Vicente E, Miracle M (1984) Distribution of photosynthetic organisms in a temporal stratified karstic pond near Cuenca (Spain). Verh Internat Verein Limnol 22:1504–1710
Viličić D, Legović T, Žutić V (1989) Vertical distribution of phytoplankton in a stratified estuary. Aquat Sci 51:31–46

Vincent WF (1983) Phytoplankton production and winter mixing: contrasting effects in two oligotrophic lakes. J Ecol 71:1–20
Vincent W, Gibbs M, Dryden S (1984) Accelerated eutrophication in a New Zealand lake: Lake Rotoiti, central North Island. NZ J Mar Freshw Res 18:431–440
Volkman JK (1986) A review of sterol markers for marine and terrigenous organic matter. Org Geochem 9:83–99
Volkman J, Burton H, Everitt D, Allen D (1988) Pigment and lipid compositions of algal and bacterial communities in Ace Lake, Vestfold Hills, Antarctica. Hydrobiologia 165:41–57
von Elbe JH, Schwartz SJ (1996) Colorants. In: Fennema OR (ed) Food chemistry. Marcel Dekker, New York, pp 651–722
Wakeham SG (1995) Lipid biomarkers for heterotrophic alteration of suspended particulate organic matter in oxygenated and anoxic water columns of the ocean. Deep Sea Res Part I 42:1749–1771
Wakeham S, Lee C (1993) Production, transport, and alteration of particulate organic matter in the marine water column. In: Engel M, Macko S (eds) Org Geochem. Plenum Press, New York, pp 145–169
Walsby AE, Hayes PK, Boje R, Stal LJ (1997) The selective advantage of buoyancy provided by gas vesicles for planktonic cyanobacteria in the Baltic Sea. New Phytol 136:407–417
Wang F, Wang X, Zhao Y, Yang Z (2012) Long-term changes of water level associated with chlorophyll *a* concentration in Lake Baiyangdian, North China. Proc Environ Sci 13:1227–1237
Watson S, McCauley E (1988) Contrasting patterns of net-and nanoplankton production and biomass among lakes. Can J Fish Aquat Sci 45:915–920
Watson S, McCauley E, Downing JA (1992) Sigmoid relationships between phosphorus, algal biomass, and algal community structure. Can J Fish Aquat Sci 49:2605–2610
Weber TS, Deutsch C (2010) Ocean nutrient ratios governed by plankton biogeography. Nature 467:550–554
Welch EB, Lindell T (1992) Ecological effects of wastewater: applied limnology and pollution effects. Chap-man and Hall, London
Welschmeyer N, Lorenzen C (1985) Role of herbivory in controlling phytoplankton abundance: annual pigment budget for a temperate marine fjord. Mar Biol 90:75–86
Wentworth AD, Jones LH, Wentworth P, Janda KD, Lerner RA (2000) Antibodies have the intrinsic capacity to destroy antigens. PNAS 97:10930–10935
Wentworth P Jr, Jones LH, Wentworth AD, Zhu X, Larsen NA, Wilson IA, Xu X, Goddard WA III, Janda KD, Eschenmoser A (2001) Antibody catalysis of the oxidation of water. Science 293:1806–1811
Weston K, Fernand L, Mills D, Delahunty R, Brown J (2005) Primary production in the deep chlorophyll maximum of the central North Sea. J Plankton Res 27:909–922
Westrich JT, Berner RA (1984) The role of sedimentary organic matter in bacterial sulfate reduction: the G model tested. Limnol Oceanogr 29:236–249
Wheeler PA, Gosselin M, Sherr E, Thibaultc D, Kirchman DL, Benner R, Whitledge TE (1996) Active cycling of organic carbon in the central Arctic Ocean. Nature 380:697–699
White EM, Kieber DJ, Sherrard J, Miller WL, Mopper K (2010) Carbon dioxide and carbon monoxide photoproduction quantum yields in the Delaware estuary. Mar Chem 118:11–21
Whitehouse M, Korb R, Atkinson A, Thorpe S, Gordon M (2008) Formation, transport and decay of an intense phytoplankton bloom within the high-nutrient low-chlorophyll belt of the Southern Ocean. J Mar Sys 70:150–167
Wilhelm C, Rudolph I, Renner W (1991) A quantitative method based on HPLC-aided pigment analysis to monitor structure and dynamics of the phytoplankton assemblage–a study from Lake Meerfelder Maar (Eifel, Germany). Arch Hydrobiol 123:21–35
Williams R, Claustre H (1991) Photosynthetic pigments as biomarkers of phytoplankton populations and processes involved in the transformation of particulate organic matter at the biotrans site (478°N, 208°W). Deep Sea Res Part I 38:347–355
Willstätter R, Stoll A (1913) Die Wirkungen der Chlorophyllase. In: Willstätter R, Stoll A (eds) Untersuchungen über Chlorophyll. Springer, Berlin, pp 172–187

Wilson CL, Hinman NW, Cooper WJ, Brown CF (2000) Hydrogen peroxide cycling in surface geothermal waters of Yellowstone National Park. Environ Sci Technol 34:2655–2662

Winder M, Cloern JE (2010) The annual cycles of phytoplankton biomass. Phil Trans R Soc B 365:3215–3226

Winder M, Reuter JE, Schladow SG (2009) Lake warming favours small-sized planktonic diatom species. Proc R Soc B 276:427–435

Windolf J, Jeppesen E, Jensen JP, Kristensen P (1996) Modelling of seasonal variation in nitrogen retention and in-lake concentration: a four-year mass balance study in 16 shallow Danish lakes. Biogeochemistry 33:25–44

Wojciechowska W (1989) Correlation between biomass, chlorophyll-a, photosynthesis and phytoplankton structure in a lake. Ekol Polska ELPLBS 37:59–82

Wolf H, Brockmann H, Biere H, Inhoffen HH (1967) Chlorophyll and hemin XIII Preparation of diastereomeric 10-methoxy(pyro)methylpheophorbide a and determination of the relative configuration at the C-10 atom. Annalen 704:208–225

Woodward RB (1961) The total synthesis of chlorophyll. Pure Appl Chem 2:383–404

Woodward R (1990) The total synthesis of chlorophyll *a*. Tetrahedron 46:7599–7659

Woodward RB, Ayer WA, Beaton JM, Bickelhaupt F, Bonnett R, Buchschacher P, Closs GL, Dutler H, Hannah J, Hauck FP, Ito S, Langeman A, Le Goff E, Leimgruber W, Lwowski W, Sauer J, Valenta Z, Volz H (1960) The total synthesis of chlorophyll. J Am Chem Soc 82:3800–3802. doi:101021/ja01499a093

Wu SM, Rebeiz C (1985) Chloroplast biogenesis. Molecular structure of chlorophyll b (E489 F666). J Biol Chem 260:3632–3634

Wurtsbaugh WA, Gross HP, Budy P, Luecke C (2001) Effects of epilimnetic versus metalimnetic fertilization on the phytoplankton and periphyton of a mountain lake with a deep chlorophyll maxima. Can J Fish Aquat Sci 58:2156–2166

Wüthrich KL, Bovet L, Hunziker PE, Donnison IS, Hörtensteiner S (2000) Molecular cloning, functional expression and characterisation of RCC reductase involved in chlorophyll catabolism. Plant J 21:189–198

Xenopoulos MA, Bird DF (1997) Effect of acute exposure to hydrogen peroxide on the production of phytoplankton and bacterioplankton in a mesohumic lake. Photochem Photobiol 66:471–478

Xia B, Ma S, Chen J, Zhao J, Chen B, Wang F (2010) Distribution of organic carbon and carbon fixed strength of phytoplankton in Enteromorpha prolifera outbreak area of the Western South Yellow Sea, 2008]. Huan Jing Ke Xue 31:1442–1449 (in Chinese with Abstract in English)

Xiu P, Liu Y, Li G, Xu Q, Zong H, Rong Z, Yin X, Chai F (2009) Deriving depths of deep chlorophyll maximum and water inherent optical properties: A regional model. Continent Shelf Res 29:2270–2279

Yacobi YZ (2006) Temporal and vertical variation of chlorophyll a concentration, phytoplankton photosynthetic activity and light attenuation in Lake Kinneret: possibilities and limitations for simulation by remote sensing. J Plankton Res 28:725–736

Yacobi YZ, Pollingher U, Gonen Y, Gerhardt V, Sukenik A (1996) HPLC analysis of phytoplankton pigments from Lake Kinneret with special reference to the bloom-forming dinoflagellate Peridinium gatunense (Dinophyceae) and chlorophyll degradatio products. J Plankton Res 18:1781–1796

Yamashita Y, Tanoue E (2008) Production of bio-refractory fluorescent dissolved organic matter in the ocean interior. Nat Geosci 1:579–582

Yamauchi N, Harada K, Watada AE (1997) In vitro chlorophyll degradation in stored broccoli (*Brassica oleracea* L. *var. italica* Plen.) florets. Postharvest Biol Technol 12:239–245

Yoshioka T (1997) Phytoplanktonic carbon isotope fractionation: equations accounting for CO_2-concentrating mechanisms. J Plankton Res 19:1455–1476

Yoshioka T, Ueda S, Khodzher T, Bashenkhaeva N, Korovyakova I, Sorokovikova L, Gorbunova L (2002) Distribution of dissolved organic carbon in Lake Baikal and its watershed. Limnology 3:159–168

Yoshioka T, Mostofa KMG, Konohira E, Tanoue E, Hayakawa K, Takahashi M, Ueda S, Katsuyama M, Khodzher T, Bashenkhaeva N (2007) Distribution and characteristics of molecular size fractions of freshwater-dissolved organic matter in watershed environments: its implication to degradation. Limnology 8:29–44

Yoshiyama K, Nakajima H (2002) Catastrophic transition in vertical distributions of phytoplankton: alternative equilibria in a water column. J Theor Biol 216:397–408

Yoshiyama K, Mellard JP, Litchman E, Klausmeier CA (2009) Phytoplankton competition for nutrients and light in a stratified water column. Am Nat 174:190–203

Yuan J, Shiller AM (2001) The distribution of hydrogen peroxide in the southern and central Atlantic ocean. Deep Sea Res Part II 48:2947–2970

Yuma M, Timoshkin OA, Melnik NG, Khanaev IV, Ambali A (2006) Biodiversity and food chains on the littoral bottoms of Lakes Baikal, Biwa, Malawi and Tanganyika: working hypotheses. Hydrobiologia 568:95–99

Zepp RG, Skurlatov Y, Pierce J (1987) Algal-induced decay and formation of hydrogen peroxide in water: its possible role in oxidation of anilines by algae. In: Zika RG and Cooper WJ (eds) Photochemistry of environmental aquatic systems, ACS Symp Ser 327, Am Chem Soc, Washington DC, pp 213–224

Zepp R, Skurlatov Y, Pierce J (1987) Algal-induced decay and formation of hydrogen peroxide in water: its possible role in oxidation of anilines by algae. In: Zika RG, Cooper WJ (eds) Photochemistry of environmental Aquatic systems. ACS Symp Ser 327, Am Chem Soc, Washington DC, pp 213–224

Zepp RG, Faust BC, Hoigne J (1992) Hydroxyl radical formation in aqueous reactions (pH 3–8) of iron (II) with hydrogen peroxide: the photo-Fenton reaction. Environ Sci Technol 26:313–319

Zepp R, Erickson D III, Paul N, Sulzberger B (2011) Effects of solar UV radiation and climate change on biogeochemical cycling: interactions and feedbacks. Photochem Photobiol Sci 10:261–279

Zhang YL, Qin BQ (2007) Variations in spectral slope in Lake Taihu, a large subtropical shallow lake in China. J Great Lakes Res 33:483–496

Zhang Y, Zhang B, Wang X, Li J, Feng S, Zhao Q, Liu M, Qin B (2007a) A study of absorption characteristics of chromophoric dissolved organic matter and particles in Lake Taihu, China. Hydrobiologia 592:105–120

Zhang YL, Zhang EL, Liu ML, Wang X, Qin BQ (2007b) Variation of chromophoric dissolved organic matter and possible attenuation depth of ultraviolet radiation in Yunnan Plateau lakes. Limnology 8:311–319

Zhang Y, van Dijk MA, Liu M, Zhu G, Qin B (2009) The contribution of phytoplankton degradation to chromophoric dissolved organic matter (CDOM) in eutrophic shallow lakes: field and experimental evidence. Water Res 43:4685–4697

Zhao J, Cao W, Wang G, Yang D, Yang Y, Sun Z, Zhou W, Liang S (2009) The variations in optical properties of CDOM throughout an algal bloom event. Estuar Coast Shelf Sci 82:225–232

Zhu ZY, Ng WM, Liu SM, Zhang J, Chen JC, Wu Y (2009) Estuarine phytoplankton dynamics and shift of limiting factors: A study in the Changjiang (Yangtze River) Estuary and adjacent area. Estuar Coast Shelf Sci 84:393–401

Ziegler R, Blaheta A, Guha N, Schönegge B (1988) Enzymatic formation of pheophorbide and pyropheophorbide during chlorophyll degradation in a mutant of *Chlorella fusca* Shihira Ee Kraus. J Plant Physiol 132:327–332

Zimmermann P, Zentgraf U (2005) The correlation between oxidative stress and leaf senescence during plant development. Cell Mol Biol Lett 10:515

Complexation of Dissolved Organic Matter with Trace Metal Ions in Natural Waters

Khan M. G. Mostofa, Cong-qiang Liu, Xinbin Feng, Takahito Yoshioka, Davide Vione, Xiangliang Pan and Fengchang Wu

1 Introduction

Complexation of the metal ions (M) with dissolved organic matter (DOM), i.e. M-DOM formation is of fundamental importance in metal ion chemistry and can control the occurrence of free toxic metals, the transport or migration of metals, acid–base balance and solubility in water, occurrence of the photo-Fenton reaction in surface water, biological effects, the bioavailability and toxicity to organisms in water, sediment and soil environments. DOM in natural waters can either enhance or decrease metal transport and affect the bioavailability of metals depending on its composition. DOM is a complex mixture of organic compounds of allochthonous

K. M. G. Mostofa (✉) · C. Q. Liu· X. B. Feng
State Key Laboratory of Environmental Geochemistry, Institute of Geochemistry, Chinese Academy of Sciences, Guiyang 550002, China
e-mail: mostofa@vip.gyig.ac.cn

T. Yoshioka
Field Science Education and Research Center, Kyoto University, Kitashirakawa Oiwake-cho, Sakyo-ku, Kyoto 606-8502, Japan

D. Vione
Dipartimento di Chimica Analitica, University of Turin, I-10125 Turin, Italy

Centro Interdipartimentale Natrisk, I-10094 Grugliasco, (TO), Italy

X. L. Pan
Xinjiang Key Laboratory of Water Cycle and Utilization in Arid Zone, Xinjiang Institute of Ecology and Geography, Chinese Academy of Sciences, Urumqi 830011, People's Republic of China

F. C. Wu
State Environmental Protection Key Laboratory of Lake Pollution Control, Chinese Research Academy of Environmental Sciences, Chaoyang 100012, China

K. M. G. Mostofa et al. (eds.), *Photobiogeochemistry of Organic Matter*, Environmental Science and Engineering, DOI: 10.1007/978-3-642-32223-5_9,

and autochthonous origin. Allochthonous organic substances are generally derived from terrestrial plant material in soil ecosystems (and may reach the aquatic environmental because of soil leaching and runoff), while autochthonous organic substances are produced mostly from algae and phytoplankton within the water column. The contribution of fulvic and humic acids (terrestrial humic substances) accounts for the most of the dissolved organic carbon (DOC), approximately 20–85 % in rivers and 14–90 % in lakes (Moran et al. 1991; Moran and Hodson 1994; Malcolm 1985; Hedges et al. 1997; Ma et al. 2001; Mostofa et al. 2009a). In contrast, they account for a relatively minor fraction in oceans, approximately 1–35 % except the shelf (see also chapter "Dissolved Organic Matter in Natural Waters") (Moran et al. 1991; Moran and Hodson 1994; Malcolm 1985; Hedges et al. 1997; Ma et al. 2001; Mostofa et al. 2009a). They play an important role on DOM physical, chemical and biological characteristics in the aquatic environments.

On the other hand, autochthonous DOM of algal, phytoplankton or bacterial origin is generally composed of autochthonous fulvic acids, carbohydrates, amino acids, proteins, lipids, organic acids and so on (Mostofa et al. 2009a; Coble 1996; Parlanti et al. 2000; Tanoue 2000; Jennings and Steinberg 1994; Yamashita and Tanoue 2003; Wada et al. 2007; Hulatt et al. 2009; Mostofa and Sakugawa 2009; Zhang et al.2009). Phytoplankton is capable of releasing 10–60 % of the carbon and 15–50 % of the nitrogen assimilated during photosynthesis in natural waters (Sundh 1992; Bronk et al. 1994; Braven et al. 1995; MalinskyRushansky and Legrand 1996; Slawyk et al. 1998, 2000). The autochthonous DOM in the surface waters is significantly high, by approximately 0–102 % in lakes and 0–194 % in oceans estimated compared to the deeper waters during the summer stratification period (Mostofa et al. 2009a; see also DOM chapter). Therefore, autochthonous DOM plays an important role in M-DOM complexation, particularly in the surface layer of lakes, estuaries, wetlands and oceans. Extracellular polymeric substances (EPS) are produced by many microorganisms in natural waters and are mainly composed of polysaccharides, proteins, uronic acids, lipids and so on (Beech and Sunner 2004; Quiroz et al. 2006; Pal and Paul 2008; Merroun and Selenska-Pobell 2008; Zhang et al. 2008, 2010).

Most of the DOM components, such as allochthonous fulvic and humic acids of vascular plant origin, autochthonous fulvic acids of algal or phytoplankton origin, proteins, aromatic amino acids (tryptophan, tyrosine and phenylalanine), extracellular polymeric substances (EPS) and so on, show fluorescence properties and can simultaneously exhibit complexing properties (Saar and Weber 1980; Ryan and Weber 1982a; Cabaniss 1992; daSilva et al. 1996, 1998a; Smith and Kramer 1998, 1999; Mounier et al. 1999; Wu and Tanoue 2001a; Wu et al. 2004a; Dudal et al. 2006; Manciulea et al. 2009, 2011). DOM components are composed of diverse functional groups (including fluorophores or chromophores) in their molecular structures, which have strong binding capacity with metals in water (Malcolm 1985; Mostofa et al. 2009a; Saar and Weber 1980; Senesi 1990; Morel and Hering 1993; Morra et al. 1997; Xia et al. 1998; Leenheer et al. 1998; Bloom et al. 2001; Steelink 2002; Leenheer and Croue 2003; Stenson et al. 2003; Schwartz et al. 2004; Klinck et al. 2005). Therefore, it has been possible to assess the complexing properties of fluorescent DOM from the fluorescence quenching upon complexation between various DOM components and metal ions.

The M-DOM interaction is generally estimated by determining the conditional stability constant, which significantly depends on several important factors in aqueous solution such as contents, nature and molecular size of DOM (Nair and Chander 1983; Kim et al. 1990; Wu et al. 2004b; Sonke and Salters 2006; Shcherbina et al. 2007; Konstantinou et al. 2009; Vidali et al. 2009); effect of pH (Zhang et al. 2009, 2010; Ryan and Weber 1982a; Sonke and Salters 2006; Shcherbina et al. 2007; Christoforidis et al. 2010; Iskrenova Tchoukova et al. 2010; Liu and Cai 2010); effect of ions (cations and anions) and ionic strength (Sonke and Salters 2006; Christoforidis et al. 2010; Iskrenova-Tchoukova et al. 2010; Smith and Martell 1987; Cabaniss and Shuman 1988; Fu et al. 2007); effect of photoinduced processes (Zhang et al. 2009; Wu et al. 2004a; Bergquist and Blum 2007; Brooks et al. 2007; Yang and Sturgeon 2009; Sanchez-Marin et al. 2010; Vidali et al. 2010; Yin et al. 2010); effect of microbial processes (Bergquist and Blum 2007; Yin et al. 2010; Tabak et al. 2005; Francis and Dodge 2008; Fletcher et al. 2010), and finally effects of freshwater and seawater (Wu et al. 2004a, c; Fu et al. 2007; Harper et al. 2008). M-DOM complexation plays a vital role in the toxicity and bioavailability of heavy metals in the environment (Christoforidis et al. 2010; Winner 1985; Stackhouse and Benson 1988; Jiang et al. 2009). Optically darker DOM (or deep lake and marine DOM) has higher metal binding capability and typically tends to decrease metal toxicity to fish and algae in natural waters (Schwartz et al. 2004; Winch et al. 2002; Luider et al. 2004). Ternary complexes between organic ligands and trace metal ions and their relative stability constants may pave the way to find out high variations in conditional stability constants of binary complexes of DOM in aqueous solution (Martin and Prados 1974; Khalil and Radalla 1998; Khalil and Attia 1999, 2000; Khalil 2000a, b; Khalil and Taha 2004; Khalil and Fazary 2004; Radalla 2010; Rosas et al. 2010). A review by Hays et al. (2004) on the determination of conditional stability constants and ligand concentrations of fulvic acid with metal ions covers the Stern–Volmer equation for non-linear relationship, the differences between linear and non-linear relationship of formed complex and fluorescence quenching, and finally the Modified Multisite Stern–Volmer Equation. A review by Mostofa et al. (2011) summarizes the studies into metal ion complexation with various organic substances, and discusses about the possible mechanisms behind the quenching or enhancement in metal-DOM complexation in waters.

This chapter will give a general overview on the complexation of metal ions (M) with DOM (M-DOM), fluorescence characteristics of M-DOM complexes, theories for binary and ternary complexes as well as their conditional stability constants in aqueous solution, and finally the binding sites (or functional groups or flurophores or chromophores) in various DOM components. A mechanism will be provided on M-DOM complexation, with use of tryptophan, a well known molecularly characterized fluorescent DOM component, as a model to understand the complexing properties. This review will discuss the key factors affecting the M-DOM complexation and the importance of the M-DOM complexation in waters.

2 Complexation of Metal Ions (M) With DOM (M-DOM) in Natural Waters

Dissolved organic matter (DOM) is composed of a variety of organic compounds, which are the most important sources of organic ligands that can form complexes with trace metal ions in natural surface water, soil, sediment pore water and groundwater (Wu et al. 2004a, b, c; Christoforidis et al. 2010; Liu and Cai 2010; Lu and Allen 2002; Filella 2008; Yamashita and Jaffe 2008; Schmeide and Bernhard 2009; Reiller and Brevet 2010; Sachs et al. 2010; Weng et al. 2010; Tserenpil and Liu 2011). The most important organic ligands in DOM that can form complexes with metal ions are allochthonous fulvic acids (Sonke and Salters 2006; Wu et al. 2002a, b, 2004c; Schmeide and Bernhard 2009; Reiller and Brevet 2010; Vlassopoulos et al. 1990; Mandal et al. 1999; 2000; Sekaly et al. 1999, 2003; Shin et al. 2001; Haitzer et al. 2003; Cao et al. 2004; Fujii et al. 2008); allochthonous humic acids (Sonke and Salters 2006; Shcherbina et al. 2007; Christoforidis et al. 2010; Liu and Cai 2010; Wu et al. 2004c; Reiller and Brevet 2010; Sachs et al. 2010; Tserenpil and Liu 2011; Haitzer et al. 2003; Cao et al. 2004; Takahashi et al. 1997; Buschmann and Sigg 2004; Filella and May 2005; Lippold and Lippmann-Pipke 2009; Pourret and Martinez 2009; Tella and Pokrovski 2009; Kim and Czerwinski 1996); hydrophobic acids extracted from natural waters (Haitzer et al. 2002, 2003); hydrophylic acids extracted from olive cake (Konstantinou et al. 2009); autochthonous DOM (ca. autochthonous fulvic acids or marine fulvic acids) of algal origin (see also chapters "Dissolved Organic Matter in Natural Waters" and "Fluorescent Dissolved Organic Matter in Natural Waters") (daSilva et al. 1996; Xue and Sigg 1993; Xue et al. 1995); tryptophan amino acid (Wu and Tanoue 2001a, b); glycine and other amino acid having O- and N-functional groups (carboxyl, alcoholic hydroxyl, phenolic hydroxyl and amine) (Tella and Pokrovski 2009); cysteine, histidine, aspartic acid, glutamic acid (Rosas et al. 2010; Santana-Casiano et al. 2000; Shoukry 2005; Shiozawa et al. 2011); picolinic acid and dipicolinic acid (Rosas et al. 2010; Shiozawa et al. 2011; Lubes et al. 2010; Da Costa et al. 2011); protein, peptide and selenoprotein P (Shoukry 2005; Sidenius et al. 1999; Motson et al. 2004); glycoprotein exopolymer produced by *Pseudoalteromonas* sp. strain (Gutierrez et al. 2008); extracellular polymeric substances (Zhang et al.2009; Harper et al. 2008; Guibaud et al. 2004, 2006; Comte et al. 2008; d'Abzac et al. 2010); melanin (Felix et al. 1978); 2,4-diiodo-6-(((2- pyridinylmethyl)amino) methyl)phenol (Frezza et al. 2009); indolo[3,2-*c*]quinolines (Primik et al. 2010); 2-[4-dimethylaminocinnamalamino]-benzoic acid (Yalcin et al. 1998); 8-hydroxyquinoline (oxine) (Xia et al. 1996; Sarmiento et al. 2010); thenoyltrifluoroacetone (HTTA) (Xia et al. 1996); bipyridyl and 2,2′-bipyridine (Sarmiento et al. 2010; Bhattacharyya et al. 2010); 1,10-phenanthroline (Xia et al. 1996; Sarmiento et al. 2010; Bhattacharyya et al. 2010); phenols (e.g., catechol, hydroquinone, resorcinol) and polyphenols (Shcherbina et al. 2007; Tella and Pokrovski 2009; Thakur et al. 2006); quinonoid compounds (Shcherbina et al. 2007); benzylmalonic acid and n-hexadecylmalonic acid (Palmer et al. 1998); bis (2,4,4-trimethylpentyl) dithiophosphinic acid (Bhattacharyya et al. 2010); 2-(aminomethyl)-benzimidazole (El-Sherif 2010); α-isosaccharinic acid and α-isosaccharinate (Vercammen et al. 2001; van Loon et al. 2004; Warwick et al. 2003, 2004); citrate (Fujii et al. 2008); acetate (Saito et al. 2010); phthalic and salicylic acids (Panak et al. 1995); salicylate and thiosalicylate (Vlassopoulos et al. 1990); ethylenediaminetetraacetic acid (EDTA) (Fu et al. 2007; Fujii et al. 2008); galacturonic acid (Harper et al. 2008); gluconic acid (Warwick et al. 2003, 2004); aminopolycarboxylic acids (Smith and Martell 1987); and finally O-bearing organic compounds such as carboxylic acids (acetic, adipic, succinic, malic, malonic, maleic, lactic, oxalic, tartaric, citric and ortophosphoric acids) (Francis and Dodge 2008; Radalla 2010; Filella and May 2005; Tella and Pokrovski 2009; Shoukry 2005; Da Costa et al. 2011; El-Sherif 2010). Besides these organic ligands, the gill membrane in fish can form complexes with metal ions such as Ag^{+}, Ca^{2+}, Cu^{2+}, Cd^{2+}, and Co^{2+} (Playle et al. 1993; Janes and Playle 1995; Richards and Playle 1998; Playle 1998; Tao et al. 2002). In addition, algae can complex or uptake trace metals either directly or in the presence of humic acid that can enhance the metals uptake substantially (Zhou and Wangersky 1985, 1989; Xue and Sigg 1990; Koukal et al. 2003; Mylon et al. 2003; Lamelas and Slaveykova 2007; Lamelas et al. 2009).

The studied trace metal ions that form M-DOM complexes are the transition metals [Sc^{3+}, Y^{3+}, V^{2+} or $VO_2{}^{+}$, Cr^{2+} or Cr^{3+}, Mn^{2+}, Fe (Fe^{2+} or Fe^{3+}), Co^{2+}, Ni^{2+}, Cu^{2+}, Au^{+}, Mo^{2+}, Zn^{2+}, Cd^{2+}, Hg^{2+}]; lanthanides [Sc, Y, La, Ce, Pr, Nd, Pm, Sm, Eu, Gd, Tb, Dy, Ho, Er, Tm, Yb, and Lu, hereafter Ln^{3+}]; actinides [Th^{4+}, U^{4+}, $UO_2{}^{2+}$, Np (Np^{4+}, Np^{5+}), $NpO_2{}^{+}$, Pu^{3+}, Am^{3+}, Cm^{3+}]; metal/metalloid [Al^{3+}, Ga^{3+}, As^{3+}, Sb^{3+}, Sb^{5+}, Tl^{3+}, Sn^{2+}, Pb^{2+}], as well as the alkali/alkaline earth elements [H^{+}, Be^{2+}, Mg^{2+}, Ca^{2+}, Sr^{2+}, Ba^{2+}] (Nair and Chander 1983; Cabaniss and Shuman 1988; Vlassopoulos et al. 1990; Cabaniss 1992; Panak et al. 1995; Xia et al. 1996; Bidoglio et al. 1997; Takahashi et al. 1997; Kaiser 1998; Murphy et al. 1999; Sekaly et al. 1999; Wu et al. 2004a, b; Kautenburger et al. 2006; Sonke and Salters 2006; de Zarruk et al. 2007; Fu et al. 2007; Reszat and Hendry 2007; Shcherbina et al. 2007; Lippold and Lippmann-Pipke 2009; Pourret and Martinez 2009; Schmeide and Bernhard 2009; Vidali et al. 2009; Bhattacharyya et al. 2010; Christoforidis et al. 2010; Reiller and Brevet 2010; Saito et al. 2010; Tserenpil and Liu 2011). Any given metal ion in the natural environment may potentially be found in many diverse forms, namely, 'free' (hydrated), complexed by 'simple' inorganic or organic ligands, complexed by ligand atoms which are part of the structure of naturally occurring macromolecules or colloids, and adsorbed on suspended organic or inorganic particles or living organisms (Huber et al. 2002; Filella et al. 2007).

Heavy metals (e.g. Cd^{2+}, Pb^{2+}, and Sr^{2+}) show strong interaction with humic acids (HA) in forming M-DOM complexes, leading to formation of covalent bonds with the radicals of humic acids (Christoforidis et al. 2010). Metal partitioning between colloidal (1 μm–1 kDa) and truly dissolved (<1 kDa) fractions is detected to match a decrease of metal toxicity (for Cd and Zn ions) in the presence of humic acid, but not in the presence of Suwannee River Fulvic Acid (Koukal et al. 2003). This suggest that metal-HA complexes are of high molecular weight and relatively stable with regard to metal-exchange reactions, thus the cited metals (Cd and Zn ions) are less bioavailable (Koukal et al. 2003). Chemodynamic modeling suggests that the enhancement of the metal uptake flux in the presence of HA originates from an increasing amount of metal bound to the internalization sites, through ternary complex formation between metal—HA complex and internalization sites (Lamelas et al. 2009). Cell wall speciation calculations indicate that the metal—humic acid complex is the predominant species in the cell wall layer in algae, while for some other metals [e.g. Cu(II) and Cd(II)] the binding to the internalization (Cu) and adsorption (Cd) sites significantly dominates over the M−HA complexes (Lamelas et al. 2009).

2.1 Fluorescence Characteristics of the M-DOM Complexation

Allochthonous fulvic acids of vascular plant origin are generally composed of two fluorescence excitation-emission (Ex/Em) peaks (or maxima) such as the peak C at the peak C-region (280–400/380–550 nm) and the peak A at the peak A-region (215–280/380–550 nm) (see chapter "Fluorescent Dissolved Organic Matter in Natural Waters" for detailed description) (Fig. 1a) (Mostofa et al. 2005, 2009a; Coble 1996; Mostofa and Sakugawa 2009; Mounier et al. 2011; Yamashita and Jaffe 2008; Coble 2007). In contrast, allochthonous humic acids of vascular plant origin are composed of three or more fluorescence Ex/Em peaks such as the peak C at the peak C-region (280–400/380–550 nm) and the peak A at the peak A-region (215–280/380–550 nm) (Fig. 1b; see also chapter "Fluorescent Dissolved Organic Matter in Natural Waters") (Mostofa et al. 2005, 2009a; Mostofa and Sakugawa 2009; Yamashita and Jaffe 2008). Similarly, the fluorescence Ex/Em peaks of autochthonous fulvic acids of algal origin resemble those of allochthonous fulvic acids (Fig. 1c, d) (Parlanti et al. 2000; Mostofa and Sakugawa 2009; Zhang et al. 2009; Stedmon et al. 2007). The tryptophan (or protein-like) component shows two fluorescence peaks such as the peak T at the peak T-region (260–285/290–380 nm) and the peak T_{UV} at the peak T_{UV}-region (215–260/280–380 nm) (Fig. 1e; see chapter "Fluorescent Dissolved Organic Matter in Natural Waters"). The fluorophores or functional groups in fluorescent DOM are susceptible to show their fluorescent properties as well as to interact with metals via complex formation. Therefore, the fluorescence intensity of M-DOM is either enhanced or quenched compared to the original fluorescent DOM (Saar and Weber 1980; Ryan and Weber 1982a; Cabaniss and Shuman 1988; Grimm et al. 1991; Cabaniss 1992; dasilva et al. 1995, 1996, 1997, 1998; Smith and Kramer 1998; Lu and Jaffe 2001; Wu and Tanoue 2001a; Wu et al. 2004a; b; Dudal et al. 2006; Plaza et al. 2006; Fu et al. 2007; Antunes et al. 2007; Ohno et al. 2008; Manciulea et al. 2009, 2011).

Many studies show that complexation of DOM with trace elements often decreases the overall fluorescence intensity of DOM (peak C, peak A, peak T and peak T_{UV}), after addition of metal ions and with increasing their concentration. On the other hand, elements such as Al^{3+}, Be^{3+}, actinides (ca. Cm^{3+}), Ca^{2+} and Mg^{2+} are responsible for an enhancement of the fluorescence intensity of organic ligands (fulvic acid, humic

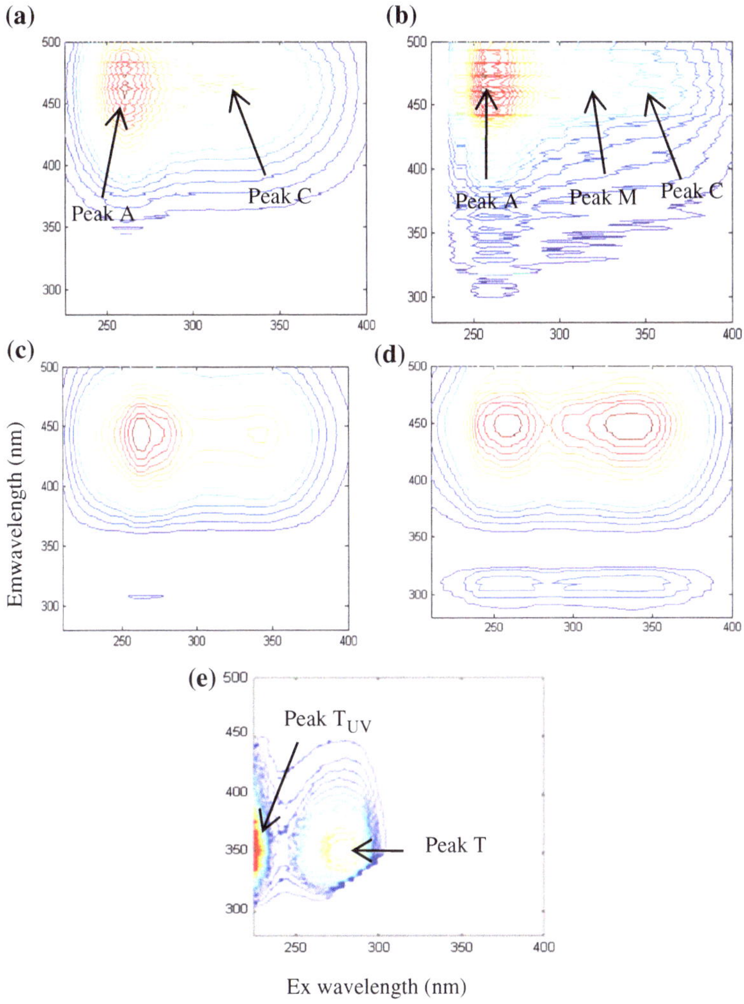

Fig. 1 The fluorescent components of allochthonous Suwannee River fulvic acid standard (C-like, **a**); allochthonous Suwannee River humic acid standard (C like, **b**), autochthonous fulvic acid (C-like) of microbial assimilation (**c**) and photoinduced assimilation (**d**) of lake algae in aqueous samples (Milli-Q and river waters), and tryptophan standard in Milli-Q waters (**e**) identified using PARAFAC modeling. *Data source* Mostofa and Sakugawa (2009); Mostofa et al. (2009b)

acid, 5-sulfosalicylic acid, and other organic compounds) that is linked with complexation (Cabaniss 1992; Smith and Kramer 1998, 1999; Cabaniss and Shuman 1988; Fu et al. 2007; Panak et al. 1995; dasilva et al. 1995; Ohno et al. 2008; Silva et al. 1994; Lakshman et al. 1993, 1996; Seritti et al. 1994). The increase in fluorescence intensity upon complexation of Cm^{3+} with 5-sulfosalicylic acid, fulvic acid, and humic acid suggests that the salicylic acid-like functional group may present in the molecular structure in allochthonous fulvic and humic acids (Panak et al. 1995). The fluorophores in fluorescent DOM (or functional groups or chromophores) are likely responsible for the formation of complexes with trace elements.

The key fluorophores in allochthonous fulvic and humic acids, tryptophan amino acid and protein in natural waters are composed of Schiff-base derivatives (–N = C–C = C–N–), –COOH, $-COOCH_3$, –OH, $-OCH_3$, –CH = O, –C = O, $-NH_2$, –NH–, –SH, –CH = CH–COOH, $-OCH_3$, $-CH_2-(NH_2)CH-COOH$, S-, O- or N-containing aromatic compounds, and so on (Malcolm 1985; Mostofa et al. 2009a; Mostofa and Sakugawa 2009; Senesi 1990; Steelink 2002; Leenheer and Croue 2003; Fu et al. 2007; Corin et al. 1996; Peña-Méndez et al. 2005; Seitzinger et al. 2005; Zhang et al. 2005). The complexation of trace elements with fluorescent substances also affects the fluorescence peak position of the respective fluorophore. Usually, both excitation and emission wavelengths of the respective peak position gradually shift toward the longer wavelength with increase in the reaction time (Wu et al. 2004a, c; Plaza et al. 2006). It has also been found that comparison of the EEM spectra before and after binding in metal-DOM complexes shows that the fast binding site in fulvic acid is responsible for 71–87 % of the total fluorescence decrease, while the remainder is associated with the slow binding site (Wu et al. 2004c).

EPSs show two fluorescence peaks (peak T and T_{UV}) in the absence and presence of trace elements such as Hg^{2+}, with the maximal fluorescence intensity at neutral pH (Fig. 2) (Zhang et al. 2010). The excitation-emission matrix (EEM)

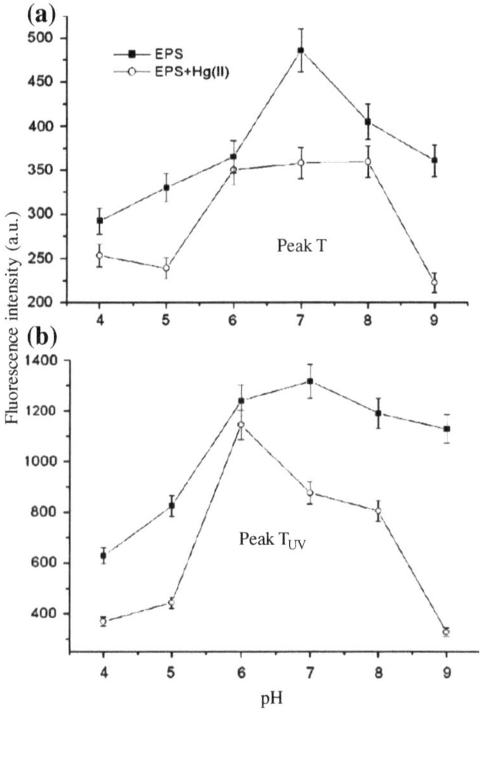

Fig. 2 Changes in the fluorescence intensities of peak T (**a**) and peak T_{UV} (**b**) for extracellular polymeric substances (EPS) with solution pH in the absence and presence of 3.0 mg L^{-1} Hg(II). The error bar indicates the standard deviation of three independent measurements. *Data source* Zhang et al. (2010)

spectra of EPS show that the fluorescence intensity of the peak T_{UV}-region is higher than the peak T-region, which is a similar behavior as tryptophan (Fig. 1e) (Mostofa et al. 2009a; Zhang et al. 2010). The fluorescence intensity of EPS and its complexation with Hg^{2+} significantly varies with pH, and is highest under neutral conditions (Fig. 2; Zhang et al. (2010)).

3 Complexation Theorem

The complexation theorem depends on the linear or non-linear relationship between formed complex and fluorescence quenching (Ryan and Weber 1982a; Cabaniss and Shuman 1988; Hays et al. 2003, 2004; Ryan and Ventry 1990; Cook and Langford 1995). The nonlinearity depends on several factors (Hays et al. 2004; Cook and Langford 1995), which can be distinguished as (i) Sample type, pretreatment, and age of the DOM; (ii) Overlap of fluorophores that typify chemically different complexation sites; (iii) The wavelengths (or scanning method) at which the fluorescence measurements are acquired; (iv) The varied physicochemical microenvironments of the fulvic acid structure as characterized by either a single fluorophore or fluorophore set; (v) Cooperational and molecular configuration changes due to metal loading or ionization state; and (vi) Shielding of binding sites or fluorescence (i.e., controlling photophysical effects). In addition (vii) the fluorescence intensity of fulvic and humic acids either at peak C-region or peak A-region might be the result of various types of fluorophores (functional groups) associated with those substances (Malcolm 1985; Senesi 1990; Steelink 2002). Sequential decomposition of the fluorophores bound at peak C- and A-regions in fulvic and humic acids by photoinduced degradation decreases the fluorescence intensity at the cited peaks during the transport from rivers to lakes and oceans (Mostofa et al. 2005a; b, 2007, 2010; Amador et al. 1989; Malcolm 1990; Allard et al. 1994; Moran et al. 2000; Xie et al. 2004; Li and Crittenden 2009; Minakata et al. 2009). Such changes may affect the nature and contents of fluorophores bound to fulvic and humic acids in various types of natural waters. Such changes are responsible for the variation of the binding capacity of metal ions, thereby exerting a significantly effect on the cited linear or non-linear relationship. The complexation theorem can therefore be expressed in two ways: (1) Theory of metal–ligand complexation based on linearity; and (2) Theory of modified Multisite Stern–Volmer (MSV) equation based on nonlinearity.

3.1 Theory of the M-DOM Complexation Based on Linearity

The fluorescence quenching of an organic ligand (e.g. fulvic acid, termed as L) by complexation with a metal ion (e.g. Cu^{2+}, termed as M) assuming a simple 1:1 metal ligand coordination ratio (Fig. 3) can be expressed in terms of the individual fluorescence of each species, as depicted by Ryan and Weber (1982a, b). The

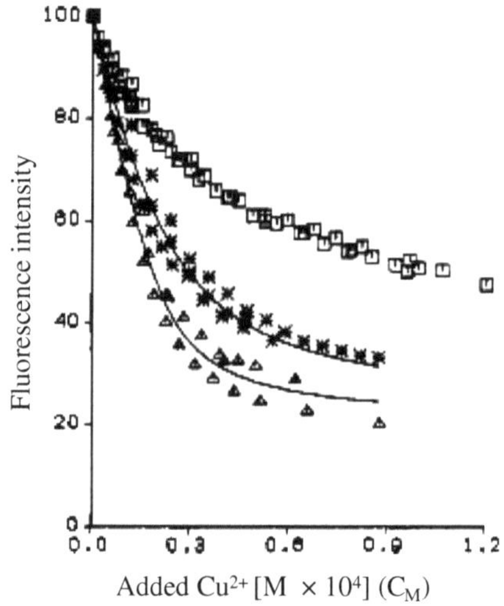

Fig. 3 Fluorescence quenching curves for 10 mg L^{-1} soil fulvic acid with Cu^{2+} in 0.1 M KNO_3 at 25 °C: (*squares*) pH 5, four replicates; (*asterisk*) pH 6, three replicates; (*triangles*) pH 7, three replicates. *Data source* Ryan and Weber (1982a)

concentration of bound metal ion with an organic ligand (ML) is the difference between total and free metal ion concentration. A titration of naturally occurring fulvic acid ligand with a metal ion can allow determination of the stability constant (K) and of the complexing capacity (C_L) of the ligand if a stoichiometry is considered (Hart 1981).

3.2 Theory of Multisite Stern–Volmer Equation for Determination of M-DOM Complexation

The multisite Stern–Volmer (MSV) equation is modified, simulated, validated for predictive capability with a suitable model compound set by Hays and his colleagues (2004), and applied to fluorescence titration data in the complexation of fulvic acid with Cu^{2+} (Fig. 4). The MSV approach assumes a simple 1:1 coordination ratio between Cu^{2+} and the fluorescent ligand components, giving the following reversible solution equilibria: $M + L1 \leftrightharpoons ML1$ and $M + L2 \leftrightharpoons ML2$ where L1 and L2 are the free ligand species (all forms of metal-free ligand) at sites 1 and 2, respectively, M is free metal, and ML1 and ML2 are the metal-bound species at these sites. Other reaction stoichiometries are possible, which rely on metal loading. For example, at low metal loadings, chelation (2:1 = ligand:metal) may be induced, contorting the fulvic acid shape and affecting fluorescence. Evidences for the effect of these molecular conformation changes on fluorescence are scarce, inconclusive, and not easily quantified.

Fig. 4 Multisite Stern–Volmer (MSV)-fitted fluorescence quenching data of soil fulvic acid (15 mg L^{-1}) titrated with Cu^{2+} at 0.1 M ionic strength: (*circles*) pH 5 data; (*squares*) pH 6 data; (*triangles*) pH 7 data; (*solid lines*) MSV predicted intensity values. *Data source* Ryan and Weber (1982a)

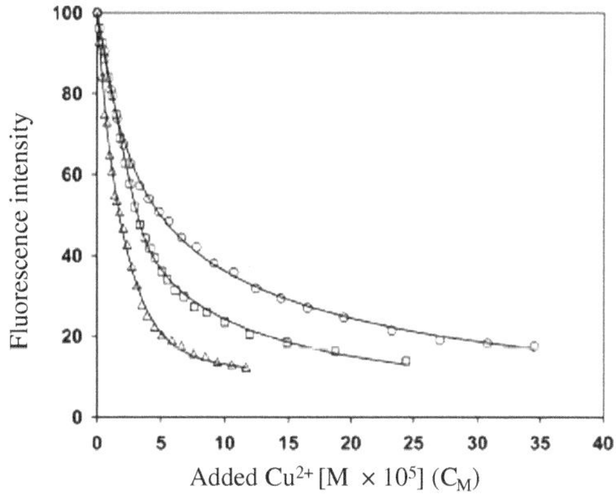

3.3 Theory for Conditional Stability Constant of M-DOM Complexation

The conditional stability constant of a M-DOM complex is operationally defined as the binding efficiency of the newly formed bond between the functional group of the DOM component (acting as an organic ligand) and the trace metal ion M, when they are mixed up under specific conditions in aqueous media. Conditional stability constants of a M-DOM complex can be useful to characterize the formed complex, to apply the strong binding capacity of organic substances to control speciation, toxicity, bioavailability and fate of toxic metals used e.g. in industries, and for predicting biological effects of metals in natural water, sediment and soil environments (Shcherbina et al. 2007; Mostofa et al. 2011; Sekaly et al. 2003; Huber et al. 2002; Filella et al. 2007; Hörnström et al. 1984; Hughes et al. 1995; Markich 2002; Matsumoto et al. 2005; Martel and Motekaitis 1988).

A conditional stability constant has been determined by Midorikawa and Tanoue (Appendix A) (Midorikawa and Tanoue 1998), adopting the relationship between measured fluorescence intensity and complexation for a divalent metal ion (M, ca. Cu^{2+}) with organic ligands, and assuming a 1:1 stoichiometry (Ryan and Weber 1982a). The complexing reactions that fit the experimental data can be described by the linear regression program. The relationship between measured fluorescence intensity and complexation can be described as follows (Eq. 3.1) (Ryan and Weber 1982a):

$$X = \frac{[ML]}{C_L} = \frac{F_0 - F}{F_0 - F_{end}} \tag{3.1}$$

where the quantity $[ML]/C$ is the fraction of the ligand bound to the metal to form the complex ML. Such a fraction can be expressed in terms of the measured fluorescence intensity, F. F_0 and F_{end} are the limiting intensities before and after metal titration. They correspond to the intensities when all ligands are entirely free and bound, respectively.

The estimated fraction $[\mathrm{ML}]/C_L$ is applied to the approach of Ruzic (1982). The conditional stability constants (K'_{ML}) for complex formation between M and the ligand, L, assuming simple 1:1 equilibrium can be written as follows (Eq. 3.2):

$$K'_{\mathrm{ML}} = \frac{[\mathrm{ML}]}{[M] \cdot [\mathrm{L}']} = \frac{[\mathrm{ML} \cdot \alpha_{\mathrm{M}}]}{(C_{\mathrm{M}} - [\mathrm{ML}])\,(C_{\mathrm{L}} - [\mathrm{ML}])} \tag{3.2}$$

where K'_{ML} is the constant with regard to the concentration of free metal ion, [M]; [M′] and [L′] are the total concentrations of all inorganic forms of M and of L unbound to M, respectively; α_{M} is the inorganic side-reaction coefficient for M that is estimated to be $\alpha_{\mathrm{M}} = [\mathrm{M}']/[\mathrm{M}] = 11$ at pH 8.15 under the same conditions (Midorikawa et al. 1990).

Substituting for $[\mathrm{ML}] = C_{\mathrm{L}}\,X$, Eq. (3.2) can be rewritten as (Eq. 3.3)

$$C_{\mathrm{M}} \cdot \left(\frac{1 - X}{X}\right) = C_{\mathrm{L}} \cdot (1 - X) + \frac{\alpha_{\mathrm{M}}}{K'_{\mathrm{ML}}} \tag{3.3}$$

By plotting of $C_{\mathrm{M}}\,(1-X)/X$ versus. $(1 - X)$, a linear regression is observed by the least-squares analysis that will give the best-fit values of C_{L} from the slope, and the conditional stability constant K'_{ML} from the intercept. For the nonlinear diagram, two 1:1 complexes by two discrete ligand classes with different stability constants are assumed, which can be treated by another model (van Den Berg 1984).

3.4 Theory for Protonation Constants of DOM in M-DOM Complexation

The protonation constants of organic ligands are estimated from the changes in fluorescence according to the changes in pH with regard to single protonation (Midorikawa and Tanoue 1998). The fluorescence (F) of the ligand (L) during the acid–base titration can be expressed by the concentration of each species of the ligand by the molar fluorescence coefficient (ε) as follows (Eqs. 3.4–3.6):

$$F_{\mathrm{H-pH}} = \varepsilon_{\mathrm{L}} C_{\mathrm{L}} \quad \text{at high pH} \tag{3.4}$$

$$F_{\mathrm{L-pH}} = \varepsilon_{\mathrm{HL}} C_{\mathrm{L}} \quad \text{at low pH} \tag{3.5}$$

$$F = \varepsilon_{\mathrm{HL}}[\mathrm{L}] + \varepsilon_{\mathrm{HL}}\,[\mathrm{HL}] \quad \text{at middle pH} \tag{3.6}$$

where the quantities $F_{\text{H-pH}}$ and $F_{\text{L-pH}}$ are the limiting fluorescence intensities at either extreme of the titration: $F_{\text{H-pH}}$ is for the free ligand (L) that is dissociated at high pH; and $F_{\text{L-pH}}$ is for HL that is protonated at low pH.

From the mass balance of the ligand, $C_{\mathrm{L}} = [\mathrm{L}] + [\mathrm{HL}]$, the above equations can be rewritten as follows:

$$F_{\mathrm{H-pH}} - F = (\varepsilon_{\mathrm{L}} - \varepsilon_{\mathrm{HL}})\,[\mathrm{HL}] \tag{3.7}$$

$$F - F_{\mathrm{L-pH}} = (\varepsilon_{\mathrm{L}} - \varepsilon_{\mathrm{HL}})\,[\mathrm{L}] \tag{3.8}$$

From Eqs. (3.7) and (3.8), the fluorescence of the ligand can be expressed in terms of the protonation constant and the proton concentration as follows:

$$\frac{F_{\mathrm{H-pH}} - F}{F - F_{\mathrm{L-pH}}} = \frac{[\mathrm{HL}]}{\mathrm{L}} = K'_{\mathrm{HL}} \cdot [\mathrm{H}] \tag{3.9}$$

Using the logarithm, Eq. (3.9) becomes

$$\log \frac{F_{\mathrm{H-pH}} - F}{F - F_{\mathrm{L-pH}}} = \log K'_{\mathrm{HL}} - \mathrm{pH} \tag{3.10}$$

The plot of $\log(F_{\mathrm{H\text{-}pH}}-F)/(F-F_{\mathrm{L\text{-}pH}})$ versus pH gives the values of $\log K'_{\mathrm{HL}}$ from the intercept. For the occurrence of a further protonation at lower pH, two protonation constants are operational that can be treated by another model (van Den Berg 1984).

3.5 *Kinetics of M-Fulvic Acid Complexation*

It has been shown that allochthonous fulvic acids are the main DOM components studied in natural waters (Moran et al. 1991; Malcolm 1990; Ma et al. 2001; Mostofa et al. 2009a). Therefore, it is vital to know how they form complexes with metal ions. Changes in the full fluorescence spectral kinetic, i.e., in both Ex/Em wavelengths of the fluorescence maxima occur in the EEM spectra of fulvic acid or DOM during their complexation with trace elements (Wu et al. 2004a, c). The complexation of a fulvic acid (extracted from Cavan Bog, Canada) with several metals (Cu^{2+}, Ni^{2+}, Co^{2+}, Cd^{2+} and Ca^{2+}) at pH 7 shows that the fulvic acid can react rapidly with all metals studied (Wu et al. 2004c). The result of pseudo-first order kinetic plots demonstrates that fulvic acid has two major kinetically distinguishable binding sites, 'fast' and 'slow', having reaction half-lives of 1.3–3.9 and 34.7–69.3 s, respectively (Wu et al. 2004c). The binding of copper to fulvic acid is found to be fairly rapid, and the reaction is virtually at equilibrium after approximately 20–30 s (Lin et al. 1995). Another study demonstrates that the three lifetimes and emission wavelength maxima for three fluorophores in fulvic acid are as follows: ~50 ps (392 nm), ~430 ps (465 nm), and 4.2 ns (512 nm) (Cook and Langford 1995). Kinetic changes of excitation-emission wavelengths of the fluorescence maxima also suggest the presence of two major binding sites. For the fast-reacting binding site, the rate constant and the site relative contribution are in the order $Cu^{2+} > Ni^{2+} > Co^{2+} > Cd^{2+} > Ca^{2+}$, which agrees with the Irving-Williams series, indicating affinity dependence of complexation kinetics (Wu et al. 2004c). For the fast-reacting binding site in fulvic acid, proteins and other organic ligands, the relative contribution of rate constant for bivalent metal complexes follows the same order (Wu et al. 2002a, 2004c; Sidenius et al. 1999; Irving and Williams 1953; Winzerling et al. 1992). This result implies that metal ions react initially with the fast reacting and strong binding site on fulvic acid; after the fast binding site is entirely occupied, the slow reacting and weak binding site starts to bind with the metal ions (Wu et al. 2004c). In each phase, the red-shifted excitation-emission spectra suggest physical structural changes (ca. molecular conformation or rigidness) of the fulvic acid-metal complex (Wu et al. 2004c). Changes in the molecular conformation and formation of aggregates of fulvic acid upon complexation are also supported by other studies (Hays et al. 2004; Filella 2008; Chakraborty et al. 2007).

Within each kinetic phase, both excitation-emission wavelengths of fluorescence maxima gradually increase with reaction time (Wu et al. 2004a, c). Such red-shifted phenomena are also observed in humic acid extracted from sewerage sludge (Plaza et al. 2006).The kinetics of metal-fulvic acid complexation depends on the concentration; chemical nature and sources of DOM components; temperature; pH; molecular size; occurrence of anions and cations; salinity; and metal affinity (da Silva et al. 1998a; Wu and Tanoue 2001a, b; Wu et al. 2004b, c; Fu et al. 2007; Mostofa et al. 2011; Lu and Jaffe 2001; Lin et al. 1995; Plankey and Patterson 1987). Recent studies show that all protonated and unprotonated forms of both the ligand and the hydrated central metal ion are involved in the formation of the precursor inner- and outer-sphere complexes, which control the kinetics of complex formation/dissociation (van Leeuwen et al. 2007; van Leeuwen and Town 2009a, b). It has also been shown that stronger the affinity of the metal, the greater proportion of strong and fast reacting fulvic acid binding sites are involved in the complexation process. This finding indicates that metal affinity affects both the thermodynamic equilibrium and the reaction kinetics (Wu et al. 2004c).

3.6 Conditional Stability Constants ($log_{10}K$) for M-DOM Complexation in Water

Conditional stability constants ($\log_{10} K$) for M-DOM complexation are presented for various DOM components in water (Table 1) (Zhang et al. 2010; Wu et al. 2001, 2004a; Nair and Chander 1983; Kim et al. 1990; Sonke and Salters 2006; Shcherbina et al. 2007; Konstantinou et al. 2009; Vidali et al. 2009; Liu and Cai 2010; Fu et al. 2007; Brooks et al. 2007; Hays et al. 2004; Cao et al. 2004; Takahashi et al. 1997; Haitzer et al. 2002; Xue and Sigg 1993; Xue et al. 1995; Wu and Tanoue 2001b; Guibaud et al. 2004, 2006; Thakur et al. 2006; van Loon et al. 2004; Plaza et al. 2006; Antunes et al. 2007; Midorikawa and Tanoue 1998; Tipping 1994; Breault et al. 1996; Mcknight et al. 1983; van Den Berg et al. 1987; Sunda and Huntsman 1991; Sunda and Hanson 1987; Sunda and Ferguson 1983; Moffett et al. 1990; Coale and Bruland 1988, 1990; van Den Berg 1984; Midorikawa and Tanoue 1996; Benoit et al. 2001; Buckau et al. 1992; Comte et al. 2006; Drexel et al. 2002; Gasper et al. 2007; Hsu and Sedlak 2003; Kim et al. 1991; Sander et al. 2005; Shank et al. 2006; Smith 1974). Fulvic acids (FA) of different origin can exhibit two major binding sites in the

Table 1 Conditional stability constants ($\log_{10} K$) of the M-DOM complexes in aqueous solution

Elements or groups	Conditional stability constants ($\log_{10} K$)					pH	References
	Fulvic acid[a]	Humic acid[a]	Tryptophan or protein or EPS[a]	Autochthonou DOM or hydrophilic or hydrophobic[a]	Other organic ligands[a]		
Model ligands							
Cu^{2+} with COOH and OH group	–	–	–	–	3.0–6.0	8.15	Martell and Smith (1974)
Cu^{2+} with NH_2 (NH_3)	–	–	–	–	2.8	8.15	Martell and Smith (1974)
Cu^{2+} with ethanol amine	–	–	–	–	4.4	8.15	Martell and Smith (1974)
Cu^{2+} with glycine (NH_2–CH_2–COOH or NH_3^+–CH_2–COO^-)	–	–	–	–	6.6	8.15	Martell and Smith (1974)
Cu^{2+} with ethylene diamine (NH2–CH_2–CH_2–NH_2)	–	–	–	–	8.5	8.15	Martell and Smith (1974)
Cu^{2+} with aspartic acid [HOOC–CH(NH_2)–CH_2–COOH]	–	–	–	–	7.2	8.15	Martell and Smith (1974)
Cu^{2+} with iminodiaacetic acid (HOOC–CH_2–NH_2–CH_2–COOH)	–	–	–	–	9.4	8.15	Martell and Smith (1974)
Cu^{2+} with diaminopropanol	–	–	–	–	8.3	8.15	Martell and Smith (1974)
Cu^{2+} with ethylenediamine-N-acetic acid	–	–	–	–	11.4	8.15	Martell and Smith (1974)
Cu^{2+} with diethylenetriamine	–	–	–	–	13.0	8.15	Martell and Smith (1974)
Cu^{2+} with fulvic acid (modeled)	9.64 (K_1); 3.26 (K_2)	–	–	–	–	?	Tipping (1994)
Cu^{2+} with fulvic acid (modeled)	4.37 (K_2)	–	–	–	–	6	Antunes et al. (2007)
Cu^{2+} with fulvic acid (extraced from soil)	5.30 (K_1)	–	–	–	–	4	Cao et al. (2004)

(continued)

Table 1 (continued)

Elements or groups	Conditional stability constants ($\log_{10} K$)					pH	References
	Fulvic acid[a]	Humic acid[a]	Tryptophan or protein or EPS[a]	Autochthonou DOM or hydrophilic or hydrophobic[a]	Other organic ligands[a]		
Cu^{2+} with fulvic acid (extracted from soil) (Ex/Em = 335/350 nm)	4.78–5.70 (K_1)	–	–	–	–	5.0–7.0	Hays et al. (2004)
Fe^{3+} with fulvic acid (modeled)	5.66 (K_1)	–	–	–	–	4	Antunes et al. (2007)
Hg^{2+} with fulvic acid (modeled)	4.44 (K_1)	–	–	–	–	6	Antunes et al. (2007)
UO_2^{2+} with fulvic acid (modeled)	5.46 (K_1)	–	–	–	–	3.5	Antunes et al. (2007)
Cu^{2+} with humic acid (modeled)	–	4.83 (K_1)	–	–	–	6	Antunes et al. (2007)
Cu^{2+} with humic acid (modeled)	–	8.55 (K_1); 4.02 (K_2)	–	–	–	?	Tipping (1994)
Cu^{2+} with humic acid (extraced from soil)	–	3.99–4.49 (K_1)	–	–	–	4	Cao et al. (2004)
Fe^{3+} with humic acid (modeled)	–	6.79 (K_1)	–	–	–	4	Antunes et al. (2007)
Hg^{2+} with humic acid (modeled)	–	5.50 (K_1)	–	–	–	6	Antunes et al. (2007)
UO_2^{2+} with humic acid (modeled)	–	4.27 (K_1)	–	–	–	3.5	Antunes et al. (2007)
As^{3+} with Aldrich humic acid (modeled)	–	5.8–7.2 (K_1); 4.5–5.3 (K_2)	–	–	–	5.2–9.3	Liu and Cai (2010)
Cu^{2+} with tryptophan (modeled) (Ex/Em = 285/360 nm)	–	–	4.88–4.90 (K_1)	–	–	6	Hays et al. (2004)
Cu^{2+} with glycyl-tryptophan (modeled) ($C_{13}H_{15}N_3O_3$) (Ex/Em = 285/360 nm)	–	–	5.81–6.02 (K_1)	–	–	6	Hays et al. (2004)
Acid/alkali/alkaline earth elements	–	–	–	–	–		

(continued)

Table 1 (continued)

Elements or groups	Conditional stability constants ($\log_{10} K$)					pH	References
	Fulvic acid[a]	Humic acid[a]	Tryptophan or protein or EPS[a]	Autochthonou DOM or hydrophilic or hydrophobic[a]	Other organic ligands[a]		
H^+ with humic acid of various sources	–	7.15–9.80 (K_1); 2.80–4.94 (K_2)	–	–	–		Vidali et al. (2009)
Be^{2+}	–	4.0	–	–	–	7	Takahashi et al. (1997)
Ca^{2+}	–	1.0	–	–	–	7	Takahashi et al. (1997)
Ca^{2+} with alpha-isosaccharinate	–	–	–	–	1.70–1.80		van Loon et al. (2004)
Sr^{2+}	–	0.5	–	–	–	7	Takahashi et al. (1997)
Ba^{2+}	–	0.5	–	–	–	7	Takahashi et al. (1997)
Ga^{3+}	–	10.0	–	–	–	7	Takahashi et al. (1997)
Transition metals							
Cu^{2+} with river DOM/humic substances	7.0–8.[illegible] (K_1); 5.4–6.1 (K_2)	–	–	–	–	?	Breault et al. (1996)
Cu^{2+} with river and canal DOM/humic substances	8.3–8.5 (K_1); 6.0–6.6 (K_2)	–	–	–	–	?	Mcknight et al. (1983)
Cu^{2+} with river fulvic acid (Ex/Em = 315–330/426–434 nm)	7.21–7.31 (K_1)	–	–	–	–	?	Wu et al. (2001)
Cu^{2+}	5.3	–	–	–	–	6	Konstantinou et al. 2009
Cu^{2+} with lake autochthonous fulvic acid (M-like: 310–320/376–386 nm) at 2.5 m	7.84–7.96 (K_1)	–	–	–	–	?	Wu et al. (2001)
Cu^{2+} with lake autochthonous fulvic acid (M-like: 250/414–438 nm) at 2.5 m	7.06–7.57 (K_2)	–	–	–	–	?	Wu et al. (2001)
Cu^{2+} with lake autochthonous fulvic acid (M-like: 310–320/376–386 nm) at 70 m	9.23 (K)	–	–	–	–	?	Wu et al. (2001)

(continued)

Table 1 (continued)

Elements or groups	Conditional stability constants ($log_{10} K$)					pH	References
	Fulvic acid[a]	Humic acid[a]	Tryptophan or protein or EPS[a]	Autochthonou DOM or hydrophilic or hydrophobic[a]	Other organic ligands[a]		
Cu^{2+} with lake autochthonous fulvic acid (M-like: 250/414–438 nm) at 70 m	8.69 (K_2)	–	–	–	–	?	Wu et al. (2001)
Cu^{2+} with lake DOM/humic substances	7.0 (K_1); 5.4 (K_2)	–	–	–	–	?	Mcknight et al. (1983)
Cu^{2+} with fulvic acid (lake)	7.05–8.78 (K_1)	–	–	–	–	?	Wu and Tanoue (2001b)
Cu^{2+} with humic acid, sewage sludge (Ex/Em = 340/438 nm)	–	4.65	–	–	–	8.1	Plaza et al. (2006)
Cu^{2+} with humic acid, soil (Ex/Em = 440/510 nm)	–	5.55	–	–	–	8.1	Plaza et al. (2006)
Cu^{2+} with humic acid, amended soil + sewage sludge (Ex/Em = 440/510 nm)	–	5.36	–	–	–	8.1	Plaza et al. (2006)
Zn^{2+} with autochthonous DOM (Lake Greifen)	–	–	–	7.8–9.6	–	8.0	Xue et al. (1995)
Cu^{2+} with autochthonous DOM (Lake Greifen)	–	–	–	13.9–14.9 (K_1); 11.8–12.9 (K_2)	–	7.8	Xue and Sigg (1993)
Cu^{2+} with autochthonous DOM (Scheldt estuary)	–	–	–	13.0–14.9 (K_1); 11.5–12.8 (K_2)	–	7.7	van Den Berg et al. (1987)
Cu^{2+} with DOM (Shelf water)	–	–	–	13.2; 10.0	–	8.1	Sunda and Huntsman (1991)
Cu^{2+} with DOM (Mississippi River plume)	–	–	–	11.1; 8.9	–	8.1	Sunda and Ferguson (1983)

(continued)

Table 1 (continued)

Elements or groups	Conditional stability constants ($\log_{10} K$)					pH	References
	Fulvic acid[a]	Humic acid[a]	Tryptophan or protein or EPS[a]	Autochthonou DOM or hydrophilic or hydrophobic[a]	Other organic ligands[a]		
Cu^{2+} with DOM (Southeastern Gulf of Mexico)	–	–	–	≥12; 9.8	–	8.2	Sunda and Ferguson (1983)
Cu^{2+} with DOM (Coast of Peru)	–	–	–	12.3; 9.2	–	8.2	Sunda and Hanson (1987)
Cu^{2+} with DOM (Sargasso Sea)	–	–	–	13.2; 9.7	–	?	Moffett et al. (1990)
Cu^{2+} with DOM (North Pacific)	–	–	–	13.0; 10.0	–	?	Coale and Bruland (1988)
Cu^{2+} with DOM (South Atlantic)	–	–	–	12.2; 10.2	–	7.7	van Den Berg (1984)
Cu^{2+} with DOM (equatorial Pacific)	–	–	–	8.92–9.26 (K_1); 6.87–7.44 (K_2)	–	8.15	Midorikawa and Tanoue (1996, 1998)
Cu^{2+} (groundwater)	–	5.60	–	–	–		Kim et al. (1990)
Hg^{2+} with isolated hydrophobic fraction of DOM (Florida Everglades surface waters)	–	–	–	11.6–12.0 (K_1); 10.5–10.9 (K_2)	–		Benoit et al. (2001)
Mn^{2+}	–	0.40	–	–	–	7	Takahashi et al. (1997)
Co^{2+}	–	0.50	–	–	–	7	Takahashi et al. (1997)
Zn^{2+} with humic acid, sewage sludge (Ex/Em = 340/438 nm)	–	4.08	–	–	–	8.1	Plaza et al. (2006)
Zn^{2+} with humic acid, soil (Ex/Em = 440/510 nm)	–	4.43	–	–	–	8.1	Plaza et al. (2006)
Zn^{2+} with humic acid, amended soil + sewage sludge (Ex/Em = 440/510 nm)	–	4.31	–	–	–	8.1	Plaza et al. (2006)

(continued)

Table 1 (continued)

Elements or groups	Conditional stability constants ($\log_{10} K$)					pH	References
	Fulvic acid[a]	Humic acid[a]	Tryptophan or protein or EPS[a]	Autochthonou DOM or hydrophilic or hydrophobic[a]	Other organic ligands[a]		
Zn^{2+}	–	0.50	–	–	–	7	Takahashi et al. (1997)
Cr^{3+}	–	6.00	–	–	–	7	Takahashi et al. (1997)
Fe^{3+}	–	10.0	–	–	–	7	Takahashi et al. (1997)
Sc^{3+}	–	15.0	–	–	–	7	Takahashi et al. (1997)
VO_2^{+}	–	2.1	–	–	–	7	Takahashi et al. (1997)
Hg^{2+} with stream humic substances (peak C: 340–355/430–455 nm)	4.34–5.20	–	–	–	–	7.92	Fu et al. (2007)
Hg^{2+} with river fulvic acid (peak A: 245/432–438 nm)	5.62–5.72		–	–	–	7.92	Fu et al. (2007)
Hg^{2+} with river fulvic acid (peak C: 330–335/426–434 nm)	5.01–5.55	–	–	–	–	7.92	Fu et al. (2007)
Hg^{2+} with stream DOM	4.34–5.2	–	–		–	4.37–7.01	Wu et al. (2004)
Hg^{2+} with surface waters	–	–	–	21.2–30.2	–	7	Hsu and Sedlak (2003)
Hg^{2+} with hydrophobic acid extracted from Florida Everglades surface waters	–	–	–	26–31	–	7	Gasper et al. (2007)
Hg^{2+} with hydrophobic acid	–	–	–	25.5	–	7	Haitzer et al. (2002)
Hg^{2+} with hydrophobic acid	–	–	–	19.8	–	6	Benoit et al. (2001)
Hg^{2+} with peat DOC (Florida Everglades peats)	–	–	–	23.2	–	6	Drexel et al. (2002)
Hg^{2+} with tryptophan (peak T: 275/340–346 nm)	–	–	4.99–5.33		–	7.92	Fu et al. (2007)

(continued)

Table 1 (continued)

Elements or groups	Conditional stability constants ($\log_{10} K$)					pH	References
	Fulvic acid[a]	Humic acid[a]	Tryptophan or protein or EPS[a]	Autochthonou DOM or hydrophilic or hydrophobic[a]	Other organic ligands[a]		
Cu^{2+} with tryptophan-like or protein-like component	–	–	7.82–9.56		–	?	Wu and Tanoue (2001b)
Pb^{2+} with humic acid, sewage sludge (Ex/Em = 340/438 nm)	–	4.95	–	–	–	8.1	Plaza et al. (2006)
Pb^{2+} with humic acid, soil (Ex/Em = 440/510 nm)	–	5.81	–	–	–	8.1	Plaza et al. (2006)
Pb^{2+} with humic acid, amended soil + sewage sludge (Ex/Em = 440/510 nm)	–	5.53	–	–	–	8.1	Plaza et al. (2006)
Cd^{2+} with humic acid, sewage sludge (Ex/Em = 340/438 nm)	–	4.24	–	–	–	8.1	Plaza et al. (2006)
Cd^{2+} with humic acid, soil (Ex/Em = 440/510 nm)	–	4.63	–	–	–	8.1	Plaza et al. (2006)
Cd^{2+} with humic acid, amended soil + sewage sludge (Ex/Em = 440/510 nm)	–	4.47	–	–	–	8.1	Plaza et al. (2006)
Hg^{2+} with extracellular polymeric substances (EPS) (peak T: 275–280/428–334 nm)	–	–	3.28–4.12 (K_1)	–	–	4.0–8.0	Zhang et al. (2010)
Hg^{2+} with EPS (peak T_{UV}: 220–230/422–336 nm)	–	–	4.28–4.49 (K_2)	–	–	4.0–8.0	Zhang et al. (2010)
Pb^{2+} with EPS	–	–	3.1–4.9 (K_1); 3.2–4.6 (K_2)	–	–	7	Comte et al. (2006)
Pb^{2+} with EPS	–	–	0.45–1.28	–	–	7	Guibaud et al. (2006)

(continued)

Table 1 (continued)

Elements or groups	Conditional stability constants ($\log_{10} K$)					pH	References
	Fulvic acid[a]	Humic acid[a]	Tryptophan or protein or EPS[a]	Autochthonou DOM or hydrophilic or hydrophobic[a]	Other organic ligands[a]		
Cd^{2+} with EPS	–	–	3.0–3.5 (K_1); 2.7–3.8 (K_2)	–	–	7	Comte et al. (2006)
Cd^{2+} with EPS	–	–	1.54–3.35	–	–	7	Guibaud et al. (2006)
Cu^{2+} with EPS	–	–	3.0–4.4	–	–	7	Guibaud et al. (2004)
Ni^{2+} with EPS	–	–	2.6–3.0	–	–	7	Guibaud et al. (2004)
Sc^{3+}	17.57	17.54–20.47	–	–	–	7–9.	Sonke and Salters (2006)
Y^{3+}	9.21–13.67	10.95–14.94	–	–	–	6–9.	Sonke and Salters (2006)
Nanthnides (14 elements)							
La^{3+}	9.15–11.65	10.64–13.29	–	–	–	6.0 –9.0	Sonke and Salters (2006)
Ce^{3+}	8.90–12.09	10.56–13.55	–	–	–	6.0 –9.0	Sonke and Salters (2006)
Pr^{3+}	8.93–12.36	10.39–13.84	–	–	–	6.0 –9.0	Sonke and Salters (2006)
Nd^{3+}	9.07–12.54	10.34–13.96	–	–	–	6.0 –9.0	Sonke and Salters (2006)
Pm^{3+}	–	–	–	–	–	6.0 –9.0	Sonke and Salters (2006)
Sm^{3+}	9.56–12.65	10.58–14.38	–	–	–	6.0 –9.0	Sonke and Salters (2006)
Eu^{3+}	9.36–11.52	10.71–14.30	–	–	–	6.0 –9.0	Sonke and Salters (2006)
Gd^{3+}	9.14–11.39	10.75–14.20	–	–	–	6.0 –9.0	Sonke and Salters (2006)

(continued)

Table 1 (continued)

Elements or groups	Conditional stability constants ($\log_{10} K$)					pH	References
	Fulvic acid[a]	Humic acid[a]	Tryptophan or protein or EPS[a]	Autochthonou DOM or hydrophilic or hydrophobic[a]	Other organic ligands[a]		
Tb^{3+}	9.39–11.75	11.19–14.80	–	–	–	6.0 –9.0	Sonke and Salters (2006)
Dy^{3+}	9.47–13.64	11.31–15.30	–	–	–	6.0 –9.0	Sonke and Salters (2006)
Ho^{3+}	9.67–13.91	11.44–15.49	–	–	–	6.0 –9.0	Sonke and Salters (2006)
Er^{3+}	9.97–14.29	12.85–15.78	–	–	–	6.0 –9.0	Sonke and Salters (2006)
Tm^{3+}	10.25–14.45	11.89–16.20	–	–	–	6.0 –9.0	Sonke and Salters (2006)
Yb^{3+}	10.34–14.32	13.26–16.23	–	–	–	6.0 –9.0	Sonke and Salters (2006)
Lu^{3+}	10.44–14.58	12.25–16.50	–	–	–	6.0 –9.0	Sonke and Salters (2006)
Eu^{3+}	–	–	–	6.3	–	6	Konstantinou et al. (2009)
Eu^{3+} with 5-Sulfosalicylic acid	–	–	–	–	6.27; 11.76	?	Nair and Chander (1983)
Actinides							
Np^{5+}	–	3.3–3.7 (2.5–3.2)	–	–	–	7.4 (4.7)	Shcherbina et al. (2007)
Cm^{3+}	5.9	6.22	–	–	–	?	Buckau et al. (1992); Kim et al. (1991)
Cm^{3+} with 5-sulfosalicylic acid	–	–	–	–	6.44; 11.99	?	Nair and Chander (1983)
Am^{3+} with 5-sulfosalicylic acid	–	–	–	–	8.06; 15.34	?	Nair and Chander (1983)

(continued)

Table 1 (continued)

Elements or groups	Conditional stability constants ($\log_{10} K$)					pH	References
	Fulvic acid[a]	Humic acid[a]	Tryptophan or protein or EPS[a]	Autochthonou DOM or hydrophilic or hydrophobic[a]	Other organic ligands[a]		
Pu^{3+} with 5-sulfosalicylic acid	–	–	–	–	8.56; 17.51	?	Nair and Chander (1983)
Th^{4+} with catechol	–	–	–	–	59.65; 14.06	?	Thakur et al. (2006)
Th^{4+} with hydroquinone	–	–	–	–	48.51; 64.86	?	Thakur et al. (2006)
Th^{4+} with resorcinol	–	–	–	–	16.98; 46.46; 59.65	?	Thakur et al. (2006)
Irradiation effect on metal-DOM complexation	–	–	–	–	–	?	
Hg^{2+} with stream DOM (before irradiation)	4.9 (K_1)	–	–	–	–	?	Wu et al. (2004)
Hg^{2+} with stream DOM (after irradiation: 1–8 d)	4.6 (1 d); 4.1 (8 d)	–	–	–	–	?	Wu et al. (2004)
Cu^{2+} with river DOM (before irradiation)	7.59 (K_1); 5.83 (K_2)	–	–	–	–	?	Brooks et al. (2007)
Cu^{2+} with river DOM (after irradiation: 72 h)	7.63 (K_1); 5.88 (K_2)	–	–	–	–	?	Brooks et al. (2007)
Cu^{2+} with wetland DOM (before irradiation)	–	–	–	–	8.02 (K_1); 6.31 (K_2)	?	Brooks et al. (2007)
Cu^{2+} with wetland DOM (after irradiation: 72 h)	–	–	–	–	7.61 (K_1); 5.85 (K_2)	?	Brooks et al. (2007)
Cu^{2+} with lake DOM (before irradiation)	–	–	–	–	16.3 (K_1)	?	Sander et al. (2005)
Cu^{2+} with lake DOM (after irradiation: 30 h and 175 h)	–	–	–	–	16.7 (30 h); 16.8 (175 h)	?	Sander et al. (2005)

(continued)

Table 1 (continued)

Elements or groups	Conditional stability constants ($\log_{10} K$)					pH	References
	Fulvic acid[a]	Humic acid[a]	Tryptophan or protein or EPS[a]	Autochthonou DOM or hydrophilic or hydrophobic[a]	Other organic ligands[a]		
Cu^{2+} with autochthonous DOM (Estuary: before irradiation)	–	–	–	–	13.5	?	Shank et al. (2006)
Cu^{2+} with autochthonous DOM (Estuary: after irradiation: 1.5 and 14.5 d)	–	–	–	–	13.5	?	Shank et al. (2006)

[a]indicates the components which are extracted or standard or under natural condition used
pH values in parentheses indicates the conditional stability constant in parentheses
Standard deviations do not mention in the results to avoid the complexiety
K_1 and K_2 values for two binding sites are mentioned using semicolon (;)
h = hours; d = days

complexation of different trace elements (Wu et al. 2004a; c; Plaza et al. 2006; Cook and Langford 1995; Kumke et al. 1998; Mcgown et al. 1995)· The conditional stability constants ($\log_{10} K_1$ and $\log_{10} K_2$) for the complexation of two major binding sites of fulvic acid with metal ions are 3.26–6.66 ($\log_{10} K_1$) and 7.0–9.64 ($\log_{10} K_2$) for Cu^{2+}-FA; 5.66 for Fe^{3+}-FA; 4.34–5.70 for Hg^{2+}-FA; 5.46 for UO_2^{2+}-FA; and 5.9 ($\log_{10} K_1$) for Cm^{3+}-FA, determined in aqueous solution except for lanthanides (Table 1) (Wu et al. 2002a, 2004; Fu et al. 2007; Hays et al. 2004; Cao et al. 2004; Sidenius et al. 1999; Antunes et al. 2007; Winzerling et al. 1992; Tipping 1994; Breault et al. 1996; Mcknight et al. 1983; Buckau et al. 1992; Kim et al. 1991). Lanthanides (Ln) are strong complexing agents for Suwannee River Fulvic Acid and show relatively high conditional stability constants that range from 8.90 to 14.58 for Ln-SRFA (Suwannee River Fulvic Acid) complexes at pH 6–9 in aqueous solution. The values obtained for SRFA are relatively lower than for humic acids extracted from various sources (Table 1; Sonke and Salters (2006)).

The conditional stability constants for Hg^{2+}-DOM complexation are 4.34–5.20 for the peak C at Ex/Em = 340–355/430–455 nm of humic substances (possibly fulvic acid) at pH 4.37–7.01 (Wu et al. 2004a). The conditional stability constants for Hg^{2+}-FA complexation are 5.01–5.55 ($\log_{10} K_1$) for peak C detected at Ex/Em = 330–335/426–434 nm and 5.62–5.72 ($\log_{10} K_2$) for peak A detected at Ex/Em = 245/432–438 nm (Fu et al. 2007). Complexation of river DOM with Hg^{2+} is 3.54–4.93 and 3.64–4.85, determined using linear and non-linear model fitting of the fluorescence maxima at Ex/Em = 304, 306/426, 430 nm (Bai et al. 2008). The fluorophores or functional groups bound for peak C at longer wavelength are presumably considered to be the fast binding sites and the fluorophores bound for peak A, correspondingly, the slow binding sites in fulvic acid. Generally, $\log_{10} K_1$ should be higher than in $\log_{10} K_2$ but the result is, in fact, the opposite. The reason behind this phenomenon is that during the complexation of the fast binding site at peak C, the decrease in fluorescence intensity occurs simultaneously at peak A of fulvic acid. Note that the fluorophores bound at a particular peak (e.g., peak C, peak A, peak T or T_{UV}) of a fluorescent molecule are the result of all fluorophores existing in the molecule. Therefore, any changes in the molecule by photoinduced or microbial degradation can alter its fluorescence properties, inducing new peak position and fluorescence intensity (Mostofa et al. 2009a, 2011; Senesi 1990). For the fast-reacting binding site of fulvic acid, the kinetic rate constant (k_1) ranges from 0.18 to 0.55 s^{-1}, whilst the k_2 values for the slow reacting binding site are similar for all metals (Cu^{2+}, Ni^{2+}, Co^{2+}, Cd^{2+}, and Ca^{2+}) at 0.01–0.02 s^{-1} (Wu et al. 2004c).

The conditional stability constants for the complexation of humic acid (HA, two binding sites) with metal ions are 3.99–5.60 ($\log_{10} K_1$) and 8.55 ($\log_{10} K_2$) for Cu^{2+}-HA; 0.40 ($\log_{10} K_1$) for Mn^{2+}-HA; 0.50 ($\log_{10} K_1$) for Co^{2+}-HA; 0.50–4.43 ($\log_{10} K_1$) for Zn^{2+}-HA; 6.79–10.0 ($\log_{10} K_1$) for Fe^{3+}-HA; 6.00 ($\log_{10} K_1$) for Cr^{3+}-HA; 15.0 ($\log_{10} K_1$) for Sc^{3+}-HA; 2.1 ($\log_{10} K_1$) for VO_2^{+}-HA; 5.50 ($\log_{10} K_1$) for Hg^{2+}-HA; 4.5–5.3 ($\log_{10} K_1$) and 5.8–7.2 ($\log_{10} K_2$) for As^{3+}-HA; 2.80–4.94 ($\log_{10} K_1$) and 7.15–9.80 ($\log_{10} K_2$) for H^{+}-HA; 4.0 ($\log_{10} K_1$) for Be^{2+}-HA; 1.0 ($\log_{10} K_1$) for Ca^{2+}-HA; 0.5 ($\log_{10} K_1$) for Sr^{2+}-HA; 0.5 ($\log_{10} K_1$) for Ba^{2+}-HA; 10.0 ($\log_{10} K_1$) for Ga^{3+}-HA; 4.95–5.81 ($\log_{10} K_1$) for Pb^{2+}-HA;

4.24–4.63 ($\log_{10} K_1$) for Cd^{3+}-HA; 4.27 ($\log_{10} K_1$) for $UO_2{}^{2+}$-HA; 3.3–3.7 ($\log_{10} K_1$) for Np^{5+}-HA; and 6.22 ($\log_{10} K_1$) for Cm^{3+}-HA in aqueous solution except lanthanides (Table 1) (Liu and Cai 2010; Cao et al. 2004; Antunes et al. 2007; Tipping 1994; Buckau et al. 1992; Kim et al. 1991). Lanthanides (Ln) are strong complexing agents for humic acid, showing relatively high conditional stability constants that range from 10.34 to 16.50 for Ln-LHA (Leonardite coal humic acid standard) and from 12.17 to 16.22 for Ln-EHA (Elliot soil humic acid standard) at pH 7–9 in aqueous media (Table 1) (Sonke and Salters 2006). The stability constants of the standard Aldrich humic acid show significantly low values (2.65–2.75 at 5 mg L^{-1}) for lanthanides. These values are greatly increased with increasing the humic acid concentration (e.g. La ranges from 2.65 to 3.85 for humic acid concentrations of 5 and 20 mg L^{-1}, respectively) for each lanthanide (Pourret and Martinez 2009). For the concentration of 20 mg L^{-1} of the standard Aldrich humic acid, the stability constants increase from La (3.85) to Eu (4.15) and then decrease from Gd (4.06) to Lu (3.95).

The stability constants of the M-HA complexes reached high values ($\log_{10} K_{Pb} = 5.81$, $\log_{10} K_{Cu} = 5.55$, $\log_{10} K_{Cd} = 4.63$ and $\log_{10} K_{Zn} = 4.43$) for the peak C (Ex/Em = 440/510 nm) of humic acid extracted from the corresponding soil (nonamended and control) compared to those of sewage sludge (4.95, 4.65, 4.24 and 4.08, respectively) and to soil amended with sewage sludge at a rate of 40 tons ha^{-1} (5.53, 5.36, 4.47 and 4.31, respectively), for metal ions such as Pb^{2+}, Cu^{2+}, Cd^{2+}, and Zn^{2+} in aqueous solution at pH 8.1 (Table 1) (Plaza et al. 2006). The overall stability constant for M-HA complexes can follow the order Pb(II) > Cu(II) > Cd(II) > Zn(II) (Plaza et al. 2006). These results suggest that the fluorophores or functional groups of humic acid of different origin are highly variable and may affect the metal-DOM complexation and transport in soil and natural water ecosystems.

The conditional stability constants for complexation between humic acid and alkaline earth metal ions follow the order Be > Ca > Sr, Ba humate. It can be hypothesized that the humates of alkaline earth metal ions with smaller ionic radii are more stable, and that humic acid is a 'hard' ligand (Takahashi et al. 1997). Correspondingly, Sc and heavy lanthanide elements (Yb and Lu) can form more stable complexes than light lanthanide elements (Ce and Eu), depending on the ionic radii (Takahashi et al. 1997). In addition, humic acid complexes of trivalent Fe and Ga are more stable than those of rare earth elements (REEs) except for Sc, while the Cr(III)-humic acid complex is less stable than REE-humic acid complexes (Takahashi et al. 1997). It has also been shown that the conditional stability constant ($\log_{10} K$) between lanthanides and humic substances (fulvic and humic acids) of varied sources significantly increases from La to Lu and increases with increasing pH and decreasing ionic strength (I) of the solution (Fig. 5a–c) (Sonke and Salters 2006). The log K values also significantly increase with decreasing ionic radius (Fig. 6) (Sonke and Salters 2006). The strength of M-DOM complexation between lanthanides and humic substances (fulvic and humic acids) thus follows the order Lu > Yb > Tm > Er > Ho > Dy > Tb > Gd > Eu > Sm > Pm > Nd > Pr > Ce > La).

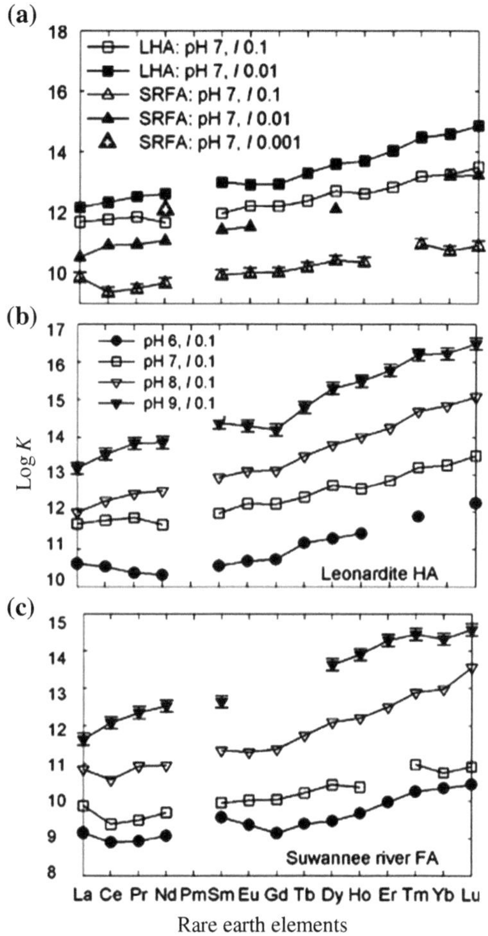

Fig. 5 Conditional binding constants (log K) for rare earth elements (REE)–Leonardite coal humic acid standard (LHA) and REE–SRFA complex formation at pH 7 with various ionic strength (**a**) and the effect pH on log K for complexation of LHA (**b**) and SRFA (**c**) with REE. Error bars (0.16 log units) are only shown for pH 7 (**a**), but apply to all log Kc values. Nominal total concentrations are 100 nM REE; 20 mg L^{-1} SRFA, 130 nM EDTA for SRFA experiments (**b** and **c**). *Data source* Sonke and Salters (2006)

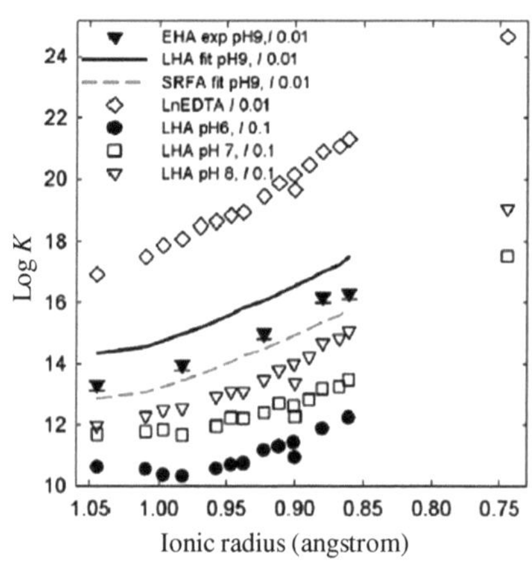

Fig. 6 Log K for lanthanides and organic ligands of various sources at pH 6, 7, and 8 as a function of ionic radius (sixfold hydrated), including Sc and Y transition metals. Log K for Y and Leonardite coal humic acid standard (LHA) values are represented by singular data points, floating just below the main rare earth elements (REE) data trends at an ionic radius of 0.900 Å. Log K for lanthanides and Elliot soil humic acid standard (EHA) values observes at pH 9, and 0.01 mol L^{-1} I and extrapolated log K values for SRFA and LHA under those conditions. EDTA conditional binding constants for 0.01 mol L^{-1} I are included for comparison. *Data source* Sonke and Salters (2006)

The conditional stability constants for the complexation of hydrophobic or hydrophilic fractions of DOM or autochthonous DOM with metal ions are 7.8–9.6 ($\log_{10} K_1$) for Zn^{2+}-DOM; 8.92–16.3 ($\log_{10} K_1$) and 6.87–12.9 ($\log_{10} K_2$) for Cu^{2+}-DOM; 11.6–31.0 ($\log_{10} K_1$) and 10.5–11.2 ($\log_{10} K_2$) for Hg^{2+}-DOM in lakes, estuaries and oceans (Table 1) (Haitzer et al. 2002; Xue and Sigg 1993; Xue et al. 1995; van Den Berg et al. 1987; Sunda and Huntsman 1991; Sunda and Hanson 1987; Sunda and Ferguson 1983; Moffett et al. 1990; Coale and Bruland 1988, 1990; van Den Berg 1984; Midorikawa and Tanoue 1996; Benoit et al. 2001; Drexel et al. 2002; Gasper et al. 2007; Hsu and Sedlak 2003; Sander et al. 2005; Shank et al. 2006). It has been shown that the conditional stability constants between autochthonous DOM of phytoplankton or algal origin and Cu^{2+} are much higher ($\log_{10} K_1 = 13.9$–15.6 and $\log_{10} K_2 = 11.8$–13.4) in lakes, and they are similar to those of organic ligands of phytoplankton or biological sources (11.1–13.2 and 9.2–10.2, respectively) in estuaries and seawater compared to those of allochthonous fulvic and humic acids (Table 1) (Xue and Sigg 1993; Xue et al. 1995; van Den Berg et al. 1987; Moffett et al. 1990; Coale and Bruland 1988; Coale and Bruland 1990; Sunda 1988). Surface water from the Irish Sea and the Atlantic Ocean contains ligand concentrations of 1.7×10^{-7} and 1.1×10^{-7} M, with conditional stability constants ($\log_{10} K$) of 9.84 ± 0.13 and 9.86 ± 0.23, respectively, at pH 8.0 (van Den Berg 1982).

The conditional stability constants of the Hg^{2+}-DOM complexation are significantly higher for the hydrophobic fraction of DOM or DOM in natural surface waters (Table 1) (Haitzer et al. 2002, 2003; Benoit et al. 2001; Drexel et al. 2002; Gasper et al. 2007; Hsu and Sedlak 2003; Ravichandran et al. 1998; 1999; Lamborg et al. 2003, 2004; Waples et al. 2005). The high affinity of the Hg^{2+}-DOM complexation is responsible for the reduced S-containing binding sites (thiol and disulfide/disulfane functional groups) bound in DOM, presumably autochthonously produced in natural waters (Haitzer et al. 2002; Benoit et al. 2001; Gasper et al. 2007; Dyrssen and Wedborg 1986; Schuster 1991; Guentzel et al. 1996; Wallschlager et al. 1996; Xia et al. 1999). The possible complexation reaction can be written as [$Hg^{2+} + R\text{-}SH^{n-} = HgR\text{-}S^{(n-1)-} + H^+$] (Benoit et al. 2001; Dyrssen and Wedborg 1991).

The conditional stability constants for the complexation of the two major binding sites of tryptophan with metal ions are 4.88–4.90 ($\log_{10} K_1$) for Cu^{2+}-tryptophan; 7.82–9.56 for Cu^{2+}-tryptophan; and 4.99–5.33 for Hg^{2+}-tryptophan for peak T at Ex/Em = 275–285/330–360 nm in aqueous solution (Table 1) (Fu et al. 2007; Hays et al. 2004; Wu and Tanoue 2001b).

The conditional stability constants ($\log_{10} K_1$ and $\log_{10} K_2$) for the complexation of two major binding sites of EPS with metal ions are 3.98–4.12 ($\log_{10} K_1$) and 4.28–4.48 ($\log_{10} K_2$) for Hg^{2+}-EPS; 0.45–4.9 ($\log_{10} K_1$) and 3.2–4.6 ($\log_{10} K_2$) for Pb^{2+}-EPS; 1.54–3.5 ($\log_{10} K_1$) and 2.7–3.8 ($\log_{10} K_2$) for Cd^{2+}-EPS; 3.0–4.4 ($\log_{10} K_1$) for Cu^{2+}-EPS; and 2.6–3.0 ($\log_{10} K_1$) for Ni^{2+}-EPS in aqueous solution (Table 1) (Zhang et al. 2010; Guibaud et al. 2004, 2006; Comte et al. 2006). The conditional stability constants of Hg^{2+}-EPS complexes are relatively low (3.98–4.12: $\log_{10} K_1$) at peak T (Ex/Em = 275–280/328–334 nm) compared to those (4.28–4.48: $\log_{10} K_2$) at peak T_{UV} (Ex/

Em = 220–230/322–336 nm) in aqueous solution (Zhang et al. 2010). Note that the fluorescence properties of EPS are similar to those of tryptophan, with higher fluorescence intensity at peak T_{UV}-region than at peak T-region and similar Ex/Em maxima.

The conditional stability constants ($\log_{10} K$) for the complexation of some standard organic and inorganic ligands with metal ions are identified in aqueous solution as 3–6 for Cu^{2+}-carboxylic or HO group; 2.8 for Cu^{2+}-amine (NH_2); 4.4 for Cu^{2+}-ethanol amine; 6.6 for Cu^{2+}-glycine (NH_2–CH_2–COOH or NH_3^+–CH_2–COO^-); 8.5 for Cu^{2+}-ethylene diamine (NH_2–CH_2–CH_2–NH_2); 7.2 for Cu^{2+}- aspartic acid [HOOC–CH(NH_2)–CH_2–COOH]; 9.4 for Cu^{2+}- iminodiaacetic acid (HOOC–CH_2–NH_2–CH_2–COOH); 8.3 for Cu^{2+}- diaminopropanol; 11.4 for Cu^{2+}- ethylenediamine-N-acetic acid; 13.0 for Cu^{2+}-diethylenetriamine; 6.27 and 11.76 for Eu^{3+}-5-sulfosalicylic acid; 6.4 and 11.99 for Cm^{3+}-5-sulfosalicylic acid; 8.06 and 15.34 for Am^{3+}-5-sulfosalicylic acid; 8.56 and 17.51 for Pu^{3+}-5-sulfosalicylic acid; 59.65 and 14.06 for Th^{4+}- catechol; 48.51 and 64.86 for Th^{4+}-hydroquinone; and 16.98, 46.46, and 59.65 for Th^{4+}- resorcinol (Table 1) (Nair and Chander 1983; Martin and Prados 1974; Thakur et al. 2006; Smith 1974). The stability constants of various standard organic substances are similar to those of allochthonous and autochthonous DOM in natural waters, indicating that carboxylic, phenolic and amino group-containing carboxylic acid bound in allochthonous and autochthonous DOM may form complexes with metal ions in aqueous solution.

The binary complexes of Cu^{2+}, Ni^{2+}, Co^{2+} and Zn^{2+} with resorcinol and some biologically important aliphatic dicarboxylic acids (adipic, succinic, malic, malonic, maleic, tartaric and oxalic) in aqueous solution at 25 °C and $I = 0.1$ mol dm^{-3} $NaNO_3$ suggest the following hypotheses (Radalla 2010): (i) Stability of the 1:1 binary complexes for all investigated ligands is higher than that of the corresponding 1:2 ones. (ii) Normal 1:1 and 1:2 complexes of resorcinol are formed with all of the metal ions studied. (iii) The order of stability constants of the 1:1 complexes of all investigated ligands with respect to the divalent transition metal ions follows the Irving-Williams series (Zn < Cu > Ni > Co) (Irving and Williams 1948). (iv) Stability of the 1:1 metal-complexes of aliphatic dicarboxylic acids follows their basicity ($pK_{a1} + pK_{a2}$), where K_{a1} and K_{a2} are the first and second dissociation constants, respectively.

Experimental studies show that photoinduced degradation can decrease the conditional stability constants of M-DOM complexes in stream waters ($\log K_1 = 4.9$ to 4.1) and in wetland waters ($\log K_1 = 8.02$ and $\log K_2 = 6.1$ to 7.61 and 5.85, respectively), whilst their values are increased in river DOM ($\log K_1 = 7.59$ and $\log K_2 = 5.83$ to 7.63 and 5.88, respectively) and in lake DOM ($\log K_1 = 16.3$ to 16.8). The values remain the same ($\log K_1 = 13.5$) before and after irradiation of estuarine waters (Wu et al. 2004a; Brooks et al. 2007; Sander et al. 2005; Shank et al. 2006). The decomposition of the functional groups in DOM is susceptible to decrease the stability constants of the M-DOM complexes in natural waters. In contrast, the increase in stability constants of the M-DOM complexes is likely caused by the formation of new photoproducts with stronger functional groups for M-DOM complexation. Irradiation of waters with high contents of DOM is not able to substantially modify the degree of M-DOM complexation, which is the most likely explanation for the cases where the same stability constants have been observed before and after irradiation. The effect of photoinduced degradation on M-DOM complexation will be explained in details later.

3.6.1 Conditional Stability Constants for Ternary Complexes in Waters

Ternary complexes are operationally defined as complexes involving two newly successive bonds between a metal ion and two different types of DOM components or organic ligands (e.g. allochthonous fulvic acid and tryptophan). It is assumed that fulvic acid (FA) acts as a primary ligand to form bonds with a metal ion (M). Therefore, one initially observes binary complex formation such as $M - FA$ ($M + FA \leftrightarrows M - FA$). Then another molecule, e.g. tryptophan (T) acts as a secondary ligand and forms ternary complexes that can be represented as such as $T - M - FA$ ($M - FA + T \leftrightarrows T - M - FA$).

Formation of ternary complexes is observed in aqueous solution (Martin and Prados 1974; Khalil and Radalla 1998; Khalil and Attia 1999; Khalil 2000a, b; Khalil and Taha 2004; Khalil and Fazary 2004; Radalla 2010; Rosas et al. 2010).

For the formation of ternary complexes of the selected bivalent metal ions (M) in the presence of resorcinol = R and aliphatic dicarboxylic acid = A, the following equilibria may be considered (Eqs. 3.11 and 3.12) (Radalla 2010):

$$M + A \rightleftharpoons MA, \quad K^{M}_{MA} = \frac{[MA]}{[M][A]} \tag{3.11}$$

$$MA + R \rightleftharpoons MAR, \quad K^{MA}_{MAR} = \frac{[MAR]}{[M][A]} \tag{3.12}$$

In the presence of both ligands, A is presumably considered to interact first with M forming a 1:1 MA binary complex. It follows interaction of R in a stepwise manner. The overall stability constant β^{M}_{MAR} can be described as below (Eq. 3.13):

$$M + A + R \rightleftharpoons MAR, \quad \beta^{M}_{MAR} = \frac{[MAR]}{[M][A][R]} = K^{MA}_{MAR} \times K^{M}_{MA} \tag{3.13}$$

The β^{M}_{MAR} constant expresses the stability of the mixed-ligand species and it does not represent the binding strength between R and M^{2+} ions directly in the presence of A. This effect is much better reflected by the equilibrium constant, K^{MA}_{MAR} calculated according to Eq. 3.14:

$$\log_{10} K^{M}_{MAR} \rightleftharpoons \log_{10} \beta^{M}_{MAR} - \log K^{M}_{MA} \tag{3.14}$$

The equilibrium constant expressed in Eq. 3.14 indicates how tightly R is bound to the simple MA binary complex in aqueous solution (Radalla 2010).

The ternary complexes of Cu^{2+}, Ni^{2+}, $Co^{2+}$and Zn^{2+} with resorcinol as primary ligand and with some biologically important aliphatic dicarboxylic acids (adipic, succinic, malic, malonic, maleic, tartaric and oxalic) as secondary ligands in aqueous solution at 25 °C and $I = 0.1$ mol dm^{-3} $NaNO_3$ show very high conditional stability constants, such as $\log_{10} \beta^{M}_{MAR} = 13.05 - 16.48$ for Cu^{2+}, 11.72–13.32 for Ni^{2+}, 10.01–11.29 for Zn^{2+}, and 8.78–10.01 for Co^{2+}, compared to those of the respective binary complexes (2.29–7.30, 2.29–6.10, 2.33–5.80 and 2.19–5.80, respectively, ranges including both $\log_{10} K_1$ and $\log_{10} K_2$) (Table 2) (Radalla 2010). The ternary complexes between vanadium(III)-picolinic acid (or dipicolinic acid) and the amino acids cysteine, histidine, aspartic acid and glutamic acid have been examined in aqueous solution at 25 °C and 3.0 mol·dm^{-3} KCl as ionic medium (Table 2) (Rosas et al. 2010; Shiozawa et al. 2011). The results show that the stability constants range from −1.96-4.8 for three ternary complexes of aspartic acid, 0.74–2.26 for three ternary complexes of glutamic acid, 1.71–8.69 for four ternary complexes of cysteine, and −0.35-10.22 for histidine (Rosas et al. 2010)· For dipicolinic acid as a primary ligand, the stability constants range from −5.98 to 8.17 for five ternary complexes of aspartic acid, −6.2 to 10.59 for five ternary complexes of glutamic acid, −1.54 to 13.91 for four ternary complexes of cysteine, and −9.7 to 14.1 for five ternary complexes of histidine (Shiozawa et al. 2011). The significantly high values of the stability constants of ternary complexes are similar to those of DOM in natural surface waters or hydrophobic acids extracted from surface waters ($K_1 = 8$–15; Table 1) (Midorikawa and Tanoue 1998; Sunda and Hanson 1987, 1991; Sunda and Ferguson 1983; Moffett et al. 1990; Coale and Bruland 1988, 1990; van Den Berg 1984; Midorikawa and Tanoue 1996). Therefore, formation of ternary complexes could be vital to understand the M-DOM complexation in natural waters. The results of a species distribution obtained for the Cu^{2+} + malonic acid + resorcinol system shows that the formation of MA starts at pH < 3, reaches a maximum concentration (70 % total Cu^{2+}) at pH 4.2 and decreases to a minimum when MAR becomes predominant. The maximum concentrations of MA_2 and MR are less than 5 % of total Cu^{2+} in solution (Radalla 2010).

These results suggest the following hypotheses (Radalla 2010): (i) Stabilities of ternary complexes with respect to the aliphatic dicarboxylic acids follow the order: adipic > succinic > malic > malonic > maleic > tartaric > oxalic; this behavior can be explained in terms of the decrease in basicity of the aliphatic dicarboxylic acids in the same direction. (ii) The complex stability of the ternary complexes with respect to the metal ion present follows the Irving-Williams series (Irving and Williams 1948). (iii) The stabilities of the ternary complexes are higher than for the 1:1 binary complexes of the corresponding aliphatic dicarboxylic acid or resorcinol for all systems studied; this behavior can be attributed to some cooperative interaction in the ternary complex between the carboxylic acid and resorcinol, such as H-bond formation. Finally, it is vital to examine the formation of ternary complexes of metal ions with fulvic acids and tryptophan in aqueous media, under laboratory conditions. This should be the focus for future studies.

Table 2 Conditional stability constants for Cu^{2+}, Ni^{2+}, Zn^{2+}, and Co^{2+} binary and ternary complexes in aqueous media at 25 °C and $I = 0.10$ mol.dm^{-3}

Elements or groups	Cu^{2+}				References
	$\text{Log}_{10}K_1$	$\text{Log}_{10}K_2$	$\log_{10}K_{MAR}^{MA}$	$\log_{10}\beta_{MAR}^{M}$	
Resorcinol (1,3-dihydroxybenzene)	7.30	5.50			Radalla (2010)
Adipic acid (HOOC–CH_2–CH_2–CH_2–CH_2–COOH)	3.84	2.29	9.17	13.81	Radalla (2010)
Succinic acid (HOOC–CH_2–CH_2–COOH)	3.2	–	9.83	13.05	Radalla (2010)
Malic acid [HOOC–CH_2–CH(OH)–COOH]	6.8	–	7.3	16.48	Radalla (2010)
Malonic acid (HOOC–CH_2–COOH)	4.82	3.37	7.3	14.31	Radalla (2010)
Maleic acid (HOOC–CH=CH–COOH)	4.08	2.82	7.3	13.5	Radalla (2010)
Tartaric acid [HOOC–CH(OH)–CH(OH)–COOH]	6.69	5.7	7.3	16.31	Radalla (2010)
Oxalic acid (HOOC–COOH)	4.66	4.06	7.3	13.75	Radalla (2010)
	Ni^{2+}				
Resorcinol (1,3-dihydroxybenzene)	6.10	5.30			Radalla (2010)
Adipic acid (HOOC–CH_2–CH_2–CH_2–CH_2–COOH)	3.54	2.29	9.17	12.71	Radalla (2010)
Succinic acid (HOOC–CH_2–CH_2–COOH)	3.17		8.84	12.01	Radalla (2010)
Malic acid [HOOC–CH_2–CH(OH)–COOH]	4.63		8.69	13.32	Radalla (2010)
Malonic acid (HOOC–CH_2–COOH)	3.93	2.92	8.58	12.51	Radalla (2010)
Maleic acid (HOOC–CH=CH–COOH)	3.68	2.82	8.42	12.10	Radalla (2010)
Tartaric acid [HOOC–CH(OH)–CH(OH)–COOH]	4.67	3.38	8.36	13.03	Radalla (2010)
Oxalic acid (HOOC–COOH)	3.46	2.96	8.26	11.72	Radalla (2010)
	Zn^{2+}				
Resorcinol (1,3-dihydroxybenzene)	5.15	3.90			Radalla (2010)
Adipic acid (HOOC–CH_2–CH_2–CH_2–CH_2–COOH)	3.41	2.35	7.88	11.29	Radalla (2010)
Succinic acid (HOOC–CH_2–CH_2–COOH)	3.06		7.79	10.83	Radalla (2010)
Malic acid [HOOC–CH_2–CH(OH)–COOH]	3.38		7.54	10.72	Radalla (2010)
Malonic acid (HOOC–CH_2–COOH)	2.98	2.33	7.48	10.82	Radalla (2010)
Maleic acid (HOOC–CH=CH–COOH)	3.40	2.45	7.37	10.77	Radalla (2010)

(continued)

Table 2 (continued)

Elements or groups	Cu^{2+}				References
	$\mathrm{Log}_{10}K_1$	$\mathrm{Log}_{10}K_2$	$\log_{10}K_{MAR}^{MA}$	$\log_{10}\beta_{MAR}^{M}$	
Tartaric acid [HOOC–CH(OH)–CH(OH)–COOH]	3.81	2.75	7.21	11.02	Radalla (2010)
Oxalic acid (HOOC–COOH)	3.03	2.47	6.98	10.01	Radalla (2010)
	Co^{2+}				
Resorcinol (1,3-dihydroxybenzene)	5.80	4.90			Radalla (2010)
Adipic acid (HOOC–CH_2–CH_2–CH_2–CH_2–COOH)	3.31	2.19	6.70	10.01	Radalla (2010)
Succinic acid (HOOC–CH_2–CH_2–COOH)	2.97		6.58	9.55	Radalla (2010)
Malic acid [HOOC–CH_2–CH(OH)–COOH]	3.12		6.39	9.51	Radalla (2010)
Malonic acid (HOOC–CH_2–COOH)	2.94	2.28	6.35	9.29	Radalla (2010)
Maleic acid (HOOC–CH=CH–COOH)	3.08	2.24	6.09	9.17	Radalla (2010)
Tartaric acid [HOOC–CH(OH)–CH(OH)–COOH]	3.15	2.37	6.00	9.15	Radalla (2010)
Oxalic acid (HOOC–COOH)	2.88	2.38	5.90	8.78	Radalla (2010)
	V^{3+}				
Picolinic acid (pyridine-2-carboxylic acid)					
Picolinic acid + aspartic acid [HOOC–CH(NH_2)–CH_2–COOH]	–	–	–	2.12–4.8	Rosas et al. (2010)*
Dipicolinic acid + glutamic acid [HOOC–CH_2–CH_2–CH(NH_2)–COOH]	–	–	–	0.74–2.26	Rosas et al. (2010)*
Dipicolinic acid + cycteine [HOOC–CH(NH_2)–CH_2–SH]	–	–	–	1.7–8.69	Rosas et al. (2010)*
Dipicolinic acid + histidine [2-amino-3-(1H-imidazol-4-yl)propanoic acid]	–	–	–	5.48–10.22	Rosas et al. (2010)*
Dipicolinic acid (pyridine-2,6-dicarboxylic acid)					
Dipicolinic acid + aspartic acid [HOOC–CH(NH_2)–CH_2–COOH]	–	–	–	3.15–8.17	Shiozawa et al. (2011)*
Dipicolinic acid + glutamic acid [HOOC–CH_2–CH_2–CH(NH_2)–COOH]	–	–	–	5.57–10.59	Shiozawa et al. (2011)*
Dipicolinic acid + cycteine [HOOC–CH(NH_2)–CH_2–SH]	–	–	–	5.30–13.91	Shiozawa et al. (2011)*
Dipicolinic acid + histidine [2-amino-3-(1H-imidazol-4-yl)propanoic acid]	–	–	–	5.7–14.1	Shiozawa et al. (2011)*

* indicates the positive stability constant values of several ternary complexes formed in the aqueous solution

3.7 *Homogeneous and Heterogeneous Complexation of Fulvic and Humic Acids to Metals*

Homogeneous complexation is operationally defined as a chemical binding equilibrium behavior (approximately in the ratio of 0.8–1:1) between metal ion and a ligand, when they are mixed up under specific conditions in aqueous media. Heterogeneous complexation is operationally defined as a chemical binding equilibrium behavior with a <0.8:1 ratio between metal ion and a ligand.

It has been shown that fulvic and humic acids (humic substances) can behave as homogeneous and heterogeneous complexants to trace metals in waters (Filella 2008; Chakraborty et al. 2007; Pinheiro et al. 1994; Town and Filella 2000; Murimboh 2002). A comparison of the heterogeneity parameter (Γ, termed Gamma: Greek Capital Letter) for Zn(II), Cd(II), Pb(II) and Cu(II) complexes in different model solutions of SRFA shows that ΓCd > ΓZn > ΓPb > ΓCu (Chakraborty et al. 2007). The results show that the value of Γ remains about the same for a given metal: the value of Γ for Cu(II) is ~0.50 in all metal to SRFA solutions with $C_M/C_{SRFA} = 0.01$, 0.001, and 0.0005, suggesting that SRFA behaves as an heterogeneous complexant for Cu(II) (Chakraborty et al. 2007). The value of Γ for Pb(II) is ~0.70 in comparable solutions, suggesting that SRFA behaved as a less heterogeneous complexant for Pb(II) compared to Cu(II). The Γ value obtained for Cd is ~0.94, suggesting that SRFA almost behaved as a homogeneous complexant for Cd, and a bit lower for Zn (Γ ~ 0.86) (Chakraborty et al. 2007). These results in combination with other studies suggest that SRFA behaves as a relatively homogeneous complexant for Zn(II) and Cd(II), as a relatively heterogeneous complexant for Pb(II) and as an even more heterogeneous complexant for Cu(II) (Chakraborty et al. 2007; Town and Filella 2000; Murimboh 2002). Coherently, it has also been shown that humic acid derived from peat performs more heterogeneous complexation of Pb(II) compared to Cd(II) (Pinheiro et al. 1994).

4 Binding Sites (or Functional Groups) in Fulvic Acid, Humic Acid and Other Ligands

Humic substances (fulvic and humic acids), autochthonous fulvic acids of phytoplankton origin (see chapters “Dissolved Organic Matter in Natural Waters” and “Fluorescent Dissolved Organic Matter in Natural Waters” for detailed discussion), polysaccharides, proteins, peptides, nucleic acids, extracellular polymeric substances and amino acids, which have properties of polyfunctionality, polyelectrolyticy, size polydispersity, physical heterogeneity and structural lability, are natural organic ligands with binding sites (or functional groups) that can form complexes with trace metals (Malcolm 1985; Mostofa et al. 2009a, b; Zhang et al. 2009, 2010; Merroun and Selenska-Pobell 2008; Filella 2008; Mandal et al. 1999; Xue et al. 1995; Hatch et al. 2009). The molecular structure of allochthonous fulvic and humic acids is composed of a variety of functional groups (or fluorophores) such as benzene-containing carboxyl, methoxylate and phenolic groups (catechol-type), carboxylic and di-carboxylic groups, alcoholic OH, carbohydrate OH, –C=C–, hydroxycoumarin-like structures, fluorophores containing Schiff-base derivatives, chromone, xanthone, quinoline, O, N, S, and P-atom-containing functional groups including aromatic carbon (17–30 %) and aliphatic carbon (47–63 %) (Malcolm 1985; Senesi 1990; Morel and Hering 1993; Leenheer et al. 1998; Steelink 2002; Leenheer and Croue 2003; Stenson et al. 2003; Fimmen et al. 2007). In humic substances, 60–90 % of the acidic groups are carboxylic and the remainder are phenolic (Morel and Hering 1993). S-XANES measures have shown that sulphur is present in humic substances in many different oxidation states: thiol, thiophene or disulfide, sulfoxide, sulfone, sulfonate and sulfate esters (Morra et al. 1997; Xia et al. 1998; Fimmen et al. 2007). A typical humic acid containing 0.2 % reduced sulphur has only 63 μmol g^{-1} of thiol sites (Bloom et al. 2001).

Fulvic and humic acids can be roughly classified into two categories: minor (approximately 1–10 %), strong sites, and major (approximately 90–99 %), weak sites (Mandal et al. 1999; Chakraborty et al. 2007; Buffle and Filella 1995).Wu and Tanoue 2001c The strong binding sites are first occupied entirely, after which the weak binding sites are occupied (Mandal et al. 1999; Chakraborty et al. 2007; Buffle and Filella 1995; Wu and Tanoue 2001c). The major sites are less diverse in type, but in number they represent the majority of the complexation sites (e.g., carboxylate and phenolate groups in humic substances). The minor sites, which are fewer in number, include all those which exhibit a wide range of binding energies, including strongly complexing nitrogen- and sulphur-containing sites (Filella 2008). Major sites represent the largest proportion of dissociable groups. They play an important role in the polyelectrolytic properties of the complexant (Filella 2008). The distinction between major and minor sites also reflects a chemical selectivity: since the major sites have oxygen-containing groups, they will preferably react with hard metals such as calcium and magnesium (Filella 2008).

EEM spectroscopy (EEMS) during the complexation between extracted fulvic acid and several metals [Cu(II), Ni(II), Co(II), Cd(II) and Ca(II)] also showed two major kinetically distinguishable binding sites on fulvic acid, 'fast' and 'slow', having reaction half-lives of 1.3–3.9 and 34.7–69.3 s, respectively (Wu et al. 2004c). The 'fast' sites in fulvic and humic acids are susceptible to be the fluorophores (or functional groups) bound at the longer wavelength peak C-region, which are rapidly complexed with metal ions. In contrast, the 'slow' sites are considered to be the fluorophores at the shorter wavelength peak A-region.

Another time-resolved fluorescence study demonstrates that fulvic acid during its compexation with metal ions has three binding sites, which can be assigned lifetimes and wavelength maxima as follows: ~50 ps at 392 nm, ~430 ps at 465 nm and 4.2 ns at 512 nm (Cook and Langford 1995). The functional groups in humic acid exhibit different proton binding properties as their concentration increases from 20 to 100 mg L^{-1}. The acidic functional group content values suggest that 40 % of the total acidity is accounted for by carboxylic-type sites in humic acids extracted from peat and soil (Vidali et al. 2009). Proton affinity constants to humic acids extracted from various sources suggest that carboxylic and phenolic-type groups in humic acids can bind to H^+, whereas carboxylic-type binding sites exhibit a smaller apparent heterogeneity than phenolic-type sites in humic acids (Vidali et al. 2009; Benedetti et al. 1996; Kinniburgh et al. 1996; Pinheiro et al. 1999; Plaza et al. 2005). Copper(II) complexation with DOM suggests that sites characterized as phenolic based on alkalimetric titration, and not carboxylic sites, account for the majority of Cu complexation under natural water conditions. Cu–DOM complexation mainly takes place through replacement of H^+ by Cu^{2+} at the phenolic binding sites (Lu and Allen 2002). Ca/Mg–Cu exchange experiments with DOM suggest that Ca and Mg are preferably bound by carboxylic sites, particularly at relatively high concentrations, which results in a weakened apparent competition effect (Lu and Allen 2002).

Conditional distribution coefficients (K'_{DOM}) for the binding of Hg(II) to dissolved organic matter (extracted hydrophobic acids) shows that very strong interactions ($K'_{DOM} = 10^{23.2\pm1.0}$ L kg^{-1} at pH = 7.0), at Hg/DOM ratios below approximately 1 μg of Hg/mg of DOM, are indicative of mercury – thiol bonds. In contrast, much weaker interactions ($K'_{DOM} = 10^{10.7\pm1.0}$ L kg^{-1} at pH = 4.9 – 5.6), at Hg/DOM ratios above approximately 10 μg of Hg/mg of DOM, are consistent with Hg binding to oxygen functional groups (Haitzer et al. 2002). Similar results have been found in another study where conditional distribution coefficients (log K'_{DOM}) for Hg(II) binding to extracted humic acids, fulvic acids and hydrophobic acids from diverse aquatic environments indicate very strong interaction, suggesting the involvement of thiol groups (Haitzer et al. 2003). It has also been shown that K'_{DOM} values decrease at low pH (4) compared to pH 7, suggesting proton competition for the strong Hg(II) binding sites that is consistent with bidentate binding of Hg(II) by one thiol group (p^{Ka} = 10.3) and by another group (p^{Ka} = 6.3) in the DOM (Haitzer et al. 2003). In addition, the hydrophobic fraction of DOM is composed of thiol functional groups with high conditional stability constant of the Hg^{2+}-DOM complexes (Benoit et al. 2001; Dyrssen and Wedborg 1986, 1991; Schuster 1991; Xia et al. 1999).

It has been shown that each SRFA molecule has approximately 3 carboxyl sites available for coordination in the pH 6–9 range (Sonke and Salters 2006). However, coordination depends on other environmental factors such as metal concentration, ligand concentration, pH and so on (Sonke and Salters 2006; Thomason et al. 1996). Lanthanide ion probe spectroscopy (LIPS) suggested that an increase in the metal/DOM ratio can result in progressively less chelated complexes, with a gradual succession from 4 to 3 to 2 to 1 carboxyl groups bound to Eu^{3+} (Thomason et al. 1996)· The LIPS study indicated that Eu^{3+} complexation by Suwannee river DOM (mixture of HA and FA) at pH 3.5 is dominated by tetra-dentate complexes at low metal to ligands ratios (100 nM Eu^{3+}, 30 mg L^{-1} DOM) (Thomason et al. 1996). The complex formation of quinonoid-enriched humic derivatives with actinides (ca. Np^{5+}) demonstrated that hydroquinone-enriched derivatives have higher stability constants than the catechol ones, and that enriched humic derivatives are more effective than the parent humic acid (Shcherbina et al. 2007). Moreover, interaction between HA and Np^{5+} in the neutral pH range is dominated by carboxylate groups in aqueous media (Sachs et al. 2005). Conditional stability constants between Cm^{3+} and 5-sulfosalicylic acid ($\log_{10} K = 6.44$), fulvic acid ($\log_{10} K = 5.90$), and humic acid ($\log_{10} K = 6.22$) are all very similar, suggesting the salicylic acid-like functional groups may be present in the molecular structure of humic substances (fulvic and humic acids) (Panak et al. 1995). This can be further highlighted by the observation of enhanced fluorescence intensity in the complexes of Cm^{3+} with 5-sulfosalicylic acid, fulvic acid, and humic acid (Panak et al. 1995). The U^{4+} complexation with humic acids with different sulfur contents (1.9, 3.9, 6.9 wt %) shows that increasing sulfur (>2 wt %) leads to an increase of the number of humic acid binding sites. This is also reflected in increased U^{4+} loading capacities and increased total humic acid ligand concentrations for U^{4+} (Sachs et al. 2010). This increase of the fraction of humic acid binding sites for U^{4+} indicates an involvement of reduced sulfur functionalities, such as thiol groups, in the complexation between U^{4+} and humic acid (Sachs et al. 2010). However, for environmentally relevant sulfur contents of humic acids (<2 wt %), compared to the oxygen functionalities and in particular to carboxylic groups, reduced sulfur functionalities play only a subordinate role in U^{4+} complexation in the acidic pH range. Notes that reduced sulfur species such as thiols, dialkylsulfides and/or disulfides are the dominating sulfur functionalities in extracted humic acids with different sulfur contents (Sachs et al. 2010). Therefore, the functional groups in fulvic and humic acids that form complexes with trace metals are phenolic OH and acidic OH groups, among which are hydroquinone-like moieties and non-quinoid phenols, O-, N- and S-containing functional groups or thiol groups (Lu and Allen 2002; Schmeide and Bernhard 2009; Haitzer et al. 2003; Zhang et al. 2004; Smith et al. 2002).

Strong organic ligands for copper (II) in seawater are likely to derive from biological sources, rather than being refractory organic materials (Wu and Tanoue 2001a, b; Midorikawa and Tanoue 1998; Moffett et al. 1990). The exudates from certain phytoplankton and bacteria, which are important sources of protein-like fluorescence, are strong Cu chelators (Zhang et al. 2009; Mcknight and Morel 1980; Determann et al. 1998). Autochthonous DOM from phytoplankton or algal biomass may contain amino and sulfidic functional groups in its molecular structure, which may form complexes with trace metals in water (Xue and Sigg 1993; Xue et al. 1995).

EEMS of tryptophan amino acid shows two fluorescence peaks: peak T for the amino carboxylic acid functional group [–CH–(NH_2)–COOH] and peak T_{UV} for the –NH group in the aromatic functionality [C_8H_5(NH)–] (Mostofa et al. 2009a, 2011). Interestingly, proteins and oligopeptides are important constituents of high molecular mass-DOM that contains primary amines in seawater (Lee and Bada 1975; Tanoue et al. 1996).

The EPS is primarily composed of polysaccharides, proteins, uronic acid, fatty acids, nucleic acids and lipids containing ionizable functional groups such as carboxyl, phosphoric, amine, acidic amino acids, hydroxyl, phenolic, sulfates, and organic phosphates (Beech and Sunner 2004; Quiroz et al. 2006; Merroun and Selenska-Pobell 2008; Zhang et al. 2008; Wingender et al. 1999; Liu and Fang 2002, 2003; Guibaud et al. 2003; Merroun et al. 2003; Sponza 2003; Guibaud et al. 2005). Such groups can form complexes with trace metals depending on the different environmental conditions. The complexation between EPS and metal ions (ca. Pu^{4+}) suggests that the carboxylic groups in EPS are the primary binding sites at pH 4 and with a ionic strength of 0.1 M NaCl (Harper et al. 2008). The amide functional groups of proteins in the EPS are susceptible to form complexes with trace metals, which have been detected using spectroscopic analysis (Guibaud et al. 2005a; b; Zhang et al. 2006). Polysaccharides can form complexes with trace metal ions, and the hydroxyl groups of neutral polysaccharides and the carboxyl groups of anionic polysaccharides are the probable binding sites for such complexes (Guibaud et al. 2005; Zhang et al. 2006; Brown and Lester 1982). The organic phosphate groups of EPS of *A. ferrooxidans* or phosphate groups in nucleic acids play a major role in the binding of U from aqueous solutions, although this bacterium contains small amounts of phosphates (Merroun and Selenska-Pobell 2008). Bacteria, algae and their exudates also consist of a mosaic of functional groups (i.e., amino, phosphoryl, sulfhydryl and carboxylic groups) and the net charge on the cell wall depends on the pH of the medium (Filella 2008). In addition, trace metals such as Th^{4+} and U form complexes with organic ligand in particulate matter that might be a nonmetal-specific chelator originating from the cell surface of microorganisms (Hirose 2004).

4.1 The Mechanism for Complex Formation Between Trace Metals and DOM in Waters

A relationship between the lability of metal–DOM complexes involving 3*d* transition metals in freshwater and their *d*-electron configuration is discussed by Sekaly and his colleagues (2003). The order of the lability of the metal complexes, Co(II) d^7 > Ni(II) d^8 > Cu(II) d^9 < Zn(II) d^{10}, follows the reverse order of the ligand field stabilization energy (LFSE). There is an exception for Cu(II), the behavior of which is due to the Jahn–Teller effect that shortens the equatorial bonds and lengthens the axial bonds of a tetragonally distorted Cu(II)-L_6 complex (Sekaly et al. 2003). Another mechanism is commonly provided by several studies (Chen et al. 1986; Dixon and Larive 1999; Sharpless and McGown 1999; Simpson et al. 2002; Wrobel et al. 2003), which hypothesize that macromolecules (humic substances) can form aggregations that can be induced by complexation with metal cations. These mechanisms do not explain how the functional groups in DOM or the organic ligands can bind to trace metal ions. Recently, Mostofa and his colleagues (2009a, 2011) firstly provided a comprehensive mechanism for M-DOM complexation using tryptophan as a model compound. A strong π-electron bond between the functional group (F:) of tryptophan and the empty *d*-orbitals of the transition metal (M^{n+}) is formed by donation of electrons (Fig. 7).

Based on the available literature data that concern M-DOM complexation, it is suggested that the functional groups in DOM (or the organic ligands) can form complexes with trace metal ions by two ways. First, donation of electrons occurs

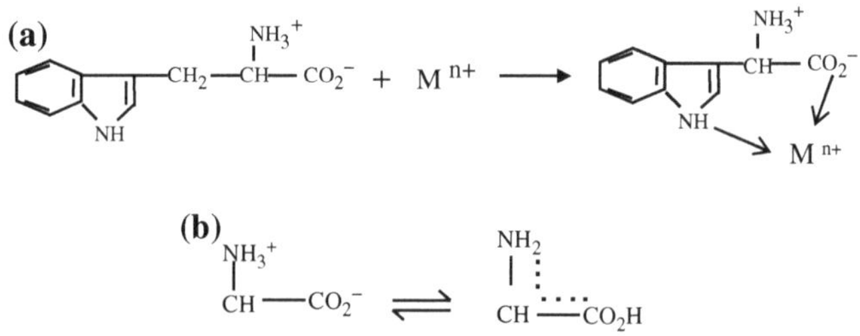

Fig. 7 The mechanism of the chemical bonding of two functional groups (amino-carboxylic and amide) in tryptophan with metal ion (M^{n+}) (**a**) and the resonance configuration of the amino-carboxylic group [–CH(NH_2)–COOH] (**b**). *Data source* with modifications Mostofa et al. (2009a; 2011)

from O-containing functional groups (carboxylic, amino-carboxylic, phenolic, alcoholic and so on) of DOM or organic ligands to metal ions. Second, donation to metal ions of non-bonding π-electrons occurs from S-, N-, and P-containing functional groups in DOM or highly unsaturated π-electrons either in aromatic or aliphatic systems. In both cases, donation of electrons takes place from the functional groups (F:) of DOM to the empty or partially filled *d*-orbitals of transition metals/lanthanides/actinides. Complex formation can also involve *s*- and *p*-orbitals in alkali/alkaline earth metals (M^{n+}) that form a strong covalent or π-electron bonding system with DOM functional groups (F:M^{n+}). This can simply be expressed as follows (Eq. 4.1):

$$F\!: + M^{n+} \rightarrow F\!: M^{n+} \quad (4.1)$$

The formation of such a bond in M-DOM complexes can greatly reduce the electron density of the functional groups (F:) in DOM. Donation of electrons from functional groups (F:) then causes the *s*-, *p*- or *d*-orbitals in metal ions (M) to be either stabilized or destabilized. Therefore, the M-DOM complexation causes the fluorescence properties of DOM to be either decreased or increased (Mostofa et al. 2009a, 2011). A stabilizing effect from a functional group in DOM lowers the energy of the interacting *s*-, *p*-, *d*-orbitals, which can considerably decrease the electron transfer probability and decrease as a consequence the fluorescence intensity of the functional groups (or fluorophores) in DOM. It is generally known that the metal ions are excellent Lewis acids and accept electron density from many molecules or ions that act as Lewis bases. On the other hand, M-DOM formation may enhance the probability of an electron transfer if a destabilizing effect from the functional groups in DOM raises the energy of the *s*-, *p*-, *d*-orbitals. This effect subsequently leads to an increase in the fluorescence intensity of the functional group (or fluorophores) in DOM.

The two types of mechanism for M-DOM complexation can be clarified by considering tryptophan amino acid as an example, because its molecular structure ($C_8H_5\,(NH)-CH_2\left(NH_3^+\right)CHCOO^-$) is composed of two functional groups such as $[-CH_2\left(NH_3^+\right)CHCOO^-]$ and [C_8H_5(NH)–] (Mostofa et al. 2009a, 2011). The two functional groups of tryptophan can electronically bind metal ions, creating strong covalent and π-electron bonds, respectively, with the metal-ion *d*-orbitals (Fig. 7). In fact, the $[-CH_2(NH_3^+)CHCOO^-]$ functional group has a strong affinity toward a resonance configuration, and the $[C_8H_5(NH)–]$ moiety has the non-bonding electrons of the NH- group in the aromatic ring (Fig. 7) that can form complexes with metal ions. The formation of the M-DOM bonding system can greatly reduce the electron density of the functional groups, i.e. the fluorophores in tryptophan, which results in a lower probability of electron transition of those fluorophores and decreases as a consequence the fluorescence intensity of tryptophan after complexation with the metal ions. On the other hand, an increase in fluorescence intensity of the DOM in M-DOM complexation may arise from enhanced probability of electron transition of the fluorophores due to M-DOM complexation that depends on the occurrences of the trace metal ions. The fluorescence peak T in the longer wavelength region is caused by the functional group $[-CH_2(NH_3^+)CHCOO^-]$ and the peak T_{UV} at shorter wavelengths is linked with the group $[C_8H_5(NH)–]$ (Mostofa et al. 2011). The conditional stability constant ($\log_{10} K_1$) for tryptophan-like or protein-like material of biological origin is 7.82–9.56 for peak T that has stronger bonding capacity than the humic-like component ($\log K_1 = 7.05–8.78$) in lake water (Wu and Tanoue 2001b). Moreover, the order of complex formation of the transition metals and other metals to DOM in natural waters depends on several associated effects (Mostofa et al. 2011). Therefore: (i) Fluorophores with high electron density will merely compete for the metal ions with stabilizing effects of *d*-orbitals.

(ii) Among the transition metals, the size or atomic radius generally decreases with increasing nuclear charge, because the electrons that experience a greater nuclear charge are pulled more strongly towards the nucleus. However, the last few elements (Cu, Zn, Ag, Cd, Pt, Au, Hg, etc.) in each row of the *d*-block are slightly larger than those preceding them because in these cases the electron–electron repulsions caused by the filling of the *d*-orbitals outweigh the increasing nuclear charge. Therefore, the two competing effects of nuclear charge and electron–electron repulsion affect the chemical binding of the *d*-orbital metals with the fluorophores. As one moves across a period, the increasing nuclear charge is usually more significant than the electron–electron repulsion. These combined effects make Cu and other metals of the same row more susceptible to complexation with DOM than the transition metals of the second and third rows. Moreover, the ground states of Sc, Ti, Fe, Co and Ni are ferromagnetic because of the presence of one unpaired electron, while V, Cr, and Mn are antiferromagnetic (Tung and Guo 2007; Iota et al. 2007). It can be proposed that the unpaired electron of ferromagnetic transition metals would easily form a bond with an electron donated by functional groups in DOM, which is not possible for antiferromagnetic metals. This hypothesis can account for the rapid complexation of the ferromagnetic transition metals compared to the antiferromagnetic elements. For example, Fe^{3+} has one electron each in its outer *d*-orbitals and no electron in its outer shell *s*-orbital (Fe^{3+}: $1s^2 2s^2 2p^6 3d^5 3s^0$). Therefore, it can accept electrons in its outer unpaired *d*-orbitals from functional groups in DOM to form M-DOM complexes. Cr^{3+} has

3 electrons in its outer d-orbitals and an empty s-orbital in its outer shell (Cr^{3+}: $1s^{2}2s^{2}2p^{6}3d^{3}3s^{0}$). It can accept electrons in its two empty and three unpaired *d*-orbitals from functional groups in DOM to form complexes. Cu^{2+} has one unpaired *d*-orbital and one empty *s*-orbital in its outer shell (Cu^{2+}: $1s^{2}2s^{2}2p^{6}3d^{9}3s^{0}$), to which the functional groups in DOM can donate electrons when forming complexes. It is assumed that donation of electrons from functional groups in DOM to unpaired *d*-orbitals can enhance the stability of the M-DOM complexes. This gives a strong bonding capacity of Cu^{2+} ions toward functional groups in DOM or organic ligands. Finally, Sc^{3+} has empty *d*- and *s*-orbitals (Sc^{3+}: $1s^{2}2s^{2}2p^{6}3s^{2}3p^{6}3d^{0}4s^{0}$) that can accept electrons when forming M-DOM complexes. Entirely empty *d*-orbitals of Sc^{3+} ions gives exceptionally strong bonding properties toward fulvic and humic acids (ca. $\log_{10} K = 17.57$ for fulvic acid and 17.54–20.47 for humic acid) among all the metal ions ($\log_{10} K = 3.26$–14.58 and 0.5–16.50, respectively) studied in aqueous solutions (Table 1).

(iii) For alkali/alkaline earth metals and metalloids, the functional groups in DOM can donate electrons to the outer empty *s*- and/or *p*-orbitals. For example, H^{+} has no electrons in its outer empty *s*-orbital (H^{+}:1s) and can accept electrons. Ca^{2+} has one empty *s*-and one *p*-orbital in its outer shell (Ca^{2+}: $1s^{2}2s^{2}2p^{6}3s^{0}3p_{x}{}^{0}3p_{y}3p_{z}$) that can accept electrons. Donation of electrons to *s*- and *p*-orbitals possibly explains the weak bonding properties of Ca^{2+} toward functional groups in DOM. Al^{3+} has an empty *s*- and two *p*-orbitals in its outer shell (Al^{3+}: $1s^{2}2s^{2}2p^{6}3d^{3}3s^{0}3p_{x}{}^{0}3p_{y}{}^{0}3p_{z}$) that can accept electrons. However, the involvement of these orbitals in complexation with functional groups in DOM would ultimately destabilize the Al^{3+}-complexes. This effect can enhance the fluorescence intensity of Al-DOM complexes in aqueous media. Sb^{3+} has empty *p*-orbitals in its outer shell (Sb^{3+}: $1s^{2}2s^{2}p^{6}3s^{2}p^{6}d^{10}4s^{2}p^{6}d^{10}5s^{2}p_{x}^{0}p_{y}^{0}p_{z}^{0}$) that can receive electrons from functional groups in DOM. Therefore, alkali and metals/metalloids can form complexes with DOM or organic ligands in aqueous media.

5 Factors Affecting the Metal-DOM Complexation in Natural Waters

The formation of M-DOM complexes is greatly affected by several factors in natural waters, which can be distinguished as (i) Quantity, nature and molecular size of DOM; (ii) Occurrence and affinity of trace metals; (iii) Effect of pH; (iv) Effects of ions (cations and anions) and their ionic strength (*I*); (v) Effects of photoinduced processes; (vi) Effects of microbial processes; and (vii) Effects of freshwater and sea waters.

5.1 *Quantity, Nature and Molecular Size of DOM*

The most common organic ligands in DOM are allochthonous fulvic acid, allochthonous humic acid, tryptophan and extracellular polymeric substances (EPS).

The complexation between DOM and trace elements depends on the nature and quantity of DOM in natural waters (Sonke and Salters 2006; Iskrenova-Tchoukova et al. 2010; Sanchez-Marin et al. 2010; Cao et al. 2004; Kaiser 1998; Reszat and Hendry 2007; Jansen et al. 2003; Naka et al. 2006).

Extracted humic acid shows a 65-fold higher affinity for lanthanide metals than SRFA, which suggests that affinity depends on the sources and nature of fulvic and humic acids (Sonke and Salters 2006). Allocthonous fulvic and humic acids have poly-carboxylic, phenolic (catechol), iminodiacetic and aminocarboxylic functionalities in their molecular structures; furthermore, the quantity of the functional groups and the aromaticity are significantly varied between them (Malcolm 1985; Senesi 1990; Morel and Hering 1993; Leenheer et al. 1998; Steelink 2002; Sonke and Salters 2006; Fimmen et al. 2007). Each SRFA molecule has ~3 carboxylic sites available for coordination that can be judged from the carboxyl site density in SRFA (Sonke and Salters 2006; Ritchie and Perdue 2003). The study of SRFA complexation with the lanthanide series hypothesizes that poly-dentate carboxylic, phenolic, and N-containing carboxylic binding sites may be involved under environmental conditions (Sonke and Salters 2006). Correspondingly, autochthonous DOM of phytoplankton or algal origin is composed of amino and sulfidic functional groups, although its exact structure remains unknown (Xue and Sigg 1993; Xue et al. 1995). The conditional stability constants of the M-DOM complexation are thus significantly variable for different DOM sources, even for the same metal ion (Table 1).

DOM in sediment elutriates and sewage-influenced water (i) was enriched by 1.4–1.7 times in DOC; (ii) absorbed and reemitted more light; and (iii) presented higher Cu complexation capacities than the natural seawater (Sanchez-Marin et al. 2010)· This suggests that differences in DOC and DOM may control the metal toxicity in natural waters (Sanchez-Marin et al. 2010). The total DOC concentrations decrease slightly when the C/metal ratio is less than 10, particularly for Al^{3+} and Fe^{3+}. Hydrophilic DOC increases and hydrophobic DOC decreases with increasing concentrations of polyvalent metal cations in the order $Ca^{2+} < Al^{3+} < Fe^{3+}$, whilst Na^{+} remains unaltered (Fig. 8) (Kaiser 1998). Such an effect is more pronounced at low DOC concentrations and high pH values (Fig. 8). This result suggests that the formation of insoluble M-DOM complexes is susceptible to reduce DOC concentrations, whilst soluble metal-DOM complexes may induce an alteration of the distribution between hydrophilic and hydrophobic DOC in natural waters (Kaiser 1998). Therefore, the polyvalent cations and their concentrations can considerably affect the distributions of the DOM fractions and their contents, determined using XAD-8 resins, particularly at low DOC and high pH (Kaiser 1998).

The interaction of several elements (Cu, Mn, Mo, Ni, Sr, U, and Zn) with high dissolved organic carbon (DOC) concentrations (21–143 mg C L^{-1}) showed that only U and Zn can form complexes with all DOC samples. The in situ association constant (K'_d) for U decreases with depth in pore waters (Reszat and Hendry 2007). It is also suggested that minor amounts of U and Zn (less than or equal to 4 % of total) can form complexes with this DOC under in situ pH conditions. Competitive complexation by other ligands may limit the importance of

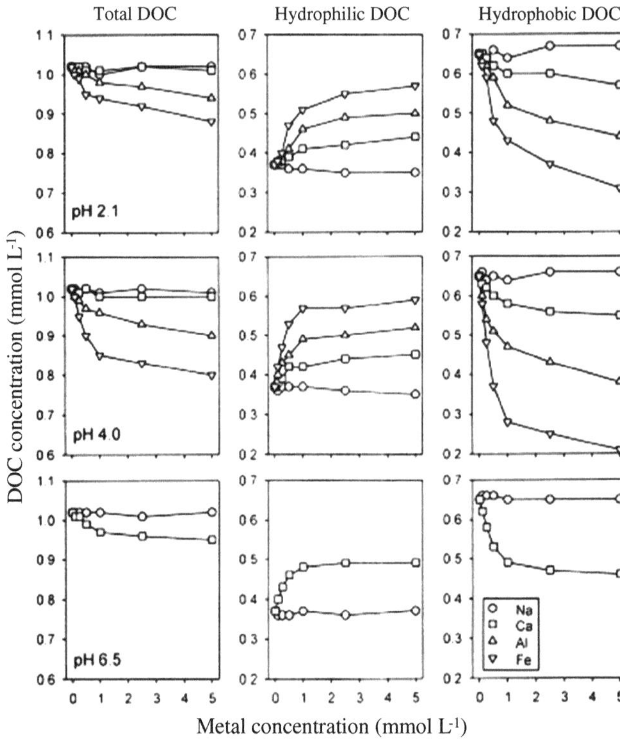

Fig. 8 Effects of increasing concentrations of Na^+, Ca^{2+}, Al^{3+} and Fe^{3+} on the total, hydrophilic and hydrophobic DOC at three pH levels (2.1, 4.0 and 6.5) and an initial total DOC concentration of 1.0 mmol L^{-1}. *Data source* Kaiser (1998)

DOC-facilitated transport of elements (Reszat and Hendry 2007). The ability of actinide ions such as uranyl to form complexes with DOC is very variable depending on the occurrence of humic fractions (fulvic and humic acids) in DOM, which undergo in turn large variations (5–80 %) in the aquatic environments (Mostofa et al. 2009a; Kim and Czerwinski 1996; Giesy et al. 1986; Kim et al. 1994; Artinger et al. 2002; Crancon and van der Lee 2003; Jackson et al. 2005). Interestingly, the complexation of U to DOM is kinetically controlled (Artinger et al. 2002).

Studies on M-DOM complexation suggest that trace metals with high binding strength are primarily distributed in the larger molecular size fractions, and metals with low binding strength are mostly distributed in the smaller molecular size fractions (Wu et al. 2004b; Shin et al. 2001; Lakshman et al. 1993; Midorikawa and Tanoue 1996, 1998; Lin et al. 1995; Rao and Choppin 1995; Sekaly et al. 1998; Sposito et al. 1982). The larger molecular size fractions have a higher affinity for metals such as Al, Cu, UO_2(II), and Np(V) than smaller molecular size fractions (Shin et al. 2001; Lakshman et al. 1993; Midorikawa and Tanoue 1998; Lin et al. 1995; Midorikawa and Tanoue 1996; Rao and Choppin 1995; Sekaly et al. 1998; Sposito et al. 1982). Kinetic studies show that strong affinity metals can form complexes with DOM more rapidly than the weak metals, and that their DOM complexes undergo a slower dissociation process (Wu et al. 2004c; Lin et al. 1995). It is also shown that the distribution of transition metals in DOM is shifted towards the larger molecular size fractions as the binding strength increases. Fe along with V and Ce is distributed mainly in the larger size fractions, but heavy metals such as U, Th and Mo are distributed mainly in the smaller molecular size fractions (Wu et al. 2004b). These studies hypothesize that the strong affinity metal ions can form complexes with the functional groups or replace the weak affinity metal ions in the larger molecular size fractions in the first binding site. Weaker metals then occupy other available sites in the small molecular size fractions.

Molecular size (or mass) distribution and levels of organic ligands suggest that new organic ligands with high molecular masses are produced during periods of high biological productivity in natural waters (Midorikawa and Tanoue 1996, 1998; Mopper et al. 1996). Such new organic ligands are considered to be the autocthonous fulvic acids (C-like) that have recently been found to originate under photoinduced or microbial assimilation of algal or phytoplankton biomass (Fig. 1c, d) (Mostofa and Sakugawa 2009; Zhang et al. 2009). The fluorescence excitation-emission maxima of autocthonous fulvic acid (C-like) are similar to those of the allochthonous fulvic acid and humic acid (C-like), showing two fluorescence peaks in the C- and A-regions (Fig. 1a–d). In the molecular mass distribution of organic ligands, the relative contribution of the fraction with <5 kDa molecular masses is dominant (67–79 %), while 17–30 % of the total organic ligands are in the 5 kDa-0.1 μm fraction, leaving 3–6 % in the 0.1 μm-GF/F fraction in lake water (Wu and Tanoue 2001a). The contribution of organic ligands in the <1 kDa fraction is 41 % of the total in estuarine water (Gordon et al. 1996). The contribution of total organic ligands in DOM accounts for 10–62 % in the case of molecular masses of >1 kDa and 50–90 % for the <1 kDa fraction in sea waters (Midorikawa and Tanoue 1998, 1996; Maurer 1976; Zsolnay 1979; Carlson et al. 1985; Benner et al. 1992; Guo et al. 1994, 1995, 1996; Buesseler et al. 1996; Guo and Santschi 1996). Contributions of total organic ligands are 0.63–4.68 % of the bulk DOM in the water of rivers, lakes and oceans (Wu and Tanoue 2001a, c; Midorikawa and Tanoue 1998; Wu et al. 2001). It has also been observed that the quantities of the weak ligands are relatively high, approximately 0.54–1.21 % of the total DOM whilst the quantities of strong ligands are low, approximately 0.06–0.21 % in stream waters (Wu and Tanoue 2001a). High-affinity nitrogenous moieties account for only 2–4 % of DOM in water (McKnight et al. 1997; Croue et al. 2003). Finally, the M-DOM complexation significantly depends on the quantities, nature and molecular size of DOM in natural waters.

5.2 *Occurrences and Affinity of Trace Metals in M-DOM Complexation*

The M-DOM complexation significantly depends on the occurrence of the trace metal ions and on their affinity toward organic ligands. Complexation shows high variation in the conditional stability constants for alkali, alkaline, heavy metals, transition metals, lanthanides and actinides (Table 1) (Nair and Chander 1983; Kim et al. 1990, 1991; Buckau et al. 1992; Xue et al. 1995; Takahashi et al. 1997; Mandal et al. 1999; Sekaly et al. 2003; van Loon et al. 2004; Wu et al. 2004b; Sonke and Salters 2006; Thakur et al. 2006; Shcherbina et al. 2007). The trace metals have different affinities toward the functional groups and the donor atoms in DOM. Affinity depends on the occurrence of trace metals and on their outer-shell electronic configuration in aqueous media (Wu et al. 2004b, c; Sonke and Salters 2006; Konstantinou et al. 2009; Fu et al. 2007; Mostofa et al. 2011; Duffus 2002; Konstantinou et al. 2007; Marang et al. 2008; Kolokassidou and Pashalidis 2006). The complexation of various trace transitional metals with different molecular size fractions of DOM in natural waters indicates that the number-averaged molecular weight follows the order: Cu > Ni > (Co, Zn, Cr) > Pb > Cd for the DOM-bound complexes. This is consistent with the Irving-Williams series. For DOM-bound complexes with other metals, the order is followed (Fe, V, Ce) > Th > U > Mo (Wu et al. 2004b). The complexation between fulvic acid and several metals (Cu^{2+}, Ni^{2+}, Co^{2+}, Cd^{2+} and Ca^{2+}) at pH 7 demonstrates that for the fast-reacting binding site, the rate constant and the site relative contribution shows an order of $Cu^{2+} > Ni^{2+} > Co^{2+} > Cd^{2+} > Ca^{2+}$ that agrees with the Irving-Williams series. This indicates that the complexation kinetics is affinity-dependent (Wu et al. 2004c). The competitive binding of Cu(II), Co(II) and Ni(II) ions by fulvic acid shows that in the absence of Cu(II) and Co(II), the strong binding sites of the fulvic acid are occupied by Ni(II), forming strong complexes that are inert. In contrast, in the presence of Cu(II) and Co(II), the strong binding sites of the fulvic acid are occupied by Cu(II). Co(II), Ni(II) and the remaining Cu(II) occupy the weak binding sites of the fulvic acid, forming weak complexes that are labile (Mandal et al. 1999). The enhanced lability of the Ni-FA complexes in the presence of Cu(II) and Co(II) suggests that Cu(II) and Co(II) successfully compete with Ni(II) for the strong binding sites of the fulvic acid (Mandal et al. 1999). The order of the lability of the metal complexes, Co(II) > Ni(II) > Cu(II) < Zn(II), follows the reverse order of the ligand field stabilization energy with the exception of Cu(II). The behavior of Cu(II) is also due to the Jahn–Teller effect, which shortens the equatorial bonds and lengthens the axial bonds of a tetragonally distorted Cu(II)-L6 complex (Sekaly et al. 2003).

It is known that the rate of octahedral ligand substitution reactions of 3*d* transition metal complexes is influenced by several factors such as effective nuclear charge, ionic radius (Douglas et al. 1994; Margerum et al. 1978) and ligand field stabilization energy (LFSE) (Morel and Hering 1993; Butler and Harrod 1989). The dissociation kinetics of the 3*d* transition metal complexes agrees with the trend predicted by the LFSE (weak field): Co(II) d^7 > Ni(II) d^8 > Cu(II) d^9 < Zn(II) d^{10} whereas the behavior of Cu(II) is also due to the Jahn–Teller effect cited above (Sekaly et al. 2003). A gradual increase in complexation strength is observed with decreasing ionic radius, an expression of the lanthanide contraction during the complexation between lanthanides and humic substances (fulvic and humic acids) in waters (Sonke and Salters 2006).

Trace metal ions have a specific interaction affinity with organic ligands or DOM that generally depends on the size and outer electronic configuration level of the metal ions. Replacement of a metal ion by a higher affinity metal ion in M-DOM complexation occurs because the number of complexing sites remains constant for a specific amount of DOM. A decrease in fluorescence intensity is often detected after further addition of Cu^{2+} to Hg–DOM complexes, which suggests that Cu^{2+} can act as a stronger quencher for DOM than Hg(II) in aqueous media (Fu et al. 2007). The complexation of humic acid, olive cake and its hydrophilic extracts with Cu^{2+} and Eu^{3+} shows that Cu^{2+} is replaced by Eu^{3+} in the aqueous solution (Konstantinou et al. 2007, 2009; Konstantinou and Pashalidis 2010). Eu^{3+} has only one electron in its outer *d*-orbital shell, which gives it strong affinity to bind to DOM compared with Cu^{2+}. Comparison between Ca–DOM and Cu–DOM complexation demonstrates that (i) Ca–DOM complexation increases of much less than an order of magnitude per pH unit and decreases at higher Ca concentration, differently from Cu–DOM complexation; and (ii) Cu–DOM complexation is highly non-linear, in contrast to the very reduced extent of non-linearity of Ca–DOM complexation (Lu and Allen 2002).

The effective distribution of affinities (Conditional Affinity Spectrum, CAS) of a metal ion binding to a humic substance under natural water conditions suggests three groups of cations (Rey-Castro et al. 2009): (a) Al, H, Pb, Hg, and Cr, which are preferentially bound to the phenolic sites of the fulvic ligand; (b) Ca, Mg, Cd, Fe(II), and Mn, which display a higher affinity for carboxylic sites, in contrast to expectations based on the individual complexation parameters; and (c) Fe(III), Cu, Zn, and Ni, for which phenolic and carboxylic distributions are overlapped.

The complexation of trace metals with the functional groups in EPS varies depending on the environmental conditions. The order is Pb > Cd > Zn for exopolysaccharides of bacterial origin (Loaec et al. 1997); Cu > Cd > Ni > Cr(III) > Cr(VI) for contaminated effluents (Bux et al. 1994); Cu > Cr > Zn > Pb for activated sludge microorganisms (Chua et al. 1999); Zn > Cu > Co > Cd > Cr(III) > CrO_4^{2-} > Ni for activated sludge (Liu et al. 2001); Pb > Cd > Cu > Zn for polluted waters of sewage sludge and paper mill waste (Lister and Line 2001); Cu > Cd > Ni for pure cultures of bacteria originated from activated sludge (Kim et al. 2002); Pb > Cu > Zn > Ni for the first site and Pb > Cu > Ni > Zn for the second site of an acidogenic thermophilic anaerobic reactor (Leighton and Forster 1997); Cd > Pb $\approx$ Cu for activated sludge originated from wastewater treatment plants (Guibaud et al. 2003); and Cu > Ni >> Zn for activated sludge (Guibaud et al. 2003).

Trace metals have highly variable affinity towards various functional groups in DOM and they show strong differences in the conditional stability constants of M-DOM complexation (Table 1) (Sonke and Salters 2006). The complexation between lanthanides and humic substances (fulvic and humic acids) suggests that a gradual increase by 2–3 orders of magnitude in the conditional stability constants from La to Lu follows the decreasing ionic radius and is an expression of the lanthanide contraction (Sonke and Salters 2006).

Overall, the M-DOM complexation greatly depends on the outer-shell electronic configuration of the metal ions in aqueous media.

5.3 Effects of pH

Complexation between DOM and trace elements is highly dependent on pH (Fig. 5) (Zhang et al. 2009, 2010; da Silva et al. 1998a, b, 2002; Sonke and Salters 2006; Shcherbina et al. 2007; Christoforidis et al. 2010; Iskrenova-Tchoukova et al. 2010; Liu and Cai 2010; Cao et al. 2004; Takahashi et al. 1997; Naka et al. 2006; Ghassemi and Christman 1968; Glaus et al. 2000). Fe can strongly form complexes with colored humic substances at low pH (Ghassemi and Christman 1968) and the iron-holding capacity of color increases with pH up to a value of 10, after which it decreases rather abruptly (Shapiro 1964). The complexation of Aldrich humic acid with As^{3+} shows that the stability constants for the first binding site are maximum under acidic conditions ($\log_{10} K_1 = 6.9$–7.2 at pH 5.2), gradually decrease under neutral conditions (6.2–7.1 at pH 7.0) and become lowest at basic pH (5.8–6.2 at pH 9.3) (Table 1) (Liu and Cai 2010). In contrast, in the case of the second binding sites the constants remain similar or increase a little, from $\log_{10} K_1 = 4.5$–5.0 (mean = 4.6) at pH 5.2 to $\log_{10} K_1 = 4.7$–5.3 (mean = 4.9) at pH 9.3 (Table 1) (Liu and Cai 2010). As^{3+} complexation to Aldrich humic acid increases with pH (particularly at and above pH 9.3), whereas the conditional stability constants decrease for the strong binding sites and remain approximately constant for weak binding sites (Liu and Cai 2010). The interaction of Cu(II), Ni(II), and Fe(III) with extracted soil fulvic acids results into quite stable soluble complexes in the acidic pH range from 3 to 6 (da Silva et al. 1998b, 2002). The pH effect can change the complexation capacity toward metals. In fact, the metal order at pH 6 is Pb > Cu > Cd and at pH 7 and 8 it is Cu > Pb >> Cd (Comte et al. 2008). The conditional stability constants (*K*) between lanthanide series (14 elements) and humic substances (standard fulvic and humic acids) increases with increasing pH in waters (Sonke and Salters 2006).

The fluorescence intensity of EPS at both peak T- and T_{UV}-regions is strongly dependent on the solution pH in the absence and presence of Hg(II), with the maximal fluorescence intensity at neutral pH (Fig. 2) (Zhang et al. 2010). EPS shows higher fluorescence intensity at the peak T_{UV}-region than at peak T, and the trend resembles that of a tryptophan standard (Mostofa et al. 2009a; Zhang et al. 2010). The effects of pH on M-DOM complexation in water imply two things. First, the pH variation (low-pH or high-pH) generates protonation-deprotonation phenomena in functional groups of DOM (or of organic ligands) that subsequently alters the complexation capacity between DOM and trace metals. An increase in pH generally increases the binding capacity between the trace metals and DOM in aqueous media. Second, the presence of cations (ca. Mg^{2+}) can significantly influence the generation of the high-pH forms of functional groups in DOM, even under circumneutral conditions, which increases M-DOM complexation (Fig. 9) (Christoforidis et al. 2010). For example, at neutral pH the resonance configuration of the amino-carboxylic group [–CH(NH_2)–COOH] in tryptophan can exist in the highest form that can require the lowest energy for the excitation of electrons (Fig. 2). This effect can result in high fluorescence intensity at

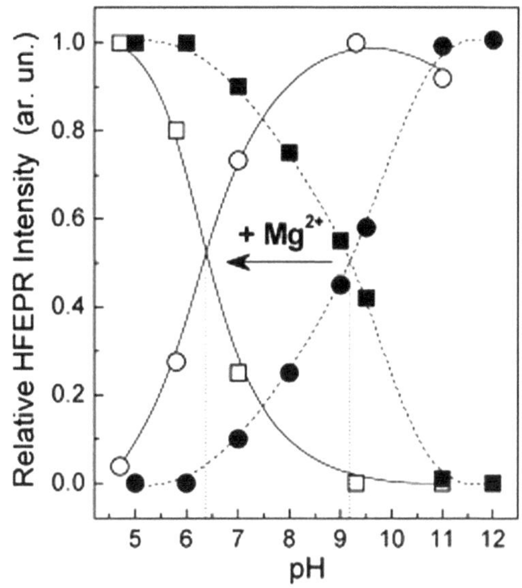

Fig. 9 Relative intensities of the HFEPR spectra of [HA-Mg^{2+}] sample (*solid lines, open symbols*) and metal-free HA (*dashed lines, solid symbols*) at different pH. The intensities are measured at $g = 2.0041$ {for [HA-Mg^{2+}] (*square*) and the metal-free HA (*filled square*)}, at $g = 2.0054$ for [HA-Mg^{2+}] (*circle*) and at $g = 2.0057$ for metal-free HA (●). *Squares* correspond to the low-pH radical forms and circles correspond to the high-pH radical forms. Data source: Christoforidis et al. (2010)

pH 7 and in donation of electrons from carboxylic functional groups to metal ions. Changes of pH (either acidic or basic) can result in protonation and deprotonation that ultimately increase the excitation energy of the functional groups, thereby decreasing the fluorescence intensity and the electron donation from the functional group to the metal ions. The changes of the fluorescence intensity of EPS and its complexation with Hg(II) as a function of pH are depicted in Fig. 2. Similarly, donation of unpaired electrons from the –NH group in the aromatic ring [C_8H_5(NH)–] of the tryptophan structure is highest under neutral condition (pH 7). Donation of electrons is significantly reduced with changes of pH (either acidic or basic), following either protonation of the unpaired electrons by H^+ in acidic conditions or deprotonation by OH^- in basic solution.

Overall, the pH effect therefore plays a very significant role in metal-DOM complexation in natural waters.

5.4 Effects of Ions (Cations and Anions) and of Ionic Strength (I)

The M-DOM complexation is significantly affected by the occurrence of ions (cations ca. Na+, $Ca^{2+}$and Mg^{2+} and anions ca. Cl^-, $CO_3{}^{2-}$, OH^-) and by the ionic strength (I) (Cabaniss 1992; Sonke and Salters 2006; Christoforidis et al. 2010; Iskrenova-Tchoukova et al. 2010; Smith and Martell 1987; Cabaniss and Shuman 1988; Fu et al. 2007; Cao et al. 2004; Takahashi et al. 1997; van Loon et al. 2004; Lu and Jaffe 2001; Glaus et al. 2000; Pinheiro et al. 2000). A computational molecular dynamics (MD) study of the interactions of Na^+, Mg^{2+}, and Ca^{2+} with the carboxylic groups of a model DOM suggests that aggregation of the DOM molecules occurs in the presence of Ca^{2+} but not of Na^+ or Mg^{2+} (Iskrenova-Tchoukova et al. 2010). These results suggest that Ca^{2+} ion bridging between NOM molecules can decrease repulsion due to the reduced net charge of the NOM–metal complexes (Iskrenova-Tchoukova et al. 2010).

An increase of Ca^{2+} can enhance the fluorescence intensity of fulvic acid involved in Hg-DOM complexation (Fu et al. 2007). In contrast, the Mg^{2+} ion has no effect on fluorescence and, therefore, on Hg-DOM complexation under natural conditions at pH 7.5 in urban river waters (Fu et al. 2007). On the other hand, the Cl^- ion is an inorganic ligand that can complex Hg(II) to form $HgCl_n^{2-n}$ depending on the concentration of Cl^- in natural waters (Lu and Jaffe 2001). It has been shown that the addition of Cl^- to Hg–DOM complexes increases the fluorescence emission intensity at both pH 7.50 and 9.50 due to the competition between DOM and Cl^- for Hg(II) (Fu et al. 2007). For divalent transition metal ions (Mn^{2+}, Co^{2+} and Zn^{2+}), carbonate complexes may be as important species as humate complexes (Takahashi et al. 1997). Hydroxide species are important for Fe^{3+}, VO^{2+} and Ga^{3+}, and significantly affect the complexation between humic acids and trace metal ions (Takahashi et al. 1997).

Mg^{2+} generates only electrostatic interaction with humic acid during complexation with heavy metals such as Cd^{2+}, Pb^{2+} and Sr^{2+} (Christoforidis et al. 2010). Furthermore, the two types of indigenous radicals that exist in all humic acids are influenced by the metal cations in a similar way, because of the presence of a unique phenolic moiety in humic acid (Christoforidis et al. 2010). Mg^{2+} ions can change the pH profile of the two radical types of humic acid, downshifting their interconversion pK_a by ca. 3 pH units (Fig. 9) (Christoforidis et al. 2010). The competition of Al^{3+} with divalent metal ions such as Pb(II) occurs primarily at carboxylic-type groups. Aluminum can reduce by a factor of 2 to 3 the amount of lead bound to humic acid (Pinheiro et al. 2000). In the case of Cd(II), the aluminum competition mainly affects the bound Cd, with only smaller changes in $[Cd^{2+}]$. Therefore, Al^{3+} competition is likely to increase Pb(II) toxicity and bioavailability and Cd(II) transport in aqueous solution (Pinheiro et al. 2000).

The conditional stability constants of the M-DOM complexation are decreased with increasing supporting electrolyte concentration in aqueous media (Sonke and Salters 2006; Cao et al. 2004; Benoit et al. 2001; Pinheiro et al. 2000; Stevenson et al. 1993; Bryan et al. 2002). It has been shown that conditional stability constants (K) values increase with decreasing ionic strength (I) (from 0.1 to 0.001 mol L^{-1} $NaNO_3$) for complexation of lanthanides with SRFA and extracted humic acids (Fig. 5a) (Sonke and Salters 2006). Similarly, the conditional stability constants of the Hg^{2+}-DOM complexes were increased by 7–12 % at $I = 0$ compared to those at $I = 0.06$ for the hydrophobic fraction of DOM extracted from the Florida Everglades surface waters (Benoit et al. 2001). An increase of ionic strength can lead to a decrease of Pb and Cd bound to humic acid in aqueous solution (Pinheiro et al. 1999, 2000). With a ionic strength of 0.1 M NaCl, the EPS from *P.*

fluorescens exhibited a much stronger affinity for the Na^+ ion compared to other EPS that originated from *S. putrefaciens* and *Clostridium* sp. This finding allows the hypothesis that the deprotonated carboxylic sites of EPS from *P. fluorescens* are mostly bound to Na^+ in solution at pH 4 (Harper et al. 2008). The mechanism behind the lower binding capacity at high ionic strength is probably that complexation at one site will decrease the tendency of a neighboring functional group to form an electrostatic complex with another metal ion. The increasing counter ion condensation in the diffuse double layer of the macromolecule will also weaken the affinity of the binding site (Stevenson et al. 1993; Bryan et al. 2002).

5.5 *Effect of Photoinduced Processes*

Photoinduced processes have a significant role in metal-DOM complexation in natural waters (Wu et al. 2004a; Bergquist and Blum 2007; Brooks et al. 2007; Yang and Sturgeon 2009; Sanchez-Marin et al. 2010; Vidali et al. 2010; Yin et al. 2010; Zheng and Hintelmann 2009). The protective effect of DOM on Cu and Pb toxicity greatly disappears when the samples are irradiated with high intensity UV-light (Sanchez-Marin et al. 2010). After UV irradiation, the bulk DOC has been found to decrease by between 60 and 75 %, whereas the decreases in fluorescence and absorbance of CDOM range from 85 to 99 % (Sanchez-Marin et al. 2010). The capacity of humic acid to bind copper appears significantly reduced for irradiated humic acid solutions in the pH range from 3 to 6 (Vidali et al. 2010). The observed apparent convergence of the percentage of copper bound to humic acid for photolytically unaltered and irradiated humic acid in the pH range from 6 to 7 is due to the precipitation of copper-soluble species and to the binding on available ionized binding sites (Vidali et al. 2010).

The photoinduced reduction of ionic Hg in natural water can result in the production of elemental Hg in presence of DOM, which is strongly affected by the Hg–DOM interaction. The subsequent reoxidation of elemental mercury to ionic mercury can occur in the presence of DOM in natural waters (Bergquist and Blum 2007; Zheng and Hintelmann 2009; Xiao et al. 1995; Costa and Liss 1999; Zhang and Lindberg 2001; Ravichandran 2004). Photochemistry can thus affect loss by volatilization and bioavailability of mercury to organisms (Ravichandran 2004).

Photoinduced degradation processes can significantly decrease the absorbance (approximately 8–100 %) and fluorescence intensity of humic substances (fulvic and humic acids) (up to 84 %), tryptophan (up to 88 %) and tyrosine (0–100 %) upon irradiation for hours to months in natural waters (Brooks et al. 2007; Mostofa et al. 2011; Stedmon et al. 2007; Moran et al. 2000; Mostofa et al. 2007, 2010; Skoog et al. 1996; Winter et al. 2007; Abboudi et al. 2008; Osburn et al. 2009; Zhang et al. 2009);Norman et al. 2011 Photoinduced degradation is able to sequentially decompose the functional groups of DOM, particularly in macromolecular fulvic and humic acids. It is induced the formation of low molecular weight photo-products, with simultaneous mineralization of dissolved organic carbon (DOC) in natural waters (Brooks et al. 2007; Corin et al. 1996; Amador et al. 1989; Allard et al. 1994; Moran et al. 2000; Mostofa et al. 2007; Backlund 1992; Kulovaara 1996; Bertilsson and Tranvik 1998; del Vecchio and Blough 2002; Vahatalo and Wetzel 2004; Lou and Xie 2006; Vione et al. 2009).

The most important functional groups or fluorophores in humic substances (fulvic and humic acids), which are responsible for complex formation with trace metals, are carboxylic and phenolic-type groups containing hydroquinone-like moieties and non-quinoid phenols, as well as non-bonding O-, N- and S-atoms contained in functional groups of DOM, such as thiols (Vidali et al. 2009; Mostofa et al. 2011; Lu and Allen 2002; Schmeide and Bernhard 2009; Haitzer et al. 2003; Benedetti et al. 1996; Kinniburgh et al. 1996; Pinheiro et al. 1999; Plaza et al. 2005; Zhang et al. 2004; Smith et al. 2002). High-affinity nitrogenous moieties that make up 2–4 % of DOM can dominate complexation at low Cu concentrations (McKnight et al. 1997; Croue et al. 2003), but model and natural organic nitrogen are readily decomposed by photoinduced processes (Langford 1973; Bushaw et al. 1996). Photoinduced losses of amide carboxyls (15 %) are responsible for a decrease of complexation between Cu and DOM in river water (Brooks et al. 2007). It is suggested that the photoinduced changes in the strong binding sites of parent organic compounds are responsible for the decrease of the stability constants of M-DOM complexes in natural waters.

On the other hand, photoirradiation can increase the conditional stability constants between Cu and DOM in river water whilst it decreases the corresponding values in wetland water (Table 1) (Brooks et al. 2007). Coherently, the ligand concentrations in DOM after photoinduced degradation are decreased by 41–45 % in rivers whilst they are increased by 65–161 % in wetland waters (Brooks et al. 2007). Cu complexation is increased by approximately 150 % in wetland DOM after photoinduced degradation, differently from riverine DOM at the same dissolved organic carbon concentrations (Brooks et al. 2007). It has similarly been found that the conditional stability constants remained the same ($\log_{10} K = 13.5$) after photoinduced degradation of estuarine DOM, but they increased a little in lake DOM after hours to days of photoirradiation. In contrast, in both cases the ligand concentrations were decreased photolytically by up to 95 and 24 %, respectively (Table 1) (Sander et al. 2005; Shank et al. 2006). DOM in wetland, lake and estuarine water is mostly composed of algal-derived material that contains comparatively high proportions of lipids, autochthonous fulvic acids and allochthonous DOM (Parlanti et al. 2000; Mostofa and Sakugawa 2009; Zhang et al. 2009; Brooks et al. 2007; Mostofa et al. 2009b; McCarthy et al. 1998; McCallister et al. 2006). Photolysis can decrease the binding affinity via scission of the allochthonous fraction of lignin-derived polyphenols in wetland DOM (Opsahl and Benner 1998; Sun et al. 1998). Simultaneously, it is hypothesized that photoirradiation can double the binding-site density of wetland DOM by polymerizing and photooxidizing the polyunsaturated lipids produced by microbial communities. This assumption is supported by the experimental observation that photooxidation of polyunsaturated fatty acids and triacylglycerides can increase their content of oxygen functional groups and produce aliphatic aldehydes (Kieber et al. 1997).

Therefore, lipid photooxidation causes a 165 % increase in ketone and aldehyde carbonyls (C-II region), which also increases the ratio of carbonyl groups to aromatic ones (Brooks et al. 2007). Correspondingly, the number of metal-binding substituents per aromatic moiety can increase, producing binding sites with weaker conditional stability constants in the residual (or photobleached) wetland DOM (Brooks et al. 2007).

Photoinduced irradiation is unable to cause complete degradation in waters having high levels of DOM, such as 2.3–32.2 mg L^{-1} in rivers, 43.3 mg L^{-1} in wetland, and 22.6–24 mg L^{-1} in estuaries. In contrast, photochemistry can degrade most of the DOM in waters having low levels of DOM, such as $\leqq 1$ mg L^{-1} in upstream rivers (Brooks et al. 2007; Moran et al. 2000; Mostofa et al. 2007). Therefore, 16–23 % DOM losses can occur in waters with high levels of DOM, whilst higher losses (32–36 %) occur in waters with low levels of DOM. As a consequence, the effects of the photoinduced degradation on M-DOM complexes are expected to be lower in high-DOM waters than in low-DOM ones.

The functional groups in protein or tryptophan are amino carboxylic acid [–CH–(NH_2)–COOH] and –NH group in an aromatic system [C_8H_5(NH)–] (Mostofa et al. 2009a, 2011). Similarly, EPS is primarily composed of several ionisable functional groups such as carboxyl, phosphoric, amine, acidic amino acids, hydroxyl, phenolic, sulfates and organic phosphates. These groups can form complexes with trace metals depending on the environmental conditions (Beech and Sunner 2004; Quiroz et al. 2006; Merroun and Selenska-Pobell 2008; Zhang et al. 2008; Merroun et al. 2003; Guibaud et al. 2005)

It has recently been shown that the functional groups of humic substances (fulvic and humic acids) and tryptophan undergo preferential photoinduced decomposition in natural waters (Xie et al. 2004; Minakata et al. 2009). Decomposition and mineralization of the functional groups of DOM by solar irradiation are responsible for the disappearance of complexation between DOM and metal ions (Sachs et al. 2010; Kulovaara 1996; Kulovaara et al. 1996; Bertilsson and Tranvik 2000). Therefore, photoinduced processes can produce a marked increase of metal toxicity in natural waters.

5.6 *Effect of Microbial Processes*

Microbial processes such as bacterial reductive precipitation, immobilization of soluble metals and M-DOM complexation can significantly affect the mobility (or transport) and toxicity of the trace metals and radionuclides in the aquatic environments (Bergquist and Blum 2007; Yin et al. 2010; Tabak et al. 2005; Francis and Dodge 2008; Fletcher et al. 2010). Key microbial processes such as biotransformation, biosorption and bioaccumulation, as well as degradation or synthesis of DOM can alter the solubility of metals and radionuclides (Fig. 10) (Bergquist and Blum 2007; Yin et al. 2010; Tabak et al. 2005; Francis and Dodge 2008; Fletcher et al. 2010). Metal-reducing or sulfate-reducing microorganisms can directly or indirectly reduce the soluble and mobile trace metals (Cr^{6+}, U^{6+}, Tc^{7+}) to metals (Cr^{3+}, U^{4+},

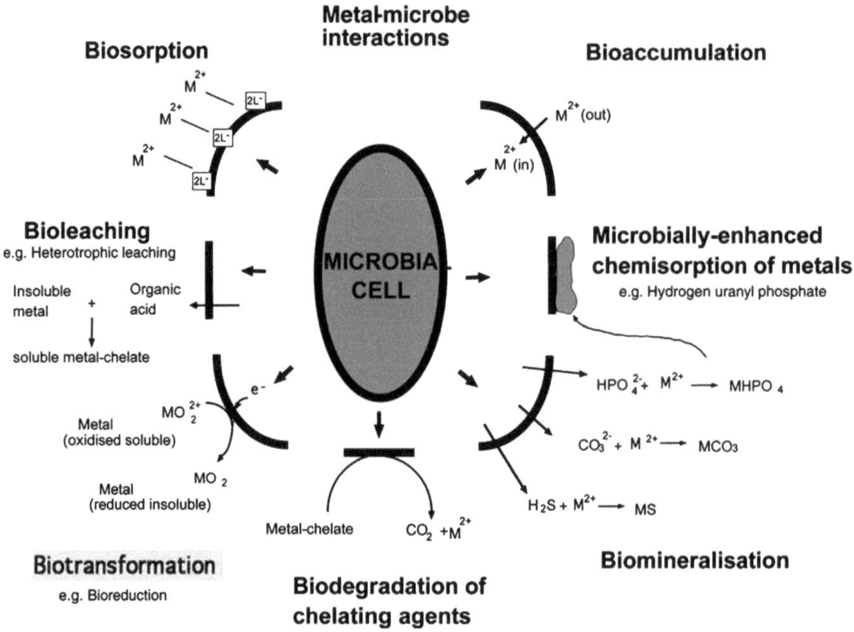

Fig. 10 Metal-microbe interactions impacting bioremediation. *Data source* Tabak et al. (2005)

Tc^{4+}) that are insoluble and less mobile in water, also affecting the M-DOM complexation (Tabak et al. 2005). Another study has shown that the anaerobic spore-forming bacteria *Clostridia*, ubiquitous in soils, sediments, and wastes, are able to reduce Fe^{3+} to Fe^{2+}, Mn^{4+} to Mn^{2+}, U^{6+} to U^{4+}, Pu^{4+} to Pu^{3+}, and Tc^{6+} to Tc^{4+}. They also reduce U^{6+} associated with citric acid in a dinuclear 2:2 U^{6+}:citric acid complex to a biligand mononuclear 1:2 U^{4+}:citric acid complex that remains in solution, in contrast to the reduction and precipitation of uranium (Francis and Dodge 2008). The bioreduction of U^{6+} to U^{4+} also occurs by environmentally relevant bacteria (Gram-positive and Gram-negative), yielding a phase or mineral composed of mononuclear U^{4+} atoms that can form inner-sphere bonds with C/N/O- or P/S-containing ligands (Fletcher et al. 2010; Bernier-Latmani et al. 2010).

Bioaccumulation and biosorption occurs in three ways (Tabak et al. 2005): (i) Sorption on surface sites: sorption of metals and actinides can take place with cell surface active sites. (ii) Surface precipitation: following initial surface sorption, additional surface complexation of metals and actinides can happen by precipitation. (iii) Precipitation with bacterial cell lysate: it occurs by both complexation and precipitation. In addition, microbes can degrade the functional groups of DOM (e.g. fulvic acid, humic acid and tryptophan amino acid) and of synthetic organic ligands (ca. EDTA), or complexes between DOM and trace metals (Tabak et al. 2005). Microbial processes can alter the functional groups (or chromophores or fluorophores) of DOM, causing significant changes (either increase or decrease) in their absorption and fluorescence properties (Moran et al. 2000; Mostofa et al. 2007; Winter et al. 2007; Helms et al. 2008). Microbial changes in the functional groups of DOM that are responsible for complex formation with trace metals are thus susceptible to alter M-DOM complexation in natural waters.

In addition, accumulation of metals by plants (phytoremediation) can reduce the toxicity and content of metals (Tabak et al. 2005; Salt et al. 1995, 1998; Raskin et al. 1997; Pulford and Watson 2003; Schwitzguébel et al. 2002) and, as a consequence, the metal complexation with DOM in the aquatic environments. The key processes are: (i) phytoextraction; the use of metal-accumulating plants to remove toxic metals from soil; (ii) rhizofiltration; the use of plant roots to remove toxic metals from polluted waters; and (iii) phytostabilization or phytoremediation; the use of plants to eliminate the bioavailability of toxic metals in soil.

5.7 Effects of Freshwater and Marine Water (Salinity)

The ionic strength and the occurrence of trace elements can affect the complexation between DOM and trace metals and can induce structural modifications of DOM in waters (Wu et al. 2004a, c; Fu et al. 2007; Harper et al. 2008). Sea water is mostly a solution of NaCl where Na and Cl contribute for more than 86 % of the salt content by mass. The order of the other cations is $Mg^{2+} > Ca^{2+} > K^+ > Sr^{2+}$ and the other main anions are SO_4^{2-}, HCO_3^-, Br^-, and F^- (Livingstone 1963; Carpente and Manella 1973; Hem 1985). Comparison of river and sea water shows that Na^+, Ca^{2+}, Mg^{2+}, K^+, HCO_3^-, Cl^- and SO_4^{2-} in the sea are typically 1670 times, 27 times, 330 times, 170 times, 2.4 times, 2,400 times and 245 times, respectively, higher than in rivers (Livingstone 1963; Hem 1985). Recently it has been shown that the affinity of EPS for complexing Pu^{4+} decreases in the order of *Clostridium* sp. > *S. putrefaciens* > *P. fluorescens*, although the concentrations of carboxylic groups in EPS isolated in laboratory cultures decrease in the order of *P. fluorescens* > *S. putrefaciens* > *Clostridium* sp. (Harper et al. 2008). The deprotonated carboxylic sites of EPS from *P. fluorescens* are mostly bound by Na^+ at a ionic strength of 0.1 M NaCl, which might be caused by a much stronger affinity of the Na^+ ion for the EPS from *P. fluorescens* compared to other EPS (Harper et al. 2008). The high quantities of cations in sea water can induce more rapid complex formation in DOM, even with metals that are relatively less effective in fresh waters. Therefore, DOM is expected to form more complexes with cations in sea waters than in fresh waters.

6 Shifts in Fluorescence Spectral Patterns Due to Metal-DOM Complexation

The complexation of trace elements with fulvic and humic acid (extracted from sewerage sludge) can shift the excitation-emission wavelengths to longer wavelength regions during the initial complexation process (Wu et al. 2004a, c; Plaza et al. 2006). During the Hg–DOM complexation, both excitation and emission wavelengths are increased by up to Ex/Em = 40/25 nm in the first 20 s, after which they remain stable over a period of 100 s (Wu et al. 2004a). The mixing of standard fluorescent organic substances with Milli-Q water and seawater shows that the excitation-emission wavelength maxima of SRFA, DAS1, tyrosine, benzoic acid, *p*-hydroxy benzoic acid, *p*-hydroxy benzaldehyde and *p*-hydroxy acetophenone in seawater are significantly shifted from shorter to longer wavelength regions compared to Milli-Q water (Nakajima 2006). For example, the fluorescence peak C of SRFA dissolved in seawater is detected Ex/Em = 345/452 nm whilst the same peak in Milli-Q water is detected at Ex/Em = 325/442 nm. The peak A remains almost the same in both aqueous media (Nakajima 2006). The fluorescence peak C of autochthonous fulvic acid (C-like) of algal origin is detected at Ex/Em = 340/442–448 nm in Milli-Q water, and at Ex/Em = 340/454–455 nm in river water during the photoinduced and microbial assimilations of algae (Mostofa et al. 2009b). In another study, the same autochthonous fulvic acid (C-like) of algal origin has been detected at Ex/Em = 365/453 nm and 270/453 nm in an isotonic solution during the microbial assimilation of lake phytoplankton (0.5 ‰ salinity) (Zhang et al. 2009). The autochthonous fulvic acid or marine humic-like material of algal origin (peak M) at the peak C-region is found to be shifted from shorter excitation wavelengths (290/400–410 nm in pure Milli-Q water) to a longer wavelength region (300–310/400–410 nm in seawater) (Parlanti et al. 2000).

The shift in excitation and emission wavelength maxima with salinity is presumably caused by the anions and cations present in sea water. Such a shift in excitation-emission from shorter to longer wavelengths is termed the red shift of fulvic acid-like fluorescence. The mechanism behind this red shift in sea water is attributed to the complex formation of the functional groups (or flurophores at peak C-region) in fulvic acid with trace elements or ions markedly present in sea water. The complexation of trace elements with the SRFA functional group (or fluorophore) can significantly enhance the electron transfer of that functional group bound at peak C from the ground state to the excited state upon absorption of longer wavelength radiation. The effect is a shift of the excitation-emission maxima of the peak C to longer wavelengths. This is evidenced by the photoinduced formation of aqueous electrons (e_{aq}^-) from organic substances, which is higher in the presence of NaCl than with organic substances alone in aqueous solution (Gopinath et al. 1972; Zepp et al. 1987; Assel et al. 1998; Richard and Canonica 2005; Fujiwara et al. 1993). Rapid excitation of electrons in ionic (saline) FDOM solution is susceptible to shift both excitation and emission maxima of fluorophores (or functional groups) associated to the peak C of allochthonous fulvic and humic acids or autochthonous fulvic acids or other autochthonous DOM. This effect is presumably responsible for the high production of hydrogen peroxide in natural waters (Mostofa and Sakugawa 2009; Fujiwara et al. 1993). Indeed, photogeneration of H_2O_2 from ultrafiltered river DOM is substantially increased with salinity, from 15 to 368 nM h^{-1} at circumneutral pH (Osburn et al. 2009). The mechanism behind this phenomenon apparently can be factors: first, irradiated CDOM can induce photoinduced production of hydrogen peroxide (H_2O_2) that is a hydroxyl radical ($HO^•$) source via photolysis or the Fenton reaction, and the photoinduced generation of H_2O_2 is enhanced by salinity. Trace metal ions (M) in salinity or sea waters can complex with DOM (M-DOM) forming a strong π–electron bonding system between metal ions and the functional groups in DOM. This π–electron in M-DOM complex is rapidly excited photolytically, which is responsible for high production of aqueous electrons (e_{aq}^-) and subsequently the high production of superoxide ion ($O_2^{\bullet -}$), H_2O_2 and $HO^{\bullet}$, respectively.

On the other hand, the mixing of some standard FDOM (e.g. DSBP, phenol, and tryptophan) with seawater show that the fluorescence excitation-emission wavelength maxima (peak C- region and peak T- region) are shifted from longer to shorter wavelength regions compared to Milli-Q water (Nakajima 2006). Such changes in fluorescence excitation-emission maxima are termed as blue-shift of the flurophores in FDOM. The blue-shift of the FDOM fluorescence peaks can be hypothesized to occur upon loss of high molecular weight fluorescent components by physico-chemical modifications such as flocculation, aggregation or precipitation when the ionic strength is increased (Sholkovitz 1976; Carlson and Mayer 1983; Sierra et al. 1997; van Heemst et al. 2000; Benner and Opsahl 2001; McCarthy et al. 1996). The mechanism behind the blue-shift phenomenon in metal-DOM complexation is presumably the fact that anions or cations can form complexes with the fluorophores (or functional groups) of few fluorescent organic components and may increase the excitation energy of the fluorophores associated to the peak C or peak A-region. The result can be a change in the excitation-emission from longer to shorter wavelengths.

7 Importance of the Metal-DOM Complexation

The complexation of DOM with trace metals is connected to several major biogeochemical phenomena that can be distinguished as: (i) Chemical speciation of the trace metals is of key importance for their biological effects and biogeochemical cycling in natural water, sediment and soil environments (Sekaly et al. 2003; Huber et al. 2002; Hughes et al. 1995; Markich 2002). (ii) DOM including humic substances can control the occurrence of free toxic metals through formation of M-DOM complexes that can significantly reduce the bioavailability and toxicity to organisms in natural waters (Shcherbina et al. 2007; Mostofa et al. 2011; Filella et al. 2007; Hörnström et al. 1984; Markich 2002; Managaki and Takada 2005; Yadav and Trivedi 2006). The bioavailability of toxic metals, their ability to bind to or traverse the cell surface of an organism, is generally dependent on the metal speciation or physicochemical form in the aquatic environments. For example, U^{4+} complexes with humic substances (fulvic and humic acids) and inorganic ligands (e.g., carbonate or phosphate) apparently reduce the bioavailability of U by reducing the activity of UO_2^{2+} and UO_2OH^+, which are the major forms of U^{4+} available to organisms, rather than U in strong complexes (e.g. uranyl fulvate) or adsorbed to colloidal and/or particulate matter (Markich 2002). (iii) Formation of M-DOM complexes can influence the transport or migration of metals, the acid–base balance and the solubility in water, sediment and soil environments (Shcherbina et al. 2007; Sekaly et al. 2003; Lippold and Lippmann-Pipke 2009). The complexation of trace metal ions with specific binding sites in fulvic acid can result in conformational folding, thereby changing the outer appearance of the molecule and affecting its mobility (Chakraborty et al. 2007). (iv) Toxic metals and radionuclides can form complexes with humic substances (fulvic and humic acids) in natural aquatic ecosystems (Tabak et al. 2005; Mantoura et al. 1978; Choppin 1988; Higgo et al. 1993; Christensen et al. 1996; Lenhart et al. 2000; Freyer et al. 2009). These trace metals can subsequently undergo facilitated transport with natural DOM via groundwater into the biosphere. As an alternative, they can remain in waters under both oxic and anoxic conditions by forming various soluble complexes with DOM (Francis and Dodge 2008; Fletcher et al. 2010; Kim and Czerwinski 1996; Kim et al. 1994; McCarthy et al. 1998; Bernier-Latmani et al. 2010; Freyer et al. 2009; Artinger et al. 1998; Schussler et al. 2001; Ranville et al. 2007). These results suggest that the complexation of DOM with toxic metals, actinides and radionuclides may affect mobility, toxicity and fate of these elements in the contaminated subsurface environments. (v) Due to the steady-state existence of DOM as M-DOM complexes in natural waters, DOM itself may substantially enhance the production of H_2O_2 by rapid charge transfer, excited-state electron transfer or intramolecular energy transfer in natural waters (Mostofa and Sakugawa 2009; Fujiwara et al. 1993; Dalrymple et al. 2010). The H_2O_2 upon irradiation of river DOM is substantially increased (from 15 to 368 nM h^{-1}) with increasing salinity at circumneutral pH values (Osburn et al. 2009). Therefore, M-DOM complexation plays a significant role in the initiation of photoinduced processes in natural waters. (vi) The M-DOM complexes are highly susceptible to inducing the photo-Fenton reaction, in the presence of the H_2O_2 produced photolytically in surface waters. Such a process is a potential pathway to produce the hydroxyl radical ($HO^{\bullet}$) (Mostofa and Sakugawa 2009; Fujiwara et al. 1993; Zepp et al. 1992; Vermilyea and Voelker 2009). The photo-Fenton reaction for iron can be depicted as follows (Zepp et al. 1992):

$$Fe^{2+} + H_2O_2 \xrightarrow{h\nu} Fe^{3+} + HO^{\bullet} + OH^- \tag{6.1}$$

$$Fe^{3+} + h\nu/H_2O^{\bullet}/O_2^{-\bullet} \rightarrow Fe^{2+} + O_2 + H^+ \tag{6.2}$$

$$\downarrow$$

Recycling of reaction 6. 1

Therefore, the M-DOM complexation can regulate the photoinduced degradation processes of DOM and other solutes in natural waters. (vii) Production of pure chemicals in petrochemical industry for extraction, separation and recycling of metals from aqueous and organic phases, where the M-DOM complexes are selectively dissolved (Mostofa et al. 2011; Bhattacharyya et al. 2010). (viii) Metal–organic ligand complexes are less bioavailable than metals and thus are less toxic to a variety of aquatic species (Koukal et al. 2003; Knezovich et al. 1981; van Leeuwen et al. 2005). These complexes can undergo partitioning between liquid–solid and solid–solid phases, thereby affecting the environmental fate and transport of metals (Hays et al. 2004). (ix) The polyvalent cations and their concentrations are considerably affected by the distribution of DOM fractions and by their quantities (which can for instance be determined using XAD-8 resins), particularly at low DOC concentrations and high pH (Kaiser 1998). It is also shown that hydrophilic DOC increases and hydrophobic DOC decreases with increasing concentrations of metal cations, in the order Ca < Al < Fe, due to formation of metal-DOM complexes (Kaiser 1998). (x) Complexation of Fe^{2+} by catechol and thiol ligands leads to the formation of aqueous species that are capable of reducing substituted nitroaromatic compounds to the corresponding anilines, and thus play an important role in the reductive transformation of persistent organic contaminants (Naka et al. 2006). (xi) Competition of Cu(II), Co(II) and Ni(II) for the few (~1–10 % of the total) strong binding sites of humic substances suggest that Ni(II) is bound to the much more numerous (~99–90 % of the total) weak binding sites, forming labile nickel complexes. The consequence is that Ni(II) is largely present as free nickel ion (Ni(II)-aquo complex), which is reported to be toxic (Mandal et al. 1999, 2000; Lavigne et al. 1987). (xii) Metal binding properties of DOM and the interaction of iron-DOM complexes with phosphate can decrease the concentration of dissolved nutrients and subsequently act as nutrient reservoirs during periods of low availability (Francko and Heath 1983; Jones 1998; Vahatalo et al. 2003). (xiii) Complexes between As and humic acids are much stronger than As-protein or As-fulvic acids ones. Therefore, intake of black soil (or dark brown soil) with high amounts of humic acid could reduce As levels in the human body. Note that while eating soil is quite unusual for humans, in some rural Bangladesh villages it is customary for pregnant women to eat *shikal* (it sounds like "chain" in English), which consists of small (2 in. × 1 in. × 1/2 in.) pellets made out of flooded soil (information source is personal experience of one of the authors). While the origin of this tradition is quite uncertain, it is noteworthy that it was observed in one of the world's regions where human exposure to As is the highest.

8 Scope of the Future Research

It is reported that autochthonous fulvic acid of algal or phytoplankton origin, typically shows fluorescence properties at peak C- and A-regions, which is a similar behavior as allochthonous fulvic and humic acids (Coble 1996, 2007; Parlanti et al. 2000; Mostofa and Sakugawa 2009; Zhang et al. 2009; Yamashita and Jaffe 2008; Stedmon et al. 2007; Mostofa et al. 2009b; Fulton et al. 2004; Yamashita and Tanoue 2004; Fu et al. 2010). Currently, PARAFAC modeling has been applied to the identification of various fluorescent components and of their characteristic changes (optical and chemical) when autochthonous fluorescent DOM originates from algae or phytoplankton and undergoes photoinduced and microbial degradation in seawater (Mostofa and Sakugawa 2009; Zhang et al. 2009; Stedmon et al. 2007). The current works do not distinguish between the photoinduced, microbial and metal-complexation processing of the autochthonous DOM and do not differentiate between e.g. autochthonous fulvic acids and terrestrial fulvic and humic acids. Such a differentiation and the elucidation of the behavior of the single components should be a key focus for future research.

An increase in water temperature due to global warming may significantly alter the biogeochemical functions in natural aquatic ecosystems. However, there is currently no study that examines the temperature effect on the metal-complexation features of fluorescent DOM in waters, which might be another key issue for future study. It may lead to understand the impacts of global warming on M-DOM complexation in the aquatic environments.

Fluorescent whitening agents (FWAs) and household detergents show fluorescence properties and are widely distributed in the waters and sediments of rivers, lakes and oceans (Mostofa and Sakugawa 2009; Mostofa et al. 2005, 2010; Managaki and Takada 2005; Stoll and Giger 1998; Poiger et al. 1999; Baker 2002; Yamaji et al. 2010). The complexation of trace elements with FWAs such as diaminostilbene-type (DAS1) and distyryl biphenyl (DSBP) and household detergents might be an interesting concern for future research.

The stability constants for the formation of ternary complexes are similar to those of DOM in surface waters (Tables 1 and 2) (Khalil and Taha 2004; Khalil and Fazary 2004; Radalla 2010; Rosas et al. 2010; Shiozawa et al. 2011). Considering the similarity of the stability constants, ternary complex formation should be taken into account in a speciation description of M-DOM complexation. It could be vital to understand the high variability of stability constants in natural waters and should be a focus for future research.

9 Nomenclature

CDOM	Chromophoric dissolved organic matter
DOM	Dissolved organic matter
EDTA	Ethylenediaminetetraacetic acid
EHA	Elliot soil humic acid standard
EPS	Extracellular polymeric substances
FA	Fulvic acid
FDOM	Fluorescent dissolved organic matter
HA	Humic acid
I	Ionic strength
$\text{Log}_{10} K$	Conditional stability constant (logarithm)
LFSE	Ligand field stabilization energy
LHA	Leonardite coal humic acid standard
SRFA	Suwannee River fulvic acid
SRHA	Suwannee River humic acid

Problems

(1) Mention the ten key organic ligands of DOM that can complex trace metal ions.
(2) Mention the key DOM components that show fluorescence properties. Explain the application of fluorescence properties to identify the complexation between metal ions (M) and DOM in aqueous media.
(3) Mention the DOM components of natural origin that can complex metal ions in aqueous solution.
(4) Mention the key possible fluorophores in the molecular structures of fulvic acid, humic acid, and tryptophan.
(5) Explain the theory of multisite Stern–Volmer equation for the determination of M-DOM complexation.
(6) Explain the kinetics of the M-fulvic acid complexation regarding the binding sites of fulvic acid.
(7) Derive the equation for the conditional stability constant of M-DOM complexes in aqueous solution.
(8) How does SRFA behave as homogeneous and heterogeneous complexant to trace metals?
(9) Explain the binding sites in fulvic acid and humic acid and how they differ from autochthonous DOM and tryptophan.
(10) Explain the mechanism for the complexation of metal ions with DOM and explain how the functional groups in tryptophan amino acid form complexes with trace metal ions.
(11) Describe the effects of the M-DOM complexation that depend on the order of complex formation of transition metals and other metals to DOM in natural waters.
(12) Explain the electronic configuration of transition metals (Cu^{2+} and Cr^{3+}), alkaline earths (Ca^{2+}), and heavy metals (Sb^{3+}) and how does it influence the interaction with functional DOM groups in M-DOM complexation.
(13) Explain why transition metals generally have higher bonding properties with DOM functional groups than alkali or alkaline earth metals.
(14) What are the controlling factors for the M-DOM complexation in aqueous solution? Describe four important factors that can regulate the M-DOM complexation in aqueous solution.
(15) Why are the fluorescence excitation-emission maxima shifted in metal-DOM complexes? Explain the mechanisms.
(16) Explain the importance of M-DOM complexation in the aquatic environment.

Acknowledgments We thank Prof. Li Siliang of Institute of Geochemistry, Chinese Academy of Sciences, China; Dr. Kazuhide Hayakawa of the Lake Biwa Environmental Research Institute, Japan; and Dr. Youhei Yamashita of Hokkaido University, Japan for their constructive comments and generous assistances. This work was financially supported by the Institute of Geochemistry, the Chinese Academy of Sciences, Guiyang, China. This work was partly supported by Field Science Education and Research Center, Kyoto University, Japan; University Turin, Italy; and Chinese

References

Abboudi M, Jeffrey WH, Ghiglione JF, Pujo-Pay M, Oriol L, Sempere R, Charriere B, Joux F (2008) Effects of photochemical transformations of dissolved organic matter on bacterial metabolism and diversity in three contrasting coastal sites in the Northwestern Mediterranean sea during summer. Microb Ecol 55:344–357

Allard B, Boren H, Pettersson C, Zhang G (1994) Degradation of humic substances by Uv irradiation. Environ Int 20:97–101

Amador JA, Alexander M, Zika RG (1989) Sequential photochemical and microbial-degradation of organic-molecules bound to humic-acid. Appl Environ Microbiol 55:2843–2849

Antunes MCG, Pereira CCC, Esteves da Silva JCG (2007) MCR of the quenching of the EEM of fluorescence of dissolved organic matter by metal ions. Anal Chim Acta 595:9–18

Artinger R, Kienzler B, Schüßler W, Kim J (1998) Effects of humic substances on the 241Am migration in a sandy aquifer: column experiments with Gorleben groundwater/sediment systems. J Contaminant Hydrol 35:261–275

Artinger R, Rabung T, Kim JI, Sachs S, Schmeide K, Heise KH, Bernhard G, Nitsche H (2002) Humic colloid-borne migration of uranium in sand columns. J Contam Hydrol 58:1–12

Assel M, Laenen R, Laubereau A (1998) Ultrafast electron trapping in an aqueous NaCl-solution. Chem Phys Lett 289:267–274

Backlund P (1992) Degradation of aquatic humic material by ultraviolet-light. Chemosphere 25:1869–1878

Bai Y, Wu F, Wan G, Liu C, Fu P, Li W (2008) Ultraviolet absorbance titration for the determination of conditional stability constants of Hg(II) and dissolved organic matter. Chinese J Geochem 27:46–52

Baker A (2002) Fluorescence properties of some farm wastes: implications for water quality monitoring. Water Res 36:189–195

Beech WB, Sunner J (2004) Biocorrosion: towards understanding interactions between biofilms and metals. Curr Opin Biotechnol 15:181–186

Benedetti MF, VanRiemsdijk WH, Koopal LK, Kinniburgh DG, Gooddy DC, Milne CJ (1996) Metal ion binding by natural organic matter: from the model to the field. Geochim Cosmochim Acta 60:2503–2513

Benner R, Opsahl S (2001) Molecular indicators of the sources and transformations of dissolved organic matter in the Mississippi river plume. Org Geochem 32:597–611

Benner R, Pakulski JD, Mccarthy M, Hedges JI, Hatcher PG (1992) Bulk chemical characteristics of dissolved organic-matter in the ocean. Science 255:1561–1564

Benoit JM, Mason RP, Gilmour CC, Aiken GR (2001) Constants for mercury binding by dissolved organic matter isolates from the Florida Everglades. Geochim Cosmochim Acta 65:4445–4451

Bergquist BA, Blum JD (2007) Mass-dependent and -independent fractionation of Hg isotopes by photoreduction in aquatic systems. Science 318:417–420

Bernier-Latmani R, Veeramani H, Vecchia ED, Junier P, Lezama-Pacheco JS, Suvorova EI, Sharp JO, Wigginton NS, Bargar JR (2010) Non-uraninite products of microbial U(VI) reduction. Environ Sci Technol 44:9456–9462

Bertilsson S, Tranvik LJ (1998) Photochemically produced carboxylic acids as substrates for freshwater bacterioplankton. Limnol Oceanogr 43:885–895

Bertilsson S, Tranvik LJ (2000) Photochemical transformation of dissolved organic matter in lakes. Limnol Oceanogr 45:753–762

Bhattacharyya A, Mohapatra P, Manchanda V (2010) Extraction of Am(III) and Eu(III) from nitrate media with Cyanex®-301 and neutral "N" donor ligands: a thermodynamic study. Radiochim Acta 98:141–147

Bidoglio G, Ferrari D, Selli E, Sena F, Tamborini G (1997) Humic acid binding of trivalent Tl and Cr studied by synchronous and time-resolved fluorescence. Environ Sci Technol 31:3536–3543

Bloom P, Bleam W, Xia K, Clapp C, Hayes M, Senesi N, Jardine P (2001) X-ray spectroscopy applications for the study of humic substances. In: Clapp CE, Hayes MHB, Senesi N, Bloom PR, Jardine PM (eds) Soil Sci Society of America Inc, Madison

Braven J, Butler EI, Chapman J, Evens R (1995) Changes in dissolved free amino-acid-composition in sea-water associated with phytoplankton populations. Sci Total Environ 172:115–150

Breault RF, Colman JA, Aiken GR, McKnight D (1996) Copper speciation and binding by organic matter in copper-contaminated streamwater. Environ Sci Technol 30:3477–3486

Bronk DA, Glibert PM, Ward BB (1994) Nitrogen Uptake, dissolved organic nitrogen release, and new production. Science 265:1843–1846

Brooks ML, Meyer JS, McKnight DM (2007) Photooxidation of wetland and riverine dissolved organic matter: altered copper complexation and organic composition. Hydrobiologia 579:95–113

Brown MJ, Lester JN (1982) Role of bacterial extracellular polymers in metal uptake in pure bacterial culture and activated-sludge.2. effects of mean cell retention time. Water Res 16:1549–1560

Bryan SE, Tipping E, Hamilton-Taylor J (2002) Comparison of measured and modelled copper binding by natural organic matter in freshwaters. Comp Biochem Physiol C: Toxicol Pharmacol 133:37–49

Buckau G, Kim JI, Klenze R, Rhee DS, Wimmer H (1992) A comparative spectroscopic study of the fulvate complexation of trivalent transuranium ions. Radiochim Acta 57:105–111

Buesseler KO, Bauer JE, Chen RF, Eglinton TI, Gustafsson O, Landing W, Mopper K, Moran SB, Santschi PH, VernonClark R, Wells ML (1996) An intercomparison of cross-flow filtration techniques used for sampling marine colloids: overview and organic carbon results. Mar Chem 55:1–31

Buffle J, Filella M (1995) Physicochemical heterogeneity of natural complexants – clarification—comment. Anal Chim Acta 313:144–150

Buschmann J, Sigg L (2004) Antimony(III) binding to humic substances: influence of pH and type of humic acid. Environ Sci Technol 38:4535–4541

Bushaw KL, Zepp RG, Tarr MA, SchulzJander D, Bourbonniere RA, Hodson RE, Miller WL, Bronk DA, Moran MA (1996) Photochemical release of biologically available nitrogen from aquatic dissolved organic matter. Nature 381:404–407

Butler IS, Harrod JF (1989) Inorg Chem: principles and applications. Benjamin/Cummings, California

Bux F, Swalaha FM, Kasan HC, Commission SAWR, Programme UoD-WWR (1994) Microbiological transformation of metal contaminated effluents. Water Res Commission

Cabaniss SE (1992) Synchronous fluorescence-spectra of metal-fulvic acid complexes. Environ Sci Technol 26:1133–1139

Cabaniss SE, Shuman MS (1988) Fluorescence quenching measurements of copper-fulvic acid binding. Anal Chem 60:2418–2421

Cao J, Lam KC, Dawson RW, Liu WX, Tao S (2004) The effect of pH, ion strength and reactant content on the complexation of Cu2 + by various natural organic ligands from water and soil in Hong Kong. Chemosphere 54:507–514

Carlson DJ, Mayer LM (1983) Relative influences of riverine and macroalgal phenolic materials on UV absorbance in temperate coastal waters. Can J Fish Aquat Sci 40:1258–1263

Carlson DJ, Brann ML, Mague TH, Mayer LM (1985) Molecular-weight distribution of dissolved organic materials in seawater determined by ultrafiltration—a re-examination. Mar Chem 16:155–171

Carpente Jh, Manella ME (1973) Magnesium to chlorinity ratios in seawater. J Geophys Res 78:3621–3626

Chakraborty P, Fasfous II, Murimboh J, Chakrabarti CL (2007) Simultaneous determination of speciation parameters of Cu, Pb, Cd and Zn in model solutions of Suwannee River fulvic acid by pseudopolarography. Anal Bioanal Chem 388:463–474

Chen Y, Stevenson F, Avnimelech Y (1986) Soil organic matter interactions with trace elements. In: Chen Y, Avnimdlch Y (eds) The role of organic matter in modern agriculture. CRC Press, Boston, pp 73–109

Choppin GR (1988) Humics Radionucl Migr. Radiochim Acta 44–5:23–28*

Christensen JB, Jensen DL, Christensen TH (1996) Effect of dissolved organic caron on the mobility of cadmium, nickel and zinc in leachate polluted groundwater. Water Res 30:3037–3049

Christoforidis KC, Un S, Deligiannakis Y (2010) Effect of metal ions on the indigenous radicals of humic acids: high field electron paramagnetic resonance study. Environ Sci Technol 44:7011–7016

Chua H, Yu P, Sin S, Cheung M (1999) Sub-lethal effects of heavy metals on activated sludge microorganisms. Chemosphere 39:2681–2692

Coale KH, Bruland KW (1988) Copper complexation in the Northeast Pacific. Limnol Oceanogr 33:1084–1101

Coale KH, Bruland KW (1990) Spatial and temporal variability in copper complexation in the North Pacific. Deep-Sea Res Part A-Oceanogr Res Pap 37:317–336

Coble PG (1996) Characterization of marine and terrestrial DOM in seawater using excitation emission matrix spectroscopy. Mar Chem 51:325–346

Coble PG (2007) Marine optical biogeochemistry: the chemistry of ocean color. Chem Rev 107:402–418

Comte S, Guibaud G, Baudu M (2006) Relations between extraction protocols for activated sludge extracellular polymeric substances (EPS) and complexation properties of Pb and Cd

with EPS Part II. Consequences of EPS extraction methods on Pb^{2+} and Cd^{2+} complexation. Enz Microb Technol 38:246–252
Comte S, Guibaud G, Baudu M (2008) Biosorption properties of extracellular polymeric substances (EPS) towards Cd, Cu and Pb for different pH values. J Hazard Mater 151:185–193
Cook RL, Langford CH (1995) Metal-ion quenching of fulvic-acid fluorescence intensities and lifetimes—nonlinearities and a possible 3-component model. Anal Chem 67:174–180
Corin N, Backlund P, Kulovaara M (1996) Degradation products formed during UV-irradiation of humic waters. Chemosphere 33:245–255
Costa M, Liss P (1999) Photoreduction of mercury in sea water and its possible implications for Hg0 air-sea fluxes. Mar Chem 68:87–95
Crancon P, van der Lee J (2003) Speciation and mobility of uranium(VI) in humic-containing soils. Radiochim Acta 91:673–679
Croue JP, Benedetti MF, Violleau D, Leenheer JA (2003) Characterization and copper binding of humic and nonhumic organic matter isolated from the South Platte River: evidence for the presence of nitrogenous binding site. Environ Sci Technol 37:328–336
Da Costa F, Lubes G, Rodriguez M, Lubes V (2011) Study of the ternary complex formation between vanadium(III), Dipicolinic acid and small blood serum bioligands. J Solut Chem 40:106–117
da Silva JCGE, Machado AASC, Oliveira CJS, Pinto MSSDS (1998a) Fluorescence quenching of anthropogenic fulvic acids by Cu(II), Fe(III) and UO(2)(2 +). Talanta 45:1155–1165
da Silva JCG, Machado AASC, Oliveira CJS (1998b) Effect of pH on complexation of Fe(III) with fulvic acids. Environ Toxicol Chem 17:1268–1273
da Silva JCGE, Herrero AI, Machado AASC, Barrado E (2002) Molecular fluorescence analysis of the effect of the pH on the complexation of Cu(II), Ni(II) and Fe(III) ions by the stronger binding sites of a soil fulvic acid. Quim Anal (Barcelona) 20:203–210
d'Abzac P, Bordas F, van Hullebusch E, Lens PNL, Guibaud G (2010) Effects of extraction procedures on metal binding properties of extracellular polymeric substances (EPS) from anaerobic granular sludges. Colloids Surf B-Biointerfaces 80:161–168
Dalrymple RM, Carfagno AK, Sharpless CM (2010) Correlations between dissolved organic matter optical properties and quantum yields of singlet oxygen and hydrogen peroxide. Environ Sci Technol 44:5824–5829
Dasilva JCGE, Machado AASC, Garcia TMO (1995) Beryllium(Ii) as a probe for study of the interactions of metals and fulvic-acids by synchronous fluorescence spectroscopy. Appl Spectr 49:1500–1506
daSilva JCGE, Machado AASC, Ramos MA, Arce F, Rey F (1996) Quantitative study of Be(II) complexation by soil fulvic acids by molecular fluorescence spectroscopy. Environ Sci Technol 30:3155–3160
daSilva JCGE, Machado AASC, Oliveira CJS (1997) Study of the interaction of Al(III) with a soil fulvic acid in the acid pH range by self-modeling mixture analysis of synchronous fluorescence spectral data. Anal Chim Acta 349:23–31
de Zarruk KK, Scholer G, Dudal Y (2007) Fluorescence fingerprints and CU2 + -complexing ability of individual molecular size fractions in soil- and waste-borne DOM. Chemosphere 69:540–548
Del Vecchio R, Blough NV (2002) Photobleaching of chromophoric dissolved organic matter in natural waters: kinetics and modeling. Mar Chem 78:231–253
Determann S, Lobbes JM, Reuter R, Rullkötter J (1998) Ultraviolet fluorescence excitation and emission spectroscopy of marine algae and bacteria. Mar Chem 62:137–156
Dixon AM, Larive CK (1999) NMR spectroscopy with spectral editing for the analysis of complex mixtures. Appl Spectr 53:426a–440a
Douglas BE, McDaniel DH, Alexander JJ (1994) Concepts and models of Inorg Chem, vol 928. Wiley, New York
Drexel RT, Haitzer M, Ryan JN, Aiken GR, Nagy KL (2002) Mercury(II) sorption to two Florida Everglades peats: evidence for strong and weak binding and competition by dissolved organic matter released from the peat. Environ Sci Technol 36:4058–4064

Dudal Y, Holgado R, Maestri G, Guillon E, Dupont L (2006) Rapid screening of DOM's metal-binding ability using a fluorescence-based microplate assay. Sci Total Environ 354:286–291
Duffus JH (2002) "Heavy metals"—A meaningless term? (IUPAC technical report). Pure Appl Chem 74:793–807
Dyrssen D, Wedborg M (1986) Titration of sulphides and thiols in natural waters. Anal Chim Acta 180:473–479
Dyrssen D, Wedborg M (1991) The Sulfur-Mercury(Ii) system in natural-waters. Water Air Soil Pollut 56:507–519
El-Sherif AA (2010) Synthesis, solution equilibria and antibacterial activity of Co(II) with 2-(Aminomethyl)-Benzimidazole and dicarboxylic acids. J Solut Chem 39:1562–1581
Felix CC, Hyde JS, Sarna T, Sealy RC (1978) Interactions of melanin with metal-ions - electron-spin resonance evidence for chelate complexes of metal-ions with free-radicals. J Am Chem Soc 100:3922–3926
Filella M (2008) NOM site binding heterogeneity in natural waters: discrete approaches. J Mol Liq 143:42–51
Filella M, May PM (2005) Critical appraisal of available thermodynamic data for the complexation of antimony(III) and antimony(V) by low molecular mass organic ligands. J Environ Monit 7:1226–1237
Filella M, Belzile N, Lett MC (2007) Antimony in the environment: a review focused on natural waters. III. Microbiota relevant interactions. Earth Sci Rev 80:195–217
Fimmen RL, Cory RM, Chin YP, Trouts TD, McKnight DM (2007) Probing the oxidation-reduction properties of terrestrially and microbially derived dissolved organic matter. Geochim Cosmochim Acta 71:3003–3015
Fletcher KE, Boyanov MI, Thomas SH, Wu QZ, Kemner KM, Loffler FE (2010) U(VI) reduction to mononuclear U(IV) by Desulfitobacterium species. Environ Sci Technol 44:4705–4709
Francis AJ, Dodge CJ (2008) Bioreduction of Uranium(VI) complexed with citric acid by clostridia affects its structure and solubility. Environ Sci Technol 42:8277–8282
Francko DA, Heath RT (1983) Abiotic uptake and photodependent release of phosphate from high-molecular-weight humic-iron complexes in bog lakes. In: Christman RF, Gjessing E (eds) Aquatic and terrestrial humic materials. Ann Arbor Scientific Publications, Ann Arbor, pp 467–480
Freyer M, Walther C, Stumpf T, Buckau G, Fanghanel T (2009) Formation of Cm humate complexes in aqueous solution at pH(c) 3 to 5.5: the role of fast interchange. Radiochim Acta 97:547–558
Frezza M, Hindo SS, Tomco D, Allard MM, Cui QC, Heeg MJ, Chen D, Dou QP, Verani CN (2009) Comparative activities of Nickel(II) and Zinc(II) complexes of Asymmetric [NN ' O] Ligands as 26S Proteasome inhibitors. Inorg Chem 48:5928–5937
Fu PQ, Wu FC, Liu CQ, Wang FY, Li W, Yue LX, Guo QJ (2007) Fluorescence characterization of dissolved organic matter in an urban river and its complexation with Hg(II). Appl Geochem 22:1668–1679
Fu PQ, Mostofa KMG, Wu FC, Liu CQ, Li W, Liao HQ, Wang LY, Wang J, Mei Y (2010) Excitation-emission matrix characterization of dissolved organic matter sources in two eutrophic lakes (Southwestern China Plateau). Geochem J 44:99–112
Fujii M, Rose AL, Waite TD, Omura T (2008) Effect of divalent cations on the kinetics of Fe(III) complexation by organic ligands in natural waters. Geochim Cosmochim Acta 72:1335–1349
Fujiwara K, Ushiroda T, Takeda K, Kumamoto Y, Tsubota H (1993) Diurnal and seasonal distribution of hydrogen-peroxide in seawater of the Seto Inland Sea. Geochem J 27:103–115
Fulton JR, McKnight DM, Foreman CM, Cory RM, Stedmon C, Blunt E (2004) Changes in fulvic acid redox state through the oxycline of a permanently ice-covered Antarctic lake. Aquat Sci 66:27–46
Gasper JD, Aiken GR, Ryan JN (2007) A critical review of three methods used for the measurement of mercury (Hg2 +)-dissolved organic matter stability constants. Appl Geochem 22:1583–1597

Ghassemi M, Christman R (1968) Properties of the yellow organic acids of natural waters. Limnol Oceanogr 13:583–597

Giesy JP, Geiger RA, Kevern NR, Alberts JJ (1986) Uo-2(2 +)-Humate Interactions in Soft, Acid, Humate-Rich Waters. J Environ Radioact 4:39–64

Glaus MA, Hummel W, van Loon LR (2000) Trace metal-humate interactions. I. Experimental determination of conditional stability constants. App Geochem 15:953–973

Gopinath C, Damle PS, Hart EJ (1972) Gamma-ray irradiated sodium-chloride as a source of hydrated electrons. J Phys Chem 76:3694

Gordon AS, Dyer BJ, Kango RA, Donat JR (1996) Copper ligands isolated from estuarine water by immobilized metal affinity chromatography: temporal variability and partial characterization. Mar Chem 53:163–172

Grimm DM, Azarraga LV, Carreira LA, Susetyo W (1991) Continuous multiligand distribution model used to predict the stability constant of Cu(Ii) metal complexation with humic material from fluorescence quenching data. Environ Sci Technol 25:1427–1431

Guentzel JL, Powell RT, Landing WM, Mason RP (1996) Mercury associated with colloidal material in an estuarine and an open-ocean environment. Mar Chem 55:177–188

Guibaud G, Tixier N, Bouju A, Baudu M (2003) Relation between extracellular polymers' composition and its ability to complex Cd, Cu and Pb. Chemosphere 52:1701–1710

Guibaud G, Tixier N, Bouju A, Baudu M (2004) Use of a polarographic method to determine copper, nickel and zinc constants of complexation by extracellular polymers extracted from activated sludge. Process Biochem 39:833–839

Guibaud G, Comte S, Bordas F, Dupuy S, Baudu M (2005a) Comparison of the complexation potential of extracellular polymeric substances (EPS), extracted from activated sludges and produced by pure bacteria strains, for cadmium, lead and nickel. Chemosphere 59:629–638

Guibaud G, Comte S, Bordas F, Baudu M (2005b) Metal removal from single and multimetallic equimolar systems by extracellular polymers extracted from activated sludges as evaluated by SMDE polarography. Process Biochem 40:661–668

Guibaud G, van Hullebusch E, Bordas F (2006) Lead and cadmium biosorption by extracellular polymeric substances (EPS) extracted from activated sludges: pH-sorption edge tests and mathematical equilibrium modelling. Chemosphere 64:1955–1962

Guo L, Santschi PH (1996) A critical evaluation of the cross-flow ultrafiltration technique for sampling colloidal organic carbon in seawater. Mar Chem 55:113–127

Guo LD, Coleman CH, Santschi PH (1994) The distribution of colloidal and dissolved organic-carbon in the Gulf-of-Mexico. Mar Chem 45:105–119

Guo LD, Santschi PH, Warnken KW (1995) Dynamics of dissolved organic carbon (DOC) in oceanic environments. Limnol Oceanogr 40:1392–1403

Guo LD, Santschi PH, Cifuentes LA, Trumbore SE, Southon J (1996) Cycling of high-molecular-weight dissolved organic matter in the middle Atlantic bight as revealed by carbon isotopic (C-13 and C-14) signatures. Limnol Oceanogr 41:1242–1252

Gutierrez T, Shimmield T, Haidon C, Black K, Green DH (2008) Emulsifying and metal ion binding activity of a glycoprotein exopolymer produced by Pseudoalteromonas sp strain TG12. Appl Environ Microbiol 74:4867–4876

Haitzer M, Aiken GR, Ryan JN (2002) Binding of mercury(II) to dissolved organic matter: the role of the mercury-to-DOM concentration ratio. Environ Sci Technol 36:3564–3570

Haitzer M, Aiken GR, Ryan JN (2003) Binding of mercury (II) to aquatic humic substances: influence of pH and source of humic substances. Environ Sci Technol 37:2436–2441

Harper RM, Kantar C, Honeyman BD (2008) Binding of Pu(IV) to galacturonic acid and extracellular polymeric substances (EPS) from Shewanella putrefaciens, Clostridium sp and Pseudomonas fluorescens. Radiochim Acta 96:753–762

Hart BT (1981) Trace-metal complexing capacity of natural-waters - a review. Environ Technol Lett 2:95–110

Hatch CD, Gierlus KM, Zahardis J, Schuttlefield J, Grassian VH (2009) Water uptake of humic and fulvic acid: measurements and modelling using single parameter Kohler theory. Environ Chem 6:380–388

Hays MD, Ryan DK, Pennell S (2003) Multi-wavelength fluorescence-quenching model for determination of Cu2 + conditional stability constants and ligand concentrations of fulvic acid. Appl Spectr 57:454–460

Hays MD, Ryan DK, Pennell S (2004) A modified multisite stern-volmer equation for the determination of conditional stability constants and ligand concentrations of soil fulvic acid with metal ions. Anal Chem 76:848–854

Hedges JI, Keil RG, Benner R (1997) What happens to terrestrial organic matter in the ocean? Org Geochem 27:195–212

Helms JR, Stubbins A, Ritchie JD, Minor EC, Kieber DJ, Mopper K (2008) Absorption spectral slopes and slope ratios as indicators of molecular weight, source, and photobleaching of chromophoric dissolved organic matter. Limnol Oceanogr 53:955–969

Hem JD (1985) Study and interpretation of the chemical characteristics of natural water, vol 2254. Dept. of the Interior, US Geological Survey

Higgo JJW, Kinniburgh D, Smith B, Tipping E (1993) Complexation of Co^{2+}, Ni^{2+}, $UO_2{}^{2+}$ and Ca^{2+} by humic substances in groundwaters. Radiochim Acta 61:91–103

Hirose K (2004) Chemical speciation of thorium in marine biogenic particulate matter. TSWJ 4:67–76

Hörnström E, Ekström C, Duraini M (1984) Effects of pH and different levels of aluminium on lake plankton in the Swedish west coast area. Rep Inst Freshw Res Drottningholm 61:115–127

Hsu H, Sedlak DL (2003) Strong Hg(II) complexation in municipal wastewater effluent and surface waters. Environ Sci Technol 37:2743–2749

Huber C, Filella M, Town RM (2002) Computer modelling of trace metal ion speciation: practical implementation of a linear continuous function for complexation by natural organic matter. Comput Geosci 28:587–596

Hughes K, Meek ME, Newhook R, Chan PKL (1995) Speciation in health risk assessments of metals: evaluation of effects associated with forms present in the environment. Regul Toxicol Pharmacol 22:213–220

Hulatt CJ, Thomas DN, Bowers DG, Norman L, Zhang C (2009) Exudation and decomposition of chromophoric dissolved organic matter (CDOM) from some temperate macroalgae. Estuar Coast Shelf Sci 84:147–153

Iota V, Klepeis JHP, Yoo CS, Lang J, Haskel D, Srajer G (2007) Electronic structure and magnetism in compressed 3d transition metals. Appl Phys Lett 90:042505–042505-3

Irving H, Williams R (1948) Reversion: a new procedure in absorptiometry. Nature 162:746

Irving H, Williams RJP (1953) The stability of transition-metal complexes. J Chem Soc 3:3192–3210

Iskrenova-Tchoukova E, Kalinichev AG, Kirkpatrick RJ (2010) Metal cation complexation with natural organic matter in aqueous solutions: molecular dynamics simulations and potentials of mean force. Langmuir 26:15909–15919

Jackson BP, Ranville JF, Bertsch PM, Sowder AG (2005) Characterization of colloidal and humic-bound Ni and U in the "dissolved" fraction of contaminated sediment extracts. Environ Sci Technol 39:2478–2485

Janes N, Playle RC (1995) Modeling silver-binding to gills of rainbow-trout (Oncorhynchus-Mykiss). Environ Toxicol Chem 14:1847–1858

Jansen B, Nierop KGJ, Verstraten JM (2003) Mobility of Fe(II), Fe(III) and Al in acidic forest soils mediated by dissolved organic matter: influence of solution pH and metal/organic carbon ratios. Geoderma 113:323–340

Jennings JG, Steinberg PD (1994) In Situ exudation of phlorotannins by the sublittoral kelp Ecklonia-Radiata. Mar Biol 121:349–354

Jiang J, Bauer I, Paul A, Kappler A (2009) Arsenic redox changes by microbially and chemically formed semiquinone radicals and hydroquinones in a humic substance model quinone. Environ Sci Technol 43:3639–3645

Jones RI (1998) Phytoplankton, primary production and nutrient cycling. In: Hessen DO, Tranvik LJ (eds) Aquatic humic substances: ecology and biogeochemistry. Springer, Berlin Heidelberg, pp 145–175

Kaiser K (1998) Fractionation of dissolved organic matter affected by polyvalent metal cations. Org Geochem 28:849–854
Kautenburger R, Nowotka K, Beck HP (2006) Online analysis of europium and gadolinium species complexed or uncomplexed with humic acid by capillary electrophoresis-inductively coupled plasma mass spectrometry. Anal Bioanal Chem 384:1416–1422
Khalil MM (2000a) Complexation equilibria and determination of stability constants of binary and ternary complexes with ribonucleotides (AMP, ADP, and ATP) and salicylhydroxamic acid as ligands. J Chem Eng Data 45:70–74
Khalil MM (2000b) Solution equilibria slid stabilities of binary and ternary complexes with N-(2-acetamido)iminodiacetic acid and ribonucleotides (AMP, ADP, and ATP). J Chem Eng Data 45:837–840
Khalil MM, Attia AE (1999) Potentiometric studies on the binary and ternary complexes of copper(II) containing dipicolinic acid and amino acids. J Chem Eng Data 44:180–184
Khalil MM, Attia AE (2000) Potentiometric studies on the formation equilibria of binary and ternary complexes of some metal ions with dipicolinic acid and amino acids. J Chem Eng Data 45:1108–1111
Khalil MM, Fazary AE (2004) Potentiometric studies on binary and ternary complexes of Di- and trivalent metal ions involving some hydroxamic acids, amino acids, and nucleic acid components. Monatshefte Fur Chemie 135:1455–1474
Khalil MM, Radalla AM (1998) Binary and ternary complexes of inosine. Talanta 46:53–61
Khalil MM, Taha M (2004) Equilibrium studies of binary and ternary complexes involving tricine and some selected alpha-amino acids. Monatshefte Fur Chemie 135:385–395
Kieber RJ, Hydro LH, Seaton PJ (1997) Photooxidation of triglycerides and fatty acids in seawater: implication toward the formation of marine humic substances. Limnol Oceanogr 42:1454–1462
Kim J, Czerwinski K (1996) Complexation of metal ions with humic acid: metal ion charge neutralization model. Radiochim Acta 73:5–10
Kim JI, Buckau G, Li GH, Duschner H, Psarros N (1990) Characterization of humic and fulvic-acids from Gorleben groundwater. Fresenius J Anal Chem 338:245–252
Kim JI, Wimmer H, Klenze R (1991) A study of Curium(Iii) humate complexation by time resolved laser fluorescence spectroscopy (Trlfs). Radiochim Acta 54:35–41
Kim J, Delakowitz B, Zeh P, Klotz D, Lazik D (1994) A column experiment for the study of colloidal radionuclide migration in Gorleben aquifer systems. Radiochim Acta 66:165–165
Kim DW, Cha DK, Wang J, Huang C (2002) Heavy metal removal by activated sludge: influence of Nocardia amarae. Chemosphere 46:137–142
Kinniburgh DG, Milne CJ, Benedetti MF, Pinheiro JP, Filius J, Koopal LK, VanRiemsdijk WH (1996) Metal ion binding by humic acid: application of the NICA-Donnan model. Environ Sci Technol 30:1687–1698
Klinck J, Dunbar M, Brown S, Nichols J, Winter A, Hughes C, Playle RC (2005) Influence of water chemistry and natural organic matter on active and passive uptake of inorganic mercury by gills of rainbow trout (Oncorhynchus mykiss). Aquat Toxicol 72:161–175
Knezovich JP, Harrison F, Tucker J (1981) The influence of organic chelators on the toxicity of copper to embryos of the Pacific oyster, Crassostrea gigas. Arch Environ Contam Toxicol 10:241–249
Kolokassidou C, Pashalidis I (2006) Potentiometric investigations on the interaction of humic acid with Cu(II) and Eu(III) ions. Radiochim Acta 94:549–552
Konstantinou M, Pashalidis I (2010) Competitive sorption of Cu (II) and Eu(III) ions on olive-cake carbon in aqueous solutions—a potentiometric study. Adsorption 16:167–171
Konstantinou M, Kolokassidou K, Pashalidis I (2007) Sorption of Cu(II) and Eu(III) ions from aqueous solution by olive cake. Adsorpt J Int Adsorpt Soc 13:33–40
Konstantinou M, Kolokassidou K, Pashalidis I (2009) Studies on the interaction of olive cake and its hydrophylic extracts with polyvalent metal ions (Cu(II), Eu(III)) in aqueous solutions. J Hazard Mater 166:1169–1173
Koukal B, Gueguen C, Pardos M, Dominik J (2003) Influence of humic substances on the toxic effects of cadmium and zinc to the green alga Pseudokirchneriella subcapitata. Chemosphere 53:953–961

Kulovaara M (1996) Light-induced degradation of aquatic humic substances by simulated sunlight. Int J Environ Anal Chem 62:85–95
Kulovaara M, Corin N, Backlund P, Tervo J (1996) Impact of UV254-radiation on aquatic humic substances. Chemosphere 33:783–790
Kumke MU, Tiseanu C, Abbt-Braun G, Frimmel FH (1998) Fluorescence decay of natural organic matter (NOM)-influence of fractionation, oxidation, and metal ion complexation. J Fluorescence 8:309–318
Lakshman S, Mills R, Patterson H, Cronan C (1993) Apparent differences in binding site distributions and aluminum (III) complexation for three molecular weight fractions of a coniferous soil fulvic acid. Anal Chim Acta 282:101–108
Lakshman S, Mills R, Fang F, Patterson H, Cronan C (1996) Use of fluorescence polarization to probe the structure and aluminum complexation of three molecular weight fractions of a soil fulvic acid. Anal Chim Acta 321:113–119
Lamborg CH, Tseng CM, Fitzgerald WF, Balcom PH, Hammerschmidt CR (2003) Determination of the mercury complexation characteristics of dissolved organic matter in natural waters with "reducible Hg" titrations. Environ Sci Technol 37:3316–3322
Lamborg CH, Fitzgerald WF, Skoog A, Visscher PT (2004) The abundance and source of mercury-binding organic ligands in long island sound. Mar Chem 90:151–163
Lamelas C, Slaveykova VI (2007) Comparison of Cd(II), Cu(II), and Pb(II) biouptake by green algae in the presence of humic acid. Environ Sci Technol 41:4172–4178
Lamelas C, Pinheiro JP, Slaveykova VI (2009) Effect of humic acid on Cd(II), Cu(II), and Pb(II) uptake by freshwater algae: kinetic and cell wall speciation considerations. Environ Sci Technol 43:730–735
Langford CH (1973) Ligand photooxidation in copper(Ii) complexes of nitrilotriacetic acid - implications for natural waters. Environ Sci Technol 7:820–822
Lavigne JA, Langford CH, Mak MKS (1987) Kinetic-study of speciation of Nickel(Ii) bound to a fulvic-acid. Anal Chem 59:2616–2620
Lee C, Bada JL (1975) Amino-acids in equatorial Pacific ocean water. Earth Planet Sci Lett 26:61–68
Leenheer JA, Croue JP (2003) Characterizing aquatic dissolved organic matter. Environ Sci Technol 37:18a–26a
Leenheer JA, Brown GK, MacCarthy P, Cabaniss SE (1998) Models of metal binding structures in fulvic acid from the Suwannee River, Georgia. Environ Sci Technol 32:2410–2416
Leighton IR, Forster CF (1997) The adsorption of heavy metals in an acidogenic thermophilic anaerobic reactor. Water Res 31:2969–2972
Lenhart JJ, Cabaniss SE, MacCarthy P, Honeyman BD (2000) Uranium(VI) complexation with citric, humic and fulvic acids. Radiochim Acta 88:345–353
Li K, Crittenden J (2009) Computerized pathway elucidation for hydroxyl radical-induced chain reaction mechanisms in aqueous phase advanced oxidation processes. Environ Sci Technol 43:2831–2837
Lin CF, Lee DY, Chen WT, Lo KS (1995) Fractionation of fulvic-acids—characteristics and complexation with copper. Environ Pollut 87:181–187
Lippold H, Lippmann-Pipke J (2009) Effect of humic matter on metal adsorption onto clay materials: testing the linear additive model. J Contam Hydrol 109:40–48
Lister SK, Line MA (2001) Potential utilisation of sewage sludge and paper mill waste for biosorption of metals from polluted waterways. Biores Technol 79:35–39
Liu GL, Cai Y (2010) Complexation of arsenite with dissolved organic matter conditional distribution coefficients and apparent stability constants. Chemosphere 81:890–896
Liu H, Fang HHP (2002) Characterization of electrostatic binding sites of extracellular polymers by linear programming analysis of titration data. Biotechnol Bioeng 80:806–811
Liu Y, Fang HHP (2003) Influences of extracellular polymeric substances (EPS) on flocculation, settling, and dewatering of activated sludge. Crit Rev Environ Sci Technol 33:237–273
Liu Y, Lam MC, Fang HHP (2001) Adsorption of heavy metals by EPS of activated sludge. Water Sci Technol 43:59–66

Livingstone DA (1963) Chemical composition of rivers and lakes. In: Fleischer M (ed) Data of Geochemistry, US Geological Survey. US Govt Printing office, Washington DC, pp 58–661
Loaec M, Olier R, Guezennec J (1997) Uptake of lead, cadmium and zinc by a novel bacterial exopolysaccharide. Water Res 31:1171–1179
Lou T, Xie HX (2006) Photochemical alteration of the molecular weight of dissolved organic matter. Chemosphere 65:2333–2342
Lu YF, Allen HE (2002) Characterization of copper complexation with natural dissolved organic matter (DOM)—link to acidic moieties of DOM and competition by Ca and Mg. Water Res 36:5083 5101
Lu XQ, Jaffe R (2001) Interaction between Hg(II) and natural dissolved organic matter: a fluorescence spectroscopy based study. Water Res 35:1793–1803
Lubes G, Rodriguez M, Lubes V (2010) Ternary complex formation between vanadium(III), Dipicolinic acid and picolinic acid in aqueous solution. J Solut Chem 39:1134–1141
Luider CD, Crusius J, Playle RC, Curtis PJ (2004) Influence of natural organic matter source on copper speciation as demonstrated by Cu binding to fish gills, by ion selective electrode, and by DGT gel sampler. Environ Sci Technol 38:2865–2872
Ma HZ, Allen HE, Yin YJ (2001) Characterization of isolated fractions of dissolved organic matter from natural waters and a wastewater effluent. Water Res 35:985–996
Malcolm R (1985) Geochemistry of stream fulvic and humic substances. In: Aiken GR, McKnight DM, Wershaw RL, MacCarthy P (eds) Humic substances in soil, sediment, and water: Geochemistry, isolation and characterization. Wiley, New York, pp 181–209
Malcolm RL (1990) The uniqueness of humic substances in each of soil, stream and marine environments. Anal Chim Acta 232:19–30
MalinskyRushansky NZ, Legrand C (1996) Excretion of dissolved organic carbon by phytoplankton of different sizes and subsequent bacterial uptake. Mar Ecol Prog Ser 132:249–255
Managaki S, Takada H (2005) Fluorescent whitening agents in Tokyo Bay sediments: molecular evidence of lateral transport of land-derived particulate matter. Mar Chem 95:113–127
Manciulea A, Baker A, Lead JR (2009) A fluorescence quenching study of the interaction of Suwannee River fulvic acid with iron oxide nanoparticles. Chemosphere 76:1023–1027
Mandal R, Sekaly ALR, Murimboh J, Hassan NM, Chakrabarti CL, Back MH, Gregoire DC, Schroeder WH (1999) Effect of the competition of copper and cobalt on the lability of Ni(II)-organic ligand complexes. Part I. In model solutions containing Ni(II) and a well-characterized fulvic acid. Anal Chim Acta 395:309–322
Mandal R, Salam MSA, Murimboh J, Hassan NM, Chakrabarti CL, Back MH, Gregoire DC (2000) Competition of Ca(II) and Mg(II) with Ni(II) for binding by a well-characterized fulvic acid in model solutions. Environ Sci Technol 34:2201–2208
Mantoura R, Dickson A, Riley J (1978) The complexation of metals with humic materials in natural waters. Estuar Coast Mar Sci 6:387–408
Marang L, Reiller PE, Eidner S, Kumke MU, Benedetti MF (2008) Combining spectroscopic and potentiometric approaches to characterize competitive binding to humic substances. Environ Sci Technol 42:5094–5098
Margerum D, Cayley G, Weatherburn D, Pagenkopf G (1978) Coordination Chemistry. Martell, AE, Ed:1-220
Markich SJ (2002) Uranium speciation and bioavailability in aquatic systems: an overview. Sci World J 2:707–729
Martel AE, Motekaitis RJ (1988) Determination and uses of stability constants. Wiley, Weinheim
Martin RB, Prados R (1974) Some factors influencing mixed complex formation. J Inorg Nucl Chem 36:1665–1670
Matsumoto ST, Rigonato A, Mantovani MS, Marin-Morales MA (2005) Evaluation of the genotoxic potential due to the action of an effluent contaminated with Chromium, by the Comet assay in CHO-K1 cultures. Caryologia 58:40–46
Maurer LG (1976) Organic polymers in seawater—changes with depth in Gulf of Mexico. Deep-Sea Res 23:1059–1064

McCallister SL, Bauer JE, Canuel EA (2006) Bioreactivity of estuarine dissolved organic matter: a combined geochemical and microbiological approach. Limnol Oceanogr 51:94–100
McCarthy MD, Hedges JI, Benner R (1996) Major biochemical composition of dissolved high molecular weight organic matter in seawater. Mar Chem 55:281–297
McCarthy JF, Czerwinski KR, Sanford WE, Jardine PM, Marsh JD (1998) Mobilization of transuranic radionuclides from disposal trenches by natural organic matter. J Contam Hydrol 30:49–77
Mcgown LB, Hemmingsen SL, Shaver JM, Geng L (1995) Total lifetime distribution analysis for fluorescence fingerprinting and characterization. Appl Spectr 49:60–66
Mcknight DM, Morel FMM (1980) Copper complexation by siderophores from filamentous blue-green-algae. Limnol Oceanogr 25:62–71
Mcknight DM, Feder GL, Thurman EM, Wershaw RL, Westall JC (1983) Complexation of copper by aquatic humic substances from different environments. Sci Total Environ 28:65–76
McKnight DM, Harnish R, Wershaw RL, Baron JS, Schiff S (1997) Chemical characteristics of particulate, colloidal, and dissolved organic material in Loch Vale Watershed, Rocky Mountain National Park. Biogeochemistry 36:99–124
Merroun ML, Selenska-Pobell S (2008) Bacterial interactions with uranium: an environmental perspective. J Contam Hydrol 102:285–295
Merroun M, Hennig C, Rossberg A, Reich T, Selenska-Pobell S (2003) Characterization of U (VI)-Acidithiobacillus ferrooxidans complexes using EXAFS, transmission electron microscopy, and energy-dispersive X-ray analysis. Radiochim Acta 91:583–592
Midorikawa T, Tanoue E (1996) Extraction and characterization of organic ligands from oceanic water columns by immobilized metal ion affinity chromatography. Mar Chem 52:157–171
Midorikawa T, Tanoue E (1998) Molecular masses and chromophoric properties of dissolved organic ligands for copper(II) in oceanic water. Mar Chem 62:219–239
Midorikawa T, Tanoue E, Sugimura Y (1990) Determination of complexing ability of natural ligands in seawater for various metal-ions using ion-selective electrodes. Anal Chem 62:1737–1746
Minakata D, Li K, Westerhoff P, Crittenden J (2009) Development of a Group contribution method to predict aqueous phase hydroxyl radical (HO center dot) reaction rate constants. Environ Sci Technol 43:6220–6227
Moffett JW, Zika RG, Brand LE (1990) Distribution and potential sources and sinks of copper chelators in the Sargasso Sea. Deep-Sea Res Part A 37:27–36
Mopper K, Feng ZM, Bentjen SB, Chen RF (1996) Effects of cross-flow filtration on the absorption and fluorescence properties of seawater. Mar Chem 55:53–74
Moran MA, Hodson RE (1994) Dissolved humic substances of vascular plant-origin in a coastal marine-environment. Limnol Oceanogr 39:762–771
Moran MA, Pomeroy LR, Sheppard ES, Atkinson LP, Hodson RE (1991) Distribution of terrestrially derived dissolved organic-matter on the Southeastern United-States Continental-Shelf. Limnol Oceanogr 36:1134–1149
Moran MA, Sheldon WM, Zepp RG (2000) Carbon loss and optical property changes during long-term photochemical and biological degradation of estuarine dissolved organic matter. Limnol Oceanogr 45:1254–1264
Morel F, Hering JG (1993) Principles and applications of aquatic chemistry. Wiley-Interscience, New York
Morra MJ, Fendorf SE, Brown PD (1997) Speciation of sulfur in humic and fulvic acids using X-ray absorption near-edge structure (XANES) spectroscopy. Geochim Cosmochim Acta 61:683–688
Mostofa KMG, Sakugawa H (2009) Spatial and temporal variations and factors controlling the concentrations of hydrogen peroxide and organic peroxides in rivers. Environ Chem 6:524–534
Mostofa KMG, Honda Y, Sakugawa H (2005a) Dynamics and optical nature of fluorescent dissolved organic matter in river waters in Hiroshima prefecture, Japan. Geochem J 39:257–271

Mostofa KMG, Yoshioka T, Konohira E, Tanoue E, Hayakawa K, Takahashi M (2005b) Three-dimensional fluorescence as a tool for investigating the dynamics of dissolved organic matter in the Lake Biwa watershed. Limnology 6:101–115
Mostofa KMG, Yoshioka T, Konohira E, Tanoue E (2007) Photodegradation of fluorescent dissolved organic matter in river waters. Geochem J 41:323–331
Mostofa KMG, Wu FC, Yoshioka T, Sakugawa H, Tanoue E (2009a) Dissolved organic matter in the aquatic environments. In: Wu FC, Xing B (eds) Natural Organic Matter and its Significance in the Environment. Science Press, Beijing, pp 3–66
Mostofa KMG, Liu CQ, Wu FC, Fu PQ, Ying WL, Yuan J (2009) Overview of key biogeochemical functions in lake ecosystem: impacts of organic matter pollution and global warming. In: Proceedings of the 13th World Lake Conference on Wuhan, China, 1–5 Nov 2009, Keynote Speech, pp 59–60
Mostofa KMG, Wu F, Liu CQ, Fang WL, Yuan J, Ying WL, Wen L, Yi M (2010) Characterization of Nanming River (southwestern China) sewerage-impacted pollution using an excitation-emission matrix and PARAFAC. Limnology 11:217–231
Mostofa KMG, Wu FC, Liu CQ, Vione D, Yoshioka T, Sakugawa H, Tanoue E (2011) Photochemical, microbial and metal complexation behavior of fluorescent dissolved organic matter in the aquatic environments. Geochem J 45:235–254
Motson GR, Fleming JS, Brooker S (2004) Potential applications for the use of lanthanide complexes as luminescent biolabels. Adv Inorg Chem: Including Bioinorg Studies 55:361–432
Mounier S, Braucher R, Benaïm J (1999) Differentiation of organic matter's properties of the Rio Negro basin by cross-flow ultra-filtration and UV-spectrofluorescence. Water Res 33:2363–2373
Mounier S, Zhao HY, Garnier C, Redon R (2011) Copper complexing properties of dissolved organic matter: PARAFAC treatment of fluorescence quenching. Biogeochemistry 106:107–116
Murimboh JD (2002) Speciation dynamics in the freshwater environment: unifying concepts in metal speciation and bioavailability. PhD Thesis, Carleton University, Ottawa, ON, Canada
Murphy RJ, Lenhart JJ, Honeyman BD (1999) The sorption of thorium (IV) and uranium (VI) to hematite in the presence of natural organic matter. Colloids Surf A-Physicochem Eng Aspects 157:47–62
Mylon SE, Twining BS, Fisher NS, Benoit G (2003) Relating the speciation of Cd, Cu, and Pb in two Connecticut rivers with their uptake in algae. Environ Sci Technol 37:1261–1267
Nair GM, Chander K (1983) Stability-constants of complexes of Plutonium(Iii) and Americium(Iii) with 5-Sulfosalicylic Acid. J Less-Common Metals 92:29–34
Naka D, Kim D, Strathmann TJ (2006) Abiotic reduction of nitroaromatic compounds by aqueous iron(II) - catechol complexes. Environ Sci Technol 40:3006–3012
Nakajima H (2006) Studies on photochemical degradation processes of dissolved organic matter in seawater. MS Thesis, Hiroshima University, pp 1–173
Norman L, Thomas DN, Stedmon CA, Granskog MA, Papadimitriou S, Krapp RH, Meiners KM, Lannuzel D, van der Merwe P, Dieckmann GS (2011) The characteristics of dissolved organic matter (DOM) and chromophoric dissolved organic matter (CDOM) in Antarctic sea ice. Deep-Sea Res Part II 58:1075–1091
Ohno T, Amirbahman A, Bro R (2008) Parallel factor analysis of excitation-emission matrix fluorescence spectra of water soluble soil organic matter as basis for the determination of conditional metal binding parameters. Environ Sci Technol 42:186–192
Opsahl S, Benner R (1998) Photochemical reactivity of dissolved lignin in river and ocean waters. Limnol Oceanogr 43:1297–1304
Osburn CL, O'Sullivan DW, Boyd TJ (2009) Increases in the longwave photobleaching of chromophoric dissolved organic matter in coastal waters. Limnol Oceanogr 54:145–159
Pal A, Paul A (2008) Microbial extracellular polymeric substances: central elements in heavy metal bioremediation. Indian J Microbiol 48:49–64
Palmer FB, Butler CA, Timperley MH, Evans CW (1998) Toxicity to embryo and adult zebrafish of copper complexes with two malonic acids as models for dissolved organic matter. Environ Toxicol Chem 17:1538–1545

Panak P, Klenze R, Kim JI, Wimmer H (1995) A Study of Intramolecular Energy-Transfer in Cm(Iii) Complexes with Aromatic Ligands by Time-Resolved Laser Fluorescence Spectroscopy. J Alloys Compd 225:261–266
Parlanti E, Worz K, Geoffroy L, Lamotte M (2000) Dissolved organic matter fluorescence spectroscopy as a tool to estimate biological activity in a coastal zone submitted to anthropogenic inputs. Org Geochem 31:1765–1781
Peña-Méndez EM, Havel J, Patočka J (2005) Humic substances–compounds of still unknown structure: applications in agriculture, industry, environment, and biomedicine. J Appl Biomed 3:13–24
Pinheiro JP, Mota AM, Goncalves MLS (1994) Complexation study of humic acids with Cadmium(Ii) and Lead(Ii). Anal Chim Acta 284:525–537
Pinheiro JP, Mota AM, Benedetti MF (1999) Lead and calcium binding to fulvic acids: salt effect and competition. Environ Sci Technol 33:3398–3404
Pinheiro JP, Mota AM, Benedetti MF (2000) Effect of aluminum competition on lead and cadmium binding to humic acids at variable ionic strength. Environ Sci Technol 34:5137–5143
Plankey BJ, Patterson HH (1987) Kinetics of aluminum-fulvic acid complexation in acidic waters. Environ Sci Technol 21:595–601
Playle RC (1998) Modelling metal interactions at fish gills. Sci Total Environ 219:147–163
Playle RC, Dixon DG, Burnison K (1993) Copper and cadmium-binding to fish gills - estimates of metal gill stability-constants and modeling of metal accumulation. Can J Fish Aquat Sci 50:2678–2687
Plaza C, Brunetti G, Senesi N, Polo A (2005) Proton binding to humic acids from organic amendments and amended soils by the NICA-Donnan model. Environ Sci Technol 39:6692–6697
Plaza C, Brunetti G, Senesi N, Polo A (2006) Molecular and quantitative analysis of metal ion binding to humic acids from sewage sludge and sludge-amended soils by fluorescence spectroscopy. Environ Sci Technol 40:917–923
Poiger T, Kari FG, Giger W (1999) Fate of fluorescent whitening agents in the River Glatt. Environ Sci Technol 33:533–539
Pourret O, Martinez RE (2009) Modeling lanthanide series binding sites on humic acid. J Colloid Interface Sci 330:45–50
Primik MF, Goschl S, Jakupec MA, Roller A, Keppler BK, Arion VB (2010) Structure-activity relationships of highly cytotoxic copper(II) complexes with modified Indolo[3,2-c]quinoline ligands. Inorg Chem 49:11084–11095
Pulford ID, Watson C (2003) Phytoremediation of heavy metal-contaminated land by trees—a review. Environ Int 29:529–540
Quiroz NGA, Hung CC, Santschi PH (2006) Binding of thorium(IV) to carboxylate, phosphate and sulfate functional groups from marine exopolymeric substances (EPS). Mar Chem 100:337–353
Radalla AM (2010) Studies on complexation of resorcinol with some divalent transition metal ions and aliphatic dicarboxylic acids in aqueous media. J Solut Chem 39:1394–1407
Ranville JF, Hendry MJ, Reszat TN, Xie QL, Honeyman BD (2007) Quantifying uranium complexation by groundwater dissolved organic carbon using asymmetrical flow field-flow fractionation. J Contaminant Hydrol 91:233–246
Rao LF, Choppin GR (1995) Thermodynamic study of the complexation of neptunium(V) with humic acids. Radiochim Acta 69:87–95
Raskin I, Smith RD, Salt DE (1997) Phytoremediation of metals: using plants to remove pollutants from the environment. Curr Opin Biotechnol 8:221–226
Ravichandran M (2004) Interactions between mercury and dissolved organic matter - a review. Chemosphere 55:319–331
Ravichandran M, Aiken GR, Reddy MM, Ryan JN (1998) Enhanced dissolution of cinnabar (mercuric sulfide) by dissolved organic matter isolated from the Florida Everglades. Environ Sci Technol 32:3305–3311
Ravichandran M, Aiken GR, Ryan JN, Reddy MM (1999) Inhibition of precipitation and aggregation of metacinnabar (mercuric sulfide) by dissolved organic matter isolated from the Florida Everglades. Environ Sci Technol 33:1418–1423

Reiller PE, Brevet J (2010) Bi-exponential decay of Eu(III) complexed by Suwannee River humic substances: spectroscopic evidence of two different excited species. Spectrochim Acta Part A-Mol Biomol Spectrosc 75:629–636
Reszat TN, Hendry MJ (2007) Complexation of aqueous elements by DOC in a clay aquitard. Ground Water 45:542–553
Rey-Castro C, Mongin S, Huidobro C, David C, Salvador J, Garces JL, Galceran J, Mas F, Puy J (2009) Effective affinity distribution for the binding of metal ions to a generic fulvic acid in natural waters. Environ Sci Technol 43:7184–7191
Richard C, Canonica S (2005) Aquatic phototransformation of organic contaminants induced by coloured dissolved natural organic matter. Environ Photochem Part II 2:299–323
Richards JG, Playle RC (1998) Cobalt binding to gills of rainbow trout (Oncorhynchus mykiss): an equilibrium model. Comp Biochem Physiol 119C:185–197
Ritchie JD, Perdue EM (2003) Proton-binding study of standard and reference fulvic acids, humic acids, and natural organic matter. Geochim Cosmochim Acta 67:85–96
Rosas H, Sarmiento LE, Rodriguez M, Lubes V (2010) Study of the ternary complex formation between Vanadium(III)-picolinic acid and the amino acids: cysteine, histidine, aspartic and glutamic acids. J Solut Chem 39:1021–1029
Ruzic I (1982) Theoretical aspects of the direct titration of natural-waters and its information yield for trace-metal speciation. Anal Chim Acta 140:99–113
Ryan DK, Ventry LS (1990) Exchange of comments on fluorescence quenching measurements of copper fulvic-acid binding. Anal Chem 62:1523–1526
Ryan DK, Weber JH (1982a) Fluorescence quenching titration for determination of complexing capacities and stability-constants of fulvic-acid. Anal Chem 54:986–990
Ryan DK, Weber JH (1982b) Copper(Ii) complexing capacities of natural-waters by fluorescence quenching. Environ Sci Technol 16:866–872
Saar RA, Weber JH (1980) Comparison of spectrofluorometry and ion-selective electrode potentiometry for determination of complexes between fulvic-acid and heavy-metal ions. Anal Chem 52:2095–2100
Sachs S, Schmeide K, Reich T, Brendler V, Heise KH, Bernhard G (2005) EXAFS study on the neptunium(V) complexation by various humic acids under neutral pH conditions. Radiochim Acta 93:17–25
Sachs S, Reich T, Bernhard G (2010) Study of the role of sulfur functionalities in humic acids for uranium(VI) complexation. Radiochim Acta 98:467–477
Saito T, Sao H, Ishida K, Aoyagi N, Kimura T, Nagasaki S, Tanaka S (2010) Application of parallel factor analysis for time-resolved Laser fluorescence spectroscopy: implication for metal speciation study. Environ Sci Technol 44:5055–5060
Salt DE, Blaylock M, Kumar NPBA, Dushenkov V, Ensley BD, Chet I, Raskin I (1995) Phytoremediation—a novel strategy for the removal of toxic metals from the environment using plants. Bio-Technology 13:468–474
Salt DE, Smith R, Raskin I (1998) Phytoremediation. Annu Rev Plant Biol 49:643–668
Sanchez-Marin P, Santos-Echeandia J, Nieto-Cid M, Alvarez-Salgado XA, Beiras R (2010) Effect of dissolved organic matter (DOM) of contrasting origins on Cu and Pb speciation and toxicity to Paracentrotus lividus larvae. Aquat Toxicol 96:90–102
Sander S, Kima JP, Anderson B, Hunter KA (2005) Effect of UVB irradiation on Cu2 + -binding organic ligands and Cu2 + speciation in alpine lake waters of New Zealand. Environ Chem 2:56–62
Santana-Casiano JM, Gonzalez-Davila M, Rodriguez MJ, Millero FJ (2000) The effect of organic compounds in the oxidation kinetics of Fe(II). Mar Chem 70:211–222
Sarmiento LE, Rodriguez M, Echevarria L, Lubes V (2010) Speciation of the Vanadium(III) Complexes with 1,10-Phenanthroline, 2,2 '-Bipyridine, and 8-Hydroxyquinoline. J Solut Chem 39:1484–1491
Schmeide K, Bernhard G (2009) Redox stability of neptunium(V) and neptunium(IV) in the presence of humic substances of varying functionality. Radiochim Acta 97:603–611
Schussler W, Artinger R, Kim JI, Bryan ND, Griffin D (2001) Numerical modeling of humic colloid borne Americium(III) migration in column experiments using the transport/speciation code K1D and the KICAM model. J Contaminant Hydrol 47:311–322

Schuster E (1991) The behavior of mercury in the soil with special emphasis on complexation and adsorption processes - a review of the literature. Water Air Soil Pollut 56:667–680

Schwartz ML, Curtis PJ, Playle RC (2004) Influence of natural organic matter source on acute copper, lead, and cadmium toxicity to rainbow trout (Oncorhynchus mykiss). Environ Toxicol Chem 23:2889–2899

Schwitzguébel JP, van der Lelie D, Baker A, Glass DJ, Vangronsveld J (2002) Phytoremediation: European and American trends successes, obstacles and needs. J Soils Sediment 2:91–99

Seitzinger S, Hartnett H, Lauck R, Mazurek M, Minegishi T, Spyres G, Styles R (2005) Molecular-level chemical characterization and bioavailability of dissolved organic matter in stream water using electrospray-ionization mass spectrometry. Limnol Oceanogr 50:1–12

Sekaly ALR, Back MH, Chakrabarti CL, Gregoire DC, Lu JY, Schroeder WH (1998) Measurements and analysis of dissociation rate constants of metal-fulvic acid complexes in aqueous solutions Part II: measurement of decay rates by inductively-coupled plasma mass spectrometry and determination of rate constants for dissociation. Spectrochim Acta Part B Atomic Spectrosc 53:847–858

Sekaly ALR, Mandal R, Hassan NM, Murimboh J, Chakrabarti CL, Back MH, Gregoire DC, Schroeder WH (1999) Effect of metal/fulvic acid mole ratios on the binding of Ni(II), Pb(II), Cu(II), Cd(II), and Al(III) by two well-characterized fulvic acids in aqueous model solutions. Anal Chim Acta 402:211–221

Sekaly ALR, Murimboh J, Hassan NM, Mandal R, Ben Younes ME, Chakrabarti CL, Back MH, Gregoire DC (2003) Kinetic speciation of Co(II), Ni(II), Cu(II), and Zn(II) in model solutions and freshwaters: lability and the d electron configuration. Environ Sci Technol 37:68–74

Senesi N (1990) Molecular and quantitative aspects of the chemistry of fulvic-acid and its interactions with metal-ions and organic-chemicals. 2. The fluorescence spectroscopy approach. Anal Chim Acta 232:77–106

Seritti A, Morelli E, Nannicini L, Giambelluca A, Scarano G (1994) Fluorescence emission characteristics of naturally-occurring organic-matter in relation to metal complexation studies. Sci Total Environ 148:73–81

Shank GC, Whitehead RF, Smith ML, Skrabal SA, Kieber RJ (2006) Photodegradation of strong copper-complexing ligands in organic-rich estuarine waters. Limnol Oceanogr 51:884–892

Shapiro J (1964) Effect of yellow organic acids on iron and other metals in water. J Am Water Works Assoc 56:1062–1082

Sharpless CM, McGown LB (1999) Effects of aluminum-induced aggregation on the fluorescence of humic substances. Environ Sci Technol 33:3264–3270

Shcherbina NS, Perminova IV, Kalmykov SN, Kovalenko AN, Haire RG, Novikov AP (2007) Redox and complexation interactions of neptunium(V) with quinonoid-enriched humic derivatives. Environ Sci Technol 41:7010–7015

Shin HS, Hong KH, Lee MH, Cho YH, Lee CW (2001) Fluorescence quenching of three molecular weight fractions of a soil fulvic acid by UO2(II). Talanta 53:791–799

Shiozawa I, Lubes G, Rodriguez M, Lubes V (2011) Speciation of the ternary complexes of vanadium(III)-Dipicolinic acid and the amino acids cysteine, histidine, aspartic and glutamic acids in 3.0 mol.dm(-3) KCl at 25 degrees C. J Solut Chem 40:17–25

Sholkovitz ER (1976) Flocculation of dissolved organic and inorganic matter during mixing of river water and seawater. Geochim Cosmochim Acta 40:831–845

Shoukry AA (2005) Complex formation reactions of (2,2 '-dipyridylamine)copper(II) with various biologically relevant ligands. The kinetics of hydrolysis of amino acid esters. Transit Metal Chem 30:814–827

Sidenius U, Farver O, Jons O, Gammelgaard B (1999) Comparison of different transition metal ions for immobilized metal affinity chromatography of selenoprotein P from human plasma. J Chromatogr B 735:85–91

Sierra MMD, Donard OFX, Lamotte M (1997) Spectral identification and behaviour of dissolved organic fluorescent material during estuarine mixing processes. Mar Chem 58:51–58

Silva CSPCO, Dasilva JCGE, Machado AASC (1994) Evolving factor-analysis of synchronous fluorescence-spectra of fulvic-acids in the presence of aluminum. Appl Spectr 48:363–372

Simpson AJ, Kingery WL, Hayes MHB, Spraul M, Humpfer E, Dvortsak P, Kerssebaum R, Godejohann M, Hofmann M (2002) Molecular structures and associations of humic substances in the terrestrial environment. Naturwissenschaften 89:84–88

Skoog A, Wedborg M, Fogelqvist E (1996) Photobleaching of fluorescence and the organic carbon concentration in a coastal environment. Mar Chem 55:333–345

Slawyk G, Raimbault P, Garcia N (1998) Measuring gross uptake of N-15-labeled nitrogen by marine phytoplankton without particulate matter collection: evidence of low N-15 losses to the dissolved organic nitrogen pool. Limnol Oceanogr 43:1734–1739

Slawyk G, Raimbault P, Garcia N (2000) Use of N-15 to measure dissolved organic nitrogen release by marine phytoplankton (reply to comment by Bronk and Ward). Limnol Oceanogr 45:1884–1886

Martell AE, Smith, RM (1974) Critical stability constants. Plenum, New York

Smith DS, Kramer JR (1998) Multi-site aluminum speciation with natural organic matter using multiresponse fluorescence data. Anal Chim Acta 363:21–29

Smith DS, Kramer JR (1999) Fluorescence analysis for multi-site aluminum binding to natural organic matter. Environ Int 25:295–306

Smith RM, Martell AE (1987) Critical stability-constants, enthalpies and entropies for the formation of metal-complexes of Aminopolycarboxylic Acids and Carboxylic-Acids. Sci Total Environ 64:125–147

Smith DS, Bell RA, Kramer JR (2002) Metal speciation in natural waters with emphasis on reduced sulfur groups as strong metal binding sites. Comp Biochem Phys C 133:65–74

Sonke JE, Salters VJM (2006) Lanthanide-humic substances complexation. I. Experimental evidence for a lanthanide contraction effect. Geochim Cosmochim Acta 70:1495–1506

Sponza DT (2003) Investigation of extracellular polymer substances (EPS) and physicochemical properties of different activated sludge flocs under steady-state conditions. Enz Microb Technol 32:375–385

Sposito G, Inouye CA, Bingham F, Yadav S (1982) Trace metal complexation by fulvic acid extracted from sewage sludge: II. Development of chemical models. Soil Sci Soc Am J 46:51–56

Stackhouse RA, Benson WH (1988) The influence of humic-acid on the toxicity and bioavailability of selected trace-metals. Aquat Toxicol 13:99–108

Stedmon CA, Markager S, Tranvik L, Kronberg L, Slatis T, Martinsen W (2007) Photochemical production of ammonium and transformation of dissolved organic matter in the Baltic Sea. Mar Chem 104:227–240

Steelink C (2002) Investigating humic acids in soils. Anal Chem 74:326a–333a

Stenson AC, Marshall AG, Cooper WT (2003) Exact masses and chemical formulas of individual Suwannee River fulvic acids from ultrahigh resolution electrospray ionization Fourier transform ion cyclotron resonance mass spectra. Anal Chem 75:1275–1284

Stevenson F, Fitch A, Brar M (1993) Stability constants of Cu (II)-humate complexes: comparison of select models. Soil Sci 155:77

Stoll JMA, Giger W (1998) Mass balance for detergent-derived fluorescent whitening agents in surface waters of Switzerland. Water Res 32:2041–2050

Sun YP, Nguyen KL, Wallis AFA (1998) Ring-opened products from reaction of lignin model compounds with UV-assisted peroxide. Holzforschung 52:61–66

Sunda W (1988) Trace metal interactions with marine phytoplankton. Biolog Oceanogr 6:411–442

Sunda W, Ferguson R (1983) Sensitivity of natural bacterial communities to additions of copper and to cupric ion activity: a bioassay of copper complexation in seawater. In: Trace metals in sea water NATO Conf Ser 4: Mar Sci V 9 Plenum, pp 871–891

Sunda WG, Hanson AK (1987) Measurement of free cupric ion concentration in seawater by a ligand competition technique involving copper sorption onto C-18 Sep-Pak cartridges. Limnol Oceanogr 32:537–551

Sunda WG, Huntsman SA (1991) The use of chemiluminescence and ligand competition with edta to measure copper concentration and speciation in seawater. Mar Chem 36:137–163

Sundh I (1992) Biochemical-composition of dissolved organic-carbon released from natural communities of lake phytoplankton. Arch Hydrobiol 125:347–369

Tabak HH, Lens P, van Hullebusch ED, Dejonghe W (2005) Developments in bioremediation of soils and sediments polluted with metals and radionuclides–1. Microbial processes and mechanisms affecting bioremediation of metal contamination and influencing metal toxicity and transport. Rev Environ Sci Biotechnol 4:115–156

Takahashi Y, Minai Y, Ambe S, Makide Y, Ambe F, Tominaga T (1997) Simultaneous determination of stability constants of humate complexes with various metal ions using multitracer technique. Sci Total Environ 198:61–71

Tanoue E (2000) Proteins in the sea—synthesis. In: Handa N, Tanoue E, Hama T (eds) Dynamics and characterization of marine organic matter. Terrapub/Kluwer, Tokyo, pp 383–463

Tanoue E, Ishii M, Midorikawa T (1996) Discrete dissolved and particulate proteins in oceanic waters. Limnol Oceanogr 41:1334–1343

Tao S, Liu GJ, Xu FL, Pan B (2002) Estimation of conditional stability constant for copper binding to fish gill surface with consideration of chemistry of the fish gill microenvironment. Comp Biochem Physiol C: Toxicol Pharmacol 133:219–226

Tella M, Pokrovski GS (2009) Antimony(III) complexing with O-bearing organic ligands in aqueous solution: an X-ray absorption fine structure spectroscopy and solubility study. Geochim Cosmochim Acta 73:268–290

Thakur UK, Shah D, Sharma RS, Sawant RM, Ramakumar KL (2006) Studies on protonation and Th(IV) complexation behaviour of dihydroxybenzenes in aqueous 1 M NaClO4 medium. Radiochim Acta 94:859–864

Thomason JW, Susetyo W, Carreira LA (1996) Fluorescence studies of metal humic complexes with the use of lanthanide ion probe spectroscopy. Appl Spectr 50:401–408

Tipping E (1994) Wham—a chemical-equilibrium model and computer code for waters, sediments, and soils incorporating a discrete site electrostatic model of ion-binding by humic substances. Comput Geosci 20:973–1023

Town RM, Filella M (2000) Determination of metal ion binding parameters for humic substances—Part 2. Utility of ASV pseudo-polarography. J Electroanal Chem 488:1–16

Tserenpil S, Liu CQ (2011) Study of antimony (III) binding to soil humic acid from an antimony smelting site. Microchem J 98:15–20

Tung JC, Guo GY (2007) Systematic ab initio study of the magnetic and electronic properties of all 3d transition metal linear and zigzag nanowires. Phys Rev B 76:094413

Vahatalo AV, Wetzel RG (2004) Photochemical and microbial decomposition of chromophoric dissolved organic matter during long (months-years) exposures. Mar Chem 89:313–326

Vahatalo AV, Salonen K, Munster U, Jarvinen M, Wetzel RG (2003) Photochemical transformation of allochthonous organic matter provides bioavailable nutrients in a humic lake. Arch Hydrobiol 156:287–314

van Den Berg CMG (1982) Determination of copper complexation with natural organic-ligands in sea-water by equilibration with Mno2.2. Experimental procedures and application to surface sea-water. Mar Chem 11:323–342

van Den Berg CMG (1984a) Determination of the complexing capacity and conditional stability-constants of complexes of copper(ii) with natural organic-ligands in seawater by cathodic stripping voltammetry of copper catechol complex-ions. Mar Chem 15:1–18

van Den Berg CMG (1984b) Determination of copper in sea-water by cathodic stripping voltammetry of complexes with catechol. Anal Chim Acta 164:195–207

van Den Berg CMG, Merks AGA, Duursma EK (1987) Organic Complexation and Its Control of the Dissolved Concentrations of Copper and Zinc in the Scheldt Estuary. Estuar Coast Shelf Sci 24:785–797

van Heemst JDH, Megens L, Hatcher PG, de Leeuw JW (2000) Nature, origin and average age of estuarine ultrafiltered dissolved organic matter as determined by molecular and carbon isotope characterization. Org Geochem 31:847–857

van Leeuwen HP, Town RM (2009a) Outer-sphere and inner-sphere ligand protonation in metal complexation kinetics: the lability of EDTA complexes. Environ Sci Technol 43:88–93

van Leeuwen HP, Town RM (2009b) Protonation effects on dynamic flux properties of aqueous metal complexes. Collect Czech Chem Commun 74:1543–1557

van Leeuwen HP, Town RM, Buffle J, Cleven RFMJ, Davison W, Puy J, van Riemsdijk WH, Sigg L (2005) Dynamic speciation analysis and bioavailability of metals in aquatic systems. Environ Sci Technol 39:8545–8556
van Leeuwen HP, Town RM, Buffle J (2007) Impact of ligand protonation on eigen-type metal complexation kinetics in aqueous systems. J Phys Chem A 111:2115–2121
van Loon LR, Glaus MA, Vercammen K (2004) Stability of the ion pair between Ca2 + and 2-(hydroxymethyl)-3-deoxy-D-erythro-pentonate (alpha-isosaccharinate). J Solut Chem 33:1573–1583
Vercammen K, Glaus M, Van Loon LR (2001) Complexation of Th(IV) and Eu(III) by α-isosaccharinic acid under alkaline conditions. Radiochim Acta 89:393
Vermilyea AW, Voelker BM (2009) Photo-Fenton reaction at near neutral pH. Environ Sci Technol 43:6927–6933
Vidali R, Remoundaki E, Tsezos M (2009) An experimental and modeling study of humic acid concentration effect on H + binding: application of the NICA-Donnan model. J Colloid Interface Sci 339:330–335
Vidali R, Remoundaki E, Tsezos M (2010) Humic acids copper binding following their photochemical alteration by simulated solar light. Aquat Geochem 16:207–218
Vione D, Lauri V, Minero C, Maurino V, Malandrino M, Carlotti ME, Olariu RI, Arsene C (2009) Photostability and photolability of dissolved organic matter upon irradiation of natural water samples under simulated sunlight. Aquat Sci 71:34–45
Vlassopoulos D, Wood SA, Mucci A (1990) Gold Speciation in Natural-Waters. 2. The importance of organic complexing—experiments with some simple-model ligands. Geochim Cosmochim Acta 54:1575–1586
Wada S, Aoki MN, Tsuchiya Y, Sato T, Shinagawa H, Hama T (2007) Quantitative and qualitative analyses of dissolved organic matter released from Ecklonia cava Kjellman, in Oura bay, Shimoda, Izu Peninsula, Japan. J Exper Mar Biol Ecol 349:344–358
Wallschlager D, Desai MVM, Wilken RD (1996) The role of humic substances in the aqueous mobilization of mercury from contaminated floodplain soils. Water Air Soil Pollut 90:507–520
Waples JS, Nagy KL, Aiken GR, Ryan JN (2005) Dissolution of cinnabar (HgS) in the presence of natural organic matter. Geochim Cosmochim Acta 69:1575–1588
Warwick P, Evans N, Hall T, Vines S (2003) Complexation of Ni (II) by α-isosaccharinic acid and gluconic acid from pH 7 to pH 13. Radiochim Acta 91:233–240
Warwick P, Evans N, Hall T, Vines S (2004) Stability constants of uranium(IV)-alpha-isosaccharinic acid and gluconic acid complexes. Radiochim Acta 92:897–902
Weng LP, van Riemsdijk WH, Temminghoff EJM (2010) Effects of lability of metal complex on free ion measurement using DMT. Environ Sci Technol 44:2529–2534
Winch S, Ridal J, Lean D (2002) Increased metal bioavailability following alteration of freshwater dissolved organic carbon by ultraviolet B radiation exposure. Environ Toxicol 17:267–274
Wingender J, Neu TR, Flemming HC (1999) Microbial extracellular polymeric substances: characterization, structure, and function. Springer, Berlin
Winner RW (1985) Bioaccumulation and toxicity of copper as affected by interactions between humic-acid and water hardness. Water Res 19:449–455
Winter AR, Fish TAE, Playle RC, Smith DS, Curtis PJ (2007) Photodegradation of natural organic matter from diverse freshwater sources. Aquat Toxicol 84:215–222
Winzerling JJ, Berna P, Porath J (1992) How to use immobilized metal ion affinity chromatography. Methods 4:4–13
Wrobel K, Sadi BBM, Wrobel K, Castillo JR, Caruso JA (2003) Effect of metal ions on the molecular weight distribution of humic substances derived from municipal compost: ultrafiltration and size exclusion chromatography with spectrophotometric and inductively coupled plasma-MS detection. Anal Chem 75:761–767
Wu F, Tanoue E (2001a) Molecular mass distribution and fluorescence characteristics of dissolved organic ligands for copper(II) in Lake Biwa, Japan. Org Geochem 32:11–20
Wu FC, Tanoue E (2001b) Geochemical characterization of organic ligands for copper(II) in different molecular size fractions in Lake Biwa, Japan. Org Geochem 32:1311–1318

Wu FC, Tanoue E (2001c) Isolation and partial characterization of dissolved copper-complexing ligands in streamwaters. Environ Sci Technol 35:3646–3652
Wu FC, Midorikawa T, Tanoue E (2001) Fluorescence properties of organic ligands for copper(II) in Lake Biwa and its rivers. Geochem J 35:333–346
Wu FC, Evans RD, Dillon PJ (2002a) High-performance liquid chromatographic fractionation and characterization of fulvic acid. Anal Chim Acta 464:47–55
Wu FC, Evans RD, Dillon PJ (2002b) Fractionation and characterization of fulvic acid by immobilized metal ion affinity chromatography. Anal Chim Acta 452:85–93
Wu FC, Cai YR, Evans D, Dillon P (2004a) Complexation between Hg(II) and dissolved organic matter in stream waters: an application of fluorescence spectroscopy. Biogeochemistry 71:339–351
Wu FC, Evans D, Dillon P, Schiff S (2004b) Molecular size distribution characteristics of the metal-DOM complexes in stream waters by high-performance size-exclusion chromatography (HPSEC) and high-resolution inductively coupled plasma mass spectrometry (ICP-MS). J Anal Atomic Spectr 19:979–983
Wu FC, Mills RB, Evans RD, Dillon PJ (2004c) Kinetics of metal-fulvic acid complexation using a stopped-flow technique and three-dimensional excitation emission fluorescence spectrophotometer. Anal Chem 76:110–113
Xia YX, Chen JF, Choppin GR (1996) Solubility, dissociation and complexation with Nd(III) and Th(IV) of oxine, thenoyltrifluoroacetone and 1,10-phenanthroline in 5.0 m NaCl. Talanta 43:2073–2081
Xia K, Weesner F, Bleam WF, Bloom PR, Skyllberg UL, Helmke PA (1998) XANES studies of oxidation states of sulfur in aquatic and soil humic substances. Soil Sci Soc Am J 62:1240–1246
Xia K, Skyllberg UL, Bleam WF, Bloom PR, Nater EA, Helmke PA (1999) X-ray absorption spectroscopic evidence for the complexation of Hg(II) by reduced sulfur in soil humic substances. Environ Sci Technol 33:257–261
Xiao ZF, Stromberg D, Lindqvist O (1995) Influence of humic substances on photolysis of divalent mercury in aqueous-solution. Water Air Soil Pollut 80:789–798
Xie HX, Zafiriou OC, Cai WJ, Zepp RG, Wang YC (2004) Photooxidation and its effects on the carboxyl content of dissolved organic matter in two coastal rivers in the Southeastern United States. Environ Sci Technol 38:4113–4119
Xue HB, Sigg L (1990) Binding of Cu(Ii) to algae in a metal buffer. Water Res 24:1129–1136
Xue HB, Sigg L (1993) Free cupric ion concentration and Cu(Ii) speciation in a eutrophic lake. Limnol Oceanogr 38:1200–1213
Xue HB, Kistler D, Sigg L (1995) Competition of copper and zinc for strong ligands in a eutrophic lake. Limnol Oceanogr 40:1142–1152
Yadav K, Trivedi S (2006) Evaluation of genotoxic potential of chromium (VI) in Channa punctata fish in terms of chromosomal aberrations. Asian Pac J Cancer Prev 7:472–476
Yalcin M, Mutluay H, Cankurtaran H (1998) Determination of the protonation constants of 2-[4-dimethylaminocinnamalamino] benzoic acid (DACAB) in dioxane - Water medium and preparation of some of its transition metal complexes. Turkish J Chem 22:209–214
Yamaji N, Hayakawa K, Takada H (2010) Role of photodegradation in the fate of fluorescent whitening agents (FWAs) in lacustrine environments. Environ Sci Technol 44:8791–8791
Yamashita Y, Jaffe R (2008) Characterizing the interactions between trace metals and dissolved organic matter using excitation-emission matrix and parallel factor analysis. Environ Sci Technol 42:7374–7379
Yamashita Y, Tanoue E (2003) Chemical characterization of protein-like fluorophores in DOM in relation to aromatic amino acids. Mar Chem 82:255–271
Yamashita Y, Tanoue E (2004) In situ production of chromophoric dissolved organic matter in coastal environments. Geophys Res Lett 31:L14302. doi:101029/2004GL019734
Yang L, Sturgeon R (2009) Isotopic fractionation of mercury induced by reduction and ethylation. Anal Bioanal Chem 393:377–385
Yin RS, Feng XB, Shi WF (2010) Application of the stable-isotope system to the study of sources and fate of Hg in the environment: a review. Appl Geochem 25:1467–1477

Zepp RG, Braun AM, Hoigne J, Leenheer JA (1987) Photoproduction of hydrated electrons from natural organic solutes in aquatic environments. Environ Sci Technol 21:485–490
Zepp RG, Faust BC, Hoigne J (1992) Hydroxyl radical formation in aqueous reactions (Ph 3–8) of Iron(Ii) with hydrogen-peroxide - the photo-Fenton reaction. Environ Sci Technol 26:313–319
Zhang H, Lindberg SE (2001) Sunlight and iron(III)-induced photochemical production of dissolved gaseous mercury in freshwater. Environ Sci Technol 35:928–935
Zhang JZ, Wang FY, House JD, Page B (2004) Thiols in wetland interstitial waters and their role in mercury and methylmercury speciation. Limnol Oceanogr 49:2276–2286
Zhang XR, Minear RA, Barrett SE (2005) Characterization of high molecular weight disinfection byproducts from chlorination of humic substances with/without coagulation pretreatment using UF-SEC-ESI-MS/MS. Environ Sci Technol 39:963–972
Zhang DY, Wang JL, Pan XL (2006) Cadmium sorption by EPSs produced by anaerobic sludge under sulfate-reducing conditions. J Hazard Mater 138:589–593
Zhang SJ, Xu C, Santschi PH (2008) Chemical composition and Th-234 (IV) binding of extracellular polymeric substances (EPS) produced by the marine diatom Amphora sp. Mar Chem 112:81–92
Zhang YL, van Dijk MA, Liu ML, Zhu GW, Qin BQ (2009a) The contribution of phytoplankton degradation to chromophoric dissolved organic matter (CDOM) in eutrophic shallow lakes: field and experimental evidence. Water Res 43:4685–4697
Zhang YL, Liu ML, Qin BQ, Feng S (2009b) Photochemical degradation of chromophoric-dissolved organic matter exposed to simulated UV-B and natural solar radiation. Hydrobiologia 627:159–168
Zhang DY, Pan XL, Mostofa KMG, Chen X, Mu GJ, Wu FC, Liu J, Song WJ, Yang JY, Liu YL, Fu QL (2010) Complexation between Hg(II) and biofilm extracellular polymeric substances: an application of fluorescence spectroscopy. J Hazard Mater 175:359–365
Zheng W, Hintelmann H (2009) Mercury isotope fractionation during photoreduction in natural water is controlled by its Hg/DOC ratio. Geochim Cosmochim Acta 73:6704–6715
Zhou X, Wangersky PJ (1985) Copper complexing capacity in cultures of phaeodactylum-tricornutum —diurnal changes. Mar Chem 17:301–312
Zhou ZL, Wangersky PJ (1989) Production of copper-complexing organic-ligands by the marine diatom phaeodactylum-tricornutum in a cage culture turbidostat. Mar Chem 26:239–259
Zsolnay A (1979) Coastal colloidal carbon: a study of its seasonal variation and the possibility of river input. Estuar Coast Mar Sci 9:559–567

Impacts of Global Warming on Biogeochemical Cycles in Natural Waters

Khan M. G. Mostofa, Cong-qiang Liu, Kunshan Gao, Shijie Li, Davide Vione and M. Abdul Mottaleb

1 Introduction

The main source of energy that drives the dynamics of Earth's outer spheres, including its climate, is unquestionably the Sun (Kandel and Viollier 2005). Therefore, electromagnetic radiation enormously dominates the energy exchange between the Earth and its cosmic environment (Kandel and Viollier 2005). At a radiative balance of 235 W m^{-2}, the Earth would have an average surface temperature of only −19 °C, resulting in a perpetually frozen planet (Ruddiman 2001). Fortunately, the planetary atmosphere traps sufficient long-wave energy that is reradiated by the warm Earth's surface (greenhouse effect) to raise the surface temperature by approximately 33 °C to a more hospitable average of 14 °C (Ferguson and Veizer 2007). The greenhouse effect is efficiently caused by the

K. M. G. Mostofa (✉) · C. Q. Liu· S. Li
State Key Laboratory of Environmental Geochemistry, Institute of Geochemistry, Chinese Academy of Sciences, Guiyang 550002, China
e-mail: mostofa@vip.gyig.ac.cn

K. Gao
State Key Laboratory of Marine Environmental Science, Xiamen University, Xiamen, Fujian, China

D. Vione
Dipartimento di Chimica Analitica, University of Turin, I-10125 Turin, Italy

Centro Interdipartimentale NatRisk, I-10095 Grugliasco, (TO), Italy

M. A. Mottaleb
Center for Innovation and Entrepreneurship (CIE), Department of Chemistry/Physics, Northwest Missouri State University, 800 University Drive, Maryville, MO 64468, USA

K. M. G. Mostofa et al. (eds.), *Photobiogeochemistry of Organic Matter*, Environmental Science and Engineering, DOI: 10.1007/978-3-642-32223-5_10,

occurrence of the atmospheric greenhouse gases (GHGs), the main one being water vapor (H_2O) followed to a lesser extent by CO_2, CH_4, N_2O, CFCs and so on (Wigley 1988, 1989; Charlson et al. 1989; Fisher et al. 1990; den Elzen et al. 1992; Kroeze and Reijnders 1992; Solomon and Daniel 1996; Kiehl and Trenberth 1997; Quaas et al. 2004; IPCC 2007a; Velders et al. 2007; May 2008; Schmidt et al. 2010; Zhang et al. 2011). A typical definition of global warming is an increase of the global average temperatures at the interface between Earth's near surface air and water. It is generally caused either by the absorption of long-wave (or thermal) infrared radiation by the GHGs and other atmospheric constituents or by high penetration of short-wave, e.g. ultraviolet (UV) radiation due to the depletion of the stratospheric ozone layer caused by ozone depleting substances.

GHGs and other atmospheric constituents are substantially released by increased soil respiration processes (Bradford et al. 2008; Bahn et al. 2010; Feng et al. 2010), high agricultural activities in soils (Mosier et al. 2004; Robertson and Grace 2004; Ambus and Robertson 2006; Smith et al. 2008), anthropogenic processes (IPCC 2007a; Sabine et al. 2004; Smith 2004; Archer 2005; Canadell et al. 2007; Hofmann et al. 2009), deforestation (IPCC 2001, 2007a; Kreileman and Bouwman 1994; van der Werf et al. 2009), photoinduced and microbial degradation of aquatic organic matter (OM) including dissolved organic matter (DOM) and particulate organic matter (POM) (Bozec et al. 2005, 2006; Schiettecatte et al. 2006, 2007; Borges et al. 2008; Omar et al. 2010; Ballaré et al. 2011; Zepp et al. 2011), and photoinduced and microbial degradation of OM in plants and soil environments (Brandt et al. 2009; Rutledge et al. 2010).

On the other hand, global warming significantly affects various biogeochemical processes of natural waters, including changes in light cycle, increase of water temperature (O'Reilly et al. 2003; Letelier et al. 2004; Huisman et al. 2006; Porcal et al. 2009), enhancement of the photoinduced activity of aquatic DOM and OM (Hiriart-Baer and Smith 2005; Molot et al. 2005; Johannessen et al. 2007; Mostofa and Sakugawa 2009; Mostofa et al. 2009a, b, 2011), changes in the microbial processing of aquatic DOM and OM (Norf et al. 2007; Vázquez-Domínguez et al. 2007; Falkowski and Oliver 2008; Peters 2008; Norf and Weitere 2010; Sarmento et al. 2010; Sawicka et al. 2010), enhancement of photosynthesis(Mostofa et al. 2009b; Marcoval et al. 2008; Zubkov and Tarran 2008; Beardall et al. 2009a, b), changes in the primary productivity (Huisman et al. 2006; Mostofa et al. 2009b; Baulch et al. 2005; Castle and Rodgers 2009; Davis et al. 2009), changes in the aquatic DOM dynamics and global carbon cycles (Zepp et al. 2011; Porcal et al. 2009; Burns et al. 2006; Vuorenmaa et al. 2006; Sobek et al. 2007; Zhang et al. 2010), and changes in the nutrients cycle (Mostofa et al. 2009b; Fu et al. 2005; Minero et al. 2007; Stedmon et al. 2007a, b; Sterner et al. 2008).

The effects of ambient levels of UV radiation (UV-B: 280–315 nm and UV-A: 315–400 nm) can alter both planktonic and benthic communities within the biota of alpine lakes (Cabrera et al. 1997; Halac et al. 1997; Sommaruga et al. 1997, 1999a; Vinebrooke and Leavitt 1998; Sommaruga and Garcia-Pichel 1999). The impact of UV radiation may interact with other important environmental changes affecting high-latitude and high-altitude lakes, such as acidification and climate warming (Vinebrooke and Leavitt 1998; Schindler et al. 1996; Yan et al. 1996; Sommaruga et al. 1999b). UV-B radiation strongly influences aquatic carbon, nitrogen, sulfur and metals cycles and affects a wide range of life processes (Epp et al. 2007). UV-B radiation changes the biological availability of dissolved organic matter (DOM) to microorganisms and accelerates DOM transformation into dissolved inorganic carbon and nitrogen, including carbon dioxide and ammonium (Epp et al. 2007). It is reported that large shifts in underwater UV-B, UV-A and photosynthetically available radiation (PAR) associated with changes in the input of colored DOM occurred into subarctic lakes during the Holocene (Pienitz and Vincent 2000). A moderate increase in UV-B also occurred in the northern hemisphere such as in the Arctic (von Der Gathen et al. 1995) and in the Swiss Alps (Blumthaler and Ambach 1990).

Global warming induces changes of climate, soil and water ecosystems (IPCC 2007a). Some 70 % of the Earth surface is covered by water containing an extremely complicated milieu of organic and inorganic chemical species (Erickson Iii et al. 2000). The photoinduced production and transformation of various greenhouse and chemically reactive gases in the ocean has been a focus of many studies over the last century (Erickson Iii et al. 2000). Increased UV radiation has implications on the biogeochemistry of the aquatic and marine boundary layer, with a focus on trace gases such as CO_2, DMS, CO, OCS, CH_4, N_2O, non-methane hydrocarbons (NMHCs) and organohalogens, which can be exchanged between the ocean and the atmosphere (Erickson Iii et al. 2000).

This chapter describes a general overview on the contributions to global warming of atmospheric constituents including GHGs, as well as their key emission processes. The aim of this review is to explain the impacts of global warming on the aquatic biogeochemical processes, including changes in light cycle and water temperature, photoinduced processes, microbial processes, photosynthesis, primary production, dissolved organic matter (DOM) dynamics and global carbon cycle, and finally the nutrients cycle in natural waters. This chapter also discusses a conceptual model for the effect of global warming on key biogeochemical processes and remedial measures for controlling algal blooms caused by global warming.

2 Global Warming

The atmospheric constituents accountable for global warming are: water vapor; clouds (condensed water in ice and liquid form); greenhouse gases (GHGs) such as carbon dioxide (CO_2), methane (CH_4), nitrous oxide (N_2O) and halocarbons including chloroflurocarbons (CFCs), hydrofluorocarbons (HFCs), perfluorocarbons (PFCs), ozone, sulphur hexafluoride (SF_6), methyl chloroform (CH_3CCl_3) and carbon tetrachloride (CCl_4) (Wigley 1988, 1989; Charlson et al. 1989; Fisher et al. 1990; den Elzen et al. 1992; Kroeze and Reijnders 1992; Solomon and Daniel 1996; Kiehl and Trenberth 1997; Quaas et al. 2004; IPCC 2007a; Velders et al. 2007; May 2008; Schmidt et al. 2010; Zhang et al. 2011; Robertson and Grace 2004; Friedlingstein et al. 2003; Jones et al. 2003a, b, c; Le Quéré et al. 2003; Archer et al. 2004; Buffett and Archer 2004; Forster and Joshi 2005; Hansen and Sato 2004; Eliseev et al. 2007). Atmospheric constituents are responsible for increasing the atmospheric temperature via two main processes. First, long-wave (or thermal) radiation emitted from the terrestrial surface is absorbed at a particular frequency and reemitted at lower frequency by greenhouse gases and clouds throughout the earth's atmosphere. The earth's surface can emit long-wave (or thermal) radiation because it is heated by sunlight. Second, gases, clouds and aerosols can absorb and scattered short-wave radiation (UV and visible) significantly. The cooling effect via short-wave reflection is dominant for clouds and aerosols. The transfer of long-wave radiation depends on both the local temperature of the gaseous absorber and the efficiency of the gases to absorb radiation at a given wavelength (Kiehl and Trenberth 1997). The absorption efficiency varies with wavelength. Note that many greenhouse gases can absorb radiation at the same wavelengths, which is called the overlap effect. In the presence of clouds, the transfer of radiation depends on the cloud amount, on the efficiency of clouds to absorb and reemit the long-wave radiation (cloud emissivity) and on the cloud top and base temperatures (Kiehl and Trenberth 1997).

It has been shown that the sulfate aerosols have a negative forcing effect that partially counterbalances the warming effect of greenhouse gases (Charlson et al. 1989; Wigley 1989; Quaas et al. 2004; IPCC 2007a; Schmidt et al. 2010; Joshi et al. 2003; Eliseev et al. 2007; Rosenfeld 2000). It is suggested that aerosols scatter sunlight and enhance the planetary short-wave albedo, an effect known as the 'aerosol direct effect' (ADE). In addition, by their ability to act as cloud condensation nuclei, (hygroscopic) aerosols change cloud properties and produce essentially an increase in cloud albedo. These processes are called 'aerosol indirect effect' (AIE).

The increase in greenhouse gas concentration could lead to a reduction of clouds at all atmospheric levels, thus decreasing the total greenhouse effect in the long-wave spectrum but increasing absorption of solar radiation upon reduction of cloud albedo (Quaas et al. 2004). Increasing anthropogenic aerosols result in a decrease of high-level cloud cover by cooling of the atmosphere, and an increase in the low-level cloud cover through the second aerosol indirect effect (Quaas et al. 2004). The decrease of the high-level cloudiness and the increase of the low-level one due to the response of cloud processes to aerosols have a contrasting impact on the short-wave radiation, and the net effect is slightly positive (Quaas et al. 2004). The total aerosol effect, including the aerosol direct and first indirect effects, remains strongly negative (Quaas et al. 2004; IPCC 2007a).

In addition, the depletion of stratospheric ozone caused by atmospheric anthropogenic GHGs can enhance penetration of harmful UV-B radiation (280–315 nm), which can have a direct influence on living organisms and also affect the global warming (IPCC 2001; Huisman et al. 2006; Kerr and McElroy 1993; Varotsos and Kondratiev 1995; Hartmann et al. 2000; Qian et al. 2001; Sarmiento et al. 2004; Schmittner 2005). The impact of UV-B radiation on global warming is of two kinds. The first is a direct heating effect of UV-B radiation penetrating in the troposphere. The second effect is the release of CO_2 to the atmosphere upon photoinduced degradation of DOM induced by UV-B radiation in natural waters (Qian et al. 2001; Sarmiento et al. 2004; Schmittner 2005). However, the observed losses in the stratospheric ozone layer over the past two decades have caused a negative climate forcing (0.15 ± 0.1 Wm^{-2}), i.e. a tendency toward cooling of the surface troposphere system (IPCC 2001). Model calculations indicate that increased penetration of ultraviolet radiation to the troposphere, as a result of stratospheric ozone depletion, leads to enhanced removal rates of gases like CH_4, with a resulting cooling effect (IPCC 2001).

In contrast, other studies suggest that stratospheric ozone depletion and GHG warming may both be producing increased meridional temperature gradients in the extratropical lower stratosphere and upper troposphere, thereby acting synergistically to produce surprisingly large trends in both surface and stratospheric climate (Hartmann et al. 2000).

2.1 Occurrence and Contribution of Atmospheric Constituents to Global Warming

The global atmospheric concentration of CO_2 has increased from a pre-industrial value of about 280–379 ppm in 2005 (IPCC 2007a). The CO_2 concentrations in 2005 exceeded by far the natural range over the last 650,000 years (Fig. 1a) (IPCC 2007a). Despite the year-to-year variation of CO_2 concentration growth rate, it is estimated that the annual rate of CO_2 concentration growth has been larger over the past 10 years (1995–2005; average: 1.9 ppm per year) than in the whole record of continuous direct atmospheric measurements (1960-2005; average: 1.4 ppm per year). The atmospheric concentrations of CH_4 in 2005 exceeded by far the natural range over the last 650,000 years (Fig. 1b) (IPCC 2007a). The global atmospheric concentration of CH_4 increased from a pre-industrial value of about 715–1732 ppb in the early 1990s, then to 1774 ppb in 2005 (IPCC 2007a). The data also suggest that the growth rates have declined since the 1990s, coherently with total emissions (sum of anthropogenic and natural sources) being nearly constant during this period. The global atmospheric N_2O concentration increased from a pre-industrial value of about 270 ppb to 319 ppb in 2005 (Fig. 1c) (IPCC 2007a).

An important greenhouse gas in both the stratosphere and the troposphere is ozone (O_3), which is formed in the atmosphere from photoinduced processes that involve both natural and human-influenced precursor species (IPCC 2001). The residence time of ozone in the atmosphere is relatively short, varying from weeks to months (IPCC 2001). The total amount of O_3 in the troposphere is estimated to have increased by 36 % since 1750, due primarily to anthropogenic emissions of several O_3-forming gases (IPCC 2001). It is also suggested that O_3 climate forcing varies considerably depending on the region and that it responds more quickly to changes in emissions than the long-lived greenhouse gases, such as CO_2.

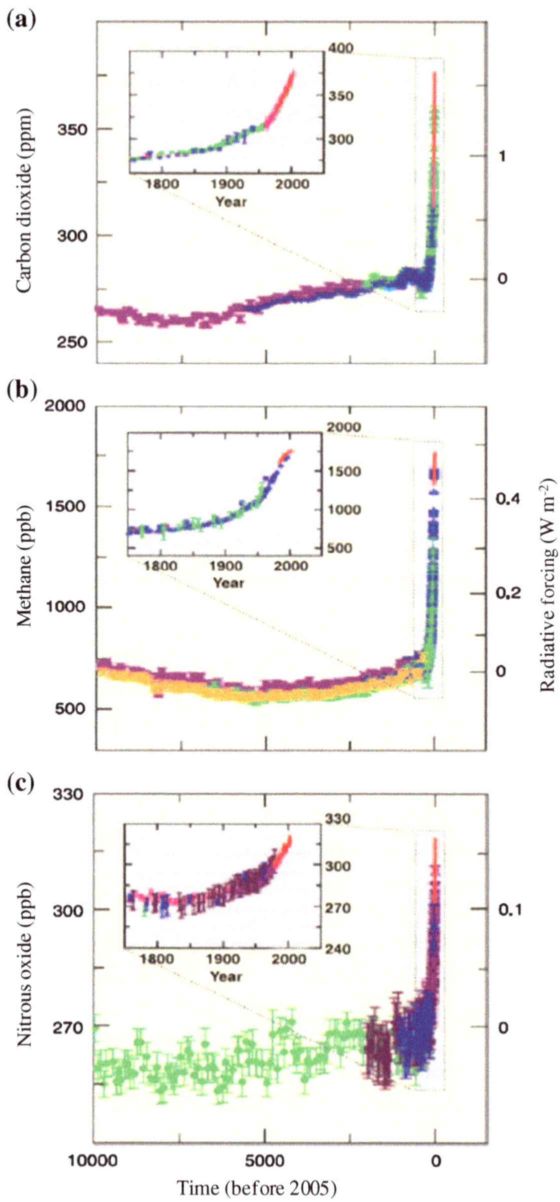

Fig. 1 Atmospheric concentrations of CO_2 (**a**), CH_4 (**b**) and N_2O (**c**) over the last 10,000 years (large panels) and since 1750 (inset panels). Measurements are shown from ice cores (symbols with different colours for different studies) and atmospheric samples (*red lines*). The corresponding radiative forcings relative to 1750 are shown on the right hand axes of the large panels. *Data source* IPCC (2007a)

Halocarbons have increased from a near-zero pre-industrial background concentration, and the increase is primarily due to human activities. The atmospheric concentrations of many halocarbon gases with ozone depleting and global warming potential (e.g. $CFCl_3$ and CF_2Cl_2) have been either increasing more slowly or even decreasing since 1995. This happened in response to reduced emissions

Table 1 Atmospheric lifetime and GWPs relative to CO_2 at different time horizon for various green house gases

Atmospheric gases	Global warming potentials (GWPs) (Time horizon in years)			Lifetime (yrs)
	20 yrs	100 yrs	500 yrs	
Carbon dioxide (CO_2)	1	1	1	~5–200[b]
Methane[a] (CH_4)	62	23	7	12[c]
Nitrous oxide (N_2O)	275	296	156	114[c]
Hydrofluorocarbons				
HFC-23 (CHF_3)	9400	12000	10000	260
HFC-32 (CH_2F_2)	1800	550	170	5.0
HFC-41 (CH_3F)	330	97	30	2.6
HFC-125 (CHF_2CF_3)	5900	3400	1100	29
HFC-134 (CHF_2CHF_2)	3200	1100	330	9.6
HFC-134a (CH_2FCF_3)	3300	1300	400	13.8
HFC-143 (CHF_2CH_2F)	1100	330	100	3.4
HFC-143a (CF_3CH_3)	5500	4300	1600	52
HFC-152 (CH_3CHF_2)	140	43	13	0.5
HFC-152a (CH_3CHF_2)	410	120	37	1.4
HFC-161 (CH_3CH_2F)	40	12	4	0.3
Fully fluorinated gases				
SF_6	15100	22200	32400	3200
CF_4	3900	5700	8900	50000
C_2F_6	8000	11900	18000	10000

GWPs are an index for estimating relative global warming contribution due to atmospheric emission of a kg of a particular greenhouse gas compared to emission of a kg of carbon dioxide. *Data source* IPCC (2001)

[a]The methane GWPs include an indirect contribution from stratospheric H_2O and O_3 production

[b]No single lifetime can be defined for CO_2 because of the different rates of uptake by different removal processes

[c]The values for methane and nitrous oxide are adjustment times, which incorporate the indirect effects of emission of each gas on its own lifetime

under the regulations of the Montreal Protocol and its Amendments (IPCC 2001). The halocarbon substitute compounds (e.g. CHF_2Cl and CF_3CH_2F) and some other synthetic compounds such as perfluorocarbons and sulphur hexafluoride, SF_6, are also greenhouse gases. Their concentrations are currently increasing (Table 1) (IPCC 2001).

Atmospheric sulfate aerosols have increased sharply during the past one-and-a-half centuries or so, with an overall increase in sulfate emissions from 1 Mt S in 1850 to 70 Mt S in the 1990s. Sulfate aerosols are mostly concentrated in the Northern Hemisphere, with distinct concentration maxima near major polluted regions (Lefohn et al. 1999; Smith et al. 2001). Carbon monoxide (CO) is identified as an important indirect greenhouse gas, which acts as a $HO^{\bullet}$ sink (thereby enhancing the lifetime of many direct greenhouse gases) and is involved in the formation of tropospheric O_3. A model study indicates that the emission of 100 Mt

CO is equivalent in terms of greenhouse gas perturbations to the emission of about 5 Mt CH_4 (IPCC 2001). The abundance of CO in the Northern Hemisphere is about twice that in the Southern Hemisphere and has increased in the second half of the twentieth century along with industrialisation and population growth (IPCC 2001). The reactive nitrogen species (NO and NO_2) and the volatile organic compounds, because of their impact over the oxidising capacity of the troposphere, may act as indirect greenhouse gases both through their influence on ozone and by impacting the lifetimes of CH_4 and other greenhouse gases via $HO^\bullet$ scavenging, although this latter effect is compensated for to a variable extent by the $HO^\bullet$ generation upon O_3 photolysis (IPCC 2001).

The contributions of different anthropogenic greenhouse gases to the 2004 total emissions in terms of CO_2 equivalents have been 56.6 % from fossil fuel use, 17.3 % from deforestation and decay of biomass, 2.8 % from other sources, 14.3 % from CH_4, 7.9 % from N_2O, and 1.1 % from fluorine gases (IPCC 2007a). The contributions of the different activity sectors to the total emissions of anthropogenic greenhouse gases in 2004, in terms of CO_2 equivalents are 25.9 % for energy supply, 19.4 % for industry, 17.4 % for forestry, 13.5 % for agriculture, 13.1 % for transport, 7.9 % for residential and commercial building purposes, and finally 2.8 % for waste and wastewater treatment (IPCC 2007a). In addition, recent studies shows that CO_2 can be significantly released to the atmosphere from other sources such as the photoinduced and microbial degradation of DOM and POM (e.g. algae or phytoplankton) in natural waters (Bozec et al. 2005, 2006; Schiettecatte et al. 2006, 2007; Borges et al. 2008; Omar et al. 2010; Kelley 1970; Kempe and Pegler 1991; Hoppema 1990, 1991; Borges and Frankignoulle 1999, 2002a, b). Also the photoinduced and microbial degradation of OM in terrestrial plant masses can release CO_2 to the atmosphere (Rutledge et al. 2010; Johannessen et al. 2007).

Among the atmospheric absorbers of long-wave radiation, H_2O vapor, clouds, CO_2, CH_4 and O_3 dominate while the aerosols and other species make small contributions to the overall effect (Schmidt et al. 2010). It has been shown that the contributions of atmospheric greenhouse gases (GHGs) to global warming are significantly variable depending on the occurrence of the atmospheric constituents and on the long-wave and short-wave fluxes under clear, cloudy or all-sky conditions (Kiehl and Trenberth 1997; IPCC 2007a; Schmidt et al. 2010).

The contributions of atmospheric GHGs to global warming are 39–70 % for H_2O vapor, 15–36 % for clouds, 14–31 % for CO_2, 8–18 % for O_3, and 6–9 % for other constituents including CH_4 and N_2O (Kiehl and Trenberth 1997; Schmidt et al. 2010; Harrison et al. 1990; IPCC 1990; Clough and Iacono 1995). In addition, the atmospheric short-wave (UV–Vis) absorbers are mostly H_2O vapor (38–43 W m^{-2}), O_3 (14–15 W m^{-2}), and O_2 (2 W m^{-2}) under both clear and cloudy conditions. In contrast, CO_2 (1 W m^{-2}) only gives a small contribution under clear-sky conditions (Kiehl and Trenberth 1997). It has also been shown that the all-sky contribution of water vapor and clouds together is approximately 72–80 % after removing all the other absorbers (Schmidt et al. 2010).

The direct emission of water vapor (a greenhouse gas) by human activities makes a negligible contribution to the radiative forcing, but an increase in global average temperature can enhance the tropospheric water vapor concentration and produce a key positive feedback for radiative forcing, thereby leading to further warming (IPCC 2007a). Interestingly, the so-called Humic-like Substances (HULIS) occurring on atmospheric aerosols can enhance water uptake and increase the role of particles as Cloud Condensation Nuclei (CCN), thereby contributing to direct and indirect climate forcing (Hatch et al. 2009). Warming also reduces terrestrial and ocean uptake of atmospheric CO_2, increasing the fraction of anthropogenic emissions that remain in the atmosphere. Such an effect is expected to lead to higher atmospheric CO_2 levels that are further involved into the global climate change (IPCC 2007a). The uptake by the oceans is approximately 25 % of the annual carbon emissions that result from fossil fuel burning and cement manufacturing (Canadell et al. 2007).

The infrared absorption cross-sections for eight commonly used CFCs (halogenated methanes and ethanes), as a function of temperature from 203 to 293 K, suggest that the combined effects of absorption by CFCs of the Earth's radiative energy in the 'window' region (700–1300 cm^{-1}) and of their O_3 depletion potential makes these compounds significant contributors to global warming (McDaniel et al. 1991). On the other hand, the sulfate aerosols can reduce global warming by about 0.1–0.4 K, depending on the scenario and on the time period. The maximum slowdown in warming (>1.5 K) is expected to occur in the Northern Hemisphere middle- and high-latitude land areas in the mid-twenty-first century (Eliseev et al. 2007). A recent study has shown that the indirect effect of stratospheric ozone depletion could have offset up to half of the predicted past increase in surface temperature that would otherwise have occurred as a result of the direct halocarbon effects (Forster and Joshi 2005). In both the troposphere and stratosphere, CFC-12 contributed most to the CFCs-related past temperature changes, and the emission projections suggest that HFC-134a could contribute most to the warming by halocarbons over the coming century (Forster and Joshi 2005).

2.2 *Global Warming Determination*

The radiative forcing (expressed in Watts per square metre, W m^{-2}) is one of the primary issues associated with potential global warming constituents (IPCC 1990, 1994, 2001). Radiative forcing is a change of the net irradiance at the top of the troposphere because of modifications in either solar or infrared radiation. Such forcing perturbs the balance between incoming and outgoing radiation (IPCC 1990, 1994). Radiative forcing is a measure of the influence of a particular factor on the balance of incoming and outgoing energy in the Earth-atmosphere system, and it is also an index for a potential climate change mechanism (IPCC 2007a). A positive radiative forcing tends to warm the climatic system while the negative forcing has a cooling effect.

The global-mean radiative forcing (ΔF) is approximately related to the equilibrium global-mean surface temperature change (ΔT) by (IPCC 1994) (Eq. 2.1):

$$\Delta T = \lambda \Delta F \tag{2.1}$$

where λ is the climate sensitivity parameter. Although there is a large discrepancy in the actual value of λ in different models, its values are assumed to be approximately independent of the agent causing the forcing. The spread in the model estimates of λ varies from about 0.4–1.2 K (W m^{-2})$^{-1}$, that is, approximately by a factor of 3 (IPCC 1990, 2001). The models also indicate generic deviations of λ from the case of global CO_2 perturbations: increases of O_3 in the upper troposphere generally produce lower values of λ, while O_3 perturbations in the lower stratosphere lead to higher values of λ (Joshi et al. 2003).

Global average radiative forcings in 2005 (best estimates with 5–95 % uncertainty ranges) with respect to 1750 for atmospheric constituents are +1.66 (range: +1.49 to +1.83) W m^{-2} for CO_2, +0.48 (+0.43 to +0.53) W m^{-2} for CH_4, +0.16 (+0.14 to +0.18) W m^{-2} for N_2O, +0.34 (+0.31 to +0.37) W m^{-2} for halocarbons, +0.35 (+0.25 to +0.65) W m^{-2} for tropospheric O_3, and +0.12 (+0.06 to +0.30) W m^{-2} for changes in solar irradiance (Fig. 1) (IPCC 2007a). On the other hand, anthropogenic contributions to aerosols (primarily sulphate aerosol, organic carbon, black carbon, nitrate and dust) produce an overall cooling effect, with a total direct radiative forcing of −0.5 (−0.9 to −0.1) W m^{-2} and an indirect cloud albedo forcing of −0.7 (−1.8 to −0.3) W m^{-2} (IPCC 2007a). The CO_2 radiative forcing increased by 20 % from 1995 to 2005, the largest change for any decade in at least the last 200 years (IPCC 2007a).

The global warming potential (GWP) is used within the Kyoto Protocol to the United Nations Framework Convention on Climate Change (UNFCCC) as a metric for weighting the climate impact of the emission of different greenhouse gases (IPCC 1990, 2001; Shine et al. 2005). The GWP is the time-integrated radiative forcing due to a pulse emission of a given gas, over some given time period (or horizon), relative to a pulse emission of carbon dioxide (IPCC 2001). GWPs are an index for estimating relative global warming contributions, due to the atmospheric emission of a kg of a particular greenhouse gas compared to the emission of a kg of carbon dioxide. For instance, CH_4 and N_2O have relatively long atmospheric residence times (12 and 114 years, respectively), which combined with their ability to efficiently absorb infrared radiation results into GWPs of 23 and 296 times, respectively, that of CO_2 on a per-kg basis and a 100 years time horizon (IPCC 2001). In addition, the perfluorocarbons (e.g. CF_4 and C_2F_6) and sulphur hexafluoride (SF_6) have really long atmospheric residence times (50000, 10000, and 3200 years, respectively) and are strong absorbers of infrared radiation. The resulting GWPs are 5700, 11900, and 22200 times, respectively, that of CO_2 on a per-kg basis for 100 years time horizon (Table 1) (IPCC 2001). Most of the halocarbons recently used (halogenated methanes and ethanes) show high GWPs ranging from 12 to 12000 times that of CO_2 on a per-kg basis for 100 years time horizon. Their atmospheric lifetimes vary from 0.3 to 260 years (Table 1) (IPCC 2001).

2.3 Key Issues that are Influenced by Global Warming

The key changes to the terrestrial and aquatic environments in response to global warming can be distinguished as: (i) increase of global average air and water temperatures (Fig. 2) (IPCC 2007a). Global surface temperatures have increased by 0.74 °C since the late nineteenth century, and 11 out of the 12 warmest years on record have occurred since 1995 (IPCC 2007a).The temperature increase is widespread over the globc and is higher at higher northern latitudes. Indeed, average Arctic temperatures have increased at almost twice the global average rate in the past 100 years (IPCC 2007a). (ii) Decreases in snow cover and in the Northern Hemisphere sea ice extent. The result is a shorter freezing season for lakes, rivers and sea ice (Fig. 2) (IPCC 2007a). Since 1978, satellite data have been showing that the annual average Arctic sea ice extent has shrunk by 2.7 % (2.1–3.3 %) per

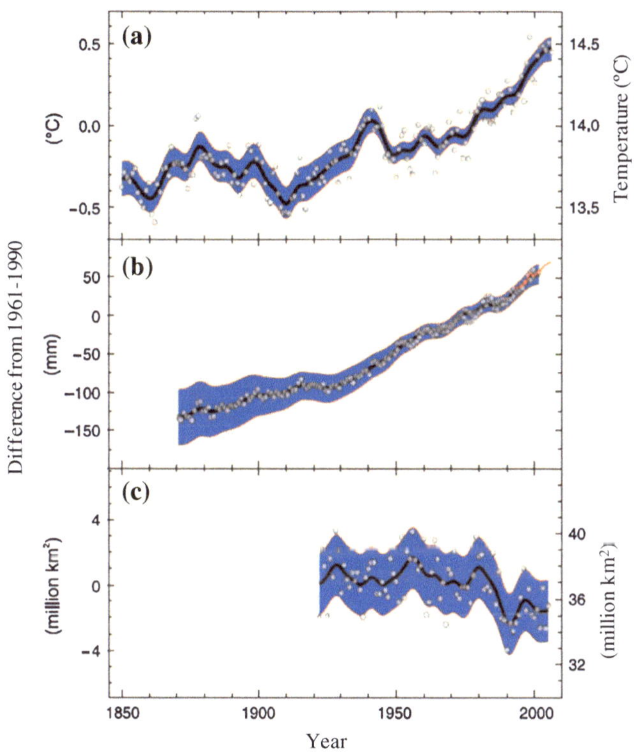

Fig. 2 Observed changes in **a** global average surface temperature; **b** global average sea level from tide gauge (*blue*) and satellite (*red*) data; and **c** Northern Hemisphere snow cover for March–April. All differences are relative to corresponding averages for the period 1961–1990. Smoothed curves represent decadal averaged values while circles show yearly values. The shaded areas are the uncertainty intervals estimated from a comprehensive analysis of known uncertainties (**a** and **b**) and from the time series (**c**). *Data source* IPCC (2007a)

decade, with summer decreases of 7.4 % (5.0–9.8 %) per decade (IPCC 2007a). The maximum surface of seasonally frozen ground has decreased by about 7 % in the Northern Hemisphere since 1900. There have been decreases in spring of up to 15 % (IPCC 2007a). (iii) Increase in soil temperature that subsequently enhances the soil respiration (Lloyd and Taylor 1994; Petersen and Klug 1994; Arnold et al. 1999; Feng and Simpson 2008, 2009; Frey et al. 2008). (iv) Weather modifications that can enhance natural disasters such as tornadoes, typhoons, storms, thunderstorms, and floods (Khalilov 2010). (v) Variations in water temperature profiles that cause changes in the euphotic zone, induce a longer summer stratification period and high photoinduced degradation of DOM and OM, make harmful algal blooms more likely, induce alteration of DOM dynamics, and change the seasonal patterns of chlorophyll or primary production and the nutrient concentrations. These effects induce as a consequence changes in species composition and in the seasonality of the water column, and finally modify food webs among phytoplankton, zooplankton, fish and birds in the aquatic environment (Huisman et al. 2006; Baulch et al. 2005; Castle and Rodgers 2009; Davis et al. 2009; Kitaysky and Golubova 2000; Hobson and McQuoid 2001; Mudie et al. 2002; Morrison et al. 2002; Johannessen and Macdonald 2009). (vi) Increases in sea level. They are consistent with warming and the global average sea level has risen at an average rate of 1.8 mm (1.3–2.3 mm) per year from 1961 to 2003. However, in the decade 1993–2003 the average rate has been of about 3.1 mm (2.4–3.8 mm) per year (IPCC 2007a). An increase of the global average temperature of about 2 ºC may cause a warming of about 2.7 ºC in the area around Greenland, possibly triggering the loss of the Greenland ice-sheet. Such a process may cause a global sea-level rise of 7 m over the next 1,000 years or more (Huybrechts et al. 1991; Gregory et al. 2004a, b). The rise of the sea level introduces vulnerability issues for agriculture, food, water resources, coral reefs, low-lying estuaries, intertidal zones, mudflats, mangrove forests, ecosystems and biodiversity (IPCC 2007a; Smith et al. 2001; Johannessen and Macdonald 2009; Doney et al. 2009; Masson and Cummins 2007; Burd et al. 2008a, b). The impacts on the coastal environments may lead to changes in the food web and affect the diversity of higher trophic levels such as marine mammals, fish and birds. (vii) A total ozone reduction of 2.5 % per decade during summer time causes a 5 % increase in UV irradiance (Varotsos and Kondratiev 1995), with a direct impact on terrestrial and aquatic environments.

3 Environmental Processes of GHGs Emission Affecting Global Warming

The key environmental processes of GHGs emission that may affect global warming can be categorized as follows: (i) soil respiration; (ii) agricultural activities in soil; (iii) anthropogenic sources of atmospheric greenhouse gases; (iv) deforestation; (v) photoinduced degradation of DOM and OM by natural sunlight; (vi) photoinduced degradation of OM in plants and soil environments.

3.1 Soil Respiration

In the soil respiration process, CO_2 fixed by terrestrial plants returns to the atmosphere.Changes in soil respiration in response to warming may contribute to the increase of CO_2 atmospheric levels (Bradford et al. 2008; Bahn et al. 2010; Feng et al. 2010; Raich and Schlesinger 1992; Oechel et al. 2000; Schlesinger and Andrews 2000; Luo et al. 2001; Melillo et al. 2002, 2004). CO_2 is produced in soils by roots, soil organisms and by chemical oxidation of carbon-containing materials (Lundegårdh 1927). Note that soil respiration through microbial activity can lead to the degradation to CO_2 of long chain ($>C_{20}$) alkanols, fatty acids (e.g. n-alkanoic acids), hydroxy fatty acids and di-acids that are major components of hydrolysable aliphatic lipids in soil organic matter (Feng et al. 2010; Nierop et al. 2003; Hajje and Jaffé 2006; Otto and Simpson 2006). These studies demonstrate that the average soil respiration rates are very variable depending on the nature of vegetation and on ambient temperature. For example, the lowest respiration rate is detected in tundra (60 ± 6 gC m^{-2} yr^{-1}), northern bogs and mires (94 ± 16 gC m^{-2} yr^{-1}), desert scrub (224 ± 38 gC m^{-2} yr^{-1}), boreal forests (322 ± 31 gC m^{-2} yr^{-1}) and marshes (413 ± 76 gC m^{-2} yr^{-1}). In contrast, respiration rates are highest in tropical moist forests (1260 ± 57 gC m^{-2} yr^{-1}), Mediterranean woodlands and heath (713 ± 88 gC m^{-2} yr^{-1}), temperate coniferous forests (681 ± 95 gC m^{-2} yr^{-1}), tropical dry forests (673 ± 134 gC m^{-2} yr^{-1}) and temperate deciduous forests (647 ± 51 gC m^{-2} yr^{-1}) (Raich and Schlesinger 1992). Temperature is the single best predictor of the annual respiration rate at a specific location, because soil respiration rates correlate significantly with average annual air temperatures and precipitation on a global scale (Raich and Schlesinger 1992).

Microbial decomposition of soil OM constituents such as lignin and hydrolysable lipids is promoted under both elevated CO_2 and N fertilization (Feng et al. 2010). Traditional tillage cultivation and rising temperature increase the flux of CO_2 from soils without increasing the stock of soil organic matter (Schlesinger and Andrews 2000). Soil warming can increase the relative abundance of Gram-positive bacteria (Frey et al. 2008; Bardgett et al. 1999; Biasi et al. 2005). It has also been shown that soil respiration is initially enhanced by warming for a few years, but that this effect is subsequently reduced over time (Frey et al. 2008; Oechel et al. 2000; Luo et al. 2001; Melillo et al. 2002, 2004). The following factors can be involved: (i) reduced plant production can lead to lower root respiration rates, decrease microbial activity because of soil drying, and to losses of labile soil organic carbon substrates such as amino acids, carbohydrates, and carboxylic acids (Frey et al. 2008; Oechel et al. 2000; Luo et al. 2001; Melillo et al. 2002). (ii) Increases in temperature can significantly change the microbial community structure that ultimately affects the soil respiration (Lloyd and Taylor 1994; Petersen and Klug 1994; Arnold et al. 1999; Feng and Simpson 2008, 2009; Frey et al. 2008).

Causes of diversity in respiration in the soil ecosystems are the variation in the decomposition factors of particulate detrital pools or vascular plant materials, which are regulated by numerous physical (temperature, moisture), chemical (redox, nutrient availability) and microbial (microfloral successional patterns, availability of microorganisms) factors (Mostofa et al. 2009a; Malcolm 1985; Wetzel 1992; Nakane et al. 1997; Uchida et al. 1998, 2000). Soil OM is typically lost upon agricultural conversion that reduces plant residue inputs, tillage-induced soil disturbance, erosion, and by the creation of more favorable conditions for microbial decomposition (Robertson and Grace 2004). The growth of typical terrestrial vegetation, rainforest, vascular plants and/or typical grassland and their degradation are significantly higher during the warm seasons than in the cold ones, because increased temperature would enhance respiration and decomposition in the soil environment (Nakane et al. 1997; Uchida et al. 1998, 2000; Duff et al. 1999; Fahey et al. 2005). The temperature increase is often found to enhance the soil CO_2 fluxes to the atmosphere (Fig. 3) (Feng and Simpson 2009).

However, the enhanced plant or litter inputs induced by warming have a stronger control on soil microbial responses than the temperature increase itself. Therefore, it is suggested that the quality of soil organic matter can control microbial responses to global warming (Feng and Simpson 2009; Rinnan et al. 2008; Zhang et al. 2005). The temperature effects outlined above may be comparatively less important at temperate latitudes, because the soil respiration rate is highly increased by temperature in areas where the soil temperatures are low (Lloyd and Taylor 1994; Biasi et al. 2005).

Warming might affect the abundance of soil microorganisms, but contrasting data are presently available. A 20–60 % increase in the fungal:bacterial ratio has been observed in a tallgrass prairie site, exposed to a ~2 °C increase in temperature over a three-year period (Zhang et al. 2005), but another study has shown that the relative abundance of fungi was significantly reduced after 15 years of soil warming (1–2 °C) in northern Sweden (Rinnan et al. 2008).

The global warming effect might be significantly different depending on the soil ecosystems. Indeed, changes in soil respiration and CO_2 fluxes are the effects of temperature and vegetation differences (Table 1) (IPCC 2001; Trumbore

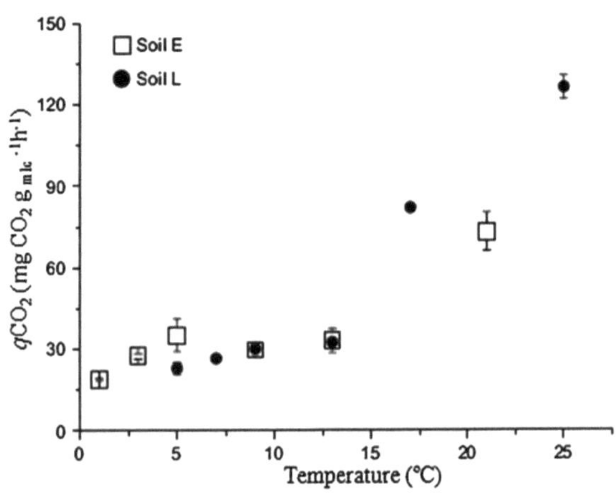

Fig. 3 Metabolic quotient (qCO_2) of both grassland soils on Day 1. Points show average values of triplicate (n = 3) and error bars represent standard error. Soil E and soil L are the two soil samples collected from the two different sites in Alberta, Canada. *Data source* Feng and Simpson (2009)

2000). Note that soil represents approximately 80 % of the carbon stocks in terrestrial ecosystems, ranging from 50 % in tropical forests to 95 % in tundra (IPCC 2002). The turnover times of OM, determined through ^{14}C tracer in well-drained boreal (Manitoba in Canada), temperate (central Massachusetts in the USA) and tropical forest soils (eastern Amazonia in Brazil), suggest that the average age of OM carbon is higher than the average age predicted from CO_2 production by OM decomposition (30, 8, and 3 yr for boreal, temperate, and tropical soil) or from total soil respiration (16, 3, and 1 yr, respectively) (Table 1) (IPCC 2001; Trumbore 2000). Most of the CO_2 produced during decomposition is derived from relatively short-lived soil organic matter (SOM) components. They do not represent a large fraction of the standing stock of soil organic matter (Trumbore 2000). Comparison of the ^{14}C in soil respiration with soil organic matter in temperate and boreal forest sites indicates a significant contribution from the decomposition of organic matter fixed >2 yr but <30 yr ago (Table 1) (IPCC 2001). Tropical soil respiration is dominated by C fixed <1 yr ago (Table 1) (IPCC 2001). Monitoring of the ^{14}C signature of CO_2 emitted from soils suggests that seasonal and interannual variability in soil respiration are the key factors in these ecosystems (IPCC 2001; Trumbore 2000) (Table 1).

These findings imply that the soil respiration is very variable in different ecosystems, with important effects on carbon sequestration and global carbon dynamics. It is estimated that on a global scale, the soil respiration in terrestrial ecosystems produces a CO_2 flux of approximately 75×10^{15} g C yr^{-1}, which is likely to increase due to changes in the Earth's condition (Schlesinger and Andrews 2000).

3.2 Agricultural Activities in Soil

The soil and the related agricultural activities can release significant amounts of CO_2, CH_4 and N_2O to the atmosphere (Mosier et al. 1989, 1991, 2004; Robertson and Grace 2004; Ambus and Robertson 2006; Smith et al. 2008; Kreileman and Bouwman 1994; IPCC 2001; Raich and Schlesinger 1992; Aselmann and Crutzen 1989; Watson et al. 1992; Bowden et al. 1993; Subak et al. 1993; Zuidema et al. 1994; Freney 1997; Tsuruta et al. 1997; Stevens and Laughlin 1998; Cole et al. 1997; Tranvik et al. 2009). CO_2 is mostly released from agricultural activities and soil disturbances (IPCC 1996, 2007a; Subak et al. 1993; Bouwman 1990; Lal et al. 1999; Schlesinger 1999; Izaurralde et al. 2000). Several processes are responsible for the production of CO_2 from such activities: (i) CO_2 is produced during the processing, transport and application of N-containing fertilizers, which cause the release of around 1.4 mol of CO_2 per mole of N applied (Schlesinger 1999; Izaurralde et al. 2000; IPCC 1996). (ii) Land limes in the form of calcium carbonate ($CaCO_3$) and dolomite [$CaMg(CO_3)_2$] can produce bicarbonate and CO_2 (Robertson and Grace 2004; Liu et al. 2010, 2011). Note that $CaCO_3$ and $CaMg(CO_3)_2$ are commonly applied to agricultural soils to counteract soil acidity and to give supplies of Ca^{2+} and Mg^{2+} for plant uptake (Robertson and Grace 2004). $CaCO_3$ can react with soil Al^{3+} to form carbonic acid, raising the soil pH by the following reaction (Robertson and Grace 2004):

$$2Al^{3+} + 3CaCO_3 + 6H_2O \rightarrow 3Ca^{2+} + 2Al\,(OH)_3 + 3H_2CO_3 \tag{3.1}$$

Similarly, carbonic acid formed in the presence of CO_2 from root and microbial respiration reacts with solid carbonates [ca. $CaMg(CO_3)_2$] to produce bicarbonate by the following reaction (Eq. 3.2) (Robertson and Grace 2004):

$$CaMg\,(CO_3)_2 + 2\,H_2CO_3 \rightarrow Ca^{2+} + Mg^{2+} + 4\,HCO_3{}^{-} \tag{3.2}$$

A strong mineral acid such as nitric acid (HNO_3) can react with carbonates [e.g. $CaMg(CO_3)_2$] to produce CO_2 by the following reaction (Robertson and Grace 2004):

$$CaMg\,(CO_3)_2 + 4\,HNO_3 \rightarrow Ca^{2+} + Mg^{2+} + 4\,NO_3{}^{-} + 2\,CO_2 + 2\,H_2O \tag{3.3}$$

Nitric acid is formed by nitrifying bacteria in most soils, including acid tropical soils (Robertson 1982; Sollins et al. 1988).

(iii) Calcium-saturated groundwater can react with soil $HCO_3{}^{-}$ to produce CO_2 by the following reaction (Robertson and Grace 2004; Schlesinger 1999):

$$Ca^{2+} + 2\,HCO_3{}^{-} \rightarrow CaCO_3 + H_2O + CO_2 \tag{3.4}$$

Carbonate reactions also occur when calcium-saturated groundwater is sprayed on calcareous surface soils (Schlesinger 1999). In arid regions groundwater often contains as much as 1 % Ca and CO_2 (Robertson and Grace 2004).

CH_4 has a microbial origin from natural (e.g. wetlands) and human-influencewd sources, such as agricultural activities (rice and crops cultivation), enteric fermentation, animal wastes and landfills (Mosier et al. 1991, 1998, 2004; Robertson and Grace 2004; Smith et al. 2008; IPCC 2001; Watson et al. 1992; Subak et al. 1993; Zuidema et al. 1994; Crutzen et al. 1986; Bingemer and Crutzen 1987; Cicerone and Oremiand 1988; Robertson et al. 2000). Methane is produced when organic materials are decomposed in oxygen-deprived conditions, including fermentative digestion by ruminant livestock, stored manures and rice grown under flooding (Mosier et al. 1998). A recent study estimates that agriculture accounts for 52 % of the global anthropogenic CH_4 emissions (Smith et al. 2008).

N_2O emission by agricultural activities in soil is accounted for by microbial nitrification, denitrification and chemo-denitrification, especially under wet conditions. N_2O is also produced by the microbial transformation of nitrogen in soil and manure (IPCC 2007a; Robertson and Grace 2004; Smith et al. 2008; Kreileman and Bouwman 1994; Mosier et al. 1989, 1991; Freney 1997; Tsuruta et al. 1997; Stevens and Laughlin 1998; Robertson et al. 2000; Cavigelli and Robertson 2000; Xing et al. 2002; Mahimairaja et al. 1994; Smith and Conen 2004; Oenema et al. 2005). Natural sources of N_2O have been estimated to be approximately 10 TgN/yr in 1990. Soils account for about 65 % of the sources, oceans for about 30 % (IPCC 2001). It is estimated that agriculture accounts for 84 % of the global anthropogenic N_2O emissions (Smith et al. 2008).

3.3 Anthropogenic Sources of Atmospheric GHGs

Anthropogenic sources are primarily responsible for the greenhouse gases (GHGs) inputs to the atmosphere. CO_2 is mostly produced anthropogenically from the burning of fossil fuels and plant litter and from oil flaring, cement manufacturing and other industrial activities (IPCC 2007a; Sabine et al. 2004; Smith 2004; Archer 2005; Canadell et al. 2007; Hofmann et al. 2009; Subak et al. 1993; Marland and Rotty 1984; Crutzen and Andreae 1990; Keeling et al. 1996). It has recently been shown that the annual emissions because of fossil fuel burning have grown by about 80 %, from 21 to 38 gigatonnes (Gt), between 1970 and 2004. The rate of growth of CO_2-eq emissions was much higher during the more recent decade 1995–2004 (0.92 $GtCO_2$-eq per year) than during the previous period 1970–1994 (0.43 $GtCO_2$-eq per year) (IPCC 2007a). CH_4 is produced from fossil fuel production and consumption including oil and gas field activities, coal mining, plant litter/wood processing, domestic sewage treatment, enteric fermentation and other biomass burning (Mosier et al. 2004; Smith 2004; Kreileman and Bouwman 1994; IPCC 2001; Subak et al. 1993; Crutzen 1991; Flessa et al. 2002). The major sources of anthropogenic aerosols are sulfur-containing fossil fuels, biomass burning and explosive volcanic eruptions (IPCC 2001, 2007a).

3.4 Deforestation

Deforestation or changes in land-surface cover can significantly affect atmospheric CO_2, CH_4 and N_2O (IPCC 2001, 2007a; Kreileman and Bouwman 1994; van der Werf et al. 2009; Raich and Schlesinger 1992; Subak et al. 1993; Bouwman 1990; Crutzen and Andreae 1990; Keller et al. 1986; Sitch et al. 2005; Detwiler and Hall 1988; Myers 1989; Houghton 1991). The above-cited processes can contribute in three ways to global warming: First, deforestation i.e. the decline of terrestrial plants can significantly reduce the uptake of CO_2 from the atmosphere by photosynthesis, which is vital for plants growth. Such a phenomenon would thus give indirect contribution to the increase of atmospheric CO_2. Second, deforestation can reduce evaporation and increases surface temperature. Third, changes in land-surface cover can enhance the degradation of soil DOM and OM by both photoinduced and microbial processes, thus uncovered land surfaces can release GHGs directly to atmosphere (Schiettecatte et al. 2006, 2007; Borges et al. 2008; Omar et al. 2010; Brandt et al. 2009; Rutledge et al. 2010; Thomas et al. 2004, 2005, 2007; Raich and Schlesinger 1992; Xie et al. 2004). Soil respiration rates are very high in the first year after the clear-cutting of plants, apparently due to the higher soil temperatures and to the decomposition of the debris that the soil incorporated during the deforestation (Raich and Schlesinger 1992). Deforestation is the second largest anthropogenic source of CO_2 to the atmosphere after fossil fuel burning. It accounted for 17.3 % of anthropogenic GHGs in the 2004 total emissions to the atmosphere (IPCC 2007a; van der Werf et al. 2009). Large-scale deforestation in the humid tropics has been identified as the main, ongoing land-surface process caused by industrialization and by growing agricultural activities because of the increasing demands of a growing population.

3.5 Photoinduced and Microbial Degradation of Organic Matter (OM) in Natural Waters

The production of CO_2 and of other dissolved inorganic carbon (DIC: generally defined as dissolved CO_2, H_2CO_3, HCO_3^-, and CO_3^{2-}) species upon photoinduced and microbial degradation of organic matter including DOM and POM (e.g. algae or phytoplankton) can potentially influence the carbon cycling and may have an impact on climate change (Bozec et al. 2005, 2006; Schiettecatte et al. 2006, 2007; Borges et al. 2008; Omar et al. 2010; Ballaré et al. 2011; Zepp et al. 2011; Kelley 1970; Kempe and Pegler 1991; Hoppema 1990, 1991; Borges and Frankignoulle 1999, 2002a, b; Brasse et al. 1999; Frankignoulle and Borges 2001; Thomas et al. 2004, 2005, 2007; Tranvik et al. 2009; Xie et al. 2004; Salonen and Vähätalo 1994; Granéli et al. 1998; Richey et al. 2002; Clark et al. 2004; Kujawinski et al. 2009; Koprivnjak et al. 2010). The water-bed in subtropical and tropical latitudes generally acts as a CO_2 source to the atmosphere, while at high and temperate latitudes it rather acts as a CO_2 sink by uptake from the atmosphere (Omar et al. 2010; Borges and Frankignoulle 2002a; Thomas et al. 2004; Sobek et al. 2005; Gattuso et al. 1993, 1997; Frankignoulle et al. 1996, 1998; Goyet et al. 1998; Tsunogai et al. 1999; Yool and Fasham 2000; Bates et al. 2001; Cai et al. 2003, 2006; Borges 2005; Ito et al. 2005; Ohde and van Woesik 1999; Wang and Cai 2004; Chen and Borges 2009; Wang et al. 2011). In temperate regions, increased temperatures and longer residence times of OM in water, which might be caused by decreased runoff could accelerate microbial respiration and photoinduced degradation of organic carbon. However, the combined effects of increased autochthonous production and increased organic carbon burial efficiency due to increased anoxia may offset increased CO_2 production (Tranvik et al. 2009). It should also be considered that increases in production, duration of stratification and sedimentation may favor the occurrence of hypolimnetic anoxia and, as a consequence, CH_4 production in temperate zones (Tranvik et al. 2009). Larger emissions of CO_2 and CH_4 may occur in Arctic regions, particularly where thermokarst erosion and ponding is occurring (Walter et al. 2006). Moreover some boreal streams, some major rivers, lakes in general or boreal lakes in particular are supersaturated in CO_2 and are considered to be net sources of CO_2 to the atmosphere (Fahey et al. 2005; Koprivnjak et al. 2010; Sobek et al. 2005; Jones and Mulholland 1998; Dawson et al. 2001; Hope et al. 2001; Finlay 2003; Öquist et al. 2009; Teodoru et al. 2009; Cole and Caraco 2001; Jones et al. 2003c; Yao et al. 2007; Huttunen et al. 2003a).

The air–sea CO_2 exchange occurs mostly in temperate regions (Borges and Frankignoulle 2002a, b; Borges 2005; Andersson et al. 2003; Andersson and Mackenzie 2004; Zhai et al. 2005). However, waters in upwelling regions act both as sinks (California and Oman coasts) and as sources (Galician and Oregon coasts) of atmospheric CO_2 (Borges and Frankignoulle 2002a, b; Goyet et al. 1998; Friederich et al. 2002; Hales et al. 2005). The global coastal zone is still a net source of CO_2 to the atmosphere, due to the combination of calcification and of net heterotrophy that is a feature of estuarine ecosystems (Frankignoulle et al. 1998; Borges 2005; Cai and Wang 1998; Raymond et al. 2000; Sarma et al. 2001; Mukhopadhyay et al. 2002; Bouillon et al. 2003; Abril et al. 2003, 2004; Mackenzie et al. 2004; Fagan and Mackenzie 2007). Indeed, when estuaries are included in the CO_2 exchange budget, the global shallow-water coastal ocean is a net source of CO_2 to the atmosphere (Borges 2005).

The production of CO_2 and its input to the atmosphere is considerably higher during the summer and fall (or dry) seasons than in winter and spring (or wet) seasons. In the latter case the waterbed actually acts as a net sink for atmospheric CO_2. The reason behind this phenomenon is that the photoinduced and microbial degradation of DOM and POM are greatly enhanced in surface waters during the summer period due to high solar radiation and longer summer day-time. CO_2 emission by boreal streams is quite high during summer and very low in spring, which might be a consequence of photoinduced processing of DOM and POM (Koprivnjak et al. 2010). Obviously, the solar intensity is significantly reduced during the winter season that also has shorter day-time. In addition, estuaries often have high contents of DOM that undergoes strong photoinduced degradation and makes these systems to be significant sources of CO_2 to the atmosphere. The concentration of dissolved organic carbon (DOC) explains the significant variation of lake pCO_2 (Sobek et al. 2005), which might be an effect of photoinduced and microbial release of CO_2 from DOM and POM in water as mentioned before. Supersaturation of CO_2 in freshwater ecosystems (streams, rivers and lakes) is possibly caused by the same photoinduced and microbial processes that degrade DOM and POM. Indeed, freshwater ecosystems generally contain high amounts of DOM and POM that are potentially important microbial or photoinduced sources of CO_2 or DIC.

The situation is much different at northern latitudes: it is estimated that the direct photo-oxidation of organic carbon to CO_2 accounted for less than 10 % of dark respiration in the epilimnion of six boreal lakes (Granéli et al. 1996). CO_2 emission is also mainly derived from in-lake respiration in the lake environments (del Giorgio et al. 1999; Jansson et al. 2000). Anyway, global warming will increase the atmospheric temperature that can enhance both the photoinduced and the microbial degradation of DOM and POM, during all seasons and at all latitudes. The consequence would obviously be a further increase of atmospheric CO_2. Warming is also expected to reduce terrestrial and ocean uptake of atmospheric CO_2, increasing the fraction of anthropogenic emissions that remain in the atmosphere. This would result into an additional increase of atmospheric CO_2 (IPCC 2007a).

A positive correlation of pCO_2 levels with CDOM and chlorophyll has been observed in the Southwest Florida Shelf, indicating that CO_2 may be produced from the photoinduced degradation of CDOM in natural waters (mostly in the dry season), from microbial respiration and from shifts in the carbonate equilibrium (Clark et al. 2004). The effect is a release of CO_2 into the atmosphere from DOM that has been formed by primary production (Thomas et al. 2009). On the other hand, microbial degradation of DOM and OM in natural waters and sediment pore waters can release CH_4 to the atmosphere (Mosier et al. 2004; Cicerone and Oremiand 1988; Pepper et al. 1992; Bastviken et al. 2004, 2008; Bergström et al. 2007). Anoxia in freshwater sediments contributes to high CH_4 emissions, and the production of CH_4 in epilimnetic sediments is the main driver of methane emission from surface waters (Bastviken et al. 2004, 2008). Methane production can also be enhanced by water temperature and lake level fluctuations. Such effects can affect carbon balances depending on the predominant plant species and sediment properties (Bergström et al. 2007). It is estimated that the contribution of CH_4 to the atmosphere is 100–200 Tg yr^{-1} from wetlands, 5–20 Tg yr^{-1} from oceans, and 1–25 Tg yr^{-1} from freshwater (Mosier et al. 2004). N_2O can be released from freshwater and oceanic environments (Watson et al. 1992; Seitzinger 1990). Increases in oxygen-deficient regions in the ocean caused by climate changecould enhance the emissions of nitrous oxide, an important greenhouse and ozone-depleting gas (Zepp et al. 2011).

The upper ocean microbial food web (mostly the autotrophs) is a huge carbon-processing machine that can remove CO_2 from the atmosphere, but part of the carbon fixed by autotrophy is actually respired in situ (Sarmento et al. 2010). The heterotrophic bacteria are responsible for the major respiration (>95 %) in the ocean (del Giorgo and Duarte 2002), and half of it (approximately 37 Gt of C per year) takes place in the euphotic layer (del Giorgio and Williams 2005). Notes that global ocean respiration is approximately as important as the oceanic primary production (del Giorgo and Duarte 2002; Karl et al. 2003; Williams PJlB et al. 2004; Riser and Johnson 2008). Increasing temperature will often increase respiration rates in natural waters (Vázquez-Domínguez et al. 2007). Increasing aquatic respiration is presumably the result of enhanced photo- and microbial products (H_2O_2, CO_2, DIC, etc.) derived from the photoinduced and microbial degradation of DOM and OM in the euphotic zone. The temperature increase accelerates the respiratory consumption of organic carbon relative to the autotrophic production, with a decrease in the biological drawdown of DIC. A decrease of up to 31 % has been observed in mesocosms warmed by 2, 4 and 6 °C (Wohlers et al. 2009). Changes in the biogenic carbon flow induced by warming have the potential to reduce the transfer of primary produced OM to higher trophic levels (Vázquez-Domínguez et al. 2007; Wohlers et al. 2009; Laws et al. 2000). This would weaken the ocean's biological carbon pump and provide a positive feedback to the rise of atmospheric CO_2 (Vázquez-Domínguez et al. 2007; Wohlers et al. 2009; Laws et al. 2000).

The photoinduced and microbial activities of DOM and POM in natural surface waters may act as sources or sinks of N_2O that is produced via nitrification and denitrification (Tranvik et al. 2009; Mengis et al. 1997; Huttunen et al. 2003b, 2004; Wang et al. 2006). N_2O may be consumed in the hypolimnion, whilst shallow sediments contribute to N_2O emissions to the atmosphere (Huttunen et al. 2003b; Wang et al. 2006).

3.6 *Photoinduced and Microbial Degradation of OM in Plants and Soil Environments*

Photoinduced and microbial processes can directly degrade the organic matter (OM) in terrestrial plants and in soil environments, releasing CO_2 to the atmosphere (Brandt et al. 2009; Rutledge et al. 2010). Photodegradation contributes 19 % of the annual CO_2 flux from peatland and almost 60 % of the dry-season CO_2 flux from grassland. The respective fractions of the summer mid-day CO_2 fluxes are up to 62 and 92 % (Rutledge et al. 2010). Photodegradation may be important in a wide range of ecosystems with exposed OM (Rutledge et al. 2010). Plant litter can be exposed outdoors to natural solar radiation. It has been shown that in clear sunny days, close to the summer solstice at mid-latitude, UV radiation (280–400 nm) accounted for 55 % of the photolytically induced CO_2 production, while shortwave visible radiation (400–500 nm) accounted for the remaining 45 % (Brandt et al. 2009). Abiotic mineralization to CO_2 is the primary mechanism by which C is lost from litter during photodegradation. It is estimated that annual CO_2 production via photodegradation could be between 1 and 4 g C m^{-2} a^{-1} in arid ecosystems in the southwestern United States (Brandt et al. 2009).

4 Impacts of Global Warming on Natural Waters

Global warming may severely affect various physical, chemical and biological processes that involve DOM in natural waters. The main effects are the following: (i) Changes in the light cycle and increase of water temperature; (ii) Increase of the photoinduced activity of natural waters; (iii) Changes in the microbial activity in natural water; (iv) Changes in photosynthetic processes in natural waters; (v) Changes in the primary production and disorders in the chlorophyll *a* maxima; (vii) Changes in the DOM dynamics and in the global carbon cycle. (viii) Changes in the nutrients cycle.

4.1 *Changes in the Light Cycle and Increase of Water Temperature*

Global warming could cause changes in the seasonal light cycle and an increase in water temperature, which affects the light distribution in the euphotic zone

(Rutledge et al. 2010; O'Reilly et al. 2003; Letelier et al. 2004; Porcal et al. 2009; Morrison et al. 2002; Ryther 1956). Notes that the depth of the euphotic zone is defined as the depth where the photon flux density equals 1 % of that measured at the air–sea interface (Ryther 1956). The temperature increase is a global effect, but it is higher northern latitudes. Average Arctic temperatures have in fact increased at almost twice the global average rate in the past 100 years (IPCC 2007a). Temperatures at the top of the permafrost layer have generally increased in the Arctic by up to 3 °C since the 1980s (IPCC 2007a). The global average surface air temperature has increased by 0.74 °C over the past century and is projected to rise by another 1.1 to 6.4 °C before 2100. The sea level could increase by 0.2 to 0.6 m or more before 2100 (Hansen and Sato 2004; IPCC 2007b). The long-term observations in European seas show that the increase of the sea-surface temperature rate is around 0.01 °C yr^{-1} since the 1860s (Wiltshire and Manly 2004; Vargas-Yañez et al. 2005; Mackensie and Schiedek 2007). The combination of temperature increase and of the decrease in water flow allow the prediction of a 10-fold increase, by the end of this century, of the number of days when the temperature of the Fraser River exceeds 20 °C. Such a phenomenon may threaten the survival of some specific fish and other aquatic microorganisms (Morrison et al. 2002). Global warming may expand the summer season and increase the water column transparency as well as the water temperature, which might accelerate the photoinduced degradation of DOM through e.g. an enhanced production of $HO^{\bullet}$ (Huisman et al. 2006; Zellner et al. 1990; Malkin et al. 2008). At the same time, there can be an increase of UV radiation during ozone hole events (Huisman et al. 2006; Kerr and McElroy 1993; Varotsos and Kondratiev 1995; Qian et al. 2001; Sarmiento et al. 2004; Schmittner 2005). Previous studies show that the incident UV-B radiation has increased at a rate of 10–20 % per decade at temperate latitudes (Kerr and McElroy 1993), and a total ozone reduction of 2.5 % per decade during summer would cause a 5 % increase in the UV irradiance (Varotsos and Kondratiev 1995).

Although the increase of the chlorine concentration in the stratosphere has slowed down, reflecting the execution of the Montreal Protocol, the time required for the recovery of the ozone layer is unconvincing and will rely on the impacts of the climate change on the stratosphere (Weatherhead and Andersen 2006). The global warming phenomenon is expected to enhance the temperature in the troposphere, but at the same time there will be cooling effects in the stratosphere that can enhance ozone depletion. An increase of UV-B radiation may greatly enhance the production of $HO^{\bullet}$ due to an increase in direct photolysis rates of NO_2^-, NO_3^- and Chromophoric Dissolved Organic Matter (CDOM), and also other redox reactions may be enhanced, in particular in the Antarctic and Arctic regions (Qian et al. 2001; Randall and Harvey 2005). The $HO^{\bullet}$ formation from nitrate, nitrite and CDOM significantly increases during ozone hole conditions (Qian et al. 2001). Two effects may derive from this scenario. First, ozone hole conditions may enhance the photoinduced degradation of aquatic DOM, which can subsequently release a large amount of CO_2 to atmosphere. Second, high production of $HO^{\bullet}$ can reduce the biological activity through oxidative damages to the living cells of biota in the aquatic environments (Berlett and Stadtman 1997; Paradies et al. 2000;

Blokhina et al. 2003). Ocean warming and acidification due to increased atmospheric CO_2 concentration may exacerbate the detrimental effects of solar UV-B radiation (Häder 2011). Finally, exposure to solar UV radiation can reduce productivity, affect reproduction and development, and increase the mutation rate in phytoplankton, macroalgae, eggs, and larval stages of fish and other aquatic animals (Häder et al. 2007). Consequences of decreased productivity are a reduced sink capacity for atmospheric carbon dioxide and negative effects on species diversity, ecosystem stability, trophic interactions and ultimately global biogeochemical cycles (Zepp et al. 2007).

Since 1993 the thermal expansion of the oceans has contributed about 57 % of the overall sea level rise, while decreases in glaciers and ice caps contributed about 28 % and the remainder was accounted for by losses from the polar ice sheets (IPCC 2007a).

4.2 Increase of Photoinduced Activity in Natural Waters

Global warming causes an increase in water temperature that can accelerate the photoinduced activity of DOM and of other chemical constituents in surface waters. Photoinduced degradation of DOM and OM can produce a number of photo-products including H_2O_2 and DIC (dissolved CO_2, H_2CO_3, HCO_3^-, and CO_3^{2-}) (Molot et al. 2005; Johannessen et al. 2007; Mostofa and Sakugawa 2009; Mostofa et al. 2009a, b, 2011; Xie et al. 2004; Clark et al. 2004; Miller and Zepp 1995; Thomas and Lara 1995; Dillon and Molot 1997; Miller 1998; Gennings et al. 2001; Johannessen and Miller 2001; Rochelle-Newall and Fisher 2002; Ma and Green 2004; Hiriart-Baer and Smith 2005). Autochthonous DOM can be released in natural waters by algae or phytoplankton upon photoinduced degradation or photorespiration (Mostofa et al. 2009a, b; Stedmon et al. 2007a, b; Thomas and Lara 1995; Rochelle-Newall and Fisher 2002; Fu et al. 2010), and it has recently been shown that the algal-derived CDOM is a more efficient photoinduced substrate than terrigenous material (Johannessen et al. 2007). In situ incubation of natural phytoplankton assemblages in Antarctic waters indicates that, under normal ozone conditions, UV-B radiation is responsible for a loss of approximately 4.9 % of the primary production in the euphotic zone. UV radiation with wavelengths between 320 and 360 nm causes a loss of approximately 6.2 % (Holm-Hansen et al. 1993a). Ambient levels of UV radiation (280–400 nm) are observed to decrease substantially the rates of carbon fixation by phytoplankton (Holm-Hansen et al. 1993a; Karentz et al. 1991; Cullen et al. 1992; Helbling et al. 1992; Smith et al. 1992; Li et al. 2011), and photoinduced release of DOM from phytoplankton can also take place. UV-B radiation accelerates the decomposition of colored DOM entering the sea via terrestrial runoff, thus having important effects on the oceanic carbon cycle (Zepp et al. 2003).

The increase of water temperature significantly enhances the efficiency of the Fenton and photo-Fenton reactions, as well as the photolysis of NO_2^-, NO_3^- and

H_2O_2. All these compounds are responsible for the production of $HO^{\bullet}$ towards the degradation of DOM or organic pollutants in aqueous solution (Zellner et al. 1990; Zafiriou and Bonneau 1987; Millero and Sotolongo 1989; Zepp et al. 1992; Farias et al. 2007). It is estimated from data of Zellner et al. (Zellner et al. 1990) that a temperature increase from 278 to 298 K can enhance the quantum yield of $HO^{\bullet}$ photoproduction (at 308 nm and at neutral pH), on average by 70 % for NO_2^- photolysis, 129 % for NO_3^- photolysis and 20 % for H_2O_2 photolysis. The efficiency of the photoinduced degradation of DOM is also significantly dependent on the wavelength, and the quantum yields (Φ_{HO}) of $HO^{\bullet}$ production decrease with increasing wavelength (Zellner et al. 1990; Zafiriou and Bonneau 1987). At pH 8 and at 298 K the Φ_{HO} for NO_2^- photolysis at 308 nm is on average 54 % higher ($\Phi_{HO} = 0.071 \pm 0.009$) than that at 351 nm ($\Phi_{HO} = 0.046 \pm 0.003$) (Zellner et al. 1990).

The formation of H_2O_2 is a key step of the photoinduced processes in surface waters (Mostofa et al. 2011) and it is as well largely dependent on the radiation wavelengths (Obernosterer et al. 2001; Richard et al. 2007). The contribution of UV-B, UV-A and photosynthetically active radiation (PAR) to H_2O_2 formation is 40, 33 and 27 %, respectively (Richard et al. 2007).

An increase in the photoinduced degradation rate of DOM may extend the water column transparency (which undergoes seasonal modifications on a variety of time scales) and the depth of the mixed layer that influences the incident UV radiation (Scully and Lean 1994; Morris et al. 1995; Morris and Hargreaves 1997). Diffuse light attenuation coefficients often undergo seasonal variations (39–81 %) in surface waters, and minimum values appear during the summer season (Morris and Hargreaves 1997). Therefore, an extension of the summer season due to global warming may enhance both photoinduced processes and photosynthesis, which could for instance increase the duration of the phytoplankton or algae productivity in lake ecosystems, particularly in the Arctic and Antarctic regions (Malkin et al. 2008). As already indicated, an increase in UV radiation due to depletion of the stratospheric ozone layer can accelerate the production of $HO^{\bullet}$, which is a key factor for the photoinduced degradation of DOM in natural waters (Huisman et al. 2006; Qian et al. 2001; Sarmiento et al. 2004; Schmittner 2005; Crutzen 1992; Stolarski et al. 1992). Therefore, global warming can enhance both the photoinduced degradation of DOM and the release of autochthonous DOM and nutrients from algae or phytoplankton. Such processes can be partially offset by the fact that the production of higher amounts of H_2O_2, CO_2, DIC and other low molecular weight substances could increase photosynthesis and enhance the primary production (Mostofa et al. 2009b).

4.3 Changes in Microbial Activity in Natural Waters

Global warming may significantly affect microbial or biological processes in natural waters, with consequences on both autotrophs (plants, algae, bacteria) and heterotrophs (animals, fungi, bacteria). An increase of temperature can change (or enhance) microbial activities including bacterial production, respiration, photosynthesis and growth efficiency, as well as bacterial–grazer trophic interactions, which can result in the rapid mineralization of organic matter in natural waters, particularly in Arctic and Antarctic ecosystems (Norf et al. 2007; Vázquez-Domínguez et al. 2007; Falkowski and Oliver 2007, 2008; Peters 2008; Norf and Weitere 2010; Sarmento et al. 2010; Sawicka et al. 2010; Nedwell and Rutter 1994; Ochs et al. 1995; Felip et al. 1996; Nedwell 1999; Reay et al. 1999; Vrede 2005; Morán et al. 2006; López-Urrutia and Morán 2007). These studies show that an increase in temperature may enhance the availability of labile substrates, which is responsible for an increase of microbial activity at elevated temperature.

The response to temperature of a species or microorganism is characterized by a number of 'cardinal temperatures': upper and lower limits of temperature for growth, and an optimum growth temperature included between the two extremes (Morita 1975). Microorganisms living near the lower temperature limit of a species can be stimulated either by higher temperature or by higher concentrations of added substrates in natural waters (Pomeroy et al. 1991; Wiebe et al. 1992, 1993). The microbial metabolism modifies organic nutrients such as glucose and the functional groups of macromolecules such as fulvic and humic acids of vascular plant origin or autochthonous fulvic acids of algal origin. The consequence of microbial processing may be the release in water of a variety of end products such as H_2O_2, CO_2, DIC, PO_4^{3-}, NH_4^+ and CH_4 (Mostofa and Sakugawa 2009; Ma and Green 2004; Fu et al. 2010; Palenik and Morel 1988; Lovley et al. 1996; Zhang et al. 2004, 2009; Kim et al. 2006; Li et al. 2008). Algae or phytoplankton biomass can release autochthonous DOM by microbial degradation or assimilation (Mostofa et al. 2009a, b, 2011; Stedmon et al. 2007a, b; Rochelle-Newall and Fisher 2002; Fu et al. 2010; Zhang et al. 2009; Biddanda and Benner 1997; Yamashita and Tanoue 2004, 2008; Stedmon and Markager 2005), and an increase in temperature can accelerate the bacterial degradation of phytoplankton-derived organic matter (Wohlers et al. 2009; Hoppe et al. 2008). Small algae carry out 40–95 % of total grazing on bacteria in the euphotic layer of the temperate North Atlantic Ocean in summer (Zubkov and Tarran 2008). A similar range (37–70 %) has been observed in the surface waters of the tropical Northeast Atlantic Ocean (Zubkov and Tarran 2008).

In Lake La Caldera it has been observed that at the lower temperature values (5.0–7.0 °C) one finds higher bacterial abundance (3.9–7.9 $\times$ 10^5 cells ml^{-1}, mean = 6.4) and higher bacterial biomass (4.0–6.7 μg C L^{-1}, mean = 5.2) compared to the higher temperature values (7.5–11.1 °C), which yielded 1.3–2.5 $\times$ 10^5 cells ml^{-1} (mean = 1.8) and 1.3–2.4 μg C L^{-1} (mean = 1.7) for bacterial abundance and biomass, respectively (Carrillo et al. 2002). The grazing on bacteria increases with increasing temperature, but the rate of the increase is maximum at temperatures lower than 2 °C, whilst bacterial production increases at higher rates at temperatures higher than 2 °C. Such a finding, obtained in a microcosm experiment with temperature manipulation (−1 to 5 °C) of Antarctic waters, suggests that bacterial production and bacterial grazing could become uncoupled processes at higher temperatures (Vaqué et al. 2009). Polar oceans at temperatures of −1 to 2–3 ºC have microbial communities, both bacterial and algal, which are physiologically stressed. In fact, the environmental temperature is well below the optimum temperature for growth of many inhabitants (Nedwell 1999). As average Arctic temperatures have increased at almost twice the global average rate in the past 100 years (IPCC 2007a), the microbial activity in the Arctic and Antarctic regions is expected to undergo a significant enhancement due to the effect of global warming.

Winter warming typically results in both stimulation (abundance and biomass) of the biofilm ciliate communities and in significant shifts in the community structure. Summer warming induces a significant decline in the ciliate biomass but does not affect the relative community composition (Norf and Weitere 2010). Gradual freeze–thaw incubation decreases the microbial activity in the frozen state to 0.25 % of the initial levels at 4 °C, but activity resumes rapidly reaching >60 % of the initial activity in the thawed state (Sawicka et al. 2010).

Uptake of nitrate by bacteria and algae is strongly dependent on temperature and consistently decreases at temperatures below the optimum. In contrast, ammonium uptake is increased at low temperatures (Reay et al. 1999). Increasing temperature can significantly accelerate the colonization speed and reduce the carrying capacity in particular seasons, e.g. during winter. At the same time, the strongest response to the temperature increase occurs during the highest DOC loadings (Norf et al. 2007). Overall, the response of microbial communities to local temperature increases strongly depends on the seasonal setting, the resource availability and the stage of succession (Norf et al. 2007).

Bacterioplankton production depends on ambient temperature, availability of nutrients and other labile substrates, and on the total DOM contents in natural waters (Ochs et al. 1995; Felip et al. 1996; Vrede 1996, 2005; Morris and Lewis 1992; Wang et al. 1992; Coveney and Wetzel 1995; Elser et al. 1995; Cotner et al. 1997; Simon and Wünsch 1998; Caron et al. 2000; Pomeroy and Wiebe 2001; Vrede et al. 1999). Bacteria in temperate lakes are temperature-dependent up to a certain threshold value, above which other factors regulate their growth (Ochs et al. 1995; Felip et al. 1996). In the mesotrophic Lake Constance it has been found that during most of the year the bacterial community is well adapted to in situ temperatures (ranging from 4 to 23 ºC) in the upper water column, whilst in the deeper strata the bacterial growth is limited by temperature (ranging between 4 and 10 ºC) (Simon and Wünsch 1998). The growth of bacteria that live at low temperatures is stimulated both by increases in temperature and by addition of organic substrates (Pomeroy et al. 1991). Bacterioplankton growth can be limited by inorganic nutrients, by phosphorus (P) and by organic carbon (C), and the limitation effect is observed either for each constituent alone or for variable constituent combinations in both freshwater and marine systems (Vrede 1996, 2005; Morris and Lewis 1992; Wang et al. 1992; Elser et al. 1995; Cotner et al. 1997; Caron et al. 2000; Vrede et al. 1999). Substrate concentrations and temperature intergo very close interactions, and the interactive effects can vary with the temperature regime (Pomeroy and Wiebe 2001). It has been shown that increased temperature can stimulate bacterioplankton production when the in situ temperatures are low, i.e., in the hypolimnion in summer and in the mixed water column in autumn (Vrede 2005). At low temperatures, both the temperature increase and the addition of P (in the hypolimnion in summer) or C (in autumn) had strong effects on bacterioplankton production (Vrede 2005). The interaction between P and temperature is only significant in the epilimnion in summer. At the same time, temperature alone had no effect whilst P alone had a strong effect on bacterioplankton production (Vrede 2005). It is hypothesized that high temperature can accelerate the photoinduced and microbial release of nutrients, labile organic substrates and other products (e.g. H_2O_2, CO_2 and DIC) from algae, phytoplankton or DOM. Such processes take place in both the epilimnion and the hypolimnion and are susceptible to enhance the bacterioplankton production in natural waters.

4.4 Changes in Photosynthetic Processes in Natural Waters

Phytoplankton cells within the euphotic zone utilize photosynthetically active radiation (PAR, 400–700 nm) to drive photosynthesis; at the same time, they are exposed to UV radiation (UVR, 280–400 nm) that can penetrate up to 60 m into the pelagic water column (Smith and Baker 1979). Short-term UV-B exposure can severely inhibit the photosynthetic capability, which can be restored quickly after transfer to low PAR conditions (Jiang and Qiu 2011). Solar UV-A radiation can act as an additional source of energy for the photosynthesis carried out by coastal marine phytoplankton assemblages in tropical areas (Li et al. 2011; Gao et al. 2007a, b), although a similar effect is not observed in pelagic water (Li et al. 2011). Global warming can significantly affect aquatic photosynthesis in different ways, by altering physical and chemical environmental conditions. First, warming of the upper ocean leads to stratification and to shoaling of the upper mixing layer. Phytoplankton cells in the upper mixing layer will be exposed to higher levels of solar UV radiation due to reduced mixing rate and depth. In this context, global warming and ozone depletion can act together to influence the primary producers. On the other hand, where higher contents of chemical constituents result in DOM-rich waters, ocean warming may stimulate photosynthesis by increasing the availability of limiting nutrients. The ongoing ocean acidification following enhanced dissolution of CO_2 may also interact with ocean warming and affect the primary production.

The photoinduced degradation of DOM and OM can produce H_2O_2, CO_2 and DIC (Molot et al. 2005; Johannessen et al. 2007; Mostofa and Sakugawa 2009; Mostofa et al. 2009b; Xie et al. 2004; Clark et al. 2004; Miller and Zepp 1995; Dillon and Molot 1997; Gennings et al. 2001; Johannessen and Miller 2001; Ma and Green 2004). Similarly, microbial degradation of DOM and OM yields for instance H_2O_2, CO_2, DIC, PO_4^{3-}, NH_4^+ and CH_4 (Mostofa and Sakugawa 2009; Ma and Green 2004; Fu et al. 2010; Palenik and Morel 1988; Lovley et al. 1996; Zhang et al. 2004, 2009; Kim et al. 2006; Li et al. 2008). The CO_2

and other compounds that are produced either photolytically (photoproducts) or microbially (microbial products), together with other environmental factors, may then influence photosynthesis (Mostofa et al. 2009a, b; Komissarov 1994, 1995, 2003):

Planktonic algae smaller than 5 μm are major fixers of inorganic carbon in the ocean (Li 1994) and dominate the phytoplankton biomass in post-bloom, stratified oceanic temperate waters (Tarran et al. 2001). Large and small phytoplankton cells have a critical and differential growth dependence on inorganic nutrients (Zubkov and Tarran 2008). UV-stimulated inorganic carbon acquisition is often observed in phytoplankton species (Beardall et al. 2009a, b; Wu and Gao 2009). Regeneration of autochthonous DOM and nutrients (NO_3^-, NO_2^-, PO_4^{3-} and NH_4^+) occurs during the photoinduced and microbial assimilation of algae or phytoplankton,and simultaneously also from the photoinduced degradation of DOM in natural waters (Mostofa et al. 2009a, b; Fu et al. 2005, 2010; Stedmon et al. 2007a, b; Ma and Green 2004; Kim et al. 2006; Li et al. 2008; Zhang et al. 2009; Carrillo et al. 2002; Mallet et al. 1998; Kopáček et al. 2004; Lehmann and Bernasconi 2004). Regeneration of nutrients significantly occurs in waters with high contents of DOM, which can control the uptake of nutrients during the photosynthetic process. Therefore, nutrients usually do not limit photosynthesis, particularly in waters with high contents of DOM. Primary production is enhanced significantly by DOM in the upper 30 m of the water column, where the production increase upon reduction of damaging UV radiation offsets the effects of attenuation of photosynthetically active radiation (PAR) (Arrigo and Brown 1996). At greater depths, where little UV radiation remains, primary production is often reduced due to removal of PAR by DOM (Arrigo and Brown 1996). When CDOM is distributed homogeneously within the euphotic zone, the depth-integrated daily primary productivity within the euphotic zone ($\int_z GPP_{ez}$) is reduced under most bio-optical (i.e. solar zenith angle, Chl *a* and DOM absorption, ozone concentration) and photophysiological (i.e. sensitivity to UV radiation) conditions. In such cases, the predicted reduction in primary production at depth is greater than the enhancement of production at the surface (Arrigo and Brown 1996). The $\int_z GPP_{ez}$ is decreased when DOM or phytoplankton is restricted to near-surface waters (−30 m) and it is enhanced when DOM or phytoplankton is restricted to a very shallow surface layer (−10 m) (Arrigo and Brown 1996). DOM effect on photosynthesis can also be justified from the observation of high primary productivity of phytoplankton biomass (Chl *a*) (Fig. 4a) and from the high photosynthetic carbon fixation rate (Fig. 4b, c) in coastal waters (usually with high contents of DOM) compared to pelagic ones (generally with low contents of DOM) (Li et al. 2011).

Another issue is the sustained photoinduced and/or microbial production of compounds such as H_2O_2, CO_2 and DIC in DOM-rich waters, which can enhance primary production (Mostofa and Sakugawa 2009; Mostofa et al. 2009b; Malkin et al. 2008; Ma and Green 2004; Meriläinen et al. 2001; Komatsu et al. 2007). Algae are significantly produced in some Chinese lakes during the summer stratification period, leading to high production of autochthonous DOM and nutrients (Mostofa et al. 2009b; Fu et al. 2010; Li et al. 2008). Such an effect is the consequence of high DOM contents that provide nutrients and simultaneously attenuate high solar radiation. Model results reveal that the progressive release of dissolved organic nitrogen (DON) in the ocean's upper layer during summer increases primary production by 30–300 %. This leads in turn to an enhancement of the dissolved organic carbon (DOC) production mainly from phytoplankton exudation in the upper layer and to solubilization of POM in the deeper layers (Druon et al. 2010).

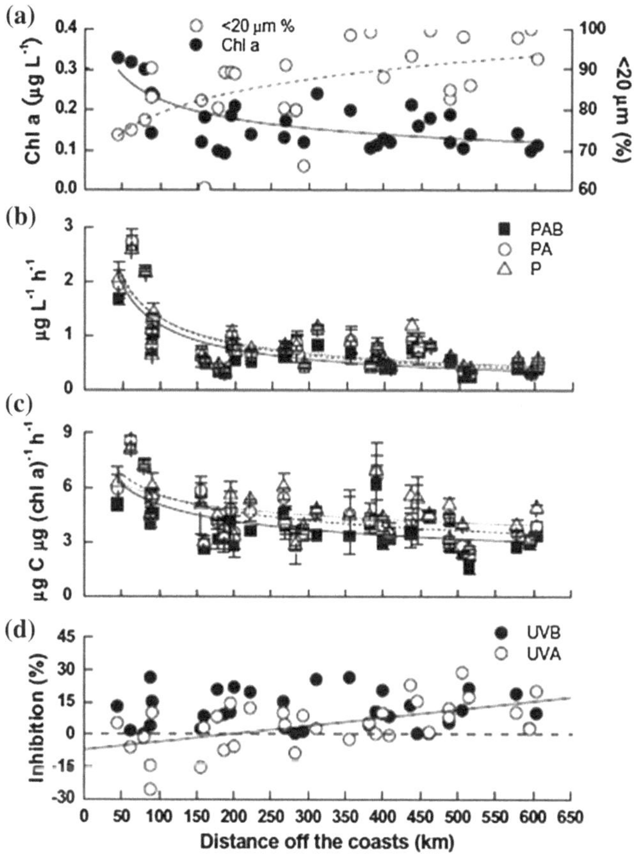

Fig. 4 Variations of biological characteristics in surface seawater from the coasts to pelagic waters. Phytoplankton biomass (Chl *a*, in µg L^{-1}) and piconanoplankton fractions (<20 µm, in %); the fitted lines are power functions ($Y = A\,x^B$, $R^2 = 0.47$ for Chl *a* and 0.32 for piconanofractions) (**a**); (**b**) and (**c**) Photosynthetic carbon fixation rates per volume of seawater (in µg C L^{-1} h^{-1}) (**b**) or based on Chl *a* [in µg C (µg Chl $a)^{-1}$ h^{-1}] (**c**) under PAR (P), PAR + UV-A (PA) or PAR + UV-A + B (PAB); the fitted lines are the power functions ($R^2 = 0.38$–0.63); Photosynthetic inhibition (in %) induced by UV-A or UV-B (**d**) where the solid line indicates the significant relationship between the UV-A-induced inhibition and the distance off the coast ($R^2 = 0.31$, $P < 0.01$, $n = 32$). Vertical bars present the standard deviations ($n = 3$). *Data source* Li et al. (2011)

UV-A does not enhance carbon fixation in pelagic (or oligotrophic) water because the picophytoplankton-dominated assemblages do not efficiently produce UV–absorbing compounds (Fig. 4) (Li et al. 2011; Garcia-Pichel 1994; Raven 1991). The UV-A related inhibition of carbon fixation increases from the coastal to pelagic waters, whereas UV-B impacts uniformly over time and space (Fig. 4d). Under reduced levels of solar radiation with heavy overcast, UV-A radiation enhances photosynthetic carbon fixation by up to 25 % in coastal waters where microplankton is abundant, but such a positive impact is not observed in offshore waters where piconanoplankton prevails (Li et al. 2011).

Water temperature, driven by solar radiation, is one of the crucial physical factors regulating photosynthesis in natural waters (Baulch et al. 2005; Mortain-Bertrand et al. 1988; Doyle et al. 2005; Yoshiyama and Sharp 2006). Primary production (approximately 67 % of variability) is mainly controlled by light availability and temperature. High nutrient concentrations do not stimulate primary production in estuary (Yoshiyama and Sharp 2006). Even the slight oceanic warming during the interglacials would result in increased affinity of active transport by algae and bacteria for nutrients (nitrate, phosphate and silicate) and would effectively increase the available pools of such nutrients in the oceans (Nedwell 1999). This increase in availability of nutrients with higher temperature would be predicted to enhance oceanic primary production and CO_2 drawdown during the interglacials (Nedwell 1999). Such a scenario is consistent with the data from the profiles of $\delta^{13}C$ isotopic ratios in benthic foraminiferan in Southern Ocean sediment cores. Such data suggest in fact increased interglacial oceanic production (Broecker and Peng 1993; Neori and Holm-Hansen 1982). It is also shown that the highest NH_4 concentrations are detected in the colder months when temperature and daily irradiance are lower, but primary production does not increase linearly with ammonium (Yoshiyama and Sharp 2006). Global warming may lengthen the summer season and enhance the water column transparency with modification of the depth of the mixing layer or euphotic zone. Such processes would influence the doses of UV radiation and PAR received by the phytoplankton cells (Malkin et al. 2008; Scully and Lean 1994; Morris et al. 1995; Morris and Hargreaves 1997). The consequence could be a photoinhibition of the cells within the upper layer in sunny days. Enhanced photosynthetic rates of polar phytoplankton are observed in response to increasing temperatures (Broecker and Peng 1993; Neori and Holm-Hansen 1982; Reay et al. 2001; Jacques 1983). This effect ultimately causes an increase in photosynthesis in natural waters, in particular in the deeper layers, because of an enhancement in its duration.

Global warming can also induce an increase in DOM contents in natural waters because of enhanced DOM leaching from terrestrial soils connected to high soil respiration following elevated atmospheric CO_2 concentrations (Porcal et al. 2009). Global warming can enhance the photosynthesis of terrestrial plants becuase of higher atmospheric CO_2 levels, which results in high primary production. The parallel increase of atmospheric temperature would also increase the soil respiration (Porcal et al. 2009; Freeman et al. 2001, 2004; Tranvik and Jasson 2002;

Evans et al. 2005, 2006; Roulet and Moore 2006; de Wit et al. 2007; Monteith et al. 2007; Dorodnikov et al. 2011). The decomposition of soil OM by microbial biomass is significantly increased under elevated atmospheric CO_2 (Dorodnikov et al. 2011; Heath et al. 2005; Lagomarsino et al. 2009; Blagodatskaya et al. 2010, 2011). High turnover rates of soil OM are apparently driven by the increasing activity of soil microorganisms under elevated CO_2 conditions (Blagodatskaya et al. 2010, 2011; Dorodnikov et al. 2009). It is also shown that the increasing activity of soil microorganisms under elevated CO_2 could accelerate the decomposition of older and fresh plant residues (Dorodnikov et al. 2009, 2011; Blagodatskaya et al. 2010; Marhan et al. 2010). In addition, the labile organic carbon released by roots stimulates microbial activity, leading to enhanced degradation of soil OM. This process is known as the 'priming mechanism' (Kuzyakov 2002). Elevated CO_2 can enhance soil organic matter mineralization by 83–218 % in a simulated wetland (Wolf et al. 2007). Therefore, elevated atmospheric CO_2 concentrations could enhance both primary production and soil respiration, inducing increased export of DOC to nearby natural waters (Porcal et al. 2009).

4.5 Changes in the Primary Production and Disorders in Chlorophyll a *Maxima*

Global warming may affect the seasonal patterns of primary production, the chlorophyll concentrations in the surface chlorophyll *a* maxima (SCM) and in the deep chlorophyll *a* maxima (DCM) in natural waters (Letelier et al. 2004; Huisman et al. 2006; Mostofa et al. 2009b; Baulch et al. 2005; Castle and Rodgers 2009; Davis et al. 2009; Hobson and McQuoid 2001). It is estimated that, depending on location, ambient UV radiation can reduce carbon fixation rates up to 65 % in surface waters of the Antarctic region, down to undetectable levels at 36 m (Boucher and Prézelin 1996). A reduction of stratospheric O_3 concentrations by 50 % would further inhibit the near-surface primary production by 8 % and the integrated primary production by 5 % (Boucher and Prézelin 1996). This effect causes the occurrence of subsurface maxima of primary production in the presence of UVR (Boucher and Prézelin 1996). About 67 % of the variability of the observed primary production indicates that estuarine production is mainly controlled by light availability and temperature. In contrast, high nutrient concentrations do not stimulate primary production (Yoshiyama and Sharp 2006).

Global warming will increase the stability of the vertical stratification in large parts of the lakes and oceans, reducing vertical mixing and suppressing the upward flux of nutrients. The effect would be a decrease in primary production (O'Reilly et al. 2003; Huisman et al. 2006). Reduced vertical mixing can generate oscillations and chaos in phytoplankton biomass and species composition of DCM, which is generated by the difference in timescale between the sinking flux of phytoplankton and the upward flux of nutrients (Huisman et al. 2006; Barbiero and Tuchman 2004). The increased stability of the water column due to global warming can thus destabilize phytoplankton dynamics in both DCM and SCM. The actual effect can be different in different types of water. Waters with low contents of DOM (apparently <100 mM C) can yield low contents of photo- and microbial products (H_2O_2, CO_2, DIC) in the euphotic zone, with limited enhancement of productivity. This effect is often found in the oligotrophic regions of the ocean where the nutrient-poor upper layer is made even poorer as a result of enhanced stratification. The phenomenon has a negative impact on net primary production and can produce oceanic 'oligotrophication' as a direct effect of global warming (Sarmento et al. 2010; Falkowski and Oliver 2007; Falkowski and Wilson 1992; Karl et al. 2001; Polovina et al. 2008; Behrenfeld et al. 2006).

A regional decrease in wind velocity in Lake Tanganyika, East Africa has contributed to reduced mixing, decreasing the deep-water nutrient upwelling and entrainment into surface waters (O'Reilly et al. 2003).

Increased stability of the water column may enhance the photoinduced degradation of DOM by combination of high temperature and longer summer season. In waters with high contents of DOM this would lead to the production of high contents of photo- and microbial products (such as H_2O_2, CO_2 and DIC). This process enhances photosynthesis and can result into in high primary production. Phytoplankton or algae productivity in DOM-rich waters would also enhance the production of autochthonous DOM and nutrients (Mostofa et al. 2009b; Stedmon et al. 2007a, b; Malkin et al. 2008; Fu et al. 2010; Li et al. 2008; Zhang et al. 2009; Carrillo et al. 2002; Kopáček et al. 2000, 2004). High production of further DOM and nutrients would severely worsten the quality of waters with high contents of DOM, particularly in lakes, reservoirs, estuaries, coastal waters and in the Arctic and Antarctic regions. Such effects of climate warming may simultaneously promote harmful algal blooms or toxic phytoplankton populations (Davis et al. 2009; Mudie et al. 2002; Richardson and Jorgensen 1996; Hallegraeff 1993; Harvell et al. 1999; Braun and Pfeiffer 2002). The occurrence of cyanobacterial blooms in freshwater has increased over the last few decades all over the world (Xu et al. 2000; Chen et al. 2003; McCarthy et al. 2007).

An increase in dissolved primary production is one of the consequences of the temperature rise in the Southern Ocean (Morán et al. 2006). Similar processes in subarctic lakes are likely to result in higher DOC concentration, bacterial production and respiration, and into emission of CO_2 to the atmosphere (Jansson et al. 2008).

The penetration to significant depths of solar UV radiation can affect arthropods, cyanobacteria, phytoplankton, macroalgae and aquatic plants in both freshwater and marine environments, including Antarctic and Arctic waters (Ballaré et al. 2011; Huisman et al. 2006; Häder et al. 2003, 2007, 2011; Karl et al. 2001; Sinha et al. 2001; Day and Neale 2002; Frenot et al. 2005; Rastogi et al. 2010). Changes in the timing of primary producers, possibly forced by UV-B radiation and temperature increase, would change connectivity in the food web among phytoplankton, zooplankton, crustaceans, amphibians, fish, corals and birds (Kitaysky and Golubova 2000; Morrison et al. 2002; Johannessen and Macdonald 2009; Häder et al. 2007, 2011; Pomeroy and Wiebe 2001).

The primary producers (e.g. phytoplankton cells) tend to be smaller in a warmer ocean (Falkowski and Oliver 2007; Daufresne et al. 2009; Morán et al. 2010). It has also been shown that UV-B influences the CO_2-concentrating mechanism of *M. aeruginosa*, and this cyanobacterium has many adaptive strategies to cope with prolonged UV-B exposure (Jiang and Qiu 2005; Song and Qiu 2007). Enhanced solar UV-A (315–400 nm) and/or UV-B radiation (280–315 nm) can reduce growth and photosynthetic rates, inhibit pigment production, increase permeability of cell membranes, damage proteins or DNA molecules, and even lead to cell death (Jiang and Qiu 2005, 2011; Behrenfeld et al. 1993; Sass et al. 1997; Helbling et al. 2001; Buma et al. 2003; Sobrino et al. 2004; Litchman and Neale 2005; Wu et al. 2005; Agustí and Llabreés 2007; Rath and Adhikary 2007; Pattanaik et al. 2008; Gao et al. 2008). At normal ozone concentrations (i.e. 344 Dobson Units), UV radiation can reduce primary productivity in surface waters by as much as 50 % (Cullen et al. 1992; Holm-Hansen et al. 1993b; Cullen and Neale 1994). A normal level of UV radiation also reduces phytoplankton production by 57 % at a depth of 1 m, while such inhibition decreases to <5 % at 30 m, at 50°S in mid December (Arrigo 1994). Such effects on aquatic organisms might be caused directly by UV radiation and indirectly through high production of $HO^{\bullet}$ in epilimnetic (upper layer) waters. Both effects are able to alter the structural configuration of organisms with release of many organic substances in epilimnetic (surface layer) waters (Mostofa et al. 2009a, b; Sinha et al. 2001; Rastogi et al. 2010; Gauslaa and McEvoy 2005; Lesser 2008; Hylander et al. 2009; Ingalls et al. 2010).

To conclude, global warming may greatly impact primary production, species composition, carbon export, and finally biological activities in the aquatic environment (Huisman et al. 2006; Häder 2011; Häder et al. 2003, 2007; Sinha et al. 2001; Rastogi et al. 2010; Petchey et al. 1999).

4.6 Changes in DOM Dynamics and the Global Carbon Cycle

The increase of DOC concentration in many catchments in Europe and North America might be the concequence of a climate effect (Zepp et al. 2011; Burns et al. 2006; Vuorenmaa et al. 2006; Sobek et al. 2007; Zhang et al. 2010; Freeman et al. 2001, 2004; Evans et al. 2005; Skjelkvåle et al. 2001; Löfgren et al. 2003; Hongve et al. 2004; Worrall et al. 2005; Larsen et al. 2011). An increase of DOC in natural waters because of global warming could be linked to the production of autochthonous DOM by phytoplankton or algae under both photoinduced and microbial-assimilation (Johannessen et al. 2007; Mostofa et al. 2009a, b; Fu et al. 2005, 2010; Stedmon et al. 2007a; Zhang et al. 2009; Biddanda and Benner 1997; Carrillo et al. 2002; Mallet et al. 1998; Lehmann and Bernasconi 2004). Indeed, increasing temperature can increase the release of organic substrates by phytoplankton (Morán et al. 2006; Watanabe 1980; Verity 1981; Zlotnik and Dubinsky 1989). Such phenomena can in turn enhance photosynthesis and primary production, as already explained, particularly in DOM-rich waters.

On the other hand, global warming can affect waters with low contents of DOM in the opposite direction, inhibiting the production of various compounds that ultimately limit photosynthesis and primary production. This effect can proceed either by gradually decreasing the total contents of DOM and nutrients or by reducing the nutrients at equal DOM. The latter scenario can occur because in waters with low contents of DOM, as can be found for instance in Lake Biwa and Lake Baikal, allochthonous DOM usually dominates than autochthonous DOM (Yoshioka et al. 2002; Mostofa et al. 2005). The removal of total organic carbon (total internal and external inputs) is accounted for by respiration (50 %), sedimentation (40 %) and photo-oxidation (10 %) in acidic lakes (Kopáček et al. 2004).

Production of algae or phytoplankton and the related photoinduced and microbial release of new DOM are greatly influenced by several factors, such as high precipitation (Zhang et al. 2010; Freeman et al. 2001; Tranvik and Jasson 2002; Hejzlar et al. 2003), land use changes that induce high transport of DOC from catchments to adjacent surface waters (Worrall et al. 2004; Raymond and Oh 2007), nitrogen deposition (Pregitzer et al. 2004; Findlay 2005), sulfate deposition (Zhang et al. 2010; Evans et al. 2006; Monteith et al. 2007), drought and alteration of hydrologic pathways (Zhang et al. 2010; Hongve et al. 2004; Knorr et al. 2005), and change in total solar UV radiation or increase in temperature due to global warming (Sobek et al. 2007; Zhang et al. 2010; Freeman et al. 2001; Sinha et al. 2001; Rastogi et al. 2010). Higher temperatures in the tropical zone may accelerate bacterial metabolism, causing a larger fraction of incoming organic carbon to be respired (Tranvik et al. 2009). In temperate zones, the additive effect of decreased water availability and increased primary production may enhance organic carbon burial through increased autochthonous production and preservation (Cotner and Biddanda 2002; Downing et al. 2008). A recent study of DOC concentrations in over 7,500 lakes in six continents suggests that DOC export is potentially enhanced by global warming in aquatic ecosystems (Sobek et al. 2007). It has also been shown that total solar radiation and precipitation can account for 49–84 % of the variation in the long-term DOC patterns in various catchments (Zhang et al. 2010). DOC concentrations in Swedish lakes and streams have substantially increased during 1970–1980, despite a reduction in temperature, most likely because of higher precipitation (Tranvik and Jasson 2002).

The uptake of inorganic nitrogen by bacteria during a phytoplankton bloom occurs particularly in lake or coastal waters where the inputs of terrestrial humic substances are much higher (Kirchman et al. 1991; Amon and Benner 1994). Photolytically produced ammonium can be assimilated by bacterial populations, which can enhance the production of autotrophic and heterotrophic biomass in planktonic environments.

Carbon storage in high-latitude peatlands is estimated to represent one-third or more of the global soil carbon pool (Post et al. 1982; Zimov et al. 2006). Warming-induced decomposition of soil organic matter, particularly in arctic and subarctic soils (Anderson 1991), can result in greater transport of allochthonous DOC to adjacent natural waters. Water bodies that usually received low DOC inputs such as alpine lakes or those resulting from glacial retreat may be strongly affected by the altered DOC quality, which causes for instance the replacement of herbs by less productive shrubs (Shaver et al. 2000).

Beneath the sea ice in the Central Basin, relatively high values of pCO_2 have been detected, ranging between 425 and 475 μatm values. Such values are larger than the mean atmospheric one in the Arctic in summertime, suggesting that CO_2 might derive from high rates of bacterial respiration (Semiletov et al. 2007). Ambient partial pressure values of CO_2 ($pCO_2 = 21$–73.5 Pa) are produced during the coral reef metabolism in Eastern Pacific reef sites. Such values are highly variable depending on depth, time, space and upwelling-nonupwelling period (Manzello 2010).

Photoinduced and microbial degradation of DOM and OM is a source of atmospheric greenhouse gases such as CO_2 and CH_4, thereby contributing to global carbon cycle and further global warming (Porcal et al. 2009; Knorr et al. 2005; Davidson and Janssens 2006). Elevated CO_2 enhances DOC supply in peat soils, an effect that is attributed to elevated net primary productivity and increased root exudation of DOC. Enhanced DOC in soil will ultimately leach into aquatic ecosystems (Freeman et al. 2004; Barbiero and Tuchman 2004; Kang et al. 2001; Pastor et al. 2003; Lavoie et al. 2005; Fenner et al. 2007a, b). Global warming could also increase soil respiration (Freeman et al. 2001, 2004; Tranvik and Jasson 2002; Evans et al. 2005, 2006; Roulet and Moore 2006; de Wit et al. 2007; Monteith et al. 2007; Dorodnikov et al. 2011).

Freshwater ecosystems that are presently located across vegetation gradients will experience significant shifts in underwater spectral irradiance. The main reasons are the effects of climate change on catchment vegetation and the export of colored DOM (Pienitz and Vincent 2000). Overall, elevated atmospheric CO_2 concentrations would increase primary production, with a consequent increase of the decomposition of soil OM and an increased export of DOC to nearby natural waters (Porcal et al. 2009). These processes can also contribute to enhance the DOM contents in natural waters.

4.7 Changes in Nutrients Cycle

The mass balance of nutrients (NO_2^-, NO_3^-, NH_4^+ and total P) is linked with the major external inputs (terrestrial and atmospheric deposition), internal sources and transformations (primary and bacterial production, secondary production, photo- and/or microbial-assimilation of algae or phytoplankton and plant debris), photoinduced transformation of both external and internal sources of nutrients, nitrification, sedimentation and outputs in natural waters (Mostofa et al. 2009b; Fu et al. 2005; Minero et al. 2007; Stedmon et al. 2007a, b; Sterner et al. 2008; Ma and Green 2004; Zhang et al. 2004; Kim et al. 2006; Li et al. 2008; Carrillo et al. 2002; Mallet et al. 1998; Kopáček et al. 1995, 2000, 2004; Lehmann and Bernasconi 2004; Schindler 1988, 1994; Carlsson et al. 1993; Urabe 1993; Bushaw et al. 1996; Goldman et al. 1996; Ramm and Scheps 1997; Mack and Bolton 1999; Sterner and Elser 2002; Demott 2003; Xie et al. 2003; Kopáček et al. 2003; Ahlgren et al. 2005). Coastal waters are generally nutrient-rich whereas open oceans are often oligotrophic, thus they are usually less productive due to nutrient limitation (Kolber et al. 1990; Shen 2001; Falkowski et al. 2004; Ho et al. 2008). Nutrient enrichment is a common feature in lakes, estuaries and coastal oceans worldwide, which can be the primary cause of eutrophication from excess algal growth (Yoshiyama and Sharp 2006; Smith 2003). Large amounts of nutrients (NO_3^-, NH_4^+ and total P or PO_4^{3-}) are produced from photoinduced and microbial-assimilations of algae or phytoplankton biomass, as well as by photoinduced degradation of DOM in natural waters (Stedmon et al. 2007a, b; Fu et al. 2010; Kim et al. 2006; Li et al. 2008; Kopáček et al. 2004; Kopáček et al. 2003). Formation of N-containing (NH_4^+, NO_3^-, and NO_2^-) and P-containing inorganic compounds (PO_4^{3-}) may take place upon degradation of dissolved organic nitrogen (DON) and dissolved organic phosphorus (DOP) in the epilimnion of natural waters (Mostofa et al. 2011; Zhang et al. 2004; Kim et al. 2006; Li et al. 2008; Bronk 2002; Vähätalo and Järvinen 2007). DIC is also produced both photolytically and microbially from DOM in natural waters (Granéli et al. 1996; Ma and Green 2004; Miller and Moran 1997; Bertilsson and Tranvik 2000). Seasonal and long-term variations of N-containing compounds in natural waters are influenced by biological processes, in which uptake by algae or phytoplankton and denitrification in bottom water play a major role (Seitzinger et al. 2006). It is therefore suggested that algae or phytoplankton might be the key controlling factors that determine the total content of nutrients in natural waters.

Global warming could cause a loop by significantly increasing the photoinduced degradation of DOM with high production of photoproducts (e.g. CO_2, DIC, H_2O_2), which can enhance photosynthesis and primary production. The effect is a higher production of nutrients, particularly in waters with high DOM contents. Therefore, global warming could lead to a considerable eutrophication of already DOM-rich waters.

The nutrients from external inputs (terrestrial and atmospheric deposition) and sedimentation sources are expected to play a less important role toward primary production. Indeed, high contents of nutrients are observed in waters with high contents of DOM or chlorophyll *a* (Stedmon et al. 2007a, b; Fu et al. 2010; Li et al. 2008; Carrillo et al. 2002; Kopáček et al. 2000, 2004). At the same time, N-compounds are strongly dependent on climatic factors in natural waters (Mitchell et al. 1996; Murdoch et al. 1998; Weyhenmeyer 2008; Hessen et al. 2009). An increase in temperature and changes in the precipitation regime can affect biological processes in soil and water, which are involved into the retention and release of N (Windolf et al. 1996; Khalili and Weyhenmeyer 2009). On the other hand, waters with low contents of either DOM or chlorophyll *a* often show low contents of nutrients that typically remain the same for long periods (Kim et al. 2006; Goldman et al. 1996; Fott et al. 1994; Vyhnálek et al. 1994; Sugiyama et al. 2005). These processes are affected by water temperature, thus climate parameters may exert a major control on nutrient variability in natural waters. Global warming could thus modify the seasonal patterns of nutrient concentrations in the water column of lakes and oceans (Letelier et al. 2004; Huisman et al. 2006). High concentrations of nutrients due to global warming could thus severely deteriorate the quality of DOM-rich waters.

5 Conceptual Model for the Impacts of Global Warming on Key Biogeochemical Processes

From the previous discussion of the 'Impacts of global warming on natural waters' section, it is possible to summarize the key biogeochemical steps that take place in natural waters. First, photoinduced transformation induced by natural sunlight is the primary step for the photo-induced generation of oxidizing species (e.g. H_2O_2 and $HO^{\bullet}$) from DOM and chemical species (such as NO_2^- and NO_3^-). Such processes can drive the photoinduced degradation of DOM in surface waters (Fig. 5), yielding a variety of photoproducts among which H_2O_2, CO_2 and DIC. Second, photosynthesis is enhanced by the occurrence of the cited photoproducts in the euphotic zone, with a subsequent increase of the primary production (algae or phytoplankton). Third, primary productivity further induces photoinduced and microbial assimilations that release autochthonous DOM and nutrients. Fourth, autochthonous DOM and nutrients undergo photoinduced and microbial degradation that further yields H_2O_2, CO_2 and DIC. Global warming can thus favor the photosynthesis and, as a consequence, the primary production.

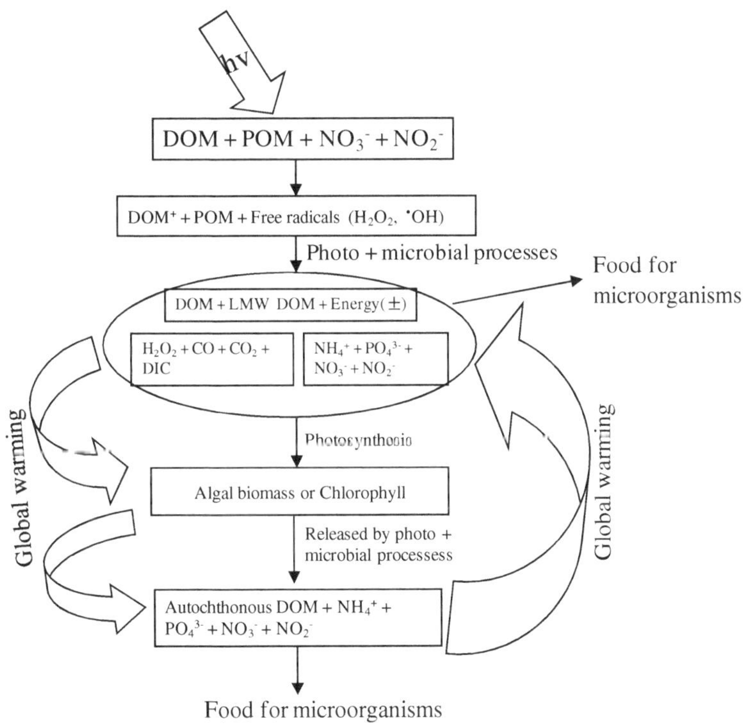

Fig. 5 Conceptual model on photochemical degradation of DOM and its possible effects on key biogeochemical processes in natural waters in response to global warming. *Data source* with few modifications Mostofa et al. (2009b)

The described processes (Fig. 5) are more important in waters having high contents of DOM, naturally or because of pollution, and can lead to further eutrophication of DOM-rich waters. The process would be enhanced by the extension of the summer season and of the euphotic zone that is expected to take place because of global warming. Total contents of DOM and global warming together thus severely affects on foodwebs, primary productivity, and nutrients cycles in freshwater environments which ultimately impact on drinking water quality, sustainable uses of agriculture and industrial purposes as well as the whole water ecosystem (Mostofa et al. 2009b; Li et al. 2008; Larsen et al. 2011; Hessen et al. 1990; Ask et al. 2009; Karlsson et al. 2009). The importance of the described processes would be much lower in DOC-poor (oligotrophic) systems. In such cases other phenomena could play a more important role, and the enhanced photoinduced DOM degradation could even lead to a further oligotrophication of DOC-poor systems.

The photoinduced degradation of DOM generally takes place through redox reactions that can lead to supply (+) or consumption (−) of energy in natural waters. Energy changes (±) also occur during photosynthesis (Komissarov 1994, 1995, 2003). DOM with its content of organic C and N is a thermodynamic anomaly that provides a major source of energy to drive aquatic and terrestrial ecosystems (Wetzel 1984, 1992; Salonen and Vähätalo 1994; Tranvik 1992; Hedges et al. 2000). Therefore, any changes in energy during the photoinduced degradation of DOM are thermodynamically vital for all the living organisms and for the aquatic environments.

The photoinduced degradation of DOM is interlinked free radical production, microbial processes, photosynthesis, autochthonous DOM, nutrients, end photoproducts and their utilization as food for microorganisms in natural waters. A conceptual schematic diagram for the global warming effects on photoinduced and microbial processes of DOM and POM, photoproducts and their importance in the aquatic environment is depicted below (Fig. 5):

Finally, it can be concluded that any changes in nature are absolutely interlinked with other changes in the natural ecosystem.

5.1 Remedial Measures for Controlling Algal Blooms due to Global Warming

The possible remedial measures for controlling the lake algal blooms that are severely affecting the water quality because of global warming are listed below: (i) The total DOM contents in lake waters should be reduced by applying coagulation processes that can reduce the regeneration of photoproducts, microbial products and nutrients. Such measures would thus reduce photosynthesis and, as a consequence, the primary production (algae or phytoplankton) in natural waters. During the algal blooms, algae or phytoplankton should be removed using fine, small-mesh nets. Such a procedure could reduce the further photoinduced and microbial release of DOM and nutrients from primary production. The consequence would be a significant reduction of further photosynthesis and, thus, of primary production. (ii) The sediments in the lake bottom should be removed, which will reduce to pore water's DOM and nutrients and their transport to surface waters, a phenomenon that generally occurs during the overturn period or because of any other physical processes. (iii) Erosion should be reduced in the surrounding soil environments, because erosion can enhance the allochthonous DOM, POM and nutrient contents of natural waters. Erosion can be reduced by proper plantation in the surrounding soils of the watershed.

6 Challenges for Future Research

Global warming is expected to enhance primary production in Chinese lakes and reservoirs, which could severely deteriorate water quality and considerably impact the sustainable use of freshwater resources. Concurrently, DOM contents have been gradually increasing in some European lakes in the last few decades, possibly because of the effect of global warming. Global warming could lead to considerable eutrophication of DOM-rich waters, by gradually increasing the occurrence of autochthonous DOM and nutrients and severely deteriorating the water quality. On the other hand it can affect DOM-poor waters, either inducing a gradual decrease of the total contents of DOM and nutrients or maintaining the same range of DOM levels but causing the nutrients to decline.

The global warming has been found to increase the average Arctic temperatures at almost twice the global average rate in the past 100 years. Therefore, warming is expected to significantly impact the biogeochemical processes of Arctic and Antarctic regions. Therefore, a number of important researches can be distinguished as: (i) Monitoring the contents of DOM and nutrients in natural waters affected by global warming. (ii) Extensively studying the photoinduced and microbial release of autochthonous DOM and nutrients from DOM and algae or phytoplankton in natural waters. (iii) Investigating the photoinduced and microbial release of CO_2, DIC and H_2O_2 from autochthonous DOM, algae or phytoplankton in natural waters. (iv) Studying the photoinduced and microbial release of autochthonous DOM from algae or phytoplankton in Arctic and Antarctic waters. (v) Development of remedial measures for high primary production caused by global warming in waters with high contents of DOM. (vi) Finally, development of remedial measures for low primary productivity caused by global warming in waters with low contents of DOM.

7 Nomenclature

CO_2	Carbon dioxide
DIC	Dissolved inorganic carbon (DIC: dissolved CO_2, H_2CO_3, HCO_3^-, and CO_3^{2-})
DOM	Dissolved organic matter
GHGs	Green house gases

GWPs	Global warming potentials
H_2O_2	Hydrogen peroxide
IPCC	Intergovernmental Panel on Climate Change
Kyr	Kilo year (1000 year)
CH_4	Methane
OM	Organic matter
NH_4^+	Ammonium
NO_2^-	Nitrite
NO_3^-	Nitrate
POM	Particulate organic matter
PO_4^{3-}	Phosphate
UNFCCC	United Nations Framework Convention on Climate Change
UV	Ultraviolet
Yrs	Years

Problems

(1) What is global warming? List the atmospheric greenhouse gases and other constituents which contribute to global warming.
(2) Explain how atmospheric greenhouse gases increase global warming.
(3) Explain how the atmospheric sulfate aerosols could reduce the global warming.
(4) Explain how the stratospheric ozone depletion affects global warming and natural water chemistry.
(5) Explain the contributions of atmospheric greenhouse gases to global warming.
(6) What is radiative forcing by greenhouse gases? Explain how global warming potentials become indicators of global warming.
(7) How are the environmental processes of greenhouse gases emission affected by global warming? Explain how global warming enhances the soil respiration that releases atmospheric greenhouse gases.
(8) Explain the various processes of CO_2 emission during agricultural activities.
(9) How does deforestation affect global warming?
(10) Explain where and why the emissions of CO_2 occur by photoinduced degradation of DOM in natural waters.
(11) What are the possible impacts of global warming on physical, chemical and biological processes in natural waters?
(12) Explain how does global warming affect photoinduced and microbial activities in natural waters.
(13) Explain how does global warming affect photosynthesis and primary productivity in natural waters.
(14) Explain why and how does global warming impact on waters with high contents of DOM.

(15) Explain the global warming impact on the DOM dynamics in natural waters.
(16) What are the sources of nutrients in natural waters and how does the global warming impact the aquatic nutrient dynamics?
(17) Explain the conceptual model for the impact of global warming on key biogeochemical processes in natural waters.

Acknowledgments We thank Dr. Benjamin Chetelat of Institute of Geochemistry, Chinese Academy of Sciences for his helpful assiatnce during the manuscript preparation. This work was financially supported by the Institute of Geochemistry, the Chinese Academy of Sciences, China. This work was partly supported by Xiamen University, China; University Turin, Itali; and Northwest Missouri State University, USA. This chapter acknowledges the reprinted from Soil Biology & Biochemistry 41(4), Feng X, Simpson MJ, Temperature and substrate controls on microbial phospholipid fatty acid composition during incubation of grassland soils contrasting in organic matter quality, 804–812, Copyright (2009), with permission from Elsevier; Li G, Gao K, Gao G, Differential Impacts of Solar UV Radiation on Photosynthetic Carbon Fixation from the Coastal to Offshore Surface Waters in the South China Sea, Photochemistry and Photobiology, Copyright (2011) by Wiley-Blackwell, Inc; Climate Change 2007: Synthesis Report. Contribution of Working Groups I, II and III to the Fourth Assessment Report of the Intergovernmental Panel on Climate Change, Figs. 1.1 and 2.3. IPCC, Geneva, Switzerland; and Climate Change 2001: The Scientific Basis. Contribution of Working Group I to the Third Assessment Report of the Intergovernmental Panel on Climate Change, Table 3. Cambridge University Press.

References

Abril G, Etcheber H, Delille B, Frankignoulle M, Borges AV (2003) Carbonate dissolution in the turbid and eutrophic Loire estuary. Marine Ecol Prog Ser 259:129–138

Abril G, Commarieu MV, Maro D, Fontugne M, Guerin F, Etcheber H (2004) A massive dissolved inorganic carbon release at spring tide in a highly turbid estuary. Geophys Res Lett 31:L09316. doi:101029/2004GL019714

Agustí S, Llabreés M (2007) Solar radiation-induced mortality of marine pico-phytoplankton in the oligotrophic ocean. Photochem Photobiol 83:793–801

Ahlgren J, Tranvik L, Gogool A, Waldebäck M, Markides K, Rydin E (2005) Depth attenuation of biogenic phosphorus compounds in lake sediment measured by ^{31}P NMR. Environ Sci Technol 39:867–872

Ambus P, Robertson G (2006) The Effect of increased N deposition on nitrous oxide, methane and carbon dioxide fluxes from unmanaged forest and grassland communities in Michigan. Biogeochemistry 79:315–337

Amon RMW, Benner R (1994) Rapid cycling of high-molecular-weight dissolved organic matter in the ocean. Nature 369:549–552

Anderson JM (1991) The effects of climate change on decomposition processes in grassland and coniferous forests. Ecol Appl 1:326–347

Andersson AJ, Mackenzie FT (2004) Shallow-water oceans: a source or sink of atmospheric CO2? Front Ecol Environ 2:348–353

Andersson AJ, Mackenzie FT, Ver LM (2003) Solution of shallow-water carbonates: an insignificant buffer against rising atmospheric CO_2. Geology 31:513–516

Archer D (2005) Fate of fossil fuel CO_2 in geologic time. J Geophys Res 110:C09S05. doi:1010 29/2004JC002625

Archer D, Martin P, Buffett B, Brovkin V, Rahmsdorf S, Ganopolski A (2004) The importance of ocean temperature to biogeochemistry. Earth Planet Sci Lett 222:333–348

Arnold SS, Fernandez IJ, Rustad LE, Zibilske LM (1999) Microbial response of an acid forest soil to experimental soil warming. Biol Fertil Soils 30:239–244
Arrigo KR (1994) The Impact of ozone depletion on phytoplankton growth In the Southern Ocean: large-scale spatial and temporal variability. Mar Ecol Prog Ser 114:1–12
Arrigo KR, Brown CW (1996) Impact of chromophoric dissolved organic matter on UV inhibition of primary productivity in the sea SERIES. Mar Ecol Prog Ser 140:207–216
Aselrnann I, Crutzen PJ (1989) Global distribution of natural freshwater wetlands and rice paddies: their net primary productivity, seasonality and possible methane emissions. J Atmos Chem 8:307–358
Ask J, Karlsson J, Persson L, Ask P, Bystrom P, Jansson M (2009) Whole-lake estimates of carbon flux through algae and bacteria in benthic and pelagic habitats of Clearwater lakes. Ecology 90:1923–1932
Bahn M, Reichstein M, Davidson EA, Grunzweig J, Jung M, Carbone MS, Epron D, Misson L, Nouvellon Y, Roupsard O, Savage K, Trumbore SE, Gimeno C, Yuste JC, Tang J, Vargas R, Janssens IA (2010) Soil respiration at mean annual temperature predicts annual total across vegetation types and biomes. Biogeosciences 7:2147–2157
Ballaré CL, Caldwell MM, Flint SD, Robinson SA, Bornman JF (2011) Effects of solar ultraviolet radiation on terrestrial ecosystems. Patterns, mechanisms, and interactions with climate change. Photochem Photobiol Sci 10:226–241
Barbiero RP, Tuchman ML (2004) The deep chlorophyll maximum in Lake Superior. J Great Lakes Res 30(Supplement 1):256–268
Bardgett RD, Kandeler E, Tscherko D, Hobbs PJ, Bezemer TM, Jones TH, Thompson LJ (1999) Below-ground microbial community development in a high temperature world. Oikos 85:193–203
Bastviken D, Cole JJ, Pace ML, Tranvik L (2004) Methane emissions from lakes: dependence of lake characteristics, two regional assessments, and a global estimate. Glob Biogeochem Cycles 18:GB4009. doi:101029/2004GB002238
Bastviken D, Cole JJ, de Bogert MCV (2008) Fates of methane from different lake habitats: connecting wholelake budgets and CH4 emissions. J Geophys Res Biogeosci 113:G02024. doi:101029/2007JG000608
Bates NR, Samuels L, Merlivat L (2001) Biogeochemical and physical factors influencing seawater fCO_2, and air–sea CO_2 exchange on the Bermuda coral reef. Limnol Oceanogr 46:833–846
Baulch HM, Schindler DW et al (2005) Effects of warming on benthic communities in a boreal lake: implications of climate change. Limnol Oceanogr 50:1377–1392
Beardall J, Sobrino C, Stojkovic S (2009a) Interactions between the impacts of ultraviolet radiation, elevated CO_2, and nutrient limitation on marine primary producers. Photochem Photobiol Sci 8:1257–1265
Beardall J, Stojkovic S, Larsen S (2009b) Living in a high CO2 world: impacts of global climate change on marine phytoplankton. Plant Ecol Divers 2:191–205
Behrenfeld M, Hardy J, Gucinski H, Hanneman A, Lee H, Wones A (1993) Effects of ultraviolet-B radiation on primary production along latitudinal transects in the south Pacific Ocean. Mar Environ Res 35:349–363
Behrenfeld MJ et al (2006) Climate-driven trends in contemporary ocean productivity. Nature 444:752–755. doi:101038/nature05317
Bergström I, Mäkelä S, Kankaala, Kortelainen P (2007) Methane efflux from littoral vegetation stands of southern boreal lakes: an upscaled regional estimate. Atmos Environ 41:339–351
Berlett BS, Stadtman ER (1997) Protein oxidation in aging, disease, and oxidative stress. J Biol Chem 272:20313–20316
Bertilsson S, Tranvik LJ (2000) Photochemical transformation of dissolved organic matter in lakes. Limnol Oceanogr 45:753–762
Biasi C, Rusalimova O, Meyer H, Kaiser C, Wanek W, Barsukov P, Junger H, Richter A (2005) Temperature-dependent shift from labile to recalcitrant carbon sources of arctic heterotrophs. Rapid Commun Mass Spectrom 19:1401–1408

Biddanda B, Benner R (1997) Carbon, nitrogen, and carbohydrate fluxes during the production of particulate and dissolved organic matter by marine phytoplankton. Limnol Oceanogr 42:506–518
Bingemer HG, Crutzen PJ (1987) The production of methane from solid wastes. J Geophys Res 92:2181–2187
Blagodatskaya E, Blagodatsky S, Dorodnikov M, Kuzyakov Y (2010) Elevated atmospheric CO_2 increases microbial growth rates in soil: results of three CO_2 enrichment experiments. Glob Chang Biol 16:836–848
Blagodatskaya E, Yuyukina T, Blagodatsky S, Kuzyakov Y, (2011) Turnover of soil organic matter and of microbial biomass under C_3–C_4 vegetation change: consideration of 13C fractionation and preferential substrate utilization. Soil Biol Biochem 43:159–166. doi:101016/jsoilbio201009028
Blokhina O, Virolainen E, Fagerstedt KV (2003) Antioxidants, oxidative damage and oxygen deprivation stress: a review. Ann Bot 91(2):179–194
Blumthaler M, Ambach W (1990) Indication of increasing solar ultraviolet-B radiation flux in alpine region. Science 248:206–208
Borges AV (2005) Do we have enough pieces of the jigsaw to integrate CO_2 fluxes in the coastal ocean? Estuaries 28:3–27
Borges AV, Frankignoulle M (1999) Daily and seasonal variations of the partial pressure of CO_2 in surface seawater along Belgian and southern Dutch coastal areas. J Mar Syst 19:251–266
Borges AV, Frankignoulle M (2002a) Distribution and air–water exchange of carbon dioxide in the Scheldt plume off the Belgian coast. Biogeochemistry 59:41–67
Borges AV, Frankignoulle M (2002b) Distribution of surface carbon dioxide and air–sea exchange in the upwelling system off the Galician coast. Glob Biogeochem Cycles 16:1020. doi:101029/2000GB001385
Borges AV, Ruddick K, Schiettecatte L-S, Delille B (2008) Net ecosystem production and carbon dioxide fluxes in the Scheldt estuarine plume. BMC Ecol 8:15. doi:101186/1472-6785-8-15
Boucher NP, Prézelin BB (1996) Spectral modeling of UV inhibition of in situ Antarctic primary production using a field-derived biological weighting function. Photochem Photobiol 64:407–418
Bouillon S, Frankignoulle M, Dehairs F, Velimirov B, Eiler A, Abril G, Etcheber H, Borges AV (2003) Inorganic and organic carbon biogeochemistry in the Gautami Godavari estuary (Andhra Pradesh, India) during pre-monsoon: the local impact of extensive mangrove forests. Glob Biogeochem Cycles 17:1114. doi:101029/2002GB002026
Bouwman AF (1990) Exchange of greenhouse gases between terrestrial ecosystems and the atmosphere. Soils and the greenhouse effect. Wiley, Chichester
Bowden R, Castro M, Melillo J, Steudler P, Aber J (1993) Fluxes of greenhouse gases between soils and the atmosphere in a temperate forest following a simulated hurricane blowdown. Biogeochemistry 21:61–71
Bozec Y, Thomas H, Elkalay K, De Baar H (2005) The continental shelf pump in the North Sea—evidence from summer observations. Mar Chem 93:131–147
Bozec Y, Thomas H, Schiettecatte L-S, Borges AV, Elkalay K, De Baar HJW (2006) Assessment of the processes controlling the seasonal variations of dissolved inorganic carbon in the North Sea. Limnol Oceanogr 51:2746–2762
Bradford MA, Davies CA, Frey SD, Maddox TR, Melillo JM, Mohan JE, Reynolds JF, Treseder KK, Wallenstein MD (2008) Thermal adaptation of soil microbial respiration to elevated temperature. Ecol Lett 11:1316–1327
Brandt LA, Bohnet C, King JY (2009) Photochemically induced carbon dioxide production as a mechanism for carbon loss from plant litter in arid ecosystems. J Geophys Res 114(G2):G02004
Brasse S, Reimer A, Seifert R, Michaelis W (1999) The influence of intertidal mudflats on the dissolved inorganic carbon and total alkalinity distribution in the German Bight, southeastern North Sea. J Sea Res 42:93–103
Braun A, Pfeiffer T (2002) Cyanobacterial blooms as the cause of a Pleistocene large mammal assemblage. Paleobiology 28:139–154

Broecker WS, Peng T-H (1993) What caused the glacial to interglacial CO_2 change? In: Heimann M (ed) The global carbon cycle. Springer, Berlin, pp 95–115
Bronk DA (2002) Dynamics of DON. In: Hansell DA, Carlson CA (eds) Biogeochemistry of marine dissolved organic matter. Academic Press, San Diego, pp 153–249
Buffett B, Archer D (2004) Global inventory of methane clathrate: sensitivity to changes in the deep ocean. Earth Planet Sci Lett 227:185–199
Buma AGJ, Boelen P, Jeffrey WH (2003) UVR-induced DNA damage in aquatic organisms. In: Helbling EW, Zagarese HE (eds) UV effects in aquatic organisms and ecosystems. The Royal Society of Chemistry, Cambridge, pp 291–327
Burd BJ, Macdonald RW, Johannessen SC, van Roodselaar A (2008a) Responses of subtidal benthos of the Strait of Georgia to ambient sediment conditions and natural and anthropogenic depositions. Mar Environ Res 66:S62–S69
Burd BJ, Barnes PAG, Wright CA, Thomson RE (2008b) A review of subtidal benthic habitats and invertebrate biota of the Strait of Georgia, British Columbia. Mar Environ Res 66:S3–S38
Burns DA, McHale MR, Driscoll CT, Roy KM (2006) Response of surface water chemistry to reduced levels of acid precipitation: comparison of trends in two regions of New York, USA. Hydrol Process 20:1611–1627
Bushaw KL, Zepp RG, Tarr MT, Schulz-Jander D, Bourbonniere RA, Hodson RE, Miller WL, Bronk DA, Moran MA (1996) Photochemical release of biologically available nitrogen from aquatic dissolved organic matter. Nature 381:404–407
Cabrera S, López M, Tartarotti B (1997) Phytoplankton and zooplankton response to ultraviolet radiation in a high altitude Andean lake: short- versus long-term effects. J Plankton Res 19:1565–1582
Cai WJ, Wang Y (1998) The chemistry, fluxes, and sources of carbon dioxide in the estuarine waters of the Satilla and Altamaha Rivers, Georgia. Limnol Oceanogr 43:657–668
Cai WJ, Wang ZA, Wang Y (2003) The role of marsh-dominated heterotrophic continental margins in transport of CO_2 between the atmosphere, the land–sea interface and the ocean. Geophys Res Lett 30:1849. doi:101029/2003GL017633
Cai W-J, Dai MH, Wang YC (2006) Air-sea exchange of carbon dioxide in ocean margins: a province-based synthesis. Geophys Res Lett 33:L12603. doi:101029/2006GL026219
Canadell JG, Le Qúeŕe C, Raupach MR, Field CB, Buitenhuis ET, Ciais P, Conway TJ, Gillett NP, Houghton RA, Marland G (2007) Contributions to accelerating atmospheric CO_2 growth from economic activity, carbon intensity, and efficiency of natural sinks. Proc Natl Acad Sci U S A 104:18353–18354
Carlsson P, Segatto AZ, Granéli E (1993) Nitrogen bound to humic matter of terrestrial origin - a nitrogen pool for coastal phytoplankton? Mar Ecol Prog Ser 97:105–116
Caron DA, Lin Lim E, Sanders RW, Dennet MR, Berninger U-G (2000) Responses of bacterioplankton and phytoplankton to organic carbon and inorganic nutrient additions in contrasting oceanic ecosystems. Aquat Microb Ecol 22:175–184
Carrillo P, Medina-Sánchez JM, Villar-Argaiz M (2002) The interaction of phytoplankton and bacteria in a high mountain lake: importance of the spectral composition of solar radiation. Limnol Oceanogr 47:1294–1306
Castle JW, Rodgers JH Jr (2009) Hypothesis for the role of toxin-producing algae in Phanerozoic mass extinctions based on evidence from the geologic record and modern environments. Environ Geosci 16:1–23
Cavigelli MA, Robertson GP (2000) The functional significance of denitrifier community composition in a terrestrial ecosystem. Ecology 81:1402–1414
Charlson RJ, Lovelock JE, Andrae MO, Warren SG (1989) Sulphate aerosols and climate. Nature 340:437–438
Chen CTA, Borges AV (2009) Reconciling opposing views on carbon cycling in the coastal ocean: continental shelves as sinks and near-shore ecosystems as sources of atmospheric CO_2. Deep Sea Res II 56:578–590
Chen YW, Qin BQ, Teubner K, Dokulil MT (2003) Long-term dynamics of phytoplankton assemblages: microcystis-domination in Lake Taihu, a large shallow lake in China. J Plankton Res 25:445–453

Cicerone RJ, Oremiand RS (1988) Biogeochemical aspects of atmospheric methane. Glob Biogeochem Cycles 2:299–327
Clark CD, Hiscock WT, Millero FJ, Hitchcock G, Brand L, Miller WL, Ziolkowski L, Chen RF, Zika RG (2004) CDOM distribution and CO_2 production on the Southwest Florida Shelf. Mar Chem 89:145–167
Clough SA, Iacono MJ (1995) Line-by-line calculations of atmospheric fluxes and cooling rates: II. Application to carbon dioxide, ozone, methane, nitrous oxide, and the halocarbons. J Geophys Res 100(D8):16519–16535. doi:101029/95JD01386
Cole JJ, Caraco NF (2001) Carbon in catchments: connecting terrestrial carbon losses with aquatic metabolism. Mar Freshw Res 52:101–110. doi:101071/MF00084
Cole CV et al (1997) Global estimates of potential mitigation of greenhouse gas emissions by agriculture. Nutr Cycl Agroecosyst 49:221–228. doi:101023/A:1009731711346
Cotner JB, Biddanda BA (2002) Small players, large role: microbial influence on biogeochemical processes in pelagic aquatic ecosystems. Ecosystems 5:105–121
Cotner JB, Ammerman JW, Peele ER, Bentzer E (1997) Phosphorus-limited bacterioplankton growth in the Sargasso Sea. Aquat Microb Ecol 13:141–149
Coveney MF, Wetzel RG (1995) Biomass, production, and specific growth rate of bacterioplankton and coupling to phytoplankton in an oligotrophic lake. Limnol Oceanogr 40:1187–1200
Crutzen PJ (1991) Methane's sinks and sources. Nature 350:380–381
Crutzen PJ (1992) Ultraviolet on the increase. Nature 356:104–105
Crutzen EJ, Andreae MO (1990) Biomass burning in the tropics: impact on atmospheric chemistry and biogeochemical cycles. Science 250:1669–1677
Crutzen PJ, Aselmann I, Seiler W (1986) Methane production by domestic animals, wild ruminants, other herbivorous fauna and humans. Tellus 38B:271–284
Cullen JC, Neale PJ (1994) Ultraviolet radiation, ozone depletion, and marine photosynthesis. Photosynth Res 39:303–320
Cullen JC, Neale PJ, Lesser MP (1992) Biological weighting function for the inhibition of phytoplankton photosynthesis by ultraviolet radiation. Science 258:646–650
Daufresne M, Lengfellner K, Sommer U (2009) Global warming benefits the small in aquatic ecosystems. Proc Natl Acad Sci U S A 106:12788–12793. doi:101073/pnas0902080106
Davidson EA, Janssens IA (2006) Temperature sensitivity of soil carbon decomposition and feedbacks to climate change. Nature 440:165–173
Davis TW, Berry DL, Boyer GL, Gobler CJ (2009) The effects of temperature and nutrients on the growth and dynamics of toxic and non-toxic strains of Microcystis during cyanobacteria blooms. Harmful Algae 8:715–725
Dawson JJC, Bakewell C, Billett MF (2001) Is in-stream processing an important control on spatial changes in carbon fluxes in headwater catchments? Sci Total Environ 265:153–167. doi:101016/S0048-9697(00)00656-2
Day TA, Neale PJ (2002) Effects of UV-B radiation on terrestrial and aquatic primary producers. Annu Rev Ecol Syst 33:371–396
de Wit H, Flory JD, Acheson A, McCloskey M, Manuck SB (2007) IQ and nonplanning impulsivity are independently associated with delay discounting in middle-aged adults. Pers Individ Dif 42:111–121
del Giorgio PA, Williams PJlB (2005) The global significance of respiration in aquatic ecosystems: from single cells to the biosphere. In: del Giorgio PA, Williams P (eds) Respiration in aquatic ecosystems. Academic Press, New York, pp 267–316
del Giorgio PA, Cole JJ, Caraco NF, Peters RH (1999) Linking planktonic biomass and metabolism to net gas fluxes in northern temperate lakes. Ecology 80:1422–1431
del Giorgo PA, Duarte CM (2002) Respiration in the open ocean. Nature 420:379–384. doi:101038/nature01165
Demott WR (2003) Implications of element deficits for zooplankton growth. Hydrobiologia 491:177–184
den Elzen MGJ, Swart RJ, Rotmans J (1992) Strengthening the Montreal protocol: does it cool down the greenhouse? Sci Total Environ 113:229–250

Detwiler RR, Hall CAS (1988) Tropical forests and the global carbon cycle. Science 239:42–47
Dillon PJ, Molot LA (1997) Dissolved organic and inorganic carbon mass balances in central Ontario lakes. Biogeochemistry 36:29–42
Doney SC, Fabry VJ, Feely RA, Kleypas JA (2009) Ocean acidification: the other CO_2 problem. Ann Rev Mar Sci 1:169–192
Dorodnikov M, Blagodatskaya E, Blagodatsky S, Marhan S, Fangmeier A, Kuzyakov Y (2009) Stimulation of microbial extracellular enzyme activities by elevated CO_2 depends on soil aggregate size. Glob Chang Biol 15:1603–1614
Dorodnikov M, Kuzyakov Y, Fangmeier A, Wiesenberg GLB (2011) C and N in soil organic matter density fractions under elevated atmospheric CO_2: turnover vs stabilization. Soil Biol Biochem 43:579–589
Downing JA, Cole JJ, Middelburg JJ, Striegl RG, Duarte CM, Kortelainen P, Prairie YT, Laube KA (2008) Sediment organic carbon burial in agriculturally eutrophic impoundments over the last century. Glob Biogeochem Cycles 22:GB1018. doi:10.1029/2006GB002854
Doyle SA, Saros JE, Williamson CE (2005) Interactive effects of temperature and nutrient limitation on the response of alpine phytoplankton growth to ultraviolet radiation. Limnol Oceanogr 50:1362–1367
Druon JN, Mannino A, Signorini S, McClain C, Friedrichs M, Wilkin J, Fennel K (2010) Modeling the dynamics and export of dissolved organic matter in the Northeastern U.S. continental shelf. Estuar Coast Shelf Sci 88:488–507
Duff KE, Laing TE, Smol JP, Lean DRS 91999) Limnological characteristics of lakes located across arctic treeline in northern Russia. Hydrobiologia 391:205–222
Eliseev A, Mokhov I, Karpenko A (2007) Influence of direct sulfate-aerosol radiative forcing on the results of numerical experiments with a climate model of intermediate complexity Izvestiya. Atmos Ocean Phys 43:544–554
Elser JJ, Stable BL, Hassett PR (1995) Nutrient limitation of bacterial growth and rates of bacterivory in lakes and oceans: a comparative study. Aquat Microb Ecol 9:105–110
Epp RG, Erickson DJ, Paul ND, Sulzberger B (2007) Interactive effects of solar UV radiation and climate change on biogeochemical cycling. Photochem Photobiol Sci 6:286–300
Erickson Iii DJ, Zepp RG, Atlas E (2000) Ozone depletion and the air-sea exchange of greenhouse and chemically reactive trace gases. Chemosph Glob Chang Sci 2:137–149
Evans CD, Monteith DT, Cooper DM (2005) Long-term increases in surface water dissolved organic carbon: observations, possible causes and environmental impacts. Environ Pollut 137:55–71
Evans CD, Chapman PJ, Clark JM, Monteith DT, Cresser MS (2006) Alternative explanations for rising dissolved organic carbon export from organic soils. Glob Chang Biol 12:2044–2053
Fagan KE, Mackenzie FT (2007) Air-sea CO_2 exchange in a subtropical estuarine-coral reef system, Kaneohe Bay, Oahu, Hawaii. Marine Chem 106:174–191
Fahey TJ, Siccama TG, Driscoll CT, Likens GE, Campbell J, Johnson CE, Battles JJ, Aber JD, Cole JJ, Fisk MC, Groffman PM, Hamburg SP, Holmes RT, Schwarz PA, Yanai RD (2005) The biogeochemistry of carbon at Hubbard Brook. Biogeochemistry 75:109–176
Falkowski PG, Oliver MJ (2007) Mix and match: how climate selects phytoplankton. Nat Rev Microbiol 5:813–819. doi:101038/nrmicro1751
Falkowski PG, Oliver MJ (2008) Diatoms in a future ocean-stirring it up: reply from Falkowski and Oliver. Nat Rev Microbiol 6:407
Falkowski PG, Wilson C (1992) Phytoplankton productivity in the North Pacific Ocean since 1900 and implications for absorption of anthropogenic CO_2. Nature 358:741–743. doi:101038/358741a0
Falkowski PG, M Koblížek M, Gorbunov M, Kolber Z (2004) Development and application of variable chlorophyll fluoresence techniques in marine ecosystems. In: Papageorgiou GC, Govindjee (eds) Chlorophyll a fluorescence: a signature of photosynthesis. Springer, The Netherlands, pp 757–778
Farias J, Rossetti GH, Albizzati ED, Alfano OM (2007) Solar degradation of formic acid: temperature effects on the photo-Fenton reaction. Ind Eng Chem Res 46:7580–7586

Felip M, Pace ML, Cole JJ (1996) Regulation of planktonic bacterial growth rates: the effects of temperature and resources. Microb Ecol 31:15–28

Feng X, Simpson MJ (2008) Temperature responses of individual soil organic matter components. J Geophys Res 113:G03036. doi:101029/2008JG000743

Feng X, Simpson MJ (2009) Temperature and substrate controls on microbial phospholipid fatty acid composition during incubation of grassland soils contrasting in organic matter quality. Soil Biol Biochem 41:804–812

Feng XJ, Simpson AJ, Schlesinger WH, Simpson MJ (2010) Altered microbial community structure and organic matter composition under elevated CO2 and N fertilization in the duke forest. Glob Chang Biol 16:2104–2116

Fenner N, Freeman C, Lock MA, Harmens H, Reynolds B, Sparks T (2007a) Interactions between elevated CO_2 and warming could amplify DOC exports from peatland catchments. Environ Sci Tech 41:3146–3152

Fenner N, Ostle NJ, McNamara N, Sparks T, Harmens H, Reynolds B, Freeman C (2007b) Elevated CO2 effects on peatland plant community carbon dynamics and DOC production. Ecosystems 10:635–647

Ferguson PR, Veizer J (2007) Coupling of water and carbon fluxes via the terrestrial biosphere and its significance to the Earth's climate system. J Geophys Res 112:D24S06. doi:10.1029/2007JD008431

Findlay SEG (2005) Increased carbon transport in the Hudson River: unexpected consequence of nitrogen deposition? Front Ecol Environ 3:133–137

Finlay JC (2003) Controls of streamwater dissolved inorganic carbon dynamics in a forested watershed. Biogeochemistry 62:231–252. doi:101023/A:1021183023963

Fisher DA, Hales CH, Wang W-C, Ko MKW, Sze ND (1990) Model calculations of the relative effects of CFCs and their replacements on global warming. Nature 344:513–516

Flessa H, Ruser R, Dörsch P, Kamp T, Jimenez M, Munch J, Beese F (2002) Integrated evaluation of greenhouse gas emissions (CO_2, CH_4, N_2O) from two farming systems in southern Germany. Agric Ecosyst Environ 91(1–3):175–189

Forster P, Joshi M (2005) The role of halocarbons in the climate change of the troposphere and stratosphere. Clim Chang 71:249–266

Fott J, Pražáková M, Stuchlík I, Stuchlíková Z (1994) Acidification of lakes in Šumava (Bohemia) and in the High Tatra Mountains (Slovakia). Hydrobiologia 274:37–47

Frankignoulle M, Borges AV (2001) The European continental shelf as a significant sink for atmospheric carbon dioxide. Glob Biogeochem Cycles 15:569–576

Frankignoulle M, Copin-Montegut G, Pichon M, Gattuso JP, Biondo R, Bourge I (1996) Carbon fluxes in coral reefs II Eulerian study of inorganic carbon dynamics and measurement of air–sea CO_2 exchanges. Mar Ecol Prog Ser 145:123–132

Frankignoulle M, Abril G, Borges A, Bourge I, Canon C, Delille B, Libert E, Theate JM (1998) Carbon dioxide emission from European estuaries. Science 282:434–436

Freeman C, Ostle N, Kang H (2001) An enzymatic 'latch' on a global carbon store. Nature 409:149

Freeman C, Fenner N, Ostle NJ, Kang H, Dowrick DJ, Reynolds B, Lock MA, Sleep D, Hughes S, Hudson J (2004) Export of dissolved organic carbon from peatlands under elevated carbon dioxide levels. Nature 430:195–198

Freney JR (1997) Emission of nitrous oxide from soils used for agriculture. Nutr Cycl Agroecosyst 49:1–6

Frenot Y, Chown SL, Whinam J, Selkirk PM, Convey P, Skotnicki M, Bergstrom DM (2005) Biological invasions in the Antarctic: extent, impacts and implications. Biol Rev 80:45–72

Frey SD, Drijber R, Smith H, Melillo J (2008) Microbial biomass, functional capacity, and community structure after 12 years of soil warming. Soil Biol Biochem 40:2904–2907

Friederich GE, Walz PM, Burczynski MG, Chavez FP (2002) Inorganic carbon in the central California upwelling system during the 1997–1999 El Nino–La Nina event. Prog Oceanogr 54:185–203

Friedlingstein P, Dufresne JJ, Cox PM, Rayner P (2003) How positive is the feedback between climate change and the carbon cycle? Tellus 55B:692–700

Fu F-X, Zhang Y, Leblanc K, Sañudo-Wilhelmy SA, Hutchins DA (2005) The biological and biogeochemical consequences of phosphate scavenging onto phytoplankton cell surfaces. Limnol Oceanogr 50:1459–1472

Fu P, Mostofa KMG, Wu FC, Liu CQ, Li W, Liao H, Wang L, Wang J, Mei Y (2010) Excitation-emission matrix characterization of dissolved organic matter sources in two eutrophic lakes (Southwestern China Plateau). Geochem J 44:99–112

Gao K, Wu YP, Li G, Wu HY, Villafañe VE, Helbling EW (2007a) Solar UV radiation drives CO_2 fixation in marine phytoplankton: a double-edged sword. Plant Physiol 144:54–59

Gao K, Li G, Helbling EW, Villafañe VE (2007b) Variability of UVR effects on photosynthesis of summer phytoplankton assemblages from a tropical coastal area of the South China Sea. Photochem Photobiol 83:802–809

Gao K, Li P, Watanabe T, Helbling EW (2008) Combined effects of ultraviolet radiation and temperature on morphology, photosynthesis, and DNA of *Arthrospira* (*spirulina*) *platensis* (*Cyanophyta*). J Phycol 44:777–786

Garcia-Pichel F (1994) A model for internal self-shading in planktonic organisms and its implications for the usefulness of ultraviolet sunscreen. Limnol Oceanogr 39:1704–1717

Gattuso JP, Pichon M, Delesalle B, Frankignoulle M (1993) Community metabolism and air–sea CO_2 fluxes in a coral-reef ecosystem (Moorea, French-Polynesia). Mar Ecol Prog Ser 96:259–267

Gattuso JP, Payri CE, Pichon M, Delesalle B, Frankignoulle M (1997) Primary production, calcification, and air–sea CO_2 fluxes of a macroalgal-dominated coral reef community (Moorea, French Polynesia). J Phycol 33:729–738

Gauslaa Y, McEvoy M (2005) Seasonal changes in solar radiation drive acclimation of the sun-screening compound parietin in the lichen Xanthoria parietina. Basic Appl Ecol 6:75–82

Gennings C, Molot LA, Dillon PJ (2001) Enhanced photochemical loss of DOC in acidic waters. Biogeochemistry 52:339–354

Goldman CR, Elser JJ, Richards RC, Reuters JE, Priscu JC, Levin AL (1996) Thermal stratification, nutrient dynamics, and phytoplankton productivity during the onset of spring phytoplankton growth in Lake Baikal, Russia. Hydrobiologia 331:9–24

Goyet C, Eischeid G, McCue SJ, Bellerby RGJ, Millero FJ O'Sullivan DW (1998) Temporal variations of pCO2 in surface seawater of the Arabian Sea in 1995. Deep Sea Res Pt 1 Oceanogr Res Pap 45:609–623

Granéli W, Lindell M, Tranvik L (1996) Photooxidative production of dissolved inorganic carbon in lakes of different humic content. Limnol Oceanogr 41:698–706

Granéli W, Lindell M, Marcal De Farria B, De Assis Esteves F (1998) Photoproduction of dissolved inorganic carbon in temperate and tropical lakes-dependence on wavelength band and dissolved organic carbon concentration. Biogeochemistry 43:175–195

Gregory JM, Ingram WJ, Palmer MA, Jones GS, Stott PA, Thorpe RB, Lowe JA, Johns TC, Williams KD (2004a) A new method for diagnosing radiative forcing and climate sensitivity. Geophys Res Lett 31:L03205. doi:101029/2003GL018747

Gregory JM, Huybrechts P, Raper SCB (2004b) Threatened loss of the Greenland ice-sheet. Nature 428:616

Häder D-P (2011) Does enhanced solar UV-B radiation affect marine primary producers in their natural habitats? Photochem Photobiol 87:263–266

Häder D-P, Kumar HD, Smith RC, Worrest RC (2003) Aquatic ecosystems: effects of solar ultraviolet radiation and interactions with other climatic change factors. Photochem Photobiol Sci 2:39–50

Häder D-P, Kumar HD, Smith RC, Worrest RC (2007) Effects of solar UV radiation on aquatic ecosystems and interactions with climate change. Photochem Photobiol Sci 6:267–285

Häder D-P, Helbling EW, Williamson CE, Worrest RC (2011) Effects of UV radiation on aquatic ecosystems and interactions with climate change. Photochem Photobiol Sci 10:242–260

Hajje N, Jaffé R (2006) Molecular characterization of Cladium peat from the Florida Everglades: biomarker associations with humic fractions. Hydrobiologia 569:99–112

Halac S, Felip M, Camarero L, Sommaruga-Wögrath S, Psenner R, Catalán J, Sommaruga R (1997) An in situ enclosure experiment to test the solar UV-B impact on microplankton in a high-altitude mountain lake I Lack of effect on phytoplankton species composition and growth. J Plankton Res 19:1671–1686

Hales B, Takahashi T, Bandstra L (2005) Atmospheric CO_2 uptake by a coastal upwelling system. Glob Biogeochem Cycles 19:GB1009. doi:101029/2004GB002295

Hallegraeff GM (1993) A review of harmful algal blooms and their apparent global increase. Phycologia 32:79–99

Hansen J, Sato M (2004) Greenhouse gas growth rates. Proc Natl Acad Sci 101:16109–16114

Harrison EF, Minnis P, Barkstrom BR, Ramanathan V, Cess RD, Gibson GG (1990) Seasonal variation of cloud radiative forcing derived from the Earth Radiation Budget Experiment. J Geophys Res 95(D11):18687–18703. doi:101029/JD095iD11p18687

Hartmann DL, Wallace JM, Limpasuvan V, Thompson DWJ, Holton JR (2000) Can ozone depletion and global warming interact to produce rapid climate change? Proc Nat Acad Sci U S A 97:1412–1417

Harvell CD et al (1999) Emerging marine diseases-climate links and anthropogenic factors. Science 285:1505–1510

Hatch CD, Gierlus KM, Zahardis J, Schuttlefield J, Grassian VH (2009) Water uptake of humic and fulvic acid: measurements and modelling using single parameter Köhler theory. Environ Chem 6:380–388

Heath J, Ayres E, Possell M, Bardgett RD, Black HIJ, Grant H, Ineson P, Kerstiens G (2005) Rising atmospheric CO2 reduces sequestration of root derived soil carbon. Science 309:1711–1713

Hedges JI, Eglinton G, Hatcher PG, Kirchman DL, Arnosti C, Dereenne S, Evershed RP, Kögel-Knabner I, de Leeuw JW, Littke R, Michaelis W, Rullkötter J (2000) The molecularly-uncharacterized component of nonliving organic matter in natural environments. Org Geochem 31:945–958

Hejzlar J, Dubrovský M, Buchtele J, Růžička M (2003) The apparent and potential effects of climate change on the inferred concentration of dissolved organic matter in a temperate stream (the Malse River, south Bohemia). Sci Total Environ 310:143–152

Helbling EW, Villafane V, Ferrario M, Holm-Hansen O (1992) Impact of natural ultraviolet radiation on rates of photosynthesis and on specific marine phytoplankton species. Mar Ecol Prog Ser 80:89–100

Helbling EW, Buma AGJ, de Boer MK, Villafañe VE (2001) In situ impact of solar ultraviolet radiation on photosynthesis and DNA in temperate marine phytoplankton. Mar Ecol Prog Ser 211:43–49

Hessen DO, Andersen T, Lyche A (1990) Carbon metabolism in a humic lake—pool sizes and cycling through zooplankton. Limnol Oceanogr 35:84–99

Hessen DO, Andersen T, Larsen S, Skjelkvale BL, de Wit HA (2009) Nitrogen deposition, catchment productivity, and climate as determinants of lake stoichiometry. Limnol Oceanogr 54:2520–2528

Hiriart-Baer VP, Smiith REH (2005) The effect of ultraviolet radiation on freshwater planktonic primary production: the role of recovery and mixing processes. Limnol Oceanogr 50(5):1352–1361

Ho AYT, Xu J, Yin K, Yuan X, He L, Jiang Y, Lee JHW, Anderson DM, Harrison PJ (2008) Seasonal and spatial dynamics of nutrients and phytoplankton biomass in Victoria Harbour and its vicinity before and after sewage abatement. Mar Pollut Bull 57:313–324

Hobson LA, McQuoid MR (2001) Pelagic diatom assemblages are good indicators of mixed water intrusions into Saanich Inlet, a stratified fjord in Vancouver Island. Mar Geol 174:125–138

Hofmann DJ, Butler JH, Tans PP (2009) A new look at atmospheric carbon dioxide. Atmos Environ 43:2084–2086

Holm-Hansen O, Helbling EW, Lubin D (1993a) Ultraviolet radiation in Antartica: inhibition of primary production. Photochem Photobiol 58:567–570

Holm-Hansen O, Lubin D, Helbling EW (1993b) Ultraviolet radiation and its effects on organisms in aquatic environments. In: Young A (ed) Environmental UV photobiology. Plenum Press, New York, pp 379–425
Hongve D, Riise G, Kristiansen JF (2004) Increased colour and organic acid concentrations in Norwegian forest lakes and drinking water—a result of increased precipitation? Aquat Sci 66:231–238
Hope D, Palmer SM, Billett MF, Dawson JJ (2001) Carbon dioxide and methane evasion from a temperate peatland stream. Limnol Oceanogr 46:847–857. doi:104319/lo20014640847
Hoppe HG, Breithaupt P, Walther K, Koppe R, Bleck S, Sommer U, Jürgens K (2008) Climate warming in winter affects the coupling between phytoplankton and bacteria during the spring bloom: a mesocosm study. Aquat Microb Ecol 51:105–115. doi:103354/ame01198
Hoppema JMJ (1990) The distribution and seasonal variation of alkalinity in the Southern Bight of the North Sea and in the western Wadden Sea. Neth J Sea Res 26:11–23
Hoppema JMJ (1991) The seasonal behaviour of carbon dioxide and oxygen in the coastal North Sea along the Netherlands. Neth J Sea Res 28:167–179
Houghton RA (1991) Tropical deforestation and atmospheric carbon dioxide. Clim Chang 19:99–118
Huisman JP, Pham Thi NN, Karl DM, Sommeijer B (2006) Reduced mixing generates oscillations and chaos in the oceanic deep chlorophyll maximum. Nature 439:322–325
Huttunen JT, Alm J, Liikanen A, Juutinen S, Larmola T, Hammar T, Silvola J, Martikainen PJ (2003a) Fluxes of methane, carbon dioxide and nitrous oxide in boreal lakes and potential anthropogenic effects on the aquatic greenhouse gas emissions. Chemosphere 52:609–621. doi:101016/S0045-6535(03)00243-1
Huttunen JT, Juutinen S, Alm J, Larmola T, Hammar T, Silvola J, Martikainen PJ (2003b) Nitrous oxide flux to the atmosphere from the littoral zone of a boreal lake. J Geophys Res Atmos 108:4421. doi:101029/2002JD002989
Huttunen JT, Hammar T, Manninen P, Servomaa K, Martikainen PJ (2004) Potential springtime greenhouse gas emissions from a small southern boreal Lake Keihasjarvi (Finland). Boreal Environ Res 9:421–427
Huybrechts P, Letreguilly A, Reeh A (1991) The Greenland ice-sheet and global warming. Glob Planet Chang 89:399–412
Hylander S, Boeing WJ, Granéli W, Karlsson J, von Einem J, Gutseit K, Hansson L-S (2009) Complementary UV protective compounds in zooplankton. Limnol Oceanogr 54:1883–1893
Ingalls AE, Whitehead K, Bridoux MC (2010) Tinted windows: the presence of the UV absorbing compounds called mycosporine-like amino acids embedded in the frustules of marine diatoms. Geochim Cosmochim Acta 74:104–115
IPCC (1990) Climate change 1990. Cambridge University Press, London
IPCC (1994) Climate change 1994 Radiative forcing of climate change. Cambridge University Press, London
IPCC (1996) Revised 1996 guidelines for national greenhouse gas inventories, reference manual, Washington, DC. Organization for economic cooperation and development
IPCC (2001) Climate change 2001. Cambridge University Press, London
IPCC (2002) IPCC special report on land use, land-use change and forestry. Cambridge University Press, Cambridge
IPCC (2007a) Climate Change 2007: synthesis report contribution of working groups I, II and III to the fourth assessment report of the intergovernmental panel on climate change IPCC. IPCC, Geneva, Switzerland
IPCC (2007b) Climate change 2007: the physical science basis contribution of working group I to the 4th assessment report of the intergovernmental panel on climate change. Cambridge University Press, Cambridge
Ito RG, Schneider B, Thomas H (2005) Distribution of surface fCO_2 and air–sea fluxes in the southwestern subtropical Atlantic and adjacent continental shelf. J Mar Syst 56:227–242
Izaurralde RC, McGill WB, Rosenberg NJ (2000) Carbon cost of applying nitrogen fertilizer. Science 288:809

Jacques G (1983) Some ecophysiological aspects of the Antarctic phytoplankton. Polar Biol 2:27–33

Jansson M, Bergström A-K, Blomqvist P, Drakare S (2000) Allochthonous organic carbon and phytoplankton/bacterioplankton production relationship in lakes. Ecology 81:3250–3255

Jansson M, Hickler T, Jonsson A, Karlsson J (2008) Links between terrestrial primary production and bacterial production and respiration in lakes in a climate gradient in subarctic Sweden. Ecosystems 11:367–376

Jiang HB, Qiu BS (2005) Photosynthetic adaptation of a bloomforming cyanobacterium Microcystis aeruginosa (Cyanophyceae) to prolonged UV-B exposure. J Phycol 41:983–992

Jiang H, Qiu B (2011) Inhibition of photosynthesis by UV-B exposure and its repair in the bloom-forming cyanobacterium *Microcystis aeruginosa*. J Appl Phycol 23:1–6. doi:101007/s10811-010-9562-2

Johannessen SC, Macdonald RW (2009) Effects of local and global change on an inland sea: the Strait of Georgia, British Columbia, Canada. Clim Res 40:1–21

Johannessen SC, Miller WL (2001) Quantum yield for the photochemical production of dissolved inorganic carbon in seawater. Mar Chem 76:271–283

Johannessen SC, Peña MA, Quenneville ML (2007) Photochemical production of carbon dioxide during a coastal phytoplankton bloom. Estuar Coast Shelf Sci 73:236–242

Jones JB, Mulholland PJ (1998) Influence of drainage basin topography and elevation on carbon dioxide and methane supersaturation of stream water. Biogeochemistry 40:57–72. doi:101023/A:1005914121280

Jones CD, Cox P, Essery RLH, Roberts DL, Woodage MJ (2003a) Strong carbon cycle feedbacks in a climate motel with interactive CO2 and sulphate aerosols. Geophys Res Lett 30:1479–1483

Jones CD, Cox P, Huntingford C (2003b) Uncertainty in climate carbon-cycle projections associated with the sensitivity of soil respiration to temperature. Tellus 55B:642–648

Jones JB, Stanley EH, Mulholland PJ (2003c) Long-term decline in carbon dioxide supersaturation in rivers across the contiguous United States. Geophys Res Lett 30:1495. doi:101029/2003GL017056

Joshi M, Shine K, Ponater M, Stuber N, Sausen R, Li L (2003) A comparison of climate response to different radiative forcings in three general circulation models: towards an improved metric of climate change. Clim Dyn 20:843–854

Kandel R, Viollier M (2005) Planetary radiation budgets. Space Sci Rev 120:1–26

Kang H, Freeman C, Ashendon TW (2001) Effects of elevated CO_2 on fen peat biogeochemistry. Sci Total Environ 279:45–50

Karentz D, Cleaver JE, Mitchell DL (1991) Cell survival characteristics and molecular responses of Antarctic phytoplankton to ultraviolet-B radiation. J Phycol 27:326–334

Karl DM, Bidigare RR, Letelier RM (2001) Longterm changes in plankton community structure and productivity in the North Pacific Subtropical Gyre: the domain shift hypothesis. Deep Sea Res Pt II 48:1449–1470

Karl DM, Laws EA, Morris P, Williams PJl, Emerson S (2003) Global carbon cycle (communication arising): metabolic balance of the open sea. Nature 426:32. doi:101038/426032a

Karlsson J, Bystrom P, Ask J, Ask P, Persson L, Jansson M (2009) Light limitation of nutrient-poor lake ecosystems. Nature 460:506–580

Keeling RF, Piper SC, Heimann M (1996) Global and hemispheric CO2 sinks deduced from changes in atmospheric O2 concentration. Nature 381:218–221

Keller M, Kaplan WA, Wofsy SC (1986) Emissions of N2O, CH4, and CO2 from tropical forest soils. J Geophys Res 91:11791–11802

Kelley JJ (1970) Carbon dioxide in the surface waters of the North Atlantic ocean and the Barents and Kara seas. Limnol Oceanogr 15:80–87

Kempe S, Pegler K (1991) Sinks and sources of CO_2 in coastal seas: the North Sea. Tellus 43B:224–235

Kerr JB, McElroy CT (1993) Evidence for large upward trends of ultraviolet B radiation linked to ozone depletion. Science 262:1032–1034

Khalili M, Weyhenmeyer G (2009) Growing season variability of nitrate along a trophic gradient-contrasting patterns between lakes and streams. Aquat Sci 71:25–33

Khalilov EN (2010) Global changes of the environment: threatening the progress of civilization (GEOCHANGE). Probl Glob Chang Geol Environ 1:1–168

Kiehl JT, Trenberth KE (1997) Earth's annual global mean energy budget. Bull Am Meteorol Soc 78:197–208. doi:101175/1520-0477(1997)078<0197:EAGMEB>20CO;2

Kim C, Nishimura Y, Nagata T (2006) Role of dissolved organic matter in hypolimnetic mineralization of carbon and nitrogen in a large, monomictic lake. Limnol Oceanogr 51:70–78

Kirchman DL, Suzuki Y, Garside C, Ducklow HW (1991) High turnover rates of dissolved organic carbon during a spring phytoplankton bloom. Nature 352:612–614

Kitaysky AS, Golubova EG (2000) Climate change causes contrasting trends in reproductive performance of planktivorous and piscivorous alcids. J Anim Ecol 69:248–262

Knorr W, Prentice IC, House JI, Holland EA (2005) Long-term sensitivity of soil carbon to warming. Nature 433:298–301

Kolber Z, Wyman KD, Falkowski PG (1990) Natural variability in photosynthetic energy conversion efficiency: a field study in the Gulf of Maine. Limnol Oceanogr 35:72–79

Komatsu E, Fukushima T, Harasawa H (2007) A modeling approach to forecast the effect of long-term climate change on lake water quality. Ecol model 209:351–366

Komissarov GG (1994) Photosynthesis: a new look. Sci Russ 5:52–55

Komissarov GG (1995) Photosynthesis as a physical process. Chem Phys Rep 14:1723–1732

Komissarov GG (2003) Photosynthesis: the physical-chemical approach. J Advan Chem Phys 2:28–61

Kopáćek J, Hejzlar J, Kaňa J, Porcal P, Klementová S (2003) Photochemical, Chemical, and biological transformations of dissolved organic carbon and its effect on alkalinity production in acidified lakes. Limnol Oceanogr 48:106–117

Kopáćek J, Procházková L, Stuchlík E, Blažka P (1995) The nitrogen-phosphorus relationship in mountain lakes: influence of atmospheric input, watershed, and pH. Limnol Oceanogr 40:930–937

Kopáćek J, Hejzlar J, Borovec J, Porcal P, Kotorová I (2000) Phosphorus inactivation by aluminum in the water column and sediments: a process lowering in-lake phosphorus availability in an acidified watershed-lake ecosystem. Limnol Oceanogr 45:212–225

Kopáćek J, Brzáková M, Hejzlar J, Nedoma J, Porcal P, Vrba J (2004) Nutrient cycling in a strongly acidified mesotrophic lake. Limnol Oceanogr 49:1202–1213

Koprivnjak J-F, Dillon PJ, Molot LA (2010) Importance of CO_2 evasion from small boreal streams. Glob Biogeochem Cycles 24:GB4003. doi:10.1029/2009GB003723

Kreileman GJJ, Bouwman AF (1994) Computing land use emissions of greenhouse gases. Water Air Soil Pollut 76:231–258

Kroeze C, Reijnders L (1992) Halocarbons and global warming. Sci Total Environ 111:1–24

Kujawinski EB, Longnecker K, Blough NV, Del Vecchio R, Finlay L, Kitner JB, Giovannoni SJ (2009) Identification of possible source markers in marine dissolved organic matter using ultrahigh resolution mass spectrometry. Geochim Cosmochim Acta 73:4384–4399

Kuzyakov Y (2002) Review: factors affecting rhizosphere priming effects. J Plant Nut Soil Sci 165:382–396

Lagomarsino A, De Angelis P, Moscatelli MC, Grego S (2009) The influence of temperature and labile C substrates on heterotrophic respiration in response to elevated CO_2 and nitrogen fertilization. Plant Soil 137:223–234

Lal R, Kimble JM, Follett RF, Cole CV (1999) The Potential of US Cropland to Sequester Carbon and Mitigate the Greenhouse Effect, Chelsea. Ann Arbor Press, Michigan

Larsen S, Andersen TOM, Hessen DO (2011) Climate change predicted to cause severe increase of organic carbon in lakes. Glob Chang Biol 17:1186–1192

Lavoie M, Paré D, Bergeron Y (2005) Impact of global change and forest management on carbon sequestration in northern forested peatlands. Environ Rev 13:199–240

Laws EA, Falkowski PG, Smith WO Jr, Ducklow H, McCarthy JJ (2000) Temperature effects on export production in the open ocean. Glob Biogeochem Cycles 14:1231–1246. doi:101029/1999GB001229

Le Quéré C, Aumont O, Bopp L, Bousquet P, Ciais P, Francey R, Heimann M, Keeling CD, Keeling RF, Kheshgi H, Peylin P, Piper SC, Prentice IC, Rayner PJ (2003) Two decades of ocean CO_2 sink and variability. Tellus B 55:649–656

Lefohn AS, Husar JD, Husar RB (1999) Estimating Historical Anthropogenic Global Sulfur Emission Patterns for the Period 1850–1990. Atmos Environ 33:3435–3444

Lehmann MF, Bernasconi SM (2004) Seasonal variation of the $\delta^{13}C$ and $\delta^{15}N$ of particulate and dissolved carbon and nitrogen in Lake Lugano: constraints on biogeochemical cycling in a eutrophic lake. Limnol Oceanogr 49:415–429

Lesser M (2008) Effects of ultraviolet radiation on productivity and nitrogen fixation in the Cyanobacterium, *Anabaena* sp. (Newton's strain). Hydrobiologia 598:1–9

Letelier RM, Karl DM, Abbott MR, Bidigare RR (2004) Light driven seasonal patterns of chlorophyll and nitrate in the lower euphotic zone of the North Pacific Subtropical Gyre. Limnol Oceanogr 49:508–519

Li WKW (1994) Primary production of prochlorophytes, cyanobacteria, and eukaryotic ultraphytoplankton—measurements from flow cytometric sorting. Limnol Oceanogr 39:169–175

Li W, Wu FC, Liu CQ, Fu PQ, Wang J, Mei Y, Wang L, Guo J (2008) Temporal and spatial distributions of dissolved organic carbon and nitrogen in two small lakes on the Southwestern China Plateau. Limnology 9:163–171

Li G, Gao K, Gao G (2011) Differential impacts of solar UV radiation on photosynthetic carbon fixation from the coastal to offshore surface waters in the South China Sea. Photochem Photobiol 87:329–334

Litchman E, Neale PJ (2005) UV effects on photosynthesis, growth and acclimation of an estuarine diatom and cryptomonad. Mar Ecol Prog Ser 300:53–62

Liu Z, Dreybrodt W, Wang H (2010) A new direction in effective accounting for the atmospheric CO_2 budget: considering the combined action of carbonate dissolution, the global water cycle and photosynthetic uptake of DIC by aquatic organisms. Earth Sci Rev 99:162–172

Liu Z, Dreybrodt W, Liu H (20110 Atmospheric CO_2 sink: silicate weathering or carbonate weathering? Appl Geochem 26:S292–S294

Lloyd J, Taylor JA (1994) On the Temperature Dependence of Soil Respiration. Funct Ecol 8:315–323

Löfgren S, Forsius M, Andersen T (2003) Climate induced water color increase in Nordic lakes and streams due to humus. Nordic Council of Ministers, Brochure, Copenhagen, Denmark, p 12

López-Urrutia A, Morán XAG (2007) Resource limitation of bacterial production distorts the temperature dependence of oceanic carbon cycling. Ecology 88:817–822. doi:101890/06-1641

Lovley DR, Coates JD, Blunt-Harris EL, Phillips EJP, Woodward JC (1996) Humic substances as electron acceptors for microbial respiration. Nature 382:445–448

Lundegårdh H (1927) Carbon dioxide evolution and crop growth. Soil Sci 23:417–453

Luo Y, Wan S, Hui D, Wallace LL (2001) Acclimatization of soil respiration to warming in a tall grass prairie. Nature 413:622–624

Ma X, Green SA (2004) Photochemical transformation of dissolved organic carbon in Lake Superior-an in situ experiment. J Great Lakes Res 30(suppl 1):97–112

Mack J, Bolton JR (1999) Photochemistry of nitrite and nitrate in aqueous solution: a review. J Photochem Photobiol A Chem 128:1–13

Mackensie BR, Schiedek D (2007) Daily ocean monitoring since the 1860s shows record warming of northern European seas. Glob Chang Biol 13:1335–1347

Mackenzie FT, Lerman A, Andersson AJ (2004) Past and present of sediment and carbon biogeochemical cycling models. Biogeosciences 1:11–32

Mahimairaja S, Bolan N, Hedley M, Macgregor A (1994) Losses and transformation of nitrogen during composting of poultry manure with different amendments: an incubation experiment. Bioresour Technol 47(3):265–273

Malcolm RL (1985) Geochemistry of stream fulvic and humic substances. In: Aiken GR, McKnight DM, Wershaw RL, MacCarthy P (eds) Humic substances in soil, sediment, and water: geochemistry, isolation and characterization. Wiley, New York, pp 181–209

Malkin SY, Guildford SJ, Hecky RE (2008) Modeling the growth response of *Cladophora* in a Laurentian Great Lake to the exotic invader *Dreissena* and to lake warming. Limnol Oceanogr 53:1111–1124
Mallet C, Charpin MF, Devaux J (1998) Nitrate reductase activity of phytoplankton populations in eutrophic Lake Aydat and mesooligotrophic Lake Pavin: a comparison. Hydrobiologia 373(374):135–148
Manzello DP (2010) Ocean acidification hotspots: spatiotemporal dynamics of the seawater CO_2 system of eastern Pacific coral reefs. Limnol Oceanogr 55:239–248
Marcoval MA, Villafañe VE, Helbling EW (2008) Combined effects of solar ultraviolet radiation and nutrients addition on growth, biomass and taxonomic composition of coastal marine phytoplankton communities of Patagonia. J Photochem Photobiol B Biol 91:157–166
Marhan S, Kandeler H, Rein S, Fangmeier A, Niklaus P (2010) Indirect effects of soil moisture reverse soil C sequestration responses of a spring wheat agroecosystem to elevated CO_2. Glob Chang Biol 16:469–483
Marland G, Rotty RM (1984) Carbon dioxide emissions from fossil fuels: a procedure for estimation and results for 1950–1982. Tellus 36B:232–261
Masson D, Cummins PF (2007) Temperature trends and interannual variability in the Strait of Georgia, British Columbia. Cont Shelf Res 27:634–649
May W (2008) Climatic changes associated with a global 2 °C-stabilization scenario simulated by the ECHAM5/MPI-OM coupled climate model. Clim Dyn 31:283–313
McCarthy MJ, Lavrentyev PJ, Yang LY, Zhang L, Chen YW, Qin BQ, Gardner WS (2007) Nitrogen dynamics and microbial food web structure during a summer cyanobacterial bloom in a subtropical, shallow, well-mixed, eutrophic lake (Lake Taihu, China). Hydrobiologia 581:195–207
McDaniel AH, Cantrell CA, Davidson JA, Shetter RE, Calvert JG (1991) The temperature dependent, infrared absorption cross-sections for the chlorofluorocarbons: CFC-11, CFC-12, CFC-13, CFC-14, CFC-22, CFC-113, CFC-114, and CFC-115. J Atmos Chem 12:211–227
Melillo JM, Steudler PA, Aber JD, Newkirk K, Lux H, Bowles FP, Catricala C, Magill A, Ahrens T, Morrisseau S (2002) Soil warming and carbon-cycle feedbacks to the climate system. Science 298:2173–2176
Melillo JM, Steudler PA, Aber JD, Newkirk K, Lux H, Bowles FP, Catricala C, Magill A, Ahrens T, Morrisseau S, Burrows E, Nadelhoffer K (2004) Soil warming—a major consequence of global climate change. In: Foster D, Aber J (eds) Forests in time. Yale University Press, Hartford, pp 280–295
Mengis M, Gachter R, Wehrli B (1997) Sources and sinks of nitrous oxide (N_2O) in deep lakes. Biogeochemistry 38:281–301
Meriläinen JJ, Hynynen J et al (2001) Pulp and paper mill pollution and subsequent ecosystem recovery of a large boreal lake in Finland: a palaeolimnological analysis. J Paleolimnol 26:11–35
Miller WL (1998) Effects of UV radiation on aquatic humus: photochemical principles and experimental considerations. Ecol Stud 133:125–143
Miller WL, Moran MA (1997) Interaction of photochemical and microbial processes in the degradation of refractory dissolved organic matter from a coastal marine environment. Limnol Oceanogr 42:1317–1324
Miller WL, Zepp RG (1995) Photochemical production of dissolved inorganic carbon from terrestrial organic matter: significance to the oceanic organic carbon cycle. Geophys Res Lett 22:417–420
Millero FJ, Sotolongo S (1989) The oxidation of Fe(II) with H_2O_2 in seawater. Geochim Cosmochim Acta 53:1867–1873
Minero C, Chiron S et al (2007) Photocheminical processes involving nitrite in surface water samples. Aquat Sci 69:71–85
Mitchell J, Driscoll CT, Kahl JS, Likens GE, Murdoch PS, Pardo LH (1996) Climatic control of nitrate losses from forested watersheds in the Northeast United States. Environ Sci Technol 30:2609–2612

Molot LA, Hudson JJ, Dillon PJ, Miller SA (2005) Effect of pH on photo-oxidation of dissolved organic carbon by hydroxyl radicals in a coloured, softwater stream. Aquat Sci 67:189–195
Monteith DT, Stoddard JL, Evans CD, de Wit H, Forsius M, Høåsen T, Wilander A, Skjelkvåle BL, Jeffries DS, Vuorenmaa J, Keller B, Kopáćek J, Veselý J (2007) Dissolved organic carbon trends resulting from changes in atmospheric deposition chemistry. Nature 450:537–541
Morán XAG, Sebastián M, Pedrós-Alió C, Estrada M (2006) Response of southern ocean phytoplankton and bacterioplankton production to short-term experimental warming. Limnol Oceanogr 51:1791–1800
Morán XAG, López-Urrutia A, Calvo-Díaz L, Li WKW (2010) Increasing importance of small phytoplankton in a warmer ocean. Glob Chang Biol 16:1137–1144. doi:101111/j1365-2486200901960x
Morita RY (1975) Psychrophilic bacteria. Bacteriol Rev 39:144–167
Morris DP, Hargreaves BR (1997) The role of photochemical degradation of dissolved organic carbon in regulating the UV transparency of three lakes on the Pocono Plateau. Limnol Oceanogr 42:239–249
Morris DP, Lewis WM Jr (1992) Nutrient limitation of bacterioplankton growth in Lake Dillon, Colorado. Limnol Oceanogr 37:1179–1192
Morris DP, Zagarese H, Williamson CE, Balseiro EG, Hargreaves BR, Modenutti B, Moeller R, Queimalinos C (1995) The attenuation of solar UV radiation in lakes and the role of dissolved organic carbon. Limnol Oceanogr 40:1381–1391
Morrison J, Quick MC, Foreman MGG (2002) Climate change in the Fraser River watershed: flow and temperature projections. J Hydrol (Amst) 263:230–244
Mortain-Bertrand A, Descolas-Gros C, Jupin H (1988) Growth, photosynthesis and carbon metabolism in the temperate marine diatom *Skeletonema costatum* adapted to low temperature and low photon-flux density. Mar Biol 100:135–141
Mosier AR, Chapman SL, Freney JR (1989) Determination of dinitrogen emission and retention in floodwater and porewater of a lowland rice field fertilized with ^{15}N-urea. Fert Res 19:127–136
Mosier AR, Schimel DS, Valentine DW, Bronson KF, Parton WJ (1991) Methane and nitrous oxide fluxes in native, fertilized, and cultivated grasslands. Nature 350:330–332
Mosier AR, Duxbury JM, Freney JR, Heinemeyer O, Minami K, Johnson DE (1998) Mitigating agricultural emissions of methane. Clim Change 40:39–80. doi:101023/A:1005338731269
Mosier A, Wassmann R, Verchot L, King J, Palm C (2004) Methane and nitrogen oxide fluxes in tropical agricultural soils: sources, sinks and mechanisms. Environ Dev Sustain 6:11–49
Mostofa KMG, Sakugawa H (2009) Spatial and temporal variations and factors controlling the concentrations of hydrogen peroxide and organic peroxides in rivers. Environ Chem 6:524–534
Mostofa KMG, Yoshioka T, Konohira E, Tanoue E, Hayakawa K, Takahashi M (2005) Three-dimensional fluorescence as a tool for investigating the dynamics of dissolved organic matter in the Lake Biwa watershed. Limnology 6:101–115
Mostofa KMG, Wu FC, Yoshioka T, Sakugawa H, Tanoue E (2009a) Dissolved organic matter in the aquatic environments. In: Wu FC, Xing B (eds) Natural organic matter and its significance in the environment. Science Press, Beijing, pp 3–66
Mostofa KMG, Liu CQ, Wu FC, Fu PQ, Ying WL, Yuan J (2009) Overview of key biogeochemical functions in lake ecosystem: impacts of organic matter pollution and global warming. In: Proceedings of the 13th World Lake Conference. Wuhan, China, 1–5 Nov 2009, Keynote Speech, pp 59–60
Mostofa KMG, Wu FC, Liu CQ, Vione D, Yoshioka T, Sakugawa H, Tanoue E (2011) Photochemical, microbial and metal complexation behavior of fluorescent dissolved organic matter in the aquatic environments. Geochem J (Invit Rev) 45:235–254
Mudie PJ, Rochon A, Levac E (2002) Palynological records of red tide-producing species in Canada: past trends and implications for the future. Palaeogeogr Palaeoclimatol Palaeoecol 180:159–186

Mukhopadhyay SK, Biswas H, De TK, Sen S, Jana TK (2002) Seasonal effects on the air–water carbon dioxide exchange in the Hooghly estuary, NE coast of Bay of Bengal, India. J Environ Monit 4:549–552

Murdoch PS, Burns DA, Lawrence GB (1998) Relation of climate change to the acidification of surface waters by nitrogen deposition. Environ Sci Technol 32:1642–1647

Myers N (1989) Deforestation Rates in Tropical Forests and their climatic implications, A Friends of the Earth report (December)

Nakane K, Kohno T Horikoshi T, Nakatsubo T (1997) Soil carbon cycling at a black spruce (Picea mariana) forest stand in Saskatchewan, Canada. J Geophy Res 102(D24):28785–28793

Nedwell DB (1999) E¡ect of low temperature on microbial growth: lowered a¤nity for substrates limits growth at low temperature. FEMS Microbiol Ecol 30:101–111

Nedwell DB, Rutter M (1994) Influence of temperature on growth rate and competition between two psychrotolerant Antarctic bacteria: low temperature diminishes affinity for substrate uptake. Appl Environ Microbiol 60:1984–1992

Neori A, Holm-Hansen O (1982) Effect of temperature on rate of photosynthesis in Antarctic phytoplankton. Polar Biol 1:33–38

Nierop GJK, Naafs DFW, Verstraten JM (2003) Occurrence and distribution of ester-bound lipids in Dutch coastal dune soils along a pH gradient. Org Geochem 34:719–729

Norf H, Weitere M (2010) Resource quantity and seasonal background alter warming effects on communities of biofilm ciliates. FEMS Microbiol Ecol 74:361–370

Norf H, Arndt H, Weitere M (2007) Impact of local temperature increase on the early development of biofilm-associated ciliate communities. Oecologia 151:341–350

Obernosterer I, Ruardij P, Herndl GJ (2001) Spatial and diurnal dynamics of dissolved organic matter (DOM) fluorescence and H_2O_2 and the photochemical oxygen demand of surface water DOM across the subtropical Atlantic Ocean. Limnol Oceanogr 46:632–643

Ochs CA, Cole JJ, Likens GE (1995) Population dynamics of bacterioplankton in an oligotrophic lake. J Plankton Res 17:365–391

Oechel WC, Vourlitis GL, Hastings SJ, Zulueta RC, Hinzman L, Kane D (2000) Acclimation of ecosystem CO2 exchange in the Alaskan Artic in response to decadal climate warming. Nature 406:978–981

Oenema O, Wrage N, Velthof GL, van Groenigen JW, Dolfing J, Kuikman PJ (2005) Trends in global nitrous oxide emissions from animal production systems. Nutr Cycl Agroecosyst 72:51–65. doi:101007/s10705-004-7354-2

Ohde S, van Woesik R (1999) Carbon dioxide flux and metabolic processes of a coral reef, Okinawa. Bull Mar Sci 65:559–576

Omar AM, Olsen A, Johannessen T, Hoppema M, Thomas H, Borges AV (2010) Spatiotemporal variations of fCO_2 in the North Sea. Ocean Sci 6:77–89

Öquist M, Wallin M, Siebert J, Bishop K, Laudon H (2009) Dissolved inorganic carbon export across the soil/stream interface and its fate in a boreal headwater stream. Environ Sci Technol 43:7364–7369. doi:101021/es900416h

O'Reilly CM, Alin SR, Plisnier P-D, Cohen AS, McKee BA (2003) Climate change decreases aquatic ecosystem productivity of Lake Tanganyika, Africa. Nature 424(6950):766–768

Otto A, Simpson MJ (2006) Sources and composition of hydrolysable aliphatic lipids and phenols in soils from western Canada. Org Geochem 37:385–407

Palenik B, Morel FMM (1988) Dark production of H_2O_2 in the Sargasso Sea. Limnol Oceanogr 33:1606–1611

Paradies G, Petrosillo G, Pistolese M, Ruggiero FM (2000) The effect of reactive oxygen species generated from mitochondrial electron transport chain on the cytochrome c oxidase activity and on the cardiolipin content in bovine heart submitochondrial particles. FEBS Lett 466:323–326

Pastor J, Solin J, Bridgham SD, Updegraff K, Harth C, Weishampel P, Dewey B (2003) Global warming and the export of dissolved organic carbon from boreal peatlands. Oikos 100:380–386

Pattanaik B, Roleda MY, Schumann R, Karsten U (2008) Isolate specific effects of ultraviolet radiation on photosynthesis, growth and mycosporine-like amino acids in the microbial mat-forming cyanobacterium Microcoleus chthonoplastes. Planta 227:907–916

Pepper W, Leggett J, Swart R, Watson J, Edmonds J, Mintzer I (1992) Emission scenarios for the IPCC: an update assumptions, methodology and results, prepared for IPCC Working Group I, May1992, p 115

Petchey OL, McPhearson PT, Casey TM, Morin PJ (1999) Environmental warming alters food-web structure and ecosystem function. Nature 402:69–72

Peters F (2008) Diatoms in a future ocean–stirring it up. Nat Rev Microbiol 6:407

Petersen SO, Klug MJ (1994) Effects of sieving, storage, and incubation temperature on the phospholipid fatty acid profile of a soil microbial community. Appl Environ Microbiol 60:2421–2430

Pienitz R, Vincent WF (2000) Effect of climate change relative to ozone depletion on UV exposure in subarctic lakes. Nature 404:484–487

Polovina JJ, Howell EA, Abecassis M (2008) Ocean's least productive waters are expanding. Geophys Res Lett 35:L03618. doi:101029/2007GL031745

Pomeroy LR, Wiebe WJ (2001) Temperature and substrates as interactive limiting factors for marine heterotrophic bacteria. Aquat Microb Ecol 23:187–204

Pomeroy LR, Wiebe WJ, Deibel D, Thompson JT, Rowe GT, Pakulaski JD (1991) Bacterial responses to temperature and substrate concentration during the Newfoundland spring bloom. Mar Ecol Prog Ser 75:143–159

Porcal P, Koprivnjak J-F, Molot LA, Dillon PJ (2009) Humic substances—part 7: the biogeochemistry of dissolved organic carbon and its interactions with climate change. Environ Sci Pollut Res 16:714–726

Post WM, Emanuel WR, Zinke PJ, Stangenberger AG (1982) Soil carbon pools and world life zones. Nature 298:156–159

Pregitzer K, Zak DR, Burton AJ, Ashby JA, MacDonald NW (2004) Chronic nitrate additions dramatically increase the export of carbon and nitrogen from northern hardwood ecosystems. Biogeochemistry 68:179–197

Qian J, Mopper K, Kieber DJ (2001) Photochemical production of the hydroxyl radical in Antarctic waters. Deep Sea Res I 48:741–759

Quaas J, Boucher O, Dufresne J-L, Le Treut H (2004) Impacts of greenhouse gases and aerosol direct and indirect effects on clouds and radiation in atmospheric GCM simulations of the 1930–1989 period. Clim Dyn 23:779–789

Raich JW, Schlesinger WH (1992) The global carbon dioxide flux in soil respiration and its relationship to vegetation and climate. Tellus B 44:81–99

Ramm K, Scheps V (1997) Phosphorus balance of a polytrophic shallow lake with consideration of phosphorus release. Hydrobiologia 342(343):43–53

Randall CE, Harvey VL (2005) Stratospheric effects of energetic particle precipitation in 2003–2004. Geophys Res Lett 32:L05082. doi:101029/2004GL022003

Rastogi RP, Richa Sinha RP, Singh SP, Häder D-P (2010) Photoprotective compounds from marine organisms. J Ind Microbiol Biotechnol 37:537–558

Rath J, Adhikary SP (2007) Response of the estuarine cyanobacterium Lyngbya aestuarii to UV-B radiation. J Appl Phycol 19:529–536

Raven JA (1991) Responses of photosynthetic organisms to increased solar UV-B. J Photochem Photobiol B Biol 9:239–244

Raymond PA, Oh NH (2007) An empirical study of climatic controls on riverine C export from three major US watersheds. Glob Biogeochem Cycles 21:GB2022

Raymond PA, Bauer JE, Cole JJ (2000) Atmospheric CO_2 evasion, dissolved inorganic carbon production, and net heterotrophy in the York River estuary. Limnol Oceanogr 45:1707–1717

Reay DS, Nedwell DB, Priddle J, Ellis-Evans JC (1999) Temperature dependence of inorganic nitrogen uptake: reduced affinity for nitrate at suboptimal temperatures in both algae and bacteria. Appl Environ Microbiol 65:2577–2584

Reay DS, Priddle J, Nedwell DB, Whitehouse MJ, Ellis-Evans JC, Deubert C, Connelly DP (2001) Regulation by low temperature of phytoplankton growth and nutrient uptake in the Southern Ocean. Mar Ecol Prog Ser 219:51–64

Richard LE, Peake BM, Rusak SA, Cooper WJ, Burritt DJ (2007) Production and decomposition dynamics of hydrogen peroxide in freshwater. Environ Chem 4:49–54. doi:101071/EN06068

Richardson K, Jorgensen BB (1996) Eutrophication: definition, history, and effects. In: Jorgensen BB, Richardson K (eds) Eutrophication in coastal marine ecosystems. American Geophysical Union, Washington, pp 1–19

Richey JE, Melack JM, Aufdenkampe AK, Ballester VM, Hess LL (2002) Outgassing from Amazonian rivers and wetlands as a large tropical source of atmospheric CO_2. Nature 416:617–620

Rinnan R, Michelsen A, Jonasson S (2008) Effects of litter addition and warming on soil carbon, nutrient pools and microbial communities in a subarctic heath ecosystem. Appl Soil Ecol 39:271–281

Riser SC, Johnson KS (2008) Net production of oxygen in the subtropical ocean. Nature 451:323–325. doi:101038/nature06441

Robertson GP (1982) Nitrification in forested ecosystems. Philos Trans R Soc Lond 296:445–457

Robertson G, Grace P (2004) Greenhouse gas fluxes in tropical and temperate agriculture: the need for a full-cost accounting of global warming potentials. Environ Dev Sustain 6:51–63

Robertson GP, Paul EA, Harwood RR (2000) Greenhouse gases in intensive agriculture: contributions of individual gases to the radiative forcing of the atmosphere. Science 289:1922–1925

Rochelle-Newall EJ, Fisher TR (2002) Production of chromophoric dissolved organic matter fluorescence in marine and estuarine environments: an investigation into the role of phytoplankton. Mar Chem 77:23–41

Rosenfeld D (2000) Suppression of rain and snow by urban and industrial air pollution. Science 287:1793–1796

Roulet N, Moore TR (2006) Browning the waters. Nature 444(2):283–284

Ruddiman WF (2001) Earth's climate: past and future. WH Freeman, New York, p 465

Rutledge S, Campbell DI, Baldocchi D, Schipper LA (2010) Photodegradation leads to increased carbon dioxide losses from terrestrial organic matter. Glob Chang Biol 16:3065–3074

Ryther JH (1956) Photosynthesis in the ocean as a function of light intensity. Limnol Oceanogr 1:61–69

Sabine CL, Feely RA, Key RM, Lee K, Bullister JL, Wanninkhof R, Wong CS, Wallace DWR, Tilbrook B, Millero FJ, Peng T-H, Kozyr A, Ono T, Rios AF (2004) The oceanic sink for anthropogenic CO_2. Science 305:367–371

Salonen K, Vähätalo A (1994) Photochemical mineralization of dissolved organic matter in Lake Skjervatjern. Environ Int 20:307–312

Sarma V, Kumar MD, Manerikar M (2001) Emission of carbon dioxide from a tropical estuarine system, Goa, India. Geophys Res Lett 28:1239–1242

Sarmento H, Montoya JM, Vázquez-Domínguez E, Vaqué D, Gasol JM (2010) Warming effects on marine microbial food web processes: how far can we go when it comes to predictions? Phil Trans R Soc B 365:2137–2149

Sarmiento JL, Slater R, Barber R, Bopp L, Doney SC, Hirst AC, Kleypas J, Matear R, Mikolajewicz U, Monfray P, Soldatov V, Spall SA, Stouffer RJ (2004) Response of ocean ecosystems to global warming. Glob Biogeochem Cycles 18:GB3003. doi:101029/2003GB002134

Sass L, Spetea C, Máté Z, Nagy F, Vass I (1997) Repair of UV-B induced damage of photosystem II via de novo synthesis of the D1 and D2 reaction centre subunits of Scynechocystis sp PCC 6803. Photosynth Res 54:55–62

Sawicka JE, Robador A, Hubert C, Jorgensen BB, Bruchert V (2010) Effects of freeze-thaw cycles on anaerobic microbial processes in an Arctic intertidal mud flat. ISME J 4:585–594

Schiettecatte L, Gazeau F, van der Zee C, Brion N, Borges AV (2006) Time series of the partial pressure of carbon dioxide (2001–2004) and preliminary inorganic carbon budget in the

Scheldt plume (Belgian coastal waters). Geochem Geophy Geosy 7:Q06009. doi:101029/2005GC001161
Schiettecatte L-S, Thomas H, Bozec Y, Borges AV (2007) High temporal coverage of carbon dioxide measurements in the Southern Bight of the North Sea. Mar Chem 106:161–173
Schindler DW (1988) Effects of acid rain on freshwater ecosystems. Science 239:149–157
Schindler DW (1994) Changes caused by acidification to the biodiversity: productivity and biogeochemical cycles of lakes. In: Steinberg CEW, Wright RF (eds) Acidification of freshwater ecosystems: implications for the future. Wiley, New York, pp 153–164
Schindler DW, Curtis PJ, Parker BR, Stainton MP (1996) Consequences of climatic warming and lake acidification for UV-B penetration in North American boreal lakes. Nature 379:705–708
Schlesinger WH (1999) Carbon sequestration in soils. Science 284:2095–2097
Schlesinger WH, Andrews JA (2000) Soil respiration and the global carbon cycle. Biogeochemistry 48:7–20
Schmidt GA, Ruedy R, Miller RL, Lacis AA (2010) The attribution of the present-day total greenhouse effect. J Geophys Res 115:D20106. doi:101029/2010JD014287
Schmittner A (2005) Decline of the marine ecosystem caused by a reduction in the Atlantic overturning circulation. Nature 434(7033):628–633
Scully NM, Lean DRS (1994) The attenuation of ultraviolet radiation in temperate lakes. Arch Hydrobiol Beih Ergebn Limnol 43:135–144
Seitzinger SP (1990) Denitrification in aquatic sediments. In: Revsbech NP, Sorensen J (eds) Denitrification in soil and sediment. Plenum Press, New York, pp 301–322
Seitzinger S, Harrison JA, Böhlke JK, Bouwman AF, Lowrance R, Peterson B, Tobias C, van Drecht G (2006) Denitrification across landscapes and waterscapes: a synthesis. Ecol Appl 6:2064–2090
Semiletov IP, Pipko II, Repina I, Shakhova NE (2007) Carbonate chemistry dynamics and carbon dioxide fluxes across the atmosphere-ice-water interfaces in the Arctic Ocean: Pacific sector of the Arctic. J Mar Syst 66:204–226
Shaver GR, Candell J, Chapin FS III, Gurevitch J, Harte J, Henry G, Ineson P, Jonasson S, Melillo J, Pitelka L, Rustad L (2000) Global warming and terrestrial ecosystems: a conceptual framework for analysis. Bioscience 50:871–882
Shen Z (2001) Historical changes in nutrient structure and its influences on phytoplankton composition in Jiaozhou Bay. Estuar Coast Shelf Sci 52:211–224
Shine K, Fuglestvedt J, Hailemariam K, Stuber N (2005) Alternatives to the Global Warming Potential for Comparing Climate Impacts of Emissions of Greenhouse Gases. Clim Chang 68:281–302
Simon M, Wünsch C (1998) Temperature control of bacterioplankton growth in a temperate large lake. Aquat Microb Ecol 16:119–130
Sinha R, Klisch M, Gröniger A, Häder D-P (2001) Responses of aquatic algae and cyanobacteria to solar UV-B. Plant Ecol 154:219–236
Sitch S, Brovkin V, von Bloh W, van Vuuren D, Eickhout B, Ganopolski A (2005) Impacts of future land cover changes on atmospheric CO_2 and climate. Global Biogeochemical Cycles 19(2):GB2013
Skjelkvåle BL, Stoddard JK, Andersen T (2001) Trends in surface water acidification in Europe and North America 1989–1998. Water Soil Air Pollut 130:787–792
Smith VH (2003) Eutrophication of freshwater and coastal marine ecosystems: a global problem. Environ Sci Pollut Res 10:1–14
Smith P (2004) Engineered biological sinks on land. In: Field CB, Raupach MR (eds) The global carbon cycle Integrating humans, climate, and the natural world. Island Press, Washington, pp 479–491
Smith RC, Baker KS (1979) Penetration of UV-B and biologically effective dose-rates in natural waters. Photochem Photobiol 29:311–323
Smith KA, Conen F (2004) Impacts of land management on fluxes of trace greenhouse gases. Soil Use Manag 20:255–263. doi:101079/SUM2004238

Smith RC, Prezelin BB, Baker KS, Bidigare RR, Boucher NP, Cooley T, Karentz D, Maclntyre S, Matlick HA, Menzies D, Ondrusek M, Wan Z, Waters KJ (1992) Ozone depletion: ultraviolet radiation and phytoplankton biology in Antarctic waters. Sclence 255:952–959

Smith JB, Schellnhuber HJ, Mirza MQM (2001) Vulnerability to climate change and reasons for concern: a synthesis. In: McCarthy JJ, Canziani OF, Leary NA, Dokken DJ, White KS (eds) Climate change 2001 Impacts, adaptation, and vulnerability. Cambridge University Press, Cambridge, pp 913–967

Smith P, Martino D, Cai Z, Gwary D, Janzen H, Kumar P, McCarl B, Ogle S, O'Mara F, Rice C, Scholes B, Sirotenko O, Howden M, McAllister T, Pan G, Romanenkov V, Schneider U, Towprayoon S, Wattenbach M, Smith J (2008) Greenhouse gas mitigation in agriculture. Phil Trans R Soc B 363:789–813

Sobek S, Tranvik LJ, Cole JJ (2005) Temperature independence of carbon dioxide supersaturation in global lakes. Glob Biogeochem Cycles 19:GB2003. doi:101029/2004GB002264

Sobek S, Tranvik LJ, Prairie YT, Kortelainen P, Cole JJ (2007) Patterns and regulation of dissolved organic carbon: an analysis of 7,500 widely distributed lakes. Limnol Oceanogr 52:1208–1219

Sobrino C, Montero O, Lubián LM (2004) UV-B radiation increases cell permeability and damages nitrogen incorporation mechanisms in Nannochloropsis gaditana. Aquat Sci 66:421–429

Sollins P, Robertson GP, Uehara G (1988) Nutrient mobility in variable- and permanent-charge soils. Biogeochemistry 6:181–199

Solomon S, Daniel JS (1996) Impact of the Montreal protocol and its amendments on the rate of change of global radiative forcing. Clim Chang 32:7–17

Sommaruga R, Garcia-Pichel (1999) UV-absorbing mycosporine-like compounds in planktonic and benthic organisms from a high-mountain lake. Arch Hydrobiol 144:255–269

Sommaruga R, Obernosterer I, Herndl G, Psenner R (1997) Inhibitory effect of solar radiation on thymidine and leucine incorporation by freshwater and marine bacterioplankton. Appl Environ Microbiol 63:4178–4184

Sommaruga R, Sattler B, Oberleiter A, Wille A, Sommaruga-Wögrath S, Psenner R, Felip M, Camarero L, Pina S, Gironés R, Catalán J (1999a) An in situ enclosure experiment to test the solar UV-B impact on microplankton in a high-altitude mountain lake II effects on the microbial food web. J Plankton Res 21:859–876

Sommaruga R, Psenner R, Schafferer E, Koinig KA, Sommaruga-Wögrath S (1999b) Dissolved organic carbon concentration and phytoplankton biomass in high-mountain lakes of the Austrian Alps: potential effects of climatic warming on UV underwater attenuation. Arct Antarct Alp Res 31:247–254

Song YF, Qiu BS (2007) The CO_2 concentrating mechanism in the bloom-forming cyanobacterium Microcystis aeruginosa (Cyanophyceae) and effects of UV-B radiation on its operation. J Phycol 43:957–964

Stedmon CA, Markager S (2005) Tracing the production and degradation of autochthonous fractions of dissolved organic matter by fluorescence analysis. Limnol Oceanogr 50:1415–1426

Stedmon CA, Markager S, Tranvik L, Kronberg L, Slätis T, Martinsen W (2007a) Photochemical production of ammonium and transformation of dissolved organic matter in the Baltic Sea. Mar Chem 104:227–240

Stedmon CA, Thomas DN, Granskog M, Kaartokallio H, Papaditriou S, Kuosa H (2007b) Characteristics of dissolved organic matter in Baltic coastal sea ice: allochthonous or autochthonous origins? Environ Sci Technol 41:7273–7279

Sterner RW, Elser JJ (2002) Ecological stoichiometry: the biology of elements from molecules to the biosphere. Princeton University Press, Princeton

Sterner RW, Andersen T, Elser JJ, Hessen DO, Hood JM, McCauley E, Urabe J (2008) Scale-dependent carbon : nitrogen : phosphorus seston stoichiometry in marine and freshwaters. Limnol Oceanogr 53:1169–1180

Stevens RJ, Laughlin RJ (1998) Measurement of nitrous oxide and di-nitrogen emissions from agricultural soils. Nutr Cycl Agroecosyst 52:131–139

Stolarski R, Bojkov R, Bishop L, Zereros C, Staehelin J, Zawodny J (1992) Measured trends in stratospheric ozone. Science 256:342–349

Subak S, Raskin P, von Hippel D (1993) National greenhouse gas accounts: current anthropogenic sources and sinks. Clim Chang 25:15–58

Sugiyama Y, Anegawa A, Inokuchi H, Kumagai T (2005) Distribution of dissolved organic carbon and dissolved fulvic acid in mesotrophic Lake Biwa, Japan. Limnology 6:161–168

Tarran GA, Zubkov MV, Sleigh MA, Burkill PH, Yallop M (2001) Microbial community structure and standing stocks in the NE Atlantic in June and July of 1996. Deep Sea Res II 48:963–985

Teodoru CR, del Giorgio PA, Prairie YT, Camire M (2009) Patterns of pCO_2 in boreal streams and rivers of northern Quebec, Canada. Glob Biogeochem Cycles 23:GB2012. doi:101029/2008GB003404

Thomas DN, Lara RJ (1995) Photodegradation of algal derived dissolved organic carbon. Mar Ecol Prog Ser 116:309–310

Thomas H, Bozec Y, Elkalay K, De Baar H (2004) Enhanced open ocean storage of CO2 from shelf sea pumping. Science 304:1005–1008

Thomas H, Bozec Y, Elkalay K, de Baar HJW, Borges AV, Schiettecatte L-S (2005) Controls of the surface water partial pressure of CO_2 in the North Sea. Biogeosciences 2:323–334

Thomas H, Prowe AEF, van Heuven S, Bozec Y, De Baar HJW, Schiettecatte L-S, Suykens K, Koñe M, Borges AV, Lima ID, Doney SC (2007) Rapid decline of the CO2 buffering capacity in the North Sea and implications for the North Atlantic Ocean. Global Biogechem Cycles 21:GB4001. doi:101029/2006GB002825

Thomas H, Schiettecatte L-S, Suykens K, Koñe YJM, Shadwick EH, Prowe AEF, Bozec Y, de Baar HJW, Borges AV (2009) Enhanced ocean carbon storage from anaerobic alkalinity generation in coastal sediments. Biogeosciences 6:267–274

Tranvik LJ (1992) Allochthonous disolved organic matter as an energy source for pelagic bacteria and the concept of the microbial loop. Hydrobiologia 229:107–114

Tranvik LJ, Jasson M (2002) Climate change—terrestrial export of organic carbon. Nature 415:861–862

Tranvik LJ, Downing JA, Cotner JB, Loiselle SA, Striegl RG, Ballatore TJ, Dillon P, Finlay K, Fortino K, Knoll LB, Kortelainen PL, Kutser T, Larsen S, Laurion I, Leech DM, McCallister SL, McKnight DM, Melack JM, Overholt E, Porter JA, Prairie Y, Renwick WH, Roland F, Sherman BS, Schindler DW, Tremblay SSA, Vanni MJ, Verschoor AM, von Wachenfeldt E, Weyhenmeyer GA (2009) Lakes and reservoirs as regulators of carbon cycling and climate. Limnol Oceanogr 54:2298–2314

Trumbore S (2000) Age of soil organic matter and soil respiration: radiocarbon constraints on below ground dynamics. Ecol Appl 10:399–411

Tsunogai S, Watanabe S, Sato T (1999) Is there a continental shelf pump for the absorption of atmospheric CO_2? Tellus 51B:701–712

Tsuruta H, Kanda K, Hirose T (1997) Nitrous oxide emission from a rice paddy field in Japan. Nutr Cycl Agroecosyst 49:51–58

Uchida M, Nakatsubo T, Horikoshi T, Nakane K (1998) Contribution of micro-organisms to the carbon dynamics in black spruce (Picea mariana) forest soil in Canada. Ecol Res 13:17–26

Uchida M, Nakatsubo T, Kasai Y, Nakane K, Horikoshi T (2000) Altitude differences in organic matter mass loss and fungal biomass in a subalpine coniferous forest, Mr Fuji, Japan. Arct Antarct Alp Res 32:262–269

Urabe J (1993) N and P cycling coupled by grazers' activities: food quality and nutrient release by zooplankton. Ecology 74:2337–2350

Vähätalo AV, Järvinen M (2007) Photochemically produced bioavailable nitrogen from biologically recalcitrant dissolved organic matter stimulates production of a nitrogen-limited microbial food web in the Baltic Sea. Limnol Oceanogr 52:132–143

van der Werf GR, Morton DC, DeFries RS, Olivier JGJ, Kasibhatla PS, Jackson RB, Collatz GJ, Randerson JT (2009) CO_2 emissions from forest loss. Nat Geosci 2:737–738

Vaqué D, Guadayol O, Peters F, Felipe J, Malits A, Pedrós-Alió C (2009) Differential response of grazing and bacterial heterotrophic production to experimental warming in Antarctic waters. Aquat Microb Ecol 54:101–112. doi:10.3354/ame01259

Vargas-Yañez M, Salat J, Fernandez de Puelles ML, Lopez-Jurado JL, Pascual J, Ramirez T, Corté D, Franco I (2005) Trends and time varibility in the Northern continental shelf of the western Mediterranean. J Geophys Res 110:1–18

Varotsos K, Kondratiev KY (1995) Changes in solar ultraviolet-radiation reaching the ground due to tropospheric and stratospheric ozone variations. Earth Observ Remote Sens 12:1–10

Vázquez-Domínguez E, Vaqué D, Gasol AM (2007) Ocean warming enhances respiration and carbon demand of coastal microbial plankton. Glob Chang Biol 13:1327–1334. doi:101111/j1365-2486200701377x

Velders GJM, Andersen SO, Daniel JS, Fahey DW, McFarland M (2007) The importance of the Montreal Protocol in protecting climate. Proc Natl Acad Sci 104:4814–4819

Verity PG (1981) Effects of temperature, irradiance, and daylength on the marine diatom Leptocylindrus danicus Cleve I photosynthesis and cellular composition. J Exp Mar Biol Ecol 55:79–91. doi:101016/0022-0981(81)90094-0

Vinebrooke RD, Leavitt PR (1998) Direct and interactive effects of allochthonous dissolved organic matter, inorganic nutrients, and ultraviolet radiation on an alpine littoral food web. Limnol Oceanogr 43:1065–1081

von Der Gathen P et al (1995) Observational evidence for chemical ozone depletion over the Arctic in winter 1991–92. Nature 375:131–134

Vrede K (1996) Regulation of bacterioplankton production and biomass in an oligotrophic clear-water lake—the importance of the phytoplankton community. J Plankton Res 18:1009–1032

Vrede K (2005) Nutrient and temperature limitation of bacterioplankton growth in temperate lakes. Microb Ecol 49:245–256

Vrede K, Vrede T, Isaksson A, Karlsson A (1999) Effects of nutrients (P, N, C) and zooplankton on bacterioplankton and phytoplankton—a seasonal study. Limnol Oceanogr 44:1616–1624

Vuorenmaa J, Forsius M, Mannio J (2006) Increasing trends of total organic carbon concentrations in small forest lakes in Finland from 1987 to 2003. Sci Total Environ 365:47–65

Vyhnálek V, Fott J, Kopáček J (1994) Chlorophyll-phosphorus relationship in acidified lakes of the High Tatra Mountains (Slovakia). Hydrobiologia 274:49–56

Walter KM, Zimov SA, Chanton JP, Verbyla D, Chapin FS (2006) Methane bubbling from Siberian thaw lakes as a positive feedback to climate warming. Nature 443:71–75

Wang ZHA, Cai WJ (2004) Carbon dioxide degassing and inorganic carbon export from a marsh-dominated estuary (the Duplin River): a marsh CO_2 pump. Limnol Oceanogr 49:341–354

Wang L, Miller TD, Priscu JC (1992) Bacterioplankton nutrient deficiency in a eutrophic lake. Arch Hydrobiol 125:423–439

Wang HJ, Wang WD, Yin CQ, Wang YC, Lu JW (2006) Littoral zones as the "hotspots" of nitrous oxide (N_2O) emission in a hyper-eutrophic lake in China. Atmos Environ 40:5522–5527

Wang FS, Wang BL, Liu CQ, Wang YC, Guan J, Liu XL, Yu YX (2011) Carbon dioxide emission from surface water in cascade reservoirs-river system on the Maotiao River, southwest of China. Atmosphric Environ 45:3827–3834

Watanabe Y (1980) A Study of the excretion and extracellular products of natural phytoplankton in Lake Nakanuma, Japan. Int Rev Ges Hydrobiol 65:809–834. doi:101002/iroh19800650606

Watson RT, Meira Filho LG, Sanhueza E, Janetos A (1992) Sources and sinks. In: Houghton JT, Callander BA, Vamey SK (eds) Climate change 1992, the supplementary report to the IPCC scientific assessment. University Press, Cambridge, pp 25–46

Weatherhead EC, Andersen SB (2006) The search for signs of recovery of the ozone layer. Nature 441:39–45. doi:10.1038/nature04746

Wetzel RG (1984) Detrital dissolved and particulate organic carbon functions in aquatic ecosystems. Bull Mar Sci 35:503–509

Wetzel RG (1992) Gradient-dominated ecosystems: sources and regulatory functions of dissolved organic matter in freshwater ecosystems. Hydrobiologia 229:181–198

Weyhenmeyer GA (2008) Water chemical changes along a latitudinal gradient in relation to climate and atmospheric deposition. Clim Chang 88:199–208
Wiebe WJ, Sheldon WM Jr, Pomeroy LR (1992) Bacterial growth in the cold: evidence for an enhanced substrate requirement. Appl Environ Microbiol 58:359–364
Wiebe WJ, Sheldon WM Jr, Pomeroy LR (1993) Evidence for enhanced substrate requirement by marine mesophilic bacterial isolates at minimal growth temperatures. Microb Ecol 25:151–159
Wigley TML (1988) Future CFC concentrations under the Montreal Protocol and their greenhouse-effect implications. Nature 335:333–335
Wigley TML (1989) Possible climate change due to SO_2-derived cloud condensation nuclei. Nature 339:365–367
Williams PJlB, Morris PJ, Karl DM (2004) Net community production and metabolic balance at the oligotrophic ocean site, station ALOHA. Deep Sea Res Pt I 51:1563–1578
Wiltshire KH, Manly BFJ (2004) The warming trend at Helgoland Roads, North Sea: phytoplankton response Helgoland. Mar Res 58:269–273. doi:101007/s10152-004-0196-0
Windolf J, Jeppesen E, Jensen JP, Kristensen P (1996) Modelling of seasonal variation in nitrogen retention and inlake concentration: a four-year mass balance study in 16 shallow Danish lakes. Biogeochemistry 33:25–44
Wohlers J, Engel A, Zöllner E, Breithaupt P, Jürgens K, Hoppe H-G, Sommer U, Riebesell U (2009) Changes in biogenic carbon flow in response to sea surface warming. Proc Natl Acad Sci U S A 106:7067–7072. doi:101073/pnas0812743106
Wolf AA, Drake BG, Erickson JE, Megonigal JP (2007) An oxygenmediated positive feedback between elevated carbon dioxide and soil organic matter decomposition in a simulated anaerobic wetland. Glob Chang Biol 13:2036–2044
Worrall F, Burt T, Adamson J (2004) Can climate change explain increases in DOC flux from upland peat catchments? Sci Total Environ 326:95–112
Worrall F, Burt T, Adamson J (2005) Fluxes of dissolved carbon dioxide and inorganic carbon from an upland peat catchment: implications for soil respiration. Biogeochemistry 73:515–539
Wu H, Gao K (2009) UV radiation-stimulated activity of extracellular carbonic anhydrase in the marine diatom Skeletonema costatum Funct. Plant Biol 36:137–143
Wu HY, Gao KS, Villafañe VE, Watanabe T, Helbling EW (2005) Effects of solar UV radiation on morphology and photosynthesis of filamentous cyanobacterium Arthrospira platensis. Appl Environ Microbiol 71:5004–5013
Xie LQ, Xie P, Tang HJ (2003) Enhancement of dissolved phosphorus release from sediment in a hyper-eutrophic, subtropical Chinese lake. Environ Pollut 122:391–399
Xie HX, Zafiriou OC, Cai WJ, Zepp RG, Wang YC (2004) Photooxidation and its effects on the carboxyl content of dissolved organic matter in two coastal rivers in the Southeastern United States. Environ Sci Technol 38:4113–4119
Xing GX, Shi SL, Shen GY, Du LJ, Xiong ZQ (2002) Nitrous oxide emissions from paddy soil in three rice-based cropping systems in China. Nutr Cycl Agroecosyst 64:135–143
Xu LH, Lam PK, Chen JP, Xu JM, Wong BS, Zhang YY, Wu RS, Harada KI (2000) Use of protein phosphatase inhibition assay to detect microcystins in Donghu Lake and a fish pond in China. Chemosphere 41:53–58
Yamashita Y, Tanoue E (2004) In situ production of chromophoric dissolved organic matter in coastal environments. Geophys Res Lett 31:L14302
Yamashita Y, Tanoue E (2008) Production of biorefractory fluorescent dissolved organic matter in the ocean interior. Nature Geosci 1:579–582. doi:101038/ngeo279
Yan ND, Keller W, Scilly NM, Lean DRS, Dillon PJ (1996) Increased UV-B penetration in lake owing to drought-induced acidification. Nature 381:141–143
Yao G, Gao Q, Wang Z, Huang X, He T, Zhang Y, Jiao S, Ding J (2007) Dynamics of CO_2 partial pressure and CO_2 outgassing in the lower reaches of the Xijiang River, a subtropical monsoon river in China. Sci Total Environ 376:255–266. doi:101016/jscitot env200701080

Yool A, Fasham MJR (2000) An examination of the "continental shelf pump" in an open ocean general circulation model. Global Biogeochem Cycles 15:831–844

Yoshioka T, Ueda S, Khodzher T, Bashenkhaeva N, Korovyakova I, Sorokovikova L, Gorbunova L (2002) Distribution of dissolved organic carbon in Lake Baikal and its watershed. Limnology 3:159–168

Yoshiyama K, Sharp JH (2006) Phytoplankton response to nutrient enrichment in an urbanized estuary: Apparent inhibition of primary production by overeutrophication. Limnol Oceanogr 51:424–434

Zafiriou OC, Bonneau R (1987) Wavelength-dependent quantum yields of OH radical formation from photolysis of nitrite in water. Photochem Photobiol 45:723–727

Zellner R, Exner M, Herrmann H (1990) Absolute OH quantum yields in the laser photolysis of nitrate, nitrite and dissolved H2O2 at 308 and 351 nm in the temperature range 278–353 K. J Atmos Chem 10:411–425

Zepp RG, Faust BC, Hoigné J (1992) Hydroxyl radical formation in aqueous reactions (pH 3–8) of iron(II) with hydrogen peroxide: the photo-Fenton reaction. Environ Sci Technol 26:313–319

Zepp RG, Callaghan TV, Erickson Iii DJ (2003) Interactive effects of ozone depletion and climate change on biogeochemical cycles. Photochem Photobiol Sci 2:51–61

Zepp RG, Erickson DJ, Paul ND, Sulzberger B (2007) Interactive effects of solar UV radiation and climate change on biogeochemical cycling. Photochem Photobiol Sci 6: . doi:10.1039/b700021a

Zepp RG, Erickson DJ, Paul ND, Sulzberger B (2011) Effects of solar UV radiation and climate change on biogeochemical cycling: interactions and feedbacks. Photochem Photobiol Sci 10:261–279

Zhai WD, Dai MH, Cai WJ, Wang YC, Hong HS (2005) The partial pressure of carbon dioxide and air–sea fluxes in the northern South China Sea in spring, summer and autumn. Mar Chem 96:87–97

Zhang Y, Zhu L, Zeng X, Lin Y (2004) The biogeochemical cycling of phosphorus in the upper ocean of the East China Sea. Est Coast Shelf Sci 60:369–379

Zhang W, Parker KM, Luo Y, Wan S, Wallace LL, Hu S (2005) Soil microbial responses to experimental warming and clipping in a tallgrass prairie. Glob Chang Biol 11:266–277

Zhang Y, van Dijk MA, Liu M, Zhu G, Qin B (2009) The contribution of phytoplankton degradation to chromophoric dissolved organic matter (CDOM) in eutrophic shallow lakes: field and experimental evidence. Water Res 43:4685–4697

Zhang J, Hudson J, Neal R, Sereda J, Clair T, Turner M, Jeffries D, Dillon P, Molot L, Somers K, Hesslein R (2010) Long-term patterns of dissolved organic carbon in lakes across eastern Canada: evidence of a pronounced climate effect. Limnol Oceanogr 55:30–42

Zhang H, Wu J, Shen ZP (2011) Radiative forcing and global warming potential of perfluorocarbons and sulfur hexafluoride. Sci China Earth Sci 1:1–9

Zimov SA, Schuur EAG, Chapin FS (2006) Permafrost and the global carbon budget. Science 312:1612–1613

Zlotnik I, Dubinsky Z (1989) The effect of light and temperature on DOC excretion by phytoplankton. Limnol Oceanogr 34:831–839

Zubkov MV, Tarran GA (2008) High bacterivory by the smallest phytoplankton in the North Atlantic Ocean. Nature 455:224–226. doi:101038/nature07236

Zuidema G, van Den Born GJ, Alcamo J, Kreileman GJJ (1994) Simulating changes in land cover as affected by economic and climate factors. Water Air Soil Pollut 76:163–198

Editors Biography

Dr. Khan M. G. Mostofa Photochemist and Geochemist, is a Research Scientist at the State Key Laboratory of Environmental Geochemistry, Institute of Geochemistry, the Chinese Academy of Sciences, PR China. He received double MS degree (Chemistry, Jahangirnagar University, Dhaka, Bangladesh in 1992 and Biogeochemistry, Institute for Hydrospheric-Atmospheric Sciences (IHAS), Nagoya University, Japan in 2001) and Ph.D degree from Hiroshima University, Japan, in 2005 in the field of Environmental Dynamics and Management. Dr. Mostofa then joined as a postdoctoral fellow the Institute of Geochemistry, Chinese Academy of Sciences, China, 2006–2008. He has then been appointed as Associate Professor in the same Institute on 2008 and presently holds that position. Dr. Mostofa also worked as key researcher on several international projects, particularly on the Montreal Protocol Project entitled 'Institutional strengthening for the Phase out of ODSs in Bangladesh', 1996–1999, Bangladesh and the global "International Geosphere-Biosphere Program (IGBP)" project, 1999–2001, Japan. Currently, Dr. Mostofa served as reviewer for several international peer-reviewed Journals. His key researches are mostly focused on the photo-bio-geochemistry of organic matter and nutrients, characterization and identification of organic substances of both allochthonous and autochthonous origins, complexation of OM with trace metal ions, photosynthesis and global warming. Approximately three years were needed to complete the primary drafts of the manuscripts for the ten chapters in this book.

K. M. G. Mostofa et al. (eds.), *Photobiogeochemistry of Organic Matter*,
Environmental Science and Engineering, DOI: 10.1007/978-3-642-32223-5,

Davide Vione Received his summa cum laude Laurea degree in Chemistry at the University of Torino (Italy) in 1998, and the Ph.D. in Chemistry in 2001. In 2002–2011 he has been University Lecturer in Torino, and he was appointed Associate Professor in 2011. His research interests focus on environmental photochemistry, particularly the sunlight-driven processes that involve transient species in surface and atmospheric waters, and on Advanced Oxidation Processes for water and wastewater decontamination. In June 2012 the ISI database reported 107 entries under his name, with 1,499 citations and h index 22 (for Scopus: 110 entries, 1,549 citations, h index 22). In 2003 Dr. Vione was presented the Young Researcher's award from the Analytical Chemistry division of the Italian Chemical Society (SCI) and the "European Young Researcher of the Year" award by the European Association of Chemistry and the Environment (ACE). He is Editorial Board Member of the E-Journal of Chemistry, and reviews around 30 articles per year for several environmental chemistry journals. Dr. Vione has contributed some chapters as a co-author as well as an editor of this book.

Dr. M. Abdul Mottaleb Received B.Sc. (Honors) and Masters of Science degree in Applied Chemistry from the University of Rajshahi, Rajshahi, Bangladesh. He earned his Ph.D. degree in Analytical and Environmental Chemistry/Sciences from the University of Strathclyde, Glasgow, UK. Over past 22 years Dr. Mottaleb got experience in teaching and research in the area of analytical and environmental sciences in different countries: Bangladesh, United Kingdom, South Korea and the United States of America. Examples are (i) Scientific officer, Bangladesh Atomic Energy Commission, (ii) Associate Professor in Chemistry, University of Rajshahi, Bangladesh, (iii) KOSFE postdoctoral research fellow, South Korea, (iv) Research Associate, National Exposure Research Laboratory, U.S. Environmental Protection Agency (EPA), Las Vegas, Nevada, and (v) Research Scientist/Professor, Baylor University, Texas. Dr. Mottaleb was a recipient of the Science Achievement Awards (SAA) in Chemistry 2005, and Science & Technological Achievement Awards (STAA) 2005, offered by U.S. EPA and American Chemical Society, for his outstanding and impressive discovery of emerging contaminants and nitro musk-haemoglobin adducts in aquatic organisms such as fish. Currently Dr. Mottaleb is working as an Analytical Chemist at the Center for Innovation and Entrepreneurship, Northwest Missouri State University, USA and is also serving as Editorial Board Member of peer-reviewed journals; Phytochemical Analysis, Maejo International Journal of Science and Technology, International Journal of Current Research, International Journal of Pure and Applied Sciences and Technology as well as a reviewer for many professional journals. Dr. Mottaleb contributes some chapters as a co-author as well as an editor of this book.

Dr. Takahito Yoshioka Received his Doctor of Science degree from Nagoya University on the mechanism of lacustrine nitrification. Dr. Yoshioka got experience in several organizations such as the Mitsubishi-Kasei Institute for Life Sciences (1984–1986), Faculty of Science in Shinshu University (1988–1992), Institute for Hydrospheric-Atmospheric Sciences in Nagoya University (1993–2000), Research Institute for Humanity and Nature (2001–2006) and Field Science Education and Research Center in Kyoto University (2007 to date). Dr. Yoshioka is distinguished as one of the Japanese pioneers of stable isotope ecology. Dr. Yoshioka was awarded a Biwako Prize for Ecology in 1999, for his outstanding contribution to stable isotope studies of lacustrine ecosystems. During the IGBP project (1997–2001), Dr. Yoshioka studied the dynamics of freshwater dissolved organic matter with Dr. Mostofa, using three-dimensional Fluorescence Spectroscopy, Ultra-filtration and Stable Isotope Techniques. Dr. Yoshioka was former Editor-In-Chief of Limnology (Springer-Verlag) and will contribute some chapters as co-author as well as an editor of this book.